Intercellular Signaling in
Development and Disease

Intercellular Signaling in Development and Disease

Editors-in-Chief

Edward A. Dennis
Department of Chemistry and Biochemistry,
Department of Pharmacology, School of Medicine,
University of California, San Diego,
La Jolla, California

Ralph A. Bradshaw
Department of Pharmaceutical Chemistry,
University of California, San Francisco,
San Francisco, California

AMSTERDAM • BOSTON • HEIDELBERG • LONDON • NEW YORK • OXFORD
PARIS • SAN DIEGO • SAN FRANCISCO • SINGAPORE • SYDNEY • TOKYO

Academic Press is an imprint of Elsevier

Academic Press is an imprint of Elsevier
525 B Street, Suite 1900, San Diego, CA 92101-4495, USA
30 Corporate Drive, Suite 400, Burlington, MA 01803, USA
32 Jamestown Road, London NW1 7BY, UK
360 Park Avenue South, New York, NY 10010-1710, USA

First edition 2011

Library of Congress Cataloging in Publication Data

Intercellular signaling in development and disease / editors-in-chief, Ralph A. Bradshaw, Edward A. Dennis.
    p. ; cm.
  Includes bibliographical references and index.
  ISBN 978-0-12-382215-4 (alk. paper)
  1. Cellular signal transduction. 2. Cell interaction. I. Bradshaw, Ralph A., 1941- II. Dennis, Edward A.
    [DNLM: 1. Signal Transduction. 2. Cell Communication. QU 375]
  QP517.C45I595 2011
  571.7'4–dc22
                        2011001763

British Library Cataloging in Publication Data
A catalog record for this book is available from the British Library

ISBN : 978-0-12-382215-4

For information on all Academic Press publications
visit our website at www.elsevierdirect.com

Printed and bound by CPI Group (UK) Ltd, Croydon, CR0 4YY

11 12 13   10 9 8 7 6 5 4 3 2 1

Contents

Since cell signaling is a major area of biomedical/biological research and continues to advance at a very rapid pace, scientists at all levels, including researchers, teachers, and advanced students, need to stay current with the latest findings, yet maintain a solid foundation and knowledge of the important developments that underpin the field. Carefully selected articles from the 2nd edition of the *Handbook of Cell Signaling* offer the reader numerous, up-to-date views of intracellular signal processing, including membrane receptors, signal transduction mechanisms, the modulation of gene expression/translation, and cellular/organotypic signal responses in both normal and disease states. In addition to material focusing on recent advances, hallmark papers from historical to cutting-edge publications are cited. These references, included in each article, allow the reader a quick navigation route to the major papers in virtually all areas of cell signaling to further enhance his/her expertise.

The Cell Signaling Collection consists of four independent volumes that focus on *Functioning of Transmembrane Receptors in Cell Signaling, Transduction Mechanisms in Cellular Signaling, Regulation of Organelle and Cell Compartment Signaling,* and *Intercellular Signaling in Development and Disease.* They can be used alone, in various combinations or as a set. In each case, an overview article, adapted from our introductory chapter for the Handbook, has been included. These articles, as they appear in each volume, are deliberately overlapping and provide both historical perspectives and brief summaries of the material in the volume in which they are found. These summary sections are not exhaustively referenced since the material to which they refer is.

The individual volumes should appeal to a wide array of researchers interested in the structural biology, biochemistry, molecular biology, pharmacology, and pathophysiology of cellular effectors. This is the ideal go-to books for individuals at every level looking for a quick reference on key aspects of cell signaling or a means for initiating a more in-depth search. Written by authoritative experts in the field, these papers were chosen by the editors as the most important articles for making the Cell Signaling Collection an easy-to-use reference and teaching tool. It should be noted that these volumes focus mainly on higher organisms, a compromise engendered by space limitations.

We wish to thank our Editorial Advisory Committee consisting of the editors of the Handbook of Cell Signaling, 2nd edition, including Marilyn Farquhar, Tony Hunter, Michael Karin, Murray Korc, Suresh Subramani, Brad Thompson, and Jim Wells, for their advice and consultation on the composition of these volumes. Most importantly, we gratefully acknowledge all of the individual authors of the articles taken from the Handbook of Cell Signaling, who are the 'experts' upon which the credibility of this more focused book rests.

Ralph A. Bradshaw, San Francisco, California
Edward A. Dennis, La Jolla, California
January, 2011

Rexford S. Ahima (45), University of Pennsylvania School of Medicine, Department of Medicine, Division of Endocrinology, Diabetes and Metabolism, Philadelphia, Pennsylvania

Daniel A. Albert (50), Dartmouth Medical School and Dartmouth-Hitchcock Medical Center, Lebanon, New Hampshire

Simon Alford (31), Department of Biological Sciences, University of Illinois at Chicago, Chicago, Illinois

William J. Arendshorst (15), Department of Cell and Molecular Physiology, University of North Carolina at Chapel Hill, Chapel Hill, North Carolina

Elsa Bello-Reuss (15), Department of Internal Medicine, Division of Nephrology and Department of Cell Physiology and Molecular Biophysics, Texas Tech University Health Science Center, Lubbock, Texas

Ramji K. Bhandari (14), Center for Reproductive Biology, School of Molecular Biosciences, Washington State University, Pullman, Washington

Trillium Blackmer (31), Department of Biological Sciences, University of Illinois at Chicago, Chicago, Illinois

Lynda F. Bonewald (13), Bone Biology Program, University of Missouri-Kansas City, Kansas City, Missouri

Kanika A. Bowen (12), Department of Surgery, The University of Texas Medical Branch, Galveston, Texas

Ralph A. Bradshaw (1, 2), Department of Pharmaceutical Chemistry, University of California, San Francisco, San Francisco, CA

Matthew H. Brush (48), Department of Pharmacology and Cancer Biology, Duke University Medical Center, Durham, North Carolina

Paul M. Campbell (38), Drug Discovery Program, H. Lee Moffitt Cancer Center & Research Institute, University of South Florida, Tampa, Florida

Catherine Carrière (42), Departments of Medicine, and Pharmacology and Toxicology, Dartmouth Medical School, Norris Cotton Cancer Center and Dartmouth Hitchcock Medical Center, Lebanon, New Hampshire

Moses V. Chao (33), Molecular Neurobiology Program, Skirball Institute of Biomolecular Medicine, New York University School of Medicine, New York

Jau-Nian Chen (25), Department of Molecular, Cell and Developmental Biology, University of California, Los Angeles, California

Lena Claesson-Welsh (28), Department of Genetics and Pathology, Rudbeck Laboratory, Uppsala University, Uppsala, Sweden

Robert Clarke (40), Department of Oncology, Lombardi Comprehensive Cancer Center, Georgetown University, Washington, DC

Nicholas O. Deakin (4), Department of Cell and Developmental Biology, State University of New York, Upstate Medical University, Syracuse, New York

Tushar B. Deb (40), Department of Oncology, Lombardi Comprehensive Cancer Center, Georgetown University, Washington, DC

Lanika DeGraffenreid (38), Departments of Surgery and Pharmacology, Lineberger Comprehensive Cancer Center, University of North Carolina at Chapel Hill, Chapel Hill, North Carolina

Katrin Deinhardt (33), Molecular Neurobiology Program, Skirball Institute of Biomolecular Medicine, New York University School of Medicine, New York

Larry Denner (20), McCoy Stem Cells and Diabetes Mass Spectrometry Research Laboratory, Stark Diabetes Center, Sealy Center for Molecular Medicine, Department of Internal Medicine, University of Texas Medical Branch, Galveston, Texas

Edward A. Dennis (1), Department of Chemistry and Biochemistry and Department of Pharmacology, School of Medicine, University of California, San Diego, La Jolla, CA, USA

Channing J. Der (38, 39), Departments of Surgery and Pharmacology, Lineberger Comprehensive Cancer Center, University of North Carolina at Chapel Hill, Chapel Hill, North Carolina, Department of Pharmacology and Curriculum in Genetics and Molecular Biology, Lineberger Comprehensive Cancer Center, University of North Carolina at Chapel Hill, Chapel Hill, North Carolina

Peter N. Devreotes (7), Department of Cell Biology, Johns Hopkins University School of Medicine, Baltimore, Maryland

K.M. Dibb (10), Unit of Cardiac Physiology, University of Manchester, Manchester, England, UK

Robert B. Dickson (40), Department of Oncology, Lombardi Comprehensive Cancer Center, Georgetown University, Washington, DC

Anna Dimberg (28), Department of Genetics and Pathology, Rudbeck Laboratory, Uppsala University, Uppsala, Sweden

Daniel J. Donoghue (30), Department of Chemistry and Biochemistry, University of California San Diego, La Jolla, California, Moores University of California San Diego Cancer Center, La Jolla, California

Thomas Dörner (49), Department of Medicine, Rheumatology, and Clinical Immunology, Charite University Medicine Berlin, Germany

Kristine A. Drafahl (30), Department of Chemistry and Biochemistry, University of California San Diego, La Jolla, California

Edward Eivers (23), Howard Hughes Medical Institute and Department of Biological Chemistry, University of California, Los Angeles, California

D.A. Eisner (10), Unit of Cardiac Physiology, University of Manchester, Manchester, England, UK

B. Mark Evers (12), Department of Surgery, The University of Texas Medical Branch, Galveston, Texas

Jonathan Franca-Koh (7), Department of Cell Biology, Johns Hopkins University School of Medicine, Baltimore, Maryland

Robert E. Friesel (8), Center for Molecular Medicine, Maine Medical Center Research Institute, Scarborough, Maine

Luis C. Fuentealba (23), Howard Hughes Medical Institute and Department of Biological Chemistry, University of California, Los Angeles, California

Tatyana Gerachshenko (31), Department of Biological Sciences, University of Illinois at Chicago, Chicago, Illinois

Nicolas Girard (44), Human Oncology and Pathogenesis Program, Memorial Sloan-Kettering Cancer Center, New York, New York

Vikas Gupta (17), Department of Medicine, Howard Hughes Medical Institute, University of California, San Francisco, California

Edaeni Hamid (31), Department of Biological Sciences, University of Illinois at Chicago, Chicago, Illinois

Sam Hanash (51), Fred Hutchinson Cancer Research Center, Seattle, Washington

Ariella B. Hanker (39), Curriculum in Genetics and Molecular Biology, Lineberger Comprehensive Cancer Center, University of North Carolina at Chapel Hill, Chapel Hill, North Carolina

Michelle L. Hermiston (17), Department of Pediatrics, Howard Hufghes Medical Institute, University of California, San Francisco, California

Margaret Howe (20), McCoy Stem Cells and Diabetes Mass Spectrometry Research Laboratory, Stark Diabetes Center, Department of Internal Medicine, University of Texas, Galveston, Texas

Marcin Iwanicki (3), Department of Microbiology, University of Virginia Health System, Charlottesville, Virginia

Sophie Jarriault (26), IGBMC (Institut de Génétique et de Biologie Moléculaire et Cellulaire), Department of Cell and Developmental Biology, Illkirch, France; and Université de Strasbourg, Strasbourg, France

Michael D. Johnson (40), Department of Oncology, Lombardi Comprehensive Cancer Center, Georgetown University, Washington, DC

Louis B. Justement (19), Department of Microbiology, University of Alabama at Birmingham, Birmingham, Alabama

Randal J. Kaufman (47), Department of Biological Chemistry, University of Michigan Medical School, Ann Arbor, Michigan, Department of Internal Medicine, University of Michigan Medical School, Ann Arbor, Michigan, Department of Howard Hughes Medical Institute, University of Michigan Medical School, Ann Arbor, Michigan

Sakari Kauppinen (52), Santaris Pharma, Hørsholm, Denmark, and Wilhelm Johannsen Centre for Functional Genome Research, Department of Cellular and Molecular Medicine, University of Copenhagen, Copenhagen, Denmark

Rolf Kemler (21), Department of Molecular Embryology, Max-Planck Institute of Immunobiology, Freiburg, Germany

Rolf König (18), Department of Microbiology and Immunology, The University of Texas Medical Branch, Galveston, Texas

Murray Korc (41, 42, 43), Departments of Medicine, Pharmacology and Toxicology, Norris Cotton Cancer Center, Dartmouth-Hitchcock Medical Center, Lebanon, New Hampshire, and Dartmouth Medical School, Hanover, New Hampshire

Chulan Kwon (9), Gladstone Institute of Cardiovascular Disease and Departments of Pediatrics and Biochemistry & Biophysics, University of California at San Francisco, San Francisco, California

Adam D. Langenbacher (25), Department of Molecular, Cell and Developmental Biology, University of California, Los Angeles, California

Hojoon X. Lee (23), Howard Hughes Medical Institute and Department of Biological Chemistry, University of California, Los Angeles, California

Lucy Liaw (8), Center for Molecular Medicine, Maine Medical Center Research Institute, Scarborough, Maine

Volkhard Lindner (8), Center for Molecular Medicine, Maine Medical Center Research Institute, Scarborough, Maine

Peter E. Lipsky (49), NIAMS, National Institutes of Health, Bethesda, MD

Karen M. Lyons (29), Department of Molecular, Cell and Developmental Biology, Department of Orthopaedic Surgery, University of California, Los Angeles, California

James P. Madigan (38), Departments of Surgery and Pharmacology, Lineberger Comprehensive Cancer Center, University of North Carolina at Chapel Hill, Chapel Hill, North Carolina

Jyoti D. Malhotra (47), Department of Biological Chemistry, University of Michigan Medical School, Ann Arbor, Michigan

Karen H. Martin (3), Department of Microbiology, University of Virginia Health System, Charlottesville, Virginia, Department of Neurobiology and Anatomy, West Virginia University, Morgantown West Virginia

Christopher W. McAndrew (30), Department of Chemistry and Biochemistry, University of California San Diego, La Jolla, California

Frank McCormick (37), Cancer Research Institute, University of California Comprehensive Cancer Center, San Francisco, California

Laurens A. van Meeteren (28), Department of Genetics and Pathology, Rudbeck Laboratory, Uppsala University, Uppsala, Sweden

Borna Mehrad (46), Division of Pulmonary and Critical Care Medicine, University of Virginia, Charlottesville, Virginia

Daniel Messerschmidt (21), Department of Molecular Embryology, Max-Planck Institute of Immunobiology, Freiburg, Germany

Guo-li Ming (34), The Salk Institute, La Jolla, California Division of Neurobiology, Department of Molecular and Cell Biology, University of California, Berkeley, California

James T. Nichols (27), Department of Biological Chemistry David Geffen School of Medicine, University of California, Los Angeles, California

Eric E. Nilsson (14), Center for Reproductive Biology, School of Molecular Biosciences, Washington State University, Pullman, Washington

Alixanna Norris (41), Departments of Medicine, Pharmacology and Toxicology, Norris Cotton Cancer Center, Dartmouth-Hitchcock Medical Center, Lebanon, New Hampshire, and Dartmouth Medical School, Hanover, New Hampshire

Stefan Offermanns (6), Department of Pharmacology, Max-Planck-Institute, Bad Nauheim, Germany

Matthew T. Pankratz (36), Gene Expression Laboratory, The Salk Institute for Biological Studies, La Jolla, California

William Pao (44), Vanderbilt-Ingram Cancer Center, Nashville, Tennessee

J. Thomas Parsons (3), Department of Microbiology, University of Virginia Health System, Charlottesville, Virginia

Samuel L. Pfaff (36), Gene Expression Laboratory, The Salk Institute for Biological Studies, La Jolla, California

Tatjana Piotrowski (22), University of Utah, Department of Neurobiology and Anatomy, Salt Lake City, Utah

Fiona J. Pixley (16), Department of Developmental and Molecular Biology, Albert Einstein College of Medicine, Bronx, New York, New York, *Current address: School of Medicine and Pharmacology, University of Western Australia, Nedlands, WA, Australia

Mu-ming Poo (34), The Salk Institute, La Jolla, California Division of Neurobiology, Department of Molecular and Cell Biology, University of California, Berkeley, California

Igor Prudovsky (8), Center for Molecular Medicine, Maine Medical Center Research Institute, Scarborough, Maine

Barbara Ranscht (5), Burnham Institute for Medical Research, La Jolla, California

Danica Ramljak (40), Department of Oncology, Lombardi Comprehensive Cancer Center, Georgetown University, Washington, DC

Kelsey N. Retting (29), Department of Biological Chemistry, University of California, Los Angeles, California

E. M. De Robertis (23), Howard Hughes Medical Institute and Department of Biological Chemistry, University of California, Los Angeles, California

Patrick J. Roberts (38), Departments of Surgery and Pharmacology, Lineberger Comprehensive Cancer Center, University of North Carolina at Chapel Hill, Chapel Hill, North Carolina

Charlotte Rolny (28), Department of Genetics and Pathology, Rudbeck Laboratory, Uppsala University, Uppsala, Sweden

Stefan Rudloff (21), Department of Molecular Embryology, Max-Planck Institute of Immunobiology, Freiburg, Germany

Frederic de Sauvage (24), Department of Molecular Biology, Genentech, Inc., South San Francisco, California

Eric F. Schmidt (35), Department of Neurology and Section of Neurobiology, Yale University School of Medicine, New Haven, Connecticut

Lorenzo F. Sempere (52), Department of Medicine, Norris Cotton Cancer Center, Lebanon, New Hampshire

Shirish Shenolikar (48), Department of Pharmacology and Cancer Biology, Duke University Medical Center, Durham, North Carolina, Signature Research Program in Neuroscience and Behavioral Disorders, Duke-NUS Graduate Medical School Singapore

Michael K. Skinner (14), Center for Reproductive Biology, School of Molecular Biosciences, Washington State University, Pullman, Washington

Jill K. Slack-Davis (3), Department of Microbiology, University of Virginia Health System, Charlottesville, Virginia

Deepak Srivastava (9), Gladstone Institute of Cardiovascular Disease and Departments of Pediatrics and Biochemistry & Biophysics, University of California at San Francisco, San Francisco, California

E. Richard Stanley (16), Department of Developmental and Molecular Biology, Albert Einstein College of Medicine, Bronx, New York, New York, *Current address: School of Medicine and Pharmacology, University of Western Australia, Nedlands, WA, Australia

Andrew W. Stoker (32), Neural Development Unit, Institute of Child Health, University College London, London, England, UK

Robert M. Strieter (46), Division of Pulmonary and Critical Care Medicine, University of Virginia, Charlottesville, Virginia

Stephen M. Strittmatter (35), Department of Neurology and Section of Neurobiology, Yale University School of Medicine, New Haven, Connecticut

David B. Talmadge (50), Dartmouth Medical School and Dartmouth-Hitchcock Medical Center, Lebanon, New Hampshire

E. Brad Thompson (2), Department of Human Biological Chemistry and Genetics, The University of Texas Medical Branch, Galveston, Texas

Robert W. Tilghman (3), Department of Microbiology, University of Virginia Health System, Charlottesville, Virginia

Hideaki Togashi (35), Department of Neurology and Section of Neurobiology, Yale University School of Medicine, New Haven, Connecticut

A.W. Trafford (10), Unit of Cardiac Physiology, University of Manchester, Manchester, England, UK

Christopher E. Turner (4), Department of Cell and Developmental Biology, State University of New York, Upstate Medical University, Syracuse, New York

Randall J. Urban (20), McCoy Stem Cells and Diabetes Mass Spectrometry Research Laboratory, Stark Diabetes Center

Gladys M. Varela (45), University of Pennsylvania School of Medicine, Department of Medicine, Division of Endocrinology, Diabetes and Metabolism, Philadelphia, Pennsylvania

Calvin Vary (8), Center for Molecular Medicine, Maine Medical Center Research Institute, Scarborough, Maine

Gerry Weinmaster (27), Department of Biological Chemistry David Geffen School of Medicine, Molecular Biology Institute, Jonsson Comprehensive Cancer Center; University of California, Los Angeles, California

Arthur Weiss (17), Department of Medicine, Howard Hughes Medical Institute, University of California, San Francisco, California

Chery A. Whipple (43), Departments of Medicine, Pharmacology and Toxicology, and the Norris Cotton Cancer Center, Dartmouth-Hitchcock Medical Center, Lebanon, New Hampshire and Dartmouth Medical School, Hanover, New Hampshire

Stacey Sedore Willard (7), Department of Cell Biology, Johns Hopkins University School of Medicine, Baltimore, Maryland

Susan Wray (11), Physiology Department, School of Biomedical Sciences, University of Liverpool, Liverpool, England, UK

Jen Jen Yeh (38), Departments of Surgery and Pharmacology, Lineberger Comprehensive Cancer Center, University of North Carolina at Chapel Hill, Chapel Hill, North Carolina

Jianxin A. Yu (4), Department of Cell and Developmental Biology, State University of New York, Upstate Medical University, Syracuse, New York

Overview

Chapter 1

Signaling in Development and Disease*

Edward A. Dennis[1] and Ralph A. Bradshaw[2]

[1]*Department of Chemistry and Biochemistry and Department of Pharmacology, School of Medicine, University of California, San Diego, La Jolla, CA*

[2]*Department of Pharmaceutical Chemistry, University of California, San Francisco, San Francisco, CA*

Cell signaling, which is also often referred to as signal transduction or, in more specialized cases, transmembrane signaling, is the process by which cells communicate with their environment and respond temporally to external cues that they sense there. All cells have the capacity to achieve this to some degree, albeit with a wide variation in purpose, mechanism, and response. At the same time, there is a remarkable degree of similarity over quite a range of species, particularly in the eukaryotic kingdom, and comparative physiology has been a useful tool in the development of this field. The central importance of this general phenomenon (sensing of external stimuli by cells) has been appreciated for a long time, but it has truly become a dominant part of cell and molecular biology research in the past three decades, in part because a description of the dynamic responses of cells to external stimuli is, in essence, a description of the life process itself. This approach lies at the core of the developing fields of proteomics and metabolomics, and its importance to human and animal health is already plainly evident.

ORIGINS OF CELL SIGNALING RESEARCH

Although cells from polycellular organisms derive substantial information from interactions with other cells and extracellular structural components, it was humoral components that first were appreciated to be intercellular messengers. This idea was certainly inherent in the 'internal secretions' initially described by Claude Bernard in 1855 and thereafter, as it became understood that ductless glands, such as the spleen, thyroid, and adrenals, secreted material into the bloodstream. However, Bernard did not directly identify hormones as such. This was left to Bayliss and Starling and their description of secretin in 1902 [1].

Recognizing that it was likely representative of a larger group of chemical messengers, the term *hormone* was introduced by Starling in a Croonian Lecture presented in 1905. The word, derived from the Greek word meaning 'to excite or arouse,' was apparently proposed by a colleague, W. B. Hardy, and was adopted, even though it did not particularly connote the messenger role but rather emphasized the positive effects exerted on target organs via cell signaling (see Wright [2] for a general description of these events). The realization that these substances could also produce inhibitory effects, gave rise to a second designation, 'chalones,' introduced by Schaefer in 1913 (see Schaefer [3]), for the inhibitory elements of these glandular secretions. The word 'autocoid' was similarly coined for the group as a whole (hormones and chalones). Although the designation chalone has occasionally been applied to some growth factors with respect to certain of their activities (e.g., transforming growth factor β), autocoid has essentially disappeared. Thus, if the description of secretin and the introduction of the term hormone are taken to mark the beginnings of molecular endocrinology and the eventual development of cell signaling, then we have passed the hundredth anniversary of this field.

The origins of endocrinology, as the study of the glands that elaborate hormones and the effect of these entities on target cells, naturally gave rise to a definition of hormones as substances produced in one tissue type that traveled systemically to another tissue type to exert a characteristic response. Of course, initially these responses were couched in organ and whole animal responses, although they increasingly were defined in terms of metabolic and other chemical changes at the cellular level. The early days of endocrinology were marked by many important discoveries, such as the discovery of insulin [4], to name one, that solidified the definition, and a well-established list of

Portions of this article were adapted from Bradshaw RA, Dennis EA. Cell signaling: yesterday, today, and tomorrow. In Bradshaw RA, Dennis EA, editors. Handbook of cell signaling. 2nd ed. San Diego, CA: Academic Press; 2008; pp 1–4.

hormones, composed primarily of three chemical classes (polypeptides, steroids, and amino acid derivatives), was eventually developed. Of course, it was appreciated even early on that the responses in the different targets were not the same, particularly with respect to time. For example, adrenalin was known to act very rapidly, while growth hormone required a much longer time frame to exert its full range of effects. However, in the absence of any molecular details of mechanism, the emphasis remained on the distinct nature of the cells of origin versus those responding and on the systemic nature of transport, and this remained the case well into the 1970s. An important shift in endocrinological thinking had its seeds well before that, however, even though it took about 25 years for these 'new' ideas that greatly expanded endocrinology to be enunciated clearly.

Although the discovery of polypeptide growth factors as a new group of biological regulators is generally associated with nerve growth factor (NGF), it can certainly be argued that other members of this broad category were known before NGF. However, NGF was the source of the designation *growth factor* and has been, in many important respects, a Rosetta stone for establishing principles that are now known to underpin much of signal transduction. Thus, its role as the progenitor of the field and the entity that keyed the expansion of endocrinology, and with it the field of cell signaling, is quite appropriate. The discovery of NGF is well documented [5] and how this led directly to identification of epidermal growth factor (EGF) [6], another regulator that has been equally important in providing novel insights into cellular endocrinology, signal transduction and, more recently, molecular oncology. However, it was not till the sequences of NGF and EGF were determined [7, 8] that the molecular phase of growth factor research began in earnest. Of particular importance was the postulate that NGF and insulin were evolutionarily related entities [9], which suggested a similar molecular action (which, indeed, turned out to be remarkably clairvoyant), and was the first indication that the identified growth factors, which at that time were quite limited in number, were like hormones. This hypothesis led quickly to the identification of receptors for NGF on target neurons, using the tracer binding technology of the time (see Raffioni *et al.* [10] for a summary of these contributions), which further confirmed their hormonal status. Over the next several years, similar observations were recorded for a number of other growth factors, which in turn, led to the redefinition of endocrine mechanisms to include paracrine, autocrine, and juxtacrine interactions [11]. These studies were followed by first isolation and molecular characterization using various biophysical methods and then cloning of their cDNAs, initially for the insulin and EGFR receptors [12–14] and then many others. Ultimately, the powerful techniques of molecular biology were applied to all aspects of cell signaling and are largely responsible for the detailed depictions we have today. They have allowed the broad understanding of the myriad of mechanisms and responses employed by cells to assess changes in their environment and to coordinate their functions to be compatible with the other parts of the organism of which they are a part.

RECEPTORS AND INTRACELLULAR SIGNALING

At the same time that the growth factor field was undergoing rapid development, major advances were also occurring in studies on hormonal mechanisms. In particular, Sutherland and colleagues [15] were redefining hormones as messengers and their ability to produce second messengers. This was, of course, based primarily on the identification of cyclic AMP (cAMP) and its production by a number of classical hormones. However, it also became clear that not all hormones produce this second messenger nor was it stimulated by any of the growth factors known at that time. This enigma remained unresolved for quite a long time until tyrosine kinases were identified [16, 17] and it was shown, first with the EGF receptor [18], that these modifications were responsible for initiating the signal transduction for many of those hormones and growth factors that did not stimulate the production of cAMP.

Aided by the tools of molecular biology, it was a fairly rapid transition to the cloning of most of the receptors for hormones and growth factors and the subsequent development of the main classes of signaling mechanisms. These data allowed the six major classes of cell surface receptors for hormones and growth factors to be defined, which included, in addition to the receptor tyrosine kinases (RTKs) described previously, the G-protein coupled receptors (GPCRs) (including the receptors that produce cAMP) that constitute the largest class of cell surface receptors; the cytokine receptors, which recruit the soluble JAK tyrosine kinases and directly activate the STAT family of transcription factors; serine/threonine kinase receptors of the TGFβ superfamily; the tumor necrosis factor (TNF) receptors that activate nuclear factor kappa B (NFκB) via TRAF molecules, among other pathways; and the guanylyl cyclase receptors. Structural biology has not maintained the same pace, and there are still both ligands and receptors for which we do not have three-dimensional information as yet.

In parallel with the development of our understanding of ligand/receptor organization at the plasma membrane, a variety of experimental approaches have also revealed the general mechanisms of transmembrane signal transduction in terms of the major intracellular events that are induced by these various receptor classes. There are three principal means by which intracellular signals are propagated: protein posttranslational modifications (PTMs), lipid messengers, and ion fluxes. There are also additional moieties that play significant roles, such as cyclic nucleotides, but their effects are generally manifested in downstream PTMs. There is considerable interplay between the three, particularly in the more complex pathways.

By far the most significant of the PTMs is phosphorylation of serine, threonine, and tyrosine residues. Indeed, there are over 500 protein kinases in the human genome with more than 100 phosphatases [19]. Many of these modifications activate various enzymes, which are designated effectors, but it has also become increasingly clear that many PTM additions were inducing new, specific ('docking') sites for protein–protein interactions. These introduced the concept of both adaptors and multisite scaffolds that bound to the sites through specific motifs and as the process is repeated, successively built up multicomponent signaling structures [20]. There has now emerged a significant number of binding motifs, recognizing, in addition to PTMs, phospholipids and proline-rich peptide segments to name a few, that are quite widely scattered through the large repertoire of signaling molecules and that are activated by different types of receptors in a variety of cell types.

Although the intracellular signaling pathways are characterized by a plethora of modifications and interactions that alter existing proteomic and metabolomic landscapes, the major biological responses, such as mitosis, differentiation, and apoptosis, require alterations in the phenotypic profile of the cell, and these require the modulation of transcription and translation. Indeed, signaling can be thought of at two levels: responses (events) that affect (or require) preexisting structures (proteins) and those that depend on generating new proteins. Temporally, rapid responses are perforce of the first type, while longer-term responses generally are of the second. Thus, it may be viewed that the importance of the complex, largely cytoplasmic, machinery, involving receptors, effectors, adaptors, and scaffolds, has two purposes: to generate immediate changes and then to ultimately reprogram the transcriptional activities for more permanent responses.

The process of gene expression in eukaryotes can be considered at several levels: the generation of the primary RNA transcript, its processing, and transport, translation of the mRNA into protein, and finally, its turnover. Since the amount of the potential activity associated with a given protein is fundamentally dependent on both its rate of synthesis and its rate of degradation, the turnover of the protein itself is also critical to signaling processes and is certainly largely, if not completely, affected by signaling events, too. In eukaryotes, transcription and mRNA processing take place in the nucleus; translation and mRNA turnover are cytoplasmic events. All of these processes are controlled or affected by signal transduction pathways.

The effects on transcription occur at a number of levels and usually involve phosphorylation, either of transcription factors or cofactors. In some cases, this occurs in the cytoplasm, and the effect of the modification is to induce transport into the nucleus; in other cases, the modifications affect binding of regulatory cofactors or to the DNA itself. One class of transcription factors, the nuclear receptor family, requires ligand binding before they are functional. Members of this family form the core of signal transduction pathways that regulate gene expression in response to steroid and thyroid hormones, fatty acids, bile acids, cholesterol metabolites, and certain xenobiotic compounds. In fact, this can be viewed as an extension of lipid signaling, as most of the ligands for these receptors are hydrophobic in character. The ligands exert their affects through allosteric regulation, which has a dramatic effect on either the DNA binding or transcriptional activation properties of the transcription factor [21].

Two biological phenomena of critical importance in all organisms are cell generation (cell division or mitosis/meiosis) and cell death (apoptosis and necrosis). Both are extensively regulated and not surprisingly, much of this control is under the aegis of cell signaling events. The progression through the cell cycle and its various checkpoints is a symphony of protein modifications coupled to programmed protein turnover. The key players are a complement of kinases, known as cyclin-dependent kinases (Cdks), whose activation and deactivation are involved in every stage of the cycle. Interaction with cyclins, required for their activity, allows them to cycle in an on–off manner, and the ubiquitin-dependent degradation of the cyclins controls the vectoral nature of the cycle. The cyclin–Cdk complexes can be further regulated by phosphorylation or complexation with other proteins, which also allows for pausing at checkpoints if the cell senses it should not continue with the division process.

There are also feed-forward mechanisms that allow early steps to regulate successive ones. Apoptosis is equally tightly regulated and its progression easily recognized by distinct phenotypic responses (membrane blebbing, cell shrinking, and chromosomal condensation) as the cell progresses to its end. It is predicated on a family of cysteine proteases, called caspases (because they cleave their substrates to the C-terminal side of aspartic acid residues) that are activated in either an extrinsic or intrinsic pathway. The ten caspases generally exist as inactive precursors (zymogens) and can be subclassified into executioner, initiator, and inflammatory types. These have different structural features and different roles in apoptosis. One apoptotic pathway is directly related to the TNF superfamily, transmembrane receptors that contain a death domain. When activated, these lead to the activation of caspases 8, which in turn, activates the executioner caspases 3. Apoptosis is also triggered by cellular stress, and this leads to the involvement of the mitochondria (as noted above). In a complex pathway involving many proteins, an apoptosome is formed which also leads to the eventual activation of the executioner caspases. Clearly, the connections between these two fundamental processes are of great importance and are closely related to a number of human diseases, notably cancer and neural degeneration.

INTERCELLULAR SIGNALING

All living cells must be able to interact with their environment if they are to remain viable, whether to sense and move to sources of nourishment or to adjust and adapt to changes that may have occurred there. In multicellular organisms, where communication can become quite complex, the effects of cell signaling extend well beyond the intracellular events triggered in the cytoplasm, and these must also be coordinated with those of sister cells to allow higher-level functions, such as exhibited by an organ (see Figure 1.1). External information can be transmitted to a recipient cell by soluble factors, by interactions with the extracellular matrix (ECM), or by cell–cell contacts that can involve a variety of specific and nonspecific

interactions, and the types of reactions initiated may be similar or different than those generated by the soluble ligands. Signals received from these sources are essential to direct developmental pathways and can play key roles in the support of some abnormal tissues such as cancers. The cues inherent in these signaling pathways can be tissue-specific or they may be of a general nature. The same signal in two different cell types may lead to very different results. The general appreciation of signaling at this level is not as well-founded as the knowledge of the more detailed events that follow the activation of intracellular signaling pathways, but it will be of great importance for understanding, for example, how stem cells differentiate and what controls their ultimate phenotype. Given the issues surrounding the use of embryonic stem cells and the apparent gains in manufacturing induced pluripotent cells, these will be important targets for signaling research in the future.

There is developing a considerable interest in the role of cell signaling in development. Genetic studies have been enormously valuable in this regard and have pointed to the

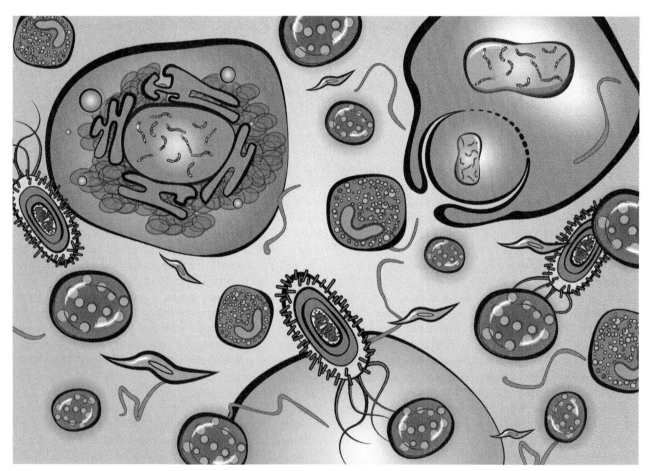

FIGURE 1.1 There are over 200 cell types in the human body, and signaling in individual cells has extracellular manifestations that result from mediators effecting surrounding cells as well as controlling cell–cell interactions. The signaling cascade can extend outward to cause pleiotropic effects on tissues and organs and can, if gone awry, result in significant disease ramifications ranging from metabolic syndrome including insulin sensitivity and obesity to cardiovascular effects, to effects on the CNS, and to numerous forms of cancer throughout the body.

role of a number of factors (and their receptors) such as the Wnt family, notch, and hedgehog in development. Several systems have direct involvement of more specialized factors like VEGF and the FGFs for angiogenesis, bone morphogenic proteins (BMP) and FGFs for skeletal growth, and a number of neurotrophic factors for neurogenesis, for example. Interestingly, most of these factors, which have defined functions in selected developmental systems, are found in the main growth factor families. Since factors like FGF and VEGF are known to activate similar intracellular pathways, there must be other means to distinguish their contributions. In bone growth, for example, FGFR1 is replaced by FGFR3 during key points in bone development, and this restricted expression becomes significant in the presence of certain mutations in the FGFR3 that lead to well-known human skeletal dysplasias [22].

Following the discovery that the EGFR was related to the oncoprotein ErbB2 [23] and that the platelet-derived growth factor was related to v-Sis [24], another oncogene product, it confirmed what had already been widely speculated for some time previously, namely that growth factors (and by association, other signaling molecules) not only controlled growth processes like hypertrophy and hyperplasticity but also aberrations in the processes they controlled lead to growth disorders like cancer. The subsequent efforts to define these relationships have occupied molecular oncologists for the succeeding 30 or so years. These investigations have provided considerable insight into what underlies cell transformations and have provided a number of drug targets. The tyrosine kinases have been a particularly rich source [25, 26] but are by no means the only ones. Signaling disorders are also not solely linked to cancer, and there are many other areas of translational research that have their basis in signaling systems. It may be expected that this area of applications will continue to expand as our understanding of cell signaling develops.

FOCUS AND SCOPE OF THIS VOLUME

The chapters of this volume have been selected from a larger collection [27] and have been organized to emphasize the role of cell–cell interactions as well as organ and tissue effects in signaling activities and their disease implications. They have been contributed by recognized experts and they are authoritative to the extent that size limitations allow. It is our intention that this survey will be useful in teaching, particularly in introductory courses, and to more seasoned investigators new to this area.

It is not possible to develop any of the areas covered in this volume in great detail, and expansion of any topic is left to the reader. The references in each chapter provide an excellent starting point, and greater coverage can also be found in the parent work [27]. It is important to realize that this volume does not cover other aspects of cell signaling such as transmembrane receptor structure, receptor organization and function, intracellular signaling mechanisms, and transcriptional activation and responses in other organelles. These can be found in other volumes in this series [28–30].

REFERENCES

1. Bayliss WM, Starling EH. The mechanism of pancreatic secretion. *J Physiol* 1902;**28**:325–53.
2. Wright RD. The origin of the term "hormone". *Trends Biochem Sci* 1978;**3**:275.
3. Schaefer EA. *The Endocrine Organs*. London: Longman & Green; 1916; p. 6.
4. Banting FG, Best CH. The internal secretion of the pancreas. *J Lab Clin Med* 1922;**7**:251–66.
5. Levi-Montalcini R. The nerve growth factor 35 years later. *Science* 1987;**237**:1154–62.
6. Cohen S. Origins of growth factors: NGF and EGF. *J Biol Chem* 2008;**283**:33793–7.
7. Angeletti RH, Bradshaw RA. Nerve growth factor from mouse submaxillary gland: amino acid sequence. *Proc Natl Acad Sci USA* 1971;**68**:2417–20.
8. Savage CR, Inagami T, Cohen S. The primary structure of epidermal growth factor. *J Biol Chem* 1972;**247**:7612–21.
9. Frazier WA, Angeletti RH, Bradshaw RA. Nerve growth factor and insulin. *Science* 1972;**176**:482–8.
10. Raffioni S, Buxser SE, Bradshaw RA. The receptors for nerve growth factor and other neurotrophins. *Annu Rev Biochem* 1993;**62**:823–50.
11. Bradshaw RA, Sporn MB. Polypeptide growth factors and the regulation of cell growth and differentiation: introduction. *Fed. Proc* 1983;**42**:2590–1.
12. Ullrich A, Bell JR, Chen EY, Herrera R, Petruzzelli LM, Dull TJ, et al. Human insulin receptor and its relationship to the tyrosine kinase family of oncogenes. *Nature* 1985;**313**:756–61.
13. Ullrich A, Coussens L, Hayflick JS, Dull TJ, Gray A, Tam AW, et al. Human epidermal growth factor receptor cDNA sequence and aberrant expression of the amplified gene in A431 epidermoid carcinoma cells. *Nature* 1985;**309**:418–25.
14. Ebina Y, Ellis L, Jarnagin K, Edery M, Graf L, Clauser E, et al. The human insulin receptor cDNA: the structural basis for hormone transmembrane signalling. *Cell* 1985;**40**:747–58.
15. Robison GA, Butcher RW, Sutherland EW. *Cyclic AMP*. San Diego, CA: Academic Press; 1971.
16. Eckert W, Hutchinson MA, Hunter T. An activity phosphorylating tyrosine in polyoma T antigen immunoprecipitates. *Cell* 1979;**18**:925–33.
17. Hunter T, Sefton BM. Transforming gene product of Rous sarcoma virus phosphorylates tyrosine. *Proc Natl Acad Sci USA* 1980;**77**:1311–5.
18. Ushiro H, Cohen S. Identification of phosphotyrosine as a product of epidermal growth factor-activated protein kinase in A-431 cell membranes. *J Biol Chem* 1980;**255**:8363–5.
19. Manning G, Whyte DB, Martinez R, Hunter T, Sudarsanam S. The protein kinase complement of the human genome. *Science* 2002;**298**:1912–34.
20. Pawson T. Regulation and targets of receptor tyrosine kinases. *Eur J Cancer* 2002;**38**(Suppl 5):S3–10.
21. Evans RM. The steroid and thyroid hormone receptor superfamily. *Science* 1988;**240**:889–95.

22. Robertson SC, Tynan J, Donoghue DJ. RTK mutations and human syndromes: when good receptors turn bad. *Trends Genet* 2000;**16**:265–71.

23. Downward J, Yarden Y, Mayes E, Scrace G, Totty N, Stockwell P, Ullrich A, Schlessinger J, Waterfield MD. Close similarity of epidermal growth factor receptor and *v-erb-B* oncogene protein sequences. *Nature* 1984;**307**:521–7.

24. Waterfield MD, Scrace GT, Whittle N, Stroobant P, Johnsson A, Wasteson A, et al. Platelet-derived growth factor is structurally related to the putative transforming protein p28^{sis} of simian sarcoma virus. *Nature* 1983;**304**:35–9.

25. Krause DS, Van Etten RA. Tyrosine kinases as targets for cancer therapy. *N Engl J Med* 2005;**353**:172–87.

26. Zwick E, Bange J, Ullrich A. Receptor tyrosine kinases as targets for anticancer drugs. *Trends Mol Med* 2002;**8**:17–23.

27. Bradshaw RA, Dennis EA, editors. *Handbook of cell signaling.* 2nd ed. San Diego, CA: Academic Press; 2008.

28. Bradshaw RA, Dennis EA, editors. *Functioning of transmembrane receptors in cell signaling mechanisms.* San Diego, CA: Academic Press; 2011.

29. Dennis EA, Bradshaw RA, editors. *Transduction mechanisms in cellular signaling.* San Diego, CA: Academic Press; 2011.

30. Bradshaw RA, Dennis EA, editors. *Regulation of organelle and cell compartment signaling.* San Diego, CA: Academic Press; 2011.

Cell-Cell Signaling

Chapter 2

Overview of Cell–Cell and Cell–Matrix Interactions

E. Brad Thompson [1] and Ralph A. Bradshaw [2]

[1] *Department of Human Biological Chemistry and Genetics, The University of Texas Medical Branch, Galveston, Texas*

[2] *Department of Pharmaceutical Chemistry, University of California, San Francisco, San Francisco, CA*

Cell signaling pathways are not simply linear, but in fact form extensive interactive networks. Indeed it is the overlapping and interconnecting nature of these that provides the distinctive features distinguishing many of the response properties of specific tissues and organs. The complexity of these networks will require a great deal of research before their organization is understood in detail, but some generalities are beginning to emerge [1]. The advent of the techniques of genomics, including microchip arrays of genes and proteomics, will stimulate much more rapid development of understanding of pathway interactions as we see how a given signal reverberates through the tissue and cellular systems. Indeed, a considerable amount is already known, and in many systems the overall patterns are beginning to be made clear. Completing this knowledge will undoubtedly be the goal of a great deal of research in the near future. Part V of this book describes how many major tissue and organ systems work as they consider the many signals that play upon them.

On reading entries in this part, it may be necessary for the uninitiated peruser to refer to earlier entries in other parts of the book, which explain in detail how particular signal pathways function, as this part deals only with the identification of those pathways that are important in the control of organ and tissue function, and not the iteration of how each pathway works. However, one concept not covered previously needs to be introduced, i.e., the idea that chemical signals may be provided locally or regionally in tissues by a group of mechanisms that have become known as paracrine, autocrine, intracrine, and juxtacrine interactions. These constitute means for regulating tissue-specific signal responses by providing the needed signals only on a local basis.

These concepts were born from the field of endocrinology. Traditionally, this discipline held that certain specific organs produce and secrete particular signaling molecules into the bloodstream, which delivered them elsewhere in the body to carry out their signaling activity. A classic example is the production of insulin by the beta cells of the pancreas, with the hormone transferred systemically to many other tissues, where it activates its receptors thus affecting glucose metabolism, among other responses. Over the past 10–20 years, it has become clear that, in addition to this classic endocrine notion, signaling molecules are also produced to function more locally. That is to say, although they may not enter the general circulation and consequently act only locally, they nevertheless work by binding receptors–either on the surface or within cells–and set off the same types of signal transduction pathways, as do traditional hormones. In fact, a general understanding has evolved that signal transduction pathways that usually have been addressed separately because of their inclusion in a particular discipline–e.g., endocrinology as contrasted with immunology–behave in much the same ways. A great unanimity in general mechanisms is seen as signals are transmitted between cells, whether they be signals from one immune cell to another, or signals from a classic endocrine target organ to another tissue. In this sense, the same types of chemical and physical behaviors ultimately carry out cell signaling universally. The specific types of mechanisms mentioned above, in addition to endocrine, are defined according to the degree of localization of effect, but the signals generated carry out their functions by the same sorts of receptor-transduced pathways as do the ubiquitous signaling molecules.

In addition to the classical endocrine mechanism, localized signaling mechanisms can be conveniently sub-grouped into four types. Paracrine interactions induce signaling activities that occur from cell to cell within a given tissue or organ, rather than through the general circulation. This takes place as locally produced hormones or other small signaling molecules exit their cell of origin, and then, by diffusion or local circulation, act only regionally on other cells of a different type within that tissue. This has been found to be important in many organs, and is a field of investigation that continues to develop rapidly. The local concentrations of paracrine signals can be quite high compared to the circulating levels, and thus can trigger effects by acting on low-affinity receptors or by supplying sufficient local signals to bind to high-affinity receptors even when the circulating level of a molecule that produces a similar signal is too low to do so. Paracrine signaling molecules sometimes are very rapidly metabolized locally to further limit the physical extent of their action. Examples are the prostaglandins and nitric oxide. Other, longer-acting signaling molecules also are employed in a paracrine fashion. Limiting their access to the blood supply and/or the total amounts produced within a tissue, so that local receptors can bind them before the general circulation is reached, keeps them acting in a paracrine fashion. Sometimes paracrine signaling molecules are moved by local physical means, so as to provide greater or lesser regional concentrations. An example is found in the development of the heart (see Chapter 309 of Handbook of Cell Signaling, Second Edition). Many growth factors function by paracrine mechanisms.

The term *autocrine* refers to entities that are released from a cell and bind to receptors on that same cell, thereby activating it. This sort of self-stimulation occurs in carefully timed phases during normal embryonic development and tissue differentiation. It is also used in inflammation and wound healing. Such localized signals help direct the concentration of appropriate cells at the wound or inflamed tissue. In addition to their importance in normal tissues, these localized signaling systems have been discovered to be quite important in understanding the autonomy achieved by cancer cells. Quite often, one of the contributing mechanisms by which a malignant cell population escapes the normal control mechanisms for regulated growth is by producing autocrine, paracrine, and other localized types of signals that stimulate cell division or other activities that favor survival and expansion of the cancer cell population.

It is interesting to note that historically the underlying concepts inherent in these mechanisms were appreciated, although not at the molecular level, as early as 1775. De Bordeu, and later Brown-Sequard in 1891 [2], proposed that every cell, not just tissues and organs, actively secreted into the circulation substances that influenced other tissues. The focus on the role of endocrine glands (thyroid, pituitary, adrenal, pancreas, etc.) in providing these "internal secretions" obscured the existence of the autocrine and paracrine messengers (as did the emphasis on the circulation) for some time, but the appreciation that all cells can and do actively secrete regulatory elements was an essential concept that was clearly recognized (although alas not considered to any great extent until relatively recently) over 200 years ago.

Both normal and pathological conditions can use the same hormone for autocrine and paracrine interactions. Autocrine regulation of a phase of keratinocyte development by NGF is well established [3], while the same factor generally acts as a paracrine regulator of sympathetic and selected sensory neurons [4]. At the same time, many tumors progress by autocrine stimulation by any one of several mitogenic factors, such EGF and FGF. These same substances also participate in numerous normal tissue situations using paracrine mechanisms (see entries on these factors in earlier parts of this book).

Two additional mechanisms have been proposed that further extend cell signaling beyond the action of circulating messengers. These are juxtacrine signals, in which the signaling entity (receptor ligand) is not soluble but is membrane-bound on one cell, and is delivered by cell–cell physical approximation to the cell bearing the receptor (usually but not necessarily different in type from the target cell) and intracrine signals, in which both receptor and ligand are expressed intracellularly and signals are generated without external stimuli. The former mechanism is now well established and is exemplified by a number of systems, such as the notch receptor [5] and the tyrosine kinase receptor family, Eph [6]. There is less compelling evidence for intracrine mechanisms, although there are a number of growth factors, such as FGF1 and 2 and interleukin 1, that are not exported in the usual manner via the endoplasmic reticulum, and that clearly exhibit both extracellular and intracellular concentrations of factor. Thus, the intracellular material could be appropriate for such signaling.

Throughout the chapters of Part V, the reader will find multiple applications of these concepts as they are used in physiologically relevant systems.

REFERENCES

1. Barolo S, Posakony JW. Three habits of highly effective signaling pathways: principles of transcriptional control by developmental cell signaling. *Genes Dev* 2002;**16**:1167–81.

2. Brown-Sequard CE, D'Absonval A. *Compt rend Soc Boil* 1891. 5s, 248.

3. Di_Marco E, Mathor M, Bondanza S, Cutuli N, Marchisio PC, Cancedda R, De_Luca M. Nerve growth factor binds to normal human keratinocytes through high and low affinity receptors and stimulates their growth by a novel autocrine loop. *J Biol Chem* 1993;**268**:22,838–22,846.

4. Huang EJ, Reichardt LF. Neurotrophins: roles in neuronal development and function. *Annu Rev Neurosci* 2001;**24**:677–736.

5. Weinmaster G. Notch signal transduction: a real rip and more. *Curr Opin Genet Dev* 2000;**10**:363–9.

6. Zisch AH, Pasquale EB. The Eph family: a multitude of receptors that mediate cell recognition signals. *Cell Tissue Res* 1997;**290**:217–26.

Chapter 3

Integrin Signaling: Cell Migration, Proliferation, and Survival

J. Thomas Parsons[1], Jill K. Slack-Davis[1], Robert W. Tilghman[1], Marcin Iwanicki[1] and Karen H. Martin[1,2]

[1]Department of Microbiology, University of Virginia Health System, Charlottesville, Virginia

[2]Department of Neurobiology and Anatomy, West Virginia University, Morgantown West Virginia

INTRODUCTION

Integrins are a family of heterodimeric, transmembrane receptors that mediate attachment of cells to the surrounding extracellular matrix (ECM) [1]. Different combinations of α and β subunits heterodimerize to form receptors with specificity for distinct extracellular ligands [2]. There are 24 identified receptors expressed in diverse tissues and cell types, allowing the selective interaction of different cell types with different ECM ligands [3]. The cytoplasmic tail of the β subunit is both necessary and sufficient to mediate the linkage of integrins to the actin cytoskeleton [4]. Although subunit cytoplasmic tails bind to cytoskeletal proteins [2], the major functional role of the α subunit is to modulate cytoskeletal interactions by directly interacting with the cytoplasmic tail region of the β subunit [5]. Thus, integrins are ligand-dependent sensors of the ECM environment. Integrins represent a unique class of bidirectional membrane receptors in that they are conduits for mechanical and chemical information providing "outside-in signals" (for example, providing a signal in response to binding to a defined ECM proteins during cell migration) as well as providing "inside-out" signals (for example, altering ligand binding activity in response to a cytoplasmic signal) [5]. As such, integrins are responsible for sensing many aspects of the microenvironment, including the structure and composition of the ECM as well as biochemical signals generated following growth factor or cytokine stimulation. Integration of these complex signals contributes to the regulation of cellular migration, growth, and survival within an organism.

INTEGRINS NUCLEATE THE FORMATION OF DYNAMIC MULTI-PROTEIN COMPLEXES

A central function of integrins is to mediate a structural linkage between the dynamic intracellular cytoskeleton and the ECM that conveys both mechanical and chemical signals. More than 150 proteins comprise the "integrin adhesome" [6], although it is more likely that only about 50 of these are a part of the integrin complex. The "adhesome" can be broken down into functional subnets comprised of structural proteins, kinases, and phosphatases that modulate phosphorylation and dephosphorylation; GTPases and their regulatory exchange factors (GEFs) and GTPase activating proteins (GAPs); lipid modifying enzymes; and regulators of proteolytic activity [6]. The dynamic interaction amongst members of these subnets is subject to yet another set of interactions involving adapter proteins and scaffolds. Thus, upon engagement with specific ligands, integrins nucleate a complex array of interactions that together play a central role in regulating dynamic intracellular complexes.

The association of integrin receptors with focal adhesion and actin binding proteins including talin, α-actinin, and vinculin serves to illustrate how integrins nucleate adhesion complexes that are linked to actin filaments and other components in cytoplasm (Figure 3.1). Talin, a major structural component of focal adhesions, binds directly to the tails of β1, β2, and β3 integrins [4]. In addition, talin binds to actin, vinculin, focal adhesion kinase (FAK), and phospholipids. Cells deficient for the expression of talin exhibit significant increases in membrane blebbing, defects

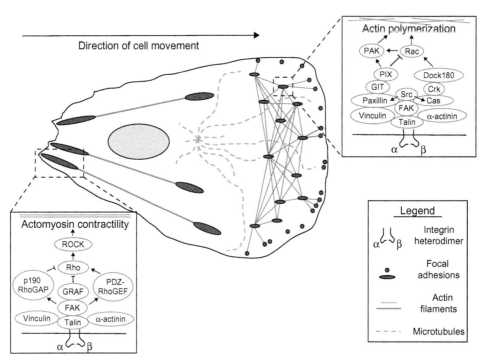

FIGURE 3.1 Integrin signaling pathways involved in the regulation of cell migration.
Integrin signaling events are regulated in a spatial-temporal manner across the cell during cell migration. At the front of the cell, integrin clustering leads to the formation of dynamic adhesions, and the recruitment of scaffold and signaling proteins such as talin, vinculin, α-actinin, FAK, and paxillin. FAK activation leads to the recruitment of Src, paxillin and Cas. Subsequent tyrosine phosphorylation of Cas provides a binding site for Crk leading to the recruitment and activation of Dock180, a GEF for Rac. Paxillin recruits GIT and PIX, both of which regulate Rac stimulating PAK activity (see text for details). At the rear of migrating cells, integrin-dependent regulation of actomyosin contractility leads to retraction of the tail of the cell. Integrins control actomyosin contractility by stimulating FAK, which regulates the activation/inactivation of RhoGTPases via different RhoGEFs and RhoGAPs, ultimately leading to modulation of ROCKs resulting in enhanced contraction (see text for details).

in cell adhesion and spreading, and a failure to assemble focal adhesions and stress fibers [7]. Alpha-actinin is an actin-binding/crosslinking protein that binds the cytoplasmic tails of β1, β2, and β3 integrins [4], as well as several additional focal adhesion proteins, including vinculin and zyxin. Localization of α-actinin to adhesion complexes occurs by a direct interaction with β-integrin cytoplasmic tails. Vinculin is a conformationally regulated protein that is recruited to adhesions upon its activation. Although vinculin does not bind integrins directly [4], once activated it binds several well characterized adhesion-associated proteins, including talin and α-actinin as well as actin and phospholipids. Thus, vinculin functions as a molecular bridge to regulate integrin dynamics and clustering and provides a link to the mechanotransduction machinery [8]. Vinculin-deficient cells exhibit decreased mechanical stiffness and increased cell motility [4, 7].

The assembly and disassembly of integrin complexes is dynamic. The integrin adhesion complexes are observed to either turnover rapidly or stabilize through a mechanism that involves the association of the complexes with the actin cytoskeleton. Fluorescent speckle microscopy of newly formed integrin complexes and actin filaments indicates that the linkage of integrin adhesion complexes to

actin is dynamic [9]. Proteins such as paxillin, zyxin, and FAK do not exhibit movements correlated with actin, and thus appear not linked to actin, whereas the dynamics of α-actinin complexes indicate a strong association with actin filaments. Interestingly, the dynamics of talin and vinculin demonstrate a partial coupling to actin [9]. The differential coupling of adhesion complex components with actin suggests the ordered recruitment and dissociation of integrin complexes with the actin cytoskeleton that is regulated by a molecular "clutch." Assembly and disassembly of adhesions is also regulated by the generation of tension on integrin adhesion complexes by the actomyosin contractility network. Signals generated from newly formed adhesions (see below) act to stimulate localized actomyosin contractility, which is required to promote adhesion maturation and stability [10]. For example, highly motile cells (e.g., macrophages) tend to have small, highly dynamic adhesions that turnover rapidly, whereas the adhesions in slower-moving, more contractile cells (e.g., fibroblasts) stabilize and grow in response to increased tension before being turned over [11, 12].

Integrin-mediated adhesion complexes contain proteins that are directly involved in regulating the formation and turnover of adhesion complexes and the promotion of intracellular signals (Figure 3.1). FAK is a focal-adhesion-associated, non-receptor protein tyrosine kinase. FAK binds *in vitro* to the cytoplasmic tails of β1 and β3 integrins, although to date this interaction has not been demonstrated *in vivo* [13]. FAK is recruited to newly formed adhesions via the C-terminal focal adhesion targeting domain [14]. This same region also directs the formation of a stable complex with the focal adhesion protein paxillin. FAK also associates with Src, talin, the cytoskeletal adaptor protein Cas, and GAPs for Rho (GRAF [15]) and ARF1 (ASAP1 [16]). In addition to binding FAK, the multidomain protein paxillin serves as a scaffold to recruit and organize a number of additional signaling molecules at the sites of adhesion (Figure 3.1). Paxillin binds Src, Crk, vinculin, actopaxin, and the serine/threonine kinase ILK, as well as the ARF GAPs GIT1 and GIT2 [17–19]. In addition, paxillin binds directly to the cytoplasmic tails of α4 integrins [20]. Cells deficient for either FAK or paxillin exhibit defects in cell spreading and cell migration; in the case of paxillin-null cells, tyrosine phosphorylation of FAK and Cas is also decreased [21, 22]. Cas is another adaptor protein that binds to both FAK and Src and serves to recruit additional signaling molecules to focal adhesions. Cas associates with the guanine nucleotide exchange factor C3G, protein phosphatases, and adaptor proteins Crk and Nck [23]. Coupling between FAK, Src, and Cas appears to be important for FAK-stimulated cell migration [24]. In lymphocyte migration, paxillin and Src appear to be important signaling intermediates downstream of the integrin, α4β1 [25].

A number of other proteins and kinases have been classified as integrin-binding or integrin-associated proteins, including (1) adaptor proteins and kinases (e.g., RACK1, Shc, Grb2, and ILK); (2) growth factor receptors (EGF receptor, ErbB2, PDGF receptor-β, insulin receptor, VEGF receptor); (3) cytoplasmic, chaperone, calcium-binding proteins (calnexin, calreticulin, CIB, endonexin); and (4) membrane-associated proteins (tetraspanins, Ig superfamily proteins, GPI-linked receptors, transmembrane proteins, and ion channels) [26]. The functional and structural diversity amongst integrin-associated proteins underscores the importance of integrins as initiators of many intracellular signaling pathways.

CELL MIGRATION: A PARADIGM FOR STUDYING INTEGRIN SIGNALING

Cell migration provides an exceptionally relevant model to study integrin signaling. Migration is a complex cellular process that involves the extension of lamellipodia; adhesion at sites within newly formed lamella, organization of force-generating adhesions, contraction and cell-body displacement, and detachment of the cell rear. These events require the coordination of multiple signaling pathways [27].

LAMELLIPODIA EXTENSION, AND ADHESION FORMATION AND STABILIZATION AT THE LEADING EDGE

The initial steps in cell migration require the formation of protrusive structures (lamellipodia) at the leading edge of the cell, and the stabilization of the protrusion by newly formed adhesion complexes. Cell protrusions are regulated by the activity of surface receptors and Rho family GTPases Cdc42 and Rac [28]. The interaction of Cdc42 and Rac with members of the Wiskott-Aldrich syndrome protein (WASP)/Scar1 superfamily regulates actin polymerization at the front of the cell [29]. Binding of Cdc42/Rac to WASP/Scar proteins activates the Arp2/3 complex [30], triggering its binding to the sides of pre-existing actin filaments and stimulating new filament formation, which results in branched actin networks [31]. The formation of branched actin fosters the growth of actin filaments in the barbed end direction. The growing filaments push the plasma membrane forward in a concerted fashion to generate lamellipodia [31–33].

The lamellipodium is stabilized by the generation of new adhesions. This process involves integrin clustering, formation of integrin complexes containing structural and signaling proteins, and linkage to the actin cytoskeleton. While it is unclear how or if the process of adhesion assembly is regulated, it is clear that the activation of FAK/Src complexes within the adhesion complex appears central to the process of adhesion turnover (breakdown) [11]. Cells deficient for FAK expression form adhesions at the same rate as wild-type cells, but in these cells the adhesions that form fail to turn over, thus compromising the migratory process. Formation of the active FAK/Src signaling complex also involves the recruitment of two adaptor proteins, Cas and paxillin, and the activation of downstream kinases, MAP kinase and PAK (p21-activated kinase) [34]. Targeted deletion of either paxillin or Cas or the inhibition of MAP kinase activity leads to the inhibition of adhesion turnover, affirming the importance of this pathway in regulating adhesion dynamics [11]. Formation of the FAK/Src signaling complex is also important to sustain activation of Rac and integrate signals for adhesion turnover and protrusion [11]. FAK/Src-mediated tyrosine phosphorylation of Cas or paxillin creates binding sites for the adaptor proteins such as Crk and Nck. Cas/Crk complexes mediate Rac activation by binding DOCK180 [35, 36]. Tyrosine phosphorylated paxillin also binds Crk and signals to Rac. However FAK mutants deficient in binding to paxillin efficiently restore migration of FAK null cells to a wild-type level [37]. Thus, in this setting, signaling to Cas appears to be sufficient to mediate adhesion turnover while tyrosine and serine phosphorylation of paxillin appears to be important in the regulation of protrusive activity at the front of the cell [38, 39].

Paxillin is an important regulator of Cdc42 and Rac through its binding to GIT family proteins and the subsequent interaction of GIT with PIX, a Cdc42/Rac

GEF [40–42]. PIX was originally reported to exhibit GEF activity for Rac and Cdc42 [43]. However this property has been questioned, raising speculation that PIX proteins might activate PAK by binding the GTP form of Cdc42/Rac rather than directly activating the GTPases [40]. While the formation of the paxillin/GIT/PIX complex has been reported to activate PAK, the interaction of paxillin with Rac also leads to the downregulation of Rac activity [44], providing a possible mechanism for Rac turnover/downregulation. Figure 3.1 illustrates one possible pathway of regulating protrusion and adhesion dynamics at the leading edge of the cell. Integrin recruitment and activation of FAK/Src leads to binding and phosphorylation of Cas, and activation of Rac via Cas/Crk/Dock180 complexes to promote lamellipodia formation and adhesion turnover. GTP-Rac may then bind to PIX proteins complexed with paxillin/GIT, resulting in PAK activation and Rac downregulation limiting protrusive activity and favoring adhesion maturation.

MATURATION, DETACHMENT, AND RELEASE OF ADHESIONS

Following lamellipodia extension and adhesion formation, cell migration ensues with maturation of adhesions, translocation of the cell body, and release of rear adhesions, all of which depend on actomyosin-dependent contractile forces regulated by Rho (Figure 3.1). Activated Rho regulates Rho kinase (ROCK) I and II, resulting in the phosphorylation of myosin light chains (MLC) [45]. In addition to phosphorylating MLC, ROCK phosphorylation of myosin phosphatase (MYPT) [46] inhibits its phosphatase activity, thereby maintaining MLCs in a highly phosphorylated (contractile) state. The resultant contractile forces are essential for the organization of actin filaments and adhesion complexes [47, 48]. ROCK also phosphorylates and activates LIM kinases, which in turn regulate the actin-depolymerizing protein, cofilin [49], possibly linking activation of ROCK to regulation of protrusive activity. Recently, distinct roles for ROCK I and II have been shown. ROCK I rather than ROCK II appears to be important for stress fiber formation, whereas ROCK II acts to regulate the microfilament bundling at sites of adhesion [50].

During cell migration, signaling by Cdc42/Rac and Rho is regulated in a reciprocal fashion, leading to the breakdown of stress fibers and focal adhesions (due to the downregulation of Rho) and the commensurate reorganization of cortical actin networks at the leading edge of the cells [51–53]. Plating cells on ECM stimulates a transient decrease in Rho activity, which is necessary for cell spreading [54, 55]. FAK contributes to the transient decrease in Rho activity; changes in Rho activity are not observed in cells deficient for FAK expression [55]. The mechanism by which FAK regulates the initial decrease in Rho activity

may involve its interaction with the Rho GTPase-activating protein GRAF [15] or its ability to activate Src, which has been shown to phosphorylate p190RhoGAP, resulting in decreased Rho activity upon integrin engagement [56, 57]. A subsequent increase in Rho activity is necessary to restore contractile forces, leading to strengthening of attachment sites, stress fiber formation, and generation of the forces necessary for continued cell movement [47].

In addition to stabilizing lamellipodia formation at the front of the cell, detachment of adhesions at the rear of the cell requires sustained contraction and disassembly of integrin complexes. Integrin signaling through FAK to Rho appears to be important in this process. The inhibition of Rho in several different cell types leads to the formation of an extended tail, possibly because actomyosin-based contractility in the body of the cell is decreased. In fibroblasts, depletion of FAK, PDZ-RhoGEF, or the Rho effector ROCK II results in formation of long tail extensions and an inhibition of adhesion turnover [58]. These observations are consistent with the requirement for tension on adhesions located at the rear of the cell. Rho may also act in the tail by stabilizing microtubules, which would then promote focal adhesion turnover [59]. Proteolytic cleavage of adhesion proteins may also play a role in rear retraction. Phosphorylation of focal-adhesion-localized calpain by active MAPK [60] is reported to stimulate calpain-mediated cleavage of adhesion proteins and cell detachment [61, 62].

GROWTH FACTOR RECEPTOR AND INTEGRIN SIGNALING-SYNERGISTIC REGULATION OF CELL PROLIFERATION AND SURVIVAL

Ligand activation of growth factor receptors and integrins activates many of the same pathways, leading to the possibility that such pathways synergize to promote cell growth and survival. Integrins activate MAPK through three different Ras-dependent pathways. Integrin-mediated activation of FAK and recruitment of Src results in phosphorylation of FAK on Tyr925 [63, 64]. Phosphorylation on Tyr925 creates a binding site for Grb2 [63], an SH2/SH3 adaptor protein that links growth factor receptor tyrosine kinases to the Ras/MEK/MAPK pathway through the Ras guanosine diphosphate (GDP)/guanosine triphosphate (GTP) exchange protein SOS (Figure 3.2). Integrins also activate SOS through caveolin-1-mediated recruitment of Shc to integrins and subsequent phosphorylation by Fyn [65] (Figure 3.1). Finally, integrin engagement results in phosphorylation and activation of the epidermal growth factor (EGF) receptor in the absence of EGF stimulation [66]. Activated EGF receptor recruits Shc to the receptor, where phosphorylation creates a binding site for Grb2/SOS [66] (Figure 3.2). Indeed, in this setting, ECM-mediated phosphorylation of Shc and activation of MAPK is blocked by

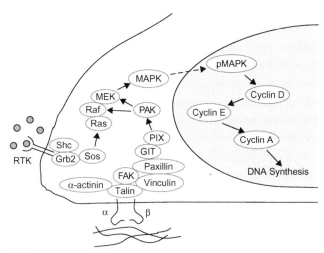

FIGURE 3.2 Integrin signaling cooperates with receptor tyrosine kinase (RTK) pathways to regulate cell proliferation.
Upon activation by growth factors, RTKs recruit adapter proteins including Shc and Grb2 and the Ras GEF Sos leading to the activation of Ras. Active Ras recruits Raf to the membrane for activation leading to MEK and MAP kinase activation, and ultimately activation of cyclin D. Integrin signaling through the paxillin/GIT/PIX pathway stimulates PAK activity, which phosphorylates Raf and MEK and stabilizes the Raf/MEK complex. This results in a prolonged activation of MAPK and the cyclins, leading to an increase in DNA synthesis (see text for details).

inhibitors of EGF receptor tyrosine kinase activity. Integrin-stimulated migration is inhibited by MAPK inhibitors and stimulated by expression of active MEK [67]. Interestingly, dominant-negative Ras expression has little effect on ECM-stimulated migration [68, 69], indicating the existence of Ras-independent mechanisms of MAPK activation.

Several studies show that Rac synergizes with Raf to stimulate MAPK-dependent migration in response to EGF [70]. Rac-dependent activation of PAK stimulates phosphorylation of MEK, resulting in an increased affinity of MEK for Raf [71]. An important consequence of Rac activation may be the enhancement of Raf–MEK interaction, leading to maximum MAPK activity in a setting of only basal Ras and Raf stimulation. This may provide a mechanism by which integrins potentiate signals from growth factor receptors, allowing cells to respond to low levels (gradients) of chemotactic signals in the environment.

INTEGRIN SIGNALS AND LINKS TO CANCER

The ECM plays a critical role in the altered growth and metastatic behavior of cancer cells. In the case of both normal and cancer cells, these signals contribute to the balance of cell growth and death by regulating the apoptotic machinery of the cell. Under appropriate circumstances, integrin signals regulate G_0-to-G_1 and G_1-to-S progression [72–75], as well as the expression of growth-related gene

products associated with these transition states. Transient MAPK activation stimulated by either growth factors or cell adhesion is sufficient to initiate G_0-to-G_1 phase transition and the coincident expression of immediate–early response genes, including c-Fos, c-Myc, and c-Jun [76–79]. Serum stimulation in the absence of adhesion to the ECM abrogates progression through G_1 into S phase by increasing the accumulation of cdk2 inhibitors $p21^{cip1}$ and $p27^{kip1}$ [80]. Cyclin D1 functions to promote G_1 progression by sequestering $p21^{cip1}$ and $p27^{kip1}$ [81]. Cyclin D1 expression requires cell adhesion to mediate sustained MAPK activity initiated by growth factors [82]. Growth factor stimulation of cells held in suspension results in a modest, transient activation of MAPK compared to the robust MAPK stimulation following serum treatment of adherent cells [83]. Maximal, sustained MAPK activation requires FAK and Rho activity [84, 85]. Indeed, FAK and Rho have both been implicated in regulating cyclin D1 expression and progression through G_1 [75, 85]. Cyclin-D-dependent downregulation of $p21^{cip1}$ and $p27^{kip1}$ is necessary for cyclin E/Cdk2 activity, which induces the expression of cyclin A [80], a key regulator of S-phase progression [86, 87]. Therefore, cell–substrate adhesion indirectly promotes cyclin A expression and S-phase progression by stimulating cyclin E/Cdk2 activity. Collectively, these data indicate that the G_0-to-G_1 transition and progression through G_1 of the cell cycle is regulated by either serum or adhesion; however, progression through S phase requires both serum and adhesion.

Cell proliferation is in dynamic balance with cell death. Shifts in this equilibrium, as a result of increasing cell proliferation or decreasing cell death, often result in tumorigenesis. Integrins provide key signals to regulate this balance. Depriving epithelial or endothelial cells of contact with the ECM rapidly induces apoptosis [88–90]. (This specialized form of cell death is referred to as *anoikis*). Normal epithelial cells acquire resistance to anoikis upon expression of certain oncogenes [91–93]. A number of studies have implicated integrin signaling to the phosphoinositide-3'-kinase (PI3K)–AKT pathway as a central regulator of anoikis [92]. FAK is thought to regulate anoikis by direct activation of PI3K and AKT, and perhaps indirectly via interactions with Cas/Crk/DOCK180/Rac [89]. Integrin-linked kinase (ILK) has also been implicated in signaling to AKT, although this pathway is poorly understood. However, ILK binds to the cytoplasmic tails of $\beta 1$ and $\beta 3$ integrin, and overexpression of ILK results in activation of AKT [94, 95].

In cancer, increasing evidence indicates that integrins synergize with growth factor receptor signals to promote cell proliferation and to stimulate the migration of tumor cells from the primary site, and function to promote growth and survival at distant metastatic sites [96, 97]. Integrins play a major role in remodeling the tumor microenvironment, and are important regulators of migration and metastatic growth [98]. In a variety of cancers, tumor progression and acquisition of metastatic behavior is often accompanied by changes in integrin expression and upregulation of expression of integrin signaling proteins, such as FAK. Also of note is that many cancers exhibit loss of the tumor suppressor gene PTEN, which leads to the upregulation of AKT and the suppression of anoikis [99]. In the case of human breast cancer cells, blocking β1 integrin binding activity results in attenuation of EGF receptor signaling and cell cycle progression [100]. In mouse models of breast cancer, the targeted deletion of β1 or β4 integrin or the deletion of FAK or c-Src inhibits tumor progression [101]. Many studies have shown a role for integrins in the metastatic spread of cancers. Metastasis of human breast cancer cells to the lung in a mouse xenograph model is blocked by systemic administration of an inhibitory anti-β1 integrin antibody [102]. In another study, delivery of a peptide designed to block the α5β1 and αvβ3 integrin receptors impaired the growth and metastasis of invasive human breast cancer cells in a mouse xenograph model [103]. Currently, inhibitors of integrin signaling are being characterized in preclinical and clinical settings. PF-573,228, a prototype ATP competitive inhibitor of FAK, inhibits FAK autophosphorylation in normal fibroblasts and human tumor cells, and inhibits cell migration by blocking adhesion turnover [104]. In preclinical animal models, the bioavailable compound, PF-562,271, inhibited FAK phosphorylation *in vivo* in a dose-dependent fashion in several human subcutaneous xenograft models, including: prostate, breast, pancreatic, colon, glioblastoma, and lung [105]. The FAK inhibitor PF-562,271 is currently being evaluated in Phase I clinical trials [106]. It is likely that additional molecules targeting components of the integrin signaling pathway will be evaluated for clinical efficacy in the near future.

CONCLUDING REMARKS

The recognition that integrins are not only adhesive receptors but also "integrators" of growth factor and ECM signaling has had a profound impact on our understanding of cellular processes, including cell migration, differentiation, and cancer. The challenge for the next decade is to understand how different ECM proteins, growth factors, and chemotactic molecules function in coordinating multiple signaling pathways in the context of individual tissues and the organism itself.

ACKNOWLEDGEMENTS

The authors acknowledge support from the NIH-NCI, CA40042, CA29243, and CA80606 to JTP.

REFERENCES

1. Hynes RO. Integrins: bidirectional, allosteric signaling machines. *Cell* 2002;**110**:673–87.
2. Liu S, Calderwood DA, Ginsberg MH. Integrin cytoplasmic domain-binding proteins. *J Cell Sci* 2000;**113**:3563–71.
3. Luo BH, Carman CV, Springer TA. Structural basis of integrin regulation and signaling. *Annu Rev Immunol* 2007;**25**:619–47.
4. Calderwood DA, Shattil SJ, Ginsberg MH. Integrins and actin filaments: reciprocal regulation of cell adhesion and signaling. *J Biol Chem* 2000;**275**:22,607–22,610.
5. Ginsberg MH, Partridge A, Shattil SJ. Integrin regulation. *Curr Opin Cell Biol* 2005;**17**:509–16.
6. Zaidel-Bar R, Itzkovitz S, Ma'ayan A, Iyengar R, Geiger B. Functional atlas of the integrin adhesome. *Nat Cell Biol* 2007;**9**:858–67.
7. Priddle H, Hemmings L, Monkley S, Woods A, Patel B, Sutton D, Dunn GA, Zicha D, Critchley DR. Disruption of the talin gene compromises focal adhesion assembly in undifferentiated but not differentiated embryonic stem cells. *J Cell Biol* 1998;**142**:1121–33.
8. Humphries JD, Wang P, Streuli C, Geiger B, Humphries MJ, Ballestrem C. Vinculin controls focal adhesion formation by direct interactions with talin and actin. *J Cell Biol* 2007;**179**:1043–57.
9. Hu K, Ji L, Applegate KT, Danuser G, Waterman-Storer CM. Differential transmission of actin motion within focal adhesions. *Science* 2007;**315**:111–15.
10. Bershadsky A, Kozlov M, Geiger B. Adhesion-mediated mechanosensitivity: a time to experiment, and a time to theorize. *Curr Opin Cell Biol* 2006;**18**:472–81.
11. Webb DJ, Donais K, Whitmore LA, Thomas SM, Turner CE, Parsons JT, Horwitz AF. FAK-Src signalling through paxillin, ERK and MLCK regulates adhesion disassembly. *Nat Cell Biol* 2004;**6**(2):154–61.
12. Webb DJ, Parsons JT, Horwitz AF. Adhesion assembly, disassembly and turnover in migrating cells-over and over and over again. *Nat Cell Biol* 2002;**4**:E97–E100.
13. Schaller MD, Otey CA, Hildebrand JD, Parsons JT. Focal adhesion kinase and paxillin bind to peptides mimicking beta integrin cytoplasmic domains. *J Cell Biol* 1995;**130**:1181–7.
14. Hildebrand JD, Schaller MD, Parsons JT. Identification of sequences required for the efficient localization of the focal adhesion kinase, pp125FAK, to cellular focal adhesions. *J Cell Biol* 1993;**123**:993–1005.
15. Hildebrand JD, Taylor JM, Parsons JT. An SH3 domain-containing GTPase-activating protein for Rho and Cdc42 associates with focal adhesion kinase. *Mol Cell Biol* 1996;**16**:3169–78.
16. Liu Y, Loijens JC, Martin KH, Karginov AV, Parsons JT. The association of ASAP1, an ADP ribosylation factor-GTPase activating protein, with focal adhesion kinase contributes to the process of focal adhesion assembly. *Mol Biol Cell* 2002;**13**:2147–56.
17. Hoefen RJ, Berk BC. The multifunctional GIT family of proteins. *J Cell Sci* 2006;**119**:1469–75.
18. Schaller MD. Paxillin: a focal adhesion-associated adaptor protein. *Oncogene* 2001;**20**:6459–72.
19. Turner CE. Paxillin interactions. *J Cell Sci* 2000;**113**:4139–40.
20. Liu S, Thomas SM, Woodside DG, Rose DM, Kiosses WB, Pfaff M, Ginsberg MH. Binding of paxillin to alpha4 integrins modifies integrin-dependent biological responses. *Nature* 1999;**402**:676–81.
21. Hagel M, George EL, Kim A, Tamimi R, Opitz SL, Turner CE, Imamoto A, Thomas SM. The adaptor protein paxillin is essential for normal development in the mouse and is a critical transducer of fibronectin signaling. *Mol Cell Biol* 2002;**22**:901–15.

22. Wade R, Bohl J, Vande Pol S. Paxillin null embryonic stem cells are impaired in cell spreading and tyrosine phosphorylation of focal adhesion kinase. *Oncogene* 2002;**21**:96–107.

23. O'Neill GM, Fashena SJ, Golemis EA. Integrin signalling: a new Cas(t) of characters enters the stage. *Trends Cell Biol* 2000;**10**:111–19.

24. Cary LA, Han DC, Polte TR, Hanks SK, Guan JL. Identification of p130Cas as a mediator of focal adhesion kinase- promoted cell migration. *J Cell Biol* 1998;**140**:211–21.

25. Rose DM, Alon R, Ginsberg MH. Integrin modulation and signaling in leukocyte adhesion and migration. *Immunol Rev* 2007;**218**:126–34.

26. van der Flier A, Sonnenberg A. Function and interactions of integrins. *Cell Tissue Res* 2001;**305**:285–98.

27. Ridley AJ, Schwartz MA, Burridge K, Firtel RA, Ginsberg MH, Borisy G, Parsons JT, Horwitz AR. Cell migration: integrating signals from front to back. *Science* 2003;**302**:1704–9.

28. Hall A. Rho GTPases and the actin cytoskeleton. *Science* 1998;**279**:509–14.

29. Mullins RD. How WASP-family proteins and the Arp2/3 complex convert intracellular signals into cytoskeletal structures. *Curr Opin Cell Biol* 2000;**12**:91–6.

30. Machesky LM, Gould KL. The Arp2/3 complex: a multifunctional actin organizer. *Curr Opin Cell Biol* 1999;**11**:117–21.

31. Mullins RD, Heuser JA, Pollard TD. The interaction of Arp2/3 complex with actin: nucleation, high affinity pointed end capping, and formation of branching networks of filaments. *Proc Natl Acad Sci USA* 1998;**95**:6181–6.

32. Machesky LM, Mullins RD, Higgs HN, Kaiser DA, Blanchoin L, May RC, Hall ME, Pollard TD. Scar, a WASp-related protein, activates nucleation of actin filaments by the Arp2/3 complex. *Proc Natl Acad Sci USA* 1999;**96**:3739–44.

33. Pollard TD, Borisy GG. Cellular motility driven by assembly and disassembly of actin filaments. *Cell* 2003;**112**:453–65.

34. Slack-Davis JK, Eblen ST, Zecevic M, Boerner SA, Tarcsafalvi A, Diaz HB, Marshall MS, Weber MJ, Parsons JT, Catling AD. PAK1 phosphorylation of MEK1 regulates fibronectin-stimulated MAPK activation. *J Cell Biol* 2003;**162**:281–91.

35. Kiyokawa E, Hashimoto Y, Kobayashi S, Sugimura H, Kurata T, Matsuda M. Activation of Rac1 by a Crk SH3-binding protein, DOCK180. *Genes Dev* 1998;**12**:3331–6.

36. Kiyokawa E, Hashimoto Y, Kurata T, Sugimura H, Matsuda M. Evidence that DOCK180 up-regulates signals from the CrkII-p130(Cas) complex. *J Biol Chem* 1998;**273**:24,479–24,484.

37. Sieg DJ, Hauck CR, Schlaepfer DD. Required role of focal adhesion kinase (FAK) for integrin-stimulated cell migration. *J Cell Sci* 1999;**112**:2677–91.

38. Nayal A, Webb DJ, Brown CM, Schaefer EM, Vicente-Manzanares M, Horwitz AR. Paxillin phosphorylation at Ser273 localizes a GIT1–PIX–PAK complex and regulates adhesion and protrusion dynamics. *J Cell Biol* 2006;**173**:587–9.

39. Zaidel-Bar R, Milo R, Kam Z, Geiger B. A paxillin tyrosine phosphorylation switch regulates the assembly and form of cell-matrix adhesions. *J Cell Sci* 2007;**120**:137–48.

40. Bagrodia S, Bailey D, Lenard Z, Hart M, Guan JL, Premont RT, Taylor SJ, Cerione RA. A tyrosine-phosphorylated protein that binds to an important regulatory region on the cool family of p21-activated kinase-binding proteins. *J Biol Chem* 1999;**274**:22,393–22,400.

41. Bagrodia S, Cerione RA. Pak to the future [see comments]. *Trends Cell Biol* 1999;**9**:350–5.

42. Turner CE, Brown MC, Perrotta JA, Riedy MC, Nikolopoulos SN, McDonald AR, Bagrodia S, Thomas S, Leventhal PS. Paxillin LD4 motif binds PAK and PIX through a novel 95-kD ankyrin repeat, ARF-GAP protein: a role in cytoskeletal remodeling. *J Cell Biol* 1999;**145**:851–63.

43. Manser E, Loo TH, Koh CG, Zhao ZS, Chen XQ, Tan L, Tan I, Leung T, Lim L. PAK kinases are directly coupled to the PIX family of nucleotide exchange factors. *Mol Cell* 1998;**1**:183–92.

44. West KA, Zhang H, Brown MC, Nikolopoulos SN, Riedy MC, Horwitz AF, Turner CE. The LD4 motif of paxillin regulates cell spreading and motility through an interaction with paxillin kinase linker (PKL). *J Cell Biol* 2001;**154**:161–76.

45. Kureishi Y, Kobayashi S, Amano M, Kimura K, Kanaide H, Nakano T, Kaibuchi K, Ito M. Rho-associated kinase directly induces smooth muscle contraction through myosin light chain phosphorylation. *J Biol Chem* 1997;**272**:12,257–12,260.

46. Kawano Y, Fukata Y, Oshiro N, Amano M, Nakamura T, Ito M, Matsumura F, Inagaki M, Kaibuchi K. Phosphorylation of myosin-binding subunit (MBS) of myosin phosphatase by Rho-kinase *in vivo*. *J Cell Biol* 1999;**147**:1023–38.

47. Burridge K, Chrzanowska-Wodnicka M. Focal adhesions, contractility, and signaling. *Annu Rev Cell Dev Biol* 1996;**12**:463–518.

48. Vicente-Manzanares M, Zareno J, Whitmore L, Choi CK, Horwitz AF. Regulation of protrusion, adhesion dynamics, and polarity by myosins IIA and IIB in migrating cells. *J Cell Biol* 2007;**176**:573–80.

49. Maekawa M, Ishizaki T, Boku S, Watanabe N, Fujita A, Iwamatsu A, Obinata T, Ohashi K, Mizuno K, Narumiya S. Signaling from Rho to the actin cytoskeleton through protein kinases ROCK and LIM-kinase. *Science* 1999;**285**:895–8.

50. Yoneda A, Multhaupt HA, Couchman JR. The Rho kinases I and II regulate different aspects of myosin II activity. *J Cell Biol* 2005;**170**:443–53.

51. Burridge K. Crosstalk between Rac and Rho [comment]. *Science* 1999;**283**:2028–9.

52. Horwitz AR, Parsons JT. Cell migration – movin' on. *Science* 1999;**286**:1102–3.

53. Rottner K, Hall A, Small JV. Interplay between Rac and Rho in the control of substrate contact dynamics. *Curr Biol* 1999;**9**:640–8.

54. Ren XD, Kiosses WB, Schwartz MA. Regulation of the small GTP-binding protein Rho by cell adhesion and the cytoskeleton. *EMBO J* 1999;**18**:578–85.

55. Ren XD, Kiosses WB, Sieg DJ, Otey CA, Schlaepfer DD, Schwartz MA. Focal adhesion kinase suppresses Rho activity to promote focal adhesion turnover. *J Cell Sci* 2000;**113**(Pt 20):3673–8.

56. Arthur WT, Burridge K. RhoA inactivation by p190RhoGAP regulates cell spreading and migration by promoting membrane protrusion and polarity. *Mol Biol Cell* 2001;**12**:2711–20.

57. Arthur WT, Petch LA, Burridge K. Integrin engagement suppresses RhoA activity via a c-src-dependent mechanism. *Curr Biol* 2000;**10**:719–22.

58. Iwanicki MP, Vomastek T, Tilghman RW, Martin KH, Banerjee J, Wedegaertner PB, Parsons JT. FAK, PDZ-RhoGEF and ROCKII cooperate to regulate adhesion movement and trailing-edge retraction in fibroblasts. *J Cell Sci* 2008;**121**:895–905.

59. Small JV, Kaverina I. Microtubules meet substrate adhesions to arrange cell polarity. *Curr Opin Cell Biol* 2003;**15**:40–7.

60. Fincham VJ, James M, Frame MC, Winder SJ. Active ERK/MAP kinase is targeted to newly forming cell-matrix adhesions by integrin engagement and v-Src. *Embo J* 2000;**19**:2911–23.

61. Glading A, Bodnar RJ, Reynolds IJ, Shiraha H, Satish L, Potter DA, Blair HC, Wells A. Epidermal growth factor activates m-calpain (calpain II), at least in part, by extracellular signal-regulated kinase-mediated phosphorylation. *Mol Cell Biol* 2004;**24**:2499–512.

62. Glading A, Chang P, Lauffenburger DA, Wells A. Epidermal growth factor receptor activation of calpain is required for fibroblast motility and occurs via an ERK/MAP kinase signaling pathway. *J Biol Chem* 2000;**275**:2390–8.

63. Schlaepfer DD, Hanks SK, Hunter T, van der Geer P. Integrin-mediated signal transduction linked to Ras pathway by GRB2 binding to focal adhesion kinase. *Nature* 1994;**372**:786–91.

64. Schlaepfer DD, Hunter T. Evidence for in vivo phosphorylation of the Grb2 SH2-domain binding site on focal adhesion kinase by Src-family protein-tyrosine kinases [published erratum appears in Mol Cell Biol 1996, 16(12):7182–4]. *Mol Cell Biol* 1996;**16**:5623–33.

65. Wary KK, Mariotti A, Zurzolo C, Giancotti FG. A requirement for caveolin-1 and associated kinase Fyn in integrin signaling and anchorage-dependent cell growth.. *Cell* 1998;**4**:625–34.

66. Moro L, Venturino M, Bozzo C, Silengo L, Altruda F, Beguinot L, Tarone G, Defilippi P. Integrins induce activation of EGF receptor: role in MAP kinase induction and adhesion-dependent cell survival. *EMBO J* 1998;**17**:6622–32.

67. Klemke RL, Cai S, Giannini AL, Gallagher PJ, de Lanerolle P, Cheresh DA. Regulation of cell motility by mitogen-activated protein kinase. *J Cell Biol* 1997;**137**:481–92.

68. Kundra V, Anand-Apte B, Feig LA, Zetter BR. The chemotactic response to PDGF-BB: evidence of a role for Ras. *J Cell Biol* 1995;**130**:725–31.

69. Slack JK, Catling AD, Eblen ST, Weber MJ, Parsons JT. c-Raf-mediated inhibition of epidermal growth factor-stimulated cell migration. *J Biol Chem* 1999;**274**:27,177–27,184.

70. Leng J, Klemke RL, Reddy AC, Cheresh DA. Potentiation of cell migration by adhesion-dependent cooperative signals from the GTPase Rac and Raf kinase. *J Biol Chem* 1999;**274**:37,855–37,861.

71. Frost JA, Steen H, Shapiro P, Lewis T, Ahn N, Shaw PE, Cobb MH. Cross-cascade activation of ERKs and ternary complex factors by Rho family proteins. *EMBO J* 1997;**16**:6426–38.

72. Guadagno TM, Ohtsubo M, Roberts JM, Assoian RK. A link between cyclin A expression and adhesion-dependent cell cycle progression. *Science* 1993;**262**:1572–5.

73. Oktay M, Wary KK, Dans M, Birge RB, Giancotti FG. Integrin-mediated activation of focal adhesion kinase is required for signaling to Jun NH2-terminal kinase and progression through the G1 phase of the cell cycle. *J Cell Biol* 1999;**145**:1461–9.

74. Schwartz MA, Ingber DE. Integrating with integrins. *Mol Biol Cell* 1994;**5**:389–93.

75. Zhao JH, Reiske H, Guan JL. Regulation of the cell cycle by focal adhesion kinase. *J Cell Biol* 1998;**143**:1997–2008.

76. Benaud CM, Dickson RB. Regulation of the expression of c-Myc by β1 integrins in epithelial cells. *Oncogene* 2001;**20**:759–68.

77. Dike LE, Farmer SR. Cell adhesion induces expression of growth-associated genes in suspension-arrested fibroblasts. *Proc Natl Acad Sci USA* 1988;**85**:6792–6.

78. McNamee HP, Ingber DE, Schwartz MA. Adhesion to fibronectin stimulates inositol lipid synthesis and enhances PDGF-induced inositol lipid breakdown. *J Cell Biol* 1993;**121**:673–8.

79. Schwartz MA, Lechene C, Ingber DE. Insoluble fibronectin activates the Na/H antiporter by clustering and immobilizing integrin alpha 5 beta 1, independent of cell shape. *Proc Natl Acad Sci USA* 1991;**88**:7849–53.

80. Fang F, Orend G, Watanabe N, Hunter T, Ruoslahti E. Dependence of cyclin E-CDK2 kinase activity on cell anchorage. *Science* 1996;**271**:499–502.

81. Assoian RK, Schwartz MA. Coordinate signaling by integrins and receptor tyrosine kinases in the regulation of G1 phase cell-cycle progression. *Curr Opin Genet Dev* 2001;**11**:48–53.

82. Roovers K, Davey G, Zhu X, Bottazzi ME, Assoian RK. α5β1 integrin controls cyclin D1 expression by sustaining mitogen-activated protein kinase activity in growth factor-treated cells. *Mol Biol Cell* 1999;**10**:3197–204.

83. Renshaw MW, Ren XD, Schwartz MA. Growth factor activation of MAP kinase requires cell adhesion. *EMBO J* 1997;**16**:5592–9.

84. Renshaw MW, Price LS, Schwartz MA. Focal adhesion kinase mediates the integrin signaling requirement for growth factor activation of MAP kinase. *J Cell Biol* 1999;**147**:611–18.

85. Welsh CF, Roovers K, Villanueva J, Liu Y, Schwartz MA, Assoian RK. Timing of cyclin D1 expression within G1 phase is controlled by Rho. *Nat Cell Biol* 2001;**3**:950–7.

86. Girard F, Strausfeld U, Fernandez A, Lamb NJ. Cyclin A is required for the onset of DNA replication in mammalian fibroblasts. *Cell* 1991;**67**:1169–79.

87. Pagano M, Pepperkok R, Verde F, Ansorge W, Draetta G. Cyclin A is required at two points in the human cell cycle. *EMBO J* 1992;**11**:961–71.

88. Frisch SM, Francis H. Disruption of epithelial cell-matrix interactions induces apoptosis. *J Cell Biol* 1994;**124**:619–26.

89. Frisch SM, Ruoslahti E. Integrins and anoikis. *Curr Opin Cell Biol* 1997;**9**:701–6.

90. Paez J, Sellers WR. PI3K/PTEN/AKT pathway. A critical mediator of oncogenic signaling. *Cancer Treat Res* 2003;**115**:145–67.

91. Coniglio SJ, Jou TS, Symons M. Rac1 protects epithelial cells against anoikis. *J Biol Chem* 2001;**276**:28,113–28,120.

92. Khwaja A, Rodriguez-Viciana P, Wennstrom S, Warne PH, Downward J. Matrix adhesion and Ras transformation both activate a phosphoinositide 3-OH kinase and protein kinase B/Akt cellular survival pathway. *EMBO J* 1997;**16**:2783–93.

93. McFall A, Ulku A, Lambert QT, Kusa A, Rogers-Graham K, Der CJ. Oncogenic Ras blocks anoikis by activation of a novel effector pathway independent of phosphatidylinositol 3-kinase. *Mol Cell Biol* 2001;**21**:5488–99.

94. Hanahan D, Weinberg RA. The hallmarks of cancer. *Cell* 2000;**100**:57–70.

95. Lynch DK, Ellis CA, Edwards PA, Hiles ID. Integrin-linked kinase regulates phosphorylation of serine 473 of protein kinase B by an indirect mechanism. *Oncogene* 1999;**18**:8024–32.

96. Hehlgans S, Haase M, Cordes N. Signalling via integrins: implications for cell survival and anticancer strategies. *Biochim Biophys Acta* 2007;**1775**:163–80.

97. Howe A, Aplin AE, Alahari SK, Juliano RL. Integrin signaling and cell growth control. *Curr Opin Cell Biol* 1998;**10**:220–31.

98. Parsons JT, Slack-Davis J, Tilghman R, Roberts WG. Focal adhesion kinase: targeting adhesion signaling pathways for therapeutic intervention. *Clin Cancer Res* 2008;**14**:627–32.

99. Yamada KM, Araki M. Tumor suppressor PTEN: modulator of cell signaling, growth, migration and apoptosis. *J Cell Sci* 2001;**114**:2375–82.

100. Bill HM, Knudsen B, Moores SL, Muthuswamy SK, Rao VR, Brugge JS, Miranti CK. Epidermal growth factor receptor-dependent regulation of integrin-mediated signaling and cell cycle entry in epithelial cells. *Mol Cell Biol* 2004;**24**:8586–99.

101. White DE, Muller WJ. Multifaceted roles of integrins in breast cancer metastasis. *J Mammary Gland Biol Neoplasia* 2007;**12**:135–42.

102. Elliott BE, Ekblom P, Pross H, Niemann A, Rubin K. Anti-beta 1 integrin IgG inhibits pulmonary macrometastasis and the size of micrometastases from a murine mammary carcinoma. *Cell Adhes Commun* 1994;**1**:319–32.

103. Khalili P, Arakelian A, Chen G, Plunkett ML, Beck I, Parry GC, Donate F, Shaw DE, Mazar AP, Rabbani SA. A non-RGD-based integrin binding peptide (ATN-161) blocks breast cancer growth and metastasis *in vivo. Mol Cancer Ther* 2006;**5**:2271–80.

104. Slack-Davis JK, Martin KH, Tilghman RW, Iwanicki M, Ung EJ, Autry C, Luzzio MJ, Cooper B, Kath JC, Roberts WG, Parsons JT. Cellular characterization of a novel focal adhesion kinase inhibitor. *J Biol Chem* 2007;**282**:14,845–14,852.

105. Roberts WG, Ung E, Whalen P, Cooper B, Hulford C, Autry C, Richter D, Emerson E, Lin J, Kath J, Coleman K, Yao L, Martinez-Alsina L, Lorenzen M, Berliner M, Luzzio M, Patel N, Schmitt E, LaGreca S, Jani J, Wessel M, Marr E, Griffor M, Vajdos F. Antitumor activity and pharmacology of a selective focal adhesion kinase inhibitor, PF-562,271. *Cancer Res* 2008;**68**:1935–44.

106. Siu LL, Burris HA, Mileshkin L, Camidge DR, Rischin D, Chen EX, Jones S, Yin D, Fingert H. ASCO annual meeting proceedings part I. *J Clin Oncol* 2007;**25**(18S):3527.

Chapter 4

The Focal Adhesion: A Network of Molecular Interactions

Jianxin A. Yu, Nicholas O. Deakin and Christopher E. Turner

Department of Cell and Developmental Biology, State University of New York, Upstate Medical University, Syracuse, New York

INTRODUCTION

Dynamic changes in cell interaction or adhesion to the surrounding extracellular matrix (ECM) is of fundamental importance for numerous physiologic processes, including cell proliferation and migration, tissue remodeling and differentiation during embryonic development, immune surveillance, and wound healing. It also underlies the progression of a broad spectrum of pathologies, including cancer cell metastasis, atherosclerosis, and cardiac hypertrophy, as well as fibrotic and neurodegenerative disorders [1]. Much of our current understanding of the molecular complexity, regulation and function of cell adhesion sites, or focal adhesions, has come from the analysis of fibroblast adhesion and migration in simple two-dimensional cell culture systems. These studies, carried out over the past 37 years, have identified upwards of 150 focal adhesion protein components [2]. Detailed structure–function analysis of individual proteins has revealed a complex network of protein–protein interactions utilizing a wide spectrum of protein- and lipid-binding domains or motifs. Precise spatial and temporal control of these interactions, and thus the composition of focal adhesions, serves to stabilize and regulate the structural links between the ECM and the cell's internal cytoskeleton to control cell shape and motility. They are also important for coordinating the long-term cellular response to changes in the external environment to regulate cell survival, proliferation, and apoptosis [3]. In this brief review we will highlight some of the unique ways that protein–protein interactions are modulated within focal adhesions to control the cell's phenotypic readout.

INTEGRIN ACTIVATION

The integrin family of heterodimeric transmembrane receptors primarily mediates cell adhesion to the ECM. To date, 24 individual integrin heterodimers have been identified in the human genome, each comprising an alpha and beta subunit [4, 5]. Each subunit consists of a short cytoplasmic domain, a transmembrane domain, and a large extracellular domain. Through its complement of cell surface expressed integrins, a cell is able to adhere to and respond to a wide variety of extracellular molecules, including fibronectin, collagens, and laminins [6]. Integrins are able to transduce signals from the ECM to the cell interior (outside-in signaling) as well as from the cell to the surrounding environment (inside-out signaling) through a complex series of conformational alterations including alpha-beta cytoplasmic domain spatial separation and extracellular domain unbending, as indicated by fluorescence resonance energy transfer (FRET), nuclear magnetic resonance (NMR), X-ray crystallography, and electron microscopy [7–10]. Integrin activation is a prerequisite for focal adhesion formation and subsequent cell migration, and has been shown to be mediated by the direct interaction of integrin cytoplasmic domains with numerous proteins, including talin-1/2, kindlin-2/3, alpha-actinin, and ICAP-1 [11–14].

ADHESION STRENGTHENING

Talin is able to activate integrins through direct interaction with the highly conserved membrane proximal Asn–Pro–x–Tyr (NPXY) motif of integrin beta subunit cytoplasmic domains, through its globular N-terminal phosphotyrosine-binding (PTB) FERM (band 4.1, ezrin, radixin, and moesin) domain [15–18]. As well as interacting with the beta integrin cytoplasmic domains, talin also binds vinculin [19] and F-actin [20], and is believed to primarily function to reinforce integrin-mediated adhesions to enable linkage to the actomyosin contractile machinery and subsequent cell spreading and migration [21]. The talin–vinculin interaction is one of the best-studied molecular associations in focal adhesions. Structural studies of talin suggest that only one of the potential vinculin binding sites (VBS) is exposed on the surface of the talin molecule [19, 22]. The other VBS domains are buried in the series of amphipathic helical bundles [23] comprising the talin rod domain, suggesting a conformational rearrangement must occur to enable increased talin–vinculin interaction. Cells devoid of talin are unable to form mature focal adhesions [21, 24], whereas cells devoid of vinculin display smaller and weaker adhesions, more sensitive to disruption by applied force [25, 26], thereby suggesting that vinculin acts to stabilize and strengthen focal adhesions. Indeed, it has been suggested that as actomyosin-driven contraction increases, conformational alterations occur in the talin rod, thereby unmasking increasing numbers of VBSs and in so doing strengthen the focal adhesion link to the cytoskeleton [27] and consequently negatively regulating focal adhesion dynamics [28]. Importantly, additional factors, such as interaction with plasma membrane phospholipids including phosphatidylinositol (4,5)-bisphosphate (PtdIns(4,5)P_2), further contribute to vinculin and talin activation and the regulation of their actin binding activity to stabilize focal adhesions. Indeed, phospholipids provide membrane-docking sites for a number of focal adhesion proteins, including alpha-actinin, WASp, ERM proteins, and Calpain, to control both their spatial distribution and activation states [29]. Vinculin, through recruitment of the Arp2/3 complex to focal adhesions, may also play a role in coordinating actin filament branching within the extending lamellipodium of migrating cells [30].

Stabilization of cell adhesion to the ECM and coupling to the actin cytoskeleton permits the generation of tension within the cell, resulting in the conversion of mechanical force into the initiation of intracellular signaling. Such mechanotransduction or mechanosensing [31, 32], which is sensitive to the deformability of the surrounding matrix, is essential for numerous physiological events, including wound healing and blood pressure maintenance [33], as well as patho-physiological conditions, including atherosclerosis [34]. Several focal adhesion proteins, in addition to talin, have been identified as mechanosensors. In the case of p130Cas (Crk-associated substrate), mechanical tension on the cell leads to molecular extension of p130Cas exposing a series of tyrosine residues that can then serve as substrates for the Src kinase [35]. These phosphotyrosine residues provide docking sites for SH2 domain-containing proteins such as Crk, thus permitting coupling to downstream signaling [35]. Conversely, mechanical stretching causes zyxin, along with VASP, to translocate from focal adhesions to the attached actin stress fibers, resulting in increased actin assembly within focal adhesions and adhesion strengthening [36, 37].

INTRACELLULAR SIGNALING AND MOLECULAR SCAFFOLDS

Cell migration is a multi-step process requiring dynamic changes in focal adhesion organization and signaling.

Extension of the plasma membrane or lamellipodium at the front of the cell is driven by the assembly of a meshwork of actin filaments, and is followed by the formation, upon contact with the ECM, of small adhesions called focal complexes. These nascent adhesions may disassemble rapidly, or mature into focal adhesions connected to robust actin filaments called stress fibers that are necessary for developing tension within the cell to facilitate translocation. Focal adhesion disassembly at the cell rear completes the migration cycle. The dynamic changes in the organization of the actin cytoskeleton, as well as the organization and composition of cell adhesions, is regulated by the activity of the Rho family of GTPases comprising Cdc42, Rac1, and RhoA [38]. Cdc42 activation promotes actin filament assembly, filopodia formation, and thus membrane extension. Cdc42 is also required for establishing cell polarity. Rac1 stimulates lamellipodia formation at the front of migrating cells, while RhoA activation stimulates focal adhesion maturation and stress-fiber formation to promote cell contractility, as well as focal adhesion disassembly at the cell rear. These molecular switches are themselves regulated by activators (guanine nucleotide exchange factors, GEFs), inhibitors (GTPase activating proteins, GAPs), and guanine nucleotide dissociating factors (GDIs) that sequester inactive GTPases in the cytosol [39]. Integrins lack any intrinsic enzymatic activity, thus the initiation and transduction of intracellular signaling necessary for controlling the functional readout from these molecules is dependent on the coordinated recruitment and regulation of multiple structural and signaling proteins to focal adhesions – a role performed by multi-domain focal adhesion adaptor proteins such as paxillin and p130Cas (Figure 4.1).

Paxillin is comprised of four carboxyl-terminal LIM domains that are essential for targeting the protein to focal adhesions. They also mediate interactions with microtubules and the tyrosine-phosphatase PTP-PEST to promote focal adhesion disassembly [1]. The amino-terminal portion of the molecule contains an abundance of protein–protein interaction modules, including five non-redundant, short alpha-helical leucine-rich LD motifs that provide docking sites for both structural actin-binding proteins including vinculin and actopaxin/parvin [1, 40], as well as signaling proteins such as the focal adhesion kinase (FAK) and Arf GAPs PKL/GIT [41]. The paxillin amino-terminus also contains a polyproline-rich motif that can bind the SH3 domain of the cytoskeletal protein ponsin as well as the Src kinase. Finally, there are numerous serine, threonine, and tyrosine residues distributed throughout the paxillin molecule which, when phosphorylated/dephosphorylated by adhesion-activated kinases or phosphatases respectively, either function as regulatable protein–protein interaction domains themselves through the creation of SH2 binding motifs, or indirectly modulate the affinity of paxillin's other binding modules [1, 40].

One of the earliest signaling events resulting from integrin-mediated cell attachment to the ECM is activation

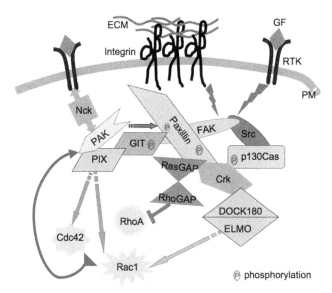

FIGURE 4.1 Focal adhesion protein networks transduce extracellular stimuli to regulate intracellular Rho family GTPase signaling.
The engagement of integrins with the extracellular matrix (ECM) as well as activation of growth factor receptors results in the localized activation of protein tyrosine kinases including Focal Adhesion Kinase (FAK) and Src. Adaptor proteins paxillin and p130Cas are recruited to nascent focal adhesions and phosphorylated by the activated FAK/Src complex. Tyrosine phosphorylated p130Cas and paxillin can then interact with the SH2/SH3 domain containing protein Crk, thereby coupling to Rac activation via the DOCK180/ELMO complex. Paxillin phosphorylation at Y31 and 118 also generates binding sites for p120RasGAP, which through p190RhoGAP can contribute to local suppression of RhoA activity. Similarly, FAK/Src phosphorylation of the Arf GAP, GIT facilitates recruitment of the GIT–PIX–PAK–Nck complex into focal adhesions via interaction with the paxillin LD4 motif. The Cdc42/Rac GEF PIX can locally stimulate Rac1 and Cdc42 and trigger cytoskeleton reorganization via activation of various effector proteins including p21-activated kinase (PAK). Active PAK may in turn, phosphorylate paxillin on S273 within the LD4 motif, to modulate paxillin's interaction with GIT and/or FAK. In addition, the association of the SH2/SH3 adaptor protein, Nck with both PAK and GIT (not shown) provides a putative mechanism for linking focal adhesion signaling nodes to growth factor receptor signaling. Recruitment of protein phosphatases PTP-PEST, SHP-2 and PTP-1B to focal adhesions via interactions with paxillin may also promote protein dephosphorylation, culminating in signal termination and focal adhesion disassembly. GF-growth factor, PM-plasma membrane, RTK-receptor tyrosine kinase

of FAK, requiring phosphorylation-dependent conformation changes to relieve autoinhibition by FAK's amino-terminal FERM domain [42, 43]. Autophosphorylation of FAK on Y397 results in the binding and activation of another tyrosine kinase, Src, and together the FAK/Src complex regulates adhesion dynamics by phosphorylating a wide spectrum of focal adhesion proteins, including paxillin and p130Cas, to generate binding sites for SH2-containing adaptor proteins such as Crk [44]. In turn, Crk binding to either protein facilitates cell migration by stimulating the local activity of Rac via Crk-mediated recruitment of the atypical Rac GEF Dock180/ELMO complex [45].

Additionally, in the case of paxillin, tyrosine phosphorylation may also promote binding of the p120RasGAP/ p190RhoGAP complex, resulting in localized suppression of Rho GTPase activity [46]. Interestingly, tyrosine phosphorylation of paxillin by FAK has also been reported to indirectly enhance the interaction between FAK and the LD 2/4 motifs of paxillin to facilitate focal adhesion maturation [47].

Paxillin can also modulate localized Rac1 activity through a tightly regulated interaction with a complex of proteins comprising the Arf GAPs GIT1/2, the Rac1/Cdc42 GEF PIX (p21-activated kinase interacting exchange factor), PAK (the p21-activated kinase), a serine/threonine kinase functioning for modulating cytoskeletal reorganization, and finally the adaptor protein Nck [48]. The adhesion-stimulated interaction of this complex with the LD4 motif of paxillin, and thus its recruitment to focal adhesions, involves a multi-step conformation-dependent activation cascade requiring both the binding of active Cdc42/Rac to PAK as well as the Src/FAK-dependent tyrosine phosphorylation of GIT [48]. It is likely that PIX GEF activity is also increased during this process, thereby amplifying local Cdc42/Rac activity. The stability of this complex in focal adhesions can be further enhanced via PAK-dependent phosphorylation of S273 within the paxillin LD4 motif [49] (Figure 4.1). The paxillin–GIT–PIX–PAK–Nck complex provides an important mechanism for restricting the spatio-temporal activity of Cdc42/Rac1 to the leading lamellipodia of migrating cells, thus contributing to cell polarization and directional migration [50, 51]. Through Nck, it may also play a role in functional cross-talk between integrins and growth factor receptors [52].

FOCAL ADHESION TURNOVER

Scaffold proteins like paxillin play an equally important role in the termination of adhesion-based signaling, thereby contributing to focal adhesion disassembly/turnover [53]. For example, the recruitment of the tyrosine phosphatase PTP-PEST to focal adhesions via interaction with the paxillin LIM domains is a prerequisite for dephosphorylation of multiple focal adhesion proteins, including FAK, p130Cas, GIT, as well as paxillin, and in turn accounts for the negative regulation of Rac signaling by PTP-PEST [54, 55]. Other phosphatases, including SHP-2 and PTP-1B, also bind to paxillin and presumably perform a similar role [1]. Interestingly, proteolytic degradation of key focal adhesion components provides an alternative mechanism for the regulation of focal adhesion organization and signaling. Both FAK/Src and PIX recruit and activate the calcium-dependent protease Calpain to promote focal adhesion disassembly and tail release via cleavage of talin, paxillin, FAK, Src, tensin, α-actinin, vinculin, and β1-integrins [44, 56], while the E3 ligase RNF5 binds to paxillin, leading to ubiquitin-mediated proteosomal degradation of paxillin, Abl, and Src [44]. These mechanisms, albeit somewhat unrefined, seem particularly important for adhesion disassembly at the rear of migrating cells, as well as for convergent extension migration in the developing embryo [57].

FOCAL ADHESIONS AND GENE EXPRESSION

In addition to controlling the dynamic relationship between the ECM and the cytoskeleton, protein interactions within focal adhesions also play a significant role in regulation of gene expression and signaling associated with cell survival, proliferation, and differentiation [58]. For example, the tyrosine kinase Abl is recruited to focal adhesions via interaction with phosphorylated paxillin, where it is activated before returning to the nucleus to regulate gene transcription. Several other focal adhesion proteins, including Hic-5, and zyxin contain nuclear localization and/or nuclear export signaling motifs and undergo nuclear–cytoplasmic shuttling [58]. Conversely, FAK activation and phosphorylation on Y925 facilitates the recruitment of the scaffold protein Grb2 to adhesion contacts, thereby coupling to Ras, and activation of the Erk-MAP kinase cascade to regulate anchorage-dependent survival. The direct binding of Erk and vinculin to paxillin may further facilitate signaling along this axis [59].

THE FUTURE

Integrin-based adhesion signaling is of paramount importance to nearly every aspect of a multicellular organism's development and tissue homeostasis. Despite the abundance of information now available regarding the molecular complexity of the focal adhesion interactome and its role in cellular communication, important questions remain. For instance, the precise mechanism(s) determining the spectrum and timing of individual protein–protein interactions within the focal adhesion and in establishing specific signaling cassettes is still largely unknown. In addition, many cells express more than one integrin type, each capable of interacting with overlapping, as well as distinct, complements of ECM and intracellular proteins, often utilizing different modalities. For example, paxillin is recruited to integrin $\beta1$ adhesions via its carboxyl-terminal LIM domains, while in contrast it binds directly via the amino terminus to the cytoplasmic tails of integrin $\alpha4$ and 9 [60, 61]. How this affects the capacity of paxillin to bind and signal through its numerous binding partners to influence the cellular response remains to be determined. Furthermore, the recent development of three-dimensional matrix model systems to evaluate adhesion signaling in a setting more analogous to the *in vivo* environment has revealed similarities, as well as important differences, in focal adhesion organization, composition, and signaling as compared to adhesions formed in two-dimensions. For example, cell adhesions formed within a three-dimensional fibronectin-rich matrix contain integrin $\alpha5\beta1$, but not $\alpha v\beta3$, and striking differences in phosphorylation of focal adhesion proteins such as FAK and ERK, as well as the relative activity of Rho GTPase family members, have been documented [62, 63]. These model systems, in combination with the use of advanced real-time microscopy such as Fluorescence Recovery After Photobleaching (FRAP), FRET, and Fluorescence Lifetime Imaging Microscopy (FLIM) [64], as well as detailed proteomic analysis [65], is expected to provide further insight into the complex, dynamic world of the focal adhesion molecular network.

ACKNOWLEDGEMENTS

Work in the author's laboratory is supported by grants from the NIH (Christopher E. Turner) and a Predoctoral Fellowship from the American Heart Association (Founders Affiliate) (Jianxin A. Yu).

REFERENCES

1. Brown MC, Turner CE. Paxillin: adapting to change. *Physiol Rev* 2004;**84**:1315–39.
2. Zaidel-Bar R, Itzkovitz S, Ma'ayan A, Iyerngar R, Geiger B. Functional atlas of the integrin adhesome. *Nat Cell Biol* 2007;**9**:858–67.
3. Schwartz MA, Schaller MD, Ginsberg MH. Integrins: emerging paradigms of signal transduction. *Annu Rev Cell Dev Biol* 1995;**11**:549–99.
4. Humphries MJ, McEwan PA, Barton SJ, Buckley PA, Bella J, Mould AP. Integrin structure: heady advances in ligand binding, but activation still makes the knees wobble. *Trends Biochem Sci* 2003;**28**:313–20.
5. Hynes RO. Integrins: bidirectional, allosteric signaling machines. *Cell* 2002;**110**:673–87.
6. Humphries JD, Byron A, Humphries MJ. Integrin ligands at a glance. *J Cell Sci* 2006;**119**:3901–3.
7. Ginsberg MH, Partridge A, Shattil SJ. Integrin regulation. *Curr Opin Cell Biol* 2005;**17**:509–16.
8. Iber D, Campbell ID. Integrin activation – the importance of a positive feedback. *Bull Math Biol* 2006;**68**:945–56.
9. Kim M, Carman CV, Springer TA. Bidirectional transmembrane signaling by cytoplasmic domain separation in integrins. *Science* 2003;**301**:1720–5.
10. Takagi J, Strokovich K, Springer TA, Walz T. Structure of integrin alpha5beta1 in complex with fibronectin. *EMBO J* 2003;**22**:4607–15.
11. Montanez E, Ussar S, Schifferer M, Bosl M, Zent R, Moser M, Fassler R. Kindlin-2 controls bidirectional signaling of integrins. *Genes Dev* 2008;**22**:1325–30.
12. Moser M, Nieswandt B, Ussar S, Pozgajova M, Fassler R. Kindlin-3 is essential for integrin activation and platelet aggregation. *Nat Med* 2008;**14**:325–30.
13. Tadokoro S, Shattil SJ, Eto K, Tai V, Liddington RC, de Pereda JM, Ginsberg MH, Calderwood DA. Talin binding to integrin beta tails: a final common step in integrin activation. *Science* 2003;**302**:103–6.

14. Millon-Fremillon A, Bouvard D, Grichine A, Manet-Dupe S, Block MR, Albiges-Rizo C. Cell adaptive response to extracellular matrix density is controlled by ICAP-1-dependent beta1-integrin affinity. *J Cell Biol* 2008;**180**:427–41.

15. Banno A, Ginsberg MH. Integrin activation. *Biochem Soc Trans* 2008;**36**:229–34.

16. Calderwood DA, Yan B, de Pereda JM, Alvarez BG, Fujioka Y, Liddington RC, Ginsberg MH. The phosphotyrosine binding-like domain of talin activates integrins. *J Biol Chem* 2002;**277**:21,749–58.

17. Garcia-Alvarez B, de Pereda JM, Calderwood DA, Ulmer TS, Critchley D, Campbell ID, Ginsberg MH, Liddington RC. Structural determinants of integrin recognition by talin. *Mol Cell* 2003;**11**:49–58.

18. Xing B, Jedsadayanmata A, Lam SC. Localization of an integrin binding site to the C terminus of talin. *J Biol Chem* 2001;**276**:44,373–8.

19. Gingras AR, Ziegler WH, Frank R, Barsukov IL, Roberts GC, Critchley DR, Emsley J. Mapping and consensus sequence identification for multiple vinculin binding sites within the talin rod. *J Biol Chem* 2005;**280**:37,217–24.

20. Critchley DR. Cytoskeleton proteins talin and vinculin in integrin-mediated adhesion. *Biochem Soc Trans* 2004;**32**:831–6.

21. Zhang X, Jiang G, Cai Y, Monkley SJ, Critchley DR, Sheetz MP. Talin depletion reveals independence of initial cell spreading from integrin activation and traction. *Nat Cell Biol* 2008;**10**:1062–8.

22. Patel B, Gingras AR, Bobkov AA, Fujimoto LM, Zhang M, Liddington RC, Mazzeo D, Emsley J, Roberts GC, Barsukov IL, Critchley DR. The activity of the vinculin binding sites in talin is influenced by the stability of the helical bundles that make up the talin rod. *J Biol Chem* 2006;**281**:7458–67.

23. Papagrigoriou E, Gingras AR, Barsukov IL, Bate N, Fillingham IJ, Patel B, Frank R, Ziegler WH, Roberts GC, Critchley DR, Emsley J. Activation of vinculin-binding site in the talin rod involves rearrangement of a five-helix bundle. *EMBO J* 2004;**23**:2942–51.

24. Priddle H, Hemmings L, Monkley S, Woods A, Patel B, Sutton D, Dunn GA, Zicha D, Critchley DR. Disruption of the talin gene compromises focal adhesion assembly in undifferentiated but not differentiated embryonic stem cells. *J Cell Biol* 1998;**142**:1121–33.

25. Alenghat FJ, Fabry B, Tsai KY, Goldmann WH, Ingber DE. Analysis of cell mechanics in single vinculin-deficient cells using a magnetic tweezer. *Biochem Biophys Res Commun* 2000;**277**:93–9.

26. Goldmann WH, Ingber DE. Intact vinculin protein is required for control of cell shape, cell mechanics, and Rac-dependent lamellipodia formation. *Biochem Biophys Res Commun* 2002;**290**:749–55.

27. Ziegler WH, Liddington RC, Critchley DR. The structure and regulation of vinculin. *Trends Cell Biol* 2006;**16**:453–60.

28. Saunders RM, Holt MR, Jennings L, Sutton DH, Barsukov IL, Bobkov A, Liddington RC, Adamson EA, Dunn GA, Critchley DR. Role of vinculin in regulating focal adhesion turnover. *Eur J Cell Biol* 2006;**85**:487–500.

29. Ling K, Schill NJ, Wagoner MP, Sun Y, Anderson RA. Movin' on up: the role of PtdIns(4,5)P(2) in cell migration. *Trends Cell Biol* 2006;**16**:276–84.

30. DeMali KA, Barlow CA, Burridge K. Recruitment of the Arp2/3 complex to vinculin: coupling membrane protrusion to matrix adhesion. *J Cell Biol* 2002;**159**:881–91.

31. Geiger B, Bershadsky A. Exploring the neighborhood: adhesion-coupled cell mechanosensors. *Cell* 2002;**110**:139–42.

32. Schwartz MA, Desimone DW. Cell adhesion receptors in mechanotransduction. *Curr Opin Cell Biol* 2008;**20**:551–6.

33. Davies PF. Flow-mediated endothelial mechanotransduction. *Physiol Rev* 1995;**75**:519–60.

34. Thubrikar MJ, Robicsek F. Pressure-induced arterial wall stress and atherosclerosis. *Ann Thorac Surg* 1995;**59**:1594–603.

35. Sawada Y, Tamada M, Dubin-Thaler BJ, Cherniavskaya O, Sakai R, Tanaka S, Sheetz MP. Force sensing by mechanical extension of the Src family kinase substrate p130Cas. *Cell* 2006;**127**:1015–26.

36. Yoshigi M, Hoffman LM, Jensen CC, Yost HJ, Beckerle MC. Mechanical force mobilizes zyxin from focal adhesions to actin filaments and regulates cytoskeletal reinforcement. *J Cell Biol* 2005;**171**:209–15.

37. Hirata H, Tatsumi H, Sokabe M. Mechanical forces facilitate actin polymerization at focal adhesions in a zyxin-dependent manner. *J Cell Sci* 2008;**121**:2795–804.

38. Ridley AJ, Schwartz MA, Burridge K, Firtel RA, Ginsberg MH, Borisy G, Parsons JT, Horwitz AR. Cell migration: integrating signals from front to back. *Science* 2003;**302**:1704–9.

39. Burridge K, Wennerberg K. Rho and Rac take center stage. *Cell* 2004;**116**:167–79.

40. Deakin N, Turner CE. Paxillin comes of age. *J Cell Sci* 2008;**121**:2435–44.

41. Turner CE, West KA, Brown MC. Paxillin-ARF GAP signaling and the cytoskeleton. *Curr Opin Cell Biol* 2001;**13**:593–9.

42. Burridge K, Turner CE, Romer LH. Tyrosine phosphorylation of paxillin and pp125FAK accompanies cell adhesion to extracellular matrix: a role in cytoskeletal assembly. *J Cell Biol* 1992;**119**:893–903.

43. Cai X, Lietha D, Ceccarelli DF, Karginov AV, Rajfur Z, Jacobson K, Hahn KM, Eck MJ, Schaller MD. Spatial and temporal regulation of focal adhesion kinase activity in living cells. *Mol Cell Biol* 2008;**28**:201–14.

44. Carragher NO, Frame MC. Focal adhesion and actin dynamics: a place where kinases and proteases meet to promote invasion. *Trends Cell Biol* 2004;**14**:241–9.

45. Cote JF, Vuori K. GEF what? Dock180 and related proteins help Rac to polarize cells in new ways. *Trends Cell Biol* 2007;**17**:383–93.

46. Tsubouchi A, Sakakura J, Yagi R, Mazaki Y, Schaefer E, Yano H, Sabe H. Localized suppression of RhoA activity by Tyr31/118-phosphorylated paxillin in cell adhesion and migration. *J Cell Biol* 2002;**159**:673–83.

47. Zaidel-Bar R, Milo R, Kam Z, Geiger B. A paxillin tyrosine phosphorylation switch regulates the assembly and form of cell-matrix adhesions. *J Cell Sci* 2007;**120**:137–48.

48. Brown MC, Cary LA, Jamieson JS, Cooper JA, Turner CE. Src and FAK kinases cooperate to phosphorylate paxillin kinase linker, stimulate its focal adhesion localization, and regulate cell spreading and protrusiveness. *Mol Biol Cell* 2005;**16**:4316–28.

49. Nayal A, Webb DJ, Brown CM, Schaefer EM, Vicente-Manzanares M, Horwitz AR. Paxillin phosphorylation at Ser273 localizes a GIT1–PIX–PAK complex and regulates adhesion and protrusion dynamics. *J Cell Biol* 2007;**173**:587–9.

50. Li Z, Hannigan M, Mo Z, Liu B, Lu W, Wu Y, Smrcka AV, Wu G, Li L, Liu M, Huang CK, Wu D. Directional sensing requires G beta gamma-mediated PAK1 and PIX alpha-dependent activation of Cdc42. *Cell* 2003;**114**:215–27.

51. Mazaki Y, Hashimoto S, Tsujimura T, Morishige M, Hashimoto A, Aritake K, Yamada A, Nam JM, Kiyonari H, Nakao K, Sabe H. Neutrophil direction sensing and superoxide production linked by the GTPase-activating protein GIT2. *Nat Immunol* 2006;**7**:724–31.

52. Bokoch GM. Biology of the p21-activated kinases. *Annu Rev Biochem* 2003;**72**:743–81.

53. Webb DJ, Donais K, Whitmore LA, Thomas SM, Turner CE, Parsons JT, Horwitz AF. FAK-Src signalling through paxillin, ERK and MLCK regulates adhesion disassembly. *Nat Cell Biol* 2004;**6**:154–61.

54. Sastry SK, Lyons PD, Schaller MD, Burridge K. PTP-PEST n of cell spreading and motility: a roel for paxillin kinase linker. *J Cell Sci* 2002;**118**:5835–47.

55. Jamieson JS, Tumbarello DA, Halle M, Brown MC, Turner CE. Paxillin is essential for PTP-PEST-dependent regulation of cell spreading and motility: a roel for paxillin kinase linker. *J Cell Sci* 2005;**118**:5835–47.

56. Franco SJ, Huttenlocher A. Regulating cell migration: calpains make the cut. *J Cell Sci* 2005;**118**:3829–38.

57. Iioka H, Iemura S, Natsume T, Kinoshita N. Wnt signaling regulates paxillin ubiquirinization essential for mesodermal cell motility. *Nat Cell Biol* 2007;**9**:813–21.

58. Kadrmas JL, Beckerle MC. The LIM domain: from the cytoskeleton to the nucleus. *Nat Rev Mol Cell Biol* 2004;**5**:920–31.

59. Mitra SK, Hanson DA, Schlaepfer DD. Focal adhesion kinase: in command and control of cell motility. *Nat Rev Mol Cell Biol* 2005;**6**:56–68.

60. Liu S, Ginsberg MH. Paxillin binding to a conserved sequence motif in the alpha 4 integrin cytoplasmic domain. *J Biol Chem* 2000;**275**:22,736–42.

61. Liu S, Slepak M, Ginsberg MH. Binding of paxillin to alpha 9 integrin cytoplasmic domain inhibits cell spreading. *J Biol Chem* 2001;**276**:37,086–92.

62. Cukierman E, Pankov R, Stevens DR, Yamada KM. Taking cell matrix adhesion to the third dimension. *Science* 2001;**294**:1708–12.

63. Green JA, Yamada KM. Three-dimensional microenvironments modulate fibroblast signaling responses. *Adv Drug Deliv Rev* 2007;**59**:1293–8.

64. Webb DJ, Brown CM, Horwitz AF. Illuminating adhesion complexes in migrating cells: moving toward a bright future. *Curr Opin Cell Biol* 2003;**15**:614–20.

65. Mann M. Functional and quantitative proteomics using SILAC. *Nat Rev Mol Cell Biol* 2006;**7**:952–8.

Chapter 5

Cadherin Regulation of Adhesive Interactions

Barbara Ranscht

Burnham Institute for Medical Research, La Jolla, California

INTRODUCTION

Adhesive interactions between neighboring cells regulate the dynamic organization and patterning of tissues in developing embryos, and are critical for tissue homeostasis [1, 2]. Cadherins are a large family of calcium-dependent cell adhesion molecules that affect diverse cellular properties *in vivo* and *in vitro* [for reviews, see [3–8]]. A multitude of studies have shown that the dynamic regulation of cadherin function affects development and physiology in vertebrate and invertebrate animals by promoting the formation of tissue structures and regulating a fine-tuned balance of adhesive interactions and signal exchange between cells. The cadherin family now includes more than a hundred members across species ranging from unicellular organisms to higher vertebrates [9]. Since their discovery as molecules promoting calcium-dependent cell adhesion and compaction [10–12], numerous studies have exposed roles beyond these functions. In developing embryos, the dynamic regulation of cadherin subtype expression controls cell segregation into tissue layers and organization of cells within organs [13]. Cadherins affect changes of tissue shape due to cellular rearrangements during morphogenetic movements such as gastrulation [14]. The formation of tissue boundaries in developing embryos requires cadherin functions [15, 16]. In the vertebrate nervous system, different cadherin subtypes contribute to the embryonic formation of specific brain subdivisions and the connectivity of neurons into functional circuits [17]. Cadherin expression is required for the initiation, elongation and pathfinding of axons to their targets [5, 18–20]. They are expressed in an isotype-specific manner at synapses [21, 22], and regulate dendritic spine formation as well as synaptic plasticity [23–26]. In epithelial cells, cadherins establish cell polarity and form adherens junctions that bond cells through a cortical actin belt within epithelial structures [27]. As cells transition from an epithelial to a migratory phenotype during development or tumorigenesis, the junctional cadherin complex disintegrates to allow for cell detachment and assembly of the migratory machinery.

Cells use diverse mechanisms to regulate cadherin expression and adhesion. The surface concentrations of cadherin molecules and differential adhesion are principal contributors of cell behavior and cell–cell communication [28, 29]. Cadherin expression and functions are controlled at multiple levels, ranging from transcriptional control, to exocytosis and endocytosis, ligand engagement, association with lateral proteins, and dynamic molecular interactions that impinge on connections with the cytoskeletal network and downstream signaling pathways. This chapter will discuss the different cadherin subclasses as regulators of intercellular communication, and highlight the principal mechanisms of cadherin function in different cellular contexts. Since the large amount of literature in the field exceeds the scope of this chapter, the reader is referred to several excellent reviews from leaders in the field for more detailed information on structure, function, and physiological roles of cadherins [3–9, 30].

THE CADHERIN FAMILY

Classical and Divergent Cadherins Interacting with Catenins

The *vertebrate classical cadherins* are typically single-span transmembrane proteins composed of five extracellular cadherin domains and a conserved cytoplasmic region (Figure 5.1). The hallmark of the cadherin family is calcium-dependent cell-to-cell binding between cadherin

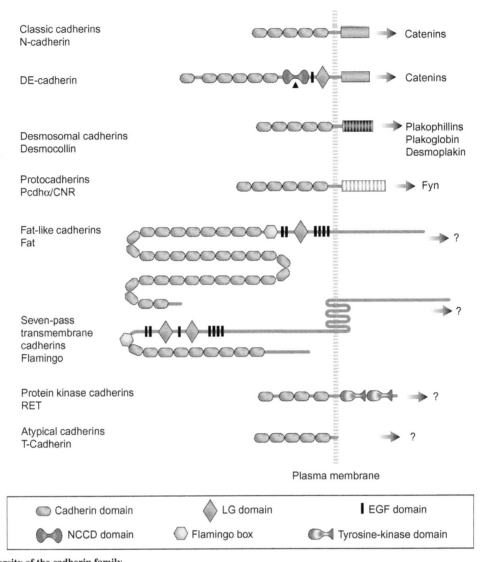

Classic cadherins
N-cadherin → Catenins

DE-cadherin → Catenins

Desmosomal cadherins
Desmocollin → Plakophillins
Plakoglobin
Desmoplakin

Protocadherins
Pcdhα/CNR → Fyn

Fat-like cadherins
Fat → ?

Seven-pass
transmembrane
cadherins
Flamingo → ?

Protein kinase cadherins
RET → ?

Atypical cadherins
T-Cadherin → ?

Plasma membrane

Cadherin domain	LG domain	EGF domain
NCCD domain	Flamingo box	Tyrosine-kinase domain

FIGURE 5.1 Diversity of the cadherin family.
The domain organization of representative members of each of the identified cadherin families is depicted. Arrows indicate the linkage with indicated intracellular molecules.

homo- or heterodimers [5, 30, 31]. Classical cadherins are subdivided into two major subgroups, type I and type II cadherins, based on phylogenetic maps [9]. Type I classical cadherins including E-, N-, P-, and R-cadherins generally mediate strong adhesive interactions. They are essential for the formation of the embryonic germ layers and assembly of organs. Deletions of E- cadherin gene expression halt mouse embryo development at the preimplantation stage [32, 33]. Disruptions in the N-cadherin gene affect cardiac development and limit mouse embryo survival to day 10 [34]. Type I cadherins contain a conserved amino terminal HAV motif that has gained attention as a potential cell-recognition sequence [35]. Although soluble HAV peptides indeed can inhibit cadherin-mediated adhesion *in vitro* [36, 37], structural studies have established that the HAV motif is necessary for cadherin dimerization rather than for cell recognition [38]. Type II cadherins, including

VE-cadherin, MN-cadherin, cadherin 8, and cadherin 11, share the overall structure with type I cadherins, but represent a phylogenetically separate group [9]. Type II cadherins lack the HAV sequence, and accordingly form dimers with a binding interface structurally different from the type I group [39, 40]. In comparison to type I cadherins, type II cadherins confer weak interactions. VE-cadherin regulates vascular remodeling and maturation rather than assembling endothelial cells into vascular plexi [41]. In the spinal cord, type II cadherins sort motor neurons into pools innervating specific muscles [42], thereby fine-tuning motor neuron patterns rather than controlling motor neuron development.

A characterizing feature of classical vertebrate and invertebrate cadherins is the association of their highly conserved cytoplasmic domains with the "amadillo repeat" family protein β-catenin [3, 7]. β-catenin serves multiple functions within cells and switches roles between the cadherin-associated state at the plasma membrane, the cytoplasmic state characterized by interactions affecting cell signaling events, and transcriptional activation in the nucleus upon Wnt-induced signaling [43]. Bound to the cadherin cytoplasmic region, β-catenin assembles a signaling complex that regulates the dynamics of the actin cytoskeleton via α-catenin [3, 4]. Thus, availability of β-catenin and its association with the cytoplasmic region of cadherins critically regulate cell adhesion.

In *Drosophila*, DE-cadherin encoded by *shotgun* shares its extracellular and intracellular domain organization and interactions with vertebrate E-cadherin [44, 45]. Mosaic analysis of mutations in the *shotgun (shg)* gene provides evidence for homophilic binding activity similar to that of classical cadherins [46]. Other invertebrate members of the cadherin family, including *Drosophila* DN-cadherin [47], *C. elegans* HMR-1 [48], and sea urchin LvG-cadherin [49], diverge in their extracellular domain organization from their vertebrate counterparts by the addition of cadherin domains, cysteine-rich EGF repeats, laminin G domains, and often a non-chordate classic cadherin domain [NCCD [50]] (Figure 5.1). These cadherins are listed under classical cadherins, as the intercellular domain of these molecules is conserved and interacts with amadillo-domain-containing β-catenin homologs. Thus, despite structural divergence of their extracellular domains, principal functions of classical cadherins are conserved between vertebrate and non-vertebrate species [51].

Desmosomal Cadherins

Desmosomal cadherins comprise two separate subfamilies, the Desmocollins (DSC) and Desmogleins (DSG), each represented by three members (DSC-1, -2, -3, and DSG-1, -2, -3) [52, 53]. Desmosomal cadherins and associated intracellular proteins orchestrate the assembly of desmosomal plaques [54–57], and are expressed in a cell type- or differentiation-specific manner [58, 59]. The extracellular domain of desmosomal cadherins is composed of five cadherin domains and confers homophilic or heterophilic binding interactions with other members of the desmosomal cadherin family (Figure 5.1). DSCs and DSGs contain characteristic intracellular domains that diverge from those of the classical cadherins and interact with either of the amadillo family proteins plakoglobin and plakophilin. The latter provide a link (via desmoplakin) to intermediate filaments. In the skin, autoimmune skin-blistering diseases such as Pemphigus vulgaris or foliaceus are caused by desmosome disruption through autoantibodies [60].

Cadherins with Divergent Structures

Bioinformatics has revealed cadherin-like domains in yeast and bacteria [61], suggesting that the cadherin motif discovered in classical vertebrate cadherins is an ancient structural motif. The number of proteins containing cadherin domains has expanded enormously during evolution, and in humans more than 180 genes encoding proteins with cadherin domains have been reported (www.pfam.wustl.edu). Most, but not all, of these genes are classified into distinct subgroups based on sequence homologies, domain organization, number of cadherin repeats, and genomic organization [7, 9, 51]. Representative members of the major divergent cadherin subclasses are shown, along with classical and desmosomal cadherins, in Figure 5.1.

Protocadherins

The largest subgroup of cadherin-related proteins, with about 70 members in vertebrates, are transmembrane protocadherins that are composed of six or seven extracellular cadherin domains and distinct cytoplasmic regions [62–64]. Genomic organization subdivides protocadherins into clustered and non-clustered groups. The clustered protocadherins Pdchα, Pdchβ, and Pdchγ are generated from a small genome locus on human chromosome 5q31, and encode 52 cadherin-related molecules [65]. The Pdchα and Pdchγ are transcribed from variable exons encoding different extracellular domains and a constant exon generating a conserved, cluster-specific carboxy terminal domain. The Pdchβ cluster lacks the constant domain. Pdchα orthologs were independently identified as cadherin-related neuronal receptors (CNR) in mouse brain by their ability to interact with Fyn [66], a Src-related intracellular protein kinase that supports synaptic plasticity [67]. The observation that CNRs demarcate specific synapse populations [66] points to possible functions of CNR cadherins in regulating synaptic function or plasticity.

Genetic deletion of the entire protocadherin γ cluster comprising 22 genes induces apoptosis of spinal interneurons and premature death of the mutant mice [68]. In an elegant experiment in which apoptosis was prevented by crossing the mutation into Bax-deficient mice, loss of protocadherin γ altered synapse number and strength of the now surviving spinal cord neurons, suggesting functions in synaptic modulation [69].

Non-clustered protocadherins comprise the Pdchδ subgroup and phylogenetically solitary protocadherins. Pdchδ protocadherins include nine different cadherins that are characterized by conserved regions, CM1 and CM2, in their cytoplasmic domains. Structural studies reveal the lack of a homophilic binding interface and a protocadherin-specific loop structure in the extracellular domain [70]. It is not clear if protocadherins exert adhesion *in vivo*, or if they modulate cell functions by other mechanisms, as recently suggested for Arcadlin (Activity-Regulated Cadherin-like Protein; also paraxial protocadherin (PAPC) or Pdch8). Arcadlin expression is induced by synaptic activity and involved in long-term potentiation [71]. Recent work demonstrates that Arcadlin affects the density of dendritic spines. Spine density is reduced by endocytosis of the synaptic N-cadherin–Arcadlin complex via a p38 MAPK pathway [24]. Outside the nervous system, PAPC controls cell movements during gastrulation in *Xenopus* embryos [72] and establishes segmental boundaries during somite formation [15]. In parallel to the findings in the central nervous system, PAPC affects morphogenetic movements by regulating C-cadherin activity in early embryos [73]. The regulation of classical cadherin expression or function through protocadherin activity is an emerging theme in the cadherin field.

FAT-Like Cadherins

Fat and Dachsous, the prototypes of the fat-like cadherins, are required for several developmental processes in *Drosophila* [74–77]. The hallmark of fat-like transmembrane cadherins is their large extracellular domain, which is composed of 17–34 tandemly arranged cadherin domains, EGF repeats, laminin G domains, and, in some subtypes, a flamingo box (Figure 5.1, [7, 75]). *Drosophila fat* and the fat-like *dachsous* gene product regulate cell growth, the proximodistal patterning of the appendages, and planar cell polarity (PCP) – for example, the coordinated orientation of cells in the fly eye and wing [78, 79]. Heterophilic binding between Fat and Dachsous does not confer strong cell adhesion, but seems to activate downstream signaling pathways that are not fully understood [80, 81]. Fat-like cadherins with functions in tissue morphogenesis are reported for the nematode *Caenorhabditis elegans* [82, 83], and mammalian homologs with yet unknown functions have been identified [84, 85].

Seven-Pass Transmembrane Cadherins

The *Drosophila melanogaster* gene product Flamingo is a cadherin family protein anchored in the membrane by a seven-pass transmembrane domain that shows similarity to those of the secretin receptor family of G-protein-coupled receptors [86]. The extracellular *Flamingo* domain consists of nine amino terminal cadherin domains, a flamingo box, EGF repeats, and Laminin A G-repeat homology domains (Figure 5.1, [86]). *Flamingo* acts in concert with frizzled and strabismus to establish planar polarity of hair cells in the *Drosophila* wing [76, 79, 86, 87]. Frizzled and strabismus relay asymmetric and intercellular signals through homotypic binding of the cadherin domains that are asymmetrically associated with core components of the planar cell polarity (PCP) pathway [88]. Flamingo-regulated PCP signaling represents a parallel pathway to Fat and Dachsous interactions discussed above. In the *Drosophila* nervous system, Flamingo regulates patterning of axon projections from R8 photoreceptor cells to the medulla [89] and dendritic tiling of peripheral neurons, the restriction of dendrites within boundaries complementary to the fields occupied by dendrites from other neurons [90]. In vertebrates, three seven-pass transmembrane cadherins, Celsr1, Celsr2, and Celsr3, are known. Celsr 2 confers dendritic maturation [91], and Celsr3 affects development of specific axon tracts [92]. The mechanism of Celsr functions remains elusive; however, Celsr 3 and frizzled 3 are co-expressed by certain neurons and positioned to act in concert similar to their functions in the *Drosophila* wing.

Protein Kinase Cadherins

The protein kinase cadherins are represented by Ret, which consists of four cadherin-like domains and an intracellular kinase region [93, 94]. Ret is part of a tripartite receptor complex that is activated by interactions with neurotrophic factors of the glial cell line-derived neurotrophic factor (GDNF) family. GDNF induces or stabilizes a complex between Ret and GPI-linked alpha receptors (GFR alpha 1–4), resulting in dimerization and activation of the Ret kinase [95]. Mutagenesis studies have shown that Ret, GFR alpha 1, and GDNF affect multiple developmental events, including development of the enteric nervous system affected in Hirschsprung's disease [96–98].

Atypical Cadherins

Several atypical cadherins, such as Ksp-cadherin (Cdh16), LI-cadherin (Cdh17), and T-cadherin (Cdh13), each represent a phylogenetically singular cadherin branch that shares the ectodomain structure with the classical cadherins but has either a short or no cytoplasmic domain [9]. T-cadherin (T for "truncated") is anchored in the membrane through a glycosylphosphatidyl inositol moiety [99]. In cellular adhesion assays, T-cadherin confers calcium-dependent homotypic interactions [100]; this interaction leads to repulsion of extending neuronal processes *in vitro* and *in vivo* [101, 102]. T-cadherin is also prominent in the vasculature, and analyses of mice with the systemic null mutation have identified pro-angiogenic *in vivo* functions for T-cadherin that seem to depend on binding the circulating adipokine adiponectin [103].

Clearly, the size and diversity of the cadherin family is extensive, and many of the principal mechanisms of nonclassical cadherin function remain to be elucidated. On the other hand, research into the regulation of classical cadherin function has brought to light complex regulation and molecular interactions. The second part of this chapter discusses principle mechanisms of classical cadherin function.

CADHERIN STRUCTURE–FUNCTION RELATIONSHIPS

Proper folding and formation of the active binding configuration of classical cadherins requires calcium [104], and mutation of one of these sites in E-cadherin, abolishes cadherin-mediated adhesion [105]. Calcium binding sites are located between adjacent cadherin domains within the same molecule. Without calcium, E-cadherin appears collapsed and disorganized by rotary shadowing and electron microscopy, while calcium at low concentrations (50 μM) enforces the formation of rigid rod-like structures [106, 107]. A further increase in calcium concentration (500 μM) supports *cis* association of adjacent cadherin extracellular domains, and concentrations above 1 mM drive *trans* interactions between opposing *cis* strand dimers [107–109]. *Trans* dimer configuration leads to strong adhesive forces between cadherin extracellular domains [110]. Cadherins undergo both homodimer (between the same cadherin type) and heterodimer formation (between different cadherin types) [111]. Thus, calcium is required on all levels of cadherin function, and enables the formation of rigid cadherin EC domains, dimerization in *cis*, and interdigitatation with dimers on opposing cell surfaces.

Structural studies of the type I cadherin amino terminal domain revealed a β-strand organization similar to the immunoglobulin fold [112, 113]. One key feature common to both type I and type II classical cadherins is the intermolecular exchange of N-terminal β-strands between corresponding EC1 domains, also referred to as "domain swap" [30, 39, 113]. The binding interface differs between type I and type II cadherins. In the type I dimerized structure, the terminal tryptophan, Trp-2, of EC1 domain interacts with Alanine in position 80 (Ala-80), within the hydrophobic acceptor pouch formed around the HAV sequence [38, 111, 114–116]. The binding interface of type II classical cadherins shows two conserved tryptophan side-chains that anchor each swapped domain strand, and a large hydrophobic pocket unique to the type II interfaces [39, 40].

Several experimental approaches provide evidence that cadherins engage multiple binding sites along the extracellular domain to confer adhesion. First, biophysical studies demonstrate that the distance between opposing cadherin-covered lipid bilayers (250 Å) corresponds to the length of the cadherin extracellular domain [117, 118]. Force application results in the stepwise increase of the intermolecular distance between cadherin molecules, and suggests additional binding sites along the molecule [117, 118]. Second, analyses of the binding properties of C-cadherin domain constructs show that the highest homophilic binding activity is conferred by the entire extracellular domain, while domain 1 polypeptides exhibit only low binding activity [119]. Third, the crystal structure of the full C-cadherin ectodomain displays *cis* and *trans* association of the CD1 domain ,with multiple sites along the cadherin extracellular domain [38].

The cumulative structural and biophysical data probing cadherin structure–function relationships lead to a model in which adhesion occurs in multiple steps (Figure 5.2): cadherins undergo *cis* dimerization, which is a prerequisite for subsequent engagement of *cis* dimers with their dimerized counterparts on opposing cell surfaces. The structural data

present a snapshot view of interactions between opposing EC1 domains. The biophysical data support a model in which cadherin-mediated adhesion is dynamic in strength and strongest upon full interdigitation of the entire EC domains. Cadherin interactions *in vivo* may be more complex than in these simplified experimental models. Their biophysical properties may differ as they traverse the lipid bilayer of the plasma membrane, and association with neighboring proteins may result in structural modifications. Indeed, cadherins at cellular adherens junctions cross-communicate with another adhesion system, the actin-linked nectin–afadin complex [120, 121]. Studies into the molecular and structural interactions between these combinatorially expressed

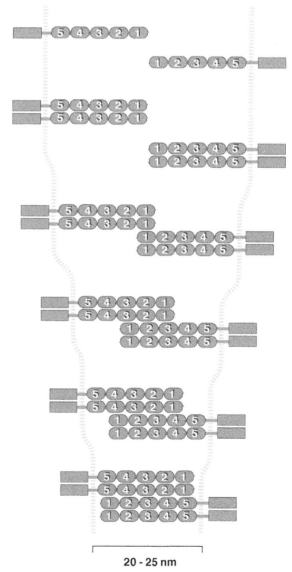

20 - 25 nm

FIGURE 5.2 Stages of cadherin-mediated adhesive interactions. Cadherin monomers associate into strand dimers that subsequently engage with *cis* dimers from opposing cell surfaces. Initial interactions may occur between the opposing amino terminal domains. Under favorable conditions, the adhesive bonds tighten and the cadherin extracellular domains fully interdigitate to bring adjacent cell surfaces into close contact.

protein complexes may reveal new insights into adherens junction assembly and function.

MULTIPLE MODES FOR REGULATING CADHERIN ADHESIVE ACTIVITY

The manifold cadherin functions in development and disease require a highly dynamic regulation of cadherin expression and function. In embryonic germlayers and epithelial structures, cadherins are engaged at adherens junctions interconnecting adjacent cells. The broadly accepted view is that cadherin clusters at cell junctions link the cadherins via associated β- and α-catenin to the filamentous actin cytoskeleton [122, 123]. Data from the Nelson and Weiss laboratories have challenged this model, and provide a new perspective on the regulation of cadherin-mediated adhesion [124, 125]. Cell migration requires the loosening of adhesive bonds, which is often accompanied by a switch in cadherin function. How do cadherins support migratory behavior? Studies show that cadherin functions are regulated at multiple levels, including gene transcription, proteolysis, endocytosis, and association with intracellular proteins. The highly conserved cytoplasmic domain of classical cadherins provides selective target sites for regulating cadherin functions and thus adhesive cell properties (Figure 5.3).

Association With Intracellular Proteins

α- and β-Catenin

Classical cadherins display their most adhesive configuration as components of adherens junction in epithelial cells [126, 127]. They directly interact with several cytoplasmic proteins, including the amadillo-repeat proteins β-catenin and its close relative plakoglobin. The availability of β-catenin, its posttranslational modifications and its intracellular associations are central to the regulation of cadherin-based adhesive activity. Classical cadherins display a highly conserved binding site for β-catenin at the extreme carboxy terminal region of the cadherin cytoplasmic domain [104]. β-catenin is a central signaling molecule that associates in a mutually exclusive manner with the cadherin cytoplasmic domain to regulate adhesion or with components of the Wnt signaling pathway to regulate transcription [43, 128–132]. Unbound β-catenin is ubiquitinated and targeted for degradation though the proteosome (Figure 5.3). β-catenin also binds to α-catenin, an actin-binding protein distantly related to the focal adhesion plaque protein vinculin. Biochemical work identified E-cadherin, β-catenin, and α-catenin in a roughly stoichiometric manner in a complex resistant to harsh detergent conditions, suggesting strong interactions [133]. Moreover, there is genetic evidence that adherens junctions and the underlying cortical actin belt are interdependent [134, 135]. Combined, these data have led to the prevailing view that transmembrane classical cadherins

bind to β-catenin, which in turn associates with α-catenin and links the complex to the actin-based cytoskeleton. However, more recent studies have challenged this model [124, 125]. Biochemical data confirm the association of cadherin, α- and β-catenin in a complex, but in association with the cadherin-β-catenin complex, α-catenin does not simultaneously bind to actin filaments [124, 125]. Instead, α-catenin seems to regulate actin dynamics [136, 137]. Monomeric α-catenin preferentially binds β-catenin, while dimerized α-catenin interacts with F-actin and suppresses actin filament branching by inhibiting the activity of the Arp2/3 complex. These findings now raise the question how cadherins connect to the actin cytoskeleton. Various possibilities are discussed, including the direct link via α-catenin-interacting proteins (such as α-actinin, vinculin, afadin, ZO1, ajuba, formin), or connection via other junction proteins such as the nectin–afadin complex. Another possibility is the concerted action with other adherens junction proteins that together achieve the connection with the cytoskeleton. Certainly, this work on cadherin functions has raised new questions that need to be addressed.

Classical cadherin function is critically regulated by the available β-catenin pool and, *vice versa*, limitation of cadherin binding sites for intracellular β-catenin leads to accumulation of free β-catenin that is rapidly degraded. β-catenin association with a large protein complex (APC complex) containing the adenomatous polyposis coli (APC) tumor suppressor gene product, axin, conductin, and the glycogen synthase kinase-3 (Figure 5.3) leads to phosphorylation, ubiquitination, and targeting for degradation by the 26S proteosome [138]. An effective way to counteract β-catenin phosphorylation and degradation is activation of the Wnt signaling pathway. Wnts play an important role in cell fate determination during embryonic development and in cancers [138, 139], but functions of many Wnt proteins remain unexplored. Wnt-binding to a receptor of the Frizzled family antagonizes β-catenin phosphorylation and degradation. Free unphosphorylated β-catenin can bind to unoccupied cadherin cytoplasmic domains and enforce adhesive interactions through regulating cytoskeletal dynamics. When the free β-catenin pool exceeds the number of available cadherin binding sites, it accumulates in the nucleus, where it interacts with DNA binding proteins of the T cell factor (TCF)/lymphoid enhancer binding factor (LEF) family. β-catenin binding to TCF/LEF transcription factors enhances expression of genes that, for example, switch cellular functions by regulating the cell cycle [138]. Thus, β-catenin both enforces the strength of cadherin-mediated adhesion and acts as a molecular switch for regulating cell functions.

Cadherin Association With Growth Factor Receptors

Classical cadherins can alternatively associate with growth factor receptors and associated signaling pathways to promote

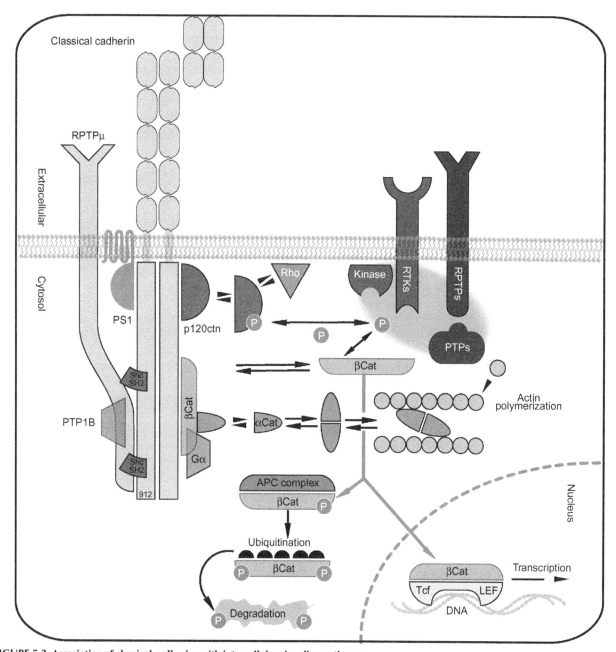

FIGURE 5.3 Association of classical cadherins with intracellular signaling pathways.
The conserved cytoplasmic region of the classical cadherins interacts with multiple proteins to regulate adhesion. The association of the cadherin cytoplasmic region with β-catenin and α-catenin is critical for regulating actin dynamics. Receptor protein kinases (RTKs), receptor tyrosine phosphatases (RPTPs), and intracellular phosphatases (PTPs) balance cadherin, β-catenin, and p120ctn phosphorylation. PS1, presenilin-1; βcat, β-catenin; α-cat, α-catenin; RPTPμ, receptor tyrosine phosphatase μ.

migratory cell behavior. During epithelial–mesenchymal transition in tumorigenesis, epithelial cells often substitute junctional E-cadherin with another cadherin that integrates different signal transduction pathways and allows loosening of adhesive interactions, detachment, and migration [140]. Similarly, downregulation of N-cadherin is required for neural crest migration [141]. Growth cone movements during N-cadherin-mediated neurite extension *in vitro* [142, 143] and axon guidance *in vivo* [144, 145] require

low-stringency adhesive interactions. Cadherin-mediated migratory behavior results from a switch in cadherin association and signaling. N-cadherin associates through binding sites in extracellular domain 4 with the HAV motif in fibroblast growth factor (FGF) receptor 1 to prevent FGF-induced receptor internalization [146]. In the presence of N-cadherin, FGF-2 binding to its receptor results in sustained activation of the MAPK–Erk pathway, which in turn leads to activation of metalloprotease transcription and cell migration [147].

p120 Catenins and Rho Family GTPases

An earlier study [148] attributed functional importance to the cadherin membrane proximal cytoplasmic domain, and it has now become clear that this site is the target for complex regulation of cadherin-mediated adhesivity [149]. The membrane proximal cadherin cytoplasmic region contains the binding site for the amadillo repeat protein p120 catenins (p120ctn), a family of four proteins – ARVCF, δ-catenin, p0071, and p120 catenin – that structurally differ from β-catenin in the number of amadillo repeats [150]. p120ctn is an important regulator of adherens junction assembly and disassembly. Like β-catenin, p120ctn undergoes diverse interactions in a pool-specific manner. At the cytoplasmic face of the cell membrane, p120ctn associates with the cadherin cyoplasmic domain and regulates Rho-GTPase activity. In the nucleus it associates with Kaiso to repress transcription. p120ctn *per se* seems dispensable for adhesive interactions, but is thought to act as a molecular switch to promote or prevent adhesion, depending on its state of phosphorylation [149, 151–154]. First, p120ctn regulates cadherin stability on the cell surface by controlling endocytosis and degradation [155–157]. It is speculated that p120ctn occupation of the cadherin membrane-proximal domain competes with proteins that target cadherins for degradation. A possible candidate for p120ctn competition is the ubiquitin E3 ligase Hakai that binds specifically to E-cadherin upon tyrosine phosphorylation by Src, and targets it to the proteosome [158]. Hakai is specific for E-cadherin, thus Hakai-like E3-ligases may regulate degradation of other classical cadherins. Second, p120ctn binding is essential for maintaining the association of β-catenin with the cadherin cytoplasmic region. The tyrosine kinase Fer associates with p120ctn to phosphorylate and activate the tyrosine phosphatase PTP1B that binds and dephosphorylates β-catenin at tyrosine 654. β-catenin associates with the cadherin cytoplasmic region only in the dephosphorylated state [159]. Third, p120ctn associates with microtubules and kinesin motor proteins to partake in regulating cadherin trafficking and thus expression on the cell surface [160, 161].

Lastly, p120ctn functions at least in part through regulation of RhoGTPases. The small GTPases Rac, Rho, and Cdc42 are well known for regulating and specifying membrane interactions with cortical actin filaments [162]. p120ctn inhibits RhoA and activates Rac and Cdc42 to increase or decrease cell motility [163–166]. IQGAP1, an effector of the Rho family GTPases Rac and Cdc42 can modify the association of the cadherin cytoplasmic region with α-catenin [167, 168]. At cell-to-cell contact sites, IQGAP1 associates with active GTP-bound Rac and Cdc42 that prohibits binding to β-catenin. In their inactive GDP-bound form, Rac and Cdc42 do not associate with IQGAP1, which is then free for binding β-catenin and for dissociating the cadherin–β-catenin complex from α-catenin and the actin cytoskeleton, thereby downregulating adhesivity [166]. The association of β-catenin with the cadherin cytoplasmic domain is also affected by the Gα subunit of heterotrimeric G proteins. Binding of activated GTP-bound Gα12/13 to the cadherin cytoplasmic region dissociates β-catenin from the complex and negates adhesive interactions [169, 170]. These and other studies provide evidence that cadherin-mediated adhesive interactions are controlled by multiple mechanisms that converge on regulating cadherin stability through β-catenin.

Phosphorylation and Dephosphorylation of Cadherins and β-Catenin

Structural studies of the interacting E-cadherin-β-catenin domains reveal that adoption of structural conformation of the cadherin cytoplasmic domain depends on β-catenin association and cadherin phosphorylation on serine residues [171]. The cadherin cytoplasmic region displays serine phosphorylation consensus sites for casein kinase II and glycogen synthase β3. Serine phosphorylation strengthens the association between the cadherin cytoplasmic region and β-catenin, and fortifies adhesion [172]. In contrast, tyrosine phosphorylation dramatically decreases binding between these proteins and weakens adhesive bonds [173–175]. Both non-receptor and receptor protein type kinases regulate phosphorylation of the cadherin cytoplasmic tail and β-catenin. Overexpression of the non-receptor tyrosine kinases Src or Fer in cultured cells promote tyrosine phosphorylation of cadherins and β-catenin [173, 176, 177] and decrease adhesive interactions in favor of a motile phenotype. Interestingly, in cancer cells, the lipid phosphatase activity of PTEN can counteract Src-induced cell scattering and invasiveness and stabilize the E-cadherin junctional complex through a yet unknown mechanism [178]. Similarly, E-cadherin and β-catenin tyrosine phosphorylation by receptor protein kinase type growth factor receptors results in cell scattering [179, 180]. Tyrosine-phosphorylated cadherin binds the adaptor protein Shc, which participates in stimulating mitogenic signaling pathways by growth factor activation of Ras [181]. Thus, cadherin function is controlled by multiple cell signaling pathways that regulate the availability of cadherins on the call surface and balance the cadherin association with β-catenin and hence its interactions with the cytoskeleton.

Formation of the cadherin–catenin complex is further fine-tuned by the balance between protein kinase and phosphatase activity. Several phosphatases associate with and stabilize the cadherin–β-catenin complex and prevent undesired phosphorylation. The non-receptor type phosphatase PTP1B is targeted to the cadherin complex, where it interacts with sequences partially overlapping with the binding site for β-catenin [182]. The receptor protein tyrosine phosphatase μ (PTPμ) interacts with multiple cadherins [183] and dynamically regulates its association with the cadherin cytoplasmic region through IQGAP1 and Cdc42 [184, 185]. Downregulation of either PTP1B or PTPμ suppresses N-cadherin-mediated neurite extension [186, 187].

Other phosphatases, including LAR, receptor tyrosine phosphatase β/ζ, and the Meprin/A5/Mu domain receptors κ and γ, do not associate directly with cadherins, but appear to regulate cadherin function through β-catenin modifications.

Regulation by Proteolytic Cleavage

An effective strategy for regulating cadherin activity is their cleavage by extracellular and intracellular proteases. Mature classical cadherins are derived from precursor proteins that are cleaved at the RKQR sequence in transit to or at the cell surface [188]. Prevention of proteolytic processing through mutation of the cleavage site abolishes the adhesive functions of E-cadherin [188]. The proprotein convertase furin is suggested to mediate the proprotein cleavage of E-cadherin [189].

Metalloproteases cleave the cadherin extracellular domain. The cleaved extracellular fragment of N-cadherin generated during retinal development is thought to partake in modulating retinal axon guidance [190, 191]. Numerous studies associate the loss of E-cadherin with tumor cell growth and metastasis [192]. E-cadherin is proteolytically cleaved in non-cancerous mammary epithelial cells by ectopically expressed metalloprotease stromelysin-1 [193]. The cleavage triggers the progressive conversion of the epithelial into a mesenchymal invasive phenotype characterized by the disappearance of E-cadherin and β-catenin from cell–cell contacts and activation of growth factors and metalloproteases [193]. Stromolysin, however, could not be detected in cancer cells or the embryo, and the tissue endogenous metalloproteases cleaving E-cadherin [194] remain to be defined. Adam family disintegrins and metalloprotease cleave cadherins in their extracellular domain causing changes in adhesion, cell migration and β-catenin translocation to the nucleus [195]. In endothelial cells, Adam 10 cleaves vascular VE-cadherin, thereby loosening endothelial cell interactions and allowing transcytosis of inflammatory cells [196].

Cellular responses to apoptotic signals are characterized by the disruption of cell-to-cell and cell-to-extracellular matrix contacts and cytoskeletal reorganization. During programmed cell death, adherens junctions disintegrate through the actions of both metalloproteases and caspases on cadherin and β-catenin/plakoglobin molecules [197, 198]. A metalloprotease activity releases most of the E-cadherin extracellular domain, while caspase-3 cleaves at an intracellular membrane proximal site [197]. These data enforce the view that cadherin structural integrity, assembly within adherens junctions and linkage to the actin filament network are critical for cell survival [41].

Lastly, in response to apoptotic stimuli, the γ-secretase activity of presenilin-1, a protein associated with Alzheimer's disease, can cleave E-cadherin at the membrane–cytoplasm interface [199]. This releases the cadherin intracellular domain, increases the intracellular pool of β-catenin, and facilitates the disassembly of adherens junctions by disconnecting cadherins from the cytoskeleton. However, under conditions that favor cell-to-cell adhesion, presenilin-1 binding stabilizes the junctional complex [200]. Such dual functions have also been reported for p120ctn that competes for the presenilin-1 binding site on E-cadherin in a mutually exclusive manner [200]. Presenilin-1 also affects β-catenin function and trafficking, thereby providing an additional mechanism for regulating cadherin functions in cell adhesion [201–203]. As presenilin-1 is recruited to sites of synaptic contact [204], and synaptic morphology and function are regulated by the cadherin–catenin system [25, 26, 205], the cadherin–catenin–presenilin-1 interaction may favor the loss of synaptic structures at an early stage of Alzheimer's disease and increase vulnerability to neuronal apoptosis.

CONCLUSIONS AND PERSPECTIVES

It is now clear that adhesive functions of classical cadherins are dynamically regulated. While beginning to grasp some of the principal mechanisms of this regulation, we are faced with new challenges. First, a large number of new cadherin-like molecules with cytoplasmic sequences different from those of the classical cadherins have been revealed. Little is known about the distribution, function, and modes of signaling of these molecules. Second, recent work suggests that cadherins may be far more promiscuous in their binding specificities than previously assumed [206]. N-cadherin-deficient mutant mice die of defects in heart development [34], but this phenotype can be rescued by the cardiac specific ectopic expression of E-cadherin [207] suggesting that cadherin-mediated adhesion but not adhesive specificity is required. Moreover, cells express multiple cadherins. Cadherin function is required for the sorting of motor neurons into specific pools which are defined by the combinatorial expression of multiple cadherins [42]. Overexpression of the type II MN cadherin disrupts pool sorting. Although the mechanism for MN-cadherin function remains to be determined, available evidence suggests that cadherin homophilic binding activity is not required [42]. Thus, other mechanisms will need to explain the functions of the type II cadherins in this context. The identification of multiple molecular interactions has made clear that regulation of cadherin functions is multifaceted. The diversity and magnitude of the cadherin family adds a new level of complexity to the cellular interactions conferred by the combinatorial cadherin expression during development and in adult organisms.

ACKNOWLEDGEMENTS

I thank Dr Rainer Stiemer for stimulating discussions on cadherin functions, and Dr Chris Kintner and Martin Denzel for comments on the manuscript. Our research on cadherins was supported by NIH grant HD 25938.

REFERENCES

1. Steinberg MS. Resconstruction of tissue by dissociated cells. *Science* 1963;**141**:401–8.

2. Townes PLaN J. Directed movements and selective adhesion of embryonic amphibian cells. *J Exp Zool* 1955;**128**:53–120.

3. Gumbiner BM. Regulation of cadherin-mediated adhesion in morphogenesis. *Nat Rev Mol Cell Biol* 2005;**6**:622–34.

4. Halbleib JM, Nelson WJ. Cadherins in development: cell adhesion, sorting, and tissue morphogenesis. *Genes Dev* 2006;**20**:3199–214.

5. Takeichi M. The cadherin superfamily in neuronal connections and interactions. *Nat Rev Neurosci* 2007;**8**:11–20.

6. Redies C. Cadherins in the central nervous system. *Prog Neurobiol* 2000;**61**:611–48.

7. Tepass U, Truong K, Godt D, Ikura M, Peifer M. Cadherins in embryonic and neural morphogenesis. *Nat Rev Mol Cell Biol* 2000;**1**:91–100.

8. Yagi T, Takeichi M. Cadherin superfamily genes: functions, genomic organization, and neurologic diversity. *Genes Dev* 2000;**14**:1169–80.

9. Nollet F, Kools P, van Roy F. Phylogenetic analysis of the cadherin superfamily allows identification of six major subfamilies besides several solitary members. *J Mol Biol* 2000;**299**:551–72.

10. Gallin WJ, Edelman GM, Cunningham BA. Characterization of L-CAM, a major cell adhesion molecule from embryonic liver cells. *Proc Natl Acad Sci USA* 1983;**80**:1038–42.

11. Peyrieras N, Hyafil F, Louvard D, Ploegh HL, Jacob F. Uvomorulin: a nonintegral membrane protein of early mouse embryo. *Proc Natl Acad Sci USA* 1983;**80**:6274–7.

12. Yoshida C, Takeichi M. Teratocarcinoma cell adhesion: identification of a cell-surface protein involved in calcium-dependent cell aggregation. *Cell* 1982;**28**:217–24.

13. Takeichi M. Morphogenetic roles of classic cadherins. *Curr Opin Cell Biol* 1995;**7**:619–27.

14. Keller R. Shaping the vertebrate body plan by polarized embryonic cell movements. *Science* 2002;**298**:1950–4.

15. Kim SH, Jen WC, De Robertis EM, Kintner C. The protocadherin PAPC establishes segmental boundaries during somitogenesis in xenopus embryos. *Curr Biol* 2000;**10**:821–30.

16. Tepass U, Godt D, Winklbauer R. Cell sorting in animal development: signalling and adhesive mechanisms in the formation of tissue boundaries. *Curr Opin Genet Dev* 2002;**12**:572–82.

17. Redies C, Puelles L. Modularity in vertebrate brain development and evolution. *Bioessays* 2001;**23**:1100–11.

18. Nern A, Zhu Y, Zipursky SL. Local N-cadherin interactions mediate distinct steps in the targeting of lamina neurons. *Neuron* 2008;**58**:34–41.

19. Piper M, Dwivedy A, Leung L, Bradley RS, Holt CE. NF-protocadherin and TAF1 regulate retinal axon initiation and elongation *in vivo*. *J Neurosci* 2008;**28**:100–5.

20. Rhee J, Buchan T, Zukerberg L, Lilien J, Balsamo J. Cables links Robo-bound Abl kinase to N-cadherin-bound beta-catenin to mediate Slit–induced modulation of adhesion and transcription. *Nat Cell Biol* 2007;**9**:883–92.

21. Uchida N, Honjo Y, Johnson KR, Wheelock MJ, Takeichi M. The catenin/cadherin adhesion system is localized in synaptic junctions bordering transmitter release zones. *J Cell Biol* 1996;**135**:767–79.

22. Fannon AM, Colman DR. A model for central synaptic junctional complex formation based on the differential adhesive specificities of the cadherins. *Neuron* 1996;**17**:423–34.

23. Saglietti L, Dequidt C, Kamieniarz K, et al. Extracellular interactions between GluR2 and N-cadherin in spine regulation. *Neuron* 2007;**54**:461–77.

24. Yasuda S, Tanaka H, Sugiura H, et al. Activity-induced protocadherin arcadlin regulates dendritic spine number by triggering N-cadherin endocytosis via TAO2beta and p38 MAP kinases. *Neuron* 2007;**56**:456–71.

25. Togashi H, Abe K, Mizoguchi A, Takaoka K, Chisaka O, Takeichi M. Cadherin regulates dendritic spine morphogenesis. *Neuron* 2002;**35**:77–89.

26. Bruses JL. Cadherin-mediated adhesion at the interneuronal synapse. *Curr Opin Cell Biol* 2000;**12**:593–7.

27. Nejsum LN, Nelson WJ. A molecular mechanism directly linking E-cadherin adhesion to initiation of epithelial cell surface polarity. *J Cell Biol* 2007;**178**:323–35.

28. Nose A, Nagafuchi A, Takeichi M. Expressed recombinant cadherins mediate cell sorting in model systems. *Cell* 1988;**54**:993–1001.

29. Foty RA, Steinberg MS. The differential adhesion hypothesis: a direct evaluation. *Dev Biol* 2005;**278**:255–63.

30. Shapiro L, Love J, Colman DR. Adhesion molecules in the nervous system: structural insights into function and diversity. *Annu Rev Neurosci* 2007;**30**:451–74.

31. Pokutta S, Weis WI. Structure and mechanism of cadherins and catenins in cell–cell contacts. *Annu Rev Cell Dev Biol* 2007;**23**:237–61.

32. Larue L, Ohsugi M, Hirchenhain J, Kemler R. E-cadherin null mutant embryos fail to form a trophectoderm epithelium. *Proc Natl Acad Sci USA* 1994;**91**:8263–7.

33. Riethmacher D, Brinkmann V, Birchmeier C. A targeted mutation in the mouse E-cadherin gene results in defective preimplantation development. *Proc Natl Acad Sci USA* 1995;**92**:855–9.

34. Radice GL, Rayburn H, Matsunami H, Knudsen KA, Takeichi M, Hynes RO. Developmental defects in mouse embryos lacking N-cadherin. *Dev Biol* 1997;**181**:64–78.

35. Blaschuk OW, Sullivan R, David S, Pouliot Y. Identification of a cadherin cell adhesion recognition sequence. *Dev Biol* 1990;**139**:227–9.

36. Doherty P, Rowett LH, Moore SE, Mann DA, Walsh FS. Neurite outgrowth in response to transfected N-CAM and N-cadherin reveals fundamental differences in neuronal responsiveness to CAMs. *Neuron* 1991;**6**:247–58.

37. Tang L, Hung CP, Schuman EM. A role for the cadherin family of cell adhesion molecules in hippocampal long-term potentiation. *Neuron* 1998;**20**:1165–75.

38. Boggon TJ, Murray J, Chappuis-Flament S, Wong E, Gumbiner BM, Shapiro L. C-cadherin ectodomain structure and implications for cell adhesion mechanisms. *Science* 2002;**296**:1308–13.

39. Patel SD, Ciatto C, Chen CP, et al. Type II cadherin ectodomain structures: implications for classical cadherin specificity. *Cell* 2006;**124**:1255–68.

40. May C, Doody JF, Abdullah R, et al. Identification of a transiently exposed VE-cadherin epitope that allows for specific targeting of an antibody to the tumor neovasculature. *Blood* 2005;**105**:4337–44.

41. Carmeliet P, Lampugnani MG, Moons L, et al. Targeted deficiency or cytosolic truncation of the VE-cadherin gene in mice impairs VEGF-mediated endothelial survival and angiogenesis. *Cell* 1999;**98**:147–57.

42. Price SR, De Marco Garcia NV, Ranscht B, Jessell TM. Regulation of motor neuron pool sorting by differential expression of type II cadherins. *Cell* 2002;**109**:205–16.

43. Nelson WJ, Nusse R. Convergence of Wnt, beta-catenin, and cadherin pathways. *Science* 2004;**303**:1483–7.

44. Tepass U, Gruszynski-DeFeo E, Haag TA, Omatyar L, Torok T, Hartenstein V. shotgun encodes Drosophila E-cadherin and is preferentially required during cell rearrangement in the neurectoderm and other morphogenetically active epithelia. *Genes Dev* 1996;**10**:672–85.

45. Uemura T, Oda H, Kraut R, Hayashi S, Kotaoka Y, Takeichi M. Zygotic *Drosophila* E-cadherin expression is required for processes of dynamic epithelial cell rearrangement in the *Drosophila* embryo. *Genes Dev* 1996;**10**:659–71.

46. Niewiadomska P, Godt D, Tepass U. DE-Cadherin is required for intercellular motility during *Drosophila* oogenesis. *J Cell Biol* 1999;**144**:533–47.

47. Iwai Y, Usui T, Hirano S, Steward R, Takeichi M, Uemura T. Axon patterning requires DN-cadherin, a novel neuronal adhesion receptor, in the *Drosophila* embryonic CNS. *Neuron* 1997;**19**:77–89.

48. Costa M, Raich W, Agbunag C, Leung B, Hardin J, Priess JR. A putative catenin-cadherin system mediates morphogenesis of the *Caenorhabditis elegans* embryo. *J Cell Biol* 1998;**141**:297–308.

49. Miller JR, McClay DR. Characterization of the role of cadherin in regulating cell adhesion during sea urchin development. *Dev Biol* 1997;**192**:323–39.

50. Oda H, Tsukita S. Nonchordate classic cadherins have a structurally and functionally unique domain that is absent from chordate classic cadherins. *Dev Biol* 1999;**216**:406–22.

51. Tepass U. Genetic analysis of cadherin function in animal morphogenesis. *Curr Opin Cell Biol* 1999;**11**:540–8.

52. Koch PJ, Walsh MJ, Schmelz M, Goldschmidt MD, Zimbelmann R, Franke WW. Identification of desmoglein, a constitutive desmosomal glycoprotein, as a member of the cadherin family of cell adhesion molecules. *Eur J Cell Biol* 1990;**53**:1–12.

53. Parker AE, Wheeler GN, Arnemann J, et al. Desmosomal glycoproteins II and III. Cadherin-like junctional molecules generated by alternative splicing. *J Biol Chem* 1991;**266**:10,438–10,445.

54. Allen E, Yu QC, Fuchs E. Mice expressing a mutant desmosomal cadherin exhibit abnormalities in desmosomes, proliferation, and epidermal differentiation. *J Cell Biol* 1996;**133**:1367–82.

55. Koch PJ, Mahoney MG, Ishikawa H, et al. Targeted disruption of the pemphigus vulgaris antigen (desmoglein 3) gene in mice causes loss of keratinocyte cell adhesion with a phenotype similar to pemphigus vulgaris. *J Cell Biol* 1997;**137**:1091–102.

56. Roberts GA, Burdett ID, Pidsley SC, King IA, Magee AI, Buxton RS. Antisense expression of a desmocollin gene in MDCK cells alters desmosome plaque assembly but does not affect desmoglein expression. *Eur J Cell Biol* 1998;**76**:192–203.

57. Serpente N, Marcozzi C, Roberts GA, et al. Extracellularly truncated desmoglein 1 compromises desmosomes in MDCK cells. *Mol Membr Biol* 2000;**17**:175–83.

58. Green KJ, Gaudry CA. Are desmosomes more than tethers for intermediate filaments? *Nat Rev Mol Cell Biol* 2000;**1**:208–16.

59. Ishii K, Green KJ. Cadherin function: breaking the barrier. *Curr Biol* 2001;**11**:R569–72.

60. Stanley JR. Autoantibodies against adhesion molecules and structures in blistering skin diseases. *J Exp Med* 1995;**181**:1–4.

61. Dickens NJ, Beatson S, Ponting CP. Cadherin-like domains in alpha-dystroglycan, alpha/varepsilon- sarcoglycan and yeast and bacterial proteins. *Curr Biol* 2002;**12**:R197–9.

62. Morishita H, Yagi T. Protocadherin family: diversity, structure, and function. *Curr Opin Cell Biol* 2007;**19**:584–92.

63. Sano K, Tanihara H, Heimark RL, et al. Protocadherins: a large family of cadherin-related molecules in central nervous system. *EMBO J* 1993;**12**:2249–56.

64. Hamada S, Yagi T. The cadherin-related neuronal receptor family: a novel diversified cadherin family at the synapse. *Neurosci Res* 2001;**41**:207–15.

65. Wu Q, Maniatis T. A striking organization of a large family of human neural cadherin-like cell adhesion genes. *Cell* 1999;**97**:779–90.

66. Kohmura N, Senzaki K, Hamada S, et al. Diversity revealed by a novel family of cadherins expressed in neurons at a synaptic complex. *Neuron* 1998;**20**:1137–51.

67. Grant SG, O'Dell TJ, Karl KA, Stein PL, Soriano P, Kandel ER. Impaired long-term potentiation, spatial learning, and hippocampal development in fyn mutant mice [see comments]. *Science* 1992;**258**:1903–10.

68. Wang X, Weiner JA, Levi S, Craig AM, Bradley A, Sanes JR. Gamma protocadherins are required for survival of spinal interneurons. *Neuron* 2002;**36**:843–54.

69. Weiner JA, Wang X, Tapia JC, Sanes JR. Gamma protocadherins are required for synaptic development in the spinal cord. *Proc Natl Acad Sci USA* 2005;**102**:8–14.

70. Morishita H, Umitsu M, Murata Y, et al. Structure of the cadherin-related neuronal receptor/protocadherin-alpha first extracellular cadherin domain reveals diversity across cadherin families. *J Biol Chem* 2006;**281**:33,650–33,663.

71. Yamagata K, Andreasson KI, Sugiura H, et al. Arcadlin is a neural activity-regulated cadherin involved in long term potentiation. *J Biol Chem* 1999;**274**:19,473–19,979.

72. Kim SH, Yamamoto A, Bouwmeester T, Agius E, Robertis EM. The role of paraxial protocadherin in selective adhesion and cell movements of the mesoderm during *Xenopus* gastrulation. *Development* 1998;**125**:4681–90.

73. Chen X, Gumbiner BM. Paraxial protocadherin mediates cell sorting and tissue morphogenesis by regulating C-cadherin adhesion activity. *J Cell Biol* 2006;**174**:301–13.

74. Bryant PJ, Huettner B, Held LI Jr, Ryerse J, Szidonya J. Mutations at the fat locus interfere with cell proliferation control and epithelial morphogenesis in *Drosophila*. *Dev Biol* 1988;**129**:541–54.

75. Mahoney PA, Weber U, Onofrechuk P, Biessmann H, Bryant PJ, Goodman CS. The fat tumor suppressor gene in *Drosophila* encodes a novel member of the cadherin gene superfamily. *Cell* 1991;**67**:853–68.

76. Clark HF, Brentrup D, Schneitz K, Bieber A, Goodman C, Noll M. Dachsous encodes a member of the cadherin superfamily that controls imaginal disc morphogenesis in *Drosophila*. *Genes Dev* 1995;**9**:1530–42.

77. Buratovich MA, Bryant PJ. Enhancement of overgrowth by gene interactions in lethal(2)giant discs imaginal discs from *Drosophila melanogaster*. *Genetics* 1997;**147**:657–70.

78. Yang CH, Axelrod JD, Simon MA. Regulation of Frizzled by fat-like cadherins during planar polarity signaling in the *Drosophila* compound eye. *Cell* 2002;**108**:675–88.

79. Adler PN, Charlton J, Liu J. Mutations in the cadherin superfamily member gene dachsous cause a tissue polarity phenotype by altering frizzled signaling. *Development* 1998;**125**:959–68.

80. Casal J, Lawrence PA, Struhl G. Two separate molecular systems, Dachsous/Fat and Starry night/Frizzled, act independently to confer planar cell polarity. *Development* 2006;**133**:4561–72.

81. Matakatsu H, Blair SS. Separating the adhesive and signaling functions of the Fat and Dachsous protocadherins. *Development* 2006;**133**:2315–24.

82. Pettit J, Wood WB, Plasterk RH. cdh-3, a gene encoding a member of the cadherin superfamily, functions in epithelial cell morphogenesis in *Caenorhabditis elegans*. *Development* 1996;**122**:4149–57.

83. Hill E, Broadbent ID, Chothia C, Pettitt J. Cadherin superfamily proteins in Caenorhabditis elegans and Drosophila melanogaster. *J Mol Biol* 2001;**305**:1011–24.

84. Dunne J, Hanby AM, Poulsom R, et al. Molecular cloning and tissue expression of FAT, the human homologue of the *Drosophila* fat

gene that is located on chromosome 4q34–q35 and encodes a putative adhesion molecule. *Genomics* 1995;**30**:207–23.

85. Ponassi M, Jacques TS, Ciani L, ffrench CC. Expression of the rat homologue of the Drosophila fat tumour suppressor gene. *Mech Dev* 1999;**80**:207–12.

86. Usui T, Shima Y, Shimada Y, et al. Flamingo, a seven-pass transmembrane cadherin, regulates planar cell polarity under the control of Frizzled. *Cell* 1999;**98**:585–95.

87. Chae J, Kim MJ, Goo JH, et al. The Drosophila tissue polarity gene starry night encodes a member of the protocadherin family. *Development* 1999;**126**:5421–9.

88. Chen WS, Antic D, Matis M, et al. Asymmetric homotypic interactions of the atypical cadherin flamingo mediate intercellular polarity signaling. *Cell* 2008;**133**:1093–105.

89. Lee RC, Clandinin TR, Lee CH, Chen PL, Meinertzhagen IA, Zipursky SL. The protocadherin Flamingo is required for axon target selection in the *Drosophila* visual system. *Nat Neurosci* 2003;**6**:557–63.

90. Gao FB, Kohwi M, Brenman JE, Jan LY, Jan YN. Control of dendritic field formation in Drosophila: the roles of flamingo and competition between homologous neurons. *Neuron* 2000;**28**:91–101.

91. Shima Y, Kengaku M, Hirano T, Takeichi M, Uemura T. Regulation of dendritic maintenance and growth by a mammalian 7-pass transmembrane cadherin. *Dev Cell* 2004;**7**:205–16.

92. Tissir F, Bar I, Jossin Y, De Backer O, Goffinet AM. Protocadherin Celsr3 is crucial in axonal tract development. *Nat Neurosci* 2005;**8**:451–7.

93. Takahashi M, Cooper GM. ret transforming gene encodes a fusion protein homologous to tyrosine kinases. *Mol Cell Biol* 1987;**7**:1378–85.

94. Anders J, Kjar S, Ibanez CF. Molecular modeling of the extracellular domain of the RET receptor tyrosine kinase reveals multiple cadherin-like domains and a calcium-binding site. *J Biol Chem* 2001;**276**:35,808–35,817.

95. Airaksinen MS, Titievsky A, Saarma M. GDNF family neurotrophic factor signaling: four masters, one servant? *Mol Cell Neurosci* 1999;**13**:313–25.

96. Schuchardt A, D'Agati V, Larsson-Blomberg L, Costantini F, Pachnis V. Defects in the kidney and enteric nervous system of mice lacking the tyrosine kinase receptor Ret. *Nature* 1994;**367**:380–3.

97. Romeo G, Ronchetto P, Luo Y, et al. Point mutations affecting the tyrosine kinase domain of the RET proto- oncogene in Hirschsprung's disease. *Nature* 1994;**367**:377–8.

98. Edery P, Lyonnet S, Mulligan LM, et al. Mutations of the RET proto-oncogene in Hirschsprung's disease. *Nature* 1994;**367**:378–80.

99. Ranscht B, Dours-Zimmermann MT. T-cadherin, a novel cadherin cell adhesion molecule in the nervous system lacks the conserved cytoplasmic region. *Neuron* 1991;**7**:391–402.

100. Vestal DJ, Ranscht B. Glycosyl phosphatidylinositol-anchored T-cadherin mediates calcium-dependent, homophilic cell adhesion. *J Cell Biol* 1992;**119**:451–61.

101. Fredette BJ, Ranscht B. T-cadherin expression delineates specific regions of the developing motor axon-hindlimb projection pathway. *J Neurosci* 1994;**14**:7331–46.

102. Fredette BJ, Miller J, Ranscht B. Inhibition of motor axon growth by T-cadherin substrata. *Development* 1996;**122**:3163–71.

103. Hebbard LW, Garlatti M, Young LJ, Cardiff RD, Oshima RG, Ranscht B. T-cadherin supports angiogenesis and adiponectin association with the vasculature in a mouse mammary tumor model. *Cancer Res* 2008;**68**:1407–16.

104. Takeichi M. Cadherins: a molecular family important in selective cell–cell adhesion. *Annu Rev Biochem* 1990;**59**:237–52.

105. Ozawa M, Engel J, Kemler R. Single amino acid substitutions in one Ca2+ binding site of uvomorulin abolish the adhesive function. *Cell* 1990;**63**:1033–8.

106. Pokutta S, Herrenknecht K, Kemler R, Engel J. Conformational changes of the recombinant extracellular domain of E-cadherin upon calcium binding. *Eur J Biochem* 1994;**223**:1019–26.

107. Pertz O, Bozic D, Koch AW, Fauser C, Brancaccio A, Engel J. A new crystal structure, Ca^{2+} dependence and mutational analysis reveal molecular details of E-cadherin homoassociation. *EMBO J* 1999;**18**:1738–47.

108. Tomschy A, Fauser C, Landwehr R, Engel J. Homophilic adhesion of E-cadherin occurs by a co-operative two-step interaction of N-terminal domains. *EMBO J* 1996;**15**:3507–14.

109. Koch AW, Pokutta S, Lustig A, Engel J. Calcium binding and homoassociation of E-cadherin domains. *Biochemistry* 1997;**36**:7697–705.

110. Brieher WM, Yap AS, Gumbiner BM. Lateral dimerization is required for the homophilic binding activity of C-cadherin. *J Cell Biol* 1996;**135**:487–96.

111. Shan WS, Tanaka H, Phillips GR, et al. Functional cis-heterodimers of N- and R-cadherins. *J Cell Biol* 2000;**148**:579–90.

112. Overduin M, Harvey TS, Bagby S, et al. Solution structure of the epithelial cadherin domain responsible for selective cell adhesion. *Science* 1995;**267**:386–9.

113. Shapiro L, Kwong PD, Fannon AM, Colman DR, Hendrickson WA. Considerations on the folding topology and evolutionary origin of cadherin domains. *Proc Natl Acad Sci USA* 1995;**92**:6793–7.

114. Shapiro L, Fannon AM, Kwong PD, et al. Structural basis of cell–cell adhesion by cadherins. *Nature* 1995;**374**:327–37.

115. Tamura K, Shan WS, Hendrickson WA, Colman DR, Shapiro L. Structure–function analysis of cell adhesion by neural (N-) cadherin. *Neuron* 1998;**20**:1153–63.

116. Kitagawa M, Natori M, Murase S, Hirano S, Taketani S, Suzuki ST. Mutation analysis of cadherin-4 reveals amino acid residues of EC1 important for the structure and function. *Biochem Biophys Res Commun* 2000;**271**:358–63.

117. Sivasankar S, Brieher W, Lavrik N, Gumbiner B, Leckband D. Direct molecular force measurements of multiple adhesive interactions between cadherin ectodomains. *Proc Natl Acad Sci USA* 1999;**96**:11,820–11,824.

118. Sivasankar S, Gumbiner B, Leckband D. Direct measurements of multiple adhesive alignments and unbinding trajectories between cadherin extracellular domains. *Biophys J* 2001;**80**:1758–68.

119. Chappuis-Flament S, Wong E, Hicks LD, Kay CM, Gumbiner BM. Multiple cadherin extracellular repeats mediate homophilic binding and adhesion. *J Cell Biol* 2001;**154**:231–43.

120. Tachibana K, Nakanishi H, Mandai K, et al. Two cell adhesion molecules, nectin and cadherin, interact through their cytoplasmic domain-associated proteins. *J Cell Biol* 2000;**150**:1161–76.

121. Pokutta S, Weis WI. The cytoplasmic face of cell contact sites. *Curr Opin Struct Biol* 2002;**12**:255–62.

122. Adams CL, Nelson WJ, Smith SJ. Quantitative analysis of cadherin–catenin–actin reorganization during development of cell–cell adhesion. *J Cell Biol* 1996;**135**:1899–911.

123. Gumbiner B, Stevenson B, Grimaldi A. The role of the cell adhesion molecule uvomorulin in the formation and maintenance of the epithelial junctional complex. *J Cell Biol* 1988;**107**:1575–87.

124. Yamada S, Pokutta S, Drees F, Weis WI, Nelson WJ. Deconstructing the cadherin–catenin–actin complex. *Cell* 2005;**123**:889–901.

125. Drees F, Pokutta S, Yamada S, Nelson WJ, Weis WI. Alpha-catenin is a molecular switch that binds E-cadherin-beta-catenin and regulates actin-filament assembly. *Cell* 2005;**123**:903–15.

126. Yap AS, Brieher WM, Gumbiner BM. Molecular and functional analysis of cadherin-based adherens junctions. *Annu Rev Cell Dev Biol* 1997;**13**:119–46.

127. Adams CL, Nelson WJ. Cytomechanics of cadherin-mediated cell–cell adhesion. *Curr Opin Cell Biol* 1998;**10**:572–7.

128. Harris TJ, Peifer M. Decisions, decisions: beta-catenin chooses between adhesion and transcription. *Trends Cell Biol* 2005;**15**:234–7.

129. Brembeck FH, Rosario M, Birchmeier W. Balancing cell adhesion and Wnt signaling, the key role of beta-catenin. *Curr Opin Genet Dev* 2006;**16**:51–9.

130. Hulsken J, Birchmeier W, Behrens J. E-cadherin and APC compete for the interaction with beta-catenin and the cytoskeleton. *J Cell Biol* 1994;**127**:2061–9.

131. von Kries JP, Winbeck G, Asbrand C, et al. Hot spots in beta-catenin for interactions with LEF-1, conductin and APC. *Nat Struct Biol* 2000;**7**:800–7.

132. Gottardi CJ, Gumbiner BM. Adhesion signaling: how beta-catenin interacts with its partners. *Curr Biol* 2001;**11**:R792–4.

133. Ozawa M, Kemler R. Molecular organization of the uvomorulin–catenin complex. *J Cell Biol* 1992;**116**:989–96.

134. Cox RT, Kirkpatrick C, Peifer M. Armadillo is required for adherens junction assembly, cell polarity, and morphogenesis during *Drosophila* embryogenesis. *J Cell Biol* 1996;**134**:133–48.

135. Quinlan MP, Hyatt JL. Establishment of the circumferential actin filament network is a prerequisite for localization of the cadherin-catenin complex in epithelial cells. *Cell Growth Differ* 1999;**10**:839–54.

136. Weis WI, Nelson WJ. Re-solving the cadherin-catenin-actin conundrum. *J Biol Chem* 2006;**281**:35,593–35,597.

137. Gates J, Peifer M. Can 1000 reviews be wrong? Actin, alpha-catenin, and adherens junctions. *Cell* 2005;**123**:769–72.

138. Polakis P. Wnt signaling and cancer. *Genes Dev* 2000;**14**:1837–51.

139. Wodarz A, Nusse R. Mechanisms of Wnt signaling in development. *Annu Rev Cell Dev Biol* 1998;**14**:59–88.

140. Wheelock MJ, Shintani Y, Maeda M, Fukumoto Y, Johnson KR. Cadherin switching. *J Cell Sci* 2008;**121**:727–35.

141. Nakagawa S, Takeichi M. Neural crest emigration from the neural tube depends on regulated cadherin expression. *Development* 1998;**125**:2963–71.

142. Bixby JL, Zhang R. Purified N-cadherin is a potent substrate for the rapid induction of neurite outgrowth. *J Cell Biol* 1990;**110**:1253–60.

143. Matsunaga M, Hatta K, Nagafuchi A, Takeichi M. Guidance of optic nerve fibres by N-cadherin adhesion molecules. *Nature* 1988;**334**:62–4.

144. Riehl R, Johnson K, Bradley R, et al. Cadherin function is required for axon outgrowth in retinal ganglion cells *in vivo*. *Neuron* 1996;**17**:837–48.

145. Lee CH, Herman T, Clandinin TR, Lee R, Zipursky SL. N-cadherin regulates target specificity in the Drosophila visual system. *Neuron* 2001;**30**:437–50.

146. Williams EJ, Williams G, Howell FV, Skaper SD, Walsh FS, Doherty P. Identification of an N-cadherin motif that can interact with the fibroblast growth factor receptor and is required for axonal growth. *J Biol Chem* 2001;**276**:43,879–43,886.

147. Suyama K, Shapiro I, Guttman M, Hazan RB. A signaling pathway leading to metastasis is controlled by N-cadherin and the FGF receptor. *Cancer Cell* 2002;**2**:301–14.

148. Kintner C. Regulation of embryonic cell adhesion by the cadherin cytoplasmic domain. *Cell* 1992;**69**:225–36.

149. Yap AS, Niessen CM, Gumbiner BM. The juxtamembrane region of the cadherin cytoplasmic tail supports lateral clustering, adhesive strengthening, and interaction with p120ctn. *J Cell Biol* 1998;**141**:779–89.

150. Hatzfeld M. The p120 family of cell adhesion molecules. *Eur J Cell Biol* 2005;**84**:205–14.

151. Alema S, Salvatore AM. p120 catenin and phosphorylation: Mechanisms and traits of an unresolved issue. *Biochim Biophys Acta* 2007;**1773**:47–58.

152. Anastasiadis PZ, Reynolds AB. The p120 catenin family: complex roles in adhesion, signaling and cancer. *J Cell Sci* 2000;**113**:1319–34.

153. Ozawa M, Kemler R. The membrane–proximal region of the E-cadherin cytoplasmic domain prevents dimerization and negatively regulates adhesion activity. *J Cell Biol* 1998;**142**:1605–13.

154. Aono S, Nakagawa S, Reynolds AB, Takeichi M. p 120 (ctn) acts as an inhibitory regulator of cadherin function in colon carcinoma cells. *J Cell Biol* 1999;**145**:551–62.

155. Chen X, Kojima S, Borisy GG, Green KJ. p120 catenin associates with kinesin and facilitates the transport of cadherin–catenin complexes to intercellular junctions. *J Cell Biol* 2003;**163**:547–57.

156. Davis MA, Ireton RC, Reynolds AB. A core function for p120-catenin in cadherin turnover. *J Cell Biol* 2003;**163**:525–34.

157. Kowalczyk AP, Reynolds AB. Protecting your tail: regulation of cadherin degradation by p120-catenin. *Curr Opin Cell Biol* 2004;**16**:522–7.

158. Fujita Y, Krause G, Scheffner M, et al. Hakai, a c-Cbl-like protein, ubiquitinates and induces endocytosis of the E-cadherin complex. *Nat Cell Biol* 2002;**4**:222–31.

159. Xu G, Craig AW, Greer P, et al. Continuous association of cadherin with beta-catenin requires the non-receptor tyrosine-kinase Fer. *J Cell Sci* 2004;**117**:3207–19.

160. Teng J, Rai T, Tanaka Y, et al. The KIF3 motor transports N-cadherin and organizes the developing neuroepithelium. *Nat Cell Biol* 2005;**7**:474–82.

161. Yanagisawa M, Kaverina IN, Wang A, Fujita Y, Reynolds AB, Anastasiadis PZ. A novel interaction between kinesin and p120 modulates p120 localization and function. *J Biol Chem* 2004;**279**:9512–21.

162. Hall A. Rho GTPases and the actin cytoskeleton. *Science* 1998;**279**:509–14.

163. Anastasiadis PZ. p120-ctn: A nexus for contextual signaling via Rho GTPases. *Biochim Biophys Acta* 2007;**1773**:34–46.

164. Grosheva I, Shtutman M, Elbaum M, Bershadsky AD. p120 catenin affects cell motility via modulation of activity of Rho-family GTPases: a link between cell–cell contact formation and regulation of cell locomotion. *J Cell Sci* 2001;**114**:695–707.

165. Anastasiadis PZ, Moon SY, Thoreson MA, et al. Inhibition of RhoA by p120 catenin. *Nat Cell Biol* 2000;**2**:637–44.

166. Fukata M, Kaibuchi K. Rho-family GTPases in cadherin-mediated cell-cell adhesion. *Nat Rev Mol Cell Biol* 2001;**2**:887–97.

167. Kuroda S, Fukata M, Nakagawa M, et al. Role of IQGAP1, a target of the small GTPases Cdc42 and Rac1, in regulation of E-cadherin- mediated cell–cell adhesion. *Science* 1998;**281**:832–5.

168. Fukata M, Kuroda S, Nakagawa M, et al. Cdc42 and Rac1 regulate the interaction of IQGAP1 with beta-catenin. *J Biol Chem* 1999;**274**:26,044–26,050.

169. Meigs TE, Fedor-Chaiken M, Kaplan DD, Brackenbury R, Casey PJ. Galpha 12 and galpha 13 negatively regulate the adhesive functions of cadherin. *J Biol Chem* 2002;**277**:24,594–24,600.

170. Kaplan DD, Meigs TE, Casey PJ. Distinct regions of the cadherin cytoplasmic domain are essential for functional interaction with Galpha 12 and beta-catenin. *J Biol Chem* 2001;**276**:44,037–44,043.

171. Huber AH, Weis WI. The structure of the beta-catenin/E-cadherin complex and the molecular basis of diverse ligand recognition by beta-catenin. *Cell* 2001;**105**:391–402.

172. Lickert H, Bauer A, Kemler R, Stappert J. Casein kinase II phosphorylation of E-cadherin increases E-cadherin/beta-catenin interaction and strengthens cell-cell adhesion. *J Biol Chem* 2000;**275**:5090–5.

173. Hamaguchi M, Matsuyoshi N, Ohnishi Y, Gotoh B, Takeichi M, Nagai Y. p60v-src causes tyrosine phosphorylation and inactivation of the N-cadherin–catenin cell adhesion system. *EMBO J* 1993;**12**:307–14.

174. Roura S, Miravet S, Piedra J, Garcia de Herreros A, Dunach M. Regulation of E-cadherin/Catenin association by tyrosine phosphorylation. *J Biol Chem* 1999;**274**:36,734–36,740.

175. Piedra J, Martinez D, Castano J, Miravet S, Dunach M, de Herreros AG. Regulation of beta-catenin structure and activity by tyrosine phosphorylation. *J Biol Chem* 2001;**276**:20,436–20,443.

176. Behrens J, Vakaet L, Friis R, et al. Loss of epithelial differentiation and gain of invasiveness correlates with tyrosine phosphorylation of the E-cadherin/beta-catenin complex in cells transformed with a temperature-sensitive v-SRC gene. *J Cell Biol* 1993;**120**:757–66.

177. Rosato R, Veltmaat JM, Groffen J, Heisterkamp N. Involvement of the tyrosine kinase fer in cell adhesion. *Mol Cell Biol* 1998;**18**:5762–70.

178. Kotelevets L, van Hengel J, Bruyneel E, Mareel M, van Roy F, Chastre E. The lipid phosphatase activity of PTEN is critical for stabilizing intercellular junctions and reverting invasiveness. *J Cell Biol* 2001;**155**:1129–35.

179. Hoschuetzky H, Aberle H, Kemler R. Beta-catenin mediates the interaction of the cadherin-catenin complex with epidermal growth factor receptor. *J Cell Biol* 1994;**127**:1375–80.

180. Hazan RB, Norton L. The epidermal growth factor receptor modulates the interaction of E-cadherin with the actin cytoskeleton. *J Biol Chem* 1998;**273**:9078–84.

181. Xu Y, Guo DF, Davidson M, Inagami T, Carpenter G. Interaction of the adaptor protein Shc and the adhesion molecule cadherin. *J Biol Chem* 1997;**272**:13,463–13,466.

182. Lilien J, Balsamo J, Arregui C, Xu G. Turn-off, drop-out: Functional state switching of cadherins. *Dev Dyn* 2002;**224**:18–29.

183. Brady-Kalnay SM, Mourton T, Nixon JP, et al. Dynamic interaction of PTPmu with multiple cadherins *in vivo*. *J Cell Biol* 1998;**141**:287–96.

184. Phillips-Mason PJ, Gates TJ, Major DL, Sacks DB, Brady-Kalnay SM. The receptor protein-tyrosine phosphatase PTPmu interacts with IQGAP1. *J Biol Chem* 2006;**281**:4903–10.

185. Rosdahl JA, Ensslen SE, Niedenthal JA, Brady-Kalnay SM. PTP mu-dependent growth cone rearrangement is regulated by Cdc42. *J Neurobiol* 2003;**56**:199–208.

186. Pathre P, Arregui C, Wampler T, et al. PTP1B regulates neurite extension mediated by cell–cell and cell–matrix adhesion molecules. *J Neurosci Res* 2001;**63**:143–50.

187. Burden-Gulley SM, Brady-Kalnay SM. PTPmu regulates N-cadherin-dependent neurite outgrowth. *J Cell Biol* 1999;**144**:1323–36.

188. Ozawa M, Kemler R. Correct proteolytic cleavage is required for the cell adhesive function of uvomorulin. *J Cell Biol* 1990;**111**:1645–50.

189. Posthaus H, Dubois CM, Laprise MH, Grondin F, Suter MM, Muller E. Proprotein cleavage of E–cadherin by furin in baculovirus over-expression system: potential role of other convertases in mammalian cells. *FEBS Letts* 1998;**438**:306–10.

190. Paradies NE, Grunwald GB. Purification and characterization of NCAD90, a soluble endogenous form of N-cadherin, which is generated by proteolysis during retinal development and retains adhesive and neurite-promoting function. *J Neurosci Res* 1993;**36**:33–45.

191. Roark EF, Paradies NE, Lagunowich LA, Grunwald GB. Evidence for endogenous proteases, mRNA level and insulin as multiple mechanisms of N-cadherin down-regulation during retinal development. *Development* 1992;**114**:973–84.

192. Birchmeier C, Birchmeier W, Brand–Saberi B. Epithelial-mesenchymal transitions in cancer progression. *Acta Anat* 1996;**156**:217–26.

193. Lochter A, Galosy S, Muschler J, Freedman N, Werb Z, Bissell MJ. Matrix metalloproteinase stromelysin-1 triggers a cascade of molecular alterations that leads to stable epithelial-to-mesenchymal conversion and a premalignant phenotype in mammary epithelial cells. *J Cell Biol* 1997;**139**:1861–72.

194. Ito K, Okamoto I, Araki N, et al. Calcium influx triggers the sequential proteolysis of extracellular and cytoplasmic domains of E-cadherin, leading to loss of beta-catenin from cell–cell contacts. *Oncogene* 1999;**18**:7080–90.

195. Maretzky T, Reiss K, Ludwig A, et al. ADAM10 mediates E-cadherin shedding and regulates epithelial cell-cell adhesion, migration, and beta-catenin translocation. *Proc Natl Acad Sci USA* 2005;**102**:9182–7.

196. Ponnuchamy B, Khalil RA. Role of ADAMs in endothelial cell permeability: cadherin shedding and leukocyte rolling. *Circ Res* 2008;**102**:1139–42.

197. Steinhusen U, Weiske J, Badock V, Tauber R, Bommert K, Huber O. Cleavage and shedding of E-cadherin after induction of apoptosis. *J Biol Chem* 2001;**276**:4972–80.

198. Herren B, Levkau B, Raines EW, Ross R. Cleavage of beta-catenin and plakoglobin and shedding of VE-cadherin during endothelial apoptosis: evidence for a role for caspases and metalloproteinases. *Mol Biol Cell* 1998;**9**:1589–601.

199. Marambaud P, Shioi J, Serban G, et al. A presenilin-1/gamma-secretase cleavage releases the E-cadherin intracellular domain and regulates disassembly of adherens junctions. *Embo J* 2002;**21**:1948–56.

200. Baki L, Marambaud P, Efthimiopoulos S, et al. Presenilin-1 binds cytoplasmic epithelial cadherin, inhibits cadherin/p120 association, and regulates stability and function of the cadherin/catenin adhesion complex. *Proc Natl Acad Sci USA* 2001;**98**:2381–6.

201. Kang DE, Soriano S, Frosch MP, et al. Presenilin 1 facilitates the constitutive turnover of beta-catenin: differential activity of Alzheimer's disease-linked PS1 mutants in the beta-catenin-signaling pathway. *J Neurosci* 1999;**19**:4229–37.

202. Nishimura M, Yu G, Levesque G, et al. Presenilin mutations associated with Alzheimer disease cause defective intracellular trafficking of beta-catenin, a component of the presenilin protein complex. *Nat Med* 1999;**5**:164–9.

203. Soriano S, Kang DE, Fu M, et al. Presenilin 1 negatively regulates beta-catenin/T cell factor/lymphoid enhancer factor-1 signaling independently of beta-amyloid precursor protein and notch processing. *J Cell Biol* 2001;**152**:785–94.

204. Georgakopoulos A, Marambaud P, Efthimiopoulos S, et al. Presenilin-1 forms complexes with the cadherin/catenin cell–cell adhesion system and is recruited to intercellular and synaptic contacts. *Mol Cell* 1999;**4**:893–902.

205. Murase S, Mosser E, Schuman EM. Depolarization drives beta-Catenin into neuronal spines promoting changes in synaptic structure and function. *Neuron* 2002;**35**:91–105.

206. Niessen CM, Gumbiner BM. Cadherin-mediated cell sorting not determined by binding or adhesion specificity. *J Cell Biol* 2002;**156**:389–99.

207. Luo Y, Ferreira-Cornwell M, Baldwin H, et al. Rescuing the N-cadherin knockout by cardiac-specific expression of N- or E-cadherin. *Development* 2001;**128**:459–69.

Chapter 6

In vivo Functions of Heterotrimeric G Proteins

Stefan Offermanns

Department of Pharmacology, Max-Planck-Institute, Bad Nauheim, Germany

INTRODUCTION

The transmembrane signaling system, which uses heterotrimeric G proteins to couple heptahelical receptors to various effectors, operates in all cells of the mammalian organism and is involved in many physiological and pathological processes [1]. The main properties of individual G proteins are determined by the identity of their α subunits. To elucidate the role of G-protein-mediated signaling processes in the intact mammalian organism, almost all known genes encoding G-protein α subunits have been inactivated by gene targeting in mice (Table 6.1). Also, several mouse lines have been reported carrying targeted mutations of Gβ- or Gγ-genes. This short review summarizes the main phenotypical changes observed in mice lacking G-protein α subunits.

DEVELOPMENT

Various Gα-deficient mouse models have pointed to the involvement of G-protein-mediated signaling pathways in certain developmental processes. For example, lack of Gα_{13} results in embryonic lethality at about mid-gestation due to a defect in angiogenesis [2], which most likely is due to the lack of Gα_{13} in embryonic endothelial cells [3]. Mice deficient in both Gα_q and Gα_{11} suffer from a defect in heart development and die *in utero* (see below). In addition, signaling through G$_q$ class members has also been implicated in the proliferation and/or migration of craniofacial neural crest cells [4–6]. The complete loss of Gα_s in mice homozygous for an inactivating Gα_s mutation leads to embryonic lethality before embryonic day 10 [7]. It is interesting that heterozygotes show varying phenotypes depending on the paternal origin of the intact allele; these are probably caused by genetic haploinsufficiency and/or tissue-specific imprinting of the maternal Gα_s allele [8, 9].

CENTRAL NERVOUS SYSTEM

In the central nervous system (CNS), many mediators and neurotransmitters function through G-protein-coupled receptors to modulate neuronal activity or morphology. Neurotransmitters that induce an inhibitory modulation typically act on receptors that are coupled to members of the G$_{i/o}$ family, whereas G$_q$ and G$_s$ family members are primarily involved in excitatory responses.

The G-protein G$_o$ is highly abundant in the mammalian nervous system, and has been shown to mediate inhibition of neuronal (N-, P/Q-, R-type) voltage-dependent Ca^{2+} channels via its $\beta\gamma$-complex, thereby reducing the excitability of the cell. Gα_o-deficient mice suffer from tremors and have occasional seizures [10, 11]. In addition, Gα_o-deficient mice appear to be hyperalgesic when tested in the hot plate assay [10]. The latter finding is consistent with the observation that opioid receptor-mediated inhibition of Ca^{2+} currents in dorsal root ganglia (DRG) from Gα_o-deficient animals was reduced by about 30 percent compared to those in wild-type DRGs [10].

G$_z$, a member of the G$_{i/o}$-family of G proteins, shares with G$_{i1}$, G$_{i2}$, and G$_{i3}$ the ability to inhibit adenylyl cyclases but has a rather limited pattern of expression, being found in brain, adrenal medulla, and platelets. Gα_z-deficient mice exhibit altered responses to a variety of psychoactive drugs. Cocaine-induced increases in locomotor activity were more pronounced, and short-term antinociceptive effects of morphine were altered [12, 13]. In addition, behavioral effects

TABLE 6.1 Phenotypical changes in mice lacking a-subunits of heterotrimeric G-proteins

Class/type	Gene	Expression	Effector(s)	Phenotype	Reference
Gα_s class					
Gα_s[1]	Gnas	Ubiquitous	AC (all types) ↑	Embryonic lethal[4]	[7]
				Osteoblast-restricted: reduced bone turnover	[44]
				Chondrocyte-restricted: epiphyseal and growth plate abnormalities	[45]
				Hepatocyte-restricted: increased glucose tolerance	[46]
				Juxtaglomerular cell restricted: reduced renin formation	[47]
				Pancreatic β cells: reduced β cell mass, diabetes	[48]
				Hematopoietic system: defective engraftment of haematopoietic stem cells in bone marrow	[49]
Gα_{sXL}	(GnasXL)	Neuroendocrine	AC ↑	Perinatal lethal	[50, 51]
Gα_{olf}	Gnal	Olfactory epithelium, brain	AC ↑	Anosmia, hyperactivity	[14]
G$\alpha_{i/o}$ class					
Gα_{i1}	Gnai1	Widely distributed	AC ↓[5]	Impaired memory formation	[52]
Gα_{i2}	Gnai2	Ubiquitous		Inflammatory bowel disease	[22, 23, 53]
Gα_{i3}	Gnai3	Widely distributed		Lack of anti-autophagic action of insulin	[54]
Gα_o[2]	Gnao	Neuronal, neuroendocrine	VDCC↓, GIRK↑[6]	Various CNS defects	[10, 11, 33, 55]
Gα_z	Gnaz	Neuronal, platelets	AC (e.g. V,VI) ↓	Viable, increased bleeding time	[12, 13]
Gα_{gust}	Gnat3	Taste cells, brush cells	PDE ↑?	Impaired bitter and sweet sensation	[34]
Gα_{t-r}	Gnat1	Retinal rods, taste cells	PDE 6 (rod) ↑	Mild retinal degeneration	[32]
Gα_{t-c}	Gnat2	Retinal cones	PDE 6 (cone) ↑	Achromatopsia	[56]
Gα_{i1}+Gα_{i3}				Viable, immunological defects	[57]
Gα_{i2}+Gα_{i3}				Lethal[9]	[54]
G$\alpha_{q/11}$ class					
Gα_q	Gnaq	Ubiquitous	PLC-β1-4 ↑	Ataxia, defective platelet activation	[18, 38]
Gα_{11}	Gna11	Almost ubiquitous	PLC-β1-4 ↑	No obvious phenotype seen so far	[6]
Gα_{14}	Gna14	Kidney, lung, spleen	PLC-β1-4 ↑	No obvious phenotype seen so far[10]	
Gα_{15}	Gna15[3]	Hematopoietic cells	PLC-β1-4 ↑	No obvious phenotype seen so far	[58]
Gα_q+Gα_{11}				Myocardial hypoplasia (lethal e11)	[6]

Class/type	Gene	Expression	Effector(s)	Phenotype	Reference
				Cardiomyocyte-restricted: pressure overload induced hypertrophy ↓	[27]
				Nervous system-restricted: perinatal lethal	[59]
				Neural crest-restricted: craniofacial defects	[4]
				Forebrain-restricted: abnormal mothering behaviors; epilepsy, impaired endo-cannabinoid formation	[60, 61]
				Parathyroid-restricted: hypercalcemia, hyperparathyroidism	[21]
				Thyrocyte-restricted: impaired thyroid function and goiter development	[62]
				Smooth muscle-restricted: hypotension, salt-dependent hypertension ↓	[63]
				Endothelial cell restricted: protection against anaphylactic shock	[30]
G$\alpha_{12/13}$ class					
Gα_{12}	*Gna12*	Ubiquitous	RhoGEF[7]	No obvious phenotype seen so far	[64]
Gα_{13}	*Gna13*	Ubiquitous	RhoGEF[8]	Defective angiogenesis (lethal e9.5)	[2]
				Platelet-restricted : activation defect	[42]
Gα_q+Gα_{13}				Severe platelet defect	[43]
Gα_{12}+Gα_{13}				Embryonic lethal (e8.5)	[64]
				Platelet-restricted : like *Gna13$^{-/-}$*	[62]
				Neural crest-restricted: cardiac defects	[4]
				B-cell-restricted: lack of marginal zone B-cells	[24, 25]
				Smooth muscle-restricted: salt-dependent hypertension ↓	[63]
				T cell-restricted: increased susceptibility towards T cell-mediated diseases	[26]

[1]*Several splice variants;*
[2]*two splice variants;*
[3]*human ortholog: GNA16;*
[4]*parent of origin specific defects in heterozygotes;*
[5]*types I, III, V, VI, VIII, IX;*
[6]*via G$\beta\gamma$;*
[7]*PDZ-RhoGEF/LARG+Btk, Gap1m, Cadherin;*
[8]*p115RhoGEF, PDZ-RhoGEF/LARG+radixin;*
[9]*L. Birnbaumer, M. Jiang, G. Boulay, K. Spicher (personal communication);*
[10]*H. Jiang and M.I. Simon (personal communication); AC, adenylyl cyclase; PDE, phosphodiesterase; PLC, phospholipase C; GIRK, G-protein regulated inward rectifier potassium channel; VDCC, voltage-dependent Ca^{2+}-channel; RhoGEF, Rho guanine nucleotide exchange factor.*

of catecholamine reuptake inhibitors were abolished in Gα_z-deficient mice [13], indicating that G$_z$ is involved in signaling processes regulated by various neurotransmitters.

Gα_{olf} is expressed in various regions of the CNS, including olfactory sensory neurons and basal ganglia. Gα_{olf}-deficient mice exhibit clear motoric abnormalities such as hypermotoric behavior [14]. Recent data indicate that G$_{olf}$ is critically involved in dopamin(D$_1$)- and adenosine(A$_{2A}$)-receptor-mediated effects in the striatum [15, 16].

The two main members of the G$_q$ family, G$_q$ and G$_{11}$, are widely expressed in the central nervous system. Mice lacking Gα_q develop an ataxia with clear signs of motor coordination deficits, and functional defects could be observed in the cerebellar cortex of Gα_q-deficient mice [17, 18]. In addition, lack of Gα_q resulted in defective cerebellar and hippocampal long-term depression [19, 20], and loss of both, Gα_q and Gα_{11} in principal neurons of the forebrain leads to a decreased on-demand formation of endocannabinoids and epilepsy [21].

IMMUNE SYSTEM

Mice lacking $G\alpha_{i2}$ develop a lethal, diffuse inflammatory bowel disease that resembles in many aspects ulcerative colitis in humans [22]. In subsequent studies, dramatic changes in the phenotype and function of intestinal lymphocytes and epithelial cells have been described that are likely to be due to defective lymphocyte homing in enteric epithelia [23]. On a cellular level, G_{i2} may be involved in the regulation of T cell function and trafficking. These processes are regulated through chemoattractant and chemokine receptors that show a predominant coupling to G_i-type G proteins. In addition to the colitis, many $G\alpha_{i2}$-deficient mice develop colonic adenocarcinomas, which are probably secondary to colonic inflammation [22].

Recently, G_{12}/G_{13} have also been found to be involved in defined immunological functions. Mice lacking the α subunits of G_{12} and G_{13} specifically in B cells lack a normal splenic marginal zone due to defects in marginal zone B cell migration [24]. This defect may be based on a role of G_{12}/G_{13} in mediating the lysophospholipid-induced integrin activation of splenic B cells [25]. Animals with T cell-specific $G\alpha_{12}/G\alpha_{13}$ deficiency show altered T cell adhesiveness and mobility, resulting in increased susceptibility towards T cell-mediated diseases [26].

HEART

The $G\alpha_q/G\alpha_{11}$-mediated signaling pathway appears to play a pivotal role in the regulation of physiological myocardial growth during embryogenesis. This is demonstrated by the phenotype of $G\alpha_q/G\alpha_{11}$ double-deficient mice that die at embryonic day 11 due to a severe thinning of the myocardial layer of the heart [6]. Adult cardiomyocytes are terminally differentiated post-mitotic cells that respond to stimulatory signals with cell growth rather than proliferation. Myocardial hypertrophy in the adult heart following mechanical stress depends on $G\alpha_q/G\alpha_{11}$-mediated signaling as demonstrated by the absence of a hypertrophic response in adult mice with cardiomyocyte-specific $G\alpha_q/G\alpha_{11}$ deficiency [27].

Inhibition of L-type Ca^{2+} channels in the heart through muscarinic M_2 receptors was found to be abrogated in hearts lacking $G\alpha_o$ as well as $G\alpha_{i2}$ [11, 28]. This unexpected finding suggests that both G proteins may regulate this downstream signaling event in a complex fashion.

Recently, G proteins of the G_q/G_{11} and G_{12}/G_{13} family have been specifically deleted in endothelial cells and smooth muscle cells. Lack of G_q/G_{11}-mediated signaling in vascular smooth muscle cells resulted in a decrease in the basal blood pressure, whereas G_{12}/G_{13} deficiency was without effect. However, salt-induced hypertension was severely affected by both $G\alpha_q/G\alpha_{11}$ and $G\alpha_{12}/G\alpha_{13}$ deficiency [29], indicating that the G_q/G_{11}-mediated signaling pathway in vascular smooth muscle cells is required for maintaining basal blood pressure, while both G_q/G_{11}- and G_{12}/G_{13}-mediated signaling in vascular smooth muscle cells is required for salt-dependent hypertension. Lack of $G\alpha_q/G\alpha_{11}$ or $G\alpha_{12}/G\alpha_{13}$ in endothelial cells did not have acute effects on the blood pressure. However, most of the acute effects of inflammatory and anaphylactic mediators on the vessel wall were absent in endothelial cell-specific $G\alpha_q/G\alpha_{11}$-deficient mice which were also protected against the deleterious effects of severel systemic anaphylactic reactions [30].

SENSORY SYSTEMS

Odors, light, and many tastants act directly on G-protein-coupled receptors. The G protein G_{olf} is centrally involved in the transduction of odorant stimuli in olfactory cilia, and $G\alpha_{olf}$-deficient mice exhibit dramatically reduced electrophysiological responses to all odors tested [14]. Since nursing and mothering behavior in rodents is mediated a great deal by the olfactory system, most $G\alpha_{olf}$-deficient pups die a few days after birth due to insufficient feeding, and rarely survive their mothers' inadequate maternal behavior. In contrast to the olfactory epithelium, the vomeronasal organ, which detects pheromones, expresses receptors that are coupled to $G_{i/o}$. Absence of $G\alpha_o$ results in apoptotic death of receptor cells that usually express $G\alpha_o$ [31].

Rod-transducin (G_{t-r}) and cone-transducin (G_{t-c}) play well-established roles in the phototransduction cascade in the outer segments of retinal rods and cones, where they couple light receptors to cGMP-phosphodiesterase. In mice lacking $G\alpha_{t-r}$, the majority of retinal rods do not respond to light any more, and these animals develop mild retinal degeneration with age [32]. The light response is transferred from the receptor cell to bipolar cells of the retina. In mice lacking $G\alpha_o$, modulation of ON bipolar cells in response to light is abrogated, indicating that G_o is critically involved in the tonic inhibition of these cells mediated by metabotropic glutamate (mGluR6) receptors [33].

Among the four taste qualities – sweet, bitter, sour, and salty – bitter and sweet tastes appear to signal through heterotrimeric G-proteins. Gustducin is a G protein mainly expressed in taste cells, and $G\alpha_{gust}$-deficient mice show impaired electrophysiological and behavioral responses to bitter and sweet agents [34]. The residual bitter- and sweet-taste responsiveness of $G\alpha_{gust}$-deficient mice could be further diminished by a dominant-negative mutant of gustducin-α, suggesting the involvement of other G proteins related to $G\alpha_{gust}$ [35]. Interestingly, recent evidence indicates that gustducin-mediated signaling is also involved in the regulation of various cells of the enteric epithelium [36, 37].

HEMOSTASIS

Hemostasis is a complex process involving platelet adhesion and aggregation as well as formation of fibrin through the coagulation cascade. Platelet activation results in a rapid shape-change reaction immediately followed by secretion of granule contents, as well as inside-out activation of the fibrinogen receptor, integrin $\alpha_{IIb}\beta_3$, leading to platelet aggregation. Most physiological platelet activators act through G-protein-coupled receptors, which in turn activate $G_{i2/3}$, G_q, G_{12}, and G_{13}. In platelets from $G\alpha_q$-deficient mice, the effect of various platelet stimuli on aggregation and degranulation was abrogated, demonstrating that $G\alpha_q$-mediated phospholipase C activation represents an essential event in platelet activation [38]. However, platelet shape change can still be induced in the absence of $G\alpha_q$, indicating that it is mediated by G proteins other than G_q, most likely G_{12}/G_{13} [39]. The defective activation of $G\alpha_q$-deficient platelets results in a primary hemostasis defect, and $G\alpha_q$ $(-/-)$ mice are protected against platelet-dependent thromboembolism.

The role of G proteins of the $G_{i/o}$ family in platelet activation has recently been elucidated. Platelets contain at least three members of this class: G_{i2}, G_{i3}, and G_z. ADP, which is released from activated platelets and functions as a positive feedback mediator during platelet activation, induces platelet activation through the G_q-coupled $P2Y_1$ receptor as well as through the G_i-coupled $P2Y_{12}$ purinergic receptor. The general importance of the G_i-mediated pathway is indicated by the fact that responses to ADP but also to thrombin were markedly reduced in platelets lacking $G\alpha_{i2}$ [40, 41]. In contrast to ADP or thrombin, epinephrine is not a full platelet activator *per se* in murine platelets. However, it is able to potentiate the effect of other platelet stimuli. In platelets from $G\alpha_z$-deficient mice, epinephrine's potentiating effects were clearly impaired, while the effects of other platelet activators appeared to be unaffected by the lack of $G\alpha_z$ [13]. Thus, members of the G-protein families G_q, G_{12}, and $G_{i/o}$ are involved in processes leading to platelet activation.

Interestingly, activation of platelets by various stimuli was severely inhibited in platelets lacking $G\alpha_{13}$, but not in $G\alpha_{12}$-deficient platelets [42]. These defects were accompanied by a reduced activation of the RhoA-mediated signaling pathway as well as by an inability to form stable platelet thrombi under high sheer stress conditions. In addition, mice carrying $G\alpha_{13}$-deficient platelets have an increased bleeding time and are protected against the formation of arterial thrombi [42]. Various studies with $G\alpha$-deficient platelets have clearly shown that three G proteins are the major mediators of platelet activation: G_q, G_{i2}, and G_{13}. While in the absence of either G_q, G_{i2}, or G_{13} some platelet activation can still be induced, in the absence of both $G\alpha_q$ and $G\alpha_{13}$, platelets are completely unresponsive to various stimuli [43].

CONCLUSIONS

Mouse models lacking almost all known genes encoding G-protein α subunits have been generated, and they provide a first insight into the biological roles of G-protein-mediated signaling pathways. To overcome embryonic lethality or complex phenotypes of some $G\alpha$-null mutations and to understand the degree of functional redundancy of closely related G proteins, researchers have begun to cross individual mutants and to generate mouse lines that allow for the conditional inactivation of genes in a time- and tissue-specific manner. These approaches will soon provide more detailed views on the functions of G-protein-mediated signaling pathways in the developing and adult mammalian organism.

REFERENCES

1. Wettschureck N, Offermanns S. Mammalian G proteins and their cell type specific functions. *Physiol Rev* 2005;**85**:1159–204.

2. Offermanns S, Mancino V, Revel JP, Simon MI. Vascular system defects and impaired cell chemokinesis as a result of Galpha13 deficiency. *Science* 1997;**275**:533–6.

3. Ruppel KM, Willison D, Kataoka H, Wang A, Zheng YW, Cornelissen I, Yin L, Xu SM, Coughlin SR. Essential role for Galpha13 in endothelial cells during embryonic development. *Proc Natl Acad Sci USA* 2005;**102**:8281–6.

4. Dettlaff-Swiercz DA, Wettschureck N, Moers A, Huber K, Offermanns S. Characteristic defects in neural crest cell-specific Galphaq/Galpha11- and Galpha12/Galpha13-deficient mice. *Dev Biol* 2005;**282**:174–82.

5. Ivey K, Tyson B, Ukidwe P, McFadden DG, Levi G, Olson EN, Srivastava D, Wilkie TM. Galphaq and Galpha11 proteins mediate endothelin-1 signaling in neural crest-derived pharyngeal arch mesenchyme. *Dev Biol* 2003;**255**:230–7.

6. Offermanns S, Zhao LP, Gohla A, Sarosi I, Simon MI, Wilkie TM. Embryonic cardiomyocyte hypoplasia and craniofacial defects in G alpha q/G alpha 11-mutant mice. *EMBO J* 1998;**17**:4304–12.

7. Yu S, Yu D, Lee E, Eckhaus M, Lee R, Corria Z, Accili D, Westphal H, Weinstein LS. Variable and tissue-specific hormone resistance in heterotrimeric Gs protein alpha-subunit (Gsalpha) knockout mice is due to tissue-specific imprinting of the gsalpha gene. *Proc Natl Acad Sci USA* 1998;**95**:8715–20.

8. Weinstein LS, Xie T, Zhang QH, Chen M. Studies of the regulation and function of the G(s)alpha gene Gnas using gene targeting technology. *Pharmacol Ther* 2007;**115**:271–91.

9. Weinstein LS, Yu S. The role of genomic imprinting of Galpha in the pathogenesis of Albright Hereditary Osteodystrophy. *Trends Endocrinol Metab* 1999;**10**:81–5.

10. Jiang M, Gold MS, Boulay G, Spicher K, Peyton M, Brabet P, Srinivasan Y, Rudolph U, Ellison G, Birnbaumer L. Multiple neurological abnormalities in mice deficient in the G protein Go. *Proc Natl Acad Sci USA* 1998;**95**:3269–74.

11. Valenzuela D, Han X, Mende U, Fankhauser C, Mashimo H, Huang P, Pfeffer J, Neer EJ, Fishman MC. G alpha(o) is necessary for muscarinic regulation of Ca2+ channels in mouse heart. *Proc Natl Acad Sci USA* 1997;**94**:1727–32.

12. Hendry IA, Kelleher KL, Bartlett SE, Leck KJ, Reynolds AJ, Heydon K, Mellick A, Megirian D, Matthaei KI. Hypertolerance to morphine in G(z alpha)-deficient mice. *Brain Res* 2000;**870**:10–19.

13. Yang J, Wu J, Kowalska MA, Dalvi A, Prevost N, O'Brien PJ, Manning D, Poncz M, Lucki I, Blendy JA, Brass LF. Loss of signaling

through the G protein, Gz, results in abnormal platelet activation and altered responses to psychoactive drugs. *Proc Natl Acad Sci USA* 2000;**97**:9984–9.

14. Belluscio L, Gold GH, Nemes A, Axel R. Mice deficient in G(olf) are anosmic. *Neuron* 1998;**20**:69–81.

15. Corvol JC, Studler JM, Schonn JS, Girault JA, Herve D. Galpha(olf) is necessary for coupling D1 and A2a receptors to adenylyl cyclase in the striatum. *J Neurochem* 2001;**76**:1585–8.

16. Zhuang X, Belluscio L, Hen R. G(olf)alpha mediates dopamine D1 receptor signaling. *J Neurosci* 2000;**20**:RC91.

17. Hartmann J, Blum R, Kovalchuk Y, Adelsberger H, Kuner R, Durand GM, Miyata M, Kano M, Offermanns S, Konnerth A. Distinct roles of Galpha(q) and Galpha11 for Purkinje cell signaling and motor behavior. *J Neurosci* 2004;**24**:5119–30.

18. Offermanns S, Hashimoto K, Watanabe M, Sun W, Kurihara H, Thompson RF, Inoue Y, Kano M, Simon MI. Impaired motor coordination and persistent multiple climbing fiber innervation of cerebellar Purkinje cells in mice lacking Galphaq. *Proc Natl Acad Sci USA* 1997;**94**:14,089–14,094.

19. Kleppisch T, Voigt V, Allmann R, Offermanns S. G(alpha)q-deficient mice lack metabotropic glutamate receptor-dependent long-term depression but show normal long-term potentiation in the hippocampal CA1 region. *J Neurosci* 2001;**21**:4943–8.

20. Miura M, Watanabe M, Offermanns S, Simon MI, Kano M. Group I metabotropic glutamate receptor signaling via Galpha q/Galpha 11 secures the induction of long-term potentiation in the hippocampal area CA1. *J Neurosci* 2002;**22**:8379–90.

21. Wettschureck N, Lee E, Libutti SK, Offermanns S, Robey PG, Spiegel AM. Parathyroid-specific double knockout of Gq and G11 alpha-subunits leads to a phenotype resembling germline knockout of the extracellular Ca^{2+}-sensing receptor. *Mol Endocrinol* 2007;**21**:274–80.

22. Rudolph U, Finegold MJ, Rich SS, Harriman GR, Srinivasan Y, Brabet P, Bradley A, Birnbaumer L. Gi2 alpha protein deficiency: a model of inflammatory bowel disease. *J Clin Immunol* 1995;**15**:101S–105S.

23. Hornquist CE, Lu X, Rogers-Fani PM, Rudolph U, Shappell S, Birnbaumer L, Harriman GR. G(alpha)i2-deficient mice with colitis exhibit a local increase in memory CD4+ T cells and proinflammatory Th1-type cytokines. *J Immunol* 1997;**158**:1068–77.

24. Rieken S, Sassmann A, Herroeder S, Wallenwein B, Moers A, Offermanns S, Wettschureck N. G12/G13 family G proteins regulate marginal zone B cell maturation, migration, and polarization. *J Immunol* 2006;**177**:2985–93.

25. Rieken S, Herroeder S, Sassmann A, Wallenwein B, Moers A, Offermanns S, Wettschureck N. Lysophospholipids control integrin-dependent adhesion in splenic B cells through G(i) and G(12)/G(13) family G-proteins but not through G(q)/G(11). *J Biol Chem* 2006;**281**:36,985–36,992.

26. Herroeder S, Reichardt P, Sassmann A, Zimmermann B, Jaeneke D, Hoeckner J, Hollmann MW, Fischer KD, Vogt S, Grosse R, Hogg N, Gunzer M, Offermanns S, Wettschureck N. Guanine nucleotide-binding proteins of the G(12) family shape immune functions by controlling CD4(+) T cell adhesiveness and motility. *Immunity* 2009;**30**:1–13.

27. Wettschureck N, Rutten H, Zywietz A, Gehring D, Wilkie TM, Chen J, Chien KR, Offermanns S. Absence of pressure overload induced myocardial hypertrophy after conditional inactivation of Galphaq/Galpha11 in cardiomyocytes. *Nat Med* 2001;**7**:1236–40.

28. Chen F, Spicher K, Jiang M, Birnbaumer L, Wetzel GT. Lack of muscarinic regulation of Ca(2+) channels in G(i2)alpha gene knockout mouse hearts. *Am J Physiol Heart Circ Physiol* 2001;**280**:H1989–95.

29. Wirth A, Benyo Z, Lukasova M, Leutgeb B, Wettschureck N, Gorbey S, Örsy P, Horvath B, Maser-Gluth C, Greiner E, Lemmer B, Schutz G, Gutkind JS, Offermanns S. G12-G13-LARG-mediated signaling in vascular smooth muscle is required for salt-induced hypertension. *Nat Med* 2008;**14**:64–8.

30. Korhonen H, Fisslthaler B, Moers A, Wirth A, Habermehl D, Wieland T, Schutz G, Wettschureck N, Fleming I, Offermanns S. Anaphylactic shock depends on endothelial Gq/G11. *J Exp Med* 2009;**206**:411–20.

31. Tanaka M, Treloar H, Kalb RG, Greer CA, Strittmatter SM. G(o) protein-dependent survival of primary accessory olfactory neurons. *Proc Natl Acad Sci USA* 1999;**96**:14,106–14,111.

32. Calvert Jr. PD, Krasnoperova NV, Lyubarsky AL, Isayama T, Nicolo M, Kosaras B, Wong G, Gannon KS, Margolskee RF, Sidman RL, Pugh EN, Makino CL, Lem J. Phototransduction in transgenic mice after targeted deletion of the rod transducin alpha -subunit. *Proc Natl Acad Sci USA* 2000;**97**:13,913–13,918.

33. Dhingra Jr. A, Lyubarsky A, Jiang M, Pugh EN, Birnbaumer L, Sterling P, Vardi N. The light response of ON bipolar neurons requires G[alpha]o. *J Neurosci* 2000;**20**:9053–8.

34. Wong GT, Gannon KS, Margolskee RF. Transduction of bitter and sweet taste by gustducin. *Nature* 1996;**381**:796–800.

35. Ruiz-Avila L, Wong GT, Damak S, Margolskee RF. Dominant loss of responsiveness to sweet and bitter compounds caused by a single mutation in alpha -gustducin. *Proc Natl Acad Sci USA* 2001;**98**:8868–73.

36. Jang HJ, Kokrashvili Z, Theodorakis MJ, Carlson OD, Kim BJ, Zhou J, Kim HH, Xu X, Chan SL, Juhaszova M, Bernier M, Mosinger B, Margolskee RF, Egan JM. Gut-expressed gustducin and taste receptors regulate secretion of glucagon-like peptide-1. *Proc Natl Acad Sci USA* 2007;**104**:15,069–15,074.

37. Margolskee RF, Dyer J, Kokrashvili Z, Salmon KS, Ilegems E, Daly K, Maillet EL, Ninomiya Y, Mosinger B, Shirazi-Beechey SP. T1R3 and gustducin in gut sense sugars to regulate expression of Na+-glucose cotransporter 1. *Proc Natl Acad Sci USA* 2007;**104**:14,887–14,888.

38. Offermanns S, Toombs CF, Hu YH, Simon MI. Defective platelet activation in G alpha(q)-deficient mice. *Nature* 1997;**389**:183–6.

39. Klages B, Brandt U, Simon MI, Schultz G, Offermanns S. Activation of G12/G13 results in shape change and Rho/Rho-kinase-mediated myosin light chain phosphorylation in mouse platelets. *J Cell Biol* 1999;**144**:745–54.

40. Jantzen HM, Milstone DS, Gousset L, Conley PB, Mortensen RM. Impaired activation of murine platelets lacking G alpha(i2). *J Clin Invest* 2001;**108**:477–83.

41. Yang J, Wu J, Jiang H, Mortensen R, Austin S, Manning DR, Woulfe D, Brass LF. Signaling through Gi family members in platelets. Redundancy and specificity in the regulation of adenylyl cyclase and other effectors. *J Biol Chem* 2002;**277**:46,035–46,042.

42. Moers A, Nieswandt B, Massberg S, Wettschureck N, Gruner S, Konrad I, Schulte V, Aktas B, Gratacap MP, Simon MI, Gawaz M, Offermanns S. G13 is an essential mediator of platelet activation in hemostasis and thrombosis. *Nat Med* 2003;**9**:1418–22.

43. Moers A, Wettschureck N, Gruner S, Nieswandt B, Offermanns S. Unresponsiveness of Platelets Lacking Both G{alpha}q and G{alpha}13: implications for collagen-induced platelet activation. *J Biol Chem* 2004;**279**:45,354–45,359.

44. Sakamoto A, Chen M, Nakamura T, Xie T, Karsenty G, Weinstein LS. Deficiency of the G-protein alpha-subunit G(s)alpha in osteoblasts leads to differential effects on trabecular and cortical bone. *J Biol Chem* 2005;**280**:21,369–21,375.

45. Sakamoto A, Chen M, Kobayashi T, Kronenberg HM, Weinstein LS. Chondrocyte-specific knockout of the G protein G(s)alpha leads to epiphyseal and growth plate abnormalities and ectopic chondrocyte formation. *J Bone Miner Res* 2005;**20**:663–71.

46. Chen M, Gavrilova O, Zhao W-Q, Nguyen A, Lorenzo J, Shen L, Nackers L, Pack S, Jou W, Weinstein LS. Increased glucose tolerance and reduced adiposity in the absence of fasting hypoglycemia in mice with liver-specific Gsalpha deficiency. *J Clin Invest* 2005;**115**:3217–27.

47. Chen L, Kim SM, Oppermann M, Faulhaber-Walter R, Huang Y, Mizel D, Chen M, Lopez ML, Weinstein LS, Gomez RA, Briggs JP, Schnermann J. Regulation of renin in mice with Cre recombinase-mediated deletion of G protein Gsalpha in juxtaglomerular cells. *Am J Physiol Renal Physiol* 2007;**292**:F27–37.

48. Xie T, Chen M, Zhang QH, Ma Z, Weinstein LS. Beta cell-specific deficiency of the stimulatory G protein alpha-subunit Gsalpha leads to reduced beta cell mass and insulin-deficient diabetes. *Proc Natl Acad Sci USA* 2007;**104**:19,601–19,606.

49. Adams GB, Alley IR, Chung UI, Chabner KT, Jeanson NT, Lo Celso C, Marsters ES, Chen M, Weinstein LS, Lin CP, Kronenberg HM, Scadden DT. Haematopoietic stem cells depend on Galpha(s)-mediated signalling to engraft bone marrow. *Nature* 2009;**459**:103–7.

50. Plagge A, Gordon E, Dean W, Boiani R, Cinti S, Peters J, Kelsey G. The imprinted signaling protein XL alpha s is required for postnatal adaptation to feeding. *Nat Genet* 2004;**36**:818–26.

51. Xie T, Plagge A, Gavrilova O, Pack S, Jou W, Lai EW, Frontera M, Kelsey G, Weinstein LS. The alternative stimulatory G protein alpha-subunit XLalphas is a critical regulator of energy and glucose metabolism and sympathetic nerve activity in adult mice. *J Biol Chem* 2006;**281**:18,989–18,999.

52. Pineda VV, Athos JI, Wang H, Celver J, Ippolito D, Boulay G, Birnbaumer L, Storm DR. Removal of G(ialpha1) constraints on adenylyl cyclase in the hippocampus enhances LTP and impairs memory formation. *Neuron* 2004;**41**:153–63.

53. He J, Gurunathan S, Iwasaki A, Ash-Shaheed B, Kelsall BL. Primary role for Gi protein signaling in the regulation of interleukin 12 production and the induction of T helper cell type 1 responses. *J Exp Med* 2000;**191**:1605–10.

54. Gohla A, Klement K, Piekorz RP, Pexa K, vom Dahl S, Spicher K, Dreval V, Haussinger D, Birnbaumer L, Nurnberg B. An obligatory requirement for the heterotrimeric G protein Gi3 in the antiautophagic action of insulin in the liver. *Proc Natl Acad Sci USA* 2007;**104**:3003–8.

55. Dhingra A, Jiang M, Wang TL, Lyubarsky A, Savchenko A, Bar-Yehuda T, Sterling P, Birnbaumer L, Vardi N. Light response of retinal ON bipolar cells requires a specific splice variant of Galpha(o). *J Neurosci* 2002;**22**:4878–84.

56. Chang B, Dacey MS, Hawes NL, Hitchcock PF, Milam AH, Atmaca-Sonmez P, Nusinowitz S, Heckenlively JR. Cone photoreceptor function loss-3, a novel mouse model of achromatopsia due to a mutation in Gnat2. *Invest Ophthalmol Vis Sci* 2006;**47**:5017–21.

57. Fan H, Zingarelli B, Peck OM, Teti G, Tempel GE, Halushka PV, Spicher K, Boulay G, Birnbaumer L, Cook JA. Lipopolysaccharide- and gram-positive bacteria-induced cellular inflammatory responses: role of heterotrimeric Galpha(i) proteins. *Am J Physiol Cell Physiol* 2005;**289**:C293–301.

58. Davignon I, Catalina MD, Smith D, Montgomery J, Swantek J, Croy J, Siegelman M, Wilkie TM. Normal hematopoiesis and inflammatory responses despite discrete signaling defects in Galpha15 knockout mice. *Mol Cell Biol* 2000;**20**:797–804.

59. Wettschureck N, Moers A, Wallenwein B, Parlow AF, Maser-Gluth C, Offermanns S. Loss of Gq/11 family G proteins in the nervous system causes pituitary somatotroph hypoplasia and dwarfism in mice. *Mol Cell Biol* 2005;**25**:1942–8.

60. Wettschureck N, Moers A, Hamalainen T, Lemberger T, Schutz G, Offermanns S. Heterotrimeric G proteins of the Gq/11 family are crucial for the induction of maternal behavior in mice. *Mol Cell Biol* 2004;**24**:8048–54.

61. Wettschureck N, van der Stelt M, Tsubokawa H, Krestel H, Moers A, Petrosino S, Schutz G, Di Marzo V, Offermanns S. Forebrain-specific inactivation of Gq/G11 family G proteins results in age-dependent epilepsy and impaired endocannabinoid formation. *Mol Cell Biol* 2006;**26**:5888–94.

62. Kero J, Ahmed K, Wettschureck N, Tunaru S, Wintermantel T, Greiner E, Schutz G, Offermanns S. Thyrocyte-specific G(q)/G(11) deficiency impairs thyroid function and prevents goiter development. *J Clin Invest* 2007;**117**:2399–407.

63. Wirth A, Benyó Z, Lukasova M, Leutgeb B, Wettschureck N, Gorbey S, Örsy P, Horváth B, Maser-Gluth C, Greiner E, Lemmer B, Schütz G, Gutkind S, Offermanns S. G12/G13-LARG-mediated signalling in vascular smooth muscle is required for salt-induced hypertension. *Nat Med* 2007;**14**:64–8.

64. Gu JL, Muller S, Mancino V, Offermanns S, Simon MI. Interaction of G alpha(12) with G alpha(13) and G alpha(q) signaling pathways. *Proc Natl Acad Sci USA* 2002;**99**:9352–7.

Chapter 7

G-Protein Signaling in Chemotaxis

Jonathan Franca-Koh, Stacey Sedore Willard and Peter N. Devreotes
Department of Cell Biology, Johns Hopkins University School of Medicine, Baltimore, Maryland

INTRODUCTION

Chemotaxis is the directed migration of cells in response to concentration gradients of extracellular signals. In unicellular organisms, such as bacteria and amoebae, chemotaxis is frequently used as a foraging mechanism [1]. In multicellular organisms, it ensures that the right cells get to the right place at the right time during development, and plays an essential role in processes such as wound healing and inflammation [2, 3]. Chemotaxis is also a contributing factor to many diseases. For example, metastatic cancer cells migrate toward stereotypic regions of the body that promote further growth, and the unregulated chemotaxis of immune cells can lead to inflammatory diseases such as asthma and arthritis.

Much of our current understanding of chemotaxis-signaling pathways through G-protein-coupled receptors (GPCRs) is derived from studies on the social amoeba, *Dictyostelium discoideum*, and mammalian neutrophils (this term will be used to refer to both primary neutrophils and HL60s, a neutrophil-like cell line). *Dictyostelium* cells feed on microorganisms that they track down by chemotaxis towards secreted metabolites such as folic acid. More dramatic, however, is the response of this organism to starvation. The individual amoebae aggregate and, through a series of morphogenetic changes and cell-fate choices, form multicellular structures containing spores that can survive starvation. The process of aggregation is directed by gradients of cAMP, and can easily be studied under physiologically relevant conditions using combined genetic, biochemical, and cell biological analyses [1]. Neutrophils are important cells of the immune system, and are most frequently studied in the context of chemotaxis to either formyl-Met-Leu-Phe (fMLP) or chemokines – chemoattractants that regulate inflammation *in vivo*. Neutrophils from knockout mice and cell lines that can be manipulated with retroviruses are available. As studies in these two systems have revealed many similarities, distinctions will only be made when differences have been observed.

CHEMOTAXIS: MEMBRANE EXTENSIONS, DIRECTIONAL SENSING, AND POLARIZATION

Chemotaxis can be thought of as the result of three separate processes: membrane extensions, directional sensing, and polarization [2, 4]. Membrane extensions are the periodic pseudopods and blebs that cells make at regular intervals, and drive cell motility [5–7]. In *Dictyostelium*, membrane extensions can occur in cells lacking functional heterotrimeric G proteins [8]. Neutrophils, though, are relatively quiescent in the absence of ligand. Directional sensing refers to the capacity of chemotactic cells to sense the direction of external gradients and localize proteins or reactions towards or away from the high concentration. This process obviously requires receptor/G-protein signaling, but can occur when cell movement is inhibited. Polarization refers to the elongated cell morphology and the stable localization of molecules to the anterior and posterior poles that is acquired by neutrophils and starved *Dictyostelium* cells during chemotaxis. Polarization depends on the cytoskeleton as well as chemoattractant receptor/G-protein signaling, but does not require a gradient.

CHEMOATTRACTANT SIGNALING REGULATES MULTIPLE DOWNSTREAM PATHWAYS

Recent advances in our understanding of the molecular mechanisms that regulate chemotaxis have revealed the important and diverse roles played by G proteins [9, 10].

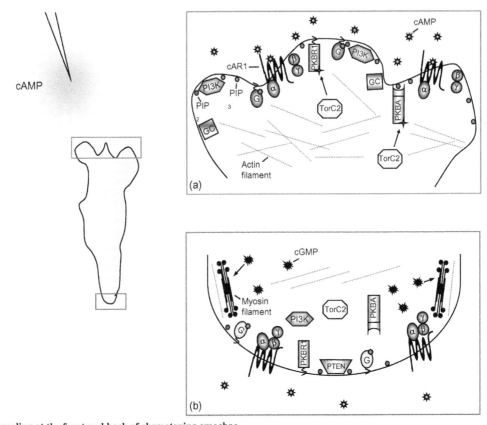

FIGURE 7.1 Signaling at the front and back of chemotaxing amoebae.
Panels (a) and (b) illustrate some key signaling components that are localized to the front and back of migrating cells in a gradient of cAMP. At the front (a), cAMP binding to cAR1 results in PI3K recruitment, production of PIP3, PKBA translocation to the membrane, GTPase (such as Ras, Rap and Rac; gray ovals) activation, PKB phosphorylation (white stars) by TorC2, sGC activation and F-actin polymerization. These signaling events, as well as others (see text for an expanded list), are required for efficient chemotaxis. At the back (b), PI3K is cytosolic and PTEN is localized to the membrane where it degrades PIP3 to PIP2. In addition, Rho is activated (light gray oval) and cGMP regulates myosin II filament formation.

These studies not only highlight the critical function of G proteins as "molecular switches," but also show how their signaling in the context of chemotactic signaling networks allows cells to translate the directional information of external concentration gradients into directional movement.

Downstream of GPCRs, many signal transduction events are initiated via heterotrimeric G proteins. *In vivo*, chemoattractant binding triggers a rapid dissociation or rearrangement of Gα and Gβγ subunits. Within seconds, this leads to activation of the small G proteins Ras, Rho, Rac, Cdc42 and Rap; the increase or decrease of the second messengers phosphatidylinositol (3′,4′,5′) trisphosphate (PIP3), arachidonic acid, diacylglycerol (DAG), inositol trisphosphate (IP3), cAMP, cGMP, Ca^{2+} and H$^+$ ions; and stimulation of the kinases protein kinase A (PKA), protein kinase C (PKC), target of rapamycin (Tor), mitogen activated protein kinase (MAPK), protein kinase B (PKB), and a PKB-related kinase (PKB-R1). Interestingly, although the heterotrimeric G-protein complex is thought to remain dissociated as long as receptors are occupied, most of the downstream pathways are only transiently activated in response to a uniform stimulus and return to

basal levels within a few minutes (see below) [11, 12]. A key breakthrough in understanding how this signaling network controls chemotactic migration was the finding that in a gradient, many responses are persistently activated and become asymmetrically localized and oriented according to the direction of the gradient (Figure 7.1) [13].

FRONT AND BACK SIGNALING

PIP3 was the first molecule found to have an asymmetric localization in a gradient, and has served as a model for understanding the temporal and spatial activation of chemotactic signal transduction pathways [14]. In *Dictyostelium*, the correct orientation of PIP3 in a gradient is achieved by the coordinated regulation of phosphatidylinositol-3′-kinase (PI3K) and phosphatase and tensin homolog deleted on chromosome 10 (PTEN) [15, 16]. PI3K produces PIP3 by phosphorylating the 3′-hydroxl group of phosphatidylinositol (4′,5′) bisphosphate (PIP2) and PTEN catalyzes the reverse reaction. In response to chemoattractant, PI3K is rapidly recruited from the cytosol to the plasma membrane, where it is likely activated by binding to Ras-GTP [17, 18]. Conversely, PTEN is bound to the plasma membrane of resting cells and with stimulation it dissociates. In a uniform stimulus, the response is transient, as PI3K and PTEN return to their original locations after a few minutes. In a gradient, however, PI3K is persistently bound to the front and PTEN is restricted to the back, resulting in steady-state PIP3 accumulation at the front (Figure 7.1). Previous studies have demonstrated that PIP3 can recruit proteins to the plasma membrane via pleckstrin homology (PH) domains, indicating that this may be a mechanism to localize downstream effectors [18–20].

Recent work has provided some insight into the mechanisms of PI3K and PTEN localization in *Dictyostelium*. The N-terminal domains from PI3K isoforms 1 and 2 are necessary and sufficient for cAMP-dependent membrane translocation [15, 17]. Furthermore, this work has shown that PI3K also appears to localize to the membrane and in a narrow band adjacent to the membrane. Treating cells with latrunculin A, an inhibitor of actin polymerization, impairs localization, suggesting that PI3K recruitment to the cell cortex may depend on the cytoskeleton [21]. For PTEN, it has been shown that the N-terminus contains an amphipathic "PIP2 binding motif," and that this stretch of about 15 amino acids is essential for membrane binding [22]. A recent study suggests that signaling through phospholipase C (PLC), which degrades PIP2, may play a role in controlling PTEN localization. In *plc⁻* cells, PTEN does not dissociate from the membrane during stimulation, whereas in cells overexpressing PLC, PTEN is not associated with the membrane [23]. Interestingly, some cAMP analogs, by coupling the receptor to different Gα proteins, can inhibit PLC and thereby act as repellents [24].

PIP3 also marks the front of neutrophils, suggesting that chemoattractant regulation of PIP3 metabolizing enzymes occurs in these cells. The recruitment of the PI3Kγ catalytic subunit is dependent on the interaction with the p101 regulatory subunit and is regulated by Gβγ (coupled to Gαᵢ) and Ras [25, 26]. In migrating neutrophils, PI3Kγ is found in a broad region at the leading edge. The requirement for binding to Gβγ and Ras-GTP may further confine PI3K activity to an even narrower region. Compared with the *Dictyostelium* enzyme, less mammalian PTEN is associated with the plasma membrane, but its binding can be detected at the single molecule level by Total Internal Reflection Fluorescence (TIRF) microscopy [27]. Membrane association is essential for activity and depends on the conserved "PIP2 binding motif." The C2 domain has also been implicated in membrane binding as mutations in this domain have been found to inhibit lipid binding *in vitro* [28]. Other evidence suggests that phosphorylation and interactions with binding proteins may be important for localization. Mutating phosphorylated residues on the C-terminus to alanine is thought to favor an "open" conformation and strongly enhances membrane recruitment. The interaction of

PTEN with several membrane proteins via its PDZ domain may also play a role [29]. However, it is somewhat controversial whether membrane binding occurs preferentially at the back and sides of migrating neutrophils [25, 30].

Although many studies have highlighted the deleterious effects of elevated PIP3, there is now general agreement that chemotaxis is less severely impaired when PIP3 production is inhibited. *Dictyostelium* amoebae lacking PTEN are defective in their ability to degrade PIP3 and chemotax poorly due to the production of numerous lateral pseudopods [16]. Chemotaxis defects due to high PIP3 levels are also seen in neutrophils, although the role of PTEN is less clear in this system. One study found that chemotaxis, PIP3 levels, and actin polymerization are normal in *pten⁻/⁻* cells, and instead suggest that SHIP1, which removes the 5′ phosphate from PIP3, is the key regulator of PIP3 in neutrophils [31]. *Ship1⁻/⁻* neutrophils were found to have a prolonged PIP3 response and a chemotaxis defect similar to that of *pten⁻* amoebae. A second study reported that *pten⁻/⁻* neutrophils have marginally elevated levels of PKB phosphorylation and actin polymerization, but do not have a strong chemotaxis defect [32]. Several groups have also looked at chemotaxis in conditions where PIP3 production is inhibited. Most recently, *Dictyostelium* cells lacking all type I PI3Ks and PI3Kγ⁻/⁻ neutrophils were found still to perform chemotaxis relatively well [33, 34]. Similar results have also been obtained in cells where PI3K activity was inhibited pharmacologically [35–37].

The limited effects of inhibiting PIP3 production clearly suggest that other pathways may act in parallel, and this has been substantiated by recent results. In *Dictyostelium*, loss of phospholipase A2 (PLA2) activity, either through inhibitors or genetic manipulation, does not have a significant effect [35, 38]. However, when combined with a loss of PI3K function, chemotaxis is severely impaired. PLA2 cleaves phospholipids to produce free fatty acids (such as arachidonic acid) and lyso-phospholipids. Additionally, it appears that the activity of this enzyme is regulated by chemoattractant [38]. It remains unclear what the downstream effects of this pathway are, and whether PLA2 enzymes play a similar role in neutrophils. There is also increasing evidence for the role of the TorC2 complex in regulating chemotaxis. *Ras interacting protein 3* (*Rip3*) and *Pianissimo* (*PiaA*), were originally isolated as chemotactic mutants in *Dictyostelium* [39, 40]. The homologs of these proteins, Sin1/Avo1 and Rictor/Avo3, respectively, were subsequently found to be part of the highly conserved TorC2 complex that is thought to play a critical role in regulating PKB activity [41–43]. This function appears to be conserved in *Dictyostelium*, which has two PKB homologs: PKB-A and PKB-R1. Furthermore, in chemotaxing cells, activation of TorC2 is localized to the leading edge (Yoichiro Kamimura, personal communication).

Small GTPases play important roles in regulating actin polymerization and myosin II function in neutrophils and

Dictyostelium [10]. For example, activated Rac and Cdc42 localize to the front of neutrophils and are thought to play a key role in initiating actin polymerization [44]. Consistent with this hypothesis, expression of dominant-negative Rac1 inhibits neutrophil migration and actin polymerization, while dominant-negative Cdc42 prevents neutrophils from maintaining a persistent leading edge [45]. *Dictyostelium* has 17 Rac isoforms which, at the sequence level, cannot be divided into specific Rac, Rho, or Cdc42 homologs, and for simplicity have been named RacA–Q. Many have been knocked out, and RacB, C, and G are reported to have defects in chemotaxis and actin polymerization [46, 47]. Additionally, cells overexpressing dominant-negative RacB have reduced pseudopod extension and migration, as do cells lacking RacGEF1 [46]. The small G proteins Rho and Rap have been implicated as regulators of myosin II function. Myosin II (a hexameric enzyme composed of two myosin heavy chains (MHC), two essential light chains (ELC), and two regulatory light chains (RLC)) is a key regulator of chemotaxis which is thought to both facilitate the retraction of the cell rear and to suppress lateral pseudopods through its actin crosslinking and motor protein functions [48]. Both of these functions depend on multiple myosin II molecules assembling into bipolar filaments that can then associate with cortical actin cytoskeleton. Rap is activated at the front of *Dictyostelium* cells, and expressing constitutively active Rap inhibits myosin II filament formation, possibly by promoting the phosphorylation of MHC [49, 50]. This may be mediated either directly or indirectly by Phg2, a Rap effector kinase that is required for chemoattractant-stimulated MHC phosphorylation [51]. In contrast, Rho is localized to the back and sides of neutrophils and is thought to promote myosin II motor activity by phosphorylating the RLC through p120 ROCK [25]. Inhibiting this kinase impairs RLC phosphorylation and leads to increased lateral pseudopod production, an indicator of reduced myosin II activity. A similar result is seen when dominant-negative Rho is expressed. In neutrophils, this is probably regulated by $G\alpha_{12/13}$ as pertussis toxin (PT), which inhibits $G\alpha_i$ but not $G\alpha_{12/13}$, does not inhibit Rho activation. Consistent with the important role of Rho at the back and not the front, PT-treated cells fail to generate pseudopods but extend uropods at the back when a chemoattractant gradient is applied. An analogous pathway may exist in *Dictyostelium* involving another Rac isoform and p21 activated kinase A (PakA), a Rac effector. This protein is reported to co-localize with myosin II at the back, and cells lacking PakA appear to have a defect in myosin II filament assembly [52, 53].

Other proteins and reactions have also been found to localize to the front of migrating amoebae, including myosin II heavy chain kinase (MHCK-A), a Na$^+$–H$^+$ exchanger (NHE), and soluble Guanylate Cyclase (sGC). MHCK-A is concentrated at the front by associating with newly polymerized F-actin, and can inhibit myosin II filament assembly by phosphorylating the heavy chain [54]. NHE mutants make increased numbers of lateral pseudopods and, given that myosin filament assembly is enhanced by an acidic pH *in vitro*, the concentration of this protein at the leading edge may inhibit filament assembly in this region by making the cytosol more alkaline [55]. The binding of sGC to the membrane is essential for guanylate cyclase activity, and although this protein is recruited to the leading edge, cGMP diffuses throughout the cell and has mainly been implicated in promoting myosin II filament assembly through cGMP binding protein C (GbpC) at the back [56, 57]. Cells lacking either sGC or GbpC have strong chemotaxis defects that can be attributed to dramatic reduction in the levels of myosin II at the actin cortex. Amoebae lacking $G\alpha_9$ have decreased cGMP levels, suggesting a regulatory role for this protein.

MECHANISMS OF DIRECTIONAL SENSING

The asymmetric localization of the molecules discussed above raises the question, How do cells orient these events based on a chemoattractant gradient? Models based on Local Excitation–Global Inhibition (LEGI) have proved very successful at explaining many features of directional sensing [4, 58]. In these models, it is proposed that receptor–ligand binding triggers at least two signals: an excitatory signal that is turned on rapidly and diffuses slowly (local excitation), and an inhibitory signal that is turned on slowly and diffuses rapidly (global inhibition). These models can explain responses to both uniform and gradient stimuli. In LEGI models, cells respond transiently to a uniform stimulus because excitation occurs more rapidly than inhibition. Thus the initial activation of downstream pathways is attenuated over time as the slower forming global inhibitor builds up (Figure 7.2a). In a gradient, the LEGI model accurately predicts that downstream signaling will be persistently activated at the front. Since diffusion of the excitatory signals is slow, the level of excitation at each point along the cell membrane reflects the receptor occupancy at that site, and is higher at the front than at the back. In contrast, since the inhibitor is freely diffusible, inhibition will be averaged across the cell. Consequently, at steady state, excitation will exceed inhibition at the front but not at the back (Figure 7.2d). In this way the cell translates the directional information of the gradient into differences between front and back, and can readily adjust to changes in the temporal–spatial pattern of stimulation.

At what point in the pathway does this asymmetry occur? By expressing CFP and YFP fusions of $G\alpha$ and $G\beta$, the dissociation state of the heterotrimeric G proteins can be monitored by FRET [12, 59]. In immobilized amoebae, the dissociation of heterotrimeric G proteins matches the

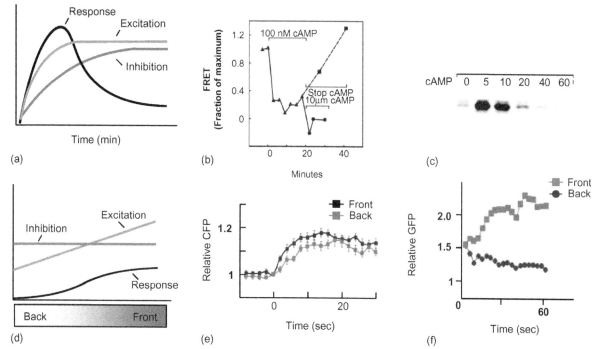

FIGURE 7.2 Temporal and spatial dynamics of chemotactic signaling.
Panels (a)–(c) illustrate the temporal dynamics of signals according to the LEGI model. In (a), the graphic representation of the LEGI model shows that many signaling responses ("response") are transient after a uniform stimulus, since build-up of the diffusible inhibitor ("inhibition") eventually dampens the quickly diffusing excitation signal ("excitation"). Not all responses are transient, however. In (b), FRET studies with fluorescently labeled Gα and Gβγ show that once the Gβγ complex dissociates, it remains dissociated as long as steady-state levels of the stimulus are present. When the stimulus is removed, the FRET response returns and additionally, increased concentrations of cAMP elicit further Gβγ dissociation (adapted from [12]). In contrast, in (c), precipitation of activated Ras shows that the GTPase remains active only transiently following cAMP stimulation (adapted from [8]). Panels (d)–(f) refer to spatial regulation of signaling. The LEGI model predicts that in a gradient of chemoattractant, responses (black line in (d)) will only be seen at the front of the cell, since excitation is greater than inhibition here. Panel (e) displays G-protein dissociation in a gradient of cAMP. An increase in Gα-CFP signal indicates Gβγ dissociation in this experiment, which can be seen at higher levels at the front of the cell than at the back (taken from [11]). Note that this response reflects the shallow gradient outside the cell. In (f), Ras binding domain-GFP localizes very strongly to the front of the cell; the intensity of this response indicates that amplification of this signal has occurred downstream of G proteins (adapted from [11]).

steepness of the gradient across the cells (Figure 7.2b, d). The earliest localized events are activation of Ras (Figure 7.2c, f) and loss of PTEN at the front [16, 21]. Thus, directional sensing must occur in between the G proteins and these downstream events, possibly by regulating the localization or activity of a RasGEF.

While LEGI models accurately describe the behaviors of proteins within immobilized cells, they cannot account for certain features of chemotaxis displayed by polarized cells. In particular, LEGI predicts that when challenged with a change in the direction of the gradient, cells should respond by establishing a new anterior–posterior axis. Polarized cells such as starved *Dictyostelium* cells and neutrophils, however, typically respond by turning, and thus maintain the same front and back regions [6, 36]. Careful analysis of pseudopod extensions has also revealed other interesting points. First, cells tend to form pseudopods by splitting existing pseudopods, indicating that one outcome of polarization is to restrict the regions in the cell that can produce pseudopod extensions. Second, once a pseudopod

splits, the cell appears to make a choice to maintain one pseudopod or the other based on which one is closer to the chemoattractant source. The mechanism cells use to make the "right" choice could be similar to the LEGI model we have described for directional sensing. These observations indicate that the interplay between directional sensing and polarization mechanisms must be accounted for in a complete description of chemotaxis.

POLARIZATION

Feedback loops are the key to establishing and maintaining polarization. Positive feedback loops amplify the absolute level of front or back signaling respectively, while negative feedback loops serve to increase the separation of these two pathways in space. In neutrophils, a positive feedback loop has been identified in the PIP3–Rac–actin polymerization pathway. First, introducing PIP3 lipid directly into neutrophils is sufficient to make neutrophils polarize and migrate [60, 61]. As this lipid is degraded over time, a positive feedback loop is required to account for the persistence of these effects. Second, inhibiting actin polymerization attenuates both PIP3 production and Rac activation, even in the presence of constant stimulus. As actin is thought to be downstream of both PIP3 and Rac, this indicates that persistent and robust activation requires an actin-dependent positive feedback loop. The ability of several molecules to localize to the leading edge by associating with newly polymerized F-actin provides a possible mechanism for this feedback loop. As F-actin is polymerized, proteins that promote PIP3 production and Rac activation are recruited to the leading edge and thereby initiate more actin polymerization. PI3K, Ras and RacGEF1 in *Dictyostelium* and PI3Kγ in neutrophils are possible candidates. There is also evidence for the existence of negative feedback pathways. Expressing constitutively active Rho in neutrophils inhibits actin polymerization and Rac activation, while expressing activated Rac inhibits GTP exchange of Rho and the assembly of contractile myosin II filaments in the rear [25]. The effect of Rho is probably mediated by myosin II filament assembly, as expressing activated myosin II RLC inhibits actin and Rac. How Rac inhibits Rho and myosin II in neutrophils is less clear. One possibility is that, as in *Dictyostelium*, actin polymerization may recruit MHCKs to the leading edge. Whatever the mechanism, two recent experiments highlight the importance of this negative feedback loop in neutrophils. First, whereas untreated cells turn when the gradient is reversed, cells treated with an inhibitor of p120 ROCK retract the original pseudopod and extend a new one towards the chemoattractant source [25]. Second, neutrophils that are treated with latrunculin B, to inhibit actin polymerization, have a reversed localization of activated Rho in a gradient [62]. These data indicate that this feedback loop is critical for maintaining polarization and for restricting Rho and myosin II activity to the rear.

CONCLUSION

Chemotaxis can be viewed as a modular process composed of membrane extensions, directional sensing, and polarization. G proteins play a central role in regulating each of these modules, and we are beginning to understand how the signal transduction pathways they regulate are controlled in space and time. Future work will need to examine how these signaling networks interact, and new models need to be developed that can account for both directional sensing and polarization.

ACKNOWLEDGEMENTS

The authors wish to acknowledge the NIH for funding and the members of the PND laboratory for useful discussions; in particular Yoichiro Kamimura, who shared unpublished data.

REFERENCES

1. Willard SS, Devreotes PN. Signaling pathways mediating chemotaxis in the social amoeba, Dictyostelium discoideum. *Eur J Cell Biol* 2006;**85**:897–904.

2. Franca-Koh J, Kamimura Y, Devreotes P. Navigating signaling networks: chemotaxis in Dictyostelium discoideum. *Curr Opin Genet Dev* 2006;**16**:333–8.

3. Rickert P, Weiner OD, Wang F, Bourne HR, Servant G. Leukocytes navigate by compass: roles of PI3Kgamma and its lipid products. *Trends Cell Biol* 2000;**10**:466–73.

4. Devreotes P, Janetopoulos C. Eukaryotic chemotaxis: distinctions between directional sensing and polarization. *J Biol Chem* 2003;**278**:20,445–20,448.

5. Weiner OD, Marganski WA, Wu LF, Altschuler SJ, Kirschner MW. An actin-based wave generator organizes cell motility. *PLoS Biol* 2007:5.

6. Andrew N, Insall RH. Chemotaxis in shallow gradients is mediated independently of PtdIns 3-kinase by biased choices between random protrusions. *Nat Cell Biol* 2007;**9**:193–200.

7. Yoshida K, Soldati T. Dissection of amoeboid movement into two mechanically distinct modes. *J Cell Sci* 2006;**119**:3833–44.

8. Sasaki AT, Janetopoulos C, Lee S, Charest PG, Takeda K, Sundheimer LW, Meili R, Devreotes PN, Firtel RA. G protein-independent Ras/PI3K/F-actin circuit regulates basic cell motility. *J Cell Biol* 2007;**178**:185–91.

9. Manahan CL, Iglesias PA, Long Y, Devreotes PN. Chemoattractant signaling in dictyostelium discoideum. *Annu Rev Cell Dev Biol* 2004;**20**:223–53.

10. Charest PG, Firtel RA. Big roles for small GTPases in the control of directed cell movement. *Biochem J* 2007;**401**:377–90.

11. Xu X, Meier-Schellersheim M, Jiao X, Nelson LE, Jin T. Quantitative imaging of single live cells reveals spatiotemporal dynamics of multistep signaling events of chemoattractant gradient sensing in *Dictyostelium*. *Mol Biol Cell* 2005;**16**:676–88.

12. Janetopoulos C, Jin T, Devreotes P. Receptor-mediated activation of heterotrimeric G-proteins in living cells. *Science* 2001;**291**:2408–11.

13. Parent CA, Devreotes PN. A cell's sense of direction. *Science* 1999;**284**:765–70.

14. Parent CA, Blacklock BJ, Froehlich WM, Murphy DB, Devreotes PN. G protein signaling events are activated at the leading edge of chemotactic cells. *Cell* 1998;**95**:81–91.

15. Funamoto S, Meili R, Lee S, Parry L, Firtel RA. Spatial and temporal regulation of 3-phosphoinositides by PI 3-kinase and PTEN mediates chemotaxis. *Cell* 2002;**109**:611–23.

16. Iijima M, Devreotes P. Tumor suppressor PTEN mediates sensing of chemoattractant gradients. *Cell* 2002;**109**:599–610.

17. Huang YE, Iijima M, Parent CA, Funamoto S, Firtel RA, Devreotes P. Receptor-mediated regulation of PI3Ks confines PI(3,4,5)P3 to the leading edge of chemotaxing cells. *Mol Biol Cell* 2003;**14**:1913–22.

18. Funamoto S, Milan K, Meili R, Firtel RA. Role of phosphatidylinositol 3′ kinase and a downstream pleckstrin homology domain-containing protein in controlling chemotaxis in dictyostelium. *J Cell Biol* 2001;**153**:795–810.

19. Comer FI. Lippincott CK, Masbad JJ, Parent CA: The PI3K-mediated activation of CRAC independently regulates adenylyl cyclase activation and chemotaxis. *Curr Biol* 2005;**15**:134–9.

20. Meili R, Ellsworth C, Lee S, Reddy TB, Ma H, Firtel RA. Chemoattractant-mediated transient activation and membrane localization of Akt/PKB is required for efficient chemotaxis to cAMP in *Dictyostelium*. *EMBO J* 1999;**18**:2092–105.

21. Sasaki AT, Chun C, Takeda K, Firtel RA. Localized Ras signaling at the leading edge regulates PI3K, cell polarity, and directional cell movement. *J Cell Biol* 2004;**167**:505–18.

22. Iijima M, Huang YE, Luo HR, Vazquez F, Devreotes PN. Novel mechanism of PTEN regulation by its phosphatidylinositol 4,5-bisphosphate binding motif is critical for chemotaxis. *J Biol Chem* 2004;**279**:16,606–16,613.

23. Korthol A, King JS, Keizer-Gunnink I, Harwood AJ, Van Haastert PJ. Phospholipase C regulation of phosphatidylinositol 3,4,5-trisphosphate-mediated chemotaxis. *Mol Biol Cell* 2007;**18**:4772–9.

24. Keizer-Gunnink I, Kortholt A, Van Haastert PJ. Chemoattractants and chemorepellents act by inducing opposite polarity in phospholipase C and PI3-kinase signaling. *J Cell Biol* 2007;**177**:579–85.

25. Xu J, Wang F, Van Keymeulen A, Herzmark P, Straight A, Kelly K, Takuwa Y, Sugimoto N, Mitchison T, Bourne HR. Divergent signals and cytoskeletal assemblies regulate self-organizing polarity in neutrophils. *Cell* 2003;**114**:201–14.

26. Suire S, Condliffe AM, Ferguson GJ, Ellson CD, Guillou H, Davidson K, Welch H, Coadwell J, Turner M, Chilvers ER, et al. Gbetagammas and the Ras binding domain of p110gamma are both important regulators of PI(3)Kgamma signalling in neutrophils. *Nat Cell Biol* 2006;**8**:1303–9.

27. Vazquez F, Matsuoka S, Sellers WR, Yanagida T, Ueda M, Devreotes PN. Tumor suppressor PTEN acts through dynamic interaction with the plasma membrane. *Proc Natl Acad Sci USA* 2006;**103**:3633–8.

28. Lee JO, Yang H, Georgescu MM, Di Cristofano A, Maehama T, Shi Y, Dixon JE, Pandolfi P, Pavletich NP. Crystal structure of the PTEN tumor suppressor: implications for its phosphoinositide phosphatase activity and membrane association. *Cell* 1999;**99**:323–34.

29. Vazquez F, Devreotes P. Regulation of PTEN function as a PIP3 gatekeeper through membrane interaction. *Cell Cycle* 2006;**5**:1523–7.

30. Li Z, Dong X, Wang Z, Liu W, Deng N, Ding Y, Tang L, Hla T, Zeng R, Li L, et al. Regulation of PTEN by Rho small GTPases. *Nat Cell Biol* 2005;**7**:399–404.

31. Nishio M, Watanabe K, Sasaki J, Taya C, Takasuga S, Iizuka R, Balla T, Yamazaki M, Watanabe H, Itoh R, et al. Control of cell polarity and motility by the PtdIns(3,4,5)P3 phosphatase SHIP1. *Nat Cell Biol* 2007;**9**:36–44.

32. Subramanian KK, Jia Y, Zhu D, Simms BT, Jo H, Hattori H, You J, Mizgerd JP, Luo HR. Tumor suppressor PTEN is a physiologic suppressor of chemoattractant-mediated neutrophil functions. *Blood* 2007;**109**:4028–37.

33. Hoeller O, Kay RR. Chemotaxis in the absence of PIP3 gradients. *Curr Biol* 2007;**17**:813–17.

34. Ferguson GJ, Milne L, Kulkarni S, Sasaki T, Walker S, Andrews S, Crabbe T, Finan P, Jones G, Jackson S, et al. PI(3)Kgamma has an important context-dependent role in neutrophil chemokinesis. *Nat Cell Biol* 2007;**9**:86–91.

35. van Haastert PJ, Keizer-Gunnink I, Kortholt A. Essential role of PI3-kinase and phospholipase A2 in Dictyostelium discoideum chemotaxis. *J Cell Biol* 2007;**177**:809–16.

36. Chen L, Janetopoulos C, Huang YE, Iijima M, Borleis J, Devreotes PN. Two phases of actin polymerization display different dependencies on PI(3,4,5)P3 accumulation and have unique roles during chemotaxis. *Mol Biol Cell* 2003;**14**:5028–37.

37. Loovers HM, Postma M, Keizer-Gunnink I, Huang YE, Devreotes PN, van Haastert PJ. Distinct roles of PI(3,4,5)P3 during chemoattractant signaling in Dictyostelium: a quantitative in vivo analysis by inhibition of PI3-kinase. *Mol Biol Cell* 2006;**17**:1503–13.

38. Chen L, Iijima M, Tang M, Landree MA, Huang YE, Xiong Y, Iglesias PA, Devreotes PN. PLA2 and PI3K/PTEN pathways act in parallel to mediate chemotaxis. *Dev Cell* 2007;**12**:603–14.

39. Lee S, Parent CA, Insall R, Firtel RA. A novel Ras-interacting protein required for chemotaxis and cyclic adenosine monophosphate signal relay in *Dictyostelium*. *Mol Biol Cell* 1999;**10**:2829–45.

40. Chen MY, Long Y, Devreotes PN. A novel cytosolic regulator, Pianissimo, is required for chemoattractant receptor and G protein-mediated activation of the 12 transmembrane domain adenylyl cyclase in *Dictyostelium*. *Genes Dev* 1997;**11**:3218–31.

41. Sarbassov DD, Ali SM, Sabatini DM. Growing roles for the mTOR pathway. *Curr Opin Cell Biol* 2005;**17**:596–603.

42. Lee 3rd S, Comer FI, Sasaki A, McLeod IX, Duong Y, Okumura K, Yates JR, Parent CA, Firtel RA. TOR complex 2 integrates cell movement during chemotaxis and signal relay in Dictyostelium. *Mol Biol Cell* 2005;**16**:4572–83.

43. Sarbassov DD, Guertin DA, Ali SM, Sabatini DM. Phosphorylation and regulation of Akt/PKB by the rictor–mTOR complex. *Science* 2005;**307**:1098–101.

44. Benard V, Bohl BP, Bokoch GM. Characterization of rac and cdc42 activation in chemoattractant-stimulated human neutrophils using a novel assay for active GTPases. *J Biol Chem* 1999;**274**:13,198–13,204.

45. Srinivasan S, Wang F, Glavas S, Ott A, Hofmann F, Aktories K, Kalman D, Bourne HR. Rac and Cdc42 play distinct roles in regulating PI(3,4,5)P3 and polarity during neutrophil chemotaxis. *J Cell Biol* 2003;**160**:375–85.

46. Park KC, Rivero F, Meili R, Lee S, Apone F, Firtel RA. Rac regulation of chemotaxis and morphogenesis in Dictyostelium. *EMBO J* 2004;**23**:4177–89.

47. Han JW, Leeper L, Rivero F, Chung CY. Role of RacC for the regulation of WASP and phosphatidylinositol 3-kinase during chemotaxis of Dictyostelium. *J Biol Chem* 2006;**281**:35,224–35,234.

48. Bosgraaf L, van Haastert PJ. The regulation of myosin II in *Dictyostelium*. *Eur J Cell Biol* 2006;**85**:969–79.

49. Jeon TJ, Lee DJ, Lee S, Weeks G, Firtel RA. Regulation of Rap1 activity by RapGAP1 controls cell adhesion at the front of chemotaxing cells. *J Cell Biol* 2007;**179**:833–43.

50. Kortholt A, Rehmann H, Kae H, Bosgraaf L, Keizer-Gunnink I, Weeks G, Wittinghofer A, Van Haastert PJ. Characterization of the GbpD-activated Rap1 pathway regulating adhesion and cell polarity in Dictyostelium discoideum. *J Biol Chem* 2006;**281**:23,367–23,376.

51. Jeon TJ, Lee DJ, Merlot S, Weeks G, Firtel RA. Rap1 controls cell adhesion and cell motility through the regulation of myosin II. *J Cell Biol* 2007;**176**:1021–33.

52. Lee S, Rivero F, Park KC, Huang E, Funamoto S, Firtel RA. *Dictyostelium* PAKc is required for proper chemotaxis. *Mol Biol Cell* 2004;**15**:5456–69.

53. Chung CY, Potikyan G, Firtel RA. Control of cell polarity and chemotaxis by Akt/PKB and PI3 kinase through the regulation of PAKa. *Mol Cell* 2001;**7**:937–47.

54. Russ M, Croft D, Ali O, Martinez R, Steimle PA. Myosin heavy-chain kinase A from Dictyostelium possesses a novel actin-binding domain that cross-links actin filaments. *Biochem J* 2006;**395**:373–83.

55. Patel H, Barber DL. A developmentally regulated Na-H exchanger in *Dictyostelium discoideum* is necessary for cell polarity during chemotaxis. *J Cell Biol* 2005;**169**:321–9.

56. Bosgraaf L, Waijer A, Engel R, Visser AJ, Wessels D, Soll D, van Haastert PJ. RasGEF-containing proteins GbpC and GbpD have differential effects on cell polarity and chemotaxis in Dictyostelium. *J Cell Sci* 2005;**118**:1899–910.

57. Veltman DM, Roelofs J, Engel R, Visser AJ, Van Haastert PJ. Activation of soluble guanylyl cyclase at the leading edge during *Dictyostelium* chemotaxis. *Mol Biol Cell* 2005;**16**:976–83.

58. Ma L, Janetopoulos C, Yang L, Devreotes PN, Iglesias PA. Two complementary, local excitation, global inhibition mechanisms acting in parallel can explain the chemoattractant-induced regulation of PI(3,4,5)P3 response in dictyostelium cells. *Biophys J* 2004;**87**:3764–74.

59. Xu X, Brzostowski JA, Jin T. Using quantitative fluorescence microscopy and FRET imaging to measure spatiotemporal signaling events in single living cells. *Methods Mol Biol* 2006;**346**:281–96.

60. Weiner OD, Neilsen PO, Prestwich GD, Kirschner MW, Cantley LC, Bourne HR. A PtdInsP(3)- and Rho GTPase-mediated positive feedback loop regulates neutrophil polarity. *Nat Cell Biol* 2002,**4**:509–13.

61. Niggli V. A membrane-permeant ester of phosphatidylinositol 3,4,5-trisphosphate (PIP(3)) is an activator of human neutrophil migration. *FEBS Letts* 2000;**473**:217–21.

62. Wong K, Pertz O, Hahn K, Bourne H. Neutrophil polarization: spatiotemporal dynamics of RhoA activity support a self-organizing mechanism. *Proc Natl Acad Sci USA* 2006;**103**:3639–44.

Chapter 8

Interactive Signaling Pathways in the Vasculature

Lucy Liaw, Igor Prudovsky, Volkhard Lindner, Calvin Vary and Robert E. Friesel

Center for Molecular Medicine, Maine Medical Center Research Institute, Scarborough, Maine

INTRODUCTION

The developmental specification of vascular cells, their formation into a circulatory network, and mature vascular function are under tight control of multiple signaling pathways. In the adult, physiological or pathological angiogenesis and vascular remodeling events are accompanied by the re-activation and interaction of these pathways. Numerous excellent reviews summarize recent progress in vascular cell differentiation and maturation [1–4], FGF [5], TGFβ [6, 7], Notch signaling [8, 9], VEGF [10], Tie/angiopoietin, and ephrins [11–13], semaphorins [14], Krüppel-like transcription factors [15], Fox and Ets transcription factors [16], and sphingosine 1-phosphate [17]. These mechanisms will not be discussed; rather, our focus is the signaling interaction of FGF, Notch, and TGFβ signaling and implications in the vasculature.

INTERACTIVE NETWORKS AS MODELS OF CELL SIGNALING

A tremendous amount of work has addressed the cellular, biochemical, and molecular mechanisms of linear signaling pathways and their phenotypic outcomes. The repertoire of second messengers active in a particular signaling pathway is reiteratively used during development and remodeling events, allowing for cross-talk between multiple pathways. Many cellular signals are activated by ligand binding to a transmembrane receptor, leading to a series of events that control nuclear functions, including gene transcription. Technological advances that provide extensive data sets describing a cellular phenotype (i.e., gene chip microarray or proteomic approaches) have led to computational modeling of signaling networks. Models of network interactions are likely to be more common in the future, and will reveal

additional unanticipated complexity [18–20]. Studying the interaction of select well-characterized pathways, however, can also provide insight into mechanisms of cross-talk. We focus on three major pathways critical in the development and maintenance of the vasculature: TGFβ/BMP, Notch, and FGF. We provide examples of cross-talk at the level of gene regulation, inhibitory molecules, and protein–protein interactions of second messengers. In addition, cell–cell interactions are considered, since these pathways are also triggered *in trans* from neighboring cells. Figure 8.1 summarizes interactions between the major components of the three signaling pathways discussed in this chapter.

CROSS-TALK BETWEEN FGF AND NOTCH SIGNALING

FGF and Notch signaling regulate vascular development and remodeling. Genetic models with mutations in the Notch signaling pathway (Notch1, Notch2, Jagged1, HRT2 signaling) [8, 21] or FGF signaling [22] all lead to deficiencies in cardiovascular development. Likewise, FGF is a major cytokine responsible for neointimal lesion formation following injury [23], and Notch participates in the maintenance and remodeling of the mature cardiovascular system [24]. An early indication of the interplay between FGF and Notch signaling was that oncogenic transformation caused by the mouse mammary tumor virus is due to multiple proviral integration events that simultaneously activate the expression of FGF2, FGF3, and Notch4 [25]. The regulation of Notch signaling by FGF occurs in several developmental situations [26–29]. During *Drosophila* tracheal development, a branching morphogenesis process similar to vascular development, Notch signaling is activated by the FGF homolog, Branchless, through the FGFR homolog, Breathless [26]. This activation is a result of Breathless-dependent stimulation of Delta

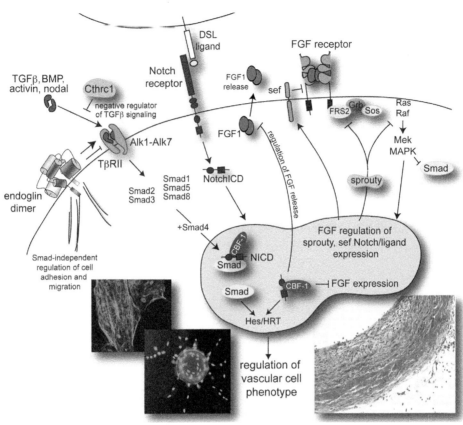

FIGURE 8.1 **Major signaling pathways TGFβ/BMP, Notch, and FGF are depicted, with potential areas of overlap or synergy.**
Shown schematically is a cell containing these receptors on the plasma membrane, with cytoplasmic signals impinging upon nuclear function (shaded). A neighboring cell (top) is shown expressing a Notch (DSL) ligand. From left: endoglin and Cthrc1 both regulate TGFβ/BMP signals, which are mediated through Smad activation and translocation into the nucleus. Notch signaling represents a cell–cell interaction signal that also leads to regulation of downstream transcriptional targets. Smads and Notch intracellular domain (ICD) interact to regulate each other. FGF signaling is regulated at the level of ligand release, activity of negative regulators including sef and sprouty, and cytoplasmic signaling cascades including the Ras/MAP kinase pathway. This cascade regulates Smad activity, and impinges in the Notch pathway by modulating expression levels of Notch ligands and receptors. Interaction of these signaling pathways is important for vascular function and remodeling. Photomicrographs (from left) show examples of phenotypic regulation of vascular cells: human primary smooth muscle cells (SMC) with immunofluorescence staining against smooth muscle actin, which is expressed in differentiated SMC; human primary endothelial cells (EC) undergoing tubulogenesis in a three-dimensional matrix; cross section of a Masson's Trichrome stained rat carotid artery 2 weeks after endothelial denudation injury. Note the thick neointimal lesion (N) in comparison to the medial layer (M) and adventitia (A).

expression at the tip of growing tracheal branches. Notch signaling is required to restrict the activation of MAPK to the tip of the branches in part through the negative regulation of Breathless expression.

FGF2 induces expression of Jagged1 during *in vitro* angiogenesis [27, 28], and stimulates Notch4 transcription in endothelial cells by enhancing AP-1 and glucocorticoid receptor binding to the Notch4 promoter [29]. Thus, Notch expression is positively regulated by FGF. However, FGF-dependent regulation of the Notch pathway can also be negative. FGF signaling induces the expression of the hes-related gene hey 13.2 [30], which is required for periodic repression of Notch-regulated hey1 and hey7 genes. Moreover, Hey13.2 protein enhances the auto-repression of hey1 gene by associating with Hey1 protein. Joint inactivation of hey13.2 and hey1 leads to a complete loss of somite borders [31]. Interestingly, Hey2 (HRT2) cooperates with Hey1 (HRT1) in embryonic vascular development [32]. It is of great interest

to understand whether FGF negatively regulates Notch signaling in the cardiovasculature.

The connection between the FGF and Notch pathways is reciprocal. Experiments with soluble Notch ligands and receptors demonstrated that the Notch pathway regulates FGF signaling [33, 34]. In human primary endothelial cells, Dll4 signaling through Notch can regulate the expression of angiogenic factors including genes of the VEGF and FGF families [35]. Due to limitations in studying primary vascular cells, regulation of FGF signaling by Notch has been explored using fibroblast cell models. NIH3T3 cells transfected with soluble Jagged1 (sJag1), which inhibits Notch signaling, constitutively release FGF1, which contains no signal sequence. The same results were demonstrated for the extracellular dominant negative forms of Notch1, Notch2, and soluble Delta1 (sDl1) transfectants [33, 34]. Inhibition of Notch signaling by sJag1 induced the transcription of FGF1, FGF4 and FGF5 [33]. Transfection of sJag1 cells with constitutively active Notch1 (intracellular domain) completely abrogated the spontaneous release of FGF1, although it did not interfere with heat shock-induced FGF1 export. sJag1 transfected cells give rise to highly angiogenic solid tumors in nude mice, and form multicellular substrate-dependent cord-like structures *in vitro*, reminiscent of angiogenesis [36]. Consistent with their tumorigenicity, sJag1 cells form colonies in soft agar, a phenotype that is inhibited by constitutively active Notch1 and a specific chemical inhibitor of FGFR. Thus, the transformed phenotype of sJag1 cells is dependent on the inhibition of Notch signaling and activation of FGF signaling. This may be explained in part by de-repression of the transcription of FGF1, FGF4, and FGF5, and by the release of FGF1. However, the situation appears to be more complex. Indeed, when exogenous recombinant FGF1 was present in the medium, the growth of sJag1 cells in soft agar was still inhibited by activation of Notch1 signaling. Also, constitutive activation of either Notch1 or Notch2 blocked soft agar colony formation by cells that express a constitutively secreted, oncogenic form of FGF1. Interestingly, neither Notch activation nor sJag1 interfere with the expression of FGFR1, FGFR2, or FGFR3 [33]. Constitutively active Notch4 also inhibits FGF2-dependent endothelial sprouting [37]. Thus, Notch interferes with the FGF pathway at the levels of FGF expression, release, and signaling.

Given that FGF and Notch components are expressed in a temporally regulated fashion during vascular injury, we suggest that these pathways cooperatively regulate vascular repair. For example, FGF release from mechanically injured vessels is immediate, and FGF signaling is required for neointimal lesion formation and endothelial regrowth [23, 38]. We hypothesize that injury-induced enhancement of FGF signaling is mediated by the attenuation of Notch signaling, which results from the loss of cell contacts. In turn, FGF induces components of the Notch signaling pathway, leading to regulation of the cell cycle, cell migration, and gene transcription. Of interest is that Notch targets/effector molecules, Hes and HRT, while mediating Notch effects, may also serve as a negative regulator of cell differentiation [39]. Notch suppression of FGF expression and export may be a negative feedback mechanism for decreasing cell proliferation.

NOVEL MODULATORS OF TGFβ SIGNALING

The broad role of transforming growth factor beta (TGFβ) signaling in vascular development, homeostasis and repair is well appreciated (reviewed in [40]). However, novel TGFβ regulatory mechanisms continue to emerge. Examples include the accessory protein endoglin, the recently identified collagen triple helix repeat containing 1 (Cthrc1), and Notch interactions with Smads. Canonical TGFβ signaling is mediated through seven type I and five type II receptors. The type I receptors are serine/threonine kinases, and include activin-like kinase 1 (ALK1) and TβRI, also known as ALK5. ALK1 and ALK5 associate with, and are activated via ligand-dependent phosphorylation by the type II TGFβ receptor, TβRII. The activated type I receptor propagates canonical Smad-dependent signals by phosphorylating Smads [7]. Mutations in endoglin [41], a TGFβ type III receptor, or ALK1 [42], cause the vascular dysplasia hereditary hemorrhagic telangiectasia. Endoglin directly associates with TGFβ receptors to potentiate ALK1 or inhibit ALK5 Smad-dependent signaling. Endoglin is also a substrate for TβRII-, ALK5-, and ALK1-mediated phosphorylation [43].

Phosphorylation of endoglin may be a mechanism by which endoglin regulates ALK1 Smad-independent effects on endothelial cell growth and adhesion [44]. Phosphorylation of endoglin affects its subcellular localization by a mechanism that may involve its interactions with focal adhesion proteins such as zyxin and zyxin-related protein-1 [45, 46], thus modifying adhesive properties of endoglin-expressing cells. Recent studies indicate that endoglin can affect both Smad-dependent and Smad-independent pathways that regulate the level and activity of cyclooxygenase-2 [47] and endothelial nitric oxide synthase [48], respectively, which are important regulators of vascular tone.

Involvement of endoglin in an alternative non-canonical TGFβ signaling pathway is suggested based on the phenotypic comparison between the *eng*$^{-/-}$ and TGFβ activated kinase-1 (*TAK1*)$^{-/-}$ mouse strains [49]. TAK1 is a non-canonical Smad-independent effector of TGFβ and bone morphogenetic protein (BMP) signaling. Similar to the *eng*$^{-/-}$ mouse, SMC development is impaired with normal endothelial cell development, thereby raising the possibility that TAK1 may mediate Smad-independent signals downstream of endoglin. More recently, genetic data obtained combining Smad4 conditional inactivation with conditional endoglin expression in cells of the embryonic neural crest indicate that endoglin operates in pathways that are separate from the canonical TGFβ receptor signaling pathways required for smooth muscle cell fate determination [50]. Recent data also show that endoglin participates in signaling independent of TGFβ receptors? The novel ligand BMP-9 binds to ALK1 and the BMP type II receptor [51]. BMP-9 and the related cytokine BMP-10 exhibit high-affinity binding to ALK1 and endoglin on endothelial cells [52] in the absence of other TGFβ signaling receptors, providing more evidence for endoglin function in non-canonical pathways. Interestingly, this BMP-9 signaling through ALK1/endoglin antagonizes FGF-2-mediated endothelial proliferation and migration [53] and functions as a mediator of vascular quiescence [54]. The aforementioned studies suggest that endoglin modulates multiple interactions between TGFβ Smad-dependent and independent signaling pathways.

Cthrc1 is a secreted 28-kDa protein identified as a gene upregulated after vascular injury [55], and now known to modulate TGFβ signaling, Cthrc1 mRNA expression levels are increased in response to TGFβ1 and BMP-4 [56]. Though disease-causing human mutations in Cthrc1 have not yet been identified, studies of transgenic mice suggest a phenotype that is predictive of connective tissue disorders [57]. More recent studies indicate that Cthrc1 expression reduces TGFβ Smad-dependent signaling in arterial smooth muscle, as demonstrated by reduced Smad phosphorylation and procollagen synthesis. These results indicate that Cthrc1 is a cell-type-specific inhibitor of TGFβ signaling that modulates collagen deposition, neointimal formation, and dedifferentiation of smooth muscle cells. Thus, endoglin and Cthrc1 exemplify the emergence of novel and unanticipated mechanism of vascular regulation.

NOTCH, FGF, AND SMAD SIGNALING INTERACTIONS

There are significant instances of interpathway cross-talk involving FGF, Notch, and TGFβ. Ras activation, acting via the Erk MAP kinases, phosphorylates Smad2 and Smad3 at sites that are separate from the TGFβ receptor phosphorylation sites. This event overcomes the growth inhibitory response to TGFβ in Ras-transformed cells [58]. Initial interaction between Notch and Smad-mediated signaling pathways was shown in embryonic neuroepithelial cells, which respond to both BMP2 and constitutively activated Notch1 (Notch1ICD) with an increase in Hes5 and Hesr1 (HRT1) promoter activity. The combination of BMP2 and Notch1ICD synergistically promoted activity of both reporters [59], and this was dependent on intact Smad and CBF-1 (RBP-Jĸ) binding sites in the Hes5 promoter. One mechanism behind this synergistic regulation is the bridging of Smad1 and Notch1ICD via p300 and P/CAF. Direct protein–protein interactions between Smad1 and NotchICD were subsequently demonstrated *in vitro* [60], in relation to BMP4/Notch cross-talk in suppression of myogenic differentiation. In endothelial cells, NotchICD interacts with Smad1, which is facilitated by P/CAF, and together they transcriptionally activate herp2 (HRT1) [61]. Thus, a major mechanism of cross-talk in which Notch/Smad signals are potentiated is via protein–protein interactions to regulate known Notch target genes. Smad3/4 complexes also regulate hey1 (HRT1) at the promoter level in a Notch-independent manner [62], and other components of these pathways may be targeted. For example, TGFβ induces expression of Jagged1, potentiating Notch signaling [62].

Bi-directional signaling through the Notch ligand Delta-like1 (Dll1) also converges on Smad signaling. The transmembrane form of Dll1 is proteolytically processed, and the Dll1 intracellular domain (Dll1-ICD) can localize in the nucleus. Dll1-ICD binds Smad2, Smad3, and Smad4, and promotes Smad3-dependent activation of CAGA-Luc reporters [63].

Conversely, antagonistic interactions may also occur through these cross-talk mechanisms. NotchICD binds to p300, and this was shown to suppress Smad3 signaling by recruiting p300 away from Smad3 [64]. Notch4ICD also binds Smad3, and to a lesser extent Smad2 and Smad4, and as a result inhibits its TGFβ signaling [65]. These data all support the interactive coordination of Notch with Smad pathways, which have implications for both TGFβ and BMP signaling.

FEEDBACK INHIBITORY MECHANISMS IN VASCULAR CELL SIGNALING

Receptor tyrosine kinases (RTKs) control many cellular processes during development and vascular remodeling, including proliferation, migration, differentiation, and survival. RTKs are tightly controlled by positive and negative regulators, which function at multiple levels. When RTK signaling is dysregulated, developmental abnormalities and pathological conditions including retinopathies, restenosis, and tumor angiogenesis can occur [66–68]. Recent studies in model organisms such as *Drosophila* and zebrafish have identified feedback inhibitory mechanisms that negatively regulate the duration and intensity of RTK signaling. In *Drosophila*, Sprouty (Spry) was identified in a genetic screen as an inhibitor of tracheal development [69]. These and subsequent studies characterized Spry as a general inhibitor of RTK-mediated Ras signaling. There are four vertebrate Spry genes encoding unique proteins that share a highly conserved cysteine-rich domain similar to that in *Drosophila* Spry. Spry proteins inhibit the Ras–ERK pathway, although the exact point of regulation is controversial. Spry interacts with multiple proteins, including Grb2, c-Cbl, Shp2, FRS2, Raf1, PTPB1, and Caveolin-1, as well forming homo- and heterodimers among themselves. These interactions and their biological significance have recently been reviewed [68]. Spry1, Spry2, and Spry4 are expressed in endothelial cells in response to mitogenic stimulation, and overexpression of Spry proteins in endothelial cells inhibits their proliferation and capillary-like tube formation *in vitro*. While little is known about the role of Spry in vascular development *in vivo*, adenoviral-mediated expression of Spry4 in E8.0 cultured mouse embryos resulted in a significant reduction in PECAM stained vessels after 24h [70]. A better understanding of the role of Spry in vascular development and remodeling awaits gain- and loss-of-function studies of *Spry* genes both individually and in combination using tissue-specific mouse models.

Another feedback inhibitor of potential importance to vascular development and homeostasis is Sef (*similar expression to FGF*). Sef was identified in zebrafish as an inhibitor of FGF signaling [71]. Sef is also expressed in epithelial tissues and endothelial cells [72], associates with FGFR1 and FGFR2 in co-immunoprecipitation experiments, and inhibits FGF-induced ERK activation, Akt phosphorylation, and

FGFR activation [73–75]. Overexpression of Sef in human umbilical vein endothelial cells inhibits FGF signaling and induces apoptosis [76]. Interestingly, deletion of the cytoplasmic domain of Sef did not affect its inhibition of FGFR activation in endothelial cells, whereas replacement of the transmembrane domain of Sef with that of the PDGFR completely abrogated the inhibitory effect of Sef on FGF signaling, suggesting an important role for the transmembrane region in Sef function. Disruption of the Sef gene by gene trap methodologies has not resulted in any discernable phenotypes [77]; however, analyses are still ongoing. It is intriguing to speculate that Sef and Spry may constitute a coordinately regulated feedback inhibitory pathway that regulates FGF signaling at multiple levels. Future studies using compound mouse mutants of Sef and Spry will likely reveal their cooperative roles on regulating vascular development and homeostasis.

IMPLICATIONS IN VASCULAR REMODELING

There are many situations in the adult vasculature where remodeling is necessary, and a convenient model for vascular cell phenotype is injury due to endothelial/smooth muscle mechanical stress. In the model of endothelial denudation, for example, signal inhibitory reagents and genetics have clarified the roles of these major pathways. Antibodies against FGF2 inhibit neointimal lesion formation [23]; soluble TGFβ receptor (pathway inhibitor) suppresses lesion formation, constrictive remodeling, and fibrosis [78]; increased expression of Cthrc1 prevents neointimal lesion formation and TGFβ signaling [79]; and we have preliminary evidence showing that Notch pathway antagonists also inhibit lesion formation (data not shown). Since FGFs are present in a moderate level in the vessel wall under normal conditions, we propose that release of FGF by injury and cellular damage is an acute response that serves to trigger cell proliferation as well as the expression of the Notch and TGFβ pathways. In addition, platelet aggregation following cellular damage also contributes TGFβ to the vessel wall, and Notch signaling and TGFβ may then serve as a sustained signal for continued remodeling. This model allows for soluble signals (FGF, TGFβ, Cthrc-1), cell membrane or intracellular signals that may act cell autonomously (endoglin, sef, sprouty), and potential heterotypic cell signaling systems activated by cell contact (Jagged or Delta/Notch). Our findings of high Jagged1 expression in regenerating endothelial cells abutting high Notch-expressing neointimal smooth muscle cells *in vivo* [80] support this model. A high priority in the future is to define the gene targets of these signaling pathways that cooperatively mediate these intercellular interactions in the vessel wall.

REFERENCES

1. Lamont RE, Childs S. Mapping out arteries and veins. *Sci STKE* 2006;**2006**. pe39.
2. Suburo AM, D'Amore PA. Development of the endothelium. *Handb Exp Pharmacol* 2006:71–105.
3. Majesky MW. Developmental basis of vascular smooth muscle diversity. *Arterioscler Thromb Vasc Biol* 2007;**27**:1248–58.
4. Armulik A, Abramsson A, Betsholtz C. Endothelial/pericyte interactions. *Circ Res* 2005;**97**:512–23.
5. Presta M, Dell'Era P, Mitola S, Moroni E, Ronca R, Rusnati M. Fibroblast growth factor/fibroblast growth factor receptor system in angiogenesis. *Cytokine Growth Factor Rev* 2005;**16**:159–78.
6. Bertolino P, Deckers M, Lebrin F, ten Dijke P. Transforming growth factor-beta signal transduction in angiogenesis and vascular disorders. *Chest* 2005;**128**:585S–590S.
7. Shi Y, Massague J. Mechanisms of TGF-beta signaling from cell membrane to the nucleus. *Cell* 2003;**113**:685–700.
8. Gridley T. Notch signaling in vascular development and physiology. *Development* 2007;**134**:2709–18.
9. Hofmann JJ, Iruela-Arispe ML. Notch signaling in blood vessels: who is talking to whom about what?. *Circ Res* 2007;**100**:1556–68.
10. Ferrara N. The role of VEGF in the regulation of physiological and pathological angiogenesis. *Exs* 2005:209–31.
11. Heroult M, Schaffner F, Augustin HG. Eph receptor and ephrin ligand-mediated interactions during angiogenesis and tumor progression. *Exp Cell Res* 2006;**312**:642–50.
12. Hofer E, Schweighofer B. Signal transduction induced in endothelial cells by growth factor receptors involved in angiogenesis. *Thromb Haemost* 2007;**97**:355–63.
13. Pfaff D, Fiedler U, Augustin HG. Emerging roles of the Angiopoietin-Tie and the ephrin-Eph systems as regulators of cell trafficking. *J Leukoc Biol* 2006;**80**:719–26.
14. Casazza A, Fazzari P, Tamagnone L. Semaphorin signals in cell adhesion and cell migration: functional role and molecular mechanisms. *Adv Exp Med Biol* 2007;**600**:90–108.
15. Atkins GB, Jain MK. Role of Kruppel-like transcription factors in endothelial biology. *Circ Res* 2007;**100**:1686–95.
16. Dejana E, Taddei A, Randi AM. Foxs and Ets in the transcriptional regulation of endothelial cell differentiation and angiogenesis. *Biochim Biophys Acta* 2007;**1775**:298–312.
17. Brinkmann V. Sphingosine 1-phosphate receptors in health and disease: mechanistic insights from gene deletion studies and reverse pharmacology. *Pharmacol Ther* 2007;**115**:84–105.
18. Hua F, Hautaniemi S, Yokoo R, Lauffenburger DA. Integrated mechanistic and data-driven modelling for multivariate analysis of signalling pathways. *J R Soc Interface* 2006;**3**:515–26.
19. Janes KA, Lauffenburger DA. A biological approach to computational models of proteomic networks. *Curr Opin Chem Biol* 2006;**10**:73–80.
20. Steffen M, Petti A, Aach J, D'Haeseleer P, Church G. Automated modelling of signal transduction networks. *BMC Bioinformatics* 2002;**3**:34.
21. Iso T, Hamamori Y, Kedes L. Notch signaling in vascular development. *Arterioscler Thromb Vasc Biol* 2003;**23**:543–53.
22. Lavine KJ, White AC, Park C, et al. Fibroblast growth factor signals regulate a wave of Hedgehog activation that is essential for coronary vascular development. *Genes Dev* 2006;**20**:1651–66.
23. Lindner V. Role of basic fibroblast growth factor and platelet-derived growth factor (B-chain) in neointima formation after arterial injury. *Z Kardiol* 1995;**84**(Suppl. 4):137–44.

24. Karsan A. The role of notch in modeling and maintaining the vasculature. *Can J Physiol Pharmacol* 2005;**83**:14–23.

25. van Leeuwen F, Nusse R. Oncogene activation and oncogene cooperation in MMTV-induced mouse mammary cancer. *Semin Cancer Biol* 1995;**6**:127–33.

26. Ikeya T, Hayashi S. Interplay of Notch and FGF signaling restricts cell fate and MAPK activation in the *Drosophila* trachea. *Development* 1999;**126**:4455–63.

27. Zimrin AB, Pepper MS, McMahon GA, Nguyen F, Montesano R, Maciag T. An antisense oligonucleotide to the notch ligand jagged enhances fibroblast growth factor-induced angiogenesis in vitro. *J Biol Chem* 1996 20;**271**:32,499–32,502.

28. Matsumoto T, Turesson I, Book M, Gerwins P, Claesson-Welsh L. p38 MAP kinase negatively regulates endothelial cell survival, proliferation, and differentiation in FGF-2-stimulated angiogenesis. *J Cell Biol* 2002;**156**:149–60.

29. Wu J, Bresnick EH. Glucocorticoid and growth factor synergism requirement for Notch4 chromatin domain activation. *Mol Cell Biol* 2007;**27**:2411–22.

30. Kawamura A, Koshida S, Hijikata H, Sakaguchi T, Kondoh H, Takada S. Zebrafish hairy/enhancer of split protein links FGF signaling to cyclic gene expression in the periodic segmentation of somites. *Genes Dev* 2005;**19**:1156–61.

31. Sieger D, Ackermann B, Winkler C, Tautz D, Gajewski M. her1 and her13.2 are jointly required for somitic border specification along the entire axis of the fish embryo. *Dev Biol* 2006;**293**:242–51.

32. Fischer A, Schumacher N, Maier M, Sendtner M, Gessler M. The Notch target genes Hey1 and Hey2 are required for embryonic vascular development. *Genes Dev* 2004;**18**:901–11.

33. Small D, Kovalenko D, Soldi R, et al. Notch activation suppresses fibroblast growth factor-dependent cellular transformation. *J Biol Chem* 2003;**278**:16,405–16,413.

34. Trifonova R, Small D, Kacer D, et al. The non-transmembrane form of Delta1, but not of Jagged1, induces normal migratory behavior accompanied by fibroblast growth factor receptor 1-dependent transformation. *J Biol Chem* 2004;**279**:13,285–13,288.

35. Harrington LS, Sainson RC, Williams CK, et al. Regulation of multiple angiogenic pathways by Dll4 and Notch in human umbilical vein endothelial cells. *Microvasc Res* 2008;**75**:144–54.

36. Wong MK, Prudovsky I, Vary C, et al. A non-transmembrane form of Jagged-1 regulates the formation of matrix-dependent chord-like structures. *Biochem Biophys Res Commun* 2000 24;**268**:853–9.

37. MacKenzie F, Duriez P, Larrivee B, et al. Notch4-induced inhibition of endothelial sprouting requires the ankyrin repeats and involves signaling through RBP-Jkappa.. *Blood* 2004;**104**:1760–8.

38. Lindner V, Majack RA, Reidy MA. Basic fibroblast growth factor stimulates endothelial regrowth and proliferation in denuded arteries. *J Clin Invest* 1990;**85**:2004–8.

39. Doi H, Iso T, Yamazaki M, et al. HERP1 inhibits myocardin-induced vascular smooth muscle cell differentiation by interfering with SRF binding to CArG box. *Arterioscler Thromb Vasc Biol* 2005;**25**:2328–34.

40. Bernabeu C, Conley BA, Vary CP. Novel biochemical pathways of endoglin in vascular cell physiology. *J Cell Biochem* 2007;**102**:1375–88.

41. McAllister KA, Grogg KM, Johnson DW, et al. Endoglin, a TGF-beta binding protein of endothelial cells, is the gene for hereditary haemorrhagic telangiectasia type 1. *Nature Genet* 1994;**8**:345–51.

42. Johnson DW, Berg JN, Baldwin MA, et al. Mutations in the activin receptor-like kinase 1 gene in hereditary haemorrhagic telangiectasia type 2. *Nat Genet* 1996;**13**:189–95.

43. Bobik A. Transforming growth factor-betas and vascular disorders. *Arterioscler Thromb Vasc Biol* 2006;**26**:1712–20.

44. Koleva RI, Conley BA, Romero D, et al. Endoglin structure and function: determinants of endoglin phosphorylation by transforming growth factor-beta receptors. *J Biol Chem* 2006;**281**:25,110–25,123.

45. Sanz-Rodriguez F, Guerrero-Esteo M, Botella LM, Banville D, Vary CP, Bernabeu C. Endoglin regulates cytoskeletal organization through binding to ZRP-1, a member of the Lim family of proteins. *J Biol Chem* 2004;**279**:32.858–32,868.

46. Conley BA, Koleva R, Smith JD, et al. Endoglin controls cell migration and composition of focal adhesions: function of the cytosolic domain. *J Biol Chem* 2004;**279**:27.440–27,449.

47. Jerkic M, Rivas-Elena JV, Santibanez JF, et al. Endoglin regulates cyclooxygenase-2 expression and activity. *Circ Res* 2006;**99**:248–56.

48. Toporsian M, Gros R, Kabir MG, et al. A role for endoglin in coupling eNOS activity and regulating vascular tone revealed in hereditary hemorrhagic telangiectasia. *Circ Res* 2005;**96**:684–92.

49. Jadrich JL, O'Connor MB, Coucouvanis E. The TGF beta activated kinase TAK1 regulates vascular development in vivo. *Development* 2006;**133**:1529–41.

50. Mancini ML, Verdi JM, Conley BA, et al. Endoglin is required for myogenic differentiation potential of neural crest stem cells. *Dev Biol* 2007;**308**:520–33.

51. Brown MA, Zhao Q, Baker KA, et al. Crystal structure of BMP-9 and functional interactions with pro-region and receptors. *J Biol Chem* 2005;**280**:25,111–18.

52. David L, Mallet C, Mazerbourg S, Feige JJ, Bailly S. Identification of BMP9 and BMP10 as functional activators of the orphan activin receptor-like kinase 1 (ALK1) in endothelial cells. *Blood* 2007;**109**:1953–61.

53. Scharpfenecker M, van Dinther M, Liu Z, et al. BMP-9 signals via ALK1 and inhibits bFGF-induced endothelial cell proliferation and VEGF-stimulated angiogenesis. *J Cell Sci* 2007;**120**:964–72.

54. David L, Mallet C, Keramidas M, et al. Bone morphogenetic protein-9 is a circulating vascular quiescence factor. *Circ Res* 2008;**102**:914–22.

55. Durmus T, LeClair RJ, Park KS, Terzic A, Yoon JK, Lindner V. Expression analysis of the novel gene collagen triple helix repeat containing-1 (Cthrc1). *Gene Expr Patterns* 2006;**6**:935–40.

56. Pyagay P, Heroult M, Wang Q, et al. Collagen triple helix repeat containing 1, a novel secreted protein in injured and diseased arteries, inhibits collagen expression and promotes cell migration. *Circ Res* 2005;**96**:261–68.

57. LeClair RJ, Durmus T, Wang Q, Pyagay P, Terzic A, Lindner V. Cthrc1 is a novel inhibitor of transforming growth factor-beta signaling and neointimal lesion formation. *Circ Res* 2007;**100**:826–33.

58. Kretzschmar M, Liu F, Hata A, Doody J, Massague J. The TGF-beta family mediator Smad1 is phosphorylated directly and activated functionally by the BMP receptor kinase. *Genes Dev* 1997 15;**11**:984–95.

59. Takizawa T, Ochiai W, Nakashima K, Taga T. Enhanced gene activation by Notch and BMP signaling cross-talk. *Nucleic Acids Res* 2003;**31**:5723–31.

60. Dahlqvist C, Blokzijl A, Chapman G, et al. Functional Notch signaling is required for BMP4-induced inhibition of myogenic differentiation. *Development* 2003;**130**:6089–99.

61. Itoh F, Itoh S, Goumans MJ, et al. Synergy and antagonism between Notch and BMP receptor signaling pathways in endothelial cells. *Embo J* 2004;**23**:541–51.

62. Zavadil J, Cermak L, Soto-Nieves N, Bottinger EP. Integration of TGF-beta/Smad and Jagged1/Notch signalling in epithelial-to-mesenchymal transition. *EMBO J* 2004;**23**:1155–65.

63. Hiratochi M, Nagase H, Kuramochi Y, Koh CS, Ohkawara T, Nakayama K. The Delta intracellular domain mediates TGF-beta/Activin signaling through binding to Smads and has an important bidirectional function in the Notch-Delta signaling pathway. *Nucleic Acids Res* 2007;**35**:912–22.

64. Masuda S, Kumano K, Shimizu K, et al. Notch1 oncoprotein antagonizes TGF-beta/Smad-mediated cell growth suppression via sequestration of coactivator p300. *Cancer Sci* 2005;**96**:274–82.

65. Sun Y, Lowther W, Kato K, et al. Notch4 intracellular domain binding to Smad3 and inhibition of the TGF-beta signaling. *Oncogene* 2005;**24**:5365–74.

66. Cabrita MA, Christofori G. Sprouty proteins: antagonists of endothelial cell signaling and more. *Thromb Haemost* 2003;**90**:586–90.

67. Bicknell R, Harris AL. Novel angiogenic signaling pathways and vascular targets. *Annu Rev Pharmacol Toxicol* 2004;**44**:219–38.

68. Mason JM, Morrison DJ, Basson MA, Licht JD. Sprouty proteins: multifaceted negative-feedback regulators of receptor tyrosine kinase signaling. *Trends Cell Biol* 2006;**16**:45–54.

69. Hacohen N, Kramer S, Sutherland D, Hiromi Y, Krasnow MA. Sprouty encodes a novel antagonist of FGF signaling that patterns apical branching of the Drosophila airways. *Cell* 1998 23;**92**:253–63.

70. Lee SH, Schloss DJ, Jarvis L, Krasnow MA, Swain JL. Inhibition of angiogenesis by a mouse sprouty protein. *J Biol Chem* 2001;**276**:4128–33.

71. Tsang M, Friesel R, Kudoh T, Dawid IB. Identification of Sef, a novel modulator of FGF signalling. *Nature Cell Biol* 2002;**4**:165–69.

72. Yang RB, Ng CK, Wasserman SM, Komuves LG, Gerritsen ME, Topper JN. A novel interleukin-17 receptor-like protein identified in human umbilical vein endothelial cells antagonizes basic fibroblast growth factor-induced signaling. *J Biol Chem* 2003;**278**:33,232–33,238.

73. Furthauer M, Lin W, Ang SL, Thisse B, Thisse C. Sef is a feedback-induced antagonist of Ras/MAPK-mediated FGF signalling. *Nat Cell Biol* 2002;**4**:170–74.

74. Kovalenko D, Yang X, Nadeau RJ, Harkins LK, Friesel R. Sef inhibits fibroblast growth factor signaling by inhibiting FGFR1 tyrosine phosphorylation and subsequent ERK activation. *J Biol Chem* 2003;**278**:14,087–14,091.

75. Yang X, Kovalenko D, Nadeau RJ, et al. Sef interacts with TAK1 and mediates JNK activation and apoptosis. *J Biol Chem* 2004;**279**:38,099–102.

76. Kovalenko D, Yang X, Chen PY, et al. A role for extracellular and transmembrane domains of Sef in Sef-mediated inhibition of FGF signaling. *Cell Signal* 2006;**18**:1958–966.

77. Nord AS, Chang PJ, Conklin BR, et al. The international gene trap consortium website: a portal to all publicly available gene trap cell lines in mouse. *Nucleic Acids Res* 2006;**34**:D642–D648.

78. Smith JD, Bryant SR, Couper LL, et al. Soluble transforming growth factor-beta type II receptor inhibits negative remodeling, fibroblast transdifferentiation, and intimal lesion formation but not endothelial growth. *Circ Res* 1999 28;**84**:1212–22.

79. Leclair R, Lindner V. The role of collagen triple helix repeat containing 1 in injured arteries, collagen expression, and transforming growth factor Beta signaling. *Trends Cardiovasc Med* 2007;**17**:202–5.

80. Lindner V, Booth C, Prudovsky I, Small D, Maciag T, Liaw L. Members of the Jagged/Notch gene families are expressed in injured arteries and regulate cell phenotype via alterations in cell matrix and cell–cell interaction. *Am J Pathol* 2001;**159**:875–83.

Chapter 9

Signaling Pathways Involved in Cardiogenesis

Deepak Srivastava and Chulan Kwon

Gladstone Institute of Cardiovascular Disease and Departments of Pediatrics and Biochemistry & Biophysics, University of California at San Francisco, San Francisco, California

INTRODUCTION

The heart is a complex organ that is derived from multiple cell types and requires extensive cell–cell signaling events which are often guided by specialized forms of extracellular matrices. As the earliest organ to form in an embryo, the heart must be functional well before it has the opportunity to shape itself into a mature organ. The combination of multifarious morphogenetic events necessary for cardiogenesis and the superimposed hemodynamic influences may contribute to the exquisite sensitivity of the heart to perturbations. This phenomenon is reflected in the estimated 10 percent incidence of severe cardiac malformations observed in early miscarriages, and in the nearly 1 percent of live births affected by cardiac developmental defects [1] An additional 1–2 percent of the population harbor more subtle cardiac developmental anomalies that only become apparent as age-dependent phenomena reveal the underlying pathology. With more than 1 million survivors of congenital heart disease (CHD) in the United States, it is becoming apparent that genetic disruptions that predispose to developmental defects can have ongoing consequences in maintenance of specific cell types and cellular processes over decades [2]. In addition, deciphering nature's secrets of heart formation might lead to novel approaches to repair or regenerate damaged heart muscle in the adult. Recent studies using stem cells have led to heightened interest in the early events involved in cardiac cell-fate decisions and cardiomyocyte differentiation, migration, and survival. Stem cells have enormous potential in regenerative medicine, and insights into cardiogenesis from progenitor cells during embryogenesis will form the basis of reprogramming cells for therapeutic use [3].

While genetic approaches have been important in understanding human CHD, detailed molecular analysis of cardiac development in humans has been difficult. The recognition that genetic pathways that dictate cardiac development are highly conserved across vastly diverse species ranging from flies to man has resulted in a rapid expansion of information from studies in more tractable biological models [4]. Despite the diversity of body plans adopted by different species, there seems to exist a common genetic program for the early formation of a circulatory system. Cardiovascular systems seem to have developed increasing complexity in order to adapt to specific environments. In a simplified view, it appears that higher organisms have retained the morphologic steps utilized by lower organisms and have built complexity into the heart as needed. In particular, the specification of chamber structures and the advent of a parallel circulation through chamber duplication and outflow tract division by neural crest derivatives have facilitated the development of larger, air-breathing organisms utilizing complex circulatory systems. In such a scheme, defects in particular regions of the heart may arise from specific genetic and environmental effects during discrete developmental windows of time. To simplify the complex events of cardiogenesis and CHD, different regions of the developing heart will be considered individually in the context described above, weaving knowledge from model systems and human genetics when available.

ORIGIN OF CARDIOMYOCYTE PRECURSORS

Despite decades of cell lineage tracings and descriptive embryology of the heart's origins, only recently has a more complete and accurate picture of cardiogenesis emerged (reviewed in [4, 5]. Two distinct mesodermal heart fields that share a common origin appear to contribute cells to the developing heart in a temporally and spatially specific manner. The well-studied "first heart field" (FHF) is derived from cells in the anterior lateral plate mesoderm that align in a crescent shape at approximately embryonic (E) day 7.5 in the mouse embryo, roughly corresponding to week 2 of human gestation (Figure 9.1). By mouse E8.0, or 3 weeks in humans, these cells coalesce along the ventral midline to form a primitive heart tube, which consists of an interior layer of endocardial cells and an exterior layer of myocardial cells separated by extracellular matrix necessary for reciprocal signaling between the two layers. The tubular heart initiates rhythmic contractions at about day 23 in humans.

Previous lineage tracings using dye-labeling techniques suggested that cells along the anterior–posterior axis of the heart tube were destined to contribute to specific chambers of the future heart (reviewed in [6]). However, such studies could not determine the clonal contributions of individual cells [7]. More recent studies using Cre-lox technologies to mark progenitor cells and all their descendents indicate – in stark contrast to previous models – that the heart tube derived from the FHF may predominantly provide a scaffold that enables a second population of cells to migrate and expand into cardiac chambers [5]. These additional cells arise from an area often referred to as the "second heart field" (SHF) or "anterior heart field," based on its location anterior and medial to the crescent-shaped primary heart field [8, 10] (Figure 9.1). Both heart fields appear to be regulated by complex positive and negative signaling networks involving members of the bone morphogenetic protein (BMP), sonic hedgehog (Shh), fibroblast growth factor (Fgf), Wnt, and Notch proteins. Such signals often arise from the adjacent endoderm, although the precise nature and role of these signals remain unknown (reviewed in [11, 12–14]. SHF cells remain in an undifferentiated progenitor state until incorporation into the heart, and this may in part be due to closer proximity to inhibitory Wnt signals emanating from the midline. Recent work has raised the possibility that the *Tbx18* may be required for the formation of a venous pole, which contributes portions of the atria and venous structures [15]. Proepicardial (Tbx18$^+$ or Wt1$^+$) progenitors give rise to the epicardium and a subset of atrial and ventricular myocytes [16, 17].

As the heart tube forms, the SHF cells migrate into the midline and position themselves dorsal to the heart tube in the pharyngeal mesoderm. Upon rightward looping of the heart tube, SHF cells cross the pharyngeal mesoderm into the anterior and posterior portions, populating a large portion of the outflow tract, future right ventricle, and atria [18] (Figure 9.1). Precursors of the left ventricle are sparsely populated by the SHF and appear to largely be derived from the FHF. In contrast to the FHF, SHF cells do not differentiate into cardiac cells until they are positioned within the heart. Once within the heart, FHF and SHF cells

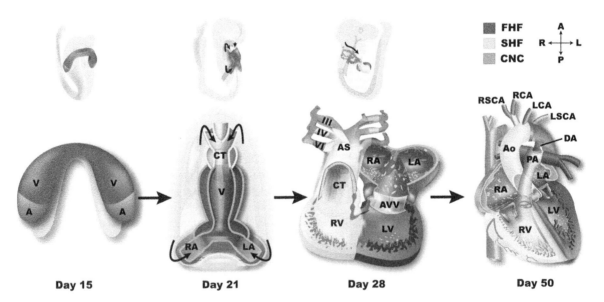

Day 15   **Day 21**   **Day 28**   **Day 50**

FIGURE 9.1 Illustration of cardiac development.
Illustrations depict cardiac development, with morphologically related regions color-coded, seen from a ventral view. (a) Two distinct cardiogenic precursor fields form a crescent that is specified to form specific regions of the heart tube (A, artery; V, ventricle), which is patterned to form the various regions and chambers of the looped and mature heart. (b) The secondary heart field (SHF) contributes to much of the right ventricle and outflow tract as the heart loops. (c) Each cardiac chamber balloons from the outer curvature of the looped heart tube in a segmental fashion. Neural crest cells populate the bilaterally symmetric aortic arch arteries (III, IV, and VI) and aortic sac that together contribute to specific segments of the mature aortic arch, also color-coded. (c, d) Mesenchymal cells form the cardiac valves from the conotruncal (CT) and atrioventricular valve (AVV) segments, which divide into separate left- and right-sided valves. Corresponding days of human embryonic development are indicated. RV, right ventricle; LV, left ventricle; RA, right atrium; LA, left atrium; PA, pulmonary artery; Ao, aorta; DA, ductus arteriosus; RSCA, right subclavian artery; RCC, right common carotid; LCC, left common carotid; LSCA, left subclavian artery. (Color shown in online version.)

appear to proliferate in response to endocardial derived signals such as neuregulin and epicardial signals dependent on retinoic acid, although the mechanisms through which these non-cell autonomous events occur remain poorly understood [19, 20].

CARDIOMYOCYTE AND HEART TUBE FORMATION

Knowledge gained from multiple model systems has begun to establish a molecular network that controls early fate decisions and subsequent morphogenetic events (Figure 9.2). The earliest discoveries arose from studies in flies. Fruitflies have a primitive heart-like structure, known as the dorsal vessel, that is analogous to the straight heart tube of the vertebrate embryo. It contracts rhythmically, and pumps hemolymph through an open circulatory system. Formation of the dorsal vessel in flies is dependent on a protein, tinman, whose name is based on the *Wizard of Oz* character that lacks a heart [21]. Tinman belongs to the homeodomain family of proteins, and was initially described as playing a role in establishing regional identity of cells and organs during embryogenesis.

In contrast to the requirement of *tinman* for heart formation in flies, its mammalian ortholog, *Nkx2.5*, is not essential for specification of the cardiac lineage in mice, suggesting either that other genes may share functions with *Nkx2.5*, or that cardiogenesis in flies and vertebrates differs with respect to its dependence on this family of homeobox

genes [22, 23]. The possibility of functional redundancy between *Nkx2.5* and other cardiac-expressed homeobox genes in vertebrates is supported by the ability of dominant negative versions of Nkx2.5 to block cardiogenesis in frog and zebrafish embryos [24, 25]. Similarly, the transcriptional co-activator, myocardin, is necessary and sufficient in frogs for cardiac gene expression, likely through activation of serum response factor (SRF)-dependent genes [26, 27]. Combinations of these transcription factors along with Mef2, Gata, Hand and Tbx family members appear to form core regulatory circuits that control early events during cardiogenesis (reviewed in [4]).

The cardiac outflow tract (conotruncus) and parts of the right ventricle are the last segments to form, and are derived from SHF cells as described above. The transcription factor Tbx1, which appears to be a cause of cardiac and craniofacial disorders in humans [28–31], is a major transcriptional regulator of the SHF, and is necessary for proper development of conotruncal myocardium and fibroblast growth factor secretion [32–34]. Islet1 (Isl1), a transcription factor involved in pancreatic development, also marks this population and is necessary for its development [18]. Interestingly, Isl1-positive cells mark niches of cardiac progenitor cells in the postnatal heart [35], suggesting that understanding the regulation of SHF-derived progenitor pools may be useful in developing approaches for cardiac repair.

Recent evidence suggests that Wnt signals play a dynamic and critical role in regulating cardiac progenitors [36].

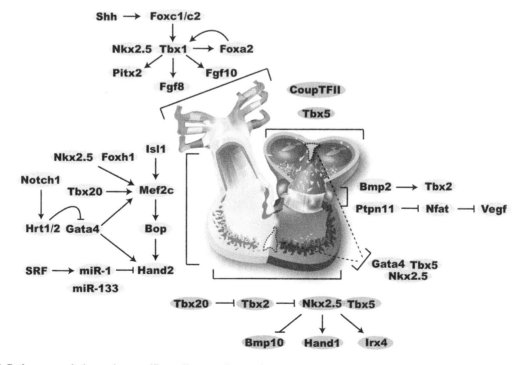

FIGURE 9.2 Pathways regulating region-specific cardiac morphogenesis.
A partial list of transcription factors, signaling proteins, and miRNAs that can be placed in pathways that influence the formation of regions of the heart. Positive influences are indicated by arrowheads, and negative effects by bars. Physical interactions are indicated by direct contact of factors.

Initially, Wnt signals are necessary to promote mesoderm formation, but subsequently need to be repressed in order for cardiac mesoderm to emerge. However, after cardiac progenitors become committed to this fate, canonical Wnt signals again promote expansion of the progenitor pool and proliferation of cardiomyocytes [37]. Such signals emanate from both the endoderm and the mesoderm.

In addition to the AP segmentation, a discrete dorsal–ventral (DV) polarity is present in the primitive heart tube. As the heart tube loops to the right, the ventral surface of the tube rotates, becoming the outer curvature of the looped heart with the dorsal surface forming the inner curvature. The outer curvature becomes the site of active growth, while remodeling of the inner curvature is essential for ultimate alignment of the inflow and outflow tracts of the heart. A model in which individual chambers "balloon" from the outer curvature in a segmental fashion has been proposed [38]. Consistent with this model, numerous genes, including the transcription factor *Hand1*, are expressed specifically on the ventral and outer curvature of the heart [39, 40]. Remodeling of the inner curvature occurs, allowing migration of the inflow tract to the right and outflow tract to the left, facilitating proper alignment and separation of right- and left-sided circulations. Defects of inner curvature remodeling may underlie a host of human congenital heart malformations that involve improper alignment of the atria, ventricles, and outflow tract, and are often observed in the setting of abnormalities of left–right asymmetry. Other cardiac defects are a result of genetic defects that cause disruption of discrete developmental events, making it useful to consider the molecular processes governing central morphogenetic aspects of cardiogenesis.

COMPLEX REGULATION OF CARDIAC MORPHOGENESIS

While the pathways regulating individual cell lineages contributing to the heart are deeply understood, the subsequent complex events involved in integrating multiple cell types, formation of chambers, and patterning of the distinct regions of the heart are also now being elucidated. Some aspects of these morphogenetic events are described below, while others have been reviewed in depth elsewhere [4, 41, 42].

Left–right asymmetry

The heart is the first organ to break the bilateral symmetry present in the early embryo, and the rightward direction of its looping reflects a more global establishment of left–right (LR) asymmetry that affects the lungs, liver, spleen, and gut. Defects in establishment of LR asymmetry in humans are associated with a wide range of cardiac alignment defects, suggesting that pathways regulating LR asymmetry dramatically affect cardiac development. The elegant pathways that control the direction of cardiac looping along the left–right axis and the general left–right body plan have been elucidated in recent years, and are summarized below [43] (Figure 9.2).

A cascade of signaling molecules regulating the establishment of embryonic LR asymmetry has been revealed from recent studies of chick embryonic development. Before the formation of organs in the developing embryo, asymmetric expression of the morphogen, Sonic hedgehog (Shh), on the left side of Hensen's node leads to left lateral mesoderm expression of nodal and lefty, members of the transforming growth factor-β (TGFβ) family [44]. Transfer of this signal from the node to the lateral mesoderm is mediated by the secreted molecule, caronte. Caronte inhibits BMP on the left side, relieving BMP-mediated repression of nodal in the left lateral plate mesoderm [45]. Left-sided expression of nodal induces rightward looping of the midline heart tube. Fibroblast growth factor and activin receptor-mediated pathways suppress caronte expression on the right side and the resulting activity of BMP signaling results in suppression of right-sided nodal expression. Conversely, the snail-related (cSnR-1) zinc finger transcription factor is expressed in the right lateral mesoderm and is repressed by Shh on the left [46]. The above signaling pathways are active in the lateral plate mesoderm, but not in the heart or other organs that actually display LR asymmetry. Ultimately, the nodal-dependent pathways result in expression of a homoedomain protein, Pitx2, on the left side of visceral organs, and repression of Pitx2 on the right [47]. Pitx2 appears to be the major factor that interprets the LR signaling cascade at the organ level. Asymmetric expression of Pitx2 is sufficient for establishing the LR asymmetry of the heart, lungs, and gut [48].

The mechanisms that control directionality of cardiac looping have also been explored by genetic analysis of mouse mutants with abnormalities in left–right asymmetry. Mice homozygous for mutation in the *left–right dynein* gene (*iv/iv*) display randomization of left–right orientation of the heart and viscera, and have bilaterally symmetric, absent or randomized expression of nodal and Pitx2 [49, 50]. Nodal and Pitx2 are expressed along the right lateral mesoderm rather than the left, displaying complete reversal of the LR signals, and have bilaterally symmetric, absent, or randomized expression of nodal and Pitx2. In contrast, in the situs inversus (*inv*) mouse, which has nearly 100 percent reversal of left–right asymmetry, nodal and Pitx2 are expressed along the right lateral mesoderm rather than the left, displaying complete reversal of the LR signals. *Pitx2* mutant mice have abnormal LR asymmetry of the lungs and a low penetrance of reversed cardiac looping, similar to *Shh* and *Fgf8* mutant mice [51, 52]. Oddly, the initial LR asymmetry and roles of Fgf and Shh are opposite in mice and chicks; however, the left–right orientation of later events involving nodal and Pitx2 are conserved [53].

While the necessity of LR asymmetric gene expression is intuitive, how the initial asymmetry of molecules is established remains in question. Initial clues came from studies of immotile cilia syndrome, also known as Kartagener's syndrome, in which individuals had situs inversus totalis, with mirror-image reversal of all organs. It was recently found that, prior to organ formation, Hensen's node contains ciliary processes that beat in a vortical fashion, pushing morphogens to the left side of the embryo [54]; concurrent establishment of a midline barrier, possibly by lefty gene expression along the left midline, may be responsible for subsequent asymmetric gene expression. Mice lacking ciliary movement in the node display abnormal LR patterning, consistent with this model.

Cardiac outflow tract regulation

Congenital cardiac defects involving the cardiac outflow tract, aortic arch, ductus arteriosus, and proximal pulmonary arteries account for 20–30 percent of all CHD. This region of the heart undergoes extensive and rather complex morphogenetic changes, with reciprocal interactions between neural crest cells and the SHF and endoderm playing critical roles.

Mesenchyme cells originating from the crest of the neural folds are essential for proper septation and remodeling of the outflow tract and aortic arch (reviewed in [55]). Such neural crest-derived cells migrate away from the neural folds and retain the ability to differentiate into multiple cell types. The migratory path and ultimate fate of these cells depends on their relative position of origin along the anterior–posterior axis, and are partly regulated by the Hox code [56]. Neural crest cells differentiate and contribute to diverse embryonic structures, including the cranial ganglia, peripheral nervous system, adrenal glands, and melanocytes. Neural crest cells that arise from the otic placode to the third somite migrate through the developing pharyngeal arches and populate the mesenchyme of each of the aortic arch arteries, and the mesenchyme necessary to septate the outflow tract septum (Figure 9.1). Because of their migratory path and role, this segment of the neural crest is often referred to as the cardiac neural crest. Mutations in many signaling cascades affect neural crest migration or development, including the endothelin and semaphorin pathways, and cause outflow tract defects similar to those observed in humans [55].

Disruption of SHF development by mutation of genes such as *Tbx1*, *Fgf8*, and *Isl1* results in defects similar to those observed with neural crest disruption, including persistent truncus arteriosus (failure of outflow septation), malalignment of the outflow tract of the heart with the ventricular chambers, and ventricular septal defects [18, 31, 57, 58]. SHF-derived myocardial cells neighbor neural crest-derived cells and secrete growth factors such as Fgf8, in a Tbx1-dependent manner [33]. Such growth factors influence neural crest cells, and reciprocal interactions between the SHF and neural crest-derived cells in the outflow tract are likely essential for normal development. Consistent with this, humans with deletion or mutation of *TBX1* [59], expressed in the SHF, appear to have cell-autonomous defects of SHF development and non-cell-autonomous anomalies of neural crest-derived tissues.

Cardiac valve formation

Appropriate placement and function of cardiac valves is essential for chamber septation and for unidirectional flow of blood through the heart. A molecular network involving BMP2 and Tbx2 defines the position of the valves relative to the chambers [60–62]. During early heart tube formation, "cushions" of extracellular matrix between the endocardium and myocardium presage valve formation at each end of the heart tube. Reciprocal signaling, mediated in part by TGFβ family members, between the myocardium and endocardium in the cushion region induces a transformation of endocardial cells into mesenchymal cells that migrate into the extracellular matrix cushion [63–65]. These mesenchymal cells differentiate into the fibrous tissue of the valves and are involved in septation of the common atrioventricular canal into right- and left-sided orifices.

The Smad proteins are intracellular transcriptional mediators of signaling initiated by TGFβ ligands. Smad6 is specifically expressed in the atrioventricular cushions and outflow tract during cardiogenesis, and is a negative regulator of TGFβ signaling. Targeted disruption of *Smad6* in mice results in thickened and gelatinous atrioventricular and semilunar valves, comparable to those observed in human aortic and pulmonary valve disease [66]. Similarly, the absence of *Ptpn11*, which encodes the protein tyrosine phosphatase Shp-2, results in dysplastic outflow valves through its involvement in a signaling pathway mediated by epidermal growth factor receptor [67]. The importance of *PTPN11* in CHD was shown by the presence of point mutations in *PTPN11* in patients with Noonan syndrome, who commonly have pulmonic valve stenosis [68]. Finally, mice lacking Ephrin B2 also have thickened valves and, although the mechanism for this remains unclear, it will be interesting to determine how these signaling pathways intersect [69].

In contrast to the thickened leaflets described above, disruption of signaling pathways converging on the transcription factor Nfatc revealed a requirement of this calcium-activated regulator. *Nfatc* is expressed specifically in the forming embryonic valves, and targeted deletion of *Nfatc* in mice results in absence of cardiac valve formation [70, 71]. Signaling via the phosphatase, calcineurin, results in nuclear translocation of Nfatc and is similarly involved in cardiac valve formation, in part through regulation of vascular endothelial growth factor (Vegf) expression in the endocardium [72].

The Notch signaling pathway is required for cell-fate and differentiation decisions throughout the embryo [73], but only recently have Notch proteins been implicated in vertebrate cardiac development. In fish and frogs, Notch appears to be involved in development of the endocardial cushions that contribute to valve tissue [74]. In humans, heterozygous *NOTCH1* mutations disrupt normal development of the aortic valve and occasionally the mitral valve [75]. While not specifically affecting valves, human mutations in *JAGGED1*, a NOTCH ligand, also cause outflow tract defects associated with the autosomal dominant disease, Alagille syndrome [76–78]. The hairy-related family of transcriptional repressors (Hrt1, Hrt2, and Hrt3) may mediate the Notch signal during valve and myocardial development; however, their targets for repression remain unknown ([79], reviewed in [80]).

MOLECULAR REGULATION OF SEPTAL FORMATION

Recent findings with the cardiac transcription factors NKX2.5, TBX5, and GATA4 exemplify the synergy between human genetics and studies of model organisms for understanding the etiology of human CHD. Numerous point mutations have been identified in *NKX2.5* in families with atrial septal defects and progressive cardiac conduction abnormalities [81]. Retrospective analysis of mice heterozygous for Nkx2.5 disruption revealed a similar phenotype and progressive apoptotic loss of conduction cells, suggesting a likely mechanism for the human phenotype [82, 83].

Humans with Holt-Oram syndrome, caused by mutations in *TBX5*, have cardiac anomalies similar to those with *NKX2.5* mutations (atrial and ventricular septal defects) as well as limb abnormalities [84, 85]. Intriguingly, mutations responsible for defects in the heart and limbs are clustered in different regions of the protein, suggesting that TBX5 engages different downstream genes or co-factors in these tissues that depend on unique structural motifs in the protein. One potential cofactor is NKX2-5, as the two physically interact and cooperate to activate common target genes [86].

Like the NKX2.5 and TBX5 mutations, mutations in the zinc-finger-containing protein GATA4 cause similar atrial and ventricular septal defects in autosomal dominant non-syndromic human pedigrees [87]. GATA4 or related proteins are essential for cardiogenesis in flies, fish, and mice [88–91]. Like NKX2.5, GATA4 and TBX5 also form a complex to regulate downstream genes, such as myosin heavy chain. Consistent with an important role for such combinatorial interactions, a familial GATA4 point mutation disrupts GATA4's ability to interact with TBX5 [75]. Conversely, several human TBX5 mutations disrupt TBX5 interaction with GATA4, suggesting that the two cooperate

in cardiac septation events [87]. GATA4, TBX5, and NKX2-5 may form a common complex that is necessary for proper cardiac septation. Disruption of any one of the three proteins or their interactions can result in atrial or ventricular septal defects.

MICRORNA REGULATION OF CARDIOMYOCYTE DIFFERENTIATION

While transcriptional and epigenetic events regulate many critical cardiac genes, translational control by small non-coding RNAs, such as microRNAs (miRNAs), has recently emerged as another mechanism to "fine-tune" dosages of key proteins during cardiogenesis [92–94]. miRNAs are genomically encoded 20–22 nucleotide RNAs that target mRNAs for translational inhibition or degradation by many of the same pathways as small interfering RNA (siRNA) [95–97]. Over 650 human microRNAs have been identified, but in only a few cases are the biological function and mRNA targets known.

The miRNA-1 family (miR-1-1 and miR-1-2) is highly conserved from worms to humans, and is specifically expressed in the developing cardiac and skeletal muscle progenitor cells as they differentiate [92]. Enrichment of miR-1-1 is initially observed in the atrial precursors before becoming ubiquitous in the heart, while miR-1-2 is specific for the ventricle throughout development, suggesting that the two may have chamber-specific effects *in vivo*. Both are highly expressed in the SHF-derived cells of the cardiac outflow tract (Figure 9.3). Interestingly, expression of these miRNAs is directly controlled by well-studied transcriptional regulatory networks that promote muscle differentiation. Cardiac expression is dependent on serum response factor (SRF), and skeletal muscle expression requires the myogenic transcription factors MyoD and Mef2. SRF recruits the potent co-activator, myocardin, to cardiac and smooth muscle-specific genes that control differentiation [26].

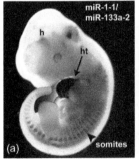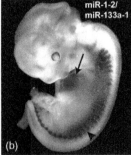

FIGURE 9.3 miR-1-1 and miR-1-2 enhancer-driven lacZ expression.
The expression patterns of miR-1-1 (a) and miR-1-2 (b) are demonstrated by the β-gal (blue) staining in embryonic day 11.5 mouse embryos. h, head; ht, heart; arrowhead indicates somites. (Color shown in online version.)

Consistent with a role in differentiation, overexpression of miR-1 in the developing mouse heart results in a decrease in ventricular myocyte expansion, with fewer proliferating cardiomyocytes remaining in the cell cycle. *In vivo* validation of Hand2, a transcription factor that regulates ventricular expansion, as an miR-1 target suggests that tight regulation of Hand2 protein levels may be involved in controlling the balance between cardiomyocyte proliferation and differentiation. Disruption of the single fly ortholog of miR-1 had catastrophic consequences, resulting in uniform lethality at embryonic or larval stages with a frequent defect in maintaining cardiac gene expression [93]. In a subset of flies lacking miR-1, a severe defect of cardiac progenitor cell differentiation provided loss-of-function evidence that miR-1 was involved in muscle differentiation events, similar to the gain-of-function findings in mice. Targeted deletion of miR-1-2 in mice results in ventricular septal defects and cardiac conduction abnormalities [94].

SUMMARY

The steps of cardiogenesis described here illustrate some of the signaling networks necessary for multiple cell types to communicate with one another in order to form a functioning organ. Reciprocal interactions between cell layers function to guide cells in the correct temporo-spatial pattern, and ultimately to adopt specific cell fates and achieve terminal differentiation. Disruption of such signaling events often underlies pathologic development of the heart, which manifests as congenital heart disease. Because fetal gene programs are often reactivated in the adult diseased heart with negative consequences, it is possible that inhibition or activation of specific signaling pathways involved in cardiogenesis may prove to have therapeutic value, even in late-onset heart disease.

ACKNOWLEDGEMENTS

The authors thank B. Taylor for editorial assistance. Chulan Kwon was supported by an American Heart Association post-doctoral fellowship, and is a post-doctoral scholar of the California Institute of Regenerative Medicine; Deepak Srivastava is supported by grants from NHLBI/NIH, California Institute of Regenerative Medicine, and is an Established Investigator of the American Heart Association.

REFERENCES

1. Hoffman JI, Kaplan S. The incidence of congenital heart disease. *J Am Coll Cardiol* 2002;**39**:1890–900.

2. Srivastava D. Heart disease: An ongoing genetic battle? *Nature* 2004;**429**:819–22.

3. Srivastava D, Ivey KN. Potential of stem cell-based therapies for heart disease. *Nature* 2006;**441**:1097–9.

4. Srivastava D. Making or breaking the heart: From lineage determination to morphogenesis. *Cell* 2006;**126**:1037–48.

5. Buckingham M, Meilhac S, Zaffran S. Building the mammalian heart from two sources of myocardial cells. *Nat Rev Genet* 2005;**6**:826–35.

6. Srivastava D, Olson EN. A genetic blueprint for cardiac development. *Nature* 2000;**407**:221–6.

7. Meilhac SM, Esner M, Kelly RG, Nicolas JF, Buckingham ME. The clonal origin of myocardial cells in different regions of the embryonic mouse heart. *Dev Cell* 2004;**6**:685–98.

8. Kelly RG, Brown NA, Buckingham ME. The arterial pole of the mouse heart forms from Fgf10-expressing cells in pharyngeal mesoderm. *Dev Cell* 2001;**1**:435–40.

9. Waldo KL, Kumiski DH, Wallis KT, et al. Conotruncal myocardium arises from a secondary heart field. *Development* 2001;**128**:3179–88.

10. Mjaatvedt CH, Nakaoka T, Moreno-Rodriguez R, et al. The outflow tract of the heart is recruited from a novel heart-forming field. *Dev Biol* 2001;**238**:97–109.

11. Zaffran S, Frasch M. Early signals in cardiac development. *Circ Res* 2002;**91**:457–69.

12. Schultheiss TM, Burch JB, Lassar AB. A role for bone morphogenetic proteins in the induction of cardiac myogenesis. *Genes Dev* 1997;**11**:451–62.

13. Schneider VA, Mercola M. Wnt antagonism initiates cardiogenesis in *Xenopus laevis*. *Genes Dev* 2001;**15**:304–15.

14. Marvin MJ, Di Rocco G, Gardiner A, Bush SM, Lassar AB. Inhibition of Wnt activity induces heart formation from posterior mesoderm. *Genes Dev* 2001;**15**:316–27.

15. Christoffels VM, Mommersteeg MT, Trowe MO, et al. Formation of the venous pole of the heart from an Nkx2-5-negative precursor population requires Tbx18. *Circ Res* 2006;**98**:1555–63.

16. Zhou B, Ma Q, Rajagopal S, et al. Epicardial progenitors contribute to the cardiomyocyte lineage in the developing heart. *Nature* 2008;**454**:109–13.

17. Cai CL, Martin JC, Sun Y, et al. A myocardial lineage derives from Tbx18 epicardial cells. *Nature* 2008;**454**:104–8.

18. Cai CL, Liang X, Shi Y, et al. Isl1 identifies a cardiac progenitor population that proliferates prior to differentiation and contributes a majority of cells to the heart. *Dev Cell* 2003;**5**:877–89.

19. Garratt AN, Ozcelik C, Birchmeier C. ErbB2 pathways in heart and neural diseases. *Trends Cardiovasc Med* 2003;**13**:80–6.

20. Stuckmann I, Evans S, Lassar AB. Erythropoietin and retinoic acid, secreted from the epicardium, are required for cardiac myocyte proliferation. *Dev Biol* 2003;**255**:334–49.

21. Bodmer R. The gene tinman is required for specification of the heart and visceral muscles in *Drosophila*. *Development* 1993;**118**:719–29.

22. Lyons I, Parsons LM, Hartley L, et al. Myogenic and morphogenetic defects in the heart tubes of murine embryos lacking the homeo box gene Nkx2-5. *Genes Dev* 1995;**9**:1654–66.

23. Tanaka M, Wechsler SB, Lee IW, Yamasaki N, Lawitts JA, Izumo S. Complex modular cis-acting elements regulate expression of the cardiac specifying homeobox gene Csx/Nkx2.5. *Development* 1999;**126**:1439–50.

24. Fu Y, Yan W, Mohun TJ, Evans SM. Vertebrate tinman homologues XNkx2-3 and XNkx2-5 are required for heart formation in a functionally redundant manner. *Development* 1998;**125**:4439–49.

25. Grow MW, Krieg PA. Tinman function is essential for vertebrate heart development: Elimination of cardiac differentiation by dominant inhibitory mutants of the tinman-related genes, XNkx2-3 and XNkx2-5. *Dev Biol* 1998;**204**:187–96.

26. Wang D, Chang PS, Wang Z, et al. Activation of cardiac gene expression by myocardin, a transcriptional cofactor for serum response factor. *Cell* 2001;**105**:851–62.

27. Small EM, Warkman AS, Wang DZ, Sutherland LB, Olson EN, Krieg PA. Myocardin is sufficient and necessary for cardiac gene expression in *Xenopus*. *Development* 2005;**132**:987–97.

28. Lindsay EA, Botta A, Jurecic V, et al. Congenital heart disease in mice deficient for the DiGeorge syndrome region. *Nature* 1999;**401**:379–83.

29. Lindsay EA, Vitelli F, Su H, et al. Tbx1 haploinsufficieny in the DiGeorge syndrome region causes aortic arch defects in mice. *Nature* 2001;**410**:97–101.

30. Merscher S, Funke B, Epstein JA, et al. TBX1 is responsible for cardiovascular defects in velo-cardio-facial/DiGeorge syndrome. *Cell* 2001;**104**:619–29.

31. Jerome LA, Papaioannou VE. DiGeorge syndrome phenotype in mice mutant for the T-box gene, Tbx1. *Nat Genet* 2001;**27**:286–91.

32. Yamagishi H, Maeda J, Hu T, et al. Tbx1 is regulated by tissue-specific forkhead proteins through a common Sonic hedgehog-responsive enhancer. *Genes Dev* 2003;**17**:269–81.

33. Hu T, Yamagishi H, Maeda J, McAnally J, Yamagishi C, Srivastava D. Tbx1 regulates fibroblast growth factors in the anterior heart field through a reinforcing autoregulatory loop involving forkhead transcription factors. *Development* 2004;**131**:5491–502.

34. Xu H, Morishima M, Wylie JN, et al. Tbx1 has a dual role in the morphogenesis of the cardiac outflow tract. *Development* 2004;**131**:3217–27.

35. Laugwitz KL, Moretti A, Lam J, et al. Postnatal isl1+ cardioblasts enter fully differentiated cardiomyocyte lineages. *Nature* 2005;**433**:647–53.

36. Kwon C, Cordes KR, Srivastava D. Wnt/β-catenin signaling acts at multiple developmental stages to promote mammalian cardiogenesis. *Cell Cycle* 2009. in press.

37. Kwon C, Arnold J, Hsiao EC, Taketo MM, Conklin BR, Srivastava D. Canonical Wnt signaling is a positive regulator of mammalian cardiac progenitors. *Proc Natl Acad Sci USA* 2007;**104**:10,894–10,899.

38. Moorman AF, Christoffels VM. Cardiac chamber formation: Development, genes, and evolution. *Physiol Rev* 2003;**83**:1223–67.

39. Thomas T, Kurihara H, Yamagishi H, et al. A signaling cascade involving endothelin-1, dHAND and msx1 regulates development of neural-crest-derived branchial arch mesenchyme. *Development* 1998;**125**:3005–14.

40. Biben C, Harvey RP. Homeodomain factor Nkx2-5 controls left/right asymmetric expression of bHLH gene eHand during murine heart development. *Genes Dev* 1997;**11**:1357–69.

41. Olson EN. A decade of discoveries in cardiac biology. *Nature Med* 2004;**10**:467–74.

42. Parmacek MS, Epstein JA. Pursuing cardiac progenitors: Regeneration redux. *Cell* 2005;**120**:295–8.

43. Palmer AR. Symmetry breaking and the evolution of development. *Science (New York NY)* 2004;**306**:828–33.

44. Levin M, Johnson RL, Stern CD, Kuehn M, Tabin C. A molecular pathway determining left–right asymmetry in chick embryogenesis. *Cell* 1995;**82**:803–14.

45. Rodriguez Esteban C, Capdevila J, Economides AN, Pascual J, Ortiz A, Izpisua Belmonte JC. The novel Cer-like protein Caronte mediates the establishment of embryonic left–right asymmetry. *Nature* 1999;**401**:243–51.

46. Isaac A, Sargent MG, Cooke J. Control of vertebrate left–right asymmetry by a snail-related zinc finger gene. *Science (New York NY)* 1997;**275**:1301–4.

47. Piedra ME, Icardo JM, Albajar M, Rodriguez-Rey JC, Ros MA. Pitx2 participates in the late phase of the pathway controlling left–right asymmetry. *Cell* 1998;**94**:319–24.

48. Logan M, Pagan-Westphal SM, Smith DM, Paganessi L, Tabin CJ. The transcription factor Pitx2 mediates situs-specific morphogenesis in response to left–right asymmetric signals. *Cell* 1998;**94**:307–17.

49. Supp DM, Witte DP, Potter SS, Brueckner M. Mutation of an axonemal dynein affects left–right asymmetry in inversus viscerum mice. *Nature* 1997;**389**:963–6.

50. Supp DM, Brueckner M, Kuehn MR, et al. Targeted deletion of the ATP binding domain of left–right dynein confirms its role in specifying development of left–right asymmetries. *Development* 1999;**126**:5495–504.

51. Lin CR, Kioussi C, O'Connell S, et al. Pitx2 regulates lung asymmetry, cardiac positioning and pituitary and tooth morphogenesis. *Nature* 1999;**401**:279–82.

52. Campione M, Steinbeisser H, Schweickert A, et al. The homeobox gene Pitx2: Mediator of asymmetric left–right signaling in vertebrate heart and gut looping. *Development* 1999;**126**:1225–34.

53. Capdevila J, Vogan KJ, Tabin CJ, Izpisua Belmonte JC. Mechanisms of left–right determination in vertebrates. *Cell* 2000;**101**:9–21.

54. Nonaka S, Tanaka Y, Okada Y, et al. Randomization of left–right asymmetry due to loss of nodal cilia generating leftward flow of extraembryonic fluid in mice lacking KIF3B motor protein. *Cell* 1998;**95**:829–37.

55. Hutson MR, Kirby ML. Neural crest and cardiovascular development: A 20-year perspective. *Birth Defects Res C Embryo Today* 2003;**69**:2–13.

56. Le Douarin NM, Creuzet S, Couly G, Dupin E. Neural crest cell plasticity and its limits. *Development* 2004;**131**:4637–50.

57. Abu-Issa R, Smyth G, Smoak I, Yamamura K, Meyers EN. Fgf8 is required for pharyngeal arch and cardiovascular development in the mouse. *Development* 2002;**129**:4613–25.

58. Frank DU, Fotheringham LK, Brewer JA, et al. An Fgf8 mouse mutant phenocopies human 22q11 deletion syndrome. *Development* 2002;**129**:4591–603.

59. Yagi H, Furutani Y, Hamada H, et al. Role of TBX1 in human del22q11.2 syndrome. *Lancet* 2003;**362**:1366–73.

60. Harrelson Z, Kelly RG, Goldin SN, et al. Tbx2 is essential for patterning the atrioventricular canal and for morphogenesis of the outflow tract during heart development. *Development* 2004;**131**:5041–52.

61. Beis D, Bartman T, Jin SW, et al. Genetic and cellular analyses of zebrafish atrioventricular cushion and valve development. *Development* 2005;**132**:4193–204.

62. Ma L, Lu MF, Schwartz RJ, Martin JF. Bmp2 is essential for cardiac cushion epithelial-mesenchymal transition and myocardial patterning. *Development* 2005;**132**:5601–11.

63. Gaussin V, Van de Putte T, Mishina Y, et al. Endocardial cushion and myocardial defects after cardiac myocyte-specific conditional deletion of the bone morphogenetic protein receptor ALK3. *Proc Natl Acad Sci USA* 2002;**99**:2878–83.

64. Kim RY, Robertson EJ, Solloway MJ. Bmp6 and Bmp7 are required for cushion formation and septation in the developing mouse heart. *Dev Biol* 2001;**235**:449–66.

65. Brown CB, Boyer AS, Runyan RB, Barnett JV. Requirement of type III TGF-beta receptor for endocardial cell transformation in the heart. *Science (New York NY)* 1999;**283**:2080–2.

66. Galvin KM, Donovan MJ, Lynch CA, et al. A role for smad6 in development and homeostasis of the cardiovascular system. *Nat Genet* 2000;**24**:171–4.

67. Chen B, Bronson RT, Klaman LD, et al. Mice mutant for Egfr and Shp2 have defective cardiac semilunar valvulogenesis. *Nat Genet* 2000;**24**:296–9.

68. Tartaglia M, Mehler EL, Goldberg R, et al. Mutations in PTPN11, encoding the protein tyrosine phosphatase SHP-2, cause Noonan syndrome. *Nat Genet* 2001;**29**:465–8.

69. Cowan CA, Yokoyama N, Saxena A, et al. Ephrin-B2 reverse signaling is required for axon pathfinding and cardiac valve formation but not early vascular development. *Dev Biol* 2004;**271**:263–71.

70. Ranger AM, Grusby MJ, Hodge MR, et al. The transcription factor NF-ATc is essential for cardiac valve formation. *Nature* 1998;**392**:186–90.

71. de la Pompa JL, Timmerman LA, Takimoto H, et al. Role of the NF-ATc transcription factor in morphogenesis of cardiac valves and septum. *Nature* 1998;**392**:182–6.

72. Chang CP, Neilson JR, Bayle JH, et al. A field of myocardial–endocardial NFAT signaling underlies heart valve morphogenesis. *Cell* 2004;**118**:649–63.

73. Artavanis-Tsakonas S, Rand MD, Lake RJ. Notch signaling: cell fate control and signal integration in development. *Science (New York NY)* 1999;**284**:770–6.

74. Timmerman LA, Grego-Bessa J, Raya A, et al. Notch promotes epithelial-mesenchymal transition during cardiac development and oncogenic transformation. *Genes Dev* 2004;**18**:99–115.

75. Garg V, Muth AN, Ransom JF, et al. Mutations in NOTCH1 cause aortic valve disease. *Nature* 2005;**437**:270–4.

76. Li L, Krantz ID, Deng Y, et al. Alagille syndrome is caused by mutations in human Jagged1, which encodes a ligand for Notch1. *Nat Genet* 1997;**16**:243–51.

77. Oda T, Elkahloun AG, Pike BL, et al. Mutations in the human Jagged1 gene are responsible for Alagille syndrome. *Nat Genet* 1997;**16**:235–42.

78. Krantz ID, Smith R, Colliton RP, et al. Jagged1 mutations in patients ascertained with isolated congenital heart defects. *Am J Med Genet* 1999;**84**:56–60.

79. Nakagawa O, Nakagawa M, Richardson JA, Olson EN, Srivastava D. HRT1, HRT2, and HRT3: A new subclass of bHLH transcription factors marking specific cardiac, somitic, and pharyngeal arch segments. *Dev Biol* 1999;**216**:72–84.

80. Kokubo H, Miyagawa-Tomita S, Johnson RL. Hesr, a mediator of the Notch signaling, functions in heart and vessel development. *Trends Cardiovasc Med* 2005;**15**:190–4.

81. Schott JJ, Benson DW, Basson CT, et al. Congenital heart disease caused by mutations in the transcription factor NKX2-5. *Science (New York NY)* 1998;**281**:108–11.

82. Biben C, Weber R, Kesteven S, et al. Cardiac septal and valvular dysmorphogenesis in mice heterozygous for mutations in the homeobox gene Nkx2-5. *Circ Res* 2000;**87**:888–95.

83. Jay PY, Harris BS, Maguire CT, et al. Nkx2-5 mutation causes anatomic hypoplasia of the cardiac conduction system. *J Clin Invest* 2004;**113**:1130–7.

84. Basson CT, Bachinsky DR, Lin RC, et al. Mutations in human TBX5 cause limb and cardiac malformation in Holt-Oram syndrome. *Nat Genet* 1997;**15**:30–5.

85. Mori AD, Bruneau BG. TBX5 mutations and congenital heart disease: Holt-Oram syndrome revealed. *Curr Opin Cardiol* 2004;**19**:211–15.

86. Hiroi Y, Kudoh S, Monzen K, et al. Tbx5 associates with Nkx2-5 and synergistically promotes cardiomyocyte differentiation. *Nat Genet* 2001;**28**:276–80.

87. Garg V, Kathiriya IS, Barnes R, et al. GATA4 mutations cause human congenital heart defects and reveal an interaction with TBX5. *Nature* 2003;**424**:443–7.

88. Molkentin JD, Lin Q, Duncan SA, Olson EN. Requirement of the transcription factor GATA4 for heart tube formation and ventral morphogenesis. *Genes Dev* 1997;**11**:1061–72.

89. Kuo CT, Morrisey EE, Anandappa R, et al. GATA4 transcription factor is required for ventral morphogenesis and heart tube formation. *Genes Dev* 1997;**11**:1048–60.

90. Reiter JF, Kikuchi Y, Stainier DY. Multiple roles for Gata5 in zebrafish endoderm formation. *Development* 2001;**128**:125–35.

91. Gajewski K, Zhang Q, Choi CY, et al. Pannier is a transcriptional target and partner of Tinman during *Drosophila* cardiogenesis. *Dev Biol* 2001;**233**:425–36.

92. Zhao Y, Samal E, Srivastava D. Serum response factor regulates a muscle-specific mircroRNA that targets *Hand2* during cardiogenesis. *Nature* 2005;**436**:214–20.

93. Kwon C, Han Z, Olson EN, Srivastava D. MicroRNA1 influences cardiac differentiation in Drosophila and regulates Notch signaling. *Proc Natl Acad Sci USA* 2005;**102**:18,986–18,991.

94. Zhao Y, Ransom JF, Li A, et al. Dysregulation of cardiogenesis, cardiac conduction, and cell cycle in mice lacking miRNA-1-2. *Cell* 2007;**129**:303–17.

95. He L, Hannon GJ. MicroRNAs: Small RNAs with a big role in gene regulation. *Nat Rev Genet* 2004;**5**:522–31.

96. Ambros V. The functions of animal microRNAs. *Nature* 2004;**431**:350–5.

97. Zhao Y, Srivastava D. A developmental view of microRNA function. *Trends Biochem Sci* 2007;**32**:189–97.

Chapter 10

Calcium Signaling in Cardiac Muscle

K.M. Dibb, A.W. Trafford and D.A. Eisner

Unit of Cardiac Physiology, University of Manchester, Manchester, England, UK

INTRODUCTION

Each heartbeat is initiated by an increase of intracellular calcium concentration, the so-called "systolic calcium transient." Changes of the amplitude of the calcium transient are the major factor controlling the force of contraction of the heart during, for example, exercise. Abnormalities in calcium signaling have been implicated in clinically important conditions such as heart failure and cardiac arrhythmias. The purpose of this chapter is to provide an overview of aspects of the state of knowledge of calcium signaling. It is impossible to cover the whole field of cardiac calcium signaling in this brief review, and interested readers are referred to recent reviews [1,2].

CALCIUM-INDUCED CALCIUM RELEASE

The systolic calcium transient has two sources (see Figure 10.1): calcium enters the cell from the extracellular fluid, and it is released from the sarcoplasmic reticulum (SR). As far as calcium entry is concerned, the major source is via the voltage-activated L-type calcium current. This channel is activated by depolarization to voltages

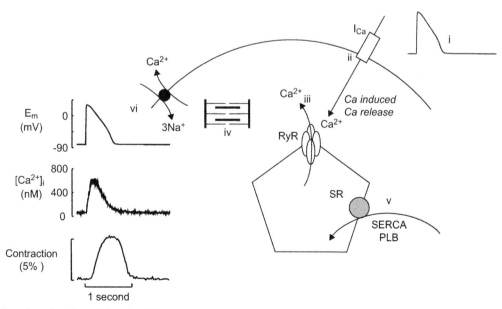

FIGURE 10.1 Overview of cardiac calcium handling.
The sequence of steps in excitation–contraction coupling. (i) The cardiac action potential results in the opening of the L-type calcium channel (ii); this produces the entry of a small amount of calcium that triggers the opening of the RyR (iii), a process known as calcium induced calcium release. This released calcium binds to the myofilaments (iv), resulting in contraction. Relaxation requires that $[Ca^{2+}]_i$ is decreased, and this occurs by (v) re-uptake into the SR by SERCA (the activity of SERCA is controlled by phospholamban, PLB) and (vi) removal from the cell by Na/Ca exchange. The inset shows typical records of (from top to bottom), membrane potential, $[Ca^{2+}]_i$, cell contraction.

positive to about -50 to $-40 \, \text{mV}$. Further depolarization increases the number of channels that open, but this is off-set by the decreased electrochemical driving force, and the relationship between membrane potential and calcium entry is therefore bell-shaped with a maximum at about $0 \, \text{mV}$ [3,4]. The major source of calcium for the systolic calcium transient is the SR. Calcium release from the SR occurs through a specialized release channel known as the ryanodine receptor (RyR); in cardiac muscle, the RyR2 isoform. Release occurs by the process of Calcium-Induced Calcium Release (CICR). The probability that the RyR is open (p_o) is increased by an increase of $[Ca^{2+}]_i$. Therefore, the calcium influx on the L-type calcium channel leads to the opening of the RyR and release of a larger amount of calcium from the SR [5]. At first sight it might appear that such a calcium-induced calcium release mechanism has the potential for positive feedback, since the calcium released from the SR would be expected to further activate RyRs and therefore the SR would be expected always to empty completely. However, experimental work shows that cal-cium release is a graded function of membrane potential, and calcium release from the SR changes in the same way as does the amplitude of the calcium current. The solution to this paradox came with the discovery of "local control of calcium release." According to this hypothesis, one L-type calcium channel activates a small group of RyRs. Calcium release from this group of RyRs cannot diffuse far enough to activate other RyRs [6], and therefore there is no ten-dency to positive feedback. The underlying calcium release events are seen as so-called "calcium sparks" [7,8]. The gradation of calcium release with changes of membrane potential results from recruitment of more of these release units as the L-type calcium current is increased [9]. One question which still remains, however, is: what terminates calcium release from the SR? The decay of the systolic cal-cium transient requires that calcium release stops in order that SERCA can return calcium to the sarcoplasmic reticu-lum. There is still considerable controversy as to which mechanisms are responsible for RyR closing. Possible explanations include: (1) an inactivation or "adaptation" of the RyR [10]; and (2) a decrease of SR calcium content resulting in a decreased opening of the RyR (see [11,12] for extensive discussion of this issue).

HOW IS SR CALCIUM CONTENT CONTROLLED?

As discussed below, the calcium content of the SR is a major factor controlling the amplitude of the calcium transient, and it is therefore important that SR calcium content be control-led precisely. Breakdown of this control can result in cal-cium waves and thence arrhythmias (see [13] for review), as well as *pulsus alternans* (a condition in which the strength of the heartbeat alternates from beat to beat) [14]. We have

investigated the factors responsible for regulating SR cal-cium content. Briefly, control results from the fact that the major pathways for calcium movement across the surface membrane are sensitive to the amplitude of the systolic cal-cium transient. An increase of the calcium transient results in more efflux from the cell (by activation of NCX) and less influx (by calcium-dependent inactivation of the L-type cal-cium current) [15-17]. This results in the following negative feedback loop to control SR calcium content: (i) an increase of SR calcium content will increase the amplitude of the systolic calcium transient; (ii) this will increase calcium efflux and decrease calcium influx across the sarcolemma; (iii) as a consequence, the cell and therefore SR calcium content will decrease towards control levels. It is important to realize that changes in the number or activity of any of the sarcolemmal or SR calcium pumps and channels will change the steady state SR calcium content. For example, an increase of SERCA activity or a decrease of NCX will increase SR calcium content. However, as reviewed above, the beat-to-beat control of SR calcium depends on the fact that SR calcium content controls the amplitude of the cal-cium transient, and this affects sarcolemmal calcium fluxes.

WHICH FACTORS CONTROL THE AMPLITUDE OF THE SYSTOLIC CALCIUM TRANSIENT?

Changes of the amplitude of the systolic calcium transient are the major means for changing contraction. There are at least three ways in which the amount of calcium released from the SR by CICR can be increased: (1) an increase of the trigger produced by the L-type calcium current and therefore the trigger for calcium release from the SR; (2) an increase in the opening of the RyR such that more cal-cium is released for a given trigger L-type current and SR calcium content; and (3) an increase of SR calcium content. We will consider these factors in turn.

1. *An increase of the trigger produced by the L-type calcium current.* It is well known that an increase in the L-type current increases the amplitude of the calcium transient and contraction [18], and this accounts for part of the positive inotropic effects of β-adrenergic stimulation.

2. *An increase in the opening of the RyR.* The question regarding the effects of changing the RyR is rather more controversial. A number of interventions affect the open probability of the RyR. For example, cyclic ADP ribose increases its open probability [19,20], as does phosphorylation [21]. Indeed, it has been sug-gested that phosphorylation of the RyR may contrib-ute to the positive inotropic effect of β-adrenergic stimulation [22]. Arguing against this, however, is the demonstration that mice in which a phosphorylation site on the RyR has been removed show quantitatively

similar inotropic responses to β-adrenergic stimulation [23]. A further issue to consider is that maneuvers that affect the open probability of the RyR do not produce maintained effects on systolic calcium or contractility. Thus, the application of low concentrations of caffeine produces an immediate increase of systolic $[Ca^{2+}]_i$. However, this increase of amplitude is not maintained and, in the steady state, the calcium amplitude of the calcium transient is exactly the same in the presence of a low concentration of caffeine as in control [24-26]. The transient nature of this response arises because the initial increase of the amplitude of the calcium transient results in more calcium efflux from the cell, thereby decreasing cell and thus SR calcium content. In the steady state, calcium efflux must equal calcium influx. If a maneuver has no effect on calcium influx, then in the steady state, calcium efflux must also be unaltered. If the properties of NCX are not altered, then this requirement for constant calcium efflux means that the calcium transient must also be unaffected and we therefore conclude that increasing the open probability of the RyR will not produce a positive inotropic effect.

3. *An increase of SR calcium content.* The remaining control point for CICR is the amount of calcium in the SR. Much work has shown that calcium release from the SR is a steep function of SR content [27], with the release being approximately proportional to the cube of SR content. It is therefore likely that increases of SR calcium content produced by interventions such as an increase of heart rate and β-adrenergic stimulation play an important role in regulating the force of contraction of the heart.

An interesting special case of an inotropic maneuver is revealed by considering the effects of increasing the calcium current. This has two effects on calcium handling: (1) an increase in the trigger for calcium release; and (2) an increase in the calcium entry into the cell and therefore into the SR. The former effect will tend to decrease SR calcium content, while the latter will increase it. Experimentally, we have found that large changes of L-type calcium current have very little effect on the SR calcium content [28]. This suggests that the two effects of altering the L-type calcium current are well balanced. As noted previously [29], the relative constancy of SR content means that positive inotropic effects of an increase of SR calcium content can occur without the delay required if SR content needs to increase.

CALCIUM SIGNALING IN HEART FAILURE

The general area of calcium handling in heart failure has been reviewed extensively [30]. The decrease of contractility is associated with a decrease in both the amplitude and rate of decay of the systolic calcium transient [31]. The decreased amplitude of the calcium transient can account

for the decreased contractility, and the decreased rate of decay will impair relaxation. Most studies find no change in the amplitude of the calcium current, and the most commonly accepted explanation of the decreased calcium transient is that it results from a decrease of SR calcium content [32]. An area of current controversy, however, is the origin of this decrease of content. There are two explanations. The first and longest-established hypothesis is that SERCA activity is decreased. This can account not only for the decrease of SR calcium content, but also for the fact that the rate of decay of the systolic calcium transient is decreased. The other hypothesis is that there is an increase of leak of calcium out of the SR through the RyR [33,34]. This leak has been attributed to hyperphosphorylation of the RyR by protein kinase A (PKA) [35]. However, recent work directly measuring the SR leak in cardiac myocytes has implicated Ca^{2+}/calmodulin-dependent protein kinase [36]. It is also worth noting that this calcium leak has also been suggested to trigger arrhythmias by initiating waves of CICR in conditions when the SR calcium is elevated or in the presence of mutations in the RyR.

REFERENCES

1. Bers DM. *Excitation–contraction coupling and cardiac contractile force.* Boston: Kluwer; 2001 p.2.
2. Dibb KM, Graham HK, Venetucci L, Eisner DA, Trafford AW. Analysis of cellular calcium fluxes in cardiac muscle to understand calcium homeostasis in the heart. *Cell Calcium* 2007;**42**:503–12.
3. Cannell MB, Berlin JR, Lederer WJ. Effect of membrane potential changes on the calcium transient in single rat cardiac muscle cells. *Science* 1987;**238**:1419–23.
4. Beuckelmann DJ, Wier WG. Mechanism of release of calcium from sarcoplasmic reticulum of guinea-pig cardiac cells. *J Physiol (Lond)* 1988;**405**:233–55.
5. Fabiato A. Time and calcium dependence of activation and inactivation of calcium-induced release of calcium from the sarcoplasmic reticulum of a skinned canine cardiac Purkinje cell. *J Gen Physiol* 1985;**85**:247–89.
6. Stern MD. Theory of excitation–contraction coupling in cardiac muscle. *Biophys J* 1992;**63**:497–517.
7. Cheng H, Lederer WJ, Cannell MB. Calcium sparks: elementary events underlying excitation-contraction coupling in heart muscle. *Science* 1993;**262**:740–4.
8. Wier WG, Egan TM, López-López JR, Balke CW. Local control of excitation–contraction coupling in rat heart cells. *J Physiol (Lond)* 1994;**474**:463–71.
9. Cannell MB, Cheng H, Lederer WJ. The control of calcium release in heart muscle. *Science* 1995;**268**:1045–9.
10. Györke S, Fill M. Ryanodine receptor adaptation: control mechanism of Ca^{2+}-induced Ca^{2+} release in heart. *Science* 1993;**260**:807–9.
11. Sobie EA, Song LS, Lederer WJ. Local recovery of Ca^{2+} release in rat ventricular myocytes. *J Physiol (Lond)* 2005;**565**:441–7.
12. Sobie EA, Dilly KW, Dos Santos CJ, Lederer WJ, Jafri MS. Termination of cardiac Ca^{2+} sparks: an investigative mathematical model of calcium-induced calcium release. *Biophys J* 2002;**83**:59–78.

13. Venetucci LA, Trafford AW, O'Neill SC, Eisner DA. The sarcoplasmic reticulum and arrhythmogenic calcium release. *Cardiovasc Res* 2008;**77**:285–92.

14. Eisner DA, Diaz ME, Li Y, O'Neill SC, Trafford AW. Stability and instability of regulation of intracellular calcium. *Exp Physiol* 2005;**90**:3–12.

15. Eisner DA, Choi HS, Díaz ME, O'Neill SC, Trafford AW. Integrative analysis of calcium cycling in cardiac muscle. *Circ Res* 2000;**87**:1087–94.

16. Eisner DA, Trafford AW, Díaz ME, Overend CL, O'Neill SC. The control of Ca release from the cardiac sarcoplasmic reticulum: regulation versus autoregulation. *Cardiovasc Res* 1998;**38**:589–604.

17. Trafford AW, Díaz ME, Negretti N, Eisner DA. Enhanced calcium current and decreased calcium efflux restore sarcoplasmic reticulum Ca content following depletion. *Circ Res* 1997;**81**:477–84.

18. Beeler GW, Reuter H. The relation between membrane potential, membrane currents and activation of contraction in ventricular myocardial fibres. *J Physiol (Lond)* 1970;**207**:211–29.

19. Sitsapesan R, McGarry SJ, Williams AJ. Cyclic ADP-ribose, the ryanodine receptor and Ca^{2+} release. *Trends Pharmacol Sci* 1995;**16**:386–91.

20. Sitsapesan R, Williams AJ. Cyclic ADP-ribose and related compounds activate sheep skeletal sarcoplasmic reticulum Ca^{2+} release channel. *Am J Physiol* 1995;**268**:C1235–40.

21. Lokuta AJ, Rogers TB, Lederer WJ, Valdivia HH. Modulation of cardiac ryanodine receptors of swine and rabbit by a phosphorylation–dephosphorylation mechanism. *J Physiol (Lond)* 1995;**487**:609–22.

22. Wehrens XH, Lehnart SE, Marks AR. Intracellular calcium release and cardiac disease. *Annu Rev Physiol* 2005;**67**:69–98.

23. MacDonnell SM, Garcia-Rivas G, Scherman JA, Kubo H, Chen X, Valdivia H, Houser SR. Adrenergic regulation of cardiac contractility does not involve phosphorylation of the cardiac ryanodine receptor at serine 2808. *Circ Res* 2008;**102**:e65–72.

24. O'Neill SC, Eisner DA. A mechanism for the effects of caffeine on Ca^{2+} release during diastole and systole in isolated rat ventricular myocytes. *J Physiol (Lond)* 1990;**430**:519–36.

25. Trafford AW, Díaz ME, Sibbring GC, Eisner DA. Modulation of CICR has no maintained effect on systolic Ca^{2+}: simultaneous measurements of sarcoplasmic reticulum and sarcolemmal Ca^{2+} fluxes in rat ventricular myocytes. *J Physiol (Lond)* 2000;**522**:259–70.

26. Trafford AW, Díaz ME, Eisner DA. Stimulation of Ca-induced Ca release only transiently increases the systolic Ca transient: measurements of Ca fluxes and SR Ca. *Cardiovasc Res* 1998;**37**:710–17.

27. Bassani JWM, Yuan W, Bers DM. Fractional SR Ca release is regulated by trigger Ca and SR Ca content in cardiac myocytes. *Am J Physiol* 1995;**268**:C1313–29.

28. Trafford AW, Díaz ME, Eisner DA. Coordinated control of cell Ca^{2+} loading and triggered release from the sarcoplasmic reticulum underlies the rapid inotropic response to increased L-type Ca^{2+} current. *Circ Res* 2001;**88**:195–201.

29. Eisner DA, Trafford AW. No role for the ryanodine receptor in regulating cardiac contraction?. *News Physiol Sci* 2000;**15**:275–9.

30. Hasenfuss G, Pieske B. Calcium cycling in congestive heart failure. *J Mol Cell Cardiol* 2002;**34**:951–69.

31. Beuckelmann DJ, Näbauer M, Erdmann E. Intracellular calcium handling in isolated ventricular myocytes from patients with terminal heart failure. *Circulation* 1992;**85**:1046–55.

32. Díaz ME, Graham HK, Trafford AW. Enhanced sarcolemmal Ca^{2+} efflux reduces sarcoplasmic reticulum Ca^{2+} content and systolic Ca^{2+} in cardiac hypertrophy. *Cardiovasc Res* 2004;**62**:538–47.

33. Marks AR. Cardiac intracellular calcium release channels role in heart failure. *Circ Res* 2000;**87**:8–11.

34. Belevych A, Kubalova Z, Terentyev D, Hamlin RL, Carnes CA, Györke S. Enhanced ryanodine receptor-mediated calcium leak determines reduced sarcoplasmic reticulum calcium content in chronic canine heart failure. *Biophys J* 2007;**93**:4083–92.

35. Reiken S, Gaburjakova M, Guatimosim S, Gomez AM, D'Armiento J, Burkhoff D, Wang J, Vassort G, Lederer WJ, Marks AR. Protein kinase A phosphorylation of the cardiac calcium release channel (ryanodine receptor) in normal and failing hearts. Role of phosphatases and response to isoproterenol. *J Biol Chem* 2003;**278**:444–53.

36. Ai X, Curran JW, Shannon TR, Bers DM, Pogwizd SM. Ca^{2+}/calmodulin-dependent protein kinase modulates cardiac ryanodine receptor phosphorylation and sarcoplasmic reticulum Ca^{2+} leak in heart failure. *Circ Res* 2005;**97**:1314–22.

Chapter 11

Calcium Signaling in Smooth Muscle

Susan Wray

Physiology Department, School of Biomedical Sciences, University of Liverpool, Liverpool, England, UK

INTRODUCTION

Calcium signaling in smooth muscle is a varied affair. The types of local and global calcium signals, the release process from the internal calcium store (SR, Sarcoplasmic Reticulum), and the calcium homeostatic control mechanisms found in this one cell type encompass most of the mechanisms only found separately in other cell types. Some of this diversity can perhaps be explained by the different roles smooth muscles play in the body, leading to specialization; compare, for example, the role of a myocyte in a large blood vessel to that of one in the gut or vas deferens. However functional diversity cannot fully explain the extent of calcium signaling mechanisms found. For example, uterine smooth muscle cells possess all three isoforms of both IP₃ receptors and ryanodine receptors on their SR [1, 2]; vascular myocytes can express L-, P-/Q-, N-, R- and T-type calcium channels [3, 4] and 10 different TRP channels [5]. The ability of smooth muscle cells to transform their phenotype from contractile to secretory or proliferative may also contribute to the plethora of calcium signaling components.

As has been demonstrated for other cell types, the ion channels and the contractile machinery in smooth muscle cells are sensitive not simply to the amount of calcium, but also the rate at which it rises or falls, its source, and how long it is present for (e.g. [6]). Calcium can, for example, both activate and inactivate channels, including voltage-gated calcium channels. An example of how rate of change [Ca] affects smooth muscle function has been elegantly showed in ileal myocytes [7]. The muscarinic response in the cells, which is a cationic current, was most responsive when [Ca] was rapidly elevated (using flash photolysis of IP₃), and poorly responsive to a rise of calcium produced in small steps, or by calcium sparks from the SR. Thus rapid, global calcium rises excite the muscarinic current in these cells, not slow or local calcium signals. There will also be intracellular heterogeneity in calcium regulation, as the distribution of plasmalemmal channels, receptors, and the SR is not uniform throughout the cell. This level of complexity presumably conveys to the smooth muscle cell mechanisms for transducing a wide range of signals impinging on their membranes from inside and out. The temporal, spatial, and quantitative aspects of the calcium signal in smooth muscle, along with microdomains (lipid rafts/caveolae) [8] in the plasma membrane and microdomains in the cell between, for example, SR and plasma membrane, all add to the complexity of understanding calcium signaling in smooth muscle.[9, 10].

A major challenge to investigators is elucidating which aspects of calcium signaling are physiologically relevant to particular smooth muscles, and what the functional consequences are of those calcium signals in health and disease. The challenge in reviewing this topic is to synthesize a cohesive account, while making some reference to aspects of specialization of mechanism found only in certain smooth muscles. In this short chapter, therefore, the focus is on calcium signals and contraction in those tissues with which I am most familiar–uterus, ureter, and vascular. The following reviews and the references cited within cover aspects of signaling in other smooth muscles [3, 11–15].

THE ROLE OF CALCIUM SIGNALING IN SMOOTH MUSCLE

The functions of smooth muscle in the body can be viewed as powering the movement of substances (blood, urine, chyme, air, sperm, babies) through hollow tubes and organs. Cross bridges between actin and myosin provide the motive force, and calcium is the usual trigger for these molecular events. However, even the most fundamental property of muscle cells, i.e. contraction, does not necessarily involve calcium signals. Some smooth muscles, especially blood vessels, affect tone (pressure) by calcium-independent sensitization mechanisms [13, 16, 17]. The relationship between calcium and force in these vascular myocytes can be changed by altering the activity of myosin phosphatase, and thus the amount of phosphorylated myosin available for force generation [18, 19]. It is notable that some blood vessels do not generate action potentials, and thus this reliance on calcium sensitization to modify force could be a consequence of the decreased role of voltage-gated calcium entry in these myocytes. In vascular myocytes, it has been calculated that the density of the peak calcium current (I_{ca}) is $<1\,\mu A\,cm^{-2}$ [20, 21], which is less than one-tenth of that in more excitable smooth muscles (see, for example, [22]), and that there are only ~ 100 open L-type calcium channels at the peak I_{ca} [23]. However, despite the fact that many blood vessels have contractile responses that are resistant to L-type channel blockers [24, 25], it remains the case that L-type voltage-gated calcium entry is a major pathway for calcium entry in most blood vessels [26, 27], and their inhibition is used to treat hypertension. Furthermore, most blood vessels produce a steady (tonic) level of force and are not required to undergo rhymic (phasic) activity in the way that visceral smooth muscles are. These differences in contractile activity may also affect how they produce calcium signals and regulate force [12]. Although mostly beyond the scope of this chapter, calcium signals in smooth muscles will underlie activities other than contraction–for example, secretion of extracellular matrix components, proliferation [28], gene regulation, [29], and protein synthesis–and are therefore also intimately connected with normal growth and development, and disease processes. Removal of any smooth muscle and placement in culture medium is associated with a rapid phenotypic change from contractile to synthetic and proliferative phenotype. The changes in phenotype, as well as being associated with loss of contractile proteins, also involve changes in calcium handling; effects on SR release [30], decrease of L-type calcium channels [31] and altered TRP channel expression [32] have all been reported. TRP proteins, STIM1, and Orai are also upregulated [28, 33]. Caveolae (discussed below) and cavolin-1 appear to play a suppressive role in proliferation, and their number/expression decreases with culturing [34].

OVERVIEW OF TYPES OF CALCIUM SIGNALS IN SMOOTH MUSCLE

Calcium signals in smooth muscles are associated with contraction and relaxation, and the control of excitability. Apart from some blood vessels, these changes in calcium are initiated and related to electrical activity (i.e., excitability), which in turn depends upon ion channel activity. The internal calcium store can then add to this event and shape the calcium signals, but, as we will see, not necessarily to augment contraction. Although not always discrete and separate entities, it is possible to describe three major types of calcium signals produced in smooth muscle: global calcium transients, global or partial calcium waves/ oscillations/spikes, and local transient signals from the SR, (calcium sparks and puffs). The calcium entry associated with action potential activity is referred to as a calcium transient. It is a global calcium signal causing a more or less simultaneous calcium rise throughout the cells in the tissue, spread via gap junctions. The large surface area to volume ratio of the long, thin, spindle-shaped smooth muscle myocytes allows a uniform rise of calcium throughout individual cells, without the need for an additional delivery system, such as t-tubules.

Oscillating waves of calcium occur in smooth muscle cells in response to agonists, and involve interactions between calcium entry on the action potential and SR calcium release. There is still uncertainty about whether SR calcium releases without calcium entry can generate such global oscillations, as well as the roles of the SR calcium release through IP$_3$-gated channels (IP$_3$Rs) and calcium-gated channels (ryanodine receptors, RyRs). In addition, just how these calcium waves relate, if at all, to local calcium releases remains an active area of research.

Local rises of calcium that do not spread throughout the cell occur as the SR releases calcium spontaneously. These release events are known as calcium sparks if they occur from RyRs, and calcium puffs if they are from IP$_3$Rs. These are brief calcium signals, and may be more related to controlling excitability via calcium sensitive ion channels than to contributing calcium to augment contraction. However global calcium signals are also important for activating calcium-sensitive ion channels, and so care is needed in making any broad generalizations. For example, calcium-activated Cl channels (Ca$_{ci}$) may not be activated by local calcium signals, but require global calcium signals in the portal vein [35]. In addition, agonists can produce substantial calcium release from the SR, leading to the discrete release events increasing in spread and number and eliding into calcium waves, which may be partial or global within the cell.

Thus, a variety of calcium signals can be produced by smooth muscle. To understand the mechanisms underlying them and their functional roles, it is necessary to understand the processes of calcium entry and efflux across both the plasma and sarcolemma that produce calcium homeostasis and the basal levels of [Ca], from which calcium signals arise [36, 37]. These processes (i.e., influx and efflux mechanisms and SR calcium release and re-uptake) will also shape the calcium signals. For example, if calcium extrusion is inhibited as part of the mechanism by which an agonist stimulates contraction, then the normal kinetics of the calcium transient, particularly its relaxation rate, will change. To maintain calcium balance inside the smooth muscle myocyte, the calcium that enters the cell in response to an action potential or agonist binding has to be removed and returned to the extracellular fluid, and any release of calcium from the SR has to be re-sequestered by its Ca-ATPase (SERCA). These homeostatic mechanisms result in resting [Ca] of a few hundred nM, compared to the 1–2 mM extracellularly and the 0.5–1.5 μM levels seen intracellularly during stimulation. These drivers of calcium signaling are affected by microdomains, both in the plasma membrane and between this membrane and the SR, and between the SR and other intracellular organelles, particularly mitochondria. These aspects of calcium signaling will therefore also be discussed.

CALCIUM ENTRY MECHANISMS

L-Type Calcium Channels

Although some basal calcium entry will occur across the smooth muscle membrane [38], the predominant mechanism of elevating [Ca] occurs via increasing the membrane permeability to calcium, allowing calcium to enter rapidly down its electrochemical gradient. The opening of voltage-gated calcium channels underlies the increased permeability, and it is the L-type, dihydropyridine-sensitive calcium channel (Ca_{v2}) that is expressed ubiquitously in smooth muscles. The fundamental importance of these channels to the activity of smooth muscles can be convincingly demonstrated by the application of L-type channel-blocking drugs, such as nifedipine; calcium transients rapidly stop and contractions are simultaneously abolished. Such is the efficacy of these drugs in decreasing smooth muscle activity that they have great clinical utility in conditions such as hypertension, premature contractions of the uterus, GIT spasm, and renal (ureteric) colic.

The relationship between electrical activity, calcium, and force has been demonstrated in some smooth muscles [39–41]. In ureteric smooth muscle, it was shown that about one-third of the change in calcium occurs on the upstroke of the action potential and the remainder during the plateau phase. In addition, the slow kinetics of force development and relaxation in smooth muscles makes it unlikely that changes of [Ca] will ever be rate-limiting. Indeed, simultaneous measurements show that more force is produced as calcium is declining, due to the lag between the two. Myosin phosphorylation has been shown to be a major contributor to the delay in the development of force [41–43].

T-Type Calcium Channels

Most smooth muscles also express the Ca_{v3} gene product, the T-type (low-voltage-activated) form of voltage-gated calcium channels (e.g., uterus [44], urethra [45], vas deferens [46], bladder, umbilical artery [4]). However, it must be noted that T-type calcium channel expression occurs at much lower levels than L-type calcium channels, and, furthermore, at the resting membrane potential in smooth muscle, they will be mostly inactivated [47], although some small calcium leak may occur [48]. T-type channel calcium

signals play a role in cell migration and proliferation of tissue [49], and thus their expression may reflect these properties of smooth muscle and not point to a role in excitation–contraction coupling.

Ligand-Gated and TRP Channels

Calcium can also enter smooth muscles when ligand-gated ion channels open. These receptor-operated channels may be gated by specific agonist (e.g., ATP, endothelin) and are often non-selective for cations, rather than being calcium selective. These channels are activated by excitatory agonists such as nor-adrenaline, and therefore physiologically important [50]. However, a lack of good inhibitors has made their study difficult. Diacyl-glycerol, formed with IP_3 from PIP_2 hydrolysis, has also been identified as being an important part of the signaling pathway associated with these channels. Recent evidence suggests receptor-operated cation channels are part of the TRP channel family, which is discussed next. For more detailed information and an excellent review of cation channels in vascular myocytes, see the [51].

In addition to receptor-operated cation channels, other channels associated with calcium entry in smooth muscle (store-operated, and stretch activation) are now viewed as being part of the large family of channels first identified in drosophila. The transient receptor potential-TRP channels are a wide-ranging and growing family of channels expressed in many tissues, including smooth muscle. For example, 10 TRP cationic channels have been reported in vascular myocytes [5] and related to vasoconstriction, proliferation, and disease. TRP channels are almost all voltage-independent, non-selective cation channels, and their properties have been well covered in several recent reviews [52, 53]. TRP channels are activated by PIP_2 hydrolysis, stretch, SR calcium depletion and agonists; some are also calcium-activated. It seems reasonable therefore to expect that there will be interactions between not just TRPs and the SR, but also between TRPs and elements of G-protein-coupled receptors and their pathways. Recently, progress has been made in understanding the mechanism linking store depletion and TRP activation, with the identification of the calcium-sensing protein STIM 1 (STromal-Interacting Molecule). It is thought that this ET/SR calcium sensor translocates to the plasma membrane as luminal calcium depletes, and mediates the pathway to calcium entry. STIM-1 preferentially locates to regions of the plasma membrane where it can couple with Orai proteins, which are in the calcium-selective, low-conductance entry channels [54, 55].

The effects of TRP channels on tone in several smooth muscles can be explained by their relationship to the non-selective cation currents. This current (Icat) is of considerable interest, as it underlies a large part of the mechanism whereby muscarinic agonists (e.g., ACh) and vasoconstrictors (e.g., Nad) affect smooth muscle tone. The molecular identify of Icat was unclear until TRP channels were identified. It now seems reasonable to conclude that in GI and vascular smooth muscles TRP channel are an essential part of the response to muscarinic agonists, and are responsible for Icat [56–59].

This cation entry will cause depolarization and the opening of L-type calcium channels, and thereby contribute to contraction in these myocytes. However, sodium rather than calcium is the dominant ion entering through these non-selective cationic TRP channels. Indeed, it has been calculated that under physiological conditions calcium contributes only 1 percent to the current induced by carbachol in gastric myocytes [60]. Therefore, these channels do not produce a discernable calcium signal in the way that SR calcium release or selective calcium channels (L, T types) do. The TRP channels may also contribute to resting membrane conductance and basal calcium influx in vascular myocytes [51]. Thus, they may be considered part of the homeostatic mechanisms that allow rises of calcium to act as signals. A role in pacemaking in GI smooth muscle has also been proposed for TRP channels [61]. The voltage-independent cation entry into pacemaking interstitial cells leads to excitation in the myocytes.

For all that we know more about TRPs, especially their molecular biology, the lack of good specific inhibitors and the existence of redundant pathways still leaves much more to be elucidated about their role in smooth muscle physiology and pathology. As a recent review concluded. "At present, the exact physiological role of the TRP superfamily in SMC is still largely unknown" [62]. For further discussion of TRP channels, readers are referred to the references given and to Chapter 113 of Handbook of Cell Signaling, Second Edition.

CALCIUM EFFLUX MECHANISMS

Plasma Membrane Ca-ATPase (PMCA)

Apart from non-regulated leakage, there are only two routes by which calcium can leave the smooth muscle cell; via Ca-ATPase and via sodium/calcium exchange. As these mechanisms have been the subject of recent reviews, only a short overview will be given here [36, 37, 63, 64]. The plasma membrane Ca-ATPase, PMCA, is a P-type enzyme with four isoforms, PMCA 1–4. It is generally considered to be a lower-capacity but higher calcium-affinity system compared to the Na/Ca exchanger. This has led to PMCA being considered initially as a fine-tuning system for maintaining basal calcium levels in myocytes. However, more recent work in uterine cells has shown that it is the predominant efflux mechanism. Thus following stimulation (a depolarizing pulse) PMCA can contribute 70 percent to the calcium efflux [65]. In experiments designed to investigate the role of different PMCA isoforms, a value of 85 percent was found [36]. A figure of 65 percent was determined for its contributions to efflux following agonist stimulation

[66]. Figures from studies of other smooth muscle vary from 25 to 100 percent (see [37]).

PMCA is regulated by calmodulin through its carboxyl tail, and the affinity for calmodulin differs between the four isoforms [67]. PMCA-4 has the lowest basal activity and greatest calmodulin stimulation, and a role in shaping myocytes' calcium signals has been suggested for it [68]. Along with PMCA-1, the housekeeping isoform, these two appear to be the predominant isoforms expressed in smooth muscles. It has been suggested that there is a caveolae (see later) localization of smooth muscle PMCA, but more work is required to test this [69]. In studies of PMCA4 knock-out mice, a contribution of ~25 percent to relaxation of contraction (calcium not measured) was reported in bladder [70].

Na/Ca Exchange

The Na/Ca exchanger uses the sodium gradient to drive calcium extrusion, in an electrogenic manner (1 Ca^{2+} for 3 Na^+ with each cycle). The exchanger has a low affinity for calcium, but has a high capacity. Mice overexpressing the Na/Ca exchanger show an increased rate of calcium decline during relaxation [71]. In smooth muscles the housekeeping form of the exchanger is NCX1, and either NCX2 or 3 have also been reported to be expressed [72]. Interest in the contribution of the Na/Ca exchanger to smooth muscle calcium signaling has been heightened following the demonstration that it is located in microdomains than overlie the SR [73] and is spatially coupled to the Na, K-ATPase (sodium pump) [74]. Thus, in vascular smooth muscle evidence suggests that, by co-localizing these transporters and the SR, microdomains of low sodium aid calcium extrusion on the exchanger and promote relaxation [37, 75], as discussed below. Therefore, both PMCA and Na/Ca exchange calcium efflux pathways are important for smooth muscle calcium signaling, as well as being responsible for lowering calcium after stimulation, and maintaining low basal calcium levels.

SR AND CALCIUM SIGNALING

Structure, Distribution, and Release Mechanisms

The SR of smooth muscle has been the subject of several recent reviews, [8, 76, 77]. Confocal images of the SR, targeted with low affinity fluorescent calcium indicators, have allowed 3D reconstructions of its structure in smooth muscle cells [77–79]. A rich reticular formation is apparent; with a dense network around the nucleus and close apposition to the plasma membrane. Recent estimates of SR volume in vascular smooth muscle cells are around 5–7 percent [78], which is somewhat greater than those obtained earlier with electron probe analysis ($\sim$2 percent [80]). Early EM studies had also reported the SR membrane running parallel to that of the plasma membrane for distances of 1μM [81]. Fluorescent probes directed at RyRs and the SR Ca-ATPase (SERCA) showed a non-homogenous distribution in uterine myocytes.

Smooth muscle cells contain IP$_3$-and calcium-induced calcium release mechanisms. As mentioned above, many smooth muscles have both IP$_3$Rs and RyRs, but substantial differences between tissues and species exist [82]. Other more recently identified SR calcium-release mechanisms, such as cADP-ribose, nicotinic acid adenine dinucleotide phosphate, sphingosine, and sphingosine-1-phosphate [83], have been little studied in smooth muscle [84–87].

SERCA, Luminal Calcium Content, and Calcium Signals

Calcium is taken back into the SR by the SR Ca-ATPase, referred to as SERCA. This P-type ATPase has three genes and several isoforms [88]. The "housekeeping" form, SERCA 2b, has been identified in smooth muscles, and SERCA 2a and 3 additionally in uterus and some vessels [37, 89–93]; calcium transport into the SR leads to SR luminal calcium levels estimated at 100–500μM [76, 94], with SR calcium-binding proteins calsequestrin and calreticulin providing the necessary calcium buffering.

Luminal SR calcium levels have been shown to regulate IP$_3$-induced calcium release in cultured vascular cells (A7r5) [95] and more recently in freshly dispersed uterine myocytes [96]. The luminal calcium ([Ca]$_L$) was directly monitored in uterine myocytes using the low affinity indicator mag-fluo-4 and cytosolic calcium simultaneously monitored with Fura-2, a high-affinity calcium indicator. Agonist application causes, in the absence of external calcium, a rapid rise of intracellular ([Ca]) and decline of [Ca]$_L$. Interestingly, re-uptake into the SR was found to begin even while agonist was still present [96]. Abolition of SERCA activity by cyclopiazonic acid or thapsigargin prevented this re-uptake. This in turn suggests an overwhelming role for SERCA as the reuptake mechanism, as opposed to the recently identified secretory pathway Ca-ATPases, SPCAs [97]. If external calcium is elevated, then [Ca]$_L$ increases [96]. However, it was found in this study of uterine myocytes that the overloaded SR did not increase the size of calcium transients elicited by agonists. As discussed below, the SR [Ca]$_L$ can also increase following depolarizing with elevated K solution; this is associated with inhibition of spontaneous action potentials and calcium transients in uterine myocytes [94]. In contrast to overloaded SR, depletion of SR [Ca]$_L$ had substantial effects on agonist-evoked calcium transient [96]; a 20 percent reduction led to the abolition of the response. Thus, at least in uterine myocytes, a steep relation between IP$_3$-induced SR calcium

signals and [Ca]$_L$ exists. Although difficult to accomplish, more studies directly measuring [Ca]$_L$ in smooth muscle cells are required to further investigate this and other aspects of SR function on calcium signaling [98].

As mentioned above, TRP channels play an important role in refilling the SR in non-excitable cells; this is known as store-operated calcium (SOC) entry or capacitative calcium entry [99]. While a role for this process in the more non-excitable vascular tissues has been demonstrated, a role in excitable smooth muscles (e.g., ureter, GIT, and uterus) is much more open to question. The current flow through SOCs has been difficult to measure, and is considered to be small [100]. Although TPP channels are expressed in most smooth muscles, this does not prove a functional role. Similarly, a rise of calcium when SERCA has been inhibited can be explained by a reduction in calcium buffering. Unfortunately, many experiments trying to show such a role have been made on cultured or transfected cells, which are so phenotypically altered that their relevance to excitation–contraction is difficult to gauge [101]. However, a recent study by Kovac and colleagues [102] in human freshly isolated colonic cells showed SERCA inhibition produced an initial non-specific cation current, and regional "hotspots" of calcium entry were observed in 70 percent of the cells. The authors suggest that these are due to store-operated calcium entry. Store-operated calcium influx in cerebral arterioles had previously been reported when the SR was inhibited [103], which, although causing a significant rise in [Ca], did not evoke vasoconstriction. This led to the authors suggesting it was a discrete subset of calcium channels allowing calcium influx into a non-contractive compartment of the myocytes. As discussed later, such a compartment may exist in the sub-sarcolemma space adjacent to SR membrane.

SERCA Inhibition and Calcium Signals

Quantitative studies of the calcium rise in myocytes in response to stimulation have been performed in single cells under voltage-clamp conditions. In a variety of different tissues, a voltage step of around 60 mV produced a rise of calcium of $\sim$100 nM (stomach [104], uterus [22], bladder [105]). Inhibition of the SR Ca-ATPase causes a rise in basal calcium (e.g., 75–170 nM in uterine myocytes [106] and $\sim$50–120 nM in gastric myocytes [104]). The same authors found that CPA also increased the size of the calcium transient from 5.3 to 12.3 nM/pC.

The inhibition of SERCA by depleting SR [Ca]$_L$ might be anticipated to lower the rise in intracellular calcium needed for agonist calcium signals as both IP$_3$- and calcium-induced SR calcium release mechanisms are prevented. This has been directly examined for IP$_3$ in isolated, freshly dissociated uterine myocytes, in the study discussed above. The role of the SR in shaping the calcium transient has been examined in several smooth muscles using SERCA inhibitors. Specifically, its role in determining the decay of the calcium transient, as calcium is taken up into the SR, has been investigated. We have found in uterine myocytes that calcium transients elicited by depolarizing voltage-clamp pluses or carbachol application had a slower rate of decay if SERCA was inhibited [65, 106]. The rate constant of decay fell from $0.6 \, s^{-1}$ to 0.3^{-1}, while the rate of the rising phase was unaffected [106]. However, we also showed that SERCA alone could not produce the decay of calcium; PMCA and Na/Ca exchange were essential for this. We concluded that in these cells the SR takes up calcium and then releases it very close to the plasma membrane extruders–i.e., it acts in series with them and facilitates calcium decay. Mention should also be made of a study by Gomez-Viquez and colleagues [107], suggesting that SERCA *per se* can influence calcium release and signaling. In urinary bladder myocytes they rapidly blocked SERCA, so that the SR still contained calcium (confirmed with direct luminal measurements), and found that both RyR and IP$_3$R calcium releases were smaller and slower than those occurring when SERCA was not blocked. They conclude that SERCA pumps are involved in sustaining agonist-evoked calcium release, although the mechanism for this is unknown. These data are consistent with ours on uterine myocytes whereby we demonstrated that blocking SERCA decreases agonist-induced calcium responses, although we attributed this entirely to the larger, agonist-induced [Ca]$_L$ depletion [108].

The SERCA accessory protein phospholamban may contribute to the kinetics of SERCA in smooth muscle [109], although a prominent role for it has not been demonstrated yet in many muscles. Phospholamban reduces SERCA activity, and its phospharylation relieves this inhibition. In a recent study of airway smooth muscle [110], it was shown that small interfering RNA to phospholamban slowed the rate of fall of calcium following ACh stimulation. Using phospholamban knockouts, an increase in frequency and amplitude of spontaneous phasic contractions of gastric antrum was reported [109], and the rise in basal tone with cyclopiazonic acid (CPA) was smaller than that in wild-type mice. In portal vein, the knockout reduced frequency of spontaneous activity while increasing force amplitude [111], and CPA had little effect on wild-type activity and increased contraction frequency in knockouts. Interestingly, these authors also measured a lower resting membrane potential in the knockouts compared to wild-types.

Calcium-Induced Calcium Release

In cardiac muscle, a clear role in calcium signaling for calcium release from ryanodine receptors (RyR$_2$) eliciting Calcium-Induced Calcium Release (CICR) is well documented. A role for CICR in smooth muscle remains somewhat controversial. Earlier studies in voltage-clamped urinary bladder and portal vein myocytes [23, 112] indicated the presence of CICR; hence, a mechanism to augment L-type calcium current entry, as occurs in cardiac muscle, was anticipated. A small amount of CICR could be demonstrated in single uterine myocytes under voltage-clamp conditions [22], but not in intact tissue [113]. In a follow-up study from their work on bladder cells, using coronary myocytes Isenberg and colleagues found no evidence for CICR, which they explained by the smaller depolarizing currents in the vascular myocytes [23]. Other more recent studies have also questioned the occurrence of CICR in other smooth muscles. It appears that when CICR can be demonstrated (for example, in bladder) it is graded and non-obligate–i.e., calcium influx does not necessary produce CICR. These differences may only partially be explained by differences in peak inward currents, as suggested above, as, for example, pregnant uterine myocytes have very large current (6.3 pA/pF [22]) but little CICR. Another explanation may reside with the mixture of RyR expressed in the different smooth muscles. In cardiac muscle the RyR2 form is expressed, whereas in smooth muscles all three forms can be expressed [2]. In addition, as discussed later, non-functional splice variants of RyR3 may also explain the lack of CICR, and of local calcium signals (calcium sparks). Kotlikoff's groups in particular have brought modern imaging techniques to bear on this issue, and suggest that the coupling between calcium flux and RyR calcium release is "loose," and a sufficient rise in calcium/calcium entry is required for it to occur [114, 115]. In a recent paper using two-photon localized calcium uncaging in urinary bladder myocytes, his group showed that CICR was not confined to a few specific subcellular sites and that the process could occur through IP$_3$Rs as well [116]. In addition, by using caged calcium release the lack of involvement of depolarization or calcium entry *per se* was also demonstrated. This study therefore also questions whether frequent discharging sites are solely due to clustering of a few specialized RyRs, as uncaging could elicit release from a homogenously distributed SR. A role for FKBP12.6 in modifying RyR2 calcium sparks in mouse bladder was also reported by the group [117], and CICR increased in FKBP12.6 knockout mice. Normal calcium sparks and CICR occurred in RyR3 knockout mice.

SR, Calcium Signals, and Ion Channels

A paradigm shift in the role played by the SR in smooth muscle has occurred with the discovery that its most important role, at least in some tissues, is to control membrane excitability, and act to dampen calcium influx and limit contraction.

The evidence for this has been elegantly demonstrated by Nelson and colleagues in vascular smooth muscle. Using confocal microscopy, he demonstrated that the local SR RyR signals (i.e., calcium sparks) did not produce CICR but rather were directed to the plasma membrane, where they activated calcium-activated K channels, in particular BK channels [118–123]. The opening of BK channels produced small, transient hyperpolarizations of the membrane (STOCs), which had previously been observed by Benham and Bolton [124] but not linked to then unknown calcium spark events. This sparks–STOCs coupling mechanism will lead to a reduced probability of L-type calcium channels opening as the membrane hyperpolarizes. In turn, if calcium falls, then force will also decline. This mechanism has been shown to be relevant to control of vascular tone, and aberrations produce hyperpolarization [119, 122]. In bladder, the feedback mechanism may help prevent hyperactivity and urinary incontinence [118].

We have shown this mechanism to be fundamental to the control of the action potential refractory period in ureteric smooth muscle [125]. Thus, when an action potential is fired, L-type calcium channels open and produce contraction. Some of this global calcium also enters the SR. The increase in $[Ca]_L$ leads to an increase in calcium sparks, which in turn activate more BK channels. The increase in STOCs produces hyperpolarization of the membrane, and terminates calcium entry through L-type channels. Consequently, calcium is reduced and dissocates from calmodulin and force falls. During this period of increased sparks and STOCs, the ureter is refractory. Only as $[Ca]_L$ returns to resting values, and sparks and STOCs decrease, can a stimulus produce an action potential and the next global calcium transient and contraction [125, 126].

Unresolved Questions

While elucidation of the sparks–STOCs mechanism has undoubtedly greatly advanced our understanding of calcium signaling and the role of the SR in smooth muscle, some major conundrums remain. One of these involves the mechanism whereby the SR regulates signaling and contraction in the uterus. It has been known for over a decade that the uterus expresses RyR1–3 and IP$_3$R1–3. However, when the SR was inhibited (for example, by cyclopiazonic acid), calcium signaling and contractions increased [113, 127]. As the sparks–STOCs mechanism was uncovered, it was assumed that this would underlie the increased calcium signaling and force when the uterine SR was inhibited–the negative feedback was prevented, and hence more calcium entered the uterine SR and promoted force production. As BK channels were also identified in the uterus [128–131], this viewpoint was reinforced. However, no reports of calcium sparks in uterine myocytes were produced, apart from in calcium overloaded transgenic mice [132]. When we recently used confocal microscopy in rat myometrium and studied *in situ* calcium signaling, we confirmed that the pregnant and non-pregnant

rat uterus does not produce calcium sparks [40]. Thus, despite having all the necessary elements–a well-developed SR, expression of RyR1–3 and BK channels–calcium sparks are not produced and cannot stimulate BK channels. Even using caffeine, a potent agonist at RyRs, calcium sparks do not occur, and in fact caffeine's predominant action on the uterus is as an inhibitor of phosphodiesterases and promoter of relaxation. Ryanodine at blocking concentrations on RyR also has no significant effect on force and calcium in non-pregnant myometrium, and only a small augmentation in pregnant tissues [113, 133].

The lack of effect of caffeine, ryanodine, and absence of calcium sparks in myometrium may be due to truncated forms of RyR3 being expressed. Jiang and colleagues [134] reported that smooth muscles express a splice variant and major dominant negative short form of RyR3. This variant inhibits RyR2. Dabertrand and colleagues [135, 136] confirmed these findings in native uterine cells, and showed them to be expressed close to the plasma membrane. Studies of RyR3 knockout mice also led to the conclusion that this isoform could inhibit calcium signaling in vascular myocytes [137]. These non-function release channels and dominant negative effects on other isoforms may exist in uterine and other smooth muscles. While these data explain the lack of calcium sparks, they do not explain why SR inhibition has such a large effect on calcium signals and contraction in the myometrium, which is particularly marked in neonatal uterus [127]. The data are suggestive of membrane depolarization occurring, but the mechanism remains to be elucidated.

SR Summary

The spatial distribution of SERCA, IP$_3$Rs, RyRs, and plasma membrane ion channels and transporters plays an important role in shaping and regulating local and global calcium signals in smooth muscle cells. The roles of the SR calcium release and SERCA activity in calcium signaling in smooth muscle are: (1) to help maintain low intracellular [Ca]; (2) to facilitate decay of the calcium transient and relaxation of force; and (3) to contribute calcium for agonist-induced calcium waves. However, it is now clear that an additional and crucial role of the SR is regulating membrane excitability, and thereby calcium signals. Furthermore, this function acts as a negative feedback control step on calcium entry and contractility, as BK channels are activated by local calcium sparks in many smooth muscles. This calcium sparks–STOCs mechanism explains the augmentation of calcium signals and contractions seen in several smooth muscles when the SR is inhibited. Other as yet unknown mechanisms must underlie the increases in calcium signaling and force seen with SR inhibition in the uterus, where calcium sparks are not produced and splice variants of RyR3 are expressed.

MITOCHONDRIAL AND OTHER ORGANELLAR CONTRIBUTION TO CALCIUM SIGNALING

Mitochondria have the capacity to take up calcium rapidly, although resting values of calcium appear to be below the affinity of their calcium transporters [138]. Until relatively recently, therefore, the view prevailed that mitochondria would only contribute to calcium accumulation and calcium signaling during pathologically high levels of calcium. When techniques suitable for reporting mitochondrial calcium levels became available [139], they led to a gradual change in how mitochondrial calcium uptake was viewed; rapid uptake and changes in mitochondrial calcium occur during cell stimulation [140, 141]. In addition, microdomains of high calcium (i.e., exceeding bulk cytoplasmic) were also demonstrated, overcoming the low-affinity problem [142]. The microdomains arise because of the close proximity of mitochondria to the SR (or ER in non-muscle cells). As Rizzuto and colleagues write, the new targeted calcium probes for measuring mitochondrial calcium have allowed the rediscovery of these organelles in calcium signaling [143].

Studies in a variety of smooth muscles have now shown that mitochondria can contribute to calcium signaling. [144–147]. Thus, elevations of mitochondrial calcium have been reported when global calcium transients occur in arterial smooth muscle and contribute to curtail the calcium rise [148, 149]. If mitochondria are disabled by loss of their membrane potential (which is usually $\sim -180\,mV$ negative with respect to the cytoplasm), then calcium transient decay is slowed and calcium-activated membrane currents are altered [150]. Cross-talk between the SR and mitochondria has been noted in some studies [145]. Close apposition between SR and mitochondria in smooth muscle cells facilitates such interchange of calcium between the two organelles [146]. More recently investigators have turned their attention to other organelles within the cell, including the Golgi, nucleus, and acidic granules (see, for example, [151, 152]). The SR is contiguous with the nuclear envelope, and thus it may contribute to calcium-regulated gene transcription, via, for example, cAMP responsive element-binding proteins (CREB) [153, 154] and nuclear factor of activated T cell [155]. However, little or no information for smooth muscles exists so far, and thus nothing further can be said regarding how these structures may influences their calcium signals.

GLOBAL CALCIUM TRANSIENTS

Spontanous Activity

The rise intracellular [Ca] upon stimulation is the best described type of calcium signal in smooth muscle. This is the calcium transient that underlies contraction. During spontaneous contractions of smooth muscles, the phasic rhythmic activity is entirely due to this calcium transient activating the contractile machinery. The entry of calcium into the myocytes is through voltage-gated L-type calcium channels. Removal of external calcium or blocking of the L-type calcium channel will abolish calcium transients and contractions, whereas blockade of T-type channels has little effect. No changes in $[Ca]_L$ were detected during spontaneous activity in isolated uterine myocytes. These direct measurements of luminal [Ca] therefore further support the view that calcium entry is the only contributor to spontaneous phasic activity in smooth muscle.

Agonist Stimulation

In several smooth muscles, especially vascular and in the presence of agonists, the calcium signal is often composed of two components, L-type calcium entry and SR calcium release, predominately though IP$_3$Rs [156]. The simplest explanation for these agonist-induced calcium signals is that there is an initial, rapid and transient part of this biphasic calcium signal due to SR calcium release, and a subsequent, more or less maintained but lower level of calcium due to calcium entry [157, 158]. Thus nifedine and other calcium channel blockers have no effect on the initial calcium rise [159] and any agonist-induced calcium sensitization [160], but abolish the sustained effects [161]. However, more recent studies have shown that the picture is probably more dynamic than this, as interactions occur between the SR release channels and calcium entry, so that repeated releases of SR calcium occur during both the sustained period ("regional switches and relays," (see [83]) and in the phasic period interactions with the SR and ion channels [162, 163]. These repetitive calcium releases from the SR spread throughout the cell and are known as calcium waves or oscillations, discussed next. Calcium entry is, however, needed to sustain the oscillations, as the SR calcium store will gradually run down as some of the released calcium leaves the cell rather than being taken up again by the SR. Agonists differ in their reliance on calcium entry vs SR calcium release in bringing about their effects. For example, 5-HT is very sensitive to calcium channel blockers, whereas U46619, a thromboxane mimetic, is insensitive [156], and calcium sensitization may play an enhanced role.

LOCAL CALCIUM SIGNALS

Calcium Sparks and Puffs

Sometimes referred to as elemental calcium signals, the calcium signals occurring as calcium is released from the SR can be characterized by their spatially local and transient nature. These release events occur spontaneously, and are also modified by stimulation. The local calcium signals are called calcium puffs and sparks, depending upon whether they arise from the activation of IP$_3$Rs or RyRs, respectively. Although IP$_3$ is a crucial part of signaling in smooth muscles, calcium puffs have been reported far less frequently that calcium sparks. This is generally explained by a lack of clustering of IP$_3$Rs compared to RyRs, and thus the size of the calcium puffs being too small for resolution with most apparatus. Emptying of the SR (for example, by blocking the Ca-ATPase with cyclopiazonic acid (CPA) or thapsigargin) abolishes puffs and sparks. Calcium sparks were first reported in smooth muscle by Nelson and colleagues [123]. They have been shown to arise from a limited number of SR sites in the myocytes; termed frequent discharging sites [164, 165], which are often close to the nucleus [78]. They have essentially similar characteristics to those described previously in striated muscle [6, 166]; duration ~100 ms, frequency 1 Hz, magnitude ~100–300 nM (causing global calcium to rise only 2 nM), and spatial spread of around 2 μM at half maximal [Ca] [167]. The rise and fall times of smooth muscles sparks were slower than those found in striated muscles. It is unlikely that a single RyR can produce a calcium spark, but small (10–50) clusters do [168]. It is thought that calcium diffusion rather than SERCA uptake largely accounts for termination calcium sparks in smooth muscle [166]. The properties of calcium sparks appear to be the same in phasic and ionic smooth muscles, although it should be appreciated that their properties can vary even within a single myocyte [6]. The coupling of calcium sparks to STOCs has now been demonstrated in several smooth muscles, and the average time course of both events is very similar [123]. It has been estimated that a single calcium spark activates 13 BK channels. As discussed by Fay [169], given what is known about the calcium required to activate BK channels (i.e., a several micromolar level), it is likely that the estimation of calcium spark amplitude is larger than that revealed by global confocal imaging, and that higher resolution imaging methods are required. A later study with modeling of data undertook this, using fluo-3 and a high-speed widefield imaging system [165]. The conclusion reached was that BK channels lie close to RyRs and experience 20–150 μM [Ca]. The STOC activity was determined by the BK channels kinetics. Global calcium rises of lower amplitude would not be capable of activating the BK channels.

Contribution to Global Calcium Signals?

It has been proposed that the local calcium signals are building-blocks of global calcium signals; hence the sobriquet "elemental" [164]. While this is the case for, say, striated muscles, there is a fundamental problem with this concept in smooth muscle. This arises from the now well-documented

evidence that calcium sparks target and activate calcium-activated K channels (K$_{ca}$), especially these with a large conductance, referred to as BK channels. Activation of these channels will increase K conductance and produce a membrane hyperpolarization via STOCs, as discussed earlier. Such hyperpolarization will curtail calcium entry through voltage-gated calcium channels. Hence, increasing calcium sparks does not seem compatible with increasing the calcium signal so that it becomes global. The fusion of calcium sparks may occasionally produce excitatory propagating calcium waves during spontaneous activity [170,171], but mostly the global rise in calcium is insignificant and thus the effects of local calcium signals on global calcium rises are indirect. There are reports of agonists producing calcium oscillations from areas where calcium spark activity was increased, leading to the proposal that calcium sparks can act as "primers" for agonist-induced global calcium signals [6]. Further studies are required to test this suggestion in *in situ* preparations. A role for calcium sparks in agonist-induced responses can perhaps be inferred from the findings that the second messengers cAMP and cGMP, produced during agonist stimulation, can affect calcium spark activity. However these actions are associated with relaxation via BK channels, rather than production of global calcium signals. For example, in vascular myocytes they have been shown to increase spark frequency [166, 172]. In cerebral arteries, PKC activation decreased spark frequency and was associated with vasoconstriction [173].

These local calcium signals may also contribute to producing microdomains of higher [Ca]. It is suggested that, in the superficial SR, release of calcium is directed to the plasma membrane or mitochondria, and that it reaches a higher level of this microdomain than in bulk cytoplasm [78]. In this way low-affinity channels or transporters can be activated, as discussed below.

In addition to BK channels, many smooth muscles have calcium-activated (CL$_{ca}$) channels which may be activated by both global and local calcium signals. These channels produce spontaneous transient inward current (STICs), and will tend to depolarize smooth muscle. As smooth muscle cells can express both Kca and Cl$_{ca}$ channels, it is possible to conceive of local calcium signals from the SR producing both STOCs and STICs. Few papers have investigated thus, but a good analysis is provided by Large and colleagues, based on channel kinetics and distribution [174].

CALCIUM OSCILLATIONS AND WAVES

Characteristics and Origins

Repetitive waves of calcium travelling the length of the myocytes, have been reported in many smooth muscle tissues, in response to agonist stimulation [175–181]. The rise of calcium may be as high as 500 nM and frequency from 5 to 3 per minute, which is usually dependent on basal calcium levels [11] and relatively independent of agonist dose in contrast to global contransients [6]. The mechanisms underlying calcium oscillations may be multiple and require further elucidation. Conclusions drawn from a particular smooth muscle and a specific agonist, may not be generalizable to others.

Oscillations of calcium often start at a limited number of sites in the myocytes, and usually at one end, perhaps reflecting a local clustering of IP$_3$Rs or RyRs or ion channel waves initiated at more than one site also occurs, and they can collide with each other. Speeds of around $100 \mu m \ s^{-1}$ for smooth muscle calcium waves have been reported [78]. The frequency of oscillations and their speed may increase as agonist concentration increases [182]. The majority of studies have been of calcium waves arising from SR, which ramifies throughout the cell, releasing and taking up calcium, with or without a crucial role for calcium influx or a particular type of IP$_3$ or RyR subtype. However calcium oscillations may also arise due to fluctuations of membrane potential, such as occur in gastrointestinal smooth muscle, leading to repetitive cycles of opening of voltage-gated calcium channels and calcium entry rather than SR calcium releases. However, it has been demonstrated (although in other cell types) that oscillations can occur without oscillations of membrane potential (or IP$_3$), and in permeabilized preparations [179, 183, 184]. Although changes in calcium entry and efflux may contribute to the time-course or frequency of calcium oscillations, in many tissues the SR is the predominant player. Elevation of luminal calcium is also associated with an increase in calcium waves [185]. SR calcium release through IP$_3$R may initiate RyR-dependent calcium oscillations in tracheal myocytes [163, 186]. It has also been reported that cADP-ribose can modulate ACh-evoked calcium oscillations of tracheal myocytes [187]. Wang and colleagues [87] showed recently that FK506 binding protein (FKBP12.6) associates with and regulates type-2 RyRs. Tracheal myocytes which had increased spontaneous calcium release in the presence of low concentrations of cADP-ribose, and global calcium releases at higher cADP-ribose concentrations, were shown to exhibit these effects via FKBP12.6 [87]. In FKBP 12.6 null mice, cADP-ribose could not influence SR calcium release (see Jude and colleagues [11] for further discussion). Thus, more than one mechanism can produce calcium oscillations in smooth muscle, and this may be a way of tailoring calcium signals to specific functions of smooth muscle. As both contraction and relaxation occur on much slower time-courses than calcium oscillations, it is expected that these outputs reflect on integration of the spatial and temporal characteristics of the oscillation.

Despite the presence of gap junctions between smooth muscle cells, the oscillations in individual cells are often found to be asynchronous with other cells in the preparation understudy [170]. Particularly in vascular smooth muscle, the view is emerging that although a smooth global calcium transient may be recorded, it is composed of many asynchronous calcium oscillations in all the different cells of the vessel [188]. Functionally, this may be associated with vasomotion. At higher agonist concentrations, a point may come where the calcium signals become synchronized and produce a powerful vasodilation. As discussed in the next section, sub-sarcolemmal membrane domains are considered to exist in smooth muscle myocytes. In their studies of the generation of asynchronous repetitive calcium waves in the inferior vena cava, Lee and colleagues [188], have suggested that non-selective, store-operated channels and reverse mode Na/Ca exchange enhance SERCA filling of the SR, and thereby permit cyclic wave activity. Application of calyculin-A, a phosphatise inhibitor, disrupts the close contact between the SR and plasma membrane and the asynchronous phenylephrine-induced calcium waves were progressively lost.

Despite the presence of calcium waves and oscillations being reported over a decade ago, and much research effort, it appears that there is still no clear consensus on both their mechanism and function. For example, is extracellular calcium entry always essential? Do RyRs or IP$_3$R play the dominant/an essential role in their generation? Can they activate/inactivate other ion channels, including calcium-activated K and Cl channels? How are they propagated? Does their occurrence correspond with increased or decreased tone? Some of the different conclusions reached may, of course, be due to the intrinsic differences between tissues; further experiments, under the same experimental conditions, are called for [90].

CAVEOLAE, MICRODOMAINS, AND CALCIUM SIGNALS

Superficial Buffer Barrier and SR

First expressed by van Breemen and colleagues [189, 190] and Blaustein [191, 192], the suggestion that spatial microdomains must be present in smooth muscle cells to explain aspects of calcium signaling has received support as molecular tools for investigating restricted spaces have advanced (see, for example, [193, 194]). As noted already, the SR is in close apposition to the plasma membrane. As proposed by van Breemen, it may take up or buffer some of the calcium that enters on depolarization. This calcium can then be vectorially released to membrane pumps and exchangers, facilitating calcium extrusion after stimulation [195]. Moore and colleagues [196] demonstrated that certain signaling elements, specifically Na/Ca exchanger, sodium pump, and SERCA, appeared grouped in gastric smooth muscle cells. Support for the SR acting as a superficial buffer barrier was also provided by studies on vascular

[197, 198] and gastric [199] myocytes. The possibility of increased concentration of ions within a restricted space either between the surface membrane and SR, or mitochondria and SR, can explain how pumps, receptors, or exchangers with high Kd for ions, can be activated under physiological conditions. Details of this have already been given for mitochondrial calcium signaling. A role for Na/Ca exchange is also apparent during calcium extrusion, indicating that the exchanger sees locally higher [Ca] than those occurring globally [200, 201]. Participation of Na/Ca exchange in contraction is discussed below.

Sodium Pump, Na/Ca Exchange, and Calcium Signaling

By locating the $\alpha 2$ isoform of the sodium pump close to the Na/Ca exchanger (spatial coupling), it has been proposed that the sodium pump contributes to calcium transport in vascular smooth muscle [202, 203]. The [Na] in the sub-sarcolemmal space adjacent to SR calcium release sites will be particularly lowered due to sodium-pump activity driving and facilitating Na/Ca exchange in these regions [73]. In mice which are two-sodium pump heterozygous, increased blood pressure and myogenic tone were noted, and cardiac muscle was hypercontractile [204]. Furthermore, endogenous cardiac glycosides (ouabains [205]) are suggested to play a role in control of blood pressure via the Na/Ca exchanger, as inhibition of the sodium pump causes calcium influx (or decreased efflux) [206]. A role for Na/Ca exchange separate from its link to the sodium pump in vascular agonist-induced contraction has also been demonstrated [90, 207, 208].

Lipid Rafts and Caveolae

It is now well accepted that there are functional microdomains within the plasma membrane. Called lipid rafts, these regions have a higher concentration of sphingolipids and cholesterol, resulting in locally decreased fluidity–hence they are "rafts" floating in the membrane sea [209, 210]. Interest in these membrane regions has grown as the view has emerged that a range of signaling components (e.g., ion channel subunits, receptors, and enzymes) are dynamically associated in the lipid rafts, or excluded from them, as part of signaling mechanisms, in a variety of cells [211, 212]. In smooth muscle, for example, PMCA, BK channel χ subunit, and oxytocin receptors [213] have all been reported to be preferentially located in caveolae, while eNOS is excluded [214]. Lipid rafts, in their caveolar form (i.e., containing the protein caveolin and thereby also invaginating the membrane), have been shown by us [215–217] and others [218, 219] to be functionally important in smooth muscles. For example, caveolar disruption by extraction of membrane cholesterol increases calcium signaling and

contractility in rat [215] and human [220] myometrium. Our recent work showing that this is associated with a decrease in outward current [221] is consistent with BK channel localizing to caveolae [8, 216, 222]; as caveolae are disrupted, so their ability to generate outward current is diminished, and excitability increases. We have postulated that elevated cholesterol in obese pregnant women may lead to enhanced outward current and increased uterine quiescence [220], which could underlie the difficult births and increased requirements for labor induction and emergency caesarean section in these women.

REFERENCES

1. Morgan JM, De Smedt H, Gillespie JI. Identification of three isoforms of the InsP$_3$ receptor in human myometrial smooth muscle. *Pflügers Arch* 1996;**431**:697–705.
2. Martin C, Hyvelin JM, Chapman KE, Marthan R, Ashley RH, Savineau JP. Pregnant rat myometrial cells show heterogeneous ryanodine- and caffeine-sensitive calcium stores. *Am J Physiol* 1999;**46**:C243–52.
3. Orallo F. Regulation of cytosolic calcium levels in vascular smooth muscle. *Pharmacol Ther* 1996;**69**:153–71.
4. Salemme S, Rebolledo A, Speroni F, Petruccelli S, Milesi V. L, P-/Q- and T-type Ca^{2+} channels in smooth muscle cells from human umbilical artery. *Cell Physiol Biochem* 2007;**20**:55–64.
5. Beech DJ. Emerging functions of 10 types of TRP cationic channel in vascular smooth muscle. *Clin Exp Pharmacol Physiol* 2005;**32**:597–603.
6. Pabelick CM, Sieck GC, Prakash AS. Significance of spacial and temporal heterogeneity of calcium transients in smooth muscle. *J Appl Physiol* 2001;**91**:488–96.
7. Gordienko DV, Zholos AV. Regulation of muscarinic cationic current in myocytes from guinea-pig ileum by intracellular Ca^{2+} release: a central role of inositol 1, 4, 5-trisphosphate receptors. *Cell Calcium* 2004;**36**:367–86.
8. Noble K, Zhang J, Wray S. Lipid rafts, the sarcoplasmic reticulum and uterine calcium signalling: an integrated approach. *J Physiol* 2006;**570**:29–35.
9. Dupont G, Combettes L, Leybaert L. Calcium dynamics: spatio-temporal organization from the subcellular to the organ level. *Intl Rev Cytol* 2007;**261**:193–245.
10. Wray S, Burdyga T, Noble K. Calcium signalling in smooth muscle. *Cell Calcium* 2005:397–407.
11. Jude JA, Wylam ME, Walseth TF, Kannan MS. Calcium signaling in airway smooth muscle. *Proc Am Thorac Soc* 2008;**5**:15–22.
12. Harnett KM, Cao W, Biancani P. Signal-transduction pathways that regulate smooth muscle function I. Signal transduction in phasic (esophageal) and tonic (gastroesophageal sphincter) smooth muscles. *Am J Physiol Gastrointest Liver Physiol* 2005;**288**:G407–16.
13. Hirano K. Current topics in the regulatory mechanism underlying the Ca^{2+} sensitization of the contractile apparatus in vascular smooth muscle. *J Pharmacol Sci* 2007;**104**:109–15.
14. Sanders KM. Regulation of smooth muscle excitation and contraction. *Neurogastroenterol Motil* 2008;**20**(Suppl. 1):39–53.
15. Bolton TB. Calcium events in smooth muscles and their interstitial cells; physiological roles of sparks. *J Physiol* 2006;**570**:5–11.
16. Sanderson MJ, Delmotte P, Bai Y, Perez-Zogbhi JF. Regulation of airway smooth muscle cell contractility by Ca^{2+} signaling and sensitivity. *Proc Am Thorac Soc* 2008;**5**:23–31.

17. Christ G, Wingard C. Calcium sensitization as a pharmacological target in vascular smooth-muscle regulation. *Curr Opin Investig Drugs* 2005;**6**:920–33.

18. Horowitz A, Menice CB, Laporte R, Morgan KG. Mechanisms of smooth muscle contraction. *Physiol Rev* 1996;**76**:967–1003.

19. Wray S, Jones K, Kupittayanant S, Matthew AJG, Monir-Bishty E, Noble K, Pierce SJ, Quenby S, Shmygol AV. Calcium signalling and uterine contractility. *J Soc Gynecol Invest* 2003;**10**:252–64.

20. Aaronson PI. Intracellular Ca^{2+} release in cerebral arteries. *Pharmacol Ther* 1994;**64**:493–507.

21. Langton PD, Standen NB. Calcium currents elicited by voltage steps and steady voltages in myocytes isolated from the rat basilar artery. *J Physiol* 1993;**469**:535–48.

22. Shmigol A, Eisner DA, Wray S. Properties of voltage-activated Ca^{2+} transients in single smooth muscle cells isolated from pregnant rat uterus. *J Physiol Lond* 1998;**511**:803–11.

23. Ganitkevich VY, Isenberg G. Efficacy of peak Ca^{2+} currents (I$_{Ca}$) as trigger of sarcoplasmic reticulum Ca^{2+} release in myocytes from the guinea-pig coronary artery. *J Physiol Lond* 1995;**484**:287–306.

24. Haws CW, Heistad DD. Effects of nimodipine on cerebral vasoconstrictor responses. *Am J Physiol* 1984;**247**:H170–6.

25. Edwards R, Trizna W. Response of isolated intracerebral arterioles to endothelins. *Pharmacology* 1990;**41**:149–52.

26. Hill MA, Zou H, Potocnik SJ, Meininger GA, Davis MJ. Invited review: arteriolar smooth muscle mechanotransduction: Ca(2+) signaling pathways underlying myogenic reactivity. *J Appl Physiol* 2001;**91**:973–83.

27. Brandt L, Andersson KE, Edvinsson l, Ljunggren B. Effects of extracellular calcium and of calcium antagonists on the contractile responses of isolated human pial and mesenteric arteries. *J Cereb Blood Flow Metab* 1981;**1**:339–47.

28. Berra-Romani R, Mazzocco-Spezzia A, Pulina MV, Golovina VA. Ca2+ handling is altered when arterial myocytes progress from a contractile to a proliferative phenotype in culture. *Am J Physiol Cell Physiol* 2008;**295**:C779–90.

29. Isenberg G. Ca^{2+} control of transcription: can we extrapolate signaling cascades from neurons to vascular smooth muscle cells? *Circ Res* 2004;**94**:1276–8.

30. Dreja K, Hellstrand P. Differential modulation of caffeine-and IP3-induced calcium release in cultured arterial tissue. *Am J Physiol* 1999;**276**:C1115–20.

31. Gollasch M, Haase H, Ried C, Lindschau C, Morano I, Luft FC, Haller H. L-type calcium channel expression depends on the differentiated state of vascular smooth muscle cells. *FASEB J* 1998;**12**:593–601.

32. Bergdahl A, Gomez MF, Wihlborg AK, Erlinge D, Eyjolfson A, Xu SZ, Beech DJ, Dreja K, Hellstrand P. Plasticity of TRPC expression in arterial smooth muscle: correlation with store-operated Ca^{2+} entry. *Am J Physiol Cell Physiol* 2005;**288**:C872–80.

33. Ng LC, Kyle BD, Lennox AR, Shen XM, Hatton WJ, Hume JR. Cell culture alters Ca^{2+} entry pathways activated by store-depletion or hypoxia in canine pulmonary arterial smooth muscle cells. *Am J Physiol Cell Physiol* 2008;**294**:C313–23.

34. Halayko AJ, Tran T, Gosens R. Phenotype and functional plasticity of airway smooth muscle: role of caveolae and caveolins. *Proc Am Thorac Soc* 2008;**5**:80–8.

35. Mironneau J, Arnaudeau S, rez-Lepretre N, Boittin FX. Ca^{2+} sparks and Ca^{2+} waves activate different Ca(2+)-dependent ion channels in single myocytes from rat portal vein. *Cell Calcium* 1996;**20**:153–60.

36. Matthew A, Shmygol A, Wray S. Ca^{2+} entry, efflux and release in smooth muscle. *Biol Res* 2004,**37**:617–24.

37. Floyd R, Wray S. Calcium transporters and signalling in smooth muscles. *Cell Calcium* 2007;**42**:467–76.

38. Poburko D, Lhote P, Szado T, Behra T, Rahimian R, McManus B, van Breemen C, Ruegg UT. Basal calcium entry in vascular smooth muscle. *Eur J Pharmacol* 2004;**505**:19–29.

39. Burdyga TV, Wray S. Simultaneous measurements of electrical activity. intracellular Ca^{2+}. and force in intact smooth muscle. *Pflügers Arch* 1997;**435**:182–4.

40. Burdyga T, Wray S, Noble K. *In situ* calcium signaling: no calcium sparks detected in rat myometrium. *Ann NY Acad Sci* 2007;**1101**:85–96.

41. Burdyga TV, Wray S. The relationship between the action potential, intracellular calcium and force in intact phasic, guinea-pig uretic smooth muscle. *J Physiol Lond* 1999;**520**:867–83.

42. Zimmermann B, Somlyo AV, Ellis-Davies GCR, Kaplan JH, Somlyo AP. Kinetics of prephosphorylation reactions and myosin light chain phosphorylation in smooth muscle. *J Biol Chem* 1995;**270**:23,966–74.

43. Burdyga TV, Wray S. On the mechanisms whereby temperature affects excitation–contraction coupling in smooth muscle. *J Gen Physiol* 2002;**19**:93–104.

44. Young RC, Smith LH, McLaren MD. T-type and L-type calcium currents in freshly dispersed human uterine smooth muscle cells. *Am J Obstet Gynecol* 1993;**169**:785–92.

45. Bradley JE, Anderson UA, Woolsey SM, Thornbury KD, McHale NG, Hollywood MA. Characterization of T-type calcium current and its contribution to electrical activity in rabbit urethra. *Am J Physiol Cell Physiol* 2004;**286**:C1078–88.

46. Park SY, Lee MY, Keum EM, Myung SC, Kim SC. Ionic currents in single smooth muscle cells of the human vas deferens. *J Urol* 2004;**172**:628–33.

47. Nilsson H. Interactions between membrane potential and intracellular calcium concentration in vascular smooth muscle. *Acta Physiol Scand* 1998;**164**:559–66.

48. Perez-Reyes E. Molecular physiology of low-voltage-activated t-type calcium channels. *Physiol Rev* 2003;**83**:117–61.

49. Rodman DM, Reese K, Harral J, Fouty B, Wu S, West J, Hoedt-Miller M, Tada Y, Li KX, Cool C, Fagan K, Cribbs L. Low-voltage-activated (T-type) calcium channels control proliferation of human pulmonary artery myocytes. *Circ Res* 2005;**96**:864–72.

50. Large WA. Receptor-operated Ca2(+)-permeable nonselective cation channels in vascular smooth muscle: a physiologic perspective. *J Cardiovasc Electrophysiol* 2002;**13**:493–501.

51. Albert AP, Large WA. Signal transduction pathways and gating mechanisms of native TRP-like cation channels in vascular myocytes. *J Physiol* 2006;**570**:45–51.

52. Minke B. TRP channels and Ca^{2+} signaling. *Cell Calcium* 2006;**40**:261–75.

53. Plant TD, Schaefer M. TRPC4 and TRPC5: receptor-operated Ca^{2+}-permeable nonselective cation channels. *Cell Calcium* 2003;**33**:441–50.

54. Wang Y, Deng X, Hewavitharana T, Soboloff J, Gill DL. Stim, orai and trpc channels in the control of calcium entry signals in smooth muscle. *Clin Exp Pharmacol Physiol* 2008;**35**:1127–33.

55. Peel SE, Liu B, Hall IP. ORAI and store-operated calcium influx in human airway smooth muscle cells. *Am J Respir Cell Mol Biol* 2008;**38**:744–9.

56. Xu S-Z, Beech DJ. TrpC1 is a membrane-spanning subunit of store-operated Ca^{2+} channels in native vascular smooth muscle cells. *Circ Res* 2001;**88**:84–7.

57. Inoue R, Okada T, Onoue H, Hara Y, Shimizu S, Naitoh S, Ito Y, Mori Y. The transient receptor potential protein homologue TRP6 is the essential component of vascular alpha(1)-adrenoceptor-activated Ca(2+)-permeable cation channel. *Circ Res* 2001;**88**:325–32.

58. Zholos AV, Zholos AA, Bolton TB. G-protein-gated TRP-like cationic channel activated by muscarinic receptors: effect of potential on single-channel gating. *J Gen Physiol* 2004;**123**:581–98.

59. Gosling M, Poll C, Li S. TRP channels in airway smooth muscle as therapeutic targets. *Naunyn Schmiedebergs Arch Pharmacol* 2005;**371**:277–84.

60. Kim SJ, Koh E-M, Kang TM, Kim YC, So I, Isenberg G, Kim KW. Ca^{2+} influx through carbachol-activated non-selective cation channels in guinea-pig gastric myocytes. *J Physiol* 1998;**513**:749–60.

61. Kim BJ, So I, Kim KW. The relationship of TRP channels to the pacemaker activity of interstitial cells of Cajal in the gastrointestinal tract. *J Smooth Muscle Res* 2006;**42**:1–7.

62. Dietrich A, Chubanov V, Kalwa H, Rost BR, Gudermann T. Cation channels of the transient receptor potential superfamily: their role in physiological and pathophysiological processes of smooth muscle cells. *Pharmacol Ther* 2006;**112**:744–60.

63. Ishida Y, Paul RJ. Ca2+ clearance in smooth muscle: lessons from gene-altered mice. *J Smooth Muscle Res* 2005;**41**:235–45.

64. Sanders KM. Mechanisms of caclium handling in smooth muscles. *J Appl Physiol* 2001;**91**:1438–49.

65. Shmigol A, Eisner DA, Wray S. Carboxyeosin decreases the rate of decay of the Ca^{2+}.$_i$ transient in uterine smooth muscle cells isolated from pregnant rats. *Pflügers Arch* 1998;**437**:158–60.

66. Taggart MJ, Wray S. Agonist mobilization of sarcoplasmic reticular calcium in smooth muscle: functional coupling to the plasmalemmal Na$^+$/Ca^{2+} exchanger. *Cell Calcium* 1997;**22**:333–41.

67. Enyedi A, Verma AK, Heim R, Adamo HP, Filoteo AG, Strehler EE, Penniston JT. The Ca^{2+} affinity of the plasma membrane Ca^{2+} pump is controlled by alternative splicing. *J Biol Chem* 1994;**269**:41–3.

68. Padanyi R, Paszty K, Penheiter AR, Filoteo AG, Penniston JT, Enyedi A. Intramolecular interactions of the regulatory region with the catalytic core in the plasma membrane calcium pump. *J Biol Chem* 2003;**278**:35,798–804.

69. Sugi H, Suzuki S. Physiological and ultrastructural studies on the intracellular calcium translocation during contraction in invertebrate smooth muscles. *Soc Gen Physiol Ser* 1982;**37**:359–70.

70. Liu L, Ishida Y, Okunade G, Shull GE, Paul RJ. Role of plasma membrane Ca^{2+}-ATPase in contraction-relaxation processes of the bladder: evidence from PMCA gene-ablated mice. *Am J Physiol Cell Physiol* 2006;**290**:C1239–47.

71. Yao A, Su Z, Nonaka A, Zubair I, Lu L, Philipson KD, Bridge JH, Barry WH. Effects of overexpression of the Na$^+$–Ca^{2+} exchanger on Ca^{2+}.$_i$ transients in murine ventricular myocytes. *Circ Res* 1998;**82**:657–65.

72. Sakai Y, Kinoshita H, Saitou K, Homma I, Nobe K, Iwamoto T. Functional differences of Na$^+$/Ca^{2+} exchanger expression in Ca2+ transport system of smooth muscle of guinea pig stomach. *Can J Physiol Pharmacol* 2005;**83**:791–7.

73. Juhaszova M, Blaustein MP. Na$^+$ pump low and high ouabain affinity a subunit isoforms are differently distributed in cells. *Proc Natl Acad Sci* 1997;**94**:1800–5.

74. Shelly DA, He S, Moseley A, Weber C, Stegemeyer M, Lynch RM, Lingrel J, Paul RJ. Na(+) pump alpha 2-isoform specifically couples to contractility in vascular smooth muscle: evidence from gene-targeted neonatal mice. *Am J Physiol Cell Physiol* 2004;**286**:C813–20.

75. Zhang J, Lee MY, Cavalli M, Chen L, Berra-Romani R, Balke CW, Bianchi G, Ferrari P, Hamlyn JM, Iwamoto T, Lingrel JB, Matteson DR,

Wier WG, Blaustein MP. Sodium pump alpha2 subunits control myogenic tone and blood pressure in mice. *J Physiol* 2005;**569**:243–56.

76. Wray S, Shmygol A. Role of the calcium store in uterine contractility. *Semin Cell Dev Biol* 2007;**18**:315–20.

77. Shmygol A, Wray S. Functional architecture of the SR calcium store in uterine smooth muscle. *Cell Calcium* 2004;**35**:501–8.

78. Gordienko DV, Greenwood IA, Bolton TB. Direct visualization of sarcoplasmic reticulum regions discharging Ca^{2+} sparks in vascular myocytes. *Cell Calcium* 2001;**29**:13–28.

79. Young RC, Mathur SP. Focal sarcoplasmic reticulum calcium stores and diffuse inositol 1, 4, 5-trisphosphate and ryanodine receptors in human myometrium. *Cell Calcium* 1999;**75**:69–75.

80. Bond M, Kitazawa T, Somlyo AP, Somlyo AV. Release and recycling of calcium by the sarcoplasmic reticulum in guinea-pig portal vein smooth muscle. *J Physiol Lond* 1984;**355**:677–95.

81. Gabella G. Hypertrophic smooth muscle. II. Sarcoplasmic reticulum, caveolae and mitochondria. *Cell Tissue Res* 1979;**201**:79–92.

82. Burdyga TV, Taggart MJ, Wray S. Major difference between rat and guinea-pig ureter in the ability of agonists and caffeine to release Ca^{2+} and influence force. *J Physiol Lond* 1995;**489**:327–35.

83. Bootman MD, Berridge MJ, Roderick HL. Calcium signalling: more messengers, more channels, more complexity. *Curr Biol* 2002;**12**:R563–5.

84. Bradley KN, Currie S, MacMillan D, Muir TC, McCarron JG. Cyclic ADP-ribose increases Ca2+ removal in smooth muscle. *J Cell Sci* 2003;**116**:4291–306.

85. Barata H, Thompson M, Zielinska W, Han YS, Mantilla CB, Prakash YS, Feitoza S, Sieck G, Chini EN. The role of cyclic-ADP-ribose-signaling pathway in oxytocin-induced Ca2+ transients in human myometrium cells. *Endocrinology* 2004;**145**:881–9.

86. Deshpande DA, Dogan S, Walseth TF, Miller SM, Amrani Y, Panettieri RA, Kannan MS. Modulation of calcium signaling by interleukin-13 in human airway smooth muscle: role of CD38/cyclic adenosine diphosphate ribose pathway. *Am J Respir Cell Mol Biol* 2004;**31**:36–42.

87. Wang YX, Zheng YM, Mei QB, Wang QS, Collier ML, Fleischer S, Xin HB, Kotlikoff MI. FKBP12.6 and cADPR regulation of Ca^{2+} release in smooth muscle cells. *Am J Physiol Cell Physiol* 2004;**286**:C538–46.

88. Moller JV, Nissen P, Sorensen TL, Le Maire M. Transport mechanism of the sarcoplasmic reticulum Ca^{2+}-ATPase pump. *Curr Opin Struct Biol* 2005;**15**:387–93.

89. Khan I, Tabb T, Garfield RE, Jones LR, Fomin VP, Samson SE, Grover AK. Expression of the internal calcium pump in pregnant rat uterus. *Cell Calcium* 1993;**14**:111–17.

90. Lagaud GJL, Randriamboavonjy V, Roul G, Stoclet J-C, Andriantsitohaina R. Mechanism of Ca release and entry during contraction elicited by norepinephrine in rat resistance arteries. *Am Physiol Soc* 1999:300–8.

91. Lompre AM, Anger M, Levitsky D. Sarco(endo)plasmic reticulum calcium pumps in the cardiovascular system: function and gene expression. *J Mol Cell Cardiol* 1994;**26**:1109–21.

92. Tribe RM, Moriarty P, Poston L. Calcium homeostatic pathways change with gestation in human myometrium. *Biol Reprod* 2000;**63**:748–55.

93. O'Reilly J, GAK. Molecular biology of calcium pumps in myometrium. In: Garfield RE, editor. *Control of uterine contractility*: CRC Press; 1994. p. 155–72.

94. Shmigol AV, Eisner DA, Wray S. Simultaneous measurements of changes in sarcoplasmic reticulum and cytosolic Ca^{2+}. in rat uterine smooth muscle cells. *J Physiol Lond* 2001;**531**:707–13.

95. Missiaen L, De Smedt H, Droogmans G, Casteels R. Ca release induced by inositol 1, 4, 5-trisphosphate is a steady-state phenomenon controlled by luminal Ca in permeabilized cells. *Nature* 1992;**357**:599–601.

96. Shmygol A, Wray S. Modulation of agonist-induced Ca2+ release by SR Ca^{2+} load: direct SR and cytosolic Ca^{2+} measurements in rat uterine myocytes. *Cell Calcium* 2005;**37**:215–23.

97. Wuytack F, Raeymaekers L, Missiaen L. Molecular physiology of the SERCA and SPCA pumps. *Cell Calcium* 2002;**32**:279–305.

98. Gomez-Viquez L, Rueda A, Garcia U, Guerrero-Hernandez A. Complex effects of ryanodine on the sarcoplasmic reticulum Ca^{2+} levels in smooth muscle cells. *Cell Calcium* 2005;**38**:121–30.

99. Ramsey IS, Delling M, Clapham DE. An introduction to TRP channels. *Annu Rev Physiol* 2006;**68**:619–47.

100. Ng LC, Gurney AM. Store-operated channels mediate Ca(2+) influx and contraction in rat pulmonary artery. *Circ Res* 2001;**89**:923–9.

101. Shlykov SG, Yang M, Alcorn JL, Sanborn BM. Capacitative cation entry in human myometrial cells and augmentation by hTrpC3 overexpression. *Biol Reprod* 2003;**69**:647–55.

102. Kovac JR, Chrones T, Sims SM. Temporal and spatial dynamics underlying capacitative calcium entry in human colonic smooth muscle. *Am J Physiol Gastrointest Liver Physiol* 2008;**294**:G88–98.

103. Flemming R, Cheong A, Dedman AM, Beech DJ. Discrete store-operated calcium influx into an intracellular compartment in rabbit arteriolar smooth muscle. *J Physiol Lond* 2002;**543**:455–64.

104. White C, McGeown JG. Ca2+ uptake by the sarcoplasmic reticulum decreases the amplitude of depolarization-dependent Ca2+.i transients in rat gastric myocytes. *Pflugers Arch* 2000;**440**:488–95.

105. Ganitkevich VY, Isenberg G. Contribution of calcium induced calcium release to Ca^{2+}.i transients in myocytes from guinea-pig urinary bladder. *J Physiol Lond* 1992;**458**:119–37.

106. Shmigol AV, Eisner DA, Wray S. The role of the sarcoplasmic reticulum as a calcium sink in uterine smooth muscle cells. *J Physiol Lond* 1999;**520**:153–63.

107. Gomez-Viquez L, Guerrero-Serna G, Garcia U, Guerrero-Hernandez A. SERCA pump optimizes Ca2+ release by a mechanism independent of store filling in smooth muscle cells. *Biophys J* 2003;**85**:370–80.

108. Shmigol AV, Eisner DA, Wray S. Simultaneous measurements of changes in sarcoplasmic reticulum and cytosolic. *J Physiol* 2001;**531**:707–13.

109. Kim M, Hennig GW, Smith TK, Perrino BA. Phospholamban knockout increases CaM kinase II activity and intracellular Ca^{2+} wave activity and alters contractile responses of murine gastric antrum. *Am J Physiol Cell Physiol* 2008;**294**:C432–41.

110. Sathish V, Leblebici F, Kip SN, Thompson MA, Pabelick CM, Prakash YS, Sieck GC. Regulation of sarcoplasmic reticulum Ca^{2+} reuptake in porcine airway smooth muscle. *Am J Physiol Lung Cell Mol Physiol* 2008;**294**:L787–96.

111. Sutliff RL, Conforti L, Weber CS, Kranias EG, Paul RJ. Regulation of the spontaneous contractile activity of the portal vein by the sarcoplasmic reticulum: evidence from the phospholamban gene-ablated mouse. *Vascul Pharmacol* 2004;**41**:197–204.

112. Gregoire G, Loirand G, Pacaud P. Ca^{2+} and Sr^{2+} entry induced Ca2+ release from the intracellular Ca^{2+} store in smooth muscle cells of rat portal vein. *J Physiol Lond* 1993;**474**:483–500.

113. Taggart MJ, Wray S. Contribution of sarcoplasmic reticular calcium to smooth muscle contractile activation: gestational dependence in isolated rat uterus. *J Physiol Lond* 1998;**511**:133–44.

114. Collier ML, Ji G, Wang Y, Kotlikoff MI. Calcium-induced calcium release in smooth muscle: loose coupling between the action potential and calcium release. *J Gen Physiol* 2000;**115**:653–62.

115. Kotlikoff MI. Calcium-induced calcium release in smooth muscle: the case for loose coupling. *Prog Biophys Mol Biol* 2003;**83**:171–91.

116. Ji G, Feldman M, Doran R, Zipfel W, Kotlikoff MI. Ca^{2+} -induced Ca^{2+} release through localized Ca^{2+} uncaging in smooth muscle. *J Gen Physiol* 2006;**127**:225–35.

117. Ji G, Feldman ME, Greene KS, Sorrentino V, Xin HB, Kotlikoff MI. RYR2 proteins contribute to the formation of Ca(2+) sparks in smooth muscle. *J Gen Physiol* 2004;**123**:377–86.

118. Heppner TJ, Herrera GM, Bonev AD, Hill-Eubanks D, Nelson MT. Ca^{2+} sparks and K(Ca) channels: novel mechanisms to relax urinary bladder smooth muscle. *Adv Exp Med Biol* 2003;**539**:347–57.

119. Amberg GC, Bonev AD, Rossow CF, Nelson MT, Santana LF. Modulation of the molecular composition of large conductance, Ca(2+) activated K(+) channels in vascular smooth muscle during hypertension. *J Clin Invest* 2003;**112**:717–24.

120. Wellman GC, Nelson MT. Signaling between SR and plasmalemma in smooth muscle: sparks and the activation of Ca^{2+}-sensitive ion channels. *Cell Calcium* 2003;**34**:211–29.

121. Perez GJ, Bonev AD, Nelson MT. Micromolar Ca^{2+} from sparks activates Ca^{2+}-sensitive K^+ channels in rat cerebral artery smooth muscle. *Am J Physiol* 2001;**281**:C1769–75.

122. Brenner R, Perez GJ, Bonev AD, Eckman DM, Kosek JC, Wiler SW, Patterson AJ, Nelson MT, Aldrich RW. Vasoregulation by the B1 subunit of the calcium-activated potassium channel. *Nature* 2000;**407**:870–6.

123. Nelson MT, Cheng H, Rubart M, Santana LF, Bonev AD, Knot HJ, Lederer WJ. Relaxation of arterial smooth muscle by calcium sparks. *Science* 1995;**270**:633–7.

124. Benham CD, Bolton TB. Spontaneous transient outward currents in single visceral and vascualr smooth muscle cells of the rabbit. *J Physiol Lond* 1986;**381**:385–406.

125. Burdyga T, Wray S. Action potential refractory period in ureter smooth muscle is set by Ca sparks and BK channels. *Nature* 2005;**436**:559–62.

126. Borisova L, Shmygol A, Wray S, Burdyga T. Evidence that a Ca(2+) sparks/STOCs coupling mechanism is responsible for the inhibitory effect of caffeine on electro-mechanical coupling in guinea pig ureteric smooth muscle. *Cell Calcium* 2007;**42**:303–11.

127. Noble K, Wray S. The role of the sarcoplasmic reticulum in neonatal uterine smooth muscle: enhanced role compared to adult rat. *J Physiol Lond* 2002;**545**:557–66.

128. Chanrachakul B, Pipkin FB, khan RN. Contribution of coupling between human myometrial beta2-adrenoreceptor and the BK(Ca) channel to uterine quiescence. *Am J Physiol Cell Physiol* 2004;**287**:C1747–52.

129. Eghbali M, Toro L, Stefani E. Diminished surface clustering and increased perinuclear accumulation of large conductance Ca^{2+}-activated K^+ channel in mouse myometrium with pregnancy. *J Biol Chem* 2003;**278**:45,311–7.

130. Khan R, Smith SK, Morrison JJ, Ashford MLJ. Properties of large-conductance K^+ channels in human myometrium during pregnancy and labour. *Proc R Soc Lond B* 1993;**251**:9–15.

131. Anwer K, Oberti C, Perez J, Perez-Reyes N, McDougall JK, Monga M, Sanborn BM, Stefani E, Toro L. Calcium-activated K^+ channels as modulators of human myometrial contractile activity. *Am J Physiol* 1993;**265**:C967–85.

132. Mironneau J, Macrez N, Morel JL, Sorrentino V, Mironneau C. Identification and function of ryanodine receptor subtype 3 in non-pregnant mouse myometrial cells. *J Physiol* 2002;**538**:707–16.

133. Kupittayanant S, Luckas MJM, Wray S. Effects of inhibiting the sarcoplasmic reticulum on spontaneous and oxytocin-induced contractions of human myometrium. *Br J Obstet Gynaecol* 2002;**109**:289–96.

134. Jiang D, Xiao B, Li X, Chen SR. Smooth muscle tissues express a major dominant negative splice variant of the type 3 Ca^{2+} release channel (ryanodine receptor). *J Biol Chem* 2003;**278**:4763–9.

135. Dabertrand F, Morel J-L, Sorrentino V, Mironneau J, Mironneau C, macrez N. Modulation of calcium signalling by dominant negative splice variant of ryanodine receptor subtype 3 in native smooth muscle cells. *Cell Calcium* 2006;**40**:11–21.

136. Dabertrand F, Fritz N, Mironneau J, Macrez N, Morel JL. Role of RYR3 splice variants in calcium signalling in mouse non-pregnant and pregnant myometrium. *Am J Physiol Cell Physiol* 2007. in press.

137. Lohn M, Jessner W, Furstenau M, Wellner M, Sorrentino V, Haller H, Luft FC, Gollasch M. Regulation of calcium sparks and spontaneous transient outward currents by RyR3 in arterial vascular smooth muscle cells. *Circ Res* 2001;**89**:1051–7.

138. Bernardi P. Mitochondrial transport of cations: channels, exchangers, and permeability transition. *Physiol Rev* 1999;**179**:127–1155.

139. Rizzuto R, Brini M, Pozzan T. Targeting recombinant aequorin to specific intracellular organelles. *Methods Cell Biol* 1994;**40**:339–58.

140. Berridge MJ, Lipp P, Bootman MD. The versatility and universality of calcium signalling. *Nat Rev Mol Cell Biol* 2000;**1**:11–21.

141. Duchen MR. Mitochondria and calcium: from cell signalling to cell death. *J Physiol* 2000;**529**:57–68.

142. Rizzuto R, Pinton P, Carrington WA, Fay FS, Fogarty KE, Lifshitz LM, Tuft RA, Pozzan T. Close contacts with the endoplasmic reticulum as determinants of mitochondrial Ca^{2+} responses. *Science* 1998;**280**:1763–6.

143. Bianchi K, Rimessi A, Prandini A, Szabadkai G, Rizzuto R. Calcium and mitochondria: mechanisms and functions of a troubled relationship. *Biochim Biophys Acta* 2004;**1742**:119–31.

144. Drummond RM, Fay FS. Mitochondria contribute to Ca^{2+} removal in smooth muscle cells. *Eur J Physiol* 1996;**431**:473–82.

145. Restini CA, Moreira JE, Bendhack LM. Cross-talk between the sarcoplasmic reticulum and the mitochondrial calcium handling systems may play an important role in the regulation of contraction in anococcygeus smooth muscle. *Mitochondrion* 2006;**6**:71–81.

146. Nixon GF, Migneri GA, Somlyo AV. Immunogold localisation of inositol 1, 4, 5-triphosphate receptors and characterisation of ultrastructural features of the sarcoplasmic reticulum in phasic and tonic smooth muscle. *J Musc Res Cell Res* 1994;**15**:682–99.

147. McCarron JG, Muir TC. Mitochondrial regulation of the cytosolic Ca^{2+} concentration and the insP$_3$-sensitive Ca^{2+} store in guinea-pig colonic smooth muscle. *J Physiol Lond* 1999;**516**:149–62.

148. Drummond RM, Tuft RA. Release of Ca^{2+} from the sarcoplasmic reticulum increases mitochondrial Ca^{2+}. in rat pulmonary artery smooth muscle cells. *J Physiol Lond* 1999;**516**:139–48.

149. Monteith GR, Blaustein MP. Heterogeneity of mitochondrial matrix free Ca^{2+}: resolution of Ca^{2+} dynamics in individual mitochondria *in situ*. *Am J Physiol* 1999;**276**:C1193–204.

150. Greenwood IA, Helliwell RM, Large WA. Modulation of Ca^{2+}- activated Cl^- currents in rabbit portal vein smooth muscle by an inhibitor of mitochondrial Ca^{2+} uptake. *J Physiol Lond* 1997;**505**:53–64.

151. Dolman NJ, Tepikin AV. Calcium gradients and the Golgi. *Cell Calcium* 2006;**512**:505–12.

152. Gerasimenko O, Gerasimenko J. New aspects of nuclear calcium signalling. *J Cell Sci* 2004;**117**:3087–94.

153. Cartin L, Lounsbury KM, Nelson MT. Coupling of Ca(2+) to CREB activation and gene expression in intact cerebral arteries from mouse: roles of ryanodine receptors and voltage-dependent Ca(2+) channels. *Circ Res* 2000;**86**:760–7.

154. Barlow CA, Rose P, Pulver-Kaste RA, Lounsbury KM. Excitation-transcription coupling in smooth muscle. *J Physiol* 2006;**570**:59–64.

155. Stevenson AS, Gomez MF, Hill-Eubanks DC, Nelson MT. NFAT4 movement in native smooth muscle. A role for differential Ca(2+) signaling. *J Biol Chem* 2001;**276**:15,018–24.

156. Wadsworth RM. Selectivity of calcium antagonist drugs on vascular smooth muscle. *Clin Exp Pharmacol Physiol* 1993;**120**:745–51.

157. Baron CB, Cunningham M, Strauss JF, III Coburn RF. Pharmacomechanical coupling in smooth muscle may involve phosphatidyli-nositol metabolism. *Proc Natl Acad Sci USA* 1984;**81**:6899–903.

158. Murray RK, Fleischmann BK, Kotlikoff MI. Receptor-activated Ca influx in human airway smooth muscle: use of Ca imaging and per-forated patch-clamp techniques. *Am J Physiol* 1993;**264**:C485–90.

159. Godfraind T, Kaba A. The role of calcium in the action of drugs on vascular smooth muscle. *Arch Intl Pharmacodyn Ther* 1972;**196**:35.

160. Hirano K, Kanaide H, Nakamura M. Effects of okadaic acid on cytosolic calcium concentration and contraction of porcine coronary artery. *Adv Second Messenger Phosphoprotein Res* 1990;**24**:461–6.

161. Karaki H, Sato K, Ozaki H. Different effects of verapamil on cytosolic Ca2+ and contraction in norepinephrine-stimulated vascular smooth muscle. *Jpn J Pharmacol* 1991;**55**:35–42.

162. Liu X, Farley JM. Acetylcholine-induced Ca^{2+}-dependent chloride current oscillations are meduated by inositol 1, 4, 5-trisphosphate in tracheal myocytes. *J Pharmacol Exp Ther* 1996;**277**:796–804.

163. Prakash YS, Kannan MS, Sieck GC. Regulation of intracellular calcium oscillations in porcine tracheal smooth muscle cells. *Am J Physiol* 1997;**272**:C966–75.

164. Gordienko DV, Bolton TB, Cannell MB. Variability in spontaneous subcellular calcium release in guinea-pig ileum smooth muscle cells. *J Physiol* 1998;**507**:707–20.

165. ZhuGe R, Tuft RA, Fogarty KE, Bellve K, Fay FS, Walsh JV. The influence of sarcoplasmic reticulum Ca^{2+} concentration on Ca^{2+} sparks and spontaneous transient outward currents in single smooth muscle cells. *J Gen Physiol* 1999;**113**:215–28.

166. Jaggar JH, Porter VA, Lederer WJ, Nelson MT. Calcium sparks in smooth muscle. *Am J Physiol Cell Physiol* 2000;**278**:C235–56.

167. Pabelick CM, Prakash YS, Kannan MS, Sieck GC. Spatial and temporal aspects of calcium sparks in porcine tracheal smooth muscle cells. *Am J Physiol* 1999;**277**:L1018–25.

168. Mejia-Alvarez R, Kettlun C, Rios E, Stern M, Fill M. Unitary Ca2+ current through cardiac ryanodine receptor channels under quasi-physiological ionic conditions. *J Gen Physiol* 1999;**113**:177–86.

169. Fay FS. Calcium sparks in vascular smooth muscle: relaxation regulators. *Science* 1995;**270**:588–9.

170. Iino M, Kasai H, Yamazawa T. Visualization of neural control of intracellular Ca^{2+} concentration in single vascular smooth muscle cells *in situ*. *EMBO J* 1994;**13**:5026–31.

171. Curtis TM, Tumelty J, Dawicki J, Scholfield CN, McGeown JG. Identification and spatiotemporal characterization of spontaneous Ca^{2+} sparks and global Ca^{2+} oscillations in retinal arteriolar smooth muscle cells. *Invest Ophthalmol Vis Sci* 2004;**45**:4409–14.

172. Porter VA, Bonev AD, Knot HJ, Heppner TJ, Stevenson AS, Kleppisch T, Lederer WJ, Nelson MT. Frequency modulation of Ca^{2+} sparks is involved in regulation of arterial diameter by cyclic nucleotides. *Am J Physiol* 1998;**274**:C1346–55.

173. Bonev AD, Jagger JH, Rubart M, Nelson MT. Activators of protein kinase C decreases Ca^{2+} spark frequency in smooth muscle cells from cerebral arteries. *Am J Physiol* 1997;**273**:C2090–5.

174. Large WA, Wang Q. Characteristics and physiological role of the Ca^{2+}-activated Cl^- conductance in smooth muscle. *Am J Physiol* 1996;**271**:C435–54.

175. Blatter LA, Wier WG. Agonist-induced $Ca^{2+}._i$ and Ca^{2+}-induced Ca^{2+} release in mammalian vascular smooth muscle cells. *Am J Physiol* 1992;**263**:H576–86.

176. Mayer EA, Kodner A, Sun XP, Wilkes J, Scott D, Sachs G. Spatial and temporal patterns of intracellular calcium in colonic smooth muscle. *J Membr Biol* 1992;**125**:107–18.

177. Karaki H, Ozaki H, Hori M, Mitsui-Saito M, Amano K, Harada K, Miyamoto S, Nakazawa H, Won K-J, Sato K. Calcium movements, distribution and function in smooth muscle. *Pharmacol Rev* 1997;**282**:157–230.

178. Lee C-H, Poburko D, Kuo K-H, Seow CY, van Breeman C. Ca^{2+} oscillations, gradients, and homeostasis in vascular smooth muscle. *Am J Physiol* 2002;**466**:H1571–83.

179. Kalthof B, Bechem M, Flocke K, Pott L, Schramm M. Kinetics of ATP-induced Ca^{2+} transients in cultured pig aortic smooth muscle cells depend on ATP concentration and stored Ca^{2+}. *J Physiol* 1993;**82**:245–62.

180. Prakash YS, van der Heijden HF, Kannan MS, Sieck GC. Effects of salbutamol on intracellular calcium oscillations in porcine airway smooth muscle. *J Appl Physiol* 1997;**182**:1836–43.

181. Fu X, Favini R, Kindahl K, Ulmsten U. Prostaglandin F2alpha-induced Ca++ oscillations in human myometrial cells and the role of RU 486. *Am J Obstet Gynecol* 2000;**49**:582–8.

182. Thomas AP, Renard DC, Rooney TA. Spatial and temporal organization of calcium signalling in hepatocytes. *Cell Calcium* 1991;**12**:111–26.

183. Wakui M, Potter BV, Petersen OH. Pulsatile intracellular calcium release does not depend on fluctuations in inositol trisphosphate concentration. *Nature* 1989;**339**:317–20.

184. Amundson J, Clapham D. Calcium waves. *Curr Opin Neurobiol* 1993;**3**:375–82.

185. Kim M, Perrino BA. CaM kinase II activation and phospholamban phosphorylation by SNP in murine gastric antrum smooth muscles. *Am J Physiol Gastrointest Liver Physiol* 2007;**292**:G1045–54.

186. Kannan MS, Prakash YS, Brenner T, Mickelson JR, Sieck GC. Role of Ryanodine receptor channels in Ca^{2+} oscillations of porcine tracheal smooth muscle. *Am J Physiol* 1997;**272**:L659–64.

187. Prakash YS, Kannan MS, Walseth TF, Sieck GC. Role of cyclic ADP-ribose in the regulation of $Ca^{2+}.i$ in porcine tracheal smooth muscle. *Am J Physiol* 1998;**274**:C1653–60.

188. Lee CH, Kuo KH, Dai J, Leo JM, Seow CY, Breemen C. Calyculin-A disrupts subplasmalemmal junction and recurring Ca^{2+} waves in vascular smooth muscle. *Cell Calcium* 2005;**37**:9–16.

189. van BC, Leijten P, Yamamoto H, aaronson P, Cauvin C. Calcium activation of vascular smooth muscle. State of the art lecture. *Hypertension* 1986;**8**:II89–95.

190. van Breemen C, Saida K. Cellular mechanisms regulating $Ca^{2+}.i$ smooth muscle. *Ann Rev Physiol* 1989;**51**:315–29.

191. Blaustein MP. Sodium ions, calcium ions, blood pressure regulation, and hypertension: a reassessment and a hypothesis. *Am J Physiol* 1977;**232**:C165–73.

192. Blaustein MP. Endogenous ouabain: role in the pathogenesis of hypertension. *Kidney Intl* 1996;**49**:1748–53.

193. Kotlikoff MI. Genetically encoded Ca^{2+} indicators: using genetics and molecular design to understand complex physiology. *J Physiol* 2007;**578**:55–67.

194. Lee MY, Song H, Nakai J, Ohkura M, Kotlikoff MI, Kinsey SP, Golovina VA, Blaustein MP. Local subplasma membrane Ca^{2+} signals detected by a tethered Ca^{2+} sensor. *Proc Natl Acad Sci USA* 2006;**103**:13,232–7.

195. Shmigol AV, Eisner DA, Wray S. The role of the sarcoplasmic reticulum as a Ca^{2+} sink in rat uterine smooth muscle cells. *J Physiol* 1999;**520**:153–63.

196. Moore EDW, Etter EF, Philipson KD, Carrington WA, Fogarty KE, Lifshitz LM, Fay FS. Coupling of the Na^+/Ca^{2+} exchanger, Na^+/K^+ pump and sarcoplasmic reticulum in smooth muscle. *Nature* 1993;**365**:657–60.

197. Nazer MA, van Breemen C. A role of the sarcoplasmic reticulum in Ca^{2+} extrusion from rabbit inferior vena cava smooth muscle. *Am J Physiol* 1998;**43**:H123–31.

198. Rembold CM, Cheng X-L. The buffer barrier hypothesis, $Ca^{2+}._i$ homogenity, and sarcoplasmic reticulum function in swine carotid artery. *J Physiol* 1998;**523**:477–92.

199. Petkov GV, Boev KK. The role of sarcoplasmic reticulum and sarcoplasmic reticulum Ca^{2+}-ATPase in the smooth muscle tone of the cat gastric fundus. *Pflügers Arch* 1996;**431**:928–35.

200. van Breemen C, Chen Q, Laher I. Superficial buffer barrier function of smooth muscle sarcoplasmic reticulum. *TiPS* 1995;**16**:98–105.

201. Iwamoto T, Kita S. Hypertension, Na^+/Ca^{2+} exchanger, and Na^+, K^+-ATPase. *Kidney Intl* 2006;**69**:2148–54.

202. Lingrel J, Moseley A, Dostanic I, Cougnon M, He S, James P, Woo A, O'Connor K, Neumann J. Functional roles of the alpha isoforms of the Na, K-ATPase. *Ann NY Acad Sci* 2003;**986**:354–9.

203. Golovina V, Song H, James P, Lingrel J, Blaustein M. Regulation of Ca2+ signaling by Na^+ pump alpha-2 subunit expression. *Ann NY Acad Sci* 2003;**986**:509–13.

204. James PF, Grupp IL, Grupp G, Woo AL, Askew GR, Croyle ML, Walsh RA, Lingrel JB. Identification of a specific role for the Na, K-ATPase alpha 2 isoform as a regulator of calcium in the heart. *Mol Cell* 1999;**3**:555–63.

205. Allen DG, Eisner DA, Wray SC. Birthday present for digitalis. *Nature* 1985;**675**:674–5.

206. Iwamoto T, Kita S, Zhang J, Blaustein MP, Arai Y, Yoshida S, Wakimoto K, Komuro I, Katsuragi T. Salt-sensitive hypertension is triggered by Ca^{2+} entry via Na^+/Ca^{2+} exchanger type-1 in vascular smooth muscle. *Nat Med* 2004;**10**:1193–9.

207. Khoyi MA, Bjur RA, Westfall DP. Time-dependent increase in Ca^{2+} influx in rabbit abdominal aorta: role of Na-Ca exchange. *Am J Physiol* 1993;**265**:C1325–31.

208. Khoyi MA, Bjur RA, Westfall DP. Norepinephrine increases Na–Ca exchange in rabbit abdominal aorta. *Am J Physiol* 1991;**261**:C685–90.

209. Simons K, Toomre D. Lipid rafts and signal transduction. *Nat Rev Mol Cell Biol* 2000;**1**:31–9.

210. Simons K, Ikonen E. How cells handle cholesterol. *Science* 2000;**290**:1721–6.

211. Quest AF, Leyton L, Parraga M. Caveolins, caveolae, and lipid rafts in cellular transport, signalling and disease. *Biochem Cell Biol* 2004;**82**:129–44.

212. O'Connell KM, Martens JR, Tamkun MM. Localization of ion channels to lipid Raft domains within the cardiovascular system. *Trends Cardiovasc Med* 2004;**14**:37–42.

213. Klein U, Gimpl G, Fahrenholz F. Alteration of the myometrial plasma membrane cholesterol content with beta-cyclodextrin modulates the binding affinity of the oxytocin receptor. *Biochemistry* 1995;**34**:13,784–93.

214. Goligorsky MS, Li H, Brodsky S, Chen J. Relationaships between caveolae and eNOS: everything in proximity and the proximity of everything. *Am J Physiol Renal Physiol* 2002;**283**:F1–10.

215. Smith RD, Babiychuk EB, Noble K, Draeger A, Wray S. Increased cholesterol decreases uterine activity: functional effects of cholesterol in pregnant rat myometrium. *Am J Physiol* 2005;**288**:C982–8.

216. Babiychuk EB, Smith RD, Burdyga TV, Babiychuk VS, Wray S, Draeger A. Membrane cholesterol selectively regulates smooth muscle phasic contraction. *J Memb Biol* 2004;**198**:95–101.

217. Zhang J, Kendrick A, Quenby S, Wray S. Contractility and calcium signalling of human myometrium are profoundly affected by cholesterol manipulation: implications for labour? *Reproductive Sci* 2007;**14**:456–66.

218. Dreja K, Voldstedlund M, Vinten J, Tranum-Jensen J, Hellstrand P, Sward K. Cholesterol depletion disrupts caveolae and differentially impairs agonist-induced arterial contraction. *Arterioscler Thromb Vasc Biol* 2002;**22**:1267–72.

219. Sampson LJ, Hayabuchi Y, Standen NB, Dart C. Caveolae localize protein kinase A signaling to arterial ATP-sensitive potassium channels. *Circ Res* 2004;**95**:1012–18.

220. Zhang J, Bricker L, Wray S, Quenby S. Poor uterine contractility in obese women. *Br J Obstet Gynaecol* 2007;**114**:343–8.

221. Shmygol A, Noble K, Wray S. Depletion of membrane cholesterol eliminates the Ca2+-activated component of outward potassium current and decreases membrane capacitance in rat uterine myocytes. *J Physiol* 2007;**581**:445–56.

222. Brainard AM, Miller AJ, Martens JR, England SK. Maxi-K channels localize to caveolae in human myometrium: a role for an actin-channel-caveolin complex in the regulation of myometrial smooth muscle K$^+$ current. *Am J Physiol Cell Physiol* 2005;**289**:C49–57.

Chapter 12

Trophic Effects of Gut Hormones in the Gastrointestinal Tract

Kanika A. Bowen and B. Mark Evers

Department of Surgery, The University of Texas Medical Branch, Galveston, Texas

INTRODUCTION

The mucosa of the gastrointestinal (GI) tract is a complex and constantly renewing tissue which is characterized by rapid proliferation, differentiation, and subsequent apoptosis, followed by extrusion into the GI lumen. These events occur as GI luminal epithelial cells ascend the vertical axis of the microfolded crypts lining the GI tract [1, 2]. This process normally takes 2–5 days to reach the upper portion of the villus, depending upon the species and the location along the GI tract [3, 4]. Numerous factors can contribute to growth of the GI mucosa. This chapter will specifically focus on the effects of GI hormones on the proliferation and repair of non-neoplastic tissues, and the receptors and signaling pathways which transmit signals from the cell surface to the nucleus.

By definition, any agent that stimulates growth can be considered as a growth factor; however, these growth-stimulating agents are usually divided into those that are produced by normal cells and are thought to act locally to control proliferation, and hormones which are thought to act at a distance. Peptide growth factors that act locally include members of the epidermal growth factor (EGF) family, the transforming growth factor-β (TGFβ) family, the insulin-like growth factor (IGF) family, the fibroblast growth factor (FGF) family, the trefoil factor (TFF) family, the colony-stimulating factor (CSF) family, and a few other unrelated regulatory peptides, such as hepatocyte growth factor (HGF), platelet-derived growth factor (PDGF), various interleukins, interferons, and tumor necrosis factor-related proteins [5].

Various gut hormones can regulate growth of the GI mucosa, usually through an endocrine effect. On occasion, an autocrine or paracrine mechanism has been postulated for the proliferative effects of these trophic peptides [6, 7].

Neuroendocrine effects also mediate the effect of the gut mucosa. The gut peptides which have been best described in their role as stimulating mucosal proliferation of the stomach, small bowel, or colon include gastrin, bombesin (BBS)/gastrin-releasing peptide (GRP), neurotensin (NT), glucagon-like peptide-2 (GLP-2), and peptide YY (PYY) [1]. These hormones are secreted by endocrine cells which are widely distributed throughout the GI mucosa and pancreas. In addition to mucosal proliferation, these gut peptides control many other functions in the GI tract, including regulation of secretion, motility, absorption, and digestion (Table 12.1).

TROPHIC EFFECTS OF GUT PEPTIDES IN THE STOMACH, SMALL BOWEL, AND COLON

Stomach

Gastrin is the GI hormone that has been best characterized for its trophic effects in the stomach. Gastrin stimulates acid secretion from gastric parietal cells, and is the single most important trophic hormone for the gastric mucosa. The trophic effect of gastrin on gastric mucosa was initially demonstrated with the synthetic gastrin analog, pentagastrin, which, when given to rats, stimulated protein synthesis and parietal cell mass [8, 9]. These results were further confirmed using the natural amidated gastrins, G17 and G34, with most pronounced effects noted in the oxyntic acid-secreting mucosa and enterochromaffin-like cells. Resection of the gastric antrum, which removes endogenous gastrin, results in gastric mucosal atrophy; this atrophy can be prevented by administration of exogenous gastrin [10]. Further confirmation of the effects of gastrin on gastric mucosal growth is

TABLE 12.1 Gut hormones contributing to GI mucosal growth

Hormone	Location	Primary effects
Gastrin	Antrum, duodenum (G cells)	• Stimulates gastric acid and pepsinogen secretion • Stimulates gastric mucosal growth
Gastrin releasing peptide (GRP) (mammalian equivalent of bombesin, BBS)	Small bowel	• Stimulates release of all GI hormones • Stimulates GI secretion and motility • Stimulates gastric acid secretion and release of antral gastrin • Stimulates growth of intestinal mucosa and pancreas
Neurotensin (NT)	Small bowel (N cells)	• Stimulates pancreatic water and bicarbonate secretion • Inhibits gastric secretion • Stimulates growth of small and large bowel mucosa
Glucagon-like peptide-2 (GLP-2)	Small bowel (L cells)	• Potent enterotrophic factor
Peptide YY (PYY)	Distal small bowel, colon	• Inhibits gastric and pancreatic secretion • Inhibits gallbladder contraction • Stimulates intestinal growth?

provided by transgenic mice which either overexpress gastrin or are gastrin-deficient. In mice overexpressing either unprocessed gastrin or the amidated gastrins (G17 and G34), there is a marked thickening of the oxyntic mucosa with increased BrdU labeling, representing an 85 percent increase in cells undergoing proliferation [11, 12]. Gastrin-deficient mice secrete acid in response to agonist stimulation, and there is a decrease in parietal cell size with an increase of parietal cell number when compared with wild-type controls [13, 14].

Another peptide that has been shown to stimulate gastric mucosal proliferation is BBS/GRP, which stimulates pancreatic, gastric, and intestinal secretion, gut motility, and smooth muscle contraction, and release of all gut hormones [15]. In addition, these peptides can stimulate growth of GI mucosa and pancreas. BBS stimulates gastric weight, fundic and antral mucosal height, and density of parietal cells in neonatal rats compared with saline-treated controls [16, 17]. These results were confirmed in adult rats given BBS for 7 days, demonstrating increased weight, RNA, and DNA contents of the oxyntic mucosa of the stomach and the duodenal mucosa; the inhibitory hormone, somatostatin, attenuated the proliferative effect of BBS [18]. In another study, the BBS receptor antagonist, RC-3095, prevented the proliferative effect of BBS on the gastric mucosa, thus providing evidence that BBS/GRP stimulates growth of stomach and duodenum, predominately due to a direct effect of hormone stimulation and not secondary to release of other gut hormones [19].

Small intestine

The intestinal hormones that have been shown to stimulate growth of small intestine mucosa include NT, BBS/GRP,

PYY, and GLP-2. Wood and colleagues [20] first noted that NT stimulated the small bowel mucosa of rats fed a normal diet. Investigators in our laboratory have shown that administration of NT prevents gut mucosal atrophy induced by feeding rats an elemental diet and stimulates mucosal growth in defunctionalized, self-emptying jejunoileal loops or isolated small bowel loops termed Thiry-Vella fistulas (TVFs), thus supporting a direct role for NT in the stimulation of gut mucosal growth [21, 22]. Assimakopoulos and colleagues [23] reported that NT restores gut mucosal integrity in rats and prevents the translocation of indigenous bacteria after partial hepatectomy. Furthermore, Izukura et al. [24] and Ryan et al. [25] demonstrated, in separate studies, that administration of NT can augment the normal adaptive hyperplasia of gut mucosa that is associated with a massive small bowel resection.

BBS also stimulates growth of the small bowel mucosa. Administration of BBS effectively prevented mucosal atrophy associated with feeding rats a liquid elemental diet [26]. Furthermore, BBS was noted to increase mucosal weight, DNA, and protein content in both jejunal and ileal TVFs compared to control animals, suggesting that the effects of BBS were directly mediated as opposed to indirect effects of stimulation of luminal pancreatic or biliary secretion [27]. In addition to its effects on gut mucosal growth, BBS exhibits protective effects on the gut after injury [28]. Using a lethal enterocolitis model in rats induced by the chemotherapeutic agent methotrexate (MTX), BBS enhanced gut mucosal growth and significantly inhibited mortality. The beneficial effect of BBS on survival was noted when BBS was given prior to or at the same time as MTX, which suggested that BBS may act through additional mechanisms other than gut mucosal growth alone. One possibility is that BBS may produce its beneficial effects through enhancement of the immune system, a known action of BBS.

Although the data are somewhat controversial, the gut peptide PYY has likewise been shown to produce a trophic effect in small bowel mucosa of both rat and mouse [29]. These effects were noted at relatively high dosages. Similarly, Chance and colleagues [30] found that PYY treatment in Sprague-Dawley rats given total parenteral nutrition (TPN) produced significant increases in jejunal, ileal, and colonic protein contents.

A trophic effect for glucagon-derived peptides in the intestinal mucosa has been postulated since the description of a glucagon-secreting tumor of the kidney associated with small bowel mucosal hypertrophy. Drucker and colleagues [31] were the first to demonstrate that the intestinal trophic factor was GLP-2, which produced a 50 percent increase in small bowel weight and a significant increase in mucosal thickness. Similarly, Ghatei et al. [32] demonstrated prominent trophic effects of GLP-2 in Wistar rats, and Litvak et al. [33] demonstrated that GLP-2 significantly increased the weight of jejunum, ileum, and colon of athymic nude mice compared to control mice. In addition to the effects of GLP-2 on normal mucosa, the effects of this agent during periods of gut injury or atrophy have also been assessed. Mice treated with indomethacin developed small bowel enteritis associated with significant mortality at 48–72 hours after administration; treatment with human [Gly^2]-GLP-2, before, during, or after indomethacin administration, resulted in reduced mortality and decreased mucosal injury [34]. The protective effects were attributed to the significantly increased crypt cell proliferation and decreased crypt compartment apoptosis. The effect of GLP-2 on chemotherapy-induced intestinal mucositis has also been assessed. Pretreatment of mice with human [Gly^2]-GLP-2 before administration of the topoisomerase inhibitor, irinotecan, resulted in reduced bacterial translocation, intestinal damage, and mortality [35]. Histological and biochemical analyses revealed significant reductions in crypt compartment apoptosis and reduced caspase-8 activation. Consistent with these reports, Tavakkolizadeh and colleagues [36] noted decreased intestinal damage in rats given GLP-2 in combination with the chemotherapeutic agent 5-fluorouracil. Finally, repeated cyclical administration of human [Gly^2]-GLP-2 resulted in significantly decreased mortality in groups of Balb/c mice given irinotecan.

Colon

Colonic mucosal growth may be affected by the gut peptides gastrin, BBS/GRP, NT, and GLP-2. Earlier reports suggested a role for amidated gastrin (i.e., G17 and G34) as a trophic factor in the colon [37]. Recent studies now suggest that glycine-extended progastrin (G-Gly) may be the responsible agent producing the effects noted with gastrin

administration. These findings have sparked renewed interest in a role for gastrin precursor products in colonic growth. Koh and colleagues [38] generated mice that overexpressed progastrin truncated at glycine-72 which demonstrate elevated serum and mucosal levels of G-Gly compared with wild-type mice. Mice overexpressing G-Gly displayed a 43 percent increase in colonic mucosal thickness and a 41 percent increase in the percentage of goblet cells per crypt. Furthermore, administration of G-Gly to gastrin-deficient mice resulted in a 10 percent increase in colonic mucosal thickness and an 81 percent increase in colonic proliferation as measured by BrdU incorporation.

Although the small bowel is significantly more sensitive to the effects of GLP-2, studies have shown that GLP-2 and analogs can stimulate the growth of colonic mucosa. Litvak and colleagues [33] demonstrated the trophic effect of GLP-2 on the colonic mucosa of athymic nude mice. Drucker et al. [39] demonstrated an increase in colonic growth using dipeptidyl peptidase IV-resistant GLP-2 analog, human [Gly^2]-GLP-2, in 6-week-old female mice. A significant increase in large bowel mass was detected in mice treated with this analog for 10 days. Furthermore, the combination of this agent with either IGF-1 or an IGF-1 analog produced a greater increase in large bowel mass than in mice treated with [Gly^2]-GLP-2 alone. Administration of GLP-2 increased colonic weight in Wistar rats with atrophic colonic mucosa induced by TPN administration and reduced colonic mucosal injury in a dextran sulfate-induced colitis model [32].

Other intestinal hormones may play a contributory role in colonic proliferation; however, the effects are relatively minimal. For example, BBS administered three times a day for 7 days stimulated rat colonic mucosal growth [40]; moreover, administration of BBS orally during the neonatal period stimulated colonic growth [17]. Investigators in our laboratory have shown that, in rats given an elemental diet, the proliferative effect of BBS was confined to the proximal colon [41]. Colonic proliferation is likewise noted with NT administration; however, the effects of NT on the colon are much less pronounced than in the small bowel [42]. NT-induced colonic proliferation appears to be dependent upon age, with hyperplasia noted in the colon of young rats given NT, whereas NT significantly increased hypertrophy in aged rats. Similarly, PYY has been shown to have a modest effect on growth of the colonic mucosa [29].

GI HORMONE RECEPTORS AND SIGNAL TRANSDUCTION PATHWAYS

GI hormone-stimulated signal transduction occurs with the binding of hormones to their cognate cell surface receptors, which are G-protein-coupled receptors (GPCRs) [43]. These receptors have the typical structural features of G-protein-binding seven-transmembrane receptors which can regulate a number of physiological processes, including proliferation, growth, and development. It was originally thought that in order for GPCR signaling to occur, specific interactions between the GI hormone and the receptor were necessary to produce conformational changes in the receptor and stimulate intercellular signal transduction pathways. However, recent studies suggest a more complex regulation of the GPCRs through (1) dimerization with themselves and other receptors, (2) activation of differing G proteins, (3) internalization and desensitization, and (4) the ability to change in conformation and interactions with empty or inactive receptors [44].

The seven-transmembrane-spanning α-helical domains function as ligand-regulated guanine nucleotide exchange factors for the intercellular heterotrimeric G proteins [43]. Heterotrimeric G proteins are composed of the products of three gene families encoding α, β, and γ subunits. The agonist-activated GPCR catalyzes the exchange of GTP for GDP bound to the Gα subunit, as well as the dissociation of GTP-Gα from its cognate Gβγ dimer. The activated GTP-Gα and Gβγ subunits, in turn, regulate the activity of various intercellular effector proteins, such as phospholipases, adenyl cyclases, protein kinases, membrane ion channels, and members of the Ras family of GTP-binding proteins. In addition, based on structural similarities, the 20 identified Gα subunits have been divided into four subfamilies: (1) the cholera toxin-sensitive (α) subunits that stimulate adenyl cyclase and increase cyclic AMP levels; (2) the pertussin toxin-sensitive ($\alpha_{i/o}$) subunits that inhibit adenyl cyclase activity; (3) the pertussin toxin-sensitive ($\alpha_{q/11/14}$) subunits which stimulate membrane phospholipases; and (4) the ($\alpha_{12/13}$) subfamily that links GPCR to the Ras-related GTP binding protein, Rho [43]. Additionally, 12 Gγ and 6 Gβ subunits have been identified. These βγ dimers have been linked to the signaling molecules, phosphatidylinositol-3 kinase (PI3K), and select forms of adenyl cyclase and receptor kinases.

Among the multiple intercellular signaling pathways that mediate the proliferative effects of GPCRs, a family of related serine-threonine kinases, collectively known as the mitogen-activated protein kinases, or MAPKs, appear to play a central role [45, 46]. Hormones act as ligands to eventually activate p42 and p44 ERKs, which occurs through the involvement of a complex interplay of several known non-receptor kinases and receptor kinases. The ability of tyrosine kinase inhibitors to reduce the activation of MAPK by GPCR and the rapid tyrosine phosphorylation of Shc (*src* homology and collagen) following GPCR stimulation with the consequent formation of Shc-Grb2 (growth factor receptor bound 2) complexes provides evidence that tyrosine kinases link GPCRs to the Ras–MAPK pathway [47, 48]. Additionally, GPCRs link to the Jun-N terminal kinase (JNK), p38, MAPK, and the big mitogen-activated kinase-1 (BMK-1) or ERK5 pathways [43].

The molecular mechanisms though which GPCRs transduce signals are complex, and likely involve multiple signaling pathways. In addition, the signaling pathways are likely cell-specific, which may explain the diverse physiologic functions controlled by gut hormones, ranging from regulation of secretion, motility, and in some instances growth, depending on the target tissue.

SIGNALING PATHWAYS MEDIATING THE EFFECTS OF INTESTINAL PEPTIDES

Figure 12.1 summarizes how signaling pathways mediate the effects of intestinal peptides. Once a trophic GI peptide binds its seven-transmembrane GPCR, signal transduction pathways are activated which ultimately can lead to cell proliferation depending upon cell type [43, 49–52]. A number of pathways and proteins have been identified that are stimulated by the trophic gut peptides. For the most part, these pathways have been identified using neoplastic cells that possess the receptor for the trophic gut hormone.

Pathways Involving Phospholipase C, Phosphatidylinositol Activation, Calcium Mobilization, and Protein Kinase C

An early event associated with binding of trophic peptides to its receptor is activation of the phospholipase C (PLC) signal transduction pathway. For example, gastrin stimulates PLC in a number of cell types, including gastric parietal cells and colonic epithelial cells, NIH-3T3 fibroblasts that express the gastrin receptor (CCK-B/gastrin), and various neoplastic cell lines (reviewed in [53, 54]). The activation of PLC, which may involve coupling of the CCK-B/gastrin receptor with selected members of the G protein superfamily, induces the breakdown of membrane phosphatidylinositol 4,5-bisphosphate (PIP_2), leading to the formation of two second messengers, IP_3 and 1,2-diacylglycerol (DAG). IP_3 binds to its intracellular receptor and triggers the release of calcium from internal stores. Gastrin stimulates inositol phosphate production in a variety of cell types, leading to IP_3 and DAG formation, protein kinase C (PKC) activation, and intracellular calcium mobilization [53, 54]. These effects can be blocked by CCK-B/gastrin receptor antagonists. In addition to gastrin, other peptides, such as NT, have been shown to stimulate the PKC pathway, IP_3 turnover, and calcium mobilization in a number of cell types, including the colon cancer cell lines HT29 and KM20, as well as the pancreatic cancer cell line MIA PaCa-2 [55–58].

The trophic GI hormones also activate downstream PKC isoenzymes. Gastrin activates classic calcium- and phospholipid-dependent PKCs, as demonstrated by translocation of the cytosolic activity of PKC to the membrane compartment of rat colonic epithelial cells [59]. The isoforms, PKCα and -β, are responsive to gastrin treatment, and mobilize from

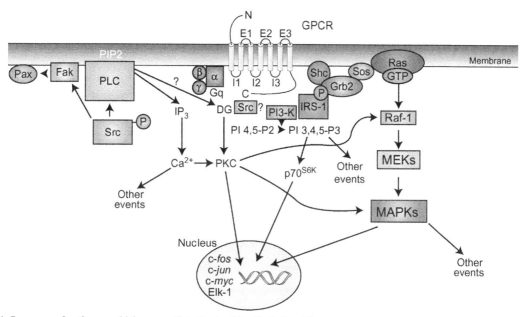

FIGURE 12.1 Summary of pathways which can mediate the trophic effect of gastrin.
Although these pathways have been best described for gastrin, other gut peptides can interact with their specific G-protein-coupled receptor (GPCR) to stimulate similar pathways in receptor-positive cells. See text for details and abbreviations. Adapted from [53] (Yassin, 1999).

the cytosol with treatment [60]. In addition, recent results demonstrate that gastrin induces the novel protein kinase D (PKD, also known as PKCμ), which has distinct structural and enzymological properties from the PKCs [61], in Rat-1 cells transfected with the human CCK-B/gastrin receptor [62]. Similarly, Guha and colleagues [63] recently reported that NT induced a rapid activation of PKD, which was linked to the mitogenic effect of NT in pancreatic cancer cells.

Tyrosine Kinases, Tyrosine Phosphorylation of Focal Adhesion Kinase, Paxillin, and CRK-Associated Substrate (CAS)

Through pairing with non-receptor tyrosine kinases, GPCRs utilize the tyrosine kinase pathway to stimulate cell growth [64]. Gastrin stimulates tyrosine kinase activity and tyrosine phosphorylation of membrane proteins in rat colonic mucosal cells, and membrane-associated protein tyrosine kinase and tyrosine phosphorylation of endogenous proteins in the IEC-6 intestinal cell line [65–67]. In addition, gastrin results in tyrosine phosphorylation of 62- and 54-kDa Src-like proteins in IEC-6 cells and pp60[c-src] kinase in rat colonic epithelial cells, which leads to tyrosine phosphorylation and activation of PLCγ1 [67, 68].

Growth factor receptors with intrinsic kinase activity and those that signal through G proteins can promote tyrosine phosphorylation of the adaptor protein Shc, which link activated growth factor receptors to the Ras signaling pathway, and its subsequent association with the Grb2–Sos (growth factor receptor binding protein-2/Son of sevenless) complex [69]. Gastrin promotes a rapid and transient

increase in tyrosine phosphorylation of the Shc proteins and association with Grb2-Sos, leading to activation of MAPKs in the human gastric cancer cell line, AGS-B, which expresses the human CCK-B/gastrin receptor [70]. The involvement of PI3K, as well as other SH2 anchoring proteins, in gastrin's mitogenic pathway has been reported. Gastrin induces tyrosine phosphorylation of the insulin receptor substrate 1 (IRS-1) and association with the 85-kDa subunit of PI3K [71]. Gastrin also stimulates the association of phosphorylated IRS-1 with the adapter Grb2, indicating that tyrosine phosphorylation of IRS-1 may be a mechanism whereby gastrin activates PI3K and other adapters [71]. Therefore, mobilization of the adapter proteins IRS-1, Shc, Grb2, and Sos could serve to link the CCK-B/gastrin receptor to the Ras-MAPK cascade, ultimately leading to transcriptional regulation [53].

Gastrin treatment results in the tyrosine phosphorylation of various proteins [67, 72]. For example, gastrin promotes tyrosine phosphorylation of focal adhesion kinase (p125[fak]), a tyrosine kinase that localizes at focal adhesions and is important for cell adhesion and transformation [72]. The focal adhesion proteins paxillin and CAS are potential downstream targets for p125[fak] and function as adapter proteins. Gastrin induces phosphorylation of p125[fak], CAS, and paxillin in Rat-1 and NIH-3T3 cells transfected with the CCK-B/gastrin receptor [72, 73].

MAPK Pathway

The MAPKs are a family of highly conserved serine-threonine kinases which are activated by a variety of extracellular signals and relay mitogenic signals to the nucleus [74]. To date, four groups of MAPKs have been identified in mammals. These include ERK1/2, JNK, p38, and ERK5. The activation of the MAPK pathway by gastrin has been demonstrated in a variety of cell types, both containing the endogenous CCK-B/gastrin receptor and cells stably transfected with this receptor [69, 70, 72]. Gastrin also stimulates the serine-threonine kinase Raf-1, the cellular homolog of the Raf oncogene and an upstream modulator of ERK1/2, JNK, and p38 [70, 72]. The mechanism for the effect of these hormones on MAPK activation is dependent upon cell type. For example, in the AR4-2J pancreatic cell line, gastrin-activated ERK was attenuated by treatment with agents that interfere with calcium mobilization or PKC activation [75]. In other cells, the activation of MAPKs by gastrin does not involve PKC. In Chinese hamster ovary (CHO) cells transfected with the CCK-B/gastrin receptor, gastrin induces ERK activation partly through the Src and PI3K pathways and partly through PKC [76]. Conversely, gastrin-stimulated Raf-1 and ERK activation in Rat-1 fibroblasts stably transfected with the CCK-B/gastrin receptor are independent of PKC [72]. Likewise, PYY stimulates ERK1/2, JNK, and p38 through binding of the Y1 receptor [77]. The stimulation of these three MAPKs can occur via multiple and diverse pathways, but in the case of PYY it has been shown that PKC plays a major role in the signaling pathway between the Y1 receptor and MAPKs, acting between the EGFR and MAPKs. In CHO cells transfected with the human Y1 receptor, both PKC and Ras are needed for the activation of the MAPK pathway [78]. Specifically, only PKCε, an isoform that has been specifically linked to mitogenic effects in gut epithelium [79], was activated by PYY in the IEC-6 intestinal cell line. NT has been shown to stimulate ERK and JNK activity in various neoplastic cell types containing endogenous NT receptors, ultimately leading to transcription factor activation [55, 56]. Therefore, activation of MAPKs appears to be an important mechanism for the mitogenic effects of intestinal hormones; this activation can occur by a variety of signal transduction mechanisms, and depends upon cell context as to which pathway is active in which cell type.

Downstream Transcription Factors

Ultimately, stimulation of various signaling pathways, such as PKC, the MAPKs, or ribosomal S6 kinase (p70^{S6K}), can lead to activation of downstream transcription factors. Gastrin stimulates the expression of early response genes, including c-*fos* and c-*jun* in AR4-2J cells, and c-*fos* and c-*myc* in NIH-3T3 cells transfected with the CCK-B/gastrin receptor [73, 80]. Gastrin-induced ERK-mediated phosphorylation and activation of these transcription factors was prevented by pharmacologic inhibition of PKC [81], suggesting an important role for the PKCs and ERKs in the activation of the AP-1 transcription factors. In addition, NT and BBS have been shown to stimulate expression of c-*jun* and c-*fos* in cell types possessing endogenous receptors for these peptides [55, 56, 82]. It is likely that the activation of these, as well as other transcription factors such as Elk-1, ultimately play a major role in the mitogenic response of these hormones.

CONCLUSIONS

The growth of GI mucosa is modulated by multiple factors, including intraluminal nutrients and the local release of growth factors and various GI hormones. These hormones bind to their specific receptors, stimulating many signal transduction pathways which ultimately lead to the mitogenic effects of these gut peptides. In this regard, the GI hormones have been suggested as potential therapeutic agents in disease states in the non-neoplastic GI tract related to gut disuse or atrophy, mucosal ulcers, or inflammatory conditions. For example, the trophic peptides, NT, BBS/GRP, and GLP-2, can augment or maintain GI mucosal growth during periods of gut disuse or atrophy [21, 22, 27, 34]. In addition, these peptides enhance adaptive hyperplasia associated with massive intestinal resection [24, 25]. In a limited clinical trial, administration of GLP-2 improved intestinal energy absorption, decreased energy excretion, increased body weight and lean body mass, and enhanced urinary creatinine excretion in patients with short bowel syndrome [83]. In addition, GI hormones may play a role in preventing the severe sequelae of chemotherapeutic agents on the intestinal mucosa. Both BBS and GLP-2 have been shown to prevent the severe mucosal inflammation associated with various chemotherapeutic agents [28, 35, 36]. In the future, it will be important to further define the signaling pathways regulated by the trophic intestinal peptides so that more effective agents can be developed which can take advantage of the gut-specific effects of these hormones on the proliferation and maintenance of the GI mucosa.

ACKNOWLEDGEMENTS

Thanks to Karen Martin for assistance with preparation of the manuscript.

REFERENCES

1. Babyatsky MW, Podolsky DK. Growth and development of the gastrointestinal tract. In: Yamada T, editor. *Textbook of gastroenterology*. 3rd edn, Vol. 1 Philadelphia, PA: Lippincott Williams & Wilkins Publishers; 1999. p. 547–84.

2. Slorach EM, Campbell FC, Dorin JR. A mouse model of intestinal stem cell function and regeneration. *J Cell Sci* 1999;**112**(18):3029–38.

3. Potten CS, Owen G, Booth D. Intestinal stem cells protect their genome by selective segregation of template DNA strands. *J Cell Sci* 2002;**115**:2381–8.

4. Stappenbeck TS, Wong MH, Saam JR, Mysorekar IU, Gordon JI. Notes from some crypt watchers: regulation of renewal in the mouse intestinal epithelium. *Curr Opin Cell Biol* 1998;**10**:702–9.

5. Dignass AU, Sturm A. Peptide growth factors in the intestine. *Eur J Gastroenterol Hepatol* 2001;**13**:763–70.

6. Aykan NF. Message-adjusted network (MAN) hypothesis in gastro-entero-pancreatic (GEP) endocrine system. *Med Hypotheses* 2007;**69**(3):571–4.

7. Walsh JH. Gastrointestinal hormones. In: Johnson LR, Alpers DH, Christensen J, Jacobson ED, Walsh JH, editors. *Physiology of the gastrointestinal tract.* 3rd edn New York, NY: Raven Press; 1994. p. 1–128.

8. Johnson LR, Aures D, Hakanson R. Effect of gastrin on the *in vivo* incorporation of 14C-leucine into protein of the digestive tract. *Proc Soc Exp Biol Med* 1969;**132**:996–8.

9. Crean GP, Marshall MW, Rumsey RD. Parietal cell hyperplasia induced by the administration of pentagastrin (ICI 50,123) to rats. *Gastroenterology* 1969;**57**:147–55.

10. Johnson LR. The trophic action of gastrointestinal hormones. *Gastroenterology* 1976;**70**:278–88.

11. Friis-Hansen L. Lessons from the gastrin knockout mice. *Regul Pept* 2007;**139**:5–22.

12. Wang TC, Koh TJ, Varro A, Cahill RJ, Dangler CA, Fox JG, Dockray GJ. Processing and proliferative effects of human progastrin in transgenic mice. *J Clin Invest* 1996;**98**:1918–29.

13. Fukushima Y, Matsui T, Saitoh T, Ichinose M, Tateishi K, Shindo T, Fujishiro M, Sakoda H, Shojima N, Kushiyama A, Fukuda S, Anai M, Ono H, Oka M, Shimizu Y, Kurihara H, Nagai R, Ishikawa T, Asano T, Omata M. Unique roles of G protein-coupled histamine H2 and gastrin receptors in growth and differentiation of gastric mucosa. *Eur J Pharmacol* 2004;**502**:243–52.

14. Hinkle KL, Bane GC, Jazayeri A, Samuelson LC. Enhanced calcium signaling and acid secretion in parietal cells isolated from gastrin-deficient mice. *Am J Physiol Gastrointest Liver Physiol* 2003;**284**:G145–53.

15. Guo Jr YS, Townsend CM. Gastrointestinal hormones and gastrointestinal cancer growth. In: Greeley Jr GH, editor. *Gastrointestinal endocrinology.* Totowa, NJ: Humana Press Inc; 1999. p. 189–214.

16. Puccio F, Lehy T. Bombesin ingestion stimulates epithelial digestive cell proliferation in suckling rats. *Am J Physiol* 1989;**256**:G328–34.

17. Lehy T, Puccio F, Chariot J, Labeille D. Stimulating effect of bombesin on the growth of gastrointestinal tract and pancreas in suckling rats. *Gastroenterology* 1986;**90**:1942–9.

18. Dembinski A, Konturek PK, Konturek SJ. Role of gastrin and cholecystokinin in the growth-promoting action of bombesin on the gastroduodenal mucosa and the pancreas. *Regul Pept* 1990;**27**:343–54.

19. Dembinski A, Warzecha Z, Konturek SJ, Banas M, Cai RZ, Schally AV. The effect of antagonist of receptors for gastrin, cholecystokinin and bombesin on growth of gastroduodenal mucosa and pancreas. *J Physiol Pharmacol* 1991;**42**:263–77.

20. Wood JG, Hoang HD, Bussjaeger LJ, Solomon TE. Neurotensin stimulates growth of small intestine in rats. *Am J Physiol* 1988;**255**:G813–17.

21. Evers Jr. BM, Izukura M, Townsend CM, Uchida T, Thompson JC Neurotensin prevents intestinal mucosal hypoplasia in rats fed an elemental diet. *Dig Dis Sci* 1992;**37**:426–31.

22. Chung Jr DH, Evers BM, Shimoda I, Townsend CM, Rajaraman S, Thompson JC. Effect of neurotensin on gut mucosal growth in rats with jejunal and ileal Thiry-Vella fistulas. *Gastroenterology* 1992;**103**:1254–9.

23. Assimakopoulos SF, Alexandris IH, Scopa CD, Mylonas PG, Thomopoulos KC, Georgiou CD, Nikolopoulou VN, Vagianos CE. Effect of bombesin and neurotensin on gut barrier function in partially hepatectomized rats. *World J Gastroenterol* 2005;**11**:6757–64.

24. Izukura Jr. M, Evers BM, Parekh D, Yoshinaga K, Uchida T, Townsend CM, Thompson JC Neurotensin augments intestinal regeneration after small bowel resection in rats. *Ann Surg* 1992;**215**:520–7.

25. Ryan CK, Miller JH, Seydel AS, de Mesy Jensen K, Sax HC. Epidermal growth factor and neurotensin induce microvillus hypertrophy following massive enterectomy. *J Gastrointest Surg* 1997;**1**:467–73.

26. Evers Jr BM, Izukura M, Townsend CM, Uchida T, Thompson JC. Differential effects of gut hormones on pancreatic and intestinal growth during administration of an elemental diet. *Ann Surg* 1990;**211**:630–8.

27. Chu Jr. KU, Higashide S, Evers BM, Ishizuka J, Townsend CM, Thompson JC Bombesin stimulates mucosal growth in jejunal and ileal Thiry-Vella fistulas. *Ann Surg* 1995;**221**:602–11.

28. Chu Jr. KU, Higashide S, Evers BM, Rajaraman S, Ishizuka J, Townsend CM, Thompson JC Bombesin improves survival from methotrexate-induced enterocolitis. *Ann Surg* 1994;**220**:570–7.

29. Gomez Jr. G, Zhang T, Rajaraman S, Thakore KN, Yanaihara N, Townsend CM, Thompson JC, Greeley GH Intestinal peptide YY: ontogeny of gene expression in rat bowel and trophic actions on rat and mouse bowel. *Am J Physiol* 1995;**268**:G71–81.

30. Chance WT, Zhang X, Balasubramaniam A, Fischer JE. Preservation of intestine protein by peptide YY during total parenteral nutrition. *Life Sci* 1996;**58**:1785–94.

31. Drucker DJ, Erlich P, Asa SL, Brubaker PL. Induction of intestinal epithelial proliferation by glucagon-like peptide 2. *Proc Natl Acad Sci USA* 1996;**93**:7911–16.

32. Ghatei MA, Goodlad RA, Taheri S, Mandir N, Brynes AE, Jordinson M, Bloom SR. Proglucagon-derived peptides in intestinal epithelial proliferation: glucagon-like peptide-2 is a major mediator of intestinal epithelial proliferation in rats. *Dig Dis Sci* 2001;**46**:1255–63.

33. Litvak Jr DA, Hellmich MR, Evers BM, Banker NA, Townsend CM. Glucagon-like peptide 2 is a potent growth factor for small intestine and colon. *J Gastrointest Surg* 1998;**2**:146–50.

34. Boushey RP, Yusta B, Drucker DJ. Glucagon-like peptide 2 decreases mortality and reduces the severity of indomethacin-induced murine enteritis. *Am J Physiol* 1999;**277**:E937–47.

35. Boushey RP, Yusta B, Drucker DJ. Glucagon-like peptide (GLP)-2 reduces chemotherapy-associated mortality and enhances cell survival in cells expressing a transfected GLP-2 receptor. *Cancer Res* 2001;**61**:687–93.

36. Tavakkolizadeh A, Shen R, Abraham P, Kormi N, Seifert P, Edelman ER, Jacobs DO, Zinner MJ, Ashley SW, Whang EE. Glucagon-like peptide 2: a new treatment for chemotherapy-induced enteritis. *J Surg Res* 2000;**91**:77–82.

37. Johnson LR. New aspects of the trophic action of gastrointestinal hormones. *Gastroenterology* 1977;**72**:788–92.

38. Koh TJ, Dockray GJ, Varro A, Cahill RJ, Dangler CA, Fox JG, Wang TC. Overexpression of glycine-extended gastrin in transgenic mice results in increased colonic proliferation. *J Clin Invest* 1999;**103**:1119–26.

39. Drucker DJ, DeForest L, Brubaker PL. Intestinal response to growth factors administered alone or in combination with human [Gly2]glucagon-like peptide 2. *Am J Physiol* 1997;**273**:G1252–62.

40. Johnson LR, Guthrie PD. Regulation of antral gastrin content. *Am J Physiol* 1983;**245**:G725–9.

41. Chu Jr. KU, Evers BM, Ishizuka J, Townsend CM, Thompson JC Role of bombesin on gut mucosal growth. *Ann Surg* 1995;**222**:94–100.

42. Evers Jr. BM, Izukura M, Chung DH, Parekh D, Yoshinaga K, Greeley GH, Uchida Jr T, Townsend CM, Thompson JC Neurotensin stimulates growth of colonic mucosa in young and aged rats. *Gastroenterology* 1992;**103**:86–91.

43. Marinissen MJ, Gutkind JS. G-protein-coupled receptors and signaling networks: emerging paradigms. *Trends Pharmacol Sci* 2001;**22**:368–76.

44. Wilkie TM. Treasures throughout the life-cycle of G-protein-coupled receptors. *Trends Pharmacol Sci* 2001;**22**:396–7.

45. Goldsmith ZG, Dhanasekaran DN. G protein regulation of MAPK networks. *Oncogene* 2007;**26**:3122–42.

46. Gutkind JS. The pathways connecting G protein-coupled receptors to the nucleus through divergent mitogen-activated protein kinase cascades. *J Biol Chem* 1998;**273**:1839–42.

47. Hordijk PL, Verlaan I, van Corven EJ, Moolenaar WH. Protein tyrosine phosphorylation induced by lysophosphatidic acid in Rat-1 fibroblasts. Evidence that phosphorylation of map kinase is mediated by the Gi-p21ras pathway. *J Biol Chem* 1994;**269**:645–51.

48. van Biesen T, Hawes BE, Luttrell DK, Krueger KM, Touhara K, Porfiri E, Sakaue M, Luttrell LM, Lefkowitz RJ. Receptor-tyrosine-kinase- and G beta gamma-mediated MAP kinase activation by a common signalling pathway. *Nature* 1995;**376**:781–4.

49. Ji TH, Grossmann M, Ji I. G protein-coupled receptors. I. Diversity of receptor-ligand interactions. *J Biol Chem* 1998;**273**:17,299–302.

50. Wank SA. G protein-coupled receptors in gastrointestinal physiology. I. CCK receptors: an exemplary family. *Am J Physiol* 1998;**274**:G607–13.

51. Rozengurt E. Early signals in the mitogenic response. *Science* 1986;**234**:161–6.

52. Rozengurt E. Signal transduction pathways in the mitogenic response to G protein-coupled neuropeptide receptor agonists. *J Cell Physiol* 1998;**177**:507–17.

53. Yassin RR. Signaling pathways mediating gastrin's growth-promoting effects. *Peptides* 1999;**20**:885–98.

54. Rozengurt E, Walsh JH. Gastrin, CCK, signaling, and cancer. *Annu Rev Physiol* 2001;**63**:49–76.

55. Ehlers 2nd RA, Bonnor RM, Wang X, Hellmich MR, Evers BM. Signal transduction mechanisms in neurotensin-mediated cellular regulation. *Surgery* 1998;**124**:239–47.

56. Ehlers RA, Zhang Y, Hellmich MR, Evers BM. Neurotensin-mediated activation of MAPK pathways and AP-1 binding in the human pancreatic cancer cell line, MIA PaCa-2. *Biochem Biophys Res Commun* 2000;**269**:704–8.

57. Slogoff MI, Evers BM. Neurotensin. In: Henry HL, Norman AW, editors. *Encyclopedia of hormones*. San Diego: Academic Press; 2003. p. 45–53.

58. Vincent JP, Mazella J, Kitabgi P. Neurotensin and neurotensin receptors. *Trends Pharmacol Sci* 1999;**20**:302–9.

59. Yassin RR, Clearfield HR, Little KM. Gastrin's trophic effect in the colon: identification of a signaling pathway mediated by protein kinase C. *Peptides* 1993;**14**:1119–24.

60. Yassin RR, Little KM. Early signalling mechanism in colonic epithelial cell response to gastrin. *Biochem J* 1995;**311**(3):945–50.

61. Auer A, von Blume J, Sturany S, von Wichert G, Van Lint J, Vandenheede J, Adler G, Seufferlein T. Role of the regulatory domain of protein kinase D2 in phorbol ester binding, catalytic activity, and nucleocytoplasmic shuttling. *Mol Biol Cell* 2005;**16**:4375–85.

62. Chiu TT, Duque J, Rozengurt E. Protein kinase D (PKD) activation is a novel early event in CCKB/gastrin receptor signaling. *Gastroenterology* 1999;**116**:A597.

63. Guha S, Rey O, Rozengurt E. Neurotensin induces protein kinase C-dependent protein kinase D activation and DNA synthesis in human pancreatic carcinoma cell line PANC-1. *Cancer Res* 2002;**62**:1632–40.

64. Malarkey K, Belham CM, Paul A, Graham A, McLees A, Scott PH, Plevin R. The regulation of tyrosine kinase signalling pathways by growth factor and G-protein-coupled receptors. *Biochem J* 1995;**309**(2):361–75.

65. Malecka-Panas E, Fligiel SE, Jaszewski R, Majumdar AP. Differential responsiveness of proximal and distal colonic mucosa to gastrin. *Peptides* 1997;**18**:559–65.

66. Malecka-Panas E, Tureaud J, Majumdar AP. Gastrin activates tyrosine kinase and phospholipase C in isolated rat colonocytes. *Acta Biochim Pol* 1996;**43**:539–46.

67. Singh P, Narayan S, Adiga RB. Phosphorylation of pp62 and pp54 src-like proteins in a rat intestinal cell line in response to gastrin. *Am J Physiol* 1994;**267**:G235–44.

68. Yassin RR, Abrams JT. Gastrin induces IP3 formation through phospholipase C gamma 1 and pp60^{c-src} kinase. *Peptides* 1998;**19**:47–55.

69. Seva C, Kowalski-Chauvel A, Blanchet JS, Vaysse N, Pradayrol L. Gastrin induces tyrosine phosphorylation of Shc proteins and their association with the Grb2/Sos complex. *FEBS Letts* 1996;**378**:74–8.

70. Hocker M, Henihan RJ, Rosewicz S, Riecken EO, Zhang Z, Koh TJ, Wang TC. Gastrin and phorbol 12-myristate 13-acetate regulate the human histidine decarboxylase promoter through Raf-dependent activation of extracellular signal-regulated kinase-related signaling pathways in gastric cancer cells. *J Biol Chem* 1997;**272**:27,015–024.

71. Kowalski-Chauvel A, Pradayrol L, Vaysse N, Seva C. Gastrin stimulates tyrosine phosphorylation of insulin receptor substrate 1 and its association with Grb2 and the phosphatidylinositol 3-kinase. *J Biol Chem* 1996;**271**:26,356–361.

72. Seufferlein T, Withers DJ, Broad S, Herget T, Walsh JH, Rozengurt E. The human CCKB/gastrin receptor transfected into rat1 fibroblasts mediates activation of MAP kinase, p74raf-1 kinase, and mitogenesis. *Cell Growth Differ* 1995;**6**:383–93.

73. Taniguchi T, Matsui T, Ito M, Murayama T, Tsukamoto T, Katakami Y, Chiba T, Chihara K. Cholecystokinin-B/gastrin receptor signaling pathway involves tyrosine phosphorylations of p125FAK and p42MAP. *Oncogene* 1994;**9**:861–7.

74. Widmann C, Gibson S, Jarpe MB, Johnson GL. Mitogen-activated protein kinase: conservation of a three-kinase module from yeast to human. *Physiol Rev* 1999;**79**:143–80.

75. Daulhac L, Kowalski-Chauvel A, Pradayrol L, Vaysse N, Seva C. Ca^{2+} and protein kinase C-dependent mechanisms involved in gastrin-induced Shc/Grb2 complex formation and P44-mitogen-activated protein kinase activation. *Biochem J* 1997;**325**(2):383–9.

76. Daulhac L, Kowalski-Chauvel A, Pradayrol L, Vaysse N, Seva C. Src-family tyrosine kinases in activation of ERK-1 and p85/p110-phosphatidylinositol 3-kinase by G/CCKB receptors. *J Biol Chem* 1999;**274**:20,657–20,663.

77. Mannon PJ. Peptide YY as a growth factor for intestinal epithelium. *Peptides* 2002;**23**:383–8.

78. Mannon PJ, Raymond JR. The neuropeptide Y/peptide YY Y1 receptor is coupled to MAP kinase via PKC and Ras in CHO cells. *Biochem Biophys Res Commun* 1998;**246**:91–4.

79. Perletti Jr. GP, Folini M, Lin HC, Mischak H, Piccinini F, Tashjian AH Overexpression of protein kinase C epsilon is oncogenic in rat colonic epithelial cells. *Oncogene* 1996;**12**:847–54.

80. Todisco A, Takeuchi Y, Seva C, Dickinson CJ, Yamada T. Gastrin and glycine-extended progastrin processing intermediates induce different programs of early gene activation. *J Biol Chem* 1995;**270**:28,337–28,341.

81. Stepan VM, Tatewaki M, Matsushima M, Dickinson CJ, del Valle J, Todisco A. Gastrin induces c-*fos* gene transcription via multiple signaling pathways. *Am J Physiol* 1999;**276**:G415–24.

82. Kim Jr. HJ, Evers BM, Guo Y, Banker NA, Hellmich MR, Townsend CM Bombesin-mediated AP-1 activation in a human gastric cancer (SIIA). *Surgery* 1996;**120**:130–7.

83. Jeppesen PB, Hartmann B, Thulesen J, Graff J, Lohmann J, Hansen BS, Tofteng F, Poulsen SS, Madsen JL, Holst JJ, Mortensen PB. Glucagon-like peptide 2 improves nutrient absorption and nutritional status in short-bowel patients with no colon. *Gastroenterology* 2001;**120**:806–15.

Chapter 13

Cell–Cell and Cell–Matrix Interactions in Bone

Lynda F. Bonewald

Bone Biology Program, University of Missouri-Kansas City, Kansas City, Missouri

INTRODUCTION TO BONE AND BONE DISEASE

Far from being the static, hard skeleton hanging in the anatomy classroom, the skeleton within the body is dynamic and constantly responding to internal and external forces. The internal forces include cytokines, growth factors, and hormones, and the external force is response to muscle and to strain placed on the skeleton. In fact, the adult skeleton undergoes greater remodeling than other organs in the body. Estrogens, androgens, parathyroid hormone (PTH), 1,25-dihydroxy vitamin D_3 (1,25 D_3), and other hormones have been shown to play important roles in the skeleton. The three major bone cell types – osteoclasts, osteoblasts, and osteocytes –are in constant communication with each other and with cells of the immune and hemopoietic systems. Bone cells are in constant communication not only with cells of other systems but also with the extracellular matrix (ECM), which is composed of osteoid, non-mineralized bone tissue and the mineralized bone matrix. Although previously viewed as mainly a support structure for bone cells, it is now clear than the bone ECM controls and directs bone cell function.

DISEASES OF BONE

Manifestation of bone disease is usually later or slower than manifestation of disease in other organs. For example, bone cancer is usually discovered after manifestation in other tissues such as breast or lung. Bone cancer such as osteosarcoma usually does not present until fracture or pain occurs, often after the cancer has become fully entrenched and difficult, if not impossible, to cure. Another example is osteoporosis, in which bone loss can occur over decades before being identified and treated. As to bone malformations, at present the only hope of treatment is surgery. Therefore, a greater understanding of normal bone function and pathology is required for the design of preventive therapy and for treatment of disease.

Osteoporosis

Osteoporosis has become a major medical problem as the world population ages. Bone strength is reduced in the postmenopausal female, and in both sexes with aging. Bone strength is a function of size, connectivity of trabecular structures, level of remodeling, and the intrinsic strength of the bone itself. Osteoporosis is defined as "the condition of generalized skeletal fragility in which bone strength is sufficiently weak that fractures occur with minimal trauma, often no more than is applied by routine daily activity" [1]. "Primary" osteoporosis is a disorder of postmenopausal women and of older men and women. "Secondary" osteoporosis occurs due to clinical disorders such as endocrinopathies, genetic diseases, or to drugs as in glucocorticoid-induced bone loss. As the US population ages, osteoporosis is taking a greater and greater toll in terms of both suffering and economic cost. Each year, osteoporosis is the underlying basis for 1.5 million fractures. These cause not only pain and morbidity, but also diminish the quality of life for these individuals, as they lose their independence. Hip fracture results in up to 24 percent mortality, 25 percent of hip fracture patients require long-term care, and only a third regain their pre-fracture level of independence [2].

Treatment for osteoporosis includes hormone therapy and the use of bone resorption inhibitors such as the bisphosphonates, and, in addition, stimulators of bone formation such as modified forms of PTH. The use of hormone (i.e., sex steroid) replacement therapy has become unclear and controversial since the Women's Health Initiative Study [3, 4] showed an associated increase in breast cancer. It has been argued that this study mainly evaluated elderly and asymptomatic woman, not symptomatic peri- and early postmenopausal women in whom hormone replacement therapy may be beneficial [5]. This continues to be an area of intense discussion and investigation. Bisphosphonates are used not only for the treatment of osteoporosis but also to prevent bone loss due to glucocorticoid use, and Paget's disease. These compounds are useful not only to treat bone loss, but also to treat and reduce bone metastasis in metastatic breast cancer and multiple myeloma, [6]. However, recently, high-dose bisphosphonate treatment has been associated with osteonecrosis of the jaw [7]. It is not known if other parts of the skeleton are affected. The newest agent developed to treat osteoporosis is anabolic, the amino terminal fragment of parathyroid hormone, PTH (1-34), called teriparatide. This promising treatment restores bone and relieves pain [8].

Skeletal Malignancies

It has become clear that tumor cells use disruption or enhancement of normal cell–cell and cell–ECM interactions to enhance their own growth and metastasis. A prime example is multiple myeloma, in which the tumor cells express an integrin complex, VLA4, that allows them to home to bone marrow [9, 10]. Myeloma is characterized by extensive bone destruction; therefore efforts are underway to prevent metastasis to bone through the use of agents that block myeloma–bone ECM interactions. Breast, lung, and prostate cancer preferentially metastasize to bone [11]. Factors that may play a role in osteoblastic bone metastasis include fibroblast growth factors, (FGFs), transforming growth factor beta (TGFβ), platelet-derived growth factor (PDGF), and, more recently, endothelin-1 [12]. Systemic syndromes can be associated with such cancers; these include leukocytosis and hypercalcemia. A well-studied factor clearly associated with hypercalcemia is parathyroid hormone-related peptide (PTHrp), which is normally produced by keratinocytes, uterus, placenta, and mammary tissue. This factor mimics PTH action by binding to the same receptor. Other factors implicated in osteolytic bone loss due to malignancy include interleukin-1, interleukin-6, tumor necrosis factor α, RANKL, and MIP-1α [13]. These factors will be discussed in more detail below.

Bone Malformations and Genetic Defects

As bone growth proceeds through the growth plate in long bones, defects or mutations in a number of factors essential in this process lead to chondrogenic dysplasias. Jansen's metaphyseal chondrodysplasia and Blomstrand's lethal chondrodysplasia are due to mutations in the PTH receptor [14]. The genetics and environmental insult responsible for other genetic disorders such as Paget's disease of bone, an autosomal dominant, are being identified. Complications of this disease include deformations of skull, face, and lower extremities, pain, degenerative arthritis, hearing loss, hypercalcemia, and hyperuricemia, due to enhanced osteoclast formation and activation [15]. Mutations in the sequestosome-1 gene occur in 30 percent of patients with familial Paget's [16]. Such mutations may make these individuals predisposed to the effects of *Paramyxioviridae* viruses on their osteoclasts [17]. Targeted expression of measles virus nucleocapsid to osteoclasts in mice results in a phenotype similar to Paget's disease [18]. The mechanism may be through induction of overexpression of TATA box-associated factor II-17, a potential co-activator of the vitamin D receptor [19].

Another serious genetic bone disease is osteogenesis imperfecta, a heritable disease of bone characterized by recurring bone fractures. It is caused by mutations affecting the structure of the collagen type I molecule, and is the most common single gene defect causing bone disease [20]. Mutations in type I collagen can cause moderate disease, or can be lethal during the perinatal period. Interestingly, bisphosphonates are being used to successfully treat this condition, but their mechanism of action is unknown [21].

BONE CELLS AND THEIR FUNCTIONS

Osteoclasts

The sole function of the osteoclast is to resorb bone. The mature osteoclast is described histologically as a multinucleated, tartrate-resistant acid phosphatase (TRAP)-positive cell. However, macrophage polykaryons can have these same characteristics, so the "gold standard" for identifying an osteoclast is the formation of resorption lacunae or "pits" on a mineralized surface. Other characteristics of the osteoclast include the expression of calcitonin receptors, enzymes such as cathepsin K and matrix metalloprotein-9 (MMP-9) that play a role in matrix degradation, and the vacuolar proton pump for the transport of protons to the resorption lacunae. For the osteoclast to resorb, it must form a "sealing zone" around the periphery of its attached area to concentrate its secreted proteases and protons into a limited area. Underneath the cell a ruffled border is formed, and in this region the pH is reduced to approximately 2–3, which enhances the degradation of mineralized matrix. [22].

Osteoclast precursors are derived from the same stem cell hematopoietic precursors that can become granulocytes and monocytes/macrophages. Cell lines such as RAW 267.4 and MOPC-5 are available that represent osteoclast precursors, as these cells can form TRAP-positive multinucleated cells that resorb bone [23]. It has been well known for the past 10–15 years in the bone field that osteoclast precursors require supporting cells for osteoclast formation. The importance of macrophage colony stimulating factor (M-CSF) as a supporter of proliferation of osteoclast precursors has been determined [24]. Critical factors and cell surface molecules involved in osteoclast formation have only recently been elucidated with the discovery of RANK ligand (RANKL) and osteoprotogerin (OPG) [25, 26]. The osteoclast precursor expresses a receptor known as RANK (receptor activator of NFκB) that signals through the NFκB pathway. The binding of the cell membrane bound ligand, RANKL, activates RANK receptor. However, a soluble factor, OPG, acting as a "decoy" receptor can bind to RANKL, preventing osteoclast formation. The expression of RANKL on the surface of supporting cells occurs when these cells are exposed to bone-resorbing cytokines, hormones, and factors such as interleukins-1, -6, -11, PTH, PTHrp, oncostatin M, leukemia inhibitory factor, prostaglandin E_2, or 1,25 D_3 [27]. These factors upregulate RANKL to a level capable of overcoming the effects of circulating OPG, thereby resulting in osteoclast formation. Efforts to generate osteoclasts without supporting cells have only recently been accomplished *in vitro* by using an artificial, soluble form of RANKL [28].

Osteoblasts

The formation of bone matrix on bone-forming surfaces has been well studied. The osteoblast is derived from a yet to be identified precursor stem cell of mesenchymal origin, and in cell culture behaves similarly to fibroblasts, except for its specialized ability to form mineralized matrix. The osteoblast undergoes three major phenotypically identifiable stages of differentiation, which Stein, Lian and co-workers have characterized as proliferation, matrix production, and maturation in which mineralization occurs [29]. During the proliferation phase there is high level expression of c-fos, histone H4 and, during matrix production, transforming growth factor-1β (TGFβ) and type I collagen. During the maturation phase these proteins decrease, and the expression of alkaline phosphatase, osteopontin, and Cbfa1 increases. During the mineralization phase these previous proteins decline in expression while proteins such as osteocalcin increase, with increases in mineralized bone formation. Osteoblast cells in each of these phases are often described as early pre-osteoblasts, proliferating osteoblasts, mature osteoblasts, and, finally, pre-osteocytes/osteocytes within the mineralized matrix.

Osteocytes

Osteocytes are terminally differentiated osteoblasts, making up the majority (over 90–95 percent) of all bone cells [30–32]. During osteocyte ontogeny, the matrix producing osteoblast becomes either a lining cell or a pre-osteocyte embedded in the newly-formed osteoid. These pre-osteocytes produce factors (such as osteocalcin) that locally inhibit mineralization, and form a lacuna around the main body of the osteocyte and canaliculi around the dendritic processes [33, 34]. A mature osteocyte is defined as a cell surrounded by mineralized bone, and is described as a stellate or star-shaped cell with a large number of slender, cytoplasmic processes radiating in all directions, but generally perpendicular to the bone surface. Osteocytes first attracted the attention of electron microscopists because of their extensive networks within the mineralized bone matrix that connect the embedded osteocytes to form an extensive network with cells on the surface of bone. Mature osteocytes are most likely coupled by GAP junctions, and appear to be linked to lining cells by the same connections [35] Recently it has been shown that osteocytes appear to also possess functional hemichannels, and that prostaglandins may be released through hemichannels in response to mechanical strain [36].

The potential functions of osteocytes include: to respond to mechanical strain and to send signals of bone formation or bone resorption to the bone surface, to modify their microenvironment, and to regulate both local and systemic mineral homeostasis. These functions are proposed to be accomplished through gap junctions, through the secretion of factors, and through the direct dendritic contact with cells on the bone surface (for reviews, see [30–32, 37]).

MECHANICAL STRAIN

Julius Wolff, in 1892, was the first to suggest that bone accommodates or responds to strain. To paraphrase Wolff's Law: the law of bone remodeling, alteration of internal and external architecture, occurs as a consequence of the stressing of bone. In general, athletes who put great stresses on their bones, such as wrestlers, and those who are chronic exercisers, such as tennis players, have higher bone mineral density and mass than matched, non-exercising controls. Astronauts subjected to long periods of weightlessness during space flight lose bone. The cells of bone that have the potential for sensing mechanical strain and translating these forces into biochemical signals include bone lining cells, osteoblasts, and osteocytes.

The skeleton adapts to mechanical usage. When the skeleton is not used, as in immobilization, bone is lost. During growth the skeleton is in "mild overload," which results in bone modeling and resulting new bone formation. When growth ceases, muscle strength is no longer in overload and bone strain is therefore reduced to the adapted level of strain. The estimated levels of micro-strain for each "window" or level of strain, have been determined by *in vivo* animal and human experiments. At less than 100 υE only resorption occurs, and this has been called the "disuse window." The "adapted window" is where remodeling (resorption followed by formation) occurs, between 100 and 1000 υE. This adapted window can be raised or lowered, depending on the hormonal environment. The "mild overload window" is where modeling (formation only) occurs between 1000 and 3000 υE, and this only occurs in growing animals. The "pathologic overload window" is where microdamage (resorption followed by formation) occurs at strains greater than 3000 υE. Microdamage can occur in race horses and in military recruits during basic training. Microdamage due to pathologic overload results in rapid resorption followed by formation at areas of microdamage. Fracture strain is approximately 25,000 υE [38].

It has been proposed that bone possesses a "mechanostat," a mechanism whereby bone can reset its response to particular levels of strain [39, 40]. Hormones have been proposed to lower the mechanostat; that is, addition of hormones such as estrogen, PTH, or 1,25,D_3 can lower the magnitude of strain necessary to induce a response [41–43]. For example, if bone normally responds to 2000 υE with an increase in modeling, in the presence of hormone the same response would occur to lower strain levels – say 1000 υE. However, it has been shown that the skeleton cannot respond optimally to mechanical strain without the presence of estrogen; therefore, estrogen is essential for the normal response of bone and not for just resetting the mechanostat [44]. Thus, for normal skeletal growth and maintenance, both biochemical signals and mechanical strain are essential. For the skeleton to optimally respond to mechanical strain, hormones must be present; conversely, bone cannot develop normally in the absence of strain.

The cell that senses mechanical strain and translates strain into biochemical signals is thought to be the osteocyte (see description above). It has been proposed that molecules travel in the bone fluid surrounding the osteocyte through a glycocalyx which attaches the dendritic processes to the caniluculi [45]. The glycocalyx acts as a sieve or "fishnet" to allow molecules below approximately 7 nm to pass [46]. Studies suggest that molecules as large as albumin can pass through the canaliculi, and that the bone fluid serves to provide nutrients to the osteocyte. Channels in osteocytes are proposed to open in response to mechanical strain induced by shear stress caused by moving bone fluid. Potential openings or channels to the extracellular bone fluid have been identified, such as calcium-, ion-, voltage-, and stretch-activated channels. Molecules proposed to be involved in osteocyte signaling include nitric oxide, ATP, prostaglandin, glutamate, and others (for reviews, see [30, 31]).

The cell processes of osteocytes are connected with each other and cells on the bone surface via gap junctions [35], thereby allowing direct cell-to-cell coupling. Recently it has been shown that connexins can compose and function in the form of un-apposed halves of gap junction channels, called hemichannels. These channels are localized at the cell surface, independent of physical contact with adjacent cells [47]. Recently, evidence of functional hemichannels formed by Cx43 has been reported in neural progenitors and neurons, astrocytes, heart, and especially, osteoblasts and osteocytes. The opening of hemichannels appears to provide a mechanism for ATP and NAD^+ release, which raises intracellular Ca^{2+} levels and promotes Ca^{2+} wave propagation in astrocytes, bone cells, epithelial cells, and outer retina. Hemichannels expressed in bone cells appear to function as essential transducers of the anti-apoptotic effects of bisphosphonates[48], and seem to allow the extracellular release of PGE_2 in osteocytes upon exposure to fluid flow shear stress [36]. Therefore, in osteocytes, gap junctions at the tip of dendrites appear to mediate a form of intracellular communication, and hemichannels along the dendrite appear to mediate a form of extracellular communication.

HORMONES RESPONSIBLE FOR BONE DEVELOPMENT, GROWTH AND MAINTENANCE

A number of hormones play a role in the maintenance of bone. These include estrogens, progesterone, aldosterone, androgens, vitamin A, and glucocorticoids. These hormones' actions are mediated by hormone-activated transcription factors belonging to the superfamily of ligand-dependent nuclear receptors. The superfamily is usually referred to as Nuclear Hormone Receptors, or simply Nuclear Receptors, a subset of which contains the ligand-dependent group (see the NURSA website at NIH). These nuclear receptors, upon binding to ligand, can form homodimers or heterodimers at specific DNA binding sites. They also interact with a wide array of other transcription factors, as well as general and specific co-regulatory proteins. The complexity of these interactions leads to intrinsic specificity of gene regulation, but the exact determinants of specificity are still under study.

Estrogen

Estrogen is clearly the major sex hormone affecting growth, remodeling, and homeostasis of the skeleton. Reduction in estrogen levels that occurs with the menopause or through ovariectomy can result in bone loss. However, estrogen also plays a role in skeletal integrity in the male. Strong support for this comes from the clinical reports of two human males, one with a mutation in the alpha isomer of the estrogen receptor (ER-α) gene, causing partial estrogen resistance; and another with aromatase p450 deficiency, causing complete estrogen deficiency. They both showed continued longitudinal bone growth due to delayed epiphyseal growth-plate ossification and osteopenia [49, 50].

Therefore, there is increased clinical interest in the use of selective estrogen receptor modulators (SERMS) in the treatment of bone loss. These synthetic ligands appear to have greater specificity for estrogen receptor in bone, and therefore the undesired side-effects of estrogen in other tissues is avoided while bone mass is maintained.

Estrogen appears to have direct effects on osteoblasts, osteocytes, and osteoclasts. Estrogen is a viability factor for osteoblasts and osteocytes, but appears to induce apoptosis of osteoclasts [51, 52]. Estrogen also appears to downregulate the production of several factors that play a role in bone resorption, such as interleukin-1, tumor necrosis factor, and interleukin-6 [53]. Therefore, considerable interest has been directed to inhibitors of these cytokines, such as their receptor antagonists.

There are two forms of the ER, designated α and β, and both forms are found in bone. The phenotypes of mice lacking either or both of these receptors are complicated, and show sexual dimorphism. One receptor isoform can partially compensate for the other. Also, it appears that estrogen can have opposite effects in mice compared to humans, especially on longitudinal bone growth, where estrogen enhances long bone growth in rodents but causes epiphyseal closure in humans. Homozygous deletion of ER-α results in reduced cortical bone formation and density in both male and female mice [54]. Female mice still lose bone with ovariectomy, and estrogen responsiveness is reduced. Studies with female estrogen receptor β (ER-β) knockout ($-/-$) mice indicate that ER-β is involved in the regulation of trabecular bone during adulthood by suppressing bone resorption, whereas this is not the case for male mice, where there is no effect [55]. Both sexes exhibit delayed growth-plate closure. Mice with both ER isoforms deleted generate a similar skeleton in the male as the ER-α knockout, but in contrast, females exhibit a more pronounced phenotype with reduced cortical thickness and trabecular bone density [56]. It is clear that ER-α and ER-β perform different functions in cortical and trabecular bone, and that these functions differ between the sexes.

1α,25-Dihydroxyvitamin D_3 (1,25,D_3)

Vitamin D is well known to prevent rickets; however, the compounds ergocalciferol (vitamin D_2) and cholecalciferol (vitamin D_3) are really pro-hormones that are converted to the biologically active form, 1α, 25(OH)$_2$vitamin D_3, which functions in a manner analogous to that of the steroid hormones [57]. Metabolism of the precursors occurs first in the liver and is completed in the kidney. The final active form then acts on the major target organs, bone, intestine, and kidney. A vitamin D binding protein (DBP) transports the hormone in the circulation for delivery to cells, where the hormone is freed from its binding protein for binding to the vitamin D receptor (VDR), a canonical nuclear receptor.

However, there is mounting evidence that 1,25,D_3 can also signal through membrane receptors [58, 59] and can have non-genomic effects [60]. These rapid responses, such as enhanced transport of Ca^{2+}, and activation of protein kinase C and of phospholipase A2, have been attributed to the interaction of the hormone with a membrane receptor and not with the canonical VDR. This membrane receptor remains to be completely characterized.

Hereditary vitamin D-resistant rickets is a rare autosomal recessive disease in which the patients exhibit defective bone mineralization and hypocalcemia due to decreased intestinal calcium absorption. These patients are unresponsive to 1,25,D_3 due to mutations in the VDR. Targeted disruption of this receptor in mice results in animals that appear normal at birth, but after weaning show growth retardation and alopecia [25, 61]. These mice are also infertile, which suggests a role for vitamin D in gonadal function. Much attention has focused on 1α, 25(OH)$_2$vitamin D_3, the major metabolite of vitamin D. The second metabolite 24R, 25 (OH)$_2$ D_3, may also be important. Deletion of the hydroxylase necessary for the generation of this metabolite results in mice that show poor viability, and bones with an accumulation of unmineralized matrix [62].

Parathyroid Hormone (PTH) and PTH-Related Protein (PTH-rp)

PTH is responsible for calcium homeostasis in the body both by its direct actions to provoke calcium release from bone and to enhance calcium reabsorption from kidney; and indirectly by actions on the gastrointestinal tract to effect conversion of 25 (OH)D_3 to 1,25 (OH)$_2D_3$. The principal form is intact PTH (1–84); however, there are several circulating cleavage forms, the functions of which are not clear [63]. It is clear that both the intact and amino terminal forms of PTH bind to the PTH type I receptor, a G-protein-coupled, seven-transmembrane receptor that signals through cAMP and potentially also through protein kinase C and calcium activation. Mutations of this receptor in humans result in various forms of chondrodysplasias that resemble the disorganized growth-plate phenotype of mice lacking this receptor [64]. Whereas both PTH and PTHrp bind to the PTH type I receptor, a second receptor, PTH type II, has been identified that responds to PTH but not to PTHrp [65]. Recently, it has been clearly demonstrated that receptors also exist that are specific for carboxy terminal fragments of PTH [66].

PTH has been shown to have both anabolic and catabolic effects on bone *in vivo*, to stimulate activation of osteoclasts, and either stimulate or inhibit osteoblast proliferation and matrix production *in vitro*. These effects of PTH have generated considerable debate on the role of PTH in bone remodeling. It appears that continuous PTH is responsible for resorption, whereas intermittent PTH can induce new bone formation [67]. Studies suggest that continuous exposure of cells results in downregulation of PTH receptor. Injections of either PTH or PTHrp into humans and animals can result in new bone formation, and intermittent application can stimulate mineralization of osteoblasts *in vitro* [68, 69].

PTHrp was first identified as the factor responsible for causing humoral hypercalcemia of malignancy, but soon afterwards was identified in many normal tissues. Nonetheless, it was undetectable in the normal circulation [70]. This molecule shares homology with PTH at the amino acid level, with 8 of the first 13 amino acids being identical, and three-dimensional homology in regions 13–34. PTHrp mRNA is alternatively spliced to yield three isoforms of 139, 141, and 173 amino acids. PTHrp is required for development of cartilage, morphogenesis of mammary gland, and tooth eruption, as demonstrated with tissue specific rescue of in PTHrp null mice that normally die at birth [71]. PTHrp also plays a role in calcium transfer across the placenta, and appears to play a role in smooth muscle contractility. Generally, bone cells appear to respond in a similar manner to PTHrp as to PTH.

GROWTH, SIGNALING, AND TRANSCRIPTION FACTORS RESPONSIBLE FOR BONE DEVELOPMENT AND GROWTH

Bone formation proceeds through two ossification processes: endochondral ossification and intramembranous ossification. The former process involves a cartilage intermediate and occurs during most of skeletal ossification, postnatal growth, bone remodeling, and fracture repair. The second process, by which bones form from mesenchymal condensations without a cartilage intermediate, only occurs in some craniofacial bones. The process of chondrogenesis and osteogenesis is tightly regulated at specific times and sites. Several transcription factors are important in this process, including Indian hedgehog and Sonic hedgehog, and the fibroblast growth factors and their receptors (see below). Indian hedgehog, expressed in prehypertrophic and hypertrophic chondrocytes [72], couples chondrogenesis to osteogenesis through PTHrP-dependent and PTHrP-independent pathways [73, 74]. The PTH signaling pathway also plays a critical role in growth-plate development (see below).

Arachidonic Acid Metabolites

One well-known group of arachidonic acid metabolites, the prostaglandins, clearly plays a role in both bone formation and bone resorption, most likely coupling the two processes. Prostaglandins are generally thought to be skeletal anabolic agents, as administration of these agents can increase bone mass [75, 76]; however prostaglandins also have catabolic effects on bone, and have been shown to stimulate osteoclast formation and activation and osteoclastic bone resorption. [77]. The differential effects of prostaglandins are thought to be mediated through multiple subtypes of specific G-protein-coupled PGE_2 receptors, designated EP_1, EP_2, EP_3, and EP_4 [78]. Recently, agonists of both the EP_2 and EP_4 have been shown to play a role in bone repair [79, 80] . However, mice lacking the genes for the cycloxygenases COX-1 or COX-2 do not appear to have any major bone developmental defects [81]. Nevertheless, in the postnatal animal, the bone formation response to mechanical strain is blocked by inhibitors of prostaglandin synthesis [82]. Mechanical strain in the form of pulsatile fluid flow shear stress raises intracellular Ca^{2+} and inositol trisphosphate, which then stimulates arachidonic acid production and PGE_2 release [83]. These observations suggest that prostaglandins are not as important in development as they are in the adult skeleton, and specifically in response to strain.

Clearly, cyclooxygenase metabolites of arachidonic acid such as the prostaglandins are important in bone formation, but it appears that the other side of the coin, the alternate pathway of arachidonic acid conversion through the lipoxygenases, results in bone loss. Metabolites of the actions of the enzyme 5-lipoxygenase (5LO) stimulate osteoclastic bone resorption [84–86] independent of RANKL [36]. 5LO metabolites also block the positive effects of prostaglandins and bone morphogenetic protein-2 on new bone formation *in vitro* [87]. Mice lacking the 5LO enzyme have greater bone mass [88], and the gene for 5LO, *Alox5*, has been shown to be a susceptibility gene for both obesity and bone traits [89]. Recently, the 12/15 lipoxygenase gene *Alox15* has been shown to also be a susceptibility gene for bone mineral density [90]. This opens new avenues for the treatment of diseases of bone loss.

Core Binding Factor 1 (Cbfa1), a Master Gene for Bone

Cbfa1, also known as Pebp2a1, Aml3, and Runx2, was originally thought to be T cell specific [91]. However, direct evidence that Cbfa1 is essential in bone and tooth development comes from Cbfa1 gene knockout experiments [92]. In these mice there is a total absence of bone, as well as arrested tooth development. The membranous bones of the skull are replaced by fibrous tissue, and endochondral bone does not replace the cartilaginous skeleton. Heterozygous Cbfa1 mice express a phenotype that is similar to the clinical manifestations of cleidocranial dysplasia, in which functional mutations of the Cbfa1 gene have been identified. Cleidocranial dysplasia is characterized by hypoplasia/aplasia of the clavicles, patent fontanelles, supernumerary teeth, short stature, and changes in skeletal patterning and growth. Cbfa1 is the earliest and most specific marker of osteogenesis identified to date [93]. Several homeodomain transcription factors, such as Msx2, Dlx5, Bapx1, and Hoxa-2 have been suggested to regulate Cbfa1 expression.

Osterix (Osx), a second transcription factor required for osteoblast differentiation during development, acts downstream of Cbfa1 [94]. Mice lacking Osx have a similar phenotype to those lacking Cbfa1. Target genes for Cbfa1 and Osx include osteocalcin, collagen type I, collagenase 3, TGFβ type II receptor, and other genes necessary for osteoblast funcion. Although Cbfa1 mRNA does not correlate with target gene regulation, phosphorylation of Cbfa1 protein does [95].

Low-density Lipoprotein Receptor-related Protein 5 (Lrp5) as a High Bone Mass Gene

Recently, a mutation in the extracellular domain of the Lrp5 gene was shown to result in extremely high bone mass in a human cohort. These individuals essentially never break their bones and have no other clinical features, suggesting solely positive effects of this mutation [96]. This was unexpected, as this protein is ubiquitously expressed and had previously only been associated with lipoproteins and liver function. However, mutations in the intracellular domain of this receptor result in a condition called osteoporosis pseudoglioma syndrome of juvenile onset. These individuals have osteoporosis and exhibit progressive blindness [97]. The extracellular mutation appears to result in constitutive activation, and the intracellular mutation results in a loss of function. These studies show the importance of Lrp5 in regulation of bone mass.

Lrp5 is a co-receptor with the seven-transmembrane receptor, frizzled, in the canonical Wnt signaling pathway [98]. These discoveries have focused considerable attention on the Wnt signaling pathway in maintenance of bone density, involving other members of this pathway such as β-catenin, glycogen synthase kinase 3B, Dishevelled, Dickkopt-1, axin, and targets of this pathway such as members of the T cell factor/lymphocyte enhancer factor family, including COX-2, c-jun, and connexin 43 [99, 100]. Studies are now in progress to fully dissect the role of this pathway in osteocyte, osteoblast, and osteoclast biology. This gene may regulate bone mass during development, but, even more exciting, may regulate bone mass through responses to mechanical strain. Particularly intriguing is the connection between the role of Wnt signaling in mechanosensation and osteocyte function as the primary cell sensing mechanical load in bone [101]. A connection between regulation of Wnt signaling in osteoblasts and regulation of osteoclastogenesis through regulation of osteoprotegerin (OPG) has been established [102]. All of these early studies implicate the Wnt signaling pathway as a major component through

which bone mass is regulated and the behavior of bone cells is orchestrated.

Transforming Growth Factor β

TGFβ-1 is the prototype and the founding molecule for the TGFβ superfamily [37]. This family has grown to include more than 40 members, including the TGFβ isoforms, the activins and inhibins, Müllerian inhibitory substance, growth differentiation factors (GDFs), and an ever-increasing number of bone morphogenetic proteins (BMPs). Members of this superfamily appear to mediate many key events in growth and development evolutionarily maintained from fruit flies to mammals. The actions of these proteins appear to be mediated through structurally similar serine/threonine kinase transmembrane receptors.

Members of the TGFβ superfamily bind to two distinct forms of serine/threonine kinase receptors, called type I and type II [103]. The constitutively active type II receptor initially binds to active TGFβ, and upon binding subsequently associates with the type I receptor and phosphorylates it. The direct substrates for the phosphorylated Type I receptor appear to be Smad2 and Smad3, also known as receptor-activated Smads (R-Smads), whereas negative regulators of the Type I receptor include Smad6 and Smad7, the inhibitory Smads (I-Smads) [104]. The discovery and naming of the Smads originated from studies of the *dpp* signaling path-way in *Drosophila* [105]. Smad2- or Smad3-heterozygous mutant mice are viable, but the compound heterozygous Smad2/Smad3 mutant is lethal, suggesting a gene dosage effect. Thus the relative expression level of Smad2 and -3 in the cell may influence the nature of the TGFβ response. Loss of Smad3 results in lower bone formation rate and osteopenia in mice [106]. Smad4, also called a common mediator (Co-Smad), appears to bring the cytoplasmic Smad2 and Smad3 into the nucleus, where together they can regulate transcription of target genes. Smad4 was found to be homologous to a gene deleted in pancreatic carcinomas called "Deleted in Pancreatic Cancer-4," or DPC-4 [107]. Smad4 is not always required for TGFβ signaling, since a number of Smad4-independent TGFβ responses have been identified [104]. These include Jun N-terminal kinase (JNK) and extracellular signal-related kinases (ERK) mitogen-activated protein (MAP) kinase pathways. The potential exists for other co-Smads to be identified.

The major function of TGFβ in bone is as an inducer of matrix formation. Mice lacking specific isoforms of TGFβ have bony defects. Injections of TGFβ can induce new bone formation or prevent bone loss, but inappropriate expression of TGFβ or its receptor can lead to bone loss. Even though TGFβ stimulates osteoid production, it actually inhibits mineralization of osteoid. Therefore, this factor must be activated in a specific time and tissue and then inactivated for normal bone remodeling to occur. TGFβ can enhance either bone formation or bone resorption, depending on the assay system and the presence of other factors. TGFβ was proposed to be a "coupling" factor, coupling bone resorption with bone formation, as this factor is released from the bone matrix where it is stored by resorbing osteoclasts [37, 108].

TGFβ is well-known among growth factors for its potent and widespread actions. Almost every cell in the body has been shown to make some form of TGFβ, and almost every cell expresses receptors for TGFβ. The largest source of TGFβ in the body is bone. This growth factor must be tightly regulated to prevent disease. Appropriately, the mechanisms of regulation of TGFβ are extensive and complex. One unique set of regulatory mechanisms centers on the fact that TGFβ is produced in a latent form that must be activated to produce biologically active TGFβ. The mechanisms of regulation include not only regulation of the latency of the molecule, but also the production of different latent forms, such as the small and large latent complexes, TGFβ targeting to matrix for storage or to cells for activation, and the various means of activation. The extracellular matrix protein, latent TGFβ binding protein (LTBP-1), appears to play a major role in the regulation of TGFβ (see below).

Bone Morphogenetic Factors

Unlike the TGFβs, which can only induce bone when injected in close proximity to existing bone, the BMPs can induce new bone formation when injected into muscle. Urist [65] was the first to describe bone regenerative capacity of bone extracts, but Celeste and co-workers were the first to identify the factors responsible through the use of peptide sequences from these mixtures and then cloning the resulting recombinant DNA for *in vivo* studies [109]. The BMPs are more closely related to proteins involved in differentiation during embryogenesis than they are to the TGFβs. In fact, while it is clear that these factors are important or essential for development, it is not clear if these factors play an important role in the adult skeleton. This will not be known until time- and tissue-specific null mice are generated.

At present, it is known that deletion of BMP2 or -4 is embryonic lethal, whereas deletion of other BMPs is not so dramatic. Deletion of BMP7 results in mice with mild limb skeleton abnormalities, BMP6$^{-/-}$ mice appear normal, and BMP5$^{-/-}$ mice exhibit the short ear phenotype. These results suggest that BMPs in some cases can compensate for deletion of one member. BMP3$^{-/-}$ mice are normal, and in fact have increased bone density, which may explain why injection of recombinant BMP3 has never induced bone formation. The growth and differentiation factors (GDFs) are also members of the BMP family. GDF5$^{-/-}$ mice exhibit brachypodism, reduction in number of digits,

and misshapen bones. Deletion of GFD11 leads to defects in skeletal patterning, and palate abnormalities. Even deletion of the some of the receptors for BMPs (see below) does not result in severe or lethal phenotypes, suggesting that the receptors can compensate for one another. Deletion of either the BMPR1B receptor or the ActRIIA receptor is not severe, whereas deletion of the BMPR1A receptor is embryonic lethal [110]. Negative regulators of the BMPs include the following; noggin for BMPs 2, 4, and 7, and GDFs 5 and 6; chordin for BMPs 2, 4, and 7; follistatin for BMPs 2, 4, 7, and 11; and gremlin for BMPs 2 and 4, and GDF5.

Like the TGFβs, the BMPs signal through type I and type II receptors; though BMPs can signal through type II receptors alone, this signal is enhanced when both receptors are engaged [111]. Seven type I receptors, called activin receptor-like kinases (ALKs), and three type II receptors have been identified. Members of the BMP family bind with different affinities to the type I and type II receptors, adding to redundancy and complexity in signaling. BMP receptors also signal through the Smads. The R-Smads for BMPs include Smads 1, 5, and 8 (for TGFβ they are 2 and 3, as described above). Smad4 is the only co-Smad that is shared by both the BMPs and TGFβ. I-Smads are Smads 6 and 7. Smad5-deficient mice have defects in angiogenesis, and Smad-6 mice exhibit cardiac defects. Transcriptional co-repressors of the Smads include TGIF, c-Ski, and SnoN. Target genes of the BMPs, such as Tlx-2, a homeobox gene related to human HoxII, Dad (Daughters against Dpp), and Id gene products, are generally responsible for patterning and development.

Recombinant BMPs 2, 4, and 7 are being used for clinical studies to induce fracture repair, and augmentation of alveolar bone, and for gene therapy. To date, none of these applications has been approved for general application, probably due to the difficulties in determining ideal doses, times for administration, and carriers. However, the potential still exists for more general therapeutic use.

Insulin-like Growth Factors (IGFs)

The IGFs were shown to be the mediators of the effects of growth hormone. IGFs I and II are 7-kDa proteins that share homology with pro-insulin. Originally it was found that IGF is made by the liver, but it has also been shown that osteoblasts produce this growth factor. Bone is the major storage organ for IGFs, and IGF II is the most abundant of all the growth factors stored in the skeleton [112]. Factors such as PTH, estrogen, prostaglandin E2, and BMP2 increase IGF expression in osteoblasts.

Animals with targeted overexpression of IGF in bone by use of the osteocalcin promoter have increased bone mineral density [113]. IGF I-deficient neonates have a marked increase in death rate compared to IGF II-deficient animals, which have normal survival rates. Mice lacking the IGF receptor die at birth. Regardless of whether ligand or receptor is deleted, the pups express normal morphogenesis [108]. This suggests that the major function of the IGFs is growth, not morphogenesis, as is true for other growth factors such as the BMPs and FGFs.

The actions of the IGFs appear to be tightly regulated by the IGF binding proteins. These are found in serum and in bone matrix. Six have been cloned and characterized, They bind with high affinity to the IGFs, preventing their interaction with the IGF receptor. IGFBP-1 can inhibit or enhance IGF action, dependent on phosphorylation state of the binding protein, and may be responsible for suppression of bone formation in malnourished individuals. IGFBP-2 is the major binding protein secreted by osteoblasts. IGFBP-3 has both inhibitory and stimulatory activity, depending on location within the cell. IGFBP-5 is not normally in the circulation, but is preferentially found in the bone matrix, where it appears to be protected from proteases and appears to potentiate IGF activity. IGFBP-6 has a selective affinity for IGF II over IGF I. [108, 114]. The binding proteins can be degraded by specific and non-specific proteases, therefore adding another level of regulation of IGF activity. For example, cathepsin D will degrade IGFBPs 1–5, whereas pregnancy-associated plasma protein-A will specifically proteolyze IGFBP-4.

The IGFs appear to stimulate new bone formation *in vivo*, with little or no preliminary resorption phase. In animal models, IGF I can enhance longitudinal growth, bone formation, and bone mass in various, but not all, models. For example, growth can be restored in hypophysectomized rats, but there is little effect on normal rats. It is clear that recombinant IGF I can enhance trabecular and cortical bone mineral density in humans with an impaired growth hormone–IGF axis, but such data are not available for normal or older adults. Therefore, at this time, use of IGF can only be justified in specific conditions, such as growth hormone-resistant short stature [115].

Fibroblast Growth Factors

The FGFs were named for their ability to stimulate the growth of 3T3 fibroblasts. The FGF family has grown to include at least 23 genes [108, 116]. The first two FGFs were called acidic and basic FGF, based on their isoelectric points, but have since been renamed FGF-1 and FGF-2. In addition to promoting cell growth, these factors can induce a mitogenic response, and stimulate cell migration, angiogenesis, vasculogenesis, transformation, morphogenesis, wound healing, and tissue repair. FGF-2 and -3 are distinguished from all other growth factors by a novel translation initiation mechanism. Four high molecular weight isoforms of FGF-2 are initiated with an unconventional CUG translation codon, whereas a smaller 18-kDa isoform

is initiated by the classical AUG codon. Another interesting feature of some of the FGFs, like interleukin 1, is their ability to be non-classically secreted even though they do not contain hydrophobic signal peptide sequences. On the cell surface, the FGFs interact with at least three types of molecules, including four high-affinity signaling receptors (FGFRs 1–4), low-affinity receptors such as perlecan and syndecan that potentiate ligand/receptor interactions, and cysteine-rich non-signaling receptors that may function to antagonize and remove ligand.

FGFs clearly stimulate new bone formation; however, injections of FGF cause serious side-effects, such as acute hypotension. Mutations in FGF receptors result in a number of human dysmorphic (dwarfism) syndromes, such as achondroplasia, thanatophoric dysplasia, Jackson-Weiss syndrome, and Pfeiffer syndrome. The clearest indications that FGFs are important in bone development is revealed through the bone phenotypes of null mice lacking the FGF receptors, rather than mice lacking a particular FGF, as compensation among this family of factors appears to occur. Disruption of FGF-2, however, results in decreased bone mass and bone formation [117]. Mice expressing activated *FGFR3* mutants reproduce the dwarfism phenotype of the chondrodysplasias, and show a marked decrease in the proliferation rate of the columnar proliferating chondrocytes and a decrease in size of the zone of hypertrophic chondrocytes [118–121]. Thus, a normal function of FGF signaling in chondrocytes is to inhibit chondrocyte proliferation.

FGFs may have additional effects on the skeleton. The newest member of the family, FGF-23, appears to play a key role in hyophosphatemic disorders [122]. FGF23 is a phosphaturic factor that prevents reabsorption of Pi by the kidney, leading to hypophosphatemia. This FGF is produced by tumors that cause osteomalacia, and when injected into mice causes hypophosphatemic rickets. In all tumors causing hypophosphatemic osteomalacia, mutations around Ser180 have been identified resulting in a non-cleaved 32-kDa protein. FGF-23 is not cleaved in autosomal dominant hypophosphatemic rickets. It is presumed that under normal conditions FGF-23 is cleaved at residue Ser180, and that these mutations may cause a gain of function for FGF-23. There are two key molecules highly expressed in osteocytes that play a role in phosphate homeostasis and the control of FGF-23 in osteocytes: dentin matrix protein 1, (DMP1), and *Pex*/Phex (phosphate regulating neutral endopeptidase on chromosome X) [123, 124]. FGF23 is also expressed in osteocytes [125], but deletion or mutation of either Pex or DMP1 results in hypophosphatemic rickets resulting from a dramatic elevation of FGF23 in osteocytes [125, 126]. Hypophosphatemic rickets in humans is caused by inactivating mutations of *Pex,* and autosomal recessive hypophosphatemia in humans is due to mutations in DMP1, both resulting in elevated circulating levels of FGF23 [126, 127]. As both DMP1 and Phex are regulated by mechanical loading [128–130], it will be important to determine if skeletal loading can play a role in mineral and phosphate metabolism. The mechanisms for the role of this FGF in hypophosphatemia are under intense investigation.

BONE EXTRACELLULAR MATRIX (ECM)

The extracellular matrix and the proteins it contains have not received the same attention as other areas of bone biology, such as cytokines, receptors, cell signaling, transcription factors, etc. This is partially due to the difficulty in determining the potential functions of large, extensively modified extracellular matrix proteins. Recently, more attention has focused on the extracellular components of bone due to the advent of new technologies. Transgenic animals and null mice have greatly assisted in determining the functions of these proteins [131]. Knockouts of the extracellular matrix proteins lead to various bone defects, such as thickened bones in osteocalcin-null mice, and an osteoporosis-like phenotype in biglycan-null mice. On the other hand, no or little bone phenotype was observed in the decorin- or osteonectin-null mice [131]. Deletion of the gene for dentin matrix protein 1, DMP1, resulted in a dramatic bone and growth-plate phenotype resembling chondrodysplasia and osteomalacia [132]. Occasionally, deleting a specific ECM protein gene, such as that for fibronectin, can result in embryonic lethality, which unfortunately does not give information concerning the specific function of the matrix protein. In these cases, the phenotype of a transgenic animal or a deletion heterozygote can be more informative.

Another new technology that has given a boost and additional insight into the function of extracellular matrix proteins is dynamic imaging, which combines fluorescent protein labeling, pH sensitive dyes, FRET, and laser confocal microscopy to obtain spatio-temporal and kinetic information. This approach is being applied to the bone ECM to study ECM molecules such as fibronectin, LTBPs, and bone-specific proteins [133, 134]. Cell and fibrillar movement in the bone ECM are considerably more dynamic than previously thought.

Although the major ECM protein in bone is collagen type 1, there are numerous non-collagenous proteins. These include proteoglycans such as decorin and biglycan, which are characterized by glycosaminoglycans attached to core protein; chondroitin sulfate proteoglycans such as aggrecan and versican; glycoproteins such as osteonectin, vitronectin and thrompospondins; proteins containing γ carboxy glutamic acid such as Matrix Gla protein and osteocalcin; and a group of proteins known as the SIBLINGS (Small Integrin-Binding Ligands with N-linked Glycosylation) [135]. Members of the SIBLINGS include osteopontin, bone sialoprotein, dentin matrix protein-1 (DMP-1), dentin

sialophosphoprotein, and MEPE. Bone proteins are proposed to have a role in the mineralization process. Whereas deletion of osteocalcin, osteonectin, and bone sialoprotein have not resulted in significant changes in the bone phenotype, deletion of DMP-1 has dramatic effects on the growth plate. DMP-1$^{-/-}$ mice display a chondrodysplastic phenotype and dwarfism [136]. Additionally, DMP-1 is highly expressed in osteocytes and may be important in osteocyte function [123].

Latent TGFβ Binding Proteins (LTBPs)

LTBPs appear to be an important mechanism whereby TGFβ is controlled. To date, four LTBP genes have been isolated: LTBPs 1, 2, 3, and 4, containing cysteine and EGF-like repeating domains. LTBPs are highly homologous to fibrillins 1 and 2, major constituents of connective tissue microfibrils. LTBP-2 does not bind latent TGFβ and therefore may be more homologous than LTBPs 1, 3, and 4 to the fibrillins. The third eight-cysteine repeat in LTBP-1 forms a covalent disulfide bond with the TGFβ1 precursor or "latency associated peptide". The major isoform of TGFβ stored in bone matrix is TGFβ-1 (80–90 percent) as part of a latent complex containing LTBP-1.

LTBP-1 does not confer latency to the TGFβ complex (the conformation of the TGFβ dimer does), but has other unique functions. The latent TGFβ complex produced by matrix-forming osteoblasts is directed by LTBP-1 to fibrillar structures known as microfibrils in bone extracellular matrix [137]. Although LTBP-1 covalently associates with small latent TGFβ1, it is also produced by osteoblasts in a free form (80 percent of total) not associated with latent TGFβ1. This molar excess suggests a function separate and distinct from its association with TGFβ. Many extracellular matrix proteins contain EGF-like repeats that mediate protein–protein interactions, suggesting that LTBP-1 may have similar functions [37].

Deletion of LTBP-2 is lethal, whereas deletion of LTBP-3 results in mice with osteopetrosis [138] and premature ossification of synchondroses at 2 weeks [138]. At 6 and 9 months, these animals develop osteosclerosis and osteoarthritis. The gene for LTBP3 has been disrupted, resulting in only trace amounts of protein in these animals [139]. These animals have defective lung and colorectal function, most likely due to a reduction in available TGFβ and an enhancement of BMP4 availability [140]. To date, the LTBP-1 gene has not been successfully deleted.

Microfibrils

Recently, the role of components of microfibrils in bone development has attracted attention. The components of microfibrils include fibronectin, fibrillins 1 and 2, elastin, microfibril-associated glycoprotein (MAGPs) 1 and 2, fibulin, and others [141, 142]. Mice lacking the genes for the microfibril proteins often have a more dramatic bone phenotype than mice lacking genes for many of the "bone-specific" matrix proteins. For example, deletion of the bone-specific gene osteocalcin results in a modest bone phenotype, where deletion of latent TGFβ binding protein-3 (LTBP-3, see below) appears to have a more dramatic effect in bone [138].

The deletion of fibronectin, a major component of ECM and microfibrils, is neonatal lethal, and there are no known human mutations of this protein [143, 144]. The fibrillin 1 knockout does not appear to have a bone phenotype, whereas fragments of overexpressed fibrillin 1 result in mice with overgrowth of ribs and long bones [145]. Based on these results, it was suggested that microfibrils control bone growth in a negative fashion. The fibrillin 2 null mouse has a bone phenotype with contracture at birth which resolves with age, rear joints that do not flex, and fusion of three toes in the hind limbs into one phalange. The elastin null mouse dies at 4.5 days *post partum* due to arterial obstruction, while the heterozygote shows functional haploinsufficiency – a model for human supravalvular aortic stenosis [146]. Proteins that are components of microfibrils could function by physical influence through alteration of the physical properties of matrix, through indirect signaling by way of retaining and releasing growth factors such as TGFβ, and through presentation and binding of protein to receptors or signaling molecules on cell surfaces.

Components of microfibrils have strong protein–protein interactions and protein–cell interactions. Cells have been shown to bind to fibrillin through integrins [147]. MAGP-2 has an RGD sequence motif that modulates cell to microfibril interactions and binds to αvβ3 integrins, as does fibrillin-1 [148]. Fibulins all have long EGF repeats and bind to the matrix glycoproteins nidogen, aggrecan, versican, fibronectin, endostatin, collagen IV, laminin α2, and perlecan [149, 150], all of which have calcium-binding EGF-like repeats. These calcium binding repeats are necessary for stabilizing their tightly folded structures. Several growth factors have also been shown to bind to components of microfibrils – for example, connective tissue growth factor (CTGF) binds to fibrillin.

Matrix Metalloproteinases (MMPs)

An intricate balance between deposition and breakdown of extracellular matrix (ECM) is critical for growth and development of bone, and significant progress has been made in understanding the roles of MMPs in the balance between osteoblasts and osteoclasts [151]. MMPs belong to a family of zinc- and calcium-dependent endopeptidases that catalyze the proteolysis of components of ECM at neutral pH. Each member has specificity for a particular subset of ECM components. The most important members are MMP-2 and MMP-9. Martignetti and colleagues found that a human disease which features osteolytic lesions in facial bones, arthritis, and subcutaneous nodules is associated with an enhanced degradation of ECM, due to the lack of a single proteolytic enzyme, matrix metalloproteinase 2 [152]. Similarly, deficiency of mouse MT1-MMP, which activates MMP2, results in a decrease of collagen breakdown by osteoblasts, a decrease in bone formation, and an increase in the number of osteoclasts [113, 153]. MMP-13 is predominantly expressed in the skeleton, and null mice have elongated growth plates and reduced bone resorption, suggesting that MMP13 directly or indirectly inhibits chondrocyte growth and stimulates osteoclastogenesis [154], while overexpression of MMP-13 leads to osteoarthritis [155].

CONCLUSION AND SUMMARY

Though this chapter merely touches on many important areas in bone research, room is not available for review of others – topics such as the bone-resorbing cytokines including interleukins 1, 6, 11, tumor necrosis factors, and regulatory factors such as nitric oxide and its regulatory enzymes. Though many of the factors involved in regulation of bone function are similar to those in other organs, a level of complexity is added due to the mineralized nature of bone. Hematopoietic and immune cells have been well-characterized because they are relatively easy to obtain, but this is not the case for bone cells, especially for osteocytes embedded in bone. Investigators in the bone field have often referred back to the areas of hematology, immunology, and development to understand the potential role of factors in bone. Determining the function of matrix proteins in bone has relied heavily on studies in other tissues, such as cartilage, skin, and other connective tissues. However, bone biologists are not able to rely on studies in other organs with regard to mineralization, the unique feature of the skeleton.

In summary, bone is a storehouse of factors ready to be released during resorption that can modify the bone coupling process or provide circulating growth factors. A number of transcription factors have been identified that are specific for bone induction and development. Clearly these growth factors and transcription factors are regulated by a number of circulating hormones, such as parathyroid hormone, estrogen, and 1, 25 (OH)2 D3. Another layer of complexity is added due to the fact that bone structure is also regulated by mechanical strain. Understanding the normal physiology of bone and its diseases should lead to prevention and treatment of disease, acceleration and initiation of repair and treatment, or reversal of abnormal development.

REFERENCES

1. Marcus R, Majumder S. The nature of osteoporsis. In: Marcus R, Feldman D, Kelsey J, editors. *Osteoporosis*, vol. 2: Academic Press; 2001. p. 3–17.

2. Foundation NO. *America's bone health: the state of osteoporosis and low bone mass in our nation Fighting osteoporosis & promoting bone health.* Washington, DC: National Osteoporosis Foundation; 2002.

3. Rossouw JE, Anderson GL, Prentice RL, LaCroix AZ, Kooperberg C, Stefanick ML, Jackson RD, Beresford SA, Howard BV, Johnson KC, Kotchen JM, Ockene J. Risks and benefits of estrogen plus progestin in healthy postmenopausal women: principal results from the women's health initiative randomized controlled trial. *J Am Med Assoc* 2002;**288**:321–33.

4. Anderson GL, Limacher M, Assaf AR, Bassford T, Beresford SA, Black H, Bonds D, Brunner R, Brzyski R, Caan B, Chlebowski R, Curb D, Gass M, Hays J, Heiss G, Hendrix S, Howard BV, Hsia J, Hubbell A, Jackson R, Johnson KC, Judd H, Kotchen JM, Kuller L, LaCroix AZ, Lane D, Langer RD, Lasser N, Lewis CE, Manson J, Margolis K, Ockene J, O'Sullivan MJ, Phillips L, Prentice RL, Ritenbaugh C, Robbins J, Rossouw JE, Sarto G, Stefanick ML, Van Horn L, Wactawski-Wende J, Wallace R, Wassertheil-Smoller S. Effects of conjugated equine estrogen in postmenopausal women with hysterectomy: the women's health initiative randomized controlled trial. *J Am Med Assoc* 2004;**291**:1701–12.

5. Birkhaeuser MH. The women's health initiative conundrum. *Arch Womens Ment Health* 2005;**8**:7–14.

6. Brown DL, Robbins R. Developments in the therapeutic applications of bisphosphonates. *J Clin Pharmacol* 1999;**39**:651–60.

7. Bamias A, Kastritis E, Bamia C, Moulopoulos LA, Melakopoulos I, Bozas G, Koutsoukou V, Gika D, Anagnostopoulos A, Papadimitriou C, Terpos E, Dimopoulos MA. Osteonecrosis of the jaw in cancer after treatment with bisphosphonates: incidence and risk factors. *J Clin Oncol* 2005;**23**:8580–7.

8. Bilezikian JP, Rubin MR, Finkelstein JS. Parathyroid hormone as an anabolic therapy for women and men. *J Endocrinol Invest* 2005;**28**:41–9.

9. Michigami T, Shimizu N, Williams PJ, Niewolna M, Dallas SL, Mundy GR, Yoneda T. Cell–cell contact between marrow stromal cells and myeloma cells and myeloma cells via VCAM-1 and alpha-4beta1 integrin enhances production of osteoclast stimulating activity. *Blood* 2000;**96**:1953–60.

10. Teoh G, Anderson KC. Interaction of tumor and host cells with adhesion and extracellular matrix molecules in the development of multiple myeloma. *Hematol Oncol Clinics North Am* 1997;**11**:27–42.

11. Mundy GR, Martin TJ. Pathophysiology of skeletal complications of cancer. In: Mundy G, Martin TJ, editors. *Physiology and pharmacology of bones: handbook of experimental pharmacology*, vol. 18. Berlin: Springer-Verlag; 1993. p. 642–7.

12. Yin JJ, Grubbs BG, Cui Y, Wu-Wong JR, Wessale J, Padley RJ, Guise TA. Endothelin a receptor blockade inhibtis osteoblastic metastases. *J Bone Miner Res* 2000;**15**:1254.

13. Mundy GR, Toshiyuki Y, Guise TA, Oyajobi B. Local factors in skeletal malignancy. In: Bilezikian JP, Raisz LG, Rodan GA, editors. *Principles of bone biology*, vol. 2. San Diego, CA: Academic Press; 2002. p. 1093–104.

14. Juppner H, Schipani E, Silve C. Jansen's metaphyseal chondrodysplasia and Blomstrand's lethal chondrodysplasia: two genetic disorders caused by PTH/PTHrP receptor mutation. In: Bilezikian JP, Raisz LG, Rodan GA, editors. *Principles of bone biology*, vol. 2. San Diego, CA: Academic Press; 2002. p. 1117–35.

15. Singer FR, Krane SM. Paget's disease of bone. In: Krane SM, Avioli LV, editors. *Metabolic bone disease*. Philadelphia, PA: Saunders; 1998. p. 546–615.

16. Laurin N, Brown JP, Morissette J, Raymond V. Recurrent mutation of the gene encoding sequestosome 1 (SQSTM1/p62) in Paget disease of bone. *Am J Hum Genet* 2002;**70**:1582–8.

17. Kurihara N, Reddy SV, Menaa C, Rodman GD. Osteoclasts formed by normal human bone marrow cells transduced with th measles virus nucleocapsid gene express a pagetic phenotype. *J Clin Invest* 2000;**105**:607–14.

18. Kurihara N, Zhou H, Reddy SV, Garcia Palacios V, Subler MA, Dempster DW, Windle JJ, Roodman GD. Expression of measles virus nucleocapsid protein in osteoclasts induces Paget's disease-like bone lesions in mice. *J Bone Miner Res* 2006;**21**:446–55.

19. Ishizuka S, Kurihara N, Reddy SV, Cornish J, Cundy T, Roodman GD. (23S)-25-Dehydro-1α-hydroxyvitamin D3-26,23-lactone, a vitamin D receptor antagonist that inhibits osteoclast formation and bone resorption in bone marrow cultures from patients with Paget's disease. *Endocrinology* 2005;**146**:2023–30.

20. Rowe DW. Osteogensis imperfecta. In: Bilezikian JP, Raisz LG, Rodan GA, editors. *Principles of bone biology*, vol. 2. San Diego, CA: Academic Press; 2002. p. 1177–93.

21. Rauch F, Plotkin H, Travers R, Zeitlin L, Glorieux FH. Osteogenesis imperfecta types I, III, and IV: effect of pamidronate therapy on bone and mineral metabolism. *J Clin Endocrinol Metab* 2003;**88**:986–92.

22. Roodman GD. Advances in bone biology: the osteoclast. *Endocrinol Rev* 1996;**17**:308–32.

23. Chen W, Li YP. Generation of mouse osteoclastogenic cell lines immortalized with SV40 large T antigen. *J Bone Miner Res* 1998;**13**:1112–23.

24. Kodama H, Nose M, Niida S, Yamasaki A. Essential role of macrophage colony-stimulating factor in the osteoclast differentiation supported by stromal cells. *J Exp Med* 1991;**173**:1291–4.

25. Yoshizawa T, Handa Y, Uematsu Y, Takeda S, Sekine K, Yoshihara Y, Kawakami T, Arioka K, Sato H, Uchiyama Y, Masushige S, Fukamizu A, Matsumoto T, Kato S. Mice lacking the vitamin D receptor exhibit impaired bone formation, uterine hypoplasia and growth retardation after weaning. *Nat Genet* 1997;**16**:391–6.

26. Simonet WS, Lacey DL, Dunstan CR, Kelley M, Chang MS, Luthy R, Nguyen HQ, Wooden S, Bennett L, Boone T, Shimamoto G, DeRose M, Elliott R, Colombero A, Tan HL, Trail G, Sullivan J, Davy E, Bucay N, Renshaw-Gegg L, Hughes TM, Hill D, Pattison W, Campbell P, Boyle WJ. Osteoprotegerin: a novel secreted protein involved in the regulation of bone density. *Cell* 1997;**89**:309–19.

27. Aubin JE, Bonnelye E. Osteoprotegerin and its ligand: a new paradigm for regulation of osteoclastogenesis and bone resorption. *Medscape Womens Heal* 2000;**5**:5.

28. Quinn JM, Elliott J, Gillespie MT, Martin TJ. A combination of osteoclast differentiation factor and macrophage-colony stimulating factor is sufficient for both human and mouse osteoclast formation in vitro. *Endocrinology* 1998;**139**:4424–7.

29. Pockwinse SM, Wilming LG, Conlon DM, Stein GS, Lian JB. Expression of cell growth and bone specific genes at single cell resolution during development of bone tissue-like organization in primary osteoblast cultures. *J Cell Biochem* 1992;**49**:310–23.

30. Bonewald LF. Mechanosensation and transduction in osteocytes. *Bonekey Osteovision* 2006;**3**:7–15.

31. Bonewald LF. Osteocytes as dynamic, multifunctional cells. *Ann NY Acad Sci.* 2007;**116**:281–90.

32. Bonewald LF. Osteocyte messages from a bony tomb. *Cell Metab* 2007;**5**:410–11.

33. Marotti G, Cane V, Palazzini S, Palumbo C. Structure–function relationships in the osteocyte. *Ital J Min Electrolyte Metab* 1990:93–106.

34. Marotti G. The structure of bone tissues and the cellular control of their desposition. *Ital J Anat Embryol* 1996:25–79.

35. Doty SB. Morphological evidence of gap junctions between bone cells. *Calc Tissue Intl* 1981;**33**:509–12.

36. Cherian PP, Siller-Jackson AJ, Gu S, Wang X, Bonewald LF, Sprague E, Jiang JX. Mechanical strain opens connexin 43 hemichannels in osteocytes: a novel mechanism for the release of prostaglandin. *Mol Biol Cell* 2005;**16**:3100–6.

37. Bonewald LF. Osteocytes: a proposed multifunctional bone cell. *J Musculoskel Neuronal Interact* 2002;**2**:239–41.

38. Frost HM. Perspectives: bone's mechanical usage windows. *Bone Miner* 1992;**19**:257–71.

39. Frost HM. Bone "mass" and the "mechanostat": a proposal. *Anat Rec* 1987;**219**:1–9.

40. Martin RB, Burr DB. *Structure, function and adaptation of compact bone*. New York, NY: Raven Press; 1989.

41. Turner CH, Riggs BL, Spelsberg TC. Skeletal effects of estrogen. *Endocrin Rev* 1994;**15**:275–300.

42. Cheng MZ, Zaman G, Lanyon LE. Estrogen enhances the stimulation of bone collagen synthesis by loading and exogenous prostacyclin, but not prostaglandin E2, in organ cultures of rat ulnae. *J Bone Miner Res* 1994;**9**:805–16.

43. Cheng MZ, Zaman G, Rawlinson SC, Suswillo RF, Lanyon LE. Mechanical loading and sex hormone interactions in organ cultures of rat ulna. *J Bone Miner Res* 1996;**11**:502–11.

44. Damien E, Price JS, Lanyon LE. Mechanical strain stimulates osteoblast proliferation through the estrogen receptor in males as well as females. *J Bone Miner Res* 2000;**15**:2169–77.

45. Cowin SC, Weinbaum S, Zeng Y. A case for bone canaliculi as the anatomical site of strain generated potentials. *J Biomech* 1995;**28**:1281–97.

46. Wang L, Ciani C, Doty SB, Fritton SP. Delineating bone's interstitial fluid pathway *in vivo*. *Bone* 2004;**34**:499–509.

47. Goodenough DA, Paul DL. Beyond the gap: functions of unpaired connexon channels. *Nat Rev Mol Cell Biol* 2003;**4**:285–94.

48. Plotkin LI, Manolagas SC, Bellido T. Transduction of cell survival signals by connexin-43 hemichannels. *J Biol Chem* 2002;**277**:8648–57.

49. Carani C, Qin K, Simoni M, Faustini-Fustini M, Serpente S, Boyd J, Korach KS, Simpson ER. Effect of testosterone and estradiol in a man with aromatase deficiency. *N Engl J Med* 1997:91–5.

50. Smith EP, Boyd J, Frank GR, Takahashi H, Cohen RM, Specker B, Williqms TC, Lubahn DB, Korach KS. Estrogen resistance caused by a mutation in the estrogen receptor gene in a man. *N Engl J Med* 1994:1056–61.

51. Hughes DE, Dai A, Tiffee JC, Li HH, Mundy GR, Boyce BF. Estrogen promotes apoptosis of murine osteoclasts mediated by TGF-beta. *Nat Med* 1996;**2**:1132–6.

52. Kousteni S, Bellido T, Plotkin LI, O'Brien CA, Bodenner DL, Han L, Han K, DiGregorio GB, Katzenellenbogen JA, Katzenellenbogen BS, Roberson PK, Weinstein RS, Jilka RL, Manolagas SC. Nongenotropic, sex-nonspecific signaling through the estrogen or androgen receptors: dissociation from transcriptional activity. *Cell* 2001;**104**:719–30.

53. Pacifici R. Estrogen, cytokines and pathogenesis of postmenopause osteoporosis. *J Bone Miner Res* 1996:1043–51.

54. Korach KS, Taki M, Kimbro KS. *The effects of estrogen receptor gene disruption on bone*. Dordrecht: Kluwer Academic and Fondazione Giovanni Lorenzini; 1997.

55. Vidal O, Lindberg MK, Hollberg K, Baylink DJ, Andersson G, Lubahn DB, Mohan S, Gustafsson JA, Ohlsson C. Estrogen receptor specificity in the regulation of skeletal growth and maturation in male mice. *Proc Natl Acad Sci USA* 2000;**97**:5474–9.

56. Sims NA, Dupont S, Resche-Rogon M, Clement-Lacroix P, Bouali Y, DaPonte F, Galien R, Gaillard-Kelly M, Baron R. *In vivo* analysis of male and female estrogen receptor a, b, and double knockouts reveals a dual role for ER beta in bone remodelling. *J Bone Miner Res* 2000:S160.

57. Norman AW. Vitamin D. In: Ziegler EE, Filer LJ, editors. *Present knowledge in nutrition (PKN7)*. Washington, DC: International Life Sciences Institute; 1996.

58. Nemere I, Yoshimoto Y, Norman AW. Studies on the model of action of calciferol. LIV. Calcium transport in perfused duodena from normal chicks: enhancement with 14 minutes of exposure to 1 alpha, 25-dihydroxvitamin D3. *Endocrinology* 1984;**115**:1476–83.

59. Pedrozo HA, Schwartz Z, Rimes S, Sylvia VL, Nemere I, Posner GH, Dean DD, Boyan BD. Physiological importance of the 1,25(OH)2D3 membrane receptor and evidence for a membrane receptor specific for 24,25(OH)2D3. *J Bone Miner Res* 1999;**14**:856–76.

60. Boyan BB, Swartz Z, Snyder SP, Dean DD, Yang F, Twardzik D, Bonewald LF. Latent transforming growth factor beta is produced by chondrocytes and activated by extracellular matrix vesicles upon exposure to 1,25(OH)2D3. *J Biol Chem* 1994;**269**:28,374–28,381.

61. Li YC, Pirro AE, Amling M, Delling G, Baroni R, Demay MB. Targeted ablation of the vitamin D receptor: an animal model of vitamin D-dependent rickets type II with alopecia. *Proc Natl Acad Sci USA* 1997;**94**:9831–5.

62. St. Arnaud R, Arabian A, Travers R, Glorieux FH. Abnormal intramembranous ossification in mice deficient for the vitamin D 24-hydroxylase. In: Norman AW, Bouillon R, Thomasset M, editors. *Vitamin D: chemistry, biology and clinical application of the steriod hormone*. Los Angeles, CA: University of California; 1997. p. 635–44.

63. Hock JM, Fitzpatric LA, Bilezikian JP. Actions of parathyriod hormone. In: Bilezikian JP, Raisz LG, Rodan GA, editors. *Principles of bone biology*, vol. 1. San Diego, CA: Academic Press; 2002. p. 463–81.

64. Lanske B. Ablation of the PTHrP gene or the PTH/PTHrP receptor gene leads to distinct abnormalities in bone development. *J Clin Invest* 1999;**104**:399–407.

65. Juppner H. Receptors for parathyroid hormone and parathyroid hormone-related peptide: from molecular cloning to definition of diseases. *Curr Opin Nephrol Hypertens* 1996;**5**:300–6.

66. Divieti P. Receptors for the carboxyl-terminal region of pth(1-84) are highly expressed in osteocytic cells. *Endocrinology* 2001;**142**:916–25.

67. Schaefer F. Pulsatile parathyroid hormone secretion in health and disease. *Novart Fdn Symp* 2000;**227**:225–39. discussion 239–43.

68. Bauer E, Aub J, Algright F. Studies of calcium and phosphorus metabolism. V. A study of the bone trabeculae as a readily available reserve supply of calcium. *J Exp Med* 1929;**49**:145–62.

69. Ishizuya T, Yokose S, Hori M, Noda T, Suda T, Yashiki S, Yamaguchi A. Parathyroid hormone exerts disparate effect on osteoblast differentiation depending on exposure time in rat osteoblastic cells. *J Clin Invest* 1997;**99**:2961–70.

70. Philbrick WM. Defining the roles of parathyroid hormone-related protein in normal physiology. *Physiol Rev* 1996;**76**:127–73.

71. Karaplis AC, Luz A, Glowacki J, Bronson RT, Tybulewicz VL, Kronenberg HM, Mulligan RC. Lethal skeletal dysplasia from targeted

disruption of the parathyroid hormone-related peptide gene. *Genes Dev* 1994;**8**:277–89.

72. Bitgood MJ, McMahon AP. Hedgehog and Bmp genes are coexpressed at many diverse sites of cell-cell interaction in the mouse embryo. *Dev Biol* 1995;**172**:126–38.

73. Karp SJ, Schipani E, St-Jacques B, Hunzelman J, Kronenberg H, McMahon AP. Indian hedgehog coordinates endochondral bone growth and morphogenesis via parathyroid hormone related-protein-dependent and -independent pathways. *Development* 2000;**127**:543–8.

74. Chung UI, Schipani E, McMahon AP, Kronenberg HM. Indian hedgehog couples chondrogenesis to osteogenesis in endochondral bone development. *J Clin Invest* 2001;**107**:295–304.

75. Jee WS, Ueno K, Deng YP, Woodbury DM. The effects of prostaglandin E2 in growing rats: increased metaphyseal hard tissue and cortico-endosteal bone formation. *Calcif Tissue Intl* 1985;**37**:148–57.

76. Baylink TM, Mohan S, Fitzsimmons RJ, Baylink DJ. Evaluation of signal transduction mechanisms for the mitogenic effects of prostaglandin E2 in normal human bone cells in vitro. *J Bone Miner Res* 1996;**11**:1413–18.

77. Raisz LG, Pilbeam CC, Fall PM. Prostaglandins: mechanisms of action and regulation of production in bone. *Osteoporos Intl* 1993;**3**(Suppl. 1):136–40.

78. Kiriyama M, Ushikubi F, Kobayashi T, Hirata M, Sugimoto Y, Narumiya S. Ligand binding specificities of the eight types and subtypes of the mouse prostanoid receptors expressed in Chinese hamster ovary cells. *Br J Pharmacol* 1997;**122**:217–24.

79. Paralkar VM, Borovecki F, Ke HZ, Cameron KO, Lefker B, Grasser WA, Owen TA, Li M, DaSilva-Jardine P, Zhou M, Dunn RL, Dumont F, Korsmeyer R, Krasney P, Brown TA, Plowchalk D, Vukicevic S, Thompson DD. An EP2 receptor-selective prostaglandin E2 agonist induces bone healing. *Proc Natl Acad Sci USA* 2003;**100**:6736–40.

80. Tanaka M, Sakai A, Uchida S, Tanaka S, Nagashima M, Katayama T, Yamaguchi K, Nakamura T. Prostaglandin E2 receptor (EP4) selective agonist (ONO-4819.CD) accelerates bone repair of femoral cortex after drill-hole injury associated with local upregulation of bone turnover in mature rats. *Bone* 2004;**34**:940–8.

81. Langenbach R, Loftin C, Lee C, Tiano H. Cyclooxygenase knockout mice: models for elucidating isoform-specific functions. *Biochem Pharmacol* 1999;**58**:1237–46.

82. Forwood MR. Inducible cyclo-oxygenase (COX-2) mediates the induction of bone formation by mechanical loading *in vivo*. *J Bone Miner Res* 1996;**11**:1688–93.

83. Ajubi NE, Klein-Nulend J, Alblas MJ, Burger EH, Nijweide PJ. Signal transduction pathways involved in fluid flow-induced PGE2 production by cultured osteocytes. *Am J Physiol* 1999;**276**:E171–8.

84. Gallwitz WE, Mundy GR, Lee CH, Qiao M, Roodman GD, Raftery M, Gaskell SJ, Bonewald LF. 5-Lipoxygenase metabolites of arachidonic acid stimulate isolated osteoclasts to resorb calcified matrices. *J Biol Chem* 1993;**268**:10,087–10,094.

85. Garcia C, Boyce BF, Gilles J, Dallas M, Qiao M, Mundy GR, Bonewald LF. Leukotriene B4 stimulates osteoclastic bone resorption both *in vitro* and *in vivo*. *J Bone Miner Res* 1996;**11**:1619–27.

86. Garcia C, Qiao M, Chen D, Kirchen M, Gallwitz W, Mundy GR, Bonewald LF. Effects of synthetic peptido-leukotrienes on bone resorption *in vitro*. *J Bone Miner Res* 1996;**11**:521–9.

87. Traianedes K, Dallas MR, Garrett IR, Mundy GR, Bonewald LF. 5-Lipoxygenase metabolites inhibit bone formation in vitro. *Endocrinology* 1998;**139**:3178–84.

88. Bonewald LF, Flynn M, Qiao M, Dallas MR, Mundy GR, Boyce BF. Mice lacking 5-lipoxygenase have increased cortical bone thickness. *Adv Exp Med Biol* 1997;**433**:299–302.

89. Mehrabian M, Allayee H, Stockton J, Lum PY, Drake TA, Castellani LW, Suh M, Armour C, Edwards S, Lamb J, Lusis AJ, Schadt EE. Integrating genotypic and expression data in a segregating mouse population to identify 5-lipoxygenase as a susceptibility gene for obesity and bone traits. *Nat Genet* 2005;**37**:1224–33.

90. Klein RF, Allard J, Avnur Z, Nikolcheva T, Rotstein D, Carlos AS, Shea M, Waters RV, Belknap JK, Peltz G, Orwoll ES. Regulation of bone mass in mice by the lipoxygenase gene Alox15. *Science* 2004;**303**:229–32.

91. Ogawa E, Maruyama M, Kagoshima H, Inuzuka M, Lu J, Satake M, Shigesada K, Ito Y. PEBP2/PEA2 represents a family of transcription factors homologous to the products of the *Drosophila* runt gene and the human AML1 gene. *Proc Natl Acad Sci USA* 1993;**90**:6859–63.

92. Ducy P, Zhang R, Geoffroy V, Ridall AL, Karsenty G. Osf2/Cbfa1: a transcriptional activator of osteoblast differentiation. *Cell* 1997;**89**:747–54.

93. Karsenty G. How many factors are required to remodel bone? *Nat Med* 2000;**6**:970–1.

94. Nakashima K, Zhou X, Kunkel G, Zhang Z, Deng JM, Behringer RR, de Crombrugghe B. The novel zinc finger-containing transcription factor osterix is required for osteoblast differentiation and bone formation. *Cell* 2002;**108**:17–29.

95. Xiao G, Jiang D, Thomas P, Benson MD, Guan K, Karsenty G, Franceschi RT. MAPK pathways activate and phosphorylate the osteoblast-specific transcription factor, Cbfa1. *J Biol Chem* 2000;**275**:4453–9.

96. Little RD, Carulli JP, Del Mastro RG, Dupuis J, Osborne M, Folz C, Manning SP, Swain PM, Zhao SC, Eustace B, Lappe MM, Spitzer L, Zweier S, Braunschweiger K, Benchekroun Y, Hu X, Adair R, Chee L, FitzGerald MG, Tulig C, Caruso A, Tzellas N, Bawa A, Franklin B, McGuire S, Nogues X, Gong G, Allen KM, Anisowicz A, Morales AJ, Lomedico PT, Recker SM, Van Eerdewegh P, Recker RR, Johnson ML. A mutation in the LDL receptor-related protein 5 gene results in the autosomal dominant high-bone-mass trait. *Am J Hum Genet* 2002;**70**:11–19.

97. Gong Y, Vikkula M, Boon L, Liu J, Beighton P, Ramesar R, Peltonen L, Somer H, Hirose T, Dallapiccola B, De Paepe A, Swoboda W, Zabel B, Superti-Furga A, Steinmann B, Brunner HG, Jans A, Boles RG, Adkins W, van den Boogaard MJ, Olsen BR, Warman ML. Osteoporosis-pseudoglioma syndrome, a disorder affecting skeletal strength and vision, is assigned to chromosome region 11q12–13. *Am J Hum Genet* 1996;**59**:146–51.

98. Wehrli M, Dougan ST, Caldwell K, O'Keefe L, Schwartz S, Vaizel-Ohayon D, Schejter E, Tomlinson A, DiNardo S. arrow encodes an LDL-receptor-related protein essential for Wingless signalling. *Nature* 2000;**407**:527–30.

99. Mann B, Gelos M, Siedow A, Hanski ML, Gratchev A, Ilyas M, Bodmer WF, Moyer MP, Riecken EO, Buhr HJ, Hanski C. Target genes of beta-catenin-T cell-factor/lymphoid-enhancer-factor signaling in human colorectal carcinomas. *Proc Natl Acad Sci USA* 1999;**96**:1603–8.

100. Mao 3rd J, Wang J, Liu B, Pan W, Farr GH, Flynn C, Yuan H, Takada S, Kimelman D, Li L, Wu D. Low-density lipoprotein receptor-related protein-5 binds to Axin and regulates the canonical Wnt signaling pathway. *Mol Cell* 2001;**7**:801–9.

101. Johnson ML, Picconi JL, Recker RR. The gene for high bone mass. *Endocrinologist* 2002;**12**:445–53.

102. Glass 2nd DA, Bialek P, Ahn JD, Starbuck M, Patel MS, Clevers H, Taketo MM, Long F, McMahon AP, Lang RA, Karsenty G. Canonical Wnt signaling in differentiated osteoblasts controls osteoclast differentiation. *Dev Cell* 2005;**8**:751–64.

103. Heldin CH, Miyazono K, ten Dijke P. TGF-beta signalling from cell membrane to nucleus through SMAD proteins. *Nature* 1997;**390**:465–71.

104. de Caestecker MP, Piek E, Roberts AB. Role of transforming growth factor-beta signaling in cancer. *J Natl Cancer Inst* 2000;**92**:1388–402.

105. Massague J. TGF-beta signal transduction. *Annu Rev Biochem* 1998:753–91.

106. Borton AJ, Frederick JP, Datto MB, Wang XF, Weinstein RS. The loss of Smad3 results in a lower rate of bone formation and osteopenia through dysregulation of osteoblast differentiation and apoptosis. *J Bone Miner Res* 2001;**16**:1754–64.

107. Hahn SA, Schmiegel WH. Recent discoveries in cancer genetics of exocrine pancreatic neoplasia. *Digestion* 1998;**59**:493–501.

108. Bonewald LF, Dallas SL. The role of growth factors in bone formation. In: . *Advances in oral biology*, vol. 5B. Stamford, CT: JAI Press Inc; 1998. p. 591–613. 108.

109. Celeste AJ, Iannazzi JA, Taylor RC, Hewick RM, Rosen V, Wang EA, Wozney JM. Identification of transforming growth factor beta family members present in bone-inductive protein purified from bovine bone. *Proc Natl Acad Sci USA* 1990;**87**:9843–7.

110. Rosen V, Wozney JM. Bone morphogenetic proteins. In: Bilezikian JP, Raisz LG, Rodan GA, editors. *Principles of bone biology*, vol. 2. San Diego, CA: Academic Press; 2002. p. 919–28.

111. Kohei M. Bone morphogenetic protein receptor and actions. In: Bilezikian JP, Raisz LG, Rodan GA, editors. *Pinciples of bone biology*, vol. 2. San Diego, CA: Academic Press; 2002. p. 929–42.

112. Mohan S, Linkhart TA, Jennings JC, Baylink DJ. Identifaction and quantification of four distinct growth factors stored in human bone matrix. *J Bone Miner Res* 1987;**2**:44–7.

113. Zhao G, Monier-Faugere MC, Langub MC, Geng Z, Nakayama T, Pike JW, Chernausek SD, Rosen CJ, Donahue LR, Malluche HH, Fagin JA, Clemens TL. Targeted overexpression of insulin-like growth factor I to osteoblasts of transgenic mice: increased trabecular bone volume without increased osteoblast proliferation. *Endocrinology* 2000;**141**:2674–82.

114. Conover CA, Rosen C. The role of insulin-like growth factors and binding proteins in bone cell biology. In: Bilezikian JP, Raisz LG, Rodan GA, editors. *Principles of bone biology*, vol. 2. San Diego, CA: Academic Press; 2002. p. 801–15.

115. Wuster C, Rosen C. Growth hormone, insulin-like growth factors: potential applications and limitations in the management of osteoporosis. In: Marcus R, Feldman D, Kelsey J, editors. *Osteoporosis*, vol. 2. San Diego, CA: Academic Press; 2001. p. 747–67.

116. Hurley MM, Marie PJ, Florkiewicz RZ. Fibroblast growth factor (FGF) and FGF receptor families in bone. In: Bilezikian JP, Raisz LG, Rodan GA, editors. *Principles of bone biology*, vol. 1. San Diego, CA: Academic Press; 2002. p. 825–51.

117. Montero A, Okada Y, Tomita M, Ito M, Tsurukami H, Nakamura T, Doetschman T, Coffin JD, Hurley MM. Disruption of the fibroblast growth factor-2 gene results in decreased bone mass and bone formation. *J Clin Invest* 2000;**105**:1085–93.

118. Naski MC, Colvin JS, Coffin JD, Ornitz DM. Repression of hedgehog signaling and BMP4 expression in growth plate cartilage by fibroblast growth factor receptor 3. *Development* 1998;**125**:4977–88.

119. Iwata T, Chen L, Li C, Ovchinnikov DA, Behringer RR, Francomano CA, Deng CX. A neonatal lethal mutation in FGFR3 uncouples proliferation and differentiation of growth plate chondrocytes in embryos. *Hum Mol Genet* 2000;**9**:1603–13.

120. Iwata T, Li CL, Deng CX, Francomano CA. Highly activated Fgfr3 with the K644M mutation causes prolonged survival in severe dwarf mice. *Hum Mol Genet* 2001;**10**:1255–64.

121. Chen L, Li C, Qiao W, Xu X, Deng C. A Ser(365) → Cys mutation of fibroblast growth factor receptor 3 in mouse downregulates Ihh/PTHrP signals and causes severe achondroplasia. *Hum Mol Genet* 2001;**10**:457–65.

122. Thakker RV. Hereditary hypophosphataemic rickets: role for a fibroblast growth factor, FGF23. *IBMS BoneKEy* 2001.

123. Toyosawa S, Shintani S, Fujiwara T, Ooshima T, Sato A, Ijuhin N, Komori T. Dentin matrix protein 1 is predominantly expressed in chicken and rat osteocytes but not in osteoblasts. *J Bone Miner Res* 2001;**16**:2017–26.

124. Westbroek I, De Rooij KE, Nijweide PJ. Osteocyte-specific monoclonal antibody MAb OB7.3 is directed against Phex protein. *J Bone Miner Res* 2002;**17**:845–53.

125. Liu S, Zhou J, Tang W, Jiang X, Rowe DW, Quarles LD. Pathogenic role of Fgf23 in hyp mice. *Am J Physiol Endocrinol Metab* 2006;**291**:E38–49.

126. Feng JQ, Ward LM, Liu S, Lu Y, Xie Y, Yuan B, Yu X, Rauch F, Davis SI, Zhang S, Rios H, Drezner MK, Quarles LD, Bonewald LF, White KE. Loss of DMP1 causes rickets and osteomalacia and identifies a role for osteocytes in mineral metabolism. *Nat Genet* 2006;**38**:1310–15.

127. Lorenz-Depiereux B, Bastepe M, Benet-Pages A, Amyere M, Wagenstaller J, Muller-Barth U, Badenhoop K, Kaiser SM, Rittmaster RS, Shlossberg AH, Olivares JL, Loris C, Ramos FJ, Glorieux F, Vikkula M, Juppner H, Strom TM. DMP1 mutations in autosomal recessive hypophosphatemia implicate a bone matrix protein in the regulation of phosphate homeostasis. *Nat Genet* 2006;**38**:1248–50.

128. Gluhak-Heinrich J, Bonewald L, Feng JQ, MacDougall M, Harris SE, Pavlin D. Mechanical loading stimulates dentin matrix protein 1 (DMP1) expression in osteocytes in vivo. *J Bone Miner Res* 2003;**18**:807–17.

129. Gluhak-Heinrich J, Pavlin D, Yang W, MacDougall M, Harris SE. MEPE expression in osteocytes during orthodontic tooth movement. *Arch Oral Biol* 2007;**52**:684–90.

130. Yang W, Lu Y, Kalajzic I, Guo D, Harris MA, Gluhak-Heinrich J, Kotha S, Bonewald LF, Feng JQ, Rowe DW, Turner CH, Robling AG, Harris SE. Dentin matrix protein 1 gene cis-regulation: use in osteocytes to characterize local responses to mechanical loading in vitro and in vivo. *J Biol Chem* 2005;**280**:20,680–20690.

131. Lian JB, Stein GS, Canalis E, Robey P, Boskey AL. Osteoblast lineage cells, growth factors, matrix proteins, and the mineralization process. In: Editor E, editor. *Primer on the metabolic bone diseases and disorders of mineral metabolism. bone formation*. 4th ed. Philadelphia, PA: Lippincott Williams & Wilkins; 1999. p. 14–19.

132. Ye L, Mishina Y, Chen D, Huang H, Dallas SI, Dallas MR, Sivakumar P, Kunieda T, Tsutsui TW, Boskey A, Bonewald LF, Feng JQ. Dmp1-deficient mice display severe defects in cartilage formation responsible for a chondrodysplasia-like phenotype. *J Biol Chem* 2005;**280**:6197–6203.

133. Dallas SL, Sivakumar P, Jones CJ, Chen Q, Peters DM, Mosher DF, Humphries MJ, Kielty CM. Fibronectin regulates latent transforming growth factor-beta (TGF beta) by controlling matrix assembly of latent TGF beta-binding protein-1. *J Biol Chem* 2005;**280**:18,871–18880.

134. Sivakumar P, Czirok A, Rongish BJ, Divakara VP, Wang YP, Dallas SL. New insights into extracellular matrix assembly and reorganization from dynamic imaging of extracellular matrix proteins in living osteoblasts. *J Cell Sci* 2006;**119**:1350–1360.

135. Robey PG. Bone matrix proteoglycans and glycoproteins. In: Bilezikian JP, Raisz LG, Rodan GA, editors. *Principles of bone biology*, vol. 1. San Diego, CA: Academic Press; 2002. p. 225–237.

136. Feng JQ, Ye L, Huang H, Lu Y, Ye L, Xie Y, Tsutsui T, Kunieda T, Castranio T, Scott G, Bonewald L, Mishina Y. The dentin matrix protein 1 (Dmp1) is specifically expressed in mineralized, but not soft tissues during development. *J Dent Res* 2003;**82**:776–780.

137. Dallas SL, Keene DR, Bruder SP, Saharinen J, Sakai LY, Mundy GR, Bonewald LF. Role of the latent transforming growth factor beta binding protein 1 in fibrillin-containing microfibrils in bone cells *in vitro* and *in vivo*. *J Bone Miner Res* 2000;**15**:68–81.

138. Dabovic B, Chen Y, Colarossi C, Obata H, Zambuto L, Perle MA, Rifkin DB. Bone abnormalities in latent TGF-β binding protein (Ltbp)-3-null mice indicate a role for Ltbp-3 in modulating TGF-β bioavailability. *J Cell Biol* 2002;**156**:227–232.

139. Sterner-Kock A, Thorey IS, Koli K, Wempe F, Otte J, Bangsow T, Kuhlmeier K, Kirchner T, Jin S, Keski-Oja J, von Melchner H. Disruption of the gene encoding the latent transforming growth factor-beta binding protein 4 (LTBP-4) causes abnormal lung development, cardiomyopathy, and colorectal cancer. *Genes Dev* 2002; **16**:2264–2273.

140. Koli K, Wempe F, Sterner-Kock A, Kantola A, Komor M, Hofmann WK, von Melchner H, Keski-Oja J. Disruption of LTBP-4 function reduces TGF-beta activation and enhances BMP-4 signaling in the lung. *J Cell Biol* 2004;**167**:123–133.

141. Christiano AM, Uitto J. Molecular pathology of the elastic fibers. *J Invest Dermatol* 1994;**103**:S53–S57.

142. Rosenbloom J, Abrams WR, Mecham R. Extracellular matrix 4: the elastic fiber. *FASEB J* 1993;**7**:1208–1218.

143. George EL, Baldwin HS, Hynes RO. Fibronectins are essential for heart and blood vessel morphogenesis but are dispensable for initial specification of precursor cells. *Blood* 1997;**90**:3073–3081.

144. George EL, Georges-Labouesse EN, Patel-King RS, Rayburn H, Hynes RO. Defects in mesoderm, neural tube and vascular development in mouse embryos lacking fibronectin. *Development* 1993;**119**:1079–1091.

145. Pereira L, Lee SY, Gayraud B, Andrikopoulos K, Shapiro SD, Bunton T, Biery NJ, Dietz HC, Sakai LY, Ramirez F. Pathogenetic sequence for aneurysm revealed in mice underexpressing fibrillin-1. *Proc Natl Acad Sci USA* 1999;**96**:3819–3823.

146. Dietz HC, Mecham RP. Mouse models of genetic diseases resulting from mutations in elastic fiber proteins. *Matrix Biol* 2000;**19**:481–488.

147. Pfaff M, Reinhardt DP, Sakai LY, Timpl R. Cell adhesion and integrin binding to recombinant human fibrillin-1. *FEBS Letts* 1996;**384**:247–250.

148. Gibson MA, Leavesley DI, Ashman LK. Microfibril-associated glycoprotein-2 specifically interacts with a range of bovine and human cell types via alphaVbeta3 integrin. *J Biol Chem* 1999; **274**:13,060–13,065.

149. Pan TC, Kluge M, Zhang RZ, Mayer U, Timpl R, Chu ML. Sequence of extracellular mouse protein BM-90/fibulin and its calcium-dependent binding to other basement-membrane ligands. *Eur J Biochem* 1993;**215**:733–740.

150. Balbona K, Tran H, Godyna S, Ingham KC, Strickland DK, Argraves WS. Fibulin binds to itself and to the carboxyl-terminal heparin-binding region of fibronectin. *J Biol Chem* 1992;**267**: 20,120–20,125.

151. Vu TH. Don't mess with the matrix. *Nat Genet* 2001;**28**:202–203.

152. Martignetti JA, Aqeel AA, Sewairi WA, Boumah CE, Kambouris M, Mayouf SA, Sheth KV, Eid WA, Dowling O, Harris J, Glucksman MJ, Bahabri S, Meyer BF, Desnick RJ. Mutation of the matrix metalloproteinase 2 gene (MMP2) causes a multicentric osteolysis and arthritis syndrome. *Nat Genet* 2001;**28**:261–265.

153. Holmbeck K, Bianco P, Caterina J, Yamada S, Kromer M, Kuznetsov SA, Mankani M, Robey PG, Poole AR, Pidoux I, Ward JM, Birkedal-Hansen H. MT1-MMP-deficient mice develop dwarfism, osteopenia, arthritis, and connective tissue disease due to inadequate collagen turnover. *Cell* 1999;**99**:81–92.

154. Krane SM. *Meeting report from the frontiers of skeletal biology: Ninth workshop on cell biology of bone and cartilage in health and disease*, Davos, Switzerland, 2002.

155. Neuhold LA, Killar L, Zhao W, Sung ML, Warner L, Kulik J, Turner J, Wu W, Billinghurst C, Meijers T, Poole AR, Babij P, DeGennaro LJ. Postnatal expression in hyaline cartilage of constitutively active human collagenase-3 (MMP-13) induces osteoarthritis in mice. *J Clin Invest* 2001;**107**:35–44.

Chapter 14

Cell–Cell Signaling in the Testis and Ovary

Michael K. Skinner, Eric E. Nilsson and Ramji K. Bhandari
Center for Reproductive Biology, School of Molecular Biosciences, Washington State University, Pullman, Washington

INTRODUCTION

The evolution of multicellular organisms was facilitated by the ability of different cells to communicate and interact. This cell–cell signaling generates a higher order functional state than possible with individual cell types. Cell–cell interactions have become an essential requirement for the physiology of any organ or tissue, and are critical in the regulation of any cell's biology. For this reason, elaborate networks of cell–cell interactions have evolved to control the development and maintenance of tissue functions. The focus of the current chapter will be on the regulatory signals that mediate cell–cell interactions in the testis and ovary.

Several previous reviews have discussed the cell–cell interactions in the testis [1–4] and ovary [5–7]. These include a focus on secretory products of the various cell types, and actions of individual regulatory molecules. The current chapter will briefly discuss the advances in cell–cell signaling in these organs.

Many different types of cell–cell interactions are required for the control of tissue physiology and cellular functions. These have been previously categorized into regulatory, nutritional, and environmental classifications [4]. Regulatory interactions are generally mediated by extracellular factors that, through receptor-mediated actions, cause a signaling event to modulate cell functions. Nutritional interactions generally involve the transport of nutritional substances, energy metabolites, or metabolic substrates between cells. Environmental interactions involve extracellular environmental factors that affect cell contacts and cytoarchitecture. The focus of the current chapter will be primarily on regulatory-type interactions that involve a receptor-mediated signaling event. It is this type of cellular signaling that actively regulates a cell's function on a molecular level. The factors involved are generally paracrine and autocrine agents such as growth factors and cytokines.

Both the testis and ovary are endocrine organs. Endocrine hormones from the pituitary (i.e., gonadotropins,

follicle stimulating hormone (FSH), and luteinizing hormone (LH)) act on various cell types to influence cellular functions and cell–cell interactions. The influence these endocrine hormones have on cell–cell signaling events is in part how hormones regulate gonadal function. The testis and ovary are also sites for the production of hormones. These gonadal hormones have an endocrine role in regulating a wide variety of tissues in the body, but also can act in a paracrine manner within the gonads to influence cell–cell signaling and cellular functions. Again, the role these gonadal steroids and peptide hormones play in the regulation of cell–cell signaling within the gonad will be discussed.

CELL–CELL SIGNALING IN THE TESTIS

Testis Cell Biology

The adult testis is a complex organ that is composed of seminiferous tubules which are enclosed by a surrounding interstitium. The seminiferous tubules are the site of spermatogenesis where germ cells develop into spermatozoa in close interaction with Sertoli cells (Figure 14.1). The Sertoli cell is an important testicular somatic cell which controls the germ cell environment by the secretion and transport of nutrients and regulatory factors. The Sertoli cells [8] form the basal and apical surface of the seminiferous tubule, and provide the cytoarchitectural framework for the developing germinal cells [3, 9]. Tight junctional complexes between the Sertoli cells contribute to the maintenance of a blood–testis barrier [10], and create a unique environment within the tubule [3, 11]. The structure of the Sertoli cell has been reviewed by several investigators [9], and a three-dimensional reconstruction has increased appreciation for the complexity of the structural relationships between cells within the seminiferous tubule [12]. The biochemical analysis of the Sertoli cell has primarily

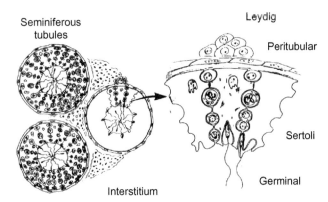

FIGURE 14.1 Testis cell biology.

focused on an examination of the components synthesized and secreted by the cell. The list of products includes steroids such as estradiol [13], metabolites such as lactate [14], and various proteins such as plasminogen activator [15], testicular transferrin [16], testicular ceruloplasmin [17], inhibin [2], and others [2]. The majority of the secretory products are hormonally regulated and provide useful markers of Sertoli cell differentiation.

Surrounding the basal surface of the Sertoli cells is a layer of peritubular myoid cells (Figure 14.1), which function in contraction of the tubule. The peritubular cells surround and form the exterior wall of the seminiferous tubule. Peritubular cells are mesenchymally derived cells that secrete fibronectin [18] and several extracellular matrix components [19]. Both the peritubular and the Sertoli cells form the basement membrane surrounding the seminiferous tubule, and their interactions are important in germ cell development.

The interstitial space around the seminiferous tubules contains another somatic cell type, the Leydig cell (Figure 14.1), which is responsible for testosterone production. Leydig cells have a major influence on spermatogenesis through the actions of testosterone on both the seminiferous tubule and the pituitary. Although the Leydig cell has numerous secretory products [4], testosterone is the most significant secretory product of these cells. Thus, interactions of all three somatic cells, Sertoli, peritubular, and Leydig, are important for regulation of normal spermatogenic function in the testis (for review, see [4]).

Testis Development

The process of fetal testis formation occurs late in embryonic development (embryonic day 13 where plug date = E0 (E13) in the rat) and is initiated by migration of primordial germ cells, first from the yolk sac to the hindgut, and then from the hindgut to the genital ridge. The first phase of migration is proposed to occur through a mechanism where transient interactions between fibronectin molecules on the extracellular matrix and corresponding receptors on the primordial germ cells cause movement of the germ cells. The second

migration is thought to occur by the release of chemoattractant factors from the genital ridge. Kit ligand and its receptor c-kit appear to be involved first in the migration to the genital ridge, and later in the proliferation of germ cells after colonization of the genital ridge. Expression of kit ligand has been localized to cells along the migratory pathway, and c-kit is expressed by primordial germ cells at this time in development (for review, see [20]). After migration, germ cell differentiation in the gonad is dependent on locally produced factors such as prostaglandins [21], growth factors [22], and the induction of specific transcription factors [23]. It is a complex network of cellular interactions that controls testis and germ cell development.

The gonad has bipotential after germ cell migration, and can be distinguished morphologically from the adjoining mesonephros (E12 in rat), but cannot yet be identified as an ovary or a testis. A variety of genes, such as SRY, SOX-9, SF1, and DMRT1, are involved in the transcriptional induction of sex determination and testis development [24–33]. Two morphological events occur early on embryonic day 13 (E13) to alter the bipotential gonad. First, Sertoli cells, which are proposed to be the first cell in the testis to differentiate, aggregate around primordial germ cells [34, 35]. Secondly, migration of mesenchymal cells occurs from the adjoining mesonephros into the developing gonad to surround the Sertoli cell–germinal cell aggregates. The migrating population of cells has been speculated to be pre-peritubular cells [36–38]. The mechanism for this migration appears to involve chemotactic factors from the Sertoli cell, such as NT3 [39] and FGF9 [22], that cause cell migration. This is postulated due to the observation that ovarian mesonephros can also be stimulated to initiate cell migration after close interaction with a developing testis [40]. In addition, using an organ culture system in which mesonephros and embryonic testis were separated by an embryonic ovary, mesonephros cells migrated through the ovary to the testis [36]. Several growth factors appear to be involved in this initial testis morphogenesis, including interactions between FGF9 and Wnt 4 [22, 41], Wnt(s) [42, 43], and Notch regulators [43]. Therefore, during early testis development Sertoli–peritubular cell interactions may allow for cord formation to occur. The cords develop neonatally into seminiferous cords and, at the onset of puberty, develop into the seminiferous tubules. Sertoli cells have been postulated to originate from stem cells in the coelomic epithelium at an early stage in gonadal development. Other cells which may potentially originate from the coelomic epithelium are interstitial or Leydig cells [44].

Seminiferous cords, precursors of adult seminiferous tubules, form as the Sertoli cell–primordial germ cell aggregates become more organized and are fully surrounded by mesenchymal cells. The formation of the seminiferous cords (E14 in rat) is a critical event in the morphogenesis of the testis, since this is the first indication of male sex differentiation [27]. During the process of cord formation Sertoli cells undergo a number of morphological changes, including: a change in expression of mesenchymal to epithelial cell markers (vimentin to cytokeratin [45]), a change in expression of cytokeratin 19 to cytokeratin 18 (cytokeratin 21 expressed in ovary [46]), and expression of Müllerian inhibiting substance (MIS), which inhibits the development of the Müllerian duct–the precursor of the female uterus, cervix, fallopian tubes and upper vagina [47, 48]. Vascular endothelial growth factor (VEGF) appears to mediate cell–cell interactions and migrations required for vascularization of the gonad [49].

Outside of the seminiferous cords, the peritubular layer of cells becomes identifiable from the interstitium or Leydig cells at E15 [50], and 3β-hydroxysteroid dehydrogenase (3βHSD) production is detected after E15 [48]. Leydig cells have been hypothesized to differentiate after cord formation and Sertoli cell differentiation is completed [51, 52]. This is important, since the production of testosterone and other androgens by the Leydig cells has been demonstrated to stabilize the Wolffian duct derivatives for normal male duct development [53, 54]. Therefore, appropriate differentiation of somatic cell types in the testis around the time of cord formation is crucial not only to the normal development of the testis, but also for the continued presence of the Wolffian duct and normal male reproductive tract development.

Testis Cell–Cell Interactions

Table 14.1 outlines a number of the factors produced locally in the testis that mediate cell–cell signaling events in the control of spermatogenesis and testis function. Several reviews address the topic of cell–cell interactions in the testis and the control of spermatogenesis [2, 4, 55, 56]. Recent observations are cited below.

Transforming growth factor-α (TGFα) is an epidermal growth factor (EGF) superfamily member, and is produced by Sertoli, peritubular, and Leydig cells. TGFα can act as a growth stimulator on all the major cell types in the testis [57–59]. In contrast, transforming growth factor-beta (TGFβ) is also produced by Sertoli, peritubular, and Leydig cells, and can act on all the major cells in the testis [60–64]. TGFβ primarily acts as a growth inhibitor, and can stimulate a variety of functions of differentiated cell types. A number of other TGFβ superfamily members have also been shown to regulate testis function and development [65], including bone morphogenic proteins (BMPs), activins, and growth differentiation factors (GDFs) [65]. Another example of a factor that is produced by all the testis cells and acts on all major cell types in the testis is insulin-like growth factor-1 (IGF-1) [66–68]. IGF-1 plays a general role in regulation of the growth cycle and homeostasis of the testis. A related family member, IGF-2, mediates paracrine interactions between Sertoli cells and germ cells [69]. These are examples of regulatory factors that mediate cell–cell signaling events between the majority of the cell types in the testis.

Several interleukins (IL-1α, IL-1β, IL-6) are produced in the testis by Sertoli cells and Leydig cells. These interleukins can regulate Sertoli, Leydig, and germ cell growth and differentiation functions [70–77]. Nitric oxide may be a mediator of interleukin actions [78]. Although further analysis is needed, interleukins appear to mediate primarily Sertoli–germ cell and Leydig–Sertoli cell interactions, as well as having autocrine roles for these factors.

Several hormonal factors produced in the testis also act locally within the testis as paracrine factors. One example is inhibin and its related peptide, activin [79–81]. Inhibin is primarily produced by Sertoli cells, and can act on germ cells and Leydig cells. Further investigation of the actions of inhibin and related compounds within the testis is needed. Another major endocrine factor produced in the testis is testosterone, which is generated by Leydig cells and can in turn act on Sertoli, peritubular, and Leydig cells [82]. Androgens have a major role in the maintenance of testis function by inducing cellular differentiated functions. The specific mechanism of action and gene products influenced by androgens remain to be elucidated. Early in prepubertal development, testosterone can also be metabolized by Sertoli cells to produce estrogen [13]. The ability of Sertoli cells to produce estrogen declines as the cells differentiate during puberty, and the role of estrogen in the testis is unclear.

Fibroblast growth factor (FGF) family members have been shown to be expressed in the testis, and regulate the growth and differentiation of a variety of cells [83–87]. FGF receptors are predominant in germ cells and Leydig cells, but are also present in others [84]. FGF-14 has recently been shown to be expressed in spermatocytes, and may influence adjacent Sertoli or peritubular cells. FGF-9 null mutants also suggest a role for FGF9 in early testis development, but this remains to be investigated in the adult [85]. Basic-FGF (bFGF) is produced by Sertoli cells and also can act on the other cells [86, 88], and appears to be influenced by androgens [88]. The role of the various FGF ligands and receptors in testis function remains to be elucidated.

Platelet-derived growth factor (PDGF) has been shown to be produced by Sertoli cells and to influence peritubular cells and Leydig cells [89–91]. Although PDGF in the adult may also be produced by the Leydig cell [92], it appears to be a factor produced within the seminiferous tubules that acts on adjacent peritubular and Leydig cells. Another factor that is only produced by Sertoli cells is stem cell factor (SCF)/kit ligand (KL), which has a direct role in regulating spermatogonial cell proliferation [93–96]. Mutations in SCF/KL block the process of spermatogenesis. A growth factor with similar activity is glial cell derived neurotropic factor (GDNF), which is produced by Sertoli cells and acts on spermatogonial stem cells [97]. These are the best examples of somatic–germ cell interactions.

TABLE 14.1 Cell–cell signaling factors in the testis

Signaling factor	Site production	Site action	Functions	Ref(s)
Transforming growth factor α (TGFα)	Sertoli	Sertoli	Growth stimulation	[57–59]
	Peritubular	Peritubular		
	Leydig	Leydig		
		Germ		
Transforming growth factor β (TGFβ)	Sertoli	Sertoli	Growth inhibition	[60–64]
	Peritubular	Peritubular	Differentiation, stimulation	
	Leydig	Leydig		
		Germ		
Insulin-like growth factor (IGF1)	Sertoli	Sertoli	Homeostasis and DNA synthesis	[66–68]
	Peritubular	Peritubular		
	Leydig	Leydig		
		Germ		
Interleukin-s	Sertoli	Sertoli	Growth regulation	[70–77]
	Leydig	Leydig	Cellular differentiation	
		Germ		
Inhibin	Sertoli	Germ	Cellular differentiation	[79–81]
		Leydig		
Androgen	Leydig	Sertoli	Cellular differentiation	[82]
		Peributular		
		Leydig		
Fibroblast growth factors	Sertoli	Germ	Growth stimulation	[83–87, 239]
	Germ	Peritubular		
	Leydig	Sertoli		
		Leydig		
Platelet-derived growth factor (PDGF)	Sertoli	Peritubular	Growth stimulation	[89–92]
		Leydig	Cellular differentiation	
Stem cell factor/kit ligand (SCF/KL)	Sertoli	Germ	Growth stimulation	[93–96]
Leukemia inhibitory factor (LIF)	Peritubular	Germ	Growth stimulation	[98, 99]
	Sertoli		Cell survival	
	Leydig			
Tumor necrosis factors	Germ	Sertoli	Cellular apoptosis	[100–102]
	Leydig	Germ	Cellular differentiation	
Hepatocyte growth factor (HGF)	Peritubular	Leydig	Growth stimulation	[105–107]
		Peritubular	Tubule formation	
		Sertoli		
Neurotropins	Germ	Sertoli	Growth stimulation	[4, 109]
	Sertoli	Peritubular	Cell migration	
			Cellular differentiation	
Glial cell derived neurotropic factor (GDNF)	Sertoli	Spermatogonia	Growth stimulation	[97]
			Cellular differentiation	

Leukemia inhibitory factor (LIF) is a pleiotropic cytokine that influences stem cell growth and survival. LIF is predominantly produced by peritubular cells, but also by Sertoli cells and Leydig cells [98]. Although LIF has been shown to influence germ cell growth and survival [99], other functions remain to be elucidated.

Tumor necrosis factors (TNFα) and related ligands (TRAIL) are produced in the testis by germ cells and Leydig cells. Both TNF and TRAIL have a role in regulating germ cells and Sertoli cells [100–103]. Germ cell apoptosis in response to hormone deficiency or environmental compound exposure is mediated in part through TNFα and TNFβ involving Sertoli cell and germ cell interactions [103, 104]. These regulatory factors for the germ cells may be more involved in apoptosis regulation, unlike in Sertoli cells, in which they may be more involved in cellular differentiated functions.

Hepatocyte growth factor (HGF) is generally a mesenchymal-derived factor that acts on adjacent epithelial cells. HGF was found to be expressed by the mesenchymal-derived peritubular cells, and its receptor (cmet) was found on both Sertoli cells and Leydig cells [105–107]. Interestingly, cmet was also found in the peritubular cells. HGF also may have a role in seminiferous tubule formation [107].

Several neurotropins have been shown to be expressed in the testis. Nerve growth factor (NGF) is produced by germ cells in the adult, and appears to act on the Sertoli cells [4, 108]. NGF can act as both an autocrine and a paracrine factor to regulate spermatogenesis [108]. In embryonic development, neurotropin-3 is expressed by Sertoli cells and acts on the migrating mesenephros cells to promote seminiferous cord formation [39, 109]. Further investigations are needed to elucidate the roles of these and other neurotropins in the testis.

Additional factors are anticipated to be identified and have critical roles in testis development. Newly identified factors such as erythropoietin (found to be expressed by Sertoli cells and peritubular cells [110]), hedgehog factors (found to affect spermatogenesis [111]), ghrelin [112] and interferon-gamma (found to act on Sertoli cells [113]), and relaxin-like factor (RLF) (expressed by Leydig cells [114]) will all likely have roles in cell–cell signaling in the testis. These and other factors [115] need to be further investigated to determine roles in testis cell biology. Clearly, a complex network of cell–cell signaling events and factors regulates testis function and spermatogenesis.

CELL–CELL SIGNALING IN THE OVARY

Ovarian Cell Biology

The ability of somatic cells in the gonad to control and maintain the process of gametogenesis is an essential requirement for reproduction. The basic functional unit in the ovary is the ovarian follicle, which is composed of somatic cells and the developing oocyte (Figure 14.2). The two primary somatic cell types in the ovarian follicle are the theca cells and granulosa cells. These two somatic cell types are the site of action and synthesis of a number of hormones which promote a complex regulation of follicular development. The proliferation of these two cell types is in part responsible for the growth of the ovarian follicle. The elucidation of factors that control ovarian somatic cell growth and development is critical to understanding ovarian physiology.

Granulosa cells are the primary cell type in the ovary that provide the physical support and microenvironment required for the developing oocyte (Figure 14.2). Granulosa cells are actively differentiating cell with several distinct populations. Alteration and progression of cellular differentiation is required during folliculogenesis from the arrested primordial stage of development through ovulation to the luteal stage of development. Regulation of granulosa cell cytodifferentiation requires the actions of a number of hormones and growth factors. Specific receptors have been demonstrated on granulosa cells for the gonadotropins FSH [116] and LH [117]. In addition, receptors have been found for factors such as EGF [118, 119], insulin-like growth factor [120], and anti-Müllerian hormone [121]. The actions of these hormones and growth factors on granulosa cells vary with the functional marker being examined and the stage of differentiation. The biosynthesis of two important ovarian steroids, estradiol (Figure 14.2) and progesterone, is a primary function of the granulosa cells in species such as cattle, humans, and rodents. Estrogen biosynthesis is controlled by the enzyme aromatase, which requires androgen (Figure 14.2) produced by the theca cells as a substrate. As the follicle develops, granulosa cells differentiate and estrogen biosynthesis increases. FSH promotes this follicular development via the actions of cAMP. As the follicle reaches the stages before ovulation, the granulosa cells develop an increased capacity to synthesize and secrete progestins under the control of LH. In contrast, the early follicle stage (e.g., primordial) granulosa cells appear to be hormone-independent and are non-steroidogenic.

Another important cell type in the ovary is the ovarian theca cell (Figure 14.2). These are differentiated stromal cells that surround the follicle and have also been termed theca interstitial cells [122]. The inner layer of cells, the theca interna, has a basement membrane separating it from the outermost layer of mural granulosa cells. One of the major functions of theca cells in species such as cattle, humans, and rodents is the secretion of androgens which are used by granulosa cells to produce estrogen [123]. Theca cells respond to LH by increasing the production of androgens from cholesterol [124] (Figure 14.2). Theca cells also produce progestins under gonadotropin control [125–128]. Other secretory products of theca cells have not been thoroughly investigated. At the primordial stage no theca

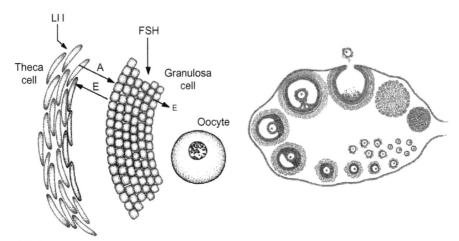

FIGURE 14.2 Ovary cell biology.

cells are present; however, during transition to the primary stage theca cells (i.e., pre-cursor cells) are recruited to the follicle [7].

Follicle Growth and Differentiation (i.e., Folliculogenesis)

The control of ovarian follicle development is complex, and involves multiple waves of growth [129]. In the initial stage of follicle development, arrested primordial follicles undergo primordial follicle transition to begin follicle growth [5]. In both the human and bovine ovary, two or three waves of follicles are initiated to develop in a single ovarian cycle [129, 130]. For both these species, follicles expand to up to 2 cm in diameter during this process. A combination of granulosa cell growth, theca cell growth, and antrum formation (i.e., formation of fluid-filled space in the developing follicle) results in the expansion of the ovarian follicle (Figure 14.2). Although a rapid stimulation of cell growth is required for the ovulatory follicle to develop, the vast majority of follicles undergo atresia, in which cell growth is arrested at various stages of follicle development. Hormones such as estrogen and FSH have been shown to promote follicle cell growth *in vivo*; however, these hormones alone do not stimulate growth of ovarian cells *in vitro* [131]. The possibility that these hormones may act indirectly through the local production of growth factors is proposed for later stages of development. Therefore, the regulation of ovarian cell growth is a complex process that requires an array of externally and locally derived regulatory agents. Interactions between theca cells, granulosa cells, and oocytes are required for follicular maturation [132]. The individual processes, such as dominant follicle selection [133] and follicle cell apoptosis/atresia [134, 135], also require integrated cell–cell interactions. A variety of specific growth factors produced in the follicle appear to mediate many of these cellular interactions in later stages of follicle development.

Ovarian Cell–Cell Interactions

Table 14.2 outlines a number of the factors produced locally in the ovary that mediate cell–cell signaling events in the control of follicle development and ovarian function. Several reviews address the topic of cell–cell interactions in the ovary and the control of follicle development [5, 7, 136–139]. Recent observations are cited below.

The epidermal growth factor (EGF) family of growth factors regulates cell–cell interactions in the ovary. TGFα has been shown to be produced by theca cells [140–143] and influence the growth of both theca and granulosa cells [143, 144]. Several *in vivo* experiments have shown that TGFα can influence follicle development [145, 146]. TGFα appears to be important for follicle development, and involves theca cell–granulosa cell interactions. TGFα has also been localized to isolated granulosa cells, but appears predominately in theca cells [147]. The primary function of TGFα is growth stimulation. Several other members of the EGF family are also involved including amphiregulin (AR), beta cellulin (Bt), and epiregulin (Ep) for granulosa cells [148, 149]. The EGF receptor is also expressed and can be regulated by hormones such as LH and GnRH [150, 151]. The EGF family also has a role in the ovarian surface epithelial cell biology [152]. Therefore, the EGF family members mediate cell–cell interactions in the ovarian follicle, with autocrine granulosa interactions being predominant.

HGF is produced by theca cells, and acts on granulosa cells to promote cell proliferation and function [153, 154]. This is an excellent example of the role HGF plays in mediating mesenchymal–epithelial interactions in tissues. Interestingly, SCF/KL produced by the granulosa cells can provide feedback to the theca cells to stimulate HGF production [155, 156]. In a similar manner, keratinocyte growth factor (KGF) is produced by theca cells and acts on granulosa cells to regulate cell growth [157, 158]. KGF can promote primordial follicle transition [159], and is also

TABLE 14.2 Cell–cell signaling factors in the ovary

Signaling factor	Site production	Site action	Functions	Ref(s)
Transforming growth factor α (TGFα)	Theca	Granulosa	Growth stimulation	[144–147, 166]
		Theca		
Transforming growth factor β (TGFβ)	Theca	Granulosa	Growth inhibition	[186–190]
	Granulosa	Theca	Cellular differentiation	
Hepatocyte growth factor (HGF)	Theca	Granulosa	Growth stimulation	[154–156, 158]
Keratocyte growth factor (KGF)	Theca	Granulosa	Growth stimulation	[153, 157, 160]
Colony stimulating factor (CSF)	Theca	Granulosa	Growth regulation	[161, 162]
		Theca		
Tumor necrosis factor (TNF)	Granulosa	Oocyte	Apoptosis	[170–173]
	Theca	Granulosa	Growth regulation	
	Oocyte	Theca		
Fas ligand	Granulosa	Oocyte	Apoptosis	[176, 177, 240, 241]
	Theca	Granulosa		
	Oocyte	Theca		
Nerve growth factor (NGF)	Theca	Granulosa	Growth stimulation	[180]
		Theca	Ovulation	
Fibroblast growth factor (bFGF)	Granulosa	Granulosa	Growth stimulation	[182–184]
	Theca	Theca		
	Oocyte			
Growth differentiation factor-9 (GDF-9)	Oocyte	Granulosa	Cellular differentiation	[192–196]
		Theca		
Bone morphogenic proteins (BMP)	Oocyte	Granulosa	Cellular differentiation	[200, 201, 203–205, 207, 108]
	Theca	Theca		
Kit ligand/stem cell factor (KL)	Granulosa	Oocyte	Growth stimulation	[209–213]
		Theca		
Leukemia inhibitory factor (LIF)	Granulosa	Oocyte	Growth stimulation	[215, 216]
		Theca	Cellular differentiation	
Vascular endothelial factor (VEGF)	Theca	Edothelium	Angiogenesis	[217, 221]
	Granulosa	Granulosa		
Interleukins	Granulosa	Granulosa	Cellular differentiation	[164–167]
	Theca	Theca		
Insulin-like growth factor (IGF-1)	Granulosa	Oocyte	Growth stimulation	[222–224]
	Theca	Granulosa	Cellular differentiation	
		Theca		
Inhibin	Granulosa	Oocyte	Cellular differentiation	[226, 227]
		Theca		
		Granulosa		
Anti-Müllerian hormone (AMH)	Granulosa	Oocyte	Cellular differentiation	[228]
		Granulosa		

expressed in the corpus luteum [160]. As was the case for HGF, SCF/KL was found to stimulate KGF expression by theca cells [156]. These factors reflect the importance of the theca cell in the regulation of follicle growth.

A number of immune-related cytokines have a potential role in the ovary. Granulocyte-macrophage colony-stimulating factor (GM-CSF) was found to be expressed primarily by theca cells in the ovary [161, 162]. The GM-CSF can influence granulosa cell growth and function. Null mice had abnormal follicle development, suggesting effects on the local cell–cell interactions [161]. Cytokines, as seen with the testis, also influence ovary function [163]. The interleukins -1, -6, and -8 have all been shown to regulate follicle development. IL-1 is expressed by the granulose, and affects granulosa function [164, 165]. IL-8 is primarily expressed by the theca, and to a lesser extent by granulose, and influences cellular function [166]. IL-6 is also expressed by granulosa cells and acts on various cells, including granulosa [167]. Further investigation of the specific roles of these and other members of the interluekin family is needed. Recent analyses of the granulosa cell transcriptome revealed that a number of immune-related cytokines are expressed, suggesting roles for these secreted factors in local cell–cell interactions that also require further investigation [168, 169].

Apoptosis is an essential aspect of follicle development and ovarian function. The vast majority of follicles undergo atresia and apoptosis. TNF has been shown to be produced by most cell types in the ovary associated with apoptosis [170–175]. TNF can act on all the cell types, and induce apoptosis or growth regulation. Another death ligand that binds death receptors to induce apoptosis is Fas ligand. Fas is also produced by all the cells associated with apoptosis, and acts to promote apoptosis in the atretic follicles [176–178]. The endocrine system can regulate the expression and action of these factors to subsequently regulate apoptosis [175, 178]. These signaling molecules are essential for ovarian function in promoting follicle atresia during folliculogenesis.

Nerve growth factor (NGF) was found to be expressed by theca cells and act on theca and granulosa cells [150]. NGF promotes the early stage of follicle growth [179]. The localization and actions suggest a potentially important role at the time of ovulation [180]. Other neurotropins (e.g., NT4) are also expressed at various stages of ovary development [181] and require further investigation.

Basic fibroblast growth factor (bFGF) has been shown to be expressed by granulosa cells, and to a lesser extent by theca cells [182]. BFGF can regulate both granulosa cell and theca cell growth and differentiated functions [183, 184]. During follicle development the expression of bFGF changes, being in the oocyte at the primordial stage and then in the granulosa at the primary stage [7]. FGF9 has been shown also to mediate ovarian cell–cell interactions, being produced by theca cells, stroma, and the CL, and acting on granulosa,

the oocyte, theca cells, and CL [185]. The role of other FGF family members has not been rigorously addressed.

The TGFβ superfamily of growth factors also has a critical role in regulating ovarian function [138]. Members of the family involved include TGFβ, GDF9, BMPs, and AMH. TGFβ is predominately produced by theca cells [186], but is also produced by isolated granulosa in selected follicle stages [187]. TGFα and TGFβ differentially regulate granulosa and theca cell differentiated functions and growth [188–190]. Although TGFβ inhibits TGFα growth stimulation, TGFβ also can influence cell functions [191].

Growth differentiation factor-9 (GDF-9) is a member of the TGFβ superfamily, and is specifically localized to the oocyte. GDF-9 can act on both granulosa cells and theca cells to regulate steroidogenesis and differentiated functions [191–197]. The actions of GDF-9 are follicle stage-specific, and appear to be expressed in a variety of species. In early follicle development in the rat, GDF9 promotes primary follicle progression [198], while in pigs GDF9 promotes primordial to primary follicle transition [199]. GDF-9 regulates the expression of other paracrine factors such as SCF/KL in the developing follicle [139, 194]. This is one of the few oocyte-specific products identified to be involved in cell–cell signaling in the ovary.

Another factor specifically expressed in the oocyte that appears to regulate granulosa cell function is BMP-15 [197, 200–202]. BMP-15 and GDF-9 may act synergistically during follicle development. Other BMP family members include BMP4 and -7, which are primarily localized in the theca cells and appear to act on the granulosa cells [203]; BMP2, which acts on granulosa cells [202, 204]; and BMP6, which also is expressed in the oocyte and acts on the granulosa cells [202, 205]. In early follicles, BMP4 promotes oocyte survival and primordial follicle transition [206]. The BMP family of growth factors are members of the TGFβ superfamily, and appear to be critical to follicle development [197, 202, 207, 208].

Stem cell factor/kit ligand (SCF/KL) is produced by the granulosa cell, and acts on the oocyte and theca cells [139, 209–213]. The null mutant suggests a critical role in oocyte viability and recruitment of primordial follicles. This role in promoting primordial follicle transition was confirmed in organ culture experiments [214]. In addition to the role in granulosa–oocyte interactions, granulosa KL also influences theca cell function and development [212]. Oocytes appear to have a regulatory role in influencing the expression of KL by granulosa cells [139, 210]. As found in the testis, this is a critical somatic–germ cell interaction. Another factor found to be expressed by granulosa cells and that regulates oocytes is LIF [215, 216]. LIF can promote primordial follicle transition [198], and is also produced by stromal cells in the ovary. This action of LIF in mediating granulosa–oocyte interactions is supported by levels of LIF that increase in follicular fluid as the follicle develops [215, 216].

Vascular endothelial factor (VEGF) has a critical role in angiogenesis. This process is important for developing follicles past the primary stage of development. VEGF is primarily expressed in theca cells, and to a reduced level by granulosa cells [217–221]. VEGF has a major role in acting on endothelial cells to promote angiogenesis, but also can influence granulosa cell functions [220]. This cell–cell signaling event controlled by VEGF is critical for follicle development.

IGF-1 also has a role in the ovarian follicle [222]. IGF-1 is expressed by granulosa and theca cells, and acts on the oocyte, granulosa, and theca cells [222–224]. Mice with null mutations in IGF-1 have impaired follicle development [224]. IGF-2 and the IGF-binding proteins also have a critical role in follicle development [223]. A related family member, relaxin, also integrates with the insulin family and may have a role in the ovary [225].

Inhibin also has a paracrine role in the developing follicle. Inhibin is primarily produced by the granulosa cells, and acts on the oocyte, theca, and granulosa cells [226, 227]. Related family members, such as the activins, are also anticipated to have similar roles. This is distinct from the roles these factors have in the endocrine system.

Additional signaling factors are anticipated to be essential for ovarian function and follicle development. One example is anti-Müllerian hormone (AMH), which is expressed by the granulosa cells [228] and may have a role as a negative regulator of oocyte viability and/or primordial follicle development [229, 230]. Local steroid production is also expected to influence the network of local cell–cell signaling events. This includes both androgen and estrogen production [231]. Newly identified developmental factors, such as Nodal, affect granulosa cell apoptosis [232, 233]; the Notch ligands (e.g., Delta) mediate oocyte and somatic cell interactions [234]; and endothelin 2 has effects on granulosa cells [235]. Platelet-derived growth factor (PDGF) also has a role in primordial follicle transition in the adult follicle and in the CL [175, 236, 237].

SUMMARY

The above descriptive discussion of cell–cell signaling in the testis and ovary demonstrates a growing complexity in the networks of cellular interactions and factors. It is anticipated some of these factors will have compensatory roles to assure growth and differentiation of the tissues. The list of factors provided is likely only partially complete, and will have more added as investigation of cell–cell interactions in the gonads expands. The advent of microarray procedures and analysis of the ovarian transcriptome have expedited this research [238]. Currently, we are primarily in the research phase of identifying the sites of production and action for these factors. The functions of some individual factors are also being analyzed. However, the next research phase of cell–cell signaling will involve a more systems biology approach to tie together all the potential interactions and gain more insight into the regulation of testis and ovary function.

The specific cell–cell signaling events identified are shown in most cases to change during development. The requirements and physiology of the embryonic testis and ovary are very different from the adult. Another research area to expand is the elucidation of cell–cell signaling at these different stages of development.

A comparison of the cell–cell signaling events between the testis and ovary is very useful. Some signaling events are the same. For example, the role SCF has in mediating direct somatic–germ cell interaction and the role HGF and KGF play in mesenchymal–epithelial cell interactions is similar. A direct correlation of the cell–cell interactions of the testis and ovary will be invaluable in elucidating the systems biology approach to understanding gonadal function.

Elucidation of cell–cell signaling events is required for the future development of therapeutic agents to control fertility and treat reproductive diseases. Through understanding the signaling events involved in testis and ovary function, basic information is provided to design more effective therapeutics. Significant advances are anticipated to be in the area of contraceptive and fertility agent development, and treatment of diseases such as polycystic ovarian disease or premature ovarian failure. Although an understanding of the intracellular signaling events is essential for understanding how a factor acts, the elucidation of the network of extracellular signaling molecules that regulates a cell's function is essential to understand how a whole tissue or organism functions.

REFERENCES

1. Berruti G. Signaling events during male germ cell differentiation: update. *Front Biosci* 2006;**11**:2144–56.
2. Griswold MD. Protein secretions of Sertoli cells. *Intl Rev Cytol* 1988;**110**:133–56.
3. Petersen C, Soder O. The sertoli cell–a hormonal target and "super" nurse for germ cells that determine testicular size. *Horm Res* 2006;**66**(4):153–61.
4. Skinner MK. Cell–cell interactions in the testis. *Endocr Rev* 1991;**12**(1):45–77.
5. Hirshfield AN. Development of follicles in the mammalian ovary. *Intl Rev Cytol* 1991;**124**:43–101.
6. Hutt KJ, Albertini DF. An oocentric view of folliculogenesis and embryogenesis. *Reprod Biomed Online* 2007;**14**(6):758–64.
7. Nilsson E, Skinner MK. Cellular interactions that control primordial follicle development and folliculogenesis. *J Soc Gynecol Investig* 2001;**8**(1 Suppl. Proc.):S17–20.
8. Sertoli E. On the existence of special branched cells in the seminiferous tubule of the human testes. *Morgangni* 1865;**7**:31–9.
9. Fawcett D. The ultrastructure and functions of the Sertoli cell. In: Greep RHE, editor.. *Handbook of physiology*, Vol. V. Washington DC: American Physiology Society; 1975. p. 22–55.

10. Setchell BWG Greep RHE, editor. *Handbook of physiology*. Washington DC: American Physiology Society; 1975. p. 143–72.

11. Waites GM, Gladwell RT. Physiological significance of fluid secretion in the testis and blood–testis barrier. *Physiol Rev* 1982;**62**(2):624–71.

12. Russell LD, Tallon-Doran M, Weber JE, Wong V, Peterson RN. Three-dimensional reconstruction of a rat stage V Sertoli cell: III. A study of specific cellular relationships. *Am J Anat* 1983;**167**(2):181–92.

13. Dorrington JH, Fritz IB, Armstrong DT. Control of testicular estrogen synthesis. *Biol Reprod* 1978;**18**(1):55–64.

14. Robinson R, Fritz IB. Metabolism of glucose by Sertoli cells in culture. *Biol Reprod* 1981;**24**(5):1032–41.

15. Lacroix M, Smith FE, Fritz IB. Secretion of plasminogen activator by Sertoli cell enriched cultures. *Mol Cell Endocrinol* 1977;**9**(2):227–36.

16. Skinner MK, Griswold MD. Sertoli cells synthesize and secrete transferrin-like protein. *J Biol Chem* 1980;**255**(20):9523–5.

17. Skinner MK, Griswold MD. Sertoli cells synthesize and secrete a ceruloplasmin-like protein. *Biol Reprod* 1983;**28**(5):1225–9.

18. Tung PS, Skinner MK, Fritz IB. Fibronectin synthesis is a marker for peritubular cell contaminants in Sertoli cell-enriched cultures. *Biol Reprod* 1984;**30**(1):199–211.

19. Skinner MK, Cosand WL, Griswold MD. Purification and characterization of testicular transferrin secreted by rat Sertoli cells. *Biochem J* 1984;**218**(2):313–20.

20. Kierszenbaum AL. Mammalian spermatogenesis in vivo and in vitro: a partnership of spermatogenic and somatic cell lineages. *Endocr Rev* 1994;**15**(1):116–34.

21. Adams IR, McLaren A. Sexually dimorphic development of mouse primordial germ cells: switching from oogenesis to spermatogenesis. *Development* 2002;**129**(5):1155–64.

22. Kim Y, Capel B. Balancing the bipotential gonad between alternative organ fates: a new perspective on an old problem. *Dev Dyn* 2006;**235**(9):2292–300.

23. Takasaki N, Rankin T, Dean J. Normal gonadal development in mice lacking GPBOX, a homeobox protein expressed in germ cells at the onset of sexual dimorphism. *Mol Cell Biol* 2001;**21**(23):8197–202.

24. Clinton M, Haines LC. An overview of factors influencing sex determination and gonadal development in birds. *Exs* 2001;**91**:97–115.

25. Drews U. Local mechanisms in sex specific morphogenesis. *Cytogenet Cell Genet* 2000;**91**(1–4):72–80.

26. Ikeda Y, Takeda Y, Shikayama T, Mukai T, Hisano S, Morohashi KI. Comparative localization of Dax-1 and Ad4BP/SF-1 during development of the hypothalamic-pituitary-gonadal axis suggests their closely related and distinct functions. *Dev Dyn* 2001;**220**(4):363–76.

27. McLaren A. Germ and somatic cell lineages in the developing gonad. *Mol Cell Endocrinol* 2000;**163**(1–2):3–9.

28. Ostrer H. Sexual differentiation. *Semin Reprod Med* 2000;**18**(1):41–9.

29. Parker KL, Schimmer BP, Schedl A. Genes essential for early events in gonadal development. *Exs* 2001;**91**:11–24.

30. Raymond CS, Murphy MW, O'Sullivan MG, Bardwell VJ, Zarkower D. Dmrt1, a gene related to worm and fly sexual regulators, is required for mammalian testis differentiation. *Genes Dev* 2000;**14**(20):2587–95.

31. Vaillant S, Magre S, Dorizzi M, Pieau C, Richard-Mercier N. Expression of AMH, SF1, and SOX9 in gonads of genetic female chickens during sex reversal induced by an aromatase inhibitor. *Dev Dyn* 2001;**222**(2):228–37.

32. Sinclair AH, Berta P, Palmer MS, Hawkins JR, Griffiths BL, Smith MJ, Foster JW, Frischauf AM, Lovell-Badge R, Goodfellow PN. A gene from the human sex-determining region encodes a protein with homology to a conserved DNA-binding motif. *Nature* 1990;**346**(6281):240–4.

33. Vidal VP, Chaboissier MC, de Rooij DG, Schedl A. Sox9 induces testis development in XX transgenic mice. *Nat Genet* 2001;**28**(3):216–17.

34. Jost A, Magre S, Agelopoulou R. Early stages of testicular differentiation in the rat. *Hum Genet* 1981;**58**(1):59–63.

35. Magre S, Jost A. The initial phases of testicular organogenesis in the rat. An electron microscopy study. *Arch Anat Microsc Morphol Exp* 1980;**69**(4):297–318.

36. Buehr M, Gu S, McLaren A. Mesonephric contribution to testis differentiation in the fetal mouse. *Development* 1993;**117**(1):273–81.

37. Merchant-Larios H, Moreno-Mendoza N, Buehr M. The role of the mesonephros in cell differentiation and morphogenesis of the mouse fetal testis. *Intl J Dev Biol* 1993;**37**(3):407–15.

38. Ricci G, Catizone A, Innocenzi A, Galdieri M. Hepatocyte growth factor (HGF) receptor expression and role of HGF during embryonic mouse testis development. *Dev Biol* 1999;**216**(1):340–7.

39. Cupp AS, Uzumcu M, Skinner MK. Chemotactic role of neurotropin 3 in the embryonic testis that facilitates male sex determination. *Biol Reprod* 2003;**68**(6):2033–7.

40. McLaren A. Development of the mammalian gonad: the fate of the supporting cell lineage. *Bioessays* 1991;**13**(4):151–6.

41. Kim Y, Kobayashi A, Sekido R, DiNapoli L, Brennan J, Chaboissier MC, Poulat F, Behringer RR, Lovell-Badge R, Capel B. Fgf9 and Wnt4 act as antagonistic signals to regulate mammalian sex determination. *PLoS Biol* 2006;**4**(6):e187.

42. Coveney D, Ross AJ, Slone JD, Capel B. A microarray analysis of the XX Wnt4 mutant gonad targeted at the identification of genes involved in testis vascular differentiation. *Gene Expr Patterns* 2007;**7**(1–2):82–92.

43. Katoh M, Katoh M. NUMB is a break of WNT–Notch signaling cycle. *Intl J Mol Med* 2006;**18**(3):517–21.

44. Karl J, Capel B. Sertoli cells of the mouse testis originate from the coelomic epithelium. *Dev Biol* 1998;**203**(2):323–33.

45. Frojdman K, Paranko J, Virtanen I, Pelliniemi LJ. Intermediate filaments and epithelial differentiation of male rat embryonic gonad. *Differentiation* 1992;**50**(2):113–23.

46. Fridmacher V, Le Bert M, Guillou F, Magre S. Switch in the expression of the K19/K18 keratin genes as a very early evidence of testicular differentiation in the rat. *Mech Dev* 1995;**52**(2–3):199–207.

47. Blanchard MG, Josso N. Source of the anti-Mullerian hormone synthesized by the fetal testis: Mullerian-inhibiting activity of fetal bovine Sertoli cells in tissue culture. *Pediatr Res* 1974;**8**(12):968–71.

48. Magre S, Jost A. Sertoli cells and testicular differentiation in the rat fetus. *J Electron Microsc Tech* 1991;**19**(2):172–88.

49. Bott RC, McFee RM, Clopton DT, Toombs C, Cupp AS. Vascular endothelial growth factor and kinase domain region receptor are involved in both seminiferous cord formation and vascular development during testis morphogenesis in the rat. *Biol Reprod* 2006;**75**(1):56–67.

50. Green R. Embryology of sexual structure and hermaphroditism. *J Clin Endocrinol* 1944;**4**:335–48.

51. Byskov AG. Differentiation of mammalian embryonic gonad. *Physiol Rev* 1986;**66**(1):71–117.

52. Greco TL, Payne AH. Ontogeny of expression of the genes for steroidogenic enzymes P450 side-chain cleavage, 3 beta-hydroxysteroid dehydrogenase, P450 17 alpha-hydroxylase/C17-20 lyase, and P450 aromatase in fetal mouse gonads. *Endocrinology* 1994;**135**(1):262–8.

53. Bloch E, Lew M, Klein M. Studies on the inhibition of fetal androgen formation. Inhibition of testosterone synthesis in rat and rabbit fetal testes with observations on reproductive tract development. *Endocrinology* 1971;**89**(1):16–31.

54. Orth JM, Weisz J, Ward OB, Ward IL. Environmental stress alters the developmental pattern of delta 5-3 beta-hydroxysteroid dehydrogenase

activity in Leydig cells of fetal rats: a quantitative cytochemical study. *Biol Reprod* 1983;**28**(3):625–31.

55. Roser JF. Endocrine and paracrine control of sperm production in stallions. *Anim Reprod Sci* 2001;**68**(3–4):139–51.

56. Weinbauer GF and Wessels J. 'Paracrine' control of spermatogenesis. *Andrologia* 1999;**31**(5):249–62.

57. Levine E, Cupp AS, Miyashiro L, Skinner MK. Role of transforming growth factor-alpha and the epidermal growth factor receptor in embryonic rat testis development. *Biol Reprod* 2000;**62**(3):477–90.

58. Mendis-Handagama SM, Ariyaratne HB. Differentiation of the adult Leydig cell population in the postnatal testis. *Biol Reprod* 2001;**65**(3):660–71.

59. Petersen C, Boitani C, Froysa B, Soder O. Transforming growth factor-alpha stimulates proliferation of rat Sertoli cells. *Mol Cell Endocrinol* 2001;**181**(1–2):221–7.

60. Avallet O, Gomez E, Vigier M, Jegou B, Saez JM. Sertoli cell–germ cell interactions and TGF beta 1 expression and secretion in vitro. *Biochem Biophys Res Commun* 1997;**238**(3):905–9.

61. Konrad L, Albrecht M, Renneberg H, Aumuller G. Transforming growth factor-beta2 mediates mesenchymal–epithelial interactions of testicular somatic cells. *Endocrinology* 2000;**141**(10):3679–86.

62. Lui WY, Lee WM, Cheng CY. Transforming growth factor-beta3 perturbs the inter-Sertoli tight junction permeability barrier *in vitro* possibly mediated via its effects on occludin, zonula occludens-1, and claudin-11. *Endocrinology* 2001;**142**(5):1865–77.

63. Olaso R, Pairault C, Habert R. Expression of type I and II receptors for transforming growth factor beta in the adult rat testis. *Histochem Cell Biol* 1998;**110**(6):613–8.

64. Wang RA, Zhao GQ. Transforming growth factor beta signal transducer Smad2 is expressed in mouse meiotic germ cells, Sertoli cells, and Leydig cells during spermatogenesis. *Biol Reprod* 1999;**61**(4):999–1004.

65. Itman C, Mendis S, Barakat B, Loveland KL. All in the family: TGF-beta family action in testis development. *Reproduction* 2006;**132**(2):233–46.

66. Le Roy C, Lejeune H, Chuzel F, Saez JM, Langlois D. Autocrine regulation of Leydig cell differentiated functions by insulin-like growth factor I and transforming growth factor beta. *J Steroid Biochem Mol Biol* 1999;**69**(1–6):379–84.

67. Rouiller-Fabre V, Lecref L, Gautier C, Saez JM, Habert R. Expression and effect of insulin-like growth factor I on rat fetal Leydig cell function and differentiation. *Endocrinology* 1998;**139**(6):2926–34.

68. Santos RL, Silva CM, Ribeiro AF, Vasconcelos AC, Pesquero JL, Coelho SG, Serakides R, Reis SR. Effect of growth hormone and induced IGF-I release on germ cell population and apoptosis in the bovine testis. *Theriogenology* 1999;**51**(5):975–84.

69. Tsuruta JK, Eddy EM, O'Brien DA. Insulin-like growth factor-II/cation-independent mannose 6-phosphate receptor mediates paracrine interactions during spermatogonial development. *Biol Reprod* 2000;**63**(4):1006–13.

70. Huleihel M, Lunenfeld E. Involvement of intratesticular IL-1 system in the regulation of Sertoli cell functions. *Mol Cell Endocrinol* 2002;**187**(1–2):125–32.

71. Jenab S, Morris PL. Interleukin-6 regulation of kappa opioid receptor gene expression in primary sertoli cells. *Endocrine* 2000;**13**(1):11–15.

72. Meroni SB, Suburo AM, Cigorraga SB. Interleukin-1beta regulates nitric oxide production and gamma-glutamyl transpeptidase activity in sertoli cells. *J Androl* 2000;**21**(6):855–61.

73. Nehar D, Mauduit C, Boussouar F, Benahmed M. Interleukin 1alpha stimulates lactate dehydrogenase A expression and lactate production in cultured porcine sertoli cells. *Biol Reprod* 1998;**59**(6):1425–32.

74. Petersen C, Boitani C, Froysa B, Soder O. Interleukin-1 is a potent growth factor for immature rat sertoli cells. *Mol Cell Endocrinol* 2002;**186**(1):37–47.

75. Soder O, Sultana T, Jonsson C, Wahlgren A, Petersen C, Holst M. The interleukin-1 system in the testis. *Andrologia* 2000;**32**(1):52–5.

76. Stephan JP, Syed V, Jegou B. Regulation of Sertoli cell IL-1 and IL-6 production in vitro. *Mol Cell Endocrinol* 1997;**134**(2):109–18.

77. Zeyse D, Lunenfeld E, Beck M, Prinsloo I, Huleihel M. Interleukin-1 receptor antagonist is produced by sertoli cells *in vitro*. *Endocrinology* 2000;**141**(4):1521–7.

78. Ishikawa T, Morris PL. Interleukin-1beta signals through a c-Jun N-terminal kinase-dependent inducible nitric oxide synthase and nitric oxide production pathway in Sertoli epithelial cells. *Endocrinology* 2006;**147**(11):5424–30.

79. de Kretser DM, Meinhardt A, Meehan T, Phillips DJ, O'Bryan MK, Loveland KA. The roles of inhibin and related peptides in gonadal function. *Mol Cell Endocrinol* 2000;**161**(1–2):43–6.

80. Ethier JF, Findlay JK. Roles of activin and its signal transduction mechanisms in reproductive tissues. *Reproduction* 2001;**121**(5):667–75.

81. Risbridger GP, Cancilla B. Role of activins in the male reproductive tract. *Rev Reprod* 2000;**5**(2):99–104.

82. Schlatt S, Meinhardt A, Nieschlag E. Paracrine regulation of cellular interactions in the testis: factors in search of a function. *Eur J Endocrinol* 1997;**137**(2):107–17.

83. Cancilla B, Davies A, Ford-Perriss M, Risbridger GP. Discrete cell- and stage-specific localisation of fibroblast growth factors and receptor expression during testis development. *J Endocrinol* 2000;**164**(2):149–59.

84. Cancilla B, Risbridger GP. Differential localization of fibroblast growth factor receptor-1, -2, -3, and -4 in fetal, immature, and adult rat testes. *Biol Reprod* 1998;**58**(5):1138–45.

85. Colvin JS, Green RP, Schmahl J, Capel B, Ornitz DM. Male-to-female sex reversal in mice lacking fibroblast growth factor 9. *Cell* 2001;**104**(6):875–89.

86. Schteingart HF, Meroni SB, Canepa DF, Pellizzari EH, Cigorraga SB. Effects of basic fibroblast growth factor and nerve growth factor on lactate production, gamma-glutamyl transpeptidase and aromatase activities in cultured Sertoli cells. *Eur J Endocrinol* 1999;**141**(5):539–45.

87. Yamamoto H, Ochiya T, Takahama Y, Ishii Y, Osumi N, Sakamoto H, Terada M. Detection of spatial localization of Hst-1/Fgf-4 gene expression in brain and testis from adult mice. *Oncogene* 2000;**19**(33):3805–10.

88. Gonzalez-Herrera IG, Prado-Lourenco L, Pileur F, Conte C, Morin A, Cabon F, Prats H, Vagner S, Bayard F, Audigier S, Prats AC. Testosterone regulates FGF-2 expression during testis maturation by an IRES-dependent translational mechanism. *FASEB J* 2006;**20**(3):476–8.

89. Basciani S, Mariani S, Arizzi M, Ulisse S, Rucci N, Jannini EA, Della Rocca C, Manicone A, Carani C, Spera G, Gnessi L. Expression of platelet-derived growth factor-A (PDGF-A), PDGF-B, and PDGF receptor-alpha and -beta during human testicular development and disease. *J Clin Endocrinol Metab* 2002;**87**(5):2310–19.

90. Chiarenza C, Filippini A, Tripiciano A, Beccari E, Palombi F. Platelet-derived growth factor-BB stimulates hypertrophy of peritubular smooth muscle cells from rat testis in primary cultures. *Endocrinology* 2000;**141**(8):2971–81.

91. Gnessi L, Basciani S, Mariani S, Arizzi M, Spera G, Wang C, Bondjers C, Karlsson L, Betsholtz C. Leydig cell loss and spermatogenic arrest in platelet-derived growth factor (PDGF)-A-deficient mice. *J Cell Biol* 2000;**149**(5):1019–26.

92. Mariani S, Basciani S, Arizzi M, Spera G, Gnessi L. PDGF and the testis. *Trends Endocrinol Metab* 2002;**13**(1):11–17.

93. Loveland KL, Schlatt S. Stem cell factor and c-kit in the mammalian testis: lessons originating from Mother Nature's gene knockouts. *J Endocrinol* 1997;**153**(3):337–44.

94. Mauduit C, Hamamah S, Benahmed M. Stem cell factor/c-kit system in spermatogenesis. *Hum Reprod Update* 1999;**5**(5):535–45.

95. Rossi P, Sette C, Dolci S, Geremia R. Role of c-kit in mammalian spermatogenesis. *J Endocrinol Invest* 2000;**23**(9):609–15.

96. Vincent S, Segretain D, Nishikawa S, Nishikawa SI, Sage J, Cuzin F, Rassoulzadegan M. Stage-specific expression of the Kit receptor and its ligand (KL) during male gametogenesis in the mouse: a Kit–KL interaction critical for meiosis. *Development* 1998;**125**(22):4585–93.

97. Naughton CK, Jain S, Strickland AM, Gupta A, Milbrandt J. Glial cell-line derived neurotrophic factor-mediated RET signaling regulates spermatogonial stem cell fate. *Biol Reprod* 2006;**74**(2):314–21.

98. Piquet-Pellorce C, Dorval-Coiffec I, Pham MD, Jegou B. Leukemia inhibitory factor expression and regulation within the testis. *Endocrinology* 2000;**141**(3):1136–41.

99. Hara T, Tamura K, de Miguel MP, Mukouyama Y, Kim H, Kogo H, Donovan PJ, Miyajima A. Distinct roles of oncostatin M and leukemia inhibitory factor in the development of primordial germ cells and sertoli cells in mice. *Dev Biol* 1998;**201**(2):144–53.

100. Boussouar F, Grataroli R, Ji J, Benahmed M. Tumor necrosis factor-alpha stimulates lactate dehydrogenase A expression in porcine cultured sertoli cells: mechanisms of action. *Endocrinology* 1999;**140**(7):3054–62.

101. Grataroli R, Vindrieux D, Gougeon A, Benahmed M. Expression of tumor necrosis factor-alpha-related apoptosis-inducing ligand and its receptors in rat testis during development. *Biol Reprod* 2002;**66**(6):1707–15.

102. Riera MF, Meroni SB, Gomez GE, Schteingart HF, Pellizzari EH, Cigorraga SB. Regulation of lactate production by FSH, iL1beta, and TNFalpha in rat Sertoli cells. *Gen Comp Endocrinol* 2001;**122**(1):88–97.

103. Yao PL, Lin YC, Sawhney P, Richburg JH. Transcriptional regulation of FasL expression and participation of sTNF-alpha in response to sertoli cell injury. *J Biol Chem* 2007;**282**(8):5420–31.

104. Maire M, Florin A, Kaszas K, Regnier D, Contard P, Tabone E, Mauduit C, Bars R, Benahmed M. Alteration of transforming growth factor-beta signaling system expression in adult rat germ cells with a chronic apoptotic cell death process after fetal androgen disruption. *Endocrinology* 2005;**146**(12):5135–43.

105. Catizone A, Ricci G, Arista V, Innocenzi A, Galdieri M. Hepatocyte growth factor and c-MET are expressed in rat prepuberal testis. *Endocrinology* 1999;**140**(7):3106–13.

106. Catizone A, Ricci G, Galdieri M. Expression and functional role of hepatocyte growth factor receptor (C-MET) during postnatal rat testis development. *Endocrinology* 2001;**142**(5):1828–34.

107. van der Wee K, Hofmann MC. An in vitro tubule assay identifies HGF as a morphogen for the formation of seminiferous tubules in the postnatal mouse testis. *Exp Cell Res* 1999;**252**(1):175–85.

108. Perrard MH, Vigier M, Damestoy A, Chapat C, Silandre D, Rudkin BB, Durand P. beta-Nerve growth factor participates in an auto/paracrine pathway of regulation of the meiotic differentiation of rat spermatocytes. *J Cell Physiol* 2007;**210**(1):51–62.

109. Cupp AS, Kim GH, Skinner MK. Expression and action of neurotropin-3 and nerve growth factor in embryonic and early postnatal rat testis development. *Biol Reprod* 2000;**63**(6):1617–28.

110. Magnanti M, Gandini O, Giuliani L, Gazzaniga P, Marti HH, Gradilone A, Frati L, Agliano AM, Gassmann M. Erythropoietin expression in primary rat Sertoli and peritubular myoid cells. *Blood* 2001;**98**(9):2872–4.

111. Szczepny A, Hime GR, Loveland KL. Expression of hedgehog signalling components in adult mouse testis. *Dev Dyn* 2006;**235**(11):3063–70.

112. Miller DW, Harrison JL, Brown YA, Doyle U, Lindsay A, Adam CL, Lea RG. Immunohistochemical evidence for an endocrine/paracrine role for ghrelin in the reproductive tissues of sheep. *Reprod Biol Endocrinol* 2005;**3**:60.

113. Kanzaki M, Morris PL. Identification and regulation of testicular interferon-gamma (IFNgamma) receptor subunits: IFNgamma enhances interferon regulatory factor-1 and interleukin-1beta converting enzyme expression. *Endocrinology* 1998;**139**(5):2636–44.

114. Ivell R. Biology of the relaxin-like factor (RLF). *Rev Reprod* 1997;**2**(3):133–8.

115. Verhoeven G, Hoeben E, De Gendt K. Peritubular cell–Sertoli cell interactions: factors involved in PmodS activity. *Andrologia* 2000;**32**(1):42–5.

116. Midgley Jr. AR Autoradiographic analysis of gonadotropin binding to rat ovarian tissue sections. *Adv Exp Med Biol* 1973;**36**(0):365–78.

117. Richards Jr. JS, Midgley AR Protein hormone action: a key to understanding ovarian follicular and luteal cell development. *Biol Reprod* 1976;**14**(1):82–94.

118. Vlodavsky I, Brown KD, Gospodarowicz D. A comparison of the binding of epidermal growth factor to cultured granulosa and luteal cells. *J Biol Chem* 1978;**253**(10):3744–50.

119. Wandji SA, Pelletier G, Sirard MA. Ontogeny and cellular localization of 125I-labeled insulin-like growth factor-I, 125I-labeled follicle-stimulating hormone, and 125I-labeled human chorionic gonadotropin binding sites in ovaries from bovine fetuses and neonatal calves. *Biol Reprod* 1992;**47**(5):814–22.

120. Adashi EY. The IGF family and folliculogenesis. *J Reprod Immunol* 1998;**39**(1–2):13–19.

121. Peng C, Ohno T, Khorasheh S, Leung PC. Activin and follistatin as local regulators in the human ovary. *Biol Signals* 1996;**5**(2):81–9.

122. Erickson GF. Primary cultures of ovarian cells in serum-free medium as models of hormone-dependent differentiation. *Mol Cell Endocrinol* 1983;**29**(1):21–49.

123. Fortune JE, Armstrong DT. Androgen production by theca and granulosa isolated from proestrous rat follicles. *Endocrinology* 1977;**100**(5):1341–7.

124. Erickson GF, Ryan KJ. Stimulation of testosterone production in isolated rabbit thecal tissue by LH/FSH, dibutyryl cyclic AMP, PGE2alpha, and PGE2. *Endocrinology* 1976;**99**(2):452–8.

125. Channing CP. Progesterone and estrogen secretion by cultured monkey ovarian cell types: influences of follicular size, serum luteinizing hormone levels, and follicular fluid estrogen levels. *Endocrinology* 1980;**107**(1):342–52.

126. Evans G, Dobias M, King GJ, Armstrong DT. Estrogen, androgen, and progesterone biosynthesis by theca and granulosa of preovulatory follicles in the pig. *Biol Reprod* 1981;**25**(4):673–82.

127. Haney AF, Schomberg DW. Estrogen and progesterone production by developing porcine follicles in vitro: evidence for estrogen formation by theca. *Endocrinology* 1981;**109**(3):971–7.

128. McNatty KP, Makris A, DeGrazia C, Osathanondh R, Ryan KJ. The production of progesterone, androgens, and estrogens by granulosa cells, thecal tissue, and stromal tissue from human ovaries *in vitro*. *J Clin Endocrinol Metab* 1979;**49**(5):687–99.

129. Fortune JE. Ovarian follicular growth and development in mammals. *Biol Reprod* 1994;**50**(2):225–32.

130. Sirois J, Fortune JE. Ovarian follicular dynamics during the estrous cycle in heifers monitored by real-time ultrasonography. *Biol Reprod* 1988;**39**(2):308–17.

131. Hsueh AJ, McGcc EA, Hayashi M, Hsu SY. Hormonal regulation of early follicle development in the rat ovary. *Mol Cell Endocrinol* 2000;**163**(1–2):95–100.

132. Yada H, Hosokawa K, Tajima K, Hasegawa Y, Kotsuji F. Role of ovarian theca and granulosa cell interaction in hormone production- and cell growth during the bovine follicular maturation process. *Biol Reprod* 1999;**61**(6):1480–6.

133. Baker SJ, Spears N. The role of intra-ovarian interactions in the regulation of follicle dominance. *Hum Reprod Update* 1999;**5**(2):153–65.

134. Chun SY, Eisenhauer KM, Minami S, Hsueh AJ. Growth factors in ovarian follicle atresia. *Semin Reprod Endocrinol* 1996;**14**(3):197–202.

135. Hsueh AJ, Eisenhauer K, Chun SY, Hsu SY, Billig H. Gonadal cell apoptosis. *Recent Prog Horm Res* 1996;**51**:433–55. discussion 455–6.

136. Einspanier R, Lauer B, Gabler C, Kamhuber M, Schams D. Egg-cumulus-oviduct interactions and fertilization. *Adv Exp Med Biol* 1997;**424**:279–89.

137. Erickson GF, Shimasaki S. The physiology of folliculogenesis: the role of novel growth factors. *Fertil Steril* 2001;**76**(5):943–9.

138. Knight PG, Glister C. TGF-beta superfamily members and ovarian follicle development. *Reproduction* 2006;**132**(2):191–206.

139. Thomas FH, Vanderhyden BC. Oocyte–granulosa cell interactions during mouse follicular development: regulation of kit ligand expression and its role in oocyte growth. *Reprod Biol Endocrinol* 2006;**4**:19.

140. Derynck R. Transforming growth factor-alpha: structure and biological activities. *J Cell Biochem* 1986;**32**(4):293–304.

141. Kudlow JE, Kobrin MS, Purchio AF, Twardzik DR, Hernandez ER, Asa SL, Adashi EY. Ovarian transforming growth factor-alpha gene expression: immunohistochemical localization to the theca-interstitial cells. *Endocrinology* 1987;**121**(4):1577–9.

142. Lobb DK, Kobrin MS, Kudlow JE, Dorrington JH. Transforming growth factor-alpha in the adult bovine ovary: identification in growing ovarian follicles. *Biol Reprod* 1989;**40**(5):1087–93.

143. Skinner Jr. MK, Coffey RJ Regulation of ovarian cell growth through the local production of transforming growth factor-alpha by theca cells. *Endocrinology* 1988;**123**(6):2632–8.

144. Skinner MK. Transforming growth factor production and actin in the ovarian follicle: theca cell and granulosa cell interactions. In: Hirschfeld A, editor. *Growth factors and the ovary*. New York: Plenum Press; 1989:141.

145. Campbell BK, Gordon BM, Scaramuzzi RJ. The effect of ovarian arterial infusion of transforming growth factor alpha on ovarian follicle populations and ovarian hormone secretion in ewes with an autotransplanted ovary. *J Endocrinol* 1994;**143**(1):13–24.

146. Ma YJ, Dissen GA, Merlino G, Coquelin A, Ojeda SR. Overexpression of a human transforming growth factor-alpha (TGF alpha) transgene reveals a dual antagonistic role of TGF alpha in female sexual development. *Endocrinology* 1994;**135**(4):1392–400.

147. Qu J, Nisolle M, Donnez J. Expression of transforming growth factor-alpha, epidermal growth factor, and epidermal growth factor receptor in follicles of human ovarian tissue before and after cryopreservation. *Fertil Steril* 2000;**74**(1):113–21.

148. Ben-Ami I, Freimann S, Armon L, Dantes A, Strassburger D, Friedler S, Raziel A, Seger R, Ron-El R, Amsterdam A. PGE2 up-regulates EGF-like growth factor biosynthesis in human granulosa cells: new insights into the coordination between PGE2 and LH in ovulation. *Mol Hum Reprod* 2006;**12**(10):593–9.

149. Shimada M, Hernandez-Gonzalez I, Gonzalez-Robayna I, Richards JS. Paracrine and autocrine regulation of epidermal growth factor-like factors in cumulus oocyte complexes and granulosa cells: key

150. Hsieh M, Lee D, Panigone S, Horner K, Chen R, Theologis A, Lee DC, Threadgill DW, Conti M. Luteinizing hormone-dependent activation of the epidermal growth factor network is essential for ovulation. *Mol Cell Biol* 2007;**27**(5):1914–24.

151. Motola S, Cao X, Ashkenazi H, Popliker M, Tsafriri A. GnRH actions on rat preovulatory follicles are mediated by paracrine EGF-like factors. *Mol Reprod Dev* 2006;**73**(10):1271–6.

152. Ahmed N, Maines-Bandiera S, Quinn MA, Unger WG, Dedhar S, Auersperg N. Molecular pathways regulating EGF-induced epithelio-mesenchymal transition in human ovarian surface epithelium. *Am J Physiol Cell Physiol* 2006;**290**(6):C1532–42.

153. Parrott JA, Skinner MK. Developmental and hormonal regulation of hepatocyte growth factor expression and action in the bovine ovarian follicle. *Biol Reprod* 1998;**59**(3):553–60.

154. Zachow RJ, Ramski BE, Lee H. Modulation of estrogen production and 17beta-hydroxysteroid dehydrogenase-type 1, cytochrome P450 aromatase, c-met, and protein kinase Balpha messenger ribonucleic acid content in rat ovarian granulosa cells by hepatocyte growth factor and follicle-stimulating hormone. *Biol Reprod* 2000;**62**(6):1851–7.

155. Ito M, Harada T, Tanikawa M, Fujii A, Shiota G, Terakawa N. Hepatocyte growth factor and stem cell factor involvement in paracrine interplays of theca and granulosa cells in the human ovary. *Fertil Steril* 2001;**75**(5):973–9.

156. Parrott JA, Skinner MK. Thecal cell–granulosa cell interactions involve a positive feedback loop among keratinocyte growth factor, hepatocyte growth factor, and Kit ligand during ovarian follicular development. *Endocrinology* 1998;**139**(5):2240–5.

157. Osuga Y, Koga K, Tsutsumi O, Yano T, Kugu K, Momoeda M, Okagaki R, Suenaga A, Fujiwara T, Fujimoto A, Matsumi H, Hiroi H, Taketani Y. Evidence for the presence of keratinocyte growth factor (KGF) in human ovarian follicles. *Endocr J* 2001;**48**(2):161–6.

158. Parrott JA, Skinner MK. Developmental and hormonal regulation of keratinocyte growth factor expression and action in the ovarian follicle. *Endocrinology* 1998;**139**(1):228–35.

159. Kezele P, Nilsson EE, Skinner MK. Keratinocyte growth factor acts as a mesenchymal factor that promotes ovarian primordial to primary follicle transition. *Biol Reprod* 2005;**73**(5):967–73.

160. Salli U, Bartol FF, Wiley AA, Tarleton BJ, Braden TD. Keratinocyte growth factor expression by the bovine corpus luteum. *Biol Reprod* 1998;**59**(1):77–83.

161. Gilchrist RB, Rowe DB, Ritter LJ, Robertson SA, Norman RJ, Armstrong DT. Effect of granulocyte-macrophage colony-stimulating factor deficiency on ovarian follicular cell function. *J Reprod Fertil* 2000;**120**(2):283–92.

162. Tamura K, Tamura H, Kumasaka K, Miyajima A, Suga T, Kogo H. Ovarian immune cells express granulocyte-macrophage colony-stimulating factor (GM-CSF) during follicular growth and luteinization in gonadotropin-primed immature rodents. *Mol Cell Endocrinol* 1998;**142**(1–2):153–63.

163. Ostanin AA, Aizikovich BI, Aizikovich IV, Kozhin AY, Chernykh ER. Role of cytokines in the regulation of reproductive function. *Bull Exp Biol Med* 2007;**143**(1):75–9.

164. Buscher U, Chen FC, Kentenich H, Schmiady H. Cytokines in the follicular fluid of stimulated and non-stimulated human ovaries; is ovulation a suppressed inflammatory reaction?. *Hum Reprod* 1999;**14**(1):162–6.

165. Ghersevich S, Isomaa V, Vihko P. Cytokine regulation of the expression of estrogenic biosynthetic enzymes in cultured rat granulosa cells. *Mol Cell Endocrinol* 2001;**172**(1–2):21–30.

166. Runesson E, Ivarsson K, Janson PO, Brannstrom M. Gonadotropin- and cytokine-regulated expression of the chemokine interleukin 8 in the human preovulatory follicle of the menstrual cycle. *J Clin Endocrinol Metab* 2000;**85**(11):4387–95.

167. Salmassi A, Lu S, Hedderich J, Oettinghaus C, Jonat W, Mettler L. Interaction of interleukin-6 on human granulosa cell steroid secretion. *J Endocrinol* 2001;**170**(2):471–8.

168. Hernandez-Gonzalez I, Gonzalez-Robayna I, Shimada M, Wayne CM, Ochsner SA, White L, Richards JS. Gene expression profiles of cumulus cell oocyte complexes during ovulation reveal cumulus cells express neuronal and immune-related genes: does this expand their role in the ovulation process?. *Mol Endocrinol* 2006;**20**(6):1300–21.

169. Skinner MK, Schmidt M, Savenkova MI, Sadler-Riggleman I, Nilsson EE. Regulation of granulosa and theca cell transcriptomes during ovarian antral follicle development. *Reprod Dev* 2008. in press.

170. Morrison LJ, Marcinkiewicz JL. Tumor necrosis factor alpha enhances oocyte/follicle apoptosis in the neonatal rat ovary. *Biol Reprod* 2002;**66**(2):450–7.

171. Prange-Kiel J, Kreutzkamm C, Wehrenberg U, Rune GM. Role of tumor necrosis factor in preovulatory follicles of swine. *Biol Reprod* 2001;**65**(3):928–35.

172. Spaczynski RZ, Arici A, Duleba AJ. Tumor necrosis factor-alpha stimulates proliferation of rat ovarian theca-interstitial cells. *Biol Reprod* 1999;**61**(4):993–8.

173. Spicer LJ. Receptors for insulin-like growth factor-I and tumor necrosis factor-alpha are hormonally regulated in bovine granulosa and thecal cells. *Anim Reprod Sci* 2001;**67**(1–2):45–58.

174. Johnson AL, Ratajczak C, Haugen MJ, Liu HK, Woods DC. Tumor necrosis factor-related apoptosis inducing ligand expression and activity in hen granulosa cells. *Reproduction* 2007;**133**(3):609–16.

175. Nilsson EE, Stanfield J, Skinner MK. Interactions between progesterone and tumor necrosis factor-alpha in the regulation of primordial follicle assembly. *Reproduction* 2006;**132**(6):877–86.

176. Bridgham JT, Johnson AL. Expression and regulation of Fas antigen and tumor necrosis factor receptor type I in hen granulosa cells. *Biol Reprod* 2001;**65**(3):733–9.

177. Quirk SM, Porter DA, Huber SC, Cowan RG. Potentiation of Fas-mediated apoptosis of murine granulosa cells by interferon-gamma, tumor necrosis factor-alpha, and cycloheximide. *Endocrinology* 1998;**139**(12):4860–9.

178. Slot KA, Voorendt M, de Boer-Brouwer M, van Vugt HH, Teerds KJ. Estrous cycle dependent changes in expression and distribution of Fas, Fas ligand, Bcl-2, Bax, and pro- and active caspase-3 in the rat ovary. *J Endocrinol* 2006;**188**(2):179–92.

179. Dissen GA, Romero C, Paredes A, Ojeda SR. Neurotrophic control of ovarian development. *Microsc Res Tech* 2002;**59**(6):509–15.

180. Dissen GA, Parrott JA, Skinner MK, Hill DF, Costa ME, Ojeda SR. Direct effects of nerve growth factor on thecal cells from antral ovarian follicles. *Endocrinology* 2000;**141**(12):4736–50.

181. Anderson RA, Robinson LL, Brooks J, Spears N. Neurotropins and their receptors are expressed in the human fetal ovary. *J Clin Endocrinol Metab* 2002;**87**(2):890–7.

182. Yamamoto S, Konishi I, Nanbu K, Komatsu T, Mandai M, Kuroda H, Matsushita K, Mori T. Immunohistochemical localization of basic fibroblast growth factor (bFGF) during folliculogenesis in the human ovary. *Gynecol Endocrinol* 1997;**11**(4):223–30.

183. Peluso JJ, Pappalardo A, Fernandez G. Basic fibroblast growth factor maintains calcium homeostasis and granulosa cell viability by stimulating calcium efflux via a PKC delta-dependent pathway. *Endocrinology* 2001;**142**(10):4203–11.

184. Puscheck EE, Patel Y, Rappolee DA. Fibroblast growth factor receptor (FGFR)-4, but not FGFR-3 is expressed in the pregnant ovary. *Mol Cell Endocrinol* 1997;**132**(1–2):169–76.

185. Drummond AE, Tellbach M, Dyson M, Findlay JK. Fibroblast growth factor-9, a local regulator of ovarian function. *Endocrinology* 2007;**148**(8):3711–21.

186. Skinner MK, Keski-Oja J, Osteen KG, Moses HL. Ovarian thecal cells produce transforming growth factor-beta which can regulate granulosa cell growth. *Endocrinology* 1987;**121**(2):786–92.

187. Christopher B. Immunolocalization of transforming growth factor-beta1 during follicular development and atresia in the mouse ovary. *Endocr J* 2000;**47**(4):475–80.

188. Feng P, Catt KJ, Knecht M. Transforming growth factor beta regulates the inhibitory actions of epidermal growth factor during granulosa cell differentiation. *J Biol Chem* 1986;**261**(30):14,167–14,170,.

189. Roberts AJ, Skinner MK. Transforming growth factor-alpha and -beta differentially regulate growth and steroidogenesis of bovine thecal cells during antral follicle development. *Endocrinology* 1991;**129**(4):2041–8.

190. Ying SY, Becker A, Ling N, Ueno N, Guillemin R. Inhibin and beta type transforming growth factor (TGF beta) have opposite modulating effects on the follicle stimulating hormone (FSH)-induced aromatase activity of cultured rat granulosa cells. *Biochem Biophys Res Commun* 1986;**136**(3):969–75.

191. Gilchrist RB, Ritter LJ, Myllymaa S, Kaivo-Oja N, Dragovic RA, Hickey TE, Ritvos O, Mottershead DG. Molecular basis of oocyte-paracrine signalling that promotes granulosa cell proliferation. *J Cell Sci* 2006;**119**(18):3811–21.

192. Elvin JA, Clark AT, Wang P, Wolfman NM, Matzuk MM. Paracrine actions of growth differentiation factor-9 in the mammalian ovary. *Mol Endocrinol* 1999;**13**(6):1035–48.

193. Fitzpatrick SL, Sindoni DM, Shughrue PJ, Lane MV, Merchenthaler IJ, Frail DE. Expression of growth differentiation factor-9 messenger ribonucleic acid in ovarian and nonovarian rodent and human tissues. *Endocrinology* 1998;**139**(5):2571–8.

194. Joyce IM, Clark AT, Pendola FL, Eppig JJ. Comparison of recombinant growth differentiation factor-9 and oocyte regulation of KIT ligand messenger ribonucleic acid expression in mouse ovarian follicles. *Biol Reprod* 2000;**63**(6):1669–75.

195. Solovyeva EV, Hayashi M, Margi K, Barkats C, Klein C, Amsterdam A, Hsueh AJ, Tsafriri A. Growth differentiation factor-9 stimulates rat theca-interstitial cell androgen biosynthesis. *Biol Reprod* 2000;**63**(4):1214–18.

196. Vitt UA, Hsueh AJ. Stage-dependent role of growth differentiation factor-9 in ovarian follicle development. *Mol Cell Endocrinol* 2001;**183**(1–2):171–7.

197. Mazerbourg S, Hsueh AJ. Genomic analyses facilitate identification of receptors and signalling pathways for growth differentiation factor 9 and related orphan bone morphogenetic protein/growth differentiation factor ligands. *Hum Reprod Update* 2006;**12**(4):373–83.

198. Nilsson EE, Skinner MK. Growth and differentiation factor-9 stimulates progression of early primary but not primordial rat ovarian follicle development. *Biol Reprod* 2002;**67**(3):1018–24.

199. Shimizu T. Promotion of ovarian follicular development by injecting vascular endothelial growth factor (VEGF) and growth differentiation factor 9 (GDF-9) genes. *J Reprod Dev* 2006;**52**(1):23–32.

200. Otsuka F, Yamamoto S, Erickson GF, Shimasaki S. Bone morphogenetic protein-15 inhibits follicle-stimulating hormone (FSH)

action by suppressing FSH receptor expression. *J Biol Chem* 2001;**276**(14):11,387–11,392,.

201. Yan C, Wang P, DeMayo J, DeMayo FJ, Elvin JA, Carino C, Prasad SV, Skinner SS, Dunbar BS, Dube JL, Celeste AJ, Matzuk MM. Synergistic roles of bone morphogenetic protein 15 and growth differentiation factor 9 in ovarian function. *Mol Endocrinol* 2001;**15**(6):854–66.

202. Brankin V, Quinn RL, Webb R, Hunter MG. BMP-2 and -6 modulate porcine theca cell function alone and co-cultured with granulosa cells. *Domest Anim Endocrinol* 2005;**29**(4):593–604.

203. Lee WS, Otsuka F, Moore RK, Shimasaki S. Effect of bone morphogenetic protein-7 on folliculogenesis and ovulation in the rat. *Biol Reprod* 2001;**65**(4):994–9.

204. Souza CJ, Campbell BK, McNeilly AS, Baird DT. Effect of bone morphogenetic protein 2 (BMP2) on oestradiol and inhibin A production by sheep granulosa cells, and localization of BMP receptors in the ovary by immunohistochemistry. *Reproduction* 2002;**123**(3):363–9.

205. Otsuka F, Moore RK, Shimasaki S. Biological function and cellular mechanism of bone morphogenetic protein-6 in the ovary. *J Biol Chem* 2001;**276**(35):32,889–32,895,.

206. Nilsson EE, Skinner MK. Bone morphogenetic protein-4 acts as an ovarian follicle survival factor and promotes primordial follicle development. *Biol Reprod* 2003;**69**(4):1265–72.

207. Elvin JA, Yan C, Matzuk MM. Oocyte-expressed TGF-beta superfamily members in female fertility. *Mol Cell Endocrinol* 2000;**159**(1–2):1–5.

208. Shimasaki S, Zachow RJ, Li D, Kim H, Iemura S, Ueno N, Sampath K, Chang RJ, Erickson GF. A functional bone morphogenetic protein system in the ovary. *Proc Natl Acad Sci USA* 1999;**96**(13):7282–7.

209. Driancourt MA, Reynaud K, Cortvrindt R, Smitz J. Roles of KIT and KIT LIGAND in ovarian function. *Rev Reprod* 2000;**5**(3):143–52.

210. Joyce IM, Pendola FL, Wigglesworth K, Eppig JJ. Oocyte regulation of kit ligand expression in mouse ovarian follicles. *Dev Biol* 1999;**214**(2):342–53.

211. Klinger FG, De Felici M. In vitro development of growing oocytes from fetal mouse oocytes: stage-specific regulation by stem cell factor and granulosa cells. *Dev Biol* 2002;**244**(1):85–95.

212. Parrott JA, Skinner MK. Kit ligand actions on ovarian stromal cells: effects on theca cell recruitment and steroid production. *Mol Reprod Dev* 2000;**55**(1):55–64.

213. Sette C, Dolci S, Geremia R, Rossi P. The role of stem cell factor and of alternative c-kit gene products in the establishment, maintenance and function of germ cells. *Intl J Dev Biol* 2000;**44**(6):599–608.

214. Parrott JA, Skinner MK. Kit-ligand/stem cell factor induces primordial follicle development and initiates folliculogenesis. *Endocrinology* 1999;**140**(9):4262–71.

215. Arici A, Oral E, Bahtiyar O, Engin O, Seli E, Jones EE. Leukaemia inhibitory factor expression in human follicular fluid and ovarian cells. *Hum Reprod* 1997;**12**(6):1233–9.

216. Coskun S, Uzumcu M, Jaroudi K, Hollanders JM, Parhar RS, al-Sedairy ST. Presence of leukemia inhibitory factor and interleukin-12 in human follicular fluid during follicular growth. *Am J Reprod Immunol* 1998;**40**(1):13–18.

217. Barboni B, Turriani M, Galeati G, Spinaci M, Bacci ML, Forni M, Mattioli M. Vascular endothelial growth factor production in growing pig antral follicles. *Biol Reprod* 2000;**63**(3):858–64.

218. Berisha B, Schams D, Kosmann M, Amselgruber W, Einspanier R. Expression and localisation of vascular endothelial growth factor and basic fibroblast growth factor during the final growth of bovine ovarian follicles. *J Endocrinol* 2000;**167**(3):371–82.

219. Garrido N, Albert C, Krussel JS, O'Connor JE, Remohi J, Simon C, Pellicer A. Expression, production, and secretion of vascular endothelial growth factor and interleukin-6 by granulosa cells is comparable in women with and without endometriosis. *Fertil Steril* 2001;**76**(3):568–75.

220. Hazzard TM, Molskness TA, Chaffin CL, Stouffer RL. Vascular endothelial growth factor (VEGF) and angiopoietin regulation by gonadotrophin and steroids in macaque granulosa cells during the peri-ovulatory interval. *Mol Hum Reprod* 1999;**5**(12):1115–21.

221. Yamamoto S, Konishi I, Tsuruta Y, Nanbu K, Mandai M, Kuroda H, Matsushita K, Hamid AA, Yura Y, Mori T. Expression of vascular endothelial growth factor (VEGF) during folliculogenesis and corpus luteum formation in the human ovary. *Gynecol Endocrinol* 1997;**11**(6):371–81.

222. Monget P, Bondy C. Importance of the IGF system in early folliculogenesis. *Mol Cell Endocrinol* 2000;**163**(1–2):89–93.

223. Giudice LC. Insulin-like growth factor family in Graafian follicle development and function. *J Soc Gynecol Investig* 2001;**8**(1 Suppl. Proc.):S26–9.

224. Kadakia R, Arraztoa JA, Bondy C, Zhou J. Granulosa cell proliferation is impaired in the Igf1 null ovary. *Growth Horm IGF Res* 2001;**11**(4):220–4.

225. Park JI, Chang CL, Hsu SY. New Insights into biological roles of relaxin and relaxin-related peptides. *Rev Endocr Metab Disord* 2005;**6**(4):291–6.

226. Findlay JK, Drummond AE, Dyson M, Baillie AJ, Robertson DM, Ethier JF. Production and actions of inhibin and activin during folliculogenesis in the rat. *Mol Cell Endocrinol* 2001;**180**(1–2):139–44.

227. Lanuza GM, Groome NP, Baranao JL, Campo S. Dimeric inhibin A and B production are differentially regulated by hormones and local factors in rat granulosa cells. *Endocrinology* 1999;**140**(6):2549–54.

228. Josso N, di Clemente N, Gouedard L. Anti-Mullerian hormone and its receptors. *Mol Cell Endocrinol* 2001;**179**(1–2):25–32.

229. Gigli I, Cushman RA, Wahl CM, Fortune JE. Evidence for a role for anti-Mullerian hormone in the suppression of follicle activation in mouse ovaries and bovine ovarian cortex grafted beneath the chick chorioallantoic membrane. *Mol Reprod Dev* 2005;**71**(4):480–8.

230. Ikeda Y, Nagai A, Ikeda MA, Hayashi S. Increased expression of Mullerian-inhibiting substance correlates with inhibition of follicular growth in the developing ovary of rats treated with E2 benzoate. *Endocrinology* 2002;**143**(1):304–12.

231. Chryssikopoulos A. The potential role of intraovarian factors on ovarian androgen production. *Ann NY Acad Sci* 2000;**900**:184–92.

232. Wang H, Jiang JY, Zhu C, Peng C, Tsang BK. Role and regulation of nodal/activin receptor-like kinase 7 signaling pathway in the control of ovarian follicular atresia. *Mol Endocrinol* 2006;**20**(10):2469–82.

233. Wang H, Tsang BK. Nodal signalling and apoptosis. *Reproduction* 2007;**133**(5):847–53.

234. Ward EJ, Shcherbata HR, Reynolds SH, Fischer KA, Hatfield SD, Ruohola-Baker H. Stem cells signal to the niche through the Notch pathway in the *Drosophila* ovary. *Curr Biol* 2006;**16**(23):2352–8.

235. Palanisamy GS, Cheon YP, Kim J, Kannan A, Li Q, Sato M, Mantena SR, Sitruk-Ware RL, Bagchi MK, Bagchi IC. A novel pathway involving progesterone receptor, endothelin-2, and endothelin receptor B controls ovulation in mice. *Mol Endocrinol* 2006;**20**(11):2784–95.

236. Sleer LS, Taylor CC. Platelet-derived growth factors and receptors in the rat corpus luteum: localization and identification of an effect on luteogenesis. *Biol Reprod* 2007;**76**(3):391–400.

237. Sleer LS, Taylor CC, Cell-type localization of platelet-derived growth factors and receptors in the postnatal rat ovary and follicle. *Biol Reprod* 2007;**76**(3):379–90.

238. Andreu-Vieyra C, Lin YN, Matzuk MM. Mining the oocyte transcriptome. *Trends Endocrinol Metab* 2006;**17**(4):136–43.

239. Yamamoto S, Mikami T, Konishi M, Itoh N. Stage-specific expression of a novel isoform of mouse FGF-14 (FHF-4) in spermatocytes. *Biochim Biophys Acta* 2000;**1490**(1–2):121–4.

240. Quirk SM, Harman RM, Cowan RG. Regulation of Fas antigen (Fas, CD95)-mediated apoptosis of bovine granulosa cells by serum and growth factors. *Biol Reprod* 2000;**63**(5):1278–84.

241. Vickers SL, Cowan RG, Harman RM, Porter DA, Quirk SM. Expression and activity of the Fas antigen in bovine ovarian follicle cells. *Biol Reprod* 2000;**62**(1):54–61.

Chapter 15

Kidney

William J. Arendshorst[1] and Elsa Bello-Reuss[2]

[1]Department of Cell and Molecular Physiology, University of North Carolina at Chapel Hill, Chapel Hill, North Carolina

[2]Department of Internal Medicine, Division of Nephrology and Department of Cell Physiology and Molecular Biophysics, Texas Tech University Health Science Center, Lubbock, Texas

OVERVIEW OF KIDNEY FUNCTION AND CELL-TO-CELL INTERACTIONS

The kidney is a complex organ comprising diverse cell types that work in coordination to perform a broad spectrum of functions, including the maintenance of body fluid and electrolyte balance, pH regulation, secretion of renin and erythropoietin, activation of vitamin D, excretion of numerous drugs and toxins, and regulation of blood pressure. Several of these functions are related and involve the process of urine formation, which takes place in the functional unit of the kidney, the nephron.

Each human kidney has about a million nephrons that operate in parallel. A nephron consists of a glomerulus and a tubule arranged in series. The glomerulus is formed by a capillary network (glomerular tuft) extending between the afferent and efferent glomerular arterioles. The capillary loops are separated by intercellular material and surrounded by a layer of specialized epithelial cells (*podocytes*) attached to the basement membrane of the capillaries. The epithelial layer reflects on itself in the vascular pole, forming the Bowman space (or urinary space) that is continued by the lumen of the proximal tubule (Figure 15.1). Urine formation is the result of filtration in the glomerulus, and reabsorption and/or secretion by the tubule. The net flux from tubule lumen to blood constitutes reabsorption, and the net flux in the opposite direction constitutes secretion. The renal-tubule epithelial cells can perform net transport between solutions of very similar or identical composition because they are polarized – that is, different transport proteins are expressed in the apical (lumen-facing) and basolateral (blood-facing) membrane domains, allowing them to carry out directional transport of specific solutes. Water transport is osmotic, and occurs in the direction of net solute transport. Further details about renal organization and function can be found in recent texts [1, 2].

Glomerular filtration, tubule transport, and excretion are finely regulated processes. The proximity of renal tubules to each other, as well as to capillaries and interstitial cells, facilitates paracrine interactions between different cell types. The glomerular urinary space is in series with the lumen of the renal tubule segments, so that cells in consecutive structures communicate with each other via the luminal fluid. The composition of this fluid can be changed by the rate of filtration (that modifies the NaCl load) or by the secretion of signaling molecules into the lumen or plasma. Glomerular release of vasoactive agents can influence the vascular resistance of the efferent arteriole and vasa recta pericytes. Both changes in luminal NaCl concentration and the presence of signaling molecules are sensed by downstream tubule segments. This signaling mode, unique to the kidney, provides functional integration at the level of the single nephron. This theme is one of the central topics of this chapter.

As in other organs, cell-to-cell communication in the kidney is largely mediated by paracrine and autocrine agents that may act in the extracellular or intracellular compartments, activating signal transduction systems. These complex interactions unify membrane transport processes and urine formation with homeostatic regulation of renal hemodynamics and glomerular filtration rate (GFR) [3, 4].

The kidney is divided in two regions: the superficial cortex and the interior medulla (Figure 15.1a). The cortex, 70 percent of the renal parenchyma, contains three classes of cells: *vascular* (large and small arteries, arterioles, capillaries), *epithelial* (proximal convoluted tubule (PCT), loop of Henle (LH) of short-looped cortical nephrons, distal convoluted tubule (DCT), cortical collecting tubule (CCT)], and *interstitial* cells. The medulla comprises 30 percent of the kidney mass, and contains vasa recta capillaries, long loops of Henle of deep (juxtamedullary) nephrons, collecting ducts (CD), and interstitial cells. Thin-descending

Handbook of Cell Signaling, Three-Volume Set 2 ed.

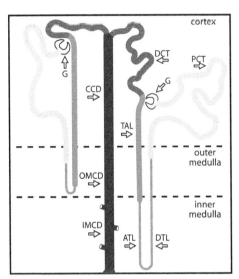
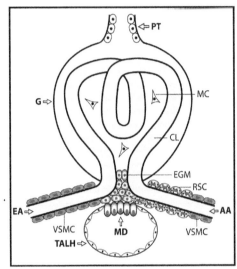

FIGURE 15.1(A) Scheme of the structure of deep and superficial nephrons.
G, Glomerulus; PTC, proximal convoluted tubule; DTL, descending thin loop of Henle; ATL, ascending thin loop of Henle; TAL, thick ascending loop of Henle; DCT, distal convoluted tubule; CCD, cortical collecting tubule; OMCD, outer medullary collecting duct; IMCD, inner medullary collecting duct.
(B) Scheme of the juxtaglomerular apparatus.
VSMCs, vascular smooth muscle cells; AA, afferent arteriole; EA, efferent arteriole; RSC, renin secreting (granular) cells; EGM, extraglomerular messangium; MD, macula densa; TALH, thick ascending loop of Henle; G, glomerulus; CL, capillary loop; MC, mesangial cell; PT, proximal tubule.

and thin-ascending loops of Henle are present in the inner medulla. The thick ascending limb of Henle's loop (TAL) resides in the outer medulla and cortex.

The vascular resistance of the kidney is lower than that of other organs, explaining the high blood flow of the organ, on average 1200 ml/min (about 25 percent of cardiac output), or 4 ml/min per g of kidney weight. The renal artery and its branches carry 85–90 percent of the total renal blood flow to the cortex, where the microvasculature consists of two highly reactive arterioles separating two capillary beds arranged in series: the glomerular capillaries (after the afferent arteriole of the glomerulus) and the peritubular capillaries in the cortex, and vasa recta in the medulla (after the efferent arteriole of the glomerulus). Afferent and efferent arterioles regulate blood flow and glomerular capillary hydrostatic pressure, the latter being the major determinant of the GFR. In the glomerulus the transcapillary hydrostatic pressure gradient exceeds the plasma colloid osmotic pressure throughout, thus causing fluid ultrafiltration all along the length of the glomerular capillaries. The glomerular capillary wall is formed, from blood to Bowman's space, by a fenestrated endothelial layer, a basement membrane, and the epithelial layer that applies over the basal membrane by interdigitating processes separated by the "slit pores." The basement membrane slit pores constitute the effective size-restrictive barrier to the glomerular filtration, with a low permeability to plasma proteins and other molecules larger than albumin. In contrast, its permeability to water, ions, and other small solutes is considerably higher than that of capillaries of other organs. After filtration, fluid flows from Bowman's space into the proximal tubule, where solutes and water are reabsorbed by transport, across both the cell membranes and the intercellular pathway, into the peritubular space and then the capillaries. The plasma colloid-osmotic pressure in the peritubular capillaries (higher than in other capillary beds because the glomerular ultrafiltrate does not contain proteins) exceeds the transcapillary hydrostatic pressure gradient, and thus there is fluid flow from peritubular interstitium into peritubular capillaries, both in the cortex and in the medulla. The capillaries in the latter are the hairpin-shaped *vasa recta* sporadically surrounded by pericytes; these smooth muscle cells may contract to act as sphincters.

Medullary blood flow is 10–15 percent of the total renal blood flow (RBF). Its control is essential for the efficient operation of the countercurrent multiplication and exchange mechanisms that accumulate solute in the medullary interstitial space, providing the driving force to concentrate or dilute the urine along the collecting ducts. In addition to upstream control by efferent arterioles, local blood-flow regulation is provided by pericytes around descending vasa recta. The medullary blood vessels may be less sensitive to vasoconstrictor agents than the cortical vessels due to protection provided by nitric oxide (NO) and eicosanoids released by endothelial and tubule cells. Abnormal regulation of the renal-medullary circulation can lead to salt retention and hypertension [5–7].

Redundant and complementary control mechanisms form the basis of structural–functional relationships that are central to two of the primary functions of the kidney, namely, the maintenance of salt and water balances and the control of blood pressure. In the steady state, the output (excretion) of fluid and electrolytes matches the input (intake = production plus administration), so that the volume and composition of the extracellular fluid remain constant, precisely controlled by homeostatic mechanisms [3, 4, 8]. In a normal hydration state, the renal tubules reabsorb more than 99 percent of the glomerular filtrate. The rate of glomerular filtration in adult humans is about 120 ml/min, or 170 l/day. About 67 percent of the reabsorption occurs in the PT, 20 percent in the loops of Henle (LH), and the rest in the distal nephron (DCT, CCD, MCD). Figure 15.1(A) outlines the main nephron segments. The luminal fluid flows from PT to the descending and ascending LH, DCT, and CD. The transport processes and permeability properties of each nephron segment are regulated by paracrine factors generated by vascular and epithelial cells. NaCl is actively reabsorbed in all nephron segments, except the thin LH. The basolateral membranes of all renal-tubule cells express Na^+,K^+-ATPase, or Na^+ pump, a primary-active transport protein that extrudes Na^+ and takes up K^+, establishing chemical and electrical gradients at the luminal membrane that favor passive Na^+ entry. The luminal membranes of the different tubule segments express diverse Na^+ transport proteins [9–13]. The main mechanisms of Na^+ entry across the luminal membrane are: Na^+/H^+ exchange in the PCT (mediated by the antiporter NHE3, sensitive to high concentrations of the drug amiloride), $Na^+–K^+–2Cl^-$ in the thick ascending LH (mediated by the symporter NKCC2, sensitive to the diuretic furosemide), $Na^+–Cl^-$ co-transport in the DCT (mediated by the symporter NCC, sensitive to diuretics of the thiazide family), and channel-mediated entry in the principal cells of the CD (via ENaC, the epithelial Na^+ channel, blocked by low concentrations of amiloride). Reabsorption of Na^+ is accompanied by anion (largely Cl^- and HCO_3^-) transport in the same direction, and/or transport of another cation (H^+ or K^+) in the opposite direction to Na^+ transport. The transport of Na^+ is coupled to transport of other solutes at the molecular level (examples given earlier) or by changing the driving force for transport of the other solute (i.e., transcellular Na^+ absorption creates an electrical potential that drives paracellular Cl^- absorption).

Changes in tubule NaCl concentration in the macula densa region of the LH determine a complex sequence of events (termed tubuloglomerular feedback (TGF), see section below and Figure 15.1b) leading to changes in the in the local secretion of renin by specialized cells in the afferent arteriole of the glomerulus and modifications of the artery arteriole resistance.

Arginine vasopressin (AVP), also known as vasopressin or antidiuretic hormone (ADH), secreted by the posterior pituitary gland, acts on CD principal cells to elicit exocytotic insertion of aquaporin 2 (AQP2) and urea transporter (UT1) in the luminal membrane, thus increasing the permeability to both urea and water. Aldosterone, secreted by the adrenal cortex, also exerts its main action on the principal cells of the CD, stimulating Na^+ reabsorption by genomic and non-genomic mechanisms that enhance the number of Na^+ channels at the luminal membrane and the number of Na^+ pumps at the basolateral membrane. Because of the differences in the salt and water permeabilities among renal-tubule segments, the fractional reabsorptions of salt and water are somewhat different in specific segments. In normal individuals, fractional Na^+ reabsorption is about 67 percent in the PT, 20 percent in the LH, 7 percent in the DCT, 5 percent in the CD, and 1 percent in the IMCD. Water reabsorption is similar in the PT, less in the LH, and variable along the DCT and CD, where the effect of vasopressin on water reabsorption is exerted, and water reabsorption depends on the hydration state and plasma concentration of vasopressin.

In summary, the kidney is a highly sophisticated organ serving multiple functions ranging from synthesis and release of the proteolytic enzyme renin to maintenance of extracellular fluid volume and composition. Several of these functions depend critically on cell-to-cell communication in a coordinated, integrated manner. Homeostasis is regulated by multiple systems featuring negative-feedback control of function characterized by coordination of hormonal, neural, paracrine, and autocrine signals as they act to regulate both the renal microcirculation and epithelial transport along the renal tubule to maintain a constant internal environment.

This chapter focuses on cell-to-cell communication within the kidney. The extrarenal neural–hormonal regulation and other important functions, such as regulation of acid–base and calcium and potassium balances, are not discussed (for recent reviews, see [3, 4, 9, 11–13]). Signal transduction among different cells is highlighted, at the expense of in-depth discussion of mechanisms underlying effector-cell responses. Due to space limitations, not all intrarenal events are discussed. The reader is referred to recent reviews on contractile mechanisms of vascular smooth muscle cells [14–20] and basic transport mechanisms in renal epithelial cells [9, 10, 21–28].

VASCULAR SMOOTH MUSCLE CELLS

Vascular smooth muscle cells (VSMCs) encircle endothelial cells and contract to regulate the blood-vessel diameter, and thereby its resistance to blood flow. Contraction and relaxation of VSMCs in interlobular arteries and in afferent and efferent glomerular arterioles are the primary determinants of total renal vascular resistance. The kidney has intrinsic autoregulatory mechanisms that maintain renal blood flow and GFR constant in the face of changes in arterial pressure. The roles of myogenic response and tubuloglomerular feedback in the control of the afferent-arteriole vascular resistance are discussed below, under "Tubulovascular interactions." In addition to circulating hormones and sympathetic nerve activity, multiple paracrine/autocrine factors regulate the contraction of VSMCs of arteries and arterioles, and of the mesangial cells (cells of smooth-muscle origin located between the glomerular capillary loops). Vasomotor tone is regulated by communication between endothelial cells and the contractile VSMCs. Just before an afferent arteriole enters a glomerulus there is a short segment of juxtaglomerular granular cells responsible for the production, storage, and release of renin and Ang II. Renin release depends on signals from the macula-densa cells of the thick ascending LH, activity of sympathetic nerve terminals, and stretch of juxtaglomerular granular cells (see Figure 15.1(B)). The juxtaglomerular apparatus (JGA) consists of four cell types: VSMCs of the terminal afferent arteriole, renin-containing granular cells, macula-densa cells, and extraglomerular mesangial cells (Figure 15.1b). Unique functions of the integrated JGA are discussed below under "Tubulovascular interactions."

Vasoconstrictor Mechanisms

The VSMCs of renal microvessels respond to local paracrine factors in addition to circulating hormones and sympathetic nerve activity [3, 4, 14, 16, 18, 19, 29-31]. Contraction of VSMCs is regulated by physical and chemical factors, with multiple signal transduction pathways coupling agonist-receptor interactions to $[Ca^{2+}]_i$, followed by activation of the actin–myosin contractile machinery. G-protein-coupled receptors (GPCRs) in the plasma membrane play a prominent role in these processes. Heterotrimeric G proteins transduce stimulatory or inhibitory signals to the cell interior. Table 15.1 shows a list of representative cell-surface GPCRs found in afferent and efferent arterioles. Ligand binding to the receptor triggers the intracellular signals leading to changes in vasomotor tone. Subsequently, a receptor is inactivated by kinase-mediated phosphorylation (GPCR serine/threonine protein kinases), thus limiting the duration of the effect of the agonist.

Most of the receptors mediating vasoconstriction couple primarily with $G\alpha_{q/11}$ to activate PLC, form inositol trisphosphate (IP_3) and diacylglycerol (DAG), and thereby elevate $[Ca^{2+}]_i$ and activate protein kinase C (PKC). Rapid increases in $[Ca^{2+}]_i$ are mediated by mobilization from sarcoplasmic-reticulum stores and more sustained elevations result from increased Ca^{2+} entry via plasma-membrane ion channels. Ca^{2+} mobilization is determined by the combined activity of two receptor/Ca^{2+} release channels on the sarcoplasmic reticulum: the IP_3 receptor and the ryanodine (Ry) receptor (RyR). There is an interaction between these two receptors/channels in Ca^{2+} release from the sarcoplasmic reticulum. The fact that RyRs are very sensitive to local increases in Ca^{2+} is the basis for the process of Ca^{2+}-induced Ca^{2+}-release [29]. In addition to responding to IP_3-mediated Ca^{2+} release, RyRs are normally sensitized by endogenous cyclic ADP ribose that is produced by a cell surface CD38 enzyme, ADP ribosyl cyclase. $G\alpha_{q/11}$-coupled receptors rapidly stimulate ADP ribosyl

cyclase by an uncertain mechanism; candidates include G proteins and physiological levels of superoxide anion.

Ca^{2+}-entry channels in the plasma membrane can be voltage-dependent or voltage-independent. Voltage-gated T- and L-type Ca^{2+} channels open upon depolarization of the plasma membrane and allow Ca^{2+} entry from the extracellular fluid. Receptor activation may lead to opening of voltage-insensitive receptor-operated channels by a poorly understood mechanism. Release of Ca^{2+} from the sarcoplasmic reticulum leads to Ca^{2+} entry by opening of store-operated cation channels in the plasma membrane. There is considerable current interest in identifying the molecular identity of store-operated and receptor-operated cation channels as combinations of TRP channel proteins [32–35]. In addition to Ca^{2+} effects on contractility, sensitivity of the contractile apparatus to a given level of $[Ca^{2+}]_i$ is increased by Rho/Rho kinase and PKC. Cell-surface receptors for vasoconstrictor agents are phosphorylated and inactivated or desensitized by an indirect process consisting of PKC-mediated phosphorylation or a family of G-protein regulated kinases.

VASCULAR ENDOTHELIAL CELLS

Endothelial cells are in contact with blood elements, and subjected to hemodynamic forces such as hydrostatic pressure and shear stress [36–40]. The lumen hydrostatic pressure tends to distend the arterioles, causing a contractile myogenic response, whereas the shear stress, depending on the flow velocity, causes release of vasodilator agents, including nitric oxide (NO), cyclooxygenase (COX) products, and endothelium-derived hyperpolarizing factor (EDHF); and vasoconstrictor factors, including endothelin-1 (ET), cytochrome P450 (Cyt-P450) products, and COX2 derivatives. The interplay of these mechanisms determines the basal tone in resistance arterioles and allows for rapid adaptations to fluctuations in flow and pressure.

Endothelial cells constitute a barrier whose permeability varies in different organs, largely depending on the presence of fenestrations and on the properties of the tight junctions that separate the endothelial cells. Endothelial cells synthesize paracrine agents (gases, fatty acids, and peptides) that act on both nearby VSMCs and epithelial cells [36–41].

The release of vasoactive paracrine/autocrine factors, both relaxing and constricting, modulates the degree of contraction of VSMCs (Table 15.1). Vasodilators produced by endothelial cells include NO, PGI_2, epoxy-eicosatrienoic acids (EETs), and EDHF. The most potent vasoconstrictor made is ET. In addition, endothelial cells express angiotensin-converting enzyme (ACE), tethered to the blood-facing membrane; this enzyme catalyzes the local production of the vasoconstrictor angiotensin II (Ang II). Under physiological conditions the actions of local dilator agents predominate, whereas constrictor systems tend to prevail in stressful and disease states. The physiological actions of vasodilator agents are important in counteracting

TABLE 15.1 Major receptors and signaling pathways in the kidney vasculature

Vasoconstrictor agents (direct actions on VSMC)			
Hormone/paracrine/autocrine Agents	Receptor	G protein	Intermediates/messengers
Adenosine	P_1–A_1	$G\alpha_i$	↓ cAMP/PKA
ATP	P_{2X}	—	Non-selective cation channel/Ca^{2+} ↓ K^+ channel activity/Depolarization/
	P_{2Y}	$G\alpha_{q/11}$	Ca^{2+}/PKC Ca^{2+}/PLA_2/Cyt P450/HETE
Angiotensin II	AT_1	$G\alpha_{q/11}$	Ca^{2+}/PKC Ca^{2+}/ADP ribosyl cyclase/cADP ribose Ca^{2+}/PLA_2/Cyt P450/HETE
Catecholamines	α_1	$G\alpha_{q/11}$	Ca^{2+}/PKC Ca^{2+}/ADP ribosyl cyclase/cADP ribose
	α_2	$G\alpha_i$	↓ cAMP/PKA
Endothelin	ET_A	$G\alpha_{q/11}$	Ca^{2+}/PKC Ca^{2+}/ADP ribosyl cyclase/cADP ribose Ca^{2+}/PLA_2/Cyt P450/HETE
	ET_B	$G\alpha_{q/11}$	Ca^{2+}/PKC Ca^{2+}/ADP ribosyl cyclase/cADP ribose Ca^{2+}/PLA_2/Cyt P450/HETE
PGE_2	EP_1	$G\alpha_{q/11}$	Ca^{2+}/PKC
	EP_3	$G\alpha_i$	↓ cAMP/PKA
$PGF_{2\alpha}$	FP	$G\alpha_{q/11}$	Ca^{2+}/PKC
Thromboxane	TP	$G\alpha_{q/11}$	Ca^{2+}/PKC Ca^{2+}/ADP ribosyl cyclase/cADP ribose
Vasopressin	V_1	$G\alpha_{q/11}$	Ca^{2+}/PKC
20-HETE	*	—	K channel (Ca^{2+} activated)/depolarization

Vasodilator agents (direct actions on VSMC)			
Hormone/paracrine/autocrine agents	Receptor	G protein	Intermediates/messengers
Adenosine	P_1–A_2	$G\alpha_s$	cAMP/PKA
Catecholamines	B	$G\alpha_s$	cAMP/PKA
Dopamine	D_1	$G\alpha_s$	cAMP/PKA
PGE_2	EP_4	$G\alpha_s$	cAMP/PKA
PGI_2	IP	$G\alpha_s$	cAMP/PKA
EDHF	*	—	K channel (Ca^{2+} activated)/hyperpolarization
11,12-/14,15-EET	*	—	K channel (Ca^{2+} activated)/hyperpolarization
Nitric oxide	sGC	—	cGMP/PKG cGMP/↓ PDE/↑ cAMP ↓ ADP ribosyl cyclase/↓ cADP ribose ↓ Cyt P450/↓ 20-HETE/hyperpolarization

(Continued)

TABLE 15.1 (Continued)

Hormone/paracrine/autocrine Agents	Receptor	G protein	Intermediates/messengers
Carbon monoxide	sGC	—	cGMP/PKG

Vasodilator agents (with preferential actions on endothelial cells)

Hormone/paracrine/autocrine Agents	Receptor	G protein	Intermediates/messengers
Endothelin	ET_B	$G\alpha_{q/11}$	Ca^{2+} NOS/NO Ca^{2+} PLA_2/COX/PGs/PLA_2/EET

Vasodilator agents (with preferential actions on endothelial cells)

Hormone/paracrine/autocrine Agents	Receptor	G protein	Intermediates/messengers
Bradykinin	B_2	$G\alpha_{q/11}$	Ca^{2+} NOS/NO Ca^{2+} PLA_2/COX/PGs/PLA_2/EET
Acetylcholine	M	$G\alpha_{q/11}$	Ca^{2+} NOS/NO Ca^{2+} PLA_2/COX/PGs/PLA_2/EET

Vasoconstrictor agents (also stimulators of vasodilators products by endothelial cells)

Hormone/paracrine/autocrine agents	Receptor	G protein	Intermediates/messengers
Adenosine	P_1–A_1	$G\alpha q/11$	Ca^{2+} NOS/NO Ca^{2+} PLA_2/COX/PGs/PLA_2/Cyt-P450/EET
Angiotensin II	AT_1	$G\alpha_{q/11}$	Ca^{2+} NOS/NO Ca^{2+} PLA_2/COX/PGs/PLA_2/Cyt-P450/EET
ATP	P_{2Y}	$G\alpha_{q/11}$	Ca^{2+} NOS/NO Ca^{2+} PLA_2/COX/PGsPLA_2/Cyt-P450/EET
Vasopressin	V_1	$G\alpha_{q/11}$	Ca^{2+} NOS/NO Ca^{2+} PLA_2/COX/PGs/PLA_2/Cyt-P450/EET

*Abbreviations: cAMP, cyclic adenosine monophosphate; ATP, adenosine triphosphate; CO, carbon dioxide; COX, cycloxygenase; cGMP, cyclic guanine monophosphate; Cyt-P450, cytochrome P450 enzyme; EET, Epoxy-eicosatrienoic acid; HETE, Hydroxy-eicosatrienoic acid; HO, heme oxgenase enzyme; NO, nitric oxide; NOS, nitric oxide synthase; PDE, phosphodiesterasePGs, prostaglandins; PLA_2, phospholipase A_2; SGC, soluble guanylyl cyclase; *, undetermined; ↓, inhibits.*

the actions of constrictor systems (for example, angiotensin II and norepinephrine) rather than producing vasodilation *per se*.

Vasodilator mechanisms

Relaxation of vascular smooth muscle involves opposite mechanisms to those eliciting contraction. Whereas an increase $[Ca^{2+}]_i$ causes contraction, a decrease in $[Ca^{2+}]_i$ elicits vasodilation. The prostanoids PGE_2 and PGI_2 and NO exert similar vasodilator effects on afferent- and efferent-arteriole VSMCs, primarily by reducing the entry of Ca^{2+} mobilization and decreasing myosin phosphorylaton. Vasodilation mechanisms can be cAMP- and/or cGMP-mediated [17, 42, 43].

The mechanisms proposed for cAMP mediated vasodilation (e.g., IP receptors for PGI_2, or EP_4 receptors for PGE_2) include the following:

1. GPCR activation of G_s, which activates adenylyl cyclase, increasing cAMP levels and thus stimulating PKA; cAMP phosphorylation of myosin kinase via a cAMP-dependent pathway decreases the affinity of the myosin kinase for the calcium-calmodulin kinase that is necessary for the phosphorylation of myosin. The result is a decreased calcium sensitivity of smooth-muscle cells to stimuli for contraction.
2. Inhibition of IP_3-mediated Ca^{2+} release from the SR.
3. Enhanced calcium pump activity in the sarcoplasmic reticulum, without increase in the plasma-membrane calcium pump activity.
4. Phosphorylation and decreasing in the opening of dihydropyridine sensitive calcium channels.

The mechanisms proposed for cGMP-mediated vasodilation (e.g., nitrovasodilators, EDRF, natriuretic peptides) include the following:

1. Activation of soluble guanylyl cyclase and production of cGMP, which can activate PKG and also inhibit cAMP breakdown via a cGMP-sensitive phosphodiesterase. Agents such as 11, 12 EET and EDHF cause renal vasodilation by relaxing VSMCs secondary to

plasma-membrane hyperpolarization following activation of K^+ channels.

2. Inhibition of L-type Ca^{2+} channels, which decreases Ca^{2+} entry into the cell.

3. PKG activation of Ca^{2+} uptake by the sarcoplasmic reticulum.

4. Increasing membrane polarization by activating K^+ channels, which will increase Ca^{2+} exit via Na–Ca exchanger.

5. Decreased Ca^{2+} mobilization by inhibition of IP3 formation or inhibition of the IP3 receptor in the sarcoplasmic reticulum.

6. Decreased Ca^{2+} mobilization mediated by ryanodine receptor due to inhibition of ADP ribosyl cyclase and cADP ribose production.

7. Activation of myosin light-chain phosphatase, conducing to dephosphorylation of myosin light chains.

Cross-talk mechanisms between cAMP and cGMP have also been proposed.

Indirect effects have also been reported. An example is the NO inhibition of Cyt-P450-mediated production of the vasoconstrictor 20-HETE.

VASOACTIVE PARACRINE/AUTOCRINE AGENTS – ACTIONS IN THE RENAL VASCULATURE

Nitric Oxide

Nitric oxide synthase (NOS) is activated, by hypoxia, shear stress, cell deformation, or vasoactive substances, to produce NO from l-arginine and molecular oxygen [41, 44–48]. NO is an important mediator of communication between endothelial cells and VSMCs, as well as between endothelial cells and tubule cells. Constitutively-active endothelial NOS (eNOS) is richly expressed in blood vessels and IMCD. Vasoactive agents such as Ang II and bradykinin stimulate eNOS, and ATP can stimulate NOS isoenzymes. The activation by these agents is mediated by G-protein-coupled receptors, and involves activation of phospholipase C (PLC), increase in cytosolic calcium concentration ($[Ca^{2+}]_i$), and calmodulin-dependent enzyme activation. Thus, many vasoconstrictor–ligand–receptor interactions leading to increased $[Ca^{2+}]_I$ elicit primarily VSMC contraction, but also stimulate endothelial production of the vasodilator NO, which limits the degree of constriction. The final balance of these two effects depends on the agonist, and the density of receptors/enzymes in endothelial cells and VSMCs. Ang II produces net vasoconstriction by virtue of the dominant action of AT_1 receptors on VSMCs, whereas bradykinin and ATP cause net vasodilation due to their dominant effects on endothelial cells and NO production.

NO rapidly permeates plasma membranes and binds to the heme moiety of soluble (no membrane tether) cyclic guanylyl cyclases to form a heterodimer of α and β subunits that catalyzes production of cGMP. NO is rapidly inactivated by hemoglobin. In health, NO may produce vasodilator effect by reducing the bioavailability of the constrictor O_2-. In disease states, the function of NOS is altered to generate vasoconstricting agents such as superoxide radicals and other reactive oxygen species. In these conditions, superoxide can scavenge NO to produce $ONOO^-$, reducing the availability of NO.

The low renal vascular resistance, compared to other vascular beds, is explained in part by a high dependence on NO. Inhibition of all three NOS isoforms (type I or neuronal, type II or inducible, and type III or endothelial) reduces RBF by about 35 percent, reduces GFR to a lesser extent, and increases arterial pressure by about 30 mmHg. Vasoconstriction occurs along both afferent and efferent glomerular arterioles, and the glomerular filtration coefficient (directly proportional to the capillary surface area per glomerulus and the hydraulic conductivity) is reduced.

Heme oxygenases/carbon monoxide system

Microsomal heme-oxygenase (HO) catalyzes the metabolism of heme to form carbon monoxide (CO), biliverdin and free iron [49–51]. Two isoforms, HO-1 and HO-2, are expressed in the kidney. Constitutively active HO-2 is ubiquitous, present in the renal vasculature and in almost all nephron segments. Under basal conditions, inducible HO-1 appears to be at low levels or absent from renal structures. In the vasculature, HO-2 is stimulated by vasoconstrictor substances such as Ang II and hypoxia. CO contributes to the renal vascular reactivity by acting as a vasodilator agent that counteracts the vasoconstriction produced by Ang II, catecholamines, 20-HETE, or pressure-induced myogenic tone. CO activates guanylate cyclase/cGMP signaling and Ca^{2+}-sensitive K^+ channels that hyperpolarize the membrane of VSMCs, therefore reducing Ca^{2+} entry via voltage-sensitive channels. There may be an interaction between CO and NO in eliciting vasodilation. Little is known about effects of endogenous CO on tubule transport and renal excretion of salt and water. Oxidative stress and cell injury, as well as chronic activation of AT_1 receptors by Ang II or V_1 receptors by vasopressin (AVP), induce HO-1 mRNA by Ca^{2+}- and perhaps PKC-dependent mechanisms. It has been reported that upregulation of the HO-1 gene attenuates oxidative stress caused by angiotensin II in primary cultured of TALH cells, a nephron segment highly susceptible to ischemic injury. In these cells Ang II causes a decrease in glutathione levels and DNA damage – a response that is respectively ameliorated and blocked by HO-1 expression. The response to HO-1 is, however, cell-type dependent: in endothelial cells HO-1 overexpression suppresses inflammation-induced apoptosis (by generation of HO-1-derived carbon monoxide production that triggers a protective vasodilatation), but in VSMCs it leads to apoptosis, an effect that can be secondary to bilirubin production [52].

Endothelin

Endothelins (ET) are potent paracrine vasoconstrictors produced by endothelial cells [53–56]. They are 21-amino-acid peptides, and three isoforms have been described, ET-1 being the most prevalent in humans. Human ET is derived by successive proteolytic steps from preproendothelin (212 amino acids), which is hydrolyzed by a specific endopeptidase to a 37–39 amino acid molecule named preproendothelin-1, which is cleaved by a neutral endopeptidase to form proendothelin-1 or big ET-1 (a non-functional peptide). This is further cleaved by the endothelium-bound endothelin converting enzyme-1 (ECE-1) to ET-1. The stimuli for ET-1 production and immediate secretion are Ang II, ATP, bradykinin, hypoxia, isoprostane, superoxide anion, shear stress, thrombin, cytokines, and high-salt diet. ET is predominantly secreted toward adjacent VSMCs. Local effects of endogenous ET are more pronounced than those elicited by the usually low concentrations of circulating ET.

Both ET_A and ET_B receptors are found in renal VSMCs, and both are vasoconstrictors, activating $G\alpha_{q/11}$, promoting an increase in $[Ca^{2+}]_i$ and triggering a contraction more pronounced and longer lasting than those elicited by norepinephrine or Ang II. This is due to a higher affinity of ET for its receptor and slower desensitization. ET-induced renal vasoconstriction is mediated by stimulation of NADPH oxidase, superoxide production, and stimulation of Ca^{2+} signaling via the ADP ribosyl cyclase and RyR pathway. Only ET_B receptors are expressed in endothelial cells, where their activation increases $[Ca^{2+}]_i$ and stimulates eNOS to produce the vasodilator NO. Production of vasodilator prostaglandins may also increase. Whether ET exerts a dilator or a constrictor action on arterioles depends on the relative abundance of ET receptors on the endothelium *versus* the VSMCs. Under normal conditions, ET-1 exerts a tonic constrictor effect. ET is thought to be primarily a vasoconstrictor agent in pathological conditions such as congestive heart failure and chronic renal failure, acting on both afferent and efferent arterioles of the glomerulus. In the renal medulla, ET-1 dilates the vasculature as a result of strong ET_B effects that increase NO production and vasodilator COX metabolites.

Gene-targeted deletion of the ET_A receptor has no effect on arterial pressure in unstressed animals. On the other hand, ET_B-receptor knockout animals have an increased arterial pressure, and high-salt diet induces salt-sensitive hypertension. Interestingly, specific deletion of ET_B receptors on endothelial cells does not result in a change in arterial pressure in animals consuming either a normal or high-sodium diet, suggesting the importance of ET_B receptors on collecting duct cells in maintaining salt balance and blood pressure [57, 58].

Arachidonic Acid Metabolites

Eicosanoid production is governed by the availability of the membrane fatty acid arachidonic acid. Its release is mediated by calcium-dependent cytosolic phospholipase A2 (PLA_2), and the activity of enzymes of the cyclooxygenase (COX), lipoxygenase, and cytochrome (Cyt) P450 families. COX1 and -2 synthesize prostaglandin H_2, from which different eicosanoids (PGE_2, PGI_2, and thromboxane, TxA_2) are produced by the action of PGE_2-isomerase, prostacyclin synthase, and thromboxane synthase, respectively. The lipoxygenase family generates leukotrienes. Very little is known about kidney lipoxygenase activity in physiological conditions. Finally, the Cyt-P450 monooxygenase family yields epoxy-eicosatrienoic acids (EETs) and hydroxy-eicosatrienoic acids (HETEs). In the vasculature, 80 percent of Cyt-P-450 activity is localized in endothelial cells.

Arachidonic Acid–COX Metabolites

Prostaglandins (PGs) and thromboxane (TxA_2) are synthesized by endothelial cells and act on vascular and tubule cells to function as autocrine or paracrine agents, contributing to the regulation of renal hemodynamics, renin release, tubule transport, and salt and water balance [59–61]. When renin levels are normal, the net effects of COX metabolites are vasodilatory and natriuretic (i.e., promoting Na^+ excretion), reflecting the predominant production and actions of PGE_2 and/or PGI_2 over $PGF_{2\alpha}$ and TxA_2. Healthy kidneys produce very little of the vasoconstrictor TxA_2. In the cortical vasculature PGI_2 is the major COX metabolite, whereas tubule cells, especially IMCD cells, produce primarily PGE_2. PGE_2 produced via COX2 in macula-densa cells is an important stimulus of renin release acting via EP_4 receptors and cAMP/PKA signal transduction in juxtaglomerular granular cells.

Prostaglandins activate G-protein-coupled receptors. In the renal vasculature, the main isotypes are EP_4 for dilator PGE_2, IP for dilator PGI_2, FP for constrictor $PGF_{2\alpha}$, and TP for constrictor TxA_2. EP_4 and IP receptors signal through $G\alpha_s$-proteins and cAMP/PKA cascade to relax VSMCs via inhibition of IP_3 mediated Ca^{2+} mobilization from sarcoplasmic-reticulum stores. Under resting conditions, vascular endothelial cells primarily produce PGI_2. Descending vasarecta have dilatory IP and EP_4 receptors. PGE_2 may exert a vasoconstricting effect via EP_1 and EP_3 receptors (see Table 15.1). TP receptors favor platelet aggregation and are vasoconstrictors, stimulating $G\alpha_{q/11}$ proteins to activate PLC and IP_3R-mediated Ca^{2+} mobilization and Ca^{2+} entry through L-type Ca^{2+} channels. In addition, ADP ribosyl cyclase, cADP-ribose and RyR signaling are important in Ca^{2+} release from sarcoplasmic reticumum and calcium-induced calcium release, resulting in renal vasoconstriction.

Vasoconstrictors such as Ang II and ET increase $[Ca^{2+}]_i$ in endothelial cells as well as in VSMCs. Some of the $[Ca^{2+}]_i$ increase in endothelial cells may originate from VSMCs, traversing via gap-junction channels. The $[Ca^{2+}]_i$ increase in endothelial cells stimulates PLA_2 to release arachidonic acid, the rate-limiting step in eicosanoid production.

Vasodilator COX metabolites (PGE_2 and PGI_2) are important in limiting the degree of vasoconstriction and thus maintaining adequate renal blood flow during low-salt diet and pathophysiological conditions such as heart failure and chronic renal failure. Production of vasodilating prostaglandins is increased by chronic salt restriction, an effect presumably mediated in part by high levels of Ang II and continuous elevation of $[Ca^{2+}]_i$ in endothelial cells and perhaps VSMCs. COX inhibition magnifies the vasoconstrictor effect of Ang II. The enhanced vasodilator effect of endogenous PGs during low sodium diet is due to a combination of Ang II stimulation of PG production and apparent upregulation of EP_4 receptors. A greater density of EP_4 receptors also may contribute to elevated renin release during extracellular-volume contraction.

Arachidonic Acid–CYT-P450 Metabolites

In renal vascular-endothelial cells, VSMCs and tubule cells, arachidonic-acid metabolism by Cyt-P450s generates EETs, diHETEs, and 20-HETE – substances that are autocrine agents and second messengers [48, 62–66]. These agents are lipophilic, bind to proteins, and partition into phospholipids. Their actions may be independent of conventional receptors, perhaps acting directly on membrane channels. VSMCs of large cortical arteries produce 20-HETE via the Cyt-P450-4 A hydroxylase gene family. Smaller arterioles produce a combination of vasodilator EETs and constrictor 20-HETE, which can have opposing effects on tubule transport. The metabolites 5, 6-EET, 11, 12-EET and 14, 15-EET are derived from the P450-2C epoxoxygenase family. Vasodilator 11, 12- and 14, 15-EETs are primarily produced by endothelial cells in the renal cortex, whereas 20-HETE is preferentially synthesized in the medulla, especially the thick ascending LH.

Vasoconstrictors such as Ang II and ET increase vascular production of 20-HETE, which reinforces agonist-induced vasoconstriction by receptor-mediated increases in $[Ca^{2+}]_i$ and activation of PLA_2, with 20-HETE formation via Cyt-P450 ω-hydroxylation of arachidonic acid. 20-HETE produces vasoconstriction due to increased $[Ca^{2+}]_i$ and reduces the open-state probability of the Ca^{2+}-activated high-conductance K^+ channel, causing depolarization and activation of voltage-gated Ca^{2+} channels.

5, 6-EET appears to elicit vasodilation secondary to formation of the COX metabolites PGI_2 and PGE_2. Bradykinin and acetylcholine are thought to produce vasodilation due to their ability to produce endothelial-derived EET and di-HETE (or EDHF) in addition to stimulating COX and NOS.

Isoprostanes are prostaglandin-like compounds formed by free radical lipid peroxidation of arachidonate, independent of COX, that are increased during high salt intake and chronic states of oxidative stress associated with diseases [67–70]. Two common isoprostanes, 8-isoPGE$_2$ and 8-isoPGF$_{2a}$, activate the thromboxane TP receptor to cause vasoconstriction and other TxA$_2$-like effects in the vasculature.

Endothelium-Dependent Hyperpolarizing Factor

In response to stimulation by bradykinin or acetylcholine, endothelial cells release a vasodilating agent distinct from prostaglandins and NO [48, 71, 72]. This elusive agent, termed Endothelium-Derived Hyperpolarizing Factor (EDHF), consists of one or more diffusible factor(s) and relaxes VSMCs by hyperpolarization caused by cAMP/PKA signaling and activation of high-conductance, Ca^{2+}-dependent K^+ channels or by propagation of the endothelial-cell hyperpolarization via myoendothelial gap junctions. Recent evidence implicates 11, 12 EET and 14, 15 EET as EDHFs, with greater relaxing effects in the microcirculation than in larger arteries.

Bradykinin

Intrarenal bradykinin has vascular actions in the cortex and medulla mediated primarily, if not exclusively, by B_2 receptors [73–77]. The main action of bradykinin is vasodilatation, mediated by a predominance of B_2 receptors on endothelial cells, which stimulate PLC production and increase $[Ca^{2+}]_i$ to stimulate eNOS activity and release EDHF/EET. Vasodilation is mediated in part by activation of K^+ channels, leading to plasma-membrane hyperpolarization. The B_2 receptor effects to increase medullary blood flow and inhibit Na^+ reabsorption appear to be primarily mediated by NO. A smaller population of constrictor B_2 receptors coupled to $G\alpha_{q/11}$ resides on VSMCs.

Purine Nucleotides and Purinoreceptors

Vasoconstrictor adenosine-sensitive A_1 receptors are the predominant type expressed in preglomerular afferent arterioles, glomerular mesangial cells, JG cells, and vasa recta [78–80]. Extracellular adenosine acts on A_1 purinergic receptors that also respond to AMP but not to ADP or ATP. P_1-A_1 receptors couple to $G\alpha_i$ proteins and decrease cAMP/PKA activity (as well as stimulating PLC_β, presumably by $G_{\beta\gamma}$) and increase $[Ca^{2+}]_i$ in VSMCs. In epithelial and endothelial cells, A_1 receptors couple to $G\alpha_{q/11}$, stimulating PLC signaling, increasing both $[Ca^{2+}]_i$ and PKC, and activating Ca^{2+}-dependent eNOS in endothelial cells. Afferent arterioles have more A_1 receptors than efferent arterioles, and adenosine produces net vasoconstriction by activating A_1 receptors. Renal P_1-A_2 receptors linked to $G\alpha_s$ proteins stimulate adenylyl cyclase to activate the cAMP/PKA pathway in VSMCs and endothelial eNOS. A_2 receptor stimulation by adenosine causes vasodilation mediated by opening of K_{ATP} channels of both afferent and efferent arterioles and a natriuresis without a change in the filtered sodium load. Adenosine has a greater affinity for A_1 than for A_2 receptors. Therefore, low adenosine concentrations elicit vasoconstriction, whereas high concentrations produce vasodilation. A_1-receptor antagonists attenuate or abolish TGF activity, as do mutations of A_1 receptors. In the medulla, adenosine activation of epithelial A_1 receptors appears to be antinatriuretic, in contrast to A_2-receptor-mediated increase in medullary blood flow and natriuresis. The balance between A_2 vasodilating receptors and A_1 vasoconstricting receptors varies with salt intake. P_1-receptor stimulation is more effective in animals maintained on a low-salt diet. In certain pathological conditions, Ang II and adenosine are synergistic – i.e., Ang II enhances the vasoconstrictor response to adenosine, and *vice versa*. The underlying mechanism is unclear.

ATP can be released from nerve terminals, endothelial cells, VSMCs, and epithelial cells, and acts locally, producing vasoconstriction. Extracellular ATP and ADP preferentially activate P_2 receptors, which have less affinity for adenosine or AMP. ATP or P_2-receptor agonists may cause renal vasodilation or vasoconstriction, depending on the predominating receptor type. ATP-responsive P_{2X} receptors are present in the VSMCs of the preglomerular vasculature but not in glomeruli and efferent arterioles, with P_{2Y} present in both glomerular arterioles. P_2 receptors are present on preglomerular vessels (endothelial cells and VSMCs), glomerular mesangial cells, and tubule cells (PCT and CD). P_{2X} receptors are ligand-gated cation channels with two membrane-spanning domains. In preglomerular VSMCs, P_{2X} receptors allow entry of Ca^{2+} and Na^+ and efflux of K^+; the plasma membrane depolarizes, triggering Ca^{2+} entry via voltage-gated L-type Ca^{2+} channels. The result is rapid, albeit transient, vasoconstriction. ATP activation of P_{2X} receptors may also stimulate Cyt-P450 production of the vasoconstrictor 20-HETE. P_{2Y} receptors are classic $G\alpha_{q/11}$-protein-coupled receptors. Endothelial P_{2Y} receptors are coupled to $G\alpha_{q/11}$-proteins that activate PLC to mobilize Ca^{2+} from sarcoplasmic-reticulum stores and stimulate PKC. Ca^{2+}-dependent eNOS and COX1 produce the vasodilators NO and PGI_2. P_{2Y} receptors on VSMCs activate the same second-messenger systems to produce contraction, which is usually weaker than endothelium-mediated relaxation.

Circulating ATP elicits variable responses involving a combination of vasodilator P_{2Y} receptors on endothelial cells and vasoconstrictor P_{2X} receptors on VSMCs. When the vasodilating component is eliminated by inhibition of NO production, the net response becomes P_{2X}-receptor-mediated vasoconstriction. ATP constricts the afferent arteriole considerably more than the efferent arteriole. ATP acting on P_{2X} receptors is thought to mediate or modulate renal autoregulatory mechanisms involving the myogenic response and TGF [4, 79, 81].

Intracellular ATP can modulate vascular resistance by regulating ATP-sensitive K^+ channels; high ATP levels lead to VSMCs hyperpolarization and vasodilatation.

Reactive Oxygen Species – Superoxide and Hydrogen Peroxide

Actions of Ang II, norepinephrine, and ET on AT_1, α_1-adrenergic, and ET_A and ET_B receptors stimulate vascular NADPH oxidase to produce superoxide anion, thereby activating ADP ribosyl cyclase and causing Ca^{2+} mobilization via cyclic ADP-ribose sensitization of RyR in the preglomerular vasculature. This results in marked renal vasoconstriction [29, 30, 56, 82–84]. Agonist-induced acute renal vasoconstriction produced by superoxide is largely independent of the presence or absence of NO. On the other hand, the pressure-induced myogenic response is markedly enhanced by superoxide in the absence, but not in the presence, of NO. Stimulation of NADPH oxidase expression and activity and O_2^- production in the vasculature is more pronounced in pathological conditions.

In non-renal arteries, generally H_2O_2 tends to produce vasodilation by activating KCa channels to produce hyperpolarization. Little is known about the effects of physiological levels of H_2O_2 on renal resistance arterioles. In the renal medulla, H_2O_2 is thought to act as a vasoconstrictor.

Summary

Vascular endothelial cells play an important role in regulating renal vascular resistance as well as ion and water transport by the renal tubules. Endothelial cells produce vasodilator and vasoconstrictor agents, whose actions are finely balanced to regulate extracellular fluid homeostasis and blood pressure. Under physiological conditions, endothelium-derived vasodilators such as PGI_2, nitric oxide, carbon monoxide, and EDHF counteract part of the renal vasoconstrictor effects produced by circulating agents such as vasopressin and norepinephrine, and local, as well as circulating, Ang II. In disease states such as congestive heart failure, chronic renal disease, and hypertension, the vascular endothelium can produce large amounts of constrictor substances such as ET and TxA_2. Our understanding of cell–cell interactions between endothelium and VSMCs, as well as the paracrine influences of vascular cells on tubule cells, is still quite limited.

VSMCs contract and relax in response to a host of vasoactive factors that change the diameter of the interlobular arteries and glomerular arterioles, the major renal resistance vessels. The balance of afferent and efferent arteriolar tone determines glomerular capillary pressure and, in turn, glomerular filtration rate. In a regulated fashion, tubular reabsorption returns between about 99 percent of the filtered fluid and solutes to peritubular capillaries and vasa recta and, eventually, the systemic circulation. Capillary uptake is favored by a low hydrostatic pressure and a high colloid-osmotic pressure in postglomerular, peritubular capillaries. Most of the constrictor agents, such as angiotensin II, vasopressin, and norepinephrine, activate specific G-protein-coupled cell surface receptors, leading to activation of a phospholipase C that elicits IP_3 and DAG signals which lead to increased intracellular $[Ca^{2+}]$ and stimulation of PKC. Another pathway for Ca^{2+} release from sarcoplasmic reticular stores involves ADP ribosyl cyclase and RyR. Vasodilator agents such as PGE_2 and PGI_2 activate specific receptors linked to the cAMP/PKA signaling pathway, which cause reductions in intracellular $[Ca^{2+}]$. On the other hand, the vasodilators nitric oxide and carbon monoxide exert their relaxing effects on vascular smooth muscle cells through a cGMP/PKC pathway, as well as mechanisms independent of cGMP. Dilator EETs and constrictor HETEs act by stimulating or inhibiting K^+ channels, respectively, changing the membrane potential and thus the activity of L-type Ca^{2+} channels.

ENDOTHELIAL CELL CONNECTIONS: CONNEXINS AND GAP JUNCTIONS

Connexins (Cx), specialized channel proteins, combine to form endothelial and myoendothelial gap junctions that transduce chemical, electrical, and mechanical signals, thereby facilitating cell–cell communication in the renal vasculature [85–87]. Gap-junction channels result from the docking of the extracellular loops of two hemichannels or connexons, each from one of the adjacent cells. Each connexon is a Cx hexamer, either homomeric or heteromeric, circling a large aqueous pore, permeable to hydrophilic molecules up to 1 kDa, including paracrine agents. Hundreds to thousands of gap-junction channels cluster to form a gap junction or gap-junction plaque. Gap-junction channels have high conductances (15–300 pS), high open-probability, and exist in multiple subconductance states. These functional characteristics are regulated by membrane voltage, intracellular pH, $[Ca^{2+}]_i$, signaling molecules, and phosphorylation events in an isoform-dependent manner. Cells with constitutive higher sensitivity to a given stimulus may act as pacemakers, initiating a response and signaling neighboring cells by intercellular permeation of ions or second-messenger molecules. Gap junctions may participate in intercellular signaling by coordinating oscillations of $[Ca^{2+}]_i$.

The physiological roles of gap junctions in the kidney are just emerging. Vasodilators such as NO and PGI_2 as well as Ca^{2+} may pass between endothelial cells and VSMCs, via gap junctions. Cx40 deficient mice are hypertensive, with impaired conduction of dilatation signals in the arterioles and irregular conduction of vasomotion that can lead to complete vascular occlusion. Endothelial deficiency of Cx43 also leads to abnormal propagation of vasodilation along arteries/arterioles [87].

The major connexins found in the renal circulation are Cx37, Cx40, and Cx43 [85–87]. Their functional roles are less well explored than those in other organs. Cx37 and Cx40 are highly expressed in endothelial cells of renal vessels and glomeruli. Endothelial cells of preglomerular arteries and arterioles also express Cx37, Cx 40, and Cx43. Cx37 is the main connexin in the media of arcuate and interlobular arteries and afferent arterioles. Endothelial-cell Cx40 is most abundant in large intrarenal arteries, including the interlobular artery and the proximal portion of the afferent arteriole, but markedly decreases as the arteriole approaches the glomerulus. Afferent arterioles have considerably more Cx isoforms than efferent arterioles. With regard to the efferent arteriole, Cx43 is present in endothelial cells, but no connexins appear to be expressed in VSMCs. Cx37 and Cx40 are expressed in juxtaglomerular cells and extraglomerular mesangial cells. Cx43 is localized to extraglomerular mesangium, and seems to be involved in renin secretion. Intraglomerular mesangial cells have primarily Cx40. NO appears to upregulate Cx40 expression in VSMCs of afferent arterioles, and downregulate Cx43 immunoreactivity in endothelial cells of efferent arterioles [88].

Gap junctions appear to contribute to conduction of endothelium-derived hyperpolarization in the renal vasculature. Peptides that are targeted to antagonize Cx function produce renal vasodilation (acetylcholine-induced, EDHF mediated) during inhibition of NOS and COX enzymes [89]. Under these conditions, putative blockade of Cx40 abolishes the increase in RBF, whereas that of Cx43 partially reduces the renal vasodilation. Ca^{2+} spreading along interlobular arterioles appears to be mediated by gap junctions and Ca^{2+} entry through L-type Ca^{2+} channels [90]. Descending vasa-recta endothelial cells express Cx40 and Cx43 and function as an electrical syncytium, whereas pericytes express Cx37 without electrical conduction [91].

Gap junctions formed by Cx40 connect endothelial cells and renin-containing juxtaglomerular cells [86]. Upregulation of Cx40 in high-renin states implicates connexins in the regulation of renin release. In Cx40-null mice, pressure-dependent control of renin synthesis and release is impaired. Normally, renin-producing cells are located in the walls of afferent arterioles before the glomerular capillary tuft, but in Cx40-null mice renin-positive cells are absent in the vessel walls and instead are found in cells of the extraglomerular mesangium, glomerular tuft, and periglomerular interstitium. Stimulation of renin secretion by severe sodium depletion fails to replicate the normal expression of renin in juxtaglomerular cells. Cx43 is critical for the large increases in renin secretion associated with renal-artery stenosis and changes in salt intake [85]. Present information indicates that uptake of Na^+, K^+, and Cl^- by the macula-densa cells results in production of adenosine that activates A1 receptors in the extraglomerular mesangial cells and triggers increase in $[Ca^{2+}]_i$; a Ca^{2+} wave propagates from extracellular mesangial cells to the affererent granular rennin-containing cells and to VSMCs via gap junctions, resulting in a decrease in renin release and afferent arteriole vasoconstriction [92, 93]. Cx37 and Cx40 may play important roles in the autoregulation of

RBF and GFR [94]. Inactivation of Cx37 and Cx43 inhibits myogenic responses in mesenteric arteries [95]; however, no expression of Cx43 has been described in the pericytes of the glomerular efferent arteriole.

Tubulovascular Interactions: the Juxtaglomerular Apparatus

The juxtaglomerular apparatus (JGA, Figure 15.1b) is a structural and functional unit in which paracrine signals are transmitted between the macula-densa cells (differentiated tubule cells at the end of the TAL), the extraglomerular mesangial cells (interposed between the glomerular arterioles and the macula densa), the arteriolar VSMCs, and the renin-producing granular cells in the afferent arteriole [3, 4, 96].

Vascular and Neural Control of Renin Secretion

Granular cells at the end of the afferent arteriole synthesize and release renin by a process regulated by local changes in arteriolar hydrostatic pressure and stretch, and β-adrenoceptor stimulation by norepinephrine released from perivascular sympathetic nerve terminals [86, 97–99]. Increased afferent arteriolar pressure inhibits renin release, an effect presumably due to stretch of JGA cells causing an increase in Ca^{2+} influx that, in contrast with the effects in most secretory cells, inhibits renin secretion. The pressure-dependent control of renin release requires an intact coupling of endothelial cells and JG cells via Cx40. Mice with specific endothelial cell knockout of Cx43 (normally expressed between endothelial cell of the efferent arterioles) have hypotension, and elevated levels of Ang I and II. The first seems to be a primary mechanism due to NO production; the latter a secondary compensatory mechanism.

β-adrenoceptors activate a $G\alpha_s$-protein linked to the cAMP–PKA messenger pathway, stimulating renin secretion.

Circulating hormones and paracrine factors also influence renin secretion, some of them acting via the cAMP/PKA signaling pathway. Endothelium-derived PGE_2 and PGI_2 and β-adrenergic receptor agonists are potent stimuli of renin secretion. Dopamine stimulates renin release from granular cells via D_1 receptors and cAMP generation. Interactions such as Ang II with AT_1 receptors, AVP with V_1 receptors, or ET with ET_A receptors, activate a $G\alpha_{q/11}$-protein to increase $[Ca^{2+}]_i$ and activate PKC, causing a decrease in renin secretion. It is not clear whether the key signal is increased $[Ca^{2+}]_i$ by itself, Ca^{2+}-activation of Cl^- channels and a fall in $[Cl^-]_i$, depolarization associated with the increase in Cl^- permeability, or increased PKC activity.

NO has biphasic effects on granular cells. Acute increases are inhibitory, whereas long-term exposure stimulates renin release. In this regard, endothelial NO appears to have a tonic permissive effect, mediated by the ability of cGMP to elevate cAMP by inhibiting phosphodiesterase and thereby cAMP breakdown. Under certain conditions, NO can inhibit renin release via cGMP-dependent protein kinase (PKG) activity. In contrast, NO generated from nNOS in TAL and macula-densa cells does not mediate changes in renin secretion in individuals on a low-sodium diet, and renin secretion is normally regulated in eNOS-deficient mice.

Macula-Densa Control of Renin Release

The macula densa participates in the regulation of renin release from juxtaglomerular granular cells. Renin secretion depends on NaCl delivery to and reabsorption by the macula-densa cells at the end of the TAL. Inhibition of renin release occurs when solute delivery to this section of the renal tubule is high, and stimulation is associated with low solute delivery. Examples include high- and low-salt diet, respectively. Renin release leads to increased concentration of Ang I and Ang II in the adjacent interstitial compartment, as well as in the systemic circulation. This is a regulatory mechanism mediated by macula-densa nNOS and COX2 metabolites [60, 61, 100]. The COX2 metabolites PGE_2 and PGI_2 act on EP_4 and IP receptors on JG cells to stimulate cAMP and PKA, thereby enhancing renin secretion. COX2 expression can be induced either by chronic sodium restriction or by inhibition of TAL NaCl reabsorption by furosemide. In contrast, the Cyt-P450 metabolite 20-HETE inhibits renin secretion, presumably by elevating $[Ca^{2+}]_i$ in juxtaglomerular granular cells.

Macula-densa cells may signal granular cells to inhibit renin release by secreting adenosine and/or ATP across the basolateral membrane in response to increased sodium delivery [96, 101, 102]. The precise mechanism of this effect is not clear. Adenosine stimulates granular-cell A_1 and A_2 receptors. Adenosine A_1-receptor activation inhibits renin secretion via stimulation of PLC, by increasing $[Ca^{2+}]_i$ and reducing cAMP/PKA. As stated above, this is an exception in secretory cells. Conversely, adenosine A_2 receptors stimulate renin secretion via cAMP/PKA signaling.

Summary

The juxtaglomerular apparatus has unique anatomical and functional properties. Macula-densa cells at the end of the thick ascending LH respond to low lumen [NaCl] by signaling to juxtaglomerular granular cells at the end of the afferent arteriole to increase renin secretion. Other stimuli for renin secretion are high sympathetic nerve activity, low systemic and afferent-arteriole pressure, and increased PGE_2. High rates of NaCl reabsorption by macula-densa cells inhibit renin secretion. The signal mediating the functional connection between macula-densa cells and afferent-arteriolar granular cells is not clear. Attractive candidates are adenosine and ATP.

Tubulovascular Interactions: Juxtaglomerular Apparatus and Tubuloglomerular Feedback

Tubuloglomerular feedback (TGF), a process mediated by the juxtaglomerular apparatus (Figure 15.1b), is the inverse relationship between the NaCl concentration in the tubule fluid at the macula densa and the capillary pressure and filtration rate of the same nephron [96, 101–105]. In functional terms, an increase in distal delivery of NaCl produces contraction of the afferent arteriole and a decrease in glomerular filtration.

The kidney regulates RBF and GFR during changes in arterial pressure by intrarenal mechanisms that adjust in preglomerular vascular resistance in proportion to perfusion pressure. This is referred to as *autoregulation*, and is mediated by two main mechanisms [3, 105]. The first is a myogenic response intrinsic to the afferent-arteriole VSMCs [106]. These cells contract in response to the increase in tension of the wall when stretched by a higher intravascular pressure. The second is TGF. Primary increases in arterial pressure initially increase glomerular filtration and fluid delivery to the TAL, where the macula-densa cells respond by sending a vasoconstricting signal to the afferent-arteriole VSMCs. The nature of the signal mediating TGF is uncertain. Current postulates focus primarily on ATP and adenosine [107]. As NaCl transport by the macula-densa cells increases, so does $[Ca^{2+}]_i$, which can stimulate PLA_2 and nNOS as well as COX2; ATP may exit across the basolateral membrane. The Cyt-P450 metabolite 20-HETE and NO may also play roles in adjusting preglomerular tone, and extraglomerular mesangial cells may be also involved. These agents may act as either mediators or modulators. For example, NO generated by macula-densa nNOS acts as a modulator of TGF responsiveness, as it inhibits macula-densa NaCl reabsorption at high tubule flow rates, compared to little or no effect when [NaCl] at the macula densa is low. NO also is thought to act directly on macula-densa cells suppressing release of a constrictor agent, rather than diffusing to the afferent arteriole. In this regard, NO appears to quench superoxide produced locally. Also, NO produced at upstream nephron sites may contribute to inhibition of TGF due to its ability to inhibit NaCl transport by macula-densa cells. Nevertheless, a chronic deficit in nNOS in gene-targeted animals does not affect TGF.

It is clear that Ang II does not mediate TGF. An increase in renin release and Ang II formation occurs when NaCl delivery to the macula densa is low and the preglomerular vessels dilate, not contract. However, Ang II may modulate TGF sensitivity and glomerular vascular reactivity by a mechanism that remains elusive. Other vasoconstrictors, such as norepinephrine and ET, have little effect on TGF sensitivity. There is an inverse relationship between TGF activity and chronic salt intake. High renin and Ang II levels are associated with strong TGF during salt restriction. The attenuated TGF during high-salt diet is attributable to enhanced NO production and low Ang II levels.

The involvement of eicosanoids in TGF is uncertain. COX2 is present in macula-densa and TAL cells; its expression is enhanced during chronic sodium restriction. COX2 metabolites such as PGE_2 diminish the vasoconstriction accompanying increased NaCl delivery. However, TGF is normal in animals null for COX2 or thromboxane TP receptors.

Role of Purinonucleotides and Purinoceptors in Tubuloglomerular Feedback

The purinergic agents adenosine and ATP can be released from epithelial cells into or formed in the interstitial compartment, providing a metabolic link between epithelial-cell NaCl transport in macula-densa cells and preglomerular vascular resistance [78, 79, 81, 97, 103, 108]. As load-dependent reabsorption of NaCl by the macula densa increases, so does ATP hydrolysis and hence adenosine production. Adenosine diffuses to the afferent arterioles to activate vascular $P_1 A_1$ (adenosine) receptors and thereby elicit vasoconstriction; adenosine may also be released from nerve terminals. ATP can be released by exocytosis from macula-densa cells and epithelial cells of the thick ascending LH nerve terminals, or may exit from endothelial cells, VSMCs, and epithelial cells, probably through membrane channels, to activate P_2 (ATP) receptors.

Modulation of TGF by Nitric Oxide and Superoxide Anion

Other metabolites produced by macula-densa cells can modulate TGF [104, 109–111]. For example, NO production catalyzed by macula-densa nNOS induces vasodilatation, buffering the vasoconstriction that follows arteriolar stretching at high tubular fluid flow rates. Macula-densa production of superoxide anion (O_2^-) has a vasoconstrictor role, with actions to quench the availability of NO (to form $ONOO^-$) and director constrictor action on afferent arterioles. Also contributing may be the HO-2 system in macula-densa cells, with production of CO and biliverdin that exert vasodilator-like actions. NO and CO may act to inhibit NaCl uptake across the luminal membrane of macula-densa cells. O_2^- may have the opposite effect.

A second Tubuloglomerular Feedback System

A recently unveiled second type of TGF system appears to link NaCl reabsorption in the cortical connecting tubule to the control of afferent arteriolar resistance [112]. Increasing NaCl delivery in a connecting tubule increases capillary pressure in its parent glomerulus as a result of afferent arteriole dilation. The signal is initiated by Na reabsorption via ENaC (sensitive to amiloride inhibition). The effector limb is not certain, but may involve the kallikrein/kinin system that is found in this nephron segment [75]; other dilator candidates are prostanoids, EETs. NO, and CO. An interaction between the primary and secondary TGF systems and the relative importance of this second TGF system await investigation.

Intrarenal Angiotensin II Production, Storage, and Actions

Circulating Ang II is formed by the action of angiotensin converting enzyme (ACE) on plasma Ang I in the lungs and the renal vasculature [30, 112–116]. Ang I can be generated from angiotensinogen by renin and other proteolitic enzymes (e.g., chymase and catepsin). Recent evidence suggests important formation, storage, and actions of Ang II in the kidney. All necessary substrates and enzymes for Ang II production are present in juxtaglomerular granular cells, and in PCT and CD principal cells. ACE lines endothelial cells and luminal and basolateral membranes of PCT cells. About 20 percent of circulating inactive Ang I is converted to biologically active Ang II in a single pass through the kidney. The PCT can produce Ang II from circulating or locally generated Ang I, concentrates Ang II by AT_1-receptor-mediated uptake, and stores it in the cells. Ang II is secreted into the lumen, where it can act locally or downstream, binding to luminal-membrane receptors in TAL, DCT, and CD. The effects are to increase Na^+, Cl^- and HCO_3^- reabsorption. Angiotensinogen also appears to be secreted into tubule fluid by the PCT.

There are two classes of Ang II receptors. AT_1 receptors, the predominant if not exclusive class under normal conditions, are present along the renal vasculature, including preglomerular arteries and arterioles, juxtaglomerular granular cells, glomeruli, efferent arterioles and vasa recta, multiple nephron segments (PCT, TAL, DCT, CD), and medullary interstitial cells. AT_1 receptors couple to $G\alpha_{q/11}$ proteins, PLC, and the classical IP_3-Ca^{2+} and DAG–PKC pathways to cause vasoconstriction and decrease salt excretion. AT_1 receptors also stimulate vascular NADPH oxidase to produce superoxide anion, activate ADP ribosyl cyclase and Ca^{2+} mobilization in glomerular arterioles, to cause marked renal vasoconstriction. Cyclic ADP-ribose sensitizes RyR on sarcoplasmic reticulum to release Ca^{2+} and promote Ca^{2+}-induced Ca^{2+} release, which may amplify Ca^{2+} signaling initiated via IP_3R and other pathways [29].

Elevated Ang II levels, as seen with chronic salt restriction, constrict both afferent and efferent arterioles, reducing RBF more than GFR. High Ang II concentrations desensitize the blood vessels by reducing the AT_1 receptor density, and thus reduce the chronic effects of increased levels of the vasoconstrictor. In contrast, high endogenous Ang II concentration may upregulate tubule AT_1 receptors, favoring Na^+ retention. AT_1 receptors on granular cells exert short-loop feedback inhibition of renin release in association with increased $[Ca^{2+}]_i$ and PKC activation. Stimulation of AT_1 receptors

activates the Ca^{2+}-dependent enzymes PLA_2 and NOS in endothelial cells, promoting production of prostanoids and NO that buffer part of the Ang II-induced vasoconstriction. AT_1 receptors stimulate Na^+-H^+ exchange in PCT and TAL.

AT_2 receptors are weakly expressed along the vasculature and nephron in a healthy kidney, and may be upregulated during low-salt diet. Their actions are primarily mediated by NO and cGMP, tending to oppose the predominant vasoconstrictor and anti-natriuretic effects of AT_1 receptor stimulation [117].

ACE-2 converts either Ang I or Ang II to Ang 1-7, which appears to act on a unique receptor to exert vasodilation and natriuresis [118–120]. The actions of Ang 1-7 are more prevalent during inhibition of ACE, which upregulates ACE-2 expression, and reduce the opposing effects of Ang II.

Pro-renin/renin Receptor and Regulation of Tissue Angiotensin II

The pro-renin/renin receptor and its actions are emerging concepts [121–123]. Recent evidence suggests inactive plasma pro-renin may bind to a pro-renin/renin receptor, a single-transmembrane-domain protein, to become an active form of renin, by a non-proteolytic process capable of catalyzing local production of Ang I from plasma angiotensinogen, leading to increased Ang II in the proximity of AT_1 receptors to signal by traditional pathways. In addition, it has been proposed that the ligand/receptor complex can activate ERK 1/2 signaling leading to proliferation and chronic inflammation, effects independent of Ang II and AT_1 receptors. In the kidneys, the pro-renin/renin receptor resides in VSMCs and glomerular mesangial cells and podocytes, and thus may impact on glomerular perfusion and filtration. Pharmacological agents have been developed to function as pro-renin/renin receptor blockers that seem beneficial in improving renal structure and function and blood pressure control in disease states characterized by high pro-renin/renin conditions such a diabetes mellitus, possibly more so than conventional inhibitors of AT_1 receptors or ACE.

Summary

Tubuloglomerular feedback is an adaptive mechanism that links the rate of glomerular filtration to the concentration of salt in the tubule fluid at the macula densa. A high [NaCl] and reabsorption rate at this site causes contraction of the afferent arteriole and a reduction in GFR. This autoregulatory mechanism is intrinsic to the kidney – i.e., it does not require neural or humoral agents. The nature of the signal is still controversial. It could well involve adenosine or ATP. In addition to the general circulation, a renin–Ang II system exists within the kidney, apparently both in the medulla as well as in the cortex, with substrates and enzymes localized in proximal-tubule and collecting-duct cells in addition to juxtaglomerular granular cells. The major effects of local and circulating Ang II are vasoconstriction, Na^+ retention, and inhibition of renin release, all of which are primarily mediated by AT_1 receptors.

Vasculotubular Communication

Receptors in the basolateral membranes of renal-tubule cells convey chemical information from the blood, vascular cells, and interstitial cells. Interactions between extracellular matrix and cell functions are beginning to be understood. In general, vasodilating agents inhibit Na^+ reabsorption by one or more nephron segments. NO, PGE_2/PGI_2, dopamine, and bradykinin conform to this oversimplified notion. Vasoconstrictors such as Ang II and norepinephrine usually stimulate Na^+ reabsorption. As a consequence of contraction of afferent and efferent arterioles, vasoconstrictor agents favor postglomerular capillary uptake of reabsorbed fluid by reducing blood flow, reducing capillary hydrostatic pressure, and increasing plasma colloid osmotic pressure. Vasodilators tend to have the opposite effects, promoting less fluid uptake from the interstitium into peritubular capillaries. Notable exceptions are the vasoconstrictors ET and 20-HETE, both of which are natriuretic. Paracrine agents produced by vascular cells exert actions on tubule cells by receptor-mediated events, as those elicited by agents produced by epithelial cells. To minimize duplication, paracrine control of carrier proteins is discussed below under the heading of "Tubule–tubule communication."

Pressure Natriuresis

Over a specific pressure range, urinary sodium excretion increases as a function of arterial pressure. The slope of the relation may vary, depending on experimental conditions and sodium balance. The mechanisms by which acute changes in arterial pressure influence tubule transport and Na^+ excretion are incompletely understood [5, 8, 124, 125]. Increased vascular hydrostatic pressure may release the endothelial-derived NO and PGE_2 that are vasodilators and natriuretic. Blood flow in the inner medulla may be more sensitive to changes in arterial pressure than cortical blood flow, and increased medullary blood flow is thought to elevate capillary and interstitial hydrostatic pressure and decrease colloid-osmotic pressure. Together, these changes lead to a reduction in net salt and water reabsorption along the LH and CD.

PARACRINE SIGNALING IN RENAL-TUBULE EPITHELIAL CELLS

Paracrine Control of Solute Transport

The major nephron segments produce and release autocrine and paracrine factors that modulate transepithelial solute and water transport [21, 24, 60, 126–128]. The main solute carrier proteins for solutes and water in the nephron are as follows:

1. In the PT, the Na^+–H^+ exchanger.
2. In the thick ascending LH, Na^+–K^+–$2Cl^-$ co-transporter, ROMK channel, and Na^+–H^+ exchanger.
3. In the DT, Na–Cl co-transporter, Na^+–H^+ exchanger, Na^+ channel (ENaC), ROMK channel, and H^+, K^+-ATPase.
4. In the principal cells of the CD, Na^+ channel, Aquaporin 2, urea transporter, H^+-ATPase and ROMK channel.
5. In the intercalated cells of the CD, H^+-ATPase, Cl^-/HCO_3^- exchanger and Na^+–H^+ exchanger.

All these segments express Na^+, K^+-ATPase in their basolateral membranes. The thick ascending LH also expresses Na^+–H^+ exchanger and a Cl^- channel in the basolateral membrane; the latter is also expressed in the basolateral membrane of the DT. The functions of these transporters are modified by a number of substances produced by the tubule cells that act upon the same cells or in neighboring ones. The PCT produces dopamine, NO, and Ang II; the thick ascending LHL synthesizes ET, 20-HETE, NO, and PGE_2; the DCT generates kallikrein 20-HETE, NO, and COX2 metabolites; and the CD produces ET, NO, and PGE_2. All renal-epithelial cells consume ATP, and produce adenosine and purine nucleotides. In this section we will discuss the main signaling molecules involved in the modulation of salt and water transport in neighboring cells from a tubule segment, as well as signaling to more distal segments.

Angiotensin II

Ang II exerts important effects on water and electrolyte transport by indirect and direct mechanisms [115, 129]. Its main indirect effect is the stimulation of secretion of the sodium-retaining hormone aldosterone by the adrenal cortex. The renin–angiotensin–aldosterone system (RAAS) is now recognized as an endocrine, paracrine, and autocrine system [3, 4, 115, 130]. Ang II directly stimulates NaCl reabsorption in PCT, TAL, DCT, and CD. Systemically generated Ang II reaches the kidney via the circulation. New evidence demonstrates that in proximal tubules there is internalization of Ang II via AT_1 receptors (receptor-mediated endocytosis). After internalization, the receptor may be recycled, degraded together with Ang II, or mediate intracellular actions of Ang II.

The non-hemodynamic effects of Ang II in the kidney (cell hypertrophy, proliferation, stimulation of extracellular matrix production, activation of inflammatory pathways) have been implicated in the pathogenesis of hypertension and the progression of kidney damage. Ang II is also synthesized in the PT (where angiotensinogen, renin, ACE, and Ang II receptors are expressed) and secreted to the lumen, reaching concentrations 100- to 1000-fold higher than in the systemic concentration ($<10^{-12}$M) [114]. In the PCT, the classical effect of Ang II is mediated by activation of AT_1 receptors expressed on the basolateral membrane. Receptor activation inhibits the cAMP/PKA pathway via $G\alpha_i$ protein, and stimulates IP3 and PKC formation via $G\alpha_{q/11}$ protein. In this segment, physiological concentrations of Ang II stimulate salt and fluid reabsorption, whereas higher concentrations are inhibitory. The stimulatory effect is mediated by increased expression in the apical membrane of the Na^+–H^+ exchanger (NHE-3); in the basolateral membrane, of the Na^+,K^+-ATPase, the Na^+/HCO_3^- co-transporter, and the K^+ channels – all transport proteins that coordinately account for Na^+ reabsorption. The inhibitory effect of high levels of Ang II is mediated by binding to AT_2 receptors, resulting in PLA_2-mediated arachidonic acid release. In the thick ascending LH, low Ang II concentrations ($\leq 10^{-12}$M) stimulate 20-HETE production and inhibit Na^+ reabsorption by inhibiting the apical-membrane K^+ channel, an effect that decreases K^+ recycling and K^+ availability to the Na^+–K^+–$2Cl^-$ co-transporter. $NaHCO_3$ transport in this segment is also inhibited by Ang II (10^{-8}M) via a mechanism that can be inhibited by Cyt-P450 blockade and by inhibitors of AA metabolism. Higher Ang II concentrations stimulate AT_1 receptors, elevating PLC and PKC activities and stimulating the Na^+–K^+–$2Cl^-$ co-transporter expressed in the apical membrane. There are additional binding sites for Ang II in the DCT and cortical and medullary CD consistent with an action of Ang II in these segments: luminal Ang II stimulates bicarbonate reabsorption in the late distal tubule by increasing expression and insertion of H^+ pumps in the α-intercalated cells, whilst in the collecting duct, basolateral exposure stimulates bicarbonate secretion by the β-intercalated cells. A significant effect of Ang in the distal segments was demonstrated by studies of knockout mice for tissue ACE; the mice exhibit a defect in urine concentration that is concomitant with a decrease in transport proteins (UT-A, CIC-K1, NKCC2/BSC1, and AQP1).

In summary, Ang II modulates water and electrolyte transport indirectly, via the stimulation of aldosterone secretion, and directly, by stimulating NaCl reabsorption in PCT, TAL, DCT, and CD. In the PCT this activation is mediated by AT_1 receptors that inhibit the cAMP/PKA pathway via $G\alpha_i$ protein and stimulate IP_3 and PKC formation via $G\alpha_{q/11}$ protein. In the PCT, physiological concentrations of Ang II stimulate salt and fluid reabsorption, whereas high concentrations are inhibitory; the latter effect is mediated by AT_2 receptors. In the TAL, low concentrations of Ang II stimulate production of 20-HETE and inhibit Na^+ reabsorption by intracellular mechanisms that seem to differ for the different Na^+ transporters in the segment. At high concentrations, Ang II elevates PLC

and PKC activities and stimulates the Na^+–K^+–$2Cl^-$ co-transporter in the apical membrane. Effects in more distal segments are suspected from the existence of Ang II receptors in the cells of these segments and studies in ACE knockout mice. Ang II is also produced by tubule cells, and circulating Ang II can be internalized in the kidney, remaining available for intracellular signaling that triggers cell growth and proliferation, and can play a pathogenic role in hypertension and in chronic damage to the kidney conducive to end-stage renal disease.

Dopamine

Dopamine is an intrarenal hormone or paracrine agent that elicits vasodilation and natriuresis [131, 132]. Endogenous renal dopamine is a major regulator of renal Na^+ excretion [131, 132]. Renal dopamine is generated mainly in the PCT (from l-dopa by the action of aromatic l-amino acid decarboxylase) and secreted across both apical and basolateral membranes, to the lumen and the blood, respectively. The renal PT lacks the dopamine β-hydroxylase that converts it to norepinephrine in neural tissue. Renal nerves are an additional, minor source of dopamine. During salt loading, dopamine production and excretion are increased in parallel with sodium excretion.

Dopamine receptors are classified as D_1-like and D_2-like receptors. Both classes have been identified in the kidney: D_1-like receptors comprise D_1 and D_5; D_2-like receptors comprise D_2, D_3, and D_4. Both D_1-like and D_2-like receptors are present in the renal tubule cells. PTC cells express predominantly D_1, but also D_5 and D_3. The mTAL expresses D_5. The CCD expresses predominantly D_5 and, to a lesser extent, D_1, D_3, and D_4. The overall effect of dopamine is inhibition of Na^+ reabsorption. Activation of D_1-like receptors stimulates both the cAMP/PKA and the PLC pathways, with downstream activation of the IP_3–Ca^{2+} and DAG–PKC signaling pathways. Both paths inhibit the Na^+,K^+-ATPase. In the PT, via the D_1 receptor, dopamine inhibits apical NHE3, Na^+-phosphate co-transporter, and the Cl^-/HCO_3^- exchanger; as well as the basolateral Na^+, K^+-ATPases and Na^+/HCO_3^- co-transporter [132–140]. D_2-like receptors inhibit cAMP/PKA production, which stimulates the luminal Na^+–H^+ exchanger and the basolateral Na^+,K^+-ATPase. Thus, the actions of D_1-like and D_2-like receptor on salt reabsorption oppose each other, but in some instances D_2 receptors enhance D_1 receptor effects, perhaps by shifting the effect of D_2 from inhibition of adenylyl cyclase to stimulation of phospholipase A_2, resulting in an increase of arachidonic acid production. Nevertheless, the dominant effect of dopamine is to reduce Na^+ reabsorption. In the TAL, D_1 receptors inhibit the Na^+–H^+ exchanger and the basolateral Na^+, K^+-ATPase, with weaker stimulation of Na^+–K^+–$2Cl^-$ co-transport via the cAMP/PKA pathway. The net effect is a decrease in Na^+ transport due to the predominance of the inhibition of the Na^+ pump. Finally, D_1 receptors may also inhibit Na^+ reabsorption in the CCD, perhaps by antagonizing the stimulatory effects of aldosterone as well as inhibiting the Na^+,K^+-ATPase.

In summary, renal dopamine is generated mainly in the PCT, and can exert effects on more distal segments of the nephron. During salt loading, dopamine production and urinary excretion increase in parallel with Na^+ excretion. Activation of D_1-like receptors in the PCT results in stimulation of both cAMP/PKA signaling and PLC activity, with downstream activation of the IP_3–Ca^{2+} and DAG–PKC signaling pathways resulting in inhibition of Na^+, K^+-ATPase. D_1 receptor stimulation also inhibits the Na^+–H^+ exchanger and the Na^+,K^+-ATPase. Inhibition of the Na^+ pump also account for the natriuretic effect of dopamine in the mTAL and the CD.

Nitric Oxide

The primary action of endogenous NO on renal-tubule cells is to inhibit Na^+ reabsorption, causing natriuresis and diuresis [45, 127, 141]. The systemic inhibition of NOS results in Na^+ retention if the resulting hypertension and pressure natriuresis are pharmacologically prevented. The l-arginine/NO system inhibits solute and water reabsorption in most nephron segments, including PCT, TAL, and cortical and medullary CD. Supporting the notion that NO is a regulator operating in physiological conditions, the natriuresis associated with a chronic high-salt diet is accompanied by high levels of NOS activity and by increases in NO production in the kidney and excretion in the urine.

All three NOS isoforms (type I or neuronal, type II or inducible, and type III or endothelial) are expressed in tubule cells. Endothelial NOS is present in PCT, TAL, and CD. Inducible NOS is found in PCT, TAL, DCT, and cortical and inner medullary CD. Neuronal NOS is limited to the TAL, macula densa, and CD. Ang II is a potent stimulus of NO production by PCT cells. In turn, NO inhibits the stimulatory action of Ang II on NaCl reabsorption.

A direct effect of NO on PCT transport is controversial, as both stimulation and inhibition of bicarbonate and fluid transport have been reported. However, evidence from knockout mice experiments indicates that type I and type II NOS reduce HCO_3^- and fluid absorption. There is agreement that NO inhibits both the luminal Na^+–H^+ exchanger and the basolateral Na^+,K^+-ATPase. NO may also influence ion transport by reducing cell ATP, an effect of potential importance in hypoxic injury. Very high NO inhibits the Na^+,K^+-ATPase via activation of PKC. In PT, NOS types I and II are responsible for these effects (reviewed in [45]). NO is produced by eNOS in the medullary thick ascending LH, where it inhibits NaCl and HCO_3^- reabsorption and thereby attenuates TGF at high tubule flow rates. The

inhibition of NaCl reabsorption is primarily by decreasing luminal-membrane Na^+–K^+–$2Cl^-$ co-transport, with a secondary inhibitory effect on the Na^+–H^+ exchanger. These effects result from stimulation of soluble guanylyl cyclase to form cGMP and consequent stimulation of phosphodiesterase II, reducing cAMP levels. NO inhibits $NaHCO_3$ absorption by activating soluble guanylate cyclase, increasing cGMP, and enhancing activity of cGMP-dependent protein kinase [127]. NOS isoforms are expressed in DCT, but there is no information regarding effects of NO on ion transport in this nephron segment. The CD produces NO, which inhibits NaCl reabsorption by principal cells by a mechanism independent of the Na^+,K^+-ATPase, probably involving mobilization of intracellular Ca^{2+} and inhibition of ENaC. At high concentrations, NO also impairs urine acidification by inhibiting the H^+-ATPase in intercalated cells. At low NO concentrations, a stimulatory pathway has been described that is mediated by small increases in cGMP that activate basolateral K^+ channels, hyperpolarizing the CD cell and increasing the driving force for Na^+ entry across the apical membrane. A stimulus of CD-NOS is AVP, acting via V_2 receptors. In turn, NO inhibits AVP-stimulated osmotic water permeability in the medullary CD via decreased intracellular cAMP secondary to activation of guanylyl cyclase and cGMP-dependent PKG.

In summary, the l-arginine/NO system inhibits solute and water reabsorption in most nephron segments, including PCT, TAL, and cortical and medullary CD. Ang II is a potent stimulus for NO production by PCT cells and inhibits the stimulatory action of Ang II on NaCl reabsorption. A direct effect of NO on PCT transport is supported by knockout-mice experiments demonstrating that type I and type II NOS are required for HCO_3^- and fluid reabsorption. It is likely that NO inhibits both the luminal Na^+–H^+ exchanger and the basolateral Na^+,K^+-ATPase. In the TAL, NO inhibits NaCl transport and may thereby contribute to attenuation of TGF at high tubule-flow rates. The CD also produces NO, which inhibits salt transport in principal cells by a mechanism independent of the Na^+,K^+-ATPase, probably involving mobilization of intracellular Ca^{2+} and inhibition of ENaC.

Endothelin

The ET system is a family of peptides with vasoconstricting properties. Of the three isoforms (ET-1, ET-2, and ET-3), only ET-1 is expressed in the kidney [56, 58]. The highest concentration of ET-1 in the body is in the renal medulla, where CD cells secrete ET-1 across the basolateral membrane. The major stimulant is AVP, acting via V_2 receptors, but production of ET-1 is also elicited by Ang II, adrenaline, insulin, cortisol, IL-1, transforming growth factor-β, low-density lipoproteins, hypoxia, and endothelin itself. ET-1 synthesis is inhibited by atrial natriuretic peptide, NO, and prostacyclins.

The ET receptors ET_A and ET_B (both of them G-protein coupled) are found in the inner medullary collecting duct cells, where ET may exert an autocrine effect. ET provides acute negative-feedback control of the AVP-induced osmotic water permeability in medullary CD. However, a chronic increase of interstitial osmolality may inhibit ET production and/or secretion. TAL and CD have ET_B receptors. ET inhibits NaCl transport in these nephron segments by increasing $[Ca^{2+}]_i$ and stimulating nNOS to produce NO, which in turn inhibits both Na^+,K^+-ATPase and ENaC at downstream sites – that is, the DCT and CD. ET is natriuretic by virtue of its action on ET_B receptors and NO production in the terminal CD. ET_B-receptor-deficient mice develop salt-dependent hypertension that is reversed by inhibition of ENaC by amiloride in the distal nephron. The role of AT-1 in human natriuresis and diuresis has not been categorically demonstrated. However, ET-1 urinary excretion and free-water clearance increase in people that climb high altitudes, an effect that is inhibited by treatment with an ET-1 receptor antagonist. The inhibition of water excretion is not accompanied by changes in proximal or distal Na^+ reabsorption [142].

In summary, CD cells are the main source of ET-1 in the renal medulla. ET production is elicited by AVP, Ang II, norepinephrine, insulin, cortisol, IL-1, TGF–, low-density lipoproteins, hypoxia, and endothelin itself. ET synthesis is inhibited by atrial natriuretic peptide, NO, and prostacyclin. ET is natriuretic and diuretic by virtue of its action on ET_B receptors. In humans, inhibition of the ET_B receptor inhibits the high-altitude diuretic response associated with urinary excretion of ET-1. TAL and CD have ET_B receptors. ET inhibits NaCl transport in these nephron segments by elevating $[Ca^{2+}]_i$ and nNOS to produce NO, which in turn inhibits both Na^+,K^+-ATPase and ENaC.

Eicosanoids

Eicosanoids are a family of biologically active, oxidized arachidonic acid metabolites. Of the three pathways of arachidonic acid metabolism, the one mediated by cyclooxygenase (COX) is the most important. The lipoxygenase pathway mediates the formation of mono-, di- and tri-hydroxyeicosatetraenoic acids (HETEs), leukotrienes and lipoxins. The cytochrome P450 (CYP450) mediates the formation of epoxyeicosatrienoic acids (EETs, diols, HETEs, and monooxygenated arachidonic acid derivatives). Arachidonic acid metabolites affect renal-tubule transport as well as renal hemodynamics [60, 62, 126, 143]. Eicosanoid synthesis is complex, and there is cross-talk between the three groups of enzymes [144].

Tubule segments containing COX are the cortical thick ascending LH, which expresses the inducible COX2, and the medullary thick ascending LH and CD, which express the constitutive COX1. Both COX1 and COX2 participate in the synthesis of PGs in the normal and diseased kidney; PGI_2 synthesis is COX2-dependent. COX2 inhibition suppresses prostanoid biosynthesis in normal kidneys, and lowers the ratio PGI_2/TXA_2 in the medulla. These findings dispelled the idea that selective inhibition of COX2 (an anti-inflammatory agent) would not produce adverse effects in the kidney. An additional COX (COX3) has been identified in canine tissues. COX3 is a splice variant of COX1; its relevance to human kidney function is unclear.

Little is known about the lipoxygenase activity along the nephron. Cyt-P450 monoxygenases are present in PCT, TAL, and CD, where both EETs and 20-HETE are produced. The Cyt-P450-2 enzyme family in PCT and CD leads to preferential production of EETs in response to stimulation by Ang II, bradykinin, or ET. Increased levels of NO and CO inhibit Cyt-P450 activity, and thus decrease production of EET and 20-HETE. A role for EETs in the CD is suggested by experiments in perfused tubules showing that addition of EETs to the lumen diminished Na^+ reabsorption and K^+ excretion. Other studies demonstrate that EETs inhibit ENaC; that the inhibition by arachidonic acid requires conversion to 11, 12-EET; that CYP2C23 epooxygenase is present in CD; that the effect of EETs is not blocked by COX inhibitors; and that 20-HETE does not mimic the EET effect [66]. A high-salt diet is associated with increased production of EET by PCT and CD and inhibition of NaCl transport. Agents that signal via cAMP/ PKA, such as dopamine and PTH, increase 20-HETE formation in PCT. 20-HETE is the exception to the proposition that vasoconstrictors favor Na^+ retention; it inhibits NaCl reabsorption in both PCT and TAL, by PKC-mediated phosphorylation and inhibition of the a-subunit of the Na^+,K^+-ATPase. Ang II stimulates 20-HETE production in the TAL by increasing $[Ca^{2+}]_i$ and stimulating PLA_2 to liberate arachidonic acid. In the TAL, 20-HETE inhibits NaCl reabsorption by acting on the luminal $Na^+–K^+–2Cl^-$ co-transporter and the basolateral Na^+,K^+-ATPase, opposing the stimulatory action of AT_1 receptors. In addition to being stimulated by paracrine and autocrine factors, PGE_2 and 20-HETE production in TAL are triggered by activation of a luminal Ca^{2+}-sensing receptor that is coupled to $G\alpha_{q/11}$-proteins and induces PLA_2 activation and an increase in $[Ca^{2+}]_i$.

High PGE_2 in the inner medulla, produced locally by epithelial and interstitial cells, antagonizes vasopressin-stimulated salt reabsorption in the thick ascending LH and water reabsorption in the CD. The major tubule-cell PGE_2 receptors are EP_1 and EP_3, with highest densities found in TAL and CD. Activation of EP_1 receptors inhibits sodium reabsorption through a $G\alpha_{q/11}$–PLC system that increases $[Ca^{2+}]_i$ and stimulates PKC, inhibiting the basolateral Na^+,K^+ -ATPase. EP_3 receptors stimulate a $G\alpha_i$ -protein and reduce cAMP/PKA activity. EP_3 receptors inhibit NaCl reabsorption in the thick ascending LH by downregulating the density of luminal-membrane Na^+–K^+–$2Cl^-$ co-transporters independently of $[Ca^{2+}]_i$, and in CD by inhibiting ENaC. Thought to be of minor functional significance during physiological conditions, EP receptors sensitive to PGF_2 predominate along the DCT. They signal through PKC, and perhaps also a PKC-independent Rho-mediated pathway. Water transport (AVP regulated) in CD is modified by prostaglandins. COX inhibition magnifies the effect of AVP. In turn, AVP stimulates endogenous PGE_2 production, creating a negative-feedback loop in which endogenous PGE_2 diminishes the antidiuretic effect of AVP.

In summary, the main effects of arachidonic acid metabolites are natriuresis and water diuresis. Tubule segments distal to the PT contain COX, leading to the production of PGE_2 and PGI_2. Cyt-P450 monoxygenases are present in PCT, TAL, and CD, where both EETs and 20-HETE are produced. In PCT and CD, preferential production of EETs occurs in response to Ang II, bradykinin, or ET. Increased levels of NO and CO inhibit Cyt-P450 activity, and thus decrease production of EET and 20-HETE. In the TAL, 20-HETE inhibits NaCl reabsorption by acting on the luminal Na^+–K^+–$2Cl^-$ co-transporter and the basolateral Na^+,K^+-ATPase. High concentrations of PGE_2 in the inner medulla, produced locally by epithelial cells and interstitial cells, antagonize vasopressin-stimulated salt reabsorption in the TAL and water reabsorption in the CD. Activation of PGE_2 EP_1 receptors inhibits Na^+ reabsorption through a $G\alpha_{q/11}$–PLC mechanism that increases $[Ca^{2+}]_i$ and stimulates PKC, inhibiting the Na^+,K^+-ATPase. EP_3 receptors inhibit NaCl reabsorption in the TAL by reducing the density of luminal Na^+–K^+–$2Cl^-$ transporters independently of $[Ca^{2+}]_i$ and in CD by inhibiting ENaC. In the CD, ENaC is also inhibited by EETs.

Bradykinin

Bradykinin is a peptide that regulates vascular tone, and water and electrolyte balance, and has a role in the control of arterial pressure [73, 145]. Bradykinin of renal origin is vasodilator and natriuretic. The enzyme kallikrein is produced and released from the distal convoluted tubule and connecting segment by exocytosis at both the apical and basolateral membranes, to the lumen and interstitial fluid, respectively. Secretion at this site makes the enzyme available at the site of substrate (kininogen) production (i.e., the principal cells of the CD). Bradykinin produced by the action of kallikrein on kininogen is present both in the lumen of the CD and in the interstitial fluid. Renal bradykinin formation is normally low, and is increased during sodium restriction and water deprivation. Bradykinin is inactivated by kininase II, the same enzyme as ACE. The B_2 receptors are expressed in mesangial cells, juxtaglomerular granular cells, TAL, cortical and medullary CD, and renomedullary

interstitial cells. In the TAL, B_2 receptor activation inhibits NaCl reabsorption. In the CD, luminal bradykinin acts on B_2 receptors and inhibits both NaCl and water transport. The mechanism of natriuresis is the inhibition of ENaC, an effect probably mediated by PGE_2, although NO may have a role by modifying the local metabolism of atrial natriuretic peptide. Bradykinin exerts a negative regulatory effect on the water reabsorption induced by AVP. BK induces release of PGs via phosphatidylinositol-3-kinase and mitogen-activated protein signaling by modifying COX activity. In some models (rabbit CD), it induces release of AA. In CD cells, BK caused a transient increase in Ca^{2+} concentration via activation of $G\alpha_q$/PLC pathway. BK decreases AQP2 translocation to the apical membrane of CD cells, an effect present in the presence of PLC inhibitors, suggesting that increases in intracellular $[Ca^{2+}]$ are not responsible for the effect. BK also induces $G\alpha_{13}$, an upstream effector of Rho protein whose activation inhibits AQP2 targeting to the plasma membrane [145]. Although absent during physiological conditions, tubule-cell B_1 receptors are induced during inflammation, primarily in efferent arterioles, PCT, TAL, and DCT.

In summary, bradykinin of renal origin has vasodilator, natriuretic and diuretic effects. Bradykinin is produced by the action of kallikrein on kininogen, and is present in both the lumen of the CD and the interstitial fluid. Renal bradykinin formation is normally low, and increases during sodium restriction and water deprivation. In the CD, luminal bradykinin acts on B_2 receptors to inhibit NaCl and water transport, by antagonizing the action of AVP.

Adenosine and ATP

Adenosine and ATP are produced by renal-tubule cells and can affect their transport functions by autocrine or paracrine mechanisms [79, 102, 146–148]. Purinergic receptors (P_1 and P_2) are responsible for these effects. P_1 receptors comprise four subtypes (A_1, A_{2a}, A_{2b} and A_3), of which A_1 is expressed in PT, medullary thick ascending LH, cortical CD, and inner medullary CD. A_2 receptors are found in thick ascending LH, cortical CD, and inner medullary CD. The A_3 receptor, identified in cortical thick ascending LH and DCT, does not have a defined function in these segments. A_1 receptor activation by adenosine produce (via pertussis-sensitive G_i and G_o-proteins) inhibition of adenylyl cyclase and stimulation of PLC. A_2 receptor activation (via cholera toxin-sensitive stimulatory G proteins) activates adenylyl cyclase and has effects on Na^+ and water excretion. The P_2 receptors respond predominantly to ATP, and can be P_{2X} (Ca^{2+}-permeable non-selective cation channels) and P_{2Y} (G-protein-coupled receptor). Block of A_1 receptors produces a reduction in Na^+ reabsorption, and hence natriuresis. In cultures of PT cells, A_1-receptor activation increases Na^+-glucose and Na^+-phosphate co-transport at the luminal membrane, and also increases $NaHCO_3$ transport at the basolateral membrane. The increase in HCO_3^- reabsorption could be secondary to the activation of NHE3 that is seen at low concentrations of adenosine ($<10^{-8}$M), and contributes to explains the natriuretic effect [101]. Adenosine can be formed from cAMP secreted in the proximal tubule by an ectophosphodiesterase. In the medullary thick ascending LH, activation of A_1 receptors decreases the transepithelial voltage, and inhibits Cl^- transport without affecting HCO_3^- absorption, suggesting that the effect occurs by a signaling mechanism other than cAMP. These two effects result in a reduction in salt reabsorption. In the DCT, adenosine has different effects on Mg^{2+} absorption, with A_1 receptor being stimulatory and A_2 receptor inhibitory. A_1 receptor also stimulates Ca^{2+} absorption. These findings suggest an effect of adenosine on the function of the transient-receptor-potential channels (TRPM6 and TRPV5) involved in divalent cation transport. In the CCD, A_1 receptor activation stimulates an apical membrane Cl^- channel, and in the inner medullary CD it inhibits the AVP stimulation of cAMP and decreases the hydraulic water permeability, increasing excretion of water in the urine, and also decreases Na^+ reabsorption. Activation of A_2 receptors also decreases the water permeability of the CD.

The PCT is a rich source of ATP that is secreted to the lumen and travels downstream to regulate transport in more distal nephron segments. Luminal P_{2X} and P_{2Y} receptors in PCT and IMCD inhibit Na^+ reabsorption by Ca^{2+}-, PKC-, and arachidonic acid-dependent pathways. Activation of P_{2Y} receptors in DCT and CD increases PLC activity and $[Ca^{2+}]_i$, resulting in increased luminal-membrane Cl^- permeability and inhibition of Na^+ reabsorption, at least in part because of the resulting apical-membrane depolarization

In summary, adenosine and ATP are produced by renal-tubule cells, and can affect their transport functions by autocrine or paracrine mechanisms. Their effects are mediated by stimulation of the purinergic receptors P_1 and P_2. Adenosine is coupled to different heterotrimeric G proteins, altering Na^+ and water excretion. The ATP produced in the PCT is secreted to the lumen, and regulates transport in more distal nephron segments. Inhibition of Na^+ transport is mediated by both P_2 receptor types.

Gap Junctions

Gap junctions are intercellular channels that, in the kidney, communicate between adjacent epithelial cells, as well as adjacent vascular cells and mesangial cells (see section on "Endothelial cell connections: connexins and gap junctions") [149, 150]. Gap junctions are high-permeability, low-ion-selectivity channels that allow for intercellular flux of ions and water-soluble molecules of up to 1000 Da. The individual gap junction is formed by two hemichannels (connexons), one from each cell, which are hexamers of the membrane proteins called connexins. Prior to docking with the neighboring cell, the hemichannels inserted in the plasma membrane remain closed because of the cell-negative

membrane potential and the presence of high extracellular $[Ca^{2+}]$. Information on the localization and function of gap junctions in renal epithelial cells is scanty. Gap junctions coordinate oscillations of intracellular $[Ca^{2+}]$ in renal epithelial monolayers stimulated with bradykinin [149]. In CD cells, which express both Cx43 and the mechanosensitive transient receptor potential channel TRPV4, mechanical stimulation causes an increase in intracellular Ca^{2+} that is propagated to neighboring cells by a Cx-dependent, sensitive to the gap-junction blocker heptanol. The rise in $[Ca^{2+}]_i$ results from both Ca^{2+} influx and release from intracellular stores. The Ca^{2+} ionophore ionomycin and a five-fold increase in extracellular glucose concentration elevate Cx43 expression. Immunohistochemistry in the mouse, rat, and rabbit kidney demonstrates localization of Cx30 hemichannels throughout the luminal membrane of the medullary thick ascending LH, DCT, and CD (in the latter, the expression is confined to the β-intercalated cells). It has been suggested that these hemichannels may be involved in ion transport. A high-salt diet upregulates Cx30 expression in the rat inner medulla, which may be related to the increase in Na^+ reabsorption associated with hypertension. The function of Cx30 in the apical membrane of β-intercalated cells of the CD remains unclear, but a role in secretion of small metabolites like ATP and NAD that participate in paracrine signaling has been suggested [151]. There is better evidence for Cx participation in intercellular signaling in tubule cells [150]. Single cells with constitutive higher sensitivity to a stimulus may act as pacemakers, initiating a response (e.g., increase in ion concentration or production of a messenger molecule) and signaling neighboring cells by intercellular permeation of ions or second-messenger molecules.

INTERSTITIAL CELL–TUBULE COMMUNICATION

Renal interstitial cells occupy the space that surrounds the blood vessels and tubules. In the renal cortical interstitium, there are stellate fibroblasts and lymphocyte-like cells. In the inner medulla, there are lipid-laden cells (located between thin descending limbs of Henle in a ladder-like appearance), cells similar to the fibroblasts and lymphocytes found in the cortex, and SMC-like pericytes that encircle descending vasa recta capillaries and act as sphincters to regulate medullary blood flow.

Cortical Interstitial Cells

Fibroblasts have the potential to differentiate to myofibroblasts containing α-smooth muscle actin and desmin, a process observed in response to inflammatory cytokines and thought to be important in the interstitial fibrosis of kidney diseases. In response to hypoxia, cortical fibroblasts close to the outer medullary border produce erythropoietin, a glycopeptide hormone that stimulates erythropoiesis in the bone marrow, as well as the synthesis of hemoglobin [152, 153]. In anemia, cortical interstitial cells also produce erythropoietin, and synthesis may also occur in proximal tubule cells [154]. It is likely that the oxygen sensor located in cells that produce erythropoietin is a heme protein which, when activated, triggers expression of several factors (hypoxia inducible factor-1, HIF-1; hepatic nuclear factor, HNF-4; p300, and hypoxia inducible factor-1 (HIF-1) alpha and – beta). Secreted erythropoietin binds to specific receptors on erythroid precursor cells. In normoxia, only the β subunit of HIF-1 is expressed. The α subunit (in the presence of iron and oxygen) undergoes proline hydroxylation, binds to the Von Hippel-Lindau protein, and is ubiquitinated.

Erythropoietin appears to have additional effects, such as acting as an anti-apoptotic cytokine to protect tissue from ischemia-reperfusion injury, and mobilizing hematopoietic progenitor cells that result in improved angiogenesis and greater density of capillaries [155, 156].

The function of lymphocyte-like cells that reside in the cortical interstitium has been studied in co-cultures with proximal tubule cells. The findings indicate that the exposure of proximal tubule cells to albumin triggers the generation of reactive oxygen species, and apical release of pro-inflammatory cytokines (IL-6, sICAM-1) and basolateral release of chemokines (CCL2, CCL5, and CXCL8) and sICAM-1. The release of cytokines recruits mononuclear cells from the microcirculation to the interstitial tissue via chemotaxis. The released chemokines from the albumin-activated PT cells upregulate the expression of chemokine receptors on infiltrating mononuclear cells. In turn, the T cells and monocytes release TNFα and IL-1β, which stimulate the PT cells, creating a positive loop that maintain the release of IL-6, sICAM-1, CCL2, and CCL5, resulting in inflammatory injury in the renal interstitium [157].

Medullary Interstitial Cells

Renal medullary interstitial cells (RMICs) participate in the regulation of Na^+ and water balance by interacting with endocrine and paracrine factors including COX2 metabolites, primarily PGE_2, in response to Ang II, AVP, ET, and bradykinin [158, 159]. These cells express an unusually high density of AT_1 and ET_A and ET_B receptors, as well as bradykinin receptor (B_2), atrial natriuretic peptide receptors (NPR_A and NPR_b), and AVP receptor (V_{1a}). The common signaling pathway involves $G\alpha_{q/11}$ activation of PLC and IP_3-induced mobilization of $[Ca^{2+}]_i$, which activates PLA_2. Locally produced PGE_2 causes relaxation of vasa recta pericytes, and inhibits NaCl reabsorption in the TAL and medullary CD. In addition, PGE_2 can inhibit the ability of AVP to mobilize aquaporin-2 water channels to the luminal membrane, and thus reduces the osmotic water permeability in the inner medullary CD. The natriuretic peptides (atrial natriuretic peptide and the kidney-produced urodilatin) are endogenous antagonists of the RAS. They increase medullary blood flow, activating glomerular receptors, and are natriuretic and diuretic by inhibiting PT transport of Na and antagonizing aldosterone and the effect of vasopressin in CCD [159, 160].

Cultured RMICs respond to vasoactive peptides with contraction, prostaglandin release, extracellular matrix synthesis, and cell proliferation, as well as with release of autocrine and paracrine factors like NO, PGs, and medullipins [159, 161, 162]. Medullipins are lipids produced by RMICs. An inactive form is released into the systemic circulation, markedly so after surgical relief from renal arterial stenosis. Medullipin becomes activated by a liver Cyt-P450 oxidase, and acts as a vasodilating and natriuretic agent. It may also suppress sympathetic nerve activity. The physiological roles of medullipins, if any, are not understood.

In summary, cortical-interstitial fibroblasts respond to hypoxia by producing erythropoietin, a hormone that stimulates erythropoiesis in the bone marrow. Medullary-interstitial cells produce COX2 metabolites, primarily PGE_2, in response to Ang II, AVP, ET, and bradykinin. The common signaling pathway involves $G\alpha_{q/11}$ activation of PLC and IP_3-induced mobilization of $[Ca^{2+}]_i$, which activates PLA_2. RMICs may have an important role in the regulation of renal medullary functions and in blood pressure homeostasis.

CONCLUSIONS

The kidney is a complex, highly sophisticated organ containing diverse cells types responsible for specialized functions, both in the renal vasculature and nephron segments. The structure of the kidney and the process of urine formation allow for cell-to-cell influences along single nephron segments that are distant in space. In addition, the proximity of parallel structures permits lateral communication between tubules, capillaries, and interstitial cells. These two kinds of cell-to-cell communication are central for functional integration at the single-nephron and whole-organ levels. Communication is mediated by paracrine and autocrine agents that act extracellularly or intracellularly by turning on signaling systems, thus eventually unifying homeostatic regulation of renal hemodynamics, glomerular filtration rate, tubule-transport processes, and urinary excretion. Endothelial cells regulate renal vascular resistance and reabsorption of salt and water by the renal tubules via a variety of messenger molecules, including nitric oxide, prostaglandins, EETs, 20-HETE, and endothelin. The vascular and epithelial cells regulate their own functions via the production of the autocrine agent 20-HETE.

Tubule cells produce dopamine, purine nucleotides, nitric oxide, prostanoids, and endothelin. These agents act locally to regulate the vasculature and transepithelial transport of salt and water. Despite a rapidly growing understanding of cell–cell signaling in the kidney, the wide variety of autocrine and paracrine systems and their actions on multiple cell types are not completely understood.

Interactions with hormones and neural control systems add to this complexity. Current research is largely reductionist. Integrative studies are also needed, in particular to discern the relative importance, cross-talk, redundancy, and compensatory effects of the various systems under physiological and pathophysiological conditions. A thorough understanding of these systems will be the foundation for rational and effective approaches to the treatment of renal diseases.

ACKNOWLEDGEMENTS

We thank Larry Starr for assistance in creating Figure 318.1.

REFERENCES

1. Giebisch G, Windhager EE. Organization of the urinary system. In: Boron WF, Boulpaep EL, editors. *Medical physiology: a cellular and molecular approach*. Philadelphia, PA: W.B. Saunders; 2003. p. 737–56.
2. Reilly Jr RF, Bulger RE, Kriz W. Structural–functional relationships in the kidney. In: Schrier RW, editor. *Diseases of the Kidney and Urinary Tract*. Philadelphia, PA: Lippincott Williams & Wilkins; 2007. p. 2–54.
3. Arendshorst WJ, Navar LG. Renal circulation and glomerular hemodynamics. In: Schrier RW, editor. *Diseases of the kidney & urinary tract*. Philadelphia, PA: Lippincott Williams & Wilkins; 2007. p. 54–95.
4. Navar LG, Inscho EW, Majid SA, Imig JD, Harrison-Bernard LM, Mitchell KD. Paracrine regulation of the renal microcirculation. *Physiol Rev* 1996;**76**:425–536 [Review] [1214 refs].
5. Cowley AW. Role of the renal medulla in volume and arterial pressure regulation. *Am J Physiol Regul Integr Comp Physiol* 1997;**273**: R1–R15 [Review] [109 refs].
6. Evans RG, Eppel GA, Anderson WP, Denton KM. Mechanisms underlying the differential control of blood flow in the renal medulla and cortex. *J Hypertens* 2004;**22**:1439–51.
7. Pallone TL, Zhang Z, Rhinehart K. Physiology of the renal medullary microcirculation. *Am J Physiol Renal Physiol* 2003;**284**:F253–66.
8. Hall JE, Brands MW, Henegar JR. Angiotensin II and long-term arterial pressure regulation: the overriding dominance of the kidney. *J Am Soc Nephrol* 2002;**13**(Suppl. 3):S173–8.
9. Malnic G, Muto S, Giebisch G. Regulation of potassium excretion. In: Alpern RJ, Hebert SC, editors. *Seldin and Giebisch's the Kidney Physiology and Pathology*. 4th edn. London: Academic Press; 2008. p. 1301–47.
10. Wang W, Hebert SC. The molecular biology of renal potassium channels. In: Alpern RJ, Hebert SC, editors. *Seldin and Giebisch's the Kidney Physiology and Pathology*. 4th edn. London: Academic Press; 2008. p. 1249–76.
11. DiBona GF. Physiology in perspective: the wisdom of the body. Neural control of the kidney. *Am J Physiol. Regul Integr Comp Physiol* 2005;**289**:R633–41.
12. Hamm LL, Alpern RJ, Preisig PA. Cellular mechanisms of renal tubular acidification. In: Alpern RJ, Hebert SC, editors. *Seldin and Giebisch's the Kidney Physiology and Pathology*. 4th edn. London: Academic Press; 2008. p. 1539–80.
13. Mount DB, Yu AS. Transport of inorganic solutes: sodium, chloride, potassium, magnesium, calcium and phosphate. In: Brenner BM,
editor. *Brenner and rector's the kidney*. 8th edn. Philadelphia, PA: WB Saunders Co; 2008. p. 156–213.
14. Abdel-Latif AA. Cross talk between cyclic nucleotides and polyphosphoinositide hydrolysis, protein kinases, and contraction in smooth muscle. *Exp Biol Med (Maywood)* 2001;**226**:153–63.
15. Earley S, Nelson MT. Central role of Ca^{2+}-dependent regulation of vascular tone *in vivo*. *J Appl Physiol* 2006;**101**:10–11.
16. Gollasch M, Nelson MT. Voltage-dependent Ca^{2+} channels in arterial smooth muscle cells. *Kidney Blood Press Res* 1997;**20**:355–71 [Review] [223 refs].
17. Lincoln TM, Dey N, Sellak H. Invited review: cGMP-dependent protein kinase signaling mechanisms in smooth muscle: from the regulation of tone to gene expression. *J Appl Physiol* 2001;**91**:1421–30.
18. Morgan KG, Gangopadhyay SS. Invited review: cross-bridge regulation by thin filament-associated proteins. *J Appl Physiol* 2001;**91**:953–62.
19. Somlyo AP, Somlyo AV. Ca2+ sensitivity of smooth muscle and nonmuscle myosin II: modulated by G proteins, kinases, and myosin phosphatase. *Physiol Rev* 2003;**83**:1325–58.
20. Thorneloe KS, Nelson MT. Ion channels in smooth muscle: regulators of intracellular calcium and contractility. *Can J Physiol Pharmacol* 2005;**83**:215–42.
21. Andreoli TE. An overview of salt absorption by the nephron. *J Nephrol* 1999;**12**(Suppl. 2):S3–S15.
22. Aronson PS, Giebisch G. Mechanisms of chloride transport in the proximal tubule. *Am J Physiol* 1997;**273**:F179–92.
23. Feraille E, Doucet A. Sodium-potassium-adenosinetriphosphatase-dependent sodium transport in the kidney: hormonal control. *Physiol Rev* 2001;**81**:345–418.
24. Knepper MA, Brooks HL. Regulation of the sodium transporters NHE3, NKCC2 and NCC in the kidney. *Curr Opin Nephrol Hypertens* 2001;**10**:655–9.
25. Giebisch G. Renal potassium channels: function, regulation, and structure. *Kidney Intl* 2001;**60**:436–45.
26. Reeves WB, Winters CJ, Andreoli TE. Chloride channels in the loop of Henle. *Annu Rev Physiol* 2001;**63**:631–45.
27. Schafer JA. Abnormal regulation of ENaC: syndromes of salt retention and salt wasting by the collecting duct. *Am J Physiol Renal Physiol* 2002;**283**:F221–35.
28. Schafer JA. Renal water reabsorption: a physiologic retrospective in a molecular era. *Kidney Intl* 2004(Suppl):S20–7.
29. Arendshorst WJ, Thai TL. Regulation of the renal microcirculation by ryanodine receptors and calcium-induced calcium release. *Cur Opin Nephrol Hypertens* 2009;**18**:40–49.
30. Just A, Olson AJ, Whitten CL, Arendshorst WJ. Superoxide mediates acute renal vasoconstriction produced by angiotensin II and catecholamines by a mechanism independent of nitric oxide. *Am J Physiol Heart Circ Physiol* 2007;**292**:H83–92.
31. Salomonsson M, Sorensen CM, Arendshorst WJ, Steendahl J, Holstein-Rathlou N-H. Calcium handling in afferent arterioles. *Acta Physiol Scand* 2004;**181**:421–9.
32. Facemire CS, Mohler PJ, Arendshorst WJ. Expression and relative abundance of short transient receptor potential channels in the rat renal microcirculation. *Am J Physiol Renal Physiol* 2004;**286**:F546–51.
33. Ma R, Du J, Sours S, Ding M. Store-operated Ca^{2+} channel in renal microcirculation and glomeruli. *Exp Biol Med (Maywood)* 2006;**231**:145–53.
34. Mene P. Transient receptor potential channels in the kidney: calcium signaling, transport and beyond. *J Nephrol* 2006;**19**:21–9.
35. Nishida M, Hara Y, Yoshida T, Inoue R, Mori Y. TRP channels: molecular diversity and physiological function. *Microcirculation* 2006;**13**:535–50.

36. Ballermann BJ. Contribution of the endothelium to the glomerular permselectivity barrier in health and disease. *Nephron Physiol* 2007;**106**:19–25.

37. Haraldsson B, Nystrom J, Deen WM. Properties of the glomerular barrier and mechanisms of proteinuria. *Physiol Rev* 2008;**88**:451–87.

38. Mehta D, Malik AB. Signaling mechanisms regulating endothelial permeability. *Physiol Rev* 2006;**86**:279–367.

39. Pries AR, Kuebler WM. Normal endothelium. *Handb Exp Pharmacol* 2006;**176**:1–40.

40. Spieker LE, Luscher TF. Protection of endothelial function. *Handb Exp Pharmacol* 2005;**176**:619–44.

41. Busse R, Fleming I. Vascular endothelium and blood flow. *Handb Exp Pharmacol* 2006;**176**:43–78.

42. Lucas KA, Pitari GM, Kazerounian S, Ruiz-Stewart I, Park J, Schulz S, Chepenik KP, Waldman SA. Guanylyl cyclases and signaling by cyclic GMP. *Pharmacol Rev* 2000;**52**:375–414.

43. Purdy KE, Arendshorst WJ. Iloprost inhibits inositol-1,4,5-trisphosphate-mediated calcium mobilization stimulated by angiotensin II in cultured preglomerular vascular smooth muscle cells. *J Am Soc Nephrol* 2001;**12**:19–28.

44. Deng A, Miracle CM, Suarez JM, Lortie M, Satriano J, Thomson SC, Munger KA, Blantz RC. Oxygen consumption in the kidney: effects of nitric oxide synthase isoforms and angiotensin II. *Kidney Intl* 2005;**68**:723–30.

45. Kone BC. Nitric oxide synthesis in the kidney: isoforms, biosynthesis, and functions in health. *Semin. Nephrol* 2004;**24**:299–315.

46. Moncada S. Adventures in vascular biology: a tale of two mediators. *Philos. Trans R Soc Lond B Biol Sci* 2006;**361**:735–59.

47. Raij L, Baylis C. Glomerular actions of nitric oxide [editorial]. *Kidney Intl* 1995;**48**:20–32 [Review] [216 refs].

48. Fleming I. Vascular cytochrome p450 enzymes: physiology and pathophysiology. *Trends Cardiovasc Med* 2008;**18**:20–5.

49. Botros FT, Navar LG. Interaction between endogenously produced carbon monoxide and nitric oxide in regulation of renal afferent arterioles. *Am J Physiol Heart Circ Physiol* 2006;**291**:H2772–8.

50. Kaide J, Zhang F, Wei Y, Wang W, Gopal VR, Falck JR, Laniado-Schwartzman M, Nasjletti A. Vascular CO counterbalances the sensitizing influence of 20-HETE on agonist-induced vasoconstriction. *Hypertension* 2004;**44**:210–16.

51. Rodriguez F, Zhang F, Dinocca S, Nasjletti A. Nitric oxide synthesis influences the renal vascular response to heme oxygenase inhibition. *Am J Physiol Renal Physiol* 2003;**284**:F1255–62.

52. Quan S, Yang L, Schnoud S, Schwatzman ML, Nasjlett A, Goodman AL, Abraham NG. Expression of human heme oxygenase-1 in the thick ascending limb attenuates angiotensinII-mediated increase in oxidative injury. *Kidney Intl* 2004;**65**:1628–39. *Renal Physiol.* **284**:F1255–1262.

53. Granger JP, Abram S, Stec D, Chandler D, LaMarca B. Endothelin, the kidney, and hypertension. *Curr Hypertens Rep* 2006;**8**:298–303.

54. Just A, Olson AJ, Falck JR, Arendshorst WJ. NO and NO-independent mechanisms mediate ETB receptor buffering of ET-1-induced renal vasoconstriction in the rat. *Am J Physiol Regul Integr Comp Physiol* 2005;**288**:R1168–77.

55. Just A, Whitten CL, Arendshorst WJ. Reactive oxygen species participate in acute renal vasoconstrictor responses induced by ETA- and ETB-receptors. *Am J Physiol Renal Physiol* 2008;**294**:F719–28.

56. Schneider MP, Boesen EI, Pollock DM. Contrasting actions of endothelin ET(A) and ET(B) receptors in cardiovascular disease. *Annu Rev Pharmacol Toxicol* 2007;**47**:731–59.

57. Bagnall AJ, Kelland NF, Gulliver-Sloan F, Davenport AP, Gray GA, Yanagisawa M, Webb DJ, Kotelevtsev YV. Deletion of endothelial

58. Kohan DE. The renal medullary endothelin system in control of sodium and water excretion and systemic blood pressure. *Curr Opin Nephrol Hypertens* 2006;**15**:34–40.

59. Cheng HF, Harris RC. Cyclooxygenases, the kidney, and hypertension. *Hypertension* 2004;**43**:525–30.

60. Hao CM, Breyer MD. Physiological regulation of prostaglandins in the kidney. *Annu Rev Physiol* 2008;**70**:357–77.

61. Harris RC. An update on cyclooxygenase-2 expression and metabolites in the kidney. *Curr Opin Nephrol Hypertens* 2008;**17**:64–9.

62. Carroll MA, McGiff JC, Ferreri NR. Products of arachidonic acid metabolism. *Methods Mol Med* 2003;**86**:385–97.

63. Imig JD. Eicosanoids and renal vascular function in diseases. *Clin Sci (Lond)* 2006;**111**:21–34.

64. Miyata N, Roman RJ. Role of 20-hydroxyeicosatetraenoic acid (20-HETE) in vascular system. *J Smooth Muscle Res* 2005;**41**:175–93.

65. Roman RJ. P-450 metabolites of arachidonic acid in the control of cardiovascular function. *Physiol Rev* 2002;**82**:131–85.

66. Capdevila JH, Falck JR, Imig JD. Roles of the cytochrome P450 arachidonic acid monooxygenases in the control of systemic blood pressure and experimental hypertension. *Kidney Intl* 2007;**72**:683–9.

67. Badr KF, Abi-Antoun TE. Isoprostanes and the kidney. *Antioxid Redox Signal* 2005;**7**:236–43.

68. Kitiyakara C, Chabrashvili T, Chen Y, Blau J, Karber A, Aslam S, Welch WJ, Wilcox CS. Salt intake, oxidative stress, and renal expression of NADPH oxidase and superoxide dismutase. *J Am Soc Nephrol* 2003;**14**:2775–82.

69. Morrow JD. The isoprostanes – unique products of arachidonate peroxidation: their role as mediators of oxidant stress. *Curr Pharm Des* 2006;**12**:895–902.

70. Wang D, Chabrashvili T, Wilcox CS. Enhanced contractility of renal afferent arterioles from angiotensin-infused rabbits: roles of oxidative stress, thromboxane prostanoid receptors, and endothelium. *Circ Res* 2004;**94**:1436–42.

71. Campbell WB, Falck JR. Arachidonic acid metabolites as endothelium-derived hyperpolarizing factors. *Hypertension* 2007;**49**:590–6.

72. Wang D, Borrego-Conde LJ, Falck JR, Sharma KK, Wilcox CS, Umans JG. Contributions of nitric oxide, EDHF, and EETs to endothelium-dependent relaxation in renal afferent arterioles. *Kidney Intl* 2003;**63**:2187–93.

73. Erdos EG. Kinins, the long march – a personal view. *Cardiovasc Res* 2002;**54**:485–91.

74. Madeddu P, Emanueli C, El-Dahr SS. Mechanisms of disease: the tissue kallikrein-kinin system in hypertension and vascular remodeling. *Nature Clin Pract Nephrol* 2007;**3**:208–21.

75. Rohrwasser A, Ishigami T, Gociman B, Lantelme P, Morgan T, Cheng T, Hillas E, Zhang S, Ward K, Bloch-Faure M, et al. Renin and kallikrein in connecting tubule of mouse. *Kidney Intl* 2003;**64**:2155–62.

76. Tornel J, Madrid MI, Garcia-Salom M, Wirth KJ, Fenoy FJ. Role of kinins in the control of renal papillary blood flow, pressure natriuresis, and arterial pressure. *Circ Res* 2000;**86**:589–95.

77. Wang X, Trottier G, Loutzenhiser RD. Determinants of renal afferent arteriolar actions of bradykinin: evidence that multiple pathways mediate responses attributed to EDHF. *Am J Physiol Renal Physiol* 2003;**285**:F540–9.

78. Burnstock G. Purinergic signalling– an overview. *Novartis Found Symp* 2006;**276**:26–48.

79. Guan Z, Osmond DA, Inscho EW. P2X receptors as regulators of the renal microvasculature. *Trends Pharmacol Sci* 2007;**28**:646–52.

80. Hansen PB, Hashimoto S, Oppermann M, Huang Y, Briggs JP, Schnermann J. Vasoconstrictor and vasodilator effects of adenosine in the mouse kidney due to preferential activation of A1 or A2 adenosine receptors. *J Pharmacol Exp Ther* 2005;**315**:1150–7.

81. Nishiyama A, Rahman M, Inscho EW. Role of interstitial ATP and adenosine in the regulation of renal hemodynamics and microvascular function. *Hypertens Res* 2004;**27**:791–804.

82. Clempus RE, Griendling KK. Reactive oxygen species signaling in vascular smooth muscle cells. *Cardiovasc Res* 2006;**71**:216–25.

83. Lyle AN, Griendling KK. Modulation of vascular smooth muscle signaling by reactive oxygen species. *Physiology (Bethesda)* 2006;**21**:269–80.

84. Majid DSA, Nishiyama A, Jackson KE, Castillo A. Superoxide scavenging attenuates renal responses to ANG II during nitric oxide synthase inhibition in anesthetized dogs. *Am J Physiol Renal Physiol* 2005;**288**:F412–19.

85. Figueroa XF, Isakson BE, Duling BR. Connexins: gaps in our knowledge of vascular function. *Physiology (Bethesda)* 2004;**19**:277–84.

86. Haefliger JA, Krattinger N, Martin D, Pedrazzini T, Capponi A, Doring B, Plum A, Charollais A, Willecke K, Meda P. Connexin43-dependent mechanism modulates renin secretion and hypertension. *J Clin Invest* 2006;**116**:405–13.

87. Wagner C. Function of connexins in the renal circulation. *Kidney Intl* 2007;**73**:547–55.

88. Zhang JH, Kawashima S, Yokoyama M, Huang P, Hill CE. Increased eNOS accounts for changes in connexin expression in renal arterioles during diabetes. *Anat Rec A Discov Mol Cell Evol Biol* 2006;**288**:1000–8.

89. De Vriese AS, Van de VJ, Lameire NH. Effects of connexin-mimetic peptides on nitric oxide synthase- and cyclooxygenase-independent renal vasodilation. *Kidney Intl* 2002;**61**:177–85.

90. Salomonsson M, Gustafsson F, Andreasen D, Jensen BL, Holstein-Rathlou NH. Local electric stimulation causes conducted calcium response in rat interlobular arteries. *Am J Physiol Renal Physiol* 2002;**283**:F473–80.

91. Zhang Q, Cao C, Mangano M, Zhang Z, Silldorff EP, Lee-Kwon W, Payne K, Pallone TL. Descending vasa recta endothelium is an electrical syncytium. *Am J Physiol Regul Integr Comp Physiol* 2006;**291**:R1688–99.

92. Peti-Peterdi J. Calcium wave of tubuloglomerular feedback. *Am J Physiol Renal Physiol* 2006;**291**:F473–80.

93. Ren Y, Carretero OA, Garvin JL. Role of mesangial cells and gap junctions in tubuloglomerular feedback. *Kidney Intl* 2002;**62**:525–31.

94. Takenaka T, Inoue T, Kanno Y, Okada H, Hill CE, Suzuki H. Connexins 37 and 40 transduce purinergic signals mediating renal autoregulation. *Am J Physiol Regul Integr Comp Physiol* 2008;**294**:R1–R11.

95. Earley S, Resta TC, Walker BR. Disruption of smooth muscle gap junctions attenuates myogenic vasoconstriction of mesenteric resistance arteries. *Am J Physiol Heart Circ Physiol* 2004;**287**:H2677–86.

96. Schnermann J, Levine DZ. Paracrine factors in tubuloglomerular feedback: adenosine, ATP, and nitric oxide. *Annu Rev Physiol* 2003;**65**:501–29.

97. Friis UG, Jensen BL, Jorgensen F, Andreasen D, Skott O. Electrophysiology of the renin-producing juxtaglomerular cells. *Nephrol Dial Transplant* 2005;**20**:1287–90.

98. Schweda F, Friis U, Wagner C, Skott O, Kurtz A. Renin release. *Physiology (Bethesda)* 2007;**22**:310–19.

99. Wagner C, de Wit C, Kurtz L, Grunberger C, Kurtz A, Schweda F. Connexin40 is essential for the pressure control of renin synthesis and secretion. Circ Res 2007;**100**:556–63.

100. Hocherl K, Kammerl MC, Schumacher K, Endemann D, Grobecker HF, Kurtz A. Role of prostanoids in regulation of the renin-angiotensin–aldosterone system by salt intake. *Am J Physiol Renal Physiol* 2002;**283**:F294–301.

101. Bell PD, Lapointe JY, Peti-Peterdi J. Macula densa cell signaling. *Annu Rev Physiol* 2003;**65**:481–500.

102. Vallon V, Muhlbauer B, Osswald H. Adenosine and kidney function. *Physiol Rev* 2006;**86**:901–40.

103. Blantz RC, Deng A. Coordination of kidney filtration and tubular reabsorption: considerations on the regulation of metabolic demand for tubular reabsorption. *Acta Physiol Hung* 2007;**94**:83–94.

104. Wilcox CS. Redox regulation of the afferent arteriole and tubuloglomerular feedback. *Acta Physiol Scand* 2003;**179**:217–23.

105. Just A. Mechanisms of renal blood flow autoregulation: dynamics and contributions. *Am J Physiol Regul Integr Comp Physiol* 2007;**292**:R1–R17.

106. Hill MA, Davis MJ, Meininger GA, Potocnik SJ, Murphy TV. Arteriolar myogenic signalling mechanisms: Implications for local vascular function. *Clin Hemorheol Microcirc* 2006;**34**:67–79.

107. Castrop H. Mediators of tubuloglomerular feedback regulation of glomerular filtration: ATP and adenosine. *Acta Physiol (Oxf)* 2007;**189**:3–14.

108. Komlosi P, Fintha A, Bell PD. Renal cell-to-cell communication via extracellular ATP. *Physiology (Bethesda)* 2005;**20**:86–90.

109. Gill PS, Wilcox CS. NADPH oxidases in the kidney. *Antioxid Redox Signal* 2006;**8**:1597–607.

110. Liu R, Ren Y, Garvin JL, Carretero OA. Superoxide enhances tubuloglomerular feedback by constricting the afferent arteriole. *Kidney Intl* 2004;**66**:268–74.

111. Thomson SC, Deng A. Cyclic GMP mediates influence of macula densa nitric oxide over tubuloglomerular feedback. *Kidney Blood Press Res* 2003;**26**:10–18.

112. Ren Y, Garvin JL, Liu R, Carretero OA. Crosstalk between the connecting tubule and the afferent arteriole regulates renal microcirculation. *Kidney Intl* 2007;**71**:1116–21.

113. Arendshorst WJ, Brannstrom K, Ruan X. Actions of angiotensin II on the renal microvasculature. *J Am Soc Nephrol* 1999;**10**:S149–61 [Review] [100 refs].

114. Ichihara A, Kobori H, Nishiyama A, Navar LG. Renal renin–angiotensin system. *Contrib Nephrol* 2004;**143**:117–30.

115. Kobori H, Nangaku M, Navar LG, Nishiyama A. The intrarenal renin–angiotensin system: from physiology to the pathobiology of hypertension and kidney disease. *Pharmacol Rev* 2007;**59**:251–87.

116. Navar LG. The intrarenal renin–angiotensin system in hypertension. *Kidney Intl* 2004;**65**:1522–32.

117. Carey RM. Update on the role of the AT2 receptor. *Curr Opin Nephrol Hypertens* 2005;**14**:67–71.

118. Ferrario CM, Jessup J, Gallagher PE, Averill DB, Brosnihan KB, Ann TE, Smith RD, Chappell MC. Effects of renin–angiotensin system blockade on renal angiotensin-(1–7) forming enzymes and receptors. *Kidney Intl* 2005;**68**:2189–96.

119. Santos RA, Ferreira AJ. Angiotensin-(1–7) and the renin–angiotensin system. *Curr Opin Nephrol Hypertens* 2007;**16**:122–8.

120. Chappell MC, Modrall JG, Diz DI, Ferrario CM. Novel aspects of the renal renin–angiotensin system: angiotensin-(1–7), ACE2 and blood pressure regulation. *Contrib Nephrol* 2004;**143**:77–89.

121. Danser AH. Novel drugs targeting hypertension: renin inhibitors. *J Cardiovasc Pharmacol* 2007;**50**:105–11.

122. Ichihara A, Kaneshiro Y, Takemitsu T, Sakoda M, Itoh H. The (pro)renin receptor and the kidney. *Semin Nephrol* 2007;**27**:524–8.

123. Nguyen G. The (pro)renin receptor: pathophysiological roles in cardiovascular and renal pathology. *Curr Opin Nephrol Hypertens* 2007;**16**:129–33.

124. Majid DSA, Navar LG. Nitric oxide in the control of renal hemodynamics and excretory function. *Am J Hypertens* 2001;**14**:S74–82.

125. Park J, Kemp BA, Howell NL, Gildea JJ, Keller SR, Carey RM. Intact microtubules are required for natriuretic responses to nitric oxide and increased renal perfusion pressure. *Hypertension* 2008;**51**:494–9.

126. Harris RC, Breyer MD. Arachidonic acid metabolites and the kidney. In: Brenner BM, editor. *Brenner and Rector's The Kidney*. 8th edn. Philadelphia, PA: WB Saunders Co; 2008. p. 363–97.

127. Herrera M, Ortiz PA, Garvin JL. Regulation of thick ascending limb transport: role of nitric oxide. *Am J Physiol Renal Physiol* 2006;**290**:F1279–84.

128. Licht C, Laghmani K, Yanagisawa M, Preisig PA, Alpern RJ. An autocrine role for endothelin-1 in the regulation of proximal tubule NHE3. *Kidney Intl* 2004;**65**:1320–6.

129. Hiranyachattada S, Harris PJ. Regulation of renal proximal fluid uptake by luminal and peritubular angiotensin II. *J Renin Angiotensin Aldosterone Syst* 2004;**5**:89–92.

130. Zhuo JL, Li XC. Novel roles of intracrine angiotensin II and signalling mechanisms in kidney cells. *J Renin Angiotensin Aldosterone Syst* 2007;**8**:23–33.

131. Carey RM. Theodore Cooper Lecture: renal dopamine system: paracrine regulator of sodium homeostasis and blood pressure. *Hypertension* 2001;**38**:297–302.

132. Zeng C, Zhang M, Asico LD, Eisner GM, Jose PA. The dopaminergic system in hypertension. *Clin Sci (Lond)* 2007;**112**:583–97.

133. Aperia AC. Intrarenal dopamine: a key signal in the interactive regulation of sodium metabolism. *Annu Rev Physiol* 2000;**62**:621–47.

134. Asghar M, Hussain T, Lokhandwala MF. Activation of dopamine D(1)-like receptor causes phosphorylation of alpha(1)-subunit of Na(+), K(+)-ATPase in rat renal proximal tubules. *Eur J Pharmacol* 2001;**411**:61–6.

135. Bacic D, Kaissling B, McLeroy P, Zou L, Baum M, Moe OW. Dopamine acutely decreases apical membrane Na/H exchanger NHE3 protein in mouse renal proximal tubule. *Kidney Int* 2003;**64**:2133–41.

136. Bacic D, Capuano P, Baum M, Zhang J, Stange G, Biber J, Kaissling B, Moe OW, Wagner CA, Murer H. Activation of dopamine D1-like receptors induces acute internalization of the renal Na+/phosphate cotransporter NaPi-IIa in mouse kidney and OK cells. *Am J Physiol Renal Physiol* 2005;**288**:F740–7.

137. Efendiev R, Cinelli AR, Leibiger IB, Bertorello AM, Pedemonte CH. FRET analysis reveals a critical conformational change within the Na, K-ATPase alpha1 subunit N-terminus during GPCR-dependent endocytosis. *FEBS Letts* 2006;**580**:5067–70.

138. Felder CC, Albrecht FE, Campbell T, Eisner GM, Jose PA. cAMP-independent, G protein-linked inhibition of Na$^+$/H$^+$ exchange in renal brush border by D$_1$ dopamine agonists. *Am J Physiol Renal Fluid Electrolyte Physiol* 1993;**264**:F1032–7.

139. Kunimi M, Seki G, Hara C, Tanaguchi S, Uwatoko S, Goto A, Kimura S, Fujita T. Dopamine inhibits Na$^+$: HCO$_3^-$ cotransport in rabbits and normotensive rats. *Kidney Intl* 2000;**57**:543.

140. Pedrosa R, Jose PA, Soares-da-Silva P. Defective D1-like receptor-mediated inhibition of the Cl$^-$HCO$_3^-$ exchanger in immortalized SHR proximal tubular epithelial cells. *Am J Physiol Renal Physiol* 2004;**286**:F1120–6.

141. Tojo A, Guzman NJ, Garg LC, Tisher CC, Madsen KM. Nitric oxide inhibits bafilomycin-sensitive H(+)-ATPase activity in rat cortical collecting duct. *Am J Physiol* 1994;**267**:F509–15.

142. Modesti PA, Vanni S, Morabito M, Modesti A, Marchetta M, Gamberi T, Sofi F, Savia G, Mancia G, Gensini GF, et al. Role of endothelin-1 in exposure to high altitude: Acute Mountain Sickness and Endothelin-1 (ACME-1) study. *Circulation* 2006;**114**:1410–16.

143. Good DW, George T, Wang DH. Angiotensin II inhibits HCO-3 absorption via a cytochrome P-450- dependent pathway in MTAL. *Am J Physiol* 1999;**276**:F726–36.

144. Nasrallah R, Clark J, Hebert RL. Prostaglandins in the kidney: developments since Y2K. *Clin Sci (Lond)* 2007;**113**:297–311.

145. Tamma G, Carmosino M, Svelto M, Valenti G. Bradykinin signaling counteracts cAMP-elicited aquaporin 2 translocation in renal cells. *J Am Soc Nephrol* 2005;**16**:2881–9.

146. Jackson EK, Raghvendra DK. The extracellular cyclic AMP–adenosine pathway in renal physiology. *Annu Rev Physiol* 2004;**66**:571–99.

147. Vallon V. P2 receptors in the regulation of renal transport mechanisms. *Am J Physiol Renal Physiol* 2008;**294**:F10–27.

148. Schwiebert EM. ATP release mechanisms, ATP receptors and purinergic signalling along the nephron. *Clin Exp Pharmacol Physiol* 2001;**28**:340–50.

149. De Blasio BF, Rottingen JA, Sand KL, Giaever I, Iversen JG. Global, synchronous oscillations in cytosolic calcium and adherence in bradykinin-stimulated Madin-Darby canine kidney cells. *Acta Physiol Scand* 2004;**180**:335–46.

150. Hills CE, Bland R, Wheelans DC, Bennett J, Ronco PM, Squires PE. Glucose-evoked alterations in connexin43-mediated cell-to-cell communication in human collecting duct: a possible role in diabetic nephropathy. *Am J Physiol Renal Physiol* 2006;**291**:F1045–51.

151. McCulloch F, Chambrey R, Eladari D, Peti-Peterdi J. Localization of connexin 30 in the luminal membrane of cells in the distal nephron. *Am J Physiol Renal Physiol* 2005;**289**:F1304–12.

152. Donnelly S. Why is erythropoietin made in the kidney? The kidney functions as a critmeter. *Am J Kidney Dis* 2001;**38**:415–25.

153. Maxwell PH, Ferguson DJ, Nicholls LG, Iredale JP, Pugh CW, Johnson MH, Ratcliffe PJ. Sites of erythropoietin production. *Kidney Intl* 1997;**51**:393–401.

154. Loya F, Yang Y, Lin H, Goldwasser E, Albitar M. Transgenic mice carrying the erythropoietin gene promoter linked to lacZ express the reporter in proximal convoluted tubule cells after hypoxia. *Blood* 1994;**84**:1831–6.

155. Arcasoy MO. The non-haematopoietic biological effects of erythropoietin. *Br J Haematol* 2008;**141**:14–31.

156. Nangaku M, Fliser D. Erythropoiesis-stimulating agents: past and future. *Kidney Intl* 2007(Suppl):S1–3.

157. Lai KN, Leung JC, Chan LY, Guo H, Tang SC. Interaction between proximal tubular epithelial cells and infiltrating monocytes/T cells in the proteinuric state. *Kidney Intl* 2007;**71**:526–38.

158. Breyer MD, Breyer RM. G protein-coupled prostanoid receptors and the kidney. *Annu Rev Physiol* 2001;**63**:579–605.

159. Zhuo JL. Renomedullary interstitial cells: a target for endocrine and paracrine actions of vasoactive peptides in the renal medulla. *Clin Exp Pharmacol Physiol* 2000;**27**:465–73.

160. Candido R, Burrell LM, Jandeleit-Dahm KM, Cooper ME. Vasoactive peptides and the kidney. In: Brenner BM, editor. *Brenner and Rector's The Kidney*. 8th edn Philadelphia, PA: WB Saunders Co; 2008. p. 363–97.

161. Cowley Jr AW. Franz Volhard Lecture. Evolution of the medullipin concept of blood pressure control: a tribute to Eric Muirhead. *J Hypertens* 1994;**12**(Suppl):S25–34.

162. Folkow B. Incretory renal functions – Tigerstedt, renin and its neglected antagonist medullipin. *Acta Physiol (Oxf)* 2007;**190**:99–102.

Chapter 16

Cytokines and Cytokine Receptors Regulating Cell Survival, Proliferation, and Differentiation in Hematopoiesis

Fiona J. Pixley* and E. Richard Stanley

Department of Developmental and Molecular Biology, Albert Einstein College of Medicine, Bronx, New York, New York
Current address: School of Medicine and Pharmacology, University of Western Australia, Nedlands, WA, Australia

GENERAL ASPECTS OF HEMATOPOIESIS

Blood contains red cells, megakaryocytes, lymphocytes, monocytes, and the various types of granulocyte. All mature blood cell types turn over rapidly, requiring active synthesis of large numbers of new cells to maintain a steady state. Differentiated blood cells are all ultimately derived from a small pool of undifferentiated, pluripotent hematopoietic stem cells. This process, involving extensive cell proliferation and differentiation, is known as *hematopoiesis*, and begins in the embryonic yolk sac (primitive hematopoiesis) before maturing into definitive hematopoiesis in the fetal liver and adult bone marrow. Hematopoiesis is regulated by a large number of cytokines that are present in the microenvironment. Specific subsets of these cytokines influence each step in the process. This chapter reviews our current knowledge of the biology of adult hematopoiesis, with an emphasis on the cytokines and their target signaling pathways important in steady-state maintenance of non-lymphoid cell lineages. Due to page limitations, only selected papers or comprehensive reviews are referenced.

Till, McCulloch and colleagues [1] introduced the concept of a hematopoietic stem cell with the capacity to (1) self-replicate; (2) proliferate to produce many progeny; and (3) differentiate to generate all the mature blood cell types. The pool of such stem cells, which are normally quiescent or cycling slowly, represents only 10^{-5} of the total nucleated bone marrow cells in the mouse. Upon division, a stem cell gives rise to an indistinguishable daughter cell to replenish stem cell stocks, and a daughter cell that proliferates extensively and differentiates to give rise to common myeloid and lymphoid progenitor cells, which proliferate and differentiate while progressively developing a more restricted capacity for differentiation, eventually giving rise to cells that are capable of forming only one mature blood or lymphoid cell type [2] (Figure 16.1).

Many different cytokines in the microenvironment of the bone marrow stimulate the development of cells of different lineages. These growth factors may be circulating, or bound to either the surface of their producing cells or to the extracellular matrix. Progenitor cells can also proliferate and differentiate in semisolid culture under the influence of specific cytokines, to form macroscopic colonies of differentiated cells – hence the term colony-stimulating factor (CSF) for the responsible growth factor [3]. The analysis of mice with targeted inactivations of the genes encoding most of these cytokines and their receptors has greatly increased our understanding of the biology of hematopoiesis. The dominant role of cytokines in the regulation of hematopoiesis has led to their rapid clinical application in maintaining normal hematopoiesis in the face of a variety of pathological conditions.

HEMATOPOIETIC CYTOKINES

The hematopoietic cytokines are glycoproteins that are either constitutively present in the circulation (e.g., CSF-1, SCF, FL, G-CSF, EPO and TPO) or appear in response to infection or inflammation (e.g., GM-CSF, IL-3, IL-5, IL-6 and IL-11), and their concentrations may be increased in response to specific triggering conditions such as hypoxia and/or anemia, which stimulate EPO production (Table 16.1).

Primitive multipotent hematopoietic cells, which co-express different lineage-specific cytokine receptors at low

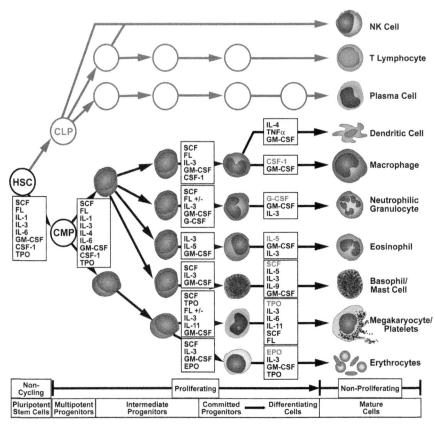

FIGURE 16.1 Hematopoietic cells, indicating the points of regulation by hematopoietic cytokines.
The primary cytokines regulating proliferation and differentiation of committed progenitors of individual lineages are colored in red. Abbreviations: HSC, hematopoietic stem cell; CLP, common lymphoid progenitor; CMP, common myeloid progenitor; SCF, stem cell factor; FL, Flt ligand; IL, interleukin; GM-CSF, granulocyte/macrophage-colony stimulating factor; CSF-1, colony stimulating factor-1; G-CSF, granulocyte-CSF; TPO, thrombopoietin; EPO, erythropoietin; TNF, tumor necrosis factor.

levels, require a combination of cytokines (e.g., SCF, IL-1, IL-3, IL-6, GM-CSF, and CSF-1) for lineage commitment. As these cells differentiate, they lose receptors for some cytokines (e.g., SCF or IL-3) while increasing expression of receptors for the late-acting cytokines (e.g., CSF-1 or EPO). When they reach the stage of committed progenitor cell, their further proliferation and differentiation is along one particular lineage and is regulated by one or more late-acting cytokines. Within specific lineages, the most primitive cells respond to cytokines by both proliferating and differentiating (e.g., committed macrophage progenitors → monoblasts → promonocytes → monocytes → macrophages), while differentiating, non-dividing cells (e.g., peritoneal macrophages) require specific cytokines for survival, activation, and function. Despite an apparent overlap in target cell specificity of several cytokines, their functions are largely non-redundant, as indicated by the distinct hematopoietic phenotypes of cytokine or receptor nullizygous mice (Table 16.1). Transcription factors such as GATA-1 and PU.1, which regulate cytokine receptor expression, are also important in hematopoietic cell commitment.

The "permissive" model of hematopoietic cell regulation by cytokines is one in which the growth factor does not have a role in multipotent progenitor cell commitment, but simply allows the survival and proliferation of committed cells. In contrast, the "instructive" model posits that specific cytokines direct multipotent progenitors to become committed to a specific lineage. It is uncertain whether cytokine regulation of differentiation is simply permissive or instructive, since there is good evidence for both mechanisms. Both may be utilized, depending on the receptors and commitment steps involved [3].

The phenomenon of synergism between predominantly late-acting, lineage-restricted cytokines, such as CSF-1, EPO and G-CSF, with predominantly early-acting cytokines, such as SCF, in stimulating the proliferation and differentiation of primitive multipotent cells provides a mechanism for coupling the changes in levels of the late-acting cytokine, which are tightly regulated by the primary stimuli, to the channeling of multipotent cells into a lineage in order to satisfy the demand for differentiated cells. The mechanisms underlying synergism between cytokines in the regulation of primitive hematopoietic cell proliferation can occur directly at the level of the receptors for the synergizing cytokines, or be due to synergistic effects between post-receptor signal transduction pathways [3].

TABLE 16.1 Hematopoietic cytokines

Cytokine	Sources	Primary hematopoietic phenotype[1]
SCF	Bone marrow stroma, fibroblasts, placenta, others	Severe macrocytic anemia
FL	Ubiquitous	NK cell and dendritic cell
CSF-1	Endothelial cells, fibroblasts, uterine epithelium macrophages	Osteopetrosis
EPO	Kidney proximal tubular cells, liver	Severe anemia
G-CSF	Activated bone marrow stroma, macrophages, fibroblasts, endothelial cells	Neutropenia
TPO	Hepatocytes, endothelial cells, fibroblasts	Thrombocytopenia
IL-3	Activated T cells	Reduced delayed hypersensitivity
IL-5	Activated helper T cells	Eosinophil deficiency
GM-CSF	Bone marrow stroma, activated T cells, endothelial cells, fibroblasts, macrophages	Pulmonary alveolar proteinosis
IL-2	Activated T cells	Autoimmune disease
IL-4	TH2 and NK1.1+ T cells, mast cells, basophils, and eosinophils	TH2 deficient
IL-7	Fetal liver, bone marrow and thymic, stromal cells, lymphoid cells, others	Reduced T and B cells
IL-9	TH2 cells, mast cells and eosinophils	Pulmonary mastocytosis and goblet cellhyperplasia
IL-15	Ubiquitous, increased by activation	Lymphopenic, deficient in NK, NK-T, CD8+ cells and gδT cells
IL-21	TH2 cells	No obvious defects
IL-6	Ubiquitous, in response to inflammatory stimuli	Reduced T cells, IgG & IgA responses, impaired neutrophil/macrophage function
IL-11	Ubiquitous, in response to inflammatory stimuli	Embryonic lethal
LIF	Monocytes, bone marrow stroma	None reported

[1]*References for these cytokine and cytokine receptor-deficient mice, except for IL-9[16] and IL-21[17], are listed in [3] (Stanley, 2001).*

SIGNALING THROUGH CYTOKINE RECEPTORS

General

Hematopoietic cytokine action on target cells is mediated by specific, high-affinity, cell-surface receptors that signal for progenitor cell survival, proliferation, and differentiation, and mature cell survival and activation. Some cytokines are recognized by a single, unique receptor, while some receptors or receptor subunits can recognize multiple cytokines. Cytokine receptors can be classified according to the presence of either intrinsic or associated tyrosine kinase activity and the structure of the extracellular domain (ECD), as well as their requirement for common shared receptor subunits (Figure 16.2). The end result of any cytokine binding to its cognate

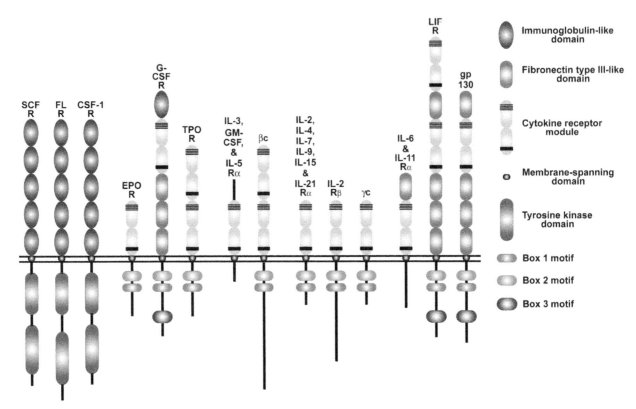

FIGURE 16.2 Schematic of selected high-affinity hematopoietic cytokine receptors showing the similarity between those containing intrinsic tyrosine kinase domains (left) and the modular nature of the cytokine receptor family that are pre-associated with cytoplasmic tyrosine kinases.

receptor is the phosphorylation of particular receptor intracellular domain (ICD) tyrosine residues that create binding sites for downstream signaling molecules. Different activated receptor cytoplasmic domains often bind a common signaling molecule or family of signaling molecules. Targeted downstream pathways include those regulating gene transcription, protein translation, actin cytoskeletal remodeling, and cell adhesion and motility [3–5].

Tyrosine Kinase Receptors

The three hematopoietic cytokines signaling through tyrosine kinase receptors, stem cell factor (SCF), Flt3 ligand (FL) and colony stimulating factor-1 (CSF-1), are members of a family of homodimeric cytokines that share some sequence and structural similarity. SCF and CSF-1 have been shown to have effects on non-hematopoietic as well as hematopoietic cells.

The SCF, CSF-1, and FL receptors, all members of the PDGF receptor family, possess ECDs comprised of five heavily glycosylated immunoglobulin-like repeats, a transmembrane domain, and an intracellular tyrosine kinase domain that is interrupted by a kinase insert domain. Binding of their cognate bivalent ligands by this class of receptors stabilizes their non-covalent dimerization, permitting receptor activation and trans-tyrosine phosphorylation of one ICD by the other. The receptor phosphotyrosines create binding sites for *src* homology region 2 (SH2) and other phosphotyrosine binding domains of signaling and adaptor proteins that bind

the receptor and may themselves become tyrosine phosphorylated (Figure 16.3a). Many of the signaling pathways activated by these receptors, including the MAP kinase (MAPK) cascade, the JAK/STAT pathway, Src family members and PI3-kinase, are shared. Two of the receptors, SCFR and CSF-1R, are encoded by the proto-oncogenes c-*kit* and c-*fms* respectively. v-*kit* and v-*fms* oncogenes are present in feline sarcomatous retroviruses, and contain mutations of the normal cellular genes that render their encoded receptors constitutively active in the absence of cytokines.

SCF and SCF Receptor (SCFR)

SCF influences development in the three different lineages involving pigmentation, hematopoiesis, and fertility. Both secreted and cell surface forms are widely expressed during embryogenesis and in a variety of adult tissues, including bone marrow stromal cells, fibroblasts, and endothelial cells, as well as yolk sac and placenta. Cells expressing the SCFR are frequently contiguous with SCF-expressing cells, and include germ cells, interstitial cells of Cajal in the gut, melanocytes, and early hematopoietic cells. Hematopoietic cell SCFR expression is low in very primitive multipotent progenitors, highest in committed progenitors, and decreases as cells mature. Because pluripotent stem cells do not appear to express the SCFR and can survive *in vitro* in the absence of SCF, it may not act on the earliest

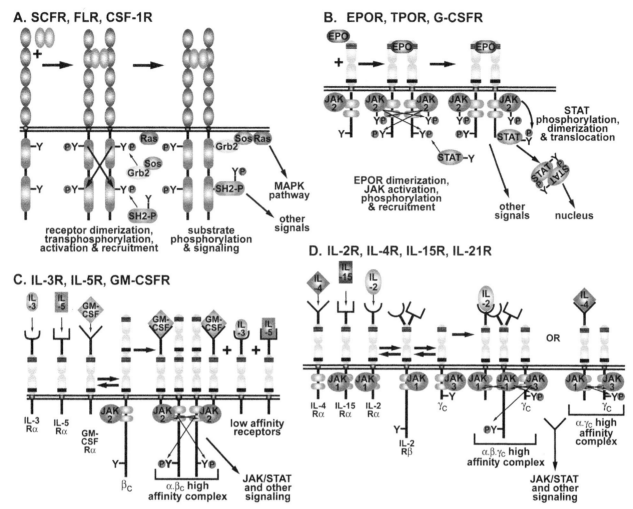

FIGURE 16.3 Models for the activation and signaling of hematopoietic growth factor receptors.
(A) Homodimeric receptors possessing intrinsic tyrosine kinase domains; SH2-P, SH2 domain- or PTB domain-containing signaling molecules. (B) Homodimeric cytokine family receptors. (C) Multimeric cytokine receptors sharing the common β subunit, βc, illustrating the concept of cross-competition between cytokines for the β subunit that is necessary for the formation of the high-affinity signaling complex. GM-CSF is shown forming the high-affinity complex. (D) Receptors sharing the common γ subunit, γc, form a high affinity α.γc receptor complex (e.g. IL-4) or a high affinity α.β.γc receptor complex utilizing a second signaling subunit, IL-2β (e.g., IL-2, IL-15).

hematopoietic precursors [3]. Although SCF has a broad spectrum of activity on hematopoietic cells, it has little activity alone and acts synergistically with many of the hematopoietic cytokines, especially IL-6 and IL-3, to increase the numbers of precursors of most if not all lineages. Spontaneously occurring SCF-null *Sl/Sl* mice are embryonic lethals, but a partially functional *Sl^d* allele allows embryonic survival, yielding severely anemic and mast-cell deficient mice, indicating that the major effects of SCF are on erythropoiesis and mast cell development. In combination with EPO, SCF enhances the number of erythroid precursor cells generated from primitive multipotent cells, and allows precursors to respond to levels of EPO that are too low to elicit a response in the absence of SCF. Mast cells and their progenitors require SCF throughout the differentiation of the lineage from early precursors to mature, primed tissue mast cells. There is also strong evidence of biologic activity for SCF on the megakaryocyte,

granulocyte/macrophage, and lymphoid lineages, yet the effects of the absence of SCF on their development are minimal, implying some redundancy in cytokine action on these lineages [6, 7].

FL and Flt3

FL, which occurs largely as a cell-surface, non-covalently associated homodimer, regulates the proliferation of primitive hematopoietic cells. It is widely expressed, while hematopoietic expression of its cognate receptor, Flt3, is predominantly restricted to the progenitor/stem cell compartment. Like SCF, FL alone cannot stimulate the proliferation of its primitive hematopoietic target cells, but synergizes with other hematopoietic cytokines. In contrast to SCF, however, there is little or no effect of FL on erythroid, or megakaryocyte progenitor cells, since Flt3-deficient mice have no defects in red cell, megakaryocyte, or platelet production. In combination with GM CSF, TNF, and IL-4, FL enhances the production of dendritic cells, and dendritic cell numbers are reduced in FL-deficient mice [7]. However, increased flt3 activity, due either to overexpression of or activating mutations in flt3, underlie most cases of acute myelogenous leukemia and many cases of acute lymphocytic leukemia [8].

CSF-1 and CSF-1 Receptor (CSF-1R)

CSF-1 regulates the survival, differentiation, and function of cells of the mononuclear phagocytic (monocyte/macrophage) lineage, osteoclasts, and epidermal Langerhans cells, as well as the function of cells of the female reproductive tract. Mature forms of the disulfide-linked CSF-1 homodimer include a secreted glycoprotein, a secreted proteoglycan, and a membrane-spanning, cell-surface glycoprotein. CSF-1 is synthesized by a variety of cell types, including fibroblasts, endothelial cells, bone marrow stromal cells, osteoblasts, keratinocytes, astrocytes, myoblasts, and breast and uterine epithelial cells. Circulating CSF-1 is elevated in response to bacterial, viral, and parasitic infections, and is primarily cleared by Kupffer cells, so the number of sinusoidally located macrophages determines the concentration of the cytokine responsible for their production – a simple feedback control. CSF-1 homodimer binding by the CSF-1R results in the formation or stabilization and activation of a receptor dimer. After activation and tyrosine phosphorylation of signaling molecules, most of the receptor–ligand complexes are ubiquitinated, internalized, and destroyed intralysosomally [9].

The osteopetrotic ($Csf-1^{op}/Csf-1^{op}$) mouse, which possesses an inactivating mutation in the CSF-1 gene, exhibits impaired bone resorption associated with a paucity of osteoclasts, no incisors, poor fertility, a lower body weight, a shortened lifespan, and deficiencies in macrophages in most tissues, indicating that CSF-1 is the primary regulator of mononuclear phagocytes. Restoration of circulating CSF-1 in newborn $Csf-1^{op}/Csf-1^{op}$ mice cures their osteopetrosis and monocytopenia, and some but not all of the tissue macrophage populations, demonstrating additional local regulation by CSF-1. CSF-1 regulates the development of macrophages with trophic and scavenger (i.e., physiologic) functions, while the development of macrophages involved in inflammatory and immunologic (i.e., pathologic) functions such as lymph node and thymic macrophages apparently depends on other cytokines. However, once developed, these macrophages are regulated by CSF-1. CSF-1 also synergizes with cytokines such as IL-1, SCF, IL-3, IL-6 to stimulate the proliferation and differentiation of multipotent cells to committed macrophage progenitors that respond to CSF-1 alone [9, 10]. A second cytokine, Il-34, that signals through the CSF-1R and has in vitro activities similar to those of CSF-1, has recently been identified (Lin et al., 2009).

Cytokines Signaling by Homodimerization of a Single, Non-Tyrosine Kinase Receptor Polypeptide Chain

Erythropoietin (EPO), granulocyte-colony stimulating factor (G-CSF), and thrombopoietin (TPO) signal by initiating homodimerization of their cognate non-tyrosine kinase receptors, although additional receptor subunits may be involved. These receptors belong to a large family of receptors, the cytokine receptor family, defined by the presence of an ECD cytokine receptor module (Figure 16.2). These modules contain conserved short amino acid sequence elements, particularly two pairs of conserved cysteine residues near the amino terminal end of the motif and a Trp-Ser–X–Trp–Ser (WSXWS) sequence near the transmembrane domain, the function of which is not clear although it is structurally important. The EPOR, G-CSFR, and TPOR each contain conserved motifs in their ICDs that mediate constitutive association with members of the cytosolic Janus kinase (JAK) family (Figures 16.2, 16.3b). The associated JAKs (JAK2 with EPOR and TPOR; JAK1 with G-CSFR) are activated by formation of the ligand/receptor complex, resulting in autophosphorylation and tyrosine phosphorylation of the receptor. This creates docking sites for molecules with protein tyrosine binding (PTB) or SH2 domains, and leads to phosphorylation and activation of these recruited signaling molecules. Recruited molecules include the Signal Transducers and Activators of Transcription (STATs), additional JAKs, other cytosolic tyrosine kinases, and SH2 domain-containing protein tyrosine phosphatases. The MAPK pathway and PI3 kinase are also activated in response to these cytokines.

EPO and the EPO Receptor (EPOR)

EPO is the primary regulator of erythropoiesis, and is synthesized primarily by the proximal convoluted tubules of the kidney in response to hypoxia. Expression and homodimerization of the EPOR is necessary and sufficient for cell responsiveness to EPO, although, to facilitate the synergistic interaction of EPO and SCF, it functionally and physically interacts with the SCFR and possibly also with the IL-3R β chain. EPOR expression, which is correlated with the proliferative and differentiation effects of EPO in erythroid lineage cells, increases from BFU-E to CFU-E and erythroblasts (Figure 16.1) before decreasing again during the terminal stages of differentiation. EPO also stimulates the release of maturing normoblasts from the bone marrow, and increases the amount of hemoglobin synthesized per erythrocyte. An activating mutation in the EPOR-associated JAK2 has recently been identified in most cases of polycythemia vera as well as other diseases in the myeloproliferative disease spectrum, essential thrombocythemia and idiopathic myelofibrosis [11].

G-CSF and the G-CSF Receptor (G-CSFR)

G-CSF, which shares sequence homology with IL-6, is synthesized by a variety of cell types, including stromal cells, fibroblasts, and endothelial cells, usually in response to inflammatory stimuli such as lipopolysaccharide (LPS), tumor necrosis factor (TNF), and IL-1. The large ECD of its receptor contains one immunoglobulin and three fibronectin domains as well as the cytokine receptor module, and it binds G-CSF with high affinity in a 2:2 stoichiometry. G-CSF is the physiological regulator of neutrophil production, stimulating the proliferation and differentiation of committed neutrophil progenitor cells without affecting other granulocytic lineages, and its levels are increased during infection. It also synergizes with IL-3 or SCF to stimulate the proliferation and differentiation of primitive multipotent hematopoietic progenitor cells. However, mice lacking G-CSF or its receptor, while neutropenic, possess some mature neutrophils and have only a mild reduction in committed granulocyte-macrophage progenitor cells, suggesting that G-CSF is not necessary for neutrophil lineage commitment. It also enhances the survival of mature neutrophils and may prime their functional responses [3]. G-CSF is routinely used in the clinic for treatment of neutropenia.

TPO and the TPO Receptor (TPOR)

Structurally, TPO can be divided into two domains: an amino terminal portion with EPO homology, and a carboxy terminal domain that is widely species divergent and lacks homology with other known proteins. It is synthesized primarily by hepatocytes, endothelial cells, fibroblasts, and the proximal tubule cells of the kidney. The TPOR gene, c-mpl, was originally identified as the cellular counterpart of v-mpl, the oncogene carried by the murine myeloproliferative leukemia virus, and is restricted in its expression to megakaryocytes, platelets, and primitive hematopoietic cells. TPO affects all aspects of megakaryocyte and platelet development, including stimulation of megakaryocyte progenitor cell proliferation and differentiation, and stimulation of platelet release. It has also been shown to have profound and non-redundant effects on survival and proliferation of hematopoietic stem cells. Mice lacking either TPO or TPOR possess an identical phenotype and exhibit an 80–90 percent decrease in platelets with no effect on numbers of other differentiated peripheral blood cells, but they display a 60 percent reduction in multipotent and committed myeloid progenitors and are deficient in stem cells capable of long-term repopulation [12].

Cytokines Signaling Through Receptors with a Common β Subunit

The three cytokines, interleukin-3 (IL-3), granulocyte/macrophage CSF (GM-CSF), and interleukin-5 (IL-5), which signal through receptors sharing a common β subunit, are members of a subfamily of cytokines that appear to share a common ancestry despite their lack of homology at the amino acid level. Their genes have a similar structure, and map closely together on chromosome 5. The cytokines signal through receptors comprised of a cytokine-specific α chain, which alone exhibits low affinity for the cytokine, and a larger, shared β chain, βc, that can interact with any of the three α chain/cytokine complexes to generate a specific, high-affinity complex (Figure 16.3c). Ligand-induced disulfide bonding between an α and β subunit and dimerization of the βc subunit are required for signaling, producing a complex consisting of two receptor α chains, two βc chains, and two ligand molecules. The shared βc leads to competition between IL-3, GM-CSF, and IL-5, although the biologic significance of this is not well understood. However, they all act synergistically to stimulate differentiation and function of myeloid cells, and are important in allergic inflammation. Their α subunit ECD cytokine receptor module structures are similar to that of the EPOR (Figure 16.2), and the divergent ICDs are required for proper receptor function. The 120-kDa β subunit contains two ECD cytokine receptor motifs and a long cytoplasmic tail that is required for proliferative signaling. Box 1 and box 2 motifs in the ICD contain the docking sites necessary for the association of JAK2 and recruitment of other signaling molecules to the activated receptor complex, resulting in rapid tyrosine phosphorylation of several cellular proteins, including the βc receptor subunit and activation of JAK/STAT, MAPK, and PI3 kinase pathways (Figure 16.3c) [13].

IL-3 and the IL-3 Receptor (IL-3R)

IL-3 is a secreted monomeric glycoprotein that is synthesized almost exclusively by T cells in response to antigen stimulation. The IL-3R consists of a unique IL-3-specific, low-affinity α subunit and the βc subunit. However, a duplication of the βc gene has occurred in the mouse to produce βIL-3, which interacts with IL-3Rα to create an additional low-affinity receptor for IL-3. IL-3 is a pleiotropic hematopoietic cytokine supporting the proliferation and differentiation of both primitive multipotent progenitor cells and committed myeloid progenitors. Working alone, it stimulates primitive hematopoietic cells to form multilineage colonies, comprised of neutrophils, basophils, eosinophils, monocytes, and megakaryocytes. In combination with SCF, IL-1, and CSF-1, IL-3 stimulates the proliferation and differentiation of even more primitive precursors and, in concert with other late-acting hematopoietic cytokines, it stimulates multipotent cells to specific lineage commitment. For example, the combination of IL-3 and CSF-1 allows the proliferation of primitive cells that do not respond to CSF-1 alone to give rise to committed, CSF-1R-expressing macrophage progenitors. Once committed, progenitor cells and their progeny lose expression of IL-3R and their ability to respond to IL-3. Since IL-3 synthesis is highly restricted and regulated and IL-3-deficient mice display only a diminished delayed hypersensitivity with no obvious steady-state hematopoietic phenotype, it seems to be important only during hematopoietic demand [3].

GM-CSF and the GM-CSF Receptor (GM-CSFR)

GM-CSF, a monomeric glycoprotein, is constitutively synthesized by macrophages, mast cells, eosinophils, basophils, endothelial cells, and fibroblasts, and inducibly expressed in a variety of cells, especially T cells. The GM-CSFR α subunit has three alternative transcripts used to produce the main form, a soluble form, or an alternative membrane-spanning subunit with an elongated C-terminus. All are functional, but their relative physiologic significance is not yet well understood. GM-CSF has been shown to have a broad range of biologic effects, acting on both progenitor cells and mature, terminally differentiated cells. The survival, proliferation, and differentiation of all stages of cells in the neutrophil, macrophage, and eosinophil lineages are supported by GM-CSF, and it appears to be important in the activation and enhancement of function of mature cells in these lineages. However, at most stages of development, each lineage also requires the presence of the other lineage-specific cytokines, G-CSF, CSF-1, and IL-5. The antigen-presenting dendritic cells appear to be derived from myeloid lineage and possibly also from lymphoid lineage cells. GM-CSF, with IL-4 and TNFα, stimulates myeloid-derived pre-dendritic cells to differentiate into mature dendritic cells. GM-CSF also synergizes with EPO and TPO on primitive hematopoietic cells to generate erythroid and megakaryocytic progeny, respectively, and in vivo GM-CSF administration increases the number of circulating neutrophils, monocytes, and eosinophils, and the number of tissue-fixed macrophages. Nevertheless, since GM-CSF-deficient mice have normal granulocyte and macrophage production in both steady-state and stressed conditions, GM-CSF apparently does not play an important role in blood cell production [14].

IL-5 and the IL-5 Receptor (IL-5R)

Biologically active IL-5 is a dimer that is secreted predominantly by antigen-stimulated T lymphocytes, but also by NK cells, mast cells, B cells, eosinophils, and endothelial cells. However, only eosinophils and basophils, precursors of both lineages, and some B cells express both subunits of the IL-5R, so IL-5 has a restricted biologic activity. It is the primary late-acting cytokine for eosinophil proliferation and differentiation, for mature eosinophil survival and activation, and, in vivo, in the development of eosinophilia. Complete abrogation of the development of eosinophilia secondary to parasitic infections is observed in IL-5-deficient mice. In combination with GM-CSF, IL-3,

and IL-4, IL-5 stimulates the survival, proliferation, and differentiation of the basophil-mast cell lineage [15].

Cytokines Signaling Through Receptors with a Common γ Subunit

Six hematopoietic cytokines, IL-2, IL-4, IL-7, IL-9, IL-15, and IL-21, signal through a high-affinity receptor comprised of a cytokine-specific α chain and a common γ chain (γc), in a manner similar to IL-3, GM-CSF, and IL-5 and their common βc, in that the cytokine binds the α subunit with low affinity and requires further binding of the γc for high-affinity α.γc complex signaling (Figure 16.3d). Additionally, IL-2R and IL-15R, which probably have a common ancestry and constitute a separate cytokine receptor subfamily, also share a common β subunit, IL-2Rβ (Figure 16.2), which can directly bind both ligands and thus replace the α chain, but is most usually included in a high-affinity α.β.γc complex (Figure 16.3d). The signaling pathways described for all these receptors are similar and, since the ICDs of each receptor lack catalytic activity, involve associated JAKs. Upon ligand binding and assembly of the high-affinity receptor–ligand complex, JAK1, constitutively associated with the α and β subunits, and JAK3, constitutively associated with the γc subunit, become activated and transphosphorylated. The JAKs phosphorylate the receptor chains, leading to recruitment and phosphorylation of phosphotyrosine-associating signaling intermediates, including STATs 1, 3, 5 A, 5B, and 6, which dimerize and move to the nucleus to direct transcription. The MAPK pathway is activated via Shc interactions with the α or β subunit, and PI3 kinase activation is brought about by recruitment of the Src-family kinases and/or insulin receptor substrates-1 and -2 to the receptors. A family of negative regulators of cytokine receptor signaling, the Suppressors of Cytokine Signaling (SOCS) family, bind and attenuate JAK/STAT signaling. In man, spontaneous mutations in the genes for γc and JAK3 produce, respectively, an X-linked and an autosomal severe combined immunodeficiency (SCID) syndrome, and mice lacking either γc or JAK3 are also severely immunodeficient, indicating the importance of these shared signaling pathways in immune function [16]. Since the primary role of IL-2, IL-7, IL-15, and IL-21 is regulation of the development and/or function of lymphoid lineage cells, only a very brief summary of their actions is outlined, while IL-4 and IL-9 have more pleiotropic effects and will be covered in more detail.

IL-2, IL-7, IL-15, IL-21, and their Receptors

IL-2 stimulates both a prompt T cell immune response to antigens and a subsequent, rapid dampening of this response. Since IL-2-deficient mice die from autoimmune disorders yet mount a normal response to infections, the primary role of IL-2 appears to be the promotion of apoptosis of excess activated T cells. IL-15 exhibits considerable functional overlap with IL-2 in their regulation of innate immunity through the two shared receptor subunits, IL-2Rβ and γc. However, in contrast to IL-2/IL-2Rα, IL-15Rα alone binds IL-15 with high affinity, and IL-15R signaling has distinct effects on activated T cells. Rather than promoting apoptosis of excess T cells, as does IL-2, IL-15 stimulates the survival of memory T cells for the secondary immune response. IL-7 plays a critical role in the early development of both B and T cells in mice, and of T cells only in man. Deletion of the gene for either IL-7 or IL-7R produces severe B and T cell lymphopenia. IL-21 is produced by activated T cells and targets a broad range of cells, including T cells, NK cells, B cells, and myeloid cells. It both positively and negatively regulates immune processes, and may play a role in autoimmune disease [17].

IL-4 and the IL-4 Receptor (IL-4R)

IL-4 is a pleiotropic cytokine that is secreted by T_H2 and NK1.1+ T cells, basophils, mast cells, and eosinophils. The high-affinity IL-4R is widely expressed in hematopoietic and non-hematopoietic cells, and its expression is upregulated by IL-4 itself. In hematopoietic cells, the IL-4R consists of the IL-4Rα and γc subunits, and requires JAK3 for STAT6 activation. Non-hematopoietic signaling by IL-4 utilizes the IL-4Rα subunit in combination with the IL-13R rather than γc. Consequently, IL-4 can act on many cell types and can modulate cytokine production by a variety of tissues. Its primary role, however, is in the regulation of T helper cell differentiation to T_H2 cells and in B cell Ig switching. Apart from its effects on lymphocytes, IL-4 inhibits CSF-1-induced macrophage colony formation and megakaryocyte colony formation, but enhances G-CSF-induced granulocyte colony formation and IL-5-induced basophil, mast cell, and eosinophil generation. Along with GM-CSF and TNFα, IL-4 induces the differentiation of myeloid lineage cells into dendritic cells. It also regulates the production of inflammatory mediators in a pattern consistent with its anti-inflammatory role.

IL-9 and the IL-9 Receptor (IL-9R)

IL-9 is a pleiotropic growth factor that appears to be important in the pathogenesis of asthma. It is predominantly secreted by T cells, especially T_H2 cells, but also by mast cells and eosinophils, particularly those derived from asthmatic subjects. IL-9R is expressed by T and B cells, eosinophils, and neutrophils, and IL-9 has been shown to stimulate early T cell development, B cell Ig production, including IgE, mast cell survival, proliferation and cytokine release, as well as upregulation of eosinophil IL-5R expression. Hence, IL-9 influences many aspects of the inflammatory process that underlies asthma.

Cytokines Signaling Through a Common gp130 Subunit

At least three hematopoietic cytokines, IL-6, IL-11, and Leukemia Inhibitory Factor (LIF), and four non-hematopoietic cytokines, oncostatin M (OSM), ciliary neurotrophic factor (CNTF), cardiotrophin-1 (CT-1), and cardiotrophin-like cytokine (CLC), comprise the IL-6-type cytokine group. All share at least one subunit, gp130, in their receptor complexes, and, as a result, functional overlap is frequent. In addition, IL-27, a recently identified member of the heterodimetic cytokine family that IL-12, IL-23 and IL-35, binds gp130, in conjunction with the T cell cytokine receptor (TCCR) IL-6R and IL-11R each contain a unique, cytokine-specific α chain, while the LIFR-specific subunit is much larger (Figure 16.2) and more promiscuous, as it is incorporated into receptor complexes for OSM, CNTF, and CT-1. In total, gp130 interacts with five different IL-6-type cytokines and functionally cooperates with five different receptor subunits. The signaling complexes for this group of cytokines vary in their stoichiometry. IL-6, IL-6Rα and gp130 bind together in a 2:2:2 molecular ratio to produce a high-affinity hexameric complex. A similar situation exists for the IL-11R, while LIF complexes in a 1:1:1 stoichiometry with LIFR and gp130. Gp130 signal transduction is mediated via activation of the constitutively associated JAKs, JAK1, JAK2, and Tyk-2, which activate STAT1, STAT3, and STAT5 in turn. The MAPK pathway is also activated through SHP-2 binding to a juxtamembrane phosphotyrosine on gp130. Gp130-deficient mice die *in utero* due to severe hematopoietic and cardiac problems [18, 19].

IL-6 and the IL-6 Receptor (IL-6R)

IL-6 is a multifunctional cytokine produced by both lymphoid and non-lymphoid cells, and its receptor is also expressed on a wide range of cells. In hematopoiesis, IL-6 acts synergistically with a number of cytokines, including SCF, IL-3, KL, CSF-1, and TPO, to stimulate primitive multipotent hematopoietic cell proliferation, myelopoiesis, and megakaryocyte production, as well as lymphopoiesis. IL-6-deficient mice have reduced primitive multipotent progenitors, megakaryocyte progenitors, and neutrophils, and are severely defective in their responses to tissue damage or infection.

IL-11 and the IL-11 Receptor (IL-11R)

Beyond its shared receptor subunit, IL-11 is similar to IL-6 in many other respects. Both IL-11 and its receptor are widely expressed in hematopoietic and non-hematopoietic tissues. Its expression is induced by pro-inflammatory and anti-inflammatory cytokines, and hematopoietic cell targets include those of the myeloid, erythroid, and megakaryocytic lineages. Although the effects of IL-11 on hematopoiesis are very similar to those of IL-6, and largely synergistic, adult IL-11Rα nullizygous mice are hematopoietically normal, highlighting the significant redundancy in this family of cytokines.

LIF and the LIF Receptor (LIFR)

LIF is also multifunctional, and is produced by monocytes and stromal cells in response to activating stimuli such as IL-1β and LPS. The hematopoietic effects of LIF are similar to those of IL-6 and IL-11 and, as for IL-11, LIF-deficient mice have no obvious hematopoietic phenotype.

CONCLUDING STATEMENTS

The maintenance of the normal complement of hematopoietic cells requires the complex interaction of a large number of cytokine signaling pathways, some of which are redundant while others are critically important. Moreover, this complexity is increased by the actions of a number of other cytokines that also regulate hematopoiesis but primarily act elsewhere, and which are therefore not covered in this chapter. They include IL-1 and IL-18, which do not signal through cytokine receptor motif-containing receptors; IFNγ, TGFβ, TNFα, and TNFβ, which mostly inhibit rather than promote hematopoiesis; several chemokines, and the mammalian Notch paralogs and their ligands, which regulate primitive hematopoietic cells.

In an extension of their physiological roles, some hematopoietic cytokines are used extensively in the clinical setting, usually to correct deficiencies of specific hematopoietic lineages such as EPO for anemia, TPO for thrombocytopenia, and GM-CSF and G-CSF for granulocytopenias.

REFERENCES

1. Till JE, McCulloch EA. Hemopoietic stem cell differentiation. *Biochim Biophys Acta* 1980;**605**:431–59.
2. Harrison DE. Evaluating functional abilities of primitive hematopoietic stem cell populations. *Curr Top Microbiol Immunol* 1992;**177**:13–30.
3. Stanley ER. The hematopoietic cytokines. In: Austen KF, Frank MM, Atkinson JP, Cantor H, editors. *Samter's Immunological Diseases*. 6th ed. Philadelphia, PA: Lippincott Williams and Wilkins; 2001. pp. 175–93.
4. Ihle JN. Signal transduction in the regulation of hematopoiesis. In. Stamatoyannopoulos G, Majerus PW, Perlmutter RM, Varmus H, editors. *The Molecular Basis of Blood Diseases*. 3rd ed. Philadelphia, PA: W. B. Saunders Company; 2001. p. 103–25.
5. Kaushansky K. Hematopoietic growth factors and receptors. In: Stamatoyannopoulos G, Majerus PW, Perlmutter RM, Varmus H, editors. *The Molecular Basis of Blood Diseases*. 3rd ed. Philadelphia, PA: W. B. Saunders Company; 2001. p. 25–54.
6. Galli SJ, Zsebo KM, Geissler EN. The kit ligand, stem cell factor. *Adv Immunol* 1994;**55**:1–96.
7. Lyman SD, Jacobsen SEW. *c-kit* ligand and flt3 ligand: Stem/progenitor cell factors with overlapping yet distinct activities. *Blood* 1998;**91**:1101–34.
8. Sternberg DW, Licht JD. Therapeutic intervention in leukemias that express the activated fms-like tyrosine kinase 3 (FLT3): opportunities and challenges. *Curr Opin Hematol* 2005;**12**:7–13.
9. Pixley FJ, Stanley ER. CSF-1 regulation of the wandering macrophage: complexity in action. *Trends Cell Biol* 2004;**14**:628–38.
10. Bourette RP, Rohrschneider LR. Early events in M-CSF receptor signaling. *Growth Factors* 2000;**17**:155–66.
11. Munugalavadla V, Kapur R. Role of c-Kit and erythropoietin receptor in erythropoiesis. *Crit Rev Oncol Hematol* 2005;**54**:63–75.
12. Kaushansky K. The molecular mechanisms that control thrombopoiesis. *J Clin Invest* 2005;**115**:3339–47.
13. Martinez-Moczygemba M, Huston DP. Biology of common β receptor-signaling cytokines: IL-3, IL-5, and GM-CSF. *J Allergy Clin Immunol* 2003;**112**:653–65.
14. Kastelein RA, Shanafelt AB. GM-CSF receptor: interactions and activation. *Oncogene* 1993;**8**:231–6.
15. Adachi T, Alam R. The mechanism of IL-5 signal transduction. *Am J Physiol* 1998;**275**:C623–33.
16. He YW, Malek TR. The structure and function of gamma c-dependent cytokines and receptors: regulation of T lymphocyte development and homeostasis. *Crit Rev Immunol* 1998;**18**:503–24.
17. Mehta DS, Wurster AL, Grusby MJ. Biology of IL-21 and the IL-21 receptor. *Immunol Rev* 2004;**202**:84–95.
18. Bravo J, Heath JK. Receptor recognition by gp130 cytokines. *EMBO J* 2000;**19**:2399–411.
19. Ernst M, Jenkins BJ. Acquiring signal specificity from the cytokine receptor gp130. *Trends Genet* 2004;**20**:23–32.
20. Lin H, Lee E, Hestir K, Leo C, Huang M, Bosch E, Halenbeck R, Wu G, Zhou A, Behrens D, Hollenbaugh D, Linnemann T, Qin M, Wong J, Chu K, Doberstein SK, Williams LT. Discovery of a cytokine and its receptor by functional screening of the extracellular proteome. *Science* 2008;**320**:807–811.

Chapter 17

CD45

Michelle L. Hermiston[1], Vikas Gupta[2] and Arthur Weiss[2]

[1]Department of Pediatrics, Howard Hughes Medical Institute, University of California, San Francisco, California

[2]Department of Medicine, Howard Hughes Medical Institute, University of California, San Francisco, California

INTRODUCTION

CD45 (also known as leukocyte common antigen (LCA), EC 3.1.3.48, T200, Ly5, PTPRC, and B220) constitutes the first and prototypic receptor-like protein tyrosine phosphatase (RPTP). Its expression is restricted to all nucleated hematopoietic cells, where it is one of the most abundant cell surface glycoproteins, constituting almost 10 percent of the cell surface, and estimated to be present at approximately $25\,\mu M$ in the plasma membrane [1–3]. CD45 functions as a central regulator of phosphotyrosine levels in hematopoietic cells by modulating the activity of Src family kinases (SFKs). Its importance is highlighted by observations, in both mice and humans, that its absence leads to severe combined immunodeficiency (SCID), while dysregulation of its activity correlates with autoimmunity [2–4]. This chapter briefly summarizes our current understanding of the structure, function, and regulation of this key phosphatase.

STRUCTURE

CD45 belongs to a large family of type I transmembrane receptor-like protein tyrosine phosphatases (RPTPs). The members of this family are defined by the presence of one or two highly conserved intracellular phosphatase domains, but are also characterized by very diverse extracellular domains [5, 6]. The large extracellular domain of CD45 is heavily glycosylated and contains three alternatively spliced exons (4, 5, and 6) that are both O-linked glycosylated and sialylated [2, 3]. Alternative splicing of these exons produces isoforms differing in size, shape, and charge [7]. The largest isoform containing all three alternatively spliced exons, CD45RABC, is approximately 235 kDa, while the smallest isoform lacking all three exons, CD45RO, is approximately

180 kDa (Figure 17.1). Additionally, evidence for alternative splicing of exon 7 has been reported at an mRNA level [3].

The expression of these isoforms depends on the cell type and activation state of the cell [2]. B cells express primarily the

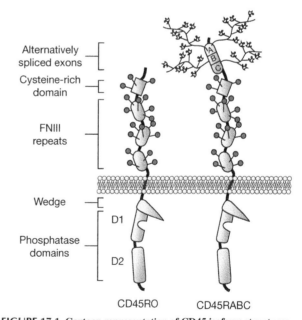

FIGURE 17.1 Cartoon representation of CD45 isoform structures.
CD45 exists as multiple isoforms due to alternative splicing of three exons (4, 5, and 6, designated A, B, and C) in the extracellular domain. The largest isoform RABC includes all three exons, while the smallest, RO, lacks all three exons. These three exons encode multiple sites of O-linked glycosylation. As a result, various isoforms differ substantially in size, shape, and charge. The remaining extracellular domain is heavily glycosylated, and contains a cysteine-rich region followed by three fibronectin type III repeats. CD45 has a single transmembrane domain and a large cytoplasmic tail containing two tandemly duplicated PTPase domains, D1 and D2. Only D1 has enzymatic activity. In addition, the CD45 crystal structure indicates that the juxtamembrane region forms a structural wedge.

full-length form of CD45, also known as B220. Naïve T cells, on the other hand, express no B220, but do express intermediate sized isoforms containing one or two of these exons. Following activation or differentiation into memory cells, as a consequence of changes in CD45 splicing, T cells replace higher-sized iso-forms with the smallest isoform, CD45RO [7]. The part of the extracellular domain common to all of the isoforms consists of a cysteine-rich region followed by three fibronectin type III repeats. CD45 homologs have been identified in various mammals, chicken, shark, and the hagfish *Eptatretus stoutii* [2, 3]. While the actual amino acid identity of the extracellular domain is low between species, the general structure is conserved.

In contrast, the transmembrane and cytoplasmic domains of CD45 are more highly conserved amongst species [2, 3]. The transmembrane domain has been shown to interact with CD45 associated-protein (CD45AP), although the function of this interaction is unclear since the phenotype of CD45AP-deficient mice is minimally affected [8, 9]. Some data suggest it may affect membrane localization of CD45 [10]. The highly conserved intracellular domain of CD45 consists of two tandemly duplicated phosphatase domains. Only the membrane proximal domain has enzymatic activity, which is necessary to rescue TCR signaling in a CD45-deficient cell line [11]. The first domain also contains a juxtamembrane helix–turn–helix motif termed the "wedge" that was first identified in the crystal structure of a related phosphatase, RPTPα [12, 13]. This motif is the most highly conserved region of CD45 between species [12]. The crystal structure of CD45 confirmed the presence of this wedge motif [13]. Interestingly, the wedge has now been shown to be present in all six of the RPTPs crystallized to date, suggesting that it may be a conserved element in this class of proteins (Figure 17.2) [3, 14, 15].

The function of the more C-terminal second phosphatase domain in CD45, as well as in other RPTPs that contain a second tandem domain, remains uncertain, since the more C-terminal domain lacks catalytic activity. Although it contains the catalytic cysteine present in all tyrosine phosphatases, the absence of a number of other key residues completely abolishes its activity. However, the second domain may contribute to CD45 activity indirectly by stabilizing the first domain. Supporting this notion, the CD45 crystal structure revealed an extensive area of contact between the two phosphatase domains. In addition, there is evidence that the second domain can bind and recruit substrates to be dephosphorylated by the first domain. The second domain also contains a unique 19 amino acid acidic insert that can be phosphorylated by casein kinase II and a long C-terminal tail; neither of these elements was included in the CD45 crystal structure. Their contribution to CD45 conformation is therefore unknown.

FUNCTION

Studies using CD45-deficient T and B cell lines demonstrate that CD45 is an obligate positive regulator of antigen receptor signaling [2, 3, 16]. Ablation of the murine CD45 gene by three independent groups reveals its critical positive role in lymphocyte development and activation [17, 18]. For example, thymocyte development is largely blocked at the positive selection developmental checkpoint, and the few mature T cells produced are refractory to TCR stimulation. Whereas B cell production is less severely affected until the most terminal stages of differentiation, B cell responses to BCR stimulation are also impaired. Similarly, loss of CD45 in humans results in a form of

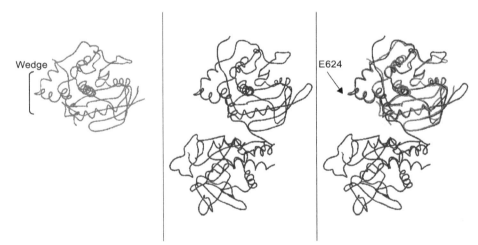

FIGURE 17.2 Three-dimensional structure of RPTPα and CD45 phosphatase domains.
Left panel: Ribbon depiction of the RPTPα domain 1 crystal structure. The wedge is highlighted. Middle panel: Ribbon depiction of the CD45 domain 1 and 2 crystal structure based upon the coordinates provided by Nam *et al.* (2005). Right panel: Overlay of RPTPα and CD45 crystal structures reveals the high degree of structural conservation, including the helix–turn–helix "wedge". The mutated wedge residue (E624 in human, E613 in mice) is highlighted.

severe combined immunodeficiency (SCID) [2, 3]. Recent studies have also shown that CD45 is essential for cytokine production, but not cell-mediated cytotoxicity, in NK cells [19, 20]. The role of CD45 in myeloid lineage cells is less clear, and may be obscured in part by functional redundancy with other RPTPs such as CD148 (J. Xhu and A. Weiss, unpublished observations).

Src family kinases (SFKs) are a primary substrate for CD45 [21, 22]. SFKs are responsible for initiating antigen receptor signaling. They also modulate signal transduction cascades emanating from growth factor, cytokine, and integrin receptors. In most CD45-deficient cells, SFKs are hyperphosphorylated at the negative regulatory tyrosine [2, 3]. Moreover, expression of a constitutively active Lck Y505F mutant in CD45-deficient mice largely rescues the block in T cell development, confirming that this is a physiologically relevant substrate of CD45 [23]. By preferentially dephosphorylating the negative regulatory tyrosine, CD45 can maintain SFKs in a primed, or signal competent, state which is capable of full activation upon receptor stimulation.

Although CD45 clearly plays a positive role in antigen receptor signaling, it can also function as a negative regulator in other settings. For example, CD45 deficient macrophages are abnormally adherent [24]. Despite hyperphosphorylation of the negative regulatory tyrosine of the SFKs, kinase activity is enhanced due to hyperphosphorylation at low stoichiometry of the autophosphorylation site, explaining the increased adhesiveness of these cells. This finding suggests that both the autophosphorylation site and the negative regulatory tyrosines can serve as CD45 substrates in some contexts. Interestingly, similar findings have been described for antigen receptor signaling in some CD45-deficient T and B cell lines, as well as CD45-deficient mice reconstituted with low levels of CD45 isoform-specific transgenes [25, 26]. The discrepancy of positive and negative effects of CD45 may be explained by its inclusion in or exclusion from clustered signaling complexes. Physical separation from the TCR during antigen recognition at the immunological synapse results in a net positive effect, while access to its substrate during integrin-mediated adhesion results in a negative effect [2, 26].

In addition to SFKs, CD45 may also negatively regulate cytokine and interferon receptor-mediated activation by dephosphorylating Janus kinases. Other possible (but controversial) substrates include ZAP-70, PAG-85, and CD3ζ [3].

REGULATION

The regulation of CD45 is complex and incompletely understood. Surprisingly, given the structural similarity between CD45 and receptor tyrosine kinases, a definitive ligand for CD45 has not been identified, although interactions with various lectins have been suggested to modulate its activity [3]. Potential and non-mutually exclusive means of regulation

include spontaneous isoform-differential homodimerization, regulated membrane localization regulated by interactions with other molecules, phosphorylation, and oxidation-induced conformational changes.

The alternative splicing of CD45 is highly conserved and tightly regulated [7]. Naive T cells predominantly express the larger isoforms, and following activation, over the course of 3–5 days, the smallest RO isoform replaces the larger isoforms on the plasma membrane. This regulated event is likely to be under the control of splicing factors that are induced in a PKC- and Ras-dependent manner after T cell activation [27]. An exonic splicing silencer (ESS) sequence in exon 4 is responsible for controlling exon usage. The ESS interacts with hnRNP L or the splicing factor PSF in response to signaling events in T cells [28]. The importance of this ESS is highlighted by a translationally silent point mutation in exon 4, which disrupts the function of the exonic splicing silencer and causes abnormally high levels and persistent expression of the larger isoform, and has been linked to the development of multiple sclerosis in a German, but not American or other European, patient cohort [3, 4]. In contrast, a point mutation in exon 6, A138G, results in enhanced production of low molecular weight isoforms and has been associated with resistance to Graves disease and hepatitis B in Japanese cohorts, where this polymorphism is prevalent [3, 4]. Together, these observations provide support for a contribution by the extracellular domain in regulating CD45 activity, and suggest differences in regulation of the various isoforms.

Unfortunately, attempts to understand the role of different isoforms by reconstituting CD45-deficient mice with isoform-specific transgenes have met with limited success due to the difficulty of reconstituting wild-type levels of expression [3, 29]. Interestingly, very low levels of any isoform of CD45 can reconstitute T, but not B, cell development. However, whether this low level expression can reconstitute an appropriate T cell repertoire has not been addressed. This is important, as positive selection has even been shown to be altered in CD45 heterozygous mice expressing a TCR transgene [30]. Recent studies have suggested that the two phosphorylation sites of SFK are differentially sensitive to different levels of CD45 transgene expression, with the C-terminal site being most sensitive to CD45 phosphatase levels [25]. Thus, the expression levels of CD45 may be critical in regulating its function, and suggest that CD45 could act as a rheostat to titrate signal transduction thresholds in the T cell lineage [2].

One possible means of regulation of CD45 is isoform-differential spontaneous homodimerization. Dimeric forms of CD45 can be detected through chemical crosslinking of cellular lysates, by using a cysteine dimer-trapping method, or by fluorescence resonance energy transfer (FRET) [31–33]. Interestingly, low molecular weight isoforms appear to dimerize more readily than larger isoforms. To test the role of dimerization, a chimeric protein containing the epidermal growth factor (EFG) receptor extracellular and transmembrane domains fused to the cytoplasmic domain of CD45 was introduced into a CD45-deficient T cell line. Although the chimera reconstituted TCR signaling in the CD45-deficient state, EGF-induced dimerization inhibited TCR-mediated signal transduction [34].

A potential molecular explanation for dimer-mediated negative regulation was suggested by the crystal structure of the membrane proximal D1 domain of RPTPα [12]. It formed a symmetrical dimer in which the catalytic site of one molecule was blocked by the juxtamembrane wedge of its partner. The phylogenetic conservation of the wedge between RPTP family members suggests a potential preserved evolutionary function. Introduction of a point mutation at the tip of this wedge (E624 in human CD45) abolished the inhibitory effect of dimerization on T cell receptor signaling in a transformed T cell line [12]. Introduction of the analogous point mutation into mice by homologous recombination (termed CD45E613R) resulted in both B cell and thymocyte hyper-responsiveness to antigen receptor stimulation [35, 36]. The phenotypic consequences of the CD45E613R mutation were development of a lymphoproliferative syndrome and severe autoimmune nephritis with autoantibody production, resulting in early death [36]. Taken together, these data validate a role for the wedge in negative regulation of CD45 function.

The presence of the wedge in the CD45 crystal structure that was subsequently solved suggests its potential role in dimerization-mediated regulation of CD45 activity [37]. At the same time, though, the crystal structure poses a significant problem to a dimerization-based model. In the crystal structure, the orientation of the first and second domains would preclude the wedge from participating in dimer formation. However, this conformation may only represent one of many possible conformations. Many other factors may influence CD45 structure and dimerization, including the extracellular domain, lipid bilayer, transmembrane segment, juxtamembrane region, acidic insert, and C-terminal tail. Alternatively, or in addition, the wedge could impact other means of CD45 regulation.

Cellular localization and access to substrate may also contribute to CD45's effect on signaling. Redistribution of an intracellular pool of CD45 upon T cell activation has been observed [2, 3]. Most studies on the localization of CD45 show that it is absent from membrane lipid rafts, and excluded from the central region of the interface between the T cell and the antigen-presenting cells, as well as regions containing stimulated TCR microclusters [2, 38]. Such exclusion is presumed to be due to the large size of CD45 and the relatively small size of molecules involved in antigen-specific recognition. Potential mechanisms regulating the segregation of enzyme and substrate have yet to be fully elucidated.

The function of CD45 may also be modulated through its interactions with other proteins. CD45 has been reported to associate at the cell surface with CD2, LFA-1, IFN receptor α chain, Thy-1, CD100, and CD26 [2, 3]. Moreover, compared to larger isoforms, CD45RO is found to preferentially associate with CD4 and TCR via the CD45 extracellular domain.

CD22, galectin-1, and glucosidase II can bind CD45 and other glycoproteins through specific sugar residues, although the functional consequences of these interactions are unclear. The transmembrane domain of CD45 mediates its interaction with lymphocyte phosphatase-associated phosphoprotein (LPAP) [39]. The cytoplasmic tail of CD45 is associated with the cytoskeletal protein fodrin [40].

CD45 conformation and activity may also be dynamically regulated by phosphorylation and/or oxidation of the second phosphatase domain. The acidic loop has been reported to be serine phosphorylated by casein kinase II, resulting in increased CD45 activity [41]. How this kinase itself is regulated during cellular activation, however, remains to be elucidated. Reversible oxidation has been shown to be an important mode of regulation for many phosphatases [42]. For example, oxidation-induced conformational changes have been observed in the oxidized crystals of the RPTP family members PTP1B and RPTPα. Interestingly, for the tandem phosphatase RPTPα, the second domain appears to be more sensitive to oxidation than the first, supporting the notion that the D2 domain can act as a redox sensor [42]. In this model, low concentrations of ROS may preferentially oxidize the second domain, leading to a conformational change allowing dimerization to occur, thereby facilitating inhibition of phosphatase activity. Intriguingly, the primary target of oxidation, the catalytic cysteine, is phylogenetically preserved in the second domain of CD45, despite the loss of catalytic activity. Moreover, a recent report has suggested that CD45 may be oxidized in B cells [43]. These authors showed that stimulation of B cells results in the generation of reactive oxygen species (ROS) by the NADPH oxidase DUOX1. CD45 was inhibited with kinetics similar to that of the ROS production, and was rescued by use of ROS scavengers. However, direct oxidization of CD45 itself was not demonstrated, and confirmation of this model will require further experimentation.

SYNOPSIS

While considerable strides have been made in our understanding of this highly abundant and critical phosphatase, significant questions remain. Why is it expressed at such high levels? What is the physiologic significance of the highly regulated isoform expression? How is membrane localization and substrate access regulated? Further research is clearly needed to fully elucidate the answers to these questions.

ACKNOWLEDGEMENTS

This works was supported in part by K08 CA098418-01 and the Campini Foundation (Michelle Hermiston) and PO1 AI35297-11 and the Rosalind Russel Medical Research Center for Arthritis (Arthur Weiss). The authors declare no financial conflict of interest.

REFERENCES

1. Trowbridge IS, Thomas ML. An emerging role as a protein tyrosine phosphatase required for lymphocyte activation and development. *Annu Rev Immunol* 1994;**12**:85–116.

2. Hermiston ML, Xu Z, Weiss A. A critical regulator of signaling thresholds in immune cells. *Annu Rev Immunol* 2003;**21**:107–38.

3. Holmes N. CD45: all is not yet crystal clear. *Immunology* 2006;**117**: 145–55.

4. Tchilian EZ, Beverley PC. Altered CD45 expression and disease. *Trends Immunol* 2006;**27**:146–53.

5. Andersen JN, Mortensen OH, Peters GH, Drake PG, Iversen LF, Olsen OH, Jansen PG, Andersen HS, Tonks NK, Moller NP. Structural and evolutionary relationships among protein tyrosine phosphatase domains. *Mol Cell Biol* 2001;**21**:7117–36.

6. Tonks NK. Protein tyrosine phosphatases: from genes, to function, to disease. *Nat Rev* 2006;**7**:833–46.

7. Lynch KW. Consequences of regulated pre-mRNA splicing in the immune system. *Nat Rev Immunol* 2004;**4**:931–40.

8. Kung C, Okumura M, Seavitt JR, Noll ME, White LS, Pingel JT, Thomas ML. CD45-associated protein is not essential for the regulation of antigen receptor-mediated signal transduction. *Eur J Immunol* 1999;**29**:3951–5.

9. Matsuda A, Motoya S, Kimura S, McInnis R, Maizel AL, Takeda A. Disruption of lymphocyte function and signaling in CD45-associated protein-null mice. *J Exp Med* 1998;**187**:1863–70.

10. Leitenberg D, Falahati R, Lu DD, Takeda A. CD45-associated protein promotes the response of primary CD4 T cells to low-potency T-cell receptor (TCR) stimulation and facilitates CD45 association with CD3/TCR and lck. *Immunology* 2007;**121**:545–54.

11. Desai DM, Sap J, Silvennoinen O, Schlessinger J, Weiss A. The catalytic activity of the CD45 membrane proximal phosphatase domain is required for TCR signaling and regulation. *EMBO J* 1994;**13**:4002–10.

12. Majeti R, Bilwes AM, Noel JP, Hunter T, Weiss A. Ligand-induced dimerization regulates receptor protein tyrosine phosphatase function via an inhibitory wedge. *Science* 1998;**279**:88–91.

13. Nam HJ, Poy F, Saito H, Frederick CA. Structural basis for the function and regulation of the receptor protein tyrosine phosphatase CD45. *J Exp Med* 2005;**201**:441–52.

14. Eswaran J, Debreczeni JE, Longman E, Barr AJ, Knapp S. The crystal structure of human receptor protein tyrosine phosphatase kappa phosphatase domain 1. *Protein Sci* 2006;**15**:1500–5.

15. Hoffmann KMV, Tonks NK, Barford D. The crystal structure of domain 1 of receptor protein-tyrosine phosphatase μ*. *J Biol Chem* 1997;**272**:27,505–608.

16. Huntington ND, Tarlinton DM. CD45: direct and indirect government of immune regulation. *Immunol Letts* 2004;**94**:167–74.

17. Byth KF, Conroy LA, Howlett S, Smith AJ, May J, Alexander DR, Holmes N. CD45-null transgenic mice reveal a positive regulatory role for CD45 in early thymocyte development, in the selection of CD4+CD8+ thymocytes, and B cell maturation. *J Exp Med* 1996;**183**:1707–18.

18. Kishihara K, Penninger J, Wallace VA, Kundig TM, Kawai K, Wakeham A, Timms E, Pfeffer K, Ohashi PS, Thomas ML, et al. Normal B lymphocyte development but impaired T cell maturation in CD45-Exon6 protein tyrosine phosphatase-deficient mice. *Cell* 1993;**74**: 143–56.

19. Hesslein DG, Takaki R, Hermiston ML, Weiss A, Lanier LL. Dysregulation of signaling pathways in CD45-deficient NK cells leads to differentially regulated cytotoxicity and cytokine production. *Proc Natl Acad Sci USA* 2006;**103**:7012–17.

20. Huntington ND, Xu Y, Nutt SL, Tarlinton DM. A requirement for CD45 distinguishes Ly49D-mediated cytokine and chemokine production from killing in primary natural killer cells. *J Exp Med* 2005;**201**:1421–33.

21. Palacios EH, Weiss A. Function of the Src-family kinases, Lck and Fyn, in T-cell development and activation. *Oncogene* 2004;**23**:7990–8000.

22. Lowell CA. Src-family kinases: rheostats of immune cell signaling. *Mol Immunol* 2004;**41**:631–43.

23. Seavitt JR, White LS, Murphy KM, Loh DY, Perlmutter RM, Thomas ML. Expression of the p56lck Y505F mutation in CD45-deficient mice rescues thymocyte development. *Mol Cell Biol* 1999;**19**:4200–8.

24. Roach T, Slater S, Koval M, White L, Cahir McFarland ED, Okumura M, Thomas M, Brown E. CD45 regulates Src family member kinase activity associated with macrophage integrin-mediated adhesion. *Curr Biol* 1997;**7**:408–17.

25. McNeill L, Salmond RJ, Cooper JC, Carret CK, Cassady-Cain RL, Roche-Molina M, Tandon P, Holmes N, Alexander DR. The differential regulation of Lck kinase phosphorylation sites by CD45 is critical for T cell receptor signaling responses. *Immunity* 2007;**27**:425–37.

26. Thomas ML. The regulation of antigen-receptor signaling by protein tyrosine phosphatases: a hole in the story. *Curr Opin Immunol* 1999;**11**:270–6.

27. Lynch KW, Weiss A. A model system for activation-induced alternative splicing of CD45 pre-mRNA in T cells implicates PKC and Ras. *Mol Cell Biol* 2000;**20**.

28. Melton AA, Jackson J, Wang J, Lynch KW. Combinatorial control of signal-induced exon repression by hnRNP L and PSF. *Mol Cell Biol* 2007;**27**:6972–84.

29. Zamoyska R. Why is there so much CD45 on T cells?. *Immunity* 2007;**27**:421–3.

30. Wallace VA, Penninger JM, Kishihara K, Timms E, Shahinian A, Pircher H, Kundig TM, Ohashi PS, Mak TW. Alterations in the level of CD45 surface expression affect the outcome of thymic selection. *J Immunol* 1997;**158**:3205–14.

31. Dornan S, Sebestyen Z, Gamble J, Nagy P, Bodnar A, Alldridge L, Doe S, Holmes N, Goff LK, Beverley P, et al. Differential association of CD45 isoforms with CD4 and CD8 regulates the actions of specific pools of p56lck tyrosine kinase in T cell antigen receptor signal transduction. *J Biol Chem* 2002;**277**:1912–18.

32. Xu Z, Weiss A. Negative regulation of CD45 by differential homodimerization of the alternatively spliced isoforms. *Nat Immunol* 2002;**3**:764–71.

33. Takeda A, Wu JJ, Maizel AL. Evidence for monomeric and dimeric forms of CD45 associated with a 30-kDa phosphorylated protein. *J Biol Chem* 1992;**267**:16,651–9.

34. Desai DM, Sap J, Schlessinger J, Weiss A. Ligand-mediated negative regulation of a chimeric transmembrane receptor tyrosine phosphatase. *Cell* 1993;**73**:541–54.

35. Hermiston ML, Tan AL, Gupta VA, Majeti R, Weiss A. The juxtamembrane wedge negatively regulates CD45 function in B cells. *Immunity* 2005;**23**:635–47.

36. Majeti R, Xu Z, Parslow TG, Olson JL, Daikh I, Killeen N, Weiss A. An inactivating point mutation in the inhibitory wedge of CD45 causes lymphoproliferation and autoimmunity. *Cell* 2000;**103**:1059–70.

37. Nam HJ, Poy F, Krueger NX, Saito H, Frederick CA. Crystal structure of the tandem phosphatase domains of RPTP LAR. *Cell* 1999;**97**:449–57.

38. Varma R, Campi G, Yokosuka T, Saito T, Dustin ML. T cell receptor-proximal signals are sustained in peripheral microclusters and terminated in the central supramolecular activation cluster. *Immunity* 2006;**25**: 117–27.

39. Schraven B, Schoenhaut D, Bruyns E, Koretzky G, Eckerskorn C, Wallich R, Kirchgessner H, Sakorafas P, Labkovsky B, Ratnofsky S, et al. LPAP, a novel 32-kDa phosphoprotein that interacts with CD45 in human lymphocytes. *J Biol Chem* 1994;**269**:29, 102–11.

40. Iida N, Lokeshwar VB, Bourguignon LY. Mapping the fodrin binding domain in CD45, a leukocyte membrane-associated tyrosine phosphatase. *J Biol Chem* 1994;**269**:28,576–83.

41. Wang Y, Guo W, Liang L, Esselman WJ. Phosphorylation of CD45 by casein kinase 2. Modulation of activity and mutational analysis. *J Biol Chem* 1999;**274**:7454–61.

42. Tonks NK. Redox redux: revisiting PTPs and the control of cell signaling. *Cell* 2005;**121**:667–70.

43. Singh DK, Kumar D, Siddiqui Z, Basu SK, Kumar V, Rao KV. The strength of receptor signaling is centrally controlled through a cooperative loop between Ca^{2+} and an oxidant signal. *Cell* 2005;**121**:281–93.

Chapter 18

Signal Transduction in T Lymphocytes

Rolf König

Department of Microbiology and Immunology, The University of Texas Medical Branch, Galveston, Texas

INTRODUCTION

The immune system provides a highly sophisticated surveillance mechanism to detect diverse antigens and protect the host organism from invading pathogens and altered cells (e.g., virus-infected and tumor cells). Adaptive immune responses depend on the recognition of antigen by specific antigen receptors that are expressed on the surface of T and B lymphocytes. To initiate effector mechanisms, binding of the antigen must induce intracellular signaling cascades that activate the lymphocyte and promote their differentiation to an effector cell type appropriate for the particular antigenic challenge. Importantly, regulatory mechanisms must also be present to safeguard against inadvertent self-reactivity, which could lead to autoimmunity, and to terminate immune responses, thereby avoiding overexposure of the organism to toxic effectors (e.g., cytotoxic T cells, cytokines).

T lymphocytes are derived from the lymphoid lineage of hematopoietic stem cells. T cell progenitors enter the thymus, where they develop into mature T lymphocytes. During thymic development, the immature thymocytes undergo rearrangement of first β and then α T cell receptor (TCR) genes [1, 2]. This process increases the diversity of available TCRs, and assures that each T cell expresses only a single type of TCR. Only those thymocytes that have successfully completed TCR gene rearrangement will be allowed to survive; unsuccessful rearrangements lead to programmed cell death. Following rearrangement, the functionality of the maturing thymocytes is tested by interactions with thymic antigen-presenting cells. A distinct lineage of T cells express $\gamma\delta$ TCR genes. Most of the information on cellular signaling in T cells has been gathered in $\alpha\beta$ T cells, and thus we focus on this lineage.

Thymocytes undergo positive selection to ensure that their TCR can engage epitopes formed by short peptides bound to molecules encoded by the major histocompatibility complex (MHC). Because of the large isotypic and allelic variability of MHC molecules within each species, not all recombination events of TCR genes lead to matches with a peptide–MHC complex potentially present in the individual organism. Two classes of MHC molecules select T cells with different functions. MHC class I molecules bind peptides derived from proteins synthesized by the presenting cell and proteolytically processed by the cell's proteosome. These peptides can combine with nascent MHC class I molecules in the endoplasmic reticulum [3]. MHC class II molecules bind peptides derived from extracellular, endocytosed proteins that are processed in lysosomes. Nascent and recycling MHC class II molecules bind these peptides while trafficking through lysosomes [4]. These same antigen processing pathways are also responsible for presenting antigenic peptides to mature T cells, and thus initiate effector functions. Naïve T cells, those T cells that have not been stimulated subsequent to thymic egress, require co-stimulation, which can only be given by specific antigen-specific cells, such as dendritic cells. Thus, the clear dichotomy between the MHC class I and MHC class II pathways, which stimulate CD8+ cytotoxic and CD4+ helper T cells, respectively, posed a problem in explaining how T cells could respond to viruses that do not infect antigen-specific cells directly or to tumor antigens that are not expressed by antigen-specific cells. The recent discovery that mannose receptor-mediated endocytosis provides an avenue for the presentation of extracellular antigens via MHC class I on dendritic cells demonstrates a constitutive mechanism whereby T cells can be stimulated by antigens that are present in non-lymphoid tissues [5].

Positive selection of thymocytes requires that the interaction between TCR and peptide–MHC complex induces a signal of sufficient strength and duration. Thymocytes that do not receive a TCR-mediated signal during positive selection die. Positive selection occurs on thymic stromal epithelial cells, which do not present a full complement of all possible antigens that the T cell might encounter during its lifetime. Thus, selected T cells have partial self-reactivity and the potential to recognize antigens that will only be encountered later in life. Maturing thymocytes express both CD4 and CD8 co-receptors. Following positive selection, the CD8 coreceptor gene is silenced. If the selection signal persists despite the absence of *Cd8* gene expression, the intermediate thymocytes differentiate into CD4+ T cells. If, however, the selection signal ceases upon *Cd8* gene silencing, the CD4+CD8-intermediate thymocytes re-initiate *Cd8* gene expression and enter the CD8+ T cell lineage [6, 7].

A separate selection step is a negative test to eliminate thymocytes that overtly respond to self-antigens. Of all the thymocytes that mature up to the double-positive (CD4+CD8+) stage, only about 3 percent complete maturation and emigrate from the thymus to peripheral lymphoid tissues (e.g., blood, lymph nodes, and spleen). Negative selection is mediated by thymic stromal cells or antigen-specific cells (mainly dendritic cells) that migrate from peripheral organs into the thymus. These cells provide a large battery, but not a complete complement, of different peptides derived from self and foreign proteins. Following interactions of TCRs with self-peptides in the thymus, cellular activation thresholds are dynamically tuned by modulating co-stimulatory and other molecules that modify cellular signaling in developing thymocytes to establish cellular activation thresholds that prevent reactivity to self [8]. Thymocytes that continue to respond vigorously to stimuli presented during the negative selection phase undergo apoptosis [9].

This chapter provides a broad overview of T cell signaling, with emphasis on the signaling molecules and pathways important for antigen-induced signal transduction and elicitation of effector functions. General principles are discussed, and open questions and areas of current research are highlighted.

SIGNALING RECEPTORS IN T CELLS FORM DYNAMIC MACROMOLECULAR SIGNALING COMPLEXES

Antigen-specific T Cell Receptors

Most mature T cells express αβ TCRs, but some T cells, especially those in mucosal tissues, express a γδ TCR. The antigen-specific TCR is a disulfide-linked heterodimer that does not have intrinsic signaling capability. The ability to transduce intracellular signals is conveyed to the TCR via its mandatory and constitutive association with a multiprotein structure, termed the CD3-ζζ complex. Neither of the individual components of the TCR/CD3-ζζ signaling machine can be transported to the cell surface without the full assembly of the complex [10].

The CD3 complex is composed of four transmembrane polypeptide chains, a δε and a γε heterodimer. These proteins have very short extracellular domains and each intracellular domain contains a conserved protein tyrosine kinase (PTK) recognition motif, termed Immunoreceptor Tyrosine-based Activation Motif (ITAM). The disulfide-linked ζζ-dimer contains three ITAMs per protein chain [11, 12]. The characteristic sequence motif of an ITAM is $(D/E)xxYxx(L/I) x_{6-8} Yxx(L/I)$, with x indicating variable amino acid residues. ITAMs are substrates for *src* family PTKs, and phosphorylation of the paired tyrosines within ITAMs is a determining initiation event for T cell signaling [11, 12] (Figure 18.1).

TCR Engagement and the Formation of Signalosomes

Binding of a peptide-MHC complex to the TCR triggers the recruitment and trans-autocatalytic activation of *src* family PTKs, such as p56^{lck} and p59^{fyn}. These *src* family PTKs rapidly phosphorylate the paired tyrosines within the ITAMs of the ζζ dimer and the CD3 subunits, thus generating binding sites for the cytosolic *syk* family PTK ζ chain-associated protein of 70kDa (ZAP-70), which contains tandem Src homology 2 (SH2) domains [13]. Recruitment of ZAP-70 to the developing TCR signaling machinery allows phosphorylation of two pairs of tyrosines by p56^{lck} (Y_{315} and Y_{319}) and by trans-autophosphorylation (Y_{492} and Y_{493}), respectively [14]. ZAP-70 in turn phosphorylates components of distinct downstream signaling pathways [15, 16]. Thus, T cell activation depends on activation of both *src* family kinases and ZAP-70.

Before we follow the signal transduction cascade further, it is worthwhile to briefly discuss the complex temporal and spatial arrangement of signaling complexes and networks activated by TCR engagement. Slightly different compositions of these signaling machines – also termed "signalosomes" [17, 18] – using essentially the same components can induce different second messenger signals and lead to drastically diverse cellular responses. In addition, the dynamic assembly and disassembly of signalosomes is likely a major factor in regulating signal transduction networks. TCR signalosomes consist of transmembrane receptors, protein kinases, phosphatases and their substrates, all of which are organized into signaling machines by anchoring, adapter, and scaffolding proteins. Signalosomes connect events on the plasma membrane to distal signaling cascades, which ultimately modulate T cell biology.

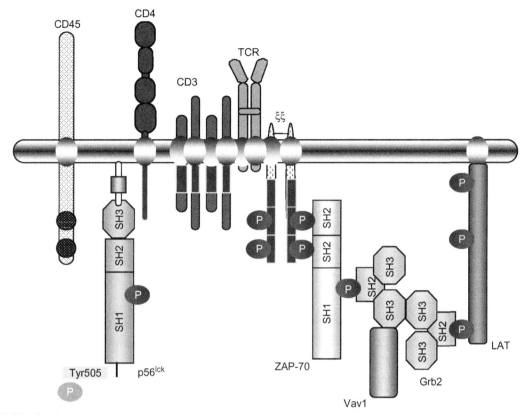

FIGURE 18.1 The signalosome.
The initial signalosome consists of the TCR, the associated CD3 and ζζ complex, the co-receptor (CD4 or CD8), the phosphatase CD45, and the *srk* kinase p56*[lck]*. The *syk* kinase ZAP-70 is recruited following phosphorylation of tyrosines located in the ITAMs of CD3 and ζζ. Adapters, such as LAT, Grb2, and Vav1 then associate with the signalsome and provide an expanding scaffold for downstream signaling molecules.

Several protein adapters, in particular Linker of Activated T cells (LAT), act as central switches that translate the quality, quantity, and duration of signals into the correct activation of specific downstream pathways [19].

Formation of signalosomes is aided by compartmentalization of the plasma membrane into detergent-insoluble, sphingolipid/cholesterol-enriched microdomains, which promote the recruitment of signal transduction molecules to the TCR signaling machine upon TCR engagement [20, 21]. These areas of the T cell surface are also known as lipid "rafts." Palmitoylation constitutively embeds several components, such as Lck and Fyn, into these lipid microdomains, whereas others, such as ZAP-70, relocalize into rafts upon TCR engagement [22–24].

While it is generally acknowledged that ITAM phosphorylation by Lck and recruitment of ZAP-70 to the T cell signaling machinery constitute crucial events for TCR-mediated signal transduction, the mechanism that causes the initial trigger is controversial [25]. Competing models that are not mutually exclusive suggest that binding of peptide–MHC induces the aggregation of TCR/CD3 complexes, or changes the conformation of TCR/CD3 complexes or the orientation of the TCR in relation to the cell membrane [26], or induces segregation and redistribution of TCR/CD3 complexes in relation to other cell membrane proteins [27].

Adapter Proteins and Macromolecular Scaffolds

Returning to the events following TCR engagement, it is now clear that the relocalization of signalosomes to receptor-associated scaffolds is crucial for effective signal transduction [28–30]. Adapter proteins with SH2 domains bind to the phosphorylated ζ chain. Among these proteins is the Src homology 2 protein of beta-cells (Shb), which recruits LAT via its central, phosphotyrosine-binding (PTB) domain-like motif. LAT contains nine tyrosine phosphorylation sites, and is a substrate for ZAP-70. Tyrosine phosphorylation of LAT leads to the recruitment of additional signaling molecules with SH2 motifs, including the adapters growth factor receptor-bound protein 2 (Grb2) and Gads, the GDP/GTP exchange factor Vav, the phospholipase Cγ1 (PLCγ1), and the p85 subunit of phosphatidylinositol 3-kinase (PI3K) [15, 31, 32] (Figures 18.1, 18.2).

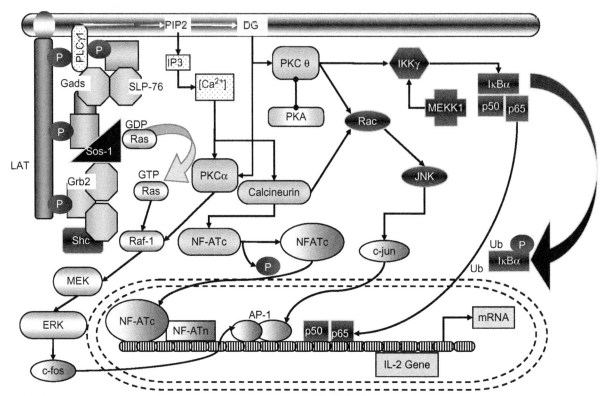

FIGURE 18.2 Intracellular signaling pathways induced by stimulation of the TCR/CD3–ζζ complex mediate the translocation of the transcription factors NFAT, c-jun, and c-fos (AP-1), and NFκB (p50/p65 dimer) to the nucleus, where they induce expression of IL-2 gene and other genes. For details, see the text in this chapter. For detailed description of the signaling pathways, see the relevant chapters.

Gads is associated with the SH2 domain containing Leukocyte Protein of 76 kDa (SLP-76), and together they form a macromolecular scaffold, which stabilizes the interaction of PLCγ1 with the TCR signalosome [32]. SLP-76 is essential in the activation of PLCγ1 and downstream signaling [33]. PLCγ1 is tyrosine phosphorylated in antigen-activated T cells, an event required for its activation. Active PLCγ1 hydrolyzes phosphotidylinositol biphosphate (PIP$_2$), producing diacylglycerol (DAG) and 1,4,5-inositol triphosphate (IP$_3$). DAG in turn activates the serine/threonine kinase family of protein kinase C (PKC), while IP$_3$ induces calcium (Ca^{2+}) mobilization in the cytosol. Thus, ZAP-70 amplifies the TCR signal by specifically phosphorylating downstream components such as LAT [34, 35] and PLCγ1 [33].

In this way, the signalosome expands in molecular complexity and amplifies the TCR initiated signal. Importantly, LAT can also bind proteins that negatively regulate TCR signaling. The SH2 domain-containing Hematopoietic Phosphotyrosine phosphatase, SHP-1, associates with LAT upon TCR stimulation and prevents further phosphorylation of the adapter by ZAP-70, suggesting a potential conversion from an "activating" to an "inhibiting" signalosome as a means for "quality control" of the stimulation event [36]. Similarly, the C-terminal *src* kinase (Csk) relocalizes

to rafts by docking to the transmembrane adapter, Csk-binding protein (Cbp), also known as Phosphoprotein Associated with Glycosphingolipid-enriched microdomains (PAG) [37]. In rafts, Csk inhibits *src* family PTKs by phosphorylating their regulatory tyrosines, and thus blocks TCR-mediated signal transduction [38].

CO-RECEPTOR AND CO-STIMULATORY PROTEINS MODULATE T CELL SIGNALING PATHWAYS

Before discussing the second messenger signals induced by TCR engagement, we need to introduce two sets of T cell surface proteins that crucially modulate TCR/CD3-ζζ-mediated signals: the co-receptors and the co-stimulators.

Co-receptors, TCRs, and the Formation of Signalosomes

Co-receptors are associated with the TCR/CD3-ζζ complex upon T cell activation. Their presence in the TCR multicomponent signaling machine amplifies or modulates the activation signal. Often, their presence is absolutely required, but not sufficient, for productive signaling – that is, signaling that results in cell cycle progression and effector functions.

CD4 and CD8

CD4 and CD8 are membrane glycoproteins associated with MHC class II and class I restriction, respectively. In mature T cells, their expression is mutually exclusive. In general, CD4+ T helper (Th) cells respond to antigenic peptides presented by MHC class II molecules, and CD8+ cytotoxic T (Tc) cells respond to MHC class I-presented antigens. CD4 and CD8 have been thought of as molecules that enhance the stability of the tripartite complex between TCR and peptide–MHC. However, in recent years it has become clear that their function is more complex and dynamic. For example, TCR/peptide–MHC interactions initiate the formation of a specialized junction between T cells and antigen-specific cells, the immunological synapse. Stimulation- and cytoskeleton-dependent processes cluster TCR/CD3 complexes in the center of the synapse, also termed the central zone of the SupraMolecular Activation Cluster (cSMAC), whereas adhesion molecules such as LFA-1 form a ring surrounding the central area called the peripheral SMAC (pSMAC) [39, 40]. After stimulation of the T cell, CD4 co-receptors are rapidly recruited into the cSMAC, but migrate towards the periphery within a few minutes, while TCR/CD3 complexes stabilize within the central area [41]. Both CD4 and CD8 associate with the PTK p56^{lck}, and the efficient transport of p56^{lck} into the cSMAC is a major function of these co-receptors [42]. In addition, co-receptor interactions with MHC molecules regulate peripheral T cell homeostasis and the survival of naïve T cells in the absence of antigenic stimulation [43]. The observation that CD4 can induce signals independent of the TCR suggests complex regulatory effects of co-receptors on T cell function [44, 45].

CD5

The CD5 lymphocyte glycoprotein is expressed on thymocytes and all mature T cells. CD5 can act as a co-stimulatory molecule for resting T cells by augmenting CD3-mediated signaling [46]. In mature, peripheral T cells, CD5 is present in lipid rafts of the T cell surface, where it promotes CD3 redistribution into rafts, thus markedly upregulating ZAP-70 and LAT activation, and Ca^{2+} influx [47]. However, CD5 is also constitutively associated with SHP-1, an interaction that increases upon TCR stimulation and negatively regulates TCR-mediated activation [48]. The differential modulatory properties of CD5 depend on the context of lymphocyte subset and their differentiation stage. A recent review discusses contradictory and complementary reports of CD5-mediated molecular intracellular signaling events [49].

CD45

CD45 is a membrane-bound tyrosine phosphatase present on hematopoietic cells. CD45 phosphatase activity is absolutely required for thymic T cell development and activation of mature T cells. CD45 dephosphorylates the negative regulatory site tyrosine-505 on the *scr* kinase p56^{lck}. However, it is much less effective in dephosphorylating the activating tyrosine-395 on p56^{lck}. Multiple exons and differential glycosylation allow the expression of different CD45 isoforms in cell- and development-specific fashion. In peripheral, human CD4+ T cells, the naïve subset (i.e., cells that have not been stimulated by antigen after thymic selection) expresses the high molecular weight form, CD45RA, whereas activated and memory CD4+ T cells express the low molecular weight form, CD45RO [50]. Importantly, TCR-dependent intracellular signaling events differ in relation to CD45 isoform expression [51]. In a rheostat-like manner, CD45 activity differentially regulates phosphorylation of p56^{lck}, thereby determining the strength of the TCR-induced signal [52]. The distribution of CD45 in activated T cells is also highly regulated in that CD45 is completely excluded from the immunological synapse. The presence of CD45 in cSMACs prevents the continuation of the TCR-mediated signal [53]. This is in contrast to the distribution of CD4 and CD8 co-receptors, which migrate into the cSMAC before relocating to the pSMAC.

Multiple Functions of Co-receptors

Views on co-receptor function have evolved with available technologies. The traditional definition identifies co-receptors as proteins that associate with the TCR upon T cell stimulation, and stabilize the TCR's interactions with its ligands. However, the ability to identify PTKs and to measure their activities has demonstrated associations between p56^{lck} and CD4 or CD8, inspiring the realization that these coreceptors enhance and regulate TCR functions by transporting Lck into the TCR/CD3-ζζ complex. Similarly, the recently identified co-receptor, CD160/BY55, which is expressed by most intestinal intraepithelial lymphocytes and by a minor subset of circulating lymphocytes including NK, γδTCR, and cytotoxic effector CD8+ T lymphocytes, associates with Lck and tyrosine phosphorylated ζζ upon TCR/CD3 cell activation [54]. In addition, CD5 is associated with SHP-1 upon T cell activation [48], and CD45 contains intrinsic phosphatase activity. Thus, physiological responses to TCR engagement at the contact site between T cells and antigen-specific cells are the result of localized and finely-tuned alterations in the balance between cellular kinases and phosphatases [55].

The ability of co-receptors to transduce signals independent of TCR stimulation suggests that they are not merely scaffolding or transport proteins, but also exert complex regulatory effects on T cell activation [44, 45].

Functionally, the distinction between co-receptors and co-stimulators blurs, and the distinguishing feature of co-receptors may be their temporal association with the TCR/CD3- $\zeta\zeta$ complex.

Co-stimulatory Receptors and TCR-mediated Activation

The concept of T cell co-stimulation was born out of the necessity to explain the phenomenon of self-tolerance within the concept of Burnet's clonal selection theory. Because not all self-proteins traffic to the thymus or are expressed at all stages of development, central tolerance induced by negative selection of auto-reactive thymocytes in the thymus cannot fully explain the unresponsiveness of mature T cells to peripheral self-antigens. Therefore, several investigators introduced the two-signal model of activation [56–58]. The model postulates that any naïve lymphocyte stimulated via engagement of only the antigen receptor will enter a state of anergy, which is characterized by unresponsiveness to future antigen-mediated stimulation. Only if a second, co-stimulatory signal is given during antigen stimulation is the lymphocyte fully activated to display effector functions and proliferate. Co-stimulatory signals for B cells and CD8+ cytotoxic T cells are mainly provided by activated CD4+ T helper cells, whereas a variety of surface molecules expressed on hematopoietic antigen-specific cells (e.g., dendritic cells, activated macrophages, activated B cells) can induce co-stimulation in T helper cells.

Co-stimulatory receptors do not have to associate with the TCR/CD3 complex to exert their function. They can transduce signals independent of the TCR/CD3. Originally, they were thought to only induce cellular effector functions in combination with TCR-transduced signals. While this paradigm has been challenged, recent experiments suggest that CD28-mediated signaling events depend on intact TCR and ZAP-70 activity [59]. Some of the co-receptors discussed in the previous chapter also have co-stimulatory functions.

CD28

Interactions of CD28 with its ligand B7 on antigen-specific cells activate the CD28-responsive element (CD28RE), contained within the interleukin-2 (IL-2) gene promoter, and thus promote induction of IL-2 gene expression. Interaction of CD28 with three intracellular proteins – PI3K, the T cell-specific Tec-family kinase Itk [60], and the complex between Grb2 and the guanine nucleotide exchange protein Son of sevenless (SOS) activates the serine/threonine kinase MAPK/ERK kinase kinase (MEKK1), which contributes to full activation of the CD28RE and regulates nuclear translocation of the transcription factors NFκB and AP-1 [61, 62].

The CD28 signal also amplifies activation of PLCγ1 and mobilization of Ca^{2+}. Further, CD28 engagement activates $p56^{lck}$ molecules that are phosphorylated at the regulatory tyrosine residue 505, generally considered to be an inactive form of $p56^{lck}$ [42]. Thus, CD28 amplifies signals from the TCR that would otherwise be too weak for T cell activation. Interestingly, the PTKs $p56^{lck}$ and $p59^{fyn}$ phosphorylate CD28, and Itk binding to CD28 is dependent on the presence of $p56^{lck}$. Thus, $p56^{lck}$ is likely to be a central switch in T cell activation, with the dual function of regulating CD28-mediated costimulation as well as TCR/CD3/CD4 signaling.

Because CD28 provides co-signals in T cell responses, a key question is whether the CD28 operates exclusively via TCR/CD3-$\zeta\zeta$ or also operates as an independent signaling unit. Recent data show that mitogenic CD28 signals depend on the expression of a functional TCR and intact ZAP-70 kinase activity. However, these CD28 signals do not phosphorylate either $\zeta\zeta$ or ZAP-70 [59]. The mitogenic CD28 signaling pathway further depends on binding of Grb2 to CD28, association with phosphorylated SLP-76, and recruitment of Vav1. SLP-76 cannot be phosphorylated independent of the TCR. The most likely pathway involves suboptimal phosphorylation of SLP-76 by ZAP-70, followed by binding of CD28 to Grb2 in the signalosome, and CD28-mediated recruitment of Vav1 and Itk. Itk then completes the phosphorylation of SLP-76 [59]. Thus, the CD28 pathway feeds into the TCR signaling pathway to amplify signals that lead to Ca^{2+} flux, IL-2 production, and proliferation.

Importantly, engagement of the TCR alone may result in an anergic state or T cell deletion, both of which can induce tolerance to antigen stimulation. Insight into the regulation of CD28 dependency comes from genetic experiments. T cells that are deficient in the adapter molecule Cbl-b do not require CD28 engagement for IL-2 production. Also, whereas B cells responding to T cell-dependent antigens cannot undergo isotype switching from IgM to IgG in CD28-deficient mice, T cell help is fully restored in CD28/Cbl-b double-deficient mice [63]. The function of Cbl-b is to selectively suppress TCR-mediated Vav activation, thus rendering T cells dependent on CD28 co-stimulation [63].

CD40L

The interaction of the tumor necrosis factor (TNF) family member CD40L on activated T cells with its receptor, the TNF receptor family member, CD40, which is expressed on macrophages, dendritic cells, and activated B cells, provides a strong signal for IL-12 production [64]. An important aspect of CD40-CD40L signaling is its synergistic relationship with the CD28-B7 signal. CD40L cell surface expression is upregulated by CD28 signaling. The subsequent interaction between CD40L and CD40 induces B7 upregulation on antigen-specific cells, enhancing the co-stimulatory activity of macrophages, dendritic cells, and B cells [65].

INTRACELLULAR SIGNALING PATHWAYS INDUCED BY ANTIGEN STIMULATION OF T CELLS

Calcium Mobilization

The TCR-induced signal transduction leads to activation of PLCγ1, which hydrolyzes PIP$_2$ to DAG and IP$_3$. Binding of IP$_3$ to its receptor in the endoplasmic reticulum membrane induces the release of Ca^{2+} into the cytosol. The subsequent increase in intracellular free Ca^{2+} opens Ca^{2+}-regulated Ca^{2+} channels in the plasma membrane, inducing additional Ca^{2+} influx. Intracellular free Ca^{2+} acts as an essential second messenger for T cell activation. Its regulatory effects on T cell activation are mediated via calmodulin, a Ca^{2+}-binding protein expressed in all eukaryotic cells. Effective T cell activation leading to IL-2 secretion requires that intracellular Ca^{2+} levels be elevated for a period of 1–2 hours. Sustained Ca^{2+} signaling is required for maintaining the transcription factor Nuclear Factor of Activated T cells (NFAT) in the nucleus in an active form. NFAT is a key transcriptional regulator of the IL-2 gene and other cytokine genes (Figure 18.2).

Ca^{2+} signaling is required for various lymphocyte activities – for example, cell mobility, change of cytoskeletal structure, cell death, differentiation, and activation. Thus, a single second messenger can elicit multiple cellular responses. The type of response induced may depend on the amplitude, duration, and temporal fluctuations of Ca^{2+} mobilization. For example, activation of NFκB is induced by high levels of Ca^{2+}, because of this transcription factor's low Ca^{2+} sensitivity. In contrast, long-lasting, low levels of Ca^{2+} selectively activate NFAT, because NFAT is highly sensitive to Ca^{2+}, but is rapidly inactivated after Ca^{2+} removal [66].

PKC

Release of DAG stimulates PKC, a family of serine/threonine kinases. In T cells, multiple PKC isoforms are expressed [67]. One of the major functions of PKC is to induce MAPKs. PKCα directly phosphorylates and activates Raf-1, another serine/threonine kinase. Activation of Raf-1 triggers a protein kinase cascade by directly phosphorylating MAPK kinase. Activation of PKC also mediates the rapid accumulation of the active, GTP-bound form of p21^{ras}. The Ca^{2+}-independent isoform PKCθ is required for the activation of mature T cells and regulates the IκB kinase (IKK)–NFκB pathway by facilitating the phosphorylation of IKKγ and subsequent degradation of IκBα and nuclear translocation of NFκB [67, 68].

Transcription Factors: Activation and Gene Expression

The multiple signaling pathways originating from T cell surface molecules initiate the expression of genes responsible for proliferation and immune functions in a cooperative manner. The most extensively studied example is the induction of the IL-2 gene. The IL-2 gene promoter contains at least seven distinct binding sites for transcription factors. Therefore, maximal transcription demands the simultaneous presence of all factors, among which AP-1, NFAT, and NFκB are the best characterized.

AP-1 is a heterodimer of c-Fos and c-Jun. Maximum activation of AP-1 requires *de novo* synthesis of c-Jun and c-Fos, and phosphorylation by MAPKs of the activation domains of both proteins, leading to translocation into the nucleus.

Activation of NFAT requires dephosphorylation by the serine/threonine protein phosphatase, calcineurin, followed by translocation into the nucleus. In the nucleus, NFAT cooperates with AP-1 in gene transactivation, resulting in a 20-fold increase in the stability of NFAT/AP-1/DNA complexes as compared with NFAT/DNA complexes.

NFκB is a homodimer or heterodimer of a family of structurally related proteins. Each member of this family contains a conserved N-terminal Rel-homology domain (RHD), which mediates dimerization and binding to DNA. The RHD contains a nuclear localization sequence that promotes NFκB translocation to the nucleus following release of NFκB from IκB. In its inactive form, NFκB is sequestered in the cytosol by non-covalent interactions with the inhibitory protein, IκB, which masks the nuclear translocation signal. Phosphorylation of IκB by IKK targets IκB for destruction by the ubiquitin-protease system. NFκB induces IL-2 gene expression, and in a negative feedback loop, promotes transcription of the IκB gene.

Role of Cyclic AMP in T Cell Activation

Cyclic AMP, Adenylyl Cyclases, and Phosphodiesterases

Cyclic AMP (cAMP) is an intracellular second messenger to a wide variety of hormones and neurotransmitters. In T cells, elevated cAMP levels antagonize T cell activation by inhibiting T cell proliferation and by suppressing the production of IL-2 and IFN-γ. TCR signaling in the absence of CD28 co-stimulation also elevates cAMP levels and adenylyl cyclase (AC) activity. The regulation of cAMP during T cell activation is mediated via other T cell surface molecules, such as CD28 and CD44 [69, 70], as well as via the co-receptor CD4 [44].

In T cells, the cAMP level is controlled by two types of enzymes: ACs and phosphodiesterases (PDE). ACs catalyze the production of cAMP from ATP, whereas PDEs control the rate of cAMP degradation to AMP. Members of the PDE families 1, 3, 4, and 7 are expressed in T cells. Different mechanisms control the activity of the different families of PDEs: Ca^{2+}/calmodulin stimulates PDE1; cyclic GMP inhibits PDE 3; p70S6 kinase and the MAPK pathway activate PDE4; and CD28-mediated signals activate PDE7 [70]. Importantly, TCR/CD28 stimulation of human T cells transiently upregulates AC and PDE activities, with different kinetics for different PDE isozymes [71]. Thus, an initial increase, followed by a rapid decrease, in intracellular cAMP is required for T cell activation, suggesting a precise kinetic regulation of cAMP production and degradation.

cAMP-Dependent Kinase

The cAMP-dependent protein kinase (PKA) is the principal intracellular cAMP receptor. In the absence of cAMP, PKA is an enzymatically inactive, tetrameric holoenzyme, consisting of two catalytic (C) subunits and two regulatory (R) subunits. The cooperative binding of four cAMP molecules to two sites on each R subunit drastically decreases the binding affinity between R and C subunits, and induces dissociation into dimeric R and two monomers of C subunits. Once freed from the R subunits, the C subunits display serine/threonine kinase activity.

PKA I, but not PKA II, mediates the inhibitory role of cAMP on T cell proliferation induced by TCR signaling by activating Csk to inhibit Lck activity [38]. Thus, PKA I can diminish T cell activation at the initiation stage. PKA I also phosphorylates Ser-43 of Raf-1 to block the MAP kinase pathway [72]. In the nucleus, activation of PKA prevents stable protein–DNA interactions at the NFκB, NFAT, and AP-1 binding sites of the IL-2 enhancer [73]. In addition, PKA I activity also inhibits cyclin D3 expression and induces the cyclin-dependent kinase inhibitor p27^{kip1} [74]. For T cells to enter the S phase of the cell cycle, D-type cylins, including cyclin D3, are synthesized during the G_1 phase [75]. These cyclins can bind to cyclin-dependent kinase (Cdk) and form an active kinase complex that phosphorylates and inactivates retinoblastoma protein (pRb). Inactivation of pRb then allows cells to pass through the late G1 phase restriction point and enter the S phase. However, cyclin D/Cdk complexes can associate with the Cdk inhibitor p27^{kip1}, thus be rendered inactive. Therefore, in addition to induction of cyclin D, downregulation of p27^{kip1} is also required for the initiation of T cell proliferation [75]. Hence, inhibition of cyclin D3 expression and induction of p27^{kip1} by PKA I both block T cell cycle progression. This suggests that PKA I activity must be regulated at all stages during T cell activation.

Interestingly, the majority of TCR-induced cAMP is generated in lipid rafts [76]. In a CD4-dependent fashion, both PDE1 and PDE4 are activated and counteract the production of cAMP [44]. CD28 co-stimulation mediates the recruitment of PDE4 activity to lipid rafts [76, 77]. In addition to regulating the level of cytosolic cAMP, T cells redistribute PKA I within minutes after TCR-stimulation to the distal cell pole opposite the aggregated TCR/CD3 complexes, thereby spatially distancing PKA I from cAMP by micro-compartmentalization [78].

CONCLUSIONS

Signal transduction research in T lymphocytes has focused on identifying the receptors and intracellular proteins affecting specific signaling pathways. It has become clear that signaling machines in T cells differ in composition depending on the extracellular signal received and the requirements of the stimulated cell. The composition and activation state of an established signalosome can also change over time in order to fine-tune or alter effector signaling pathways. For example, a combination of computer modeling of early T cell activation events and quantitative cell-based experiments revealed a rapid SHP-1-mediated negative feedback loop followed by a slower, but much more sensitive, positive feedback loop mediated by ERK [79, 80]. In addition, subcellular compartmentalization of signaling molecules is an important mechanism for regulating cellular signal transduction. For example, using a biosensor for ZAP-70 activity demonstrated an unexpected bipolar distribution of ZAP-70 activity upon T cell stimulation both at the immunological synapse and at the distal cell pole [81]. In the future, it will be important to determine the kinetics of assembly, the dynamics of composition, and the compartmentalization for all signalosomes induced in T cells, as well as to define the interactions between different signaling pathways. Future research will increasingly utilize proteomics [82], live-cell video imaging, and computer modeling to characterize the signalome – the entire complement of a T cell's signaling molecules and their interactions in a temporal and spatial context.

REFERENCES

1. Niederberger N, Holmberg K, Alam S, et al. Allelic exclusion of the TCR alpha-chain is an active process requiring TCR-mediated signaling and c-Cbl. *J Immunol* 2003;**170**:4557–63.

2. Agata Y, Tamaki N, Sakamoto S, et al. Regulation of T cell receptor beta gene rearrangements and allelic exclusion by the helix–loop–helix protein, E47. *Immunity* 2007;**27**:871–84.

3. Cresswell P, Ackerman A, Giodini A, Peaper D, Wearsch P. Mechanisms of MHC class I-restricted antigen processing and cross-presentation. *Immunol Rev* 2005;**207**:145–57.

4. Pieters J. MHC class II-restricted antigen processing and presentation. *Adv Immunol* 2000;**75**:159–208.

5. Burgdorf S, Kautz A, Böhnert V, Knolle P, Kurts C. Distinct pathways of antigen uptake and intracellular routing in CD4 and CD8 T cell activation. *Science* 2007;**316**:612–6.

6. Liu X, Bosselut R. Duration of TCR signaling controls CD4–CD8 lineage differentiation *in vivo*. *Nat Immunol* 2004;**5**:280–8.

7. Erman B, Alag A, Dahle O, et al. Coreceptor signal strength regulates positive selection but does not determine CD4/CD8 lineage choice in a physiologic *in vivo* model. *J Immunol* 2006;**177**:6613–25.

8. Nambiar M, Enyedy E, Warke V, et al. T cell signaling abnormalities in systemic lupus erythematosus are associated with increased mutations/polymorphisms and splice variants of T cell receptor zeta chain messenger RNA. *Arthritis Rheum* 2001;**44**:1336–50.

9. Palmer E. Negative selection– clearing out the bad apples from the T-cell repertoire. *Nat Rev Immunol* 2003;**3**:383–91.

10. Delgado P, Alarcón B. An orderly inactivation of intracellular retention signals controls surface expression of the T cell antigen receptor. *J Exp Med* 2005;**201**:555–66.

11. Pitcher L, van Oers N. T-cell receptor signal transmission: who gives an ITAM? *Trends Immunol* 2003;**24**:554–60.

12. Amon M, Manolios N. Hypothesis: TCR signal transduction – a novel tri-modular signaling system. *Mol Immunol* 2008;**45**:876–80.

13. Deindl S, Kadlecek T, Brdicka T, Cao X, Weiss A, Kuriyan J. Structural basis for the inhibition of tyrosine kinase activity of ZAP-70. *Cell* 2007;**129**:735–46.

14. Brdicka T, Kadlecek T, Roose J, Pastuszak A, Weiss A. Intramolecular regulatory switch in ZAP-70: analogy with receptor tyrosine kinases. *Mol Cell Biol* 2005;**25**:4924–33.

15. Horejsí V, Zhang W, Schraven B. Transmembrane adaptor proteins: organizers of immunoreceptor signalling. *Nature Rev Immunol* 2004;**4**:603–16.

16. Salvador J, Mittelstadt P, Guszczynski T, et al. Alternative p38 activation pathway mediated by T cell receptor-proximal tyrosine kinases. *Nat Immunol* 2005;**6**:390–5.

17. Werlen G, Hausmann B, Palmer E. A motif in the alphabeta T-cell receptor controls positive selection by modulating ERK activity. *Nature* 2000;**406**:422–6.

18. Werlen G, Palmer E. The T-cell receptor signalosome: a dynamic structure with expanding complexity. *Curr Opin Immunol* 2002;**14**:299–305.

19. Malissen B, Aguado E, Malissen M. Role of the LAT adaptor in T-cell development and Th2 differentiation. *Adv Immunol* 2005;**87**:1–25.

20. He H, Lellouch A, Marguet D. Lipid rafts and the initiation of T cell receptor signaling. *Semin Immunol* 2005;**17**:23–33.

21. Magee A, Adler J, Parmryd I. Cold-induced coalescence of T-cell plasma membrane microdomains activates signalling pathways. *J Cell Sci* 2005;**118**:3141–51.

22. Kabouridis P. Lipid rafts in T cell receptor signalling. *Mol Membr Biol* 2006;**23**:49–57.

23. Harder T, Rentero C, Zech T, Gaus K. Plasma membrane segregation during T cell activation: probing the order of domains. *Curr Opin Immunol* 2007;**19**:470–5.

24. Jury E, Flores-Borja F, Kabouridis P. Lipid rafts in T cell signalling and disease. *Semin Cell Dev Biol* 2007;**18**:608–15.

25. Choudhuri K, van der Merwe P. Molecular mechanisms involved in T cell receptor triggering. *Semin Immunol* 2007;**19**:255–61.

26. Kuhns M, Davis M, Garcia K. Deconstructing the form and function of the TCR/CD3 complex. *Immunity* 2006;**24**:133–9.

27. Burroughs N, van der Merwe P. Stochasticity and spatial heterogeneity in T-cell activation. *Immunol Rev* 2007;**216**:69–80.

28. Harder T, Kuhn M. Selective accumulation of raft-associated membrane protein LAT in T cell receptor signaling assemblies. *J Cell Biol* 2000;**151**:199–208.

29. Yokosuka T, Sakata-Sogawa K, Kobayashi W, et al. Newly generated T cell receptor microclusters initiate and sustain T cell activation by recruitment of Zap70 and SLP-76. *Nat Immunol* 2005;**6**:1253–62.

30. Saito T, Yokosuka T. Immunological synapse and microclusters: the site for recognition and activation of T cells. *Curr Opin Immunol* 2006;**18**:305–13.

31. Tybulewicz V. Vav-family proteins in T-cell signalling. *Curr Opin Immunol* 2005;**17**:267–74.

32. Moran O, Roessle M, Mariuzza R, Dimasi N. Structural features of the full-length adaptor protein GADS in solution determined using small angle X-ray scattering. *Biophys J* 2008.

33. Yablonski D, Kadlecek T, Weiss A. Identification of a phospholipase C-gamma1 (PLC-gamma1) SH3 domain-binding site in SLP-76 required for T-cell receptor-mediated activation of PLC-gamma1 and NFAT. *Mol Cell Biol* 2001;**21**:4208–18.

34. Salojin K, Zhang J, Madrenas J, Delovitch T. T-cell anergy and altered T-cell receptor signaling: effects on autoimmune disease. *Immunol Today* 1998;**19**:468–73.

35. Paz P, Wang S, Clarke H, Lu X, Stokoe D, Abo A. Mapping the Zap-70 phosphorylation sites on LAT (linker for activation of T cells) required for recruitment and activation of signalling proteins in T cells. *Biochem J* 2001;**356**:461–71.

36. Stefanová I, Hemmer B, Vergelli M, Martin R, Biddison W, Germain R. TCR ligand discrimination is enforced by competing ERK positive and SHP-1 negative feedback pathways. *Nat Immunol* 2003;**4**:248–54.

37. Brdicka T, Pavlistová D, Leo A, et al. Phosphoprotein associated with glycosphingolipid-enriched microdomains (PAG), a novel ubiquitously expressed transmembrane adaptor protein, binds the protein tyrosine kinase csk and is involved in regulation of T cell activation. *J Exp Med* 2000;**191**:1591–604.

38. Vang T, Abrahamsen H, Myklebust S, Horejsí V, Taskén K. Combined spatial and enzymatic regulation of Csk by cAMP and protein kinase a inhibits T cell receptor signaling. *J Biol Chem* 2003;**278**:17,597–17,600.

39. Grakoui A, Bromley S, Sumen C, et al. The immunological synapse: a molecular machine controlling T cell activation. *Science* 1999;**285**:221–7.

40. Delon J, Germain R. Information transfer at the immunological synapse. *Curr Biol* 2000;**10**:R923–33.

41. Krummel M, Sjaastad M, Wülfing C, Davis M. Differential clustering of CD4 and CD3zeta during T cell recognition. *Science* 2000;**289**:1349–52.

42. Lee K, Holdorf A, Dustin M, Chan A, Allen P, Shaw A. T cell receptor signaling precedes immunological synapse formation. *Science* 2002;**295**:1539–42.

43. König R, Shen X, Maroto R, Denning T. The role of CD4 in regulating homeostasis of T helper cells. *Immunol Res* 2002;**25**:115–30.

44. Zhou W, König R. T cell receptor-independent CD4 signalling: CD4–MHC class II interactions regulate intracellular calcium and cyclic AMP. *Cell Signal* 2003;**15**:751–62.

45. König R, Zhou W. Signal transduction in T helper cells: CD4 coreceptors exert complex regulatory effects on T cell activation and function. *Curr Issues Mol Biol* 2004;**6**:1–15.

46. Berney S, Schaan T, Wolf R, Kimpel D, van der Heyde H, Atkinson T. CD5 (OKT1) augments CD3-mediated intracellular signaling events in human T lymphocytes. *Inflammation* 2001;**25**:215–21.

47. Yashiro-Ohtani Y, Zhou X, Toyo-Oka K, et al. Non-CD28 costimulatory molecules present in T cell rafts induce T cell costimulation by enhancing the association of TCR with rafts. *J Immunol* 2000;**164**:1251–9.

48. Perez-Villar J, Whitney G, Bowen M, Hewgill D, Aruffo A, Kanner S. CD5 negatively regulates the T-cell antigen receptor signal transduction pathway: involvement of SH2-containing phosphotyrosine phosphatase SHP-1. *Mol Cell Biol* 1999;**19**:2903–12.

49. Lozano F, Simarro M, Calvo J, et al. CD5 signal transduction: positive or negative modulation of antigen receptor signaling. *Crit Rev Immunol* 2000;**20**:347–58.

50. Holmes N. CD45: all is not yet crystal clear. *Immunology* 2006;**117**:145–55.

51. Dornan S, Sebestyen Z, Gamble J, et al. Differential association of CD45 isoforms with CD4 and CD8 regulates the actions of specific pools of p56lck tyrosine kinase in T cell antigen receptor signal transduction. *J Biol Chem* 2002;**277**:1912–8.

52. McNeill L, Salmond R, Cooper J, et al. The differential regulation of Lck kinase phosphorylation sites by CD45 is critical for T cell receptor signaling responses. *Immunity* 2007;**27**:421–3.

53. Varma R, Campi G, Yokosuka T, Saito T, Dustin M. T cell receptor-proximal signals are sustained in peripheral microclusters and terminated in the central supramolecular activation cluster. *Immunity* 2006;**25**:117–27.

54. Nikolova M, Marie-Cardine A, Boumsell L, Bensussan A. BY55/CD160 acts as a co-receptor in TCR signal transduction of a human circulating cytotoxic effector T lymphocyte subset lacking CD28 expression. *Intl Immunol* 2002;**14**:445–51.

55. Leupin O, Zaru R, Laroche T, Müller S, Valitutti S. Exclusion of CD45 from the T-cell receptor signaling area in antigen-stimulated T lymphocytes. *Curr Biol* 2000;**10**:277–80.

56. Bretscher P, Cohn M. A theory of self-nonself discrimination. *Science* 1970;**169**:1042–9.

57. Lafferty K, Cunningham A. A new analysis of allogeneic interactions. *Aust J Exp Biol Med Sci* 1975;**53**:27–42.

58. Jenkins M, Pardoll D, Mizuguchi J, Quill H, Schwartz R. T-cell unresponsiveness *in vivo* and *in vitro*: fine specificity of induction and molecular characterization of the unresponsive state. *Immunol Rev* 1987;**95**:113–35.

59. Dennehy K, Elias F, Na S, Fischer K, Hünig T, Lühder F. Mitogenic CD28 signals require the exchange factor Vav1 to enhance TCR signaling at the SLP-76-Vav-Itk signalosome. *J Immunol* 2007;**178**:1363–71.

60. Berg L, Finkelstein L, Lucas J, Schwartzberg P. Tec family kinases in T lymphocyte development and function. *Annu Rev Immunol* 2005;**23**:549–600.

61. Marinari B, Costanzo A, Viola A, et al. Vav cooperates with CD28 to induce NF-kappaB activation via a pathway involving Rac-1 and mitogen-activated kinase kinase 1. *Eur J Immunol* 2002;**32**:447–56.

62. Marinari B, Costanzo A, Marzano V, Piccolella E, Tuosto L. CD28 delivers a unique signal leading to the selective recruitment of RelA and p52 NF-kappaB subunits on IL-8 and Bcl-xL gene promoters. *Proc Natl Acad Sci USA* 2004;**101**:6098–103.

63. Chiang Y, Kole H, Brown K, et al. Cbl-b regulates the CD28 dependence of T-cell activation. *Nature* 2000;**403**:216–20.

64. Bullens D, Kasran A, Thielemans K, Bakkus M, Ceuppens J. CD40L-induced IL-12 production is further enhanced by the Th2 cytokines IL-4 and IL-13. *Scand J Immunol* 2001;**53**:455–63.

65. Grewal I, Flavell R. A central role of CD40 ligand in the regulation of CD4+ T-cell responses. *Immunol Today* 1996;**17**:410–14.

66. Dolmetsch R, Lewis R, Goodnow C, Healy J. Differential activation of transcription factors induced by Ca^{2+} response amplitude and duration. *Nature* 1997;**386**:855–8.

67. Matthews S, Cantrell D. The role of serine/threonine kinases in T-cell activation. *Curr Opin Immunol* 2006;**18**:314–20.

68. Hayashi K, Altman A. Protein kinase C theta (PKCtheta): a key player in T cell life and death. *Pharmacol Res* 2007;**55**:537–44.

69. Rothman B, Kennure N, Kelley K, Katz M, Aune T. Elevation of intracellular cAMP in human T lymphocytes by an anti-CD44 mAb. *J Immunol* 1993;**151**:6036–42.

70. Li L, Yee C, Beavo J. CD3- and CD28-dependent induction of PDE7 required for T cell activation. *Science* 1999;**283**:848–51.

71. Kanda N, Watanabe S. Regulatory roles of adenylate cyclase and cyclic nucleotide phosphodiesterases 1 and 4 in interleukin-13 production by activated human T cells. *Biochem Pharmacol* 2001;**62**:495–507.

72. Ramstad C, Sundvold V, Johansen H, Lea T. cAMP-dependent protein kinase (PKA) inhibits T cell activation by phosphorylating ser-43 of raf-1 in the MAPK/ERK pathway. *Cell Signal* 2000;**12**:557–63.

73. Chen D, Rothenberg E. Interleukin 2 transcription factors as molecular targets of cAMP inhibition: delayed inhibition kinetics and combinatorial transcription roles. *J Exp Med* 1994;**179**:931–42.

74. van Oirschot B, Stahl M, Lens S, Medema R. Protein kinase A regulates expression of p27(kip1) and cyclin D3 to suppress proliferation of leukemic T cell lines. *J Biol Chem* 2001;**276**:33,854–33,860.

75. Boonen G, van Dijk A, Verdonck L, van Lier R, Rijksen G, Medema R. CD28 induces cell cycle progression by IL-2-independent down-regulation of p27kip1 expression in human peripheral T lymphocytes. *Eur J Immunol* 1999;**29**:789–98.

76. Abrahamsen H, Baillie G, Ngai J, et al. TCR- and CD28-mediated recruitment of phosphodiesterase 4 to lipid rafts potentiates TCR signaling. *J Immunol* 2004;**173**:4847–58.

77. Arp J, Kirchhof M, Baroja M, et al. Regulation of T-cell activation by phosphodiesterase 4B2 requires its dynamic redistribution during immunological synapse formation. *Mol Cell Biol* 2003;**23**:8042–57.

78. Zhou W, Vergara L, König R. T cell receptor induced intracellular redistribution of type I protein kinase A. *Immunology* 2004;**113**:453–9.

79. Chan C, Stark J, George AJT. Feedback control of T-cell receptor activation. *Proc R Soc Lond B Biol Sci* 2004;**271**:931–9.

80. Altan-Bonnet G, Germain R. Modeling T cell antigen discrimination based on feedback control of digital ERK responses. *PLoS Biol* 2005;**3**:e356.

81. Randriamampita C, Mouchacca P, Malissen B, Marguet D, Trautmann A, Lellouch AC. A novel ZAP-70 dependent FRET based biosensor reveals kinase activity at both the immunological synapse and the antisynapse. *PLoS ONE* 2008;**3**:e1521.

82. Grant M, Scheel-Toellner D, Griffiths H. Contributions to our understanding of T cell physiology through unveiling the T cell proteome. *Clin Exp Immunol* 2007;**149**:9–15.

Chapter 19

Signal Transduction via the B Cell Antigen Receptor: A Crucial Regulator of B Cell Biology

Louis B. Justement

Department of Microbiology, University of Alabama at Birmingham, Birmingham, Alabama

INTRODUCTION

B lymphocytes express antigen receptors (BCR) on their surface that impart the ability to detect foreign antigens. Binding of antigen to the BCR initiates a signal transduction cascade that in turn leads to cellular activation, proliferation, and, ultimately, differentiation into antibody-secreting plasma cells. Additionally, the BCR plays an important role in internalization of antigen, resulting in its processing and presentation in the context of MHC class II to helper T cells, which promote clonal expansion and differentiation of the antigen-specific B cells [1, 2]. Although signal transduction via the BCR is central to the generation of a productive humoral immune response, its role in regulating B cell biology is more complex and multifaceted. Indeed the BCR is involved in determining the fate of the B cell throughout its development and differentiation [3–5]. During B cell development in the bone marrow, the pre-BCR is crucial for transducing signals that ensure the formation and expression of a functional, mature BCR on the surface of the B cell. Additionally, the pre-BCR is involved in regulating allelic exclusion to ensure that only BCRs with a single antigenic specificity are expressed on the surface of any given B cell. Once a mature BCR is expressed on the surface of an immature B cell, signals delivered through it can either positively select those cells that will ultimately enter into the periphery to patrol for foreign antigen, or negatively select those cells that recognize self-antigens with a high affinity. Negative selection mediated by signals delivered via the BCR results in editing of the BCR to change its specificity, or in deletion of the self-reactive clone [6–8]. Thus the BCR is directly involved in determining the fate of the cell presumably by virtue of its ability to transduce signals that vary both quantitatively as well as qualitatively. Additionally, the response of the B cell to such signals is likely to be regulated in a developmental manner [9, 10]. In conclusion, the BCR serves a central role in regulating B cell biology by virtue of its ability to selectively transduce signals in response to antigen binding. The discussion that follows will primarily deal with the basic signal transduction pathways that are activated in response to binding of antigen to BCRs expressed on mature, quiescent B cells.

INITIATION OF SIGNAL TRANSDUCTION THROUGH THE BCR

The BCR complex is comprised of an antigen recognition structure, membrane immunoglobulin (mIg), and an associated signal transducing heterodimer. Membrane Ig consists of two heavy chains and two light chains that are disulfide bonded to one another to form the mature antigen recognition structure. Membrane Ig is non-covalently associated with one transmembrane heterodimer consisting of disulfide-linked Igα (CD79a) and Igβ (CD79b) polypeptides [11, 12]. The Igα/β heterodimer functions both as a transporter and a signal transducing structure [12]. Initiation of signal transduction through the BCR was originally thought to occur in response to BCR crosslinking mediated by binding of bivalent or multivalent antigen to the bivalent mIg molecule. This was thought to induce BCR dimerization and/or multimerization, which in turn facilitated the ability of associated Src family protein tyrosine kinases (PTK) (e.g., Lyn, Fyn) to phosphorylate tyrosine residues in the cytoplasmic domains of Igα/β [13]. Although the BCR in resting B cells is constitutively associated with Src family PTKs at a low stoichiometry, the net level of Igα/β phosphorylation is presumably minimal due to the fact that individual receptor complexes are distributed throughout the membrane and the phosphorylation of

Igα/β is counterbalanced by one or more protein tyrosine phosphatases (PTP). In contrast, BCR aggregation presumably favors a net increase in the total level of protein tyrosine phosphorylation associated with the activation complex due to the physical co-localization of Src family PTKs and their substrates (i.e., Igα/β)[13].

More recently, however, it has been proposed that the BCR exists in the membrane of unstimulated B cells in preformed oligomers, and that initiation of signaling occurs when antigen binds to one or more mIg molecules within an oligomer, thereby causing a conformational change or reorganization of the individual subunits that promote signal transduction [11, 12, 14, 15]. In this regard, the transmembrane region of mIg heavy chains is thought to assume an α-helical conformation in the membrane in which one face of the α-helix is comprised of highly conserved hydrophilic amino acid residues that mediate the interaction with the Igα/β heterodimer [14]. The other face of the α-helix is specific to each of the mIg isotypes, and contains several hydrophilic amino acid residues with large polar side groups. Because such hydrogen-binding, hydrophilic moieties exhibit a strong bias against being located within a lipid membrane, it is possible that this face of the helix constitutes an interaction surface that is involved in BCR oligomerization [14]. Although the molecular composition of the resting BCR oligomers has yet to be elucidated, it has been hypothesized that the oligomers are associated with the PTK Syk [16]. This PTK presumably promotes phosphorylation of Igα/β heterodimers within the oligomer, which in turn facilitates binding of Syk to phosphotyrosine residues via its tandem Src homology 2 (SH2) domains. It is further hypothesized that a low basal level of tyrosine phosphorylation is maintained by the action of a PTP called SHP-1, which acts to constitutively downregulate Syk kinase activity, and perhaps to dephosphorylate Igα/β [13]. Binding of antigen to the complex is thought to cause a conformational change in the resting BCR oligomer such that SHP-1 activity is attenuated, possibly by causing the PTP to become physically dissociated from the BCR complex. Dissociation of SHP-1 would presumably favor a net increase in tyrosine phosphorylation of Igα/β leading to initiation of signal transduction and cellular activation. It has been reported that SHP-1 constitutively interacts with the BCR isolated from resting B cells and is induced to dissociate from the BCR upon activation in agreement with this proposed mechanism [17].

Another recent finding is that binding of antigen causes a rapid translocation of the BCR to glycosphingolipid-enriched microdomains (GEMs) within the plasma membrane. These microdomains are enriched for Src family PTKs including Lyn, and it has been hypothesized that translocation of the BCR into GEMs physically localizes the complex with Src PTKs, thereby promoting tyrosine phosphorylation of Igα/β by Lyn [18–20]. Interestingly, translocation of the BCR into GEMs appears to consti-

tute a novel step that precedes signal transduction because translocation does not require the Igα/β heterodimer and is resistant to blockade by PTK inhibitors [21]. Thus, it is possible that binding of antigen to BCR oligomers on resting B cells leads to a reorganization/conformational change in the oligomer that promotes its translocation to GEMs. As stated, localization of the BCR to GEMs promotes Lyn-mediated phosphorylation of Igα/β, as well as phosphorylation of Syk, resulting in potentiation of its catalytic activity. Simultaneously, translocation of the BCR to GEMs may lead to dissociation of SHP-1, thereby causing a net increase in tyrosine phosphorylation of Igα/β and initiation of signal transduction. Alternatively, it is possible that Lyn activity is relatively resistant to the inhibitory action of SHP-1, in which case co-localization of Lyn with the BCR may be sufficient to favor a net increase in Igα/β phosphorylation despite the presence of SHP-1 [14].

Regardless of whether antigen binding promotes BCR aggregation or reorganization of pre-existing oligomers, it is clear that the BCR translocates to GEMs where there is a net increase in the tyrosine phosphorylation of Igα and Igβ. Both of these polypeptides contain immunoreceptor tyrosine-based activation motifs (ITAMs) within their cytoplasmic domains that function as docking sites for SH2 domain-containing proteins including Syk, Shc, and BLNK [13]. The net increase in tyrosine phosphorylation of the Igα/β ITAMs promotes the formation of multimolecular complexes that are targeted to the BCR and are critical for propagation of signal transduction. The formation of these complexes enables the BCR to activate several signaling pathways that are distinct yet interrelated, including the phospholipase C (PLCγ) pathway, the phosphatidylinositol 3-kinase (PI3K) pathway, and signaling processes that are controlled by the Ras, Rac1, and Rap1 GTPases (for reviews, see [22–26]).

PROPAGATION OF SIGNAL TRANSDUCTION VIA THE BCR

Numerous studies have demonstrated that ligand binding to the BCR results in inducible tyrosine phosphorylation and activation of the PTK Syk [27–29]. The critical role that Syk serves during BCR-mediated signaling has been demonstrated by the analysis of Syk-deficient B cells in which abrogation of Syk expression results in the loss of IP_3 generation and Ca^{2+} mobilization [30]. Moreover, studies using bone marrow cells derived from Syk-deficient mice have confirmed that Syk is essential for proper signal transduction via the BCR based on the inability of these cells to develop normally into mature B cells in chimeric mice [31, 32]. Recruitment of Syk to the BCR complex is mediated by tyrosine phosphorylation of the ITAMs in Igα and Igβ, creating docking sites that are recognized by the dual SH2 domains of Syk [33–35]. Efficient recruitment of Syk requires that both conserved tyrosines within a given ITAM be present, and that they be phosphorylated. Additionally, both SH2 domains of Syk must be present and functional in order to observe optimal association of the kinase with Igα and Igβ and its subsequent activation through the physical interaction with these polypeptides [13]. Binding of Syk to the ITAMs of Igα and Igβ may also play a crucial role in its activation by focusing Syk to the BCR activation complex within GEMs, which effectively brings Syk into close proximity to Lyn. Studies support the conclusion that Lyn phosphorylates key tyrosine residues on Syk, resulting in potentiation of its kinase activity [13].

The ability of Syk to phosphorylate downstream signaling effector proteins is dependent on the recruitment of an adaptor protein called BLNK (SLP65, BASH) to the BCR activation complex [25, 26, 36]. Recent studies have determined that BLNK is recruited to the BCR complex by virtue of its ability to bind to phosphorylated tyrosine residues in the cytoplasmic tail of Igα that lie outside the ITAM [37, 38]. It has been shown that mutation of tyrosines 176 and 204, which flank the ITAM, abrogates BLNK-dependent signaling [38]. Moreover, it was shown that the SH2 domain of BLNK binds directly to tyrosine 204 of Igα [37, 38]. Thus, it appears that Syk and BLNK are co-localized to the BCR complex through their interaction with Igα. This facilitates the ability of Syk to phosphorylate BLNK, which in turn generates phosphotyrosine-dependent binding sites for recruitment of additional signal transducing proteins [16, 39]. The formation of the BCR–Syk–BLNK complex constitutes the basic unit that is required for propagation of signal transduction leading to the activation of several interrelated downstream pathways.

Activation of PLCγ2-Dependent Signaling

The initial activation of PTKs is responsible for mediating the production of second messengers that subsequently regulate intermediate signaling processes leading to transcription factor activation. Studies have elegantly demonstrated that PLCγ2 is activated in response to BCR ligation and is responsible for the production of diacylglycerol (DAG) and inositol 1,4,5-trisphosphate (IP_3) [40–42]. These in turn promote PKC activation and mobilization of Ca^{2+}, respectively [22–26]. It has been demonstrated, using a non-lymphoid cell reconstitution system, that expression of Igα/β, Syk, and PLCγ2 is not sufficient to mediate activation of PLCγ2 and Ca^{2+} mobilization in response to BCR ligation, suggesting that one or more key components were missing [43]. Subsequent work has clearly shown that PLCγ2 is recruited to tyrosine phosphorylated BLNK via its tandem SH2 domains. In cells that lack BLNK, PLCγ2 is not observed to translocate from the cytoplasm to GEMs, and exhibits decreased activation [36]. Additionally, mutation of the amino-terminal SH2 domain of PLCγ2 blocks its

binding to phosphorylated BLNK and blocks localization to GEMs [44]. Therefore, it is apparent that BLNK provides a scaffold to localize PLCγ2 to the BCR activation complex within GEMs, where it can be activated.

Based on numerous findings, it has been concluded that Syk and the PTK Btk act in concert to regulate the phosphorylation and activity of PLCγ2 [45–47]. In addition to the Src family PTKs and Syk, Btk is inducibly activated in response to BCR ligation and plays an important role in signal transduction [46, 48]. Like the Src family PTKs, Btk contains contiguous SH3, SH2, and SH1 domains, although it does not possess a carboxyl-terminal negative regulatory tyrosine residue or a myristylation site. In addition to the SH domains, Btk contains an amino-terminal pleckstrin homology (PH) domain and an adjacent proline- and cysteine-rich Tec homology (TH) domain [46]. It has been shown that the SH2 domain of Btk is required for PLCγ2 phosphorylation and activation, and that it exhibits restricted binding specificity for tyrosine phosphorylated BLNK. Thus, PLCγ2 and Btk are co-localized to the BCR activation complex by virtue of their interaction with BLNK.

Studies in the mouse and human have clearly demonstrated that Btk expression is essential for BCR-mediated Ca^{2+} mobilization [49, 50]. Moreover, expression of Btk has been shown to restore calcium signaling in Btk-deficient XLA B cells. Although expression of Syk is required for normal mobilization of Ca^{2+} in B cells, overexpression of this PTK was not observed to restore calcium signaling in XLA B cells, demonstrating that it cannot compensate for the loss of Btk. This finding indicates that these PTKs must act in concert in a non-redundant manner to regulate PLCγ2 function [36, 47]. The ability of Btk to reconstitute calcium signaling is dependent on its catalytic function and the Btk activation loop tyrosine (Tyr551), which is phosphorylated by Syk. Mutation of the Syk transphosphorylation site or the ATP-binding site abrogates Btk-dependent phosphorylation of PLCγ2 [45, 47]. Recent studies have demonstrated that specific tyrosine residues in PLCγ2 (tyrosines 753, 759, 1197, and 1217) are direct targets for Btk, and that their phosphorylation is essential for optimal activation of its catalytic function [51, 52]. In conclusion, it appears that Syk is required for PLCγ2 activation by virtue of its role in phosphorylating BLNK to generate docking sites for Btk and PLCγ2, and through its role in phosphorylating Tyr551 on Btk leading to its catalytic activation. Btk appears to be specifically involved in phosphorylating multiple tyrosine residues on PLCγ2 that are required for optimal catalytic activation leading to hydrolysis of phosphatidylinositol 4,5-bisphosphate to produce IP3 and DAG.

The production of IP3 in response to activation of PLCγ2 results in mobilization of Ca^{2+} from the endoplasmic reticulum (ER) [53]. IP3-mediated release of Ca^{2+} involves IP3 receptors (IP3R) located in the ER, which form heterotetrameric channels. It has been shown that the nature of the calcium mobilization response in B cells may be controlled by the combinatorial function of three IP3R receptor subtypes [54]. Additional evidence suggests that the function of IP3Rs may be regulated by phosphorylation, as well as in response to binding of IP3. A B cell restricted scaffold protein with ankyrin repeats (BANK) has recently been cloned that is inducibly phosphorylated on tyrosine residues in response to BCR ligation [55]. Overexpression of BANK results in enhanced BCR-induced calcium mobilization. Interestingly, BANK has been shown to interact with Lyn and IP3Rs, suggesting that this adaptor functions to bring Lyn and IP3Rs into close proximity, thereby promoting tyrosine phosphorylation of IP3Rs by Lyn [55]. This may play a key role in potentiating calcium mobilization from the ER, as it has been shown that tyrosine phosphorylation of IP3Rs upregulates their channel activity. The release of Ca^{2+} from intracellular stores has been shown to promote Ca^{2+} influx across the plasma membrane, via the opening of store-operated Ca^{2+} channels (SOCs) [56–59]. Numerous studies suggest that the main type of SOC operating in immune cells produces a distinct Ca^{2+} release activated current (I_{CRAC}) [56–59]. Recent studies have identified two important effector proteins that appear to play a role in mediating the I_{CRAC} response to BCR crosslinking; stromal interaction molecule 1 (STIM1) and ORAI (which is also referred to as CRAC modulator 1) [60–65]. Based on studies of various mammalian cell types, STIM1 has been shown to sense Ca^{2+} levels in the lumen of the ER, and when the Ca^{2+} levels in the ER drop, STIM1 aggregates and translocates to the plasma membrane, where it interacts with ORAI and thereby promotes activation of SOCs. In the DT40 B cell line, gene targeting of STIM1 was shown to eliminate Ca^{2+} entry across the plasma membrane exhibiting electrophysiological characteristics of I_{CRAC} [66]. Although deletion of ORAI in T cells has been shown to abrogate I_{CRAC}, similar results were not observed in B cells, suggesting that other plasma membrane proteins may be critical for SOC-dependent Ca^{2+} entry. Moreover, recent studies support the conclusion that, in contrast to T cells, second messenger-operated Ca^{2+} channels (SMOCs) are important for Ca^{2+} entry in response to BCR crosslinking. Members of the transient receptor potential (TRP) proteins and IP3Rs have been shown to function as potential SMOCs in B cells [67–71]. Thus, the relative importance of SOC versus SMOC in mediating Ca^{2+} entry across the plasma membrane of B cells in response to BCR ligation remains to be completely elucidated.

One of the most important downstream targets of the calcium signaling pathway in B cells is the transcription factor NFAT [72]. Members of the NFAT family are retained in the cytosol due to constitutive phosphorylation on serine residues. NFAT activation leading to translocation to the nucleus is promoted by dephosphorylation mediated by the Ca^{2+} sensitive serine/threonine phosphatase calcineurin. The calcineurin complex is comprised of three subunits, including the catalytic A subunit (Aα, Aβ, or Aγ),

a regulatory B subunit (B1 or B2), and the Ca^{2+}-sensitive calmodulin subunit bound to Ca^{2+}. Recent work has demonstrated an important role for the B1 regulatory subunit of calcineurin in regulation of NFAT activation in B cells [73]. B cells express three NFAT family members (c1, c2, and c3) that have been shown to be activated in a Ca^{2+}/calmodulin-dependent manner in response to BCR ligation [74–76].

Activation of Protein Kinase C-Dependent Signaling and the CARMA1 Signalosome

The production of DAG by -PLCγ 2 leads to the activation of members of the PKC family. Both conventional (PKCα, -β, -βII, -γ) and novel (PKCδ, -ε, -η, -θ, -μ) members of the PKC family require DAG for their activation, whereas the former require Ca^{2+} as well [77]. Studies have detected PKCα, -β, -γ, -δ, -ε, -η, -θ and -μ expression in B cells, and have also shown that several of these kinases (PKCα, -βII, -δ, -ε) translocate from the cytoplasm to the membrane in response to BCR ligation [77, 78]. It has further been shown that PKCμ interacts with the BCR and exhibits increased specific activity in response to BCR ligation [79]. The activation of members of the PKC family has been shown to play a role in regulating activation of several transcription factors in B cells [22–26]. PKC activation leads to activation of the mitogen-activated protein (MAP) kinase ERK2 via the classical Ras/Raf-1/Mek/ERK2 pathway [22–26]. ERK2 translocates to the nucleus, where it regulates phosphorylation and activation of transcription factors including Elk-1 and members of the Ets family [80, 81]. Additionally, it has been shown that novel PKC members PKCθ and -δ may play a critical role in B cell activation by virtue of their ability to regulate the activation of the transcriptional regulator NFκB and the MAP kinase JNK [82].

Recent studies have demonstrated that PKCβ is activated in response to BCR crosslinking in a DAG and Ca^{2+}-dependent manner, and that its recruitment to the plasma membrane plays a critical role in regulating activation of the transcriptional regulator NFκB [83, 84]. Subsequent work revealed that PKCβ binds to and phosphorylates CARMA1 (caspase-recruitment domain (CARD)-membrane-associated guanylate kinase (MAGUK) protein 1), resulting in its activation, which involves a conformational reorganization of the protein thereby promoting its recruitment to lipid rafts, where it undergoes oligomerization [83, 85, 86]. CARMA1 then recruits B bell lymphoma 10 (Bcl-10), and Bcl-10 in turn binds to MALT1 (mucosa-associated lymphoid tissue lymphoma-translocation gene 1) [83]. MALT1 then directly interacts with either TRAF2 (tumor-necroisis factor receptor (TNFR)-associated factor 2 or TRAF6, resulting in activation of their E3 ubiquitin-ligase activity [83, 87]. TRAF6, once activated undergoes auto-ubiquitination promoting recruitment of the IKKγ (inhibitor of NFκB (IκB) kinase gamma regulatory subunit) and TAK1 complexes [83]. In contrast, TRAF 2 recruits RIP1, which becomes poly-ubiquitinated, thereby recruiting IKKγ and Tak1 complexes [83]. Ultimately, IKKγ is polyubiquitinated by TRAF6 or TRAF2, resulting in translocation and activation of TAK1, which in turn phosphorylates IKKβ, leading to activation of the IKK complex. The IKK complex phosphorylates IκB, thereby targeting it for degradation promoting the nuclear translocation of NFκB. The importance of this pathway for BCR signal transduction leading to B cell activation has been borne out by gene-targeting studies resulting in loss of expression of the key effectors that promote formation of the CARMA1 signalosome (i.e., PKCβ), as well as components of the signalosome itself, including CARMA1, Bcl-10, MALT1, and Tak1 [88–92].

Activation of Phosphoinositide 3-Kinase-Dependent Signaling

Another major signaling pathway that is activated upon BCR ligation involves the lipid kinase PI3K. Mice lacking the p85 subunit of PI3K exhibit severe defects in B cell development as well as impaired proliferative responses to stimulation through the BCR [93, 94]. Thus, it is apparent that PI3K plays a critical role in signaling via the BCR. Recruitment and activation of PI3K to the BCR complex has been shown to occur via the transmembrane protein CD19 [95]. BCR ligation leads to tyrosine phosphorylation of several tyrosine residues in the cytoplasmic domain of CD19. Two of these residues, Tyr 482 and 513, have been shown to mediate binding of the p85 subunit of PI3K via its tandem SH2 domains [96]. Because CD19 localizes to GEMs in response to B cell activation, its interaction with PI3K provides a mechanism to focus this kinase in regions of the cell that are enriched for its substrate phosphoinositide 4,5-bisphosphate (PIP2) [97]. It is important to note, however, that loss of CD19 expression causes a less severe defect in B cell development and function than is observed in B cells that lack the p85 subunit of PI3K. This suggests that additional adaptor proteins other than CD19 may mediate PI3K recruitment and activation [25].

Recent studies have identified a protein called B cell adaptor for phosphoinositide 3-kinase (BCAP) that is inducibly phosphorylated by Syk and Btk in response to BCR ligation [98]. Of the 31 potential tyrosine residues that can be phosphorylated on BCAP, four are contained within consensus motifs for binding to the SH2 domains of the PI3K p85 subunit. Mutation of these tyrosine residues ablates the ability of BCAP to interact with p85 or to restore Akt activation in cells that lack BCAP [98]. Additionally, loss of BCAP expression has been shown to attenuate the recruitment of PI3K to GEMs, suggesting that this adaptor plays an important role in targeting PI3K to the BCR activation

complex. Nevertheless, loss of BCAP expression does not entirely abrogate the production of phosphatidylinositol 3,4,5-trisphosphate (PIP3) or the activation of the downstream target Akt [98]. Therefore, it is clear that CD19 as well as other potential adaptors may be able to promote PI3K recruitment to GEMs, where it is activated. A candidate for such an adaptor is Gab1, which is inducibly phosphorylated on tyrosine residues in response to BCR ligation mediating its direct interaction with the SH2 domains of PI3K, Shc, and the PTP SHP-2 in a phosphotyrosine-dependent manner [99]. Overexpression of Gab1 in B cells was observed to potentiate BCR-mediated phosphorylation of Akt, which is a PI3K-dependent response. Importantly, it was observed that the pleckstrin homology domain of Gab1 is required for its ability to translocate from the cytoplasm to the plasma membrane, and for its ability to potentiate activation of PI3K [100]. PH domains have been shown to bind to PIP3, which is produced in response to activation of PI3K. Thus, it is likely that Gab1 functions as an amplifier of PI3K-dependent signaling due to the fact that its recruitment to the membrane and function require the initial production of PIP3 by PI3K [100].

Activation of PI3K is mediated by virtue of its ability to bind to tyrosine-phosphorylated proteins such as CD19 and BCAP via the SH2 domains of the p85 subunit. This in turn promotes translocation of PI3K to GEMs, where it is able to access its substrate PIP2, which is enriched in these microdomains of the plasma membrane. Although PI3K binding to adaptor proteins and translocation is required for activation, it is not sufficient for full activation of PI3K catalytic function. Studies have indicated that Vav family members may play a role in potentiating the activation of PI3K [25]. Generation of B cells lacking Vav3 resulted in significant decreases in PIP3 production and activation of Akt [101]. These defects could be corrected by expressing Vav3 or Vav2 in Vav3-deficient B cells. However, the guanine nucleotide exchange factor (GEF) mutant of Vav3 was not able to restore pI3K function, indicating that Vav3 may regulate PI3K activity through its target Rac1 [101]. This was indeed found to be the case, based on several experimental strategies. In conclusion, Vav appears to be important for potentiating PI3K activity in response to BCR ligation. Nevertheless, it has yet to be formally determined whether all members of the Vav family share the ability to potentiate PI3K activity through activation of Rac1, although it seems likely, based on the available experimental evidence.

Activation of PI3K results in the phosphorylation of PIP2 to form PIP3, which regulates the activation of numerous downstream signaling proteins that contain PH domains [102]. It has been shown that activation of PLCγ2, Btk, Rac1, and the kinase network including PDK1, Akt, GSK-3, and mTOR is regulated by the ability of these proteins to bind to PIP3 via their PH domains [22–26]. Presumably, the production of PIP3 by PI3K plays an important role in promoting recruitment of PH domain-containing proteins such as PLCγ2 and Btk to the membrane, where they are co-localized through their interaction with BLNK. Additionally, it is possible that the production of PIP3 functions to maintain activated PLCγ2 and/or Btk at the membrane once they have dissociated from BLNK, thereby prolonging the signal transduction response. It is now well documented that PI3K-dependent production of PIP3 plays a crucial role in activation of the PDK1/Akt/GSK-3 kinase cascade, which in turn promotes cell survival [102]. PIP3-dependent recruitment of PDK1 and Akt to the membrane facilitates the ability of PDK1 to phosphorylate Akt on serine/threonine residues, which in turn leads to activation of Akt [102]. Numerous studies have documented activation of Akt in response to BCR ligation [103–106]. Kinetic studies have demonstrated that Akt transiently translocates to the plasma membrane in B cells, where it is activated and then migrates to the cytoplasm and nucleus, where it presumably can interact with potential substrates [107]. Akt is a serine/threonine kinase that phosphorylates numerous downstream substrates, including Bad and GSK-3. Phosphorylation of the apoptosis-inducing Bad protein creates a binding site for 14-3-3 proteins, preventing Bad from binding to Bcl2 and Bcl-X$_L$ and thereby releasing them to mediate cell survival [102]. Phosphorylation of GSK-3, which is a serine/threonine kinase, inactivates its catalytic function. GSK-3 is constitutively active in resting cells and phosphorylates numerous proteins, including c-myc and cyclin D, maintaining them in an inactive state [102]. GSK-3 has also been shown to phosphorylate NF-AT, causing a change in its conformation that reveals a nuclear export signal [108]. Thus, inhibition of GSK-3 activity by Akt promotes retention of NF-AT in the nucleus. Therefore, phosphorylation of GSK-3 by Akt promotes the activation of proteins that regulate cell cycling and cell survival.

Activation of Small Molecular Weight G Proteins and their Signaling Pathways

BCR ligation promotes the activation of small molecular weight G proteins that in turn regulate signal transduction pathways that control transcription factor activation and cytoskeletal reorganization, which affects cell morphology and motility. In many instances it appears that BLNK plays a role in co-localizing critical signal effector proteins leading to activation of G proteins. It has been shown that BLNK interacts with the adaptor protein Grb2, which appears to bind constitutively to a proline-rich region on BLNK via one of its SH3 domains [109, 110]. The interaction between Grb2 and BLNK can also be potentiated in a tyrosine phosphorylation-dependent manner in which the SH2 domain of Grb2 binds to phosphotyrosine on BLNK. In either case, Grb2 recruits the guanine nucleotide

exchange factor (GEF) Sos to the complex. Another potential mechanism whereby Sos is recruited to the membrane involves the formation of a Shc/Grb2/Sos complex in which Shc is inducibly phosphorylated in response to BCR ligation promoting binding of Grb2 via its SH2 domain. It has been proposed that Shc in turn binds via its SH2 domain to tyrosine residues within the cytoplasmic domain of Igα/β [13, 22, 23]. An alternative mechanism by which the Shc/Grb2/Sos complex may be recruited to the membrane is through binding of Shc to tyrosine phosphorylated Gab1 [99]. Regardless of the specific mechanism by which Sos is recruited to the plasma membrane, its localization leads to direct activation of the small molecular weight G protein Ras by virtue of its GEF activity. GTP-bound Ras controls the activation of a kinase cascade that culminates in the activation of the MAP kinases ERK1 and ERK2 [22, 23, 77]. This is mediated by binding of activated Ras to the Raf-1 kinase, which then phosphorylates and activates MEK1 and 2, and these phosphorylate the ERKs. Activated ERK1 and ERK2 translocate to the nucleus, where they phosphorylate and regulate the activity of transcription factors including Elk-1 and Sap1a. The Ras/Raf-1/ERK pathway also functions to regulate cell cycle progression by virtue of its ability to upregulate the expression of cyclins D and E1 while at the same time inhibiting the expression of the cell cycle inhibitor p27^{Kip1} [111–113]. The function of the Ras/Raf-1/ERK pathway is negatively regulated by the Rap1 G protein, which is inducibly activated in response to BCR ligation via a DAG-dependent mechanism [114]. Rap1 possesses the same effector binding domain as Ras, suggesting that it can compete with Ras for binding to downstream proteins in the Ras/Raf-1/ERK cascade. Indeed, Rap1 has been shown to bind to Raf-1, but does not activate it [115]. Thus Rap1 may sequester components of the Ras/Raf-1/ERK pathway, thereby blocking activation of ERK1 and 2. Alternatively, it has been proposed that Rap1 activates a distinct pathway that may antagonize the function of the Ras pathway [116].

Activation of the Rac1 GTPase is mediated through the GEF activity of Vav [117]. Vav is effectively recruited to the plasma membrane by virtue of its PH domain, which binds to PIP3, as well as its ability to interact with adaptor proteins via SH2/phosphotyrosine-dependent interactions. It has been shown that Vav is recruited to tyrosine phosphorylated BLNK, where it is co-localized with the PTK Syk [36]. Phosphorylation of Vav by Syk potentiates its GEF activity leading to activation of Rac1. Additionally, it has been shown that Vav interacts with CD19, suggesting that this may constitute another mechanism for targeting it to the membrane. Rac1 is important for regulation of receptor-induced actin polymerization and cytoskeletal reorganization [118]. Such processes are likely to play an important role in organization of signaling proteins into effective complexes that promote B cell activation. Rac1 activation also plays an important role in linking BCR signaling to activation of the downstream MAP kinases JNK and p38 [22, 23, 77]. Studies have shown that activation of these kinases is abrogated in cells that lack BLNK, and that activation cannot be restored by reconstitution of PLCγ2 signaling alone [36]. This indicates that binding of Vav to BLNK and its activation by Syk is crucial for subsequent activation of Rac1, and the JNK and p38 MAP kinases. Activated forms of these MAP kinases translocate to the nucleus, where they play an important role in regulating the function of transcription factors through phosphorylation. JNK can activate Elk-1 and Sap1a transcription factors as well as c-Jun. It has been shown that p38 can regulate the activation of ATF-2, Sap1a, CHOP, and MEF2C transcription factors [77].

CONCLUSION

It is clear that the BCR is able to access several distinct yet interrelated signaling pathways in response to binding of antigen. The initiation of signaling is dependent on antigen-driven changes in the organization of the BCR complexes expressed on the surface of a quiescent B cell. Although the exact nature of the changes that are elicited in response to antigen binding has yet to be elucidated, it is apparent that perturbation of the BCR complexes, whether they exist as individual monomeric structures or in preformed oligomers, leads to translocation to GEMs. This is critical for promoting a net increase in tyrosine phosphorylation of the BCR-associated Igα/β heterodimer and for recruitment of Syk and BLNK, which form the central initiation complex. Formation of this initiation complex lead to propagation of signaling transduction via pathways that are regulated by activation of PLCγ2, PI3K, PKC, Ras, and Rac1. These pathways in turn lead to activation of numerous transcription factors that regulate gene expression. The ability of the BCR to control the various biological outcomes associated with B cell development, selection, and activation is ultimately regulated by both intrinsic and extrinsic factors that affect the qualitative and quantitative nature of the overall signal transduced via the BCR. Extrinsic factors include the physical nature of the antigen, the duration of exposure to antigen, and the B cell's previous exposure to that antigen. Intrinsic factors include the developmental state of the B cell, which may affect the response of the cell to a given antigenic signal at the genetic level. Additionally, it is clear that the maturational and differentiative state of the B cell dramatically alter its response to antigenic challenge. This can occur through changes in the expression of proximal BCR signaling proteins such as PTKs and PTPs, as well as through changes in the expression of other transmembrane receptors that function as co-receptors for the BCR. Examples of these include CD19, CD22, FcγRIIb, and PIR-B. These co-receptors have the ability to provide contextual information to the B cell through their ability to detect extracellular ligands that affect their ability to engage the BCR and to recruit signal transducing proteins that modify the nature of the signal transduced via the BCR [119–122]. Thus, it is clear that numerous mechanisms exist for modulating the basic signal transduced via the BCR and this in turn alters the complement of transcriptional regulators that are activated, as well as the genes that may be accessible to them. This in turn dramatically alters the expression of key regulators that determine whether the B cell undergoes apoptosis, or mounts an active immune response.

REFERENCES

1. Wagle NM, Cheng P, Kim J, Sproul TW, Kausch KD, Pierce SK. B-Lymphocyte signaling receptors and the control of class-II antigen processing. *Curr Top Microbiol Immunol* 2000;**245**:101–26.
2. Bishop GA, Hostager BS. B lymphocyte activation by contact-mediated interactions with T lymphocytes. *Curr Opin Immunol* 2001;**13**:278–85.
3. Cariappa A, Pillai S. Antigen-dependent B-cell development. *Curr Opin Immunol* 2002;**14**:241–9.
4. Kurosaki T. Regulation of B cell fates by BCR signaling components. *Curr Opin Immunol* 2002;**14**:341–7.
5. Benschop RJ, Cambier JC. B cell development: signal transduction by antigen receptors and their surrogates. *Curr Opin Immunol* 1999;**11**:143–51.
6. Nemazee D. Receptor editing in B cells. *Adv Immunol* 2000;**74**:89–126.
7. Glynne R, Ghandour G, Rayner J, Mack DH, Goodnow CC. B-lymphocyte quiescence, tolerance and activation as viewed by global gene expression profiling on microarrays. *Immunol Rev* 2000;**176**:216–46.
8. Monroe JG. Molecular mechanisms regulating B-cell negative selection. *Biochem Soc Trans* 1997;**25**:643–7.
9. King LB, Monroe JG. Immunobiology of the immature B cell: plasticity in the B-cell antigen receptor-induced response fine tunes negative selection. *Immunol Rev* 2000;**176**:86–104.
10. Gauld SB, Dal Porto JM, Cambier JC. B cell antigen receptor signaling: roles in cell development and disease. *Science* 2002;**296**:1641–2.
11. Schamel WW, Reth M. Monomeric and oligomeric complexes of the B cell antigen receptor. *Immunity* 2000;**13**:5–14.
12. Reth M, Wienands J, Schamel WW. An unsolved problem of the clonal selection theory and the model of an oligomeric B-cell antigen receptor. *Immunol Rev* 2000;**176**:10–18.
13. Justement LB. Signal transduction via the B-cell antigen receptor: the role of protein tyrosine kinases and protein tyrosine phosphatases. *Curr Top Microbiol Immunol* 2000;**245**:2–51.
14. Reth M. Oligomeric antigen receptors: a new view on signaling for the selection of lymphocytes. *Trends Immunol* 2001;**22**:356–60.
15. Matsuuchi L, Gold MR. New views of BCR structure and organization. *Curr Opin Immunol* 2001;**13**:270–7.
16. Zhang Y, Wienands J, Zurn C, Reth M. Induction of the antigen receptor expression on B lymphocytes results in rapid competence for signaling of SLP-65 and Syk. *EMBO J* 1998;**17**:7304–10.
17. Pani G, Kozlowski M, Cambier JC, Mills GB, Siminovitch KA. Identification of the tyrosine phosphatase PTP1C as a B cell antigen receptor-associated protein involved in the regulation of B cell signaling. *J Exp Med* 1995;**181**:2077–84.

18. Cheng PC, Dykstra ML, Mitchell RN, Pierce SK. A role for lipid rafts in B cell antigen receptor signaling and antigen targeting. *J Exp Med* 1999;**190**:1549–60.

19. Cherukuri A, Dykstra M, Pierce SK. Floating the raft hypothesis: lipid rafts play a role in immune cell activation. *Immunity* 2001;**14**:657–60.

20. Pierce SK. Lipid rafts and B-cell activation. *Nat Rev Immunol* 2002;**2**:96–105.

21. Cheng PC, Brown BK, Song W, Pierce SK. Translocation of the B cell antigen receptor into lipid rafts reveals a novel step in signaling. *J Immunol* 2001;**166**:3693–701.

22. DeFranco AL. The complexity of signaling pathways activated by the BCR. *Curr Opin Immunol* 1997;**9**:296–308.

23. Campbell KS. Signal transduction from the B cell antigen-receptor. *Curr Opin Immunol* 1999;**11**:256–64.

24. Kurosaki T. Functional dissection of BCR signaling pathways. *Curr Opin Immunol* 2000;**12**:276–81.

25. Kelly ME, Chan AC. Regulation of B cell function by linker proteins. *Curr Opin Immunol* 2000;**12**:267–75.

26. Kurosaki T. Regulation of B-cell signal transduction by adaptor proteins. *Nat Rev Immunol* 2002;**2**:354–63.

27. Hutchcroft JE, Harrison ML, Geahlen RL. B lymphocyte activation is accompanied by phosphorylation of a 72-kDa protein-tyrosine kinase. *J Biol Chem* 1991;**266**:14,846–14,849.

28. Hutchcroft JE, Harrison ML, Geahlen RL. Association of the 72-kDa protein tyrosine kinase PTK 72 with the B cell antigen receptor. *J Biol Chem* 1992;**267**:8613–19.

29. Saouaf SJ, Mahajan S, Rowley RB, Kut SA, Fargnoli J, Burkhardt AL, Tsukada S, Witte ON, Bolen JB. Temporal differences in the activation of three classes of non-transmembrane protein tyrosine kinases following B-cell antigen receptor surface engagement. *Proc Natl Acad Sci USA* 1994;**91**:9524–8.

30. Takata M, Sabe H, Hata A, Inazu T, Homma Y, Nukada T, Yamamura H, Kurosaki T. Tyrosine kinases lyn and Syk regulate B cell receptor-coupled Ca2+ mobilization through distinct pathways. *EMBO J* 1994;**13**:1341–9.

31. Turner M, Mee JP, Costello PS, Williams O, Price AA, Duddy LP, Furlong MT, Geahlen RL, Tybulewicz VLJ. Perinatal lethality and blocked B-cell development in mice lacking the tyrosine kinase Syk. *Nature* 1995;**378**:298–302.

32. Cheng AM, Rowley B, Pao W, Hayday A, Bolen JB, Pawson T. Syk tyrosine kinase required for mouse viability and B-cell development. *Nature* 1995;**378**:303–6.

33. Rowley RB, Burkhardt AL, Chao H-G, Matsueda GR, Bolen JB. Syk protein-tyrosine kinase is regulated by tyrosine-phosphorylated Igα/Igβ immunoreceptor tyrosine activation motif binding and autophosphorylation. *J Biol Chem* 1995;**270**:11,590–11,594.

34. Kurosaki T, Johnson SA, Pao L, Sada K, Yamamura H, Cambier JC. Role of the Syk autophosphorylation site and SH2 domains in B cell antigen receptor signaling. *J Exp Med* 1995;**182**:1815–23.

35. Turner M, Schweighoffer E, Colucci F, Di Santo JP, Tybulewicz VL. Tyrosine kinase SYK: essential functions for immunoreceptor signaling. *Immunol Today* 2000;**21**:148–54.

36. Ishiai M, Kurosaki M, Pappu R, Okawa K, Ronko I, Fu C, Shibata M, Iwamatsu A, Chan AC, Kurosaki T. BLNK required for coupling Syk to PLCγ2 and Rac1-JNK in B cells.. *Immunity* 1999;**10**:117–25.

37. Su Y-W, Zhang Y, Schweikert J, Koretsky GA, Reth M, Wienands J. Interaction of SLP adaptors with the SH2 domain of Tec family kinases. *Eur J Immunol* 1999;**19**:3702–11.

38. Kabak S, Skaggs BJ, Gold MR, Affolter M, West KL, Foster MS, Slemasko K, Chan AC, Aebersold R, Clark MR. The direct recruitment of BLNK to immunoglobin-α couples the B-cell antigen receptor to distal signaling pathways. *Mol Cell Biol* 2002;**22**:2524–35.

39. Wienands J, Schweikert J, Wollscheid B, Jumaa H, Nielsen PJ, Reth M. A new signaling component in B lymphocytes which requires expression of the antigen receptor for phosphorylation. *J Exp Med* 1998;**188**:791–5.

40. Coggeshall MK. Inhibitory signaling by B cell FcγRIIb. *Curr Opin Immunol* 1998;**10**:306–12.

41. Hempel WM, Schatzman RC, DeFranco AL. Tyrosine phosphorylation of phospholipase C-?2 upon crosslinking of membrane Ig on murine B lymphocytes. *J Immunol* 1992;**148**:3021–7.

42. Carter RH, Park DJ, Rhee SG, Fearon DT. Tyrosine phosphorylation of phospholipase C induced by membrane immunoglobulin crosslinking in B lymphocytes. *Proc Natl Acad Sci USA* 1991;**88**:2745–9.

43. Richards JD, Gold MR, Hourihane SL, DeFranco AL, Matsuuchi L. Reconstitution of B cell antigen receptor-induced signaling events in a nonlymphoid cell line by expressing the Syk protein-tyrosine kinase. *J Biol Chem* 1996;**271**:6458–66.

44. Ishiai M, Sugawara H, Kurosaki M, Kurosaki T. Cutting edge: association of phospholipase C-γ2 Src homology 2 domains with BLNK is critical for B-cell antigen receptor signaling. *J Immunol* 1999;**163**:1746–9.

45. Takata M, Kurosaki T. A role for bruton's tyrosine kinase in B cell antigen receptor-mediated activation of phospholipase C-γ2. *J Exp Med* 1996;**184**:31–40.

46. Desiderio S. Role of Btk in B cell development and signaling. *Curr Opin Immunol* 1997;**9**:534–40.

47. Fluckiger A-C, Li Z, Kato RM, Wahl MI, Ochs HD, Longnecker R, Kinet J-P, Witte ON, Scharenberg AM, Rawlings DJ. Btk/Tec kinases regulate sustained increases in intracellular Ca2+ following B cell receptor activation. *EMBO J* 1998;**17**:1973–85.

48. Satterthwaite AB, Cheroutre H, Khan WN, Sideras P, Witte ON. Btk dosage determines sensitivity to B cell antigen receptor cross-linking. *Proc Natl Acad Sci USA* 1997;**94**:13,152–13,157.

49. Rigley KP, Harnett MM, Phillips RJ, Klaus GGB. Analysis of signaling via surface immunoglobulin receptors on B cells from CBS/N mice. *Eur J Immunol* 1989;**19**:2081–6.

50. Wickler LS, Scher I. X-linked immune deficiency (xid) of CBA/N mice. *Curr Top Microbiol Immunol* 1986;**124**:87–101.

51. Rodriguez R, Matsuda M, Perisic O, Bravo J, Paul A, Jones NP, Light Y, Swann K, Williams RL, Katan M. Tyrosine residues in phospholipase Cγ2 essential for the enzyme function in B-cell signaling. *J Biol Chem* 2001;**276**:47,982–47,992.

52. Watanabe D, Hashimoto S, Ishiai M, Matsushita M, Baba Y, Kishimoto T, Kurosaki T, Tsukada S. Four tyrosine residues in phospholipase C-γ2, identified as Btk-dependent phosphorylation sites, are required for B cell antigen receptor-coupled calcium signaling. *J Biol Chem* 2001;**276**:38,595–38,601.

53. Berridge MJ. The versatility and complexity of calcium signaling. *Novartis Found Symp* 2001;**239**:64–7.

54. Miyakawa T, Maeda A, Yamazawa T, Hirose K, Kurosaki T, Iino M. Encoding of Ca2+ signals by differential expression of IP3 receptor subtypes. *EMBO J* 1999;**18**:1303–8.

55. Yokoyama K, Su I-H, Tezuka T, Yasuda T, Mikoshiba K, Tarakhovsky A, Yamamoto T. BANK regulates BCR-induced calcium mobilization by promoting tyrosine phosphorylation of IP3 receptor. *EMBO J* 2002;**21**:83–92.

56. Clapham DE. Calcium signaling. *Cell* 1995;**80**:259–68.

57. Lewis RS, Cahalan MD. Potassium and calcium channels in lymphocytes. *Annu Rev Immunol* 1995;**13**:623–53.

58. Engelke M, Engels N, Dittmann K, Stork B, Wienands J. Ca^{2+} signaling in antigen receptor-activated B lymphocytes. *Immunol Rev* 2007;**218**:235–46.

59. Scharenberg A, Humphries LA, Rawlings DJ. Calcium signalling and cell-fate choice in B cells. *Nat Rev Immunol* 2007;**7**:778–89.

60. Liou Jr. J, Kim ML, Heo WD, Jones JT, Myers JW, Ferrell JE, Meyer T STIM is a Ca^{2+} sensor essential for Ca^{2+}-store-depletion-triggered Ca^{2+} influx. *Curr Biol* 2005;**15**:1235–41.

61. Roos J, DiGregorio PJ, Yeromin AV, Ohlsen K, Lioudyno M, Zhang S, Safrina O, Kozak JA, Wagner SL, Cahalan MD, Veliçelebi G, Stauderman KA. STIM1, an essential and conserved component of store-operated Ca^{2+} channel function. *J Cell Biol* 2005;**169**:435–45.

62. Zhang SL, Yu Y, Roos J, Kozak JA, Deerinck TJ, Ellisman MH, Stauderman KA, Cahalan MD. STIM1 is a Ca^{2+} sensor that activates CRAC channels and migrates from the Ca^{2+} store to the plasma membrane. *Nature* 2005;**437**:902–5.

63. Feske S, Gwack Y, Prakriya M, Srikanth S, Puppel SH, Tanasa B, Hogan PG, Lewis RS, Daly M, Rao A. A mutation in orai1 causes immune deficiency by abrogating CRAC channel function. *Nature* 2006;**441**:179–85.

64. Prakriya M, Feske S, Gwack Y, Srikanth S, Rao A, Hogan PG. Orai1 is an essential pore subunit of the CRAC channel. *Nature* 2006;**443**:230–3.

65. Vig M, Peinelt C, Beck A, Koomoa DL, Rabah D, Koblan-Huberson M, Kraft S, Turner H, Fleig A, Penner R, Kinet JP. CRACM1 is a plasma membrane protein essential for store-operated Ca^{2+} entry. *Science* 2006;**312**:1220–3.

66. Baba Y, Hayashi K, Fujii Y, Mizushima A, Watarai H, Wakamori M, Numaga T, Mori Y, Iino M, Hikida M, Kurosaki T. Coupling of STIM1 to store-operated Ca^{2+} entry through its constitutive and inducible movement in the endoplasmic reticulum. *Proc Natl Acad Sci USA* 2006;**103**:16704–9.

67. Vazquez Jr G, Bird GS, Mori Y, Putney JW. Native TRPC7 channel activation by an inositol trisphosphate receptor-dependent mechanism. *J Biol Chem* 2006;**281**:25,250–25,258.

68. Lievremont Jr. JP, Numaga T, Vazquez G, Lemonnier L, Hara Y, Mori E, Trebak M, Moss SE, Bird GS, Mori Y, Putney JW The role of canonical transient receptor potential 7 in B-cell receptor-activated channels. *J Biol Chem* 2005;**280**:35,346–35,351.

69. Venkatachalam K, Zheng F, Gill DL. Regulation of canonical transient receptor potential (TRPC) channel function by diacylglycerol and protein kinase C. *J Biol Chem* 2003;**278**:29,031–29,040.

70. Vazquez G, Wedel BJ, Bird GS, Joseph SK, Putney JW. An inositol 1,4,5-trisphosphate receptor-dependent cation entry pathway in DT40 B lymphocytes. *EMBO J* 2002;**21**:4531–8.

71. Dellis O, Dedos SG, Tovey SC, Taufiq-Ur-Rahman., Dubel SJ, Taylor CW. Ca^{2+} entry through plasma membrane IP3 receptors. *Science* 2006;**313**:229–33.

72. Crabtree GR, Olson EN. NFAT signaling: choreographing the social lives of cells. *Cell* 2002;**109**:S67–79.

73. Winslow MM, Gallo EM, Neilson JR, Crabtree GR. The calcineurin phosphatase complex modulates immunogenic B cell responses. *Immunity* 2006;**24**:141–52.

74. Choi MS, Brines RD, Holman MJ, Klaus GG. Induction of NF-AT in normal B lymphocytes by anti-immunoglobulin or CD40 ligand in conjunction with IL-4. *Immunity* 1994;**1**:179–87.

75. Venkataraman L, Francis DA, Wang Z, Liu J, Rothstein TL, Sen R. Cyclosporin–a sensitive induction of NF-AT in murine B cells. *Immunity* 1994;**1**:189–96.

76. Glynne R, Akkaraju S, Healy JI, Rayner J, Goodnow CC, Mack DH. How self-tolerance and the immunosuppressive drug FK506 prevent B cell mitogenesis. *Nature* 2000;**403**:672–6.

77. Gold MR. Intermediary signaling effectors coupling the B-cell receptor to the nucleus. *Curr Top Microbiol Immunol* 2000;**245**:78–134.

78. Mischak H, Kolch W, Goodnight J, Davidson WF, Rapp U, Rose-John S, Mushinski JF. Expression of protein kinase C genes in hemopoietic cells is cell-type- and B cell-differentiation stage specific. *J Immunol* 1991;**147**:3981–7.

79. Sidorenko SP, Law C-L, Klaus SJ, Chandran KA, Takata M, Kurosaki T, Clark EA. Protein kinase C μ (PKCμ) associates with The B cell antigen receptor complex and regulates lymphocyte signaling. *Immunity* 1996;**5**:353–63.

80. Janknecht R, Hunter T. Convergence of MAP kinase pathways on the ternary complex factor sapla. *EMBO J* 1997;**16**:1620–7.

81. Treisman R. Regulation of transcription by MAP kinase cascades. *Curr Opin Cell Biol* 1996;**8**:205–15.

82. Krappmann D, Patke A, Heissmeyer V, Scheidereit C. B-cell receptor– and phorbol ester-induced NF-κB and c-Jun N-terminal kinase activation in B cells requires novel protein kinase C's. *Mol Cell Biol* 2001;**21**:6640–50.

83. Rawlings DJ, Sommer K, Moreno-Garcia ME. The CARMA1 signalosome links the signalling machinery of adaptive and innate immunity in lymphocytes. *Nat Rev Immunol* 2006;**6**:799–812.

84. Su TT, Guo B, Kawakami Y, Sommer K, Chae K, Humphries LA, Kato RM, Kang S, Patrone L, Wall R, Teitell M, Leitges M, Kawakami T, Rawlings DJ. PKC-beta controls I Kappa B kinase lipid raft recruitment and activation in response to BCR signaling. *Nat Immunol* 2002;**3**:780–6.

85. Shinohara H, Maeda S, Watarai H, Kurosaki T. IkappaB kinase beta-induced phosphorylation of CARMA1 contributes to CARMA1 Bcl10 MALT1 complex formation in B cells. *J Exp Med* 2007;**204**:3285–93.

86. Sommer K, Guo B, Pomerantz JL, Bandaranayake AD, Moreno-García ME, Ovechkina YL, Rawlings DJ. Phosphorylation of the CARMA1 linker controls NF-kappaB activation. *Immunity* 2005;**23**:561–74.

87. Sun L, Deng L, Ea CK, Xia ZP, Chen ZJ. The TRAF6 ubiquitin ligase and TAK1 kinase mediate IKK activation by BCL10 and MALT1 in T lymphocytes. *Mol Cell* 2004;**14**:289–301.

88. Leitges M, Schmedt C, Guinamard R, Davoust J, Schaal S, Stabel S, Tarakhovsky A. Immunodeficiency in protein kinase C beta-deficient mice.. *Science* 1996;**273**:788–91.

89. Newton K, Dixit VM. Mice lacking the CARD of CARMA1 exhibit defective B lymphocyte development and impaired proliferation of their B and T lymphocytes. *Curr Biol* 2003;**13**:1247–51.

90. Xue L, Morris SW, Orihuela C, Tuomanen E, Cui X, Wen R, Wang D. Defective development and function of Bcl10-deficient follicular, marginal zone and B1 B cells. *Nat Immunol* 2003;**4**:857–65.

91. Ruland J, Duncan GS, Wakeham A, Mak TW. Differential requirement for Malt1 in T and B cell antigen receptor signaling. *Immunity* 2003;**19**:749–58.

92. Sato S, Sanjo H, Takeda K, Ninomiya-Tsuji J, Yamamoto M, Kawai T, Matsumoto K, Takeuchi O, Akira S. Essential function for the kinase TAK1 in innate and adaptive immune responses. *Nat Immunol* 2005;**6**:1087–95.

93. Suzuki H, Terauchi Y, Fujiwara M, Aizawa S, Yazaki Y, Kadowaki T, Koyasu S. Xid-like immunodeficiency in mice with disruption of the p85α subunit of phosphoinositide 3-kinase.. *Science* 1999;**283**:390–2.

94. Fruman DA, Snapper SB, Yballe CM, Davidson L, Yu JY, Alt FW, Cantley LC. Impaired B cell development and proliferation in absence of phosphoinositide 3-kinase p85α. *Science* 1999;**283**:393–7.

95. Poe JC, Hasegawa M, Tedder TF. CD19, CD21, and CD22: multifaceted response regulators of B lymphocyte signal transduction. *Intl Rev Immunol* 2001;**20**:739–62.

96. Tuveson DA, Carter RH, Soltoff SP, Fearon DT. CD19 of B cells as a surrogate kinase insert region to bind to phosphatidylinositol 3-kinase. *Science* 1993;**260**:986–9.

97. Cherukuri A, Cheng PC, Sohn HW, Pierce SK. The CD19/CD21 complex functions to prolong B cell antigen receptor signaling from lipid rafts. *Immunity* 2001;**14**:169–79.

98. Okada T, Maeda A, Iwamatsu A, Gotoh K, Kurosaki T. BCAP: the tyrosine kinase substrate that connects B cell receptor to phosphoinositide 3-kinase activation. *Immunity* 2000;**13**:817–27.

99. Ingham RJ, Holgado-Madrug M, Siu C, Wong AJ. The Gab1 protein is a docking site for multiple proteins involved in signaling by the B cell antigen receptor. *J Biol Chem* 1998;**273**:30,630–30,637.

100. Ingham RJ, Santos L, Dang-Lawson M, Holgado-Madruga M, Dudek P, Maroun CR, Wong AJ, Matsuuchi L, Gold MR. The Gab1 docking protein links the B cell antigen receptor to the phosphatidylinositol 3-kinase/Akt signaling pathway and to the SHP2 tyrosine phosphatase. *J Biol Chem* 2001;**276**:12,257–12,265.

101. Inabe K, Ishiai M, Scharenberg AM, Freshney N, Downward J, Kurosaki T. Vav3 modulates B cell receptor responses by regulating phosphoinositide 3-kinase activation. *J Exp Med* 2002;**195**:189–200.

102. Cantley LC. The phosphoinositide 3-kinase pathway. *Science* 2002;**296**:1655–7.

103. Li H-L, Davis WW, Whiteman EL, Birnbaum MJ, Pure E. The tyrosine kinases Syk and lyn exert opposing effects on the activation of protein kinase Akt/PKB in B lymphocytes. *Proc Natl Acad Sci USA* 1999;**96**:6890–5.

104. Gold MR, Scheid MP, Santos L, Dang-Lawson M, Roth RA, Matsuuchi L, Duronio V, Krebs DL. The B cell antigen receptor activates the Akt (Protein Kinase B)/glycogen synthase kinase-3 signaling pathway via phosphatidylinositol 3-kinase. *J Immunol* 1999;**163**:1894–905.

105. Pogue SL, Kurosaki T, Bolen J, Herbst R. B cell antigen receptor-induced activation of Akt promotes B cell survival and is dependent on Syk kinase. *J Immunol* 2000;**165**:1300–6.

106. Craxton A, Jiang A, Kurosaki T, Clark EA. Syk and bruton's tyrosine kinase are required for B cell antigen receptor-mediated activation of the kinase Akt. *J Biol Chem* 1999;**274**:30,644–30,650.

107. Astoul E, Watton S, Cantrell D. The dynamics of protein kinase B regulation during B cell antigen receptor engagement. *J Cell Biol* 1999;**145**:1511–20.

108. Beals CR, Sheridan CM, Turck CW, Gardner P, Crabtree GR. Nuclear export of NF-ATc enhanced by glycogen synthase kinase-3. *Science* 1997;**275**:1930–4.

109. Fu C, Turck CW, Kurosaki T, Chan AC. BLNK: a central linker protein in B cell activation. *Immunity* 1998;**22**:267–72.

110. Wienands J, Schweikert J, Wollscheid B, Jumaa H, Nielsen PJ, Reth M. SLP-65: a new signaling component in B lymphocytes which requires expression of the antigen receptor for phosphorylation. *J Exp Med* 1998;**188**:791–5.

111. Ekholm SV, Reed SI. Regulation of G(1) cyclin-dependent kinases in the mammalian cell cycle. *Curr Opin Cell Biol* 2000;**12**:676–84.

112. Piatelli MJ, Doughty C, Chiles TC. Requirement for a Hsp90 chaperone-dependent MEK1/2-ERK pathway for B cell antigen receptor-induced cyclin D2 expression in mature B lymphocytes. *J Biol Chem* 2002;**277**:12,144–12,150.

113. Sherr CJ, Roberts JM. CDK inhibitors: positive and negative regulators of G1-phase progression. *Genes Dev* 1999;**13**:1501–12.

114. McLeod SJ, Ingham RJ, Bos JL, Kurosaki T, Gold MR. Activation of the Rap1 GTPase by the B cell antigen receptor. *J Biol Chem* 1998;**273**:29,218–29,223.

115. Zwartkruis FJ, Bos JL. Ras and Rap1: two highly related small GTPases with distinct function. *Exp Cell Res* 1999;**253**:157–65.

116. Bos JL. All in the family? new insights and questions regarding interconnectivity of Ras, Rap1, and Ral. *EMBO J* 1998;**17**:6776–82.

117. Crespo P, Schuebel KE, Ostrom AA, Gutkind JS, Bustelo XR. Phosphotyrosine-dependent activation of Rac-1 GDP/GTP exchange by the Vav proto-oncogene product. *Nature* 1997;**385**:169–72.

118. Kaibuchi K, Kuroda S, Amano M. Regulation of the cytoskeleton and cell adhesion by the Rho family GTPases in mammalian cells. *Annu Rev Biochem.* 1999;**68**:459–86.

119. Cyster JG, Goodnow CC. Tuning antigen receptor signaling by CD22: integrating cues from antigens and the microenvironment. *Immunity* 1997;**6**:509–17.

120. Fearon DT, Carroll MC. Regulation of B lymphocyte responses to foreign and self-antigens by the CD19/CD21 complex. *Annu Rev Immunol* 2000;**18**:393–422.

121. Coggeshall KM. Positive and negative signaling in B lymphocytes. *Curr Top Microbiol Immunol* 2000;**245**:213–60.

122. Takai T, Ono M. Activating and inhibitory nature of the murine paired immunoglobulin-like receptor family. *Immunol Rev* 2001;**181**:215–22.

Chapter 20

Signaling Pathways Regulating Growth and Differentiation of Adult Stem Cells

Larry Denner[1–4], Margaret Howe[1,2,4] and Randall J. Urban[1,2]

[1]McCoy Stem Cells and Diabetes Mass Spectrometry Research Laboratory
[2]Stark Diabetes Center
[3]Sealy Center for Molecular Medicine
[4]Department of Internal Medicine, University of Texas Medical Branch, Galveston, Texas

INTRODUCTION

Stem cells have remarkable capabilities of self-renewal and the potency to differentiate into many different cell types. Through these properties, adult stem cells can serve as a repair system for the body. Some stem cells can divide indefinitely to replenish other cells for the lifespan of an individual. Many types of stem cells show great promise for clinical applications due to these unique and powerful capabilities to replenish other cells. However, clinical applications have lagged due to the difficulties in isolating pure populations of adult stem cells, and in growing sufficient quantities required for transplantation indications. Consequently, it is critical to understand the mechanisms of self-renewal, differentiation, and multipotency to develop new strategies to isolate and to grow adult stem cells.

STEM CELL PROPERTIES

Self-Renewal and Differentiation

Self-renewal and differentiation are two unique properties that distinguish stem cells from somatic cells. During division, a stem cell either self-renews (remains a stem cell after cell division) or differentiates (becomes a more mature cell), resulting in the following three possibilities [1]: (1) self-renewal to produce two daughter cells identical to the parental cell; (2) self-renewal and differentiation to produce one identical and one differentiated daughter cell; (3) differentiation to produce two daughter cells that differ from the parent.

Multipotency

The number of lineages that a stem cell can form defines its differentiation potential. A stem cell is pluripotent if it can differentiate into each of the three developmental germ layers. Thus, pluripotent embryonic stem cells (ESCs) form cells of the endoderm, ectoderm, and mesoderm. A stem cell is multipotent if it differentiates into many, but not all, lineages of cells. For example, adult cord blood stem cells described below can be engineered to produce insulin [2, 3].

Adult Stem Cells

Through self-renewal and differentiation, adult stem cells are capable of tissue repair to maintain tissue homeostasis in a niche-specific controlled microenvironment [4, 5] as well as organize tissue regeneration and repair upon stress [6]. Self-renewal generates an identical cell to maintain the stem cell population [1]. Differentiation generates non-identical cells that can then differentiate into many cell types, but cannot self-renew. Although adult stem cells have been recently characterized in many tissues (including neural, muscular, hepatic, and cardiovascular tissues), hematopoietic stem cells remain one of the best-characterized adult stem cell populations with evidence of clinical applications.

Hematopoietic Stem Cells (HSCs)

Hematopoietic cells derived from bone marrow are composed of multiple subpopulations. The stem cell subpopulation is classified by its duration of self-renewal, either as long-term hematopoietic stem cells or as short-term hematopoietic stem cells. Both of these subpopulations are multipotent, but differ in the duration they maintain their self-renewal properties. They give rise to the committed myeloid and lymphoid progenitors. These progenitor subpopulations differentiate into their respective lineages, but do not self-renew. As HSCs mature, these populations multiply to maintain all the necessary cellular and functional features of blood.

Umbilical Cord Blood Stem Cells

Stem cells from umbilical cord blood (UCB), bone marrow (BM), and mobilized peripheral blood have many properties that are advantageous clinically. For example, these sources were successfully used in cellular therapy of myocardially infarcted cardiac tissues for BM and in hematotherapy using UCB [7–9]. McGuckin and colleagues previously reported [10–13] the isolation, expansion, and engineering of cord blood stem cell lineages of various degrees of multipotency into many types of functional progenitor phenotypes, including blood, neural, and hepatic. Using multiparametric phenotyping, it was demonstrated that UCB contained a complex succession of heterogeneous stem and progenitor cell groups up- and downregulating a range of surface antigens, including CD133, CD34, CD38, CD7, and CD90, as they differentiated [11, 13].

We recently reported the identification, isolation, expansion, and engineering of cord blood-derived, embryonic-like stem cells (CBEs) from the lineage negative compartment that express ESC-specific sialoprotein antigens SSEA-3 and -4 but not SSEA-1 [13], consistent with the human ESC pattern and confirming their undifferentiated phenotype [14]. These CBEs also expressed tumour rejection antigens TRA 1-60 and TRA 1-81, ESC-specific sialylated keratin sulphate proteoglycans. In addition, they formed clusters similar to embryoid bodies [15] that were quite fragile and may reflect a very undifferentiated state. CBEs were also positive for the pluripotency transcription factor Oct-4 involved in inhibition of differentiation and support of self-renewal of ESC [16, 17]. The multipotency of these clusters was demonstrated by propagation with standard cytokines used for other stem cell lineages and differentiation to the hepatic lineage phenotype.

CD133 cord blood stem cells have the capacity for self-renewal, proliferation, and multipotency. Freshly isolated cells can self-renew to grow and produce identical CD133+ daughters, and are multipotent in ability to form all three developmental germ layers (Figure 20.1, type 1). Under normal cell culture conditions cells undergo asymmetric growth, differentiate with loss of the CD133 marker, stop proliferating, and have decreased potency (Figure 20.1, type 2). As described below, inhibition of glycogen synthase kinase 3β, a key mediator of the Wnt signaling axis, causes loss of the CD133 marker, but maintenance of multipotency and, surprisingly, proliferative capacity (Figure 20.1, type 3). It will be important to identify signaling mechanisms under conditions of growth with or without CD133 (type 1 vs type 3), or lack of CD133 with or without growth (type 2 vs type 3). Using this experimental

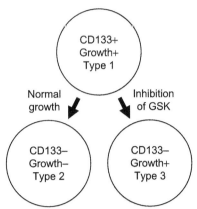

FIGURE 20.1 Characteristics of CD133 cord blood stem cell growth. Freshly isolated CD133+ cells grow to produce identical progeny through self-renewal (CD133+, Growth+, type 1). Under normal culture conditions, type 1 cells also differentiate by losing the CD133 marker and stop growing (CD133−, Growth−, type 2). When cultured with BIO to inhibit GSK, type 1 cells differentiate but continue growing (CD133−, Growth+, type 3).

paradigm, we can now distinguish signaling events important for growth independent of differentiation in addition to those important for differentiation independent of growth.

SIGNALING INTERMEDIATES AND PATHWAYS IN CD133 STEM CELLS

Oct-4 Signaling Pathway

The POU domain containing transcription factor Oct-4, previously known as Oct-3/4, confers ESC self-renewal and pluripotency [18]. As ESCs differentiate and lose pluripotency, expression of Oct-4 is downregulated [19–26]. Subsequent overexpression allows cells to regain the ESC primitive phenotype, indicating this marker alone can reprogram self-renewal mechanisms [16, 23–26]. In addition, Oct-4 deficient embryos and Oct-4 knockdown ESCs exhibit a loss of pluripotency and differentiation [16, 27], while repression of Oct-4 expression results in the loss of self-renewal in ESCs [28]. Also, in the absence of Oct-4 the inner cell mass was restricted to the trophoblast lineage and lacked ESCs [27]. Finally, overexpression of Oct-4 in combination with three other transcription factors (Sox-2, c-Myc, and Klf4) results in fibroblast transformation into pluripotent cells that have the ability to self-renew [22].

Oct-4 is comprised of two isoforms, Oct-4 A and Oct-4B. These have identical central POU DNA binding domains and C-terminal domains, but differ in the N-terminal domains. Further, the A isoform resides in the nucleus, possesses a functional N-terminal transactivation domain, and induces target gene expression, while the B isoform is cytoplasmic and has no transactivation domain. Finally, only the A isoform is responsible for stemness properties [29] and sustains stem cell renewal [18].

Previous studies had reported that subsets of cord blood stem cells, lineage negative cells and CD133$^+$ cells, expressed Oct-4 [13, 30]. However, these studies did not specify the isomer of Oct-4 detected. We recently reported that lineage negative stem cells (CD133-, CD34-, CD45-, CD33- and CD7-) isolated from human umbilical cord blood expressed many ESC markers, including Oct-4, thus raising the possibility that this transcription factor might confer ESC-like pluripotency and indefinite self-renewal to this type of stem cell [13]. However, while lineage negative cells have the potential to generate all three developmental lineages, they did not demonstrate indefinite self-renewal [31]. Furthermore, Oct-4 is expressed on hematopoietic, mesenchymal, follicular, breast, liver, pancreatic, kidney, and gastric adult stem cells [13, 20, 30, 32, 33]. These observations continue to leave open the potential for Oct-4 to confer self-renewal in some adult stem cells. However, Lengner and colleagues recently demonstrated that Oct-4 is not even essential for self-renewal and maintenance of mouse somatic stem cells [34]. Thus, the ability of Oct-4 to confer self-renewal has been under intense scrutiny [13, 18, 20, 30, 34–38].

Studies on Oct-4 have been complicated by subtleties posed by the two isoforms. Protein analysis was performed by several laboratories, using antibodies that recognize regions common to both isoforms [18, 22] and that are unable to distinguish which isoform is present in any given cell. This complexity provided the opportunity that Oct-4A might confer self-renewal. However, a goat anti-human Oct-4A antibody was raised to an undisclosed 10 amino acid peptide from within residues 10 to 60 in the 134 amino acid N-terminal domain of Oct-4A that does not exist in Oct-4B.

Detection of the Oct-4 mRNA has been equally problematic. Several primers have been reported for the analysis of Oct-4 mRNA expression, but many code for Oct-4 pseudogenes [37]. Primers specific for Oct-4A mRNA coding sequences have been recently reported wherein the 5' primer only anneals to sequences coding for the N-terminal domain of Oct-4A that do not exist in Oct-4B. Thus, tools are available for specific detection of Oct-4A mRNA that does not identify Oct-4B mRNA.

We used these new approaches to study Oct-4A expression in CD133 cord blood stem cells [38]. Oct-4A protein and mRNA were expressed in freshly isolated CD133 cells (Figure 20.1, type 1 cells) that had high proliferative potential. However, expression of Oct-4A was maintained during proliferation in culture during which CD133 and other adult stem cell markers were downregulated, while differentiated hematopoetic markers were induced, all the while maintaining multipotent capacity to give rise to hematopoetic lineages (Figure 20.1, type 2). Therefore, expression of Oct-4A mRNA or protein neither implies nor confers ESC-like properties in human umbilical cord blood CD133 stem cells or their differentiated progeny. It

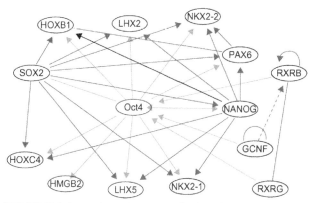

FIGURE 20.2 Ingenuity pathway analysis of the Oct-4 signaling network.
Light-gray arrows indicated relationships between Oct-4 and some of the proteins in the Oct-4 network. Black lines indicate relationships among other members of the Oct-4 network. The connectivity is an oversimplification demonstrating the complex communication in among elements of signaling networks.

will be important to use the specific tools described here to understand the role of Oct-4A in other adult stem cells to better define the emergent, but undiscovered, properties of Oct-4A.

Bioinformatics approaches such as Ingenuity Pathway Analysis (www.ingenuity.com) illuminate the complexity of the Oct-4 signaling pathway (Figure 20.2). This analysis is based on the most extensive literature-based annotation of the relationships between molecular intermediates (signaling nodes) and the signaling pathways they subserve. As has been long established, Nanog contributes to the maintenance of pluripotency, clonal expansion, and Oct-4 levels [23]. Sox-2 is critical for establishment of early cell fate decisions and activated by Oct-4 [21]. Recently, Oct-4 and Sox-2 have been shown in several cell types, thus bringing into question the universal role of Oct-4 and Sox-2 in identification of pluripotent, self renewing stem cells [20, 32, 33, 39, 40]. The nuclear orphan receptor, GCNF, also has important regulatory interactions with Oct-4 itself, as well as regulation of Oct-4 gene expression [41, 42]. Oct-4 actions are further mediated by other nuclear receptors such as retinoid X receptors (RXRB and RXRG) and the HOX family of transcription factors (HOXB1, HOXC4).

There are many potential possibilities for the inability of Oct-4 expression to confer self-renewal and multipotency in adult stem cells as in ESCs. First, expression levels of Oct-4 govern its functional consequences, and levels in CD133 cells may just be too low [28]. Second, we showed in human cord blood CD133 stem cells [38] that Oct-4A is present in both the nucleus and cytoplasm, clearly indicating some level of dysfunctional coupling of Oct-4A from its role as a nuclear transcription factor where it binds with co-regulators such as GCNF. Sox2 showed a similar subcellular redistribution. Third, a recent report from Lengner and colleagues showed in somatic stem cells that Oct-4 is not even essential for self-renewal [34]. Fourth, it is implicit that appropriate functional regulation of Oct-4 is necessary for the manifestation of self-renewal and multipotency. Indeed, Oct-4 and Sox-2 undergo many posttranslational modifications, such as phosphorylation, acetylation [43], and sumoylation [44], that influence function. These studies in different species with stem cells of different origins directly point to a more complex role of Oct-4 in stem cell biology than previously considered. Measurement of Oct-4 may be only a sentinel marker for the critical signaling pathways in stem cell self-renewal.

Wnt Signaling Pathway

Signaling of the Wnt pathway, shown in Figure 20.3, is regulated by interactions between several key proteins, including glycogen synthase kinase-3β (GSK-3β), β-catenin, and T cell factor-4 (Tcf-4). The rate-limiting signaling node in this pathway is GSK-3β, a serine-threonine kinase with two isoforms. The alpha isoform (51 kDa) is a key regulator of glucose metabolism, while the beta form (47 kDa) mediates Wnt signaling through β-catenin. In the absence of Wnt signaling (Figure 20.3b), active GSK-3β forms a multiprotein complex with Axin-1 and adenomatosis polyposis coli (APC). Casein kinase I-α (CKI-α) phosphorylates β-catenin on Ser9 [45], a "priming" site required for subsequent recognition by active GSK-3β, which can then phosphorylate β-catenin at Ser33, Ser37, and Thr41 [45, 46]. The F-box protein beta-TrCP binds to phosphorylated β-catenin, targeting it for proteosomal degradation by the E3 ubiquitin ligase (Figure 20.1) [46, 47]. Conversely, activation of the Wnt pathway (Figure 20.3a) results in inhibition of GSK-3β, thereby preventing phosphorylation-dependent degradation of β-catenin. β-catenin then translocates to the nucleus, where it binds and activates Tcf-4 inducing transcription of target genes such as cyclin D1, c-myc, PPAR-γ, MMP-7, and Axin-1.

The kinase activity of GSK-3β is regulated by phosphorylation on Tyr216 and Ser9. Tyr216 phosphorylation increases the kinase catalytic activity of GSK-3β [48, 49], while phosphorylation of Ser9 decreases GSK-3β activity [50]. Because GSK-3β has a role in a wide variety of disease processes, it has been a key target for drug discovery efforts that have produced more than 30 small molecule inhibitors of GSK-3β with different ranges of specificity [51]. Pharmacological inhibition of phosphorylation of Tyr216 on GSK-3β directly inhibits kinase activity, thereby resulting in β-catenin activation of Wnt pathway signaling. (2′Z,3′E)-6-bromoindirubine-3′-oxime (BIO), a small organic molecule, specifically inhibits the kinase activity of GSK-3β by preventing the activating phosphorylation of Tyr216 [52].

The canonical Wnt pathway has long been recognized to have an important role in development. Wnt signaling is required at four distinct developmental stages: formation of

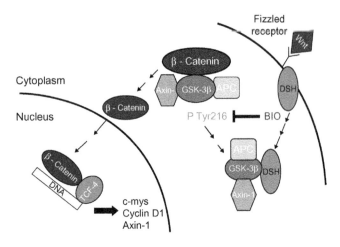

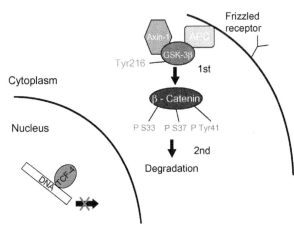

FIGURE 20.3 The Role of GSK-3β in the Wnt signaling network.
(a) Phosphorylation of Tyr216 inhibits GSK-3β, thereby activating the Wnt signaling pathway. (b) Activation of GSK leads to phosphorylation of β-catenin, targeting it for degradation and resulting in inhibition of Wnt signaling. See text for detailed abbreviations.

the primitive streak, subsequent induction of the mesoderm and endoderm, formation of hematopoietic progenitor cells, and differentiation of erythroid cells [53–55]. The pathway was first recognized for its function in the development of body axes and subsequent patterning during gastrulation [53]. Later, the role of the Wnt pathway in the development of the inner cell mass into both the mesoderm and HSCs was demonstrated [54, 55]. Since then, many reports support the hypothesis that the Wnt pathway is important in proliferation, differentiation, and self-renewal of HSCs [56–59]. More recently, the importance of the canonical Wnt pathway was demonstrated in studies on the tails of flatworms [60]. Expression of the Wnt protein, β-catenin, caused regeneration of the tail, while silencing β-catenin resulted in the formation of a head instead of a tail. Many additional studies have since shown that co-incubation of HSCs with stromal cells transfected with genes coding for Wnt ligands (Wnt1, Wnt5a, or Wnt10b) increased the percentage of primitive stem cells [56, 58, 59]. Furthermore, HSCs could be regulated by adding extracellular factors such as lipid-modified Wnt3a [56, 59, 61]. In addition to extracellular stimulation, intracellular modulation of the Wnt axis has been performed by transfection with constitutively active β-catenin, or inhibition of GSK-3β with pharmacological compounds such as LiCl, CHIR-911, or BIO in HSCs and ESCs [56, 57, 59, 62]. However, the effect of activating the Wnt pathway on proliferation, self-renewal, and differentiation of adult blood stem cells, in particular CD133 cells, remains relatively unexplored.

The specific role of GSK-3β kinase activity in self-renewal, differentiation, and multipotency of CD133 stem cells was, until recently, also unknown. However, we showed [63] that phosphorylation of Tyr216 on GSK-3β increased specifically in the CD133- differentiated population (Figure 20.1, type 2). This finding is the first to correlate GSK-3β kinase activity with stem cell differentiation. Since β-catenin was undetectable in fresh CD133 cells, GSK-3β kinase activity was not correlated with β-catenin, as previously published literature suggested [46, 64, 65]. However, since neither β- nor γ-catenin are essential for normal hematopoiesis in mice [66], these may not be essential for self-renewal of CD133 stem cell development. Another possibility is that GSK-3β is fully active in CD133+ and CD133− cells to the extent that β- and γ-catenin are produced, but continuously degraded. This latter theory seems the most plausible, since BIO treatment decreased GSK-3β kinase activity, leading to increased β-catenin levels from previously undetectable levels.

The downstream consequences of Wnt pathway activation provide important insight as well. While little is known about the role of the Wnt transcriptional mediator Tcf-4 in hematopoietic stem cells, it has been described in intestinal stem cells, where the loss of Tcf-4 resulted in differentiated, non-proliferative cells in intestinal crypts [67]. As shown in Figure 20.3, Tcf-4 is also known to co-activate the transcription of many target genes, including cyclin D1, through interactions with β- and γ-catenin [68, 69]. Tcf-4 expression and activity dramatically decreased as CD133 cells differentiated and cyclin D1 was downregulated (Howe, submitted data). Interestingly, Tcf-4 levels were unchanged by inhibition of GSK-3β even though cyclin D1 mRNA increased. Since cyclin D1 promotes progression through the G_1-S phase of the cell cycle via inactivation of the retinoblastoma protein [70], cyclin D1 could play an essential role in proliferation of CD133− BIO-treated cells. Thus, the Wnt pathway is activated in cells that maintain the CD133 marker and deactivated in cells that lose the CD133 marker and gain the differentiation markers CD14, CD15, and CD56.

Inhibition of GSK-3β has been shown to promote both self-renewal and differentiation of CD133 cord blood stem cells [57, 62]. As expected, BIO treatment activated the Wnt pathway by inhibiting phosphorylation of Tyr216 on GSK-3β and increased expression of both cyclin D1 and previously undetected β-catenin. BIO activation of the Wnt pathway led to an increased rate of loss of CD133 and

other adult stem cell markers while, surprisingly, maintaining proliferative capacity (Figure 20.1, type 3). Holmes and colleagues also found a similar phenomenon after treatment of CD34+ cord blood cells with BIO [71], suggesting potential common signaling pathways in various lineages of adult cord blood stem cells.

Although further studies are required to delineate comprehensive effects of BIO inhibition of GSK-3β, we speculate that treatment would expand the CFU-GEMM population in cord blood stem cells, thereby decreasing the time to repopulate the bone marrow. In addition, BIO-treated cells had a significantly increased erythroid potential. This is similar to a recent study showing that activation of the Wnt pathway through either addition of the Wnt3 ligand, or β-catenin overexpression in mouse Flk1+ mesodermal progenitors, increased the frequency of primitive erythroid colony formation, and inhibition of the Wnt pathway completely blocked primitive erythroid colony formation [54]. Since this is somewhat different than studies on CD133 cells, our findings highlight the importance of GSK-3β inhibition in directing cells down the hematopoietic pathway.

Proliferation of differentiated cells is commonly noted in many cancers. GSK-3β inhibition provides an additional example of how activation of the Wnt pathway contributes to cancer. Through GSK-3β inhibition, increased expression of two oncogenes, β-catenin and cyclin D1, have been shown to predict poor prognosis in patients with acute and chronic myeloid leukemia [72–74]. For example, GSK-3β inhibition may lead to major alterations in the Hedgehog and Notch pathways giving rise to different phenotypes [75, 76]. Therefore, activation of the Wnt pathway could result in proliferation of differentiated hematopoietic cells in hematological malignancies, and provide insight into potential intervention strategies in the future.

Wnt pathway signaling is critical to understanding regulation of self-renewal and differentiation in cord blood stem cells for clinical applications in regenerative medicine. For example, activation of the Wnt pathway through inhibition of GSK-3β could result in increased self-renewal of CD133+ cells. This would be a key discovery for hematopoetic transplantation. On the other hand, activation of the Wnt pathway through inhibition of GSK-3β could predispose CD133+ cells to differentiate into one or more hematopoetic lineages (erythroid, granulocyte, monocyte, or megakaryocyte). This knowledge would be relevant to the development of hematopoietic cells and hematopoietic diseases.

Mass Spectrometry and Bioinformatics Identification of Signaling Pathways and Networks

While typical studies have used reductionist approaches to focus on a single signaling intermediate or closely related members, unbiased mass spectrometry allows the discovery of new entities not constrained by prior knowledge or preconceived ideas. A stable isotope, quantitative mass spectrometry approach [77–79] was used to identify differentially expressed proteins in various CD133 populations and their differentiated progeny. Data were analyzed using bioinformatics tools (Ingenuity Pathways Analysis, www.ingenuity.com) to identify signaling pathways and networks critical for integration of convergent and divergent information processing in growth and differentiation [80, 81].

The experimental approach was rather straightforward. After 1 week in culture, the CD133+ and CD133– populations (Figure 20.1, types 1 and 2) were separated by immunomagnetic bead selection. Cell extracts from the two populations were separately digested with trypsin and differentially labeled with stable isotopes using trypsin-mediated carboxyl-terminal exchange of oxygen molecules from normal (^{16}O-H$_2$O) or heavy (^{18}O-H$_2$O) water [82, 83]. Following mixing of the samples, off-line strong cation exchange chromatography allowed deeper interrogation of the proteome. Reversed-phase tandem mass spectrometry was performed in the data-dependent mode to fragment peptides for sequence identification and to perform zoom scans to quantify differential expression ratios of tens of thousands of tryptic peptides corresponding to over 1100 proteins. Extensive filtering and validation then confirmed 263 differentially expressed proteins that changed more than ±30 percent – the accuracy limit of the method.

Bioinformatics tools are critical to the integration of large data sets into comprehensible models of pathways and related networks. Table 20.1 shows Ingenuity Pathway Analysis classification of differentially expressed proteins in the CD133+ cells compared to the CD133– cells into Molecular and Cellular Functions (top) and Signaling Network Functions (bottom).

The highest-probability Molecular and Cellular Function (P = 2.33E-06) with the most molecules (n = 60) is Cellular Growth and Proliferation (Table 20.1, top). This is excellent independent validation of the approach, since CD133+ cells do grow while CD133– cells do not grow [38]. Further validation derived from the calculated enhanced relative expression of CD34 in this functional group in CD133+ cells, an observation consistent with those that we and others have made by flow cytometry showing this marker is present in CD133+ cells and nearly absent from CD133– cells. This group also contains the growth regulators cyclins, histones, and RNA polymerase II. Other important Molecular and Cellular Functions include Cell-To-Cell Signaling, DNA Replication, and Cellular Assembly and Organization. All of these are critical to cell division in CD133+ cells.

The highest scoring (score = 42) Signaling Network Function with the most focus molecules (n = 25) was also Cellular Growth and Proliferation (Table 20.1, bottom). This confirms the central connections between molecular and cellular functions with signaling network functions. Other

TABLE 20.1 Ingenuity Pathway Analysis of CD133 cord blood stem cells[1]

Molecular and Cellular Functions		
Name	P value	# Molecules
Cellular Growth and Proliferation	2.33E-06	60
RNA Post-transcriptional Modification	3.62E-06	19
Cell-to-Cell Signaling and Interaction	2.79E-05	40
DNA Replication, Recombination, and Repair	2.79E-05	34
Cellular Assembly and Organization	3.63E-05	50

Signaling Network Functions		
Name	Score	Focus molecules
Cellular Growth and Proliferation	42	25
Cancer, Molecular Transport	39	24
Cell Morphology, Assembly, and Organization	37	23
Cell Signaling, DNA Replication	29	19
Cell Morphology, Cell Death	26	18

[1]CD133 cells were grown for 1 week in culture, and CD133+ cells separated from CD133– cells. Stable isotope labeling and quantitative mass spectrometry identified differentially expressed proteins, which were submitted to IPA. Values represent functions enriched in CD133+ cells compared to CD133– cells.

high-scoring networks included Cell Morphology, Cell Death, with members like poly-ADP ribose polymerase (PARP) and annexins that execute apoptotic death. With recent reports of the central causal role of caspase-mediated processes in ESC self-renewal, it will be important to assess engagement of the apoptotic machinery in CD133 cell growth and potency.

Thus, a mass spectrometry/bioinformatics approach has led to several emergent properties of CD133+ growth and differentiation to produce CD133-progeny. This approach will also be invaluable to identify signaling pathways that underlie these processes in normal maturation in culture as well as under conditions of Wnt pathway activation with BIO treatment that induces the differentiated CD133 cells that continue to proliferate.

CONCLUSIONS

Early attempts to expand cord blood stem cells *ex-vivo* before transplantation were disappointing, mainly due to predominant expansion of mature, rather than immature, cells [84]. It is now clear that understanding the signaling pathways of self-renewal, differentiation, and proliferation is essential to leverage the special properties of adult cord blood stem cells for regenerative medicine.

The principles that regulate these properties in ESCs are not operative in CD133 stem cells. For example, the Oct-4 pathway does not recapitulate the same properties identified in other stem cell populations such as ESCs. Further, Oct-4 is expressed in many differentiated cell types where the function of Oct-4 remains to be discovered.

Activity of the Wnt signaling pathway is also different in CD133 cord blood cells compared to ESCs. In freshly isolated CD133 cells, the Wnt pathway is active. In culture, these cells undergo a combination of self-renewal and differentiation, the latter of which is correlated with inactivation of the Wnt pathway and loss of proliferative potential. Surprisingly, activation of Wnt signaling by BIO-induced inhibition of GSK accelerated differentiation but maintained the proliferative capacity. Signaling mechanisms that maintained expression of cyclin D1 may hold the key to this apparent conundrum. Clearly, there are important underlying signaling pathways that allow the generation of differentiated cells that continue to proliferate. Unbiased quantitative mass spectrometry to identify differentially expressed proteins combined with bioinformatics approaches to identifying signaling nodes, pathways, and networks holds tremendous promise to unveil the subtleties of signaling pathways in regulation of CD133 stem cell growth, differentiation, and potency.

REFERENCES

1. NIH (2001). Stem Cells: Scientific Progress and Future Research D. (2001). Bethesda, MD.
2. Denner L, Urban RJ. Critical issues for engineering cord blood stem cells to produce insulin. *Expert Opin Biol Ther* 2008;**8**:1251–4.
3. Denner L, Bodenburg Y, Zhao JG, et al. Directed engineering of umbilical cord blood stem cells to produce C-peptide and insulin. *Cell Prolif* 2007;**40**:367–80.
4. Li L, Xie T. Stem Cell Niche: Structure and Function. *Annu Rev Cell Dev Biol* 2005;**21**:605–31.
5. Heissig B, Ohki Y, Sato Y, Rafii S, Werb Z, Hattori K. A role for niches in hematopoietic cell development. *Hematology* 2005;**10**:247–53.
6. Quesenberry PJ, Dooner G, Colvin G, Abedi M. Stem cell biology and the plasticity polemic. *Exp Hematol* 2005;**33**:389–94.
7. Stamm C, Westphal B, Kleine HD, et al. Autologous bone-marrow stem-cell transplantation for myocardial regeneration. *Lancet* 2003;**361**:45–6.
8. Perin EC, Silva GV. Stem cell therapy for cardiac diseases. *Curr Opin Hematol* 2004;**11**:399–403.
9. Cohen Y, Nagler A. Umbilical cord blood transplantation – how, when and for whom? *Blood Rev* 2004;**18**:167–79.
10. McGuckin CP, Forraz N, Allouard Q, Pettengell R. Umbilical cord blood stem cells can expand hematopoietic and neuroglial progenitors *in vitro*. *Exp Cell Res* 2004;**295**:350–9.
11. McGuckin CP, Forraz N, Baradez MO, et al. Colocalization analysis of sialomucins CD34 and CD164. *Stem Cells* 2003;**21**:162–70.
12. Forraz N, Pettengell R, Deglesne PA, McGuckin CP. AC133+ umbilical cord blood progenitors demonstrate rapid self-renewal and low apoptosis. *Br J Haematol* 2002;**119**:516–24.
13. McGuckin CP, Forraz N, Baradez MO, et al. Production of stem cells with embryonic characteristics from human umbilical cord blood. *Cell Prolif* 2005;**38**:245–55.
14. Thomson JA, Itskovitz-Eldor J, Shapiro SS, et al. Embryonic stem cell lines derived from human blastocysts. *Science* 1998;**282**:1145–7.
15. Hoffman LM, Carpenter MK. Characterization and culture of human embryonic stem cells. *Nat Biotechnol* 2005;**23**:699–708.
16. Gerrard L, Zhao D, Clark AJ, Cui W. Stably transfected human embryonic stem cell clones express OCT4-specific green fluorescent protein and maintain self-renewal and pluripotency. *Stem Cells* 2005;**23**:124–33.
17. Matin MM, Walsh JR, Gokhale PJ, et al. Specific knockdown of Oct4 and beta2-microglobulin expression by RNA interference in human embryonic stem cells and embryonic carcinoma cells. *Stem Cells* 2004;**22**:659–68.
18. Lee J, Kim HK, Rho JY, Han YM, Kim J. The human OCT-4 isoforms differ in their ability to confer self-renewal. *J Biol Chem* 2006;**281**:33,554–33,565.
19. Pesce M, Scholer HR. Oct-4: gatekeeper in the beginnings of mammalian development. *Stem Cells* 2001;**19**:271–8.
20. Zangrossi S, Marabese M, Broggini M, et al. Oct-4 expression in adult human differentiated cells challenges its role as a pure stem cell marker. *Stem Cells* 2007;**25**:1675–80.
21. Avilion AA, Nicolis SK, Pevny LH, Perez L, Vivian N, Lovell-Badge R. Multipotent cell lineages in early mouse development depend on SOX2 function. *Genes Dev* 2003;**17**:126–40.
22. Takahashi K, Yamanaka S. Induction of pluripotent stem cells from mouse embryonic and adult fibroblast cultures by defined factors. *Cell* 2006;**126**:663–76.
23. Chambers I, Colby D, Robertson M, et al. Functional expression cloning of Nanog, a pluripotency sustaining factor in embryonic stem cells. *Cell* 2003;**113**:643–55.
24. Loh YH, Wu Q, Chew JL, et al. The Oct4 and Nanog transcription network regulates pluripotency in mouse embryonic stem cells. *Nat Genet* 2006;**38**:431–40.
25. Mitsui K, Tokuzawa Y, Itoh H, et al. The homeoprotein Nanog is required for maintenance of pluripotency in mouse epiblast and ES cells. *Cell* 2003;**113**:631–42.
26. Silva J, Chambers I, Pollard S, Smith A. Nanog promotes transfer of pluripotency after cell fusion. *Nature* 2006;**441**:997–1001.
27. Nichols J, Zevnik B, Anastassiadis K, et al. Formation of pluripotent stem cells in the mammalian embryo depends on the POU transcription factor Oct4. *Cell* 1998;**95**:379–91.
28. Niwa H, Miyazaki J, Smith AG. Quantitative expression of Oct-3/4 defines differentiation, dedifferentiation or self-renewal of ES cells. *Nat Genet* 2000;**24**:372–6.
29. Cauffman G, Van de Velde H, Liebaers I, Van Steirteghem A. Oct-4 mRNA and protein expression during human preimplantation development. *Mol Hum Reprod* 2005;**11**:173–81.

30. Baal N, Reisinger K, Jahr H, et al. Expression of transcription factor Oct-4 and other embryonic genes in CD133 positive cells from human umbilical cord blood. *Thromb Haemost* 2004;**92**:767–75.

31. McGuckin CP, Forraz N. Potential for access to embryonic-like cells from human umbilical cord blood. *Cell Prolif* 2008;**41**(Suppl. 1):31–40.

32. Izadpanah R, Joswig T, Tsien F, Dufour J, Kirijan JC, Bunnell BA. Characterization of multipotent mesenchymal stem cells from the bone marrow of rhesus macaques. *Stem Cells Dev* 2005;**14**:440–51.

33. Lamoury FM, Croitoru-Lamoury J, Brew BJ. Undifferentiated mouse mesenchymal stem cells spontaneously express neural and stem cell markers Oct-4 and Rex-1. *Cytotherapy* 2006;**8**:228–42.

34. Lengner CJ, Camargo FD, Hochedlinger K, et al. Oct4 expression is not required for mouse somatic stem cell self-renewal. *Cell Stem Cell* 2007;**1**:403–15.

35. Berg JS, Goodell MA. An argument against a role for Oct4 in somatic stem cells. *Cell Stem Cell* 2007:359–60.

36. Kotoula V, Papamichos SI, Lambropoulos AF. Revisiting OCT4 expression in peripheral blood mononuclear cells. *Stem Cells* 2008;**26**:290–1.

37. Liedtke B, Enczmann J, Waclawczyk S, Wernet P, Kogler G. Oct4 and its pseudogenes confuse stem cell research. *Cell Stem Cell* 2008;**2**:364–6.

38. Howe M, Zhao J, Bodenburg Y, McGuckin CP, Forraz N, Tilton RG, Urban RJ, Denner L. Oct4-A isoform is expressed in human cord blood-derived CD133 stem cells and differentiated progeny. *Cell Prolif* 2008 in press.

39. Roche S, Richard MJ, Favrot MC. Oct-4, Rex-1, and Gata-4 expression in human MSC increase the differentiation efficiency but not hTERT expression. *J Cell Biochem* 2007;**101**:271–80.

40. Yu H, Fang D, Kumar SM, et al. Isolation of a novel population of multipotent adult stem cells from human hair follicles. *Am J Pathol* 2006;**168**:1879–88.

41. Fuhrmann G, Chung AC, Jackson KJ, et al. Mouse germline restriction of Oct4 expression by germ cell nuclear factor. *Dev Cell* 2001;**1**:377–87.

42. Liang J, Wan M, Zhang Y, et al. Nanog and Oct4 associate with unique transcriptional repression complexes in embryonic stem cells. *Nat Cell Biol* 2008;**10**:731–9.

43. Nagano K, Taoka M, Yamauchi Y, et al. Large-scale identification of proteins expressed in mouse embryonic stem cells. *Proteomics* 2005;**5**:1346–61.

44. Zhang Z, Liao B, Xu M, Jin Y. Post-translational modification of POU domain transcription factor Oct-4 by SUMO-1. *FASEB J* 2007; **21**:3042–51.

45. Liu C, Li Y, Semenov M, et al. Control of beta-catenin phosphorylation/degradation by a dual-kinase mechanism. *Cell* 2002;**108**:837–47.

46. Behrens J, Jerchow BA, Wurtele M, et al. Functional interaction of an axin homolog, conductin, with beta-catenin, APC, and GSK3beta. *Science* 1998;**280**:596–9.

47. Hinoi T, Yamamoto H, Kishida M, Takada S, Kishida S, Kikuchi A. Complex formation of adenomatous polyposis coli gene product and axin facilitates glycogen synthase kinase-3 beta-dependent phosphorylation of beta-catenin and downregulates beta-catenin. *J Biol Chem* 2000;**275**:34,399–34,406.

48. Wang QM, Fiol CJ, DePaoli-Roach AA, Roach PJ. Glycogen synthase kinase-3 beta is a dual specificity kinase differentially regulated by tyrosine and serine/threonine phosphorylation. *J Biol Chem* 1994; **269**:14,566–14,574.

49. Hagen T, Di Daniel E, Culbert AA, Reith AD. Expression and characterization of GSK-3 mutants and their effect on beta-catenin phosphorylation in intact cells. *J Biol Chem* 2002;**277**:23,330–23,335.

50. Cross DA, Alessi DR, Cohen P, Andjelkovich M, Hemmings BA. Inhibition of glycogen synthase kinase-3 by insulin mediated by protein kinase B. *Nature* 1995;**378**:785–9.

51. Kockeritz L, Doble B, Patel S, Woodgett JR. Glycogen synthase kinase-3 – an overview of an over-achieving protein kinase. *Curr Drug Targets* 2006;**7**:1377–88.

52. Meijer L, Skaltsounis AL, Magiatis P, et al. GSK-3-selective inhibitors derived from Tyrian purple indirubins. *Chem Biol* 2003;**10**:1255–66.

53. Baker JC, Beddington RS, Harland RM. Wnt signaling in Xenopus embryos inhibits bmp4 expression and activates neural development. *Genes Dev* 1999;**13**:3149–59.

54. Nostro MC, Cheng X, Keller GM, Gadue P. Wnt, activin, and BMP signaling regulate distinct stages in the developmental pathway from embryonic stem cells to blood. *Cell Stem Cell* 2008;**2**:60–71.

55. Lengerke C, Schmitt S, Bowman TV, et al. BMP and Wnt specify hematopoietic fate by activation of the Cdx–Hox pathway. *Cell Stem Cell* 2008;**2**:72–82.

56. Reya T, Duncan AW, Ailles L, et al. A role for Wnt signalling in self-renewal of haematopoietic stem cells. *Nature* 2003;**423**:409–14.

57. Trowbridge JJ, Xenocostas A, Moon RT, Bhatia M. Glycogen synthase kinase-3 is an *in vivo* regulator of hematopoietic stem cell repopulation. *Nat Med* 2006;**12**:89–98.

58. Van Den Berg DJ, Sharma AK, Bruno E, Hoffman R. Role of members of the Wnt gene family in human hematopoiesis. *Blood* 1998;**92**:3189–202.

59. Willert K, Brown JD, Danenberg E, et al. Wnt proteins are lipid-modified and can act as stem cell growth factors. *Nature* 2003;**423**:448–52.

60. Fountain H. *The gene that tells Planaria worms which end is up*: New York Times; 2007.

61. Vertino AM, Taylor-Jones JM, Longo KA, et al. Wnt10b deficiency promotes coexpression of myogenic and adipogenic programs in myoblasts. *Mol Biol Cell* 2005;**16**:2039–48.

62. Sato N, Meijer L, Skaltsounis L, Greengard P, Brivanlou AH. Maintenance of pluripotency in human and mouse embryonic stem cells through activation of Wnt signaling by a pharmacological GSK-3-specific inhibitor. *Nat Med* 2004;**10**:55–63.

63. Howe M, Zhao J, Bodenburg Y, et al. Oct-4A isoform is expressed in human cord blood-derived CD133 stem cells and differentiated progeny. *Cell Prolif.* 2009;**42**:265–75.

64. Rubinfeld B, Albert I, Porfiri E, Fiol C, Munemitsu S, Polakis P. Binding of GSK3beta to the APC-beta-catenin complex and regulation of complex assembly. *Science* 1996;**272**:1023–6.

65. Salic A, Lee E, Mayer L, Kirschner MW. Control of beta-catenin stability: reconstitution of the cytoplasmic steps of the wnt pathway in Xenopus egg extracts. *Mol Cell* 2000;**5**:523–32.

66. Jeannet G, Scheller M, Scarpellino L, et al. Long-term, multilineage hematopoiesis occurs in the combined absence of -catenin and gamma-catenin. *Blood* 2008;**111**:142–9.

67. Korinek V, Barker N, Moerer P, et al. Depletion of epithelial stem-cell compartments in the small intestine of mice lacking Tcf-4. *Nat Genet* 1998;**19**:379–83.

68. Tetsu O, McCormick F. Beta-catenin regulates expression of cyclin D1 in colon carcinoma cells. *Nature* 1999;**398**:422–6.

69. van de Wetering M, Sancho E, Verweij C, et al. The beta-catenin/TCF-4 complex imposes a crypt progenitor phenotype on colorectal cancer cells. *Cell* 2002;**111**:241–50.

70. Fu M, Wang C, Li Z, Sakamaki T, Pestell RG. Minireview: Cyclin D1: normal and abnormal functions. *Endocrinology* 2004;**145**: 5439–47.

71. Holmes T, O'Brien TA, Knight R, et al. Glycogen synthase kinase-3beta inhibition preserves hematopoietic stem cell activity and inhibits leukemic cell growth. *Stem Cells* 2008;**26**:1288–97.

72. Simon M, Grandage VL, Linch DC, Khwaja A. Constitutive activation of the Wnt/beta-catenin signalling pathway in acute myeloid leukaemia. *Oncogene* 2005;**24**:2410–20.

73. Jamieson CH, Ailles LE, Dylla SJ, et al. Granulocyte-macrophage progenitors as candidate leukemic stem cells in blast-crisis CML. *N Engl J Med* 2004;**351**:657–67.

74. Ysebaert L, Chicanne G, Demur C, et al. Expression of beta-catenin by acute myeloid leukemia cells predicts enhanced clonogenic capacities and poor prognosis. *Leukemia* 2006;**20**:1211–16.

75. Jia J, Amanai K, Wang G, Tang J, Wang B, Jiang J. Shaggy/GSK3 antagonizes Hedgehog signalling by regulating Cubitus interruptus. *Nature* 2002;**416**:548–52.

76. Foltz DR, Santiago MC, Berechid BE, Nye JS. Glycogen synthase kinase-3beta modulates notch signaling and stability. *Curr Biol* 2002;**12**:1006–11.

77. Ong SE, Mann M. Mass spectrometry-based proteomics turns quantitative. *Nat Chem Biol* 2005;**1**:252–62.

78. Goshe MB, Blonder J, Smith RD. Affinity labeling of highly hydrophobic integral membrane proteins for proteome-wide analysis. *J Proteome Res* 2003;**2**:153–61.

79. Tao WA, Aebersold R. Advances in quantitative proteomics via stable isotope tagging and mass spectrometry. *Curr Opin Biotechnol* 2003;**14**:110–18.

80. Zhao Y, Haidacher SJ, Howe M et al. Stem cell proteomics: epigenetic regulation of growth and differentiation. 4th Annual US HUPO Conference, 2008 Bethesda, MD.

81. Howe M, Zhao J, Bodenburg Y, Tilton RG, Urban RJ, Denner, L. The role of the Wnt pathway in self-renewal and differentiation of umbilical cord blood-derived stem cells Keystone Symposium: Stem Cell Interaction with their Microenvironmental Niche, Keystone, CO, (2007), p. 203.

82. Heller M, Mattou H, Menzel C, Yao X. Trypsin catalyzed 16O-to-18O exchange for comparative proteomics: tandem mass spectrometry comparison using MALDI-TOF, ESI-QTOF, and ESI-ion trap mass spectrometers. *J Am Soc Mass Spectrom* 2003;**14**:704–18.

83. Yao X, Freas A, Ramirez J, Demirev PA, Fenselau C. Proteolytic 18O labeling for comparative proteomics: model studies with two serotypes of adenovirus. *Anal Chem* 2001;**73**:2836–42.

84. Gluckman E. Ex vivo expansion of cord blood cells. *Exp Hematol* 2004;**32**:410–12.

Signaling In Development

Chapter 21

Wnt Signaling in Development

Stefan Rudloff, Daniel Messerschmidt and Rolf Kemler

Department of Molecular Embryology, Max-Planck Institute of Immunobiology, Freiburg, Germany

INTRODUCTION

The term Wnt is an acronym of *wingless (wg)* and *INT-1*, which are the the two founding member genes of this signaling pathway. *Wg* was found to be indispensable for segment polarity during *Drosophila melanogaster* larval development [1]. *Int-1* was first described as a preferential integration site in virally induced mammary tumors in mice [2]. *Wnts* are highly conserved throughout evolution, with a multiplicity of genes found in species ranging from hydra to man. They encode secreted messenger molecules that share a common palmitoylation modification that is necessary for proper function [3]. To date, at least three major pathways downstream of Wnt can be distinguished: the Wnt/Ca^{2+} pathways [4], the Wnt/PCP (planar cell polarity) pathway and the Wnt/β-catenin pathway. The latter is also referred to as the canonical Wnt signaling pathway. Only here is β-catenin the central player, and, upon Wnt binding to its receptor, it translocates to the nucleus where it functions as a transcriptional co-activator. In this chapter, we will give an overview of the present knowledge regarding Wnt signaling, focusing primarily on the Wnt/β-catenin pathway and its role in invertebrate and vertebrate development. Further, we briefly discuss recent progress concerning the non-canonical Wnt/PCP pathway.

CANONICAL WNT SIGNALING

In general, the canonical Wnt pathway involves the fine-tuned interaction of activators, inhibitors, and many co-factors at the cell membrane, in the cytoplasm, and in the nucleus. This complex molecular machinery is put into place to control cytoplasmic turnover of β-catenin, thus regulating the transcription of Wnt target genes. The Wnt homepage (www.stanford.edu/~rnusse/wntwindow.html) provides more detailed information about this signaling cascade.

In the absence of Wnt signaling, cytoplasmic β-catenin is bound to a multimeric protein complex. This so-called destruction complex, composed of the scaffolding protein axin, APC (adenomatous polyposis coli), and the kinases CK1 (casein kinase 1) and GSK3β (glycogen synthase kinase 3β), phosphorylates β-catenin at N-terminal serine/threonine residues, subjecting the protein to ubiquitination and proteasome dependent degradation [5–15]. Upon binding of Wnt to the membrane receptors Frizzled (Fz) and LRP5/6 (Low-density lipoprotein Receptor-related Protein 5/6), intracellular signaling events impede β-catenin breakdown. Wnt signaling promotes the phosphorylation of the intracellular domain of LRP6, as well as the interaction of Dishevelled (Dsh) with Fz. Together they sequester axin to the membrane, which in turn is followed by GSK3b and CK1 [16, 17]. The translocation of these proteins blocks phosphorylation of β-catenin, and thereby amplifies the Wnt signal. Hypo-phosphorylated β-catenin accumulates and translocates into the nucleus, where it binds to TCF/LEF (T cell factor/lymphoid enhancer factor) transcription factors [18]. In the present model, TCF/LEF proteins are bound to sites in the promoters of canonical Wnt target genes and act as transcriptional repressors by associating with other factors including Groucho homologs (TLEs) and histone deacetylases [19, 20]. β-catenin displaces Groucho (TLE) from TCF/LEF, other co-activators including p300/CBP are summoned, and target gene transcription is activated [21]. Taking into account that most Wnt target genes are developmentally and cell-type-specifically regulated, it is not well understood how TCF/LEF and β-catenin activate a distinct subset of target genes in a given cell. Recent evidence suggests that epigenetic mechanisms may come into play here [22].

WNT SIGNALING IN INVERTEBRATE DEVELOPMENT

The first *wg* mutation in *D. melanogaster* was induced by X-irradiation, and resulted, as the name suggests, in missing adult body structures, including the wings [23]. Yet this phenotype is misleading, as the genetic lesion did not disrupt the coding region but rather a small regulatory element 9 kb downstream of the *wg* transcription unit. This element is believed to contribute to *wg* expression in the imaginal discs [24, 25]. Major contributions to the understanding of canonical Wnt signaling came from genetic screens for zygotic mutations disrupting the embryonic body plan [1]. The ventral cuticular exoskeleton of the larva displays a segmental repetitive pattern of denticle belts and naked smooth cuticle. In *wg* loss-of-function mutants, the abdominal exoskeleton is covered with denticles. Other mutants, such as *Armadillo* (β-catenin), *Arrow* (LRP5/6) and, using the dominant female sterile mutation *ovoD*, *Dsh* and *Zestewhite3* (*Zw3*, GSK3ß) were identified [26]. Hence, the central mechanism of canonical Wnt signaling was recognized in *D. melanogaster*. The discrepancy between Armadillo transcript and protein localization in the cuticle, the earlier showing uniform distribution and the latter a striped pattern, led to the hypothesis that *wg* regulates Armadillo protein stability and that this involves *Zw3* (GSK3β) [27, 28]. However, the analysis in *D. melanogaster* did not reveal further downstream components of this signaling pathway. This was achieved in the mammalian system, with the identification of the transcription factors TCF/LEF as binding partners of β-catenin [29–31]. Subsequently, the TCF ortholog *Pangolin* was identified in *D. melanogaster* [32].

The *Caenorhabditis elegans* genome encodes multiple genes for Wnt ligands (five) and Fz receptors (four). However, unlike vertebrates and *D. melanogaster*, the nematode genome possesses three β-catenin genes, which fulfill different functions in Wnt signaling and adhesion [33]. Whilst bar-1 most likely adopts the role of canonical β-catenin in the classical signal transduction cascade, Hmp-1 only functions in cell adhesion. Remarkably, Wrm-1, the third nematode β-catenin, acts in a non-canonical form of Wnt signaling, possibly unique to nematodes. In the four-cell stage embryo the Wnt signal is thought to align the mitotic spindle along the AP body axis. The outcome of this non-canonical variant of Wnt signaling is the polarization of the bi-potent precursor cell (EMS) [34, 35]. Subsequent cell division results in a posterior daughter cell that will give rise to the future endoderm, whereas the anterior daughter will form mesoderm. Disturbances in this important fate decision can be observed in several loss-of-function mutants, which lack of endodermal structures. Genetic analysis of these mutant worms revealed a high frequency of genes for components of the Wnt signaling pathway. Among these are *mom-2* (more mesoderm) encoding one of the Wnt ligands, *mom-5*, a Fz receptor homolog, and *wrm-1*.

Canonical Wnt signaling in *C. elegans* is required for vulval induction [36]. The vulva develops post-embryonically from six vulva precursor cells (VPCs). Each of these cells can adopt one out of three fates, 1°, 2°, or 3°, respectively. Here, Wnt signaling is transduced by the β-catenin ortholog bar-1 inducing the primary fate of the central VPC (P6.p) via the activation of the homeobox gene *lin-39* [37]. Loss of *bar-1* results in the adoption of the 3° fate of all VPCs, and thus in a vulva-less phenotype. Further genetic screens have shown that many Wnt components have effects on vulva induction. The mutants *mom-3* (Wnt), *mig-14* (not identified), *pry-1* (axin), and *apr-1* (APC) all exhibit vulva phenotypes [38, 39]. As expected, the loss of the negative Wnt regulator *pry-1* results in the generation of multiple ectopic vulvae.

WNT SIGNALING IN VERTEBRATE DEVELOPMENT

Various processes during chordate embryogenesis depend on canonical Wnt signaling. Although the significance of this pathway has been described in virtually every vertebrate species, including human, we will discuss some major developmental processes in *Xenopus laevis* and in mouse. Wnt signaling controls early patterning of the body plan, when the dorsal–ventral (DV) and anterior–posterior (AP) body axes are laid down. In addition, later in development Wnt signaling is involved in many morphogenetic events leading to organogenesis.

Fertilization of the amphibian egg sets in motion a cascade of events that leads to movements of the vegetal cortical cytoplasm away from the point of sperm entry [40]. Thereby, maternal factors are relocated asymmetrically to the future dorsal side of the embryo [41]. The result of this reorganization is the establishment of the Spemann organizer that further directs *X. laevis* development. A role for Wnt signaling in this process has long been anticipated. While β-catenin depletion leads to ventralization and concomitant lack of axial structures [42], ectopic overexpression induces axial duplication [43]. In 2005, Tao and colleagues demonstrated that maternal wnt11 mRNA is necessary and sufficient to activate the canonical Wnt pathway on the dorsal side of the embryo [44]. Even though Wnt11 is considered a non-canonical member of the Wnt family [45], extracellular membrane-associated co-factors direct its potential towards a canonical signaling function. For example, posttranslational modifications of heparan sulfate proteoglycans are thought to be involved in Wnt stabilization, presentation to receptors, and extracellular transport [46]. The extent to which the composition of the extracellular environment is able to regulate the signaling property of a given Wnt molecule still remains to be investigated.

In mouse pre-implantation embryos, a cytoplasmic pool of β-catenin and other Wnt signaling components can be detected. Functionally, these molecules do not affect the key lineage decisions made at early blastocyst stages, but rather prepare the embryo for implantation [47, 48]. However, patterning of the epiblast, especially the development of the primitive streak, is highly dependent on intact canonical Wnt signaling. Monitoring canonical Wnt signaling activity during mouse development is possible with the help of TCF-reporter transgenic mice expressing β-galactosidase under the transcriptional control of seven fused TCF/LEF binding sites from the minimal promoter-TATA box of the gene *siamois* [49]. In these BATgal mice, the activity of the reporter is first observed at day 6 of embryonic development in a subpopulation of epiblast cells at the boundary between embryonic and extra-embryonic ectoderm. With the onset of gastrulation, β-galactosidase activity becomes enriched in the primitive streak, where it remains present until the end of gastrulation. The candidate Wnt ligand upstream of these events is Wnt3, having an expression pattern similar to the β-galactosidase activity [50]. *Wnt3*-null mutant embryos show a similar phenotype to β-catenin knock-out mice: failure of primitive streak formation, and no gastrulation [51]. Unlike in *X. laevis*, expression of constitutively active β-catenin does not lead axis duplication, but to disorganization of the epiblast. Analyses of such embryos point towards a precocious mesoderm differentiation of the embryonic ectoderm [48]. Despite molecular differences between *X. laevis* and mouse during initial stages of development, both species share conserved mechanisms centered on canonical Wnt signaling, necessary for body axes formation.

After the initial establishment of the major axes, organs are formed at precise positions along the body plan. Development of these highly specialized structures often depends on cross-talk between an inductive tissue and a responding cell population.

Three distinct renal organs evolve in a defined temporal–spatial order; the primary nephric duct in the intermediate mesoderm represents the vertebrate excretory system during development. In mammals, pronephros and mesonephros are transient, partially functional excretory organs of the anterior portion of the duct. They degenerate as the metanephrous, or definite, kidney develops. The generation of the adult kidney makes use of several Wnt signaling events. The key inducer of metanephric development is the ureteric bud (UB), a caudal outgrowth of the primary nephric duct [52]. A promising factor expressed in the UB to initiate nephrogenesis is Wnt9b [53]. *Wnt9b* mutants only develop rudimentary kidneys, reflecting a failure of the UB to induce the metanephric mesenchyme. In response to the signal emanating from the invading UB, the mesenchymal cells further condense, aggregate into pretubular clusters, and undergo an epithelial transition forming a tubule. *Wnt4* was found to be the factor expressed in the metanephric mesenchyme required for the tubulogenesis [54, 55]. Besides Wnt4, other Wnt ligands are expressed in highly specific patterns in the developing metanephros [56, 57]. *Wnt2b* transcripts are found in perinephric mesenchymal cells, whereas *Wnt11* is exclusively detected in the tips of the ureter; both function in ureteric bud branching morphogenesis [58]. In *X. leavis*, the pronephros is a fully functional filtration unit consisting of a glomus, tubules, and a duct. Its development partially depends on the same mechanisms as described in mice. Morphological changes in the pronephros anlage coincide with changes in gene expression – for example, upregulation of *wnt4*. Overexpression of *wnt4* results in fused pronephric tubules, while morpholino-mediated *wnt4* knockdown leads to a complete loss of pronephric tubules [59]. *X. leavis Fz8* (*Xfz8*) knockdown leads to a significant reduction in pronephric duct differentiation [60]. Wnt11, transcripts of which are found in the vicinity of the pronephric duct, could activate Xfz8. However, to date a direct association of these two proteins has not been shown.

Another well-studied example of a Wnt/β-catenin-dependent tissue specification is the hair follicle. Follicle development is guided by reciprocal exchange of signals between ectoderm and underlying dermis. At day 14 of embryonic development, an epidermal thickening or placode marks the beginning of the hair-follicle formation. *In situ* hybridization reveals an upregulation of *β-catenin* and *LEF1* in the epidermal part of these structures. Cre-mediated skin-specific knockout of *β-catenin* results in the loss of placodes and later follicle formation [61]. A similar phenotype is observed in *LEF1*-deficient mice [62]. On the other hand, the overexpression of *LEF1* or constitutively active β-catenin induces extra hair follicles [63, 64]. Several Wnts are expressed in the early placode (Wnt10a, 10b) or within its close proximity (Wnt3, 3a, 4, 5a); nevertheless, to date it remains elusive which Wnt ligands are truly involved in hair follicle formation [65].

These are examples of the requirement for Wnt signaling during development, but a detailed description of all of the processes where Wnts are involved would go beyond the scope of this chapter. Over recent years, stem cell research has evolved into one of the most prominent topics in biological science. Given the importance of Wnt signaling during development, it is not surprising that many reports have connected β-catenin and other components of the Wnt signaling machinery to embryonic and somatic stem cell renewal and cancer [66, 67].

WNT/PLANAR CELL POLARITY

Planar cell polarity (PCP) is distinct from apical–basal cell polarity. Whereas the latter comprises the functional barrier property across a tissue, PCP describes the uniform alignment of cells within a tissue. Readily visible examples are the ordered array of distally pointing bristles on

D. melanogaster wing cells, or the chevron-shaped alignment of stereocilia protruding from sensory hair cells in the mammalian cochlea. PCP, as a variant of non-canonical Wnt signaling, does not rely on β-catenin. This implies that only the membrane components of the canonical Wnt machinery including Dsh are used in this pathway. In *D. melanogaster*, Wnts seem not to be involved in the direct initiation of the PCP signaling cascade. Rather they are proposed to regulate the graded expression of protocadherins upstream of PCP signaling [68]. Activation of the PCP pathway is thought to arise from differential interaction of these cadherin superfamily members [69]. In vertebrates, non-canonical Wnt ligands have been shown to activate PCP signaling. For example, gastrulation and neural tube closure in *X. laevis* or the zebrafish *Danio rerio* embryos depend on the presence of Wnt11 and Wnt5 [70, 71]. Furthermore, the complex construction of the mammalian cochlea could be orchestrated by Wnt7a [72]. A key event in PCP is the asymmetric assembly of membrane-bound proteins (core components) into distinct complexes at the proximal or distal side of the cell. Through antagonizing and stabilizing feedback mechanisms, these complexes maintain their polarized distribution and further propagate this pattern to neighboring cells. This uniform patterning builds the basis for the execution of cellular events at one side of the cells [73, 74]. One important outcome of PCP in vertebrates is convergent extension. Activated cells preferentially extend filopodia from their proximal sides, along which they move towards the center, thereby concomitantly elongating and thinning a tissue. Recent reports have identified non-redundant functions for Wnt5a and Wnt11 in convergent extension of gastrulating *X. laevis* embryos. Wnt5a upregulates the expression of paraxial protocadherin (PAPC) in dorsal mesoderm [75]. PAPC interacts with Fz7 and direct the separation of involuting mesoderm and ectoderm [76]. Whereas Wnt5a seems to establish directionality by sorting cells, Wnt11 is necessary to stimulate cell motility, as shown in depletion experiments. Only the combined function of Wnt5a and Wnt11 is sufficient to bring about convergent extension. The propagation of the PCP signal depends on distinctive domains of core players, dispensable for canonical Wnt signaling [77]. In particular, the DEP domain of Dsh is involved in the activation of the cytoskeleton regulators JNK, Rho, Rac, and Cdc42 [78, 79].

Different responses to Wnt/PCP signaling are the generation of the invariant mosaic of sensory cells and supporting cells in the fly eye based on Delta-Notch signaling [80] or mitotic spindle orientation [81]. The involvement of PCP in a wide variety of developmental processes is increasingly appreciated. However, future research needs to better define the signaling events occurring upstream and downstream of the core cassette. Greater comprehension of the mechanisms underlying PCP will help us to understand how aberrant tissue polarity affects development and organ physiology.

ACKNOWLEDGEMENTS

We thank Dr Verdon Taylor for critically reading the manuscript. Daniel Messerschmidt and Stefan Rudloff are PhD students of the Faculty of Biology, University of Freiburg, Germany, and contributed equally to this work.

REFERENCES

1. Nusslein-Volhard C, Wieschaus E. Mutations affecting segment number and polarity in *Drosophila*. *Nature* 1980;**287**(5785):795–801.
2. Nusse R, Varmus HE. Many tumors induced by the mouse mammary tumor virus contain a provirus integrated in the same region of the host genome. *Cell* 1982;**31**(1):99–109.
3. Willert K, et al. Wnt proteins are lipid-modified and can act as stem cell growth factors. *Nature* 2003;**423**(6938):448–52.
4. Kohn AD, Moon RT. Wnt and calcium signaling: beta-catenin-independent pathways. *Cell Calcium* 2005;**38**(3–4):439–46.
5. Aberle H, et al. beta-catenin is a target for the ubiquitin-proteasome pathway. *Embo J* 1997;**16**(13):3797–83804.
6. Behrens J, et al. Functional interaction of an axin homolog, conductin, with beta-catenin, APC, and GSK3beta. *Science* 1998;**280**(5363):596–9.
7. Dominguez I, Itoh K, Sokol SY. Role of glycogen synthase kinase 3 beta as a negative regulator of dorsoventral axis formation in Xenopus embryos. *Proc Natl Acad Sci U S A* 1995;**92**(18):8498–502.
8. Hart MJ, et al. Downregulation of beta-catenin by human Axin and its association with the APC tumor suppressor, beta-catenin and GSK3 beta. *Curr Biol* 1998;**8**(10):573–81.
9. Ikeda S, et al. Axin, a negative regulator of the Wnt signaling pathway, forms a complex with GSK-3beta and beta-catenin and promotes GSK-3beta-dependent phosphorylation of beta-catenin. *Embo J* 1998;**17**(5):1371–84.
10. Morin PJ, et al. Activation of beta-catenin-Tcf signaling in colon cancer by mutations in beta-catenin or APC. *Science* 1997;**275**(5307):1787–90.
11. Polakis P. The oncogenic activation of beta-catenin. *Curr Opin Genet Dev* 1999;**9**(1):15–21.
12. Polakis P. Wnt signaling and cancer. *Genes Dev* 2000;**14**(15):1837–51.
13. Rubinfeld B, et al. Stabilization of beta-catenin by genetic defects in melanoma cell lines. *Science* 1997;**275**(5307):1790–2.
14. Rubinfeld B, et al. Association of the APC gene product with beta-catenin. *Science* 1993;**262**(5140):1731–4.
15. Sakanaka C, et al. Casein kinase iepsilon in the wnt pathway: regulation of beta-catenin function. *Proc Natl Acad Sci U S A* 1999;**96**(22):12548–52.
16. Zeng X, et al. Initiation of Wnt signaling: control of Wnt coreceptor Lrp6 phosphorylation/activation via frizzled, dishevelled and axin functions. *Development* 2008;**135**(2):367–75.
17. Zeng X, et al. A dual-kinase mechanism for Wnt co-receptor phosphorylation and activation. *Nature* 2005;**438**(7069):873–7.
18. van de Wetering M, et al. Armadillo coactivates transcription driven by the product of the Drosophila segment polarity gene dTCF. *Cell* 1997;**88**(6):789–99.
19. Cavallo RA, et al. Drosophila Tcf and Groucho interact to repress Wingless signalling activity. *Nature* 1998;**395**(6702):604–8.
20. Chen G, et al. A functional interaction between the histone deacetylase Rpd3 and the corepressor groucho in Drosophila development. *Genes Dev* 1999;**13**(17):2218–30.

21. Hecht A, et al. The p300/CBP acetyltransferases function as transcriptional coactivators of beta-catenin in vertebrates. *Embo J* 2000;**19**(8): 1839–50.

22. Wohrle S, Wallmen B, Hecht A. Differential control of Wnt target genes involves epigenetic mechanisms and selective promoter occupancy by T-cell factors. *Mol Cell Biol* 2007;**27**(23):8164–77.

23. Sharma RP, Chopra VL. Effect of the Wingless (wg1) mutation on wing and haltere development in Drosophila melanogaster. *Dev Biol* 1976;**48**(2):461–565.

24. Baker NE. Molecular cloning of sequences from wingless, a segment polarity gene in Drosophila: the spatial distribution of a transcript in embryos. *Embo J* 1987;**6**(6):1765–73.

25. van den Heuvel M, et al. Cell patterning in the Drosophila segment: engrailed and wingless antigen distributions in segment polarity mutant embryos. *Dev Suppl* 1993:105–14.

26. Perrimon N, Engstrom L, Mahowald AP. Zygotic lethals with specific maternal effect phenotypes in Drosophila melanogaster. I. Loci on the X chromosome. *Genetics* 1989;**121**(2):333–52.

27. Riggleman B, Schedl P, Wieschaus E. Spatial expression of the Drosophila segment polarity gene armadillo is posttranscriptionally regulated by wingless. *Cell* 1990;**63**(3):549–60.

28. Riggleman B, Wieschaus E, Schedl P. Molecular analysis of the armadillo locus: uniformly distributed transcripts and a protein with novel internal repeats are associated with a Drosophila segment polarity gene. *Genes Dev* 1989;**3**(1):96–113.

29. Castrop J, van Norren K, Clevers H. A gene family of HMG-box transcription factors with homology to TCF-1. *Nucleic Acids Res* 1992;**20**(3):611.

30. Huber O, et al. Nuclear localization of beta-catenin by interaction with transcription factor LEF-1. *Mech Dev* 1996;**59**(1):3–10.

31. Oosterwegel M, et al. Cloning of murine TCF-1, a T cell-specific transcription factor interacting with functional motifs in the CD3-epsilon and T cell receptor alpha enhancers. *J Exp Med* 1991;**173**(5): 1133–42.

32. Brunner E, et al. pangolin encodes a Lef-1 homologue that acts downstream of Armadillo to transduce the Wingless signal in Drosophila. *Nature* 1997;**385**(6619):829–33.

33. Korswagen HC, Herman MA, Clevers HC. Distinct beta-catenins mediate adhesion and signalling functions in C. elegans. *Nature* 2000;**406**(6795):527–32.

34. Bowerman B, Shelton CA. Cell polarity in the early Caenorhabditis elegans embryo. *Curr Opin Genet Dev* 1999;**9**(4):390–5.

35. Thorpe CJ, Schlesinger A, Bowerman B. Wnt signalling in Caenorhabditis elegans: regulating repressors and polarizing the cytoskeleton. *Trends Cell Biol* 2000;**10**(1):10–107.

36. Eisenmann DM, Kim SK. Protruding vulva mutants identify novel loci and Wnt signaling factors that function during Caenorhabditis elegans vulva development. *Genetics* 2000;**156**(3):1097–116.

37. Eisenmann DM, et al. The beta-catenin homolog BAR-1 and LET-60 Ras coordinately regulate the Hox gene lin-39 during Caenorhabditis elegans vulval development. *Development* 1998;**125**(18):3667–80.

38. Maloof JN, et al. A Wnt signaling pathway controls hox gene expression and neuroblast migration in C. elegans. *Development* 1999;**126**(1):37–49.

39. Hoier EF, et al. The Caenorhabditis elegans APC-related gene apr-1 is required for epithelial cell migration and Hox gene expression. *Genes Dev* 2000;**14**(7):874–86.

40. Vincent JP, Gerhart JC. Subcortical rotation in Xenopus eggs: an early step in embryonic axis specification. *Dev Biol* 1987; **123**(2):526–39.

41. Marikawa Y, Li Y, Elinson RP. Dorsal determinants in the Xenopus egg are firmly associated with the vegetal cortex and behave like activators of the Wnt pathway. *Dev Biol* 1997;**191**(1):69–79.

42. Heasman J, et al. Overexpression of cadherins and underexpression of beta-catenin inhibit dorsal mesoderm induction in early Xenopus embryos. *Cell* 1994;**79**(5):791–803.

43. Funayama N, et al. Embryonic axis induction by the armadillo repeat domain of beta-catenin: evidence for intracellular signaling. *J Cell Biol* 1995;**128**(5):959–68.

44. Tao Q, et al. Maternal wnt11 activates the canonical wnt signaling pathway required for axis formation in Xenopus embryos. *Cell* 2005;**120**(6):857–71.

45. Marlow F, et al. Zebrafish Rho kinase 2 acts downstream of Wnt11 to mediate cell polarity and effective convergence and extension movements. *Curr Biol* 2002;**12**(11):876–84.

46. Lin X. Functions of heparan sulfate proteoglycans in cell signaling during development. *Development* 2004;**131**(24):6009–21.

47. Haegel H, et al. Lack of beta-catenin affects mouse development at gastrulation. *Development* 1995;**121**(11):3529–37.

48. Kemler R, et al. Stabilization of beta-catenin in the mouse zygote leads to premature epithelial-mesenchymal transition in the epiblast. *Development* 2004;**131**(23):5817–24.

49. Maretto S, et al. Mapping Wnt/beta-catenin signaling during mouse development and in colorectal tumors. *Proc Natl Acad Sci U S A* 2003;**100**(6):3299–304.

50. Liu P, et al. Requirement for Wnt3 in vertebrate axis formation. *Nat Genet* 1999;**22**(4):361–5.

51. Huelsken J, et al. Requirement for beta-catenin in anterior-posterior axis formation in mice. *J Cell Biol* 2000;**148**(3):567–78.

52. Saxen L, Sariola H. Early organogenesis of the kidney. *Pediatr Nephrol* 1987;**1**(3):385–92.

53. Carroll TJ, et al. Wnt9b plays a central role in the regulation of mesenchymal to epithelial transitions underlying organogenesis of the mammalian urogenital system. *Dev Cell* 2005;**9**(2):283–92.

54. Kispert A, Vainio S, McMahon AP. Wnt-4 is a mesenchymal signal for epithelial transformation of metanephric mesenchyme in the developing kidney. *Development* 1998;**125**(21):4225–34.

55. Stark K, et al. Epithelial transformation of metanephric mesenchyme in the developing kidney regulated by Wnt-4. *Nature* 1994;**372**(6507):679–83.

56. Kispert A, et al. Proteoglycans are required for maintenance of Wnt-11 expression in the ureter tips. *Development* 1996;**122**(11):3627–37.

57. Lin Y, et al. Induction of ureter branching as a response to Wnt-2b signaling during early kidney organogenesis. *Dev Dyn* 2001; **222**(1):26–39.

58. Majumdar A, et al. Wnt11 and Ret/Gdnf pathways cooperate in regulating ureteric branching during metanephric kidney development. *Development* 2003;**130**(14):3175–85.

59. Saulnier DM, Ghanbari H, Brandli AW. Essential function of Wnt-4 for tubulogenesis in the Xenopus pronephric kidney. *Dev Biol* 2002;**248**(1):13–28.

60. Satow R, Chan TC, Asashima M. The role of Xenopus frizzled-8 in pronephric development. *Biochem Biophys Res Commun* 2004;**321**(2):487–94.

61. Huelsken J, et al. beta-Catenin controls hair follicle morphogenesis and stem cell differentiation in the skin. *Cell* 2001;**105**(4):533–45.

62. van Genderen C, et al. Development of several organs that require inductive epithelial-mesenchymal interactions is impaired in LEF-1-deficient mice. *Genes Dev* 1994;**8**(22): 2691–703.

63. Gat U, et al. De Novo hair follicle morphogenesis and hair tumors in mice expressing a truncated beta-catenin in skin. *Cell* 1998;**95**(5):605–14.

64. Zhou P, et al. Lymphoid enhancer factor 1 directs hair follicle patterning and epithelial cell fate. *Genes Dev* 1995;**9**(6):700–13.

65. Alonso L, Fuchs E. Stem cells in the skin: waste not, Wnt not. *Genes Dev* 2003;**17**(10):1189–200.

66. Dreesen O, Brivanlou AH. Signaling pathways in cancer and embryonic stem cells. *Stem Cell Rev* 2007;**3**(1):7–17.

67. Lindvall C, et al. Wnt signaling, stem cells, and the cellular origin of breast cancer. *Stem Cell Rev* 2007;**3**(2):157–68.

68. Yang CH, Axelrod JD, Simon MA. Regulation of Frizzled by fat-like cadherins during planar polarity signaling in the Drosophila compound eye. *Cell* 2002;**108**(5):675–88.

69. Lawrence PA, Struhl G, Casal J. Planar cell polarity: one or two pathways? *Nat Rev Genet* 2007;**8**(7):555–63.

70. Kilian B, et al. The role of Ppt/Wnt5 in regulating cell shape and movement during zebrafish gastrulation. *Mech Dev* 2003;**120**(4):467–76.

71. Tada M, Smith JC. Xwnt11 is a target of Xenopus Brachyury: regulation of gastrulation movements via Dishevelled, but not through the canonical Wnt pathway. *Development* 2000;**127**(10):2227–38.

72. Dabdoub A, Kelley MW. Planar cell polarity and a potential role for a Wnt morphogen gradient in stereociliary bundle orientation in the mammalian inner ear. *J Neurobiol* 2005;**64**(4):446–57.

73. Barrow JR. Wnt/PCP signaling: a veritable polar star in establishing patterns of polarity in embryonic tissues. *Semin Cell Dev Biol* 2006;**17**(2):185–93.

74. Karner Jr C, Wharton KA, Carroll TJ. Planar cell polarity and vertebrate organogenesis. *Semin Cell Dev Biol* 2006;**17**(2):194–203.

75. Schambony A, Wedlich D. Wnt-5A/Ror2 regulate expression of XPAPC through an alternative noncanonical signaling pathway. *Dev Cell* 2007;**12**(5):779–92.

76. Medina A, et al. Xenopus paraxial protocadherin has signaling functions and is involved in tissue separation. *Embo J* 2004;**23**(16):3249–58.

77. Boutros M, et al. Dishevelled activates JNK and discriminates between JNK pathways in planar polarity and wingless signaling. *Cell* 1998;**94**(1):109–18.

78. Habas R, Kato Y, He X. Wnt/Frizzled activation of Rho regulates vertebrate gastrulation and requires a novel Formin homology protein Daam1. *Cell* 2001;**107**(7):843–54.

79. Yamanaka H, et al. JNK functions in the non-canonical Wnt pathway to regulate convergent extension movements in vertebrates. *EMBO Rep* 2002;**3**(1):69–75.

80. Lanford PJ, et al. Notch signalling pathway mediates hair cell development in mammalian cochlea. *Nat Genet* 1999;**21**(3):289–92.

81. Bellaiche Y, et al. Frizzled regulates localization of cell-fate determinants and mitotic spindle rotation during asymmetric cell division. *Nat Cell Biol* 2001;**3**(1):50–7.

Chapter 22

Interactions between Wnt/β-Catenin/Fgf and Chemokine Signaling in Lateral Line Morphogenesis

Tatjana Piotrowski

University of Utah, Department of Neurobiology and Anatomy, Salt Lake City, Utah

INTRODUCTION

During embryonic development, cells organize themselves into distinct and often complex three-dimensional organs or tissues. In order for proper morphogenesis to occur, a variety of cellular behaviors have to be tightly coordinated (e.g., cell migration, cell–cell adhesion, cell proliferation, cell death, interactions with the environment and changes in cell morphology). Because of this complexity, the molecular and cell biological mechanisms that regulate embryonic morphogenesis remain poorly understood. Recent work has demonstrated that zebrafish lateral line development is an excellent model to study cell signaling pathways underlying these events *in vivo*.

The lateral line is a sensory system present in all aquatic vertebrates for the detection of water movements, which initiates the appropriate behavioral responses for capturing prey, avoiding predators and schooling. The lateral line consists of mechanosensory organs (neuromasts) distributed in lines on the head and along the flanks of the animal. Neuromasts contain hair cells that are very similar to the hair cells of the inner ear of vertebrates [1]. Despite the unusual location of the hair cells in the skin (Figure 22.1a), lateral line and ear hair cells develop by similar stereotyped mechanisms and are derived from cephalic placodes (Figure 22.1b) [2–5]. However, in contrast to the otic placode, the lateral line placode (also called primordium) undergoes a remarkable posterior migration toward

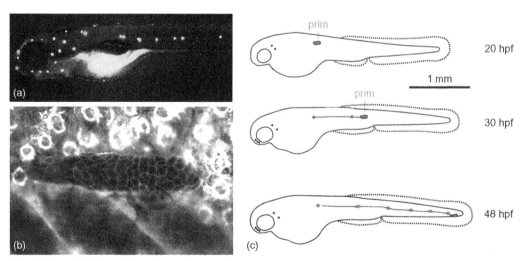

FIGURE 22.1 Morphology of the lateral line system.
(a) Live 5-day-old larva stained with Daspei (Invitrogen) which labels the hair cells yellow (shown white). Individual neuromasts are represented by a bright spot. Signal in the yolk is autofluorescence. (b) Lateral view of the posterior lateral line primordium on the trunk of a 30-hours post-fertilization (hpf) larva stained with the vital dye Bodipy (Molecular Probes). Anterior to the left. Primordium is migrating to the right. (c) Schematic drawing of neuromast deposition. Between 20 and 48 hpf the primordium (prim) gives rise to 7–9 primary neuromasts (shown as dots).

the tail. This migration is a dynamic event that involves the primordium depositing neuromasts every three to five somites until it reaches the tail tip, patterning the future lateral line (Figure 22.1c) [6–8].

Classically, the development of the lateral line system has been studied in amphibian embryos because of the relatively large size of this organ system in these animals [2, 4, 9, 10]. However, these studies have been restricted to embryological manipulations, such as grafting and skin preparations, because of the general intractability of amphibians to genetic approaches. Thus, a major advantage of using zebrafish as a model is that, in addition to being able to manipulate the embryo, genetics and molecular methods can be employed to study developmental problems.

Analyses of zebrafish mutants and experimental manipulations of signaling pathways have begun to provide important insights into how primordium migration is guided, and how collective cell movement and neurogenesis are regulated.

FGF SIGNALING CONTROLS SENSORY ORGAN FORMATION IN THE MIGRATING PRIMORDIUM

Neuromasts consist of centrally located hair cells that are surrounded by support cells. All cell types within a deposited neuromast are derived from the migrating primordium, raising the question of when these cells acquire their fates

and which signaling pathways are involved. Morphological and molecular analyses reveal that proto-neuromasts already begin to form within the migrating primordium. Two to three proto-neuromasts can be morphologically detected in the trailing region of the primordium as cells organize themselves into epithelial rosettes with their apical poles facing toward a central cell (Figure 22.2a, arrows). At this stage, these centrally located cells within proto-neuromasts can also be identified by the expression of *fgf10*, *delta A, B and C*, *neurod*, and *atoh1* (Figure 22.2d) [11–13].

The leading region of the primordium is unpatterned, and lacks *delta* A, B and C, *neurod*, and *atoh1* expression (Figure 22.2a). Interestingly, proto-neuromasts are formed and deposited in a conveyer-belt-like fashion. A new proto-neuromast is induced close to the leading region of the primordium as the most trailing proto-neuromast is deposited (Figure 22.2b). As the primordium is constantly "losing" whole clusters of cells from its trailing region and new proto-neuromasts form at the same rate, new cells have to be produced in the primordium as it migrates.

The undifferentiated leading region of the primordium serves this purpose, and acts as a progenitor zone [12]. When progenitor cells in the leading zone were fluorescently labeled, these cells were shown to proliferate and contribute to all cell types within deposited proto-neuromasts. The fact that proto-neuromasts can be studied from birth until deposition makes the lateral line primordium an excellent model to study the genetic cascades underlying

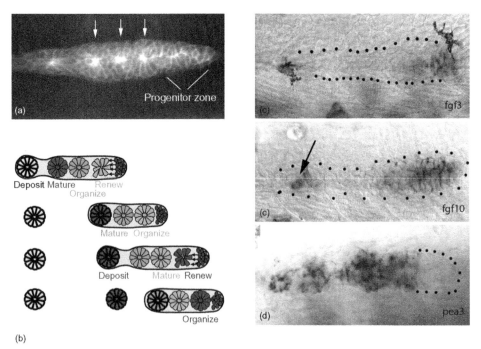

(a)

(b)

FIGURE 22.2 Fgf signaling controls neuromast formation.
(a) Primordium in a transgenic embryo in which the lateral line system is GFP-positive (*Tg(claudinb:gfp)*; [23]). Proto-neuromasts form epithelial rosettes in the trailing region of the migrating primordium (arrows). Anterior to the left. (b) Schematic drawing of neuromast formation and deposition over time. From [12] (Nechiporuk and Raible, 2008); reprinted with permission from AAAS. (c, d, e) *In situ* hybridization with *fgf3* (c), *fgf10* (d) and the Fgf target *pea3* (d). Arrow in (e) points at central cell in a forming neuromast.

cell type specification, morphogenesis, and differentiation in a vertebrate.

Several elegant studies have begun to shed light on the signaling pathways involved in rosette formation and specification of hair cell precursors. Fgf signaling has recently been shown to be required for hair cell specification within the ear and lateral line by regulating *atoh1*, a gene responsible for hair cell differentiation [14]. However, loss of Fgf signaling in the lateral line primordium not only causes loss of hair cells, but also prevents rosette formation [11, 12, 17]. Fgf ligands are expressed in an interesting pattern in the primordium. *fgf3* and *fgf10* are expressed in all cells of the leading region, whereas in the trailing two-thirds of the primordium only *fgf10* is expressed in single central cells within proto-neuromasts (Figure 22.2c, d). However, Fgf pathway activation is inhibited in the leading region, as evidenced by the absence of the Fgf target gene *pea3* (see below), but is activated by diffusion of Fgf ligands in all trailing cells (Figure 22.2e).

By inhibiting or hyperactivating Fgf signaling at different time points during primordium migration, it was demonstrated that Fgf signaling in the trailing two-thirds of the primordium is required for transitioning cells from their mesenchymal character in the leading zone to an epithelial character as the proto-neuromast forms [11, 12]. Specifically, a central *fgf10* expressing cell serves as a nucleation center around which future support cells form a rosette (Figure 22.2d, arrow). Fgf ligand expression by this central cell causes radial epithelialization of surrounding support cells. During this process, support cells elongate along the basil/apical axis while their apical poles all constrict above the central cell [11]. These Fgf-dependent cell shape changes cause the proto-neuromast to acquire a rosette shape, resembling a garlic bulb. This effect of Fgf signaling on cytoskeletal remodeling appears to be applicable to many other epithelial tissues that undergo morphogenesis, such as the inner ear and dorsal neural tube closure [15, 16].

As mentioned above, Fgf signaling is also required for hair cell differentiation. To investigate whether proto-neuromast morphogenesis depends on hair cell differentiation, Nechiporuk and Raible (12) and Lecaudey and colleagues (11) inhibited hair cell differentiation by knockdown of Atoh1 using antisense morpholino nucleotide injections. Neither rosettogenesis nor proto-neuromast deposition were affected, demonstrating that Fgf signaling is independently required for rosettogenesis and later for hair cell differentiation.

WNT/β-CATENIN SIGNALING RESTRICTS NEUROGENESIS TO TRAILING CELLS AND MAINTAINS THE PROGENITOR ZONE

The analyses of the roles of Fgf signaling in neuromast development left several questions unanswered. It was not

known how Fgf signaling is activated in the primordium, or by which mechanism Fgf signaling is restricted to the trailing region. The analysis of Wnt/β-catenin pathway mutants revealed that the Wnt/β-catenin pathway plays crucial roles in these processes [17]. Wnt/β-catenin pathway genes, such as *lef1* and *axin2*, are expressed in leading cells of the primordium overlapping with *fgf3* and *fgf10* (schematically shown in Figure 22.3a, right-hand domain). Experimental inhibition of the Wnt/β-catenin pathway leads to the loss of all four genes. On the other hand, ectopic activation of the Wnt/β-catenin pathway in trailing cells causes strong upregulation of *fgf3* and *fgf10* in these cells. These results demonstrated that Wnt/β-catenin signaling is activating the Fgf pathway. However, even though *fgf3* and *fgf10* are highly expressed in leading cells, they are not able to activate the Fgf pathway in that domain; this is reflected by the absence of the Fgf target *pea3* (Figures 22.2e, 22.3a, left-hand domain). This repression can be attributed to the fact that Wnt/β-catenin signaling simultaneously induces in leading cells the expression of *sef* (*il17rd*), a well-described cytoplasmic membrane associated Fgf signaling inhibitor [18, 19]. Thus, even though Wnt/β-catenin signaling is inducing Fgf ligand expression in the leading zone, it restricts Fgf pathway activation to the trailing cells. This is of biological importance for the primordium, as Wnt/β-catenin signaling thereby restricts neurogenesis and differentiation to the trailing region of the primordium. Inhibiting Fgf signaling in the leading cells also maintains the progenitor zone, enabling the primordium to form new proto-neuromasts as the older, trailing proto-neuromasts are deposited during migration [17].

THE FGF PATHWAY RESTRICTS WNT/β-CATENIN SIGNALING TO THE LEADING EDGE ENSURING NORMAL MIGRATION

Strikingly, abrogation of the Fgf signaling pathway and ectopic activation of the Wnt/β-catenin pathway cause migration defects, suggesting that these pathways not only regulate neurogenesis but also control migration of the primordium [17]. The similarity in phenotype also suggests that Wnt/β-catenin and Fgf possibly act in a negative-feedback loop. Indeed Wnt/β-catenin signaling does not only restrict Fgf signaling to trailing cells, but Fgf signaling in turn restricts Wnt/β-catenin signaling to leading cells. Inhibition of the Fgf pathway using a pharmacological inhibitor or heat shock induction of dominant-negative Fgfr1 leads to ectopic expression of Wnt/β-catenin target genes in trailing cells. The inhibition of the Wnt/β-catenin pathway in trailing cells is achieved by *dkk1* [20], which is expressed adjacent to the Wnt/β-catenin target gene expressing cells, but within the domain where the Fgf pathway is activated. Because of this spatial overlap, it was tested whether *dkk1* depends on Fgf signaling. Indeed, even

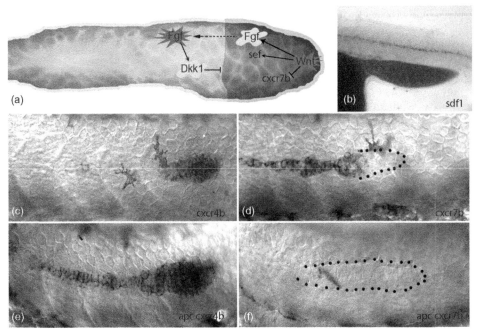

FIGURE 22.3 A feedback-loop between Fgf and Wnt/β-catenin signaling controls neurogenesis and migration.
(a) Schematic drawing of signaling pathway interactions in the migrating primordium. (b) *sdf1* is expressed along the midline of the trunk and guides the primordium. (c, d) The *sdf1* receptors *cxcr4b* (c) and *cxcr7b* (d) are expressed in the leading and trailing regions of the primordium, respectively. Constitutive activation of Wnt/β-catenin signaling in trailing cells leads to ectopic expression of *cxcr4b* in trailing cells (e) and complete loss of *cxcr7b* disrupting directed migration (f).

though *dkk1* is commonly thought to be a Wnt/β-catenin target, in the primordium *dkk1* is induced by Fgf signaling (Figure 22.3a).

All these results revealed that the Wnt/β-catenin and Fgf pathways interact in a feedback loop, which restricts each pathway to mutually exclusive domains. Because of these feedback interactions, both loss of Fgf signaling and constitutive activation of the Wnt/β-catenin pathway lead to ectopic expression of Wnt/β-catenin target genes in the trailing cells. This ectopic expression of Wnt/β-catenin target genes in trailing cells is disrupting directed cell migration (see below).

CHEMOKINE SIGNALING GUIDES THE MIGRATING PRIMORDIUM

Guided primordium migration is a fascinating process, as the primordium migrates along the embryonic axis, thereby encountering many different positional cues. Nevertheless, the primordium does not stray on its way, but migrates straight along the horizontal myoseptum (a connective tissue along the midline of the trunk) toward the tail tip. Several studies have demonstrated that chemokine signaling is the major guidance system employed by the primordium. The chemokine ligand Sdf1 is expressed in cells along the horizontal myoseptum, prefiguring the track on which the primordium will migrate (Figure 22.3b). The two Sdf1 receptors Cxcr4b and Cxcr7b, on the other hand,

are expressed in complementary regions of the primordium (Figure 22.3b, c; [21–25]).

Initially it was thought that Sdf1 was expressed in a gradient that guides the primordium to the tail tip. However, in some zebrafish mutants the primordium turns around on itself midway down the trunk and migrates back toward the head [23, 26, 27]. This finding led to the hypothesis that Sdf1 protein is not expressed as a gradient along the horizontal myoseptum, but rather that the primordium itself is polarized or is able to create an Sdf1 gradient within the primordium. This hypothesis is supported by the finding that the two chemokine receptors *cxcr4b* and *cxcr7b* are expressed in opposing poles of the primordium (Figure 22.3c, d). Both receptors are necessary for directed migration, as loss of either one disrupts migration. This finding initially seemed puzzling, since both receptors bind to Sdf1. However, given that neither receptor is dispensable, the two chemokine receptors either bind Sdf1 with different affinities, or they elicit distinct intracellular responses upon ligand binding. A beautiful study of the roles of chemokine signaling in primordial germ cell (PGC) migration in zebrafish discovered that Cxcr7b can also function as an Sdf1 sink without activating an intracellular signaling cascade [28]. Therefore, Cxcr7b modulates the amount of Sdf1 available to bind to Cxcr4b, which enables PGCs to detect an Sdf1 gradient guiding the cells toward the developing gonads. If Cxcr7b acts as an Sdf1 sink in the trailing region of the primordium as well, the regionally distinct combinations of *cxcr4b* and *cxcr7b* expression in the primordium might enable individual cells to detect a Sdf1 gradient necessary for directed migration.

A FEEDBACK LOOP BETWEEN FGF AND WNT/β-CATENIN SIGNALING CONTROLS MIGRATION BY LOCALIZING CHEMOKINE RECEPTOR EXPRESSION

Ectopic activation of Wnt/β-catenin signaling in trailing cells, or disruption of Fgf signaling, cause primordium stalling. The migration defect is not caused by an inability of cells to move, as time-lapse analyses showed that all cells are still highly motile but fail to migrate directionally [17]. This phenotype is reminiscent of the defects observed in mutants in which the chemokine guidance receptors cxcr4b or cxcr7b are disrupted [21, 23, 25], suggesting that the Fgf and Wnt/β-catenin pathways are also involved in regulating the chemokine guidance system. Indeed, chemokine receptor expression is highly aberrant in Fgf-depleted embryos, or in embryos in which the Wnt/β-catenin pathway is constitutively active. cxcr4b, which is normally more restricted to the leading cells, is ectopically expressed in trailing and deposited cells, and cxcr7b, which is normally restricted to trailing cells, is completely absent (Figure 22.3e, f). Thus, the migration defects observed in Fgf-depleted embryos, and in embryos in which the Wnt/β-catenin pathway is constitutively active, are likely due to misregulation of the two chemokine receptors cxcr4b and cxcr7b [17]. This study showed for the first time that Fgf/Wnt/β-catenin and chemokine signaling are functionally linked, which is significant not only for our understanding of developmental events but also for cancer biology.

CELL MIGRATION AND ROSETTOGENESIS ARE INDEPENDENTLY REGULATED

Since rosettes are epithelial structures within the migrating primordium, it would be intuitive that rosettes or lack thereof would influence cell migration. Surprisingly, primordia stall whether they contain rosettes, as in mutant embryos in which the Wnt/β-catenin pathway is constitutively active, or whether they lack rosettes, as in Fgf-depleted embryos. On the other hand, primordia in which Wnt/β-catenin signaling is blocked migrate normally toward the tail tip, even though they do not form or deposit any neuromasts. Therefore, migration and neurogenesis are independently influenced by Wnt/β-catenin signaling. Rosettogenesis is inhibited if loss of Wnt/β-catenin signaling leads to the loss of Fgf signaling, whereas migration is compromised if Wnt/β-catenin signaling is ectopically expressed in trailing cells, irrespective of the state of Fgf signaling in these cells.

SUMMARY

Studies of lateral line development have revealed how different signaling pathways interact with each other to control morphogenesis. On top of this cascade is Wnt/β-catenin signaling in leading cells, which initiates a feedback loop between itself and Fgf signaling in trailing cells. The restriction of these pathways in regionally distinct domains ensures the coupling of neurogenesis in trailing cells, and directed cell migration of all cells within the primordium. An interesting aspect of how these two pathways interact is that each pathway induces the expression of inhibitors for the other pathway. This novel finding justifies the reinvestigation of the hierarchy of these signaling interactions in other organ systems.

Even though the exact nature of the molecules might not be exactly recapitulated in other organ systems, it is likely that the overall characteristics of these signaling interactions are conserved. Many of the genes that have been identified to be important for lateral line development are also major players in cancer in humans. We therefore believe that, because of its relative simplicity and accessibility, the migrating lateral line primordium is a very valuable model for understanding signaling interactions in development and disease.

REFERENCES

1. Nicolson T. The genetics of hearing and balance in zebrafish. *Annu Rev Genet* 2005;**39**:9–22.
2. Harrison RG. Experimentelle Untersuchungen ueber die Entwicklung der Sinnesorgane der Seitenlinie bei den Amphibien. *Arch fuer mikr Anat und Ent* 1904;**63**.
3. Northcutt RG, Brandle K, Fritzsch B. Electroreceptors and mechanosensory lateral line organs arise from single placodes in axolotls. *Dev Biol* 1995;**168**:358–73.
4. Platt JB. Ontogenetic differentiations of the ectoderm in necturus. *Q J Microscopic Vacules* 1896;**38**:486–547.
5. Stone LS. Experiments on the development of the cranial ganglia and the lateral line sense organs in Amblystoma punctatum. *J Exp Zool* 1922;**35**:421–96.
6. Gompel N, Cubedo N, Thisse C, Thisse B, Dambly-Chaudiere C, Ghysen A. Pattern formation in the lateral line of zebrafish. *Mech Dev* 2001;**105**:69–77.
7. Ledent V. Postembryonic development of the posterior lateral line in zebrafish. *Development* 2002;**129**:597–604.
8. Metcalfe WK, Kimmel CB, Schabtach E. Anatomy of the posterior lateral line system in young larvae of the zebrafish. *J Comp Neurol* 1985;**233**:377–89.
9. Cotanche DA, Lee KH, Stone JS, Picard DA. Hair cell regeneration in the bird cochlea following noise damage or ototoxic drug damage. *Anat Embryol (Berl)* 1994;**189**:1–18.
10. Northcutt RG. Distribution and innervation of lateral line organs in the axolotl. *J Comp Neurol* 1992;**325**:95–123.
11. Lecaudey V, Cakan-Akdogan G, Norton WH, Gilmour D. Dynamic Fgf signaling couples morphogenesis and migration in the zebrafish lateral line primordium. *Development* 2008 in press.

12. Nechiporuk A, Raible DW. FGF-dependent mechanosensory organ patterning in zebrafish. *Science* 2008;**320**:1774–7.

13. Sarrazin AF, Villablanca EJ, Nunez VA, Sandoval PC, Ghysen A, Allende ML. Proneural gene requirement for hair cell differentiation in the zebrafish lateral line. *Dev Biol* 2006;**295**:534–45.

14. Millimaki BB, Sweet EM, Dhason MS, Riley BB. Zebrafish atoh1 genes: classic proneural activity in the inner ear and regulation by Fgf and Notch. *Development* 2007;**134**:295–305.

15. Hoch RV, Soriano P. Context-specific requirements for Fgfr1 signaling through Frs2 and Frs3 during mouse development. *Development* 2006;**133**:663–73.

16. Sai X, Ladher RK. FGF signaling regulates cytoskeletal remodeling during epithelial morphogenesis. *Curr Biol* 2008;**18**:976–81.

17. Aman A, Piotrowski T. Wnt/beta-catenin and Fgf signaling control collective cell migration by restricting chemokine receptor expression. *Dev Cell* 2008;**15**:749–61.

18. Furthauer M, Lin W, Ang SL, Thisse B, Thisse C. Sef is a feedback-induced antagonist of Ras/MAPK-mediated FGF signalling. *Nat Cell Biol* 2002;**4**:170–4.

19. Tsang M, Friesel R, Kudoh T, Dawid IB. Identification of Sef, a novel modulator of FGF signalling. *Nat Cell Biol* 2002;**4**:165–9.

20. Niehrs C. Function and biological roles of the Dickkopf family of Wnt modulators. *Oncogene* 2006;**25**:7469–81.

21. Dambly-Chaudiere C, Cubedo N, Ghysen A. Control of cell migration in the development of the posterior lateral line: antagonistic interactions between the chemokine receptors CXCR4 and CXCR7/RDC1. *BMC Dev Biol* 2007;**7**:23.

22. David NB, Sapede D, Saint-Etienne L, Thisse C, Thisse B, Dambly-Chaudiere C, Rosa FM, Ghysen A. Molecular basis of cell migration in the fish lateral line: role of the chemokine receptor CXCR4 and of its ligand, SDF1. *Proc Natl Acad Sci USA* 2002;**99**(16):297–302.

23. Haas P, Gilmour D. Chemokine signaling mediates self-organizing tissue migration in the zebrafish lateral line. *Dev Cell* 2006;**10**:673–80.

24. Li Q, Shirabe K, Kuwada JY. Chemokine signaling regulates sensory cell migration in zebrafish. *Dev Biol* 2004;**269**:123–36.

25. Valentin G, Haas P, Gilmour D. The chemokine SDF1a coordinates tissue migration through the spatially restricted activation of Cxcr7 and Cxcr4b. *Curr Biol* 2007;**17**:1026–31.

26. Kawasaki Y, Sato R, Akiyama T. Mutated APC and Asef are involved in the migration of colorectal tumour cells. *Nat Cell Biol* 2003;**5**:211–15.

27. Wilson AL, Shen YC, Babb-Clendenon SG, Rostedt J, Liu B, Barald KF, Marrs JA, Liu Q. Cadherin-4 plays a role in the development of zebrafish cranial ganglia and lateral line system. *Dev Dyn* 2007;**236**:893–902.

28. Boldajipour B, Mahabaleshwar H, Kardash E, Reichman-Fried M, Blaser H, Minina S, Wilson D, Xu Q, Raz E. Control of chemokine-guided cell migration by ligand sequestration. *Cell* 2008;**132**:463–73.

Chapter 23

Integration of BMP, RTK, and Wnt Signaling Through Smad1 Phosphorylations

Luis C. Fuentealba, Edward Eivers, Hojoon X. Lee and E. M. De Robertis

Howard Hughes Medical Institute and Department of Biological Chemistry, University of California, Los Angeles, California

INTRODUCTION

Understanding how multiple signaling pathways are integrated to generate simple cellular decisions is a major challenge in cellular and developmental biology. *Xenopus* ectoderm differentiation provides an excellent experimental system to study signal integration because at gastrula its cells must choose between two fates: epidermis or neural tissue. Epidermis is formed at high Bone Morphogenetic Protein (BMP) signaling levels, whereas the inhibition of BMPs by antagonists such as Noggin, Chordin, and Follistatin in the Spemann organizer leads to neuralization of the ectoderm [1–3]. BMP signals through activation of its cognate receptors, which are serine/threonine kinases that phosphorylate the transcription factors Smad1/5/8 at the carboxyl-terminal region, causing their activation and nuclear translocation [4, 5]. Recent studies have identified additional signaling inputs which converge on Smads. In this chapter, we review how three different pathways–BMP, RTK, and Wnt–are integrated at the level of a single transcription factor, Smad1, in the control of the differentiation of the ectoderm into neural or epidermal fates (Figure 23.1). We will also discuss other situations in which Smads are utilized as a platform for the integration of different signaling pathways.

NEURAL INDUCTION: LINKING RTKS AND ANTI-BMP SIGNALS

In the search for the molecular mechanisms that lead to neural induction, it was found that in amphibians, particularly in *Xenopus*, ectodermal cells acquire neural fate in response to BMP antagonists secreted by the Spemann's organizer region [6, 7]. However, neural tissue can also be induced through the activation of Receptor Tyrosine

Kinases (RTKs) by ligands such as Fibroblast Growth Factors (FGF) and Insulin-like Growth Factors (IGF). Experiments in the chick embryo have shown that FGF signals are required for CNS formation [8, 9], and overexpression of IGF in

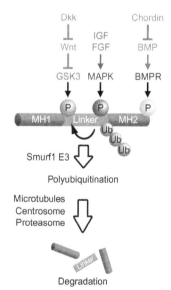

FIGURE 23.1 Biochemical pathway of Smad1 transcription factor signaling illustrating the different levels of signaling integration.
Upon binding of the BMP ligands, BMP receptor (BMPR) phosphorylates Smad1 at its C-terminal residues, resulting in its translocation into the nucleus and initiation of the transcription of target genes. Conversely, inhibitory MAPK and GSK3 phosphorylations of Smad1 at conserved sites in the linker (middle) region result in its polyubiquitinylation (mediated by the E3 ubiquitin ligase, Smurf1) and degradation in the proteasomal machinery [12, 13]. For neural induction to be achieved, the Smad1 transcriptional activity is required to be decreased by: (1) low BMP levels through the inhibition by antagonists such as Chordin and Noggin; (2) high MAPK activity through the activation of RTKs by growth factors such as FGF and IGF; and (3) high GSK3 activity through the inhibition of Wnt signals by antagonist such as Dkk1 and sFRPs. Reproduced from [13] (Fuentealba *et al.*, 2007; *Cell* **131**, 980–993) with permission.

Xenopus induced ectopic brain-like structures [10]. How do such distinct signaling pathways achieve similar outputs?

A key discovery was made in human cultured cell lines, in which the activation of the Mitogen-Activated Protein Kinase (MAPK) by Epidermal Growth Factor (EGF) resulted in the phosphorylation of Smad1 at its linker (middle) region, causing inhibition of its activity [11]. There are four conserved consensus MAPK sites (PXS[PO₃]P) in Smad1. Phosphorylation of these sites promotes polyubiquitinylation by Smurf1 E3-ubiquitin ligase, and subsequent degradation in the proteasomal machinery (Figure 23.1) [12, 13]. Thus, MAPK signals oppose BMP signaling by inhibiting the activity of the Smad1 transcription factor. Overexpression of MAPK phosphorylation-resistant mutants of Smad1 in *Xenopus* embryos resulted in increased expression of *Sizzled*, a BMP target gene, and reduced neural tissue, marked by *NCAM*, when compared to embryos injected with a wild-type Smad1 construct which had little effect [14, 15]. In other words, the Smad1 transcription factor is hyperactive in the absence of MAPK phosphorylation. This also suggests that neural genes are expressed only at very low levels of Smad1 activity, which are attained by having both low BMP levels and high MAPK signals. In agreement with this view, neural induction by the BMP antagonist Chordin can be blocked by agents that inhibit FGF or IGF signaling [14].

Signal integration during neural induction is also illustrated by dissociated *Xenopus* ectodermal cells experiments. Embryonic animal cap cells dissociated for 3 or more hours differentiate into neural tissue instead of adopting their normal epidermal fate. This default type of neural induction occurs in the absence of Spemann's organizer signals, and was thought to be caused by the dilution of endogenous BMPs into the culture medium [16]. However, analysis of the carboxyl-terminal phosphorylation of Smad1 showed that BMP ligands continue to signal in dissociated cells. Instead, cell dissociation induces a sustained activation of the MAPK pathway [17]. This sustained MAPK activity induced phosphorylation of Smad1 linker sites and inhibition of Smad1 transcriptional activity. Conversely, the inhibition of MAPK activation by various reagents resulted in epidermal differentiation [17]. These results showed that the inhibition of the BMP pathway during brain tissue formation in dissociated ectoderm was in fact caused by the activation of the MAPK pathway.

MAPK ACTIVATION EXPLAINS HETEROLOGOUS NEURAL INDUCERS

From an embryological point of view, the activation of MAPK in dissociated ectodermal cells is very interesting. Research in neural induction almost came to a halt in the 1930s, when it was found that dead organizer tissue and many heterologous non-specific substances such as fatty acids, sterols, methylene blue, and even sand particles could induce neural tissue [18]. Furthermore, Barth discovered and Holtfreter confirmed that ectoderm of the American salamander *Ambystoma maculatum* could differentiate into neural tissue by culturing them in sub-optimal saline solutions even in the absence of inducing substances [19, 20]. This was the final nail in the coffin of the once vibrant field of embryonic induction [3, 21]. We obtained *Ambystoma maculatum* embryos and reproduced these 60-year-old results [22]. It was found that MAPK/Erk is strongly induced in *Ambystoma maculatum* ectodermal cultures. A chemical inhibitor of the Ras/MAPK pathway, U0126, caused the explants to differentiate into epidermis instead of neural tissue [22]. Using sand particles as an example of a heterologous inducer produced similar results. Thus, the inhibition of Smad1 activity by MAPK activation provided the molecular explanation for the effects of the heterologous neural inducers described in the classical embryological literature [22].

EPIDERMAL DIFFERENTIATION: INTEGRATION OF WNT AND BMP SIGNALS

The choice between neural and epidermal fates is also regulated by canonical Wnt signals. In *Xenopus* and chicken embryos, Wnt promotes epidermal differentiation while its inhibition by antagonists such as Dickkopf-1 (Dkk1) and secreted Frizzled-related proteins (sFRPs) facilitates neural differentiation [8, 23]. In mouse embryonic stem cells (ESCs), sFRP2 has been isolated during overexpression screens for molecules causing differentiation into neural lineages [24]. Is there a common molecular mechanism by which BMP and Wnt signals regulate the choice between neural and epidermal fates?

The finding that Smad1 is phosphorylated by glycogen synthase kinase (GSK3) sheds light onto a mechanism integrating Smad and Wnt signals [12, 13]. GSK3 is a serine/threonine kinase that usually phosphorylates "primed," or prephosphorylated, substrates as in the well-known case of β-catenin. The canonical Wnt pathway signals through the inhibition of GSK3, resulting in the stabilization of β-catenin and translocation into the nucleus, where it functions as a transcriptional co-activator [25]. Coupled phosphorylations of β-catenin by casein kinase I followed by GSK3 result in its polyubiquitinylation and degradation in the proteasome [26]. A similar molecular mechanism was recently shown for the control of Smad1 stability. GSK3 phosphorylates Smad1 primed by MAPK in the linker region, and both phosphorylations are required for Smad1 polyubiquitinylation and degradation in the proteasome [13]. Importantly, Wnt signaling decreased this GSK3-dependent phosphorylation, resulting in the stabilization of Smad1, and prolonged the duration of the BMP signal

(Figure 23.1). In dissociated ectodermal cells, Wnt3a caused epidermal induction, which could be blocked by a dominant-negative form of Smad1 [13]. In the presence of Wnt, the activity of Smad1 is stronger (Figure 23.1). For as yet unknown reasons, the regulation of Smad1 by Wnt requires the presence of active β-catenin, as do most other canonical Wnt signaling effects [13].

The regulation of the Smad activity by both BMP and Wnt signals may also have important consequences during the formation of the body plan. During gastrulation in *Xenopus*, the embryonic patterning is regulated by gradients of growth factors and their antagonists, with BMPs controlling the dorsal–ventral (D-V) and Wnt signals the anterior–posterior (A-P) body axes [27]. This positional information must be seamlessly integrated, for when a blastula is cut in half the embryo can self-regulate, forming perfect identical twins [3]. This tendency of the embryo to regulate towards the whole is what embryologists call a self-regulating morphogenetic field. Such fields are found not only in the whole gastrula, but also during the formation of many organs, such as the developing organ fields of limb buds, brain, lens, heart, ear, and hypophysis [28].

A recent study using zebrafish as a model system proposed that the A-P pattern is also controlled by BMP signaling in a temporal progressive manner, with an early specification of anterior ventrolateral cells followed by a later specification of more posterior ventrolateral cells at critical intervals [29]. These observations are congruent with a mechanism in which the D-V and A-P axes may be integrated at the level of the Smad1 signal integration platform (Figure 23.2). We proposed that a conserved extracellular mechanism that controls BMP signaling regulates the intensity of Smad1 transcriptional activity in the D-V axis, while Wnt signals would control the stability and therefore the duration of the signal of this transcription factor in the nucleus [13]. The model of transcriptional and biochemical interactions shown in Figure 23.2 could provide the first molecular explanation for the self-regulating nature of embryonic morphogenetic fields. This biochemical patterning pathway is entirely composed of secreted extracellular signaling regulators, except for Smad1/5/8 which are transcription factors.

SMAD1 AS A PLATFORM FOR MAPK INTEGRATION

During development, there are multiple situations in which the FGF/MAPK pathway cross-talks with BMP signals. The finding that MAPK signaling can cause inhibition of signaling by BMP Smads via this hard-wired mechanism may help explain other situations in which FGF and BMP signals oppose each other during organogenesis. A classical example is the antagonism between FGF4 and BMP2 in limb development [30]. Similarly, opposing effects of FGFs

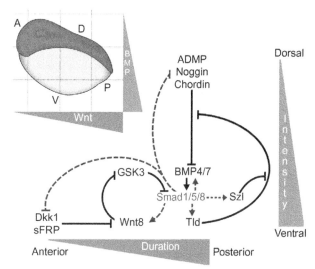

FIGURE 23.2 Model of a system Cartesian coordinates for the formation of the amphibian embryo body plan, in which the A-P (Wnt) and D-V (BMP) gradients are integrated at the level of Smad1/5/8 phosphorylations during gastrulation.
An elaborate biochemical pathway of extracellular protein–protein interactions mediates the formation of a DV gradient of BMP signaling, as well as an A-P gradient of Wnt signaling. In this model, BMP regulates the intensity, while Wnt controls the duration, of the Smad1/5/8 signal. ADMP is a dorsally expressed BMP [7]. Tolloid (Tld) is a zinc metalloprotease [52] that cleaves Chordin [53], and Sizzled (Szl) functions as a competitive inhibitor of Tld [54]. Direct protein–protein interactions are shown as a solid line, and transcriptional activity of Smad1/5/8 as a dashed line.

and BMPs have been reported in lung morphogenesis, cranial suture fusion, tooth development, and the maintenance of undifferentiated human ESCs [31–34]. In the mouse, knock-in of Smad1 forms that are insensitive to MAPK phosphorylation in the linker region exhibit phenotypes in gastrointestinal epithelium and the reproductive tract [35]. Cultured fibroblasts of these knock-in mice with Smad1 resistant to phosphorylation by MAPK have been prepared [35]. Experiments using luciferase BMP responsive reporter construct have shown that the BMP signal is no longer inhibited by the addition of FGF in mutant cells, demonstrating that in these conditions FGF signals through Smad1 [12]. The effects on the male reproductive tract may be explained by the recent discovery of a cross-talk between BMP and the androgen receptor [36]. It was found that the MAPK-dependent phosphorylations of Smad1 are required for the inhibitory effects of BMP/Smads on androgen receptor function, and there are indications that the integration of MAPK signals on Smad1 has a critical role in the androgen regulation of prostate cancer progression [36].

SMAD1 AS A PLATFORM FOR WNT SIGNALS

The potential role of Wnt signals in the regulation of the BMP pathway has also been found to include involvement in a number of developmental processes. In particular, recent studies have underscored the importance of the Wnt pathway on the pathogenesis of hereditary bone diseases that were previously thought to be caused by abnormal BMP signaling. BMPs were isolated as potent inducers of bone morphogenesis [37], yet recent genetic studies in humans and mice have revealed that the Wnt signaling pathway is the one that is mutated in most hereditary bone-formation diseases. Reduced bone formation or osteoporosis is observed in loss-of-function mutations in the Wnt co-receptor LRP5 [38], as well as in transgenic mice overexpressing the BMP antagonist Noggin [39, 40] or the Smad1-specific E3 ubiquitin ligase Smurf1 [41]. Conversely, excessive bone-mass formation or osteopetrosis is caused by loss-of-function mutations of the antagonists Dkk1 [42] and Sclerostin (SOST) [43, 44] which bind to LRP5/6 and inhibit Wnt signaling. The similarity of these bone phenotypes suggests that Wnt signals may regulate the bone formation through the mechanism involving Smad1 as a platform for signaling integration indicated in Figure 23.1.

A CONSERVED MECHANISM OF SIGNAL INTEGRATION

Is this mechanism of signal integration unique for the BMP-regulated transcription factors Smad1/5/8? In *Drosophila*, the Smad ortholog Mad also contains a single conserved MAPK phosphorylation site preceded by two GSK3 phosphorylation sites [13]. The conservation of these phosphorylation sites suggests that a similar mechanism of integration of MAPK and Wnt signals may control Mad signaling in *Drosophila*. Interestingly, *Drosophila* Mad is also a substrate for a MAPK-related kinase, Nemo, a member of the Nemo-like kinase (Nlk) family of serine/threonine kinases involved in the Wnt signal pathway [45, 46]. The Nlk phosphorylation occurs in a conserved serine at the N-terminal MH1 domain, and was shown to promote nuclear export of Mad and inhibition of the BMP signaling [47].

Finally, an analogous mechanism of signal integration has been found for the transcription factors Smad2 and Smad3, which transduce signals of Transforming Growth Factor β (TGFβ). The TGFβ growth factor pathway is very important tumor suppressor in mammals. In *Xenopus*, the TGFβ molecule Activin induces the expression of mesodermal genes in ectodermal cells, but this ability is lost after a certain point of development. This "loss of competence" correlated with the inability of Smad2 to accumulate in the nucleus, and mutation of conserved MAPK phosphorylation sites (SP) at the linker region of Smad2 prolonged the competence of cells to respond to Activin by inducing mesodermal genes [48].

TGFβs are also potent inhibitors of cell proliferation, and have crucial roles in the control of tumorigenesis. However, certain types of cancer are resistant to the growth-inhibitory effects of TGFβ. Several studies have highlighted the role of linker-phosphorylation in TGFβ resistance in cancer. In transformed cell lines with activated oncogenic Ras/MAPK signals, Smad2/3 protein displays elevated phosphorylation at their MAPK sites, which inhibited TGFβ-induced nuclear accumulation [49]. Similarly, Smad2/3 is also a substrate target for G1 cyclin-dependent kinases. CDK4 and CDK2 phosphorylate Smad2/3 in the linker region and repress its capacity to inhibit cell cycle progression from G1 to S phase [50]. Since cancer cells often contain high levels of CDK activity, inactivation of Smad2/3 by CDK may contribute to tumorigenesis and TFGβ resistance in cancer cells. Furthermore, the stability of Smad3, but not Smad2, is regulated by GSK3 phosphorylation in the MH1 domain, which triggers its degradation [51]. The use of the different Smads as unique platforms for the integration of signals suggests that these are ancient molecular mechanisms that different cellular signaling pathways utilize to regulate the cross-talk between BMP/TGFβ signals and receptor tyrosine kinases and canonical Wnt/GSK3 signals.

CONCLUDING REMARKS

It is important to emphasize that these regulations constitute hard-wired circuits that will affect the activity of many genes. The Smad1/5/8 and Smad2/3 transcription factors affect the activity of hundreds of target genes and presumably thousands of DNA enhancers that bind these proteins. All of them will be affected coordinately in this mode of signaling integration, which regulates the intensity (BMP makes the Smad1/5/8 signal stronger) and the duration of the signal (RTK/MAPK makes the signal shorter and Wnt makes the BMP signal longer). The activity of Smad1 at any one time represents the summation of these three (or more) inputs of signaling information. Hundreds of target genes are co-regulated simultaneously. Perhaps the need for coordinate control resides in the self-regulation nature of morphogenetic fields. In the case of the gastrula embryonic field, the Cartesian coordinates on which the various organs will develop at later stages are preordained in the phosphorylation sites of Smad1/5/8 (Figure 23.2). It might be simpler to integrate positional information in a few transcription factors rather than in hundreds of them, so as to generate a perfect embryo time after time. The human genome contains 1500 transcription factors, and a great many of them are phosphorylated. We can expect that other DNA-binding proteins will also serve as platforms for signaling integration once they are studied in similar detail as Smad1.

REFERENCES

1. Harland R. Neural induction. *Curr Opin Genet Dev* 2000;**10**:357–62.

2. De Robertis EM, Kuroda H. Dorsal-ventral patterning and neural induction in *Xenopus* embryos. *Annu Rev Cell Dev Biol* 2004;**20**: 285–308.

3. De Robertis EM. Spemann's organizer and self-regulation in amphibian embryos. *Nat Rev Mol Cell Bio* 2006;**7**:296–302.

4. Shi Y, Massagué J. Mechanisms of TGF-beta signaling from cell membrane to the nucleus. *Cell* 2003;**113**:685–700.

5. Feng XH, Derynck R. Specificity and versatility in TGF-β signaling through Smads. *Annu Rev Cell Dev Biol* 2005;**21**:659–93.

6. Khokha MK, Yeh J, Grammer TC, Harland RM. Depletion of three BMP antagonists from Spemann's organizer leads to a catastrophic loss of dorsal structures. *Dev Cell* 2005;**8**:401–11.

7. Reversade B, De Robertis EM. Regulation of ADMP and BMP2/4/7 at opposite embryonic poles generates a self-regulating morphogenetic field. *Cell* 2005;**123**:1147–60.

8. Wilson S, Edlund T. Neural induction: toward a unifying mechanism. *Nat Neurosci* 2001;**4**:1161–8.

9. Stern CD. Induction and initial patterning of the nervous system–the chick embryo enters the scene. *Curr Opin Genet Dev* 2002;**12**:447–51.

10. Pera EM, Wessely O, Li SY, De Robertis EM. Neural and head induction by insulin-like growth factor signals. *Dev Cell* 2001;**1**:655–65.

11. Kretzschmar M, Doody J, Massagué J. Opposing BMP and EGF signaling pathways converge on the TGF-β family mediator Smad1. *Nature* 1997;**389**:618–22.

12. Sapkota G, Alarcon C, Spagnoli FM, Brivanlou AH, Massagué J. Balancing BMP signaling through integrated outputs into the Smad linker. *Mol Cell* 2007;**25**:441–54.

13. Fuentealba LC, Eivers E, Ikeda A, Hurtado C, Kuroda H, Pera EM, De Robertis EM. Integrating patterning signals: Wnt/GSK3 regulates the duration of the BMP/Smad1 signal. *Cell* 2007;**131**:980–93.

14. Pera EM, Ikeda A, Eivers E, De Robertis EM. Integration of IGF, FGF, and anti-BMP signals via Smad1 phosphorylation in neural induction. *Genes Dev* 2003;**17**:3023–8.

15. Sater AK, El-Hodiri HM, Goswami M, Alexander TB, Al-Sheikh O, Etkin LD, Akif Uzman J. Evidence for antagonism of BMP-4 signals by MAP kinase during *Xenopus* axis determination and neural specification. *Differentiation* 2003;**71**:434–44.

16. Wilson PA, Hemmati-Brivanlou A. Induction of epidermis and inhibition of neural fate by BMP-4. *Nature* 1995;**376**:331–3.

17. Kuroda H, Fuentealba L, Ikeda A, Reversade B, De Robertis EM. Default neural induction: neuralization of dissociated *Xenopus* cells is mediated by Ras/MAPK activation. *Genes Dev* 2005;**19**:1022–7.

18. Holtfreter J, Hamburger V. Embryogenesis: progressive differentiation–Amphibians. In: Willier BH, Weiss PA, Hamburger V, editors. *Analysis of development*. Philadelphia, PA: Saunders WB Company; 1955. p. 230–96.

19. Barth LG. Neural differentiation without organizer. *J Exp Zool* 1941;**87**:371–84.

20. Holtfreter J. Neural differentiation of ectoderm through exposure to saline solution. *J Exp Zool* 1944;**95**:307–43.

21. Spemann HEmbryonic development and induction. New Haven, CT: Yale University Press; 1938.

22. Hurtado C, De Robertis EM. Neural induction in the absence of organizer in salamanders is mediated by MAPK. *Dev Biol* 2007;**307**:282–9.

23. Glinka A, Wu W, Delius H, Monaghan AP, Blumenstock C, Niehrs C. Dickkopf-1 is a member of a new family of secreted proteins and functions in head induction. *Nature* 1998;**391**:357–62.

24. Aubert J, Dunstan H, Chambers I, Smith A. Functional gene screening in embryonic stem cells implicates Wnt antagonism in neural differentiation. *Nat Biotechnol* 2002;**20**:1240–5.

25. Logan CY, Nusse R. The Wnt signaling pathway in development and disease. *Ann Rev Cell Dev Biol* 2004;**20**:781–810.

26. Cohen P, Frame S. The renaissance of GSK3. *Nat Rev Mol Cell Biol* 2001;**2**:769–76.

27. Niehrs C. Regionally specific induction by the Spemann-Mangold organizer. *Nat Rev Genet* 2004;**6**:425–34.

28. Huxley J, De Beer GRThe elements of experimental embryology. Cambridge: Cambridge University Press; 1934.

29. Tucker JA, Mintzer KA, Mullins MC. The BMP signaling gradient patterns dorsoventral tissues in a temporally progressive manner along the anteroposterior axis. *Dev Cell* 2008;**14**:108–19.

30. Niswander L, Martin GR. FGF-4 and BMP-2 have opposite effects on limb growth. *Nature* 1993;**361**:68–71.

31. Weaver M, Dunn NR, Hogan BL. Bmp4 and Fgf10 play opposing roles during lung bud morphogenesis. *Development* 2000;**127**:2695–704.

32. Warren SM, Brunet LJ, Harland RM, Economides AN, Longaker MT. The BMP antagonist noggin regulates cranial suture fusion. *Nature* 2003;**422**:625–9.

33. Thesleff I, Mikkola M. The role of growth factors in tooth development. *Intl Rev Cytol* 2002;**217**:93–135.

34. Xu RH, Peck RM, Li DS, Feng X, Ludwig T, Thomson JA. Basic FGF and suppression of BMP signaling sustain undifferentiated proliferation of human ES cells. *Nat Methods* 2005;**2**:185–90.

35. Aubin J, Davy A, Soriano P. *In vivo* convergence of BMP and MAPK signaling pathways: impact of differential Smad1 phosphorylation on development and homeostasis. *Genes Dev* 2004;**18**:1482–94.

36. Qiu T, Grizzle WE, Oelschlager DK, Shen X, Cao X. Control of prostate cell growth: BMP antagonizes androgen mitogenic activity with incorporation of MAPK signals in Smad1. *EMBO J* 2007;**26**:346–57.

37. Urist MR. Bone: formation by autoinduction. *Science* 1965;**150**:893–9.

38. Koay MA, Brown MA. Genetic disorders of the LRP5-Wnt signalling pathway affecting the skeleton. *Trends Mol Med* 2005;**11**:129–37.

39. Devlin RD, Du Z, Pereira RC, Kimble RB, Economides AN, Jorgetti V, Canalis E. Skeletal overexpression of noggin results in osteopenia and reduced bone formation. *Endocrinology* 2003;**144**:1972–8.

40. Wu XB, Li Y, Schneider A, Yu W, Rajendren G, Iqbal J, Yamamoto M, Alam M, Brunet LJ, Blair HC, Zaidi M, Abe E. Impaired osteoblastic differentiation, reduced bone formation, and severe osteoporosis in noggin-overexpressing mice. *J Clin Invest* 2003;**112**:924–34.

41. Zhao M, Qiao M, Harris SE, Oyajobi BO, Mundy GR, Chen D. Smurf1 inhibits osteoblast differentiation and bone formation *in vitro* and *in vivo*. *J Biol Chem* 2004;**279**:12,854–12,859.

42. Morvan F, Boulukos K, Clément-Lacroix P, Roman Roman S, Suc-Royer I, Vayssière B, Ammann P, Martin P, Pinho S, Pognonec P, Mollat P, Niehrs C, Baron R, Rawadi G. Deletion of a single allele of the *Dkk1* gene leads to an increase in bone formation and bone mass. *J Bone Min Res* 2006;**21**:934–45.

43. van Bezooijen RL, Roelen BA, Visser A, van der Wee-Pals L, de Wilt E, Karperien M, Hamersma H, Papapoulos SE, ten Dijke P, Löwik CW. Sclerostin is an osteocyte-expressed negative regulator of bone formation, but not a classical BMP antagonist. *J Exp Med* 2004;**199**:805–14.

44. Semenov M, Tamai K, He X. SOST is a ligand for LRP5/LRP6 and a Wnt signaling inhibitor. *J Biol Chem* 2005;**280**:26,770–26,775.

45. Ishitani T, Ninomiya-Tsuji J, Nagai S, Nishita M, Meneghini M, Barker N, Waterman M, Bowerman B, Clevers H, Shibuya H, Matsumoto K. The TAK1-NLK-MAPK-related pathway antagonizes signalling between beta-catenin and transcription factor TCF. *Nature* 1999;**399**:798–802.

46. Meneghini MD, Ishitani T, Carter JC, Hisamoto N, Ninomiya-Tsuji J, Thorpe CJ, Hamill DR, Matsumoto K, Bowerman B. MAP kinase and Wnt pathways converge to downregulate an HMG-domain repressor in *Caenorhabditis elegans*. *Nature* 1999;**399**:793–7.

47. Zeng YA, Rahnama M, Wang S, Sosu-Sedzorme W, Verheyen FM *Drosophila* Nemo antagonizes BMP signaling by phosphorylation of Mad and inhibition of its nuclear accumulation. *Development* 2007;**134**:2061–71.

48. Grimm OH, Gurdon JB. Nuclear exclusion of Smad2 is a mechanism leading to loss of competence. *Nat Cell Biol* 2002; **4**:519–22.

49. Kretzschmar M, Doody J, Timokhina I, Massagué J. A mechanism of repression of TGFbeta/ Smad signaling by oncogenic Ras. *Genes Dev* 1999;**13**:804–16.

50. Matsuura I, Denissova NG, Wang G, He D, Long J, Liu F. Cyclin-dependent kinases regulate the antiproliferative function of Smads. *Nature* 2004;**430**:226–31.

51. Guo X, Ramirez A, Waddell DS, Li Z, Liu X, Wang XF. Axin and GSK3- control Smad3 protein stability and modulate TGF-signaling. *Genes Dev* 2008;**22**:106–20.

52. Piccolo S, Agius E, Lu B, Goodman S, Dale L, De Robertis EM. Cleavage of Chordin by Xolloid metalloprotease suggests a role for proteolytic processing in the regulation of Spemann organizer activity. *Cell* 1997;**91**:407–16.

53. Sasai Y, Lu B, Steinbeisser H, Geissert D, Gont LK, De Robertis EM. *Xenopus* chordin: a novel dorsalizing factor activated by organizer-specific homeobox genes. *Cell* 1994;**79**:779–90.

54. Lee HX, Ambrosio AL, Reversade B, De Robertis EM. Embryonic dorsal–ventral signaling: secreted frizzled-related proteins as inhibitors of tolloid proteinases. *Cell* 2006;**124**:147–59.

Chapter 24

Hedgehog Signaling in Development and Disease

Frederic de Sauvage

Department of Molecular Biology, Genentech, Inc., South San Francisco, California

THE HEDGEHOG PROTEINS: GENERATION AND DISTRIBUTION

The hedgehog mutation was originally identified in a *Drosophila* genetic screen as one of the segment-polarity genes important in fly development [1]. It is now clear that Hh plays a vital role in the development of multiple organ systems in the fly and vertebrates (for review, see [2]). In mammals there are three Hh proteins, named Sonic Hh (Shh), Desert Hh (Dhh), and Indian Hh (Ihh). Dhh appears to be most closely related to *Drosophila* Hh, while Shh and Ihh are more closely related to one another. Production and diffusion of these factors in different tissues determines proper development of multiple organ systems.

Hh proteins are synthesized as precursor proteins of about 45 kD. The C-terminal portion of the Hh precursor has autoproteolytic activity, and cleaves Hh into a C-terminal peptide of about 25 kD with no known function and an N-terminal fragment (Hh-N) which constitutes the biologically active portion of Hh [3]. During autoprocessing, a cholesterol moiety is coupled to the C-terminus of Hh-N, a form which is further denoted Hh-Np [4]. It is thought that the addition of cholesterol helps to retain Hh-Np to cell membranes, thus limiting the range of action of Hh activity. However, in mice engineered to express a form of Shh lacking cholesterol modification (N-Shh), short-range Hh signaling was maintained while long-range signaling was defective, resulting in loss of digits and proper patterning in the developing limb, and suggesting differential requirements for cholesterol in Hh signaling [5]. Additional proteins involved in the secretion and diffusion of cholesterol-modified forms of Hh have been identified, such as Dispatched (Disp), which is required for release of Hh-Np from Hh-producing cells, and Tout velu (TTV), which is involved in the biosynthesis of a putative Hh-interacting proteoglycan [6–8].

Hh proteins are further modified by palmitoylation on a highly conserved N-terminal cysteine residue [9]. Mutation of the *sightless/skinny Hh* (*sig/ski*) gene, encoding a transmembrane (TM) acyl transferase, abrogates palmitoylation of Hh-N and results in a Hh-like phenotype, indicating that palmitoylation of Hh is required for some aspect of Hh function [10, 11]. In some systems both modified and unmodified forms of Hh show equivalence, indicating that the importance of fatty acid modification may be context-dependent [9, 12]. Together these data indicate that the biological activity of the Hh proteins is finely tuned through posttranslational modification, affecting its activity and capacity to diffuse.

TRANSMITTING THE HH SIGNAL

Both *Drosophila* and mouse genetics indicate that the 7-transmembrane protein Smoothened (Smo) is required to transmit the Hh signal [13–15], while Patched (Ptc), a 12-transmembrane protein, negatively regulates Smo in the absence of Hh [16, 17]. This inhibition has been shown to occur in a sub-stoichiometric or "catalytic" manner [18]. Binding of Hh to Ptc is also regulated by additional cell surface proteins, such as Hh interacting protein (Hip [19]), which sequester Hh. In contrast, growth arrest-specific gene (Gas1 [20]) and two newly identified single-transmembrane proteins, cell adhesion molecule-related/downregulated by oncogenes (Cdon/Cdo) and brother of Cdo (Boc), appear to facilitate Hh signaling [21–23].

Although much remains to be clarified, some insights into the mechanistic details describing the interplay between these two proteins have started to emerge. Work in the fly initially showed that Smo activity requires cell surface localization, and that Smo access to the plasma membrane is controlled by Ptc via regulation of vesicular trafficking [24].

Binding of Hh to Ptc leads to its internalization, and allows Smo to translocate to the cell surface and initiate signaling (Figure 24.1). Signal transduction by Smo appears to involve a conformational change in the carboxy-terminal tail of Smo homodimers [25]. Very elegant mouse genetic work has recently shown that the primary cilium, a nonmotile projection present on most vertebrate cells, appears to be required for Hh signaling in higher organisms [26, 27]. Cell-localization studies have suggested a model of reciprocal movement between Ptc and Smo similar to that observed in flies, but where Smo access to the primary cilium instead of merely the plasma membrane is critical in order to initiate signaling [28–30]. Interestingly, Ptc may control the movement of Smo to the cilium by regulating the levels of oxysterols which are able to activate the pathway and promote Smo translocation to the cilia [31, 32].

What makes the cilium special in order to allow for signaling downstream of Smo remains to be determined. Other important components of the pathway, including SuFu and the Gli family of zinc finger transcription factors known as cubitus interruptus (Ci) in the fly, are also found in cilia, where important steps in their activation may be occurring [33]. Gli activity is regulated at multiple levels, including nuclear export, proteosome-mediated degradation, and subcellular localization. In vertebrates there are three Ci orthologs, Gli1, Gli2, and Gli3, which may have retained aspects of Ci-155 and Ci-75 function (reviewed in [2, 34]). Knockout or transgenic mice of each Gli isoform have been generated, and the phenotypes observed support the idea that Gli2 and Gli3 are critical for normal development, while Gli1 may be redundant for Gli2 and Gli3 function. The interplay between the activities of these isoforms increases the complexity of vertebrate systems dramatically, making it difficult conclusively to delineate all of the functions for the different Glis during normal development (reviewed in [2]).

HH IN DEVELOPMENT AND DISEASE

Studies of the normal functions for the Hhs in animal models have helped in our understanding of Hh-related diseases. Many studies have shown that Shh acts as a morphogen in the nervous system, where it is secreted from the notochord and later from the floor plate, patterning neurons along the dorsoventral axis of the neural tube in a dose-dependent manner (reviewed in [2]) (Figure 24.2). In the spinal cord and part of the hindbrain, a fine gradient of SHH with two- to three-fold incremental changes from the source (floor plate and notochord) delineates the ventral neural tube into six distinct domains along the dorsal-ventral axis. The expression of a set of homeodomain and bHLH transcription factors is tightly controlled by different SHH protein concentrations, thus generating six intricate combinatorial transcription factor codes in these domains [35, 36]. These codes specify the identities of neural progenitors, which ultimately give rise to six cell types, including the floor plate (FP) cells, motor neurons (MN), and four classes of interneurons (V0, V1, V2, V3) (reviewed in [37]; Figure 24.2a, c). In the brains of mouse embryos, SHH acts in concert with another organizer molecule FGF8 to create information grids, which serve as spatial cues for the specification of ventral cell types. Ventral neurons

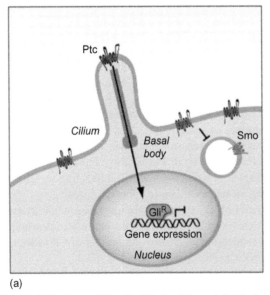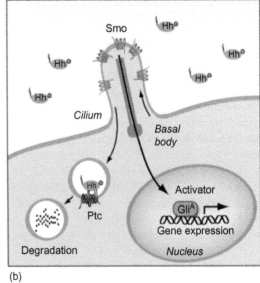

(a)   (b)

FIGURE 24.1 Model for transmitting the Hh signal through Patched and Smoothened via the primary cilium.
(a) In the absence of Hh, Ptc acts to repress Smo translocation to the cell surface and the primary cilium, leading to Gli phosphorylation and partial degradation by the proteasome in a transcriptional repressor form (Gli^R). (b) Upon Hh binding to Ptc, the Ptc–Hh complex is internalized via endocytosis resulting in the degradation of Ptc. This allows for the translocation of Smo to the cell surface and the primary cilium where signaling events downstream of Smo are activated and leads to Gli stabilization and translocation to the nucleus in an activated form (Gli^A).

such as dopaminergic (DA) neurons and serotonergic (5HT) neurons are specified in locations where Shh and FGF8 intersect (Figure 24.2a, b) [38, 39].

Shh controls cell fate not only by induction, but also by repression. It is found that SHH in the forebrain is required to repress Pax6 expression, thus resulting in the separation of a single eye field into two retinal primordia [40, 41]. This explains the cyclopic phenotype in the Shh-deficient embryos [42]. Late in development, Shh elicits cellular responses other than fate specification. For example, in the fetal cerebella, SHH induces proliferation of granule cells while inhibiting their differentiation [43–45].

Shh also acts to determine anterior–posterior (A–P) patterning in the developing skeleton, limb bud, and gut tube (reviewed in [2, 34]). Shh has also been shown to act as an angiogenic factor leading to neovascularization and the proliferation of blood cells [46]. In some tissue types, such as the pancreas, Shh acts as both a positive and a negative regulator, as its activity is needed for inhibition of pancreatic

anlagen formation, but is also needed for specification of the pancreatic β cells [47, 48]. Shh is also involved in the morphology of branching structures such as the lung [49]. The importance of Shh is highlighted by the phenotype of Shh knockout mice, which die at birth due to multiple defects, including cyclopia and holoprosencephaly (HPE), as well as other defects in limbs, brain, spinal cord, axial skeleton, and midline structures [42]. Overlapping roles of Shh and Ihh have been identified in heart development as well as specification of left/right (L–R) asymmetry, as observed in Shh/Ihh double knockout mice and Smo knockout mice [50]. The major impact of Ihh is in radial patterning of the gut as well as bone morphogenensis [51, 52]. Loss of Ihh function results in a lack of chondrocyte proliferation and differentiation [52]. Dhh is predominantly involved in peripheral nerve sheath and germ cell development, particularly in the development of the male germ line and maturation of the testes [53, 54].

The patterning functions of Hh are also highlighted in humans where mutations have clearly been linked to developmental disorders, including spina bifida, neural tube defects, and skeletal deformations. For example, mutations in human SHH result in cyclopia and HPE [55]. Downstream of Hh, mutations within GLI3 are found in Grieg's cephalopolycyndactyly [56] and Pallister-Hall syndrome (PHS) [34, 57], disorders associated with various abnormalities including polydactyly.

The first indication that the Hh pathway might be involved in tumor formation stemmed from the observation that human Patched (*PTCH*) gene was mutated in individuals with Gorlin's syndrome (also known as Basal-Cell Nevus Syndrome (BCNS)) [58], a familial inherited predisposition to the development of basal cell carcinomas (BCCs), medulloblastomas, and rhabdomyocarcomas [59, 60]. Most sporadic BCCs have been associated with mutation within PTCH [61]. Loss of PTCH function leads to constitutive SMO signaling and Hh pathway activation. Point mutations in Smo, which result in constitutively active PTCH-insensitive forms of SMO, have also been identified in sporadic BCC [62].

In addition to tumors where mutations in components of the Hh pathway lead to its activation, Hh ligand-driven pathway activation has more recently been implicated in a large number of cancers, such as small cell lung cancer, various types of upper GI tract tumors such as pancreatic ductal adenocarcinoma, prostate cancer, and, more recently, multiple myeloma or other B cell derived malignancies (see Table 24.1). However, the exact contribution of the Hh signal to tumorigenesis in these tissues remains to be defined. Evidence for a causative role is lacking, since oncogenic mutations in Hh pathway components have not been identified to date in these tumor types, and introduction of constitutively active SMO (SMOM2) in mice only leads to the formation of tumors where the pathway is known to be mutated in humans [63]. Most of the work performed *in vitro* on cell lines using high concentrations

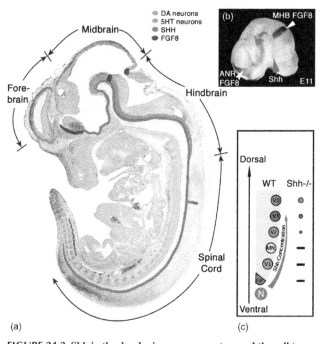

(a)

(c)

FIGURE 24.2 Shh in the developing nervous system and the cell types controlled by Shh.
(a) Shh, FGF8, dopaminergic (DA) neurons, serotonergic (5HT) neurons, and six ventral spinal cord neurons are diagrammed on a sagittal section of a 12-day-old mouse embryo. Shh in the notochord and floorplate is represented by a single line. (b) Wholemount *in situ* hybridization of Shh and FGF8 in the brain of an 11-day-old embryo. Shh is expressed in the ventral midline (dark arrow), FGF8 is expressed in two organizers: the mid-hindbrain junction (MHB) and the anterior neural ridge (ANR) (white arrow heads). (c) The position of six Shh dependent cell types (filled circles) in the ventral spinal cord of wild-type (WT) and Shh mutant embryos (Shh$^{-/-}$), and their relationship to Shh concentration (dark arrow). Different cell types are induced by different threshold concentrations of Shh. Higher concentration of SHH induces cells in progressively more ventral locations. FP, floor plate; N, notochord; MN, motor neuron; V0, V1, V2 and V3, four classes of interneurons.

TABLE 24.1 Tumor types where Hedgehog pathway activation has been described via mutation of pathway components or Hh ligand overexpression

	Tissue	Reference
Mutation-driven	BCC	59, 60, 69, 70
(PTCH, SMO, SUFU)	Medulloblastoma	70, 71
	Rhabdomyocarcoma	72
Ligand-driven	Lung (SCLC and NSCLC)	73, 74
(SHH, IHH, DHH)	Esophagus	75
	Pancreas	75, 76
	Stomach	75
	Biliary Tract	75
	Prostate	77–80
	Colon	81–83
	Liver	84
	B-cells (Multiple myeloma	85, 86

of cyclopamine posit an autocrine role where epithelial tumor cells both produce and respond to Hh ligands. In a variation of this model, the responding cell population could be a subset of tumor cells, identified as cancer stem cells or cancer initiating cells [64]. Yet another alternative model in line with the more typical role of the pathway in these tissues during development is one of paracrine signaling, where Hh ligand made by tumor cells activate the pathway in the infiltrating stroma. The activated stroma would in return provide an environment more favorable to tumor growth/survival.

The connection between the Hh pathway and human cancer drove the effort to identify small molecule Hh antagonists [65]. In addition to cyclopamine, a natural Smo antagonist isolated from corn lilies, different classes of small-molecule Hh antagonists have been identified through cell-based screens using Hh reporter assays [66, 67]. One of these small-molecule antagonists in preclinical development has been used successfully to treat endogenous medulloblastoma tumors in the $Ptch1^{+/-}P53^{-/-}$ genetic mouse model [68], suggesting that drugs specifically inhibiting of the Hh pathway may be effective in the treatment of Hh-associated tumors.

ACKNOWLEDGEMENTS

The authors would like to thank A. Bruce for help in generating figures; they also apologize to the many authors whose work was not referenced in this chapter due to space constraints.

REFERENCES

1. Nusslein-Volhard C, Wieschaus E. Mutations affecting segment number and polarity in Drosophila. *Nature* 1980;**287**:795–801.
2. Ingham PW, McMahon AP. Hedgehog signaling in animal development: paradigms and principles. *Genes Dev* 2001;**15**:3059–87.
3. Lee JJ, Ekker SC, von Kessler DP, Porter JA, Sun BI, Beachy PA. Autoproteolysis in hedgehog protein biogenesis. *Science* 1994;**266**:1528–37.
4. Porter JA, Young KE, Beachy PA. Cholesterol modification of hedgehog signaling proteins in animal development. *Science* 1996;**274**:255–9.
5. Lewis PM, Dunn MP, McMahon JA, et al. Cholesterol modification of Sonic Hedgehog is required for long-range signaling activity and effective modulation of signaling by Ptc1. *Cell* 2001;**105**:599–612.
6. Bellaiche Y, The I, Perrimon N. Tout-velu is a Drosophila homologue of the putative tumour suppressor EXT-1 and is needed for Hh diffusion. *Nature* 1998;**394**:85–8.
7. Thé I, Bellaiche Y, Perrimon N. Hedgehog movement is regulated through tout velu-dependent synthesis of a heparan sulfate proteoglycan. *Mol Cell* 1999;**4**:633–9.
8. Burke R, Nellen D, Bellotto M, et al. Dispatched, a novel sterol-sensing domain protein dedicated to the release of cholesterol-modified hedgehog from signaling cells. *Cell* 1999;**99**:803–15.
9. Pepinsky RB, Zeng C, Wen D, et al. Identification of a palmitic acid-modified form of human Sonic hedgehog. *J Biol Chem* 1998;**273**:14,037–14,045.
10. Chamoun Z, Mann RK, Nellen D, et al. Skinny hedgehog, an acyltransferase required for palmitoylation and activity of the hedgehog signal. *Science* 2001;**293**:2080–4.
11. Amanai K, Jiang J. Distinct roles of Central missing and Dispatched in sending the Hedgehog signal. *Development (Suppl)* 2001;**128**:5119–27.
12. Kohtz J, Lee H, Gaiano N, et al. N-terminal fatty-acylation of sonic hedgehog enhances the induction of rodent ventral forebrain neurons. *Development* 2001;**128**:2351–63.
13. Kalderon D. Transducing the Hedgehog signal (minireview). *Cell* 2000;**103**:371–4.
14. Alcedo J, Ayzenzon M, Von Ohlen T, Noll M, Hooper JE. The *Drosophila* smoothened gene encodes a seven-pass membrane protein, a putative receptor for the hedgehog signal. *Cell* 1996;**86**:221–32.
15. van den Heuvel M, Ingham PW. smoothened encodes a receptor-like serpentine protein required for hedgehog signalling. *Nature* 1996;**382**:547–51.
16. Ingham PW, Taylor AM, Nakano Y. Role of the *Drosophila* patched gene in positional signalling. *Nature* 1991;**353**:184–7.
17. Chen Y, Struhl G. *In vivo* evidence that Patched and Smoothened constitute distinct binding and transducing components of a Hedgehog receptor complex. *Development* 1998;**125**:4943–8.
18. Taipale J, Cooper MK, Maiti T, Beachy PA. Patched acts catalytically to suppress the activity of Smoothened. *Nature* 2002;**418**:892–7.

19. Chuang PT, McMahon AP. Vertebrate Hedgehog signalling modulated by induction of a Hedgehog-binding protein. *Nature* 1999;**397**:617–21.

20. Lee CS, Buttitta L, Fan CM. Evidence that the WNT-inducible growth arrest-specific gene 1 encodes an antagonist of sonic hedgehog signaling in the somite. *Proc Natl Acad Sci USA* 2001;**98**:11,347–11,352.

21. Yao S, Lum L, Beachy P. The ihog cell-surface proteins bind hedgehog and mediate pathway activation. *Cell* 2006;**125**:343–57.

22. Tenzen T, Allen BL, Cole F, Kang JS, Krauss RS, McMahon AP. The cell surface membrane proteins Cdo and Boc are components and targets of the Hedgehog signaling pathway and feedback network in mice. *Dev Cell* 2006;**10**:647–56.

23. Zhang W, Kang JS, Cole F, Yi MJ, Krauss RS. Cdo functions at multiple points in the Sonic Hedgehog pathway, and Cdo-deficient mice accurately model human holoprosencephaly. *Dev Cell* 2006;**10**:657–65.

24. Denef N, Neubuser D, Perez L, Cohen SM. Hedgehog induces opposite changes in turnover and subcellular localization of patched and smoothened. *Cell* 2000;**102**:521–31.

25. Zhao Y, Tong C, Jiang J. Hedgehog regulates smoothened activity by inducing a conformational switch. *Nature* 2007;**450**:252–8.

26. Huangfu D, Liu A, Rakeman AS, Murcia NS, Niswander L, Anderson KV. Hedgehog signalling in the mouse requires intraflagellar transport proteins. *Nature* 2003;**426**:83–7.

27. Huangfu D, Anderson KV. Cilia and Hedgehog responsiveness in the mouse. *Proc Natl Acad Sci USA* 2005;**102**:11,325–11,330.

28. Corbit KC, Aanstad P, Singla V, Norman AR, Stainier DY, Reiter JF. Vertebrate Smoothened functions at the primary cilium. *Nature* 2005;**437**:1018–21.

29. Rohatgi R, Scott MP. Patching the gaps in Hedgehog signalling. *Nat Cell Biol* 2007;**9**:1005–9.

30. Rohatgi R, Milenkovic L, Scott MP. Patched1 regulates hedgehog signaling at the primary cilium. *Science* 2007;**317**:372–6.

31. Dwyer JR, Sever N, Carlson M, Nelson SF, Beachy PA, Parhami F. Oxysterols are novel activators of the hedgehog signaling pathway in pluripotent mesenchymal cells. *J Biol Chem* 2007;**282**:8959–68.

32. Corcoran RB, Scott MP. Oxysterols stimulate Sonic hedgehog signal transduction and proliferation of medulloblastoma cells. *Proc Natl Acad Sci USA* 2006;**103**:8408–13.

33. Haycraft CJ, Banizs B, Aydin-Son Y, Zhang Q, Michaud EJ, Yoder BK. Gli2 and Gli3 localize to cilia and require the intraflagellar transport protein polaris for processing and function. *PLoS Genet* 2005;**1**:e53.

34. Theil T, Kaesler S, Grotewold L, Bose J, Ruther U. Gli genes and limb development. *Cell Tissue Res* 1999;**296**:75–83.

35. Briscoe J, Pierani A, Jessell TM, Ericson J. A homeodomain protein code specifies progenitor cell identity and neuronal fate in the ventral neural tube. *Cell* 2000;**101**:435–45.

36. Zhou Q, Anderson DJ. The bHLH transcription factors OLIG2 and OLIG1 couple neuronal and glial subtype specification. *Cell* 2002;**109**:61–73.

37. Stone D, Rosenthal A. Achieving neuronal patterning by repression. *Nat Neurosci* 2000;**3**:967–9.

38. Ye W, Shimamura K, Rubenstein J, Hynes M, Rosenthal A. FGF and Shh signals control dopaminergic and serotonergic cell fate in the anterior neural plate. *Cell* 1998;**93**:755–66.

39. Matise MP, Epstein DJ, Park HL, Platt KA, Joyner AL. Gli2 is required for induction of floor plate and adjacent cells, but not most ventral neurons in the mouse central nervous system. *Development* 1998;**125**:2759–70.

40. Macdonald R, Barth KA, Xu Q, Holder N, Mikkola I, Wilson SW. Midline signalling is required for Pax gene regulation and patterning of the eyes. *Development* 1995;**121**:3267–78.

41. Hallonet M, Hollemann T, Pieler T, Gruss P. Vax1, a novel homeobox-containing gene, directs development of the basal forebrain and visual system. *Genes Dev* 1999;**13**:3106–14.

42. Chiang C, Litingtung Y, Lee E, et al. Cyclopia and defective axial patterning in mice lacking Sonic hedgehog gene function. *Nature* 1996;**383**:407–13.

43. Dahmane N, Ruiz i Altaba A. Sonic hedgehog regulates the growth and patterning of the cerebellum. *Development* 1999;**126**:3089–100.

44. Wallace V. Purkinje-cell-derived Sonic hedgehog regulates granule neuron precursor cell proliferation in the developing mouse cerebellum. *Curr Biol* 1999;**9**:445–8.

45. Wechsler-Reya RJ, Scott MP. Control of neuronal precursor proliferation in the cerebellum by Sonic Hedgehog. *Neuron* 1999;**22**:103–14.

46. Pola R, Ling LE, Silver M, et al. The morphogen Sonic hedgehog is an indirect angiogenic agent upregulating two families of angiogenic growth factors. *Nat Med* 2001;**7**:706–11.

47. Hebrok M, Kim S, St Jacques B, McMahon AP, Melton D. Regulation of pancreas development by hedgehog signaling. *Development* 2000;**127**:4905–13.

48. diIorio PJ, Moss JB, Sbrogna JL, Karlstrom RO, Moss LG. Sonic hedgehog is required early in pancreatic islet development. *Dev Biol* 2002;**244**:75–84.

49. Bellusci S, Furuta Y, Rush M, Henderson R, Winnier G, Hogan B. Involvement of Sonic hedgehog (Shh) in mouse embryonic lung growth and morphogenesis. *Development* 1997;**124**:53–63.

50. Zhang XM, Ramalho-Santos M, McMahon AP. Smoothened mutants reveal redundant roles for Shh and Ihh signaling including regulation of L/R asymmetry by the mouse node. *Cell* 2001;**105**:781–92.

51. Lanske B, Karaplis AC, Lee K, et al. PTH/PTHrP receptor in early development and Indian hedgehog-regulated bone growth. *Science* 1996;**273**:663–6.

52. St-Jacques B, Hammerschmidt M, McMahon AP. Indian hedgehog signaling regulates proliferation and differentiation of chondrocytes and is essential for bone formation. *Genes Dev* 1999;**13**:2072–86.

53. Bitgood MJ, Shen L, McMahon AP. Sertoli cell signaling by Desert hedgehog regulates the male germline. *Curr Biol* 1996;**6**:298–304.

54. Yao HH, Whoriskey W, Capel B. Desert Hedgehog/Patched 1 signaling specifies fetal Leydig cell fate in testis organogenesis. *Genes Dev* 2002;**16**:1433–40.

55. Belloni E, Muenke M, Roessler E, et al. Identification of Sonic hedgehog as a candidate gene responsible for holoprosencephaly. *Nat Genet* 1996;**14**:353–6.

56. Hui CC, Joyner AL. A mouse model of Greig cephalopolysyndactyly syndrome:The extra toes mutation contains an intragenic deletion of the Gli3 gene. *Nat Genet* 1993;**3**:241–6.

57. Shin SH, Kogerman P, Lindstrom E, Toftgard R, Biesecker LG. GLI3 mutations in human disorders mimic Drosophila cubitus interruptus protein functions and localization. *Proc Natl Acad Sci USA* 1999;**96**:2880–4.

58. Gorlin RJ. Nevoid basal cell carcinoma syndrome. *Dermatol Clin* 1995;**13**:113–25.

59. Johnson RL, Rothman AL, Xie J, et al. Human homolog of patched, a candidate gene for the basal cell nevus syndrome. *Science* 1996;**272**:1668–71.

60. Hahn H, Wicking C, Zaphiropoulous PG, et al. Mutations of the human homolog of Drosophila patched in the nevoid basal cell carcinoma syndrome. *Cell* 1996;**85**:841–51.

61. Xie J, Johnson RL, Zhang X, et al. Mutations of the PATCHED gene in several types of sporadic extracutaneous tumors. *Cancer Res* 1997;**57**:2369–72.

62. Xie J, Murone M, Luoh SM, et al. Activating Smoothened mutations in sporadic basal-cell carcinoma. *Nature* 1998;**391**:90–2.

63. Mao J, Ligon KL, Rakhlin EY, et al. A novel somatic mouse model to survey tumorigenic potential applied to the Hedgehog pathway. *Cancer Res* 2006;**66**:10,171–10,178.

64. Liu S, Dontu G, Wicha MS. Mammary stem cells, self-renewal pathways, and carcinogenesis. *Breast Cancer Res* 2005;**7**:86–95.

65. Rubin LL, de Sauvage FJ. Targeting the Hedgehog pathway in cancer. *Nat Rev Drug Discov* 2006;**5**:1026–33.

66. Chen JK, Taipale J, Young KE, Maiti T, Beachy PA. Small molecule modulation of Smoothened activity. *Proc Natl Acad Sci USA* 2002;**99**:14,071–14,076.

67. Frank-Kamenetsky M, Zhang XM, Bottega S, et al. Small-molecule modulators of Hedgehog signaling: identification and characterization of Smoothened agonists and antagonists. *J Biol* 2002;**1**:10.

68. Romer JT, Kimura H, Magdaleno S, et al. Suppression of the Shh pathway using a small molecule inhibitor eliminates medulloblastoma in Ptc1(+/−)p53(−/−) mice. *Cancer Cell* 2004;**6**:229–40.

69. Xie J, Murone M, Luoh SM, et al. Activating Smoothened mutations in sporadic basal-cell carcinoma. *Nature* 1998;**391**:90–2.

70. Raffel C, Jenkins RB, Frederick L, et al. Sporadic medulloblastomas contain PTCH mutations. *Cancer Res* 1997;**57**:842–5.

71. Taylor MD, Liu L, Raffel C, et al. Mutations in SUFU predispose to medulloblastoma. *Nat Genet* 2002;**31**:306–10.

72. Tostar U, Malm CJ, Meis-Kindblom JM, Kindblom LG, Toftgard R, Unden AB. Deregulation of the hedgehog signalling pathway: a possible role for the PTCH and SUFU genes in human rhabdomyoma and rhabdomyosarcoma development. *J Pathol* 2006;**208**:17–25.

73. Watkins DN, Berman DM, Burkholder SG, Wang B, Beachy PA, Baylin SB. Hedgehog signalling within airway epithelial progenitors and in small-cell lung cancer. *Nature* 2003;**422**:313–17.

74. Yuan Z, Goetz JA, Singh S, et al. Frequent requirement of hedgehog signaling in non-small cell lung carcinoma. *Oncogene* 2007;**26**:1046–55.

75. Berman DM, Karhadkar SS, Maitra A, et al. Widespread requirement for Hedgehog ligand stimulation in growth of digestive tract tumours. *Nature* 2003;**425**:846–51.

76. Thayer SP, di Magliano MP, Heiser PW, et al. Hedgehog is an early and late mediator of pancreatic cancer tumorigenesis. *Nature* 2003;**425**:851–6.

77. Karhadkar SS, Bova GS, Abdallah N, et al. Hedgehog signalling in prostate regeneration, neoplasia and metastasis. *Nature* 2004;**431**:707–12.

78. Fan L, Pepicelli CV, Dibble CC, et al. Hedgehog signaling promotes prostate xenograft tumor growth. *Endocrinology* 2004;**145**:3961–70.

79. Sheng T, Li C, Zhang X, et al. Activation of the hedgehog pathway in advanced prostate cancer. *Mol Cancer* 2004;**3**:29.

80. Sanchez P, Hernandez AM, Stecca B, et al. Inhibition of prostate cancer proliferation by interference with SONIC HEDGEHOG-GLI1 signaling. *Proc Natl Acad Sci USA* 2004;**101**:12,561–12,566.

81. Monzo M, Moreno I, Artells R, et al. Sonic hedgehog mRNA expression by real-time quantitative PCR in normal and tumor tissues from colorectal cancer patients. *Cancer Letts* 2006;**233**:117–23.

82. Oniscu A, James RM, Morris RG, Bader S, Malcomson RD, Harrison DJ. Expression of Sonic hedgehog pathway genes is altered in colonic neoplasia. *J Pathol* 2004;**203**:909–17.

83. Qualtrough D, Buda A, Gaffield W, Williams AC, Paraskeva C. Hedgehog signalling in colorectal tumour cells: induction of apoptosis with cyclopamine treatment. *Intl J Cancer* 2004;**110**:831–7.

84. Huang S, He J, Zhang X, et al. Activation of the hedgehog pathway in human hepatocellular carcinomas. *Carcinogenesis* 2006;**27**:1334–40.

85. Dierks C, Grbic J, Zirlik K, et al. Essential role of stromally induced hedgehog signaling in B-cell malignancies. *Nat Med* 2007;**13**:944–51.

86. Peacock CD, Wang Q, Gesell GS, et al. Hedgehog signaling maintains a tumor stem cell compartment in multiple myeloma. *Proc Natl Acad Sci USA* 2007;**104**:4048–53.

Chapter 25

Regulation of Vertebrate Left-Right Axis Development by Calcium

Adam D. Langenbacher and Jau-Nian Chen

Department of Molecular, Cell and Developmental Biology, University of California, Los Angeles, California

INTRODUCTION

Vertebrates exhibit an extensive asymmetry of the internal organs with respect to the left–right axis that is established during embryogenesis and persists in the mature organism. During the past decade, numerous studies have demonstrated that organ laterality is influenced by a left–right asymmetric signaling cascade in the lateral plate mesoderm that is evolutionarily conserved from fish to mammals. In this signaling cascade, expression of the TGFβ superfamily members *nodal* and *lefty* and the transcription factor *pitx2* are upregulated in the left lateral plate mesoderm. The mechanism by which left–right symmetry is broken appears to be less conserved, and is still not fully understood in any vertebrate model organism.

In several vertebrate models, calcium has been found to plan an essential role in left–right patterning. Left–sided expression of *nodal*, *lefty*, and *pitx2* is preceded by an elevation in intracellular calcium levels on the left side of the mouse node and zebrafish Kupffer's vesicle (KV) [1, 2], or by an elevation in extracellular calcium levels on the left side of Hensen's node in the chick [3]. In the mouse, cilia in the node appear to be involved in elevating intracellular calcium on its left side. In the zebrafish, a direct link between nodal cilia and asymmetric calcium signaling has not yet been established, but intracellular calcium levels have recently been shown to influence left–right patterning by modulating KV cilia motility [4]. The increase in extracellular calcium on the left side of Hensen's node is known to depend on left–right asymmetric activity of the H^+/K^+-ATPase ion transporter and has been proposed to directly enhance Notch signaling [3]. This review will discuss the conserved and divergent mechanisms regulating left–right axis development in vertebrate model organisms, and the role of calcium at different steps in this process.

CONSERVED MOLECULAR PATHWAYS REGULATING LR ASYMMETRY

Expression of *nodal* genes in the left lateral plate mesoderm is the first molecular asymmetry that is conserved between fish, *Xenopus*, chick, and mouse. Abnormal expression of *nodal* genes in mouse and zebrafish results in organ laterality defects and disrupts asymmetric gene expression [5]. For example, inhibition of the expression of *southpaw*, the first *nodal* homolog expressed during zebrafish development, causes defects in cardiac and pancreas laterality, and an absence of *lefty1*, *cyclops*, and *pitx2* expression in the lateral plate mesoderm and forebrain [6]. Nodal protein binds to two activin type II receptors, ActRIIA and ActRIIB, and two activin type I receptors, ALK4 and ALK7, and initiates both positive- and negative-feedback loops (reviewed in [7, 8]). Transduction of Nodal signaling also requires the presence of membrane-anchored EGF-CFC co-receptors that were first identified in zebrafish (*one-eyed pinhead*) and in mouse (*Cripto, Cryptic*) [9, 10]. Activation of Nodal signaling is inhibited in the right lateral plate mesoderm by proteins of the Cerberus/DAN family, and increased right-sided expression of these Cerberus/DAN genes has been detected in mouse (*Cerberus*) and fish (*charon*) [11, 12]. Knockdown of *charon* in zebrafish using morpholino oligonucleotides results in a large percentage of embryos acquiring bilateral *southpaw* (*nodal*) expression and abnormal organ laterality, demonstrating that Charon negatively regulates Nodal signaling in the right lateral plate mesoderm and is essential for left–right patterning [13].

Lefty genes are upregulated by Nodal signaling in the left lateral plate mesoderm, and Lefty protein functions as a competitive inhibitor for Nodal, potentially restricting the activity of Nodal both spatially and temporally

[13, 14]. *Lefty* genes are also expressed along the embryonic midline, and are thought to create a barrier preventing Nodal signaling from extending over to the right side. In accordance with this model, mouse embryos lacking *lefty1* have left-sided *nodal* expression at first, but bilateral expression at later developmental stages [15]. Nodal signaling in the left lateral plate mesoderm induces expression of the homeobox transcription factor *Pitx2*, and it is currently thought that Pitx2 coordinates the situs-specific morphogenesis of the visceral organs. Complex organ laterality defects, including right isomerism of the lungs and abnormal gut rotation, have been reported in mice lacking an enhancer element required for asymmetric *Pitx2* expression. Interestingly, looping of the heart was unaffected, indicating that there may be more downstream *nodal* signaling targets that remain to be identified [16].

INITIATING A BREAK IN SYMMETRY

Nodal Cilia

A role for cilia in left–right development was first identified in the mouse, where monocilia in the node were shown to generate a leftward fluid flow that is required for initiating the break in left–right symmetry [17, 18]. Left/right-dynein (Lrd), an axonemal dynein heavy chain, is localized to some nodal cilia and is required for their motility [17, 19].

Populations of "nodal" cilia marked by expression of *lrd* have also been identified in zebrafish, *Xenopus*, and chick [20], but among non-mammalian vertebrates a function for nodal cilia as the initial symmetry-breaking mechanism has only been demonstrated so far in the zebrafish (Figure 25.1) [21, 22].

Identification of *lrd* as the relevant gene deleted by the mouse *inversus viscerum* (*iv*) mutation implied that motility of nodal cilia was essential for left–right patterning [17, 19]. The laterality of the internal organs is reversed in roughly half of *iv* mutant mice, indicating that without nodal fluid flow, establishment of the left–right axis is randomized. The expression patterns of asymmetric markers like *nodal* and *lefty* are randomized as well, making nodal flow the earliest known step in mammalian left–right development [23, 24]. Nodal flow is also required as the earliest known step in zebrafish left–right patterning [21, 22]. In zebrafish embryos injected with morpholino oligonucleotides targeting *lrd*, and in embryos in which KV cilia biogenesis is disrupted by injection of morpholinos targeting *polaris/Ift88* and *hippi/Ift57*, fluid flow in KV is inhibited and the expression of *nodal* and/or *lefty* genes in the left lateral plate mesoderm is disrupted. Exactly how nodal flow induces the asymmetric *nodal* cascade is not yet fully understood in zebrafish or mouse.

Two prevailing models have been developed to explain how nodal flow initiates the break in left–right symmetry in mouse. The first model, known as the "two-cilia model,"

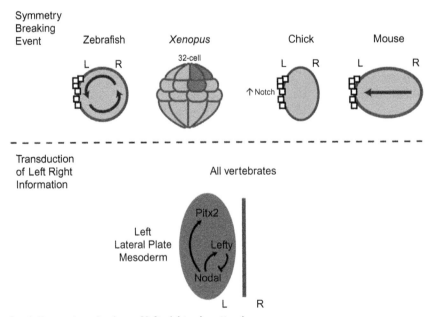

FIGURE 25.1 Conserved and divergent mechanisms of left–right axis patterning.
In zebrafish and mouse, nodal flow (gray arrows) breaks left–right symmetry by leading to a left-sided increase in intracellular calcium levels (squares). In chick, an H^+,K^+-ATPase-dependent increase in extracellular calcium levels (squares) on the left side of Hensen's node is thought to potentiate Notch signaling and trigger asymmetric gene expression. *Xenopus* left–right axis formation occurs within the first few cell divisions after fertilization, and left–right determinants (shaded dark gray) are asymmetrically localized at least as early as the 32-cell stage. Following the symmetry breaking event, all vertebrates share a similar cascade of *nodal, lefty, and pitx2* expression in the left lateral plate mesoderm that transmits left–right information to the developing organ primordia. L, left; R, right.

suggests that the mouse node contains two distinct populations of cilia: motile Lrd-containing monocilia that are restricted to more medial regions of the node and generate a leftward flow, and non-motile Polycystin-2 (Pkd2) containing cilia that sense the nodal flow [1]. Pkd2 is a Ca^{2+}-permeable channel that had previously been shown to localize to the primary cilium of kidney cells. In kidney cells, Pkd2 senses fluid flow and causes an increase in intracellular calcium levels [25]. The two-cilia model proposes that sensory cilia on the left periphery of the node respond to fluid flow leading to an accumulation of intracellular calcium on the left side of the node. In agreement with this model, embryos lacking Pkd2 have no asymmetric calcium signal on the left node border [1].

The second model of how nodal flow breaks left–right symmetry involves what have been termed "nodal vesicular parcels" (NVPs). Imaging of DiI-labeled nodes has revealed that membrane-sheathed objects 0.3- to 5-μm in diameter are released from the floor of the node and are transported to the left by nodal flow. These NVPs then fragment near the left wall of the node [26]. Fgf signaling is required for production of NVPs and asymmetric calcium signaling on the left side of the node, but has been shown to be dispensable for the generation of leftward nodal flow. In the absence of Fgf signaling, NVP production and asymmetric calcium signaling can be restored by exogenous application of Sonic hedgehog (Shh) protein or retinoic acid (RA), and immunostaining demonstrated that Shh and RA are contained within the NVPs. The NVP model is further supported by the observation that *lrd* mutant embryos tend to have bilaterally elevated intracellular calcium levels around the node, suggesting that NVPs may float indiscriminately in either direction in the absence of directed nodal flow [26]. How the fusion of Shh and RA-containing NVPs to the left edge of the node triggers an increase in intracellular calcium levels is not currently known.

Divergent Mechanisms of Symmetry Breaking

The mechanisms by which LR symmetry is broken in *Xenopus* and chick appear to differ from the mammalian and fish models. In *Xenopus*, monocilia present on the gastrocoel roof plate during neurulation have been shown to generate a leftward fluid flow that is required for normal asymmetric gene expression and organ situs [27], but molecular LR asymmetries are evident much earlier in *Xenopus* development than in any other vertebrate model organism (Figure 25.1). Within the first few cell divisions, H^+,K^+-ATPase and 14-3-3 mRNAs are asymmetrically localized with respect to the LR axis [28, 29]. The signaling molecule serotonin is also asymmetrically localized as early as the 32-cell stage [30].

In both chick and *Xenopus*, H^+,K^+-ATPase activity is greater on the right side of the embryo and is important at a very early step in left–right patterning [29]. H^+,K^+-ATPase

pumps protons out of the cell, resulting in a more negative membrane potential on the right side of the node than on the left. Chemical inhibition of H^+,K^+-ATPase activity in *Xenopus* and chick leads to a disruption of the asymmetric expression of *nodal* and *pitx2*, and also disrupts asymmetric expression of *shh* around Hensen's node, which precedes asymmetric *nodal* expression in the chick. Gap junctions are also required at a very early step in chick and *Xenopus* LR patterning, and disrupting gap junctions results in a loss of asymmetric *shh* in chick and heterotaxic organ situs in *Xenopus* [31, 32]. Taken together, the requirement for both H^+,K^+-ATPase and gap junctions at a very early step in LR patterning suggests that H^+,K^+-ATPase might maintain an asymmetric cellular voltage potential, and that charged, low molecular weight signaling molecules can travel through gap junctions and accumulate preferentially on one side of the embryo. Alternatively, these two molecules may be functioning in parallel to influence the secretion and transport of LR determinants.

CONSERVED ROLE OF CALCIUM IN LEFT – RIGHT ASYMMETRY DETERMINATION

LR asymmetric gene expression is preceded by an elevation in intracellular calcium levels on the left side of the mouse node and zebrafish KV [1, 2]. Asymmetric intracellular calcium was first demonstrated to be important for LR patterning in mouse, where loss of Pkd2 abolishes the increase in intracellular calcium on the left side of the node and leads to defects in organ laterality [1, 33]. In zebrafish, Ipk1, a kinase that converts inositol 1,3,4,5,6-pentakisphosphate (IP$_5$) to inositol hexakisphosphate (IP$_6$), is essential for LR axis determination [2]. Inositol polyphosphates function as second messengers that can affect multiple cellular processes including intracellular calcium levels. Zebrafish embryos injected with *ipk1* morpholinos exhibit reduced intracellular calcium levels in cells surrounding KV, and defects in LR patterning, suggesting that asymmetric calcium signaling is critical for LR axis determination in the zebrafish model.

In the chick, elevated extracellular calcium levels have been detected on the left side of Hensen's node, and how these elevated calcium levels influence gene expression is better understood than in mouse or zebrafish (Figure 25.1) [3]. The increased extracellular calcium on the left side is thought to modulate Notch signaling, resulting in an increase in the expression level of the Notch ligand *Delta-like 1* (*Dll1*) on the left. When Notch signaling is downregulated by the gamma-secretase inhibitor DAPT, *Dll1* expression does not increase on the left side of Hensen's node. Furthermore, overexpression of a dominant-negative form of *Dll1* led to a loss of left-sided *nodal* expression. Symmetric *Dll1* expression and a loss of left-sided *nodal* were also observed in embryos treated with the calcium chelator BAPTA, placing asymmetric calcium signaling above Notch signaling in the pathway of LR axis determination.

Recently, another role for calcium in LR patterning has been identified in the zebrafish model. Injection of morpholinos targeting one zebrafish sodium–calcium exchanger (NCX) isoform, NCX4a, or Na^+,K^+-ATPase $\alpha2$ results in a disruption of asymmetric gene expression and organ laterality [4, 34]. NCX function is one of the primary mechanisms by which calcium is extruded from cells, and the Na^+ gradient established by Na^+,K^+-ATPase helps to drive calcium extrusion via NCX (reviewed in [35]). In blastula-stage NCX4a and Na^+,K^+-ATPase $\alpha2$ morphants (morpholino injected embryos), intracellular calcium levels are globally elevated, directly demonstrating a requirement for these proteins in intracellular calcium homeostasis [4]. Surprisingly, although KV and the number and length of KV cilia appear normal in NCX4a and Na^+,K^+-ATPase $\alpha2$ morphants, injection of fluorescent beads into KV demonstrated that fluid flow in KV is completely abolished. High-speed video imaging showed that cilia within the KV of these embryos are immotile, implying that calcium signaling can regulate cilia motility and thus functions both upstream and downstream of nodal flow in zebrafish LR patterning (Figure 25.2). Furthermore, increasing calcium signaling by injecting a constitutively active form of calcium/calmodulin-dependent protein kinase II (CaMKII) led to defects in organ laterality, and decreasing CaMKII activity with the chemical inhibitor

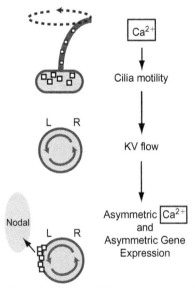

FIGURE 25.2 Roles of calcium at different steps in zebrafish left–right axis patterning.
In zebrafish, calcium (squares) functions at multiple steps in left–right patterning. Proper calcium homeostasis is required for Kupffer's vesicle (KV) cilia motility and the generation of counterclockwise fluid flow in KV. KV flow triggers a left-sided increase in intracellular calcium levels that is required for asymmetric gene expression in the left lateral plate mesoderm. L, left; R, right.

KN-62 was able to rescue the defects in organ laterality and KV flow of NCX4a and Na^+,K^+-ATPase $\alpha2$ morphants. Thus, intracellular calcium signals through CaMKII to regulate the motility of cilia in KV [4].

CONCLUSIONS

Many questions still remain in the field of LR axis determination, and the realization that calcium signaling can regulate the motility of KV cilia in zebrafish suggests that a re-examination of the role of calcium in LR patterning in other model organisms may be in order. For example, in mouse *pkd2* mutants, calcium levels appear to be reduced both in and around the node [1]. If calcium regulates nodal cilia motility in the mouse, then nodal flow might be affected and present a new interpretation of the role of Pkd2 in the two-cilia model. Direct observations of nodal cilia motility in *pkd2* knockout embryos may be required to distinguish between the roles of calcium at different steps in LR patterning.

The mechanism by which calcium regulates nodal cilia motility also requires further investigation. Calcium levels have been proposed to modulate dynein-regulated microtubule sliding, and thereby affect the waveform of *Clamydomonas* flagella and the motility of sea-urchin sperm flagella [36–39]. Since CaMKII is known to be an important component of calcium signaling in zebrafish LR development, it is possible that CaMKII directly or indirectly regulates dynein activity in KV cilia. Alternatively, proper calcium homeostasis and signaling might be required for the biogenesis of ultrastructurally normal motile KV cilia.

How the asymmetric calcium signals generated by nodal flow are translated into asymmetric gene expression in the lateral plate mesoderm in mouse and zebrafish is not yet understood. It is possible that increased intracellular calcium might stimulate production of a currently unknown left-sided determinant, or enhance the activity of a particular signaling pathway. Given the ease of forward and reverse genetic techniques in the zebrafish, it will serve as an excellent model for future studies aiming to tease out the intermediates between asymmetric calcium and asymmetric gene expression, and also between calcium signaling and cilia motility.

The question still remains of precisely how asymmetric intracellular calcium levels are triggered by fluid flow in the mouse node and zebrafish KV. A combination of the two-cilia model and the NVP hypothesis may be necessary to fully explain the LR symmetry-breaking event in mouse. For example, the non-motile Pkd2-containing cilia on the sides of the node may be required to sense NVPs and in response, Pkd2 channels trigger the release of large amounts of calcium from intracellular stores, elevating calcium levels in cells on the left side of the node. NVPs have not yet been identified in the zebrafish model, but, given the other similarities of mouse and zebrafish

LR determination, it will be interesting to determine when in evolution this symmetry-breaking mechanism arose. An examination of LR axis development in primitive chordates like amphioxus and urochordates will also reveal how early in evolution the different LR patterning mechanisms were developed, and will expand our understanding of their conservation and divergence.

ACKNOWLEDGEMENTS

Adam D. Langenbacher is supported by an NSF Graduate Research Fellowship. This work is supported by grants from NIH to Jau-Nian Chen.

REFERENCES

1. McGrath J, Somlo S, Makova S, Tian X, Brueckner M. Two populations of node monocilia initiate left–right asymmetry in the mouse. *Cell* 2003;**114**:61–73.

2. Sarmah B, Latimer AJ, Appel B, Wente SR. Inositol polyphosphates regulate zebrafish left–right asymmetry. *Dev Cell* 2005;**9**:133–45.

3. Raya A, Kawakami Y, Rodriguez-Esteban C, Ibanes M, Rassidin-Gutman D, Rodriguez-Leon J, Buscher D, Feijo J, Izpisua Belmonte J. Notch activity acts as a sensor for extracellular calcium during vertebrate left-right determination. *Nature* 2004;**427**:121–8.

4. Shu X, Huang J, Dong Y, Choi J, Langenbacher A, Chen JN. Na,K-ATPase alpha2 and Ncx4a regulate zebrafish left–right patterning. *Development* 2007;**134**:1921–30.

5. Burdine RD, Schier AF. Conserved and divergent mechanisms in left–right axis formation. *Genes Dev* 2000;**14**:763–76.

6. Long S, Ahmad N, Rebagliati M. The zebrafish nodal-related gene southpaw is required for visceral and diencephalic left-right asymmetry. *Development* 2003;**130**:2303–16.

7. Schier AF, Shen MM. Nodal signalling in vertebrate development. *Nature* 2000;**403**:385–9.

8. Tian T, Meng AM. Nodal signals pattern vertebrate embryos. *Cell Mol Life Sci* 2006;**63**:672–85.

9. Gritsman K, Zhang J, Cheng S, Heckscher E, Talbot WS, Schier AF. The EGF-CFC protein one-eyed pinhead is essential for nodal signaling. *Cell* 1999;**97**:121–32.

10. Yan YT, Gritsman K, Ding J, Burdine RD, Corrales JD, Price SM, Talbot WS, Schier AF, Shen MM. Conserved requirement for EGF-CFC genes in vertebrate left-right axis formation. *Genes Dev* 1999;**13**:2527–37.

11. Pearce JJ, Penny G, Rossant J. A mouse cerberus/Dan-related gene family. *Dev Biol* 1999;**209**:98–110.

12. Hojo M, Takashima S, Kobayashi D, Sumeragi A, Shimada A, Tsukahara T, Yokoi H, Narita T, Jindo T, Kage T, Kitagawa T, Kimura T, Sekimizu K, Miyake A, Setiamarga D, Murakami R, Tsuda S, Ooki S, Kakihara K, Naruse K, Takeda H. Right-elevated expression of charon is regulated by fluid flow in medaka Kupffer's vesicle. *Dev Growth Differ* 2007;**49**:395–405.

13. Hashimoto H, Rebagliati M, Ahmad N, Muraoka O, Kurokawa T, Hibi M, Suzuki T. The Cerberus/Dan-family protein Charon is a negative regulator of Nodal signaling during left–right patterning in zebrafish. *Development* 2004;**131**:1741–53.

14. Chen C, Shen MM. Two modes by which Lefty proteins inhibit nodal signaling. *Curr Biol* 2004;**14**:618–24.

15. Meno C, Shimono A, Saijoh Y, Yashiro K, Mochida K, Ohishi S, Noji S, Kondoh H, Hamada H. lefty-1 is required for left-right determination as a regulator of lefty-2 and nodal. *Cell* 1998;**94**:287–97.

16. Shiratori H, Yashiro K, Shen MM, Hamada H. Conserved regulation and role of Pitx2 in situs-specific morphogenesis of visceral organs. *Development* 2006;**133**:3015–25.

17. Supp D, Witte D, Potter S, Brueckner M. Mutation of an axonemal dynein affects left–right asymmetry in inversus viscerum mice. *Nature* 1997;**389**:963–6.

18. Nonaka S, Tanaka Y, Okada Y, Takeda S, Harada A, Kanai Y, Kido M, Hirokawa N. Randomization of left–right asymmetry due to loss of nodal cilia generating leftward flow of extraembryonic fluid in mice lacking KIF3B motor protein. *Cell* 1998;**95**:829–37.

19. Okada Y, Nonaka S, Tanaka Y, Saijoh Y, Hamada H, Hirokawa N. Abnormal nodal flow precedes situs inversus in iv and inv mice. *Mol Cell* 1999;**4**:459–68.

20. Essner JJ, Vogan KJ, Wagner MK, Tabin CJ, Yost HJ, Brueckner M. Conserved function for embryonic nodal cilia. *Nature* 2002;**418**:37–8.

21. Kramer-Zucker AG, Olale F, Haycraft CJ, Yoder BK, Schier AF, Drummond IA. Cilia-driven fluid flow in the zebrafish pronephros, brain and Kupffer's vesicle is required for normal organogenesis. *Development* 2005;**132**:1907–21.

22. Essner JJ, Amack JD, Nyholm MK, Harris EB, Yost HJ. Kupffer's vesicle is a ciliated organ of asymmetry in the zebrafish embryo that initiates left–right development of the brain, heart and gut. *Development* 2005;**132**:1247–60.

23. Lowe LA, Supp DM, Sampath K, Yokoyama T, Wright CV, Potter SS, Overbeek P, Kuehn MR. Conserved left–right asymmetry of nodal expression and alterations in murine situs inversus. *Nature* 1996;**381**:158–61.

24. Meno C, Saijoh Y, Fujii H, Ikeda M, Yokoyama T, Yokoyama M, Toyoda Y, Hamada H. Left–right asymmetric expression of the TGF beta-family member lefty in mouse embryos. *Nature* 1996;**381**:151–5.

25. Nauli SM, Alenghat FJ, Luo Y, Williams E, Vassilev P, Li X, Elia AE, Lu W, Brown EM, Quinn SJ, Ingber DE, Zhou J. Polycystins 1 and 2 mediate mechanosensation in the primary cilium of kidney cells. *Nature Genet* 2003;**33**:129–37.

26. Tanaka Y, Okada Y, Hirokawa N. FGF-induced vesicular release of Sonic hedgehog and retinoic acid in leftward nodal flow is critical for left–right determination. *Nature* 2005;**435**:172–7.

27. Schweickert A, Weber T, Beyer T, Vick P, Bogusch S, Feistel K, Blum M. Cilia-driven leftward flow determines laterality in Xenopus. *Curr Biol* 2007;**17**:60–6.

28. Bunney TD, De Boer AH, Levin M. Fusicoccin signaling reveals 14-3-3 protein function as a novel step in left–right patterning during amphibian embryogenesis. *Development* 2003;**130**:4847–58.

29. Levin M, Thorlin T, Robinson KR, Nogi T, Mercola M. Asymmetries in H^+/K^+-ATPase and cell membrane potentials comprise a very early step in left-right patterning. *Cell* 2002;**111**:77–89.

30. Fukumoto T, Kema IP, Levin M. Serotonin signaling is a very early step in patterning of the left–right axis in chick and frog embryos. *Curr Biol* 2005;**15**:794–803.

31. Levin M, Mercola M. Gap junctions are involved in the early generation of left–right asymmetry. *Dev Biol* 1998;**203**:90–105.

32. Levin M, Mercola M. Gap junction-mediated transfer of left-right patterning signals in the early chick blastoderm is upstream of Shh asymmetry in the node. *Development* 1999;**126**:4703–14.

33. Pennekamp P, Karcher C, Fischer A, Schweickert A, Skryabin B, Horst J, Blum M, Dworniczak B. The ion channel polycystin-2 is required for left–right axis determination in mice. *Curr Biol* 2002;**12**:938–43.

34. Shu X, Cheng K, Patel N, Chen F, Joseph E, Tsai H, Chen J. Na,K-ATPase is essential for embryonic heart development in the zebrafish. *Development* 2003;**130**:6165–73.

35. Blaustein M, Lederer W. Sodium/calcium exchange: its physiological implications. *Physiol Rev* 1999:736–854.

36. Brokaw CJ. Calcium-induced asymmetrical beating of triton-demembranated sea urchin sperm flagella. *J Cell Biol* 1979;**82**:401–11.

37. Gibbons BH, Gibbons IR. Calcium-induced quiescence in reactivated sea urchin sperm. *J Cell Biol* 1980;**84**:13–27.

38. Nakano I, Kobayashi T, Yoshimura M, Shingyoji C. Central-pair-linked regulation of microtubule sliding by calcium in flagellar axonemes. *J Cell Sci* 2003;**116**:1627–36.

39. Wargo MJ, McPeek MA, Smith EF. Analysis of microtubule sliding patterns in Chlamydomonas flagellar axonemes reveals dynein activity on specific doublet microtubules. *J Cell Sci* 2004;**117**:2533–44.

Chapter 26

LIN-12/Notch Signaling: Induction, Lateral Specification, and Interaction with the EGF/Ras Pathway

Sophie Jarriault

IGBMC (Institut de Génétique et de Biologie Moléculaire et Cellulaire), Department of Cell and Developmental Biology, Illkirch, France; and Université de Strasbourg, Strasbourg, France

THE LIN-12/NOTCH PATHWAY

LIN-12/Notch receptors are evolutionary conserved type I transmembrane proteins involved in many developmental events. Genetic and biochemical experiments from studies on invertebrates, mainly *D. melanogaster* and *C. elegans*, and more recently on vertebrates, have identified members of the LIN-12/Notch signal transduction pathway (Table 26.1). The canonical DSL LIN-12/Notch ligands, named after the *Drosophila* Delta and Serrate, and *C. elegans* Lag-2 factors are also evolutionary conserved. Most are transmembrane proteins, although secreted forms have been described, and they contain notably a DSL domain and EGF repeats in their extracellular region [1]. In addition to the N-terminal DSL domain, an adjacent "DOS" domain, which encompasses the two first EGF repeats, might be important for DSL ligand function [2]. The DOS domain is found in most known LIN-12/Notch ligands, but not in the *C. elegans* DSL proteins. However, a secreted protein containing a DOS domain, but lacking the canonical DSL domain, probably acts as a co-factor of the *C. elegans* DSL ligands during a *lin-12*-dependent specification [2]. Cell contact is the main mode to trigger activation of LIN-12/Notch signaling pathway.

LIN-12/Notch receptors undergo a proteolytic maturation (site 1 cleavage, see Figure 26.1) during their transportation to the cell surface, which produces the ligand-sensitive form. An original model for signal transduction has been proposed [3]. Upon activation of a LIN-12/Notch receptor by its ligand, the receptor is cleaved both in its extracellular domain (site 2 cleavage) and underneath its transmembrane domain (site 3 cleavage) to produce an active intracellular fragment which translocates to the nucleus (Figure 26.1).

A number of membrane-associated factors are necessary for these proteolysis: the metalloprotease SUP-17/Kuzbanian [4] and its close homolog TACE/ADM-4 [5, 6] are necessary for site 2 cleavage, while presenilin, in a complex with the transmembrane proteins APH-1, APH-2/Nicastrin, and PEN-2, is necessary for site 3 cleavage [7].

The intracellular part of the LIN-12/Notch receptor then associates with the CSL DNA binding protein (for *C*BF1/*Su*(H)/*LAG*-1, see Table 26.1] on regulatory regions of target genes [8]. Activation of the target genes is believed to be the sum of two events: release of a co-repressor complex previously associated with CSL [9, 10], and formation of an activator complex comprising LIN-12/Notch intra and LAG-1, probably associated with a number of co-factors, such as SEL-8 and SKIP [8, 11-13]. Known targets genes include *lin-12*, and EGF-MAPK negative regulators such as the *lst* genes and *lip-1* in *C. elegans* [14–16], the bHLH encoding *E(spl)* in *D. melanogaster* [17], or *HES1* and *HES1*-related genes in vertebrates [8, 18].

Finally, a number of negative modulators of LIN-12/Notch signaling have been described, which influence the ability of a ligand to signal [19], the ability of the receptor to transmit the signal [20], the stability of the receptor [21] or its ability to relieve repression of target genes [22, 23].

INDUCTIVE SIGNALING VERSUS LATERAL SPECIFICATION

Both instructive and permissive roles have been described for LIN-12/Notch signaling. These processes, called inductive signaling (between two different cells) or lateral specification

TABLE 26.1 Members of the LIN-12/Notch signal transduction pathway identified by studies on invertebrates (mainly *D. melanogaster* and *C. elegans*) and vertebrates

	C. elegans	*D. melanogaster*	*Vertebrates*
Emission of the signal	LAG-2, APX-1, DSL-1 Others* OSM-11	DELTA SERRATE	DELTA[DELTA-LIKE] -1 to -4 JAGGED-1, -2 (EGFL19)
Reception of the signal	LIN-12, GLP-1	*NOTCH*	NOTCH-1 to -4
Maturation/activation of the receptor	(FURINE) SUP-17 ADM-4 SEL-12, HOP-1 APH-1 APH-2 PEN-2	(FURINE) KUZBANIAN (TACE) PRESENILIN (APH-1) NICASTRIN (PEN-2)	FURINE (KUZBANIAN/ADAM10) TACE *PRESENILIN-1, -2* APH-1 NICASTRIN PEN-2
Nuclear effectors	LAG-1 SEL-8 (SKIP)	SU(H) "MASTERMIND" # –	*RBP-J/CBF1* "MASTERMIND" # SKIP
Direct modulators of the activity or stability of the LIN-12/Notch receptors	(NUMB) – SEL-10	NUMB FRINGE –	(NUMB) LUNATIC FRINGE SEL-10
Target genes	– *lip-1* *lin-12*	genes of the *E(spl)* complex – Notch?	HES-1, -5 and HESR-1, -2 – –

Homologs are shown on the same line for each species.
*Genes coding for proteins with similar characteristics exist in the C. elegans genome [35] (Chen and Greenwald, 2004)
–, so far no known homologs have been described or implicated; ( ), a homolog is known in the organism, but its function in the LIN-12/Notch pathway has not been shown; #, sel-8 and mastermind do not share much homology in terms of sequence, but rather behave as functional homologs.

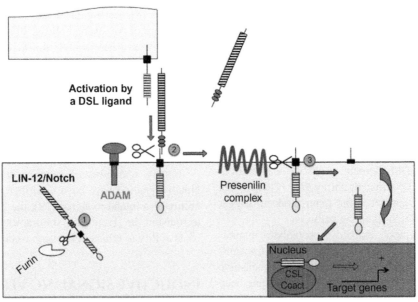

FIGURE 26.1 LIN-12/Notch activation and signal transduction.
The LIN-12/Notch receptor is cleaved by the furin protease at site 1 while in transit to the cell surface [44] (Logeat *et al.*, 1998). The resulting heterodimeric receptor, after activation by a ligand, is cleaved by a metalloprotease from the ADAM family (see Table 26.1) at site 2. The membrane-anchored intracellular domain left is then constitutively cleaved by a complex containing Presenilin, at site 3. The released intracellular part of the receptor migrates to the nucleus and associates with the CSL DNA-binding protein (see Table 26.1) and other co-activators to activate target genes transcription.

(between initially equivalent cells), respectively, are illustrated here using *C. elegans* cell-fate decisions as a model.

Inductive Signaling: Differentiation of the Germ Cells in *C. elegans*

The LIN-12/Notch signaling can take place between nonequivalent cells. In this case, only one of the cells expresses the ligand whereas one or more receptive cells express the receptor. Its activation leads to adoption of a particular fate for the receptive cells, a process called inductive signaling. Such a process is observed during differentiation of the female germ cells in *C. elegans*. The *C. elegans* gonads comprise somatic and germinal components. The germinal part is organized into two U-shaped symmetrical arms extending from the vulva, and the distal part contains the female germ cells. As the germ cell nuclei migrate proximally they enter meiosis and start to differentiate into oocytess, whereas the distal nuclei remain proliferative. Two somatic cells at the distal extremity of each arm, the *distal cells* or DTCs, control the differentiation of the germ cells (Figure 26.2a). If the DTCs are ablated, the distal germ cell nuclei enter meiosis precociously [24]. *glp-1*, a *lin-12/Notch* gene, and *lag-2*, a DSL ligand, are important for the interactions between the distal cells and the germ cells. *glp-1* loss-of-function mutants have germ-cell proliferation defects similar to those observed in animals where the DTCs have been ablated: the distal germ cells enter meiosis [25, 26]. The DTCs express *lag-2* [27] and the germ cells express *glp-1* [28]. The analysis of mosaic animals showed that *glp-1* activity is required in the germinal cells to transmit the *lag-2* signal coming from the DTCs and allow the maintenance of the germ cells in mitosis [25].

Lateral specification mediated by the LIN-12/Notch signaling pathway allows two (or more) initially equivalent cells, after interactions between them, to take on different fates. This cell-fate choice can be biased (i.e., the same cell of the equivalent group always adopt the same fate) or not (i.e., all cells in the equivalence group have an equiprobability to adopt each cell fate).

The AC/VU Decision During Gonadal Development, an Example of Unbiased Lateral Specification

During formation of the somatic gonad of the nematode, two neighboring cells, named Z1.ppp and Z4.aaa, adopt with the same probability the *anchor* (AC) or the *ventral uterine* (VU) fates [29] (Figure 26.2b). Cell adhesion mutants and cell ablation experiments have shown that cell contacts between Z1.ppp and Z4.aaa are necessary for final fate acquisition, and that the VU fate depends on a signal emanating from the future AC cell [29]. The LIN-12/Notch signaling pathway is necessary for the correct specification of the AC/VU fates, and the level of *lin-12* activity determines which fate is adopted: study of loss-of-function and gain-of-function alleles of *lin-12* showed that high levels *of lin-12* activity lead to the VU fate, whereas low levels of *lin-12* cause the cell to adopt the AC fate [30]. Both the LIN-12 receptor and its ligand LAG-2 are initially expressed at comparable levels in Z1.ppp and Z4.aaa [15]. Through continuous cell–cell interactions, a small difference in the level of LIN-12 activation eventually arises which, amplified by feedback loops, results in one cell expressing LIN-12 and the other LAG-2 only [15, 30]. Thus, these initially equivalent precursor cells eventually acquire distinct fates through a unidirectional LIN-12/Notch signaling.

Biased Lateral Signaling: Determination of the Ventral Precursor Cells (or VPC)

At a later stage during *C. elegans* larval development, a group of six equivalent precursors in the ventral hypodermis (named the *Ventral Precursor Cells*, or VPCs) can all adopt a vulval fate. However, in wild-type animals, although the six VPCs are equipotent, the cell closest to the AC, named P6.p, always adopts a primary vulval fate (noted 1°) and its two neighbors a secondary vulval fate (or 2°), and their descendants will form the vulva [31]. The other VPCs adopt a non-vulval fate (or 3°) and fuse with the underlying hypodermis, thus giving an invariant 3° 3° 2° 1° 2° 3° pattern. Two signals are necessary for the specification of the vulval cells: one inductive signal, mediated by the EGF pathway and coming from the AC cell, is necessary for the establishment of the 1° fate [32]. A lateral signal, mediated by LIN-12/Notch, takes place between the VPCs to specify the 2° fate. For example, in gain-of-function *lin-12(d)* mutants, all six VPCs adopt a secondary vulval fate. Inversely, in loss-of-function *lin-12(0)* mutants, the VPC cells adopt only the primary or non-vulval fates [33]. All six VPCs express *lin-12* and its ligand *lag-2*, and no change in *lin-12* transcription is observed during VPC determination [34]. However, a decrease in LIN-12 protein level, and a concomitant increase in expression of the three *lin-12* ligands *lag-2*, *apx-1*, and *dsl-1*, has been observed in the P6.p cell in response to the AC signal [34–36]. Thus the AC signal, transmitted through the Ras pathway, biases the direction in which the *lin-12* signal is sent between the vulval VPCs by decreasing *lin-12* activity in P6.p and increasing, simultaneously, the expression of its ligands in the same cell. In addition, the future 2° cells receive low levels of EGF/Ras signaling from the AC. Target genes of the LIN-12/Notch pathway in the future 2° cells include negative regulators of the RAS pathway, antagonizing it in these cells, thereby insuring that they do not adopt a 1° fate [14, 16].

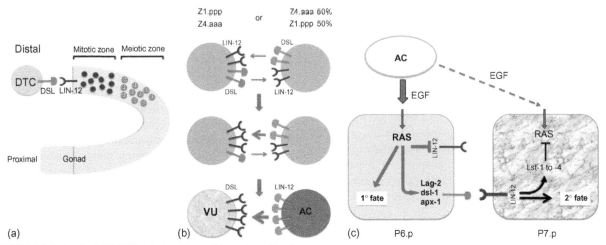

FIGURE 26.2 Inductive LIN-12/Notch signaling and lateral specification.
(a) Germ cell proliferation. The DSL ligand LAG-2 activates GLP-1 in the distal part of the gonad. This inductive signal maintains the germ cells in a mitotic state. As the germ nuclei migrate more proximally, GLP-1 signaling levels drop, resulting in the germ cell nuclei entering meiosis. (b) The AC/ VU decision, an unbiased lateral specification. Two cells, named Z1.ppp and Z4.aaa, have an equipotential to adopt the AC or the VU fate. Initially, both express comparable levels of LIN-12 and its ligand LAG-2. A stochastic event is amplified by feedback loops so that one of these cells expresses more ligand and signals to the other, which expresses more LIN-12, to adopt the VU fate. (c), Biased lateral signaling during VPC specification. In wild type animals, six equivalent cells, named P3.p to P8.p, are specified so that they invariably adopt non-vulval (3°), secondary vulval (2°), and primary vulval (1°) fates, in a 3° 3° 2° 1° 2° 3° pattern. Adoption of the 2° depends on the interplay between the LIN-12/Notch and EGF/Ras pathways. The AC cell sends a graded EGF signal to the underlying VPCs. The closest one, P6.p, receives a high EGF signal and adopts the 1° fate, upregulating DSL ligand expression (*lag-2*, *dsl-1* and *apx-1*) while downregulating LIN-12 protein levels. Its neighbors, P5.p (not represented) and P7.p, receive a high LIN-12 signal and low EGF signal. The efficient activation of LIN-12/Notch signaling in these two cells results in activation of negative regulators of the EGF/ Ras pathway and acquisition of the 2° fate.

Further examples of crosstalk between the LIN-12/ Notch signaling pathway and others, such as the frizzled pathway in Drosophila [37], have been described.

CELLULAR OUTCOME OF THE ACTIVATION OF THE LIN-12/NOTCH PATHWAY

Notch activity via cell–cell contacts generates molecular differences between adjacent cells. This pathway can mediate both instructive and lateral signaling. The latter has also been coined "lateral inhibition." Characterization of LIN-12/Notch loss-of-function phenotype in both flies and mice showed, notably, hypertrophy of neural tissue, a phenotype called "neurogenic" [38]. The additional neural cells reflect a failure of neuroepithelial progenitors to segregate both epidermal and neural cell lineages in flies, or precocious neural differentiation in mice. These data led to the interpretation that activation of the LIN-12/Notch signaling pathway results in inhibition of cell differentiation. Such an interpretation was reinforced by the apparent block in cell differentiation resulting from ectopic expression of an activated intracellular fragment of the receptor [39]. Further studies, however, have shown that the consequences of LIN-12/Notch activation are varied, ranging from keeping progenitors in a proliferative state, or orchestrating a molecular oscillator during somitogenesis, to instructing

cells to adopt a specific fate or diversifying a progenitor pool [38, 40–42]. Lateral inhibition between two equivalent cells, resulting in one adopting a fate "A" and the other a fate "B," implies that the only outcome of LIN-12/Notch signaling is to repress one of these fates–for example, "A." It is, however, likely that the outcome of LIN-12/Notch signaling in most cases is to specify fate "B" while it may, at the same time, repress fate "A" specification, as exemplified during the *C. elegans* VPC determination. In addition, it may be that activation of LIN-12/Notch signaling in certain instances results in maintenance of a larger developmental potential [43].

It remains to be understood how LIN-12/Notch signaling activates distinct targets in different cell types or at different developmental times. The answer will likely involve a combination of the ability of various LIN-12/Notch levels to displace repression complexes from target gene regulatory regions (in relationship with the number of CSL binding sites), the identity of the nuclear co-factors present in each cell type, and the integration of other signaling pathways inputs.

ACKNOWLEDGEMENT

I am grateful to Michel Labouesse for a critical reading of this manuscript.

REFERENCES

1. D'Souza B, Miyamoto A, Weinmaster G. The many facets of Notch ligands. *Oncogene* 2008;**38**(27):5148–67.

2. Komatsu H, Chao MY, Larkins-Ford J, Corkins ME, Somers GA, Tucey T, Dionne HM, White JQ, Wani K, Boxem M, Hart AC. OSM-11 facilitates LIN-12 Notch signaling during *Caenorhabditis elegans* vulval development. *PLoS Biol* 2008;**8**(6):e196.

3. Mumm JS, Kopan R. Notch signaling: from the outside in. *Dev Biol* 2000;**2**(228):151–65.

4. Zolkiewska A. ADAM proteases: ligand processing and modulation of the Notch pathway. *Cell Mol Life Sci* 2008;**13**(65):2056–68.

5. Brou C, Logeat F, Gupta N, Bessia C, Lebail O, Doedens JR, Cumano A, Roux P, Black R, Israel A. A novel proteolytic cleavage involved in Notch signaling; role of the disintegrin-metalloprotease TACE. *Mol Cell* 2000;**5**(2):207–16.

6. Jarriault S, Greenwald I. Evidence for functional redundancy between *C. elegans* ADAM proteins SUP-17/Kuzbanian and ADM-4/TACE. *Dev Biol* 2005;**1**(287):1–10.

7. Spasic D, Annaert W. Building gamma-secretase: the bits and pieces. *J Cell Sci* 2008;**121**(Pt 4):413–20.

8. Jarriault S, Brou C, Logeat F, Schroeter EH, Kopan R, Israel A. Signalling downstream of activated mammalian Notch. *Nature* 1995;**6547**(377):355–8.

9. Hsieh JJ, Henkel T, Salmon P, Robey E, Peterson MG, Hayward SD. Truncated mammalian Notch1 activates CBF1/RBPJk-repressed genes by a mechanism resembling that of Epstein-Barr virus EBNA2. *Mol Cell Biol* 1996;**3**(16):952–9.

10. Lai EC. Keeping a good pathway down: transcriptional repression of Notch pathway target genes by CSL proteins. *EMBO Rep* 2002;**9**(3):840–5.

11. Doyle TG, Wen C, Greenwald I. SEL-8, a nuclear protein required for LIN-12 and GLP-1 signaling in *Caenorhabditis elegans*. *Proc Natl Acad Sci USA* 2000;**14**(97):7877–81.

12. Petcherski AG, Kimble J. LAG-3 is a putative transcriptional activator in the *C. elegans* Notch pathway. *Nature* 2000;**6784**(405):364–8.

13. Zhou S, Fujimuro M, Hsieh JJ, Chen L, Miyamoto A, Weinmaster G, Hayward SD. SKIP, a CBF1-associated protein, interacts with the ankyrin repeat domain of NotchIC To facilitate NotchIC function. *Mol Cell Biol* 2000;**7**(20):2400–10.

14. Berset T, Hoier EF, Battu G, Canevascini S, Hajnal A. Notch inhibition of RAS signaling through MAP kinase phosphatase LIP-1 during *C. elegans* vulval development. *Science* 2001;**5506**(291):1055–8.

15. Wilkinson HA, Fitzgerald K, Greenwald I. Reciprocal changes in expression of the receptor *lin-12* and its ligand *lag-2* prior to commitment in a *C. elegans* cell fate decision. *Cell* 1994;**7**(79):1187–98.

16. Yoo AS, Bais C, Greenwald I. Crosstalk between the EGFR and LIN-12/Notch pathways in *C. elegans* vulval development. *Science* 2004;**5658**(303):663–6.

17. Lecourtois M, Schweisguth F. The neurogenic suppressor of hairless DNA-binding protein mediates the transcriptional activation of the enhancer of split complex genes triggered by Notch signaling. *Genes Dev* 1995;**21**(9):2598–608.

18. Kageyama R, Ohtsuka T, Hatakeyama J, Ohsawa R. Roles of bHLH genes in neural stem cell differentiation. *Exp Cell Res* 2005;**2**(306):343–8.

19. Irvine KD. Fringe, Notch, and making developmental boundaries. *Curr Opin Genet Dev* 1999;**4**(9):434–41.

20. Guo M, Jan LY, Jan YN. Control of daughter cell fates during asymmetric division: interaction of Numb and Notch. *Neuron* 1996;**1**(17):27–41.

21. Justice NJ, Jan YN. Variations on the Notch pathway in neural development. *Curr Opin Neurobiol* 2002;**1**(12):64–70.

22. Eimer S, Lakowski B, Donhauser R, Baumeister R. Loss of spr-5 bypasses the requirement for the *C. elegans* presenilin sel-12 by derepressing hop-1. *EMBO J* 2002;**21**(21):5787–96.

23. Jarriault S, Greenwald I. Suppressors of the egg-laying defective phenotype of sel-12 presenilin mutants implicate the CoREST corepressor complex in LIN-12/Notch signaling in *C. elegans*. *Genes Dev* 2002;**20**(16):2713–28.

24. Kimble JE, White JG. On the control of germ cell development in *Caenorhabditis elegans*. *Dev Biol* 1981;**2**(81):208–19.

25. Austin J, Kimble J. glp-1 is required in the germ line for regulation of the decision between mitosis and meiosis in *C. elegans*. *Cell* 1987;**4**(51):589–99.

26. Berry LW, Westlund B, Schedl T. Germ-line tumor formation caused by activation of glp-1, a *Caenorhabditis elegans* member of the Notch family of receptors. *Development* 1997;**4**(124):925–36.

27. Henderson ST, Gao D, Lambie EJ, Kimble J. lag-2 may encode a signaling ligand for the GLP-1 and LIN-12 receptors of *C. elegans*. *Development* 1994;**10**(120):2913–24.

28. Crittenden SL, Troemel ER, Evans TC, Kimble J. GLP-1 is localized to the mitotic region of the *C. elegans* germ line. *Development* 1994;**10**(120):2901–11.

29. Greenwald I, Rubin GM. Making a difference: the role of cell–cell interactions in establishing separate identities for equivalent cells. *Cell* 1992;**2**(68):271–81.

30. Seydoux G, Greenwald I. Cell autonomy of lin-12 function in a cell fate decision in *C elegans*. *Cell* 1989;**7**(57):1237–45.

31. Sternberg PW. Vulval development. *WormBook* 2005, 25 Jun:1–28.

32. Kenyon C. A perfect vulva every time: gradients and signaling cascades in *C. elegans*. *Cell* 1995;**2**(82):171–4.

33. Greenwald IS, Sternberg PW, Horvitz HR. The *lin-12* locus specifies cell fates in *Caenorhabditis elegans*. *Cell* 1983;**2**(34):435–44.

34. Levitan D, Greenwald I. LIN-12 protein expression and localization during vulval development in *C. elegans*. *Development* 1998;**16**(125):3101–9.

35. Chen N, Greenwald I. The lateral signal for LIN-12/Notch in *C. elegans* vulval development comprises redundant secreted and transmembrane DSL proteins. *Dev Cell* 2004;**2**(6):183–92.

36. Shaye DD, Greenwald I. Endocytosis-mediated downregulation of LIN-12/Notch upon Ras activation in *Caenorhabditis elegans*. *Nature* 2002;**6916**(420):686–90.

37. Blair SS. Eye development: Notch lends a handedness. *Curr Biol* 1999;**10**(9):R356–60.

38. Bolos V, Grego-Bessa J, de la Pompa JL. Notch signaling in development and cancer. *Endocr Rev* 2007;**3**(28):339–63.

39. Artavanis-Tsakonas S, Matsuno K, Fortini ME. Notch signaling. *Science* 1995;**5208**(268):225–32.

40. Blanpain C, Horsley V, Fuchs E. Epithelial stem cells: turning over new leaves. *Cell* 2007;**3**(128):445–58.

41. Pourquie O. Building the spine: the vertebrate segmentation clock. *Cold Spring Harb Symp Quant Biol* 2007;**72**:445–9.

42. Tomlinson A, Struhl G. Delta/Notch and Boss/Sevenless signals act combinatorially to specify the Drosophila R7 photoreceptor. *Mol Cell* 2001;**3**(7):487–95.

43. Jarriault S, Schwab Y, Greenwald I. A Caenorhabditis elegans model for epithelial-neuronal transdifferentiation. *Proc Natl Acad Sci USA* 2008;**10**(105):3790–5.

44. Logeat F, Bessia C, Brou C, LeBail O, Jarriault S, Seidah NG, Israel A. The Notch1 receptor is cleaved constitutively by a furin-like convertase. *Proc Natl Acad Sci USA* 1998;**14**(95):8108–12.

Chapter 27

Proteolytic Activation of Notch Signaling: Roles for Ligand Endocytosis and Mechanotransduction

James T. Nichols[1,4] and Gerry Weinmaster[1,2,3,4*]

[1]Department of Biological Chemistry David Geffen School of Medicine
[2]Molecular Biology Institute
[3]Jonsson Comprehensive Cancer Center
[4]University of California, Los Angeles, California

INTRODUCTION

The Notch pathway is one of a relatively few signaling systems that is absolutely required for normal invertebrate and vertebrate development; however, Notch signaling also functions in the maintenance of adult cells and tissues [1]. Signaling via the Notch pathway influences a wide range of cell types and cellular processes through effects on cell-type specification, differentiation, survival, proliferation, apoptosis, and morphogenetic events. The varied and often opposite effects of Notch signaling are strongly dependent on the cellular context as well as interactions with other signaling systems [2]. Although initially identified as important in restricting cellular differentiation, Notch signaling has also been reported to induce the differentiation of a number of different cell types [3]. Through a process known as lateral inhibition, Notch signaling restricts the selection of specific cell fates from a field of equivalent progenitors. However, Notch signaling induced between compartments of non-equivalent cells directs cell type specification and boundary formation required for tissue patterning and morphogenesis. Moreover, Notch signaling plays a critical role in coordinating binary cell-fate decisions obtained through asymmetric cell division, and losses in Notch signaling produce cell fate transformations. Finally, Notch signaling induced through interactions between stem cells and their adjacent niche cells has been implicated in stem cell survival, maintenance, and self-renewal, possibly reflecting a role for Notch in regeneration and renewal of tissues following injury and/or during aging in the adult. Given the extensive list of functions attributed to Notch signaling, it is not surprising that mutations in genes encoding components of this pathway have been found associated with a number of inherited human syndromes as well as cancer.

The Notch receptors and DSL ligands are widely expressed during development, and in many cases interacting cells express both ligands and receptors [3, 4]. Cells take on distinct fates because Notch signaling is consistently activated in only one of the two interacting cells, indicating that the signaling polarity must be highly regulated. Studies in flies and worms have identified positive and negative transcriptional feedback mechanisms that amplify small differences in Notch and DSL ligand expression that could introduce a bias for which of the interacting cells sends or receives the Notch signal [5, 6]. If this were the case, cells competent to induce a signal would be expected to display higher DSL ligand levels then cells receiving the Notch signal; however, Delta expression appears uniform among cells undergoing lateral inhibition during selection of the neural fate [7, 8]. Therefore, mechanisms in addition to transcription must exist to ensure fidelity in cell-fate decisions regulated by Notch signaling. In this regard, the localization of DSL ligands to intracellular vesicles has been used to identify the signal-sending cell [9–11], which is surprising, given that ligands need to be on the cell surface to bind and activate Notch on adjacent cells. Numerous lines of evidence have indicated that ligand must be internalized from the surface of signal-sending cells to activate signaling in Notch receiving cells (reviewed in [12–16]), underscoring the importance of endocytosis and membrane trafficking of DSL ligands in generating cell types regulated by Notch signaling. This chapter will discuss potential roles for DSL ligand endocytosis in Notch signaling, with a focus on the generation of a physical force to mechanically dissociate and activate Notch.

DSL LIGAND ENDOCYTOSIS IS REQUIRED FOR ACTIVATION OF NOTCH SIGNALING

DSL ligands need to be expressed on the cell surface to engage Notch receptors on adjacent cells, however, DSL ligands are detected in intracellular vesicles [7, 9, 17–19], and display overlap with proteins involved in sorting and recycling, such as Hrs and Rab-11 [20–22], suggesting that internalized ligands may undergo complex membrane trafficking. Antibody uptake experiments clearly demonstrate the potential of DSL ligands to be internalized from the cell surface into endosomal vesicles [21, 23, 24], but not all internalized ligands appear competent to signal [19, 25]. Nonetheless, several lines of evidence have indicated an absolute requirement for DSL ligand endocytosis in Notch signaling (reviewed in [12, 14, 16]). Genetic mosaic studies with the *Drosophila* homolog of dynamin, a key regulator of endocytosis, were the first to suggest an important role for endocytosis by the DSL ligand cell in activation of Notch signaling [26]. Consistent with this, specific inhibition of dynamin in the DSL ligand-presenting cell, in both *Drosophila* and mammalian systems, produces defects in Notch signaling [9, 24]; however, the exact function(s) of ligand endocytosis and membrane trafficking in activation of Notch have remained controversial.

Cell surface proteins are targeted for internalization through the presence of either intrinsic endocytic signals or the covalent modification of ubiquitin within their cytoplasmic domains [27]. The DSL ligand intracellular domains contain multiple lysine residues that represent potential sites for ubiquitination, and intracellular truncations yield ligands that accumulate on the cell surface, where they bind Notch but are unable to activate signaling [24]. Two structurally unrelated E3 ubiquitin ligases, Neuralized (Neur) and Mindbomb, ubiquitinate DSL ligands, and this modification is absolutely required for ligands to activate Notch signaling [23, 28–35]. Defects in these E3 ligases cause DSL ligands to accumulate on the cell surface, indicating that ubiquitination is required for ligand internalization. Importantly, in the absence of ubiquitination, the DSL ligands cannot activate Notch signaling, indicating that ubiquitination and/or endocytosis is critical for ligand activity. Interestingly, E3 ligase overexpression promotes efficient removal of ligands from the cell surface, conditions that counterintuitively enhance signaling activity. Although the mechanism by which ubiquitination promotes ligand signaling is currently unknown, cell surface proteins gain access to particular endocytic routes as well as undergo endosomal sorting and trafficking through ubiquitination [36]. In fact, Neur2 is not required for internalization,

but rather regulates membrane trafficking of DSL ligands downstream of endocytosis [22].

Internalization of ubiquitinated cell surface proteins and endosomal trafficking are regulated by a number of different ubiquitin-binding proteins [36]. Epsin is one such endocytic adaptor and sorting protein that contains ubiquitin interaction motifs (UIM), and studies in flies and worms have indicated a strict requirement for epsin homologs in the ligand signal-sending cell during several processes that involve Notch signaling [19, 25, 35, 37–39]. It seems likely that epsin directly binds ubiquitinated DSL ligands and recruits them to a specific endocytic pathway necessary for signaling activity; however, both the endocytic route and specific function provided by epsin in generating an active ligand have remained unclear. Genetic studies in flies have identified requirements for clathrin heavy chain [40] and auxilin for DSL ligand activity [40, 41], suggesting that DSL ligand endocytosis is clathrin dependent; however, recent findings supporting internalization of ubiquitinated cargo via the caveolar pathway [42, 43], raise the possibility that DSL ligands may also undergo clathrin-independent endocytosis. In fact, dynamin and epsin are absolute requirements for ligand activity, and both have been reported to function in clathrin-dependent and -independent endocytosis, and thus the specific endocytic route involved in activating DSL ligands is still unknown. Moreover, studies in flies suggest that DSL ligands use a number of different endocytic routes, but only ligands internalized in a ubiquitin- and epsin-dependent manner are competent to signal [19, 25, 35]. The challenge now is to determine the specific property and/or mechanism conferred by ubiquitin and epsin for ligand signaling potential.

UBIQUITIN AND EPSIN-DEPENDENT RECYCLING TO PRODUCE AN ACTIVE DSL LIGAND

The findings that not all internalized DSL ligands are competent to signal have suggested that a subpopulation of ligands may gain access to a particular endosomal compartment to acquire signaling activity, and this requires both ubiquitination and interactions with epsin [19, 25, 35, 37]. It is possible that ligand ubiquitination followed by epsin-dependent endocytosis and/or membrane trafficking regulates more than one event in Notch signaling, and roles both before and after binding to Notch have been proposed (reviewed in [12, 14, 16]). For example, it has been suggested that internalized ligands are trafficked through the endosome for conversion into an active ligand that is recycled back to the cell surface [25]. The idea that DSL ligands might need to recycle to activate Notch initially came from studies using a form of Delta in which the intracellular domain was replaced with a portion of the Low Density Lipoprotein Receptor (LDLR) containing a known internalization motif (FDNPXY) [25].

This chimeric ligand is surprisingly active in cells lacking epsin, suggesting that the LDLR endocytic signal allows DSL ligands to bypass the need for ubiquitination and epsin in Notch signaling, perhaps through recruiting other adaptors such as Dab2 or ARH that normally assist LDLR endocytosis [27, 36]. However, Serrate has a dileucine internalization motif that is active in endocytosis but is not sufficient for signaling [19], indicating that not all intrinsic endocytic signals confer DSL activity.

Given that internalized LDLR is rapidly recycled, the signaling activity detected with the Delta–LDLR chimera may reflect epsin-dependent recycling of endogenous DSL ligands; however, epsin is not known to function in recycling, and losses in epsin do not perturb recycling of the transferrin receptor [44]. Even though the activating properties conferred through recycling are undefined, endocytosis followed by recycling has been suggested to produce an active DSL ligand through facilitating a posttranslational modification during endosomal trafficking [25]. Recycling could also promote ligand clustering [45], or target ligand to a particular cell surface microdomain, or simply replenish ligand at the cell surface to maintain high levels for active signaling. Although it is still unclear if recycling is required for ligand activity, losses in Rab11 or Sec15, which function together in recycling endosomes, produce cell-fate transformations indicative of losses in DSL activity [20, 46–48]. However, not all Notch-dependent processes appear to require Sec 15 activity [47], as one might expect if recycling is an absolute requirement for signaling activity. It is important to note that even though Delta and Rab11 co-localize in endocytic vesicles, direct evidence that DSL ligands actually recycle and that recycling positively affects either Notch binding or activation is lacking.

NOTCH SIGNALING REQUIRES PROTEOLYSIS AND NUCLEAR TRANSLOCATION

Notch signaling depends on a series of proteolytic cleavage events that serve to release the Notch intracellular domain (NICD) from the membrane and allow it to function as a signal transducer [1, 49]. Ligand-induced proteolytic activation is dependent on furin proteolytic processing of the primary Notch translational product during its trafficking to the cell surface [50–52]. The furin-cleavage fragments remain associated through intramolecular, non-covalent interactions, which maintain the mature cell surface receptor in a heterodimeric form that is necessary to keep Notch inactive in the absence of ligand [53–55]. Binding of DSL ligands to this heterodimeric Notch receptor activates signaling by inducing additional proteolysis, first within the Notch extracellular domain (NECD) by a disintegrin and metalloprotease (ADAM). Cleavage by ADAMs allows for efficient proteolytic processing within the membrane-spanning

region of Notch by γ-secretase to generate the biologically active NICD [56, 57]. Membrane release and trafficking of NICD to the nucleus permits interactions with the DNA binding protein CSL (CBF1, SuH, LAG-1) and recruitment of co-activators, which together allows Notch to directly participate in the transcriptional activation of Notch target genes [1, 58]. Although activating proteases have been identified, the events prior to and following ligand binding required for Notch proteolysis in the generation of NICD are not well defined.

DSL LIGAND ENDOCYTOSIS TO PRODUCE A FORCE FOR NOTCH PROTEOLYTIC ACTIVATION

The requirement for DSL endocytosis could reflect a more "active" role beyond presentation of a functional cell surface ligand. Although DSL ligands are required to produce NICD from full-length Notch, it is still unclear how ligand binding promotes ADAM cleavage, a necessary step in activating ©-secretase proteolysis [49]. Studies in flies have suggested that endocytosis of DSL ligand bound to Notch on adjacent cells induces a molecular strain in Notch that effects conformational changes and facilitates ADAM cleavage within the NECD [9]. The ADAM-cleaved Notch ectodomain may be shed from the Notch cell; however, evidence initially provided by studies in flies [9, 59], and more recently from mammalian cells [24], indicates that released NECD is taken up by DSL ligand cells. In either event, these findings suggest that a critical step in Notch activation is NECD removal from intact Notch, which is also supported by previous findings, where engineered NECD truncations or deletions produce constitutively active forms of Notch [60–63]. Additionally, dissociation of the NECD subunit from heterodimeric Notch via calcium chelators [53] or mutations within the heterodimerization (HD) domain [54] also mimic signaling induced by DSL ligands. Moreover, mutations within the HD domain of human Notch1, associated with aberrant Notch signaling causative for T cell acute lymphoblastic leukemia, induce both NECD shedding and Notch signaling independent of ligand [64]. Biochemical and structural studies have delineated a negative regulatory region (NRR) within the NECD that includes the HD domain, which is responsible for keeping the Notch receptor inactive in the absence of ligand [54, 55, 65]. Structural analysis has identified multiple intramolecular interactions within the NRR, which function in maintaining the Notch heterodimeric structure and appear to mask the ADAM cleavage site [55]. Based on this analysis, the authors propose that DSL ligand binding would need to induce substantial movement within the NECD to expose the ADAM cleavage site and allow proteolytic activation. Whether the required conformational movement is mediated through allosteric changes induced by ligand binding or a mechanical force generated by DSL ligand endocytosis of bound Notch is currently unknown.

Recent findings with mammalian cells have provided additional support for ligand endocytosis in proteolytic activation of Notch [24]. Importantly, this study clearly demonstrates that Notch dissociation is driven not by ligand binding but by ligand endocytosis, and provides further insight into the role and underlying mechanism of NECD removal in Notch activation. At odds with conventional models of ligand-induced proteolytic activation of Notch, NECD release and uptake by DSL ligand cells does not require ADAM proteolysis. Specifically, ADAM inhibitors do not block NECD trans-endocytosis by DSL ligand cells, demonstrating for the first time non-enzymatic dissociation of Notch. Moreover, Notch signaling but not NECD removal by DSL ligand cells is dependent on ADAM proteolysis, suggesting that separation of the NECD heterodimeric subunit from the membrane-bound portion, rather than conformational changes within the intact Notch receptor, unmasks the ADAM cleavage site and facilitates activating Notch proteolysis. Consistent with this idea, inhibition of Notch furin processing, and thus heterodimeric formation, prevents Notch dissociation, trans-endocytosis of NECD by DSL ligand cells, and activation of Notch signaling [24]. The dependence of Notch heterodimer dissociation and NECD removal on ligand endocytosis has suggested a model in which endocytosis of ligand-bound Notch provides a physical force to mechanically remove the NECD subunit from the intact Notch heterodimer to allow activating Notch proteolysis [24]. That the critical event in Notch signaling is non-enzymatic dissociation of Notch brings its activation closer to the realm of mechanotransduction than previously proposed proteolytic cleavage models.

CONVERTING DSL LIGAND ENDOCYTOSIS INTO A FORCE-GENERATING PROCESS

Endocytosis of DSL ligand bound to Notch on adjacent cells is likely mechanistically different from constitutive or bulk ligand endocytosis in the absence of Notch (Figure 27.1). In fact, bulk Delta endocytosis occurs in the absence of epsin [25] and a Serrate mutant defective in Notch binding traffics normally [19], indicating that most endocytosis is not connected with signaling. Then how could bound Notch alter ligand endocytosis, and why is there an absolute dependence on E3 ubiquitin ligases and epsin for ligand signaling activity? Notch binding may induce ubiquitination and/or clustering of DSL ligands, which could amass multiple ubiquitin-binding sites for epsin (see Figure 27.1). By assembling multiple low-affinity mono-ubiquitin interactions, strong epsin-UIM/ubiquitinated-DSL interactions could be generated [66, 67], and this may be necessary to stabilize DSL ligands within

endocytic vesicles and overcome any resistance to internalization when bound to Notch. In fact, replacement of the Delta intracellular domain with a single ubiquitin motif that can undergo polyubiquitination promotes internalization and signaling activity in zebrafish [23]. However, a non-extendable ubiquitin only weakly signals even though it promotes endocytosis [25], supporting the idea that multiple ubiquitin interaction sites are required for DSL ligands to activate Notch, possibly through providing stable associations with epsin-containing endocytic vesicles.

Recent studies in flies have suggested that Neur may play additional roles in DSL ligand endocytosis in potentiating ligand signaling activity beyond ubiquitination [34, 68]. A phosphoinositide-binding domain identified in Neur functions to direct Neur to the plasma membrane; however, this is not required for Neur to interact with or ubiquitinate Delta, but membrane localization is required for Delta endocytosis and thus Notch signaling [68]. Epsin also binds phosphoinositides, an activity proposed to function in membrane curvature during endocytic vesicle formation [69]; however, epsin–phosphoinositide interactions also function in endosomal sorting and trafficking of internalized proteins [36]. Therefore, both epsin and Neur perform multiple functions during endocytosis and membrane trafficking of cell surface proteins. Since Neur directly binds Delta and epsin has the potential to interact with ubiquitinated Delta, it seems possible that they could work together to recruit and/or stabilize the clustering of Delta bound to Notch into endocytic vesicles. The association of epsin and Neur with Delta-Notch containing endocytic vesicles may help to generate a force during ligand internalization that could mechanically dissociate the Notch heterodimer (see Figure 27.1).

Although it is not known if ubiquitin-dependent multimerization of epsin during endocytosis of Delta–Notch complexes could stabilize endocytic vesicles, epsin is a multidomain protein that, in addition to binding ubiquitinated cargo and membrane phospholipids, also directly binds clathrin and other accessory endocytic components [69]. Recently, epsin has been reported to bind to Cdc42 GTPase-activating proteins, suggesting a role for epsin in regulating actin dynamics [70]. Given that both the actin cytoskeleton and dynamin are implicated in inducing membrane constriction and tension during the process of endocytosis [71, 72], it is tempting to speculate that the epsin requirement in ligand activity is the creation of a force-generating endocytic vesicle, capable of mechanically pulling the Notch heterodimer apart to expose the ADAM cleavage site and allow proteolytic activation (Figure 27.1). This may explain why ubiquitinated ligands internalized in cells lacking epsin are unable to signal, and why only DSL ligands internalized in an epsin-dependent manner are competent to signal. Since the FDNPXY endocytic motif allows Notch signaling in the absence of espin, clathrin adaptors recruited by this intrinsic signal might also confer endocytic vesicles with force-generating properties.

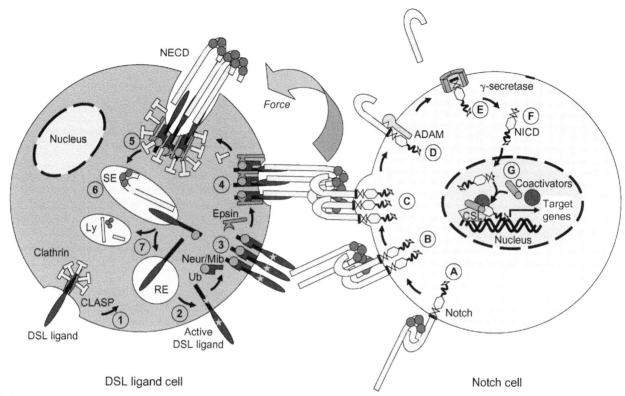

FIGURE 27.1 Potential roles for DSL ligand endocytosis in Notch signaling.
In the DSL ligand cell, surface ligands may undergo constitutive endocytosis via clathrin-mediated endocytosis facilitated by an unidentified CLASP (1). Trafficking of internalized ligands through the endosome may convert them into active ligands (indicated by a star) that are returned to the cell surface (2) via the recycling endosome (RE), where they are available to interact with heterodimeric Notch receptors on adjacent cells (A). DSL ligands are covalently modified by the addition of ubiquitin (Ub) by the E3 ligases Neuralized (Neur) or Mind bomb (Mib), an event that may or may not follow interactions with Notch (3). Cell–cell interactions stimulate binding and mutual clustering of receptors (B) and ubiquitinated DSL ligands (3). Ligand clustering could produce multiple ubiquitin binding sites for epsin that stabilize ligand–epsin interactions within endocytic pits (4). Additional interactions of epsin with clathrin and plasma membrane phospholipids may contribute to the production of a force-generating endocytic vesicle capable of mechanically removing the Notch extracellular domain (NECD) from the Notch heterodimer (5) (see text for details). The freed NECD is trans-endocytozed by the DSL ligand cell (6), where it may dissociate from the ligand in the sorting endosome (SE) (7) and be targeted for degradation in the lysosome (Ly), while the ligand is recycled back to the cell surface (2). Removal of the NECD by the DSL ligand cell exposes the ADAM cleavage site within the extracellular domain of the membrane-bound Notch heterodimeric subunit on the Notch cell (C). ADAM cleavage (D) facilitates a subsequent cleavage by γ-secretase within the membrane-spanning region of Notch (E) that allows release of the biologically active Notch intracellular domain (NICD) from the plasma membrane (F). Translocation of NICD to the nucleus (G) allows NICD to interact with the DNA-binding protein CSL and recruit co-activators required to drive transcription of target genes.

CONCLUSIONS AND FUTURE DIRECTIONS

At this point there is no consensus as to how DSL ligand endocytosis contributes to Notch activation. It is also currently unclear what role(s) ligand recycling might play in Notch activation, but recycling could provide a mechanism to replenish ligand to the cell surface following its removal through NECD trans-endocytosis during Notch activation (see Figure 27.1). High DSL ligand cell surface density may be required for continued Notch activation and sustained signaling, especially given that each activated Notch generates only one signaling molecule (NICD) that is not amplified and turns over rapidly, and once Notch is activated it cannot be reactivated. Therefore, it is possible that DSL ligands recycle following both constitutive and Notch-associated endocytosis. Understanding whether differences in endocytic pathways

and membrane trafficking routes underlie differences in constitutive and Notch-associated DSL ligand activities is likely to shed light on the molecular mechanisms underlying ligand-induced activation of Notch. Future studies are also needed to determine whether a force can indeed pull the Notch heterodimer apart, and if ligand recycling alters interactions with Notch it will be important to define the molecular bases of such ligand modifications and determine how they contribute to Notch activation.

REFERENCES

1. Bray SJ. Notch signalling: a simple pathway becomes complex. *Nat Rev Mol Cell Biol* 2006;**7**(9):678–89.
2. Hurlbut GD, Kankel MW, Lake RJ, Artavanis-Tsakonas S. Crossing paths with notch in the hyper-network. *Curr Opin Cell Biol* 2007;**19**(2):166–75.

3. Artavanis-Tsakonas S, Rand MD, Lake RI Notch signaling: cell fate control and signal integration in development. *Science* 1999;**284**(5415):770–6.

4. Weinmaster G. The ins and outs of notch signaling. *Mol Cell Neurosci* 1997;**9**(2):91–102.

5. Greenwald I, Rubin GM. Making a difference: the role of cell–cell interactions in establishing separate identities for equivalent cells. *Cell* 1992;**68**(2):271–81.

6. Seugnet L, Simpson P, Haenlin M. Transcriptional regulation of notch and Delta: requirement for neuroblast segregation in drosophila. *Development* 1997;**124**(10):2015–25.

7. Kooh PJ, Fehon RG, Muskavitch MA. Implications of dynamic patterns of delta and notch expression for cellular interactions during *Drosophila* development. *Development* 1993;**117**(2):493–507.

8. Kopczynski CC, Muskavitch MA. Complex spatio-temporal accumulation of alternative transcripts from the neurogenic gene delta during drosophila embryogenesis. *Development* 1989;**107**(3):623–36.

9. Parks AL, Klueg KM, Stout JR, Muskavitch MA. Ligand endocytosis drives receptor dissociation and activation in the notch pathway. *Development* 2000;**127**(7):1373–85.

10. Parks AL, Turner FR, Muskavitch MA. Relationships between complex delta expression and the specification of retinal cell fates during drosophila eye development. *Mech Dev* 1995;**50**(2–3):201–16.

11. Ohlstein B, Spradling A. Multipotent drosophila intestinal stem cells specify daughter cell fates by differential notch signaling. *Science* 2007;**315**(5814):988–92.

12. Chitnis A. Why is delta endocytosis required for effective activation of notch?. *Dev Dyn* 2006;**235**(4):886–94.

13. Fischer JA, Eun SH, Doolan BT. Endocytosis, endosome trafficking, and the regulation of drosophila development. *Annu Rev Cell Dev Biol* 2006;**22**:181–206.

14. Le Borgne R. Regulation of notch signalling by endocytosis and endosomal sorting. *Curr Opin Cell Biol* 2006;**18**(2):213–22.

15. Le Borgne R, Schweisguth F. Notch signaling: endocytosis makes delta signal better. *Curr Biol* 2003;**13**(7):R273–5.

16. Nichols JT, Miyamoto A, Weinmaster G. Notch signaling – constantly on the move. *Traffic* 2007;**8**(8):959–69.

17. Klueg KM, Muskavitch MA. Ligand–receptor interactions and trans-endocytosis of delta, serrate and notch: members of the notch signalling pathway in *Drosophila*. *J Cell Sci* 1999;**112**(19):3289–97.

18. Klueg KM, Parody TR, Muskavitch MA. Complex proteolytic processing acts on delta, a transmembrane ligand for notch, during *Drosophila* development. *Mol Biol Cell* 1998;**9**(7):1709–23.

19. Glittenberg M, Pitsouli C, Garvey C, Delidakis C, Bray S. Role of conserved intracellular motifs in serrate signalling, cis-inhibition and endocytosis. *EMBO J* 2006;**25**(20):4697–706.

20. Emery G, Hutterer A, Berdnik D, Mayer B, Wirtz-Peitz F, Gaitan MG, Knoblich JA. Asymmetric rab 11 endosomes regulate delta recycling and specify cell fate in the *Drosophila* nervous system. *Cell* 2005;**122**(5):763–73.

21. Le Borgne R, Schweisguth F. Unequal segregation of neuralized biases notch activation during asymmetric cell division. *Dev Cell* 2003;**5**(1):139–48.

22. Song R, Koo BK, Yoon KJ, Yoon MJ, Yoo KW, Kim HT, Oh HJ, Kim YY, Han JK, Kim CH, Kong YY. Neuralized-2 regulates a notch ligand in cooperation with Mind bomb-1. *J Biol Chem* 2006;**281**(47):36391–400.

23. Itoh M, Kim CH, Palardy G, Oda T, Jiang YJ, Maust D, Yeo SY, Lorick K, Wright GJ, Ariza-McNaughton L, Weissman AM, Lewis J, Chandrasekharappa SC, Chitnis AB. Mind bomb is a ubiquitin ligase that is essential for efficient activation of notch signaling by delta. *Dev Cell* 2003;**4**(1):67–82.

24. Nichols JT, Miyamoto A, Olsen SL, D'Souza B, Yao C, Weinmaster G. DSL ligand endocytosis physically dissociates notch1 heterodimers before activating proteolysis can occur. *J Cell Biol* 2007;**176**(4):445–58.

25. Wang W, Struhl G. *Drosophila* Epsin mediates a select endocytic pathway that DSL ligands must enter to activate notch. *Development* 2004;**131**(21):5367–80.

26. Seugnet L, Simpson P, Haenlin M. Requirement for dynamin during notch signaling in drosophila neurogenesis. *Dev Biol* 1997;**192**(2):585–98.

27. Szymkiewicz I, Shupliakov O, Dikic I. Cargo- and compartment-selective endocytic scaffold proteins. *Biochem J* 2004;**383**(1):1–11.

28. Chen W, Casey Corliss D. Three modules of zebrafish mind bomb work cooperatively to promote delta ubiquitination and endocytosis. *Dev Biol* 2004;**267**(2):361–73.

29. Deblandre GA, Lai EC, Kintner C. Xenopus neuralized is a ubiquitin ligase that interacts with XDelta1 and regulates Notch signaling. *Dev Cell* 2001;**1**(6):795–806.

30. Koo BK, Lim HS, Song R, Yoon MJ, Yoon KJ, Moon JS, Kim YW, Kwon MC, Yoo KW, Kong MP, Lee J, Chitnis AB, Kim CH, Kong YY. Mind bomb 1 is essential for generating functional notch ligands to activate notch. *Development* 2005;**132**(15):3459–70.

31. Lai EC, Deblandre GA, Kintner C, Rubin GM. Drosophila neuralized is a ubiquitin ligase that promotes the internalization and degradation of delta. *Dev Cell* 2001;**1**(6):783–94.

32. Pavlopoulos E, Pitsouli C, Klueg KM, Muskavitch MA, Moschonas NK, Delidakis C. Neuralized Encodes a peripheral membrane protein involved in delta signaling and endocytosis. *Dev Cell* 2001;**1**(6):807–16.

33. Yeh E, Dermer M, Commisso C, Zhou L, McGlade CJ, Boulianne GL. Neuralized functions as an E3 ubiquitin ligase during drosophila development. *Curr Biol* 2001;**11**(21):1675–9.

34. Pitsouli C, Delidakis C. The interplay between DSL proteins and ubiquitin ligases in notch signaling. *Development* 2005;**132**(18):4041–50.

35. Wang W, Struhl G. Distinct roles for Mind bomb, neuralized and epsin in mediating DSL endocytosis and signaling in *Drosophila*. *Development* 2005;**132**(12):2883–94.

36. Traub LM, Lukacs GL. Decoding ubiquitin sorting signals for clathrin-dependent endocytosis by CLASPs. *J Cell Sci* 2007;**120** (Pt 4):543–53.

37. Overstreet E, Chen X, Wendland B, Fischer JA. Either part of a drosophila epsin protein, divided after the ENTH domain, functions in endocytosis of delta in the developing eye. *Curr Biol* 2003;**13**(10):854–60.

38. Overstreet E, Fitch E, Fischer JA. Fat facets and Liquid facets promote delta endocytosis and delta signaling in the signaling cells. *Development* 2004;**131**(21):5355–66.

39. Tian X, Hansen D, Schedl T, Skeath JB. Epsin potentiates Notch pathway activity in drosophila and C. elegans. *Development* 2004;**131**(23):5807–15.

40. Eun SH, Lea K, Overstreet E, Stevens S, Lee JH, Fischer JA. Identification of genes that interact with drosophila liquid facets. *Genetics* 2006;**175**:1163–74.

41. Hagedorn EJ, Bayraktar JL, Kandachar VR, Bai T, Englert DM, Chang HC. *Drosophila melanogaster* auxilin regulates the internalization of delta to control activity of the notch signaling pathway. *J Cell Biol* 2006;**173**(3):443–52.

42. Chen H, De Camilli P. The association of epsin with ubiquitinated cargo along the endocytic pathway is negatively regulated by its interaction with clathrin. *Proc Natl Acad Sci USA* 2005;**102**(8):2766–71.

43. Sigismund S, Woelk T, Puri C, Maspero E, Tacchetti C, Transidico P, Di Fiore PP, Polo S. Clathrin-independent endocytosis of ubiquitinated cargos. *Proc Natl Acad Sci USA* 2005;**102**(8):2760–5.

44. Vanden Broeck D, De Wolf MJ. Selective blocking of clathrin-mediated endocytosis by RNA interference: epsin as target protein. *Biotechniques* 2006;**41**(4):475–84.

45. Hicks C, Ladi E, Lindsell C, Hsieh J, Hayward S, Collazo A, Weinmaster GA. Secreted Delta1-Fc fusion protein functions both as an activator and inhibitor of notch1 signaling. *J Neurosci Res* 2002;**69**:60–71.

46. Langevin J, Morgan MJ, Sibarita JB, Aresta S, Murthy M, Schwarz T, Camonis J, Bellaiche Y. Drosophila exocyst components Sec5, Sec6, and Sec15 regulate DE-Cadherin trafficking from recycling endosomes to the plasma membrane. *Dev Cell* 2005;**9**(3):355–76.

47. Jafar-Nejad H, Andrews HK, Acar M, Bayat V, Wirtz-Peitz F, Mehta SQ, Knoblich JA, Bellen HJ. Sec15, a component of the exocyst, promotes notch signaling during the asymmetric division of *Drosophila* sensory organ precursors. *Dev Cell* 2005;**9**(3):351–63.

48. Wu S, Mehta SQ, Pichaud F, Bellen HJ, Quiocho FA. Sec15 interacts with Rab11 via a novel domain and affects Rab11 localization *in vivo*. *Nat Struct Mol Biol* 2005;**12**(10):879–85.

49. Mumm JS, Kopan R. Notch signaling: from the outside in. *Dev Biol* 2000;**228**(2):151–65.

50. Blaumueller CM, Qi H, Zagouras P, Artavanis-Tsakonas S. Intracellular cleavage of notch leads to a heterodimeric receptor on the plasma membrane. *Cell* 1997;**90**:281–91.

51. Bush G, diSibio G, Miyamoto A, Denault JB, Leduc R, Weinmaster G. Ligand-induced signaling in the absence of furin processing of notch1. *Dev Biol* 2001;**229**(2):494–502.

52. Logeat F, Bessia C, Brou C, LeBail O, Jarriault S, Seiday N, Israel A. The notch1 receptor is cleaved constitutively by a furin-like convertase. *Proc Natl Acad Sci USA* 1998;**95**:8108–12.

53. Rand MD, Grimm LM, Artavanis-Tsakonas S, Patriub V, Blacklow SC, Sklar J, Aster JC. Calcium depletion dissociates and activates heterodimeric notch receptors. *Mol Cell Biol* 2000;**20**(5):1825–35.

54. Sanchez-Irizarry C, Carpenter AC, Weng AP, Pear WS, Aster JC, Blacklow SC. Notch subunit heterodimerization and prevention of ligand-independent proteolytic activation depend, respectively, on a novel domain and the LNR repeats. *Mol Cell Biol* 2004;**24**(21):9265–73.

55. Gordon WR, Vardar-Ulu D, Histen G, Sanchez-Irizarry C, Aster JC, Blacklow SC. Structural basis for autoinhibition of notch. *Nat Struct Mol Biol* 2007;**14**(4):295–300.

56. Brou C, Logeat F, Gupta N, Bessia C, LeBail O, Doedens JR, Cumano A, Roux P, Black RA, Israel A. A novel proteolytic cleavage involved in Notch signaling: the role of the Disintegrin-Metalloprotease TACE. *Mol. Cell* 2000;**5**:207–16.

57. Mumm JS, Schroeter EH, Saxena MT, Griesemer A, Tian X, Pan DJ, Ray WJ, Kopan R. A ligand-induced extracellular cleavage regulates g-secretase-like proteolytic activation of notch1. *Mol. Cell* 2000;**5**:197–206.

58. Wilkin MB, Baron M. Endocytic regulation of notch activation and down-regulation (review). *Mol Membr Biol* 2005;**22**(4):279–89.

59. Morel V, Le Borgne R, Schweisguth F. Snail is required for delta endocytosis and Notch-dependent activation of single-minded expression. *Dev Genes Evol* 2003;**213**(2):65–72.

60. Lieber T, Kidd S, Alcamo E, Corbin V, Young MW. Antineurogenic phenotypes induced by truncated Notch proteins indicate a role in signal transduction and may point to a novel function for Notch in nuclei. *Genes Dev* 1993;**7**(10):1949–65.

61. Rebay I, Fehon RG, Artavanis-Tsakonas S. Specific truncations of *Drosophila* notch define dominant activated and dominant negative forms of the receptor. *Cell* 1993;**74**(2):319–29.

62. Schroeter E, Kisslinger J, Kopan R. Notch1 signalling requires ligand-induced proteolytic release of the intracellular domain. *Nature* 1998;**393**(6683):382–6.

63. Struhl G, Fitzgerald K, Greenwald I. Intrinsic activity of the Lin-12 and notch intracellular domains *in vivo*. *Cell* 1993;**74**(2):331–45.

64. Weng AP, Ferrando AA, Lee W, Morris JPt, Silverman LB, Sanchez-Irizarry C, Blacklow SC, Look AT, Aster JC. Activating mutations of NOTCH1 in human T cell acute lymphoblastic leukemia. *Science* 2004;**306**(5694):269–71.

65. Malecki MJ, Sanchez-Irizarry C, Mitchell JL, Histen G, Xu ML, Aster JC, Blacklow SC. Leukemia-associated mutations within the NOTCH1 heterodimerization domain fall into at least two distinct mechanistic classes. *Mol Cell Biol* 2006;**26**(12):4642–51.

66. Barriere H, Nemes C, Lechardeur D, Khan-Mohammad M, Fruh K, Lukacs GL. Molecular basis of oligoubiquitin-dependent internalization of membrane proteins in Mammalian cells. *Traffic* 2006;**7**(3):282–97.

67. Hawryluk MJ, Keyel PA, Mishra SK, Watkins SC, Heuser JE, Traub LM. Epsin 1 is a polyubiquitin-selective clathrin-associated sorting protein. *Traffic* 2006;**7**(3):262–81.

68. Skwarek LC, Garroni MK, Commisso C, Boulianne GL. Neuralized contains a phosphoinositide-binding motif required downstream of ubiquitination for delta endocytosis and notch signaling. *Dev Cell* 2007;**13**(6):783–95.

69. Horvath CA, Vanden Broeck D, Boulet GA, Bogers J, De Wolf MJ. Epsin: inducing membrane curvature. *Intl J Biochem Cell Biol* 2007;**39**:1765–70.

70. Aguilar RC, Longhi SA, Shaw JD, Yeh LY, Kim S, Schon A, Freire E, Hsu A, McCormick WK, Watson HA, Wendland B. Epsin N-terminal homology domains perform an essential function regulating Cdc42 through binding Cdc42 GTPase-activating proteins. *Proc Natl Acad Sci USA* 2006;**103**(11):4116–21.

71. Itoh T, Erdmann KS, Roux A, Habermann B, Werner H, De Camilli P. Dynamin and the actin cytoskeleton cooperatively regulate plasma membrane invagination by BAR and F-BAR proteins. *Dev Cell* 2005;**9**(6):791–804.

72. Roux A, Uyhazi K, Frost A, De Camilli P. GTP-dependent twisting of dynamin implicates constriction and tension in membrane fission. *Nature* 2006;**441**(7092):528–31.

Chapter 28

Vascular Endothelial Growth Factors and Receptors: Signaling in Vascular Development

Anna Dimberg, Charlotte Rolny, Laurens A. van Meeteren and Lena Claesson-Welsh
Department of Genetics and Pathology, Rudbeck Laboratory, Uppsala University, Uppsala, Sweden

INTRODUCTION TO VEGFs AND VEGF RECEPTORS

The vascular endothelial growth factor (VEGF) family of ligands consists of five mammalian polypeptides and several structurally related factors of invertebrate origin. The VEGFs exist as homodimeric glycoproteins of about 45 kDa. They belong to the cysteine-knot family of ligands, which have a tight disulfide-bonded structure [1]. The prototype VEGF, now denoted VEGFA, becomes alternatively spliced and exists in isoforms denoted according to the length of the mature polypeptide; VEGFA121, -145, -165 and -189 in the human. The corresponding mouse VEGFA isoforms are each a single amino acid residue shorter (i.e., VEGFA120, etc). The splice variants are able, to different extents, to bind to co-receptors such as heparan sulfate proteoglycans (HSPGs) and neuropilins (NRPs); such interactions will retain the ligands by the producer cell and thereby restrict effects on endothelial cells located at a distance. VEGFA is expressed by most cell types; low oxygen tension, hypoxia, is an important regulator of VEGF expression [2]. Other VEGF family members, placenta growth factor (PlGF) and VEGFB, also exist as alternatively spliced isoforms with distinct biology. The biology of VEGFC and VEGFD, on the other hand, is regulated through proteolytic processing. There are related invertebrate VEGFs, such as VEGFE (parapox virus open reading frame, [3]) and VEGFF (snake venom-derived, [4]), which have served as tools to dissect VEGF biology, since they bind to VEGFRs in a different pattern than the mammalian VEGFs.

The diversity of the VEGFR family (VEGFR1, VEGFR2, and VEGFR3) is thought to have been established through gene duplication from a single common ancestral receptor, exemplified by the *Drosophila melanogaster* receptor tyrosine kinase D-VEGFR (also denoted PVR, as there is a high similarity to the platelet-derived growth factor receptors, PDGFRs) [5]. The VEGFRs are classical receptor tyrosine kinases that are organized in four regions: the extracellular ligand-binding domain, the transmembrane (TM) domain, the tyrosine kinase (TK) domain which is interrupted by a "kinase insert," and the carboxy-terminal domain (see Figure 28.1).

Ligand-binding leads to receptor dimerization and activation. The receptor dimers are held together not only by binding of the growth factor but also by direct interaction between the two receptor molecules in the dimer [6]. Dimerization creates both homo- and heterodimers of VEGF receptors [7, 8], which may have distinct functions.

The VEGFs and their receptors function in different compartments of the vascular system; in embryonic development of blood and lymphatic endothelial cells, in formation of vessels from already established vasculature (angiogenesis), and in migration of monocytes and endothelial cell progenitors. A very important function of VEGF is in regulation of vascular permeability, as inferred from its alternative designation, vascular permeability factor [9]. In addition, neuronal stem cells express VEGF receptors and are to some extent regulated by VEGF, indicative of the close relationship between vascular and neuronal development (for a review, see [10]).

Gene targeting has demonstrated a strict requirement for several of the VEGF ligand/receptor members for vascular development during embryogenesis (vasculogenesis) (see Table 28.1 for a summary of the phenotypes of all gene targeted VEGF/VEGFR mouse models described in this review). Certain VEGFs are dispensable for embryonic development but have a crucial function in regulation of blood vessel formation in pathological processes.

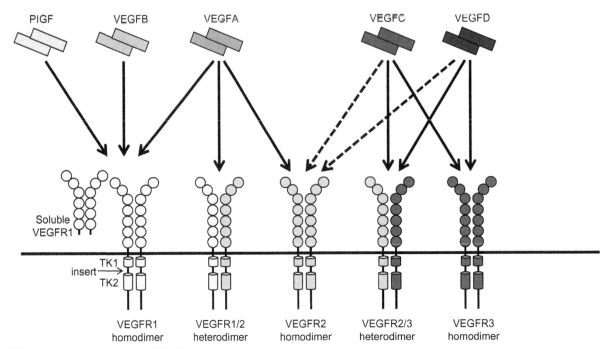

FIGURE 28.1 Schematic outline of VEGF ligands and receptors.
The specific binding of mammalian VEGFs (-A, -B, -C, -D and PlGF) to the three VEGFR tyrosine kinases results in formation of VEGFR homo- and heterodimers. Proteolytic processing regulates binding of VEGFC and -D to VEGFR2 (dashed line).

DEVELOPMENTAL PROCESSES; VASCULOGENESIS AND ANGIOGENESIS

Migration of endothelial precursors (angioblasts) and their organization into so-called blood islands allow establishment of the primitive vascular plexus [11]. By pruning and splitting, finer vessels develop (Figure 28.2). This process is denoted vasculogenesis, and is strictly dependent on VEGFA/VEGFR2. To a large extent, however, vasculogenesis remains uncharacterized in molecular terms, and very little is known about VEGFR signaling at this stage. Differentiation of endothelial cells from precursors may also occur in the adult, through homing of bone-marrow derived circulating precursors to sites of active vascular regeneration in a VEGFA/VEGFR2-dependent manner [12].

Formation of new vessels from already established vasculature is denoted angiogenesis. The main mechanisms for creation of new vessels in angiogenesis are sprouting and splitting (intussusception) of vessels (Figure 28.2). In vessel sprouting, VEGF-activated endothelial cells start to secrete proteases, which digest the vessel basement membrane. Selected cells extend "sprouts" into the surrounding tissue, followed by formation of a stalk of dividing cells headed by a non-dividing tip cell [13]. Sprouting is negatively regulated by the Notch pathway [14]; otherwise, little is known about the signaling that regulates formation of the angiogenic sprout. VEGF-driven angiogenic sprouting has been documented in the developing retina in newborn mice, and in tumors. In vessel splitting, growth of transvascular

tissue pillars allows rapid remodeling of the vasculature while maintaining the circulation [15].

Below is an outline of the current understanding of VEGFR signaling pathways. We start with an account of VEGFR2, which has been intensely studied, and which may serve as a template.

VEGFR2 AND ITS LIGANDS IN VASCULAR DEVELOPMENT

VEGFR2 (also denoted KDR in humans and Flk1 in mouse) is the first vascular marker to appear during development, and is central in endothelial cell function. It binds VEGFA and the processed forms of human VEGFC and VEGFD. It is expressed on endothelial and hematopoietic precursor cells, mature capillary endothelial cells, and neuronal cells. Although not required for hemangioblast formation, VEGFR2 is necessary for endothelial cell maturation, migration, proliferation, and vascular organization during development [16, 17]. Gene targeting of mouse *Vegfr2* is lethal at embryonic day (E)8.5, associated with defective blood island formation, and arrested endothelial and hematopoietic development [18]. These effects are phenocopied by targeting of the *Vegfa* gene [19]. VEGFA/VEGFR2 signaling has been shown to control all aspects of endothelial cell function, such as migration, survival, proliferation, and differentiation to create lumen-containing

TABLE 28.1 VEGF/VEGFR and co-receptor function assessed by gene targeting in mice

Genotype	Phenotype	Ref(s)
Vegfa$^{+/-}$	Lethal at E11–12, defect vascular development	19, 21
Vegfa$^{-/-}$	More severe defects in vascular development than heterozygote, lethal at E9.5–10.5	19
Vegfa$^{120/120}$	50% die at birth due to bleeding in multiple organs, remaining mice die before postnatal day 14 due to cardiac failure. Impaired myocardial angiogenesis, ischemic cardiomyopathy, skeletal defects, defects in vascular outgrowth and patterning in the retina	22–25
Vegfa$^{188/188}$	Impaired retinal arterial development, dwarfism, defect epiphysal vascularization, impaired development of growth plates and secondary ossification centers, knee joint dysplasia	22, 26
Vegfb$^{-/-}$	Reduced heart size, dysfunctional coronary vasculature, impaired recovery from cardiac ischemia	65, 96
Vegfc$^{-/-}$	Prenatal death due to edema, no lymphatic vessels	69
Vegfd$^{-/-}$	Normal development, slight reduction of lymphatic vessels adjacent to lung bronchiole	71
Plgf$^{-/-}$	Impaired angiogenesis during ischemia, inflammation, wound healing and cancer	64
Vegfr1$^{-/-}$	Lethal at E8.5–9.0, increased hemangioblast commitment, vascular disorganization due to endothelial cell overgrowth	57, 97
Vegfr1 (TK)$^{-/-}$	Normal development, suppressed VEGF-induced macrophage migration, decreased tumor angiogenesis	60, 98
Vegfr1 (TM-TK)$^{-/-}$	50% of mice die during embryonic development due to vascular defects	59
Vegfr2$^{-/-}$	Lethal at E8.5–9.5, defect blood-island formation and vasculogenesis	18
Vegfr3$^{-/-}$	Lethal before formation of lymphatics due to cardiovascular failure. Embryos show vascular remodeling defects and pericardial fluid accumulation	68
Nrp1$^{-/-}$	Embryonically lethal, defect neural patterning, vascular regression	84, 85
Nrp1 overexpression	Cardiovascular defects, heart malformation, excess blood vessel formation, dilated blood vessels, hemorrhage, anomalies in nervous system and limb	86
Nrp2$^{-/-}$	40% show perinatal death. Survivors are smaller than littermates. Defects in neuronal patterning, severe reduction of small lymphatic vessels and capillaries	87–89
Nrp1$^{-/-}$, Nrp2$^{-/-}$	Lethal at E8.5, defect vascular development	90

vascular structures. The critical role of VEGFA/VEGFR2 in regulation of the vasculature has promoted efforts to block its function through anti-angiogenic therapies for treatment of, for example, cancer and retinopathy. Currently, therapy with neutralizing antibodies against VEGFA is used to arrest vascularization in different malignancies such as renal, colorectal, and breast cancer [20].

There is a strict requirement for a certain level of signaling through VEGFA/VEGFR2 during development, demonstrated by the fact that mice lacking one VEGFA allele *(Vegfa$^{+/-}$)* die at E11.5 due to severe cardiovascular defects [19, 21]. Moreover, engagement of co-receptors, which is differentially regulated via alternative splicing of *Vegfa*, is critical

in VEGFA/VEGFR2 biology. Mice expressing only the VEGFA165 isoform, which binds to both HSPGs and NRP1 (see further below), show no phenotypic deviation from the wild-type animal [22]. The VEGFA120 isoform does not bind HSPGs or NRPs, and is therefore freely diffusable. Expression of VEGFA120 alone in the *Vegfa$^{120/120}$* mouse, however, leads to death shortly after birth in 50 percent of the mice, due to bleeding and cardiovascular distress [22, 23]. Remaining mice have impaired postnatal myocardial angiogenesis and die from cardiac failure within 2 weeks of birth [22, 23]. Several VEGF isoforms also plays a role in bone formation. *Vegfa$^{120/120}$* mice show skeletal defects and impaired angiogenesis and endochondral bone formation [24, 25].

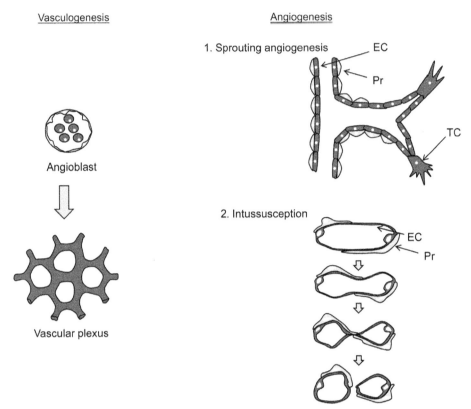

FIGURE 28.2 Endothelial cell development and formation of new vessels.
Vascular development is initiated when angioblasts differentiate to endothelial cells and assemble to form a vascular plexus through a process denoted vasculogenesis. Angiogenesis, the formation of new vessels form pre-existing ones, occurs either through formation of a vascular sprout guided by a leading tip cell (TC) or through splitting of vessels (intussusception). EC, endothelial cell; Pr, pericyte.

Mice expressing only VEGFA188, and thus lacking expression of soluble VEGFA isoforms, suffer from dwarfism, at least partially due to defects in vascularization of the epiphysis and impaired cartilage development [26].

VEGFR2 tyrosine phosphorylation sites have been identified and, apart from two positive regulatory tyrosine residues (Y1054 and Y1059), required for maximal kinase activity [27], there are three major phosphorylation sites involved in downstream signaling (see Figure 28.3). One is in the kinase insert sequence (the stretch of about 70 amino acid "non-kinase" residues that divides the VEGFR kinase domains in two parts; see Figure 28.1), at position Y951 (Y949 in the mouse). This site binds the adaptor molecule T cell-specific adaptor (TSAd), which becomes tyrosine phosphorylated in VEGF-treated cells [28, 29]. $Tsad^{-/-}$ mice survive development, but show decreased tumor growth and vascularization [29]. Tyrosine phosphorylated TSAd presents a binding site for the cytoplasmic tyrosine kinase Src, which regulates the organization of the actin cytoskeleton and cell migration. Mice deficient in expression of Src as well as the related tyrosine kinase Yes ($Src/Yes^{-/-}$ mice) survive to adulthood, but are unable to respond to VEGF with increased vascular permeability due to defects in regulation of endothelial cell junctions [30, 31]. This indicates that Src and Yes are critically involved

in the cross-talk between VEGFR2 and specialized endothelial junction proteins such as vascular endothelial (VE)-cadherin, to mediate opening of cell–cell junctions and the subsequent increase in vascular permeability.

A second main tyrosine phosphorylation site in VEGFR2 is at Y1175 (Y1173 in the mouse), in the C-terminal tail. Phosphorylation at this site allows binding and activation of phosphatidylinositol-specific phospholipase C (PLC)γ [32]. PLCγ hydrolyzes phosphatidylinositol 4,5 bisphosphate (PI-4,5-P2), a plasma membrane lipid. This results in generation of inositol 1,4,5-P3 and diacylglycerol, which leads to release of Ca^{2+} from intracellular stores and activation of protein kinase C (PKC), respectively. Many growth factor receptors transduce signals to activation of Erk and to cell proliferation via the small GTPase Ras. However, in the case of VEGFR-2, PLCγ activation leads to a Ras-independent activation of extracellular-regulated kinase (Erk)1/2 via PKC, critical for transducing VEGF signals for proliferation [33]. Elimination of the Y1173 phosphorylation site in mice ($Vegfr2^{Y1173F}$) leads to embryonic lethality at E8.5, with a phenotype very similar to that of the complete $Vegfr2$ knockout [34]. This is compatible with a role for Y1173 and PLCγ in endothelial cell proliferation. However, Y1173 in VEGFR2 is also a binding site for the adaptor molecules Shb [35] and Sck/ShcB [36], which may contribute to the

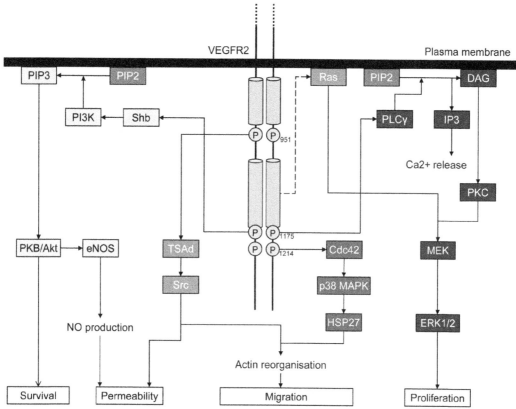

FIGURE 28.3 VEGFR2 signal transduction.
The intracellular domains of VEGF-bound, dimerized VEGFR2 are shown. Receptor dimerization leads to activation of the tyrosine kinase and subsequent autophosphorylation of major phosphorylation sites Y951, Y1175, and Y1214. This results in activation of specific downstream signaling molecules, which ultimately leads to the indicated biological responses in endothelial cells. VEGF-induced proliferation may involve activation of Ras, but this pathway has not been mapped (indicated by the dashed line) and it is unclear which phosphorylation sites are involved. See text for further information on signaling pathways.

phenotype of the $Vegfr2^{Y1173F}$ mouse. The central role of the PLCγ pathway in endothelial cell function is corroborated by the phenotype of $Plcg^{-/-}$ mice. These animals show lack of erythropoiesis and diminished vasculogenesis, and die during early embryogenesis at E9.5–10.5 [37]. Inactivation of the zebrafish (*Danio rerio*) *plcg* gene, on the other hand, leads to loss in arterial specification [38].

Another critical signal transduction pathway activated downstream of Y1173 involves activation of phosphoinositol 3′ kinase (PI3K). There are several PI3K isoforms, of which p110a has been shown to be critical for vascular development [39]. PI3K in turn mediates activation of the serine/threonine kinase Akt/PKB, which is essential in VEGFA-induced endothelial cell survival. Akt has also been implicated in regulation of endothelial permeability through activation of endothelial nitric oxide (eNOS) synthase and subsequent nitric oxide production [40–42]. $Akt1^{-/-}$ mice are viable, but show impaired pathological angiogenesis, characterized by defects in vessel maturation and increased vascular permeability [43].

Interestingly, a balance in the activities of the PI3K and PLCγ pathways appears to direct the development of arteries. Activation of Erk1/2 downstream of PLCγ is required for arterial development, and this pathway is opposed by activation of PI3K [44]. Erk1/2 activity may be essential in VEGF-mediated induction of Notch signaling pathways, in an as yet unidentified circuit (for reviews, see [45, 46]). Notch ligands and receptors have also been implicated in regulating arterial development downstream of VEGF/VEGFR-2 (see [46]).

The third autophosphorylation site, at Y1214 (Y1212 in the mouse), has been shown to regulate sequential activation of CDC42, p38 MAPK, and Hsp27 involved in VEGF-induced actin reorganization [47–49]. However, mice expressing a mutation at this site ($Vegfr2^{Y1212F}$) develop normally without vascular defects [34]. Changes in adult physiological or pathological angiogenesis in these mice have not yet been reported.

The central role of VEGFR2 signaling in vascular and hematopoietic development has been confirmed in small animal models of angiogenesis. Zebrafish is a tropical fish that survives the first day of development in the absence of blood circulation due to passive diffusion of oxygen, making it an excellent model for studying genes critically involved in vascular development. Furthermore, by injection in the zebrafish oocyte of "gene-specific morpholinos," the role of these genes in vascular development can easily and quickly be evaluated. See [10] for a review on the use of small animal models in vascular biology.

Morpholino-mediated knockdown of one VEGFR2 ortholog (Kdra) in zebrafish led to impaired intrasegmental sprouting angiogenesis [50]. Subsequently another VEGFR2 ortholog (Kdrb) was found, and a combined knockdown of both genes resulted in a more severe vascular phenotype, suggesting that Kdra and Kdrb cooperate in regulating vascular development in the zebrafish [51]. Hematopoietic development, however, was not perturbed. The fruit fly *Drosophila melanogaster* lacks a proper vascular system, but represents a useful model to study hematopoietic development. The common VEGF/PDGF receptor, PVR, is involved in guidance of cell migration during *Drosophila* development. Mutation in PVR leads to increased apoptosis of embryonic hematopoietic cells, implying that PVR regulates survival of blood cells in the embryo [52].

VEGFR1 AND ITS LIGANDS IN VASCULAR DEVELOPMENT AND INFLAMMATORY RESPONSES

VEGFR1 (also denoted Flt1 in mouse) binds three ligands: VEGFA, VEGFB, and PlGF. VEGFR1 stands out from the other VEGFRs by being expressed in two variants; a full-length transmembrane form and a soluble form (sFlt1). sFlt1 encompasses the extracellular ligand-binding domain of VEGFR1, but is devoid of the transmembrane and intracellular part, including the kinase domain (see Figure 28.1). sFlt1 acts as a decoy by sequestering VEGF. It is expressed during development in the placenta. Expression of sFlt1 increases during progression of the pregnancy and is upregulated further in pre-eclampsia [53], a condition characterized by high blood pressure, proteinurea and, finally, renal failure. Inflammatory cells infiltrating the placenta may be a source of sFlt during progression of pre-eclampsia [54]. The full-length VEGFR1 variant is expressed on endothelial and hematopoietic precursor cells, monocytes/macrophages, and vascular endothelial cells (for references, see [55]).

Signal transduction by VEGFR1 has been challenging to dissect due to its poor kinase activity compared to the other VEGFRs. Mapping of VEGFR1 tyrosine phosphorylation sites shows lack of phosphorylation at conserved sites traditionally serving in positive regulation of kinase activity [56], which may explain the difficulty in inducing the VEGFR1 kinase.

Vegfr1 gene targeting leads to increased endothelial cell proliferation and embryonic lethality at E8.5–9 due to vessel occlusion. This phenotype implicates VEGFR1 as a negative regulator in vascular development [57, 58]. A plausible mechanism for the expansion of the endothelial cell pool may be the increased availability of VEGF to bind and activate VEGFR2.

To better understand the negative regulatory role of VEGFR1, recombinant mice expressing only sFlt1 (i.e., encompassing only the soluble extracellular VEGFR1 domain) were generated. The $Vegfr1^{TK/TM-/-}$ mice show a complex fate, with 50 percent of the embryos dying at E8.5–9 due to vascular malformations [59]. The remaining 50 percent survive to adulthood and develop normally.

Mice have also been created that express a tyrosine-kinase (TK)-deleted but membrane-anchored VEGFR1 variant, $Vegfr1^{TK-/-}$. These mice survive embryogenesis and show apparently normal vascular development [60]. These results strongly suggest that the membrane-fixed, ligand-binding region of VEGFR1 traps VEGF for the appropriate regulation of VEGF signaling in vascular endothelial cells during early embryogenesis.

VEGFR1 is expressed not only in endothelial cells but also on hematopoietic precursors, bone marrow-derived progenitors, and mature monocytes/macrophages. Even though VEGFR1 kinase activity appears dispensable for both endothelial and hematopoietic cell development, loss of TK activity leads to reduced monocyte/macrophage migration, which dampens inflammatory reactions in a number of pathological conditions, such as chronic inflammatory joint diseases, atherosclerosis, and cancer [61, 62].

The ligands for VEGFR1, VEGFA, VEGFB, and PlGF all transduce distinct biological responses. This may be attributable to recruitment of different co-receptors, to induction of different autophosphorylation sites in VEGFR1, or, in the case of PlGF, to induction of intramolecular cross-talk between VEGFR1 and -2 [63].

Mice lacking PlGF also survive embryonic development, but show reduced ability to cope with pathological conditions such as myocardial infarction, ischemia, and tumor growth [64]. The reduced capacity to regulate endothelial function during pathologies indicates distinct molecular mechanisms in regulation of endothelial cell function during vasculogenesis and angiogenesis.

Mice lacking expression of the VEGFR1-specific ligand VEGFB survive development but show cardiac abnormalities and pathologically enhanced growth of the cardiac muscle [65], independent of effects on the vasculature.

VEGFR3 AND ITS LIGANDS IN LYMPHATIC DEVELOPMENT

VEGFR3 (Flt4 in the mouse) is an important regulator of lymphatic development and function. The lymphatic vascular system is critical for reabsorption of extravasated fluid and dietary fat into the circulation; in addition, it contributes to the host immune defense [66]. VEGFR3 binds VEGFC and -D. It is expressed in lymphatic endothelial precursors and mature lymphendothelial capillaries, but also in blood vascular endothelial cells – for example, in tumor vessels and on monocytes/macrophages.

The positions of phosphorylated tyrosine residues in VEGFR3 have been mapped [7]; however, their role in signal transduction has not yet been addressed using *in vivo* strategies. Phosphorylation sites in VEGFR3 are used differently, dependent on whether receptor activation occurs in homodimers or in heterodimers with VEGFR2. The biology of such differences in phosphorylation patterns also remains to be deduced.

Lymphatic endothelial cells (LECs) that express VEGFR3 bud off and migrate away from the embryonic cardinal vein at E10.5 in response to a gradient of VEGFC, which is produced by nearby mesenchymal cells. The migrating LECs subsequently assemble into lymph sacs that extend through sprouting, to lay down the framework of the lymphatic system [67]. However, targeting of *Vegfr3* leads to embryonic death at E 9.5, prior to the establishment of the lymphatics, and is accompanied by defective remodeling and maturation of the primitive blood vascular plexus into larger vessels [68]. Targeted *Vegfc*$^{-/-}$ embryos lack lymphatic vessels and die prior to birth due to severe tissue edema [69]. Interestingly, targeting of the corresponding *Vegfc/Vegfr3* genes in zebrafish leads to loss of lymphatic development [70]. Targeting of mouse *Vegfd*, on the other hand, shows that it is dispensable for lymphatic endothelial cell development [71]. Overexpression of zVEGFD in zebrafish leads to misguidance of sprouting intrasegmental vessels, implicating a role for VEGFD in vascular development in zebrafish [72].

In *Xenopus laevis* tadpoles, morpholino-targeting of *vegfc* or *vegfr3* leads to lymphedema. Moreover, VEGFD is implicated in lymphatic endothelial cell migration during development of the tadpole lymphatic system [73]. The different roles of the VEGFR3 ligands in different model organisms may be related to their processing, and different abilities to interact also with VEGFR2.

Interestingly, VEGFR3 is the only VEGFR for which naturally occurring mutations have been identified [74]. Such mutations invariably lead to loss of function and lymphedema.

VEGF AND ITS CORECEPTORS IN MODULATION OF SIGNAL TRANSDUCTION

A co-receptor implies a molecule that binds to the ligand and supports or stabilizes the formation of a signaling complex. Thereby, downstream signal transduction may be modulated quantitatively and qualitatively. Co-receptors lack intrinsic enzymatic activity but may transmit signals independently of the receptor tyrosine kinase, through participation in other molecular complexes.

There are at least two co-receptors for the VEGF ligand–receptor complex, namely HSPGs and NRPs. HSPGs are composed of a protein backbone with an attachment of repeated units of sulfated glycosaminoglycans (GAGs) that form long, linear sugar chains. GAG sulfation confers a net negative charge, which allows binding to many different growth modulatory factors [75]. In cells that lack heparan sulfate completely, or express defective heparan sulfate with reduced degree of sulfation, there is no response to VEGF even though VEGF receptors are expressed [76]. The mode of presentation of HSPGs (on both endothelial cells and perivascular cells, or only on perivascular cells) determines the level and longevity of receptor activation [77].

NRPs (1 and 2) are transmembrane molecules with a short cytoplasmic tail, which lack intrinsic enzymatic activity [78]. NRPs were first identified as negative regulators of neuronal axon guidance through binding of members of the class 3 Semaphorin (Sema) family [79, 80]. NRP1 was subsequently shown to bind exon 7-containing VEGFA isoforms such as VEGFA165 [81]. It is noteworthy that NRP1 may be modified by chondroitin and heparan sulfation [82], potentially allowing binding of VEGF to the protein core as well as to the HS side-chains. Through these interactions, NRPs become an integral part of the VEGF/VEGFR signaling complex. Whereas NRP1 is engaged in VEGFR2 signaling, NRP2 interacts with the VEGFC/VEGFR3 signaling complex in lymphendothelial cells [83].

NRPs play an essential role in vascular development. *Nrp1* knockout mice die around embryonic day E12.5 to E13.5, displaying abnormal axonal networks and defects in embryonic and yolk sac vascular formation [84, 85]. The knockout embryos have abnormal cardiovascular development, such as agenesis of the brachial arch-related great vessels, dorsal aorta, and transposition of aortic arches. Overexpression of *Nrp1* in mice also confers vascular defects [86]. These mice die *in utero* with morphological deformities, an excess of capillaries and blood vessels, dilation of blood vessels, hemorrhage, and malformed hearts.

Nrp2-deficient mice display normal development of blood vessels [87, 88], as well as of larger collecting lymphatic vessels, but exhibit a severe reduction in small lymphatic vessels and capillaries [89]. Double *Nrp1Nrp2* knockout mice have a more severe vascular phenotype than the individual knockouts, and die at E8.5 with defects reminiscent of those in *Vegfa/Vegfr2* knockout mice [90].

Endothelial-specific *Nrp1* knockout mice emphasize the essential role for NRP1 in cardiovascular development. These mice die at mid-to-late embryonic development, due to a poorly branched vasculature and multiple defects in the major arteries and failure of septation of the major cardiac outflow tract [91]. Knock-in mice expressing a mutant NRP1 that has lost the ability to bind semaphorins but still can bind VEGFA survive to birth and show no cardiovascular defects [91].

The zebrafish has two nrp*1* (*nrp1a* and *b*) and two nrp*2* (*nrp2a* and *b*) genes [92]. During formation of the vascular system, *nrp1a* is first expressed by tail angioblasts toward the end of somitogenesis, and by 48 hours post-fertilization (hpf) expression becomes more widespread to include the major trunk vessels as well as intersomitic vessels. Morpholino-induced silencing of *nrp1* in the zebrafish does not disturb vasculogenesis, but leads to loss, or anarchic sprouting, of new capillaries from pre-existing intersomitic vessels [93].

Another important binding partner of the VEGFR2, although by definition and function not a genuine co-receptor, is VE-cadherin. VE-cadherin is a strictly endothelial-specific adhesion molecule located at endothelial cell adherens junctions. VE-cadherin-mediated adhesion is essential for the control of vascular permeability and leukocyte extravasation [94]. Deficiency or truncation of VE-cadherin induces endothelial apoptosis and abolished transmission of the endothelial survival signal by VEGFA to the Akt kinase and Bcl2, via reduced complex formation with VEGFR2, beta-catenin, and PI3K, leading to embryonic death at day E9.5 [95].

CONCLUSIONS

Dissection of the biology of the VEGF/VEGFR family of ligands and receptors in genetic models such as mouse, zebrafish, and tadpoles shows a remarkable degree of biological conservation. In most cases, gene targeting leads to arrested or impaired vascular development and embryonic lethality. Interesting exceptions are mice lacking PlGF and VEGFB, which are viable but show defects in pathological angiogenesis. This implies that developmental angiogenesis and pathological angiogenesis occur, at least partially, through distinct mechanisms. Vascular development is likely to result from a pre-determined genetic program that ensures an optimal blood vessel density in each developing organ. Pathological angiogenesis, on the other hand, appears to be chaotic and largely determined by the bioavailability of VEGFs in the microenvironment.

Another fascinating aspect of research performed to date is the fact that VEGFA isoforms VEGFA121, VEGFA165, and VEGFA189, acting on the same receptor, VEGFR2, have different biology. It is possible that different co-receptors recruited into the signaling complex qualitatively modify the output signal. Since co-receptor silencing in many respects phenocopies those seen as a result of VEGF family ligand and receptor silencing, these pathways are likely to be linked.

It is vital to validate VEGFR phosphorylation sites and their functions *in vivo*, both for understanding and controlling the biology. Pertinent questions include: which signaling pathways regulate developmental processes such as precursor migration, plexus formation, and three-dimensional organization of endothelial cells to form the vascular tube: Also, what are the principal differences between neonatal and adult angiogenesis? Such knowledge will be the key to the development of high-quality novel therapies aimed to create blood vessels for, for example, tissue regeneration as well as anti-angiogenic strategies to treat cancer and other diseases characterized by excess angiogenesis.

ACKNOWLEDGEMENTS

The authors acknowledge funding for their work from the following sources: The Swedish Cancer Society, the Swedish Research Council, the Novo Nordisk foundation, Svenska Sällskapet för Medicinsk Forskning, the Magnus Bergvall Foundation and the EU6th Framework Integrated Project Lymphangiogenomics. Laurens A. van Meeteren was supported by a fellowship from the Dutch Cancer Society.

REFERENCES

1. Muller YA, Christinger HW, Keyt BA, de Vos AM. The crystal structure of vascular endothelial growth factor (VEGF) refined to 1.93 A resolution: multiple copy flexibility and receptor binding. *Structure* 1997;**5**:1325–38.

2. Shweiki D, Itin A, Soffer D, Keshet E. Vascular endothelial growth factor induced by hypoxia may mediate hypoxia-initiated angiogenesis. *Nature* 1992;**359**:843–5.

3. Takahashi H, Shibuya M. The vascular endothelial growth factor (VEGF)/VEGF receptor system and its role under physiological and pathological conditions. *Clin Sci (Lond)* 2005;**109**:227–41.

4. Suto K, Yamazaki Y, Morita T, Mizuno H. Crystal structures of novel vascular endothelial growth factors (VEGF) from snake venoms: insight into selective VEGF binding to kinase insert domain-containing receptor but not to fms-like tyrosine kinase-1. *J Biol Chem* 2005;**280**:2126–31.

5. Wood W, Faria C, Jacinto A. Distinct mechanisms regulate hemocyte chemotaxis during development and wound healing in *Drosophila melanogaster*. *J Cell Biol* 2006;**173**:405–16.

6. Ruch C, Skiniotis G, Steinmetz MO, Walz T, Ballmer-Hofer K. Structure of a VEGF-VEGF receptor complex determined by electron microscopy. *Nat Struct Mol Biol* 2007;**14**:249–50.

7. Dixelius J, Makinen T, Wirzenius M, Karkkainen MJ, Wernstedt C, Alitalo K, Claesson-Welsh L. Ligand-induced vascular endothelial growth factor receptor-3 (VEGFR-3) heterodimerization with VEGFR-2 in primary lymphatic endothelial cells regulates tyrosine phosphorylation sites. *J Biol Chem* 2003;**278**:40,973–40,979.

8. Huang K, Andersson C, Roomans GM, Ito N, Claesson-Welsh L. Signaling properties of VEGF receptor-1 and -2 homo- and heterodimers. *Intl J Biochem Cell Biol* 2001;**33**:315–24.

9. Nagy JA, Benjamin L, Zeng H, Dvorak AM, Dvorak HF. Vascular permeability, vascular hyperpermeability and angiogenesis. *Angiogenesis* 2008. in press.

10. Zacchigna S, Ruiz de Almodovar C, Carmeliet P. Similarities between angiogenesis and neural development: what small animal models can tell us. *Curr Topics Dev Biol* 2008;**80**:1–55.

11. Schmidt A, Brixius K, Bloch W. Endothelial precursor cell migration during vasculogenesis. *Circ Res* 2007;**101**:125–36.

12. Bertolini F, Shaked Y, Mancuso P, Kerbel RS. The multifaceted circulating endothelial cell in cancer: towards marker and target identification. *Nat Rev Cancer* 2006;**6**:835–45.

13. Gerhardt H, Golding M, Fruttiger M, Ruhrberg C, Lundkvist A, Abramsson A, Jeltsch M, Mitchell C, Alitalo K, Shima D, Betsholtz C. VEGF guides angiogenic sprouting utilizing endothelial tip cell filopodia. *J Cell Biol* 2003;**161**:1163–77.

14. Siekmann AF, Covassin L, Lawson ND. Modulation of VEGF signalling output by the Notch pathway. *Bioessays* 2008;**30**:303–13.

15. Burri PH, Hlushchuk R, Djonov V. Intussusceptive angiogenesis: its emergence, its characteristics, and its significance. *Dev Dyn* 2004;**231**:474–88.

16. Schuh AC, Faloon P, Hu QL, Bhimani M, Choi K. In vitro hematopoietic and endothelial potential of flk-1(−/−) embryonic stem cells and embryos. *Proc Natl Acad Sci USA* 1999;**96**:2159–64.

17. Li X, Edholm D, Lanner F, Breier G, Farnebo F, Dimberg A, Claesson-Welsh L. Lentiviral rescue of vascular endothelial growth factor receptor-2 expression in flk1−/− embryonic stem cells shows early priming of endothelial precursors. *Stem Cells* 2007;**25**:2987–95.

18. Shalaby F, Rossant J, Yamaguchi TP, Gertsenstein M, Wu XF, Breitman ML, Schuh AC. Failure of blood-island formation and vasculogenesis in Flk-1-deficient mice. *Nature* 1995;**376**:62–6.

19. Carmeliet P, Ferreira V, Breier G, Pollefeyt S, Kieckens L, Gertsenstein M, Fahrig M, Vandenhoeck A, Harpal K, Eberhardt C, Declercq C, Pawling J, Moons L, Collen D, Risau W, Nagy A. Abnormal blood vessel development and lethality in embryos lacking a single VEGF allele. *Nature* 1996;**380**:435–9.

20. Jain RK, Duda DG, Clark JW, Loeffler JS. Lessons from phase III clinical trials on anti-VEGF therapy for cancer. *Nat Clin Pract Oncol* 2006;**3**:24–40.

21. Ferrara N, Carver-Moore K, Chen H, Dowd M, Lu L, O'Shea KS, Powell-Braxton L, Hillan KJ, Moore MW. Heterozygous embryonic lethality induced by targeted inactivation of the VEGF gene. *Nature* 1996;**380**:439–42.

22. Stalmans I, Ng YS, Rohan R, Fruttiger M, Bouche A, Yuce A, Fujisawa H, Hermans B, Shani M, Jansen S, Hicklin D, Anderson DJ, Gardiner T, Hammes HP, Moons L, Dewerchin M, Collen D, Carmeliet P, D'Amore PA. Arteriolar and venular patterning in retinas of mice selectively expressing VEGF isoforms. *J Clin Invest* 2002;**109**:327–36.

23. Carmeliet P, Ng YS, Nuyens D, Theilmeier G, Brusselmans K, Cornelissen I, Ehler E, Kakkar VV, Stalmans I, Mattot V, Perriard JC, Dewerchin M, Flameng W, Nagy A, Lupu F, Moons L, Collen D, D'Amore PA, Shima DT. Impaired myocardial angiogenesis and ischemic cardiomyopathy in mice lacking the vascular endothelial growth factor isoforms VEGF164 and VEGF188. *Nat Med* 1999;**5**:495–502.

24. Zelzer E, McLean W, Ng YS, Fukai N, Reginato AM, Lovejoy S, D'Amore PA, Olsen BR. Skeletal defects in VEGF(120/120) mice reveal multiple roles for VEGF in skeletogenesis. *Development* 2002;**129**:1893–904.

25. Maes C, Carmeliet P, Moermans K, Stockmans I, Smets N, Collen D, Bouillon R, Carmeliet G. Impaired angiogenesis and endochondral bone formation in mice lacking the vascular endothelial growth factor isoforms VEGF164 and VEGF188. *Mech Dev* 2002;**111**:61–73.

26. Maes C, Stockmans I, Moermans K, Van Looveren R, Smets N, Carmeliet P, Bouillon R, Carmeliet G. Soluble VEGF isoforms are essential for establishing epiphyseal vascularization and regulating chondrocyte development and survival. *J Clin Invest* 2004;**113**:188–99.

27. Dougher M, Terman BI. Autophosphorylation of KDR in the kinase domain is required for maximal VEGF-stimulated kinase activity and receptor internalization. *Oncogene* 1999;**18**:1619–27.

28. Wu LW, Mayo LD, Dunbar JD, Kessler KM, Ozes ON, Warren RS, Donner DB. VRAP is an adaptor protein that binds KDR, a receptor for vascular endothelial cell growth factor. *J Biol Chem* 2000;**275**:6059–62.

29. Matsumoto T, Bohman S, Dixelius J, Berge T, Dimberg A, Magnusson P, Wang L, Wikner C, Qi JH, Wernstedt C, Wu J, Bruheim S, Mugishima H, Mukhopadhyay D, Spurkland A, Claesson-Welsh L. VEGF receptor-2 Y951 signaling and a role for the adapter molecule TSAd in tumor angiogenesis. *EMBO J* 2005;**24**:2342–53.

30. Eliceiri BP, Paul R, Schwartzberg PL, Hood JD, Leng J, Cheresh DA. Selective requirement for Src kinases during VEGF-induced angiogenesis and vascular permeability. *Mol Cell* 1999;**4**:915–24.

31. Weis S, Shintani S, Weber A, Kirchmair R, Wood M, Cravens A, McSharry H, Iwakura A, Yoon YS, Himes N, Burstein D, Doukas J, Soll R, Losordo D, Cheresh D. Src blockade stabilizes a Flk/cadherin complex, reducing edema and tissue injury following myocardial infarction. *J Clin Invest* 2004;**113**:885–94.

32. Takahashi T, Yamaguchi S, Chida K, Shibuya M. A single autophosphorylation site on KDR/Flk-1 is essential for VEGF-A-dependent activation of PLC-gamma and DNA synthesis in vascular endothelial cells. *EMBO J* 2001;**20**:2768–78.

33. Takahashi T, Ueno H, Shibuya M. VEGF activates protein kinase C-dependent, but Ras-independent Raf-MEK-MAP kinase pathway for DNA synthesis in primary endothelial cells. *Oncogene* 1999;**18**:2221–31.

34. Sakurai Y, Ohgimoto K, Kataoka Y, Yoshida N, Shibuya M. Essential role of Flk-1 (VEGF receptor 2) tyrosine residue 1173 in vasculogenesis in mice. *Proc Natl Acad Sci USA* 2005;**102**:1076–81.

35. Holmqvist K, Cross MJ, Rolny C, Hagerkvist R, Rahimi N, Matsumoto T, Claesson-Welsh L, Welsh M. The adaptor protein shb binds to tyrosine 1175 in vascular endothelial growth factor (VEGF) receptor-2 and regulates VEGF-dependent cellular migration. *J Biol Chem* 2004;**279**:22,267–22,275.

36. Warner AJ, Lopez-Dee J, Knight EL, Feramisco JR, Prigent SA. The Shc-related adaptor protein, Sck, forms a complex with the vascular-endothelial-growth-factor receptor KDR in transfected cells. *Biochem J* 2000;**347**:501–9.

37. Liao HJ, Kume T, McKay C, Xu MJ, Ihle JN, Carpenter G. Absence of erythrogenesis and vasculogenesis in Plcg1-deficient mice. *J Biol Chem* 2002;**277**:9335–41.

38. Lawson ND, Mugford JW, Diamond BA, Weinstein BM. phospholipase C gamma-1 is required downstream of vascular endothelial growth factor during arterial development. *Genes Dev* 2003;**17**:1346–51.

39. Graupera M, Guillermet-Guibert J, Foukas LC, Phng LK, Cain RJ, Salpekar A, Pearce W, Meek S, Millan J, Cutillas PR, Smith AJ, Ridley AJ, Ruhrberg C, Gerhardt H, Vanhaesebroeck B. Angiogenesis selectively requires the p110alpha isoform of PI3K to control endothelial cell migration. *Nature* 2008;**453**:662–6.

40. Fujio Y, Walsh K. Akt mediates cytoprotection of endothelial cells by vascular endothelial growth factor in an anchorage-dependent manner. *J Biol Chem* 1999;**274**:16,349–16,354.

41. Dimmeler S, Fleming I, Fisslthaler B, Hermann C, Busse R, Zeiher AM. Activation of nitric oxide synthase in endothelial cells by Akt-dependent phosphorylation. *Nature* 1999;**399**:601–5.

42. Fulton D, Gratton JP, McCabe TJ, Fontana J, Fujio Y, Walsh K, Franke TF, Papapetropoulos A, Sessa WC. Regulation of endothelium-derived nitric oxide production by the protein kinase Akt. *Nature* 1999;**399**:597–601.

43. Chen J, Somanath PR, Razorenova O, Chen WS, Hay N, Bornstein P, Byzova TV. Akt1 regulates pathological angiogenesis, vascular maturation and permeability *in vivo*. *Nat Med* 2005;**11**:1188–96.

44. Hong CC, Peterson QP, Hong JY, Peterson RT. Artery/vein specification is governed by opposing phosphatidylinositol-3 kinase and MAP kinase/ERK signaling. *Curr Biol* 2006;**16**:1366–72.

45. Lamont RE, Childs S. MAPping out arteries and veins. *Sci STKE* 2006;**2006**:pe39.

46. Weinstein BM, Lawson ND. Arteries, veins, Notch, and VEGF. *Cold Spring Harb Symp Quant Biol* 2002;**67**:155–62.

47. Lamalice L, Houle F, Jourdan G, Huot J. Phosphorylation of tyrosine 1214 on VEGFR2 is required for VEGF-induced activation of Cdc42 upstream of SAPK2/p38. *Oncogene* 2004;**23**:434–5.

48. McMullen ME, Bryant PW, Glembotski CC, Vincent PA, Pumiglia KM. Activation of p38 has opposing effects on the proliferation and migration of endothelial cells. *J Biol Chem* 2005;**280**:20,995–21,003.

49. Rousseau S, Houle F, Landry J, Huot J. p 38 MAP kinase activation by vascular endothelial growth factor mediates actin reorganization and cell migration in human endothelial cells. *Oncogene* 1997;**15**:2169–77.

50. Habeck H, Odenthal J, Walderich B, Maischein H, Schulte-Merker S. Analysis of a zebrafish VEGF receptor mutant reveals specific disruption of angiogenesis. *Curr Biol* 2002;**12**:1405–12.

51. Bahary N, Goishi K, Stuckenholz C, Weber G, Leblanc J, Schafer CA, Berman SS, Klagsbrun M, Zon LI. Duplicate VegfA genes and orthologues of the KDR receptor tyrosine kinase family mediate vascular development in the zebrafish. *Blood* 2007;**110**:3627–36.

52. Bruckner K, Kockel L, Duchek P, Luque CM, Rorth P, Perrimon N. The PDGF/VEGF receptor controls blood cell survival in *Drosophila*. *Dev Cell* 2004;**7**:73–84.

53. Maynard S, Epstein FH, Karumanchi SA. Preeclampsia and angiogenic imbalance. *Annu Rev Med* 2008;**59**:61–78.

54. Girardi G, Bulla R, Salmon JE, Tedesco F. The complement system in the pathophysiology of pregnancy. *Mol Immunol* 2006;**43**:68–77.

55. Olsson AK, Dimberg A, Kreuger J, Claesson-Welsh L. VEGF receptor signalling-in control of vascular function. *Nat Rev Mol Cell Biol* 2006;**7**:359–71.

56. Ito N, Wernstedt C, Engstrom U, Claesson-Welsh L. Identification of vascular endothelial growth factor receptor-1 tyrosine phosphorylation sites and binding of SH2 domain-containing molecules. *J Biol Chem* 1998;**273**:23,410–23,418.

57. Fong GH, Rossant J, Gertsenstein M, Breitman ML. Role of the Flt-1 receptor tyrosine kinase in regulating the assembly of vascular endothelium. *Nature* 1995;**376**:66–70.

58. Kearney JB, Ambler CA, Monaco KA, Johnson N, Rapoport RG, Bautch VL. Vascular endothelial growth factor receptor Flt-1 negatively regulates developmental blood vessel formation by modulating endothelial cell division. *Blood* 2002;**99**:2397–407.

59. Hiratsuka S, Nakao K, Nakamura K, Katsuki M, Maru Y, Shibuya M. Membrane fixation of vascular endothelial growth factor receptor 1 ligand-binding domain is important for vasculogenesis and angiogenesis in mice. *Mol Cell Biol* 2005;**25**:346–54.

60. Hiratsuka S, Minowa O, Kuno J, Noda T, Shibuya M. Flt-1 lacking the tyrosine kinase domain is sufficient for normal development and angiogenesis in mice. *Proc Natl Acad Sci USA* 1998;**95**:9349–54.

61. Lyden D, Hattori K, Dias S, Costa C, Blaikie P, Butros L, Chadburn A, Heissig B, Marks W, Witte L, Wu Y, Hicklin D, Zhu Z, Hackett NR, Crystal RG, Moore MA, Hajjar KA, Manova K, Benezra R, Rafii S. Impaired recruitment of bone-marrow-derived endothelial and hematopoietic precursor cells blocks tumor angiogenesis and growth. *Nat Med* 2001;**7**:1194–201.

62. Luttun A, Tjwa M, Moons L, Wu Y, Angelillo-Scherrer A, Liao F, Nagy JA, Hooper A, Priller J, De Klerck B, Compernolle V, Daci E, Bohlen P, Dewerchin M, Herbert JM, Fava R, Matthys P, Carmeliet G, Collen D, Dvorak HF, Hicklin DJ, Carmeliet P. Revascularization of ischemic tissues by PlGF treatment, and inhibition of tumor angiogenesis, arthritis and atherosclerosis by anti-Flt1. *Nat Med* 2002;**8**:831–40.

63. Autiero M, Waltenberger J, Communi D, Kranz A, Moons L, Lambrechts D, Kroll J, Plaisance S, De Mol M, Bono F, Kliche S, Fellbrich G, Ballmer-Hofer K, Maglione D, Mayr-Beyrle U, Dewerchin M, Dombrowski S, Stanimirovic D, Van Hummelen P, Dehio C, Hicklin DJ, Persico G, Herbert JM, Shibuya M, Collen D, Conway EM, Carmeliet P. Role of PlGF in the intra- and intermolecular cross talk between the VEGF receptors Flt1 and Flk1. *Nat Med* 2003;**9**:936–43.

64. Carmeliet P, Moons L, Luttun A, Vincenti V, Compernolle V, De Mol M, Wu Y, Bono F, Devy L, Beck H, Scholz D, Acker T, DiPalma T, Dewerchin M, Noel A, Stalmans I, Barra A, Blacher S, Vandendriessche T, Ponten A, Eriksson U, Plate KH, Foidart JM, Schaper W, Charnock-Jones DS, Hicklin DJ, Herbert JM, Collen D, Persico MG. Synergism between vascular endothelial growth factor and placental growth factor contributes to angiogenesis and plasma extravasation in pathological conditions. *Nat Med* 2001;**7**:575–83.

65. Aase K, von Euler G, Li X, Ponten A, Thoren P, Cao R, Cao Y, Olofsson B, Gebre-Medhin S, Pekny M, Alitalo K, Betsholtz C, Eriksson U. Vascular endothelial growth factor-B-deficient mice display an atrial conduction defect. *Circulation* 2001;**104**:358–64.

66. Saharinen P, Tammela T, Karkkainen MJ, Alitalo K. Lymphatic vasculature: development, molecular regulation and role in tumor metastasis and inflammation. *Trends Immunol* 2004;**25**:387–95.

67. Tammela T, Petrova TV, Alitalo K. Molecular lymphangiogenesis: new players. *Trends Cell Biol* 2005;**15**:434–41.

68. Dumont DJ, Jussila L, Taipale J, Lymboussaki A, Mustonen T, Pajusola K, Breitman M, Alitalo K. Cardiovascular failure in mouse embryos deficient in VEGF receptor-3. *Science* 1998;**282**:946–9.

69. Karkkainen MJ, Haiko P, Sainio K, Partanen J, Taipale J, Petrova TV, Jeltsch M, Jackson DG, Talikka M, Rauvala H, Betsholtz C, Alitalo K. Vascular endothelial growth factor C is required for sprouting of the first lymphatic vessels from embryonic veins. *Nat Immunol* 2004;**5**:74–80.

70. Kuchler AM, Gjini E, Peterson-Maduro J, Cancilla B, Wolburg H, Schulte-Merker S. Development of the zebrafish lymphatic system requires VEGFC signaling. *Curr Biol* 2006;**16**:1244–8.

71. Baldwin ME, Halford MM, Roufail S, Williams RA, Hibbs ML, Grail D, Kubo H, Stacker SA, Achen MG. Vascular endothelial growth factor D is dispensable for development of the lymphatic system. *Mol Cell Biol* 2005;**25**:2441–9.

72. Song M, Yang H, Yao S, Ma F, Li Z, Deng Y, Deng H, Zhou Q, Lin S, Wei Y. A critical role of vascular endothelial growth factor D in zebrafish embryonic vasculogenesis and angiogenesis. *Biochem Biophys Res Commun* 2007;**357**:924–30.

73. Ny A, Koch M, Vandevelde W, Schneider M, Fischer C, Diez-Juan A, Neven E, Geudens I, Maity S, Moons L, Plaisance S, Lambrechts D, Carmeliet P, Dewerchin M. Role of VEGF-D and VEGFR-3 in developmental lymphangiogenesis, a chemicogenetic study in Xenopus tadpoles. *Blood* 2008.

74. Irrthum A, Karkkainen MJ, Devriendt K, Alitalo K, Vikkula M. Congenital hereditary lymphedema caused by a mutation that inactivates VEGFR3 tyrosine kinase. *Am J Hum Genet* 2000;**67**:295–301.

75. Esko JD, Selleck SB. Order out of chaos: assembly of ligand binding sites in heparan sulfate. *Annu Rev Biochem* 2002;**71**:435–71.

76. Dougher AM, Wasserstrom H, Torley L, Shridaran L, Westdock P, Hileman RE, Fromm JR, Anderberg R, Lyman S, Linhardt RJ, Kaplan J, Terman BI. Identification of a heparin binding peptide on the extracellular domain of the KDR VEGF receptor. *Growth Factors* 1997;**14**:257–68.

77. Jakobsson L, Kreuger J, Holmborn K, Lundin L, Eriksson I, Kjellen L, Claesson-Welsh L. Heparan sulfate in trans potentiates VEGFR-mediated angiogenesis. *Dev Cell* 2006;**10**:625–34.

78. Fujisawa H, Kitsukawa T, Kawakami A, Takagi S, Shimizu M, Hirata T. Roles of a neuronal cell-surface molecule, neuropilin, in nerve fiber fasciculation and guidance. *Cell Tissue Res* 1997;**290**:465–70.

79. Kolodkin AL, Levengood DV, Rowe EG, Tai YT, Giger RJ, Ginty DD. Neuropilin is a semaphorin III receptor. *Cell* 1997;**90**:753–62.

80. He Z, Tessier-Lavigne M. Neuropilin is a receptor for the axonal chemorepellent Semaphorin III. *Cell* 1997;**90**:739–51.

81. Soker S, Takashima S, Miao HQ, Neufeld G, Klagsbrun M. Neuropilin-1 is expressed by endothelial and tumor cells as an isoform-specific receptor for vascular endothelial growth factor. *Cell* 1998;**92**:735–45.

82. Shintani Y, Takashima S, Asano Y, Kato H, Liao Y, Yamazaki S, Tsukamoto O, Seguchi O, Yamamoto H, Fukushima T, Sugahara K, Kitakaze M, Hori M. Glycosaminoglycan modification of neuropilin-1 modulates VEGFR2 signaling. *Embo J* 2006;**25**:3045–55.

83. Karpanen T, Heckman CA, Keskitalo S, Jeltsch M, Ollila H, Neufeld G, Tamagnone L, Alitalo K. Functional interaction of VEGF-C and VEGF-D with neuropilin receptors. *FASEB J* 2006;**20**:1462–72.

84. Kawasaki T, Kitsukawa T, Bekku Y, Matsuda Y, Sanbo M, Yagi T, Fujisawa H. A requirement for neuropilin-1 in embryonic vessel formation. *Development* 1999;**126**:4895–902.

85. Kitsukawa T, Shimizu M, Sanbo M, Hirata T, Taniguchi M, Bekku Y, Yagi T, Fujisawa H. Neuropilin-semaphorin III/D-mediated chemorepulsive signals play a crucial role in peripheral nerve projection in mice. *Neuron* 1997;**19**:995–1005.

86. Kitsukawa T, Shimono A, Kawakami A, Kondoh H, Fujisawa H. Overexpression of a membrane protein, neuropilin, in chimeric mice causes anomalies in the cardiovascular system, nervous system and limbs. *Development* 1995;**121**:4309–18.

87. Giger RJ, Cloutier JF, Sahay A, Prinjha RK, Levengood DV, Moore SE, Pickering S, Simmons D, Rastan S, Walsh FS, Kolodkin AL, Ginty DD, Geppert M. Neuropilin-2 is required *in vivo* for selective axon guidance responses to secreted semaphorins. *Neuron* 2000;**25**:29–41.

88. Chen H, Bagri A, Zupicich JA, Zou Y, Stoeckli E, Pleasure SJ, Lowenstein DH, Skarnes WC, Chedotal A, Tessier-Lavigne M. Neuropilin-2 regulates the development of selective cranial and sensory nerves and hippocampal mossy fiber projections. *Neuron* 2000;**25**:43–56.

89. Yuan L, Moyon D, Pardanaud L, Breant C, Karkkainen MJ, Alitalo K, Eichmann A. Abnormal lymphatic vessel development in neuropilin 2 mutant mice. *Development* 2002;**129**:4797–806.

90. Takashima S, Kitakaze M, Asakura M, Asanuma H, Sanada S, Tashiro F, Niwa H, Miyazaki Ji J, Hirota S, Kitamura Y, Kitsukawa T, Fujisawa H, Klagsbrun M, Hori M. Targeting of both mouse neuropilin-1 and neuropilin-2 genes severely impairs developmental yolk sac and embryonic angiogenesis. *Proc Natl Acad Sci USA* 2002;**99**:3657–62.

91. Gu C, Rodriguez ER, Reimert DV, Shu T, Fritzsch B, Richards LJ, Kolodkin AL, Ginty DD. Neuropilin-1 conveys semaphorin and VEGF signaling during neural and cardiovascular development. *Dev Cell* 2003;**5**:45–57.

92. Yu HH, Houart C, Moens CB. Cloning and embryonic expression of zebrafish neuropilin genes. *Gene Expr Patterns* 2004;**4**:371–8.

93. Lee P, Goishi K, Davidson AJ, Mannix R, Zon L, Klagsbrun M. Neuropilin-1 is required for vascular development and is a mediator of VEGF-dependent angiogenesis in zebrafish. *Proc Natl Acad Sci USA* 2002;**99**:10,470–10,475.

94. Weber C, Fraemohs L, Dejana E. The role of junctional adhesion molecules in vascular inflammation. *Nat Rev Immunol* 2007;**7**:467–77.

95. Carmeliet P, Lampugnani MG, Moons L, Breviario F, Compernolle V, Bono F, Balconi G, Spagnuolo R, Oostuyse B, Dewerchin M, Zanetti A, Angellilo A, Mattot V, Nuyens D, Lutgens E, Clotman F, de Ruiter MC, Gittenberger-de Groot A, Poelmann R, Lupu F, Herbert JM, Collen D, Dejana E. Targeted deficiency or cytosolic truncation of the VE-cadherin gene in mice impairs VEGF-mediated endothelial survival and angiogenesis. *Cell* 1999;**98**:147–57.

96. Bellomo D, Headrick JP, Silins GU, Paterson CA, Thomas PS, Gartside M, Mould A, Cahill MM, Tonks ID, Grimmond SM, Townson S, Wells C, Little M, Cummings MC, Hayward NK, Kay GF. Mice lacking the vascular endothelial growth factor-B gene (Vegfb) have smaller hearts, dysfunctional coronary vasculature, and impaired recovery from cardiac ischemia. *Circ Res* 2000;**86**:E29–35.

97. Fong GH, Zhang L, Bryce DM, Peng J. Increased hemangioblast commitment, not vascular disorganization, is the primary defect in flt-1 knock-out mice. *Development* 1999;**126**:3015–25.

98. Hiratsuka S, Maru Y, Okada A, Seiki M, Noda T, Shibuya M. Involvement of Flt-1 tyrosine kinase (vascular endothelial growth factor receptor-1) in pathological angiogenesis. *Cancer Res* 2001;**61**:1207–13.

Chapter 29

BMPs in Development

Kelsey N. Retting[1] and Karen M. Lyons[2]

[1]*Department of Biological Chemistry, University of California, Los Angeles, California*

[2]*Department of Molecular, Cell and Developmental Biology, Department of Orthopaedic Surgery, University of California, Los Angeles, California*

INTRODUCTION

Bone morphogenetic proteins (BMPs) are members of the transforming growth factor beta (TGFβ) superfamily, whose other members include nodals, activins, anti-Mullarian hormone, and myostatins. In all organisms studied, BMPs are expressed at the earliest stages of embryogenesis and throughout adulthood, with diverse functions in cell fate, differentiation, proliferation, and survival in many tissues. BMPs are secreted proteins that in some circumstances act as morphogens because they can be expressed in gradients and can function in a concentration-dependent manner to direct gene expression. This chapter will provide an overview of BMP signal transduction and some of the conserved functions of BMP pathways in invertebrate and vertebrate development.

BMP SIGNAL TRANSDUCTION

Like all members of the TGFβ superfamily, BMPs transduce signals by binding to heteromeric complexes of serine/threonine kinase type I and type II receptors. BMP ligands are secreted and activated by proteolytic cleavage. Once activated, BMPs form homo- and heterodimers linked by disulfide bonds. Although differences in signaling activity between heterodimeric and homodimeric ligand complexes have been demonstrated [1, 2], it has not been possible to perform genetic studies to test whether homodimers and heterodimers have differing roles *in vivo* due to extensive functional redundancy and compensatory upregulation. BMPs can be structurally subdivided into three groups, BMP2/4, BMP5/6/7, and growth and differentiation factor (GDF)5/6/7. Although originally classified as a BMP family member, the structurally divergent BMP3

is an exception, and appears to antagonize both BMP and TGFβ/activin signaling [3, 4].

BMPs bind to type I receptors BMPRIA (ALK3), BMPRIB (ALK6), and ActRI (ALK2), and type II receptors BMPRII, ActRII, and ActRIIB. Upon ligand binding, type II receptors activate type I receptors via phosphorylation of serine/threonine residues on type I receptors, initiating signal transduction (reviewed in [5, 6]). *In vivo* studies have demonstrated that the loss of *Bmpr1a* leads to lethality at gastrulation [7]. *Bmpr1b*$^{-/-}$ mice, however, are viable [8]. Moreover, in the context of skeletal development, *Bmpr1b*$^{-/-}$ mice display defects in chondrogenesis restricted to distal phalanges [8, 9], while the cartilage specific knockout of *Bmpr1a* displays a generalized skeletal dysplasia [10]. These different effects may be due in part to distinct expression patterns in the developing condensations [11, 12]. Although these studies raise the possibility that type I receptor functions are unique, redundancy in cartilage is also evident. Overexpression of *Bmpr1a* can rescue the differentiation defect of chondrocytes in *Bmpr1b*$^{-/-}$ phalanges [13], and the combined loss of *Bmpr1a* and *Bmpr1b* in skeletal elements results in embryonic lethality due to the severe loss of endochondral bone formation [10, 14]. These studies highlight the fact that there are many factors regulating the specific functions of type I receptors, which include differential expression, ligand–receptor combinations, and the level of signal activation.

Canonical Smad signaling is believed to be the major transduction pathway of the TGFβ superfamily (Figure 29.1). Smad proteins can be divided into three groups: receptor Smads (R-Smads), co-Smad (Smad4), and inhibitory Smads (I-Smads 6 and 7). Canonical BMP signaling is defined by phosphorylation of receptor Smads 1, 5, and 8, whereas TGFβs and activins signal through receptor Smads 2 and 3 [5, 15]. Smads 1 and 5 are structurally related and

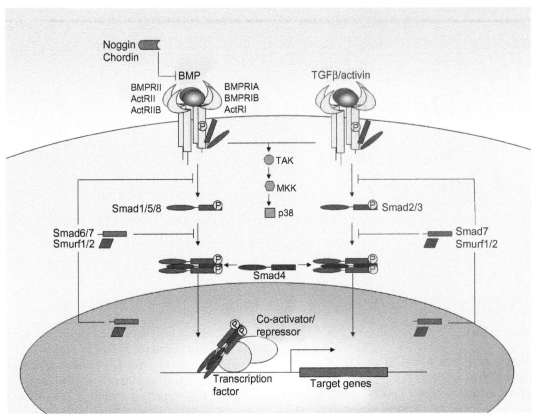

FIGURE 29.1 BMP signal transduction.
Heteromeric receptor complexes consisting of type I and type II receptors bind ligand dimers at the cell surface. Extracellular antagonists such as Noggin and Chordin can inhibit complex formation. Activated type I receptors phosphorylate receptor Smads at the carboxy terminus initiating the canonical pathway. Phosphorylated R-Smads can then complex with Co-Smad4 and translocate to the nucleus to initiate transcription of target genes. The BMP ligand–receptor complex can also activate p38 through TAK1. Inhibitory Smads 6 and 7 and Smurfs antagonize BMP signaling at the intracellular level.

transduce equivalent signals *in vitro* [6]. Both Smads 1 and 5 are essential for normal development in the early embryo [16–18], and the phenotypic similarity shared between *Smad1$^{+/-}$:Smad5$^{+/-}$* double heterozygous mutants and those lacking *Smad1* or *Smad5* further suggests equivalent signaling functions by *Smad1* and *Smad5* [19]. Smad8, however, is more divergent in structure and expression patterns, and is not required for normal development [19, 20]. The extent of functional redundancy among R-Smads *in vivo* is currently unknown, and, due to embryonic lethality of null mutations, resolution of this question will require the use of conditional knockout models.

R-Smads, which contain two globular Mad homology domains (MH1 and MH2) attached by a linker region, are phosphorylated by activated type I receptors on conserved serine residues at the carboxy terminus. Mice exhibiting a disruption in the C-terminal residues of Smad1 have severe developmental defects similar to several phenotypes of *Smad1*-null mice [21], confirming that the C-terminus contains the critical functional domain for transcriptional activation by BMP signaling, and initiates the majority of Smad1 functions during early embryonic development.

R-Smads subsequently complex with co-Smad4 and translocate to the nucleus, forming transcription complexes. It is generally believed that most canonical signaling events require complex formation [5]. Many studies have supported this view, including the finding that *Smad4$^{-/-}$* mice exhibit early embryonic lethality [22]. On the other hand, it has been demonstrated in both *Drosophila* and mammalian systems that several BMP-dependent developmental processes can occur in the absence of embryonic Smad4 [23–25]. Previous studies have suggested that Smad4 is required for R-Smad nuclear translocation [26]. However, there are also studies that indicate that Smad4 may not always be required for this activity. Smad1, located in the cytoplasm in the basal state, accumulates in the nucleus upon BMP stimulation in Smad4-null human colon carcinoma cells [27]. Moreover, the disruption of the Smad4 nuclear export signal results in normal development in transgenic mice [28], indicating that BMP signal transduction does not depend on Smad4 nucleocytoplasmic shuttling for all developmental processes.

Subcellular localization of Smad proteins can mediate BMP signal transduction. At the cytoplasmic membrane, endofin, an endosome-associated FYVE domain protein, has recently been demonstrated to anchor BMP R-Smads for phosphorylation by activated receptors [29]. Smad phosphorylation and nuclear accumulation are also mediated by actin binding proteins such as filamins and calponin 3, which can also bind Smads [30, 31]. The interaction of Smads and filamins has been developmentally implicated in several skeletal dysplasias [32]. Activity is also mediated by cytoplasmic membrane protein SANE and nuclear envelop protein MAN1, which share significant sequence similarity, can inhibit Smad phorphorylation, and can antagonize BMP signaling during *Xenopus* embryogenesis [33–35].

Mutating the linker region of Smad1 results in retention of the protein in the plasma membrane and cytoskeleton reorganization in primary cells of transgenic mice [21], suggesting that the linker region is also involved in Smad localization. R-Smads can be phosphorylated at the linker region by ERK/MAPK pathways, leading to the inhibition of R-Smad activation by C-terminal phosphorylation [36, 37]. The linker phosphorylation may restrict Smad activity by inducing polyubiquitination by Smurf1 [38]. These data suggest that BMP signaling can be antagonized at the level of R-Smad1/5/8 phosphorylation. The physiological relevance of this model was demonstrated by Pera and colleagues [39], who showed that FGF-mediated phosphorylation of the Smad1 linker region is essential to inhibit BMP activity during neural induction. This induction may be a major role in developmental systems in which opposing activities for BMP and FGF pathways have been demonstrated, including limb bud development, skeletal growth, and lung bud morphogenesis [10, 40, 41].

Although canonical Smad signaling is believed to be the predominant form of BMP signal transduction, BMP signaling has also been shown to activate alternative pathways. BMP ligands can signal via TGFβ activated kinase (TAK1), which induces a p38 mitogen-activated (MAPK) pathway. Little is known about how BMP-activated canonical and non-canonical pathways interact, and it is unclear whether these pathways cooperate or antagonize each other. Studies have indicated that the balance between Smad and p38 signaling may be influenced by ligand–receptor oligomerization complexes [42, 43], and by complex internalization through association with either clathrin-coated vesicles or caveolae [44]. *In vitro*, overexpression of *TAK1* in chondrocytes can mimic the stimulatory effects of BMPs, while dominant-negative *TAK1* expression inhibits this induction [45], suggesting cooperation between canonical Smad and TAK1 pathways. However, the nature of the interactions remains to be clarified, as TAK1 has been found to negatively interact with R-Smads, Smad4, and I-Smads in mesenchymal progenitor cells [46]. Similarly, transgenic mice constitutively active for MKK6, an upstream activator of p38, exhibit skeletal abnormalities including dwarfism and a reduction in chondrocyte proliferation and differentiation [47], consistent with the antagonism of BMP by p38 pathways. Taken together, these studies indicate that Smad

and p38-MAPK pathways can exhibit positive and negative interactions, although the extent of BMP-activated p38 signaling *in vivo* remains to be determined.

Cross-talk may also occur between TGFβ and BMP signaling. It has been shown that endogenous Smads1 and 5 can be activated in a dose- and time dependent manner by TGFβ as well as BMP stimulation [48–51]. TGFβs can complex with type I receptors ALK1 and ALK5 to transduce a signal through Smad2/3. TGFβ signaling through ALK1 has also been shown to activate both Smad2/3 and Smad1/5/8 pathways in endothelial cells [51, 52]. Embryonic null mutations in either ALK1, ALK5, or Smad5 have similar defects in angiogenesis, demonstrating that TGFβ exerts essential effects by directly activating BMP pathways in endothelial cells [53]. In addition to different heteromeric receptor combinations, TGFβ-induced activation of BMP receptor Smads may involve cross-talk with other pathways such Ras/MEK [6, 50].

EXTRACELLULAR AND INTRACELLULAR BMP ANTAGONISTS AND THE ESTABLISHMENT OF MORPHOGEN GRADIENTS

The threshold levels of morphogens give positional information to cells and activate differential gene expression patterns (reviewed in [54]). Extracellular antagonists appear to be the key architects of BMP gradients. In addition to the negative interactions of BMP and FGF pathways discussed above, regulation of the intensity and duration of the BMP signal is tightly controlled in the extracellular space by secreted antagonists including noggin, chordin, twisted gastrulation (Tsg), and members of the Dan family, such as gremlin (reviewed in [55]). These antagonists are grouped based on the presence of evolutionarily conserved cysteine-rich domains, and they inhibit BMP signaling by binding to ligands and preventing ligand–receptor interaction [56].

Much of what is known about BMP signaling components and the establishment of morphogen gradients was determined from studies of *Drosophila* development. In *Drosophila*, mutations in decapentaplegic (Dpp), which is homologous to BMP2/4, cause abnormal dorso-ventral (DV) patterning in the embryo [57]. Dpp is expressed uniformly in the dorsal blastomere embryo, and specifies the dorsal fate. Dpp activity is restricted into the dorsal-most region by a ventral to dorsal gradient of the antagonistic Sog, a chordin homolog expressed ventrally. The extracellular transport of ligands regulated by antagonists such as Sog and Tsg, may influence spatial domains of the gradient and signal stability [58, 59].

At the intracellular level, the BMP signal creates a step gradient of Smad responses by sharp transitions in phosphorylated Smad levels, which correlate to gene expression boundaries. This Smad activity gradient has been shown to pattern the dorsovental axis at the onset of grastrulation, and later to pattern the wing primordium [60, 61]. BMP canonical Smad signaling can be inhibited by inhibitory Smads 6 and 7 [5, 62]. These I-Smads can compete with R-Smads to bind to type I receptors, inhibiting R-Smad phosphorylation at the carboxy terminus. I-Smads can also prevent complex formation between R-Smads and Smad4, recruit E3 ubiquitin ligases (Smurfs 1 and 2), and target receptors for degradation [63–66]. I-Smad activity forms a feedback loop with BMP signaling, as both Smads 6 and 7 are activated by BMP signaling through R-Smad binding to the *Smad6* and *7* promoters [67–69]. I-Smads may be also be regulated by other pathways involved in development such as Wnts [70]. Thus, the activity of BMPs is tightly regulated at multiple levels, suggesting the strength of the signal is very important.

BMPS IN VERTEBRATE EMBRYO PATTERNING

The requirement for a gradient of BMP activity for dorsoventral patterning is conserved across a wide phylogenetic distance. During vertebrate embryo patterning, polarity is established similarly by the interaction of BMP and chordin to form the DV axis; however, the axis is inversed relative to that of invertebrates. *Xenopus* has been used as a model for many of these studies (reviewed in [71]) which demonstrate that BMPs promote ventralization of the mesoderm, repressing development of dorsal tissues including the neural tube and notochord. During gastrulation in *Xenopus*, BMP4 is found on the ventral side of the embryo in an opposing gradient to the dorsal expression of secreted BMP antagonists noggin and chordin [72–76]. BMP activity is reflected by the presence of C-terminal phosphorylated Smad1 (pSmad1), which is also found in the same ventral to dorsal gradient [77]. BMP4 inhibits neuralization and induces epidermis in ectodermal explant assays [72, 78], while morpholino knockdown of Chordin expression leads to a reduction in neural plate and CNS tissue with an expansion in ventral mesoderm [79]. Upon the loss of function of the three antagonists chordin, noggin, and follistatin, dorsal structures such as the neural plate fail to form at all, demonstrating cooperative and redundant functions among BMP antagonists [35]. In contrast, loss-of-function experiments simultaneously inactivating BMP2/4/7 and ADMP (a BMP-like anti-dorsalizing morphogenetic protein) show that the loss of BMP signals in both ventral and dorsal poles cause the entire ectoderm to become neural tissue [80]. This establishment of dorsal and ventral cell types may be mediated by both canonical and non-canonical pathways; overexpression of Smads 1 and 5, induce epidermis [81, 82], while dominant-negative TAK1 can block SMAD1/5-induced ventralization [83]. Inhibitory Smads 6 and 7 antagonize BMP-induced ventralization, and can induce formation of a secondary dorsal axis when misexpressed ventrally in *Xenopus* embryos [65, 84–86]. The

BMP gradient is further modulated by the level of pSmad1 in the nucleus, which regulates the transcription of dorsal and ventral genes such as Sizzled and Tolloid, altering antagonistic activity of Chordin via degradation [87].

Genetic studies in mammals also support a role for BMP signaling in early development (reviewed in [88]). In the mouse epiblast, or embryo proper, mesoderm cells arise from the primitive streak. BMPs regulate patterning of the epiblast, the position of the primitive streak as a posterior organizing center, and formation of mesoderm. For example, the BMPRIA knockout is embryonic lethal, resulting in reduced epiblast proliferation and no mesoderm formation [7]. Likewise, the loss of BMP4, which signals through BMPRIA, results in embryonic lethality at the egg cylinder stage, and little to no mesodermal differentiation [89]. Furthermore, embryos that lack Smad4 in both embryonic and extra-embryonic tissues fail to gastrulate [22].

More recent studies have utilized conditional gene ablation or overexpression models to examine BMP function during organogenesis. For example, conditional deletion of BMPR1A in the limb mesenchyme demonstrates that BMPs are essential for pattern formation in the limb. BMPR1A is required for apical ectodermal ridge (AER) development [90], distal outgrowth, antero-posterior, and dorsoventral patterning [91]. The same Prx1-Cre promoter-driven excision used to ablate BMPR1A in early limb bud mesenchyme has also been used to examine consequences of ligand deletion to limb patterning. A severe reduction in limb growth occurs in the absence of BMP2/4; however, only a mild phenotype is observed in the absence of BMP2/7 [92]. Although these results indicate different functions in limb growth, mesenchymal condensations and chondrocyte differentiation occur in both instances, indicating functional redundancy with remaining ligands at least at early stages.

BMP signaling is also required along the antero-posterior axis for somite differentiation. Compound mutants of *Bmp4* and *noggin* have axial patterning defects, demonstrating the interaction of BMP4 and noggin in patterning the mesoderm bilateral to the neural tube [93, 94]. The compound *Tsg;Bmp7* mutant displays a fusion of the distal limbs known as sirenomelia, indicating that posterior patterning of the mesoderm requires the interaction of BMP7 and Tsg [95]. Within the neural tube, overexpression of a constitutively active form of BMPRIA or BMPRIB induces an increase in dorsal cell types in neural precursor cells of the spinal cord [96], indicating BMPs are necessary for dorsoventral patterning of the neural tube.

BMPS AND BONE DEVELOPMENT

BMP function in development is well characterized in the skeletal system. BMPs were originally discovered by Marshall Urist in 1965, who found that subcutaneously implanted decalcified bone could induce ectopic cartilage and bone formation [97]. Subsequent purification of osteo-conductive factors led to the discovery of BMPs [98, 99]. The majority of the vertebrate skeleton is composed of endochondral bone. Endochondral ossification occurs when mesenchymal cells differentiate into chondrocytes, which form a highly organized template known as the growth plate [100]. This structure, in which cells are stratified in layers based on their stage of differentiation, permits assessments of the roles of BMP signaling pathways in specific aspects of cell commitment, survival, proliferation, withdrawal from the cell cycle, differentiation, and apoptosis. Several studies have demonstrated that BMP signaling is required for multiple aspects of endochondral bone formation (reviewed in [101, 102]). BMP signaling promotes the formation of early prechondrogenic condensations by sustaining the expression of Sox9, a transcription factor required for commitment of mesenchymal cells to the chondrogenic lineage, adhesion molecules such as N-cadherin, and matrix production [103–105]. BMPs can also induce proliferation and differentiation of chondrocytes by induction of target genes including *type II* and *type X collagen* [104, 106, 107]. In addition, promoter activity of type X collagen can be induced in prehypertrophic chondrocytes by overexpressing Smads1 or 5 [108, 109]. In the growth plate, chondrocytes align into distinct zones of small rounded resting, flattened disk-like proliferating, and large hypertrophic chondrocytes. Many BMP ligands are expressed within the growth plate and in the surrounding perichondrium and periosteum, while BMP antagonists are expressed in opposing patterns [12, 110]. Furthermore, the level of nuclear phosphorylated Smads1, 5, and 8 increases in chondrocytes as they progress from the resting to the prehypertrophic zone [10]. These studies raise the possibility that BMPs may act as a gradient within the growth plate to regulate the balance between proliferation and differentiation.

In vivo, overexpression of BMPs leads to enlarged, misshapen, and fused skeletal elements as a result of increased matrix production and chondrocyte proliferation [94, 111, 112]. In contrast, low levels of BMPs induced by exposure to noggin prevent mesenchymal cell aggregation and differentiation, and at later stages lead to osteopenia and reduced bone formation [113–115]. Likewise, at the Smad level, transgenic mice overexpressing the inhibitory Smad6 exhibit delayed chondrocyte hypertrophy, postnatal dwarfism, and osteopenia [116]. Chondrogenic differentiation is severely impaired in transgenic mice lacking both *Bmpr1a* and *Bmpr1b* in cartilage, leading to a virtual lack of endochondral skeletal formation [14]. It remains to be determined whether the loss of intracellular R-Smads can mimic the effects of the BMP receptor knockout on endochondral bone formation *in vivo*.

PERSPECTIVES

Recent studies have highlighted the importance of BMP signaling in embryonic stem cells (reviewed in [117]). BMPs can be used in the place of serum and feeder cells in combination with leukemia inhibitory factor (LIF) to maintain mouse ES cells in an undifferentiated state [118]. However, it is important to note that the undifferentiated state is not maintained in human ES cells under these conditions. A balance between transcriptional complexes formed by STAT3 or Smad activity mediates the BMP signal and thus the maintenance of pluripotency [119]. Upon removal of LIF, BMPs can block the neural pathway and drive mesodermal differentiation [120], demonstrating that in addition to controlling whether stem cells self-renew, they also influence ES cell differentiation. Pluripotent hematopoetic stem cells (HSCs) require BMP4 for differentiation from the mesoderm [121]. BMP signaling through BMPR1A also controls the size of the regulatory microenvironment, or niche, and differentiation of HSCs in the bone marrow [122].

An important area of future BMP research will be to understand how stem cell niches are maintained in postnatal development. In addition to HSCs, BMP signaling is essential for the control of the intestinal [123] and neural stem cell niches [124]. Mechanisms controlling the BMP signaling cascade at the extracellular and intracellular levels, as well as interactions with other pathways, require further study. It is interesting to note that although ES cells cannot be derived from embryonic lethal BMPRIA$^{-/-}$ embryos, ES cells can be derived from Smad4$^{-/-}$ embryos [22, 125]. These data further suggest that Smad4 may not be needed for R-Smad function in BMP-induced stem cell differentiation, and perhaps another protein is aiding in R-Smad nuclear translocation. Recently, it was demonstrated that Transcriptional Intermediary Factory 1γ (TIF1γ) can bind to phosphorylated Smad2/3 in competition with Smad4 [126]. Furthermore, the TIF1γ/Smad2/3 complex stimulated HSC differentiation in response to TGFβ stimulation in contrast to the inhibitory effect of Smad2/3/4 complexes [126]. These results indicate that there may be another stage of regulation in R-Smad signaling at the level of complex formation, although it remains to be determined if and how TIF1γ mediates BMP signaling. In summary, the regulation of BMP pathways and their role in the directing of cell fate must be further explored. Studies of developmental and postnatal BMP function in the manipulation of cell proliferation and regeneration will be important for future therapeutic approaches in tissue engineering and treatment of genetic diseases.

REFERENCES

1. Israel DI, Nove J, Kerns KM, et al. Heterodimeric bone morphogenetic proteins show enhanced activity *in vitro* and *in vivo*. *Growth Factors* 1996;**13**:291–300.

2. Zhu W, Kim J, Cheng C, et al. Noggin regulation of bone morphogenetic protein (BMP) 2/7 heterodimer activity *in vitro*. *Bone* 2006;**39**:61–71.

3. Daluiski A, Engstrand T, Bahamonde ME, et al. Bone morphogenetic protein-3 is a negative regulator of bone density. *Nat Genet* 2001;**27**:84–8.

4. Gamer LW, Nove J, Levin M, Rosen V. BMP-3 is a novel inhibitor of both activin and BMP-4 signaling in *Xenopus* embryos. *Dev Biol* 2005;**285**:156–68.

5. Massague J, Seoane J, Wotton D. Smad transcription factors. *Genes Dev* 2005;**19**:2783–810.

6. Derynck R, Zhang YE. Smad-dependent and Smad-independent pathways in TGF-beta family signalling. *Nature* 2003;**425**:577–84.

7. Mishina Y, Suzuki A, Ueno N, Behringer RR. Bmpr encodes a type I bone morphogenetic protein receptor that is essential for gastrulation during mouse embryogenesis. *Genes Dev* 1995;**9**:3027–37.

8. Yi SE, Daluiski A, Pederson R, Rosen V, Lyons KM. The type I BMP receptor BMPRIB is required for chondrogenesis in the mouse limb. *Development* 2000;**127**:621–30.

9. Baur ST, Mai JJ, Dymecki SM. Combinatorial signaling through BMP receptor IB and GDF5: shaping of the distal mouse limb and the genetics of distal limb diversity. *Development* 2000;**127**:605–19.

10. Yoon BS, Pogue R, Ovchinnikov DA, et al. BMPs regulate multiple aspects of growth-plate chondrogenesis through opposing actions on FGF pathways. *Development* 2006;**133**:4667–78.

11. Zou H, Wieser R, Massague J, Niswander L. Distinct roles of type I bone morphogenetic protein receptors in the formation and differentiation of cartilage. *Genes Dev* 1997;**11**:2191–203.

12. Minina E, Schneider S, Rosowski M, Lauster R, Vortkamp A. Expression of Fgf and Tgfbeta signaling related genes during embryonic endochondral ossification. *Gene Expr Patterns* 2005;**6**:102–9.

13. Kobayashi T, Lyons KM, McMahon AP, Kronenberg HM. BMP signaling stimulates cellular differentiation at multiple steps during cartilage development. *Proc Natl Acad Sci USA* 2005;**102**:18,023–18,027.

14. Yoon BS, Ovchinnikov DA, Yoshii I, Mishina Y, Behringer RR, Lyons KM. Bmpr1a and Bmpr1b have overlapping functions and are essential for chondrogenesis *in vivo*. *Proc Natl Acad Sci USA* 2005;**102**:5062–7.

15. Feng XH, Derynck R. Specificity and versatility in tgf-beta signaling through Smads. *Annu Rev Cell Dev Biol* 2005;**21**:659–93.

16. Lechleider RJ, Ryan JL, Garrett L, et al. Targeted mutagenesis of Smad1 reveals an essential role in chorioallantoic fusion. *Dev Biol* 2001;**240**:157–67.

17. Tremblay KD, Dunn NR, Robertson EJ. Mouse embryos lacking Smad1 signals display defects in extra-embryonic tissues and germ cell formation. *Development* 2001;**128**:3609–21.

18. Chang H, Huylebroeck D, Verschueren K, Guo Q, Matzuk MM, Zwijsen A. Smad5 knockout mice die at mid-gestation due to multiple embryonic and extraembryonic defects. *Development* 1999;**126**:1631–42.

19. Arnold SJ, Maretto S, Islam A, Bikoff EK, Robertson EJ. Dose-dependent Smad1, Smad5 and Smad8 signaling in the early mouse embryo. *Dev Biol* 2006;**296**:104–18.

20. Newfeld SJ, Wisotzkey RG, Kumar S. Molecular evolution of a developmental pathway: phylogenetic analyses of transforming growth factor-beta family ligands, receptors and Smad signal transducers. *Genetics* 1999;**152**:783–95.

21. Aubin J, Davy A, Soriano P. *In vivo* convergence of BMP and MAPK signaling pathways: impact of differential Smad1 phosphorylation on development and homeostasis. *Genes Dev* 2004;**18**:1482–94.

22. Sirard C, de la Pompa JL, Elia A, et al. The tumor suppressor gene Smad4/Dpc4 is required for gastrulation and later for anterior development of the mouse embryo. *Genes Dev* 1998;**12**:107–19.

23. Wisotzkey RG, Mehra A, Sutherland DJ, et al. Medea is a *Drosophila* Smad4 homolog that is differentially required to potentiate DPP responses. *Development* 1998;**125**:1433–45.

24. Chu GC, Dunn NR, Anderson DC, Oxburgh L, Robertson EJ. Differential requirements for Smad4 in TGFbeta-dependent patterning of the early mouse embryo. *Development* 2004;**131**:3501–12.

25. Zhang J, Tan X, Li W, et al. Smad4 is required for the normal organization of the cartilage growth plate. *Dev Biol* 2005;**284**:311–22.

26. Shioda T, Lechleider RJ, Dunwoodie SL, et al. Transcriptional activating activity of Smad4: roles of SMAD hetero-oligomerization and enhancement by an associating transactivator. *Proc Natl Acad Sci USA* 1998;**95**:9785–90.

27. Liu F, Pouponnot C, Massague J. Dual role of the Smad4/DPC4 tumor suppressor in TGFbeta-inducible transcriptional complexes. *Genes Dev* 1997;**11**:3157–67.

28. Biondi CA, Das D, Howell M, et al. Mice develop normally in the absence of Smad4 nucleocytoplasmic shuttling. *Biochem J* 2007;**404**:235–45.

29. Shi W, Chang C, Nie S, Xie S, Wan M, Cao X. Endofin acts as a Smad anchor for receptor activation in BMP signaling. *J Cell Sci* 2007;**120**:1216–24.

30. Sasaki A, Masuda Y, Ohta Y, Ikeda K, Watanabe K. Filamin associates with Smads and regulates transforming growth factor-beta signaling. *J Biol Chem* 2001;**276**:17,871–17,877.

31. Haag J, Aigner T. Identification of calponin 3 as a novel Smad-binding modulator of BMP signaling expressed in cartilage. *Exp Cell Res* 2007;**313**:3386–94.

32. Zheng L, Baek HJ, Karsenty G, Justice MJ. Filamin B represses chondrocyte hypertrophy in a Runx2/Smad3-dependent manner. *J Cell Biol* 2007;**178**:121–8.

33. Raju GP, Dimova N, Klein PS, Huang HC. SANE, a novel LEM domain protein, regulates bone morphogenetic protein signaling through interaction with Smad1. *J Biol Chem* 2003;**278**:428–37.

34. Osada S, Ohmori SY, Taira M. XMAN1, an inner nuclear membrane protein, antagonizes BMP signaling by interacting with Smad1 in *Xenopus* embryos. *Development* 2003;**130**:1783–94.

35. Khokha MK, Yeh J, Grammer TC, Harland RM. Depletion of three BMP antagonists from Spemann's organizer leads to a catastrophic loss of dorsal structures. *Dev Cell* 2005;**8**:401–11.

36. Kretzschmar M, Doody J, Massague J. Opposing BMP and EGF signalling pathways converge on the TGF-beta family mediator Smad1. *Nature* 1997;**389**:618–22.

37. Kretzschmar M, Doody J, Timokhina I, Massague J. A mechanism of repression of TGFbeta/ Smad signaling by oncogenic Ras. *Genes Dev* 1999;**13**:804–16.

38. Sapkota G, Alarcon C, Spagnoli FM, Brivanlou AH, Massague J. Balancing BMP signaling through integrated inputs into the Smad1 linker. *Mol Cell* 2007;**25**:441–54.

39. Pera EM, Ikeda A, Eivers E, De Robertis EM. Integration of IGF, FGF, and anti-BMP signals via Smad1 phosphorylation in neural induction. *Genes Dev* 2003;**17**:3023–8.

40. Pajni-Underwood S, Wilson CP, Elder C, Mishina Y, Lewandoski M. BMP signals control limb bud interdigital programmed cell death by regulating FGF signaling. *Development* 2007;**134**:2359–68.

41. Weaver M, Dunn NR, Hogan BL. Bmp4 and Fgf10 play opposing roles during lung bud morphogenesis. *Development* 2000;**127**:2695–704.

42. Nohe A, Hassel S, Ehrlich M, et al. The mode of bone morphogenetic protein (BMP) receptor oligomerization determines different BMP-2 signaling pathways. *J Biol Chem* 2002;**277**:5330–8.

43. Hassel S, Schmitt S, Hartung A, et al. Initiation of Smad-dependent and Smad-independent signaling via distinct BMP-receptor complexes. *J Bone Joint Surg Am* 2003;**85-A**(Suppl. 3):44–51.

44. Hartung A, Bitton-Worms K, Rechtman MM, et al. Different routes of bone morphogenic protein (BMP) receptor endocytosis influence BMP signaling. *Mol Cell Biol* 2006;**26**:7791–805.

45. Qiao B, Padilla SR, Benya PD. Transforming growth factor (TGF)-beta-activated kinase 1 mimics and mediates TGF-beta-induced stimulation of

type II collagen synthesis in chondrocytes independent of Col2a1 transcription and Smad3 signaling. *J Biol Chem* 2005;**280**:17,562–17,571.

46. Hoffmann A, Preobrazhenska O, Wodarczyk C, et al. Transforming growth factor-beta-activated kinase-1 (TAK1), a MAP3K, interacts with Smad proteins and interferes with osteogenesis in murine mesenchymal progenitors. *J Biol Chem* 2005;**280**:27,271–27,283.

47. Zhang R, Murakami S, Coustry F, Wang Y, de Crombrugghe B. Constitutive activation of MKK6 in chondrocytes of transgenic mice inhibits proliferation and delays endochondral bone formation. *Proc Natl Acad Sci USA* 2006;**103**:365–70.

48. Yingling JM, Das P, Savage C, Zhang M, Padgett RW, Wang XF. Mammalian dwarfins are phosphorylated in response to transforming growth factor beta and are implicated in control of cell growth. *Proc Natl Acad Sci USA* 1996;**93**:8940–4.

49. Lechleider RJ, de Caestecker MP, Dehejia A, Polymeropoulos MH, Roberts AB. Serine phosphorylation, chromosomal localization, and transforming growth factor-beta signal transduction by human bsp-1. *J Biol Chem* 1996;**271**:17,617–17,620.

50. Yue J, Frey RS, Mulder KM. Cross-talk between the Smad1 and Ras/MEK signaling pathways for TGFbeta. *Oncogene* 1999;**18**:2033–7.

51. Goumans MJ, Valdimarsdottir G, Itoh S, Rosendahl A, Sideras P, ten Dijke P. Balancing the activation state of the endothelium via two distinct TGF-beta type I receptors. *Embo J* 2002;**21**:1743–53.

52. Pannu J, Nakerakanti S, Smith E, ten Dijke P, Trojanowska M. Transforming growth factor-beta receptor type I-dependent fibrogenic gene program is mediated via activation of Smad1 and ERK1/2 pathways. *J Biol Chem* 2007;**282**:10,405–10,413.

53. Goumans MJ, Valdimarsdottir G, Itoh S, et al. Activin receptor-like kinase (ALK)1 is an antagonistic mediator of lateral TGFbeta/ALK5 signaling. *Mol Cell* 2003;**12**:817–28.

54. Podos SD, Ferguson EL. Morphogen gradients: new insights from DPP. *Trends Genet* 1999;**15**:396–402.

55. Gazzerro E, Canalis E. Bone morphogenetic proteins and their antagonists. *Rev Endocr Metab Disord* 2006;**7**:51–65.

56. Avsian-Kretchmer O, Hsueh AJ. Comparative genomic analysis of the eight-membered ring cystine knot-containing bone morphogenetic protein antagonists. *Mol Endocrinol* 2004;**18**:1–12.

57. Ray RP, Arora K, Nusslein-Volhard C, Gelbart WM. The control of cell fate along the dorsal-ventral axis of the *Drosophila* embryo. *Development* 1991;**113**:35–54.

58. Shimmi O, Umulis D, Othmer H, O'Connor MB. Facilitated transport of a Dpp/Scw heterodimer by Sog/Tsg leads to robust patterning of the *Drosophila* blastoderm embryo. *Cell* 2005;**120**:873–86.

59. Wang YC, Ferguson EL. Spatial bistability of Dpp-receptor interactions during Drosophila dorsal-ventral patterning. *Nature* 2005;**434**:229–34.

60. Sutherland DJ, Li M, Liu XQ, Stefancsik R, Raftery LA. Stepwise formation of a SMAD activity gradient during dorsal-ventral patterning of the *Drosophila* embryo. *Development* 2003;**130**:5705–16.

61. Teleman AA, Cohen SM. Dpp gradient formation in the *Drosophila* wing imaginal disc. *Cell* 2000;**103**:971–80.

62. Itoh S, ten Dijke P. Negative regulation of TGF-beta receptor/Smad signal transduction. *Curr Opin Cell Biol* 2007;**19**:176–84.

63. Nakao A, Afrakhte M, Moren A, et al. Identification of Smad7, a TGFbeta-inducible antagonist of TGF-beta signalling. *Nature* 1997;**389**:631–5.

64. Souchelnytskyi S, Nakayama T, Nakao A, et al. Physical and functional interaction of murine and Xenopus Smad7 with bone morphogenetic protein receptors and transforming growth factor-beta receptors. *J Biol Chem* 1998;**273**:25,364–25,370.

65. Hata A, Lagna G, Massague J, Hemmati-Brivanlou A. Smad6 inhibits BMP/Smad1 signaling by specifically competing with the Smad4 tumor suppressor. *Genes Dev* 1998;**12**:186–97.

66. Murakami G, Watabe T, Takaoka K, Miyazono K, Imamura T. Cooperative inhibition of bone morphogenetic protein signaling by Smurf1 and inhibitory Smads. *Mol Biol Cell* 2003;**14**:2809–17.

67. Benchabane H, Wrana JL. GATA- and Smad1-dependent enhancers in the Smad7 gene differentially interpret bone morphogenetic protein concentrations. *Mol Cell Biol* 2003;**23**:6646–61.

68. Afrakhte M, Moren A, Jossan S, et al. Induction of inhibitory Smad6 and Smad7 mRNA by TGF-beta family members. *Biochem Biophys Res Commun* 1998;**249**:505–11.

69. Ishida W, Hamamoto T, Kusanagi K, et al. Smad6 is a Smad1/5-induced smad inhibitor. Characterization of bone morphogenetic protein-responsive element in the mouse Smad6 promoter. *J Biol Chem* 2000;**275**:6075–9.

70. Han G, Li AG, Liang YY, et al. Smad7-induced beta-catenin degradation alters epidermal appendage development. *Dev Cell* 2006;**11**:301–12.

71. De Robertis EM, Kuroda H. Dorsal-ventral patterning and neural induction in Xenopus embryos. *Annu Rev Cell Dev Biol* 2004;**20**:285–308.

72. Wilson PA, Hemmati-Brivanlou A. Induction of epidermis and inhibition of neural fate by Bmp-4. *Nature* 1995;**376**:331–3.

73. Sasai Y, Lu B, Steinbeisser H, De Robertis EM. Regulation of neural induction by the Chd and Bmp-4 antagonistic patterning signals in Xenopus. *Nature* 1995;**377**:757.

74. Hawley SH, Wunnenberg-Stapleton K, Hashimoto C, et al. Disruption of BMP signals in embryonic Xenopus ectoderm leads to direct neural induction. *Genes Dev* 1995;**9**:2923–35.

75. Piccolo S, Sasai Y, Lu B, De Robertis EM. Dorsoventral patterning in Xenopus: inhibition of ventral signals by direct binding of chordin to BMP-4. *Cell* 1996;**86**:589–98.

76. Piccolo S, Agius E, Lu B, Goodman S, Dale L, De Robertis EM. Cleavage of Chordin by Xolloid metalloprotease suggests a role for proteolytic processing in the regulation of Spemann organizer activity. *Cell* 1997;**91**:407–16.

77. Schohl A, Fagotto F. Beta-catenin, MAPK and Smad signaling during early Xenopus development. *Development* 2002;**129**:37–52.

78. Suzuki A, Ueno N, Hemmati-Brivanlou A. *Xenopus* msx1 mediates epidermal induction and neural inhibition by BMP4. *Development* 1997;**124**:3037–44.

79. Oelgeschlager M, Kuroda H, Reversade B, De Robertis EM. Chordin is required for the Spemann organizer transplantation phenomenon in Xenopus embryos. *Dev Cell* 2003;**4**:219–30.

80. Reversade B, De Robertis EM. Regulation of ADMP and BMP2/4/7 at opposite embryonic poles generates a self-regulating morphogenetic field. *Cell* 2005;**123**:1147–60.

81. Wilson PA, Lagna G, Suzuki A, Hemmati-Brivanlou A. Concentration-dependent patterning of the Xenopus ectoderm by BMP4 and its signal transducer Smad1. *Development* 1997;**124**:3177–84.

82. Suzuki A, Kaneko E, Maeda J, Ueno N. Mesoderm induction by BMP-4 and -7 heterodimers. *Biochem Biophys Res Commun* 1997;**232**:153–6.

83. Shibuya H, Iwata H, Masuyama N, et al. Role of TAK1 and TAB1 in BMP signaling in early Xenopus development. *Embo J* 1998;**17**:1019–28.

84. Casellas R, Brivanlou AH. Xenopus Smad7 inhibits both the activin and BMP pathways and acts as a neural inducer. *Dev Biol* 1998;**198**:1–12.

85. Nakayama T, Gardner H, Berg LK, Christian JL. Smad6 functions as an intracellular antagonist of some TGF-beta family members during *Xenopus* embryogenesis. *Genes Cells* 1998;**3**:387–94.

86. Nakayama T, Berg LK, Christian JL. Dissection of inhibitory Smad proteins: both N- and C-terminal domains are necessary for full activities of *Xenopus* Smad6 and Smad7. *Mech Dev* 2001;**100**:251–62.

87. Lee HX, Ambrosio AL, Reversade B, De Robertis EM. Embryonic dorsal-ventral signaling. secreted frizzled-related proteins as inhibitors of tolloid proteinases. *Cell* 2006;**124**:147–59.

88. Zhao GQ. Consequences of knocking out BMP signaling in the mouse. *Genesis* 2003;**35**:43–56.

89. Winnier G, Blessing M, Labosky PA, Hogan BL. Bone morphogenetic protein-4 is required for mesoderm formation and patterning in the mouse. *Genes Dev* 1995;**9**:2105–16.

90. Ahn III K, Mishina Y, Hanks MC, Behringer RR, Crenshaw EB. BMPR-IA signaling is required for the formation of the apical ectodermal ridge and dorsal-ventral patterning of the limb. *Development* 2001;**128**:4449–61.

91. Ovchinnikov DA, Selever J, Wang Y, et al. BMP receptor type IA in limb bud mesenchyme regulates distal outgrowth and patterning. *Dev Biol* 2006;**295**:103–15.

92. Bandyopadhyay A, Tsuji K, Cox K, Harfe BD, Rosen V, Tabin CJ. Genetic analysis of the roles of BMP2, BMP4, and BMP7 in limb patterning and skeletogenesis. *PLoS Genet* 2006;**2**:e216.

93. McMahon JA, Takada S, Zimmerman LB, Fan CM, Harland RM, McMahon AP. Noggin-mediated antagonism of BMP signaling is required for growth and patterning of the neural tube and somite. *Genes Dev* 1998;**12**:1438–52.

94. Wijgerde M, Karp S, McMahon J, McMahon AP. Noggin antagonism of BMP4 signaling controls development of the axial skeleton in the mouse. *Dev Biol* 2005;**286**:149–57.

95. Zakin L, Reversade B, Kuroda H, Lyons KM, De Robertis EM. Sirenomelia in Bmp7 and Tsg compound mutant mice: requirement for Bmp signaling in the development of ventral posterior mesoderm. *Development* 2005;**132**:2489–99.

96. Panchision DM, Pickel JM, Studer L, et al. Sequential actions of BMP receptors control neural precursor cell production and fate. *Genes Dev* 2001;**15**:2094–110.

97. Urist MR. Bone: formation by autoinduction. *Science* 1965;**150**:893–9.

98. Sampath TK, Reddi AH. Dissociative extraction and reconstitution of extracellular matrix components involved in local bone differentiation. *Proc Natl Acad Sci USA* 1981;**78**:7599–603.

99. Sampath TK, Reddi AH. Homology of bone-inductive proteins from human, monkey, bovine, and rat extracellular matrix. *Proc Natl Acad Sci USA* 1983;**80**:6591–5.

100. Kronenberg HM. Developmental regulation of the growth plate. *Nature* 2003;**423**:332–6.

101. Retting KN, Lyons KM. BMP signaling and the regulation of endochondral bone formation. *Cellscience Rev* 2006;**3**:67–88.

102. Yoon BS, Lyons KM. Multiple functions of BMPs in chondrogenesis. *J Cell Biochem* 2004;**93**:93–103.

103. Denker AE, Haas AR, Nicoll SB, Tuan RS. Chondrogenic differentiation of murine C3H10T1/2 multipotential mesenchymal cells: I. Stimulation by bone morphogenetic protein-2 in high-density micromass cultures. *Differentiation* 1999;**64**:67–76.

104. Hatakeyama Y, Tuan RS, Shum L. Distinct functions of BMP4 and GDF5 in the regulation of chondrogenesis. *J Cell Biochem* 2004;**91**:1204–17.

105. Haas AR, Tuan RS. Chondrogenic differentiation of murine C3H10T1/2 multipotential mesenchymal cells: II. Stimulation by bone morphogenetic protein-2 requires modulation of N-cadherin expression and function. *Differentiation* 1999;**64**:77–89.

106. Shukunami C, Akiyama H, Nakamura T, Hiraki Y. Requirement of autocrine signaling by bone morphogenetic protein-4 for chondrogenic differentiation of ATDC5 cells. *FEBS Letts* 2000;**469**:83–7.

107. Fujii M, Takeda K, Imamura T, et al. Roles of bone morphogenetic protein type I receptors and Smad proteins in osteoblast and chondroblast differentiation. *Mol Biol Cell* 1999;**10**:3801–13.

108. Leboy P, Grasso-Knight G, D'Angelo M, et al. Smad-Runx interactions during chondrocyte maturation. *J Bone Joint Surg Am* 2001;**83-A**(Suppl. 1):S15–22.

109. Drissi MH, Li X, Sheu TJ, et al. Runx2/Cbfa1 stimulation by retinoic acid is potentiated by BMP2 signaling through interaction with Smad1 on the collagen X promoter in chondrocytes. *J Cell Biochem* 2003;**90**:1287–98.

110. Nilsson O, Parker EA, Hegde A, Chau M, Barnes KM, Baron J. Gradients in bone morphogenetic protein-related gene expression across the growth plate. *J Endocrinol* 2007;**193**:75–84.

111. Duprez D, Bell EJ, Richardson MK, et al. Overexpression of BMP-2 and BMP-4 alters the size and shape of developing skeletal elements in the chick limb. *Mech Dev* 1996;**57**:145–57.

112. Brunet LJ, McMahon JA, McMahon AP, Harland RM. Noggin, cartilage morphogenesis, and joint formation in the mammalian skeleton. *Science* 1998;**280**:1455–7.

113. Pizette S, Niswander L. BMPs are required at two steps of limb chondrogenesis: formation of prechondrogenic condensations and their differentiation into chondrocytes. *Dev Biol* 2000;**219**:237–49.

114. Pathi S, Rutenberg JB, Johnson RL, Vortkamp A. Interaction of Ihh and BMP/Noggin signaling during cartilage differentiation. *Dev Biol* 1999;**209**:239–53.

115. Devlin RD, Du Z, Pereira RC, et al. Skeletal overexpression of noggin results in osteopenia and reduced bone formation. *Endocrinology* 2003;**144**:1972–8.

116. Horiki M, Imamura T, Okamoto M, et al. Smad6/Smurf1 overexpression in cartilage delays chondrocyte hypertrophy and causes dwarfism with osteopenia. *J Cell Biol* 2004;**165**:433–45.

117. Wagner TU. Bone morphogenetic protein signaling in stem cells–one signal, many consequences. *FEBS J* 2007;**274**:2968–76.

118. Ying QL, Nichols J, Chambers I, Smith A. BMP induction of Id proteins suppresses differentiation and sustains embryonic stem cell self-renewal in collaboration with STAT3. *Cell* 2003;**115**:281–92.

119. Suzuki A, Raya A, Kawakami Y, et al. Nanog binds to Smad1 and blocks bone morphogenetic protein-induced differentiation of embryonic stem cells. *Proc Natl Acad Sci USA* 2006;**103**:10,294–10,299.

120. Xu RH, Peck RM, Li DS, Feng X, Ludwig T, Thomson JA. Basic FGF and suppression of BMP signaling sustain undifferentiated proliferation of human ES cells. *Nat Methods* 2005;**2**:185–90.

121. Park C, Afrikanova I, Chung YS, et al. A hierarchical order of factors in the generation of FLK1- and SCL-expressing hematopoietic and endothelial progenitors from embryonic stem cells. *Development* 2004;**131**:2749–62.

122. Zhang J, Niu C, Ye L, et al. Identification of the haematopoietic stem cell niche and control of the niche size. *Nature* 2003;**425**:836–41.

123. He XC, Zhang J, Tong WG, et al. BMP signaling inhibits intestinal stem cell self-renewal through suppression of Wnt-beta-catenin signaling. *Nat Genet* 2004;**36**:1117–21.

124. Chen HL, Panchision DM. Concise review: bone morphogenetic protein pleiotropism in neural stem cells and their derivatives–alternative pathways, convergent signals. *Stem Cells* 2007;**25**:63–8.

125. Qi X, Li TG, Hao J, et al. BMP4 supports self-renewal of embryonic stem cells by inhibiting mitogen-activated protein kinase pathways. *Proc Natl Acad Sci USA* 2004;**101**:6027–32.

126. He W, Dorn DC, Erdjument-Bromage H, Tempst P, Moore MA, Massague J. Hematopoiesis controlled by distinct TIF1gamma and Smad4 branches of the TGFbeta pathway. *Cell* 2006;**125**:929–41.

Chapter 30

Signaling from Fibroblast Growth Factor Receptors in Development and Disease

Kristine A. Drafahl[*,1], Christopher W. McAndrew[1] and Daniel J. Donoghue[1,2]

[1]Department of Chemistry and Biochemistry, University of California San Diego, La Jolla, California

[2]Moores University of California San Diego Cancer Center, La Jolla, California

[*]Authors contributed equally to this work

INTRODUCTION

Fibroblast growth factor receptors (FGFRs) constitute a family of four (FGFR1–4) [1–3] structurally related, cell surface receptor tyrosine kinases (RTKs), with 55–72 percent homology [4]. FGFRs are involved in a variety of biological processes, including cell growth, migration, differentiation, survival, and apoptosis, and are essential for embryonic and neural development, skeletal and organ formation, and adult tissue homoeostasis [5, 6]. Alternative splicing of *Fgfr* transcripts generates up to 15 isoforms, which transmit the signals of at least 22 fibroblast growth factors (FGF1–22) [7]. Each receptor is comprised of an extracellular ligand binding domain consisting of three immunoglobulin (Ig)-like domains, an acidic box between IgI and IgII [4], a transmembrane domain, and a split intracellular tyrosine kinase domain composed of an ATP binding site and catalytic site. FGFR activation is achieved upon ligand binding [8, 9], resulting in receptor dimerization and transautophosphorylation of multiple conserved intracellular tyrosine residues [10], which stimulate the receptor's intrinsic kinase activity and recruit downstream adaptor and signaling proteins [11–13]. Heparan sulfate proteoglycans (HSPGs) facilitate ligand binding and are obligate cofactors for FGFR activation by FGFs [14–17]. The three main signaling pathways associated with FGFR activation include the Ras/MAPK, PI 3-kinase, and PLCg pathways. All but one of the mutations known for the *Fgfr* genes are gain-of-function mutations, and activation of these receptors is associated with many developmental and skeletal disorders [18, 19]. Additionally, FGFR and FGF overexpression has been observed in many tumor samples, and mutations are also likely to be involved in carcinogenesis.

FGFR EXPRESSION AND ROLE DURING DEVELOPMENT

During embryonic development, FGFR signaling is essential for organ growth and patterning of the embryo. All FGFRs are widely expressed in distinct spatial patterns during development and in adult tissues [20–24]. FGFR1 expression is found mainly in the mesenchyme in the central nervous system and limbs, and targeted inactivation of *Fgfr1* in mice severely impairs growth and results in recessive embryonic lethality [25]. During early neurogenesis, FGFR1 expression is upregulated in the ventricular zone of the neural tube and mesenchyme of developing limbs [26, 27], and at later stages is expressed in spinal cord motor neurons and maturing neurons in the brain [26, 28]. Although required for correct axial organization and embryonic cell proliferation, FGFR1 is not directly required for mesoderm formation [25, 29]. FGFR1 was also shown to play a role in neurulation, as chimeric mouse embryos, created by injection of FGFR1 deficient (R1$^{-/-}$) embryonic stem (ES) cells into wild-type blastocytes, showed limb bud and tail distortion, partial neural tube duplication, and spina bifida [30]. FGFR2 is highly expressed in epithelial lineages during early gastrulation, and in both epithelial and mesenchymal cells during later development and organogenesis [26, 27]. Like *Fgfr1*, targeted disruption of *Fgfr2* results in an embryonic lethal phenotype [31]. Its expression is essential for limb outgrowth, mammalian lung branching morphogenesis [32], and keratinocyte differentiation [33]. FGFR3 expression primarily occurs in the central nervous system and bone rudiments, specifically the developing brain, spinal cord, cochlea, and hypertrophic zone of the growth plate [32]. Targeted disruption of *Fgfr3* in mice is not embryonic lethal, but leads to severe skeletal and inner ear defects, and mouse models indicate FGFR3 negatively regulates bone growth and development [34, 35]. FGFR3 also cooperates with FGFR4 to mediate liver functions and lung development. FGFR4 expression occurs in the definitive endoderm, somatic myotome, and the ventricular zone of developing dorsal root ganglia and spinal cord [36–38]. Although, *Fgfr4* null mice appear normal, they exhibit elevated liver bile acids, enhanced cholesterol biosynthesis, and depleted gall bladders [39].

SIGNALING PATHWAYS MEDIATED BY FGFRS

Activation of FGFRs can result in a variety of outcomes by initiating various intracellular signaling pathways. In many cases, the pathways activated depend on the cell type or stage of differentiation, leading to specific activation of downstream targets [40]. Specificity is also achieved through the binding of different FGFs, of which many have unique and cell-specific roles. Splice variants of FGFRs also contribute to diverse cell signaling [40]. Despite the varied outcomes of FGFR signaling, several key pathways are commonly activated in most cell types. FGFR activation results in tyrosine autophosphorylation, and these phosphorylated tyrosines serve as high-affinity binding sites for proteins containing Src-homology 2 (SH2) domains or phosphotyrosine binding (PTB) domains [41]. These intracellular proteins then transduce the activation signal from the receptor through signaling cascades which eventually lead to changes in gene transcription and a biological response [42, 43].

The membrane-associated docking protein FGF receptor substrate 2 (FRS2) binds to the FGFR juxtamembrane domain (JM) through its PTB domain, and is phosphorylated by the receptor [44]. This leads to recruitment of a variety of adaptor proteins, including growth factor receptor bound protein 2 (Grb2), which then binds the guanine nucleotide exchange factor Son of sevenless (Sos) [45]. Recruitment of this complex to the plasma membrane activates the G-protein Ras, which stimulates the mitogen-activated protein kinase (MAPK) pathway [45]. MAPK pathway activation results in a variety of outcomes depending on cell type or state, including DNA synthesis, proliferation, and/or differentiation. The adaptor molecule Shc is also phosphorylated by FGFR, leading to Grb2 recruitment and activation of the Ras/MAPK pathway [46].

FRS2 activation also signals through the PI 3-kinase pathway. The SH2 domain of Grb2 binds to a phosphorylated tyrosine residue on FRS2 while the C-terminal SH3 domain of Grb2 forms a complex with the proline-rich region of Grb2 associated binding protein 1 (Gab1) [47]. Gab1 recruitment in close proximity to the receptor results in its tyrosine phosphorylation. Recruitment of PI 3-kinase

and activation of AKT follows, leading to cell survival [48]. FGFR binding and phosphorylation of FRS2 is essential for Gab1 recruitment and eventual activation of the PI 3-kinase cascade [47], indicating that FGFR activation of FRS2 plays a prominent role in promoting cell survival. The N-terminal SH2 domain of SH2 tyrosine phosphatase 2 (Shp2) interacts with a phosphotyrosine on FRS2 and leads to phosphorylation of Shp2 itself. Phosphorylated Shp2 interacts with the Grb2/Sos complex and forms a ternary complex with FRS2 [49]. Shb also interacts with Shp2, and potentiates its FGF-mediated phosphorylation and FRS2 interaction. Interaction of phosphorylated Shp2 with FRS2 is essential for MAPK activation, indicating an important role for the adaptor Shb [50]. FRS2 has also been shown to associate with Src, a non-receptor tyrosine kinase, which phosphorylates cortactin to affect cell migration [51, 52].

Autophosphorylation of Tyr766 in the carboxy-terminal tail of FGFR1 creates a specific binding site for the SH2 domain of PLCg [48]. Activation of PLCg by tyrosine phosphorylation results in hydrolysis of phosphatidylinositol, generating diacylglycerol (DAG) and Ins(1,4,5)P3 (IP3) [53]. Generation of these second messengers results in Ca^{2+} release and activation of PKC [53]. Shb also interacts with FGFR1 through Y766, although it does not seem to compete for binding with PLCg [50].

Other adaptor molecules link FGFR activation to various biological activities. Crk interacts with Tyr463 on FGFR1 and results in cellular proliferation in certain cell types [54, 55]. The adaptor protein Nck also binds to phosphorylated FGFR, facilitating the interaction between Pak and Rac, and may link FGFR signaling to the actin cytoskeleton [56]. Activated FGFR1, 3, and 4 also promote Stat1 and Stat3 activation [57], and FGFR3 can activate STAT5 through the adaptor protein SH2-B [58]. Many of the interactions and signaling pathways activated by FGFRs described above are shown in Figure 30.1.

FGFRS AND DEVELOPMENTAL DISORDERS

Specific mutations in the *Fgfr1–3* genes lead to congenital bone diseases classified as chondrodysplasia and craniosynostosis syndromes, which cause dwarfism, deafness, and abnormalities of the skeleton, skin, and eye [59, 60]. Almost all of these are activating, gain-of-function mutations, and many occur in the IgII and IgIII domains, which mediate FGF binding [61, 62]. Over 60 mutations have been found to be associated with craniosynostosis syndromes, with a majority in FGFR2, including Antley-Bixler-like syndrome (ABS), Apert syndrome (AS), Beare-Stevenson syndrome (BSS), Crouzon syndrome (CS), Jackson-Weiss syndrome (JWS), Muenke-like syndrome (MS), Saethre-Chotzen syndrome (SCS), as well as the FGFR1-associated craniofacial dysplasia with hypophosphatemia (CFDH) and Pfeiffer

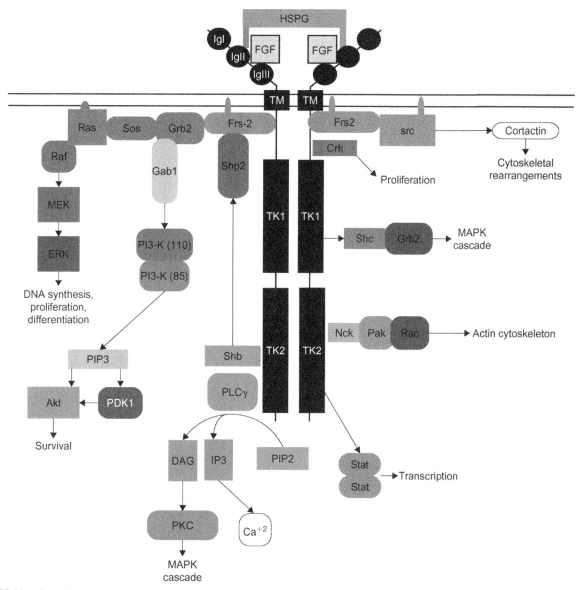

FIGURE 30.1 Signaling pathways activated by FGFRs.
FGFRs dimerize and ungergo autophosphorylation upon ligand stimulation, creating docking sites for various signaling molecules. Additional proteins are recruited to the membrane through modular domain interactions involving SH2, PTB, and other domains. Once at the membrane, these proteins activate multiple cellular signaling pathways, most notably the MAPK, PI3K, and PLCγ pathways.

syndrome (PS) [59, 60, 63–68]. All of these mutations are dominant, and craniofacial abnormalities varying in severity result from these syndromes. Missense mutations in FGFR3 result in skeletal dysplasia syndromes and short-limbed dwarfisms, including achondroplasia (ACH), Crouzon syndrome with acanthosis nigricans (CAN), hypochondroplasia (HCH), severe achondroplasia with developmental delay, and acanthosis nigricans (SADDAN), and the platyspondylic lethal skeletal dysplasias (PLSDs), including thanatophoric dysplasia (TD) types I and II [69–83]. Additionally, two syndromes caused by loss-of-function mutations in FGFRs have been described, including the FGFR1-associated type 2 Kallmann syndrome (KS) [60, 84, 85] and

the FGFR3-associated camptodactyly, tall stature, and hearing loss (CATSHL) syndrome [86]. To date, no mutations in FGFR4 are associated with any known chondrodysplasia or craniosynostosis syndromes. A list of the mutations and syndromes associated with their respective FGFR can be seen in Table 30.1.

ROLE OF FGFRS IN HUMAN CANCER

All four members of the FGFR family and many of their ligands have been implicated in human cancers as well. They play roles in cancer progression by inducing

TABLE 30.1 FGFR mutations associated with developmental syndromes

Syndrome	Missense Mutations	Receptor (Domain)
ACH	Y278C, S279C, G346E, G375C, G380R	FGFR3 (IgIII and TM)
ABS	S267P, W290C, C342R/S, S351C	FGFR2 (IgIII)
AS	S252W/F, P253R	FGFR2 (IgII-IgIII linker)
BSS	S372C, Y375C	FGFR2 (JM)
CFDH	Y372C	FGFR1 (TM)
CS	Y105C, S252L, S267P/F, C278F/Y, Y281C, Q289P, W290R/G, Y308C, Y328C, N331I, A337P, G338R/E, Y340H/S, C342S/F/Y/W/R, A344G, S347C, S354C, S355V, L357S, V359F, A362S, K526F, 549H, K659N, R678G	FGFR2 (IgI, IgIII, and KD)
CAN	A391E	FGFR3 (TM)
CATSHL	R621H	FGFR3 (KD)
HCH	S84L, G268C, R200C, N262H, V381E, I538V, N540K/T/S, K650N/Q	FGFR3 (KD)
JWS	C278F, Q289P, C342S/R, A344G	FGFR2 (IgIII)
KS	G97D, Y99C, V102I, S107X, D129A, A167S, V273M, C277Y, A520T, V607M, R622X, W666R, G687R, E692G, M719R, Y730X, P745S P772S	FGFR1 (IgI, IgII, IgIII, AB, KD, and C-term)
MS	P250R	FGFR3 (IgII-IgIII linker)
PS	P252R A172F, S252F, P253S, S267P, F276V, C278F, W290C, A314S, D321A, Y340C, T341P, C342G/S/Y/W/R, A344P, S351C, V359F, Y375C, N549H, E565G, K641R, G663E	FGFR1 (IgII-IgIII linker) FGFR2 (IgII, IgIII, IgIII-TM linker, and KD)
TDI	R248C, S249C, G370C, S371C, Y373C, x807L/G/C/R/W	FGFR3 (IgII-IgIII linker and IgIII-TM linker)
TDII	K650E	FGFR3 (KD)
SCS	Q289P	FGFR2 (IgII-IgIII linker)
SADDAN	K650M	FGFR3 (KD)

angiogenesis [87], changes in cell morphology, increased motility, and tumor cell proliferation [43]. FGFRs are over-expressed or have altered activity in cancers of the colon [88], prostate [89, 90], breast [91], kidneys [92], ovaries [93, 94], central nervous system [95], gastrointestinal system [96], thyroid [97], pituitary [98, 99], brain [100, 101], liver [102, 103], pancreas [104], skin [105], and lung [106], as well as in leukemia [107], multiple myeloma, urological cancers [108], soft tissue sarcomas [109], head and neck squamous cell carcinoma [110], and lymphoma [111]. Recent evidence indicates FGFRs may be used to target tumors for growth inhibition [87, 112, 113], and targeted inhibition of FGFRs may provide a therapeutic approach in the fight against cancer.

FGFR1 was recently found to be amplified in a small percentage of breast cancers, and contributes to the survival of lobular breast carcinomas [114]. In estrogen-receptor positive breast cancer cells, FGFR1 amplification is a prognostic of poor outcome [115]. Recent research found that activation of FGFR1 plays a role in the initiation of angiogenesis in prostate cancer [116] and may be a new marker for prostate cancer progression, as it was shown to be upregulated in late-stage prostate tumors [117]. The role of FGFR1 is most widely described in chronic myeloproliferative disorders (CMPDs). One rare CMPD, known as 8p11 myeloproliferative syndrome (EMS) or stem cell leukemia lymphoma (SCLL), is caused by an 8p11 translocation of *Fgfr1* [118]. This leads to fusion of *Fgfr1* to other genes, and constitutive activation of the receptor. The first fusion identified was to a zinc finger gene, *ZNF198*, and subsequently many *Fgfr1* rearrangements involved with a variety of partners have been demonstrated [119].

In two recent genome-wide association studies, FGFR2 was implicated in increased susceptibility to breast cancer [120, 121]. It is believed that a splice variant of FGFR2 or possibly an unwarranted estrogen receptor binding site may be the cause for the associated risk of breast cancer [121]. Also, certain types of gastric cancers overexpress FGFR2, and recent research has discovered that an inhibitor, AZD2171, exerts potent anti-tumor activity against gastric cancer xenografts overexpressing FGFR2 [122].

A frequent translocation observed in multiple myeloma, t(4;14)(p16.3;q32.3), involves the *Fgfr3* gene, and results

in increased expression of FGFR3 alleles that contain activating mutations [123, 124], including Y373C and K650E, which cause the lethal skeletal syndromes TDI and TDII [72, 79]. The splice variant FGFRIIIb is expressed in a wide range of bladder and cervical carcinoma cell lines [125], and these cancers exhibited expression of mutant alleles of FGFRIIIb, including R248C, S249C, G372C and K652E [125]. These and other FGFR mutations are shown in Figure 30.2.

Although FGFR4 is not associated with any known syndromes, it is associated with the widest range of cancers. Of recent debate is the significance of the G388R polymorphism. This polymorphism exists in approximately half the population, and appears to have no effect on cancer susceptibility. However, evidence suggests that the polymorphism leads to reduced disease-free survival in cancer patients, and correlates with a poor prognosis compared to the Gly388 allele in head and neck squamous cell carcinoma [110, 126], breast cancer [127–129], melanoma [105], lung adenocarcinoma [106], prostate cancer [130], and high-grade soft tissue sarcomas [109]. Opposing evidence suggests there is no correlation between the G388R polymorphism and cancer prognosis [101, 131–134]. Continued research into the significance of this polymorphism is needed to conclude if it is a valuable marker for cancer prognosis.

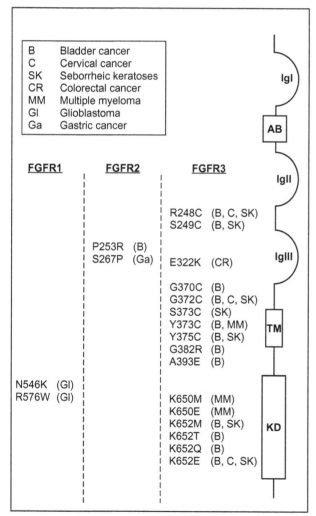

FIGURE 30.2 **FGFR mutations associated with human cancer.** FGFR mutations associated with various human cancers are indicated. The abbreviations are as follows: bladder cancer (B) [135–138], cervical cancer (C) [125], seborrheic keratoses (SK) [139], colorectal cancer (CR) [140], multiple myeloma (MM) [123, 124, 141, 142], glioblastoma (GI) [143], and gastric cancer (Ga) [140]. The mutations are placed at their approximate location in the FGF receptor.

REFERENCES

1. Dionne CA, Crumley G, Bellot F, et al. Cloning and expression of two distinct high-affinity receptors cross-reacting with acidic and basic fibroblast growth factors. *EMBO J* 1990;**9**:2685–92.

2. Partanen J, Makela TP, Alitalo R, Lehvaslaiho H, Alitalo K. Putative tyrosine kinases expressed in K-562 human leukemia cells. *Proc Natl Acad Sci USA* 1990;**87**:8913–17.

3. Keegan K, Johnson DE, Williams LT, Hayman MJ. Isolation of an additional member of the fibroblast growth factor receptor family, FGFR-3. *Proc Natl Acad Sci USA* 1991;**88**:1095–9.

4. Johnson DE, Williams LT. Structural and functional diversity in the FGF receptor multigene family. *Adv Cancer Res* 1993;**60**:1–41.

5. Szebenyi G, Fallon JF. Fibroblast growth factors as multifunctional signaling factors. *Intl Rev Cytol* 1999;**185**:45–106.

6. Powers CJ, McLeskey SW, Wellstein A. Fibroblast growth factors, their receptors and signaling. *Endocr Relat Cancer* 2000;**7**:165–97.

7. Ornitz DM, Itoh N. Fibroblast growth factors. *Genome Biol* 2001;**2**. REVIEWS3005.

8. Plotnikov AN, Schlessinger J, Hubbard SR, Mohammadi M. Structural basis for FGF receptor dimerization and activation. *Cell* 1999;**98**:641–50.

9. Xu J, Nakahara M, Crabb JW, et al. Expression and immunochemical analysis of rat and human fibroblast growth factor receptor (flg) isoforms. *J Biol Chem* 1992;**267**:17,792–17,803.

10. Bellot F, Crumley G, Kaplow JM, Schlessinger J, Jaye M, Dionne CA. Ligand-induced transphosphorylation between different FGF receptors. *EMBO J* 1991;**10**:2849–54.

11. Schlessinger J. Cell signaling by receptor tyrosine kinases. *Cell* 2000;**103**:211–25.

12. Hunter T. Signaling–2000 and beyond. *Cell* 2000;**100**:113–27.

13. McKeehan WL, Wang F, Kan M. The heparan sulfate-fibroblast growth factor family: diversity of structure and function. *Prog Nucleic Acid Res Mol Biol* 1998;**59**:135–76.

14. Harmer NJ. Insights into the role of heparan sulphate in fibroblast growth factor signalling. *Biochem Soc Trans* 2006;**34**:442–5.

15. Ibrahimi OA, Zhang F, Hrstka SC, Mohammadi M, Linhardt RJ. Kinetic model for FGF, FGFR, and proteoglycan signal transduction complex assembly. *Biochemistry* 2004;**43**:4724–30.

16. Pantoliano MW, Horlick RA, Springer BA, et al. Multivalent ligand–receptor binding interactions in the fibroblast growth factor system produce a cooperative growth factor and heparin mechanism for receptor dimerization. *Biochemistry* 1994;**33**:10,229–10,248.

17. Yayon A, Klagsbrun M, Esko JD, Leder P, Ornitz DM. Cell surface, heparin-like molecules are required for binding of basic fibroblast growth factor to its high affinity receptor. *Cell* 1991;**64**:841–8.

18. Webster MK, Donoghue DJ. FGFR activation in skeletal disorders: too much of a good thing. *Trends Genet* 1997;**13**:178–82.

19. Wilkie AO. Craniosynostosis: genes and mechanisms. *Hum Mol Genet* 1997;**6**:1647–56.

20. Hughes SE. Differential expression of the fibroblast growth factor receptor (FGFR) multigene family in normal human adult tissues. *J Histochem Cytochem* 1997;**45**:1005–19.

21. Korhonen J, Partanen J, Eerola E, et al. Novel human FGF receptors with distinct expression patterns. *Ann NY Acad Sci* 1991;**638**:403–5.

22. Partanen J, Makela TP, Eerola E, et al. FGFR-4, a novel acidic fibroblast growth factor receptor with a distinct expression pattern. *EMBO J* 1991;**10**:1347–54.

23. Katoh M, Hattori Y, Sasaki H, et al. K-sam gene encodes secreted as well as transmembrane receptor tyrosine kinase. *Proc Natl Acad Sci USA* 1992;**89**:2960–4.

24. Luqmani YA, Graham M, Coombes RC. Expression of basic fibroblast growth factor, FGFR1 and FGFR2 in normal and malignant human breast, and comparison with other normal tissues. *Br J Cancer* 1992;**66**:273–80.

25. Deng CX, Wynshaw-Boris A, Shen MM, Daugherty C, Ornitz DM, Leder P. Murine FGFR-1 is required for early postimplantation growth and axial organization. *Genes Dev* 1994;**8**:3045–57.

26. Peters KG, Werner S, Chen G, Williams LT. Two FGF receptor genes are differentially expressed in epithelial and mesenchymal tissues during limb formation and organogenesis in the mouse. *Development* 1992;**114**:233–43.

27. Wilke TA, Gubbels S, Schwartz J, Richman JM. Expression of fibroblast growth factor receptors (FGFR1, FGFR2, FGFR3) in the developing head and face. *Dev Dyn* 1997;**210**:41–52.

28. Heuer JG, von Bartheld CS, Kinoshita Y, Evers PC, Bothwell M. Alternating phases of FGF receptor and NGF receptor expression in the developing chicken nervous system. *Neuron* 1990;**5**:283–96.

29. Yamaguchi TP, Harpal K, Henkemeyer M, Rossant J. fgfr-1 is required for embryonic growth and mesodermal patterning during mouse gastrulation. *Genes Dev* 1994;**8**:3032–44.

30. Deng C, Bedford M, Li C, et al. Fibroblast growth factor receptor-1 (FGFR-1) is essential for normal neural tube and limb development. *Dev Biol* 1997;**185**:42–54.

31. Arman E, Haffner-Krausz R, Chen Y, Heath JK, Lonai P. Targeted disruption of fibroblast growth factor (FGF) receptor 2 suggests a role for FGF signaling in pregastrulation mammalian development. *Proc Natl Acad Sci USA* 1998;**95**:5082–7.

32. Arman E, Haffner-Krausz R, Gorivodsky M, Lonai P. Fgfr2 is required for limb outgrowth and lung-branching morphogenesis. *Proc Natl Acad Sci USA* 1999;**96**:11,895–11,899.

33. Werner S, Weinberg W, Liao X, et al. Targeted expression of a dominant-negative FGF receptor mutant in the epidermis of transgenic mice reveals a role of FGF in keratinocyte organization and differentiation. *EMBO J* 1993;**12**:2635–43.

34. Colvin JS, Bohne BA, Harding GW, McEwen DG, Ornitz DM. Skeletal overgrowth and deafness in mice lacking fibroblast growth factor receptor 3. *Nature Genet* 1996;**12**:390–7.

35. Deng C, Wynshaw-Boris A, Zhou F, Kuo A, Leder P. Fibroblast growth factor receptor 3 is a negative regulator of bone growth. *Cell* 1996;**84**:911–21.

36. Marcelle C, Eichmann A, Halevy O, Breant C, Le Douarin NM. Distinct developmental expression of a new avian fibroblast growth factor receptor. *Development* 1994;**120**:683–94.

37. Ozawa K, Uruno T, Miyakawa K, Seo M, Imamura T. Expression of the fibroblast growth factor family and their receptor family genes during mouse brain development. *Brain Res Mol Brain Res* 1996;**41**:279–88.

38. Stark KL, McMahon JA, McMahon AP. FGFR-4, a new member of the fibroblast growth factor receptor family, expressed in the definitive endoderm and skeletal muscle lineages of the mouse. *Development* 1991;**113**:641–51.

39. Yu C, Wang F, Kan M, et al. Elevated cholesterol metabolism and bile acid synthesis in mice lacking membrane tyrosine kinase receptor FGFR4. *J Biol Chem* 2000;**275**:15,482–15,489.

40. Dailey L, Ambrosetti D, Mansukhani A, Basilico C. Mechanisms underlying differential responses to FGF signaling. *Cytokine Growth Factor Rev* 2005;**16**:233–47.

41. Lemmon MA, Schlessinger J. Regulation of signal transduction and signal diversity by receptor oligomerization. *Trends Biochem Sci* 1994;**19**:459–63.

42. Pawson T. Protein modules and signalling networks. *Nature* 1995;**373**:573–80.

43. Klint P, Claesson-Welsh L. Signal transduction by fibroblast growth factor receptors. *Front Biosci* 1999;**4**:D165–77.

44. Xu H, Lee KW, Goldfarb M. Novel recognition motif on fibroblast growth factor receptor mediates direct association and activation of SNT adapter proteins. *J Biol Chem* 1998;**273**:17,987–17,990.

45. Ong SH, Guy GR, Hadari YR, et al. FRS2 proteins recruit intracellular signaling pathways by binding to diverse targets on fibroblast growth factor and nerve growth factor receptors. *Mol Cell Biol* 2000;**20**:979–89.

46. Browaeys-Poly E, Cailliau K, Vilain JP. Transduction cascades initiated by fibroblast growth factor 1 on Xenopus oocytes expressing MDA-MB-231 mRNAs. Role of Grb2, phosphatidylinositol 3-kinase, Src tyrosine kinase, and phospholipase Cgamma. *Cell Signal* 2001;**13**:363–8.

47. Hadari YR, Gotoh N, Kouhara H, Lax I, Schlessinger J. Critical role for the docking-protein FRS2 alpha in FGF receptor-mediated signal transduction pathways. *Proc Natl Acad Sci USA* 2001;**98**:8578–83.

48. Eswarakumar VP, Lax I, Schlessinger J. Cellular signaling by fibroblast growth factor receptors. *Cytokine Growth Factor Rev* 2005;**16**:139–49.

49. Hadari YR, Kouhara H, Lax I, Schlessinger J. Binding of Shp2 tyrosine phosphatase to FRS2 is essential for fibroblast growth factor-induced PC12 cell differentiation. *Mol Cell Biol* 1998;**18**:3966–73.

50. Cross MJ, Lu L, Magnusson P, et al. The Shb adaptor protein binds to tyrosine 766 in the FGFR-1 and regulates the Ras/MEK/MAPK pathway via FRS2 phosphorylation in endothelial cells. *Mol Biol Cell* 2002;**13**:2881–93.

51. Sandilands E, Akbarzadeh S, Vecchione A, McEwan DG, Frame MC, Heath JK. Src kinase modulates the activation, transport and

signalling dynamics of fibroblast growth factor receptors. *EMBO Rep* 2007;**8**(12):1162–9.

52. LaVallee TM, Prudovsky IA, McMahon GA, Hu X, Maciag T. Activation of the MAP kinase pathway by FGF-1 correlates with cell proliferation induction while activation of the Src pathway correlates with migration. *J Cell Biol* 1998;**141**:1647–58.

53. Carpenter G, Ji Q. Phospholipase C-gamma as a signal-transducing element. *Exp Cell Res* 1999;**253**:15–24.

54. Larsson H, Klint P, Landgren E, Claesson-Welsh L. Fibroblast growth factor receptor-1-mediated endothelial cell proliferation is dependent on the Src homology (SH) 2/SH3 domain-containing adaptor protein Crk. *J Biol Chem* 1999;**274**:25,726–25,734.

55. Reuss B, von Bohlenund und Halbach O. Fibroblast growth factors and their receptors in the central nervous system. *Cell Tissue Res* 2003;**313**:139–57.

56. Li W, Fan J, Woodley DT. Nck/Dock: an adapter between cell surface receptors and the actin cytoskeleton. *Oncogene* 2001;**20**:6403–17.

57. Hart KC, Robertson SC, Kanemitsu MY, Meyer AN, Tynan JA, Donoghue DJ. Transformation and Stat activation by derivatives of FGFR1, FGFR3, and FGFR4. *Oncogene* 2000;**19**:3309–20.

58. Kong M, Wang CS, Donoghue DJ. Interaction of fibroblast growth factor receptor 3 and the adapter protein SH2-B. A role in STAT5 activation. *J Biol Chem* 2002;**277**:15,962–15,970.

59. McIntosh I, Bellus GA, Jab EW. The pleiotropic effects of fibroblast growth factor receptors in mammalian development. *Cell Struct Funct* 2000;**25**:85–96.

60. Passos-Bueno MR, Wilcox WR, Jabs EW, Sertie AL, Alonso LG, Kitoh H. Clinical spectrum of fibroblast growth factor receptor mutations. *Hum Mutat* 1999;**14**:115–25.

61. Plotnikov AN, Hubbard SR, Schlessinger J, Mohammadi M. Crystal structures of two FGF–FGFR complexes reveal the determinants of ligand-receptor specificity. *Cell* 2000;**101**:413–24.

62. Schlessinger J, Plotnikov AN, Ibrahimi OA, et al. Crystal structure of a ternary FGF–FGFR-heparin complex reveals a dual role for heparin in FGFR binding and dimerization. *Mol Cell* 2000;**6**:743–50.

63. Hehr U, Muenke M. Craniosynostosis syndromes: from genes to premature fusion of skull bones. *Mol Genet Metab* 1999;**68**:139–51.

64. Katzen JT, McCarthy JG. Syndromes involving craniosynostosis and midface hypoplasia. *Otolaryngol Clin North Am* 2000;**33**:1257–84. vi.

65. Park WJ, Meyers GA, Li X, et al. Novel FGFR2 mutations in Crouzon and Jackson-Weiss syndromes show allelic heterogeneity and phenotypic variability. *Hum Mol Genet* 1995;**4**:1229–33.

66. Freitas EC, Nascimento SR, de Mello MP, Gil-da-Silva-Lopes VL. Q289P mutation in FGFR2 gene causes Saethre-Chotzen syndrome: some considerations about familial heterogeneity. *Cleft Palate Craniofac J* 2006;**43**:142–7.

67. Lajeunie E, Heuertz S, El Ghouzzi V, et al. Mutation screening in patients with syndromic craniosynostoses indicates that a limited number of recurrent FGFR2 mutations accounts for severe forms of Pfeiffer syndrome. *Eur J Hum Genet* 2006;**14**:289–98.

68. Wilkie AO, Bochukova EG, Hansen RM, et al. Clinical dividends from the molecular genetic diagnosis of craniosynostosis. *Am J Med Genet A* 2007;**143**:1941–9.

69. Oberklaid F, Danks DM, Jensen F, Stace L, Rosshandler S. Achondroplasia and hypochondroplasia. Comments on frequency, mutation rate, and radiological features in skull and spine. *J Med Genet* 1979;**16**:140–6.

70. Rousseau F, Bonaventure J, Legeai-Mallet L, et al. Mutations in the gene encoding fibroblast growth factor receptor-3 in achondroplasia. *Nature* 1994;**371**:252–4.

71. Shiang R, Thompson LM, Zhu YZ, et al. Mutations in the transmembrane domain of FGFR3 cause the most common genetic form of dwarfism, achondroplasia. *Cell* 1994;**78**:335–42.

72. Tavormina PL, Shiang R, Thompson LM, et al. Thanatophoric dysplasia (types I and II) caused by distinct mutations in fibroblast growth factor receptor 3. *Nature Genet* 1995;**9**:321–8.

73. Tavormina PL, Bellus GA, Webster MK, et al. A novel skeletal dysplasia with developmental delay and acanthosis nigricans is caused by a Lys650Met mutation in the fibroblast growth factor receptor 3 gene. *Am J Hum Genet* 1999;**64**:722–31.

74. Brodie SG, Kitoh H, Lachman RS, Nolasco LM, Mekikian PB, Wilcox WR. Platyspondylic lethal skeletal dysplasia, San Diego type, is caused by FGFR3 mutations. *Am J Med Genet* 1999;**84**:476–80.

75. Orioli IM, Castilla EE, Barbosa-Neto JG. The birth prevalence rates for the skeletal dysplasias. *J Med Genet* 1986;**23**:328–32.

76. Rousseau F, Bonaventure J, Le Merrer M, Maroteaux P, Munnich A. [Mutations of FGFR3 gene cause 3 types of nanisms with variably severity: achondroplasia, thanatophoric nanism and hypochondroplasia]. *Ann Endocrinol (Paris)* 1996;**57**:153.

77. Rousseau F, Bonaventure J, Le Merrer M, Munnich A. [Association of achondroplasia to a mutation in the transmembrane domain of fibroblastic growth factor receptor 3 (FGFR3)]. *Ann Endocrinol (Paris)* 1996;**57**:151–2.

78. Rousseau F, Bonaventure J, Legeai-Mallet L, et al. Clinical and genetic heterogeneity of hypochondroplasia. *J Med Genet* 1996;**33**:749–52.

79. Rousseau F, el Ghouzzi V, Delezoide AL, et al. Missense FGFR3 mutations create cysteine residues in thanatophoric dwarfism type I (TD1). *Hum Mol Genet* 1996;**5**:509–12.

80. Rousseau F, Bonaventure J, Legeai-Mallet L, et al. Mutations of the fibroblast growth factor receptor-3 gene in achondroplasia. *Horm Res* 1996;**45**:108–10.

81. Rousseau F, Saugier P, Le Merrer M, et al. Stop codon FGFR3 mutations in thanatophoric dwarfism type 1. *Nature Genet* 1995;**10**:11–12.

82. L'Hote CG, Knowles MA. Cell responses to FGFR3 signalling: growth, differentiation and apoptosis. *Exp Cell Res* 2005;**304**:417–31.

83. Cohen Jr. MM The new bone biology: pathologic, molecular, and clinical correlates. *Am J Med Genet A* 2006;**140**:2646–706.

84. Dode C, Levilliers J, Dupont JM, et al. Loss-of-function mutations in FGFR1 cause autosomal dominant Kallmann syndrome. *Nature Genet* 2003;**33**:463–5.

85. Cadman SM, Kim SH, Hu Y, Gonzalez-Martinez D, Bouloux PM. Molecular pathogenesis of Kallmann's syndrome. *Horm Res* 2007;**67**:231–42.

86. Toydemir RM, Brassington AE, Bayrak-Toydemir P, et al. A novel mutation in FGFR3 causes camptodactyly, tall stature, and hearing loss (CATSHL) syndrome. *Am J Hum Genet* 2006;**79**:935–41.

87. Rusnati M, Presta M. Fibroblast growth factors/fibroblast growth factor receptors as targets for the development of anti-angiogenesis strategies. *Curr Pharm Des* 2007;**13**:2025–44.

88. Tassi E, Wellstein A. The angiogenic switch molecule, secreted FGF-binding protein, an indicator of early stages of pancreatic and colorectal adenocarcinoma. *Semin Oncol* 2006;**33**:S50–6.

89. Giri D, Ropiquet F, Ittmann M. Alterations in expression of basic fibroblast growth factor (FGF) 2 and its receptor FGFR-1 in human prostate cancer. *Clin Cancer Res* 1999;**5**:1063–71.

90. Kwabi-Addo B, Ozen M, Ittmann M. The role of fibroblast growth factors and their receptors in prostate cancer. *Endocr Relat Cancer* 2004;**11**:709–24.

91. Penault-Llorca F, Bertucci F, Adelaide J, et al. Expression of FGF and FGF receptor genes in human breast cancer. *Intl J Cancer* 1995;**61**:170–6.

92. Zhong H, Deng F, Kong X. [Expression of basic fibroblast growth factor and its receptor in renal cell carcinoma]. *Zhonghua Wai Ke Za Zhi* 1996;**34**:651–4.

93. Jaakkola S, Salmikangas P, Nylund S, et al. Amplification of fgfr4 gene in human breast and gynecological cancers. *Int J Cancer* 1993;**54**:378–82.

94. Valve E, Martikainen P, Seppanen J, et al. Expression of fibroblast growth factor (FGF)-8 isoforms and FGF receptors in human ovarian tumors. *Intl J Cancer* 2000;**88**:718–25.

95. Chandler LA, Sosnowski BA, Greenlees L, Aukerman SL, Baird A, Pierce GF. Prevalent expression of fibroblast growth factor (FGF) receptors and FGF2 in human tumor cell lines. *Intl J Cancer* 1999;**81**:451–8.

96. Takaishi S, Sawada M, Morita Y, Seno H, Fukuzawa H, Chiba T. Identification of a novel alternative splicing of human FGF receptor 4: soluble-form splice variant expressed in human gastrointestinal epithelial cells. *Biochem Biophys Res Commun* 2000;**267**:658–62.

97. St Bernard R, Zheng L, Liu W, Winer D, Asa SL, Ezzat S. Fibroblast growth factor receptors as molecular targets in thyroid carcinoma. *Endocrinology* 2005;**146**:1145–53.

98. Abbass SA, Asa SL, Ezzat S. Altered expression of fibroblast growth factor receptors in human pituitary adenomas. *J Clin Endocrinol Metab* 1997;**82**:1160–6.

99. Ezzat S, Zheng L, Zhu XF, Wu GE, Asa SL. Targeted expression of a human pituitary tumor-derived isoform of FGF receptor-4 recapitulates pituitary tumorigenesis. *J Clin Invest* 2002;**109**:69–78.

100. Yamada SM, Yamada S, Hayashi Y, Takahashi H, Teramoto A, Matsumoto K. Fibroblast growth factor receptor (FGFR) 4 correlated with the malignancy of human astrocytomas. *Neurol Res* 2002;**24**:244–8.

101. Mawrin C, Kirches E, Diete S, et al. Analysis of a single nucleotide polymorphism in codon 388 of the FGFR4 gene in malignant gliomas. *Cancer Letts* 2006;**239**:239–45.

102. Tsou AP, Wu KM, Tsen TY, et al. Parallel hybridization analysis of multiple protein kinase genes: identification of gene expression patterns characteristic of human hepatocellular carcinoma. *Genomics* 1998;**50**:331–40.

103. Qiu WH, Zhou BS, Chu PG, et al. Over-expression of fibroblast growth factor receptor 3 in human hepatocellular carcinoma. *World J Gastroenterol* 2005;**11**:5266–72.

104. Kornmann M, Beger HG, Korc M. Role of fibroblast growth factors and their receptors in pancreatic cancer and chronic pancreatitis. *Pancreas* 1998;**17**:169–75.

105. Streit S, Mestel DS, Schmidt M, Ullrich A, Berking C. FGFR4 Arg388 allele correlates with tumour thickness and FGFR4 protein expression with survival of melanoma patients. *Br J Cancer* 2006;**94**:1879–86.

106. Spinola M, Leoni V, Pignatiello C, et al. Functional FGFR4 Gly388Arg polymorphism predicts prognosis in lung adenocarcinoma patients. *J Clin Oncol* 2005;**23**:7307–11.

107. Armstrong E, Vainikka S, Partanen J, Korhonen J, Alitalo R. Expression of fibroblast growth factor receptors in human leukemia cells. *Cancer Res* 1992;**52**:2004–7.

108. Cronauer MV, Schulz WA, Seifert HH, Ackermann R, Burchardt M. Fibroblast growth factors and their receptors in urological cancers: basic research and clinical implications. *Eur Urol* 2003;**43**:309–19.

109. Morimoto Y, Ozaki T, Ouchida M, et al. Single nucleotide polymorphism in fibroblast growth factor receptor 4 at codon 388 is associated with prognosis in high-grade soft tissue sarcoma. *Cancer* 2003;**98**:2245–50.

110. Streit S, Bange J, Fichtner A, Ihrler S, Issing W, Ullrich A. Involvement of the FGFR4 Arg388 allele in head and neck squamous cell carcinoma. *Intl J Cancer* 2004;**111**:213–17.

111. Khnykin D, Troen G, Berner JM, Delabie J. The expression of fibroblast growth factors and their receptors in Hodgkin's lymphoma. *J Pathol* 2006;**208**:431–8.

112. Lappi DA. Tumor targeting through fibroblast growth factor receptors. *Semin Cancer Biol* 1995;**6**:279–88.

113. Wang F, McKeehan K, Yu C, McKeehan WL. Fibroblast growth factor receptor 1 phosphotyrosine 766: molecular target for prevention of progression of prostate tumors to malignancy. *Cancer Res* 2002;**62**:1898–903.

114. Reis-Filho JS, Simpson PT, Turner NC, et al. FGFR1 emerges as a potential therapeutic target for lobular breast carcinomas. *Clin Cancer Res* 2006;**12**:6652–62.

115. Elbauomy Elsheikh S, Green AR, Lambros MB, et al. FGFR1 amplification in breast carcinomas: a chromogenic in situ hybridisation analysis. *Breast Cancer Res* 2007;**9**:R23.

116. Winter SF, Acevedo VD, Gangula RD, Freeman KW, Spencer DM, Greenberg NM. Conditional activation of FGFR1 in the prostate epithelium induces angiogenesis with concomitant differential regulation of Ang-1 and Ang-2. *Oncogene* 2007;**26**:4897–907.

117. Devilard E, Bladou F, Ramuz O, et al. FGFR1 and WT1 are markers of human prostate cancer progression. *BMC Cancer* 2006;**6**:272.

118. Macdonald D, Cross NC. Chronic myeloproliferative disorders: the role of tyrosine kinases in pathogenesis, diagnosis and therapy. *Pathobiology* 2007;**74**:81–8.

119. Delhommeau F, Pisani DF, James C, Casadevall N, Constantinescu S, Vainchenker W. Oncogenic mechanisms in myeloproliferative disorders. *Cell Mol Life Sci* 2006;**63**:2939–53.

120. Hunter DJ, Kraft P, Jacobs KB, et al. A genome-wide association study identifies alleles in FGFR2 associated with risk of sporadic postmenopausal breast cancer. *Nature Genet* 2007;**39**:870–4.

121. Easton DF, Pooley KA, Dunning AM, et al. Genome-wide association study identifies novel breast cancer susceptibility loci. *Nature* 2007;**447**:1087–93.

122. Takeda M, Arao T, Yokote H, et al. AZD2171 shows potent antitumor activity against gastric cancer over-expressing fibroblast growth factor receptor 2/keratinocyte growth factor receptor. *Clin Cancer Res* 2007;**13**:3051–7.

123. Chesi M, Nardini E, Brents LA, et al. Frequent translocation t(4;14)(p16.3;q32.3) in multiple myeloma is associated with increased expression and activating mutations of fibroblast growth factor receptor 3. *Nature Genet* 1997;**16**:260–4.

124. Richelda R, Ronchetti D, Baldini L, et al. A novel chromosomal translocation t(4; 14)(p16.3; q32) in multiple myeloma involves the fibroblast growth-factor receptor 3 gene. *Blood* 1997;**90**:4062–70.

125. Cappellen D, De Oliveira C, Ricol D, et al. Frequent activating mutations of FGFR3 in human bladder and cervix carcinomas. *Nature Genet* 1999;**23**:18–20.

126. da Costa Andrade Jr. VC, Parise O, Hors CP, de Melo Martins PC, Silva AP, Garicochea B The fibroblast growth factor receptor 4 (FGFR4) Arg388 allele correlates with survival in head and neck squamous cell carcinoma. *Exp Mol Pathol* 2007;**82**:53–7.

127. Bange J, Prechtl D, Cheburkin Y, et al. Cancer progression and tumor cell motility are associated with the FGFR4 Arg(388) allele. *Cancer Res* 2002;**62**:840–7.

128. Stadler CR, Knyazev P, Bange J, Ullrich A. FGFR4 GLY388 isotype suppresses motility of MDA-MB-231 breast cancer cells by EDG-2 gene repression. *Cell Signal* 2006;**18**:783–94.

129. Thussbas C, Nahrig J, Streit S, et al. FGFR4 Arg388 allele is associated with resistance to adjuvant therapy in primary breast cancer. *J Clin Oncol* 2006;**24**:3747–55.

130. Wang J, Stockton DW, Ittmann M. The fibroblast growth factor receptor-4 Arg388 allele is associated with prostate cancer initiation and progression. *Clin Cancer Res* 2004;**10**:6169–78.

131. Matakidou A, El Galta R, Rudd MF, et al. Further observations on the relationship between the FGFR4 Gly388Arg polymorphism and lung cancer prognosis. *Br J Cancer* 2007;**96**:1904–7.

132. Mawrin C, Schneider T, Firsching R, et al. Assessment of tumor cell invasion factors in gliomatosis cerebri. *J Neurooncol* 2005;**73**:109–15.

133. Spinola M, Leoni VP, Tanuma J, et al. FGFR4 Gly388Arg polymorphism and prognosis of breast and colorectal cancer. *Oncol Rep* 2005;**14**:415–19.

134. Jezequel P, Campion L, Joalland MP, et al. G388R mutation of the FGFR4 gene is not relevant to breast cancer prognosis. *Br J Cancer* 2004;**90**:189–93.

135. Van Rhijn BW, van Tilborg AA, Lurkin I, et al. Novel fibroblast growth factor receptor 3 (FGFR3) mutations in bladder cancer previously identified in non-lethal skeletal disorders. *Eur J Hum Genet* 2002;**10**:819–24.

136. Billerey C, Chopin D, Aubriot-Lorton MH, et al. Frequent FGFR3 mutations in papillary non-invasive bladder (pTa) tumors. *Am J Pathol* 2001;**158**:1955–9.

137. Andreou A, Lamy A, Layet V, et al. Early-onset low-grade papillary carcinoma of the bladder associated with Apert syndrome and a germline FGFR2 mutation (Pro253Arg). *Am J Med Genet A* 2006;**140**:2245–7.

138. Tomlinson DC, Baldo O, Harnden P, Knowles MA. FGFR3 protein expression and its relationship to mutation status and prognostic variables in bladder cancer. *J Pathol* 2007;**213**:91–8.

139. Hafner C, Hartmann A, van Oers JM, et al. FGFR3 mutations in seborrheic keratoses are already present in flat lesions and associated with age and localization. *Mod Pathol* 2007;**20**:895–903.

140. Jang JH, Shin KH, Park JG. Mutations in fibroblast growth factor receptor 2 and fibroblast growth factor receptor 3 genes associated with human gastric and colorectal cancers. *Cancer Res* 2001;**61**:3541–3.

141. Chesi M, Nardini E, Lim RS, Smith KD, Kuehl WM, Bergsagel PL. The t(4;14) translocation in myeloma dysregulates both FGFR3 and a novel gene, MMSET, resulting in IgH/MMSET hybrid transcripts. *Blood* 1998;**92**:3025–34.

142. Fracchiolla NS, Luminari S, Baldini L, Lombardi L, Maiolo AT, Neri A. FGFR3 gene mutations associated with human skeletal disorders occur rarely in multiple myeloma. *Blood* 1998;**92**:2987–9.

143. Rand V, Huang J, Stockwell T, et al. Sequence survey of receptor tyrosine kinases reveals mutations in glioblastomas. *Proc Natl Acad Sci USA* 2005;**102**:14,344–14,349.

Chapter 31

Regulation of Synaptic Fusion by Heterotrimeric G Proteins

Simon Alford, Edaeni Hamid, Trillium Blackmer and Tatyana Gerachshenko

Department of Biological Sciences, University of Illinois at Chicago, Chicago, Illinois

INTRODUCTION

Chemical synaptic transmission is central to neural communication. Alterations in its efficacy underlie our identity; physiological and pathophysiological. A component of this synaptic plasticity is controlled by presynaptic G-protein-coupled-receptors (GPCRs), which universally modify neurotransmitter release at all synapses in which their effect has been studied. GPCRs act either as autoreceptors following neurotransmitter release from that particular terminal, or as heteroreceptors activated by axo-axonic or paracrine projections to the synapse. In neurons, exocytosis, or the fusion of a synaptic vesicle with a specialized area in the plasma membrane to cause neurotransmitter release, is a tightly regulated process that must be activated with strict orchestration. Ca^{2+} entry to the terminal, the trigger for exocytosis and neurotransmitter release, might be considered to represent the conductor. Thus, GPCRs have been found to alter the entry of Ca^{2+} to the presynaptic terminal, its storage and release in the terminal, and the outcome of its entry to the presynaptic terminal.

THE VESICLE FUSION MACHINERY

To understand how G proteins modulate the release of neurotransmitter, we must understand some basic principles of exocytosis. Synaptic transmission requires regulated exocytosis; the fusion of a synaptic vesicle with a specialized area in the plasma membrane [1, 2] that utilizes a group of presynaptic proteins constituting the SNARE complex. The core complex, or SNARE (soluble NSF (N-ethylmaleimide-sensitive factor) attachment protein receptor), is a bundle of four α-helices, approximately 65 amino acids in length, which is thought to bridge the synaptic vesicle and plasma membranes [3]. These α-helices are donated by three different proteins; a family member from the syntaxin and the SNAP-25 families, located in the synaptic "active zone," and a VAMP (Vesicle Associated Membrane Protein, also known as synaptobrevin) family member, located in the synaptic vesicular membrane. The syntaxin family consists of integral membrane proteins, around 300 amino acids in length, that have been shown to bind to many regulatory proteins [4–6]. It is now clear that among these regulatory proteins, synaptotagmin represents the Ca^{2+} sensor in synchronous release of neurotransmitter. Indeed, synaptotagmin interacts with the SNARE complex C-terminal region [7–9] as well as vesicle and cell membranes [10, 11] during Ca^{2+}-dependent fusion. The core complex is sufficient to mediate fusion of lipid micelles *in vitro* [12], and fusion of the synaptic vesicle with the plasma membrane requires the interaction of syntaxin, SNAP-25, and VAMP.

Vesicles containing neurotransmitter must be able to fuse with the plasma membrane in microsecond timescales. Consequently, synaptic vesicles are located very near to the point of fusion, the active zone, at the presynaptic terminal. This organization is termed *docking*. Vesicular recruitment and docking requires ATP [13]. Multi-step fusion reactions are unlikely due to the speed of release. For this reason, it is believed that there is a pool of ready-to-fuse synaptic vesicles that have undergone a further maturation step, referred to as *priming* [14]. Priming in large dense-core vesicles requires ATP, submicromolar Ca^{2+} concentrations, and alterations in membrane lipids by lipid transferases and kinases [15]. Furthermore, the possible role in priming of the N-ethylmaleimide-sensitive factor (NSF) in synaptic vesicles may indicate similar requirements [16]. The priming reaction also provides a target for modification of release by G-protein-coupled receptors. During priming, syntaxin undergoes a conformational modification facilitated by UNC 13, which allows syntaxin to contribute to the SNARE complex [17–20]. At the final stage of evoked release, synaptic vesicle fusion is thought to require high (hundreds of μM) local concentrations of Ca^{2+} [21] following action potential invasion of the nerve terminal, although more recent work suggests that low μM increases in Ca^{2+} concentrations ($\sim$10 μM) may activate fusion in some neurons [22].

MODES OF SYNAPTIC VESICLE FUSION

Until recently, it was almost universally accepted that transmitters are released as quanta. This premise, first based on recordings of postjunctional potentials at the neuromuscular junction [23], was later combined with the vesicular theory of transmitter release [2, 24]. Consequently, presynaptic modification of neurotransmitter release has been envisaged as a means of altering the probability of exocytosis occurring during presynaptic action potentials. Synaptic neurotransmitter release may, however, be more subtle and complex. Quantal transmission is believed to result from complete vesicle fusion with the presynaptic membrane, but complete fusion may not always occur. Three hypotheses have been proposed to account for the recycling of vesicle membrane after exocytosis. Heuser and Reese [2] proposed that vesicles are recycled through an endosome. Concern that this process is too slow to account for recycling observed experimentally [25] led De Camilli and colleagues [26] to propose a model by which the vesicle completely fuses with the presynaptic membrane to be recovered into vesicles through clathrin-coated intermediaries. However, it is apparent that large, dense-core vesicles can fuse with the membrane and then return to a pool of vesicles available for release without being recycled by clathrin-mediated pathways, and this may lead to variability in transmitter release through a process that has been named "kiss and run" [27–29]. Measurements of catecholamine release from adrenal chromaffin vesicles provide direct evidence that variable amounts of hormone may be released during vesicle fusion [30–31]. More recently, it has been proposed that changes in interactions between synaptotagmin (the Ca^{2+} sensor) and SNAP-25 may govern a mode shift of fusion. It has been suggested that synaptotagmin-Imay govern complete fusion [32] and synaptotagmin-IV incomplete fusion (kiss and run) [33].

In the mammalian CNS, incomplete fusion events also occur [34–36]. Use of vesicle-staining techniques has demonstrated that various fusion modes occur at central synapses [37–34], and that incomplete fusion can alter neurotransmitter release [38, 41–43]. This modification of vesicle fusion mode represents another point at which G-protein-coupled receptors may intervene in synaptic transmission.

G-PROTEIN-COUPLED RECEPTOR MEDIATED MODULATION AT THE PRESYNAPTIC TERMINAL

Regulation of neurotransmitter release at the presynaptic terminal plays an important part in the plasticity of the nervous system [44]. Various neurotransmitters modulate release from presynaptic terminals, and many of these interactions involve the activation of a G-protein-coupled receptor (GPCR) [45]. Modulation of exocytotic release by GPCRs is an important mechanism by which neurons are able to respond and adapt to changes in secretory requirements. Some GPCRs may couple to more than one G protein, while others show a great deal of specificity. G$\beta\gamma$ binding to Gα involves widespread contacts at two distinct interfaces. Following activation by a GPCR, the heterotrimeric G protein dissociates into an activated Gα-GTP subunit and a free G$\beta\gamma$ subunit [46] Active Gα-GTP and free G$\beta\gamma$ may then activate many different signaling pathways [47].

Uncertainty over the mechanisms by which G proteins alter neurotransmitter release in part reflects the variety of G-protein effector targets and the difficulties in gaining experimental access to these small structures. Thus, most molecular studies of the detailed mechanisms come from either transfection of the relevant proteins into cultured cell lines and *Xenopus* oocytes, or from electrophysiological measurements from neuronal cell bodies.

POSSIBLE MECHANISMS OF PRESYNAPTIC INHIBITION BY G PROTEINS

GPCRs that inhibit neurotransmitter release have perhaps been the mostly widely studied modulators of synaptic transmission. A consensus mechanism by which these transmitters may modulate synaptic transmitter release has been hypothesized to involve an alteration in action potential-evoked Ca^{2+} entry to the presynaptic terminal.

1. That this reduction in Ca^{2+} entry may occur by a modification of voltage gated Ca^{2+} channels (VGCCs) was first demonstrated for adrenergic and serotonergic receptors 30 years ago [48] (Figure 31.1), but this result has since been expanded upon to include many GPCRs [49, 50]. This effect was since demonstrated to be mediated by a direct membrane-delimited [51] action of G$\beta\gamma$ [52, 53]. If G$\beta\gamma$ inhibits VGCCs, less Ca^{2+} will enter the presynaptic terminal and, since neurotransmitter release is Ca^{2+}-dependent, less neurotransmitter will be released. GPCR-mediated inhibition of release via direct inhibition of VGCCs has been demonstrated at one presynaptic terminal [54] through multiple GPCRs [55, 56]. Given the large number of neurotransmitters that inhibit release, many of which have very little effect

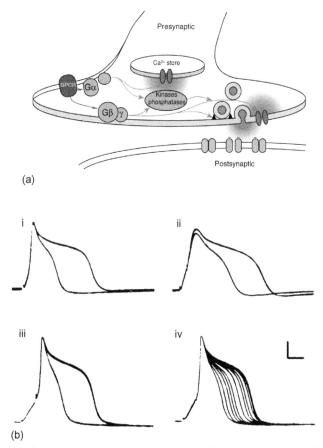

FIGURE 31.1(A) GPCRs at the presynaptic terminal may modulate transmission through either Gα or Gβγ. These pathways may in turn alter the phosphorylation state of presynaptic proteins, cause the release of Ca^{2+} from presynaptic internal stores, or alter release through interactions with the release machinery or through actions at presynaptic ion channels.

(b) This demonstrates the most common explanation for GPCR-mediated presynaptic inhibition in which activation of numerous GPCRs can inhibit Ca^{2+} channels activation by a membrane-delimited effect of Gβγ. The resultant reduction in Ca^{2+} entry may then cause less neurotransmitter release. The effect of (i), 10^{-5} M 5-HT; (ii) 10^{-4} M GABA; (iii, iv) 10^{-4} M noradrenaline on dorsal root ganglion neuron soma action potential. The two successive sweeps show the spike before (longer duration) and after drug application. Return to control shown in d with 10-s interval between sweeps. Calibration, 20 mV and 2 ms.

Figure 31.1B is reprinted from [48] (Dunlap and Fischbach, 1978) by permission from MacMillan Publishers Ltd, *Nature* (Dunlap and Fischbach, 1978; 276: 837) ©1978.

on Ca^{2+} entry through VGCCs [55], it is unlikely that this is the only pathway involved in GPCR-mediated inhibition at this terminal. Nevertheless, there is compelling evidence that Gβγ-mediated modification of Ca^{2+} channels plays an important role in presynaptic inhibition [57]. A great deal is now understood about the mechanisms by which Gβγ modifies Ca^{2+} channel function, and this has been covered in depth in a recent review [58]. However, these mechanisms are clearly important to presynaptic function. The strongest evidence for Gβγ-mediated modulation of Ca^{2+} channel function is at the $C_{av}2.x$ class of channels, which includes N, P/Q, and R type α subunits [50, 52, 53, 59, 60]. These are precisely the channels that cause neurotransmitter release. Gβγ binding to the Ca^{2+} channel is complex. Ca^{2+} channel α subunits that form the

channel pore comprise four homologous domains linked to each other with intracellular region. Each domain comprises six transmembrane regions and a re-entrant loop that contribute to the ion channel. Gβγ binds to the N-terminal of the α subunit [61] as well as the I/II domain linker region [62]. In addition, this binding may be modified by interactions with the channel C-terminus [63] and by interaction with Ca^{2+}-channel β subunits [64]. Gβγ-mediated inhibition of Ca^{2+} channels may also be modified by interaction of the channel and Gβγ with the SNARE complex protein syntaxin. In this respect, syntaxin is likely to act as a Ca^{2+} channel–Gβγ chaperone, increasing the probability of interaction [65, 66]– although, as discussed below, direct interactions between Gβγ and the SNARE complex may modify neurotransmitter release.

The direct, membrane-delimited action by which Gβγ modifies Ca^{2+} channel function is also, to some extent, voltage-dependent. A strong (to $+100$-mV) pre-pulse depolarization of the channels relieves GPCR-mediated Ca^{2+} channel inhibition [67]. Clearly, neurons never undergo such a profound stimulus under physiological circumstances; however, extended intense (200-Hz) stimulation of the channel to more physiological voltages may imitate this effect [68]. It remains unclear to what extent this might play a role in functional synapses.

2. If G-protein-coupled inwardly rectifying K^+ channels (GIRKs) were located at presynaptic terminals, activation by Gβγ could modulate action potential amplitudes, allowing fewer Ca^{2+} channels to open. Gβγ activates GIRKs in neuronal cell bodies, transfected cell lines, and *Xenopus* oocytes [69, 70]. While GIRKs have been histochemically localized to presynaptic terminals [71], physiological evidence for their presynaptic action is lacking. These channels are believed to be more important for postsynaptic modulation [72]. GPCR-mediated inhibition of voltage-gated K^+ channels at an autaptic presynaptic terminal has also been shown [73] to occur through activation of dopamine receptors, although it is not clear which G-protein subunit, Gα or Gβγ, is responsible. At the reticulospinal-motoneuron synapse of the lamprey, both glutamate and 5-HT activate GPCRs which modulate a K^+ current, although the channel subtype and G-protein subunit involved are unknown [74–75].

3. It has been suggested that G proteins may modulate voltage-gated Na^+ channels at the presynaptic terminal; however, no direct evidence has yet been presented [76]. Modulation of Na^+ channels could also indirectly affect the entry of Ca^{2+} into the presynaptic terminal.

4. Presynaptic Ca^{2+} signaling may also be modified by less direct pathways. Often termed voltage-independent G-protein inhibition, because GPCR-mediated inhibition is not relaxed by application of a pre-pulse depolarization, these effects are mediated by cytosolic signaling rather than a direct membrane-delimited action of Gβγ. Free Gβγ and activated Gα-GTP signal by numerous mechanisms in all cells. Thus, GPCRs can inhibit Ca^{2+} channels via a number of cytosolic messengers. An example of such mechanisms is the effect of dopamine D1 receptor signaling using Gαs and PKA and protein phosphatase 1 to dephosphorylate N and P/Q channels [77]. This example perhaps highlights the sensitivity of Ca^{2+} channel function to its phosphorylation state.

These studies led to the idea that Gβγ-mediated inhibition of neurotransmission at the presynaptic terminal was through a direct or indirect effect on the amount of Ca^{2+} that enters the terminal during the action potential. However, there is growing evidence that Gβγ may also inhibit synaptic transmission by modulation distal to the point of Ca^{2+} entry. The ability of G proteins to inhibit neurotransmitter release by directly targeting the release apparatus was first demonstrated by Silinsky [78] in the neuromuscular junction. Spontaneous exocytotic events, where exocytosis occurs independently of Ca^{2+} entry, can be detected by recording miniature excitatory/inhibitory postsynaptic currents (mE/IPSCs). Measurements of mE/IPSCs allow the exocytotic modulatory processes that occur independently of Ca^{2+} entry to be isolated and studied, unlike evoked EPSCs. These mE/IPSCs have been shown to be regulated by many GPCRs [79].

In support of the hypothesis that they inhibit vesicle fusion directly, G proteins can inhibit exocytosis after cell permeabilization, suggesting a role late in the exocytotic event [80]. Additionally, exocytotic processes in pancreatic β cells, peritoneal mast cells, chromaffin cells, PC12 cells, and secretory granules are regulated independently of Ca^{2+} entry by G proteins [81]. Gβγ has been shown to interact directly with the fusion machinery in rat mast cells [82]. In the lamprey giant synapse, 5-HT-mediated synaptic inhibition does not cause a reduction in Ca^{2+} entry to the synapse [75]. Furthermore, the actions of 5-HT at a GPCR are abolished by intracellular block of activated Gβγ [83]. A mechanism for a direct interaction between Gβγ and the core vesicle fusion machinery was suggested by the finding that Gβγ directly binds SNARE proteins syntaxin and SNAP-25 [65–83], as well as the cysteine string protein (CSP) [84]. It is now clear that Gβγ can inhibit neurotransmitter release by a direct interaction with the SNARE complex [85]. However, it is also clear that this modification represents an entirely different modification than that identified by effects on mini-frequency, because mini-frequency is unaffected by this mechanism [86–87]. This interaction occurs at a late phase after priming in vesicles whose SNARE complex is formed and not available for cleavage by botulinum toxin B [88]. Furthermore, Gβγ binds to the C-terminal regions of the formed SNARE complex and competes with Ca^{2+}-dependent synaptotagmin binding to the SNARE complex [89]. In this way a Ca^{2+} dependency is conferred on this Gβγ-mediated presynaptic inhibition, because high presynaptic Ca^{2+} concentrations allow synaptotagmin to compete more effectively with Gβγ.

The final outcome of Gβγ competition with synaptotagmin at the SNARE complex allows a very fine-tuned control of neurotransmitter release. In both chromaffin-cell large, dense-core vesicle fusion [90] and at lamprey giant synapses [91], Gβγ causes kiss and run fusion of the exocytosing vesicle. This in turn reduces neurotransmitter release, in allowing subquantal postsynaptic events and a reduction in the peak synaptic cleft concentration of neurotransmitter release [87].

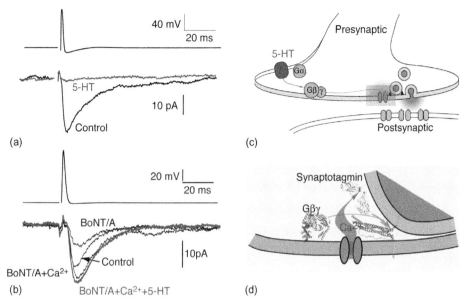

FIGURE 31.2 Serotonin reduces synaptic transmission through a direct action on the vesicle fusion machinery.

(a) Presynaptic serotonin GPCR reduces synaptic transmission in a paired recording between lamprey reticulospinal axons and motoneuron. The presynaptic neuron was held under current clamp with a microelectrode and action potential was evoked by a depolarizing current pulse (2 ms). The postsynaptic motoneuron was held under voltage clamp with a patch electrode, action potential evoked EPSCs were reduced by serotonin (gray).

(b) EPSCs before BoNT/A microinjection into the reticulospinal presynaptic terminal (control), after microinjection and subsequent 1-Hz stimulation (800 sweeps) the EPSC was reduced leaving only the BoNT/A-resistant component (BoNT/A). Increasing the extracellular Ca^{2+} concentration to 10 mM returned synaptic transmission to the level of control (BoNT/A + Ca^{2+}), which was not responsive to the 5-HT mediated inhibition (BoNT/AP + Ca^{2+} + 5-HT).

(c) Activation of presynaptic 5-HT receptors targets the SNARE complex downstream of Ca^{2+} entry into the presynaptic terminal through a direct interaction of G$\beta\gamma$ with the release machinery.

(d) At the molecular level, G$\beta\gamma$ and synaptotagmin share the same binding site on the ternary complex. At low and moderate Ca^{2+} concentrations, G$\beta\gamma$ is inhibitory. At high Ca^{2+} concentration, the binding affinity of synaptotagmin for the SNARE complex increases. This Ca^{2+}-dependent increased affinity of synaptotagmin to the SNARE complex out-competes the G$\beta\gamma$-mediated modification of fusion. Data from [88] (Gerachshenko *et al.*, 2005).

Gα_Q SIGNALING CA^{2+} STORES, DAG AND MODULATION OF NEUROTRANSMITTER RELEASE

Ca^{2+} release from internal stores located at the presynaptic terminal may lead to enhancement of transmitter release. However, this Ca^{2+} may originate either from Ca^{2+}-activated channels (CICRs) [92] or following the activation of a presynaptic GPCR leading to IP$_3$ production and activation of presynaptic IP$_3$ receptors [74, 93]. It is important to note that the activation of presynaptic receptors leading to IP$_3$ production will also produce diacylglycerol in the nerve terminal, leading to the possible activation of PKC or direct effects on proteins associated with the release machinery – for example, UNC13 [94]. Indeed, it has been proposed that effects that were previously ascribed solely to PKC activation in the presynaptic terminal (see below) can be accounted for entirely by DAG interactions with (m)UNC13 [95], which in turn leads to an enhancement in vesicle priming [96, 97].

G PROTEINS AND PHOSPHORYLATION

G proteins may alter the efficacy of synaptic transmission either through phosphorylation or dephosphorylation of presynaptic components. G$\beta\gamma$ has been suggested to activate the calcineurin phosphatase pathway to inhibit release independent of Ca^{2+} entry [98] in neuroendocrine cells. In addition, the activation of both PKA and PKC has been implicated in the enhancement of synaptic transmission in the central nervous system. Indeed, tonic activation of PKA may be necessary for vesicle fusion to occur [99–100], and PKA phosphorylates CSP to alter its binding to syntaxin [101]. Metabotropic glutamate receptors in the mammalian CNS may activate either of these latter pathways [102, 103].

Although relatively difficult to study, the presynaptic terminal may contain as rich an array of receptor-mediated mechanisms that modify information flow as has been identified at the postsynaptic side of the synapse.

REFERENCES

1. Katz B. Quantal mechanism of neural transmitter release. *Science* 1971;**173**:123–6.
2. Heuser JE, Reese TS, Dennis MJ, Jan Y, Jan L, Evans L. Synaptic vesicle exocytosis captured by quick freezing and correlated with quantal transmitter release. *J Cell Biol* 1979;**81**:275–300.
3. Sutton RB, Fasshauer D, Jahn R, Brunger AT. Crystal structure of a SNARE complex involved in synaptic exocytosis at 2.4 A resolution. *Nature* 1998;**395**:347–53.

4. Wu MN, Fergestad T, Lloyd TE, He Y, Broadie K, Bellen HJ. Syntaxin 1A interacts with multiple exocytic proteins to regulate neurotransmitter release *in vivo*. *Neuron* 1999;**23**:593–605.

5. Lin RC, Scheller RH. Mechanisms of synaptic vesicle exocytosis. *Annu Rev Cell Dev Biol* 2000;**16**:19–49.

6. Jahn R, Sudhof TC. Membrane fusion and exocytosis. *Annu Rev Biochem* 1999;**68**:863–911.

7. Sugita S, Han W, Butz S, Liu X, Fernandez-Chacon R, Lao Y, Sudhof TC. Synaptotagmin VII as a plasma membrane Ca(2+) sensor in exocytosis. *Neuron* 2001;**30**:459–73.

8. Davis AF, Bai J, Fasshauer D, Wolowick MJ, Lewis JL, Chapman ER. Kinetics of synaptotagmin responses to Ca^{2+} and assembly with the core SNARE complex onto membranes. *Neuron* 1999;**24**:363–76.

9. Chapman ER. Synaptotagmin: a Ca(2+) sensor that triggers exocytosis? *Nature Rev Mol Cell Biol* 2002;**3**:498–508.

10. Bai J, Wang P, Chapman ER. C2A activates a cryptic Ca(2+)-triggered membrane penetration activity within the C2B domain of synaptotagmin I. *Proc Natl Acad Sci USA* 2002;**99**:1665–70.

11. Martens S, Kozlov MM, McMahon HT. How synaptotagmin promotes membrane fusion. *Science* 2007;**316**:1205–8.

12. Weber T, Zemelman BV, McNew JA, Westermann B, Gmachl M, Parlati F, Sollner TH, Rothman JE. SNAREpins: minimal machinery for membrane fusion. *Cell* 1998;**92**:759–72.

13. Iida Y, Senda T, Matsukawa Y, Onoda K, Miyazaki JI, Sakaguchi H, Nimura Y, Hidaka H, Niki I. Myosin light-chain phosphorylation controls insulin secretion at a proximal step in the secretory cascade. *Am J Physiol* 1997;**273**:E782–9.

14. Sudhof TC. The synaptic vesicle cycle: a cascade of protein–protein interactions. *Nature* 1995;**375**:645–53.

15. Hay JC, Fisette PL, Jenkins GH, Fukami K, Takenawa T, Anderson RA, Martin TF. ATP-dependent inositide phosphorylation required for Ca(2+)-activated secretion. *Nature* 1995;**374**:173–7.

16. Xu J, Xu Y, Ellis-Davies GC, Augustine GJ, Tse FW. Differential regulation of exocytosis by alpha- and beta-SNAPs. *J Neurosci* 2002;**22**:53–61.

17. Brose, N, Rosenmund C, Rettig, J. Regulation of transmitter release by Unc-13 and its homologues. *Curr Opin Neurobiol* 2000;**10**:303–11.

18. Madison J. M, Nurrish S, Kaplan J M. UNC-13 interaction with syntaxin is required for synaptic transmission. *Curr Biol* 2005;**15**:2236–42.

19. Stevens, D. R., Wu, Z. X., Matti, U., Junge, H. J., Schirra, C., Becherer, U., Wojcik, S. M., Brose, N. and Rettig, J. (2005) Identification of the minimal protein domain required for priming activity of Munc13-1. *Curr Biol* 15:2243–8.

20. Basu, J., Shen, N., Dulubova, I., Lu, J., Guan, R., Guryev, O., Grishin, N. V., Rosenmund, C. and Rizo, J. (2005) A minimal domain responsible for Munc13 activity. *Nat Struct Mol Biol* 12:1017–8.

21. Augustine GJ, Adler EM, Charlton MP. The calcium signal for transmitter secretion from presynaptic nerve terminals. *Ann NY Acad Sci* 1991;**635**:365–81.

22. Bollmann JH, Sakmann B, Borst JG. Calcium sensitivity of glutamate release in a calyx-type terminal. *Science* 2000;**289**:953–7.

23. Del Castillo J, Katz B. Quantal components of the end-plate potential. *J Physiol* 1954;**124**:560–73.

24. Heuser JE, Reese TS. Evidence for recycling of synaptic vesicle membrane during transmitter release at the frog neuromuscular junction. *J Cell Biol* 1973;**57**:315–44.

25. Betz WJ, Bewick GS. Optical analysis of synaptic vesicle recycling at the frog neuromuscular junction. *Science* 1992;**255**:200–3.

26. Takei K, Mundigl O, Daniell L, De Camilli P. The synaptic vesicle cycle: a single vesicle budding step involving clathrin and dynamin. *J Cell Biol* 1996;**133**:1237–50.

27. Ceccarelli B, Hurlbut WP, Mauro A. Turnover of transmitter and synaptic vesicles at the frog neuromuscular junction. *J Cell Biol* 1973;**57**:499–524.

28. Fesce R, Meldolesi J. Peeping at the vesicle kiss. *Nature Cell Biol* 1999;**1**:E3–4.

29. Henkel AW, Meiri H, Horstmann H, Lindau M, Almers W. Rhythmic opening and closing of vesicles during constitutive exo- and endocytosis in chromaffin cells. *EMBO J* 2000;**19**:84–93.

30. Graham ME, Fisher RJ, Burgoyne RD. Measurement of exocytosis by amperometry in adrenal chromaffin cells: effects of clostridial neurotoxins and activation of protein kinase C on fusion pore kinetics. *Biochimie* 2000;**82**:469–79.

31. Elhamdani A, Palfrey HC, Artalejo CR. Quantal size is dependent on stimulation frequency and calcium entry in calf chromaffin cells. *Neuron* 2001;**31**:819–30.

32. Nishiki T, Augustine GJ. Synaptotagmin I synchronizes transmitter release in mouse hippocampal neurons. *J Neurosci* 2004;**24**:6127–32.

33. Wang CT, Lu JC, Bai J, Chang PY, Martin TF, Chapman ER, Jackson MB. Different domains of synaptotagmin control the choice between kiss-and-run and full fusion. *Nature* 2003;**424**:943–7.

34. Stevens CF, Williams JH. "Kiss and run" exocytosis at hippocampal synapses. *Proc Natl Acad Sci USA* 2000;**97**(12):828–33.

35. Kavalali ET, Klingauf J, Tsien RW. Properties of fast endocytosis at hippocampal synapses.. *Philos Trans R Soc Lond B Biol Sci* 1999;**354**:337–46.

36. Pyle JL, Kavalali ET, Piedras-Renteria ES, Tsien RW. Rapid reuse of readily releasable pool vesicles at hippocampal synapses. *Neuron* 2000;**28**:221–31.

37. Aravanis AM, Pyle JL, Tsien RW. Single synaptic vesicles fusing transiently and successively without loss of identity. *Nature* 2003;**423**:643–7.

38. Zakharenko SS, Zablow L, Siegelbaum SA. Altered presynaptic vesicle release and cycling during mGluR-dependent LTD. *Neuron* 2002;**35**:1099–110.

39. Gandhi SP, Stevens CF. Three modes of synaptic vesicular recycling revealed by single-vesicle imaging. *Nature* 2003;**423**:607–13.

40. Klyachko VA, Jackson MB. Capacitance steps and fusion pores of small and large-dense-core vesicles in nerve terminals. *Nature* 2002;**418**:89–92.

41. Richards DA, Bai J, Chapman ER. Two modes of exocytosis at hippocampal synapses revealed by rate of FM1-43 efflux from individual vesicles. *J Cell Biol* 2005;**168**:929–39.

42. Choi S, Klingauf J, Tsien RW. Fusion pore modulation as a presynaptic mechanism contributing to expression of long-term potentiation. *Philos Trans R Soc Lond B Biol Sci* 2003;**358**:695–705.

43. Choi S, Klingauf J, Tsien RW. Postfusional regulation of cleft glutamate concentration during LTP at "silent synapses." *Nature Neurosci* 2000;**3**:330–6.

44. Alford S, Grillner S. The involvement of GABAB receptors and coupled G-proteins in spinal GABAergic presynaptic inhibition. *J Neurosci* 1991;**11**:3718–26.

45. Dutar P, Nicoll RA. A physiological role for GABAB receptors in the central nervous system. *Nature* 1988;**332**:156–8.

46. Stryer L, Bourne HR. G proteins: a family of signal transducers. *Annu Rev Cell Biol* 1986;**2**:391–419.

47. Hamm HE. The many faces of G protein signaling. *J Biol Chem* 1998;**273**:669–72.

48. Dunlap K, Fischbach GD. Neurotransmitters decrease the calcium component of sensory neurone action potentials. *Nature* 1978;**276**:837–9.

49. Dolphin AC, Pearson HA, Menon-Johansson AS, Sweeney MI, Sutton K, Huston E, Cullen GP, Scott RH. G protein modulation of voltage-dependent calcium channels and transmitter release. *Biochem Soc Trans* 1993;**21**:391–5.

50. Qin N, Platano D, Olcese R, Stefani E, Birnbaumer L. Direct interaction of gbetagamma with a C-terminal gbetagamma-binding domain of the Ca^{2+} channel alpha1 subunit is responsible for channel inhibition by G protein-coupled receptors. *Proc Natl Acad Sci USA* 1997;**94**:8866–71.

51. Forscher P, Oxford GS, Schulz D. Noradrenaline modulates calcium channels in avian dorsal root ganglion cells through tight receptor-channel coupling. *J Physiol* 1986;**379**:131–44.

52. Herlitze S, Garcia DE, Mackie K, Hille B, Scheuer T, Catterall WA. Modulation of Ca2+ channels by G-protein beta gamma subunits. *Nature* 1996;**380**:258–62.

53. Ikeda SR. Voltage-dependent modulation of N-type calcium channels by G-protein beta gamma subunits. *Nature* 1996;**380**:255–8.

54. Takahashi T, Forsythe ID, Tsujimoto T, Barnes-Davies M, Onodera K. Presynaptic calcium current modulation by a metabotropic glutamate receptor. *Science* 1996;**274**:594–7.

55. Mirotznik RR, Zheng X, Stanley EF. G-Protein types involved in calcium channel inhibition at a presynaptic nerve terminal. *J Neurosci* 2000;**20**:7614–21.

56. Takahashi T, Kajikawa Y, Tsujimoto T. G-Protein-coupled modulation of presynaptic calcium currents and transmitter release by a GABAB receptor. *J Neurosci* 1998;**18**:3138–46.

57. Mizutani H, Hori T, Takahashi T. 5-HT1B receptor-mediated presynaptic inhibition at the calyx of Held of immature rats. *Eur J Neurosci* 2006;**24**:1946–54.

58. Tedford HW, Zamponi GW. Direct G protein modulation of Cav2 calcium channels. *Pharmacol Rev* 2006;**58**:837–62.

59. Currie KP, Fox AP. Comparison of N- and P/Q-type voltage-gated calcium channel current inhibition. *J Neurosci* 1997;**17**:4570–9.

60. Mehrke G, Pereverzev A, Grabsch H, Hescheler J, Schneider T. Receptor-mediated modulation of recombinant neuronal class E calcium channels. *FEBS Letts* 1997;**408**:261–70.

61. Page KM, Canti C, Stephens GJ, Berrow NS, Dolphin AC. Identification of the amino terminus of neuronal Ca^{2+} channel alpha1 subunits alpha1B and alpha1E as an essential determinant of G-protein modulation. *J Neurosci* 1998;**18**:4815–24.

62. De Waard M, Liu H, Walker D, Scott VE, Gurnett CA, Campbell KP. Direct binding of G-protein betagamma complex to voltage-dependent calcium channels. *Nature* 1997;**385**:446–50.

63. Li B, Zhong H, Scheuer T, Catterall WA. Functional role of a C-terminal Gbetagamma-binding domain of Ca(v)2.2 channels. *Mol Pharmacol* 2004;**66**:761–9.

64. Campbell V, Berrow NS, Fitzgerald EM, Brickley K, Dolphin AC. Inhibition of the interaction of G protein G(o) with calcium channels by the calcium channel beta-subunit in rat neurones. *J Physiol* 1995;**485**(2):365–3672.

65. Jarvis SE, Magga JM, Beedle AM, Braun JE, Zamponi GW. G protein modulation of N-type calcium channels is facilitated by physical interactions between syntaxin 1 A and Gbetagamma. *J Biol Chem* 2000;**275**:6388–94.

66. Stanley EF, Mirotznik RR. Cleavage of syntaxin prevents G-protein regulation of presynaptic calcium channels. *Nature* 1997;**385**:340–3.

67. Bean BP. Neurotransmitter inhibition of neuronal calcium currents by changes in channel voltage dependence. *Nature* 1989;**340**:153–6.

68. Williams S, Serafin M, Muhlethaler M, Bernheim L. Facilitation of N-type calcium current is dependent on the frequency of action potential-like depolarizations in dissociated cholinergic basal forebrain neurons of the guinea pig. *J Neurosci* 1997;**17**:1625–32.

69. Huang CL, Slesinger PA, Casey PJ, Jan YN, Jan LY. Evidence that direct binding of G beta gamma to the GIRK1 G protein-gated inwardly rectifying K$^+$ channel is important for channel activation. *Neuron* 1995;**15**:1133–43.

70. Reuveny E, Slesinger PA, Inglese J, Morales JM, Iniguez-Lluhi JA, Lefkowitz RJ, Bourne HR, Jan YN, Jan LY. Activation of the cloned muscarinic potassium channel by G protein beta gamma subunits. *Nature* 1994;**370**:143–6.

71. Ponce A, Bueno E, Kentros C, Vega-Saenz de Miera E, Chow A, Hillman D, Chen S, Zhu L, Wu MB, Wu X, Rudy B, Thornhill WB. G-protein-gated inward rectifier K$^+$ channel proteins (GIRK1) are present in the soma and dendrites as well as in nerve terminals of specific neurons in the brain. *J Neurosci* 1996;**16**:1990–2001.

72. Dutar P, Petrozzino JJ, Vu HM, Schmidt MF, Perkel DJ. Slow synaptic inhibition mediated by metabotropic glutamate receptor activation of GIRK channels. *J Neurophysiol* 2000;**84**:2284–90.

73. Congar P, Bergevin A, Trudeau LE. D2 receptors inhibit the secretory process downstream from calcium influx in dopaminergic neurons: implication of K(+) channels. *J Neurophysiol* 2002;**87**:1046–56.

74. Cochilla AJ, Alford S. Metabotropic glutamate receptor-mediated control of neurotransmitter release. *Neuron* 1998;**20**:1007–16.

75. Takahashi M, Freed R, Blackmer T, Alford S. Calcium influx-independent depression of transmitter release by 5-HT at lamprey spinal cord synapses. *J Physiol (Lond)* 2001;**532**:323–36.

76. Ma JY, Catterall WA, Scheuer T. Persistent sodium currents through brain sodium channels induced by G protein betagamma subunits. *Neuron* 1997;**19**:443–52.

77. Surmeier Jr DJ, Bargas J, Hemmings HC, Nairn AC, Greengard P. Modulation of calcium currents by a D1 dopaminergic protein kinase/phosphatase cascade in rat neostriatal neurons. *Neuron* 1995;**14**:385–97.

78. Silinsky EM. On the mechanism by which adenosine receptor activation inhibits the release of acetylcholine from motor nerve endings. *J Physiol* 1984;**346**:243–56.

79. Scanziani M, Gahwiler BH, Thompson SM. Presynaptic inhibition of excitatory synaptic transmission by muscarinic and metabotropic glutamate receptor activation in the hippocampus: are Ca^{2+} channels involved? *Neuropharmacology* 1995;**34**:1549–57.

80. Luini A, De Matteis MA. Evidence that receptor-linked G protein inhibits exocytosis by a post- second-messenger mechanism in AtT-20 cells. *J Neurochem* 1990;**54**:30–8.

81. Lang J. Molecular mechanisms and regulation of insulin exocytosis as a paradigm of endocrine secretion. *Eur J Biochem* 1999;**259**:3–17.

82. Pinxteren JA, O'Sullivan AJ, Tatham PE, Gomperts BD. Regulation of exocytosis from rat peritoneal mast cells by G protein beta gamma-subunits. *EMBO J* 1998;**17**:6210–18.

83. Blackmer T, Larsen EC, Takahashi M, Martin TF, Alford S, Hamm HE. G protein betagamma subunit-mediated presynaptic inhibition: regulation of exocytotic fusion downstream of Ca^{2+} entry. *Science* 2001;**292**:293–7.

84. Magga JM, Jarvis SE, Arnot MI, Zamponi GW, Braun JE. Cysteine string protein regulates G protein modulation of N-type calcium channels.. *Neuron* 2000;**28**:195–204.

85. Blackmer T, Larsen EC, Bartleson C, Kowalchyk JA, Yoon EJ, Preininger AM, Alford S, Hamm HE, Martin TF. G protein betagamma

directly regulates SNARE protein fusion machinery for secretory granule exocytosis. *Nature Neurosci* 2005;**8**:421–5.

86. Delaney AJ, Crane JW, Sah P. Noradrenaline modulates transmission at a central synapse by a presynaptic mechanism. *Neuron* 2007;**56**:880–92.

87. Schwartz EJ, Blackmer T, Gerachshenko T, Alford S. Presynaptic G-protein-coupled receptors regulate synaptic cleft glutamate via transient vesicle fusion. *J Neurosci* 2007;**27**:5857–68.

88. Gerachshenko T, Blackmer T, Yoon EJ, Bartleson C, Hamm HE, Alford S. Gbetagamma acts at the C terminus of SNAP-25 to mediate presynaptic inhibition. *Nature Neurosci* 2005;**8**:597–605.

89. Yoon EJ, Gerachshenko T, Spiegelberg BD, Alford S, Hamm HE. Gbetagamma interferes with Ca2+-dependent binding of synaptotagmin to the soluble N-ethylmaleimide-sensitive factor attachment protein receptor (SNARE) complex. *Mol Pharmacol* 2007;**72**:1210–19.

90. Chen XK, Wang LC, Zhou Y, Cai Q, Prakriya M, Duan KL, Sheng ZH, Lingle C, Zhou Z. Activation of GPCRs modulates quantal size in chromaffin cells through G(betagamma) and PKC. *Nature Neurosci* 2005;**8**:1160–8.

91. Photowala H, Blackmer T, Schwartz E, Hamm HE, Alford S. G protein betagamma-subunits activated by serotonin mediate presynaptic inhibition by regulating vesicle fusion properties. *Proc Natl Acad Sci USA* 2006;**103**:4281–6.

92. Peng Y. Ryanodine-sensitive component of calcium transients evoked by nerve firing at presynaptic nerve terminals. *J Neurosci* 1996;**16**:6703–12.

93. Schwartz NE, Alford S. Physiological activation of presynaptic metabotropic glutamate receptors increases intracellular calcium and glutamate release. *J Neurophysiol* 2000;**84**:415–27.

94. Nurrish S, Segalat L, Kaplan JM. Serotonin inhibition of synaptic transmission: Galpha(0) decreases the abundance of UNC-13 at release sites. *Neuron* 1999;**24**:231–42.

95. Rhee JS, Betz A, Pyott S, Reim K, Varoqueaux F, Augustin I, Hesse D, Sudhof TC, Takahashi M, Rosenmund C, Brose N. Beta phorbol ester- and diacylglycerol-induced augmentation of transmitter release is mediated by Munc13s and not by PKCs. *Cell* 2002;**108**:121–33.

96. Richmond JE, Davis WS, Jorgensen EM. UNC-13 is required for synaptic vesicle fusion in. *C. elegans. Nature Neurosci* 1999;**2**:959–64.

97. Weimer RM, Gracheva EO, Meyrignac O, Miller KG, Richmond JE, Bessereau JL. UNC-13 and UNC-10/rim localize synaptic vesicles to specific membrane domains. *J Neurosci* 2006;**26**:8040–7.

98. Renstrom E, Ding WG, Bokvist K, Rorsman P. Neurotransmitter-induced inhibition of exocytosis in insulin-secreting beta cells by activation of calcineurin. *Neuron* 1996;**17**:513–22.

99. Hilfiker S, Czernik AJ, Greengard P, Augustine GJ. Tonically active protein kinase A regulates neurotransmitter release at the squid giant synapse. *J Physiol* 2001;**531**:141–6.

100. Trudeau LE, Fang Y, Haydon PG. Modulation of an early step in the secretory machinery in hippocampal nerve terminals. *Proc Natl Acad Sci USA* 1998;**95**:7163–8.

101. Evans GJ, Wilkinson MC, Graham ME, Turner KM, Chamberlain LH, Burgoyne RD, Morgan A. Phosphorylation of cysteine string protein by protein kinase A. Implications for the modulation of exocytosis. *J Biol Chem* 2001;**276**(47):877–85.

102. Kondo S. A. Marty Protein kinase A-mediated enhancement of miniature IPSC frequency by noradrenaline in rat cerebellar stellate cells. *J Physiol (Lond)* 1997;**498**:165–76.

103. Trudeau LE, Emery DG, Haydon PG. Direct modulation of the secretory machinery underlies PKA-dependent synaptic facilitation in hippocampal neurons. *Neuron* 1996;**17**:789–97.

Chapter 32

The Role of Receptor Protein Tyrosine Phosphatases in Axonal Pathfinding

Andrew W. Stoker

Neural Development Unit, Institute of Child Health, University College London, London, England, UK

INTRODUCTION

One of the most impressive processes that occurs during development is the establishment of countless connections between neurons and their targets. Such precise connectivity requires long-distance growth and accurate pathfinding by axons, short-range detection of target cells, and in most cases the establishment of permanent synaptic connections. This chapter reviews one family of molecules, the receptor-like protein tyrosine phosphatases (RPTPs), which direct axons in this astonishing feat. Most RPTPs are found to be expressed in developing nervous systems, in several cases within axons and their motile, pathfinding growth cones [1–3]. Evidence is reviewed here for RPTP roles in axon growth and guidance. Their potential signaling mechanisms are also briefly discussed. Due to space limitations, readers will in most cases be referred to the above reviews and the references therein. Figure 32.1 summarizes the basic axon growth and guidance events discussed below, and the RPTPs currently implicated in each. Readers should always be mindful that this reductionist focus on RPTPs must be viewed in the more holistic context of all the other axonal guidance mechanisms [4].

The human genome contains 21 RPTP genes, most of which have either orthologs or homologs in other species. The encoded proteins are subgrouped according to their extracellular domain homologies and the number of cytoplasmic catalytic domains, and this has been described in detail elsewhere [5]. The structure and biological roles of these genes have been conserved across species, and because of this scientists have been able to make use of several excellent model systems to study RPTP functions. Much of the discussion below focuses on two anatomical areas, the visual and neuromuscular systems, but other axon growth and guidance models are also highlighted briefly.

RPTPS AND THE VISUAL SYSTEM

Drosophila

The compound eye of the fly contains about 800 light-receiving ommatidia, each with photoreceptor neurons named R1 through R8. Axons from these photoreceptors project from the eye to the optic lobe in the brain, where each terminates either in the lamina layer (R1 to R6) or in proximal layers (R8) or distal layers (R7) of the medulla [6]. DPTP69D and DLAR influence these axonal termination events [2, 7, 8]. For example, if axons of R1 to R6 are made DPTP69D-deficient, they will overshoot their target and terminate in the medulla. In addition, loss of DPTP69D in the R7 photoreceptor causes its axon to stop short in the R8 termination zone of the medulla. DPTP69D appears to control the ability of growth cones to de-adhere (defasciculate) from the R8 axon at correct navigational decision points (step 2, Figure 32.1). Interestingly, whereas DLAR-deficient axons from R1 to R6 terminate normally, R7 axons that lack DLAR reach, but then retract from, their medulla targets [7, 8]. This suggests a failure to establish stable adhesive or synaptic contacts with medulla target cells, indicating that DLAR is involved in this process. DLAR mutants and cadherin mutants have similar phenotypes, suggesting that they may regulate similar adhesive signaling pathways [7], which is of interest given recent evidence that a related mammalian RPTP PTPσ influences cadherin adhesion during neurite outgrowth [9]. The collective data also indicate that DPTP69D and DLAR function cell autonomously, although DLAR also shows evidence of non-autonomous function in R8, suggesting that it may "send" signals through its extracellular domain. In contrast to their guidance roles, DRPTPs do not appear to be

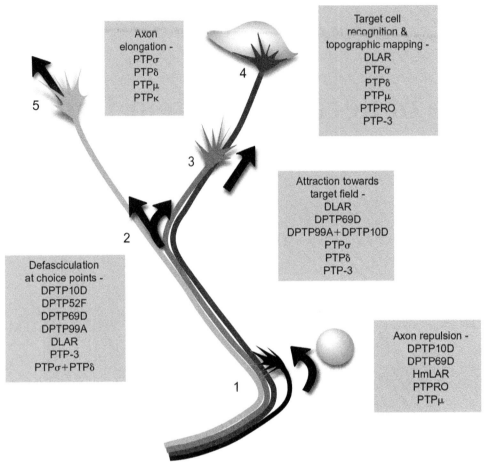

FIGURE 32.1 Schematic diagram showing the RPTPs implicated at different stages of axon growth and guidance.
Step 1 shows repulsive signaling to make growth cones avoid a cue; step 2 shows defasciculation of axons, typically seen at choice points; step 3 shows positive attraction towards a target tissue or specific cellular target; step 4 shows direct recognition of targets, halting at that point and formation of stable synaptic contacts (a "target" can also be a cellular environment without direct synaptic contact with other cells, e.g. sensory nerve endings); step 5 shows the basic process of a growth cone's forward migration. See text for further details.

necessary for axon elongation in the visual system, unlike their vertebrate counterparts.

Vertebrate Retinotectal System

In vertebrate eyes, retinal ganglion cell (rgc) axons relay visual signals from the eye to the brain. Neighboring rgc axons establish precise topographic connectivity with neighboring neurons in the optic tectum or superior colliculus, so setting up an accurate visual map. Several studies in cell culture demonstrate a role for RPTPs in rgc neurite growth. Signaling from PTPμ enhances cadherin-dependent retinal axon outgrowth [2]. Furthermore, PTPμ expression levels vary topographically across the retina and tectum, and PTPμ has a selective, growth cone collapsing function [10]. PTPμ may therefore differentially influence axon growth within the retinotectal projection, and is thought to be a dominant, inhibitory guidance cue within a restricted, but critical, time window of retinal axon development [1, 10, 11]. *Xenopus* PTPδ promotes both rgc axon growth along the optic

tract *in vivo* and neurite growth on basement membranes in culture [12]. A guidance role for PTPδ has been suggested from forebrain cultures, where soluble ectodomains of human PTPδ can attract growth cones [13]. Since both PTPδ and PTPμ bind homophilically, they may trigger signals directly between axons. Another retinal RPTP, PTPRO (initially known as CRYP-2), is anti-adhesive in retinal ganglion cell cultures [14]. PTPRO was also shown to have an axon navigation capability in culture, since its ectodomain in the form of a PTPRO-Fc fusion protein induces growth cone collapse and repulsive growth cone turning [14] (step 1, Figure 32.1). More recent studies *in ovo* have revealed that PTPRO does indeed play a role in guiding retinal axons to their topographic targets in the optic tectum [15] (steps 3 and 4, Figure 32.1). In this case it was shown that PTPRO acts through the negative regulation of another influential retinal guidance receptor, the tyrosine kinase EphB2. This elegantly confirms the concept that some RPTPs may act in growth cones by setting thresholds of sensitivity to other signals, including guidance cues [16].

Another RPTP implicated in retinal axon growth and guidance is PTPσ. The interaction between chick PTPσ and a ligand(s) on basement membranes and glial endfeet maintains optimal retinal neurite outgrowth [2] (step 5, Figure 32.1). Perhaps counterintuitively, interference with intracellular signaling of *Xenopus* PTPσ causes faster neurite outgrowth in culture [12], suggesting a possible signaling model in which PTPσ signals are inhibitory to axon growth. The first evidence for RPTP function in vertebrate axon targeting *in vivo* came from studies of chick PTPσ. Perturbation of the interactions between PTPσ and its ligands in the optic tectum causes retinal axon stalling and rostral mistargeting [17] (possible failure of steps 1, 3, or 4, Figure 32.1). PTPσ may therefore function by maintaining retinal axon growth over the tectum and facilitating the recognition of correct target sites. Whether it acts through the regulation of Eph receptors, like PTPRO, remains to be determined.

NEUROMUSCULAR SYSTEM

Drosophila genetics has highlighted key RPTP functions during motor axon guidance. The segmental and intersegmental motor nerves ISN, ISNb and SNa of the fly larva innervate body wall muscles in a highly stereotyped manner. Nerve defects arise after loss of function in *DLAR*, *DPTP69D*, *DPTP99A*, *DPTP10D*, and *DPTP52F* [3, 18]. DPTP69D and DPTP99A are required for ISNb axons to defasciculate from the ISN at the correct choice point. Gene-deficiency causes a "bypass" phenotype where axons fail to leave the ISN and thus travel past their targets (failure of step 2, Figure 32.1). DLAR influences not only this defasciculation step, but also both the entry of axons into the muscle target field, and synapse formation [19, 20] (steps 3 and 4, Figure 32.1). The DLAR-related PTP-3 protein in *C. elegans* has also been implicated in synapse formation [21]. DPTP10D collaborates with other DRPTPs in guiding SNa, but antagonizes them during navigation of the ISN. Similarly, DLAR and DPTP99A antagonize each other within SNb axons. There is therefore a complex pattern of interaction between these RPTPs with "partial redundancy, competition and collaboration," as described by Sun and co-workers [22].

The LAR-related PTP-3 protein in *C. elegans* has been implicated in the control of guidance of motor axons in the ventral nerve cord of the worm. Mutations induce defects such as incorrect choice-point decisions, defasciculations, targeting errors, and growth-cone stalling [21] (steps 1, 2, 3, and 4).

In vertebrates, there is also some evidence for the role of RPTPs in controlling motor axon growth, guidance, and targeting. In the chick model, acute knockdown of RPTP transcripts using RNAi has shown that PTPσ, PTPδ, and PTPRO are involved in the control of specific motor axon outgrowth from the spinal cord, as well as nerve fasciculation [23]. In the mouse, germline disruption of *Ptprs*

and *Ptprd* has shown that phrenic nerve targeting to the diaphragm is aberrant, with failure either to form stereotypical nerve branching patterns or to maintain diaphragm innervation [24] (steps 2, 3, and 4, Figure 32.1). These studies also suggest once again that there is both redundancy and competition between different RPTP members during motor nerve development.

FURTHER AXON GROWTH AND GUIDANCE ROLES

In the leech, LAR homolog HmLAR2 is expressed in growth cones of neurite-like processes of comb cells, where it controls the orderly outgrowth of these processes [2]. Evidence supports a homophilic interaction between HmLAR2 molecules, signaling a mutual repulsion between growth cones and neighboring processes.

In the ventral nerve cord of *Drosophila*, axon guidance across the midline is influenced by DLAR, DPTP99A, DPTP69D, and DPTP10D. The latter two in particular cooperate with Robo receptors to transduce repulsive signals from midline Slit protein [2]. How these receptors cooperate biochemically remains unclear.

As well as those described above, several RPTP gene-deficiency models have been developed in mice. Loss of PTPσ function causes motor function deficits and hyposmia, as well as defects in sciatic nerve myelination and maturation [2]. Deficiency in PTPδ also causes milder motor defects as well as memory alterations, while loss of LAR causes a reduction in forebrain cholinergic neuron numbers and some mild defects in hippocampal innervation [25]. The developmental bases for all these neuronal and axonal defects have yet to be characterized.

AXONAL SIGNALING BY RPTPS.

Instructive or Permissive?

Do axonal RPTPs send permissive or instructive signals during axon guidance? With DLAR, the fact that R7 growth cones reach targets, but then retract, supports an instructive role in securing adhesion to targets. DLAR may also control the instructive process of muscle cell recognition by motor axons [22]. For DPTP69D, the consensus is more in favor of a permissive role in allowing defasciculation, rather than active target recognition, although this remains unresolved [2]. In fact, the many complex interactions between *Drosophila* RPTPs may ultimately make simple instructive/permissive distinctions untenable. Similarly in vertebrates, the instructive/permissive distinction is difficult to assess. Vertebrate PTPδ and PTPRO may have instructive signaling roles given that they can force growth cone turning on otherwise permissive substrates [2, 14]. For PTPσ and PTPRO in the visual system, however, it is not known exactly how these phosphatases are controlled by extracellular ligands, and so they could still have either active or more passive roles in regulating graded Eph signaling, for example [15, 17].

Ligands

As already indicated above, one problem with understanding how RPTPs act is that we have relatively little knowledge of how RPTPs are themselves regulated at a molecular level. In particular, the identity of RPTP ligands and their action in cells and tissues remains an area of keen interest. PTPμ and PTPδ bind homophilically, and ectodomains of these RPTPs can act as neurite growth-promoting substrates in culture, suggesting that axon fasciculation *in vivo* may be promoted by their homophilic action. Interestingly, while PTPμ ectodomains can also have negative effects on growth cones, specifically in temporal retinal axons [10], studies with catalytically-inactive PTPμ indicate that both its permissive and repulsive signaling in retinal neurites is transduced directly by homophilic regulation of PTPμ phosphatase activity [26]. Ligands for PTPζ include the heparin-binding chemokine pleiotrophin. Pleiotrophin can inhibit PTPζ, probably by causing inactive PTPζ dimers to form [27], and this leads to increased tyrosine phosphorylation of potential targets such as β-catenin [28] (see below). The heparan sulfate proteoglycans agrin (HSPGs) and collagen XVIII are binding partners for PTPσ [29], and recent studies have shown that peptides that block the heparin-binding site can also interfere with retinal neurite outgrowth in culture (Stoker and Hawadle, unpublished work). One of several models proposed also suggests that chick PTPσ ligands may inactivate the phosphatase, thereby facilitating neurite growth [12]. Interestingly, the *Drosophila* relative of PTPσ, DLAR, also binds to HSPGs, and two of these, Dallylike

and Syndecan, regulate DLAR function in opposite ways during synaptogenesis [20].

Although RPTP ectodomains have adhesive capacities, it is their catalytic functions that are crucial for most of their signaling roles. For example, enzymatically active PTPμ is required for neurite outgrowth on cadherins [2]. Furthermore, genetic rescue studies with *Drosophila* RPTPs indicate that the rescuing genes must encode active phosphatases [2, 7, 8, 22]. There are exceptions, however, since PTPσ has been shown to have non-cell-autonomous functions both in retinal axon growth in culture [30] and in sciatic nerve guidance *in vivo* [31], while PTPRO ectodomains can clearly send repulsive signals to other axons [14].

Downstream Signals

Figure 32.2 contains a summary of some of the known substrates and binding partners of neuronal RPTPs. Most of these impinge ultimately on the actin cytoskeleton, providing a logical handle on growth cone dynamics. DLAR interacts with several molecules, including the tyrosine kinase Abl and its substrate Enabled (Ena, a VASP family member). Ena can be dephosphorylated by DLAR [32]. Dephosphorylation of Ena activates downstream signals that pass through profilin and on to actin. DLAR also interacts genetically with Trio, a large protein with two exchange factor domains for Rho family GTPases. *Drosophila* Trio also signals through the SH2–SH3 adaptor Dock and the p21-activated kinase Pak, again impinging on the cytoskeleton. Mammalian Trio binds directly to LAR family members and signals through Rho family GTPases. Human Trio promotes neurite outgrowth in PC12 neurons, although it is not known yet if LAR RPTPs are involved in this event [33]. Catenins and cadherins are a key target of several axonal RPTPs, including PTPμ, PTPκ, PTPζ, and LAR members [9, 28, 34–36]. These RPTPs may well antagonize cadherin/

RPTPs	Interacting protein	Downstream effector	Refs.
LAR	> Trio >	Rac, Rho	39
LAR, PTP-3	liprins		20, 40
LAR, PTPσ	< Trk receptors <¶		41, 42
PTPRO	EphB		14
PTPα	> grb2 >		43
PTPα, LAR	> c-src* >	TrkB	41, 44
PTPμ	> RACK1 >	PKCδ	37
PTPμ	IQGAP1	Rho GTPases	38
PTPμ, PTPκ, LAR family PTPζ	< cadherins* & catenins* <		37, 33 34–36
DLAR	< abl* <	ena*	31
DPTP10D, DPTP99A	gp150*		45

FIGURE 32.2 Table showing some of the proteins that a thought to be important interactors with RPTPs in axons, with some of the predicted downstream effectors.

Symbols > and < indicate either stimulation or repression of function by the RPTP, respectively. Note the exception that LAR causes indirect activation of TrkB via pp60^{c-src}. Asterisks indicate that the protein is a phosphatase substrate.

catenin-regulated cell adhesion by dephosphorylating β-catenin, p120 catenin, or cadherin, thereby directly influencing growth cone adhesion. PTPμ also binds to the adaptor protein RACK and in turn requires PKCδ to promote neurite outgrowth [37]. Furthermore, PTPμ binds to IQGAP1 and in so doing transmits signals downstream through Rho GTPAses [38]. Other RPTP targets include the cell adhesion molecule-like gp150 and the tyrosine kinase pp60^{c-src}. Finally, adaptor proteins of the liprin family bind to LAR family RPTPs and may be important for localizing these RPTPs in membranes and in forming complexes with roles in signaling, adhesionb and synapse function [3, 19].

REFERENCES

1. Ensslen-Craig SE, Brady-Kalnay SM. Receptor protein tyrosine phosphatases regulate neural development and axon guidance. *Dev Biol* 2004;**275**:12–22.
2. Stoker AW. Receptor tyrosine phosphatases in axon growth and guidance. *Curr Opin Neurobiol* 2001;**11**:95–102.
3. Van Vactor D. Protein tyrosine phosphatases in the developing nervous system. *Curr Opin Cell Biol* 1998;**10**:174–81.
4. Huber AB, Kolodkin AL, Ginty DD, Cloutier JF. Signaling at the growth cone: ligand–receptor complexes and the control of axon growth and guidance. *Annu Rev Neurosci* 2003;**26**:509–63.
5. Alonso A, Sasin J, Bottini N, Friedberg I, Osterman A, Godzik A, Hunter T, Dixon J, Mustelin T. Protein tyrosine phosphatases in the human genome. *Cell* 2004;**117**:699–711.
6. Mast JD, Prakash S, Chen PL, Clandinin TR. The mechanisms and molecules that connect photoreceptor axons to their targets in *Drosophila*. *Semin Cell Dev Biol* 2006;**17**:42–9.
7. Clandinin TR, Lee CH, Herman T, Lee RC, Yang AY, Ovasapyan S, Zipursky SL. Drosophila LAR regulates R1–R6 and R7 target specificity in the visual system. *Neuron* 2001;**32**:237–48.
8. Maurel-Zaffran C, Suzuki T, Gahmon G, Treisman JE, Dickson BJ. Cell-autonomous and –nonautonomous functions of LAR in R7 photoreceptor axon targeting. *Neuron* 2001;**32**:225–35.
9. Siu R, Fladd C, Rotin D. N-cadherin is an *in vivo* substrate for protein tyrosine phosphatase sigma (PTPσ) and participates in PTPσ-mediated inhibition of axon growth. *l Cell Biol* 2007;**27**:208–19.
10. Burden-Gulley SM, Ensslen SE, Brady-Kalnay SM. Protein tyrosine phosphatase-mu differentially regulates neurite outgrowth of nasal and temporal neurons in the retina. *J Neurosci* 2002;**22**:3615–27.
11. Ensslen SE, Brady-Kalnay SM. PTPmicro signaling via PKCdelta is instructive for retinal ganglion cell guidance. *Mol Cell Neurosci* 2004;**25**:558–71.
12. Johnson KG, McKinnell IW, Stoker AW, Holt CE. Receptor protein tyrosine phosphatases regulate retinal ganglion cell axon outgrowth in the developing *Xenopus* visual system. *J Neurobiol* 2001;**49**:99–117.
13. Sun QL, Wang J, Bookman RJ, Bixby JL. Growth cone steering by receptor tyrosine phosphatase delta defines a distinct class of guidance Cue. *Mol Cell Neurosci* 2000;**16**:686–95.
14. Stepanek L, Sun QL, Wang J, Wang C, Bixby JL. CRYP-2/cPTPRO is a neurite inhibitory repulsive guidance cue for retinal neurons in vitro. *J Cell Biol* 2001;**154**:867–78.
15. Shintani T, Ihara M, Sakuta H, Takahashi H, Watakabe I, Noda M. Eph receptors are negatively controlled by protein tyrosine phosphatase receptor type O. *Nat. Neurosci* 2006;**9**:761–9.
16. Stoker A, Dutta R. Protein tyrosine phosphatases and neural development. *Bioessays* 1998;**20**:463–72.
17. Rashid-Doubell F, McKinnell I, Aricescu AR, Sajnani G, Stoker AW. Chick PTPsigma regulates the targeting of retinal axons within the optic tectum. *J Neurosci* 2002;**22**:5024–33.
18. Schindelholz B, Knirr M, Warrior R, Zinn K. Regulation of CNS and motor axon guidance in Drosophila by the receptor tyrosine phosphatase DPTP52F. *Development* 2001;**128**:4371–82.
19. Kaufmann N, DeProto J, Ranjan R, Wan H, Van Vactor D. *Drosophila* liprin-alpha and the receptor phosphatase Dlar control synapse morphogenesis. *Neuron* 2002;**34**:27–38.
20. Johnson KG, Tenney AP, Ghose A, Duckworth AM, Higashi ME, Parfitt K, Marcu O, Heslip TR, Marsh JL, Schwarz TL, Flanagan JG, Van Vactor D. The HSPGs Syndecan and Dallylike bind the receptor phosphatase LAR and exert distinct effects on synaptic development. *Neuron* 2006;**49**:517–31.
21. Ackley BD, Harrington RJ, Hudson ML, Williams L, Kenyon CJ, Chisholm AD, Jin Y. The two isoforms of the *Caenorhabditis elegans* leukocyte-common antigen related receptor tyrosine phosphatase PTP-3 function independently in axon guidance and synapse formation. *J Neurosci* 2005;**25**:7517–28.
22. Sun Q, Schindelholz B, Knirr M, Schmid A, Zinn K. Complex genetic interactions among four receptor tyrosine phosphatases regulate axon guidance in Drosophila. *Mol Cell Neurosci* 2001;**17**:274–91.
23. Stepanek L, Stoker AW, Stoeckli E, Bixby JL. Receptor tyrosine phosphatases guide vertebrate motor axons during development. *J Neurosci* 2005;**25**:3813–23.
24. Uetani N, Chagnon MJ, Kennedy TE, Iwakura Y, Tremblay ML. Mammalian motoneuron axon targeting requires receptor protein tyrosine phosphatases sigma and delta. *J Neurosci* 2006;**26**:5872–80.
25. Yeo TT, Yang T, Massa SM, Zhang JS, Honkaniemi J, Butcher LL, Longo FM. Deficient LAR expression decreases basal forebrain cholinergic neuronal size and hippocampal cholinergic innervation. *J Neurosci Res* 1997;**47**:348–60.
26. Ensslen-Craig SE, Brady-Kalnay SM. PTP mu expression and catalytic activity are required for PTP mu-mediated neurite outgrowth and repulsion. *Mol Cell Neurosci* 2005;**28**:177–88.
27. Fukada M, Fujikawa A, Chow JP, Ikematsu S, Sakuma S, Noda M. Protein tyrosine phosphatase receptor type Z is inactivated by ligand-induced oligomerization. *FEBS Letts* 2006;**580**:4051–6.
28. Meng K, Rodriguez-Pena A, Dimitrov T, Chen W, Yamin M, Noda M, Deuel TF. Pleiotrophin signals increased tyrosine phosphorylation of beta-catenin through inactivation of the intrinsic catalytic activity of the receptor-type protein tyrosine phosphatase beta/zeta. *Proc Natl Acad Sci USA* 2000;**97**:2603–8.
29. Aricescu AR, McKinnell IW, Halfter W, Stoker AW. Heparan sulfate proteoglycans are ligands for receptor protein tyrosine phosphatase sigma. *Mol Cell Biol* 2002;**22**:1881–92.
30. Sajnani G, Aricescu AR, Jones EY, Gallagher J, Alete D, Stoker A. PTPsigma promotes retinal neurite outgrowth non-cell-autonomously. *J Neurobiol* 2005;**65**:59–71.
31. McLean J, Batt J, Doering LC, Rotin D, Bain JR. Enhanced rate of nerve regeneration and directional errors after sciatic nerve injury in receptor protein tyrosine phosphatase sigma knock-out mice. *J Neurosci* 2002;**22**:5481–91.
32. Wills Z, Bateman J, Korey CA, Comer A, Van Vactor D. The tyrosine kinase Abl and its substrate enabled collaborate with the receptor phosphatase Dlar to control motor axon guidance. *Neuron* 1999;**22**:301–12.

33. Estrach S, Schmidt S, Diriong S, Penna A, Blangy A, Fort P, Debant A. The human rho-GEF Trio and its target GTPase rhoG are involved in the NGF pathway, leading to neurite outgrowth. *Curr Biol* 2002;**12**:307–12.

34. Zondag GC, Reynolds AB, Moolenaar WH. Receptor protein-tyrosine phosphatase RPTPmu binds to and dephosphorylates the cat-enin p120(ctn). *J Biol Chem* 2000;**275**:11,264–11,269,.

35. Anders L, Mertins P, Lammich S, Murgia M, Hartmann D, Saftig P, Haass C, Ullrich A. Furin-, ADAM 10-, and γ-secretase-mediated cleav-age of a receptor tyrosine phosphatase and regulation of β-catenin's transcriptional activity. *Cell Biol* 2006;**26**:3917–34.

36. Stoker AW. Protein tyrosine phosphatases and signalling. *J Endocrinol* 2005;**185**:19–33.

37. Rosdahl JA, Mourton TL, Brady-Kalnay SM. Protein kinase C delta (PKCdelta) is required for protein tyrosine phosphatase mu (PTPmu)-dependent neurite outgrowth. *Mol Cell Neurosci* 2002;**19**:292–306.

38. Phillips-Mason PJ, Gates TJ, Major DL, Sacks DB, Brady-Kalnay SM. The receptor protein-tyrosine phosphatase PTPmu interacts with IQGAP1. *J Biol Chem* 2006;**281**:4903–10.

39. Debant A, Serra-Pages C, Seipel K, O'Brien S, Tang M, Park SH, Streuli M. The multidomain protein Trio binds the LAR transmem-brane tyrosine phosphatase, contains a protein kinase domain, and has separate rac-specific and rho-specific guanine nucleotide exchange factor domains. *Proc Natl Acad Sci USA* 1996;**93**:5466–71.

40. Serra-Pages C, Medley QG, Tang M, Hart A, Streuli M. Liprins, a family of LAR transmembrane protein-tyrosine phosphatase-interacting proteins. *J Biol Chem* 1998;**273**:15,611–15,620.

41. Yang T, Massa SM, Longo FM. LAR protein tyrosine phosphatase receptor associates with TrkB and modulates neurotrophic signaling pathways. *J Neurobiol* 2006;**66**:1420–36.

42. Faux C, Hawadle M, Nixon J, Wallace A, Lee S, Murray S, Stoker A. PTPsigma binds and dephosphorylates neurotrophin receptors and can suppress NGF-dependent neurite outgrowth from sensory neu-rons. *Biochim Biophys Acta* 2007;**1773**:1689–700.

43. Den Hertog J, Hunter T. Tight association of GRB2 with receptor protein-tyrosine phosphatase alpha is mediated by the SH2 and C-terminal SH3 domains. *EMBO J* 1996;**15**:3016–27.

44. Su J, Muranjan M, Sap J. Receptor protein tyrosine phosphatase alpha activates Src-family kinases and controls integrin-mediated responses in fibroblasts. *Curr Biol* 1999;**9**:505–11.

45. Fashena SJ, Zinn K. Transmembrane glycoprotein gp150 is a sub-strate for receptor tyrosine phosphatase DPTP10D in Drosophila cells. *Mol Cell Biol* 1997;**17**:6859–67.

Chapter 33

Neurotrophin Signaling in Development

Katrin Deinhardt and Moses V. Chao

Molecular Neurobiology Program, Skirball Institute of Biomolecular Medicine, New York University School of Medicine, New York

INTRODUCTION

The formation of the vertebrate nervous system is characterized by widespread programmed cell death, which determines cell number and appropriate target innervation during development. The neurotrophins, which include nerve growth factor (NGF), brain-derived growth factor (BDNF), NT-3 and NT-4, represent an important family of trophic factors that are essential for survival of selective populations of neurons during different developmental periods. The neurotrophic hypothesis postulates that during nervous system development, neurons approaching the same final target compete for limited amounts of target-derived trophic factors [1]. In this way, the nervous system moulds itself to maintain only the most competitive and appropriate connections. Competition among neurons for limiting amounts of neurotrophin molecules produced by target cells accounts for selective cell survival. Two predictions emanate from this hypothesis. First, the efficacy of neuronal survival will depend upon the amounts of trophic factors produced during development. Second, specific receptor expression in responsive cell populations will dictate neuronal responsiveness.

Neurotrophins exert their cellular effects through the actions of two different receptors: the tropomyosin-related kinase (Trk) receptor tyrosine kinase and the p75 neurotrophin receptor (p75^{NTR}), a member of the tumor necrosis factor (TNF) receptor superfamily. On one level, neurotrophins fit well with the neurotrophic hypothesis, as many peripheral neuronal subpopulations exhibit a predominant dependence on a specific neurotrophin during the period of naturally occurring cell death (Figure 33.1). However, the biological reality appears much more complex. In the central nervous system, the overlapping expression of multiple neurotrophin receptors and their cognate ligands allows for the creation of diverse connectivity, which extends well into adulthood. And even in the periphery there are additional activities, such as the molecular mechanisms underlying the retrograde signal, a pathway that must efficiently transmit

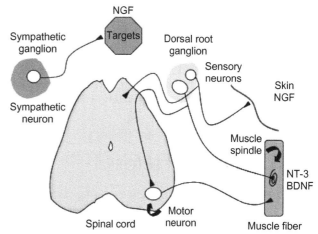

FIGURE 33.1 Neurotrophins serve as target-derived survival factors. NGF, BDNF and NT-3 display specific survival and differentiative effects upon sympathetic and sensory neuron populations in the peripheral nervous system.

information over long distances – at times more than a meter. Moreover, it has recently become clear that, in addition to mature neurotrophins, the pro-neurotrophins also play important roles in the development and modulation of the nervous system. Finally, the role of neurotrophins and their receptors is not confined to the nervous system, but is instead increasingly recognized in non-neuronal tissues such as the vasculature [2]. The role of the neurotrophin system in development has been reviewed [3]. This chapter will focus upon new views concerning ligand–receptor interactions, signal transduction, and retrograde transport in the nervous system, and discuss the roles of both mature and pro-neurotrophins.

THE NEUROTROPHIN LIGANDS

The neurotrophins are initially synthesized as precursors or pro-neurotrophins that are cleaved to release the mature, active proteins. The mature proteins form stable, non-covalent dimers, and are normally expressed at very low levels during development. Pro-neurotrophins are cleaved intracellularly by furin or pro-convertases recognizing a highly conserved dibasic amino acid cleavage site to release carboxy-terminal mature proteins of approximately 13 kDa. These extensively studied mature proteins mediate neurotrophin actions by selectively binding to members of the Trk family of receptor tyrosine kinases to regulate neuronal survival, differentiation and synaptic plasticity. In addition, all mature neurotrophins interact with p75^{NTR} which can modulate the affinity of Trk : neurotrophin associations.

Neurotrophins promote cell survival and differentiation during neural development. These effects are primarily conveyed through Trk receptor activation. Paradoxically, neurotrophins can also induce cell death. p75^{NTR} serves as a pro-apoptotic receptor during developmental cell death and after injury to the nervous system. In the context of neurotrophin processing, pro-neurotrophins are more effective than mature NGF in inducing p75^{NTR}-dependent apoptosis [4]. This suggests that the biological action of the neurotrophins can be regulated by proteolytic cleavage, with proforms preferentially activating p75^{NTR} to mediate apoptosis and mature forms selectively activating Trk receptors to promote survival.

NEUROTROPHIN RECEPTORS

One way of generating more specificity during development is by imparting greater discrimination of ligands for the Trk receptors. NGF binds most specifically to TrkA; BDNF and NT-4 to TrkB; and NT-3 to TrkC receptors. The p75^{NTR} receptor can bind to each neurotrophin, but has the additional capability of regulating a Trk's affinity for its cognate ligand. Trk and p75^{NTR} receptors have been referred to as high- and low-affinity receptors, respectively. However, this is not correct, since TrkA and TrkB actually bind their ligands with an affinity of 10^{-9}–10^{-10} M, which is lower than the high-affinity site ($K_d = 10^{-11}$ M). Also, pro-NGF displays high affinity binding to p75^{NTR}. Trk-mediated responsiveness to low concentrations of NGF is dependent upon the relative levels of p75^{NTR} and TrkA receptors and their combined ability to form high-affinity sites.

Although p75^{NTR} and Trk receptors do not appear to bind to each other directly, there is evidence that complexes form between the two receptors. As a result of these interactions, increased ligand selectivity can be conferred onto Trks by p75^{NTR}. NGF and NT-3 can both bind to TrkA, but p75^{NTR} restricts signaling of TrkA to NGF and not to NT-3 [5]. In addition to increasing ligand–receptor affinity and selectivity, p75^{NTR} contains an intracellular death domain sequence that can recruit pro-apoptotic adaptor molecules [6]. The ability of p75^{NTR} to promote apoptosis in response to pro-neurotrophins requires sortilin, a member of the Vps10p family, as a co-receptor [7, 8].

What are the reasons for having a Trk receptor that mediates neuronal survival and a p75^{NTR} receptor that mediates apoptosis? One reason neurotrophins use a death receptor may be to prune neurons efficiently during periods of developmental cell death. In addition to competing for trophic support from the target, neurons must establish connections with the proper target. In the event of mistargeting of axons, neurons may undergo apoptosis if the appropriate set of trophic factors is not encountered. In this case, a neurotrophin may not only fail to activate Trks, but also bind to p75^{NTR} and eliminate cells by an active killing process. For example, BDNF causes sympathetic cell death by binding to p75^{NTR} when TrkB is absent. Likewise, NT-4 causes p75^{NTR}-mediated cell death in BDNF-dependent trigeminal neurons due presumably to preferential p75^{NTR} rather than TrkB stimulation [9]. Therefore, Trk and p75^{NTR} receptors can give opposite outcomes in the same cells. Cell death mediated by p75^{NTR} may be important for the refinement of correct target innervation during development.

SIGNALING SPECIFICITY DURING DEVELOPMENT

Specific Trk receptor expression patterns determine the development of peripheral neuron populations. In the dorsal root ganglion, small-sized unmyelinated neurons predominantly express TrkA, whereas larger-sized neurons express TrkC receptors. Many of the small-diameter neurons are nociceptive and frequently terminate in the epidermis (Figure 33.1). NGF is important for the development of these neurons during early postnatal periods. The large-diameter neurons are proprioceptive, and are most responsive to NT-3. Consistent with the receptor expression, a lack of NGF leads to a lack of responsiveness to nociceptive stimuli, and a lack of NT-3 leads to a loss of muscle spindle afferents [10]. Furthermore, neurotrophin receptors play an instructive role in the sub-specification of dorsal root ganglia neurons into nociceptive or proprioceptive neurons as expression of TrkC from the *TrkA* locus leads these neurons to switch fate [11].

Trk receptors exhibit very high conservation in their intracellular domains, including the catalytic tyrosine kinase and the juxtamembrane NPXY motif that serves as the Shc binding site. However, several pronounced differences among the Trks exist. In a sympathetic neuronal background, TrkA relies predominantly upon phosphoinositide 3-kinase (PI3K) activation for survival, whereas TrkB uses both PI3K and ERK pathways. Thus, each Trk receptor carries distinctive signaling properties. For example, TrkB may contain sequences that bind to factors that favor alternative pathways. Since there are now a number of different adaptor proteins and enzymatic functions associated with Trk receptors (Figure 33.2), preferential interactions with these proteins must take place. Receptor

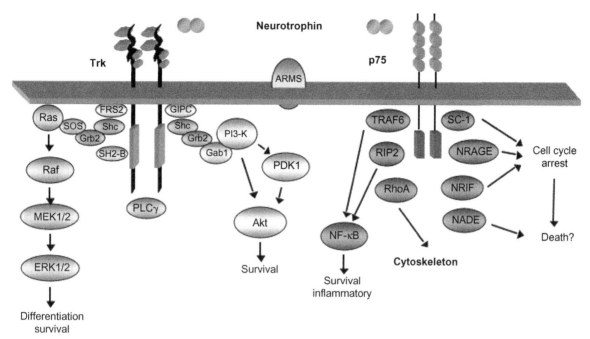

FIGURE 33.2 Neurotrophin receptors use multiple adaptor proteins.
Each receptor undergoes ligand-induced dimerization that results in the recruitment of multiple cytoplasmic proteins, which serve to increase the activities of phospholipase Cγ, PI3-, and MAP kinases. The p75 receptor is capable of initiating a cell death program in selected cells and signaling that leads to ceramide production and NFκB or JNK activation.

utilization of substrates with differential association/dissociation kinetics, competition for binding among different substrates, or recruitment of unique target proteins, such as FRS-2, rAPS and SH2 for the Trk receptors, represent mechanisms by which each receptor may differentially employ common substrates for signaling.

Alternatively, receptor processing or targeting into different membrane compartments may dictate their function. A comparison of TrkA and TrkB receptors in neuronal cell lines has revealed a difference in turnover of each receptor. While NGF binding to TrkA leads primarily to receptor recycling, BDNF binding to TrkB results in rapid turnover of TrkB receptors at the cell surface [12, 13]. In sensory neurons, activated TrkA is ubiquitinated by the E3 ubiquitin ligase NEDD4-2 and subsequently downregulated. NEDD4-2 does not bind to or ubiquitinate TrkB, further strengthening the notion that Trk receptors are differentially regulated [14]. Additionally, the number of surface TrkB receptors is highly influenced by depolarization and levels of cAMP. These observations hint at yet other receptor mechanisms that confer greater signaling specificity to the neurotrophins.

INTERACTING PROTEINS

Neurotrophin receptors undergo ligand-induced dimerization that activates multiple signal transduction pathways. Neurotrophin binding to Trk family members produces biological responses through rapid increases in the phosphorylation of ERK, phospholipase Cγ and PI3K [15]. Increased

Ras activity, a common signal from all tyrosine kinase receptors, results from the stimulation of guanine nucleotide exchange factors coupled to adaptor proteins which directly interact with Trk after ligand binding. These adaptor proteins include Shc, Grb2, SH2B, and FRS-2 (Figure 33.2).

A number of adaptor proteins also bind to p75[NTR] (Figure 33.2). Three different proteins, NRIF, NADE, and NRAGE, contribute to apoptosis in immortalized cell lines or are correlated with neurotrophin-dependent cell death. Each protein binds to separate sequence in the cytoplasmic domain of the p75[NTR] [16]. Another protein that interacts with both p75[NTR] and Trk receptors is ARMS, an ankyrin-rich transmembrane protein [17]. ARMS is rapidly tyrosine phosphorylated after binding of neurotrophins to Trk receptors, and leads to sustained ERK signaling [18]. This protein may act as a scaffold to cluster proteins essential to neurotrophin signaling. Other proteins, including RhoA GTPase, SC-1, and NRAGE, exert non-apoptotic activities, such as neurite elongation and growth arrest. These proteins expand the functional scope of neurotrophins [19]. Still other proteins, such as cytoplasmic dynein, myosin VI, and the PDZ-containing GIPC protein, serve to target neurotrophin receptors intracellularly during important cellular processes such as axonal and dendritic localization, and synaptic transmission [20]. Given the wide number of activities of neurotrophins and the small number of neurotrophins and neurotrophin receptor genes, it is likely other signaling systems are used. This includes the neurotrophin-induced modulation of ion channels such as TRP and glutamate receptors.

RETROGRADE AXONAL TRANSPORT

The biological effects of neurotrophins require that signals are conveyed over long distances from the nerve terminal to the cell body. Therefore, a central theme of the neurotrophic hypothesis is that neuronal survival and differentiation depend upon retrograde signaling of trophic factors produced at the target tissue.

During development, neurotrophins are produced and released from the target tissues and become internalized into vesicles, which are then transported to the cell body to convey the survival signal. The membrane vesicles carrying Trk, p75^{NTR}, and the neurotrophin have been termed "signaling endosomes." Increasing evidence suggests that these endocytic vesicles indeed retain their signaling capacity over long distances. A complex of NGF–TrkA has been found in clathrin-coated vesicles and endosomes, giving rise to the model that NGF and Trk are components of the retrograde signal [21]. Internalization of NGF from axon terminals is necessary for phosphorylation and activation of the CREB transcription factor, which leads to changes in gene expression and increased neuronal cell survival [22].

Measurements of ^{125}I-NGF transport from distal axons to the cell body in mice indicate a rate from 3 to 10 mm/h *in vivo* [23]. Recently, *in vitro* assays directly visualizing retrograde axonal transport of NGF, BDNF, or TrkB in cultured dorsal root ganglia and motor neurons have replicated these transport rates and identified proteins controlling this process, such as dynein and endocytic Rab GTPases. In the future, these assays will allow a more detailed assessment of the molecules that regulate retrograde axonal transport in neuronal development, health, and disease [24].

NEUROTROPHIN SIGNALING IN THE ADULT NERVOUS SYSTEM

Neurotrophin signaling is not only required during development, but also plays important roles in higher order function in the adult, such as behavior, learning, and memory. For example, knockout animals with lower levels of BDNF or its receptor TrkB, as well as mice harboring a common polymorphism in the *bdnf* gene leading to decreased BDNF secretion, show eating disorders and an increase in anxiety-related behavior [25, 26]. Moreover, decreases in BDNF have been correlated with depression, while increases in BDNF seem to have an antidepressant effect. Mature BDNF facilitates long-term potentiation and dendritic spine formation in the hippocampus, a process that has been implicated in learning and memory. This effect is specific to the action of TrkB, as reduction of TrkB but not TrkC levels reduces spine density in the hippocampus of aged mice [27]. In contrast, proBDNF signaling via p75^{NTR} has been involved in reduced spine density and long-term depression, which in turn may contribute to depressive

behavior [28]. The discovery that pro- and mature neurotrophins have opposite biological actions has been termed the "yin and yang hypothesis of neurotrophin action" [29]. Together, these findings strengthen the notion that neurotrophins, in addition to their potent properties on the cellular level, possess abilities to influence cognitive functions.

REFERENCES

1. Levi-Montalcini R. The nerve growth factor: thirty-five years later. *Biosci Rep* 1987;**7**:681–99.
2. Kraemer R, Hempstead BL. Neurotrophins: novel mediators of angiogenesis. *Front Biosci* 2003;**8**:s1181–6.
3. Huang EJ, Reichardt LF. Neurotrophins: roles in neuronal development and function. *Annu Rev Neurosci* 2001;**24**:677–736.
4. Lee R, Kermani P, Teng KK, Hempstead BL. Regulation of cell survival by secreted proneurotrophins. *Science* 2001;**294**:1945–8.
5. Benedetti M, Levi A, Chao MV. Differential expression of nerve growth factor receptors leads to altered binding affinity and neurotrophin responsiveness. *Proc Natl Acad Sci USA* 1993;**90**:7859–63.
6. Liepinsh E, Ilag L, Otting G, Ibanez C. NMR structure of the death domain of the p75 neurotrophin receptor. *EMBO J.* 1997;**16**:4999–5005.
7. Domeniconi M, Hempstead BL, Chao MV. Pro-NGF secreted by astrocytes promotes motor neuron cell death. *Mol Cell Neurosci* 2007;**34**:271–9.
8. Teng HK, Teng KK, Lee R, Wright S, Tevar S, Almeida RD, Kermani P, Torkin R, Chen ZY, Lee FS, Kraemer RT, Nykjaer A, Hempstead BL. ProBDNF induces neuronal apoptosis via activation of a receptor complex of p75NTR and sortilin. *J Neurosci* 2005;**25**:5455–63.
9. Agerman K, Baudet C, Fundin B, Willson C, Ernfors P. Attenuation of a caspase-3 dependent cell death in NT-4 and p75- deficient embryonic sensory neurons. *Mol Cell Neurosci* 2000;**16**:258–68.
10. Snider WD. Functions of the neurotrophins during nervous system development: what the knockouts are teaching us. *Cell* 1994;**77**:627–38.
11. Moqrich A, Earley TJ, Watson J, Andahazy M, Backus C, Martin-Zanca D, Wright DE, Reichardt LF, Patapoutian A. Expressing TrkC from the TrkA locus causes a subset of dorsal root ganglia neurons to switch fate. *Nature Neurosci* 2004;**7**:812–18.
12. Chen ZY, Ieraci A, Tanowitz M, Lee FS. A novel endocytic recycling signal distinguishes biological responses of Trk neurotrophin receptors. *Mol Biol Cell* 2005;**16**:5761–72.
13. Sommerfeld M, Schweigreiter R, Barde Y, Hoppe E. Down-regulation of the neurotrophin TrkB following ligand binding. *J Biol Chem* 2000;**275**:8982–90.
14. Arevalo JC, Waite J, Rajagopal R, Beyna M, Chen ZY, Lee FS, Chao MV. Cell survival through Trk neurotrophin receptors is differentially regulated by ubiquitination. *Neuron* 2006;**50**:549–59.
15. Reichardt LF. Neurotrophin-regulated signalling pathways. *Philos Trans R Soc Lond B Biol Sci* 2006;**361**:1545–64.
16. Roux P, Barker P. Neurotrophin signaling through the p75 neurotrophin receptor. *Prog Neurobiol* 2002;**67**:203–33.
17. Kong H, Boulter J, Weber J, Lai C, Chao M. An evolutionarily conserved transmembrane protein that is a novel downstream target of neurotrophin and ephrin receptors. *J Neurosci* 2001;**21**:176–85.
18. Arevalo J, Yano H, Teng K, Chao M. A unique pathway for sustained neurotrophin signaling through an ankyrin-rich membrane-spanning protein. *EMBO J* 2004;**23**:2358–68.
19. Lee F, Kim A, Khursigara G, Chao M. The uniqueness of being a neurotrophin receptor. *Curr Opin Neurobiol* 2001;**11**:281–6.

20. Yano H, Ninan I, Zhang H, Milner TA, Arancio O, Chao MV. BDNF-mediated neurotransmission relies upon a myosin VI motor complex. *Nature Neurosci* 2006;**9**:1009–18.

21. Grimes ML, Zhou J, Beattie EC, Yuen EC, Hall DE, Valletta JS, Topp KS, LaVail JH, Bunnett NW, Mobley WC. Endocytosis of activated TrkA: evidence that nerve growth factor induces formation of signaling endosomes. *J Neurosci* 1996;**16**:7950–64.

22. Riccio A, Pierchala B, Ciarallo C, Ginty D. An NGF-TrkA-mediated retrograde signal to transcription factor CREB in sympathetic neurons. *Science* 1997;**277**:1097–100.

23. Stockel K, Schwab M, Thoenen H. Comparison between the retrograde axonal transport of nerve growth factor and tetanus toxin in motor, sensory and adrenergic neurons. *Brain Res* 1975;**99**:1–16.

24. Ibanez CF. Message in a bottle: long-range retrograde signaling in the nervous system. *Trends Cell Biol* 2007;**17**:519–28.

25. Chen ZY, Jing D, Bath KG, Ieraci A, Khan T, Siao CJ, Herrera DG, Toth M, Yang C, McEwen BS, Hempstead BL, Lee FS. Genetic variant BDNF (Val66Met) polymorphism alters anxiety-related behavior. *Science* 2006;**314**:140–3.

26. Lyons WE, Mamounas LA, Ricaurte GA, Coppola V, Reid SW, Bora SH, Wihler C, Koliatsos VE, Tessarollo L. Brain-derived neurotrophic factor-deficient mice develop aggressiveness and hyperphagia in conjunction with brain serotonergic abnormalities. *Proc Natl Acad Sci USA* 1999;**96**:15,239–15,244.

27. von Bohlen Und Halbach O, Minichiello L, Unsicker K. TrkB but not trkC receptors are necessary for postnatal maintenance of hippocampal spines. *Neurobiol Aging* 2007;**29**:1247–55.

28. Martinowich K, Manji H, Lu B. New insights into BDNF function in depression and anxiety. *Nature Neurosci* 2007;**10**:1089–93.

29. Lu B, Pang PT, Woo NH. The yin and yang of neurotrophin action. *Nat Rev Neurosci* 2005;**6**:603–14.

Chapter 34

Attractive and Repulsive Signaling in Nerve Growth Cone Navigation

Guo-li Ming and Mu-ming Poo

The Salk Institute, La Jolla, California Division of Neurobiology, Department of Molecular and Cell Biology, University of California, Berkeley, California

INTRODUCTION

The function of the nervous system depends on complex and precise connections between nerve cells [1]. The formation of specific connections during development often requires the growing axons to navigate over considerable distances to reach their final target cells. This long-range navigation is achieved by guidance factors within the developing tissue that regulate the motility or directionality of the growing tip of the axon, the growth cone [2]. During the past decade several families of guidance factors have been identified, including netrins, semaphorins, ephrins, and slits [3, 4]. In addition, inhibitory factors associated with the myelin that exert repulsive actions on the navigation of regenerating axons have also been discovered. Different classes of membrane receptors for these factors have been identified, and their intracellular signal transduction mechanisms are beginning to be elucidated.

How does a guidance factor affect the navigation of the growing axon? A general scheme of signal transduction cascades from the receptor activation to cytoskeletal rearrangements is shown in Figure 34.1 [4]. It starts with the binding of the guidance factor with the receptor protein or protein complexes at the cell surface. Ligand–receptor binding in general stimulates the activities of the cytoplasmic domain of the receptor, which in turn interacts specifically with cytoplasmic adaptor proteins. These adaptors may then recruit or activate their downstream effectors to further mediate the guidance signal. The effectors (or mediators) can be enzymes or second messengers that activate or inhibit cytoskeleton-associated proteins, leading to polymerization or depolymerization of cytoskeletal structures and steering of the growth cone.

Two types of guidance signals may be distinguished: signals that convey a "stop or go" command regulating growth cone motility, and signals that provide directional instructions to the growing axon, triggering turning responses of the growth cone. For non-directional signals, mediators may simply alter the global cytoskeletal activity at the growth cone. For directional guidance signals, however, a gradient of cytoskeletal rearrangements must be created in order to induce directional motility. In the latter case, mediators must be activated or distributed in a gradient across the growth cone, and such a gradient may also need to be amplified in the cytoplasm in order to achieve a reliable directional response [4]. Although a number of cytoplasmic components have been implicated in such a scheme of signal transduction, definitive identification of signaling pathways is yet to be established for any one of the major families of guidance factors. This chapter summarizes some of the putative signaling pathways that have been shown to participate in growth cone navigation.

NETRIN SIGNALING

Netrins are a family of secreted proteins, and their receptors were identified to be DCC (deleted in colorectal cancer) and UNC-5 [5], two interacting transmembrane proteins that set the polarity of growth cone responses. Ectopic expression of UNC-5 in neurons converted netrin (UNC-6)-dependent chemoattraction to chemorepulsion in both *Caenorhabditis elegans* [6] and in dissociated *Xenopus* spinal neurons [7], a conversion that involves a netrin-dependent interaction between the cytoplasmic domain of DCC and UNC-5 [7]. The level of DCC can be actively regulated through degradation mediated by Sina/Siah protein [8] or through metalloprotease-mediated shedding [9], resulting in changes in the growth cone's sensitivity to netrin-1. Extracellular signal-regulated kinase1/2

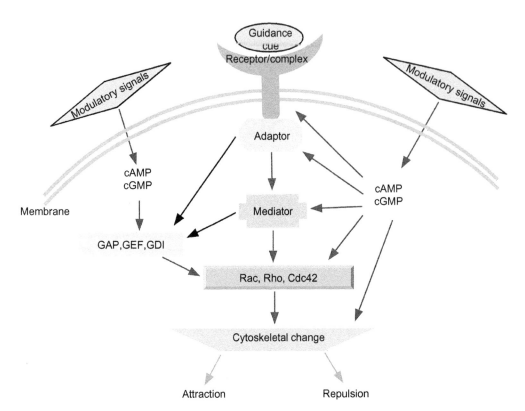

FIGURE 34.1 Signal transduction mechanism for nerve growth cone guidance.

(Erk 1/2) was found to be recruited to DCC receptor complex in rat commissural neurons [10] and activation of phosphoinositide 3-kinase (PI3K), and phospholipase Cγ appears to mediate attractive turning of *Xenopus* spinal neurons induced by a gradient of netrin-1 [11]. More recently, it has been shown that activation of MAPK, local protein synthesis, and protein degradation are involved in the netrin-induced chemoattraction of *Xenopus* spinal and retinal neurons [13, 14]. These downstream events were shown to be critical for adaptive changes of growth cone sensitivity to netrin-1 as the extracellular concentration of netrin-1 is increased [13]. How these cytoplasmic factors are linked to the cytoskeleton changes remains largely unclear. Two members of the Rho GTPase family, Rac1 and Cdc42, appear to be involved in netrin signaling [15, 16], thus providing potential links to cytoskeletal regulation [17]. To serve for directional guidance signals, a mediator is not only required for the guidance responses, but must also be activated in a gradient across the growth cone. Furthermore, such a gradient should be sufficient to induce a turning response of the growth cone. None of the putative signaling components described above fulfill these criteria. Interestingly, the well-known second messenger Ca^{2+} appears to satisfy these criteria for netrin-1 signaling. Elevation of cytoplasmic Ca^{2+}, through both Ca^{2+} influx and release from internal stores, is required for netrin-1-induced turning responses, and a netrin-1 gradient can trigger Ca^{2+} elevation and transient Ca^{2+} gradients across the growth cone [7, 12]. Experimentally creating a gradient of Ca^{2+} across the growth cone in the absence of netrin-1 signals is sufficient to induce the turning of the growth cone [18]. However, it remains unclear how Ca^{2+} signals are linked to receptor activation upstream and cytoskeletal rearrangements downstream.

SEMAPHORIN SIGNALING

The semaphorin family includes both membrane-bound and secreted molecules, thus it may work for both short- and long-range guidance [19, 20]. Neuropilins were identified as semaphorin receptors and the plexin family of receptors was shown to be a co-receptor that transduces the signal. Several proteins have been shown to bind to neuropilins or plexins. These include a transmembrane protein OTK (off-track) [21], cytoplasmic protein NIP–a PSD-95/Dlg/ZO-1 domain-containing protein that may be involved in membrane trafficking [22]–and MICAL, a flavoprotein oxidoreductase [23]. There is also evidence for the involvement of heterotrimeric G proteins [24]. The precise role of these receptor-interacting proteins and whether they mediate or modulate the signaling process remain to be determined. More is known about the downstream cascades that mediate cytoskeleton changes induced by semaphorins. Of particular interest is Rac1, a small Rho family GTPase. Introduction of dominant-negative Rac1 [25] or an inhibitory peptide for Rac1 or C3 transferase, a Rho GTPase inhibitor, blocks Sema3A-induced growth cone collapse in sensory neurons [26, 27]. A major downstream mediator of Rac1 and Cdc42 is P21-associated kinase (PAK), and LIM-domain-containing kinase (LIM-kinase), a direct substrate of PAK, is necessary for Sema3A-induced growth cone collapse [28]. LIM-kinase is a serine-threonine kinase that inhibits cofilin's actin-severing function. Thus, at least one of the mechanisms for semaphorin-mediated cytoskeletal changes is mediated through the activation of small GTPases and their target PAK, which then regulates actin dynamics through a LIM-kinase- and cofilin-dependent pathway. Other small GTPases such as Rho and Rnd1 [29, 30] and kinases such as GSK-3 (glycogen synthase kinase) [31] and Fes/Fps tyrosine kinase [32] have also been implicated in semaphorin signaling.

SLIT SIGNALING

Slits are a family of secreted proteins that can exert short- and long-range guidance functions by activating their receptors, the roundabout (Robo) family of proteins [33]. Slits appear to act not only as directional guidance factors but also as stop signals through activation in a combinatorial manner of different Robo receptors expressed on the growth cone surface [34–36]. Both Abelson tyrosine kinase (Abl) and its substrate Enabled (Ena) can bind directly to the cytoplasmic domain of Robo and modulate its function [37]. Interfering with the binding between Ena and Robo partially impairs the Robo function, while a mutation in a conserved tyrosine residue that can be phosphorylated by Abl generates a hyperactive Robo. Small GTPases and their regulators also affect slit-Robo signaling. A slit-Robo-GTPase activating protein 1 (srGAP1) can bind to Robo and inactivate Cdc42, resulting in repulsion of growth cones [38], while GEF64C, a Dbl family guanine nucleotide exchange factor (GEF), can activate Rho and block Robo-induced repulsion [39]. Thus, Robo-mediated cytoskeleton changes also appear to be mediated by activation of GAP or GEF of small GTPases.

EPHRIN SIGNALING

Ephrins and the Eph family of tyrosine kinase receptors are membrane-bound molecules that mediate short-range axon guidance via cell–cell contacts [40]. Ephrin-A ligands are attached to the plasma membrane via a glycophosphatidylinositol (GPI) linkage, whereas the ephrin-B ligand contains a transmembrane domain and a cytoplasmic tail [41]. Similar to slits, ephrins can function as either directional or non-directional guidance factors. In addition, the signaling activated by ephrin-Eph binding is bidirectional [41], so that cytoplasmic activities are triggered in both interacting

cells. A GEF, ephexin, binds to the kinase domain of EphA constitutively through its Dbl homology-pleckstrin homology (DH/PH) domain and activates both RhoA and Cdc42, thus regulating cytoskeletal structures [42]. In addition, focal adhesion kinase (FAK) and its downstream factor P130(cas), which are two proteins involved in actin reorganization, are also implicated in EphA-induced cytoskeletal changes [43].

NOGO AND MYELIN-ASSOCIATED GLYCOPROTEIN SIGNALING

Two proteins associated with myelin, Nogo and myelin-associated glycoprotein (MAG), have been identified as the major inhibitory factors that prevent axon regeneration after CNS injury [44]. Although the full lengths of these proteins are membrane-anchored, they can be released in a truncated form and function in repelling and inhibiting axon growth [45, 46]. The receptors for Nogo (NogoR) [46] and for MAG (GD1a and GT1b) [47] have been identified. Interestingly, MAG also binds to NogoR [48]. NogoR is a GPI-anchored protein at the cell surface [46], and an as yet unidentified co-receptor(s) is required for transducing the cytoplasmic signal. The downstream signal cascade for NogoR signaling is largely unknown, although Ca^{2+} and PI3K are required for MAG-induced repulsion of *Xenopus* spinal neurons [11, 49]. Rho is activated by MAG through receptor GD1a and GT1b [47]. Inhibition of Rho activity can promote CNS axon regeneration, suggesting that Rho may also be involved in MAG-induced cytoskeletal rearrangement [50].

CRITICAL ROLES OF MODULATORY SIGNALS

For a growth cone to make its navigational decisions, it must integrate information provided not only by the guidance factors but also by other modulatory signals. *In vitro* studies have shown that cytoplasmic cyclic nucleotides play key roles in modulating signal transduction events triggered by most guidance factors identified thus far [4]. For example, the growth cone responses to netrin and MAG are modulated by a cAMP-dependent pathway, whereas Sema3A and slit signaling is modulated by a cGMP-dependent pathway. Elevating the cytoplasmic level of cyclic nucleotides favors attraction/growth, while lowering their level favors repulsion/collapse [4]. Many extracellular ligands, including neuromodulators, adhesion molecules, and extracellular matrix (ECM) components, may change the level of cyclic nucleotides within the cell, thus altering the growth cone behavior when they are present concurrently with the guidance signal [51]. For example, laminin, an abundant ECM protein, reduces the cAMP level in *Xenopus* retinal ganglion neurons and converts the growth cone response to a netrin-1 gradient from attraction to repulsion both *in vitro* and *in vivo* [52]. Conversely, the repulsive response of DRG and cortical axons induced by Sema3A can be converted to attraction by exposure to soluble LI-Fc chimeric molecules; activation of guanylate cyclase activity is required for the conversion [53]. The targets of PKA/PKG that are involved in regulating the polarity of growth cone turning responses remain to be identified.

CONCLUDING REMARKS

An emerging view of axon guidance factors is that they are multifunctional molecules capable of conferring attractive, repulsive, or stop signals. The precise behavior of a growth cone is determined by the nature of specific receptors and the status of cytoplasmic signal cascades, which are under the influence of a variety of extrinsic and intrinsic factors (Figure 34.1). The combinatorial expression pattern of various receptors at the surface of a growth cone may trigger a differential downstream event [54]. In addition, the efficacy of receptor signaling across the plasma membrane can be modulated by various factors both extra- and intracellularly [4]. Recruitment of different adaptors and mediators in the cytoplasm can result in different growth cone behaviors. An area of interest for future studies is determining the mechanisms that control or modulate the recruitment and activation of these cytoplasmic factors, and that amplify the signals conveyed by the receptors. Signal cascades triggered by all known guidance factors appear eventually to converge upon different members of the Rho family GTPases or their activators/inhibitors. It is of interest now to determine how spatio-temporal patterns of GTPase activation account for distinct navigational behaviors of the growth cone.

REFERENCES

1. Kandel ER, Schwartz JH, Jessell TMPrinciples of neural science. New York: McGraw-Hill; 2000.
2. Tessier-Lavigne M, Goodman CS. The molecular biology of axon guidance. *Science* 1996;**274**:1123–33.
3. Mueller BK. Growth cone guidance: first steps towards a deeper understanding. *Annu Rev Neurosci* 1999;**22**:351–88.
4. Song H, Poo M. The cell biology of neuronal navigation. *Nat Cell Biol* 2001;**3**:E81–8.
5. Livesey FJ. Netrins and netrin receptors. *Cell Mol Life Sci* 1999;**56**:62–8.
6. Hamelin M, Zhou Y, Su MW, Scott IM, Culotti JG. Expression of the UNC-5 guidance receptor in the touch neurons of *C. elegans* steers their axons dorsally. *Nature* 1993;**364**:327–30.
7. Hong K, Nishiyama M, Henley J, Tessier-Lavigne M, Poo M. Calcium signalling in the guidance of nerve growth by netrin-1. *Nature* 2000;**403**:93–8.

8. Hu G, Fearon ER. Siah-1 N-terminal RING domain is required for proteolysis function, and C-terminal sequences regulate oligomerization and binding to target proteins. *Mol Cell Biol* 1999;**19**:724–32.
9. Galko MJ, Tessier-Lavigne M. Function of an axonal chemoattractant modulated by metalloprotease activity. *Science* 2000;**289**:1365–7.
10. Forcet C, Stein E, Pays L, Corset V, Llambi F, Tessier-Lavigne M, Mehlen P. Netrin-1-mediated axon outgrowth requires deleted in colorectal cancer-dependent MAPK activation. *Nature* 2002;**417**:443–7.
11. Ming G, Song H, Berninger B, Inagaki N, Tessier-Lavigne M, Poo M. Phospholipase C-gamma and phosphoinositide 3-kinase mediate cytoplasmic signaling in nerve growth cone guidance. *Neuron* 1999;**23**:139–48.
12. Ming G-L, Song H-J, Berninger B, Holt CE, Tessier-Lavigne M, Poo M. cAMP-dependent growth cone guidance by netrin-1. *Neuron* 1997;**19**:1225–35.
13. Ming GL, Wong ST, Henley J, Yuan XB, Song HJ, Spitzer NC, Poo M-M. Adaptation in the chemotactic guidance of nerve growth cones. *Nature* 2002;**417**:411–18.
14. Campbell DS, Holt CE. Chemotropic responses of retinal growth cones mediated by rapid local protein synthesis and degradation. *Neuron* 2001;**32**:1013–26.
15. Li X, Saint-Cyr-Proulx E, Aktories K, Lamarche-Vane N. Rac1 and Cdc42 but not RhoA or Rho kinase activities are required for neurite outgrowth induced by the Netrin-1 receptor DCC (deleted in colorectal cancer) in N1E-115 neuroblastoma cells. *J Biol Chem* 2002;**277**:15,207–15,214.
16. Shekarabi M, Kennedy TE. The netrin-1 receptor DCC promotes filopodia formation and cell spreading by activating Cdc42 and Rac1. *Mol Cell Neurosci* 2002;**19**:1–17.
17. Tapon N, Hall A. Rho, Rac and Cdc42 GTPases regulate the organization of the actin cytoskeleton. *Curr Opin Cell Biol* 1997;**9**:86–92.
18. Zheng JQ. Turning of nerve growth cones induced by localized increases in intracellular calcium ions. *Nature* 2000;**403**:89–93.
19. He Z, Wang KC, Koprivica V, Ming G, Song H-J. Functions of semaphorins in the nervous system. *Sci STKE* 2002;**119**:RE1.
20. Nakamura F, Kalb RG, Strittmatter SM. Molecular basis of semaphorin-mediated axon guidance. *J Neurobiol* 2000;**44**:219–29.
21. Winberg ML, Tamagnone L, Bai J, Comoglio PM, Montell D, Goodman CS. The transmembrane protein Off-track associates with Plexins and functions downstream of Semaphorin signaling during axon guidance. *Neuron* 2001;**32**:53–62.
22. Cai H, Reed RR. Cloning and characterization of neuropilin-1-interacting protein: a PSD-95/Dlg/ZO-1 domain-containing protein that interacts with the cytoplasmic domain of neuropilin-1. *J Neurosci* 1999;**19**:6519–27.
23. Terman JR, Mao T, Pasterkamp RJ, Yu HH, Kolodkin AL. MICALs, a amily of conserved flavoprotein oxidoreductases, function in plexin-mediated axonal repulsion. *Cell* 2002;**109**:887–900.
24. Igarashi M, Strittmatter SM, Vartanian T, Fishman MC. Mediation by G proteins of signals that cause collapse of growth cones. *Science* 1993;**259**:77–9.
25. Jin Z, Strittmatter SM. Rac1 mediates collapsin-1-induced growth cone collapse. *J Neurosci* 1997;**17**:6256–63.
26. Kuhn TB, Brown MD, Wilcox CL, Raper JA, Bamburg JR. Myelin and collapsin-1 induce motor neuron growth cone collapse through different pathways: Inhibition of collapse by opposing mutants of rac1. *J Neurosci* 1999;**19**:1965–75.
27. Liu BP, Strittmatter SM. Semaphorin-mediated axonal guidance via Rho-related G proteins. *Curr Opin Cell Biol* 2001;**13**:619–26.

28. Aizawa H, Wakatsuki S, Ishii A, Moriyama K, Sasaki Y, Ohashi K, Sekine-Aizawa Y, Sehara-Fujisawa A, Mizuno K, Goshima Y, Yahara I. Phosphorylation of cofilin by LIM-kinase is necessary for semaphorin3A-induced growth cone collapse. *Nat Neurosci* 2001;**4**:367–73.

29. Arimura N, et al. Phosphorylation of collapsin response mediator protein-2 by Rho-kinase. Evidence for two separate signaling pathways for growth cone collapse. *J Biol Chem* 2000;**275**:23,973–23,980.

30. Zanata SM, Hovatta I, Rohm B, Puschel AW. Antagonistic effects of Rnd1 and RhoD GTPases regulate receptor activity in Semaphorin3A-induced cytoskeletal collapse. *J Neurosci* 2002;**22**:471–7.

31. Eickholt BJ, Walsh FS, Doherty P. An inactive pool of GSK-3 at the leading edge of growth cones is implicated in Semaphorin3A signaling. *J Cell Biol* 2002;**157**:211–17.

32. Mitsui N, Inatome R, Takahashi S, Goshima Y, Yamamura H, Yanagi S. Involvement of Fes/Fps tyrosine kinase in semaphorin3A signaling. *EMBO J* 2002;**21**:3274–85.

33. Guthrie S. Axon guidance: Robos make the rules. *Curr Biol* 2001;**11**:R300–3.

34. Rajagopalan S, Vivancos V, Nicolas E, Dickson BJ. Selecting a longitudinal pathway: Robo receptors specify the lateral position of axons in the *Drosophila* CNS. *Cell* 2000;**103**:1033–45.

35. Simpson JH, Bland KS, Fetter RD, Goodman CS. Short-range and long-range guidance by Slit and its Robo receptors: a combinatorial code of Robo receptors controls lateral position. *Cell* 2000;**103**:1019–32.

36. Bagri A, Marin O, Plump AS, Mak J, Pleasure SJ, Rubenstein JL, Tessier-Lavigne M. Slit proteins prevent midline crossing and determine the dorsoventral position of major axonal pathways in the mammalian forebrain. *Neuron* 2002;**33**:233–48.

37. Bashaw GJ, Kidd T, Murray D, Pawson T, Goodman CS. Repulsive axon guidance: Abelson and Enabled play opposing roles downstream of the roundabout receptor. *Cell* 2000;**101**:703–15.

38. Wong K, Ren X-R, Huang Y-Z, Xie Y, Liu G, Saito H, et al. Signal transduction in neuronal migration: roles of GTPase activating proteins and the small GTPase Cdc42 in the Slit-Robo pathway. *Cell* 2001;**107**:209–21.

39. Bashaw GJ, Hu H, Nobes CD, Goodman CS. A novel Dbl family RhoGEF promotes Rho-dependent axon attraction to the central nervous system midline in *Drosophila* and overcomes Robo repulsion. *J Cell Biol* 2001;**155**:1117–22.

40. Wilkinson DG. Multiple roles of EPH receptors and ephrins in neural development. *Nat Rev Neurosci* 2001;**2**:155–64.

41. Flanagan JG, Vanderhaeghen P. The ephrins and Eph receptors in neural development. *Annu Rev Neurosci* 1998;**21**:309–45.

42. Shamah SM, Lin MZ, Goldberg JL, Estrach S, Sahin M, Hu L, Bazalakova M, Neve RL, Corfas G, Debant A, Greenberg ME. EphA receptors regulate growth cone dynamics through the novel guanine nucleotide exchange factor ephexin. *Cell* 2001;**105**:233–44.

43. Carter N, Nakamoto T, Hirai H, Hunter T. EphrinA1-induced cytoskeletal re-organization requires FAK and p130 (cas). *Nat Cell Biol* 2002;**22**:565–73.

44. Fouad K, Dietz V, Schwab ME. Improving axonal growth and functional recovery after experimental spinal cord injury by neutralizing myelin associated inhibitors. *Brain Res Brain Res Rev* 2001;**36**:204–12.

45. Tang S, Qiu J, Nikulina E, Filbin MT. Soluble myelin-associated glycoprotein released from damaged white matter inhibits axonal regeneration. *Mol Cell Neurosci* 2001;**18**:259–69.

46. Brittis PA, Flanagan JG. Nogo domains and a Nogo receptor: implications for axon regeneration. *Neuron* 2001;**30**:11–14.

47. Vyas AA, Patel HV, Fromholt SE, Heffer-Lauc M, Vyas KA, Dang J, Schachner M, Schnaar RL. Gangliosides are functional nerve cell ligands for myelin-associated glycoprotein (MAG), an inhibitor of nerve regeneration. *Proc Natl Acad Sci USA* 2002;**99**:8412–17.

48. Liu BP, Fournier A, GrandPre T, Strittmatter SM. Myelin-associated glycoprotein as a functional ligand for the Nogo-66 receptor. *Science* 2002;**297**:1190–3.

49. Song H, Ming G, He Z, Lehmann M, McKerracher L, Tessier-Lavigne M, Poo M. Conversion of neuronal growth cone responses from repulsion to attraction by cyclic nucleotides. *Science* 1998;**281**:1515–18.

50. Lehmann M, Fournier A, Selles-Navarro I, Dergham P, Sebok A, Leclerc N, Tigyi G, McKerracher L. Inactivation of Rho signaling pathway promotes CNS axon regeneration. *J Neurosci* 1999;**19**:7537–47.

51. Song HJ, Poo MM. Signal transduction underlying growth cone guidance by diffusible factors. *Curr Opin Neurobiol* 1999;**9**:355–63.

52. Hopker VH, Shewan D, Tessier-Lavigne M, Poo M, Holt C. Growth-cone attraction to netrin-1 is converted to repulsion by laminin-1. *Nature* 1999;**401**:69–73.

53. Castellani V, Chedotal A, Schachner M, Faivre-Sarrailh C, Rougon G. Analysis of the L1-deficient mouse phenotype reveals cross-talk between Sema3A and L1 signaling pathways in axonal guidance. *Neuron* 2000;**27**:237–49.

54. Yu TW, Bargmann CI. Dynamic regulation of axon guidance. *Nat Neurosci* 2001;**4**(Suppl.):1169–76.

Chapter 35

Semaphorins and their Receptors in Vertebrates and Invertebrates

Eric F. Schmidt, Hideaki Togashi and Stephen M. Strittmatter

Department of Neurology and Section of Neurobiology, Yale University School of Medicine, New Haven, Connecticut

THE SEMAPHORIN FAMILY

Semaphorins are a large family of proteins originally identified as axon guidance factors of the developing nervous system. Over 30 family members are grouped into 8 classes based on structural and phylogenetic relationships (reviewed in [1]). Classes 1 and 2 are expressed in invertebrates, classes 3 through 7 are vertebrate semaphorins, and Class V is expressed in non-neurotropic DNA viruses. All semaphorins share a highly conserved 500 amino acid "Sema" domain at their amino terminus, but different classes possess divergent sequences in their carboxyl regions. Classes 1, 4, 5, and 6 are transmembrane proteins, class 7 has a GPI-anchor, and classes 2, 3, and V are secreted proteins. The presence of both membrane-bound and secreted semaphorins suggests that semaphorins act as both short- and long-range cues. In addition, the diversity of structural properties between the classes implies roles in a diversity of biological processes.

The best-documented function of semaphorins is their role in central nervous system (CNS) development. Semaphorins act as both repellents and attractants for growing axons. The first identified vertebrate semaphorin, Sema3A, causes retraction of axons and the collapse of the growth cone, a specialization at the tip of growing axons [2]. While repellents for certain neurons, Sema3C and 3F also serve as attractants for cortical and olfactory neurons, respectively. Within the large semaphorin family, certain semaphorins can exert antagonistic activity by competitively blocking the activity of other family members at certain receptors [3]. Semaphorins may guide growing dendrites as well as axons. Specifically, Sema3A attracts the apical dendrite of pyramidal neurons in the cerebral cortex toward the pial surface [4]. Many types of neurons are responsive to semaphorins, including dorsal root ganglion, sensory, motor, hippocampal, cortical, cerebellar, and olfactory. In addition to guiding axons and dendrites, semaphorins appear to play a role in fasciculation of nerve bundles, neuronal cell migration, axoplasmic transport, and apoptosis [5]. Like developing neurons, adult neurons of the regenerating CNS are responsive to semaphorins, and semaphorin expression is upregulated after nerve injury [6].

Semaphorin signaling is not restricted to the CNS, as evidenced by widespread expression throughout the embryo and adult tissue. Migrating non-neuronal cells are responsive to semaphorins, and cardiovascular abnormalities are observed when semaphorin signaling is disrupted [7]. In the immune system, expression of Sema4D (CD100) is regulated upon B- and T-lymphocyte activation, and migrating monocytes are responsive to Sema3A and Sema4D [8]. Malignant lung cells show reduced levels and a cytoplasmic localization of semaphorins [9]. Taken together, it can be concluded that semaphorins act as guidance cues for many types of migrating cells in developmental, adult, and pathological tissue.

RECEPTORS FOR SEMAPHORINS

Neuropilins

Neuropilins are high-affinity transmembrane receptors for the secreted class 3 semaphorins in the CNS, but play no role in the activity of other semaphorins. A neuropilin family is composed of neuropilin-1 and several splice variants of neuropilin-2 [1]. Neuropilin-1 and -2 contain a number of conserved motifs on their extracellular domain, including two CUB domains, FV/FVIII, and a MAM domain. The CUB domains are required for ligand binding, and the MAM domain mediates neuropilin oligomerization [10]. Neuropilins have a short intracellular domain containing a PDZ binding motif that targets receptor localization to signaling components in the membrane of the cell [11]; however, the intracellular domain is not required to transduce the semaphorin signal [10]. The neuropilin isoforms bind differentially to various class 3 semaphorins [12], and the specificity of binding is determined by the CUB domains [10]. Although neuropilins are sufficient to bind class 3 semaphorins, the fact that the intracellular domain is not required for signaling suggests that a co-receptor transmits the semaphorin signal into the cell.

Outside of the CNS, neuropilins are found in the mesenchyme surrounding blood vessels and act as co-receptors for vascular endothelial growth factor (VEGF). Upon binding VEGF, neuropilins potentiate the kinase activity of the VEGF receptors flt-1 and KDR, resulting in endothelial cell migration [13]. There is some evidence to suggest that semaphorins and VEGF compete for neuropilin binding, and a dysregulation of the competition may lead to pathological conditions [9].

Plexins

Plexins are the predominant receptors for membrane-bound, GPI-linked and viral semaphorins, and they bind to neuropilins to act as signaling co-receptors for the secreted class 3 semaphorins [14]. The initial discovery of plexins as semaphorin receptors occurred with the identification of virus-encoded semaphorin protein receptor (VESPR; plexinC1) as a binding site for a class V semaphorin [15]. Currently, at least 10 plexins have been identified and are classified into 4 groups, plexins A–D, which have different specificities for different semaphorins [1]. Plexins are distantly related to semaphorins, since they possess the conserved Sema domain on their extracellular surface [16] and also share some sequence homology with the HGF receptor Met on their extracellular surface [17]. The intracellular domain of plexin is highly conserved among family members, but is not significantly homologous to any known signaling motif. In their native state plexins are autoinhibited by their Sema domain, and binding to semaphorin–neuropilin complexes or cleavage of the sema domain leads to activation of the protein and growth cone collapse in sensory neurons [18].

INTRACELLULAR SIGNALING PATHWAYS

Actin Cytoskeleton and Monomeric GTPases

The actin cytoskeleton in growth cones undergoes dramatic rearrangement upon exposure to Sema3A. There is a relative decrease in F-actin within the lamellipodia [19], and actin co-localizes with neuropilin-1/plexinA1 receptor complexes [20]. The actin reorganization is linked to increased endocytosis [20]. It was thought that semaphorins

might regulate the actin cytoskeleton through monomeric G proteins due to the weak similarity of the conserved intracellular domain of plexins to an R-Ras-GAP. However, no plexin protein has been shown to possess GAP activity, and semaphorin responses are not dependent on R-Ras. Instead, Rho family G proteins, namely Rac and Rho, seem to mediate the semaphorin response [21] (also reviewed in [22]). Active Rac binds directly to the intracellular domain of vertebrate and invertebrate plexinB1, and this interaction is enhanced by the presence of ligand binding [23]. Activation of plexinB1 appears to sequester active Rac from its endogenous substrate, p21-associated kinase, PAK [24–26]. RhoA is also activated as a result of plexinB1, although it is not clear whether this is due to direct or indirect action of plexinB1 on RhoA [26], or is downstream of Rac–plexin interactions [24]. Together, Rac sequestration and RhoA activation appear to mediate axon repulsion by plexinB receptors.

Although the intracellular domain is highly conserved among all plexin family members, it is not clear whether plexinA functions in a similar fashion as plexinB. PlexinA1 binds to both RhoD and Rnd1, and Rnd1 binding has been suggested to induce growth cone collapse [27, 28], perhaps due to Rnd1-dependent inhibition of Rac [29]. Direct Rac–plexinA interactions have not been demonstrated. It is possible that plexins regulate monomeric GTPases indirectly by regulating Rho family GEFs and GAPs, factors that activate or inactivate monomeric GTPases, respectively.

A direct link between Sema3A and actin dynamics was recently demonstrated. Activated complexes of NP1 and plexinA2 lead to the phosphorylation and deactivation of cofilin by LIM kinase [30]. Cofilin leads to F-actin turnover, and plays a role in protrusion of lamellipodia and filopodia [31]. Further, LIM kinase is a substrate for both PAK and Rho kinase, which is consistent with the requirement for Rac and Rho, respectively, for Sema3A-induced collapse [31].

CRMP

Collapsin-response-mediator protein (CRMP) was identified in a *Xenopus* oocyte expression screen of mRNAs required for Sema3A responses [32]. The protein sequence of CRMP shares sequence homology with the *Caenorhabditis elegans* unc-33, a protein required for proper axonal pathfinding [33]. At least five isoforms of CRMP have been identified, and they form heterotetramers *in vivo* [34]. Function blocking antibodies to CRMP block Sema3A-mediated growth cone collapse in chick DRG neurons [32], and CRMP is upregulated after axotomy of the sciatic [35] and olfactory [36] nerves. The mechanism of CRMP action is still unclear. Studies have shown that it is phosphorylated by Rho kinase, but this is not required for Sema3A-induced collapse [37]. The microtubule abnormalities of unc-33 mutants and the observation that CRMP co-localizes with microtubules at certain stages of the cell cycle [38] suggest that plexin/CRMP signaling may regulate microtubule dynamics. There is also evidence demonstrating that CRMP binds to and inactivates phospholipase D2, an enzyme implicated in a variety of cell processes, including actin dynamics, vesicle trafficking, and mitogenesis [39]. Finally, CRMP may act to mediate the cytoskeleton through Rho GTPases. Neuroblastoma cells overexpressing CRMP and constitutively active RhoA showed a Rac1-like morphology, whereas cells co-expressing CRMP and active Rac1 showed a RhoA-like morphology [40]. Thus, it seems likely that CRMP enhances the function of a plexin–Rho family G-protein axis in axon repulsion.

Protein Phosphorylation

Receptors for several other axon guidance molecules, such as ephrins and neurotrophins, act via kinase cascades. Although semaphorin receptors themselves show no kinase activity, indirect evidence suggests that protein phosphorylation occurs and is required [41]. The involvement of PAK and LIM kinase downstream of plexins is mentioned above. A recent study has shown that *Drosophila* plexinA associates with the membrane-bound receptor tyrosine kinase-related protein Off-Track (Otk; [42]). In addition, two proteins with kinase activity have been co-purified with CRMP [38, 43]. The serine/threonine kinase, glycogen synthase kinase (GSK)-3, is activated as a result of Sema3A in both neuronal cells and human breast cancer cells, and GSK-3 inhibitors prevent Sema3A-induced growth cone collapse [44].

Other Signaling Mechanisms

Another pathway that has been implicated in semaphorin signaling may utilize heterotrimeric G proteins. Much of the evidence for this has come from experiments with pertussis toxin (PTX), which blocks G-protein function [21, 45]. Indeed, neuropilins were found to bind to a Gα-interacting protein (GIPC, SEMPCAP-1) that associates with a regulator of G-protein signaling (RGS) protein via its PDZ domains [11]. Interestingly, some transmembrane semaphorins interact with SEMCAP-1 as well, suggesting that semaphorins may act as receptors to transduce signals into the cell [46]. Semaphorin reverse signaling is further supported by the fact that Sema6B binds to Src both *in vitro* and *in vivo* [47], Sema4D interacts with a serine kinase [48], and Sema6A binds to the actin binding protein EVL [49].

Semaphorin-mediated signaling can be modulated by other pathways. Cyclic nucleotides can alter the response of growth cones to various signaling molecules [50, 51]. Increasing levels of cGMP switch Sema3A responses from repulsion to attraction, and decreasing cGMP potentiates the repulsive activity of Sema3A. Apical dendrites of cerebral cortical neurons are attracted to Sema3A while the axons of the same cells are repelled [4]. Remarkably, soluble guanylate cyclase (SGC) is asymmetrically localized to the dendrites of these cells, implicating an endogenous regulation of cGMP *in vivo*. The cell adhesion molecule, L1, is also able to modulate growth cone responses to Sema3A [52]. DRG neurons from L1-deficient mice show no response to Sema3A, and soluble L1 protein switched repulsion to attraction. Finally, one or more of these pathways may impinge on protein synthesis and degradation within axons. Evidence indicates that local regulation of protein levels participates in multiple growth cone responses [53].

Semaphorin Signaling in the Immune System

Semaphorin signaling in activated lymphocytes does not rely on neuropilins and plexins, but utilizes a different receptor called CD72 [54]. Under normal conditions, CD72 is phosphorylated on its intracellular domain by a Src tyrosine kinase. This phosphorylation leads to the recruitment of an SH2-domain-containing tyrosine phosphatase SHP-1, which then dephosphorylates and inactivates signaling proteins involved in lymphocyte activation [8]. Sema4D binding prevents the phosphorylation of CD72, therefore potentiating lymphocyte activation [54]. The migration of monocytes is inhibited by Sema4D and Sema3A; this effect is likely to be mediated via plexins and neuropilins, since monocytes do not express CD72 [55].

SUMMARY AND FUTURE DIRECTIONS

Many biological systems in the developing embryo and adult animal are dependent on semaphorin signaling. Although the importance of semaphorins in the developing CNS is well documented, their involvement in the immune response, the cardiovascular system, and in pathology is still being clarified. Most of the work to date has focused on the identification and classification of the various semaphorin families and their receptors, with less clarification of downstream signaling mechanisms. Regulation of the cytoskeleton is the most obvious effect of semaphorin signaling, and a number of studies have demonstrated a signaling connection of semaphorin receptors with actin filaments and microtubules. In particular, Rho family G proteins and CRMP appear to play major roles in this connection (see Figure 35.1).

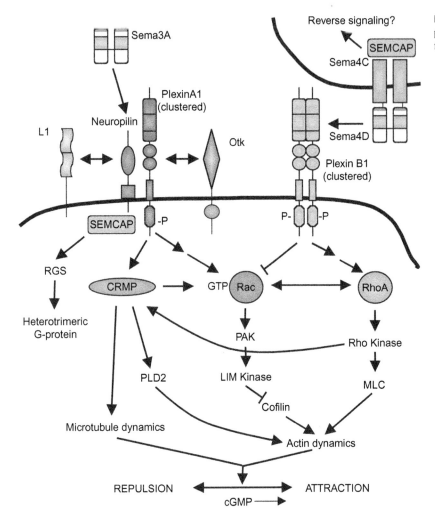

FIGURE 35.1 A schematic representation of semaphorin signaling in neuronal development. See text for details.

ACKNOWLEDGEMENTS

This work was supported by grants to Stephen M. Strittmatter from the NIH. Eric F. Schmidt is supported by an institutional NIH predoctoral training grant. Stephen M. Strittmatter is an Investigator of the Patrick and Catherine Weldon Donaghue Medical Research Foundation.

REFERENCES

1. Nakamura F, Kalb RG, Strittmatter SM. Molecular basis of semaphorin-mediated axon guidance. *J Neurobiol* 2000;**44**:219–29.
2. Luo Y, Raible D, Raper JA. Collapsin: a protein in brain that induces the collapse and paralysis of neuronal growth cones. *Cell* 1993;**75**:217–27.
3. Takahashi T, Nakamura F, Jin Z, Kalb RG, Strittmatter SM. Semaphorins A and E act as antagonists of neuropilin-1 and agonists of neuropilin-2 receptors. *Nat Neurosci* 1998;**1**:487–93.
4. Polleux F, Morrow T, Ghosh A. Semaphorin 3A is a chemoattractant for cortical apical dendrites. *Nature* 2000;**404**:567–73.
5. He Z, Wang KC, Koprivica V, Ming G, Song HJ. Knowing how to navigate: mechanisms of semaphorin signaling in the nervous system. *Sci STKE* 2002;**2002**:RE1.

6. Pasterkamp RJ, Verhaagen J. Emerging roles for semaphorins in neural regeneration. *Brain Res Brain Res Rev* 2001;**35**:36–54.
7. Brown CB, et al. PlexinA2 and semaphorin signaling during cardiac neural crest development. *Development* 2001;**128**:3071–80.
8. Bismuth G, Boumsell L. Controlling the immune system through semaphorins. *Sci STKE* 2002;**2002**:RE4.
9. Brambilla E, Constantin B, Drabkin H, Roche J. Semaphorin SEMA3F localization in malignant human lung and cell lines: a suggested role in cell adhesion and cell migration. *Am J Pathol* 2000;**156**:939–50.
10. Nakamura F, Tanaka M, Takahashi T, Kalb RG, Strittmatter SM. Neuropilin-1 extracellular domains mediate semaphorin D/III-induced growth cone collapse. *Neuron* 1998;**21**:1093–100.
11. Cai H, Reed RR. Cloning and characterization of neuropilin-1-interacting protein: A PSD-95/Dlg/ZO-1 domain-containing protein that interacts with the cytoplasmic domain of neuropilin-1. *J Neurosci* 1999;**19**:6519–27.
12. Feiner L, Koppel AM, Kobayashi H, Raper JA. Secreted chick semaphorins bind recombinant neuropilin with similar affinities but bind different subsets of neurons *in situ*. *Neuron* 1997;**19**:539–45.
13. Tamagnone L, Comoglio PM. Signalling by semaphorin receptors: cell guidance and beyond. *Trends Cell Biol* 2000;**10**:377–83.

14. Takahashi T, Fournier A, Nakamura F, Wang LH, Murakami Y, Kalb RG, Fujisawa H, Strittmatter SM. Plexin–neuropilin-1 complexes form functional semaphorin-3A receptors. *Cell* 1999;**99**:59–69.

15. Comeau MR, et al. A poxvirus-encoded semaphorin induces cytokine production from monocytes and binds to a novel cellular semaphorin receptor, VESPR. *Immunity* 1998;**8**:473–82.

16. Winberg ML, Noordermeer JN, Tamagnone L, Comoglio PM, Spriggs MK, Tessier-Lavigne M, Goodman CS. Plexin A is a neuronal semaphorin receptor that controls axon guidance. *Cell* 1998;**95**:903–16.

17. Bork P, Doerks T, Springer TA, Snel B. Domains in plexins: links to integrins and transcription factors. *Trends Biochem Sci* 1999;**24**:261–3.

18. Takahashi T, Strittmatter SM. Plexina1 autoinhibition by the plexin sema domain. *Neuron* 2001;**29**:429–39.

19. Fan J, Mansfield SG, Redmond T, Gordon-Weeks PR, Raper JA. The organization of F-actin and microtubules in growth cones exposed to a brain-derived collapsing factor. *J Cell Biol* 1993;**121**:867–78.

20. Fournier AE, Nakamura F, Kawamoto S, Goshima Y, Kalb RG, Strittmatter SM. Semaphorin3A enhances endocytosis at sites of receptor-F-actin colocalization during growth cone collapse. *J Cell Biol* 2000;**149**:411–22.

21. Jin Z, Strittmatter SM. Rac1 mediates collapsin-1-induced growth cone collapse. *J Neurosci* 1997;**17**:6256–63.

22. Liu BP, Strittmatter SM. Semaphorin-mediated axonal guidance via Rho-related G proteins. *Curr Opin Cell Biol* 2001;**13**:619–26.

23. Vikis HG, Li W, He Z, Guan KL. The semaphorin receptor plexin-B1 specifically interacts with active Rac in a ligand-dependent manner. *Proc Natl Acad Sci USA* 2000;**97**:12,457–12,462.

24. Driessens MH, Hu H, Nobes CD, Self A, Jordens I, Goodman CS, Hall A. Plexin-B semaphorin receptors interact directly with active Rac and regulate the actin cytoskeleton by activating Rho. *Curr Biol* 2001;**11**:339–44.

25. Vikis HG, Li W, Guan KL. The plexin-B1/Rac interaction inhibits PAK activation and enhances Sema4D ligand binding. *Genes Dev* 2002;**16**:836–45.

26. Hu H, Marton TF, Goodman CS. Plexin B mediates axon guidance in Drosophila by simultaneously inhibiting active Rac and enhancing RhoA signaling. *Neuron* 2001;**32**:39–51.

27. Zanata SM, Hovatta I, Rohm B, Puschel AW. Antagonistic effects of Rnd1 and RhoD GTPases regulate receptor activity in Semaphorin 3A-induced cytoskeletal collapse. *J Neurosci* 2002;**22**:471–7.

28. Rohm B, Rahim B, Kleiber B, Hovatta I, Puschel AW. The semaphorin 3A receptor may directly regulate the activity of small GTPases. *FEBS Letts* 2000;**486**:68–72.

29. Nobes CD, Lauritzen I, Mattei MG, Paris S, Hall A, Chardin P. A new member of the Rho family, Rnd1, promotes disassembly of actin filament structures and loss of cell adhesion. *J Cell Biol* 1998;**141**:187–97.

30. Aizawa H, et al. Phosphorylation of cofilin by LIM-kinase is necessary for semaphorin 3A-induced growth cone collapse. *Nat Neurosci* 2001;**4**:367–73.

31. Kuhn TB, et al. Regulating actin dynamics in neuronal growth cones by ADF/cofilin and rho family GTPases. *J Neurobiol* 2000;**44**:126–44.

32. Goshima Y, Nakamura F, Strittmatter P, Strittmatter SM. Collapsin-induced growth cone collapse mediated by an intracellular protein related to UNC-33. *Nature* 1995;**376**:509–14.

33. Li W, Herman RK, Shaw JE. Analysis of the *Caenorhabditis elegans* axonal guidance and outgrowth gene unc-33. *Genetics* 1992;**132**:675–89.

34. Wang LH, Strittmatter SM. Brain CRMP forms heterotetramers similar to liver dihydropyrimidinase. *J Neurochem* 1997;**69**:2261–9.

35. Minturn JE, Fryer HJ, Geschwind DH, Hockfield S. TOAD-64, a gene expressed early in neuronal differentiation in the rat, is related to unc-33, a *C. elegans* gene involved in axon outgrowth. *J Neurosci* 1995;**15**:6757–66.

36. Pasterkamp RJ, De Winter F, Holtmaat AJ, Verhaagen J. Evidence for a role of the chemorepellent semaphorin III and its receptor neuropilin-1 in the regeneration of primary olfactory axons. *J Neurosci* 1998;**18**:9962–76.

37. Arimura N, et al. Phosphorylation of collapsin response mediator protein-2 by Rho-kinase. Evidence for two separate signaling pathways for growth cone collapse. *J Biol Chem* 2000;**275**:23,973–23,980.

38. Gu Y, Ihara Y. Evidence that collapsin response mediator protein-2 is involved in the dynamics of microtubules. *J Biol Chem* 2000;**275**:17,917–17,920.

39. Lee S, et al. Collapsin response mediator protein-2 inhibits neuronal phospholipase D(2) activity by direct interaction. *J Biol Chem* 2002;**277**:6542–9.

40. Hall C, Brown M, Jacobs T, Ferrari G, Cann N, Teo M, Monfries C, Lim L. Collapsin response mediator protein switches RhoA and Rac1 morphology in N1E-115 neuroblastoma cells and is regulated by Rho kinase. *J Biol Chem* 2001;**276**:43,482–43,486.

41. Tamagnone L, et al. Plexins are a large family of receptors for transmembrane, secreted, and GPI-anchored semaphorins in vertebrates. *Cell* 1999;**99**:71–80.

42. Winberg ML, Tamagnone L, Bai J, Comoglio PM, Montell D, Goodman CS. The transmembrane protein Off-track associates with Plexins and functions downstream of Semaphorin signaling during axon guidance. *Neuron* 2001;**32**:53–62.

43. Inatome R, Tsujimura T, Hitomi T, Mitsui N, Hermann P, Kuroda S, Yamamura H, Yanagi S. Identification of CRAM, a novel unc-33 gene family protein that associates with CRMP3 and protein-tyrosine kinase(s) in the developing rat brain. *J Biol Chem* 2000;**275**:27,291–27,302.

44. Eickholt BJ, Walsh FS, Doherty P. An inactive pool of GSK-3 at the leading edge of growth cones is implicated in Semaphorin 3A signaling. *J Cell Biol* 2002;**157**:211–17.

45. Igarashi M, Strittmatter SM, Vartanian T, Fishman MC. Mediation by G proteins of signals that cause collapse of growth cones. *Science* 1993;**259**:77–9.

46. Wang LH, Kalb RG, Strittmatter SM. A PDZ protein regulates the distribution of the transmembrane semaphorin, M-SemF. *J Biol Chem* 1999;**274**:14,137–14,146.

47. Eckhardt F, Behar O, Calautti E, Yonezawa K, Nishimoto I, Fishman MC. A novel transmembrane semaphorin can bind c-src. *Mol Cell Neurosci* 1997;**9**:409–19.

48. Elhabazi A, Lang V, Herold C, Freeman GJ, Bensussan A, Boumsell L, Bismuth G. The human semaphorin-like leukocyte cell surface molecule CD100 associates with a serine kinase activity. *J Biol Chem* 1997;**272**:23,515–23,520.

49. Klostermann A, Lutz B, Gertler F, Behl C. The orthologous human and murine semaphorin 6A-1 proteins (SEMA6A-1/ Sema6a-1) bind to the enabled/vasodilator-stimulated phosphoprotein-like protein (EVL) via a novel carboxyl-terminal zyxin-like domain. *J Biol Chem* 2000;**275**:39,647–39,653.

50. Song H, Ming G, He Z, Lehmann M, McKerracher L, Tessier-Lavigne M, Poo M. Conversion of neuronal growth cone responses from repulsion to attraction by cyclic nucleotides. *Science* 1998;**281**:1515–18.

51. Ming G, Henley J, Tessier-Lavigne M, Song H, Poo M. Electrical activity modulates growth cone guidance by diffusible factors. *Neuron* 2001;**29**:441–52.

52. Castellani V, Chedotal A, Schachner M, Faivre-Sarrailh C, Rougon G. Analysis of the L1-deficient mouse phenotype reveals cross-talk between Sema3A and L1 signaling pathways in axonal guidance. *Neuron* 2000;**27**:237–49.

53. Campbell DS, Holt CE. Chemotropic responses of retinal growth cones mediated by rapid local protein synthesis and degradation. *Neuron* 2001;**32**:1013–26.

54. Kumanogoh A, et al. Identification of CD72 as a lymphocyte receptor for the class IV semaphorin CD100: a novel mechanism for regulating B cell signaling. *Immunity* 2000;**13**:621–31.

55. Delaire S, Billard C, Tordjman R, Chedotal A, Elhabazi A, Bensussan A, Boumsell L. Biological activity of soluble CD100. II. Soluble CD100, similarly to H-SemaIII, inhibits immune cell migration. *J Immunol* 2001;**166**:4348–54.

Chapter 36

Signaling Pathways that Regulate Cell Fate in the Embryonic Spinal Cord

Matthew T. Pankratz and Samuel L. Pfaff

Gene Expression Laboratory, The Salk Institute for Biological Studies, La Jolla, California

INTRODUCTION

A cascade of signaling events triggers the differentiation of specific neuronal and glial cell populations that comprise the central nervous system (CNS). Ectoderm deprived of bone morphogenic protein (BMP) signaling differentiates into neuroepithelia that subsequently fold to form the neural tube. The neural tube is oriented along the rostrocaudal axis of the embryo, and is the precursor of the entire CNS [1]. Neural precursors that make up the neural tube are multipotential, and respond to signals in their environment in order to generate the appropriate types of neurons and glia at the correct positions. In this chapter, we focus on the spinal cord, the most caudal region of the CNS, since it has served as a useful model in which to investigate signaling events that give rise to neuronal and glial populations within the developing neural tube.

Emerging from a combination of modern molecular studies and classical cellular studies, a central theme in spinal cord development is one in which inductive factors signal along the dorsoventral and rostrocaudal axes of the developing spinal cord to specify cell fate in a Cartesian-coordinate-like manner [2]. This signaling leads to the generation of dorsal spinal cord interneurons that process sensory information and relay it to the brain, while the ventral spinal cord forms interneurons and motor neurons involved in locomotor control (Figure 36.1a). Along the rostrocaudal axis, discontinuous subclasses of motor neurons are generated in register with the peripheral targets that they innervate. In addition, numerous glial cell types are formed, including the roof plate and floor plate, which act as organizing centers within the spinal cord, and astrocytes and oligodendrocytes, which support neuronal function and myelinate neurons, respectively.

PATTERNING ALONG THE DORSOVENTRAL AXIS

Two classes of factors play prominent roles in specifying distinct cell types along the dorsoventral axis of the spinal cord: members of the transforming growth factor β (TGFβ) superfamily acting dorsally, and Sonic hedgehog (Shh) ventrally (Figure 36.1a) [3, 4]. TGFβ signaling from the epidermal ectoderm flanking the dorsal neural tube leads to the differentiation of the roof plate [5], and Shh expression from the notochord below the neural tube triggers the formation of the floor plate [6]. These two glial structures in the spinal cord then express TGFβs dorsally and Shh ventrally. In this way, signals from the periphery are propagated into the spinal cord to control cell differentiation locally.

The dividing progenitor cells within the ventricular (medial) region of the spinal cord monitor the types and concentrations of TGFβs and Shh in order to determine their position, and consequently their fate, as they become postmitotic and migrate laterally into the mantle region (Figure 36.1a). The signaling pathways triggered by these inductive factors lead to the activation of transcriptional networks that first define distinct domains along the dorsoventral axis of the ventricular zone, and ultimately lead to the expression of genes involved in controlling cell identity and that mediate specialized functions (Table 36.1) [7–10].

DORSAL SPINAL CORD DEVELOPMENT

The embryonic dorsal spinal cord is made up of interneurons (INs) organized into six distinct classes (dorsal interneurons dI1–6), which arise in an orderly fashion from specific regions of the progenitor zone (Figure 36.1a) [11, 12].

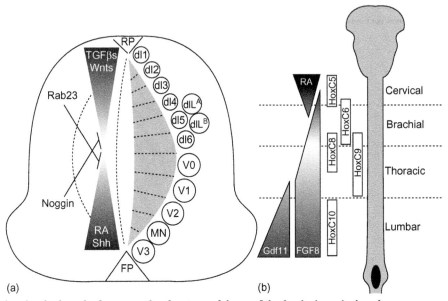

FIGURE 36.1 Patterning signals along the dorsoventral and rostrocaudal axes of the developing spinal cord.
(a) Transverse view of the spinal cord in which the gray area represents the ventricular zone containing progenitor cells and the white area represents the mantle layer with mature cell types. FP, floor plate; RP, roof plate. (b) Rostrocaudal view of the spinal cord and the graded pattern of signals expressed by paraxial and axial mesoderm adjacent to the neural tube [51].

These INs, while not absolutely defined, consist of association and commissural cells that process and relay sensory information. Dorsal INs can be further subdivided into two subclasses; those that are dependent on the roof plate for formation (dI1–3), and those that are independent (dI4–6, which gives rise to two additional sets of dorsal INs at a later stage of development, dIL^A and dIL^B) [5, 13, 14]. Each of the six classes of INs expresses a unique set of transcription factors that changes as progenitor cells in the ventricular zone mature and migrate to the mantle. For example, dI1, which is characterized postmitotically by the markers Lhx2/9, arises from progenitor cells that express the bHLH transcription factor mATH1 [9, 10]. Likewise, dI2 cells marked by Lhx1/5 arise from progenitors that express the bHLH protein Ngn1 (for a more complete listing, see Figure 36.1a and Table 36.1). The precise set of markers that labels each class of dorsal INs has allowed researchers to work backward to characterize the signal transduction pathways that activate gene expression, and the factors provide a precise readout in experiments aimed at altering cell fate by overstimulating or knocking out signaling pathways.

An important source of patterning signals in the developing spinal cord is the roof plate and adjacent neural epithelial cells, which express overlapping and nested combinations of several TGFβ members, including BMP4/5/7, Gdf6/7, and Dsl1 [5, 9]. How might the TGFβs produce different cell types in the spinal cord? Several strategies are likely to be involved, including quantitative, qualitative, and timing differences in TGFβ activity. In gain-of-function studies, expression of activated BMP receptors (BMPR1A

or BMPR1B) in transgenic mice and developing chicks showed an expansion of dorsal cell populations in the developing spinal cord, demonstrating that BMP-mediated TGFβ signaling is capable of directly inducing a dorsal fate [15, 16]. The clearest example of how TGFβs trigger the differentiation of specific IN types is based on the finding that individual members of this family have different qualitative activities [5]. The most convincing evidence for such a mechanism is found in the Gdf 7 mutant mice where dI1 cells fail to be generated [13]. These results suggest that the specificity of dorsal IN patterning is mediated, at least in part, by various TGFβ signaling molecules, some of which act directly to render class specificity (Table 36.1).

The expression pattern of the TGFβs suggests that a high-dorsal to low-ventral gradient of these proteins is present within the neural tube (Figure 36.1a). *In vitro* experiments with neural explants have detected concentration-dependent activities for Activin A in the induction of dI1 and dI3 cells [5]. The sensitivity of intermediate progenitors to BMP signaling is illustrated in studies of BMP mutants in zebrafish. In swirl mutants lacking BMP2b, there is an expansion of Lim1 + INs. If BMP signaling was normally in place to inhibit this class of INs, then we would expect that further knockdown of BMP signaling as a result of injection of mRNA encoding the secreted BMP antagonist Chordin would result in more Lim1 + INs. Instead there was a reduction in Lim1 + cells, suggesting that the intermediate progenitors are responding to a specific level or isoform of BMP [17].

An additional mechanism for generating cellular diversity in the dorsal spinal cord involves a temporal switch in

TABLE 36.1 Dorsal/ventral positioning of progenitors dictates the exposure of cells to secreted patterning factors released from the roof and floor plate. A complex and continually evolving list of transcription factors and proteins labels each population of cells at both the progenitor (ventricular zone) and mature (mantle zone) cell stage; rostral/caudal signals pattern the posterior portion of the CNS into distinct regions defined by their Hox gene expression

	Dorsal Patterning lateral ectoderm → roof plate	Ventral Patterning notochord → floor plate	Rostral/Caudal Patterning Hensen's node, paraxial mesoderm
Signal	TGFβ Superfamily BMP 4, 5, 6, 7 Gdf 6, 7 Activin B ; Dsl1	Sonic Hedgehog	Retinoic Acid FGF 8 Gdf 11
Inhibitors	Follistatin, Noggin, Chordin	Hedgehog-interacting Protein (Hip) Rab23 (vesicle transporter)	Cellular Retinoic Acid Binding Protein Follistatin, Noggin, Chordin
Receptor(s)	Type I/II TGFβ receptors (serine/threonine receptor kinases)	Patched : ligand binding Smoothened : transducing	Nuclear Receptors RAR RXR FGFR (receptor tyrosine kinase) TGFβR (receptor serine/ threonine kinase)
Signal transduction components	SMAD transcription factor family	Gli transcription factor family Gli 1, 2, 3	RAR RXR Cdx transcription factor family Cdx transcription factor family SMADs

(Continued)

TABLE 36.1 (Continued)

Downstream transcription factors (ventricular zone)

Dorsal Patterning lateral ectoderm → roof plate		Ventral Patterning notochord → floor plate			Rostral/Caudal Patterning Hensen's node, paraxial mesoderm	
Progenitor Domain	Transcription Factors	Progenitor Domain	Class I	Class II	Rostral/Caudal Level	Transcription Factors
dP1	Math1	pV0	Dbx 1/2, Irx3,	—	hindbrain	HoxB1/2/3/4
dP2	Ngn1/2	pV1	Pax6		cervical	HoxC5
dP3	Mash1, Pax7(low)	pV2	Dbx, Irx3, Pax6	Nkx6.2	brachial	HoxC6/8
dP4	Mash1, Pax7(high)	pMN	Irx3, Pax6	Nkx6.1	thoracic	HoxC8/9
dP5	Mash1, Pax7(high)	pV3	Pax6	Olig2, MNR2, Nkx6.1 Nkx2.2, Nkx6.1	lumbar	HoxC10
dP6	Pax7, Dbx2, Ngn1/2	pOlig	—	Nkx2.2, Olig2		

Differentiated cell type (mantal zone)

Dorsal Patterning		Ventral Patterning			Rostral/Caudal Patterning	
Cell Type	Functional Markers	Cell Type	Functional Markers		MN Subtypes	Functional Markers
dI1	Lhx2/9, BarH1, Brn3a	V0 IN	Evx1/2		columnar	LIM transcription family
dI2	Lhx1/5, Brn3a, Foxd3	V1 IN	En1		pool	ETS transcription family
dI3	Isl1/2, Brn3a, Rnx	V2 IN	Lhx3, Chx10		hindbrain	Phox transcription family
dI4	Lbx1, Lhx1/5, Pax2	MN	Isl1/2, HB9		visceral types	
dI5	Lbx1, Brn3a, Lmx1b, Rnx, Tlx1	V3 IN	Sim1			
dI6	Lbx1, Lhx1/5, Pax2	floor plate	HNF3β			
dIL^A	Lbx1, Lhx1/5, Pax2	oligodendrocytes	PDGFRα			
dIL^B	Lbx1, Lmx1b, Rnx, Brn3a	astrocytes	GFAP			
roof plate	Gdf7					
astrocytes	GFAP					

the way neuroepithelial cells respond to TGFβ signals. Early in development, progenitor cells produce neural crest cells when exposed to BMP4 or Activin A, but later they give rise to dorsal INs in response to the same signals [5]. The basis for qualitative differences in TGFβ signaling and the mechanisms underlying the developmental switch in TGFβ responsiveness by neural epithelial cells remain important questions. Shh signaling for ventral cell differentiation is attenuated by TGFβ signaling [18]. What limits the range of TGFβ activity to the appropriate areas of the developing spinal cord? Several TGFβ antagonists have been identified, including noggin, chordin, and follistatin, which bind directly to and sequester specific TGFβs [19]. These antagonists are expressed by the somites and notochord near the ventral surface of the neural tube, and therefore are expected to limit the exposure of ventral cells to certain TGFβs. In noggin mutant mice, TGFβ signaling in the ventral neural tube is unmasked (Figure 36.1a), which leads to a progressive loss of ventral cell differentiation [20].

Cell fate diversity might also be modulated by downstream signaling elements activated by TGFβs. The receptors of the TGFβs are serine/threonine kinases comprised of type I and type II dimers. These receptor complexes have not been well characterized in the spinal cord, but may select for different ligands and serve as the basis for the qualitative differences in cell differentiation induced by different TGFβ family members. Along these lines, the loss of Type I BMP receptors results in a loss of dI1 and reduction of dI2 cells [21]. The best known transducers of TGFβ signaling are the SMAD transcription factors [22]; however, the role of SMADs in spinal cord development also requires further characterization.

Another class of secreted morphogens present in the dorsal region of the developing spinal cord that potentially controls progenitor fate decisions consists of the Wnt proteins. Mutant mouse embryos lacking Wnt1 and Wnt3a showed a reduction in dI1 and dI2 cells, and an expansion of dI3 INs [23]. However subsequent studies have suggested that Wnt signaling may control cell identity decisions indirectly by promoting proliferation of dorsal cells specified by TGFβ signaling [24] and/or maintaining the dorsal progenitor pool by delaying differentiation [25]. Highlighting the possibility that Wnt signaling can control both patterning and proliferation, a study in zebrafish demonstrated that two separate transcriptional activators downstream of Wnt signaling were responsible for dorsal patterning (Tcf7) and proliferation (Tcf3) in the spinal cord [26]. BMPs also appear to control expression of Wnt signaling elements in the dorsal spinal cord [21, 24], which leads to further coupling of patterning and proliferation rates in the developing cord. As pointed out in a recent review by Ulloa and Briscoe [27], this coupling of growth and patterning could be important for scalability, so that the same set of morphogens can generate spinal cords of different sizes with relatively similar proportions of progenitor pools.

VENTRAL SPINAL CORD DEVELOPMENT

Genetic studies as well as *in vitro* explant experiments have implicated Shh in the differentiation of ventral spinal cord cell types involved in locomotor control (V0–V3 INs and MNs, motor neurons), as well as oligodendrocytes (Figure 36.1a) [3, 8]. Unlike the nested combinations of TGFβ molecules in the dorsal spinal cord; however, only one hedgehog member appears to be involved in ventral spinal cord patterning in higher vertebrates. This raises the question of how different ventral cell types are induced by a single factor. The active Shh signaling molecule is autoprocessed and cholesterol-modified, and binds to the patched/smoothened receptor complex [28]. In the absence of Shh, patched is thought to inhibit smoothened from signaling, and this inhibition is relieved when Shh binds patched. Extensive studies with *in vitro* explants have shown that Shh concentration differences of approximately two- to threefold dramatically influence the types of cells that are triggered to differentiate. Part of the machinery responsible for detecting differences in Shh concentration are the non-motile, microtubule-based cilia on the surface of neural progenitors. Disruption of intraflagellar transport (IFT) proteins, which are required for assembly and maintenance of cilia, results in defects in ventral neural patterning consistent with a role for these proteins downstream of smoothened in mice but, interestingly, not *Drosophila* [29].

Decreasing concentrations of Shh progressively induce cell types found further from the ventral midline, recapitulating the normal organization of cells in the ventral spinal cord [30]. As with TGFβ signaling, there are also important temporal mechanisms that modify progenitor cell responses to Shh signaling during development. At early stages Shh acts on progenitor cells to trigger MN differentiation, but later in development oligodendrocytes are produced instead of MNs. The basis for this switch is not well understood, but seems to involve the regulation of the transcription factor Nkx2.2 [31, 32]. Studies of chick neural tube explants have shown that cells are capable of interpreting both the concentration and duration of Shh signaling. Cells become gradually desensitized to ongoing Shh as a result of Shh-mediated upregulation of patched 1 [33]. Many additional components of the Shh signaling pathway have been identified through genetic studies in *Drosophila*, including the downstream Gli family of zinc finger transcription factors [28]. The gradient of Shh signaling present in the ventral spinal cord seems to be faithfully translated without amplification to intracellular mediators such as Gli3, which has a gradient of activity that matches Shh [34]. However Gli proteins can switch from transcriptional repressors to activators, adding further complexity to this signaling system [35]. For example, in the absence of Shh, Gli3 is cleaved and the resulting truncated protein directly represses Shh target genes. The rescue of some defects seen in $Shh^{-/-}$ mutant mice in the double mutant $Shh^{-/-}$, $Gli3^{-/-}$ mouse suggests that Shh signaling is important for downregulating the repressive form of Gli3 [36].

How might small gradations in the level of Shh signaling produce sharp progenitor cell domains that serve as the precursors for different ventral cell types? Studies of the factors regulated by Shh in the ventricular zone have uncovered a network of homeodomain proteins that mark distinct progenitor domains (Table 36.1) [7]. The expression of these factors is controlled at two levels. First, Shh either represses (class I) or activates (class II) the expression of the homeodomain factors. If this were the only mechanism operating to control these factors, it might be expected that the interpretation of the fine Shh gradient would lead to imprecise boundaries of gene expression. However, the domains appear to be further refined by cross-repressive transcriptional interactions between factors from different domains. In this two-step manner, graded Shh leads to the activation of unique combinations of homeodomain transcription factors in precise progenitor cell domains [3, 8]. The combinatorial activities of these homeodomain factors lead to the activation of downstream transcriptional regulators involved in cell specification and function (Table 36.1) [37].

The opposing nature of the ventral Shh gradient meeting the dorsal TGFβ factors suggests that inhibitors of Shh activity might constrain its activity, much like the inhibitors of TGFβs. Hedgehog interacting protein (Hip) is a surface membrane protein that binds Shh and attenuates its activity [38]. In addition, characterization of the mouse *open brain* (*opb*) mutant, in which ventral cell types form inappropriately in the dorsal region of the spinal cord, has led to the identification of a member of the Rab family of vesicular transporters, Rab23, important in limiting the activity of Shh dorsally (Figure 36.1a) [39]. Interestingly, mice deficient in both Shh and Rab23 regain many of the ventral cell types lost in Shh mutants. This, together with the observation that *Gli3/Shh* double mutants also regain many ventral cell types [36], suggests additional Shh-independent pathways might contribute to ventral spinal cord development. Studies to understand the basis for Shh-independent signaling have uncovered a parallel pathway involving retinoic acid (RA) expressed by paraxial mesoderm beside the neural tube [40]. RA can induce many of the class I genes expressed in the ventral spinal cord [40, 41] and null mutant mice lacking a key RA synthesizing enzyme retinaldehyde dehydrogenase-2 (Raldh2) fail to induce Pax6 and Olig2 in the spinal cord, a phenotype that is reversed by maternal dietary supplement with RA [42]. To summarize, ventral patterning in the spinal cord is accomplished via RA activation of class I transcription factors and Shh repression (class I) and activation (class II) of transcription factors that in turn cross-repress one another to demarcate discrete progenitor pools.

As in the case of Wnt/BMP signaling in the dorsal spinal cord, Shh signaling in the ventral cord also plays a mitogenic role to control the growth of neural progenitors [43, 44], and thus patterning and proliferation are coupled. Furthermore, not all patterning in the spinal cord is controlled by secreted factors. For example, recent work on interneuron subtypes generated from the p2 domain show that Notch activation drives inhibitory V2b interneuron development, whereas p2 progenitors that fail to activate Notch become excitatory V2a interneurons [45, 46].

ROSTROCAUDAL SPECIFICATION

The spinal cord can be subdivided into four broad, functional regions along the rostrocaudal axis: cervical, brachial, thoracic, and lumbosacral. The IN classes of the spinal cord extend continuously throughout these regions, while specific MN subclasses are found at each level [47]. Individual MN subclasses form discontinuous columns in register with their targets, such that MNs of the cervical region innervate axial muscles, brachial region MNs innervate the forelimb, MNs of the thoracic region innervate body wall muscle, and lumbar MNs innervate hindlimbs. Much like the initiation of dorsoventral patterning in the spinal cord, embryonic manipulations and *in vitro* explant studies suggest that members of several families of signaling molecules originating initially from sources outside the spinal cord contribute to the diversification process that leads to the generation of specific classes of MNs along the rostrocaudal axis [48–51].

Studies of the signals that control segmental identity along the rostrocaudal axis have used *Hox* gene expression patterns as downstream molecular correlates of the regional specification of cell identity (Figure 36.1b, Table 36.1). Furthermore, there are increasing functional data to suggest that *Hox* genes contribute to the proper development of MN subclasses [47, 52, 53]. As neuroepithelial cell identity is first established, it is thought to have a rostral identity which is then modified by "caudalizing" signals [4]. Hindbrain studies have found that increasing levels of RA activate more caudal-type *Hox* genes [54–56]. Likewise, the pattern of *Hox* gene expression in the cervical spinal cord is regulated by RA synthesized by the cervical paraxial mesoderm flanking the neural tube (Figure 36.1b) [51].

However, at more caudal regions of the spinal cord, RA is insufficient to confer positional identity. A major source of additional regionalizing signals is detected in Hensen's node (HN), a precursor of the axial mesoderm that moves in a caudal direction below the nascent neural tube as development progresses. Interestingly, HN tissue taken from different stages (i.e., different rostrocaudal levels) is able to specify different regional values in neural explants (51). Studies utilizing fibroblast growth factor (FGF) receptor antagonist SU5402 and expression of constitutively active FGF receptors are found to alter the Hox coding in neural cells, implicating FGF signaling as a mediator of HN activity. FGF8 is expressed by the HN, and *in vitro* studies have found that this factor can act in a concentration-dependent manner to induce progressively more caudal positional values in neural explants. The Cdx family of transcription factors represents possible downstream mediators of both RA and FGF8 signaling in the regulation of Hox expression [56].

Taken together, these findings suggest that the signaling activity of FGF8 increases as the HN moves caudally (Figure 36.1b) [51]. An additional mechanism that appears to contribute to the increased activity of FGF8 at more caudal positions is the involvement of accessory factors that enhance FGF signaling. One such example is the TGFβ superfamily member Gdf11. This factor is expressed in HN as it progresses through lumbosacral levels, where FGF signaling is expected to be highest (Figure 36.1b). Unlike other TGFβ members, Gdf11 does not influence the dorsoventral pattern of the spinal cord, but rather acts to enhance FGF8 signaling activity. In this way, progressively more caudal regions of the spinal cord are defined by the composite functions of FGF8 and Gdf11 through the regulation of Hox codes involved in establishing regional levels of the spinal cord that will generate different MN subclasses (Table 36.1).

The development of the spinal cord is a highly conserved process that utilizes the same classes of signaling molecules to induce families of transcription factors, which define progenitor pools in everything from flies to people. While the progenitor pools produced appear to be conserved across species, variation and complexity arises from the number, timing, and combination of transcription factors induced [57, 58]. Secreted patterning signals control cell identity but also division vs differentiation, which determines overall cell number and allows scalability. While partitioning of the embryonic spinal cord into distinct dorsal/ventral and rostral/caudal groups of neural precursors is a well-studied process, the subsequent stages of development, including how transcription factors control positioning of mature neurons in the cord and eventually the proper outgrowth and connection to other neurons and muscles, are exciting new areas of research. Perhaps not surprisingly, but only recently appreciated, the same mechanisms that control the fate of neurons in the developing cord also serve to specify astrocytes that emerge from the progenitor pools after neurons [59]. Understanding of the signaling pathways that are used to establish cell fate in the spinal cord has allowed researchers to recapitulate some of these events *in vitro* to create motor neurons from mouse and human embryonic stem cells [60, 61]. A more detailed understanding of how individual interneuron populations are generated and how developing neurons make axon path finding and connectivity decisions might ultimately allow researchers and clinicians to repair or even replace cells damaged by injury or disease.

REFERENCES

1. Weinstein DC, Hemmati-Brivanlou A. Neural induction. *Annu Rev Cell Dev Biol* 1999;**15**:411–33.
2. Jessell TM, Lumsden A. Inductive signals and the assignment of cell fate in the spinal cord and hindbrain: An axial coordinate system for neural patterning. In: Cowan WM, Jessell TM, Zipursky SL, editors. *Molecular and Cellular Approaches to Neural Development.* New York: Oxford University Press; 1997. p. 290–333.
3. Jessell TM. Neuronal specification in the spinal cord: inductive signals and transcriptional codes. *Nat Rev Genet* 2000;**1**:20–9.
4. Tanabe Y, Jessell TM. Diversity and pattern in the developing spinal cord. *Science* 1996;**274**:1115–23.
5. Liem Jr KF, Tremml G, Jessell TM. A role for the roof plate and its resident TGFbeta-related proteins in neuronal patterning in the dorsal spinal cord. *Cell* 1997;**91**:127–38.
6. Placzek M, Dodd J, Jessell TM. Discussion point. The case for floor plate induction by the notochord. *Curr Opin Neurobiol* 2000;**10**:15–22.
7. Briscoe J, Pierani A, Jessell TM, Ericson J. A homeodomain protein code specifies progenitor cell identity and neuronal fate in the ventral neural tube. *Cell* 2000;**101**:435–45.
8. Briscoe J, Ericson J. Specification of neuronal fates in the ventral neural tube. *Curr Opin Neurobiol* 2001;**11**:43–9.
9. Lee KJ, Jessell TM. The specification of dorsal cell fates in the vertebrate central nervous system. *Annu Rev Neurosci* 1999;**22**:261–94.
10. Gowan K, Helms AW, Hunsaker TL, et al. Crossinhibitory activities of Ngn1 and Math1 allow specification of distinct dorsal interneurons. *Neuron* 2001;**31**:219–32.
11. Chizhikov VV, Millen KJ. Roof plate-dependent patterning of the vertebrate dorsal central nervous system. *Dev Biol* 2005;**277**:287–95.
12. Helms AW, Johnson JE. Specification of dorsal spinal cord interneurons. *Curr Opin Neurobiol* 2003;**13**:42–9.
13. Lee KJ, Mendelsohn M, Jessell TM. Neuronal patterning by BMPs: a requirement for GDF7 in the generation of a discrete class of commissural interneurons in the mouse spinal cord. *Genes Dev* 1998;**12**:3394–407.
14. Muller T, Brohmann H, Pierani A, et al. The homeodomain factor lbx1 distinguishes two major programs of neuronal differentiation in the dorsal spinal cord. *Neuron* 2002;**34**:551–62.
15. Panchision DM, Pickel JM, Studer L, et al. Sequential actions of BMP receptors control neural precursor cell production and fate. *Genes Dev* 2001;**15**:2094–110.
16. Timmer JR, Wang C, Niswander L. BMP signaling patterns the dorsal and intermediate neural tube via regulation of homeobox and helix–loop–helix transcription factors. *Development* 2002;**129**:2459–72.
17. Nguyen VH, Trout J, Connors SA, Andermann P, Weinberg E, Mullins MC. Dorsal and intermediate neuronal cell types of the spinal cord are established by a BMP signaling pathway. *Development* 2000;**127**:1209–20.
18. Liem Jr KF, Jessell TM, Briscoe J. Regulation of the neural patterning activity of sonic hedgehog by secreted BMP inhibitors expressed by notochord and somites. *Development* 2000;**127**:4855–66.
19. Harland R. Neural induction. *Curr Opin Genet Dev* 2000;**10**:357–62.
20. McMahon JA, Takada S, Zimmerman LB, Fan CM, Harland RM, McMahon AP. Noggin-mediated antagonism of BMP signaling is required for growth and patterning of the neural tube and somite. *Genes Dev* 1998;**12**:1438–52.
21. Wine-Lee 3rd L, Ahn KJ, Richardson RD, Mishina Y, Lyons KM, Crenshaw EB. Signaling through BMP type 1 receptors is required for development of interneuron cell types in the dorsal spinal cord. *Development* 2004;**131**:5393–403.

22. Massague J. How cells read TGF-beta signals. *Nat Rev Mol Cell Biol* 2000;**1**:169–78.

23. Muroyama Y, Fujihara M, Ikeya M, Kondoh H, Takada S. Wnt signaling plays an essential role in neuronal specification of the dorsal spinal cord. *Genes Dev* 2002;**16**:548–53.

24. Chesnutt C, Burrus LW, Brown AM, Niswander L. Coordinate regulation of neural tube patterning and proliferation by TGFbeta and WNT activity. *Dev Biol* 2004;**274**:334–47.

25. Ille F, Atanasoski S, Falk S, et al. Wnt/BMP signal integration regulates the balance between proliferation and differentiation of neuroepithelial cells in the dorsal spinal cord. *Dev Biol* 2007;**304**:394–408.

26. Bonner J, Gribble SL, Veien ES, Nikolaus OB, Weidinger G, Dorsky RI. Proliferation and patterning are mediated independently in the dorsal spinal cord downstream of canonical Wnt signaling. *Dev Biol* 2008;**313**:398–407.

27. Ulloa F, Briscoe J. Morphogens and the control of cell proliferation and patterning in the spinal cord. *Cell Cycle* 2007;**6**:2640–9.

28. Ruiz IAA, Palma V, Dahmane N. Hedgehog-Gli signalling and the growth of the brain. *Nat Rev Neurosci* 2002;**3**:24–33.

29. Huangfu D, Anderson KV. Cilia and Hedgehog responsiveness in the mouse. *Proc Natl Acad Sci USA* 2005;**102**:11,325–11,330.

30. Ericson J, Rashbass P, Schedl A, et al. Pax6 controls progenitor cell identity and neuronal fate in response to graded Shh signaling. *Cell* 1997;**90**:169–80.

31. Zhou Q, Choi G, Anderson DJ. The bHLH transcription factor Olig2 promotes oligodendrocyte differentiation in collaboration with Nkx2.2. *Neuron* 2001;**31**:791–807.

32. Marquardt T, Pfaff SL. Cracking the transcriptional code for cell specification in the neural tube. *Cell* 2001;**106**:651–4.

33. Dessaud E, Yang LL, Hill K, et al. Interpretation of the sonic hedgehog morphogen gradient by a temporal adaptation mechanism. *Nature* 2007;**450**:717–20.

34. Stamataki D, Ulloa F, Tsoni SV, Mynett A, Briscoe J. A gradient of Gli activity mediates graded Sonic Hedgehog signaling in the neural tube. *Genes Dev* 2005;**19**:626–41.

35. Ruiz i Altaba A, Mas C, Stecca B. The Gli code: an information nexus regulating cell fate, stemness and cancer. *Trends Cell Biol* 2007;**17**:438–47.

36. Litingtung Y, Chiang C. Specification of ventral neuron types is mediated by an antagonistic interaction between Shh and Gli3. *Nat Neurosci* 2000;**3**:979–85.

37. Lee SK, Pfaff SL. Transcriptional networks regulating neuronal identity in the developing spinal cord. *Nat Neurosci* 2001;**4**(Suppl.):1183–91.

38. Chuang PT, McMahon AP. Vertebrate Hedgehog signalling modulated by induction of a Hedgehog-binding protein. *Nature* 1999;**397**:617–21.

39. Eggenschwiler JT, Espinoza E, Anderson KV. Rab23 is an essential negative regulator of the mouse Sonic hedgehog signalling pathway. *Nature* 2001;**412**:194–8.

40. Pierani A, Brenner-Morton S, Chiang C, Jessell TM. A sonic hedgehog-independent, retinoid-activated pathway of neurogenesis in the ventral spinal cord. *Cell* 1999;**97**:903–15.

41. Novitch BG, Wichterle H, Jessell TM, Sockanathan S. A requirement for retinoic acid-mediated transcriptional activation in ventral neural patterning and motor neuron specification. *Neuron* 2003;**40**:81–95.

42. Molotkova N, Molotkov A, Sirbu IO, Duester G. Requirement of mesodermal retinoic acid generated by Raldh2 for posterior neural transformation. *Mech Dev* 2005;**122**:145–55.

43. Jeong J, McMahon AP. Growth and pattern of the mammalian neural tube are governed by partially overlapping feedback activities of the hedgehog antagonists patched 1 and Hhip1. *Development* 2005;**132**:143–54.

44. Machold R, Hayashi S, Rutlin M, et al. Sonic hedgehog is required for progenitor cell maintenance in telencephalic stem cell niches. *Neuron* 2003;**39**:937–50.

45. Peng CY, Yajima H, Burns CE, et al. Notch and MAML signaling drives Scl-dependent interneuron diversity in the spinal cord. *Neuron* 2007;**53**:813–27.

46. Del Barrio MG, Taveira-Marques R, Muroyama Y, et al. A regulatory network involving Foxn4, Mash1 and delta-like 4/Notch1 generates V2a and V2b spinal interneurons from a common progenitor pool. *Development* 2007;**134**:3427–36.

47. Pfaff S, Kintner C. Neuronal diversification: development of motor neuron subtypes. *Curr Opin Neurobiol* 1998;**8**:27–36.

48. Lance-Jones C, Omelchenko N, Bailis A, Lynch S, Sharma K. Hoxd10 induction and regionalization in the developing lumbosacral spinal cord. *Development* 2001;**128**:2255–68.

49. Matise MP, Lance-Jones C. A critical period for the specification of motor pools in the chick lumbosacral spinal cord. *Development* 1996;**122**:659–69.

50. Ensini M, Tsuchida TN, Belting HG, Jessell TM. The control of rostrocaudal pattern in the developing spinal cord: specification of motor neuron subtype identity is initiated by signals from paraxial mesoderm. *Development* 1998;**125**:969–82.

51. Liu JP, Laufer E, Jessell TM. Assigning the positional identity of spinal motor neurons. Rostrocaudal patterning of Hox-c expression by FGFs, Gdf11, and retinoids. *Neuron* 2001;**32**:997–1012.

52. Tiret L, Le Mouellic H, Maury M, Brulet P. Increased apoptosis of motoneurons and altered somatotopic maps in the brachial spinal cord of Hoxc-8-deficient mice. *Development* 1998;**125**:279–91.

53. Studer M, Lumsden A, Ariza-McNaughton L, Bradley A, Krumlauf R. Altered segmental identity and abnormal migration of motor neurons in mice lacking Hoxb-1. *Nature* 1996;**384**:630–4.

54. Itasaki N, Sharpe J, Morrison A, Krumlauf R. Reprogramming Hox expression in the vertebrate hindbrain: influence of paraxial mesoderm and rhombomere transposition. *Neuron* 1996;**16**:487–500.

55. Marshall H, Nonchev S, Sham MH, Muchamore I, Lumsden A, Krumlauf R. Retinoic acid alters hindbrain Hox code and induces transformation of rhombomeres 2/3 into a 4/5 identity. *Nature* 1992;**360**:737–41.

56. Gavalas A, Krumlauf R. Retinoid signalling and hindbrain patterning. *Curr Opin Genet Dev* 2000;**10**:380–6.

57. Hobert O, Westphal H. Functions of LIM-homeobox genes. *Trends Genet* 2000;**16**:75–83.

58. Deschamps J. Ancestral and recently recruited global control of the Hox genes in development. *Curr Opin Genet Dev* 2007;**17**:422–7.

59. Hochstim C, Deneen B, Lukaszewicz A, Zhou Q, Anderson DJ. Identification of positionally distinct astrocyte subtypes whose identities are specified by a homeodomain code. *Cell* 2008;**133**:510–22.

60. Wichterle H, Lieberam I, Porter JA, Jessell TM. Directed differentiation of embryonic stem cells into motor neurons. *Cell* 2002;**110**:385–97.

61. Li XJ, Du ZW, Zarnowska ED, et al. Specification of motoneurons from human embryonic stem cells. *Nat Biotechnol* 2005;**23**:215–21.

Signaling In Disease

Chapter 37

Ras and Cancer

Frank McCormick

Cancer Research Institute, University of California Comprehensive Cancer Center, San Francisco, California

INTRODUCTION: RAS ACTIVATION IN CANCER

It has been 20 years since H-Ras mutations were identified in DNA from the bladder cancer cell line T-24. Since this seminal observation, rates of mutation in H-Ras, N-Ras, and K-Ras have been measured in most types of human cancers [1]. The clonal nature of these mutations in tumors strongly suggests a causal role, a suggestion that has been amply verified by mouse models of Ras-induced cancer. A striking result of this comprehensive survey is the considerable variation in frequency in Ras mutation between different types of cancer. In pancreatic carcinoma, K-Ras is activated by point mutation in almost every case, whereas Ras mutations are hardly ever detected in mammary carcinomas, to cite two extreme examples. The biological or molecular basis of these observations is not yet understood. One interpretation is that alternative mechanisms of activating the Ras pathway (receptor amplification, activation of downstream pathways) also occur at varying frequencies. Another interpretation is that different types of cancer vary in their dependence on the Ras pathway. Another unresolved issue is the predominance of K-Ras mutations over N-Ras and H-Ras: this may reflect different levels of expression of these genes in different tissues, and different levels of dependence on each type. Mouse knockout experiments show that K-Ras is essential [2], whereas N-Ras and H-Ras are not, consistent with K-Ras being the most important form and therefore the most likely to be directly involved in carcinogenesis. However, other models must be considered: for example, each type of Ras may signal through a different set of downstream effectors, and K-Ras happens to provide a repertoire of signals that is consistent with malignant progression. Although most evidence points toward shared effectors among all three types of Ras, evidence for discrimination among effectors also exists [3].

In addition to mutations in Ras genes, gains and losses of Ras genes have been reported in human tumors [4–9]. In mouse tumors, double minute chromosomes encoding H-Ras have been identified [10]. Also in mouse models, progress increase of copy numbers of H-Ras mutants appears to drive malignant progression, along with selective loss of the wild-type allele [11].

PATHWAYS DOWNSTREAM OF RAS

Figure 37.1 shows pathways regulated by Ras. In addition to the well-established pathways that Ras activates, the Raf-MAP kinase cascade and the PI3′ kinase pathway, Ras activates RalGDS and possibly other effector pathways that are not well characterized [12]. Raf and PI3′ kinase pathways act synergistically to mediate Ras transformation, suggesting that inhibitors of either pathway have profound effects on Ras transformation [13]. This is an important issue in the context of drug development based on Ras pathways.

The precise molecular basis of synergy between effector pathways is not fully understood. However, there are multiple elements of these pathways that intersect and could contribute to synergistic interaction. For example, the cyclin D1 gene is a transcriptional target of the Raf-MAP kinase pathway, and cyclin D1 protein is stabilized by the PI3′ kinase pathway [14].

Until recently, there was little genetic evidence that the Raf–MAP kinase pathway is activated by mutation of gene copy number change in human cancer. However, the recent discovery that B-Raf is activated by mutation has changed this view dramatically. Two types of mutant have been described: one that renders B-Ras independent of Ras and occurs in tumors in which Ras is wild-type; another one that requires Ras interaction for full activity and occurs in tumors containing mutant Ras. It is conceivable

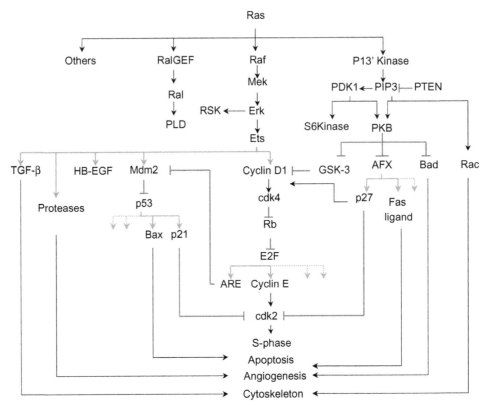

FIGURE 37.1 Pathways regulated by Ras.

that the high-throughput sequencing approach that identified these mutations may yet reveal other activating events in the Raf–MAP kinase pathway. Nonetheless, these new data imply that activation of the Raf effector pathway is the major selection for Ras mutation in these diseases.

In contrast, genetic changes activating the PI3′ kinase have been well documented and are considered of major importance in human cancer. Loss of PTEN is by far the most frequent event that activates this pathway, but increases in copy number of Akt/PKB have been documented and implicate this arm of the pathway in PTEN-deficient tumors. In endometrial and cutaneous melanoma cancers, loss of PTEN and Ras activation are mutually exclusive, suggesting that in these conditions, the major selection for Ras mutation is activation of the PI3′ kinase effector pathway [15, 16].

MOUSE MODELS OF CANCER

Mouse models of cancer provide important clues relating to the role of Ras in cancer. Many have involved forced expression of mutant Ras proteins under tissue-specific promoters, revealing transforming power of the Ras oncogene in different physiological settings. An interesting aspect of these models is the sustained requirement for Ras expression even in advanced cancers: withdrawal of Ras

expression causes complete regression of such tumors [17]. Recently a model has been developed in which mutant K-Ras is activated sporadically: this appears to be an excellent model for sporadic human lung cancer [18].

Other informative rodent models have used mutagens to initiate cancers, followed by analysis of Ras activation and progression. The classic studies of Sukamar and Barbacid and coworkers proved conclusively that Ras mutation can be the initiating event in cancer, and showed that mutations caused by an early chemical insult can persist in latent forms before progressing to cancer [19]. The skin cancer models of Balmain and coworkers have revealed a step-wise activation of H-Ras during initiation and progression: mutant H-Ras alleles created by exposure to carcinogen are selectively amplified in a step-wise manner as the tumors evolve. In parallel to increased Ras activity, levels of cyclin D1 increase during progression. A role of cyclin D1 in Ras transformation in this model was confirmed by demonstration that tumors' progression is strongly retarded in mice lacking the cyclin D1 gene. Even more striking effects of cyclin D1 were demonstrated recently in a model of mammary carcinogenesis driven by Ras, erbB, wnt, or myc: the former two oncogenes were completely dependent on cyclin D1, whereas the latter were not [20]. This clear role of cyclin D1 points toward the importance of the Raf–MAP kinase effector pathway in Ras transformation, since this pathway activates transcription of cyclin D1 directly.

However, a role of the PI kinase pathway cannot be ruled out, as this pathway stabilizes cyclin D1 through inhibition of GSK-3-mediated degradation.

PROSPECTS FOR CANCER THERAPY BASED ON RAS

Attempts to block Ras signaling in human cancers by inhibiting posttranslational farnesylation have been stalled by the fact that K-Ras, the major form of Ras involved in human cancer, can also be modified by geranylgeranylation. This allows continued K-Ras activity in the presence of farnesyl transferase inhibitors. Such inhibitors may have clinical value through their action on other cellular targets, however [21]. More recent approaches to blocking Ras activity have targeted enzymes downstream of Ras. A Raf kinase inhibitor entered clinical trials recently [22], and a MEK inhibitor followed soon afterwards [23]. Attempts to block other enzymes downstream of Ras are also under way [24].

REFERENCES

1. Bos JL. Ras oncogenes in human cancer: a review. *Cancer Res* 1989;**49**:4682–9.
2. Johnson L, Greenbaum D, Cichowski K, Mercer K, Murphy E, Schmitt E, Bronson RT, Umanoff H, Edelmann W, Kucherlapati R, Jacks T. K-ras is an essential gene in the mouse with partial functional overlap with N-ras. *Genes Dev* 1997;**11**:2468–81.
3. Wolfman A. Ras isoform-specific signaling: location, location, location. *Sci STKE* 2001;**2001**:E2.
4. Kimura E, Armelin HA. Role of proto-oncogene c-Ki-ras amplification and overexpression in the malignancy of Y-1 adrenocortical tumor cells. *Brazil J Med Biol Res* 1988;**21**:189–201.
5. Filmus J, Trent JM, Pullano R, Buick RN. A cell line from a human ovarian carcinoma with amplification of the K-ras gene. *Cancer Res* 1986;**46**:5179–82.
6. George DL, Scott AF, Trusko S, Glick B, Ford E, Dorney DJ. Structure and expression of amplified cKi-ras gene sequences in Y1 mouse adrenal tumor cells. *EMBO J* 1985;**4**:1199–203.
7. George DL, Scott AF, de Martinville B, Francke U. Amplified DNA in Y1 mouse adrenal tumor cells: isolation of cDNAs complementary to an amplified c-Ki-ras gene and localization of homologous sequences to mouse chromosome 6. *Nucleic Acids Res* 1984;**12**:2731–43.
8. Schwab M, Alitalo K, Klempnauer KH, Varmus HE, Bishop JM, Gilbert F, Brodeur G, Goldstein M, Trent J. Amplified DNA with limited homology to myc cellular oncogene is shared by human neuroblastoma cell lines and a neuroblastoma tumour. *Nature* 1983;**305**:245–8.
9. Schwab M, Alitalo K, Varmus HE, Bishop JM, George D. A cellular oncogene (c-Ki-ras) is amplified, overexpressed, and located within karyotypic abnormalities in mouse adrenocortical tumour cells. *Nature* 1983;**303**:497–501.
10. Tanaka K, Takechi M, Nishimura S, Oguma N, Kamada N. Amplification of c-MYC oncogene and point mutation of N-RAS oncogene point mutation in acute myelocytic leukemias with double minute chromosomes. *Leukemia* 1993;**7**:469–71.
11. Buchmann A, Ruggeri B, Klein-Szanto AJ, Balmain A. Progression of squamous carcinoma cells to spindle carcinomas of mouse skin is associated with an imbalance of H-ras alleles on chromosome 7. *Cancer Res* 1991;**51**:4097–101.
12. Campbell SL, Khosravi-Far R, Rossman KL, Clark GJ, Der CJ. Increasing complexity of Ras signaling. *Oncogene* 1998;**17**:1395–413.
13. Gille H, Downward J. Multiple ras effector pathways contribute to G(1) cell cycle progression. *J Biol Chem* 1999;**274**:22,033–22,040.
14. Diehl JA, Cheng M, Roussel MF, Sherr CJ. Glycogen synthase kinase-3beta regulates cyclin D1 proteolysis and subcellular localization. *Genes Dev* 1998;**12**:3499–511.
15. Ikeda T, Yoshinaga K, Suzuki A, Sakurada A, Ohmori H, Horii A. Anticorresponding mutations of the KRAS and PTEN genes in human endometrial cancer. *Oncol Rep* 2000;**7**:567–70.
16. Tsao H, Zhang X, Fowlkes K, Haluska FG. Relative reciprocity of NRAS and PTEN/MMAC1 alterations in cutaneous melanoma cell lines. *Cancer Res* 2000;**60**:1800–4.
17. Chin 2nd L, Tam A, Pomerantz J, Wong M, Holash J, Bardeesy N, Shen Q, O'Hagan R, Pantginis J, Zhou H, Horner JW, Cordon-Cardo C, Yancopoulos GD, DePinho RA. Essential role for oncogenic Ras in tumour maintenance. *Nature* 1999;**400**:468–72.
18. Johnson L, Mercer K, Greenbaum D, Bronson RT, Crowley D, Tuveson DA, Jacks T. Somatic activation of the K-ras oncogene causes early onset lung cancer in mice. *Nature* 2001;**410**:1111–16.
19. Sukumar S, Notario V, Martin-Zanca D, Barbacid M. Induction of mammary carcinomas in rats by nitroso-methylurea involves malignant activation of H-ras-1 locus by single point mutations. *Nature* 1983;**306**:658–61.
20. Yu Q, Geng Y, Sicinski P. Specific protection against breast cancers by cyclin D1 ablation. *Nature* 2001;**411**:1017–21.
21. Prendergast GC, Rane N. Farnesyltransferase inhibitors: mechanism and applications. *Expert Opin Investig Drugs* 2001;**10**:2105–16.
22. Lyons JF, Wilhelm S, Hibner B, Bollag G. Discovery of a novel Raf kinase inhibitor. *Endocr Relat Cancer* 2001;**8**:219–25.
23. Sebolt-Leopold JS, Dudley DT, Herrera R, Van Becelaere K, Wiland A, Gowan RC, Tecle H, Barrett SD, Bridges A, Przybranowski S, Leopold WR, Saltiel AR. Blockade of the MAP kinase pathway suppresses growth of colon tumors *in vivo*. *Nat Med* 1999;**5**:810–16.
24. McCormick F. Small-molecule inhibitors of cell signaling. *Curr Opin Biotechnol* 2000;**11**:593–7.

Chapter 38

Targeting Ras for Anticancer Drug Discovery

Jen Jen Yeh[1], James P. Madigan[1], Paul M. Campbell[2], Patrick J. Roberts[1], Lanika DeGraffenreid[1] and Channing J. Der[1]

[1]*Departments of Surgery and Pharmacology, Lineberger Comprehensive Cancer Center, University of North Carolina at Chapel Hill, Chapel Hill, North Carolina*

[2]*Drug Discovery Program, H. Lee Moffitt Cancer Center & Research Institute, University of South Florida, Tampa, Florida*

INTRODUCTION

Mutational activation of *RAS* genes is associated with 33 percent of human cancers [1]. The three human *RAS* genes (*HRAS*, *NRAS*, and *KRAS*) encode small GTPases that function as binary switches in signal transduction [2, 3] (Figure 38.1a). The lack of current success in the development of candidate "anti-Ras" inhibitors has dampened enthusiasm for such efforts [4, 5]. In particular, the disappointing results with farnesyltransferase inhibitors (FTIs) in clinical trial analyses prompted the misleading perception that Ras is not a clinically useful target for pancreatic cancer treatment. Instead, the correct conclusion, surprisingly not understood by many researchers, is that FTIs are not effective Ras inhibitors. When coupled with the poor "druggability" of GTPases, an outcome of the FTI saga has been greatly diminished interest in the development of anti-Ras inhibitors.

Several recent research observations have rekindled interest in targeting Ras for cancer treatment. First, genome-wide cancer genome studies have revealed, perhaps somewhat disappointingly, that the genes most commonly mutated in cancers were already identified in more systematic studies [6–9]. For example, prior to the sequencing of pancreatic cancer (PDAC), the most frequently mutated genes known to be associated with the progression of this cancer were the *KRAS* oncogene and the *TP53*, *CDKN2A*, and *SMAD4* tumor suppressors [10, 11]. The outcome of sequence analyses of 20,661 genes was that these same four genes remained the top four most frequently mutated genes, with *KRAS* mutation in 113 of 114 PDAC tumors [9]. A similar picture emerged from genome-wide sequencing of colorectal cancers (CRC), with

KRAS the second most prevalent mutated gene. For colon and pancreatic cancer, the most frequent gain-of-function genetic event was *KRAS* activation, and, consequently, the most attractive target for therapeutic intervention.

The second observation involved the association of *KRAS* mutations with resistance to treatment with inhibitors of the epidermal growth factor receptor (EGFR) [12]. Whereas *EGFR* mutation was a positive prognostic indicator, *KRAS* was a negative indicator for non-small cell lung cancer (NSCLC) [13–15]. A similar association between *KRAS* mutation and resistance to EGFR therapy has been seen for CRC [16–18]. Overall, the combined results from 20 clinical studies found that the response rate to anti-EGFR therapy was less than 3 percent for *KRAS* mutant tumors, and in contrast, 35 percent and 20 percent response rates, respectively, for *KRAS* wild-type CRC or NSCLCs. Taken together with observations made from the genome sequencing studies, a rebirth in the appreciation of Ras as a clinically important target for cancer therapy has occurred.

RAS PROTEINS FUNCTION AS SIGNALING NODES

The three *RAS* genes encode 188-189 amino acid proteins, with alternative splicing accounting for the K-Ras4A and K-Ras4B proteins that differ in their C-terminal 12 residues. The N-terminal 164 residues correspond to the G domain (Figure 38.1a). The G domain includes consensus GTP binding motifs shared with other GTP binding proteins. Ras proteins bind GDP and GTP with high affinity ($K_d = 10\,pM$) and possess intrinsic GTP hydrolysis and GDP/GTP exchange activities. These activities are accelerated by

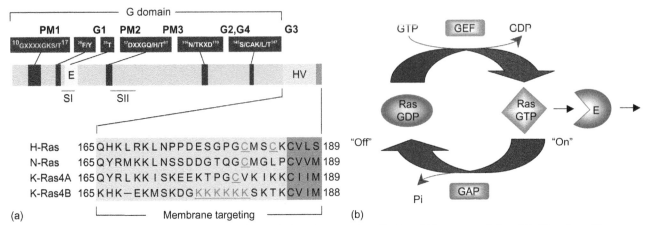

FIGURE 38.1 Ras structure and biochemistry. (a) Domain organization of Ras proteins. The three *RAS* genes encode four 188-189 amino acid proteins that share 82–90 percent sequence identity; *KRAS* encodes two splice variants that differ in their C-terminal 25 amino acids (designated 4A and 4B). Residues 1–164 comprise the G domain that contains six conserved sequence motifs shared with other Ras superfamily and GTP binding proteins and involved in either binding phosphate/Mg^{2+} (PM) or the guanine base (G) of GDP and GTP. Switch I (30–38) and II (60–76) residues change in conformation during GDP/GTP cycling. The core effector binding domain (E; residues 32–40) and flanking sequences are involved in effector binding specificity. The C-terminal 24–25 residues comprise the membrane targeting sequence and is composed of the C-terminal CAAX box, required for posttranslational lipid modification, and the hypervariable (HV) domain that includes a second membrane targeting sequence element (palmytoylated cysteine or polybasic stretches). (b) The Ras GDP/GTP cycle. Ras proteins cycle between a GTP-bound "on" state and a GDP bound "off" states. Ras proteins are positively regulated by Ras specific GEFs and negatively regulated by GAPs. When bound to GTP, Ras interacts with a variety of downstream effectors (E) to control cellular processes.

guanine nucleotide exchange factors (RasGEFs) that stimulate formation of the active GTP bound protein and by GTPase activating proteins (RasGAPs) that stimulate GTP hydrolysis and return Ras to the inactive GDP bound state [19] (Figure 38.1b).

A diverse spectrum of extracellular stimuli (acting on receptor tyrosine kinases, G protein coupled receptors, cytokine and antigen receptors, integrins, etc.) causes activation of Ras. This occurs most commonly through activation of RasGEFs. Ras-GDP and Ras-GTP differ in conformation in two regions, referred to as switch 1 (30–38) and switch 2 (60–76) (Figure 38.1a). The active conformation displays preferential affinity and binding to multiple, catalytically distinct downstream effectors [3]. Hence, Ras proteins serve as points of signaling convergence and activators of divergent effective signaling networks.

The classical Ras signaling pathway comprises EGF stimulation of the EGFR (Figure 38.2a). Activated EGFR recruits the Sos RasGEF, in complex with the Grb2 adaptor, leading to increased association of Sos with membrane bound Ras, leading to transient formation of Ras-GTP. Ras-GTP binds to and facilitates activation of the Raf serine/threonine kinases (A-Raf, B-Raf, c-Raf-1). Activated Raf phosphorylates and activates the MEK1 and MEK2 dual specificity kinases, which then phosphorylate and activate the ERK1 and ERK2 mitogen activated protein kinases (MAPKs). Activated ERK translocates into the nucleus where it phosphorylates and activates Ets family transcription factors (e.g., Elk-1). The EGF-EGFR-Grb2-Sos-Ras-Raf-MEK-ERK cascade represents the first signaling

pathway where all components from the cell surface to the nucleus were defined. However, the subsequent identification of additional Ras effectors revealed a more complex signaling network regulated by Ras activation [3]. Of these, currently five have been implicated in Ras mediated oncogenesis (Figure 38.2b). The validation of these effectors in Ras transformation has involved studies in cell culture and mouse models of Ras mediated oncogenesis. For cell cultures studies, the use of dominant negative mutants, interfering RNA or pharmacologic inhibitors has demonstrated the necessity of an effector pathway for Ras transformation [20]. For mouse model studies, an important approach has been the use of mice deficient in effector function. For example, a deficiency in Tiam1, the RalGDS RalGEF, or phospholipase C epsilon did not perturb mouse development, but did impair carcinogen activated H-Ras induced skin tumor formation [21–23]. Similarly, mice harboring a mutant p110 alpha catalytic subunit of phosphotidylinositol 3-kinase (PI3K) that cannot couple to Ras also showed impaired Ras induced tumor formation [23].

RAS ACTIVATION IN HUMAN CANCERS: VALIDATION AND DRUGGABILITY

Missense mutations in *RAS* genes are present in cancers that arise from many tissues [1]. Mutations in *KRAS* are the most common (21 percent), followed by *NRAS* (9 percent), with *HRAS* mutations the least common (3 percent). Most commonly, this results in single amino acid substitutions at residues 12, 13, or 61 (Figure 38.3). Less frequently, mis-

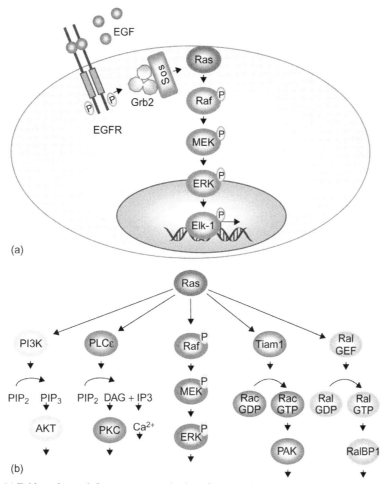

FIGURE 38.2 Ras signaling. (a) Epidermal growth factor receptor activation of Ras and the ERK MAPK cascade. The activated EGFR recruits the Grb2 adaptor protein in complex with the RasGEF, Sos, to the plasma membrane, where Ras resides, leading to Ras activation. Activated Ras then binds and activates Raf and activations the ERK MAPK cascade. (b) Effector signaling networks that promote Ras mediated oncogenesis.

sense mutations are also seen affecting other residues, typically associated with regions involved in GTP binding and/or hydrolysis. Mutant Ras proteins are refractory to GAP stimulation, rendering Ras persistently GTP bound and active.

The structural and biochemical distinctions between normal and mutant Ras are well delineated. However, these differences have not been exploited successfully to develop small molecule inhibitors that direct target mutant Ras. Unlike the successful generation of small molecule antagonists of ATP binding to protein kinases (low micromolar binding affinity), the picomolar binding affinity of Ras for GTP has made this an unrealistic approach to reduce the GTP bound state of Ras. Finally, while considerable effort was made to utilize GAP as a foundation for approaches to reactivate the GTPase activity of mutant Ras, these efforts have also not yielded promising leads. Therefore, while there remains hope that approaches can still be identified to target Ras directly, the general feeling is that indirect approaches will provide the most promising leads.

Indirect approaches for blocking Ras have focused on either inhibition of Ras membrane association or downstream effector signaling (Figure 38.4). Below we summarize the specific targets under evaluation for each of these directions. We have limited this list to those where (1) target validation has been done in cell culture or mouse models and (2) small molecule, cell permeable inhibitors have been identified and shown to exhibit target based and anti-tumor activity in cell culture and/or mouse models of cancer.

TARGETING RAS MEMBRANE ASSOCIATION

All Ras proteins terminate with a CAAX tetrapeptide motif (where C = cysteine, A = aliphatic amino acid, X = serine, methionine) (Figure 38.1a) that is recognized by farnesyltransferase (FTase), which catalyzes the covalent addition of a C15 farnesyl isoprenoid lipid [24] (Figure 38.4a). Two subsequent modifications signaled

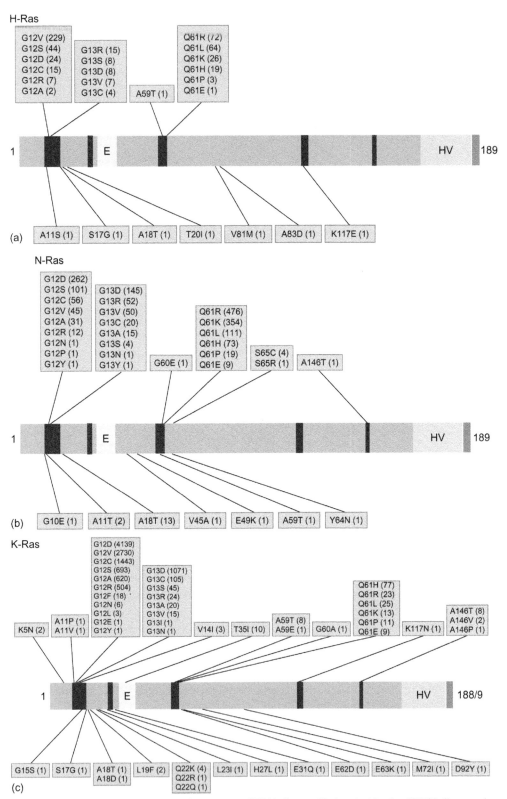

FIGURE 38.3 Ras mutations in human cancers. Shown are the consensus GTP binding motifs shared with other GTP binding proteins as described in the legend to Figure 38.1. Shown are the missense mutations found in (a) H-Ras, (b) N-Ras, and (c) K-Ras. The most frequent amino acid substitutions are found at residues 12, 13, or 61 that impair the intrinsic and GAP stimulated GTP hydrolysis activity of Ras. The occurrence of each mutation is indicated in parenthesis. Data were compiled from the COSMIC database.

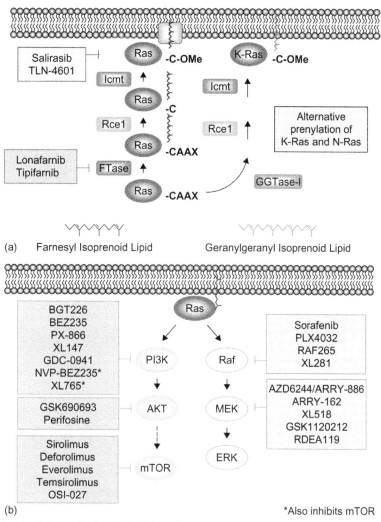

FIGURE 38.4 Anti-Ras inhibitors in clinical evaluation. (a) Inhibitors of Ras membrane association. The C-terminal CAAX motif is necessary and sufficient to signal for three posttranslational modifications: Ftase catalyzed covalent addition of a farnesyl isoprenoid to the cysteine residue, Rce1 catalyzed endoproteolytic cleavage of the AAX residues, and Icmt catalyzed carboxylmethylation of the now terminal farnesylated cysteine residue. Two FTase inhibitors are currently in clinical trial analyses. The second class of inhibitors is the farnesyl group-containing small molecules that function, in part, to block the ability of farnesylated Ras to associate with membrane receptors. (b) Inhibitors of Ras effector signaling. Inhibitors of Raf and MEK activation of ERK or inhibitors of PI3K activation of AKT and mTOR are currently in Phase I–II clinical trial. See www.clinicaltrials.gov for specific completed, active, or recruiting clinical trials for each inhibitor.

by the farnesylated CAAX motif are endoproteolytic cleavage of the AAX residues by Ras converting enzyme 1 (Rce1) and carboxyl methylation of the isoprenylated cysteine residue by isoprenylcysteine carboxyl methyltransferase (Icmt). The CAAX modifications are critical, but not sufficient, to promote Ras association with the inner face of the plasma membrane. Most Ras proteins (except K-Ras4B) undergo an additional covalent modification, the addition of palmitate fatty acid to cysteine residues upstream of the CAAX motif that serves as a second signal to facilitate plasma membrane association (Figure 38.1a). K-Ras4B contains polybasic amino acid sequences that serve as a second signal for its association with the plasma membrane. Inhibitors of Ras membrane association involve either inhibitors of CAAX motif signaled

posttranslational modifications or farnesyl moiety containing molecules that are proposed to function as antagonists of Ras membrane association.

Farnesyltransferase Inhibitors

The first evidence that farnesylation was critical for Ras transforming activity came from structure function mutagenesis studies of the CAAX motif. Mutation of the cysteine residue of the CAAX motif (e.g., H-Ras(C186S)) prevented farnesylation and all subsequent C-terminal modifications, rendering Ras cytosolic and nontransforming [25–27]. Further support for targeting farnesylation came from the isolation of farnesyltransferase the demonstration that the CAAX tetrapeptide can serve

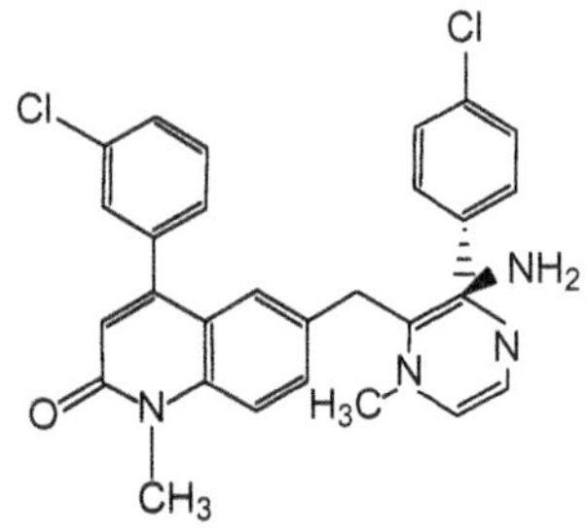

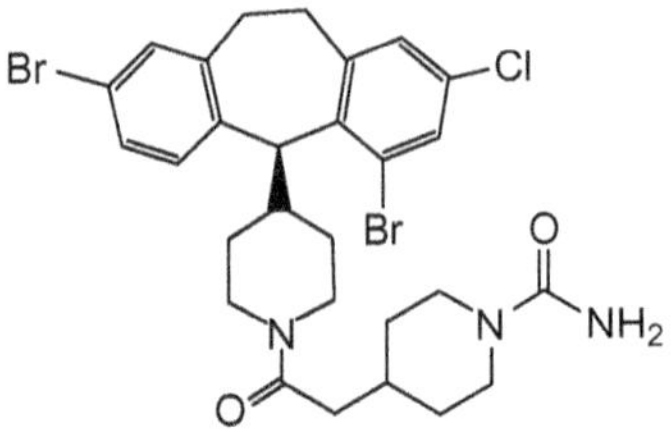

(a) Tipifarnib

Lonafarnib

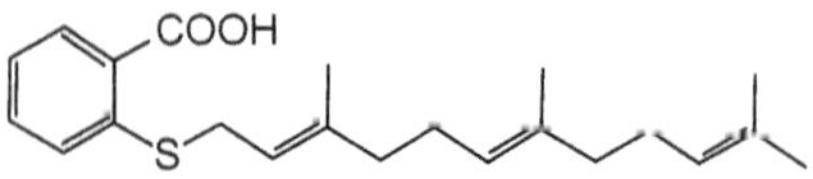

Salirasib (FTS)

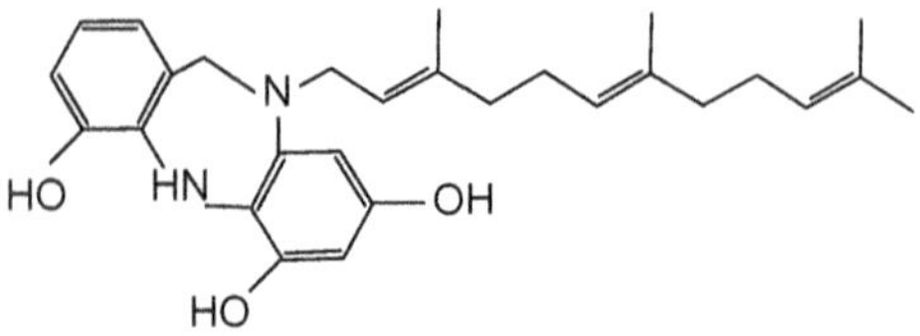

(b) TLN-4601

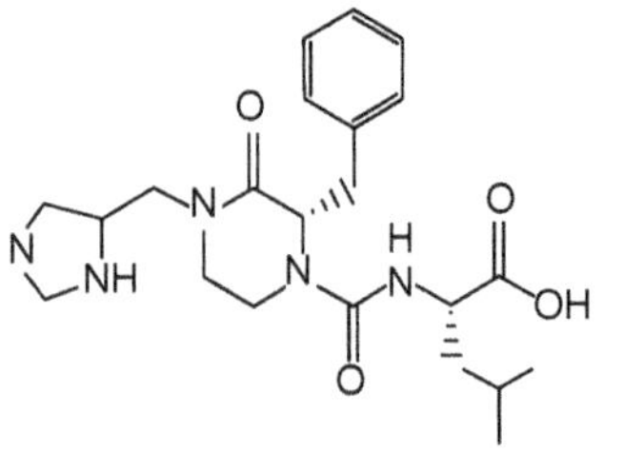

(c) GGTI-2418

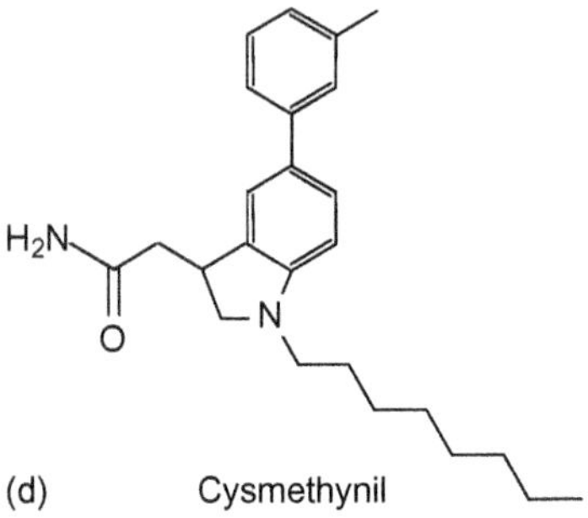

(d) Cysmethynil

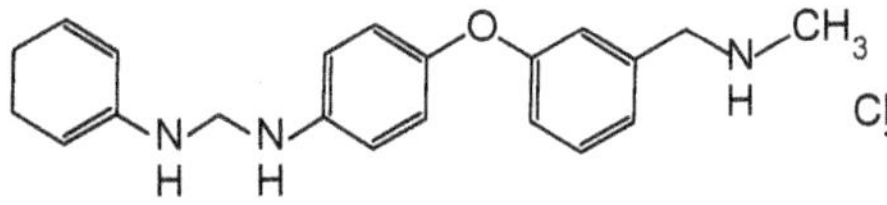

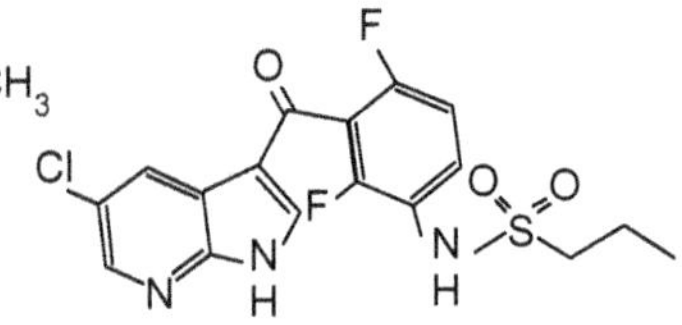

(e) Sorafenib

PLX4720

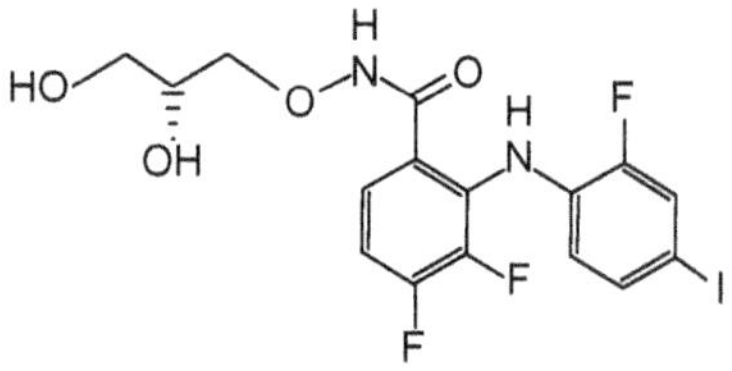

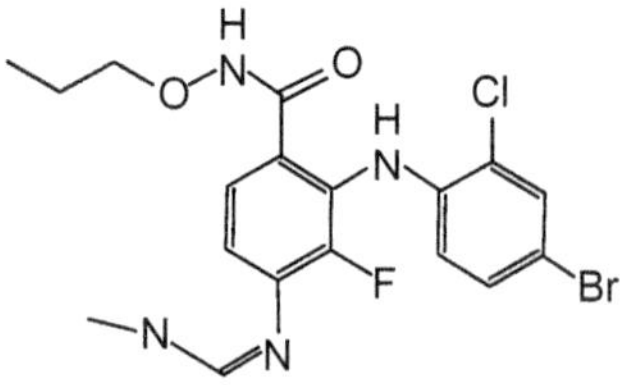

(f) PD0325901

AZD6244/ARRY-886

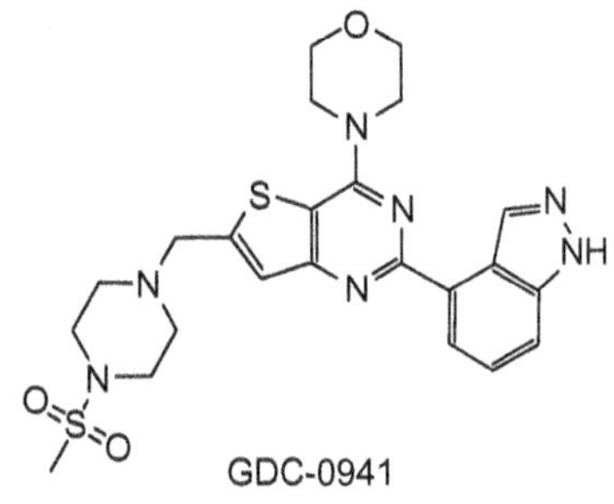

(g) GDC-0941

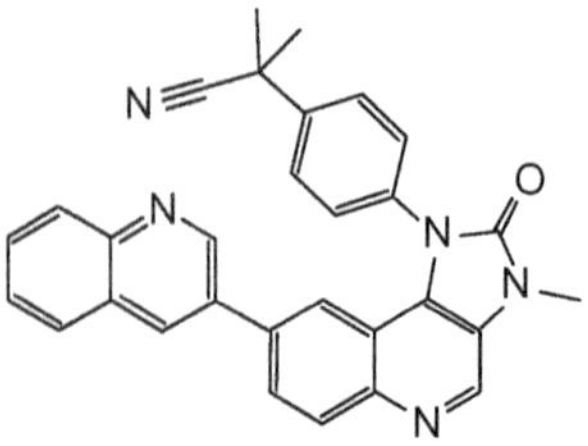

(h) NV-BEZ235

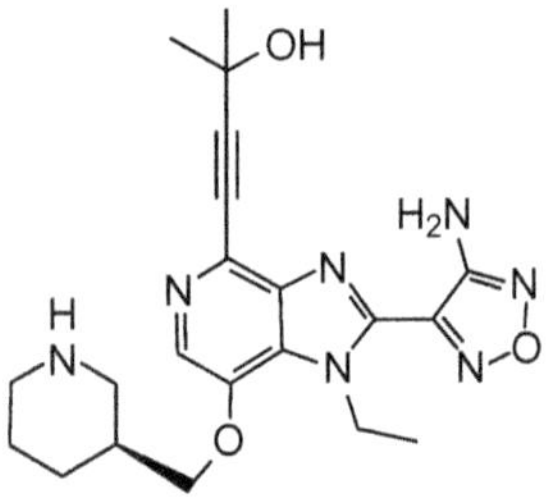

(i) GSK690693

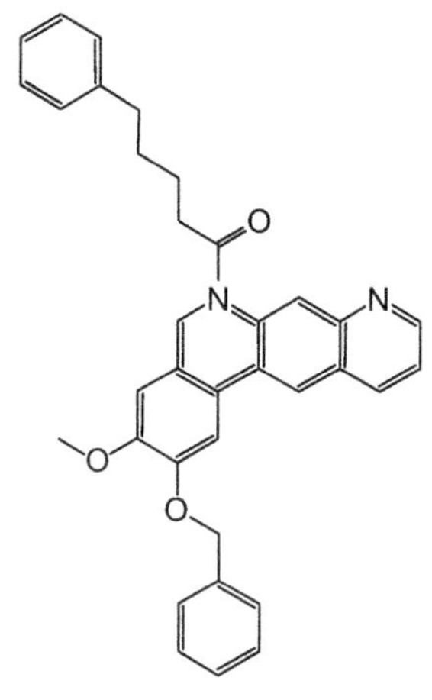

FIGURE 38.5 Structures of inhibitors of Ras membrane association or effector signaling that are currently undergoing preclinical or clinical evaluation. (a) Farnesyltransferase inhibitors. (b) Farnesyl group-containing small molecule inhibitors. (c) Geranylgeranyltransferase-I inhibitor. (d) Isoprenylcysteine carboxyl methyltransferase inhibitor. (e) Raf serine/threonine kinase inhibitors. (f) MEK1/2 inhibitors. (g) Phosphotidylinositol 3-kinase inhibitor. (h) Duel phosphotidylinositol 3-kinase and mTOR inhibitor. (i) AKT serine/threonine kinase inhibitor. (j) Ras–Raf interaction inhibitor. (k) Rac small GTPase inhibitors.

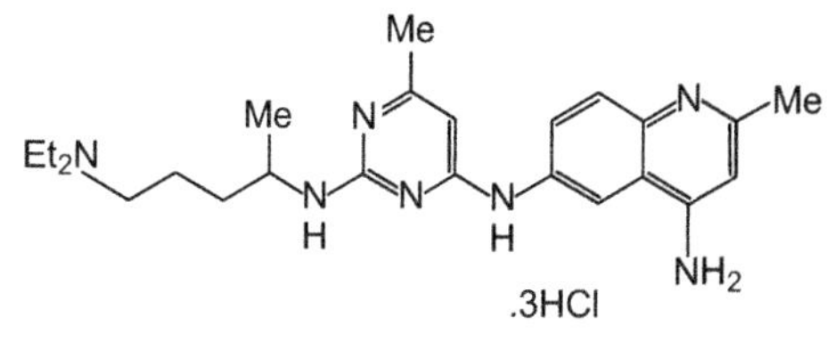

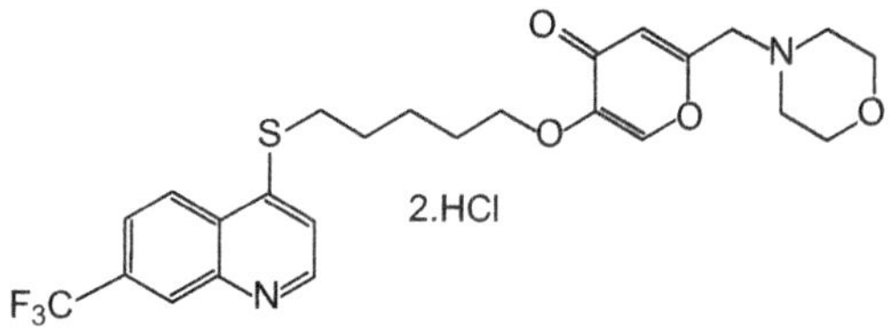

as an inhibitor of FTase [28]. These findings set the stage for the intensive development of CAAX peptidomimetics, as well as high throughput compound library screens, to develop FTase inhibitors [29]. With many pharmaceutical companies involved, it turned out to be relatively easy to identify potent and selective FTase inhibitors (FTIs) [30]. Numerous structurally distinct FTIs were identified that selectively inhibited FTase and not the closely related geranylgeranyltransferase type I (GGTase-I) enzyme. GGTase-I typically recognizes CAAX motifs where X = leucine, and catalyzes the addition of the more hydrophobic C20 geranylgeranyl isoprenoid (Figure 38.5a). The majority of FTase substrates terminate with X = methionine, serine or glutamine.

Although FTIs showed impressive anti-Ras and anti-tumor activity in preclinical cell culture and mouse models, two key issues led to the demise of FTIs as anti-Ras inhibitors [5, 30]. First, many of the early studies focused on H-Ras driven oncogenesis. While FTIs indeed effectively block H-Ras farnesylation and membrane association, and transformation, FTIs turned out to be ineffective against the N-Ras and K-Ras proteins. This was due to an unexpected biochemical difference among Ras proteins. When FTase activity is blocked, then K-Ras4B and N-Ras can serve as substrates for GGTase-I and addition of a geranylgeranyl isoprenoid [31, 32], which can effectively substitute for the farnesyl group and support Ras membrane association and transforming activity [33, 34]. Therefore, it was not surprising that clinical trial analyses with pancreatic cancer, where *KRAS* is mutated, resulted in negative findings [35–37]. Thus, the two most commonly mutated *RAS* genes encode proteins can bypass FTI mediated inhibition of function. Second, although commonly and mistakenly referred to as

"anti-Ras inhibitors," Ras proteins are not the only substrates for FTase [38]. Other farnesylated proteins with established roles in growth regulation may also be targeted by FTIs, for example the Rheb small GTPase [39], an activator of mammalian target of rapamycin (mTOR), a pathway commonly deregulated in cancer. Thus, the anti-tumor activities of FTIs very likely involve inhibition of function of other farnesylated proteins. The therapeutic value of FTIs may also be complicated by inhibiting the function of some farnesylated Ras family GTPases that function as tumor suppressors (e.g., Di-Ras1/Rig, ARHI/NOEY2, RRP22/RasL10A) [40–42]. Thus, while two FTIs continue to be evaluated for anticancer treatment (Figure 38.5a), the mystery of the key targets involved continues to hamper their development. While inhibitors of GGTase-I have been developed and considered for combination treatment with FTIs (Figure 38.5c), there is concern that this combination approach may suffer from significant off-target toxicity, since there are over 50 known putative substrates for GGTase-I .

Inhibitors of Rce1 and Icmt

Although much less explored, the other two CAAX signaled modifications have also been considered for anti-Ras inhibitors [43]. Initial studies suggested that, unlike blocking FTase activity, blocking Rce1 and Icmt modification of Ras resulted in only a 50 percent reduction in membrane association and transforming activity [44]. Hence, blocking these two enzymatic functions was not expected to significantly impair Ras mediated oncogenesis. However, more recent studies support the potential usefulness of inhibitors of Rce1 and Icmt for blocking Ras oncogenicity.

Studies in mice deficient in Rce1 or Icmt function argue that inhibitors of these enzymes can also have a significant impact on Ras mediated tumor growth. Mouse embryo fibroblasts deficient in Rce1 revealed that Ras proteins were incompletely processed and membrane associated [45, 46]. Cre mediated loss of Rce1 in fibroblasts generated from mice with a conditional Rce1 allele resulted in a loss of endoproteolytic processing and methylation of Ras. Additionally, soft agar assays revealed that excision of Rce1 reduced Ras mediated transformation. Finally, excision of Rce1 in a skin carcinoma cell line greatly reduced their growth [47]. Loss of Icmt resulted in inhibition of K-Ras mediated growth in soft agar assays and tumor growth in nude mice. Unexpectedly, loss of Icmt activity also blocked transformation by an oncogenic form of B-Raf, which is not an Icmt substrate [48]. This inhibition may be due to inhibition of other Icmt substrates. In support of this possibility, this study found evidence that RhoA inhibition may contribution to inhibition of K-Ras mediated transformation. Multiple Rho family GTPases have been implicated in Ras transformation. Our recent study found that RhoA and other Rho family members are dependent on Rce1 and Icmt mediated post-prenyl processing [49]. In a recent study, an Icmt deficiency reduced lung tumor development in a mouse model of KRAS induced cancer [50]. However, this issue may be highly context dependent, since an Rce1 deficiency was found to accelerate mutant KRAS induced myeloproliferative disease [51].

With regards to inhibitors of post-prenyl processing, Casey and colleagues described the identification of a small molecule inhibitor of Icmt [52]. Identified in a chemical library screen, cysmethynil (2-[5-(3-methylphenyl)-1-octyl-1H-indol-3-yl]acetamide) (Figure 38.5d) treatment inhibited cell growth in an Icmt dependent fashion. They also showed that cysmethynil treatment of cancer cells resulted in mislocalization of Ras. Cysmethynil treatment blocked the anchorage independent growth of a colon cancer cell line, and this effect was reversed by ectopic overexpression of Icmt, indicating that the inhibition was target based. Additionally, treatment of PC3 human prostate cell derived xenograft tumors with cysmethynil resulted in markedly reduced tumor size [53].

Other small molecules with Icmt inhibitory activity have also been described. Inhibition of Icmt function has been shown to be a critical component of the antiproliferative effect of the anti-folate methotrexate. In a colon cancer cell line, methotrexate treatment resulted in a decrease in Ras methylation by nearly 90 percent, mislocalization of Ras to the cytoplasm [54]. Finally, several natural product inhibitors of Icmt were discovered in a high throughput screen campaign [55–57].

To date, only limited development of Rce1 inhibitors has been described. In one compound library screen, several compounds were found to be effective at a low micromolar range for both yeast and human Rce1 and

were identified as possible tools for design of future Rce1 inhibitors [58]. An additional study showed that peptidyl (acyloxy)methyl ketones can inhibit Rce1 enzyme activity in vitro [59].

Finally, inhibitors of Ras palmitoylation have also been considered. However, the enzymology of Ras palmitoylation is complex and a better understanding of the specificity of the DHHC domain proteins that function as S-palmitoyltransferases remains to be achieved.

Farnesyl Containing Small Molecule Inhibitors of Ras Membrane Association

As detailed above, carboxyl terminal farnesylation of Ras proteins is critical for localization to the plasma membrane, and this localization is necessary for Ras binding to effector molecules in the various downstream signaling pathways. There is evidence that the insertion of the lipophilic prenyl moiety into the plasma (or other lipid bi-layer) membrane is not simply random, but that specific "prenyl receptors" facilitate prenylated protein binding [60]. Indeed, the data showing that inappropriately prenylated H- or K-Ras is mislocalized in the cell [61, 62] and the necessary presence of the farnesyl group for the binding of Ras to the effector c-Raf-1 [63–65] lend support to the hypothesis that prenylation also provides specificity for interaction partners. Therefore, ongoing work is focused on inhibiting the binding of farnesylated Ras to sites on the inner surface of the plasma membrane.

Two farnesyl isoprenoid containing small molecules have been described that are proposed to antagonize Ras function by competition for farnesyl binding (Figure 38.5b). Salirasib (FTS; S-trans,trans-farnesylthiosalicylic acid) is a farnesylcysteine mimetic, which selectively disrupts the association of chronically active Ras proteins with the plasma membrane [66]. Salirasib is proposed to compete with Ras for binding to membrane associated Ras escort proteins (galectins), which possess putative farnesyl binding domains, thereby dislodging Ras from the plasma membrane and disrupting effector signaling.

Galectin-1 has been shown to interact with mutant H-Ras and K-Ras, and that this interaction was required for membrane localization of the GTPases and subsequent transforming activity in human and rat epithelial cells [67, 68]. Kloog and colleagues demonstrated that the prenyl moiety is necessary for galectin-1 regulation of Ras activation by using site directed mutagenesis to create an L11A mutant in the putative prenyl binding domain of galectin-1 that was incapable of augmenting GTP loading of H-Ras and transforming oncogenic H-Ras transfected Rat-1 fibroblasts, and disrupted membrane association of mutant H-Ras [69]. The importance of the interaction in Ras driven transformation is illustrated by data showing that galectin-1 binding directs the signaling downstream of Ras from PI3K to c-Raf-1, and that inhibition of galectin-1 expression by antisense reversed this preference for c-Raf-1 activation [68]. There appears to be a preferential binding of galectins to the GTP bound form of Ras compared to Ras-GDP, and greater affinity of galectin-1 to activated H-Ras than K-Ras [68, 70]. Since there is evidence of specific proteins that both recruit activated forms of Ras proteins and guide the activation of transforming effector cascades, it is desirable to investigate potential inhibitors of this binding.

Salirasib blocks the membrane association of H-, K-, and N-Ras proteins in both transformed cells and cancer cells with oncogenic mutant Ras or hyperactivated wild-type Ras, including pancreatic, melanoma, glioblastoma, neuroblastoma, and neurofibromatosis cancer cells [71–78]. Bolstering the hypothesis that salirasib was specifically inhibiting activated Ras function, Kloog and colleagues demonstrated that signaling from three of the most studied effector pathways downstream of Ras, Raf-MEK-ERK [79], RalGEF-Ral [78], and PI3K-AKT [76–78, 80] could be blunted by treatment with salirasib. Several of the phenotypic changes attributed to aberrant Ras activation in cells were similarly inhibited by salirasib, including cell survival [81], proliferation [71, 73], and migration [80, 82]. In pancreatic and neurofibromatosis xenograft tumor models, tumor growth was inhibited by salirasib and was associated with a reduction of the abundance of Ras in the tumor tissue [78, 83]. Phase I clinical trials have shown that salirasib was well tolerated and several Phase I/II trials are ongoing.

Another putative Ras membrane binding inhibitor is also currently being developed. TLN-4601 (4,6,8-tri-hydroxy-10-(3,7,11-trimethyldodeca-2,6,10-trienyl)-5,10-dihydrodibenzo[b,e]; formerly ECO-4601) is a structurally novel farnesylated dibenzodiazepinone discovered from a *Micromonospora* sp., 046Eco-11 bacterium [84]. It was isolated independently in an antibacterial activity guided fractionation of fermentation extracts of a *Micromonospora* sp., DPJ12 and called diazepinomicin [85].

TLN-4601 was shown to have broad cytotoxic activity in micromolar concentrations when evaluated in the NCI 60 human tumor cell line panel (http://dtp.nci.nih.gov/docs/misc/common_files/cell_list.html). In contrast to salirasib, TLN-4601 has a benzodiazepinone moiety that binds peripheral and not central nervous system benzodiazepine receptors (PBR, CBR) *in vitro* [84]. PBR expression is upregulated in several tumor subtypes compared to normal tissue [86], with increased PBR ligand binding in several solid tumors including colon [87–89], brain [90], breast [91–93], prostate [86], ovary [94], and liver [95]. These studies suggest that the PBR might facilitate binding and uptake of TLN-4601 into tumor cells.

Like salirasib, it is speculated that TLN-4601 also competes with Ras for farnesyl binding proteins on membranes, and the displaced Ras is therefore susceptible to degradation. Preclinical cell culture studies support this mechanism.

TLN-4601 exhibited anti-tumor activity in rat C6 and human U-87MG glioma tumor xenograft models, as well as in MDA-MB-231 breast and PC-3 prostate human tumor xenograft models [84].

A Phase I/II trial for TLN-4601 was recently completed that examined the compound's safety, pharmacologic profile, and anti-tumor efficacy for the treatment of advanced solid tumors. TLN-4601 was found to be safe and well tolerated with no maximum tolerated dose attained. As with the experimental xenograft models, pharmacokinetic data indicated rapid clearance of TLN-4601 from the plasma upon termination of infusion, and that no accumulation of the drug occurred for subsequent dosing cycles. Based on promising results from this study and from preclinical glioma model experiments a Phase II trial has recently been initiated that will evaluate TLN-4601 as a monotherapy treatment for patients with recurrent/refractory glioblastoma multiforme.

TARGETING RAS EFFECTOR SIGNALING

Activated Ras binds preferentially to a spectrum of functionally diverse downstream effectors. Most are characterized by Ras binding (RBD) or Ras association (RA) domains that directly interact with Ras [3]. As described above, the Raf kinases are the best characterized effectors of Ras [96]. However, there exist at least 10 functionally distinct classes of Ras effectors, with evidence for Raf and four non-Raf effectors in Ras transformation (Figure 38.2b). Genetic experiments in mice argue that inhibition of any one effector pathway alone is sufficient to effectively block Ras driven tumor development. However, genetic ablation of a target is perhaps too blunt of a tool that may not accurately mimic pharmacologic inhibition of effector function. Furthermore, these analyses demonstrate tumor prevention rather than inhibition of an already developed tumor. Below we summarize the status of the development of inhibitors of Ras effector signaling.

Inhibitors of the Raf-MEK-ERK MAPK Cascade

The Raf serine/threonine kinases are the initiating signaling component of the ERK kinase cascade. The three Raf kinases (A-Raf, B-Raf, c-Raf-1) are highly conserved structurally and share the same substrate specificity, with MEK1 and MEK2 the only known substrates. Despite their similarities with one another, the highly related Raf isoforms exhibit distinct differences in their regulation and biological function [97, 98]. Furthermore, MEK independent functions have been described, although these activities remain poorly characterized. A number of structurally distinct classes of compounds have been developed as Raf and

TABLE 38.1 Small molecule Raf and MEK Inhibitors[1]

Agent	Company	Target	Status	Indication /target population
Sorafenib (BAY 43-09006; Nexavar®)	Bayer/Onyx	B-Raf, Raf-1[2]	Approved	Advanced RCC, unresectable HCC
			Phase II-III	CRC, NSCLC, pancreatic cancer, biliary tract cancer, HCC, glioblastoma, melanoma, thyroid, breast, neurofibromas, cervical cancer, prostate, Barrett's esophagus cancer, ovarian cancer, AML, multiple myeloma, myelodysplastic syndrome, as well as other solid and hematological malignances, Karposi's sarcoma, nasopharyngeal carcinoma, SCCHN
Raf265 (CHIR-265)	Novartis	A-Raf, B-Raf, c-Raf-1[3]	Phase I	Locally advanced or metastatic melanoma
PLX4032	Roche/Plexxikon	Mutant B-Raf(V600E)	Phase I	Melanoma
XL281	Exelixis	Raf	Phase I	NSCLC, CRC, papillary thyroid cancer
RO5126766	Hoffman-La Roche	Raf, MEK	Phase I	Solid tumors
PD0325901	Pfizer	MEK1, MEK2	Phase I–II	CRC, melanoma, breast
AZD6244 (ARRY-142886)	AstraZeneca/Array BioPharma	MEK1, MEK2	Phase II	Hepatocellular, AML, thyroid, ovarian
ARRY-438162	Array BioPharma	MEK1, MEK2	Phase II	Rheumatoid arthritis
XL518	Exelixis	MEK1, MEK2	Phase I	Solid tumors
GSK1120212	GlaxoSmithKline	MEK	Phase I	Solid tumors or lymphoma
RDEA119	Ardea Biosciences	MEK1, MEK2	Phase I	Advanced cancer
AZD8330 (ARRY-424704)	AstraZeneca/Array BioPharma	MEK1	Phase I	Advanced cancer

[1]Compiled from studies cited in reviews that we have referenced in the text, from company websites, and from www.clinicaltrials.gov; RCC, renal cell cancer; HCC, hepatocellular carcinoma; CRC, colorectal cancer, NSCLC, non-small cell lung cancer; AML, acute myelogenous leukemia; SCCHN, squamous cell carcinomas of the head and neck.
[2]Also activity for VEGFR-2, PDGFR-β, Flt-3, c-Kit, and FGFR-1.
[3]Also activity for VEGFR-2.

MEK kinase inhibitors, many of which have now entered clinical development [99] (Figures 38.4b, 38.5e, and 38.5f) (Table 38.1).

Raf Inhibitors

Sorafenib (tosylate salt of BAY 43-9006; Nexavar®), an orally available compound, is the first anti-Raf inhibitor to gain FDA approval in 2005 for advanced renal cell carcinoma (RCC) and later in 2007 it received approval for use in patients with unresectable hepatocellular carcinoma (HCC) (Figure 38.5e). Originally developed as an inhibitor of Raf-1 [100], sorafenib is a potent inhibitor of both wild-type and mutant B-Raf kinases *in vitro*. Crystallographic analyses determined that the inhibitor bound to the ATP binding pocket and prevented kinase activation, preventing substrate binding and phosphorylation [101].

In addition to its anti-Raf activity, sorafenib potently inhibits a diverse spectrum of additional protein kinases [102]. *BRAF* mutations are not common in RCC or HCC and it is currently unclear whether Raf is a key target of sorafenib in HCC [103]. Thus, it is believed that the antitumor activity of sorafenib is due largely to its inhibition of receptor tyrosine kinases involved in tumor angiogenesis (e.g., VEGFR-2, VEGFR-3, Flt-3, c-Kit, and FGFR-1).

In addition to RCC and HCC, several single agent and combination clinical studies are ongoing in NSCLC, prostate cancer, breast cancer, ovarian cancer, pancreatic cancer, melanoma, and hematological malignancies [104, 105].

RAF265 (formerly CHIR-265) is another orally bioavailable Raf inhibitor that is being investigated in Phase I clinical trials in locally advanced or metastatic melanoma. RAF265 inhibits all three Raf isoforms as well as mutated B-Raf (Table 38.1). Like sorafenib, RAF265 may also have anti-angiogenic activity through inhibition of VEGFR2.

PLX4032 is a potent and selective inhibitor of mutant B-Raf that is currently in Phase I clinical evaluation (Figure 38.5e). *In vitro* analysis against a panel of 65 non-Raf kinase showed PLX4032 is a highly selective inhibitor of B-Raf kinase activity, with an IC_{50} of 44 nM against V600E-mutant B-Raf [106], while only one kinase, BRK, showed inhibition in the nanomolar range (IC50 = 240 nM). Most of the kinases tested showed >100-fold higher IC_{50} than mutant Raf. In addition, cell culture experiments showed PLX4032 potently inhibited cell proliferation and MEK activation in melanoma and thyroid carcinoma cell lines harboring mutant B-Raf.

A fourth Raf kinase inhibitor, XL281, has also recently entered Phase I clinical testing (Table 38.1). XL281 is reported to be a potent and highly selective inhibitor of Raf kinases [107]. Data presented showed XL281 was generally well tolerated and pharmacodynamic analyses found that substantial modulation of Raf signaling was observed in tumor tissue, skin, and hair as indicated by decreases in the phosphorylation of MEK and ERK. Other Raf inhibitors are also in preclinical evaluation and should be entering clinical evaluation in the near future.

MEK Inhibitors

MEK1 and MEK2 are closely related dual specificity kinases, capable of phosphorylating both serine/threonine and tyrosine residues of their substrates, p44 ERK1 and p42 ERK2. They are the only known catalytic substrates of Raf kinases. The fact that ERK1/2 are the only known substrates of MEK1/2, when coupled with the observation that ERK is commonly activated in both tumor cell lines and patient tumors, has contributed to strong interest in developing pharmacological inhibitors of MEK as a means to block ERK activation [108].

In contrast to sorafenib and RAF265, small molecule inhibitors of MEK1/2 are highly specific protein kinase inhibitors. Although the first two MEK inhibitors, PD98059 and U0126, were highly specific [109] they lacked the pharmaceutical properties needed to be successful clinical candidates. Nonetheless, these compounds have been invaluable academic research tools for dissecting the MEK-ERK pathway and have provided enormous insight into the importance of ERK MAPK signaling in cancer [4, 110].

The first MEK inhibitor to enter clinical trials was CI-1040 (PD184352), an orally active, highly potent and selective inhibitor of MEK1 and MEK2 [111]. Preclinical evaluation found that CI-1040 inhibited the growth of human colon cancer cells and human melanoma cells in athymic nude mice [111,112]. Subsequent Phase I and II clinical trials reported the most common toxicities were mild skin rash, diarrhea, and fatigue [113,114]. During the Phase I trial, a partial response was seen in one patient with pancreatic cancer and 25 percent of patients with a variety of tumors had stable disease for greater than 3 months [113]. Tumor tissues from treated patients showed significant reduction in activated phosphorylated ERK, indicating that the target was inhibited. These promising results prompted a Phase II study in patients with advanced NSCLC, breast cancer, CRC, and PDAC. Unfortunately, the results of this trial were negative and CI-1040 was determined to have poor pharmacokinetic properties [114].

In contrast to the majority of protein kinase inhibitors, MEK inhibitors are non-ATP competitive inhibitors, which may account for their highly selective properties. Structural studies with an analog of CI-1040 in complex with MEK1 or MEK2 showed inhibitor binding did not perturb ATP binding, and, instead, bound to a unique inhibitor binding pocket adjacent to the ATP binding site [115]. Inhibitor binding locked MEK in a catalytically inactive conformation.

PD0325901 is a derivative of CI-1040 where several slight modifications to the chemical structure have resulted in more than a 50-fold increase in potency against MEK1/2, improved bioavailability, and longer duration of target suppression compared to CI-1040 [110] (Figure 38.5f). Anti-tumor activity for PD0325901 was demonstrated for a variety of tumor xenografts and this inhibitor has been evaluated in Phase I/II clinical trials with a focus on tumors expected to have activated ERK MAPK signaling [116, 117]. However, current clinical development has been suspended likely due to poor side effect profile.

AZD6244 (ARRY-142886) is an orally bioavailable benzimidazole derivative known to potently inhibit MEK1/2 *in vitro* and in cell based assays [118, 119] (Figure 38.5f). Like other MEK inhibitors, AZD6244 is ATP non-competitive. Preclinical evaluation of AZD6244 showed anti-tumor activity in several human xenograft models including colon, pancreas, breast, NSCLC, and melanoma [120]. Additionally, AZD6244 anti-tumor activity was found to correlate with suppression of ERK activation, which further validates that its mechanism of action is MEK dependent. Results from preclinical analysis have been extremely promising and thus AZD6244 has moved into clinical development. Recently, initial results of a first in human dose ranging study to assess the pharmacokinetics, pharmacodynamics, and toxicities of AZD6244 in patients with advanced solid tumors, concluded that

AZD6244 is well tolerated, and the most common treatment related adverse events were rash, diarrhea, nausea, fatigue, peripheral edema, and vomiting [121].

RDEA119 is a potent, non-ATP competitive, highly selective inhibitor of MEK1/2 and preclinical testing showed RDEA119 inhibited ERK phosphorylation, anchorage dependent growth, and anchorage independent growth in a variety of cancer cell lines at nanomolar concentrations. Additionally, analyses in tumor xenograft models indicated significant tumor growth delay and even tumor regression in some models treated orally with REA119. RDEA119 is currently being evaluated alone in a Phase I trial for advanced solid tumors and in Phase I/II clinical trial in combination with Nexavar in advanced cancer patients.

XL518 is a potent, selective, orally bioavailable inhibitor of MEK1 that inhibits proliferation and stimulates apoptosis in a variety of human tumor cell lines. *In vitro* analyses showed potent activity for MEK1 (0.95 nM), weak activity for MEK2 (199 nM) and no activity for a panel of 103 serine/threonine or tyrosine protein kinases [122]. In preclinical xenograft models, oral administration of XL518 resulted in sustained inhibition of ERK in tumor tissue, but not brain tissue, leading to tumor growth inhibition and regression at well tolerated doses.

Several other MEK inhibitors are also entering early phase clinical trials (Table 38.1). Patients with solid tumors and lymphoma are now being recruited for evaluation of the GSK1120212 MEK inhibitor. A similar Phase I study is planned for RO5126766, a dual Raf and MEK inhibitor.

Finally, one novel class of small molecule inhibitors of the Raf effector pathway are MCP1 and derivatives that have been described and shown to have anti-tumor activity in preclinical cell culture studies [123] (Figure 38.5j). MCP1 was identified as an inhibitor of Ras interaction with Raf in a yeast two-hybrid based screen and additional more potent analogs were identified (e.g., MCP110). Mouse model evaluation of MCP110 found anti-tumor activity against various human tumor xenografts [124]. While the exact mechanism by which MCP compounds function is currently unclear, the inability of MCP to block a second Ras effector pathway [123] and the ability to block the growth of *BRAF* mutation positive melanomas [125] support a mechanism by which it antagonizes Raf function.

INHIBITORS OF PI3K-AKT-MTOR SIGNALING

The p110 catalytic subunit of class I PI3Ks (p110α, β, δ, and γ) were found to be effectors of Ras and shown to be required for Ras transformation [126, 127]. A major activity of PI3K is the conversion of phosphoinositide(4,5) bisphosphate to phosphoinositide(3,4,5)bisphosphate (PIP3). Membrane associated PIP3 promotes the activation of the AKT serine/threonine kinase as well as other signaling proteins that include GEFs for Rho family GTPases. One downstream activity of AKT involves activation of the mTOR serine/threonine kinase. In addition to activation by Ras, the PI3K-AKT-mTOR pathway is deregulated by a variety of mechanisms in human cancers. Hence, these three components have been attractive targets for anticancer drug discovery, with many inhibitors recently entering clinical trial analyses [128, 129] (Table 38.2).

PI3K Inhibitors

The competitive ATP binding PI3K inhibitors, wortmannin and LY294002, have been used widely to study the mechanisms of the PI3K and other signal transduction pathways [130]. Improved and more potent derivatives of wortmannin are currently in clinical trials. PX-866, an orally bioavailable semisynthetic derivative of wortmannin is currently in Phase I studies for advanced solid tumors (Table 38.2). PX-866 was selected as part of a screen of semisynthetic viridans for its ability to inhibit PI3K compared to wortmannin and anchorage independent growth in HT-29 colorectal cancer cells [131]. Initial preclinical studies also demonstrated *in vivo* activity against human tumor cell line xenografts [131]. Further studies have demonstrated that PX-866 in combination with gefitinib, a small molecule inhibitor of EGFR, was able to further inhibit growth of A549 cell line xenografts compared to either one alone [132]. In addition, PX-866 has been shown to inhibit anchorage independent growth of U87 glioblastoma cells in a three-dimensional spheroid model [133]. Finally, PX-866 inhibits the number of lung nodules mutant *KRAS* driven mouse model of lung tumorigenesis [134].

XL147 is also an orally bioavailable PI3K inhibitor currently in Phase I clinical trials as monotherapy or in combination with erlotinib or paclitaxel and carboplatin in patients with solid tumors. Although the structure has not been disclosed, it is a selective reversible ATP competitive inhibitor of PI3K. XL147 demonstrated growth inhibition of MCF-7, A549, ovarian cancer, and glioma cell line xenografts [135]. In addition, it was found to also exhibit anti-angiogenic properties in a vascular endothelial growth factor driven tubule formation assay [136]. Combination treatment with rapamycin, an EGFR inhibitor or paclitaxel and carboplatin, increased apoptosis [136], leading to the Phase I trial designs mentioned above. Finally, GDC-0941 is another ATP competitive inhibitor of class I PI3Ks that entered Phase I studies (Figure 38.5g); however, currently no active or recruiting clinical trials are listed (www.clinicaltrials.gov).

XL765 [137, 138], BGT226 (http://www.novartisoncology.com/research innovation/pipeline/BGT226.jsp) and NVP-BEZ235 are dual inhibitors of PI3K and mTOR (Figure 38.5g and Table 38.2). Preclinical studies of

TABLE 38.2 Small molecule PI3K and AKT Inhibitors[1]

Agent	Company	Target	Status	Indication /target population[2]
BGT226	Novartis	PI3K	Phase I–II	Solid tumors
BEZ235	Novartis	PI3K	Phase I–II	Solid tumors
PX-866	ProlX Pharmaceuticals	PI3K	Phase I	Solid tumors
XL765	Exelixis	PI3K and mTOR	Phase I	Solid tumors, glioma
XL147	Exelixis	PI3K	Phase I	Solid tumors
NVP-BEZ235	Novartis	PI3K and mTOR	Phase I–II	
GDC-0941	Genentech/Piramed	PI3K	Not in trial	
SF1126	Semafore	PI3K and mTOR	Phase I (completed)	Solid tumors
GSK690693	GlaxoSmithKline	AKT	Phase I	Lymphoma (terminated); hematologic malignancies (not yet open)
Perifosine (KRX-04010)	Keryx Biopharmaceuticals	AKT[3]	Phase I–II	Solid tumors, hematologic malignancies
Sirolimus (rapamycin; Rapamune®)	Wyeth	mTOR	Phase II	Neurofibromatosis type I and plexiform neurofibromas, solid tumors, NSCLC, RCC, tuberous sclerosis and lymphangioleiomyomatosis, ALL, CML, Burkitt's Lymphoma, pancreatic cancer
Deforolimus (AP23573; MK-8669)	Ariad Pharmaceuticals/ Merck	mTOR	Phase II–III	Sarcomas, hematologic cancers, endometrial cancer, breast cancer
Everolimus (RAD001; Certican®)	Novartis	mTOR	Phase I–II	Breast cancer, CRC, RCC, HCC, prostate, endometrial, NSCLC, mesotheliomas, lymphoma, pancreatic, tuberous sclerosis and giant cell astrocytoma or lymphangioleiomyomatosis, gastric, CML, AML
Temsirolimus (CCI-779; Torisel®)	Wyeth	mTOR	Approved	RCC
			Phase II	NSCLC, solid tumors, chronic lymphocytic leukemia, prostate, RCC, glioblastoma, multiple myeloma
OSI-027	OSI Pharmaceuticals	mTOR	Phase I	Solid tumors, lymphoma

[1]Compiled from studies cited in reviews that we have referenced in the text, from company websites, and from www.clinicaltrials.gov.
[2]Abbreviations: NSCLC, non-small cell lung cancer; RCC, renal cell cancer, ALL, acute lymphocytic leukemia; CML, chronic myelogenous leukemia; CRC, colorectal cancer; HCC, hepatocellular carcinoma; AML, acute myelogenous leukemia.
[3]Also activity for ERK inhibition and JNK activation.

XL765 demonstrated continued tumor regression after cessation of treatment compared to rapamycin alone [139]. NVP-BEZ235 has been studied in a genetically engineered mouse model of PTEN deletion and *KRAS* activation driven ovarian cancer model [140] as well as primary pancreatic cancer xenografts [141] demonstrating activity as a single agent. XL765 is currently in Phase I trials as monotherapy and in combination with erlotinib for solid tumors. In addition, it is being tested in combination with temozolamide for malignant gliomas. NVP-BEZ235 and BGT226 are in Phase I trails for advanced solid tumors.

AKT Inhibitors

Two inhibitors of AKT are currently in clinical trials. GSK690693 is a novel ATP competitive pan-AKT kinase inhibitor [142] (Figure 38.5i). Currently in Phase II clinical trials for relapsed or refractory hematologic malignancies, GSK690693 requires parenteral administration. Preclinical studies of GSK690693 demonstrated anti-tumor activity against breast, ovarian, and prostate cancer cell line xenografts [143].

Perifosine (KRX-0401, NSC 639966) is an orally available alkylphospholipid that inhibits the translocation of AKT to the plasma membrane and its subsequent phosphorylation [144]. This lipid based inhibitor has been studied extensively for its anti-neoplastic properties both as a single agent and in combination with other targeted or chemotherapeutic agents in Phase I and II clinical trials [128] with 24 currently active trials. Unfortunately, despite promising results from preclinical studies, no study of perifosine as monotherapy demonstrated significant efficacy for a wide range of tumor types [145–153]. Therefore it is being evaluated in combination with traditional chemotherapeutics as well as tyrosine kinase inhibitors for activity against most solid tumors.

mTOR Inhibitors

Four mTOR inhibitors are currently in clinical trials. Sirolimus (rapamycin; Rapamune®), the prototype is the best studied. It was originally isolated as an antifungal metabolite produced by the bacterium *Streptomyces hygroscopicus* in a soil sample from Easter Island (Rapa Nui) [154] and early on was found to have anti-tumor effects in colorectal xenografts [155]. Although FDA approved for use as an immunosuppressing agent in transplantation, there has been a resurgence of interest and excitement in the use of sirolimus and its derivatives as targeted therapies in cancer. Several Phase I and II studies of sirolimus as monotherapy and combination therapy are currently underway for solid tumors and hematological malignancies. In addition, sirolimus is being studied as a preventative agent for post-transplant related skin cancers.

Temsirolimus (CCI-779; Torisel®), the only mTOR inhibitor with FDA approval as anticancer therapy, has demonstrated a median overall survival benefit of 3.6 months for patients with poor prognostic RCC compared to interferon therapy [156, 157]. As a result, temsirolimus is now recommended as first-line therapy for patients with RCC and poor prognostic risk factors [158]. In preclinical studies of mantle cell lymphoma, temsirolimus induced cell cycle arrest and autophagy [159], prompting its evaluation in Phase II studies. A Phase II clinical trial of mantle cell lymphoma demonstrated an overall response rate of 41 percent with 50 percent of the patients having been refractory to previous therapies [160], leading to its current evaluation

in Phase III studies. Results of temsirolimus in advanced or metastatic breast cancer have been less favorable leading to early termination of the Phase III trial. Temsirolimus continues to be studied as a single agent for head and neck squamous cell and prostate carcinomas, NSCLC, gynecologic malignancies and B cell lymphoma, and chronic lymphocytic leukemia. Studies in small cell lung cancer have not been promising and are no longer active [161]. Preclinical studies have suggested temsirolimus sensitivity in EGFR resistant squamous cell carcinomas of the head and neck [162]. In addition, temsirolimus demonstrated activity against prostate cancer cell line xenografts [163]. Finally, many trials with temsirolimus in combination with either other targeted therapeutics or traditional chemotherapeutics are underway.

In Phase III studies of metastatic renal cell carcinoma, everolimus (RAD001; Certican®) demonstrated significant improvement in progression free survival [164]. This study is ongoing and awaiting analysis of its secondary end point overall survival [165]. Phase II studies of everolimus in breast cancer have not demonstrated responses as impressive with an overall response rate of 68 percent compared to 59 percent (p = 0.062) in patients treated with everolimus plus the aromatose inhibitor letrozole, compared to letrozole alone [166]. However these early results are promising and further studies of everolimus in combination with other agents is being studied for breast cancer. Early preclinical studies of NSCLC using temsirolimus suggested that mTOR inhibition may restore cisplatin sensitivity in cisplatin resistant cell lines [167, 168]. However, in Phase II studies of NSCLC using everolimus as monotherapy, this has not been confirmed [169]. Phase II studies of everolimus in combination with gefitinib in smokers have shown promising results with a 17 percent response rate, which is higher than expected for this cohort [170]. Trials of combination therapy for NSCLC are currently ongoing. Results of everolimus in small cell lung cancer have been less promising, although Phase II studies are still ongoing [171]. Phase I–III studies are ongoing with everolimus as monotherapy or combination therapy for a wide range of solid tumors and hematologic malignancies.

Studies of deferolimus (AP23573; MK-8669), a non-pro drug derivative of rapamycin, have been more focused thus far. Phase II studies in hematologic malignancies have shown promise with 40 percent of heavily pretreated patients achieving stable disease, but only 10 percent had a partial response [172]. Preclinical studies of deferolimus as monotherapy in leiomyosarcoma xenografts demonstrated 67 percent tumor growth inhibition [173]. Studies in endometrial and sarcoma cancer cell lines demonstrated an additive effect of deferolimus with doxorubicin, carboplatin, or paclitaxel [174]. Deferolimus is reported to have promising activity in endometrial cancer and sarcomas with Phase II trials of endometrial cancer and a Phase III trial of sarcoma ongoing [173].

Unlike its predecessors, sirolimus, temsirolimus, everolimus, and deferolimus, OSI-027 is an mTOR inhibitor that inhibits both the TORC1 and TORC2 complexes of mTOR. mTORC2 is a second and distinct mTOR protein complex that is less understood [175, 176]. However recent studies have shown that mTORC2 phosphorylates and activates AKT [177], suggesting that the dual targeting of mTORC1 and mTORC2 will block additional signaling pathways important in cancer. A Phase I study of OSI-027 in patients with solid tumors and lymphoma is underway.

INHIBITORS OF OTHER RAS EFFECTOR PATHWAYS

Tiam1-Rac Pathway

Tiam1 was identified originally as a T cell invasion and metastasis gene [178]. Tiam1 is a member of the Dbl family of RhoGEFs and specifically activates the Ras related Rac small GTPase. Subsequent studies identified a Ras binding domain in Tiam1 that facilitated its function as a Ras effector [179]. The importance of Tiam1 in Ras mediated oncogenesis was demonstrated by the reduced skin tumor induction seen by carcinogen induced H-Ras activation in Tiam1 deficient mice. The importance of Rac in Ras transformation has been demonstrated in cell culture studies where dominant-negative Rac impaired Ras transformation of rodent fibroblasts [180, 181]. Additionally, conditional loss of Rac1 impaired tumor formation in a mutant *KRAS* driven lung cancer mouse model [182].

To date, two small molecule inhibitors of Rac have been described and evaluated in preclinical studies (Figure 38.5k). First, Zheng and colleagues took a rational design approach to identify a Rac inhibitor [183]. NSC23766 was identified by a structure based virtual screening of compounds that fit into a surface groove of Rac1 known to be critical for GEF specification. NSC23766 inhibited Rac1 binding and activation by the Rac specific GEF Trio or Tiam1 *in vitro*. In cell based analyses, it potently blocked serum or platelet derived growth factor induced Rac1 activation and lamellipodia formation without affecting the activity of endogenous Cdc42 or RhoA. NSC23766 did not reduce the activity of another RacGEF, Vav, or the activity of GEFs for other Rho family GTPases. When applied to human prostate cancer PC-3 cells, it was able to inhibit the proliferation, anchorage independent growth, and invasion phenotypes that require the endogenous Rac1 activity. Finally, continuous NSC23766 treatment was used in mice transplanted with BCR-Abl expressing primary mouse or human CML cells and inhibition of leukemogenesis was seen, suggesting that Rac is a therapeutic target for BCR-Abl positive leukemias [184].

EHT 1864 is another Rac inhibitor that works by a mechanism distinct from that of NSC23766 [185]. Our studies determined that EHT 1864 specifically inhibited Rac1 dependent platelet derived growth factor induced lamellipodia formation [186]. We completed biochemical analyses that showed that EHT 1864 possesses high affinity binding to Rac1, as well as the related Rac1b, Rac2, and Rac3 isoforms, and this association promoted the loss of bound nucleotide, inhibiting both guanine nucleotide association and Tiam1 RacGEF stimulated exchange factor activity *in vitro*. EHT 1864 therefore places Rac in an inert and inactive state, preventing its engagement with downstream effectors. Finally, we evaluated the ability of EHT 1864 to block Rac dependent growth transformation, and we determined that EHT 1864 potently blocked transformation caused by constitutively activated Rac1, as well as Rac dependent transformation caused by Tiam1 or Ras. No *in vivo* evaluation of EHT 1864 anti-tumor activity in mouse models of cancer has been reported.

RalGEF-Ral Pathway

There are four human RalGEFs that serve as effectors of Ras. RalGEFs are activators of the highly related Ras-like RalA and RalB small GTPases (82 percent sequence identity) [187]. Similar with Ras, Ral GTPases function as GDP/GTP regulated switches in signal transduction. Activated RalGTP can interact with multiple downstream effectors, with two components of the exocyst complex (Sec5 and Exo84) and RalBP1/RLIP (a GAP for the Rac and Cdc42 Rho family GTPases) the best characterized, and involved in regulation of exocytosis and endocytosis, respectively. Additionally, RalGEF-Ral signaling can activate various transcription factors that include the ternary complex factor, Jun, AFX (FOXO4), STAT3, and the NFκB.

The RalGEF-Ral pathway was characterized initially to play a relatively minor role in Ras transformation of rodent fibroblasts [188]. However, subsequent studies by Counter and colleagues establish a very significant role for this effector pathway in Ras transformation of human cells [189]. In particular, a significant role for Ral GTPases in pancreatic cancer has been established [190, 191]. Additionally, studies of bladder and prostate cancer support the role of RalGEF-Ral signaling in tumor invasion and metastasis [192, 193]. Finally mouse model studies showed that homozygous deletion of RalGDS (a RalGEF) caused resistance to Ras induced skin tumor formation [23]. Consequently, there is increasing interest in targeting this pathway for novel anti-Ras strategies for cancer treatment [194, 195]. Our recent studies support the possibility that inhibitors of GGTase-I (GGTI) can be effective inhibitors of Ral GTPases in oncogenesis [196] (Figure 38.5c). However, as with FTIs, since other GGTase-I substrates (e.g., RhoA, RhoC, Rac) are involved in oncogenesis, GTTI anti-tumor activity may also involve inhibition of non-Ral targets. Finally, a recent study identified RalA as a substrate for Aurora A [197]. Since Aurora A phosphorylation of RalA was important for Aurora A induced cellular motility and transformation perhaps inhibitors of Aurora A, currently in Phase I clinical trial analyses, may be effective inhibitors of RalA function.

CONCLUSIONS AND FUTURE DIRECTIONS

Despite the lack of success of past efforts to develop anti-Ras inhibitors for cancer treatment, ongoing studies support promising new directions for these efforts. One potential limitation of the current targets of Ras is the fact that none are linked exclusively to Ras function. Hence, as with FTIs, off-target activities of these inhibitors will be a potential concern. An additional potential limitation regarding inhibitors of effector signaling is fact that multiple effector pathways promote Ras transformation. Hence, whether inhibition of any one effector pathway alone will be effective for blocking Ras mediated oncogenesis is a logical concern. Future efforts will need to consider concurrent inhibition of multiple effector pathways. An additional complication is that each effector pathway is not a simple linear pathway, so effective inhibition of any one effector signaling network may also require inhibition at multiple points. In summary, with renewed appreciation and interest in Ras as a therapeutic target for cancer treatment, and with promising leads for indirect approaches for blocking Ras function, there remains strong optimism that anti-Ras therapies will finally reach the clinic.

ACKNOWLEDGEMENTS

We apologize to all colleagues whose work could not be cited due to space limitations. Research in the authors' laboratory was supported by grants from the National Institutes of Health (CA042978, CA106991, and CA67771) and from the Emerald Foundation.

REFERENCES

1. Karnoub AE, Weinberg RA. Ras oncogenes: split personalities. *Nat Rev Mol Cell Biol* 2008;**9**:517–31.
2. Vetter IR, Wittinghofer A. The guanine nucleotide-binding switch in three dimensions. *Science* 2001;**294**:1299–304.
3. Repasky GA, Chenette EJ, Der CJ. Renewing the conspiracy theory debate: does Raf function alone to mediate Ras oncogenesis? *Trends Cell Biol* 2004;**14**:639–47.
4. Cox AD, Der CJ. Ras family signaling: therapeutic targeting. *Cancer Biol Ther* 2002;**1**:599–606.
5. Rowinsky EK. Lately, it occurs to me what a long, strange trip it's been for the farnesyltransferase inhibitors. *J Clin Oncol* 2006;**24**:2981–4.
6. Sjoblom T, Jones S, Wood LD, et al. The consensus coding sequences of human breast and colorectal cancers. *Science* 2006;**314**:268–74.
7. Wood LD, Parsons DW, Jones S, et al. The genomic landscapes of human breast and colorectal cancers. *Science* 2007;**318**:1108–13.
8. Parsons DW, Jones S, Zhang X, et al. An integrated genomic analysis of human glioblastoma multiforme. *Science* 2008;**321**:1807–12.
9. Jones S, Zhang X, Parsons DW, et al. Core signaling pathways in human pancreatic cancers revealed by global genomic analyses. *Science* 2008;**321**:1801–6.
10. Hezel AF, Kimmelman AC, Stanger BZ, et al. Genetics and biology of pancreatic ductal adenocarcinoma. *Genes Dev* 2006;**20**:1218–49.
11. Yeh JJ, Der CJ. Targeting signal transduction in pancreatic cancer treatment. *Expert Opin Ther Targets* 2007;**11**:673–94.
12. Raponi M, Winkler H, Dracopoli NC. KRAS mutations predict response to EGFR inhibitors. *Curr Opin Pharmacol* 2008;**8**:413–18.
13. Eberhard DA, Johnson BE, Amler LC, et al. Mutations in the epidermal growth factor receptor and in KRAS are predictive and prognostic indicators in patients with non-small-cell lung cancer treated with chemotherapy alone and in combination with erlotinib. *J Clin Oncol* 2005;**23**:5900–9.
14. Massarelli E, Varella-Garcia M, Tang X, et al. KRAS mutation is an important predictor of resistance to therapy with epidermal growth factor receptor tyrosine kinase inhibitors in non-small-cell lung cancer. *Clin Cancer Res* 2007;**13**:2890–6.
15. Zhu CQ, da Cunha Santos G, Ding K, et al. Role of KRAS and EGFR as biomarkers of response to erlotinib in National Cancer Institute of Canada Clinical Trials Group Study BR.21. *J Clin Oncol* 2008;**26**:4268–75.
16. Lievre A, Bachet JB, Le Corre D, et al. KRAS mutation status is predictive of response to cetuximab therapy in colorectal cancer. *Cancer Res* 2006;**66**:3992–5.
17. Amado RG, Wolf M, Peeters M, et al. Wild-type KRAS is required for panitumumab efficacy in patients with metastatic colorectal cancer. *J Clin Oncol* 2008;**26**:1626–34.
18. Lievre A, Bachet JB, Boige V, et al. KRAS mutations as an independent prognostic factor in patients with advanced colorectal cancer treated with cetuximab. *J Clin Oncol* 2008;**26**:374–9.
19. Mitin N, Rossman KL, Der CJ. Signaling interplay in Ras superfamily function. *Curr Biol* 2005;**15**:R563–74.
20. Campbell PM, Singh A, Williams FJ, et al. Genetic and pharmacologic dissection of Ras effector utilization in oncogenesis. *Methods Enzymol* 2006;**407**:195–217.
21. Malliri A, van der Kammen RA, Clark K, et al. Mice deficient in the Rac activator Tiam1 are resistant to Ras-induced skin tumours. *Nature* 2002;**417**:867–71.
22. Bai Y, Edamatsu H, Maeda S, et al. Crucial role of phospholipase Cepsilon in chemical carcinogen-induced skin tumor development. *Cancer Res* 2004;**64**:8808–10.
23. Gonzalez-Garcia A, Pritchard CA, Paterson HF, et al. RalGDS is required for tumor formation in a model of skin carcinogenesis. *Cancer Cell* 2005;**7**:219–26.
24. Sebti SM, Der CJ. Opinion: searching for the elusive targets of farnesyltransferase inhibitors. *Nat Rev Cancer* 2003;**3**:945–51.
25. Willumsen BM, Christensen A, Hubbert NL, et al. The p21 ras C-terminus is required for transformation and membrane association. *Nature* 1984;**310**:583–6.
26. Hancock JF, Magee AI, Childs JE, Marshall CJ. All ras proteins are polyisoprenylated but only some are palmitoylated. *Cell* 1989;**57**:1167–77.
27. Jackson JH, Cochrane CG, Bourne JR, et al. Farnesol modification of Kirsten-ras exon 4B protein is essential for transformation. *Proc Natl Acad Sci U S A* 1990;**87**:3042–6.

28. Reiss Y, Goldstein JL, Seabra MC, et al. Inhibition of purified p21ras farnesyl:protein transferase by Cys-AAX tetrapeptides. *Cell* 1990;**62**:81–8.

29. Cox AD, Der CJ. Farnesyltransferase inhibitors: promises and realities. *Curr Opin Pharmacol* 2002;**2**:388–93.

30. Van Buskirk R, Dowling JE. Isolated horizontal cells from carp retina demonstrate dopamine-dependent accumulation of cyclic AMP. *Proc Natl Acad Sci U S A* 1981;**78**:7825–9.

31. Whyte DB, Kirschmeier P, Hockenberry TN, et al. K- and N-Ras are geranylgeranylated in cells treated with farnesyl protein transferase inhibitors. *J Biol Chem* 1997;**272**:14,459–14,464.

32. Rowell CA, Kowalczyk JJ, Lewis MD, Garcia AM. Direct demonstration of geranylgeranylation and farnesylation of Ki-Ras in vivo. *J Biol Chem* 1997;**272**:14,093–14,097.

33. Cox AD, Hisaka MM, Buss JE, Der CJ. Specific isoprenoid modification is required for function of normal, but not oncogenic, Ras protein. *Mol Cell Biol* 1992;**12**:2606–15.

34. Hancock JF, Cadwallader K, Paterson H, Marshall CJ. A CAAX or a CAAL motif and a second signal are sufficient for plasma membrane targeting of ras proteins. *EMBO J* 1991;**10**:4033–9.

35. Cohen SJ, Ho L, Ranganathan S, et al. Phase II and pharmacodynamic study of the farnesyltransferase inhibitor R115777 as initial therapy in patients with metastatic pancreatic adenocarcinoma. *J Clin Oncol* 2003;**21**:1301–6.

36. Van Cutsem E, van de Velde H, Karasek P, et al. Phase III trial of gemcitabine plus tipifarnib compared with gemcitabine plus placebo in advanced pancreatic cancer. *J Clin Oncol* 2004;**22**:1430–8.

37. Macdonald JS, McCoy S, Whitehead RP, et al. A phase II study of farnesyl transferase inhibitor R115777 in pancreatic cancer: a Southwest oncology group (SWOG 9924) study. *Invest New Drugs* 2005;**23**:485–7.

38. Reid TS, Terry KL, Casey PJ, Beese LS. Crystallographic analysis of CaaX prenyltransferases complexed with substrates defines rules of protein substrate selectivity. *J Mol Biol* 2004;**343**:417–33.

39. Basso AD, Mirza A, Liu G, et al. The farnesyl transferase inhibitor (FTI) SCH66336 (lonafarnib) inhibits Rheb farnesylation and mTOR signaling. Role in FTI enhancement of taxane and tamoxifen antitumor activity. *J Biol Chem* 2005;**280**:31,101–31,108.

40. Ellis CA, Vos MD, Howell H, et al. Rig is a novel Ras-related protein and potential neural tumor suppressor. *Proc Natl Acad Sci U S A* 2002;**99**:9876–81.

41. Luo RZ, Fang X, Marquez R, et al. ARHI is a Ras-related small G-protein with a novel N-terminal extension that inhibits growth of ovarian and breast cancers. *Oncogene* 2003;**22**:2897–909.

42. Elam C, Hesson L, Vos MD, et al. RRP22 is a farnesylated, nucleolar, Ras-related protein with tumor suppressor potential. *Cancer Res* 2005;**65**:3117–25.

43. Winter-Vann AM, Casey PJ. Post-prenylation-processing enzymes as new targets in oncogenesis. *Nat Rev Cancer* 2005;**5**:405–12.

44. Kato K, Cox AD, Hisaka MM, et al. Isoprenoid addition to Ras protein is the critical modification for its membrane association and transforming activity. *Proc Natl Acad Sci U S A* 1992;**89**:6403–7.

45. Kim E, Ambroziak P, Otto JC, et al. Disruption of the mouse Rce1 gene results in defective Ras processing and mislocalization of Ras within cells. *J Biol Chem* 1999;**274**:8383–90.

46. Bergo MO, Leung GK, Ambroziak P, et al. Targeted inactivation of the isoprenylcysteine carboxyl methyltransferase gene causes mislocalization of K-Ras in mammalian cells. *J Biol Chem* 2000;**275**:17,605–17,610.

47. Bergo MO, Ambroziak P, Gregory C, et al. Absence of the CAAX endoprotease Rce1: effects on cell growth and transformation. *Mol Cell Biol* 2002;**22**:171–81.

48. Bergo MO, Gavino BJ, Hong C, et al. Inactivation of Icmt inhibits transformation by oncogenic K-Ras and B-Raf. *J Clin Invest* 2004;**113**:539–50.

49. Roberts PJ, Mitin N, Keller PJ, et al. Rho Family GTPase modification and dependence on CAAX motif-signaled posttranslational modification. *J Biol Chem* 2008;**283**:25,150–25,163.

50. Wahlstrom AM, Cutts BA, Liu M, et al. Inactivating Icmt ameliorates K-RAS-induced myeloproliferative disease. *Blood* 2008;**112**:1357–65.

51. Wahlstrom AM, Cutts BA, Karlsson C, et al. Rce1 deficiency accelerates the development of K-RAS-induced myeloproliferative disease. *Blood* 2007;**109**:763–8.

52. Winter-Vann AM, Baron RA, Wong W, et al. A small-molecule inhibitor of isoprenylcysteine carboxyl methyltransferase with antitumor activity in cancer cells. *Proc Natl Acad Sci U S A* 2005;**102**:4336–41.

53. Wang M, Tan W, Zhou J, et al. A small molecule inhibitor of isoprenylcysteine carboxymethyltransferase induces autophagic cell death in PC3 prostate cancer cells. *J Biol Chem* 2008;**283**:18,678–18,684.

54. Winter-Vann AM, Kamen BA, Bergo MO, et al. Targeting Ras signaling through inhibition of carboxyl methylation: an unexpected property of methotrexate. *Proc Natl Acad Sci U S A* 2003;**100**:6529–34.

55. Buchanan MS, Carroll AR, Fechner GA, et al. Spermatinamine, the first natural product inhibitor of isoprenylcysteine carboxyl methyltransferase, a new cancer target. *Bioorg Med Chem Lett* 2007;**17**:6860–3.

56. Buchanan MS, Carroll AR, Fechner GA, et al. Aplysamine 6, an alkaloidal inhibitor of Isoprenylcysteine carboxyl methyltransferase from the sponge Pseudoceratina sp. *J Nat Prod* 2008;**71**:1066–7.

57. Buchanan MS, Carroll AR, Fechner GA, et al. Small-molecule inhibitors of the cancer target, isoprenylcysteine carboxyl methyltransferase, from Hovea parvicalyx. *Phytochemistry* 2008;**69**:1886–9.

58. Manandhar SP, Hildebrandt ER, Schmidt WK. Small-molecule inhibitors of the Rce1p CaaX protease. *J Biomol Screen* 2007;**12**:983–93.

59. Porter SB, Hildebrandt ER, Breevoort SR, et al. Inhibition of the CaaX proteases Rce1p and Ste24p by peptidyl (acyloxy)methyl ketones. *Biochim Biophys Acta* 2007;**1773**:853–62.

60. Marshall CJ. Protein prenylation: a mediator of protein–protein interactions. *Science* 1993;**259**:1865–6.

61. Cox AD, Der CJ. The ras/cholesterol connection: implications for ras oncogenicity. *Crit Rev Oncog* 1992;**3**:365–400.

62. Niv H, Gutman O, Henis YI, Kloog Y. Membrane interactions of a constitutively active GFP-Ki-Ras 4B and their role in signaling. Evidence from lateral mobility studies. *J Biol Chem* 1999;**274**:1606–13.

63. Drugan JK, Khosravi-Far R, White MA, et al. Ras interaction with two distinct binding domains in Raf-1 may be required for Ras transformation. *J Biol Chem* 1996;**271**:233–7.

64. Luo Z, Diaz B, Marshall MS, Avruch J. An intact Raf zinc finger is required for optimal binding to processed Ras and for ras-dependent Raf activation in situ. *Mol Cell Biol* 1997;**17**:46–53.

65. Williams JG, Drugan JK, Yi GS, et al. Elucidation of binding determinants and functional consequences of Ras/Raf-cysteine-rich domain interactions. *J Biol Chem* 2000;**275**:22,172–22,179.

66. Blum R, Cox AD, Kloog Y. Inhibitors of chronically active ras: potential for treatment of human malignancies. *Recent Patents Anticancer Drug Discov* 2008;**3**:31–47.

67. Paz A, Haklai R, Elad-Sfadia G, et al. Galectin-1 binds oncogenic H-Ras to mediate Ras membrane anchorage and cell transformation. *Oncogene* 2001;**20**:7486–93.

68. Elad-Sfadia G, Haklai R, Ballan E, et al. Galectin-1 augments Ras activation and diverts Ras signals to Raf-1 at the expense of phosphoinositide 3-kinase. *J Biol Chem* 2002;**277**:37,169–37,175.

69. Rotblat B, Niv H, Andre S, et al. Galectin-1(L11A) predicted from a computed galectin-1 farnesyl-binding pocket selectively inhibits Ras-GTP. *Cancer Res* 2004;**64**:3112–18.

70. Elad-Sfadia G, Haklai R, Balan E, Kloog Y. Galectin-3 augments K-Ras activation and triggers a Ras signal that attenuates ERK but not phosphoinositide 3-kinase activity. *J Biol Chem* 2004;**279**:34,922–34,930.

71. Haklai R, Weisz MG, Elad G, et al. Dislodgment and accelerated degradation of Ras. *Biochemistry* 1998;**37**:1306–14.

72. Jansen B, Schlagbauer-Wadl H, Kahr H, et al. Novel Ras antagonist blocks human melanoma growth. *Proc Natl Acad Sci USA* 1999;**96**:14,019–14,024,.

73. Marom M, Haklai R, Ben-Baruch G, et al. Selective inhibition of Ras-dependent cell growth by farnesylthiosalisylic acid. *J Biol Chem* 1995;**270**:22,263–22,270.

74. Gana-Weisz M, Halaschek-Wiener J, Jansen B, et al. The Ras inhibitor S-trans,trans-farnesylthiosalicylic acid chemosensitizes human tumor cells without causing resistance. *Clin Cancer Res* 2002;**8**:555–65.

75. Weisz B, Giehl K, Gana-Weisz M, et al. A new functional Ras antagonist inhibits human pancreatic tumor growth in nude mice. *Oncogene* 1999;**18**:2579–88.

76. Blum R, Jacob-Hirsch J, Amariglio N, et al. Ras inhibition in glioblastoma down-regulates hypoxia-inducible factor-1alpha, causing glycolysis shutdown and cell death. *Cancer Res* 2005;**65**:999–1006.

77. Yaari S, Jacob-Hirsch J, Amariglio N, et al. Disruption of cooperation between Ras and MycN in human neuroblastoma cells promotes growth arrest. *Clin Cancer Res* 2005;**11**:4321–30.

78. Barkan B, Starinsky S, Friedman E, et al. The Ras inhibitor farnesylthiosalicylic acid as a potential therapy for neurofibromatosis type 1. *Clin Cancer Res* 2006;**12**:5533–42.

79. Gana-Weisz M, Haklai R, Marciano D, et al. The Ras antagonist S-farnesylthiosalicylic acid induces inhibition of MAPK activation. *Biochem Biophys Res Commun* 1997;**239**:900–4.

80. Goldberg L, Kloog Y. A Ras inhibitor tilts the balance between Rac and Rho and blocks phosphatidylinositol 3-kinase-dependent glioblastoma cell migration. *Cancer Res* 2006;**66**:11,709–11,717,.

81. Shalom-Feuerstein R, Lindenboim L, Stein R, et al. Restoration of sensitivity to anoikis in Ras-transformed rat intestinal epithelial cells by a Ras inhibitor. *Cell Death Differ* 2004;**11**:244–7.

82. Reif S, Weis B, Aeed H, et al. The Ras antagonist, farnesylthiosalicylic acid (FTS), inhibits experimentally induced liver cirrhosis in rats. *J Hepatol* 1999;**31**:1053–61.

83. Haklai R, Elad-Sfadia G, Egozi Y, Kloog Y. Orally administered FTS (salirasib) inhibits human pancreatic tumor growth in nude mice. *Cancer Chemother Pharmacol* 2008;**61**:89–96.

84. Gourdeau H, McAlpine JB, Ranger M, et al. Identification, characterization and potent antitumor activity of ECO-4601, a novel peripheral benzodiazepine receptor ligand. *Cancer Chemother Pharmacol* 2008;**61**:911–21.

85. Charan RD, Schlingmann G, Janso J, et al. Diazepinomicin, a new antimicrobial alkaloid from a marine Micromonospora sp. *J Nat Prod* 2004;**67**:1431–3.

86. Han Z, Slack RS, Li W, Papadopoulos V. Expression of peripheral benzodiazepine receptor (PBR) in human tumors: relationship to

breast, colorectal, and prostate tumor progression. *J Recept Signal Transduct Res* 2003;**23**:225–38.

87. Katz Y, Eitan A, Amiri Z, Gavish M. Dramatic increase in peripheral benzodiazepine binding sites in human colonic adenocarcinoma as compared to normal colon. *Eur J Pharmacol* 1988;**148**:483–4.

88. Katz Y, Eitan A, Gavish M. Increase in peripheral benzodiazepine binding sites in colonic adenocarcinoma. *Oncology* 1990;**47**:139–42.

89. Maaser K, Hopfner M, Jansen A, et al. Specific ligands of the peripheral benzodiazepine receptor induce apoptosis and cell cycle arrest in human colorectal cancer cells. *Br J Cancer* 2001;**85**:1771–80.

90. Cornu P, Benavides J, Scatton B, et al. Increase in omega 3 (peripheral-type benzodiazepine) binding site densities in different types of human brain tumours. A quantitative autoradiography study. *Acta Neurochir (Wien)* 1992;**119**:146–52.

91. Beinlich A, Strohmeier R, Kaufmann M, Kuhl H. Specific binding of benzodiazepines to human breast cancer cell lines. *Life Sci* 1999;**65**:2099–108.

92. Carmel I, Fares FA, Leschiner S, et al. Peripheral-type benzodiazepine receptors in the regulation of proliferation of MCF-7 human breast carcinoma cell line. *Biochem Pharmacol* 1999;**58**:273–8.

93. Hardwick M, Fertikh D, Culty M, et al. Peripheral-type benzodiazepine receptor (PBR) in human breast cancer: correlation of breast cancer cell aggressive phenotype with PBR expression, nuclear localization, and PBR-mediated cell proliferation and nuclear transport of cholesterol. *Cancer Res* 1999;**59**:831–42.

94. Batra S, Larsson I, Boven E. Mitochondrial and microsomal peripheral benzodiazepine receptors in human ovarian cancer xenografts. *Int J Mol Med* 2000;**5**:619–23.

95. Venturini I, Zeneroli ML, Corsi L, et al. Up-regulation of peripheral benzodiazepine receptor system in hepatocellular carcinoma. *Life Sci* 1998;**63**:1269–80.

96. Roberts PJ, Der CJ. Targeting the Raf-MEK-ERK mitogen-activated protein kinase cascade for the treatment of cancer. *Oncogene* 2007;**26**:3291–310.

97. Wellbrock C, Karasarides M, Marais R. The RAF proteins take centre stage. *Nat Rev Mol Cell Biol* 2004;**5**:875–85.

98. Schreck R, Rapp UR. Raf kinases: oncogenesis and drug discovery. *Int J Cancer* 2006;**119**:2261–71.

99. Smith RA, Dumas J, Adnane L, Wilhelm SM. Recent advances in the research and development of RAF kinase inhibitors. *Curr Top Med Chem* 2006;**6**:1071–89.

100. Lyons JF, Wilhelm S, Hibner B, Bollag G. Discovery of a novel Raf kinase inhibitor. *Endocr Relat Cancer* 2001;**8**:219–25.

101. Wan PT, Garnett MJ, Roe SM, et al. Mechanism of activation of the RAF-ERK signaling pathway by oncogenic mutations of B-RAF. *Cell* 2004;**116**:855–67.

102. Wilhelm SM, Carter C, Tang L, et al. BAY 43-9006 exhibits broad spectrum oral antitumor activity and targets the RAF/MEK/ERK pathway and receptor tyrosine kinases involved in tumor progression and angiogenesis. *Cancer Res* 2004;**64**:7099–109.

103. Tannapfel A, Sommerer F, Benicke M, et al. Mutations of the BRAF gene in cholangiocarcinoma but not in hepatocellular carcinoma. *Gut* 2003;**52**:706–12.

104. Hahn O, Stadler W. Sorafenib. *Curr Opin Oncol* 2006;**18**:615–21.

105. Rini BI. Sorafenib. *Expert Opin Pharmacother* 2006;**7**:453–61.

106. Sala E, Mologni L, Truffa S, et al. BRAF silencing by short hairpin RNA or chemical blockade by PLX4032 leads to different responses in melanoma and thyroid carcinoma cells. *Mol Cancer Res* 2008;**6**:751–9.

107. Lankelma J, Dekker H, Groeningen C, Hoekman K. VEGF excretion in urine of colorectal cancer patients after bevacizumab administration. *AACR Meet Abstr* 2008;**383**.

108. Hoshino R, Chatani Y, Yamori T, et al. Constitutive activation of the 41-/43-kDa mitogen-activated protein kinase signaling pathway in human tumors. *Oncogene* 1999;**18**:813–22.

109. Davies SP, Reddy H, Caivano M, Cohen P. Specificity and mechanism of action of some commonly used protein kinase inhibitors. *Biochem J* 2000;**351**:95–105.

110. Sebolt-Leopold JS, Herrera R. Targeting the mitogen-activated protein kinase cascade to treat cancer. *Nat Rev Cancer* 2004;**4**:937–47.

111. Sebolt-Leopold JS, Dudley DT, Herrera R, et al. Blockade of the MAP kinase pathway suppresses growth of colon tumors in vivo. *Nat Med* 1999;**5**:810–16.

112. Collisson EA, De A, Suzuki H, et al. Treatment of metastatic melanoma with an orally available inhibitor of the Ras-Raf-MAPK cascade. *Cancer Res* 2003;**63**:5669–73.

113. Lorusso PM, Adjei AA, Varterasian M, et al. Phase I and pharmacodynamic study of the oral MEK inhibitor CI-1040 in patients with advanced malignancies. *J Clin Oncol* 2005;**23**:5281–93.

114. Rinehart J, Adjei AA, Lorusso PM, et al. Multicenter phase II study of the oral MEK inhibitor, CI-1040, in patients with advanced non-small-cell lung, breast, colon, and pancreatic cancer. *J Clin Oncol* 2004;**22**:4456–62.

115. Ohren JF, Chen H, Pavlovsky A, et al. Structures of human MAP kinase kinase 1 (MEK1) and MEK2 describe novel noncompetitive kinase inhibition. *Nat Struct Mol Biol* 2004;**11**:1192–7.

116. Thompson N, Lyons J. Recent progress in targeting the Raf/MEK/ERK pathway with inhibitors in cancer drug discovery. *Curr Opin Pharmacol* 2005;**5**:350–6.

117. Solit DB, Garraway LA, Pratilas CA, et al. BRAF mutation predicts sensitivity to MEK inhibition. *Nature* 2006;**439**:358–62.

118. Lyssikatos J, Yeh T, Wallace E, et al. ARRY-142886, a potent and selective MEK inhibitor: I) ATP-independent inhibition results in high enzymatic and cellular selectivity. *AACR Meet Abstr* 2004;**2004**:896b.

119. Yeh T, Wallace E, Lyssikatos J, Winkler J. ARRY-142886, a potent and selective MEK inhibitor: II) Potency against cellular MEK leads to inhibition of cellular proliferation and induction of apoptosis in cell lines with mutant Ras or B-Raf. *AACR Meet Abstr* 2004;**2004**:896c–897c.

120. Lee P, Wallace E, Yeh T, et al. ARRY-142886, a potent and selective MEK inhibitor: III) Efficacy in murine xenograft models correlates with decreased ERK phosphorylation. *AACR Meet Abstr* 2004;**2004**:897.

121. Adjei AA, Cohen RB, Franklin W, et al. Phase I pharmacokinetic and pharmacodynamic study of the oral, small-molecule mitogen-activated protein kinase kinase 1/2 inhibitor AZD6244 (ARRY-142886) in patients with advanced cancers. *J Clin Oncol* 2008;**26**:2139–2146.

122. Johnston S. XL518, a potent selective orally bioavailable MEK1 inhibitor, downregulates the RAS/RAF/MEK/ERK pathway in vivo, resulting in tumor growth inhibition and regression in pre-clinical models. *AACR Meet Abstr* 2007;**C209**.

123. Kato-Stankiewicz J, Hakimi I, Zhi G, et al. Inhibitors of Ras/Raf-1 interaction identified by two-hybrid screening revert Ras-dependent transformation phenotypes in human cancer cells. *Proc Natl Acad Sci U S A* 2002;**99**:14,398–14,403,.

124. Skobeleva N, Menon S, Weber L, et al. In vitro and in vivo synergy of MCP compounds with mitogen-activated protein kinase pathway- and microtubule-targeting inhibitors. *Mol Cancer Ther* 2007;**6**:898–906.

125. Hao H, Muniz-Medina VM, Mehta H, et al. Context-dependent roles of mutant B-Raf signaling in melanoma and colorectal carcinoma cell growth. *Mol Cancer Ther* 2007;**6**:2220–2229.

126. Rodriguez-Viciana P, Warne PH, Dhand R, et al. Phosphatidylinositol-3-OH kinase as a direct target of Ras. *Nature* 1994;**370**:527–532.

127. Rodriguez-Viciana P, Warne PH, Khwaja A, et al. Role of phosphoinositide 3-OH kinase in cell transformation and control of the actin cytoskeleton by Ras. *Cell* 1997;**89**:457–467.

128. Garcia-Echeverria C, Sellers WR. Drug discovery approaches targeting the PI3K/Akt pathway in cancer. *Oncogene* 2008;**27**:5511–5526.

129. Fasolo A, Sessa C. mTOR inhibitors in the treatment of cancer. *Expert Opin Investig Drugs* 2008;**17**:1717–1734.

130. Yuan TL, Cantley LC. PI3K pathway alterations in cancer: variations on a theme. *Oncogene* 2008;**27**:5497–5510.

131. Ihle NT, Williams R, Chow S, et al. Molecular pharmacology and antitumor activity of PX-866, a novel inhibitor of phosphoinositide-3-kinase signaling. *Mol Cancer Ther* 2004;**3**:763–772.

132. Ihle NT, Paine-Murrieta G, Berggren MI, et al. The phosphatidylinositol-3-kinase inhibitor PX-866 overcomes resistance to the epidermal growth factor receptor inhibitor gefitinib in A-549 human non-small cell lung cancer xenografts. *Mol Cancer Ther* 2005;**4**:1349–1357.

133. Howes AL, Chiang GG, Lang ES, et al. The phosphatidylinositol 3-kinase inhibitor, PX-866, is a potent inhibitor of cancer cell motility and growth in three-dimensional cultures. *Mol Cancer Ther* 2007;**6**:2505–2514.

134. Yang Y, Iwanaga K, Raso MG, et al. Phosphatidylinositol 3-kinase mediates bronchioalveolar stem cell expansion in mouse models of oncogenic K-ras-induced lung cancer. *PLoS ONE* 2008;**3**:e2220.

135. (2007). *AACR Meet Abstr* C205.

136. (2007). *AACR Meet Abstr* C199.

137. (2007). *AACR Meet Abstr* B265.

138. Molckovsky A, Siu LL. First-in-class, first-in-human phase I results of targeted agents: Highlights of the 2008 American Society of Clinical Oncology meeting. *J Hematol Oncol* 2008;**1**:20.

139. (2007). *AACR Meet Abstr* B250.

140. (2007). *AACR Meet Abstr* 4970.

141. (2007). *AACR Meet Abstr* C195.

142. Heerding DA, Rhodes N, Leber JD, et al. Identification of 4-(2-(4-amino-1,2,5-oxadiazol-3-yl)-1-ethyl-7-{[(3S)-3-piperidinylmethyl]oxy}-1H-imidazo[4,5-c]pyridin-4-yl)-2-methyl-3-butyn-2-ol (GSK690693), a novel inhibitor of AKT kinase. *J Med Chem* 2008;**51**:5663–5679.

143. Rhodes N, Heerding DA, Duckett DR, et al. Characterization of an Akt kinase inhibitor with potent pharmacodynamic and antitumor activity. *Cancer Res* 2008;**68**:2366–2374.

144. Kondapaka SB, Singh SS, Dasmahapatra GP, et al. Perifosine, a novel alkylphospholipid, inhibits protein kinase B activation. *Mol Cancer Ther* 2003;**2**:1093–1103.

145. Leighl NB, Dent S, Clemons M, et al. A Phase 2 study of perifosine in advanced or metastatic breast cancer. *Breast Cancer Res Treat* 2008;**108**:87–92.

146. Marsh Rde W, Rocha Lima CM, Levy DE, et al. A phase II trial of perifosine in locally advanced, unresectable, or metastatic pancreatic adenocarcinoma. *Am J Clin Oncol* 2007;**30**:26–31.

147. Bailey HH, Mahoney MR, Ettinger DS, et al. Phase II study of daily oral perifosine in patients with advanced soft tissue sarcoma. *Cancer* 2006;**107**:2462–2467.

148. Vink SR, Schellens JH, Beijnen JH, et al. Phase I and pharmacokinetic study of combined treatment with perifosine and radiation in patients with advanced solid tumours. *Radiother Oncol* 2006;**80**:207–213.

149. Argiris A, Cohen E, Karrison T, et al. A phase II trial of perifosine, an oral alkylphospholipid, in recurrent or metastatic head and neck cancer. *Cancer Biol Ther* 2006;**5**:766–770.

150. Knowling M, Blackstein M, Tozer R, et al. A phase II study of perifosine (D-21226) in patients with previously untreated metastatic or locally advanced soft tissue sarcoma: A National Cancer Institute of Canada Clinical Trials Group trial. *Invest New Drugs* 2006;**24**:435–439.

151. Posadas EM, Gulley J, Arlen PM, et al. A phase II study of perifosine in androgen independent prostate cancer. *Cancer Biol Ther* 2005;**4**:1133–1137.

152. Ernst DS, Eisenhauer E, Wainman N, et al. Phase II study of perifosine in previously untreated patients with metastatic melanoma. *Invest New Drugs* 2005;**23**:569–575.

153. Van Ummersen L, Binger K, Volkman J, et al. A phase I trial of perifosine (NSC 639966) on a loading dose/maintenance dose schedule in patients with advanced cancer. *Clin Cancer Res* 2004;**10**:7450–7456.

154. Vezina C, Kudelski A, Sehgal SN. Rapamycin (AY-22,989), a new antifungal antibiotic. I. Taxonomy of the producing streptomycete and isolation of the active principle. *J Antibiot (Tokyo)* 1975;**28**:721–726.

155. Eng CP, Sehgal SN, Vezina C. Activity of rapamycin (AY-22,989) against transplanted tumors. *J Antibiot (Tokyo)* 1984;**37**:1231–1237.

156. Atkins MB, Hidalgo M, Stadler WM, et al. Randomized phase II study of multiple dose levels of CCI-779, a novel mammalian target of rapamycin kinase inhibitor, in patients with advanced refractory renal cell carcinoma. *J Clin Oncol* 2004;**22**:909–918.

157. Hudes G, Carducci M, Tomczak P, et al. Temsirolimus, interferon alfa, or both for advanced renal-cell carcinoma. *N Engl J Med* 2007;**356**:2271–2281.

158. Hanna SC, Heathcote SA, Kim WY. mTOR pathway in renal cell carcinoma. *Expert Rev Anticancer Ther* 2008;**8**:283–292.

159. Yazbeck VY, Buglio D, Georgakis GV, et al. Temsirolimus downregulates p21 without altering cyclin D1 expression and induces autophagy and synergizes with vorinostat in mantle cell lymphoma. *Exp Hematol* 2008;**36**:443–450.

160. Ansell Jr. SM, Inwards DJ, Rowland KM, et al Low-dose, single-agent temsirolimus for relapsed mantle cell lymphoma: a phase 2 trial in the North Central Cancer Treatment Group. *Cancer* 2008;**113**:508–514.

161. Pandya KJ, Dahlberg S, Hidalgo M, et al. A randomized, phase II trial of two dose levels of temsirolimus (CCI-779) in patients with extensive-stage small-cell lung cancer who have responding or stable disease after induction chemotherapy: a trial of the Eastern Cooperative Oncology Group (E1500). *J Thorac Oncol* 2007;**2**:1036–1041.

162. Jimeno A, Kulesza P, Wheelhouse J, et al. Dual EGFR and mTOR targeting in squamous cell carcinoma models, and development of early markers of efficacy. *Br J Cancer* 2007;**96**:952–959.

163. Wu L, Birle DC, Tannock IF. Effects of the mammalian target of rapamycin inhibitor CCI-779 used alone or with chemotherapy on human prostate cancer cells and xenografts. *Cancer Res* 2005;**65**:2825–2831.

164. (May 20, 2008). *J Clin Oncol* 26 (Suppl.): LBA5026.

165. Figlin RA, Brown E, Armstrong AJ, et al. NCCN Task Force Report: mTOR Inhibition in Solid Tumors. *J Natl Compr Canc Netw* 2008;**6**:S1–S22.

166. (May 20, 2008). *J Clin Oncol* 26 (Suppl.): 530.

167. Wu C, Wangpaichitr M, Feun L, et al. Overcoming cisplatin resistance by mTOR inhibitor in lung cancer. *Mol Cancer* 2005;**4**:25

168. Wangpaichitr M, Wu C, You M, et al. Inhibition of mTOR restores cisplatin sensitivity through down-regulation of growth and anti-apoptotic proteins. *Eur J Pharmacol* 2008;**591**:124–127.

169. (June 20, 2007). 2007 ASCO Annual meeting Proceedings Part 1. *J Clin Oncol* 25 (Suppl.): 7589.

170. (June 20, 2007). 2007 ASCO Annual Meeting *J Clin Oncol* 25 (Suppl.): 7575.

171. (May 20, 2008). *J Clin Oncol* 26: (Suppl.): 19,017.

172. Rizzieri DA, Feldman E, Dipersio JF, et al. A phase 2 clinical trial of deforolimus (AP23573, MK-8669), a novel mammalian target of rapamycin inhibitor, in patients with relapsed or refractory hematologic malignancies. *Clin Cancer Res* 2008;**14**:2756–2762.

173. (2008). *AACR Meet Abstr* 1482.

174. (2008). *AACR Meet Abstr* 4006.

175. Sabatini DM. mTOR and cancer: insights into a complex relationship. *Nat Rev Cancer* 2006;**6**:729–734.

176. Chiang GG, Abraham RT. Targeting the mTOR signaling network in cancer. *Trends Mol Med* 2007;**13**:433–442.

177. Sarbassov DD, Guertin DA, Ali SM, Sabatini DM. Phosphorylation and regulation of Akt/PKB by the rictor-mTOR complex. *Science* 2005;**307**:1098–1101.

178. Habets GG, Scholtes EH, Zuydgeest D, et al. Identification of an invasion-inducing gene, Tiam-1, that encodes a protein with homology to GDP-GTP exchangers for Rho-like proteins. *Cell* 1994;**77**:537–549.

179. Lambert JM, Lambert QT, Reuther GW, et al. Tiam1 mediates Ras activation of Rac by a PI(3)K-independent mechanism. *Nat Cell Biol* 2002;**4**:621–625.

180. Qiu RG, Chen J, Kirn D, et al. An essential role for Rac in Ras transformation. *Nature* 1995;**374**:457–459.

181. Khosravi-Far R, Solski PA, Clark GJ, et al. Activation of Rac1, RhoA, and mitogen-activated protein kinases is required for Ras transformation. *Mol Cell Biol* 1995;**15**:6443–6453.

182. Kissil JL, Walmsley MJ, Hanlon L, et al. Requirement for Rac1 in a K-ras induced lung cancer in the mouse. *Cancer Res* 2007;**67**:8089–8094.

183. Gao Y, Dickerson JB, Guo F, et al. Rational design and characterization of a Rac GTPase-specific small molecule inhibitor. *Proc Natl Acad Sci U S A* 2004;**101**:7618–7623.

184. Thomas EK, Cancelas JA, Chae HD, et al. Rac guanosine triphosphatases represent integrating molecular therapeutic targets for BCR-ABL-induced myeloproliferative disease. *Cancer Cell* 2007;**12**:467–478.

185. Desire L, Bourdin J, Loiseau N, et al. RAC1 inhibition targets amyloid precursor protein processing by gamma-secretase and decreases Abeta production in vitro and in vivo. *J Biol Chem* 2005;**280**:37,516–37,525.

186. Shutes A, Onesto C, Picard V, et al. Specificity and mechanism of action of EHT 1864, a novel small molecule inhibitor of Rac family small GTPases. *J Biol Chem* 2007;**282**:35,666–35,678,.

187. Bodemann BO, White MA. Ral GTPases and cancer: linchpin support of the tumorigenic platform. *Nat Rev Cancer* 2008;**8**:133–140.

188. Urano T, Emkey R, Feig LA. Ral-GTPases mediate a distinct downstream signaling pathway from Ras that facilitates cellular transformation. *EMBO J* 1996;**15**:810–816.

189. Hamad NM, Elconin JH, Karnoub AE, et al. Distinct requirements for Ras oncogenesis in human versus mouse cells. *Genes Dev* 2002;**16**:2045–2057.

190. Lim KH, Baines AT, Fiordalisi JJ, et al. Activation of RalA is critical for Ras-induced tumorigenesis of human cells. *Cancer Cell* 2005;**7**:533–545.

191. Lim KH, O'Hayer K, Adam SJ, et al. Divergent roles for RalA and RalB in malignant growth of human pancreatic carcinoma cells. *Curr Biol* 2006;**16**:2385–2394.

192. Smith SC, Oxford G, Wu Z, et al. The metastasis-associated gene CD24 is regulated by Ral GTPase and is a mediator of cell proliferation and survival in human cancer. *Cancer Res* 2006;**66**:1917–1922.

193. Yin J, Pollock C, Tracy K, et al. Activation of the RalGEF/Ral pathway promotes prostate cancer metastasis to bone. *Mol Cell Biol* 2007;**27**:7538–7550.

194. Rodriguez-Viciana P, McCormick F. RalGDS comes of age. *Cancer Cell* 2005;**7**:205–206.

195. Feig LA. Ral-GTPases: approaching their 15 minutes of fame. *Trends Cell Biol* 2003;**13**:419–425.

196. Falsetti SC, Wang DA, Peng H, et al. Geranylgeranyltransferase I inhibitors target RalB to inhibit anchorage-dependent growth and induce apoptosis and RalA to inhibit anchorage-independent growth. *Mol Cell Biol* 2007;**27**:8003–8014.

197. Wu JC, Chen TY, Yu CT, et al. Identification of V23RalA-Ser194 as a critical mediator for Aurora-A-induced cellular motility and transformation by small pool expression screening. *J Biol Chem* 2005;**280**:9013–9022.

Chapter 39

The Roles of Ras Family Small GTPases in Breast Cancer

Ariella B. Hanker[1] and Channing J. Der[2]

[1]*Curriculum in Genetics and Molecular Biology, Lineberger Comprehensive Cancer Center, University of North Carolina at Chapel Hill, Chapel Hill, North Carolina*

[2]*Department of Pharmacology and Curriculum in Genetics and Molecular Biology, Lineberger Comprehensive Cancer Center, University of North Carolina at Chapel Hill, Chapel Hill, North Carolina*

INTRODUCTION

Approximately 180,000 new cases of breast cancer will be diagnosed in the United States in 2008 (www.cancer.gov), and despite significant advances in understanding the molecular pathways driving this disease, breast cancer remains the second leading cause of cancer related deaths among American women, with more than 40,000 deaths per year. Nearly 70 percent of breast cancers express the estrogen receptor alpha (ER) and depend on estrogen for growth. Current therapies for ER-positive breast cancers exploit this dependency on estrogen. These therapies include the anti-estrogen tamoxifen (Nolvadex®), which competes with estrogen for binding to the ER, and aromatase inhibitors (anastrazole (Arimidex®), exemestane (Aromasin®), and letrozole (Femara®)), which prevent the synthesis of estrogen [1, 2]. However, many tumors develop resistance to these anti-estrogen therapies [1–3]. ER-negative breast cancers tend to be more aggressive and metastatic than ER-positive tumors and therapeutic options for many of these patients are much more limited [4]. A subset of ER-positive and ER-negative breast cancers overexpress the human epidermal growth factor receptor HER2/ErbB2/Neu, and while the advent of HER2 inhibitors, including the HER2 monoclonal antibody trastuzumab (Herceptin®) and the lapatinib (Tykerb®) kinase inhibitor, have improved the survival of some of these patients, again the development of resistance to this targeted therapy is a major problem [5]. While a number of other targeted therapies are in clinical trials either alone or in combination with anti-estrogen or HER2 therapies, none have proven to be the "magic bullet" that selectively targets the breast cancer cell. Therefore, new therapies that block other molecules driving breast tumor growth are still required for breast cancer treatment.

Members of the Ras branch (36 genes encoding 39 proteins) of the Ras superfamily of small GTPases (156 members) regulate a multitude of cellular processes, including cell adhesion, differentiation, survival, and proliferation [6] (Figure 39.1a). The three Ras isoforms (H-, K-, and N-Ras) relay signals from outside the cell to the cytoplasm and the nucleus (Figure 39.1b). Ras proteins cycle between a GTP-bound active state and an inactive, GDP-bound conformation. Ras specific guanine nucleotide exchange factors (RasGEFs; e.g., Sos, RasGRF, RasGRP) promote the exchange of GDP for GTP and thereby activate Ras-GTP, whereas GTPase activating proteins (RasGAPs; e.g., NF1/neurofibromin) negatively regulate Ras GTPases by catalyzing the hydrolysis of GTP to GDP, thereby forming the inactive Ras-GDP. The conformation of Ras differs in these two nucleotide-bound states in amino acid residues referred to as switch 1 (30–38) and switch 2 (60–76) [7]. When bound to GTP, Ras proteins interact with multiple, functionally distinct effectors to initiate downstream cytoplasmic signaling networks that induce a plethora of cellular responses.

The domain organization of Ras family GTPases is summarized in Figure 39.2. The amino acids G12 and Q61 of Ras are conserved in the majority of Ras superfamily proteins and are essential for intrinsic and GAP stimulated GTP hydrolysis. Missense mutations of these residues are found in human cancers. These mutations render Ras insensitive to GAPs and thus constitutively GTP bound and active. Ras-GTP preferentially binds to and activates its downstream effectors via its core effector domain (residues 32–40) and flanking switch 1 and 2 sequences. Although this core effector binding domain is conserved in Ras fam-

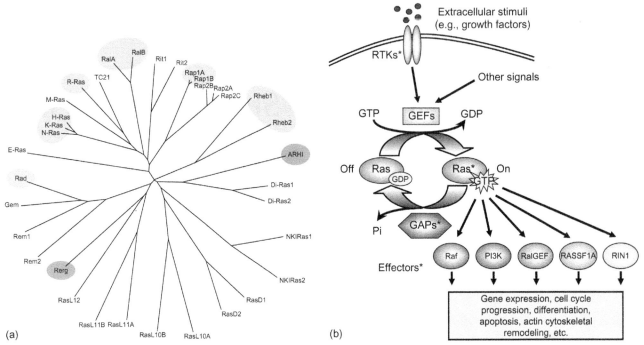

(a)   (b)

FIGURE 39.1 The Ras branch of the human Ras superfamily of small GTPases and Ras signal transduction.
(a) The Ras branch of the human Ras superfamily of small GTPases. ClustalX was used to generate a dendrogram of the GTP binding domains of human Ras subfamily members. Bubbles indicate proteins with established or putative roles in breast cancer, discussed in the text. Pink (dark gray) indicates proteins that suppress breast cancer growth (tumor suppressors), whereas green (medium gray) indicates proteins that promote tumor growth. While Rad (yellow/light gray) has also been implicated in breast cancer, whether it promotes or suppresses tumor growth is unclear. The K-Ras2B isoform of K-Ras was used. (b) Ras signal transduction. Ras GTPases cycle between a GTP-bound "on" state and a GDP-bound "off" state. Ras proteins are positively regulated by GEFs and negatively regulated by GAPs. Ras is activated by growth factor stimulated receptor tyrosine kinases (RTKs), such as EGFR and HER2. When bound to GTP, Ras interacts with a variety of downstream effectors to control cellular processes. Asterisks denote proteins that are aberrantly expressed or mutated in breast cancer. The RTKs EGFR and HER2 are frequently overexpressed in breast cancer and loss of the RasGAP NF1 leads to an increased risk of breast cancer. While Ras itself is infrequently mutated in breast cancer, mutations in some of its effectors, such as PI3K, are more common. In addition, loss of the Ras effector RIN1 has been observed in breast cancer.

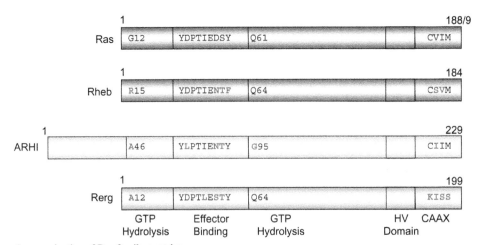

FIGURE 39.2 Domain organization of Ras family proteins.
ARHI contains a 35-residue N-terminal extension (N-extension) that may be important for its tumor suppressor activity. The amino acid residues G12 and Q61 in Ras are important for GAP stimulation of GTP hydrolysis; mutation of these residues in cancer leads to GAP insensitivity and therefore Ras is constitutively GTP bound and active. Since these residues are not conserved in ARHI, ARHI may be constitutively GTP bound. Rheb is regulated by the GAP TSC2, but whether the glutamine at position 64 is required for this regulation is unclear. Whether Rerg is regulated by GAPs is not known. Ras residues 32–40 comprise the core sequences important for effector binding. Although this core effector sequence is conserved between most Ras sub-family members, many Ras related GTPases, such as Rheb, use distinct effectors. The residues in Rheb, ARHI, and Rerg that are not conserved in Ras are shown in bold. The hypervariable (HV) domain in Ras dictates subcellular localization, as does the C-terminal CAAX box, required for lipid modification and membrane targeting. Ras, Rheb, and ARHI are farnesylated. Rerg is not farnesylated and does not localize to the membrane.

ily members, many members of this family signal through distinct effectors. Finally, most Ras family members terminate in a CAAX (C, cysteine; A, aliphatic amino acid; X, terminal amino acid) tetrapeptide sequence, a substrate for C15 farnesyl or C20 geranylgeranyl lipid modification responsible for membrane localization and critical for Ras biological activity (Figure 39.3) [8].

Ras itself is an oncoprotein and is mutationally activated in ~30 percent of all cancers but only 5 percent of breast cancers [9]. Despite the fact that Ras is rarely mutated in breast cancer, Ras is activated in many breast cancers by various upstream regulators, including the epidermal growth factor receptor family, in particular EGFR/HER1/ErbB1 and HER2/ErbB2/Neu [10, 11]. Therefore, blocking Ras signaling in breast cancer may be therapeutically beneficial. Recently, several Ras family proteins have been implicated in breast cancer and may play significant roles in the initiation and progression of this disease. These proteins include Rheb (Ras homolog enriched in brain), an activator of the mTOR signaling pathway involved in estrogen regulated breast cancer, and Rerg (Ras related estrogen regulated growth inhibitor) and ARHI (Ras homolog member I, also known as Di-Ras3 or Noey2), two Ras family proteins that, in contrast to Ras and Rheb, negatively regulate breast cancer growth and may function as tumor suppressors rather than oncogenes (Figure 39.1). In this chapter, we focus on these Ras family proteins involved in breast cancer and on possible strategies to correct the cancer related defects in these proteins for breast cancer treatment.

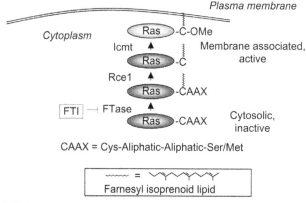

FIGURE 39.3 Ras posttranslational processing and membrane association.
Most members of the Ras subfamily, including Ras, Rheb, and ARHI/Noey2, are processed by a series of posttranslational modifications that allow insertion into the plasma membrane or endomembranes. The first step in this process is the addition of a farnesyl isoprenoid lipid to the cysteine residue of the CAAX motif by FTase; FTIs block this step. Next, the endonuclease Rce1 cleaves the -AAX tripeptide and the methyltransferase Icmt methylates the now terminal farnesylated cysteine (O-Me). Complete processing is required for proper Ras localization and function. While FTIs have shown promise in breast cancer clinical trials, they do not inhibit K-Ras, and the FTase substrates responsible FTI inhibition of breast cancer growth are not known.

RAS IN BREAST CANCER

Ectopic expression of mutationally activated Ras is capable of transforming immortalized human mammary epithelial cells and transgenic Ras expression drives mammary tumor development in mice [12]. Furthermore, ectopic expression of Ras in the estrogen dependent MCF-7 cell line promotes estrogen independent growth [13–15]. However, evidence for activation of endogenous Ras in human breast tumors is more limited. The growth factor receptors EGFR and HER2 are aberrantly activated by overexpression in breast cancers and can cause upstream activation of Ras. Studies have found that the levels of activated, GTP-bound Ras are increased in breast tumors and cell lines overexpressing EGFR or HER2 [10, 11]. Therefore, Ras may be activated in breast tumors in the absence of direct mutational activation of Ras. Ras proteins are also overexpressed in 20–50 percent of breast cancers. Another mechanism by which Ras may be activated in breast cancer could be decreased expression of a negative regulator of Ras, such as a RasGAP. Women with the genetic disease neurofibromatosis, caused by inactivating mutations of the RasGAP NF1, have an increased risk of developing breast cancer [16, 17], but whether the NF1 tumor suppressor is downregulated in spontaneous breast cancers is not known.

Recent work by Song and colleagues has revealed a mechanism by which H-Ras expression is elevated in breast tumor initiating cells, which may represent breast cancer "stem cells" [18]. They showed that expression of the microRNA *let-7*, a negative regulator of H-Ras protein expression, is reduced in breast tumor initiating cells (cancer stem cells) and in clinical samples. They further showed that restoring *let-7* expression reduced H-Ras expression, cell proliferation, mammosphere formation, tumorigenicity, and metastasis, and that silencing H-Ras alone was sufficient to reduce mammosphere formation, tumor formation, and metastasis [18]. Thus, H-Ras inhibition may represent a promising therapeutic approach for targeting breast cancer stem cells. Further work is needed to determine whether H-Ras levels are increased in independently derived populations of breast cancer stem cells.

GTP-bound Ras associates with a variety of effectors to elicit a diverse array of cellular responses. The best characterized Ras effectors implicated in oncogenesis include the Raf serine/threonine kinases, the phosphatidylinositol 3-kinases (PI3K), and GEFs for the Ral GTPases (RalGEFs) [19]. Mutations in the genes encoding B-Raf and the p110α catalytic subunit of PI3K (*PIK3CA*) have been identified in many cancers, supporting a role for these Ras effectors in oncogenesis. While *BRAF* mutations are rare in breast cancer, the gene encoding the p110α catalytic subunit of PI3K (*PIK3CA*) is mutated in 20–40 percent of human breast cancers [20] and its occurrence is mutually exclusive with Ras mutations in breast cancer cell lines [21]. However, Ras also activates some effectors that negatively regulate cancer growth; silencing of these effectors, such as the RASSF1 tumor suppressor and its family members, is a common event in cancer [22, 23]. Hyper-methylation of the *RASSF1A* promoter has been observed in 80–90 percent of breast cancers [24, 25], but a definitive role for RASSF1A in breast cancer has not yet been established. The Ras effector RIN1 was recently implicated as a tumor suppressor in breast cancer. RIN1 regulates epithelial cell properties such as cytoskeletal remodeling through Abl tyrosine kinases [26]. Colicelli and colleagues found that RIN1 levels are reduced in primary breast tumors and in breast cancer cell lines, and that restoring RIN1 expression suppressed tumor growth [27]. Whether loss of RIN1 correlates with Ras activation, and whether its loss is necessary for Ras induced tumor growth, is not known.

Due to the unequivocal importance of Ras in human cancer development and progression, there have been considerable efforts to develop pharmacologic agents that block Ras function for cancer therapy. One major approach has been the development of inhibitors of effector signaling, in particular inhibitors of Raf induced activation of the ERK mitogen activated protein kinase cascade [28]. A second major focus has been inhibition of Ras membrane association. In order for Ras to function properly, it must be targeted to the plasma membrane. This proper localization is accomplished via posttranslational addition of a farnesyl isoprenoid lipid to the C-terminus of Ras [8]. Farnesylation is catalyzed by the enzyme farnesyltransferase (FTase). FTase inhibitors (FTIs) were originally developed to block Ras localization and thereby block Ras function in cancer. However, it is now appreciated that FTIs do not in fact block localization of the most commonly mutated Ras isoforms in cancer, K- and N-Ras. K- and N-Ras escape FTI inhibition by a process known as alternative prenylation, whereby they are modified by addition of another isoprenoid lipid, geranylgeranol; geranylgeranylated Ras proteins remain functional and transforming in the presence of FTIs [29]. Despite an inability to effectively block Ras, FTIs have shown some anti-tumor activity, particularly in breast cancer [30]. Since other substrates of FTase do have known roles in cell proliferation (e.g., Rheb) [31], the anti-tumor activity of FTIs may be due to inhibiting the function of these proteins. FTIs have also demonstrated clinical activity in combination with hormonal therapies in breast cancer [32, 33]. Elucidating the FTase substrates responsible for the anti-tumor activity of FTIs in breast cancer could be very beneficial in selecting patients that may respond to this targeted therapy.

RHEB

Rheb1 and Rheb2 (51 percent identity) are also members of the Ras family of small GTPases [34] that may have increased activity in breast cancer. Rheb is a critical component of the Akt-TSC-mTOR pathway that regulates

protein translation, nutrient sensing, cell size, and cell proliferation (Figure 39.4a) [35]. The tumor suppressor TSC2 (tuberin), in complex with TSC1 (hamartin), functions as a GAP toward Rheb and thereby suppresses Rheb-GTP formation and signaling. The role of TSC2 in Rheb

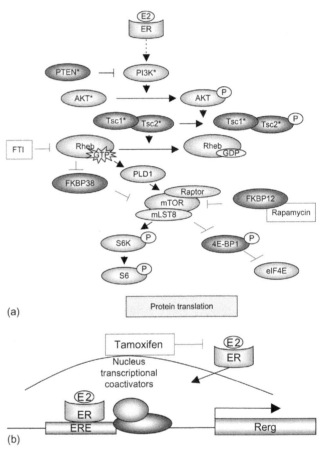

(a)

(b)

FIGURE 39.4 Estrogen regulation of Rheb signaling and estrogen regulation of Rerg expression.
(a) Estrogen regulation of Rheb signaling. When bound to estrogen (E2), the estrogen receptor (ER) can initiate rapid, "non-genomic" signaling through PI3K. PI3K is a lipid kinase that converts phosphatidylinositol (4,5) bisphosphate (PIP$_2$) into phosphatidylinositol (3,4,5) triphosphate (PIP$_3$). The reverse reaction is catalyzed by the lipid phosphatase PTEN, a negative regulator of the pathway. PIP$_3$ leads to activation of Akt, which then phosphorylates a variety of substrates involved in cell proliferation and survival, including TSC2. Akt mediated phosphorylation of TSC2 inactivates its RhebGAP activity, leading to increased levels of Rheb-GTP. Rheb then activates mTOR by displacing the mTOR inhibitor FKBP38 and by activating PLD1. mTOR, in complex with raptor and mLST8, phosphorylates substrates involved in protein translation, including S6K and 4E-BP1. Rheb is inhibited by FTIs and mTOR is inhibited by rapamycin. Green (light gray) indicates positive regulators of the pathway; red (dark gray) indicates negative regulators. Asterisks denote proteins that are known to be mutated or aberrantly expressed in breast cancer. (b) Estrogen regulation of Rerg expression. When bound to estrogen (E2), the ER can function as a transcription factor in the nucleus. The *RERG* gene contains an upstream estrogen response element (ERE). When the ER binds to EREs, it recruits transcriptional coactivators to activate transcription of estrogen responsive genes such as Rerg. Rerg expression correlates with ER expression in breast cancer; ER-negative breast cancers lack Rerg expression.

signaling is analogous to that of the RasGAP NF1: loss of NF1 in neurofibromatosis type I leads to hyperactivation of Ras. Similarly, loss-of-function mutations in TSC1 or TSC2 cause the genetic disease tuberous sclerosis complex, which is characterized by the formation of hamartomas in a variety of organs [36, 37]. In this disease, loss of TSC2 or TSC1 results in hyperactivation of Rheb and promotes Rheb-GTP association with its two known effectors, FKBP38 and phospholipase D1 (PLD1) [38–40]. Rheb-GTP activates PLD1, which is required for Rheb activation of the mTOR pathway. FKBP38 associates with and inactivates the mTOR (mammalian target of rapamycin) complex 1 (mTORC1). The mTORC1 complex, which consists of mTOR, raptor, and mLST8 (also known as GβL), regulates protein translation initiation, metabolism, cell size, and cell proliferation [41, 42]. Rheb association with FKBP38 relieves this negative regulation, leading to mTORC1 activation, which causes inactivation of the translational inhibitor eIF4E binding protein 1 (4E-BP1) and activation of p70 S6 kinase (S6K). Hypo-phosphorylated 4E-BP1 binds and sequesters eIF4E. Phosphorylation of 4E-BP1 by mTOR allows the release of eIF4E, which promotes cap dependent mRNA translation and tumorigenesis [43]. S6K phosphorylates the S6 ribosome subunit, also promoting increased protein translation and the development of benign tumors.

Although mutations of TSC1 and TSC2 are not found in breast cancer, other mechanisms may activate Rheb in breast cancer. Studies have found reduced expression of TSC1 and TSC2 in breast cancer [44]. The mRNA levels of both Rheb1 and Rheb2 are elevated in breast cancer cells [45]. The TSC2-Rheb-mTOR pathway can also be activated by genetic alterations that cause activation of PI3K and its target, the serine-threonine kinase Akt (Figure 39.4a). Akt phosphorylates and inactivates TSC2, leading to an increase in GTP-bound Rheb. The PI3K-Akt pathway is one of the most commonly activated pathways in breast cancer [20]. In particular, mutational activation of the gene encoding p110α is one of the most frequent gene mutations found in breast cancer [46].

Like Ras, Rheb can be hyperactivated in breast cancer by upstream signals. HER2 and EGFR overexpression can cause activation of PI3K, and thus Rheb. Furthermore, estrogen can trigger Rheb signaling in the MCF-7 ER-positive breast cancer cell line. Estrogen activates PI3K and Akt through non-genomic mechanisms and estrogen treatment was recently found to increase Rheb-GTP levels and to stimulate downstream signaling through mTOR [47, 48]. Rheb was also shown to be required for estrogen induced DNA synthesis and cell cycle progression in MCF-7 cells [48]. In addition, mTOR is required for estrogen induced MCF-7 cell proliferation [49]. Rheb and mTOR have also been implicated in tamoxifen resistance: Rheb inhibition increased sensitivity to tamoxifen [45] and mTOR inhibition prevented Akt induced tamoxifen resistance in MCF-7 cells [50, 51]. Rapamycin analogs, which inhibit mTOR,

are currently in clinical trials and have shown promise in combination with endocrine therapies for breast cancer treatment [32, 52]. Therefore, inhibiting Rheb signaling may be beneficial in ER-positive and tamoxifen resistant breast cancers. Further research is needed to determine the role of Rheb in tamoxifen resistance and to validate Rheb as a therapeutic target for the treatment of breast cancer.

Like Ras, Rheb terminates in a CAAX motif and is farnesylated. Farnesylation is essential for its proper localization and function [53]. FTIs block Rheb signaling and function [45]. Therefore, Rheb inhibition may be partially responsible for the anti-breast cancer effects of FTIs. Basso et al. showed that the FTI lonafarnib enhanced tamoxifen induced apoptosis, and this enhancement was prevented by an FTI resistant version of Rheb [45]. Whether this FTI resistant Rheb version prevents FTIs from decreasing breast cancer growth will be of considerable interest for the selection of patients that may benefit from FTIs.

ARHI/DI-RAS3/NOEY2

Whereas Ras is an oncogene, and many Ras subfamily members, such as Rheb, promote growth, there are also several Ras related proteins that have been implicated as tumor suppressors in breast cancer. One such protein, ARHI, was identified originally by Bast and colleagues by using differential display PCR in ovarian cancer cells to identify genes whose expression is lost in tumor cells [54]. The *ARHI* gene is maternally imprinted and expressed monoallelically. Loss of heterozygosity, in which the non-imprinted allele was deleted, was reported in 41 percent of breast and ovarian cancers [54]. ARHI expression is also negatively regulated by promoter hyper-methylation, decreased mRNA stability, and transcriptional regulation by histone deacetylase (HDAC)-E2F complexes [55–57]. ARHI expression was found to be reduced in 70 percent of invasive breast carcinoma tissues and decreased expression was associated with breast cancer progression and lymph node metastases [58, 59]. ARHI expression is also downregulated in several other cancers, including pancreatic cancer, thyroid cancer, and non-small cell lung cancer, suggesting that loss of ARHI may be an important step in oncogenesis [60].

Ectopic re-expression of ARHI was found to reduce the clonogenic growth of breast and ovarian cancer cells, reduce cyclin D1 promoter activity, and increase p21^{CIP1} expression [54], supporting a tumor suppressor function. Bast and colleagues further showed that ARHI re-expression inhibited breast cancer cell proliferation and tumor xenograft formation and induced calpain dependent apoptosis in breast and ovarian cancer cells [61]. Since ARHI is divergent at two key residues important for intrinsic GTPase activity (A46 and G95), ARHI may be constitutively active and not regulated by GDP/GTP cycling (Figure 39.2). The mechanism of ARHI induced tumor suppression was shown to depend on its unique N-terminal extension [62], absent in other Ras family members (Figure 39.2), and to involve inhibition of the signal transducer and activator of transcription STAT3 [63]. Strategies to cause re-expression of ARHI may represent an attractive therapy for ARHI-negative breast cancer patients.

Like Ras and Rheb, ARHI terminates in a CAAX motif and is farnesylated. The farnesylation of ARHI is essential for its membrane association and growth inhibitory function [62]. Therefore, it may be an important "off-target" for farnesyltransferase inhibitors that results in promotion, rather than inhibition, of growth.

RERG

Like ARHI, Rerg is another Ras related small GTPase that is downregulated in breast cancers and is thought to play a growth inhibitory role. Rerg was initially identified in a microarray screen, where its expression was decreased in the most aggressive, ER-negative subtypes [64, 65]. Rerg was found to be a transcriptional target of the ER, which functions as a transcription factor when bound to estrogen (Figure 39.4b). Rerg expression was stimulated by estrogen treatment and reduced by the anti-estrogen tamoxifen [64]. Other microarray studies have found decreased Rerg expression in several different types of cancer and in metastatic cancers relative to their non-metastatic counterparts [66], suggesting that loss of Rerg expression may play a role in tumor progression and metasasis.

While Rerg shares substantial sequence identity with Ras, particularly in the core effector domain (Figure 39.1b), Rerg does not bind to canonical Ras effectors and no Rerg effectors have been identified to date. Unlike most Ras family members, Rerg lacks the C-terminal CAAX motif and is localized primarily in the cytosol [64, 67]. Overexpression of Rerg in MCF-7 cells decreased cell proliferation and xenograft formation [64], suggesting that Rerg is a growth inhibitor in breast cancer. However, ectopic re-expression of Rerg in ER-negative, Rerg-negative breast cancer cell lines did not affect anchorage dependent or independent growth or invasion [66], casting doubt on the role of Rerg as a tumor suppressor in breast cancer. More studies are needed to confirm the role of Rerg as a growth inhibitor in breast cancer *in vivo* and the identification of Rerg binding partners may shed light on its function in breast cancer.

THERAPEUTIC RE-EXPRESSION OF ARHI OR RERG IN BREAST CANCER

Due to the putative roles of ARHI and Rerg as tumor suppressors or growth inhibitors in breast cancer, therapeutic re-expression of these proteins in tumors that have lost their expression may be beneficial. Although a gene therapy approach to simply restore expression may be the most direct strategy, this feat remains difficult to accomplish. Instead, targeting the epigenetic regulation of ARHI and Rerg may be more promising. Expression of both genes is negatively regulated by HDACs (Figure 39.5). The HDAC inhibitor trichostatin A restored expression of ARHI and Rerg in breast cancer cells [68, 69]. Several HDAC inhibitors are currently in clinical trials to treat cancer, including breast cancer, and the HDAC inhibitor vorinostat (Zolinza®) has been approved by the FDA for the treatment of cutaneous T cell lymphoma [70]. It will be important to determine whether these HDAC inhibitors restore expression of ARHI or Rerg in primary breast cancers lacking expression of these proteins and whether restoration of ARHI or Rerg expression correlates with patient response.

Many tumor suppressors, including ARHI, are epigenetically silenced by promoter hyper-methylation (Figure 39.5). DNA methyltransferases (DNMTs) have thus attracted interest as therapeutic targets [71]. The DNMT inhibitor 5-aza-2'-deoxycytidine (5-aza-dC) reactivated ARHI expression in breast cancer cells [56, 69]. Another way to restore ARHI expression could be E2F inhibitors, since RNA interference mediated suppression of E2F family member expression was shown to increase ARHI expression in SKBR3 breast cancer cells [55]. Whether the Rerg promoter is hyper-methylated or whether Rerg expression can be restored by 5-aza-dC treatment is not known,

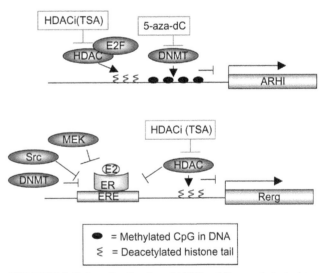

FIGURE 39.5 Epigenetic silencing of ARHI and Rerg and strategies to restore their expression.
The *ARHI* gene can be epigenetically silenced by promoter hyper-methylation, catalyzed by DNMTs, and by HDAC/E2F complexes. The DNMT inhibitor 5-aza-dC and the HDAC inhibitor (HDACi) TSA have been shown to restore ARHI expression in some breast cancer cells. TSA can also restore Rerg expression in an ER-negative cell line. Other methods to restore Rerg expression in ER-negative breast cancers could include strategies to upregulate or to stabilize the ER, such as with Src inhibitors, MEK inhibitors, or DNMT inhibitors.

but 5-aza-dC can restore ER expression in ER-negative breast cancer cells [72]. It will be important to determine if ER expression is sufficient to restore Rerg expression. If this is the case, then Rerg expression may also be restored by inhibitors of the Src tyrosine kinase or MEK1/2 dual specificity kinases, which have been shown to increase ER levels in breast cancer cells [73, 74]. Since many of these therapies are in preclinical or clinical studies, understanding the effects of ARHI or Rerg re-expression in breast cancer may help to predict patient response to these therapies.

OTHER RAS FAMILY PROTEINS IN BREAST CANCER

There are several other members of the Ras subfamily that have been implicated in breast cancer, but more studies are needed to determine the importance of these proteins in breast cancer. Rap1, a Ras related protein involved in integrin and cadherin signaling, was shown to be involved in breast cancer lumen formation [75]. Rap1 activity was elevated in a malignant breast cancer cell line compared to its non-malignant counterpart when cultured in three-dimensional laminin-rich extracellular matrix [75]. Dominant-negative Rap1 restored tissue polarity and induced lumen formation in breast cancer cells. Dominant-negative Rap1 also reduced tumor incidence in mice, whereas constitutively active Rap1 increased invasiveness and tumorigenicity [75]. Determining whether GTP-bound Rap1 is increased in primary breast cancer tissue and whether RNAi mediated silencing of Rap1 in multiple breast cancer cell lines reduces tumorigenicity and invasion will strengthen the evidence for the importance of Rap1 in breast cancer tumorigenesis.

Rad (R-Rad/Rem3) is a Ras related GTPase that is normally expressed in heart, skeletal muscle, and lung tissues [76]. Multiple groups have reported reduced expression of Rad in breast cancers [77, 78], suggesting a role for Rad as a tumor suppressor. However, ectopic expression of Rad was shown to increase anchorage independent growth and tumor formation in mice, implicating Rad as an oncogene instead of a tumor suppressor [77]. Coexpression of the metastasis suppressor nm23, which interacts with Rad, inhibited Rad induced breast cancer cell growth [77]. More work is needed to resolve whether Rad is a tumor suppressor or an oncogene in breast cancer.

The Ral GTPases (RalA and RalB) have recently garnered attention for their roles in promoting oncogenesis, tumor growth, and metastasis [79]. Ral GTPases function downstream of the Ras effector, RalGEF, and the RalGEF-Ral pathway is required for Ras induced transformation in human cells [80]. However, whether RalA and RalB are involved in breast cancer is not known. Ral was shown to be required for EGFR induced estrogen independent growth in MCF-7 cells, but constitutively active RalA was not sufficient to promote estrogen independent growth [81]. Missense mutations in the RalGEF Rgl1 have been found in breast cancer [82], but whether these mutations activate Ral and drive tumorigenesis has not been determined. Clearly, further research is warranted in order to determine whether targeting Ral or Rgl1 would be beneficial in breast cancer.

The GEF BCAR3 was found in a screen to identify proteins that promote tamoxifen resistance [83]. While BCAR3 overexpression does indeed promote anti-estrogen resistance, the question of which downstream GTPase it utilizes to do so is controversial. Studies originally implicated the mouse homolog of BCAR3, AND-34, as a GEF for RalA, Rap1A, and R-Ras [84]. However, overexpression of AND-34 failed to activate RalA or R-Ras in breast cancer cells [81, 85], and AND-34 induced anti-estrogen resistance was found to be mediated by PI3K and Rac1 [85, 86]. Nevertheless, constitutively active R-Ras was sufficient to induce estrogen independent MCF-7 cell proliferation [81]. Future studies to address whether R-Ras activity is elevated in tamoxifen resistant breast cancer tissues will help to evaluate its role in breast cancer.

In summary, members of the Ras subfamily of small GTPases may represent attractive targets for the treatment of breast cancer, but more research is needed to fully understand the roles these proteins play in breast cancer. In addition, other small GTPases, including members of the Rho and Rab subfamilies, have been implicated in breast cancer and have been reviewed elsewhere [87–89]. Currently, GTPases are not considered promising druggable targets. Since GTPases bind GTP with nanomolar to picomolar affinities, small molecule approaches to block GTP binding are not feasible. As strategies to therapeutically inhibit small GTPase signaling improve, it will be imperative to understand the contributions of each of these proteins to breast cancer development and progression.

REFERENCES

1. Ali S, Coombes RC. Endocrine-responsive breast cancer and strategies for combating resistance. *Nat Rev Cancer* 2002;**2**:101–12.
2. Johnston SR, Dowsett M. Aromatase inhibitors for breast cancer: lessons from the laboratory. *Nat Rev Cancer* 2003;**3**:821–31.
3. Massarweh S, Schiff R. Unraveling the mechanisms of endocrine resistance in breast cancer: new therapeutic opportunities. *Clin Cancer Res* 2007;**13**:1950–4.
4. Rochefort H, Glondu M, Sahla ME, Platet N, Garcia M. How to target estrogen receptor-negative breast cancer?. *Endocr Relat Cancer* 2003;**10**:261–6.
5. Nahta R, Esteva FJ. Trastuzumab: triumphs and tribulations. *Oncogene* 2007;**26**:3637–43.
6. Wennerberg K, Rossman KL, Der CJ. The Ras superfamily at a glance. *J Cell Sci* 2005;**118**:843–6.
7. Vetter IR, Wittinghofer A. The guanine nucleotide-binding switch in three dimensions. *Science* 2001;**294**:1299–304.
8. Cox AD, Der CJ. Ras family signaling: therapeutic targeting. *Cancer Biol Ther* 2002;**1**:599–606.

9. Bos JL. ras oncogenes in human cancer: a review. *Cancer Res* 1989;**49**:4682–9.

10. Eckert LB, Repasky GA, Ulku AS, McFall A, Zhou H, Sartor CI, Der CJ. Involvement of Ras activation in human breast cancer cell signaling, invasion, and anoikis. *Cancer Res* 2004;**64**:4585–92.

11. von Lintig FC, Dreilinger AD, Varki NM, Wallace AM, Casteel DE, Boss GR. Ras activation in human breast cancer. *Breast Cancer Res Treat* 2000;**62**:51–62.

12. Dimri G, Band H, Band V. Mammary epithelial cell transformation: insights from cell culture and mouse models. *Breast Cancer Res* 2005;**7**:171–9.

13. Kasid A, Lippman ME, Papageorge AG, Lowy DR, Gelmann EP. Transfection of v-rasH DNA into MCF-7 human breast cancer cells bypasses dependence on estrogen for tumorigenicity. *Science* 1985;**228**:725–8.

14. Kasid A, Lippman ME. Estrogen and oncogene mediated growth regulation of human breast cancer cells. *J Steroid Biochem* 1987;**27**:465–70.

15. Sommers CL, Papageorge A, Wilding G, Gelmann EP. Growth properties and tumorigenesis of MCF-7 cells transfected with isogenic mutants of rasH. *Cancer Res* 1990;**50**:67–71.

16. Sharif S, Moran A, Huson SM, Iddenden R, Shenton A, Howard E, Evans DG. Women with neurofibromatosis 1 are at a moderately increased risk of developing breast cancer and should be considered for early screening. *J Med Genet* 2007;**44**:481–4.

17. Guran S, Safali M. A case of neurofibromatosis and breast cancer: loss of heterozygosity of NF1 in breast cancer. *Cancer Genet Cytogenet* 2005;**156**:86–8.

18. Yu F, Yao H, Zhu P, Zhang X, Pan Q, Gong C, Huang Y, Hu X, Su F, Lieberman J, Song E. let-7 Regulates Self Renewal and Tumorigenicity of Breast Cancer Cells. *Cell* 2007;**131**:1109–23.

19. Repasky GA, Chenette EJ, Der CJ. Renewing the conspiracy theory debate: does Raf function alone to mediate Ras oncogenesis? *Trends Cell Biol* 2004;**14**:639–47.

20. Samuels Y, Ericson K. Oncogenic PI3K and its role in cancer. *Curr Opin Oncol* 2006;**18**:77–82.

21. Hollestelle A, Elstrodt F, Nagel JH, Kallemeijn WW, Schutte M. Phosphatidylinositol-3-OH Kinase or RAS Pathway Mutations in Human Breast Cancer Cell Lines. *Mol Cancer Res* 2007;**5**:195–201.

22. van der Weyden L, Adams DJ. The Ras-association domain family (RASSF) members and their role in human tumourigenesis. *Biochim Biophys Acta* 2007;**1776**:58–85.

23. Donninger H, Vos MD, Clark GJ. The RASSF1A tumor suppressor. *J Cell Sci* 2007;**120**:3163–72.

24. Shinozaki M, Hoon DS, Giuliano AE, Hansen NM, Wang HJ, Turner R, Taback B. Distinct hypermethylation profile of primary breast cancer is associated with sentinel lymph node metastasis. *Clin Cancer Res* 2005;**11**:2156–62.

25. Yeo W, Wong WL, Wong N, Law BK, Tse GM, Zhong S. High frequency of promoter hypermethylation of RASSF1A in tumorous and non-tumourous tissue of breast cancer. *Pathology* 2005;**37**:125–30.

26. Hu H, Bliss JM, Wang Y, Colicelli J. RIN1 is an ABL tyrosine kinase activator and a regulator of epithelial-cell adhesion and migration. *Curr Biol* 2005;**15**:815–23.

27. Milstein M, Mooser CK, Hu H, Fejzo M, Slamon D, Goodglick L, Dry S, Colicelli J. RIN1 is a breast tumor suppressor gene. *Cancer Res* 2007;**67**:11,510–16.

28. Roberts PJ, Der CJ. Targeting the Raf-MEK-ERK mitogen-activated protein kinase cascade for the treatment of cancer. *Oncogene* 2007;**26**:3291–310.

29. Cox AD, Der CJ. Farnesyltransferase inhibitors: promises and realities. *Curr Opin Pharmacol* 2002;**2**:388–93.

30. Kelland LR, Smith V, Valenti M, Patterson L, Clarke PA, Detre S, End D, Howes AJ, Dowsett M, Workman P, Johnston SR. Preclinical antitumor activity and pharmacodynamic studies with the farnesyl protein transferase inhibitor R115777 in human breast cancer. *Clin Cancer Res* 2001;**7**:3544–50.

31. Reid TS, Terry KL, Casey PJ, Beese LS. Crystallographic analysis of CaaX prenyltransferases complexed with substrates defines rules of protein substrate selectivity. *J Mol Biol* 2004;**343**:417–33.

32. Johnston SR. Targeting downstream effectors of epidermal growth factor receptor/HER2 in breast cancer with either farnesyltransferase inhibitors or mTOR antagonists. *Int J Gynecol Cancer* 2006;**16**:543–8.

33. O'Regan RM, Khuri FR. Farnesyl transferase inhibitors: the next targeted therapies for breast cancer? *Endocr Relat Cancer* 2004;**11**:191–205.

34. Aspuria PJ, Tamanoi F. The Rheb family of GTP-binding proteins. *Cell Signal* 2004;**16**:1105–12.

35. Arsham AM, Neufeld TP. Thinking globally and acting locally with TOR. *Curr Opin Cell Biol* 2006;**18**:589–97.

36. Inoki K, Corradetti MN, Guan KL. Dysregulation of the TSC-mTOR pathway in human disease. *Nat Genet* 2005;**37**:19–24.

37. Astrinidis A, Henske EP. Tuberous sclerosis complex: linking growth and energy signaling pathways with human disease. *Oncogene* 2005;**24**:7475–81.

38. Manning BD, Tee AR, Logsdon MN, Blenis J, Cantley LC. Identification of the tuberous sclerosis complex-2 tumor suppressor gene product tuberin as a target of the phosphoinositide 3-kinase/akt pathway. *Mol Cell* 2002;**10**:151–62.

39. Bai X, Ma D, Liu A, Shen X, Wang QJ, Liu Y, Jiang Y. Rheb activates mTOR by antagonizing its endogenous inhibitor, FKBP38. *Science* 2007;**318**:977–80.

40. Sun Y, Fang Y, Yoon MS, Zhang C, Roccio M, Zwartkruis FJ, Armstrong M, Brown HA, Chen J. Phospholipase D1 is an effector of Rheb in the mTOR pathway. *Proc Natl Acad Sci USA* 2008;**105**:8286–91.

41. Guertin DA, Sabatini DM. Defining the Role of mTOR in Cancer. *Cancer Cell* 2007;**12**:9–22.

42. Mamane Y, Petroulakis E, LeBacquer O, Sonenberg N. mTOR, translation initiation and cancer. *Oncogene* 2006;**25**:6416–22.

43. Averous J, Proud CG. When translation meets transformation: the mTOR story. *Oncogene* 2006;**25**:6423–35.

44. Jiang WG, Sampson J, Martin TA, Lee-Jones L, Watkins G, Douglas-Jones A, Mokbel K, Mansel RE. Tuberin and hamartin are aberrantly expressed and linked to clinical outcome in human breast cancer: the role of promoter methylation of TSC genes. *Eur J Cancer* 2005;**41**:1628–36.

45. Basso AD, Mirza A, Liu G, Long BJ, Bishop WR, Kirschmeier P. The farnesyl transferase inhibitor (FTI) SCH66336 (lonafarnib) inhibits Rheb farnesylation and mTOR signaling. Role in FTI enhancement of taxane and tamoxifen anti-tumor activity. *J Biol Chem* 2005;**280**:31,101–8.

46. Wood LD, Parsons DW, Jones S, Lin J, Sjoblom T, Leary RJ, Shen D, Boca SM, Barber T, Ptak J, Silliman N, Szabo S, Dezso Z, Ustyanksky V, Nikolskaya T, Nikolsky Y, Karchin R, Wilson PA, Kaminker JS, Zhang Z, Croshaw R, Willis J, Dawson D, Shipitsin M, Willson JK, Sukumar S, Polyak K, Park BH, Pethiyagoda CL, Pant PV, Ballinger DG, Sparks AB, Hartigan J, Smith DR, Suh E, Papadopoulos N, Buckhaults P, Markowitz SD, Parmigiani G, Kinzler KW,

Velculescu VE, Vogelstein B. The genomic landscapes of human breast and colorectal cancers. *Science* 2007;**318**:1108–13.

47. Stoica GE, Franke TF, Wellstein A, Czubayko F, List HJ, Reiter R, Morgan E, Martin MB, Stoica A. Estradiol rapidly activates Akt via the ErbB2 signaling pathway. *Mol Endocrinol* 2003;**17**:818–30.

48. Yu J, Henske EP. Estrogen-Induced Activation of Mammalian Target of Rapamycin Is Mediated via Tuberin and the Small GTPase Ras Homologue Enriched in Brain. *Cancer Res* 2006;**66**:9461–6.

49. Boulay A, Rudloff J, Ye J, Zumstein-Mecker S, O'Reilly T, Evans DB, Chen S, Lane HA. Dual inhibition of mTOR and estrogen receptor signaling in vitro induces cell death in models of breast cancer. *Clin Cancer Res* 2005;**11**:5319–28.

50. DeGraffenried LA, Friedrichs WE, Russell DH, Donzis EJ, Middleton AK, Silva JM, Roth RA, Hidalgo M. Inhibition of mTOR activity restores tamoxifen response in breast cancer cells with aberrant Akt Activity. *Clin Cancer Res* 2004;**10**:8059–67.

51. Beeram M, Tan QT, Tekmal R, Russell D, Middleton A, DeGraffenried L. Akt-induced endocrine therapy resistance is reversed by inhibition of mTOR signaling. *Ann Oncol* 2007;**18**:1323–8.

52. Johnston SR. Clinical efforts to combine endocrine agents with targeted therapies against epidermal growth factor receptor/human epidermal growth factor receptor 2 and mammalian target of rapamycin in breast cancer. *Clin Cancer Res* 2006;**12**:1061s–1068s.

53. Clark GJ, Kinch MS, Rogers-Graham K, Sebti SM, Hamilton AD, Der CJ. The Ras-related protein Rheb is farnesylated and antagonizes Ras signaling and transformation. *J Biol Chem* 1997;**272**:10,608–15.

54. Yu Jr Y, Xu F, Peng H, Fang X, Zhao S, Li Y, Cuevas B, Kuo WL, Gray JW, Siciliano M, Mills GB, Bast RC. NOEY2 (ARHI), an imprinted putative tumor suppressor gene in ovarian and breast carcinomas. *Proc Natl Acad Sci U S A* 1999;**96**:214–9.

55. Lu Z, Luo RZ, Peng H, Huang M, Nishimoto A, Hunt KK, Helin K, Liao WS, Yu Y. E2F-HDAC complexes negatively regulate the tumor suppressor gene ARHI in breast cancer. *Oncogene* 2006;**25**:230–9.

56. Yuan Jr J, Luo RZ, Fujii S, Wang L, Hu W, Andreeff M, Pan Y, Kadota M, Oshimura M, Sahin AA, Issa JP, Bast RC, Yu Y. Aberrant methylation and silencing of ARHI, an imprinted tumor suppressor gene in which the function is lost in breast cancers. *Cancer Res* 2003;**63**:4174–80.

57. Feng W, Lu Z, Luo RZ, Zhang X, Seto E, Liao WS, Yu Y. Multiple histone deacetylases repress tumor suppressor gene ARHI in breast cancer. *Int J Cancer* 2007;**120**:1664–8.

58. Wang Jr L, Hoque A, Luo RZ, Yuan J, Lu Z, Nishimoto A, Liu J, Sahin AA, Lippman SM, Bast RC, Yu Y. Loss of the expression of the tumor suppressor gene ARHI is associated with progression of breast cancer. *Clin Cancer Res* 2003;**9**:3660–6.

59. Shi Z, Zhou X, Xu L, Zhang T, Hou Y, Zhu W, Zhang T. [NOEY2 gene mRNA expression in breast cancer tissue and its relation to clinicopathological parameters]. *Zhonghua Zhong Liu Za Zhi* 2002;**24**:475–8.

60. Yu Jr Y, Luo R, Lu Z, Wei Feng W, Badgwell D, Issa JP, Rosen DG, Liu J, Bast RC. Biochemistry and Biology of ARHI (DIRAS3), an Imprinted Tumor Suppressor Gene Whose Expression Is Lost in Ovarian and Breast Cancers. *Methods Enzymol* 2005;**407**:455–8.

61. Bao Jr JJ, Le XF, Wang RY, Yuan J, Wang L, Atkinson EN, LaPushin R, Andreeff M, Fang B, Yu Y, Bast RC. Reexpression of the tumor suppressor gene ARHI induces apoptosis in ovarian and breast cancer cells through a caspase-independent calpain-dependent pathway. *Cancer Res* 2002;**62**:7264–72.

62. Luo RZ, Fang X, Marquez R, Liu SY, Mills GB, Liao WS, Yu Y, Bast RC. ARHI is a Ras-related small G-protein with a novel N-terminal extension that inhibits growth of ovarian and breast cancers. *Oncogene* 2003;**22**:2897–2909.

63. Nishimoto Jr A, Yu Y, Lu Z, Mao X, Ren Z, Watowich SS, Mills GB, Liao WS, Chen X, Bast RC, Luo RZ. A Ras homologue member I directly inhibits signal transducers and activators of transcription 3 translocation and activity in human breast and ovarian cancer cells. *Cancer Res* 2005;**65**:6701–10.

64. Finlin BS, Gau CL, Murphy GA, Shao H, Kimel T, Seitz RS, Chiu YF, Botstein D, Brown PO, Der CJ, Tamanoi F, Andres DA, Perou CM. RERG is a novel ras-related, estrogen-regulated and growth-inhibitory gene in breast cancer. *J Biol Chem* 2001;**276**:42,259–67.

65. Sorlie T, Perou CM, Tibshirani R, Aas T, Geisler S, Johnsen H, Hastie T, Eisen MB, van de Rijn M, Jeffrey SS, Thorsen T, Quist H, Matese JC, Brown PO, Botstein D, Eystein Lonning P, Borresen-Dale AL. Gene expression patterns of breast carcinomas distinguish tumor subclasses with clinical implications. *Proc Natl Acad Sci USA* 2001; **98**:10,869–74.

66. Hanker AB, Morita S, Repasky GA, Ross DT, Seitz RS, Der CJ. Tools to study the function of the ras-related, estrogen-regulated growth inhibitor in breast cancer. *Methods Enzymol* 2008;**439**:53–72.

67. Key MD, Andres DA, Der CJ, Repasky GA. Characterization of RERG: An Estrogen-Regulated Tumor Suppressor Gene. *Methods Enzymol* 2005;**407**:513–27.

68. Wang AG, Fang W, Han YH, Cho SM, Choi JY, Lee KH, Kim WH, Kim JM, Park MG, Yu DY, Kim NS, Lee DS. Expression of the RERG gene is gender-dependent in hepatocellular carcinoma and regulated by histone deacetyltransferases. *J Korean Med Sci* 2006; **21**:891–6.

69. Fujii Jr S, Luo RZ, Yuan J, Kadota M, Oshimura M, Dent SR, Kondo Y, Issa JP, Bast RC, Yu Y. Reactivation of the silenced and imprinted alleles of ARHI is associated with increased histone H3 acetylation and decreased histone H3 lysine 9 methylation. *Hum Mol Genet* 2003;**12**:1791–1800.

70. Glaser KB. HDAC inhibitors: Clinical update and mechanism-based potential. *Biochem Pharmacol* 2007;**74**:659–71.

71. Yoo CB, Jones PA. Epigenetic therapy of cancer: past, present and future. *Nat Rev Drug Discov* 2006;**5**:37–50.

72. Sharma D, Saxena NK, Davidson NE, Vertino PM. Restoration of Tamoxifen Sensitivity in Estrogen Receptor-Negative Breast Cancer Cells: Tamoxifen-Bound Reactivated ER Recruits Distinctive Corepressor Complexes. *Cancer Res* 2006;**66**:6370–8.

73. Bayliss J, Hilger A, Vishnu P, Diehl K, El-Ashry D. Reversal of the Estrogen Receptor Negative Phenotype in Breast Cancer and Restoration of Antiestrogen Response. *Clin Cancer Res* 2007; **13**:7029–36.

74. Chu I, Arnaout A, Loiseau S, Sun J, Seth A, McMahon C, Chun K, Hennessy B, Mills GB, Nawaz Z, Slingerland JM. Src promotes estrogen-dependent estrogen receptor alpha proteolysis in human breast cancer. *J Clin Invest* 2007;**117**:2205–15.

75. Itoh M, Nelson CM, Myers CA, Bissell MJ. Rap1 integrates tissue polarity, lumen formation, and tumorigenic potential in human breast epithelial cells. *Cancer Res* 2007;**67**:4759–66.

76. Reynet C, Kahn CR. Rad: a member of the Ras family overexpressed in muscle of type II diabetic humans. *Science* 1993;**262**:1441–4.

77. Tseng YH, Vicent D, Zhu J, Niu Y, Adeyinka A, Moyers JS, Watson PH, Kahn CR. Regulation of growth and tumorigenicity of breast cancer cells by the low molecular weight GTPase Rad and nm23. *Cancer Res* 2001;**61**:2071–9.

78. Suzuki M, Shigematsu H, Shames DS, Sunaga N, Takahashi T, Shivapurkar N, Iizasa T, Minna JD, Fujisawa T, Gazdar AF.

Methylation and gene silencing of the Ras-related GTPase gene in lung and breast cancers. *Ann Surg Oncol* 2007;**14**:1397–404.

79. Bodemann BO, White MA. Ral GTPases and cancer: linchpin support of the tumorigenic platform. *Nat Rev Cancer* 2008;**8**:133–40.

80. Lim KH, Baines AT, Fiordalisi JJ, Shipitsin M, Feig LA, Cox AD, Der CJ, Counter CM. Activation of RalA is critical for Ras-induced tumorigenesis of human cells. *Cancer Cell* 2005;**7**:533–45.

81. Yu Y, Feig LA. Involvement of R-Ras and Ral GTPases in estrogen-independent proliferation of breast cancer cells. *Oncogene* 2002;**21**:7557–68.

82. Sjoblom T, Jones S, Wood LD, Parsons DW, Lin J, Barber TD, Mandelker D, Leary RJ, Ptak J, Silliman N, Szabo S, Buckhaults P, Farrell C, Meeh P, Markowitz SD, Willis J, Dawson D, Willson JK, Gazdar AF, Hartigan J, Wu L, Liu C, Parmigiani G, Park BH, Bachman KE, Papadopoulos N, Vogelstein B, Kinzler KW, Velculescu VE. The consensus coding sequences of human breast and colorectal cancers. *Science* 2006;**314**:268–274.

83. van Agthoven T, van Agthoven TL, Dekker A, van der Spek PJ, Vreede L, Dorssers LC. Identification of BCAR3 by a random search for genes involved in antiestrogen resistance of human breast cancer cells. *Embo J* 1998;**17**:2799–808.

84. Gotoh T, Cai D, Tian X, Feig LA, Lerner A. p130Cas regulates the activity of AND-34, a novel Ral, Rap1, and R-Ras guanine nucleotide exchange factor. *J Biol Chem* 2000;**275**:30,118–23.

85. Felekkis KN, Narsimhan RP, Near R, Castro AF, Zheng Y, Quilliam LA, Lerner A. AND-34 activates phosphatidylinositol 3-kinase and induces anti-estrogen resistance in a SH2 and GDP exchange factor-like domain-dependent manner. *Mol Cancer Res* 2005;**3**:32–41.

86. Cai D, Iyer A, Felekkis KN, Near RI, Luo Z, Chernoff J, Albanese C, Pestell RG, Lerner A. AND-34/BCAR3, a GDP exchange factor whose overexpression confers antiestrogen resistance, activates Rac, PAK1, and the cyclin D1 promoter. *Cancer Res* 2003;**63**:6802–8.

87. Tang Y, Olufemi L, Wang MT, Nie D. Role of Rho GTPases in breast cancer. *Front Biosci* 2008;**13**:759–776.

88. Burbelo P, Wellstein A, Pestell RG. Altered Rho GTPase signaling pathways in breast cancer cells. *Breast Cancer Res Treat* 2004;**84**:43–48.

89. Cheng KW, Lahad JP, Gray JW, Mills GB. Emerging role of RAB GTPases in cancer and human disease. *Cancer Res* 2005;**65**:2516–9.

Chapter 40

Signaling Pathways in the Normal and Neoplastic Breast

Tushar B. Deb, Danica Ramljak, Robert B. Dickson, Michael D. Johnson and Robert Clarke

Department of Oncology, Lombardi Comprehensive Cancer Center, Georgetown University, Washington, DC

INTRODUCTION

A thorough discussion of the signaling pathways operant in the normal mammary gland would easily fill an entire volume, before even considering the derangement of those pathways in mammary cancer. This chapter is therefore, of necessity, highly selective in nature, and examines in detail a limited selection of pathways in a limited number of contexts. Where we have touched upon a topic without exploring it in depth, we have tried to direct the reader to one of the many excellent review articles written by leaders in those particular fields.

In their broadest sense, signal transduction pathways include all biochemical cellular pathways that modulate or alter cellular behavior or function. During normal embryonic development and in adult life signaling needs to be precisely coordinated and integrated at all times, because properly regulated differentiation signals are critical for preventing oncogenesis. Cancers, therefore, do not necessarily arise as a result of an increased rate of cellular proliferation. Rather, carcinogenesis is a combination of defects in cell cycle progression (cellular division), immortalization, genomic instability, cell survival regulation (apoptosis, autophagy, senescence, necrosis), cell–cell and cell–substrate adhesion, and angiogenesis. Deregulated cell and tissue growth is a defining feature of all neoplasms, both benign and malignant. Malignant neoplasms have the capacity to invade normal tissues, to induce the development of a local vasculature, to metastasize, and to grow at distant body sites. All of the behaviors of malignant mammary cancers can be found at some stage and in some context during normal mammary development and mature function, and so, to a large extent, mammary carcinogenesis and malignant progression can be viewed as the inappropriate manifestation of normal functions as a result of perturbed signaling pathways. Thus, understanding the role of signaling in normal mammary processes is invaluable to any description of lesions present in mammary cancers.

During the latter part of the last century, much effort was devoted to the identification of the mechanisms of action of estrogen and progesterone at the local tissue level, in the normal breast, and during initiation, early promotion, and later progression to malignancy. The role of estrogens in breast cancer initiation remains somewhat controversial, whereas a role in promotion and progression is relatively well established. Breast tissue regulation by these hormones is modulated in a rather complex fashion. Function of the cognate nuclear hormone receptors is regulated by a series of co-activators and co-repressors. A consequence of receptor activation regulates the expression of various transcription factors and other genes, including those encoding autocrine and paracrine growth factors. These hormone-regulated factors control epithelial cellular differentiation, and epithelial cell–cell and cell–stromal adhesion.

A large body of breast cancer research has been focused on understanding the complex interactions among growth factors and deregulated growth regulatory genes (oncogenes or proto-oncogenes and suppressor genes) in mediating or modulating endocrine steroid action in breast cancer. Another important area of investigation has been studied, of the involvement of growth factors in facilitating malignant progression of the disease. One area of research has examined defective tumor host interactions, resulting in aberrant stromal collagen synthesis (desmoplasia), epithelial cell invasion, and vascular infiltration (angiogenesis) to promote distant metastasis. Furthermore, several studies have found that certain growth factors may suppress the host immune response to the tumor and may influence a tumor's response to therapy. Over the past few years, much progress has been made toward developing an appreciation for the plethora of signaling pathways that are involved in normal mammary gland biology and in elucidating the role of specific components of those pathways. However, it is becoming evident that several major pathways may be of particular importance both in the normal and malignant gland; this chapter will focus on a selection of these pathways. Other important pathways that have been the subject of other reviews we will only mention in passing. These include signaling from adhesion molecules [1, 2], the TGFβ pathway [3, 4], and steroid hormones.

The mammary gland is a fascinating organ from a developmental standpoint, going through several distinct phases of growth during which the relationships between, and importance of, various signaling pathways are dramatically altered. Formation of the mammary gland, which originates in the ectodermal layer during embryogenesis, involves complex interactions between epithelial and mesenchymal components. These interactions result in the coordinated activation of an array of signaling pathways. Since these processes have recently been the subject of an excellent review [5], they will not be examined further here.

During pre-pubertal life, the mammary gland grows at essentially the same rate as the other organs, and undergoes little additional development until the onset of puberty. At puberty, the influence of ovarian hormones, in concert with an array of signaling pathways, stimulates the gland to begin to develop into its adult form. During reproductive life, the gland undergoes cyclical re-modeling throughout the ovarian cycle. Other pathways are invoked to prepare the organ for lactation during pregnancy. At weaning, a coordinated series of signals triggers several waves of apoptosis, initiating the regression of the mother's gland to the resting adult state. Not surprisingly, many of the key regulators of these pathways are the locus of what are thought to be key lesions that can lead to transformation and malignant progression of cells within the mammary gland. The role of several of these pathways in this process will be considered below; specifically the epidermal growth factor (EGF) family, and signaling through the phosphatidylinositol 3-kinase (PI3K)/phosphatase and tensin homolog (PTEN)/AKT axis.

THE EPIDERMAL GROWTH FACTOR FAMILY

Members of the Epidermal Growth Factor (EGF) family play a critical role in the mammary gland, and work over the past few years has revealed an increasingly complex regulatory network involving these proteins and the family of receptors on which they act. Two major classes of structurally and functionally distinct TGF families were initially

TABLE 40.1 Members of the EGFR family and their ligands[1]

Receptor	Ligands
EGFR (c-Erb-B1)	Epidermal growth factor (EGF), transforming growth factor-α, amphiregulin, heparin-binding EGF-like growth factor (Hb EGF), betacellulin (BC), epiregulin (EP)
c-ErbB2 (HER2)	Not established
c-ErbB3 (HER3)	Neuregulins
c-ErbB4 (HER4)	Neuregulins, Hb EGF, BC

[1]The four ErbB receptors are shown along with their ligands. Note that ErbB2 does not have a direct binding ligand.

characterized by the prototypic transforming growth factor α (TGFα) and transforming growth factor β (TGFβ). The TGFβs and their receptors are structurally distinct from TGFα/EGFR, and are briefly discussed below. TGFα is closely homologous to epidermal growth factor (EGF), and both ligands bind to and activate the EGF receptor (EGFR) [6]. EGFR is a tyrosine-kinase which, on activation, stimulates several cellular responses, including survival, proliferation, motility, and differentiation [6], and which is the prototype for a family of four receptors collectively called the ErbB family (EGFR/HER-1, ErbB2/HER-2, ErbB3/HER-3, and ErbB4/HER-4; see Table 40.1) [7].

The EGF family members that bind to EGFR are EGF, TGFα, amphiregulin (AR, a heparin-binding factor), heparin-binding EGF (HbEGF), epiregulin, and β-cellulin [8]. Cripto (CR-1) is an EGF family member that also plays an important role in embryogenesis and mammary gland development. CR-1 is overexpressed in several human tumors [9, 10]. However, CR-1 binds to a type I serine/threonine kinase receptor for activin (ALK4), which is expressed on the cell surface of mammary epithelial cells [11], rather than binding to EGFR or one of the other ErbB receptors. A related class of molecules known as the heregulins (human) and NDFs (neu differentiation factor, from rat) [12, 13], now commonly known simply as the neuregulins, bind to ErbB heterodimers. Whereas c-ErbB4 is a tyrosine kinase, c-ErbB3 is kinase defective. A fourth member, c-ErbB2 (HER2/neu), binds no secreted ligand and is activated through heterodimeric interaction with other family members. c-ErbB2 is particularly important as an oncogene in breast cancer [14].

ErbB receptors undergo homo- or heterodimerization following ligand binding. After ligand binding, EGFR/ErbB1/HER-1 undergoes either endocytosis and degradation by both proteasomal and lysosomal pathways, or the receptor is recycled to plasma membrane. ErbB2/HER-2 is considered to be endocytosis-impaired, and is the most stable protein among ErbBs in the plasma membrane. Largely as a consequence of its stability, ErbB2/HER-2 is the preferred heterodimerization partner among the ErbBs.

All four members of the EGFR family (Table 40.1), and most of the EGFR-related growth factors, play a role in tumor growth and development, and in the progression of human breast cancer. The availability of targeted mutant mice, in which various members of the ligand and receptor families have been knocked out, has proven to be invaluable to elucidating the role that these proteins play in mammary gland development and function. Stromal EGFR is activated by amphiregulin (AR) in a paracrine manner, following AR's shedding from neighboring mammary epithelial cells [15]. Shedding is mediated by the ADAM17 protease, which cleaves the epithelial-bound AR precursor on the outside of the plasma membrane. This mechanism revealed how the combined effect of EGFR and its ligand AR, through an epithelial–mesenchymal interaction, regulates the process of mammary ductal morphogenesis [15, 16].

Ductal growth in the mammary epithelium is defective when ErbB2 is disrupted in the mammary gland [17], implying a prominent role for ErbB2 in mammary ductal growth. Cooperation between signaling molecules such as Akt and ErbB2 was observed in mammary specific bitransgenic mice, resulting in accelerated mammary tumor formation [18]. ErbB2 also plays a significant role in ductal morphogenesis [19, 20]. While kinase-dead, ErbB3/HER-3 frequently forms a high-affinity co-receptor for heregulin by heterodimerization with ErbB2/HER-2. ErbB3/HER-3 possesses several docking sites for PI3K, and can initiate PI3K/Akt signaling when transactivated by ErbB2/HER-2. Heterodimerized ErbB-2/HER-2-ErbB3/HER-3 acts as an oncogenic unit for breast tumor cell proliferation [21].

While the basic mechanism of ErbB signaling is identical in both normal and malignant mammary gland, it is frequently deregulated and uncontrolled in breast cancer. This is often the result of gene amplification or overexpression, and/or activating mutations in the ErbB receptors and ligands. Steroid-growth factor interactions have been studied in human mammary tissue most frequently in the context of malignant epithelium; these interactions are almost certainly also crucial in the regulation of the normal gland and in the development of cancer. In hormone-responsive human breast cancer cells, estrogen-induced proliferation is accompanied by an increase in growth stimulatory TGFα, AR, and IGF-II, modulation of IGF-binding proteins, induction of EGF and IGF-I receptors, inhibition of IGF-II and c-ErbB2 receptors, and inhibition of TGFβ [22].

The role of EGF family members in tumor onset and progression is well established [8]. TGFα has been shown, both in cell lines *in vitro* and in experimental animal models *in vivo*, to be a positive modulator of cellular transformation. These effects of TGFα are due to its regulatory effects on proliferation, survival, and motility, as well as its modulation of differentiation [23]. Several laboratories have utilized mouse models in which the EGFR

ligand, TGFα, was overexpressed in the mammary gland under control of either the MMTV-LTR or WAP promoter [24–26]. Expression of the TGFα transgene in the mammary gland caused anomalous development. For example, alveoli appeared precociously and postlactational involution was impaired, resulting in the persistence of epithelial structures, termed hyperplastic alveolar nodules. The mice also developed focal mammary tumors with a high incidence and short latency [24], indicating that EGFR signaling leads to neoplastic progression in this tissue.

Humphreys and Hennighausen [25, 27], compared TGFα-induced mammary tumorigenesis in wild-type mice with those lacking a functional Stat5a gene. These investigators found that the absence of Stat5a delayed hyperplasia and tumor development and, coincidentally, promoted more epithelial regression. However, these effects were not observed in TGFα transgenic mice containing Stat5a.

Much effort has been focused on understanding the cooperation and synergy between TGFα and c-Myc. c-Myc, a downstream effector of the c-ErbB2 oncogene, is frequently amplified and overexpressed in human breast cancer [28]. Bitransgenic c-Myc/TGFα mice develop multiple aggressive mammary tumors with a dramatically shorter latency compared to either single transgenic lineage [24]. These results indicate a strong synergy between TGFα and c-Myc, which could reflect the ability of c-Myc to amplify autocrine growth circuits, including those involving EGFR.

Recent data suggest the possibility of cooperation between cyclin D1 and TGFα [29]. Early upregulation of cyclin D1 by TGFα might circumvent the normal ability of c-Myc to repress this cell cycle regulator, perhaps obviating the need for genetic alterations that would otherwise uncouple c-Myc and cyclin D1 during neoplastic progression. Work by Shroeder and colleagues [26] focused on identifying genes that cooperate with TGFα in mammary tumorigenesis. Based on their results, TGFα appears to upregulate the Wnt3 gene. Thus, the apparent synergy between the EGFR and Frizzled signaling pathways might contribute to neoplastic progression. Both Wnt and ErbB receptors can regulate β-catenin activity leading to cellular disaggregation. Therefore, it is reasonable to speculate that the observed synergy involves this pathway.

Besides the role of TGFα in mammary tumorigenesis, other EGFR ligands regulate proliferation. The closely related factor, EGF, can act as an oncogene-like molecule when transfected and overexpressed in immortalized rodent fibroblasts [30]. Furthermore, it has been shown that both AR and CR-1 may be important in early stages of the onset of breast cancer [30–35]. Forced overexpression of EGFR in the mammary gland, under the control of the MMTV or β-lactoglobulin (BLG) promoters, resulted in abnormal mammary gland development and the production of epithelial hyperplasias. With multiple pregnancies, dysplasias and tubular adenocarcinomas were also observed. Differentiation of the mammary epithelium was perturbed in response to deregulated EGFR, as reflected by fewer alveoli developing in whole-mount organ cultures. Similar to EGFR, overexpression of c-ErbB2 in the transgenic mouse, or in the transgenic mammary gland after retroviral transfer, also leads to pregnancy-induced mammary tumors [7].

The roles of c-ErbB3 and c-ErbB4 in mammary tumorigenesis are still incompletely understood. Increasing evidence suggests that whereas overexpression of c-ErbB3 like the EGFR and c-ErbB2 is associated with more aggressive cancer, c-ErbB4 tends to be lost in these tumors. c-ErbB4 signaling appears to inhibit proliferation and is pro-apoptotic. It is clear that complete elucidation of this pathway will require significant effort from researchers for many years to come.

OTHER GROWTH FACTOR FAMILIES

The TGFβ family consists of several related gene products, each forming 25-kDa homodimeric or heterodimeric species, and is expressed in both normal and neoplastic mammary epithelium. Three membrane-binding proteins interact with this family of growth factors. These were initially termed receptors (type I, II, and III). Type III receptors appear to be non-signaling proteins, whereas the other receptor types (type I and II) are serine/threonine kinases and have been shown to deliver intracellular signals [36–40]. Four different type II receptors have been cloned, and they each may associate with one of several type I receptors. The function of type II receptors is determined on the basis of which type I receptor is recruited into each heterodimer. TGFβ ligands only directly bind to type II receptors.

The role of this signaling system in mammary gland development and carcinogenesis is complex. The various components of the system are expressed, under tight special and temporal control, in both the epithelial and mesenchymal components of the gland during development. Recent advances in the role of TGFβ signaling in mammary development and carcinogenesis have been reviewed by several investigators [3, 4]. In addition to the EGFR and TGFβ growth factor families, at least five other growth factor stimulatory molecules likely play a role in breast cancer. These molecules are summarized in Table 40.2: insulin-like growth factors (type I and II) [41], members of the Wnt (wingless) growth factor receptor family (Wnt-2, Wnt-3, Wnt-4, Wnt-5a, and Wnt-7b) [42], platelet-derived growth factors A and B [43], and the fibroblast growth factor (FGF) family [44].

Each of these growth factor classes binds to one or more specific tyrosine kinase-encoding receptors. Vascular endothelial growth factor (VEGF, a member of a different family of tyrosine kinase receptor-binding factors) [45], pleiotrophin (a developmental, neurotropic factor) [46], and hepatocyte growth factor (HGF; also called scatter factor) and its tyrosine kinase encoding receptor, c-Met [47], are all

TABLE 40.2 Diverse group of growth factors and other molecules thought to play roles in breast cancer[1]

Signaling molecules	Cellular response
Insulin-like growth factors (type I and II)	Stimulatory/tumor cells
Wnt growth factor family	Stimulatory/tumor cells
Platelet-derived growth factors A and B	Stimulatory/stromal cells[2]
Fibroblast growth factors	Stimulatory/tumor cells
Vascular endothelial growth factor	Angiogenesis/vascular cells[2]
Hepatocyte growth factor, receptor c-Met	Stimulatory/tumor cells
Prolactin	Stimulatory/tumor cells
Mammary-derived growth factor 1	Stimulatory/tumor cells
Pleiotrophin (developmental neurotropic factor)	Stimulatory/tumor cells[2]

[1]Proposed autocrine and paracrine factors in breast cancer. Tumor cells release variety of growth factors that might play autocrine roles in vivo (this is based mostly on their activity in vitro). Several of the same factors also are known to play paracrine and endocrine roles as well.
[2]Additional, angiogenic effects of these factors.

produced by breast cancer. In addition, breast cancer cells also produce the hormone prolactin [48] and mammary-derived growth factor 1 (MDGF-1) [49]. While the roles of some of these molecules are well established in breast signaling, much more research is needed to understand fully their role in both mammary gland development and human breast cancer.

THE PI 3-KINASE–AKT AND PTEN AXIS

One of the most exciting advances over the past few years has been the growing appreciation of the importance of phosphatidylinositol 3-kinase (PI3K) mediated signaling in the mammary gland and its dysregulation in breast cancer. The PI3K/Akt/PTEN axis is involved in some aspect of the downstream signaling of all of the systems described above. Survival signals from the EGFR pathway, for example, are transmitted both by the extracellular signal regulated kinase (Erk1/2) pathway and the PI3K/Akt system [50, 51], though the PI3K pathway seems to predominate.

Akt is a serine/threonine kinase, downstream of PI3K activation, which delivers strong survival signals in many cell types [52–54]. Both growth factors and integrins activate one of the three isoforms of Akt (Akt1, Akt2, and Akt3), each of which is expressed at different levels in various tissues [55]. Akt then modulates the activity and expression of multiple downstream targets involved in the regulation of cell growth, metabolism, and apoptosis. Akt downstream targets include the BCL2 family of proteins [56], caspase-9 [57], the Forkhead transcription factor [58], BCL-xL [59], MCL-1 [60] and NFκB activity [61]. Akt signaling has been the subject of multiple reviews [62–64].

The past few years have revealed the important role that aberrant PI3K signaling can play in mammary carcinogenesis. Deregulation of PI3K-Akt activation either by constitutive activation of PI3K or by deletion of PTEN, an endogenous Akt inhibitor, enhances the oncogenic potential of mammary gland. Constitutively active PI3K was shown to transform human mammary epithelial cells [65], thereby underscoring the potential role for PI3K in breast neoplasia. Several PI3KCA mutations have been reported in human breast tumor samples and breast cancer cell lines [66–71]. These mutations were associated with PTEN loss, ErbB2 overexpression, and estrogen-progesterone receptor expression, thus activating several components of the carcinogenesis machinery.

Consistent with its inhibitory role in breast tumorigenesis, PTEN overexpression in a transgenic mouse model enhanced apoptosis in the mammary gland [72]. In addition to its well-established role as tumor suppressor, PTEN has been recently shown to increase the efficacy of Tratsuzumab (Hereceptin) therapy [73]. This beneficial interaction appears to be the result of a previously unknown mechanism for increased PTEN recruitment and phosphatase activity by Src-mediated tyrosine phosphorylation. This work identifies a locus of cross-talk between signaling pathways, and indicates that parameters such as PTEN status should be considered during Herceptin therapy.

Akt is one of the best characterized of the pro-survival kinases. As in other organs, Akt has a major effect in regulating both the normal and neoplastic mammary gland. Mammary-specific transgene expression of Akt delays mammary gland involution [74, 75]. Recently, three groups have simultaneously reported that Akt phosphorylates the p27 cell cycle inhibitor, leading to the latter's mislocalization from the nucleus [76–78] and a resultant enhanced cell cycle progression. It was further shown that Akt-mediated phosphorylation of p27 at Thr-157 impairs its association with importin, a component of transport machinery, thus preventing p27 re-entry to nucleus [79]. These observations present another mechanism by which Akt positively regulates biochemical mechanisms that are pro-tumorigenic in nature.

Contrary to its pro-survival nature, Akt has also been found to have an unexpected additional role in breast cancer. In a recent report, Yoeli-Lerner and colleagues have shown that Akt activation inhibits breast cancer cell migration by indirectly regulating ubiquitination and degradation of the pro-invasive NFAT transcription factor [80]. In another study, downregulation of Akt1 enhanced cell migration by EGF and IGF-1 in MCF-10A cells [81]. These contrasting roles for Akt, when compared with its pro-tumorigenic role, demand a vigorous examination. Such additional examination is particularly timely in light of the ongoing clinical trials targeting the expected pro-tumorigenic role of Akt.

Regarding the regulation of Akt activity, two recent reports described two putative PDK-2s responsible for phosphorylating the hydrophobic S473 site on Akt. One of these PDK2 molecules is a DNA-dependent protein kinase (DNA-PK) [82], the other being Rictor, in association with companion mTOR complex [83, 84]. A new protein tyrosine phosphatase, called PHLPP, has been shown to dephosphorylate Akt at the S473 site in colon cancer and glioblastoma [85]. While the status of PHLPP is currently unknown in breast cancer, this would present an interesting therapeutic target, since PTEN is frequently mutated in breast cancer. PHLPP is the only phosphatase in the human genome with a Pleckstrin homology (PH) domain, and so potentially PHLPP has the ability to be membrane targeted in a PIP3 (product of PI3K)-dependent manner.

Calmodulin, a protein long known to play multiple roles in cellular signaling and previously shown to be involved in estrogen receptor signaling and stabilization [86–89], has recently also been demonstrated to participate in Akt-mediated survival signaling in the mammary gland. In c-Myc overexpressing mouse mammary carcinoma cells, calmodulin is required for EGF, insulin, and serum induced Akt activation. This effect is mediated downstream of PI3K, and appears to involve a direct interaction between Akt and calmodulin [90–92]. Pharmacologic inhibition of calmodulin results in the suppression of Akt signaling in a manner independent of PI3K activity. In addition to a pro-survival role for calmodulin, calmodulin kinase I was recently shown to control the G0–G1 cell cycle checkpoint in MCF-7 breast cancer cells [93]. Using pharmacologic and siRNA based approaches, Rodriguez-Mora and colleagues have shown that inhibition of calmodulin kinase I and its upstream activator calmodulin kinase kinase led to cell cycle arrest at G0–G1 due to inhibition of cyclin D1 synthesis and Rb hypophosphorylation. In contrast to the pro-survival and pro-proliferative nature of calmodulin and calmodulin kinase I, Death Associated Protein kinase (DAP) [94], a calmodulin kinase and a death-promoting kinase, is expressed in normal breast epithelium. However, DAP expression is often lost in human breast cancer [95]. Another study reported significant promoter hypermethylation of the DAP kinase gene in human breast cancer samples [96], suggesting that this pathway warrants further study.

CONCLUSIONS

In this chapter we have attempted to highlight some of the more important aspects of signaling in the normal and neoplastic mammary gland. The past few years have seem major advances in our understanding of the key aspects of the control of mammary gland development, function, neoplastic transformation, and tumor progression. Increased understanding of the importance of survival signaling in all of these processes, and the recognition of the critical role that derangement of PI3K pathway plays in breast cancer, hold the promise that the next few years will be an exciting time for the development of novel therapeutic strategies for the prevention and treatment of breast cancer.

ACKNOWLEDGEMENTS

This work was supported in part by US Public Health Service awards R01-CA096483, U54-CA100970, and P30-CA51008, and research awards from the US Department of Defense Breast Cancer Research program BC030280, BC073877.

DEDICATION

This chapter is dedicated to the memory of Dr Robert Dickson.

REFERENCES

1. Schatzmann F, Marlow R, Streuli CH. Integrin signaling and mammary cell function. *J Mamm Gland Biol Neoplasia* 2003;**8**:395–408.
2. Katz E, Streuli CH. The extracellular matrix as an adhesion checkpoint for mammary epithelial function. *Intl J Biochem Cell Biol* 2007;**39**:715–26.
3. Chang CF, Westbrook R, Ma J, Cao D. Transforming growth factor-beta signaling in breast cancer. *Front Biosci* 2007;**12**:4393–401.
4. Buck MB, Knabbe C. TGF-beta signaling in breast cancer. *Ann NY Acad Sci* 2006;**1089**:119–26.
5. Robinson GW. Cooperation of signalling pathways in embryonic mammary gland development. *Nature Rev Genet* 2007;**8**:963–72.
6. Kim ES, Khuri FR, Herbst RS. Epidermal growth factor receptor biology (IMC-C225). *Curr Opin Oncol* 2001;**13**:506–13.
7. Troyer KL, Lee DC. Regulation of mouse mammary gland development and tumorigenesis by the ERBB signaling network. *J Mamm Gland Biol Neoplasia* 2001;**6**:7–21.
8. Prenzel N, Zwick E, Leserer M, Ullrich A. Tyrosine kinase signalling in breast cancer. Epidermal growth factor receptor: convergence point for signal integration and diversification. *Breast Cancer Res* 2000;**2**:184–90.
9. Adamson ED, Minchiotti G, Salomon DS. Cripto: a tumor growth factor and more. *J Cell Physiol* 2002;**190**:267–78.
10. Salomon DS, Bianco C, Khan NI, DeSantis M, Normanno N, Wechselberger C, Seno M, Williams K, Sanicola M, Foley S, Gullick WJ, Persico G. The EGF-CFC family: novel epidermal growth factor-related proteins in development and cancer. *Endocr Relat Cancer* 2000;**7**:199–226.
11. Bianco C, Adkins HB, Wechselberger C, Seno M, Normanno N, De Luca A, Sun Y, Khan N, Kenney N, Ebert A, Williams KP, Sanicola M, Salomon DS. Cripto-1 activates nodal- and ALK4-dependent and -independent signaling pathways in mammary epithelial Cells. *Mol Cell Biol* 2002;**22**:2586–97.
12. Holmes WE, Sliwkowski MX, Akita RW, Henzel WJ, Lee J, Park JW, Yansura D, Abadi N, Raab H, Lewis GD. Identification of heregulin, a specific activator of p185erbB2. *Science* 1992;**256**:1205–10.
13. Peles E, Bacus SS, Koski RA, Lu HS, Wen D, Ogden SG, Levy RB, Yarden Y. Isolation of the neu/HER-2 stimulatory ligand: a 44-kD glycoprotein that induces differentiation of mammary tumor cells. *Cell* 1992;**69**:205–16.
14. Eccles SA. The role of c-erbB-2/HER2/neu in breast cancer progression and metastasis. *J Mamm Gland Biol Neoplasia* 2001;**6**:393–406.
15. Sternlicht MD, Sunnarborg SW, Kouros-Mehr H, Yu Y, Lee DC, Werb Z. Mammary ductal morphogenesis requires paracrine activation of stromal EGFR via ADAM17-dependent shedding of epithelial amphiregulin. *Development* 2005;**132**:3923–33.
16. Wiesen JF, Young P, Werb Z, Cunha GR. Signaling through the stromal epidermal growth factor receptor is necessary for mammary ductal development. *Development* 1999;**126**:335–44.
17. Andrechek ER, White D, Muller WJ. Targeted disruption of ErbB2/Neu in the mammary epithelium results in impaired ductal outgrowth. *Oncogene* 2005;**24**:932–7.
18. Hutchinson JN, Jin J, Cardiff RD, Woodgett JR, Muller WJ. Activation of Akt-1 (PKB-alpha) can accelerate ErbB-2-mediated mammary tumorigenesis but suppresses tumor invasion. *Cancer Res* 2004;**64**:3171–8.
19. Jackson-Fisher AJ, Bellinger G, Ramabhadran R, Morris JK, Lee KF, Stern DF. ErbB2 is required for ductal morphogenesis of the mammary gland. *Proc Natl Acad Sci USA* 2004;**101**:17,138–17,143.
20. Stern DF. ErbBs in mammary development. *Exp Cell Res* 2003;**284**:89–98.
21. Holbro III T, Beerli RR, Maurer F, Koziczak M, Barbas CF, Hynes NE. The ErbB2/ErbB3 heterodimer functions as an oncogenic unit: ErbB2 requires ErbB3 to drive breast tumor cell proliferation. *Proc Natl Acad Sci USA* 2003;**100**:8933–8.
22. Dickson RB, Lippman ME. Autocrine and paracrine growth factors in the normal and neoplastic breast. In: Harris JR, editor. *Diseases of the breast.* Philadelphia, PA: Lippincott Williams and Wilkins; 2000. p. 303–17.
23. Amundadottir LT, Leder P. Signal transduction pathways activated and required for mammary carcinogenesis in response to specific oncogenes. *Oncogene* 1998;**16**:737–46.
24. Amundadottir LT, Nass SJ, Berchem GJ, Johnson MD, Dickson RB. Cooperation of TGF alpha and c-Myc in mouse mammary tumorigenesis: coordinated stimulation of growth and suppression of apoptosis. *Oncogene* 1996;**13**:757–65.
25. Humphreys RC, Hennighausen L. Transforming growth factor alpha and mouse models of human breast cancer. *Oncogene* 2000;**19**:1085–91.
26. Schroeder JA, Troyer KL, Lee DC. Cooperative induction of mammary tumorigenesis by TGFalpha and Wnts. *Oncogene* 2000;**19**:3193–9.
27. Humphreys RC, Hennighausen L. Signal transducer and activator of transcription 5a influences mammary epithelial cell survival and tumorigenesis. *Cell Growth Differ* 1999;**10**:685–94.
28. Nass SJ, Dickson RB. Defining a role for c-Myc in breast tumorigenesis. *Breast Cancer Res Treat* 1997;**44**:1–22.
29. Liao DJ, Natarajan G, Deming SL, Jamerson MH, Johnson M, Chepko G, Dickson RB. Cell cycle basis for the onset and progression of c-Myc-induced, TGFalpha-enhanced mouse mammary gland carcinogenesis. *Oncogene* 2000;**19**:1307–17.
30. Schroeder JA, Lee DC. Transgenic mice reveal roles for TGFalpha and EGF receptor in mammary gland development and neoplasia. *J Mamm Gland Biol Neoplasia* 1997;**2**:119–29.
31. Schroeder JA, Lee DC. Dynamic expression and activation of ERBB receptors in the developing mouse mammary gland. *Cell Growth Differ* 1998;**9**:451–64.

32. Kenney NJ, Smith GH, Johnson MD, Rosenberg K, Gullick WJ, Salomon DS, Dickson RB. Cripto-1 activity in the intact and ovariectomized virgin mouse mammary gland. *Pathogenesis* 1997;**1**:157–71.

33. Kenney NJ, Smith GH, Rosenberg K, Cutler ML, Dickson RB. Induction of ductal morphogenesis and lobular hyperplasia by amphiregulin in the mouse mammary gland. *Cell Growth Differ* 1996;**7**:1769–81.

34. Amundadottir LT, Merlino G, Dickson RB. Transgenic mouse models of breast cancer. *Breast Cancer Res Treat* 1996;**39**:119–35.

35. Edwards paw. Control of three dimensional growth pattern of mammary epithelium: Role of genes of the WNT and erbB families studied using reconstituted epithelium. In: Rudland PS, Fernig DG, Leinster S, Lunt GG, editors. *Mammary development and cancer*. London: Portland Press; 1998. p. 21–6.

36. Heldin CH, Miyazono K, ten Dijke P. TGF-beta signalling from cell membrane to nucleus through SMAD proteins. *Nature* 1997;**390**:465–71.

37. Massague J. Receptors for the TGF-beta family. *Cell* 1992;**69**:1067–70.

38. Ebner R, Chen RH, Shum L, Lawler S, Zioncheck TF, Lee A, Lopez AR, Derynck R. Cloning of a type I TGF-beta receptor and its effect on TGF-beta binding to the type II receptor. *Science* 1993;**260**:1344–8.

39. Attisano L, Carcamo J, Ventura F, Weis FM, Massague J, Wrana JL. Identification of human activin and TGF beta type I receptors that form heteromeric kinase complexes with type II receptors. *Cell* 1993;**75**:671–80.

40. Liu X, Yue J, Frey RS, Zhu Q, Mulder KM. Transforming growth factor beta signaling through Smad1 in human breast cancer cells. *Cancer Res* 1998;**58**:4752–7.

41. Ellis mj. The insulin-like growth factor network and breast cancer. In: Bowcock A, editor. *Breast cancer: molecular genetics, pathogenesis, and therapeutics*. Totowa, NJ: Humana Press; 1999:121.

42. Bergstein I, Brown AM. WNT genes and breast cancer. In: Bowcock A, editor. *Breast cancer: molecular genetics, pathogenesis, and therapeutics*. Totowa, NJ: Humana Press; 1999:181.

43. Bronzert DA, Pantazis P, Antoniades HN, Kasid A, Davidson N, Dickson RB, Lippman ME. Synthesis and secretion of platelet-derived growth factor by human breast cancer cell lines. *Proc Natl Acad Sci USA* 1987;**84**:5763–7.

44. Kern FG. The role of fibroblast growth factors in breast cancer pathology and progression. In: Bowcock A, editor. *Breast cancer: molecular genetics, pathogenesis, and therapeutics*. Totowa, NJ: Humana Press; 1999:59.

45. Ferrara N, Houck K, Jakeman L, Leung DW. Molecular and biological properties of the vascular endothelial growth factor family of proteins. *Endocr Rev* 1992;**13**:18–32.

46. Wellstein A, Fang WJ, Khatri A, Lu Y, Swain SS, Dickson RB, Sasse J, Riegel AT, Lippman ME. A heparin-binding growth factor secreted from breast cancer cells homologous to a developmentally regulated cytokine. *J Biol Chem* 1992;**267**:2582–7.

47. Rong S, Bodescot M, Blair D, Dunn J, Nakamura T, Mizuno K, Park M, Chan A, Aaronson S, Vande Woude GF. Tumorigenicity of the met proto-oncogene and the gene for hepatocyte growth factor. *Mol Cell Biol* 1992;**12**:5152–8.

48. Clevenger CV, Chang WP, Ngo W, Pasha TL, Montone KT, Tomaszewski JE. Expression of prolactin and prolactin receptor in human breast carcinoma. Evidence for an autocrine/paracrine loop. *Am J Pathol* 1995;**146**:695–705.

49. Bano M, Kidwell WR, Dickson RB. MDGF-1: a multifunctional growth factor in human milk and human breast cancer. In: Dickson RB, Lippman ME, editors. *Mammary tumorigenesis and malignant progression*. Boston, MA: Kluwer; 1994:193.

50. Roberts RA, James NH, Cosulich SC. The role of protein kinase B and mitogen-activated protein kinase in epidermal growth factor and tumor necrosis factor alpha-mediated rat hepatocyte survival and apoptosis. *Hepatology* 2000;**31**:420–7.

51. Sibilia M, Fleischmann A, Behrens A, Stingl L, Carroll J, Watt FM, Schlessinger J, Wagner EF. The EGF receptor provides an essential survival signal for SOS-dependent skin tumor development. *Cell* 2000;**102**:211–20.

52. Downward J. Mechanisms and consequences of activation of protein kinase B/Akt. *Curr Opin Cell Biol* 1998;**10**:262–7.

53. Kandel ES, Hay N. The regulation and activities of the multifunctional serine/threonine kinase Akt/PKB. *Exp Cell Res* 1999;**253**:210–29.

54. Datta SR, Brunet A, Greenberg ME. Cellular survival: a play in three Akts. *Genes Dev* 1999;**13**:2905–27.

55. Okano J, Gaslightwala I, Birnbaum MJ, Rustgi AK, Nakagawa H. Akt/protein kinase B isoforms are differentially regulated by epidermal growth factor stimulation. *J Biol Chem* 2000;**275**:30,934–30,942.

56. Datta SR, Dudek H, Tao X, Masters S, Fu H, Gotoh Y, Greenberg ME. Akt phosphorylation of BAD couples survival signals to the cell-intrinsic death machinery. *Cell* 1997;**91**:231–41.

57. Cardone MH, Roy N, Stennicke HR, Salvesen GS, Franke TF, Stanbridge E, Frisch S, Reed JC. Regulation of cell death protease caspase-9 by phosphorylation. *Science* 1998;**282**:1318–21.

58. Brunet A, Bonni A, Zigmond MJ, Lin MZ, Juo P, Hu LS, Anderson MJ, Arden KC, Blenis J, Greenberg ME. Akt promotes cell survival by phosphorylating and inhibiting a Forkhead transcription factor. *Cell* 1999;**96**:857–68.

59. Jones RG, Parsons M, Bonnard M, Chan VS, Yeh WC, Woodgett JR, Ohashi PS. Protein kinase B regulates T lymphocyte survival, nuclear factor kappaB activation, and Bcl-X(L) levels invivo. *J Exp Med* 2000;**191**:1721–34.

60. Kuo ML, Chuang SE, Lin MT, Yang SY. The involvement of PI 3-K/Akt-dependent up-regulation of Mcl-1 in the prevention of apoptosis of Hep3B cells by interleukin-6. *Oncogene* 2001;**20**:677–85.

61. Pianetti S, Arsura M, Romieu-Mourez R, Coffey RJ, Sonenshein GE. Her-2/neu overexpression induces NF-kappaB via a PI3-kinase/Akt pathway involving calpain-mediated degradation of IkappaB-alpha that can be inhibited by the tumor suppressor PTEN. *Oncogene* 2001;**20**:1287–99.

62. Manning BD, Cantley LC. AKT/PKB signaling: navigating downstream. *Cell* 2007;**129**:1261–74.

63. Liu W, Bagaitkar J, Watabe K. Roles of AKT signal in breast cancer. *Front Biosci* 2007;**12**:4011–19.

64. Dillon RL, White DE, Muller WJ. The phosphatidyl inositol 3-kinase signaling network: implications for human breast cancer. *Oncogene* 2007;**26**:1338–45.

65. Zhao JJ, Gjoerup OV, Subramanian RR, Cheng Y, Chen W, Roberts TM, Hahn WC. Human mammary epithelial cell transformation through the activation of phosphatidylinositol 3-kinase. *Cancer Cell* 2003;**3**:483–95.

66. Samuels Y, Wang Z, Bardelli A, Silliman N, Ptak J, Szabo S, Yan H, Gazdar A, Powell SM, Riggins GJ, Willson JK, Markowitz S, Kinzler KW, Vogelstein B, Velculescu VE. High frequency of mutations of the PIK3CA gene in human cancers. *Science* 2004;**304**:554.

67. Saal LH, Holm K, Maurer M, Memeo L, Su T, Wang X, Yu JS, Malmstrom PO, Mansukhani M, Enoksson J, Hibshoosh H, Borg A, Parsons R. PIK3CA mutations correlate with hormone receptors, node metastasis, and ERBB2, and are mutually exclusive with PTEN loss in human breast carcinoma. *Cancer Res* 2005;**65**:2554–9.

68. Levine DA, Bogomolniy F, Yee CJ, Lash A, Barakat RR, Borgen PI, Boyd J. Frequent mutation of the PIK3CA gene in ovarian and breast cancers. *Clin Cancer Res* 2005;**11**:2875–8.

69. Lee JW, Soung YH, Kim SY, Lee HW, Park WS, Nam SW, Kim SH, Lee JY, Yoo NJ, Lee SH. PIK3CA gene is frequently mutated in breast carcinomas and hepatocellular carcinomas. *Oncogene* 2005;**24**:1477–80.

70. Campbell IG, Russell SE, Choong DY, Montgomery KG, Ciavarella ML, Hooi CS, Cristiano BE, Pearson RB, Phillips WA. Mutation of the PIK3CA gene in ovarian and breast cancer. *Cancer Res* 2004;**64**:7678–81.

71. Bachman KE, Argani P, Samuels Y, Silliman N, Ptak J, Szabo S, Konishi H, Karakas B, Blair BG, Lin C, Peters BA, Velculescu VE, Park BH. The PIK3CA gene is mutated with high frequency in human breast cancers. *Cancer Biol Ther* 2004;**3**:772–5.

72. Dupont J, Renou JP, Shani M, Hennighausen L, LeRoith D. PTEN overexpression suppresses proliferation and differentiation and enhances apoptosis of the mouse mammary epithelium. *J Clin Invest* 2002;**110**:815–25.

73. Nagata Y, Lan KH, Zhou X, Tan M, Esteva FJ, Sahin AA, Klos KS, Li P, Monia BP, Nguyen NT, Hortobagyi GN, Hung MC, Yu D. PTEN activation contributes to tumor inhibition by trastuzumab, and loss of PTEN predicts trastuzumab resistance in patients. *Cancer Cell* 2004;**6**:117–27.

74. Schwertfeger KL, Richert MM, Anderson SM. Mammary gland involution is delayed by activated Akt in transgenic mice. *Mol Endocrinol* 2001;**15**:867–81.

75. Ackler S, Ahmad S, Tobias C, Johnson MD, Glazer RI. Delayed mammary gland involution in MMTV-AKT1 transgenic mice. *Oncogene* 2002;**21**:198–206.

76. Viglietto G, Motti ML, Bruni P, Melillo RM, D'Alessio A, Califano D, Vinci F, Chiappetta G, Tsichlis P, Bellacosa A, Fusco A, Santoro M. Cytoplasmic relocalization and inhibition of the cyclin-dependent kinase inhibitor p27(Kip1) by PKB/Akt-mediated phosphorylation in breast cancer. *Nature Med* 2002;**8**:1136–44.

77. Liang J, Zubovitz J, Petrocelli T, Kotchetkov R, Connor MK, Han K, Lee JH, Ciarallo S, Catzavelos C, Beniston R, Franssen E, Slingerland JM. PKB/Akt phosphorylates p27, impairs nuclear import of p27 and opposes p27-mediated G1 arrest. *Nature Med* 2002;**8**:1153–60.

78. Shin I, Yakes FM, Rojo F, Shin NY, Bakin AV, Baselga J, Arteaga CL. PKB/Akt mediates cell-cycle progression by phosphorylation of p27(Kip1) at threonine 157 and modulation of its cellular localization. *Nature Med* 2002;**8**:1145–52.

79. Shin I, Rotty J, Wu FY, Arteaga CL. Phosphorylation of p27Kip1 at Thr-157 interferes with its association with importin alpha during G1 and prevents nuclear re-entry. *J Biol Chem* 2005;**280**:6055–63.

80. Yoeli-Lerner M, Yiu GK, Rabinovitz I, Erhardt P, Jauliac S, Toker A. Akt blocks breast cancer cell motility and invasion through the transcription factor NFAT. *Mol Cell* 2005;**20**:539–50.

81. Irie HY, Pearline RV, Grueneberg D, Hsia M, Ravichandran P, Kothari N, Natesan S, Brugge JS. Distinct roles of Akt1 and Akt2 in regulating cell migration and epithelial-mesenchymal transition. *J Cell Biol* 2005;**171**:1023–34.

82. Feng J, Park J, Cron P, Hess D, Hemmings BA. Identification of a PKB/Akt hydrophobic motif Ser-473 kinase as DNA-dependent protein kinase. *J Biol Chem* 2004;**279**:41,189–41,196.

83. Hresko RC, Mueckler M. mTOR.RICTOR is the Ser473 kinase for Akt/protein kinase B in 3T3-L1 adipocytes. *J Biol Chem* 2005;**280**:40,406–40,416,.

84. Sarbassov DD, Guertin DA, Ali SM, Sabatini DM. Phosphorylation and regulation of Akt/PKB by the rictor–mTOR complex. *Science* 2005;**307**:1098–101.

85. Gao T, Furnari F, Newton AC. PHLPP: a phosphatase that directly dephosphorylates Akt, promotes apoptosis, and suppresses tumor growth. *Mol Cell* 2005;**18**:13–24.

86. Biswas DK, Reddy PV, Pickard M, Makkad B, Pettit N, Pardee AB. Calmodulin is essential for estrogen receptor interaction with its motif and activation of responsive promoter. *J Biol Chem* 1998;**273**:33,817–33,824.

87. Li L, Li Z, Sacks DB. The transcriptional activity of estrogen receptor-alpha is dependent on Ca^{2+}/calmodulin. *J Biol Chem* 2005;**280**:13,097–13,104.

88. Li L, Li Z, Sacks DB. Calmodulin regulates the transcriptional activity of estrogen receptors. Selective inhibition of calmodulin function in subcellular compartments. *J Biol Chem* 2003;**278**:1195–200.

89. Li Z, Joyal JL, Sacks DB. Calmodulin enhances the stability of the estrogen receptor. *J Biol Chem* 2001;**276**:17,354–17,360.

90. Deb TB, Coticchia CM, Dickson RB. Calmodulin-mediated activation of Akt regulates survival of c-Myc-overexpressing mouse mammary carcinoma cells. *J Biol Chem* 2004;**279**:38,903–38,911.

91. Shen X, Valencia CA, Szostak JW, Dong B, Liu R. Scanning the human proteome for calmodulin-binding proteins. *Proc Natl Acad Sci USA* 2005;**102**:5969–74.

92. Dong B, Valencia CA, Liu R. $Ca^{(2+)}$/calmodulin directly interacts with the pleckstrin homology domain of AKT1. *J Biol Chem* 2007;**282**:25,131–25,140.

93. Rodriguez-Mora OG, LaHair MM, McCubrey JA, Franklin RA. Calcium/calmodulin-dependent kinase I and calcium/calmodulin-dependent kinase kinase participate in the control of cell cycle progression in MCF-7 human breast cancer cells. *Cancer Res* 2005;**65**:5408–16.

94. Deiss LP, Feinstein E, Berissi H, Cohen O, Kimchi A. Identification of a novel serine/threonine kinase and a novel 15-kD protein as potential mediators of the gamma interferon-induced cell death. *Genes Dev* 1995;**9**:15–30.

95. Levy D, Plu-Bureau G, Decroix Y, Hugol D, Rostene W, Kimchi A, Gompel A. Death-associated protein kinase loss of expression is a new marker for breast cancer prognosis. *Clin Cancer Res* 2004;**10**:3124–30.

96. Dulaimi E, Hillinck J, Ibanez de Caceres I, Al Saleem T, Cairns P. Tumor suppressor gene promoter hypermethylation in serum of breast cancer patients. *Clin Cancer Res* 2004;**10**:6189–93.

Chapter 41

Aberrant Signaling Pathways in Pancreatic Cancer: Opportunities for Targeted Therapeutics

Alixanna Norris and Murray Korc

Departments of Medicine, Pharmacology and Toxicology, Norris Cotton Cancer Center, Dartmouth-Hitchcock Medical Center, Lebanon, New Hampshire, and Dartmouth Medical School, Hanover, New Hampshire

INTRODUCTION

The pancreas is a small organ located just behind the stomach. It measures approximately 6 inches long and 2 inches wide, situated horizontally in the abdomen. It is a compound gland, containing both exocrine and endocrine cells. The exocrine pancreas produces pancreatic juice that contains digestive enzymes that are synthesized in the acinar cells and bicarbonate-rich fluid that is secreted by the ductal cells. The majority (95 percent) of the cells of the pancreas are acinar cells, whereas only 2 percent are ductal cells, and an even smaller percentage of cells are endocrine cells, which are arranged into small clusters called the islets of Langerhans. These islets release insulin and glucagon into the bloodstream, thereby regulating glucose homeostasis.

To date, the cell type that gives rise to pancreatic ductal adenocarcinoma (PDAC) has not been clearly delineated. Candidate cell types include the ductal epithelial cell, the acinar cell, the centro-acinar cell, and progenitor-stem cells. The first evidence of pancreatic dysplasia appears histologically as pancreatic intraepithelial neoplasias (PanINs). As depicted in Figure 41.1, PanIN-1 A lesions exhibit only minimal molecular alterations and microscopically appear as tall columnar cells with some crowding. PanIN-1B are similar to PanIN-1 A, but with enhanced crowding of the columnar cells. PanIN-2 lesions exhibit nuclear atypia and papillary projections, whereas PanIN-3 lesions are highly dysplastic and have atypical ductal hyperplasia with severe atypia. The PanIN-3 lesions represent carcinomas *in situ*, and are the immediate precursors to invasive carcinoma [1].

Less common types of exocrine cancers include adenosquamous carcinomas, squamous cell carcinomas, cystadocarcinomas, and giant cell carcinomas, to name a few.

Tumors may also arise in the endocrine portion of the pancreas. These are known as neuroendocrine tumors, and include such tumors as insulinomas and glucanomas.

PDAC, the most common pancreatic tumor, is currently the fourth leading cause of cancer death in the United States, with just over 37,000 new cases diagnosed each year. It is one of the deadliest forms of cancer with a median survival of 5 months, an overall 1-year survival rate of 15 percent, and a 5-year survival rate of 3–5 percent [2, 3]. The majority of PDAC patients have extensive local and/or metastatic disease upon presentation, which excludes them from surgical resection. Yet, non-surgical treatment modalities for PDAC are generally ineffective [4, 5], often leaving palliative care as the only other option in these patients.

Specific molecular alterations have been shown to correlate with discrete pathological stages of pancreatic tumorigenesis (Figure 41.1) and contribute to specific features of PDAC (Table 41.1). Based on the multi-hit hypothesis that was proposed for colorectal adenocarcinoma, a similar hypothesis has been put forth for PDAC based on a series of gene array and immunohistochemical studies [6]. The first detectable alteration in pancreatic cancer appears to be point mutations in the K-*ras* oncogene, followed by the overexpression of the *HER-2/neu* gene. The second "hit" is then thought to be inactivation of the *p16^{INK4a}* gene, followed by the loss of the *p53*, *DPC4* (*Smad4*), and *BRCA2* genes [7–11].

Additional alterations suggested to play a role in PDAC include the overexpression of a number of important growth factors and their corresponding high affinity tyrosine kinase receptors, transforming growth factor beta (TGF-β) isoforms, Hedgehog signaling components, the Wnt signaling pathway, the cell cycle regulatory pathway, and nuclear effectors such as Notch, NFκB, and Stat3. In general, the somatic

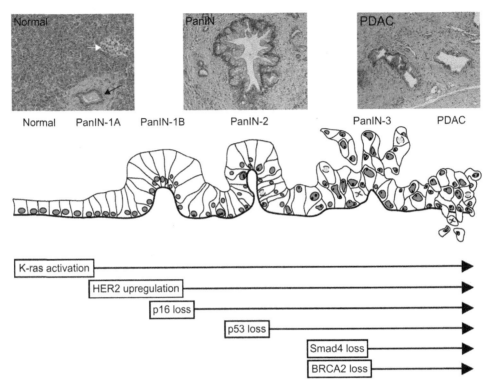

FIGURE 41.1 Molecular alterations in pancreatic cancer progression.
Pancreatic ductal adenocarcinoma (PDAC) arises as a consequence of a series of histological and molecular alterations, as depicted. This progression from histologically normal cells through various stages of pancreatic intraepithelial neoplasia (PanIN) lesions is now taken to represent the most common pathway leading to the genesis of PDAC. Pathological sections are pictured at top. A sample from a normal pancreas (left) includes numerous acinar cells, one duct consisting of uniformly sized ductal cells (black arrow) surrounded by a small amount of stroma, and one endocrine islet (white arrow) consisting of many small endocrine cells. A PanIN-1B lesion is shown in the middle panel. A PDAC sample is shown in the right panel, exhibits ductal-like structures that represent the pancreatic cancer cells, embedded in an extensive stroma that has a readily evident micro-vasculature. Original magnifications for all three hematoxylin-eosin sections: X 20.

alterations that are found in PDAC lead to increased proliferation, decreased apoptosis, and aberrant angiogenesis.

ONCOGENIC ACTIVATION IN PDAC

Growth Receptor Pathways

Pancreatic cancer cells, like several other cancers [12], have a high mitotic index and proliferative capacity due, in part, to the excessive activation of a number of growth factor receptors (Figure 41.2). One pathway that is particularly important in PDAC is dependent on signaling via the epidermal growth factor (EGF) receptor (EGFR). This transmembrane tyrosine-kinase receptor is also known as human epidermal growth factor receptor 1 (HER1; ErbB-1). Human pancreatic cancer cell lines express high levels of EGFR, as well as at least five of the seven factors that bind to and activate EGFR [12, 13]. These factors include EGF, transforming growth factor-alpha (TGF-α), amphiregulin, betacellulin, and heparin binding EGF-like growth factor (HB-EGF). EGFR expression correlated with a more advanced clinical stage of PDAC and shorter survival of patients following tumor resection due to more rapid disease reactivation, even though

EGFR is also expressed at high levels in chronic pancreatitis [13, 14]. Additionally, c-Erb-B2 (HER2) and c-Erb-B3 (HER3) are overexpressed in PDAC [13, 15, 16]. These two EGFR related receptors further potentiate mitogenic signaling through their heterodimerization with EGFR. Moreover, the heregulin ligands that bind to HER3 are also abundant in PDAC, with heregulin-beta exerting growth promoting effects on pancreatic cancer cells [17, 18].

A number of other mitogenic factors are overexpressed in pancreatic cancers, including fibroblast growth factor-1 (FGF-1), FGF-2, FGF-5, keratinocyte growth factor-1 (FGF-1), platelet derived growth factor (PDGF) B chain, insulin-like growth factor (IGF), vascular endothelial growth factor (VEGF), and hepatocyte growth factor (HGF). Additional overexpressed receptors include the PDGF receptors α and β, FGF receptors, IGF-I, VEGF receptors, and the MET receptor. The elevation of these factors directly contributes to the highly active intracellular signaling (and subsequent proliferation and invasion) observed in cultured human pancreatic cells. In addition, high insulin levels deriving from the islet cells can stimulate the overexpressed IGF-I receptors on pancreatic cancer cells, thereby contributing to their excessive proliferation.

TABLE 41.1 Molecular Alterations in Pancreatic Cancer

	Type of alteration	Specific example
1	Altered oncogenes and proto-oncogenes	Mutated K-ras
2	Loss of tumor suppressor gene functions	Mutated p53
3	Aberrant growth factor and receptor expression	EGF overexpression
4	Aberrant activation of signaling pathways	BMP-2 mediated mitogenesis
5	Enhanced angiogenesis	Increased VEGF expression
6	Resistance to apoptosis	Expression of Fas and FasL
7	Altered epithelial–mesenchymal interactions	Increased glypican-1 and CTGF
8	Loss of negative growth constraints	Mutated Smad4
9	Loss of metastasis suppressing genes	Loss of Kai1

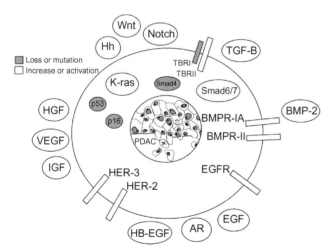

FIGURE 41.2 Altered signaling components in pancreatic cancer. A number of key signaling proteins are known to contribute to the progression of pancreatic cancer. As depicted, items in white are elevated or constitutively active in the cancer cells, while items in black are either not expressed or harbor inactivating mutations.

K-ras

The K-*ras* oncogene is mutated in 75 to 95 percent of PDACs [19–21]. K-ras is a 21 kDa membrane associated guanyl nucleotide exchange factor with intrinsic and extrinsic GTPase activity. K-*ras* mutations on codons 12 and 13,

of the type observed in PDAC, often lead to attenuated GTPase activity and constitutive mitogenic signaling. K-*ras* mutations are present in 40 percent of PanIN lesions, indicating it may be one of the earliest alterations in pancreatic tumorigenesis (Figure 41.1). It is postulated that signaling from mutated K-ras provides a strong selective pressure for cells to escape normal growth restraints and continuously proliferate, thereby setting the stage for cancer development. In support of this hypothesis, several mouse models of PDAC have been produced in which a key component has been the expression of oncogenic K-ras in the pancreas through its own endogenous promoter [22].

Hedgehog Signaling Pathway

The mammalian Hedgehog family comprises three members: Sonic (Shh), Indian (Ihh), and Desert (Dhh). These proteins regulate the growth and development of many organs in embryogenesis [23] and are vital for gastrointestinal patterning early in development [23, 24]. The two Hedgehog receptors are called Patched1 (Ptch1) and Patched2 (Ptch2). In the absence of bound Hedgehog ligand, these receptors are bound to and repress Smoothened (Smo). After ligand binding, Smo is released and subsequently signals for the translocation and activation of Gli family proteins [25, 26]. Glis are transcriptional regulators that ultimately regulate the expression of Hedgehog responsive genes, including cyclin D1, N-Myc, p21, Wnt, Ptch, and Gli [25, 27–30].

Deregulation of the Hedgehog pathway is found in many types of human cancer, including cancers of the lung, brain, skin, and pancreas [26, 31–34]. The activation of Hedgehog signaling has been demonstrated in a majority of PDAC cases [26, 33]. Activation in pancreatic cancer can occur by a number of mechanisms including loss of *PTCH*, constitutive activation of Smo, and overexpression of Gli and Shh [26]. Hedgehog signaling blockade, both *in vivo* and *in vitro*, was shown to inhibit proliferation, induce apoptosis, and suppress pancreatic cancer cell metastasis [26, 35]. Oncogenic K-ras suppresses Gli1 protein degradation, thereby activating Hedgehog signaling [36].

Transgenic mouse models that overexpress Shh in a pancreas specific fashion have yielded conflicting results. While Thayer *et al.* reported that Shh overexpression leads to the development of pancreatic intraepithelial neoplasia (PanIN) lesions, a recent consensus report concluded that this model developed only ductal-intestinal metaplasia without PanIN lesions [26, 37]. Similarly, a mouse model with inducible activation of *Gli2* developed pancreatic tumors, but the lesions did not resemble those found in humans [38]. However, when combined with mutated K-ras (*KrasG12D*), *Gli2* activation exacerbated PanIN formation [38], suggesting that Hedgehog signaling may only be important after dysplasia has already been initiated. Other transgenic animal and *in vitro* studies support the notion that activation of Hedgehog signaling may be secondary to K-ras activation [39]. Nonetheless, this pathway has been shown to be a target for novel therapies, such as cyclopamine and IPI-269609 [40].

Notch Signaling Pathway

The Notch signaling pathway is another developmental pathway that is implicated in PDAC. This pathway maintains the balance between cell proliferation, apoptosis, and differentiation through cell–cell interactions involving membrane bound Notch receptors [41, 42]. The Notch receptors (Notch 1–4) are activated by five ligands named Delta-like (DLL) 1, 3, and 4 and Jagged 1 and 2 in a ligand–receptor interaction between neighboring cells. This activation initiates proteolytic cleavage events, resulting in the release and nuclear translocation of the receptor cytoplasmic domain (NICD) [43–46]. Once in the nucleus, NICD interacts with a number of transcriptional repressors and activators [47–51] to affect gene expression. Notch target genes include the Hairy enhance of split (Hes) family genes, Cdkn1a, cyclin D1, the ubiquitin ligase SKIP2, and the c-myc proto-oncogene [52–56].

The proliferative consequences of Notch activation are complex and context dependent. Notch signaling is an important regulator of embryogenesis and pancreatic development by serving to maintain a pool of undifferentiated precursor cells [57–60]. Notch signaling is also required in normal pancreatic tissue regeneration after acute pancreatitis, as a consequence of its ability to promote acinar cell maturation by a crosstalk mechanism with Wnt signaling [61]. Notch activation in PDAC is oncogenic [62], and alterations in Notch signaling have been observed in early mouse and human PanIN lesions [63], suggesting that Notch signaling may contribute to pancreatic tumorigenesis by promoting the growth of undifferentiated cells. The Notch pathway is therefore a potential therapeutic target in PDAC [64]. Moreover, blockade of Dll4 interferes with tumor angiogenesis, leading to a significant decrease in tumor growth both alone and in combination with anti-VEGF therapies [65, 66]. Notch signaling can be targeted by sequestration of Notch ligands, by antisense Notch therapy, or by RNA interference and have been used in experimental anticancer approaches with varying degrees of efficacy [67–70]. However, Notch 1 and 2 exert opposite effects on tumor growth (67), indicating that such therapies may have potential pitfalls.

LOSS OF TUMOR SUPPRESSOR FUNCTION IN PDAC

p53

The p53 tumor suppressor is mutated in approximately 85 percent of PDAC cases [71–73]. p53 is a key regulator of cellular proliferation, DNA repair, and the cell stress response. It is a nuclear protein that regulates the cellular response by modifying gene expression and triggering the G1/S cell cycle checkpoint in unfavorable conditions. The spectrum of p53 mutations in PDAC has been analyzed indirectly through immunohistochemical studies and direct sequence analysis [73, 74]. p53 mutations become prevalent in later stages of PanIN formation and thus appear to be an early event in pancreatic tumorigenesis, but not an initiating event. Cells containing mutated p53 are less likely to undergo apoptosis or cell cycle arrest in response to stress stimuli and have reduced genomic stability [75]. p53 deficient cells are more sensitive to oncogene induced transformation and more resistant to standard anti-tumor treatment regimens (e.g., chemotherapy and radiation therapy).

p16^{INK4a}

$p16^{INK4a}$ is a tumor suppressor commonly altered in human adenocarcinoma. The p16 protein is an inhibitor of the cyclin D-Cdk4 and cyclin D-Cdk6 complexes, which themselves are regulators of the cell cycle. Loss of p16 results in increased activity of the CDKs and subsequent Rb protein activation, leading to unchecked cellular proliferation. The $p16^{INK4a}$ gene is mutated in 85 percent of PDAC cases and is epigenetically silenced in the remaining 15 percent of cases [19, 76]. $p16^{INK4a}$ mutations include somatic point mutations (in ~40 percent of cases) and deletion of both alleles (~40 percent). This alteration is found early in PanIN development and precedes loss of p53 in pancreatic cancer progression (Figure 41.2) [77–79]. Clinical studies suggest that PDAC patients carrying $p16^{INK4a}$ mutations have more aggressive tumors and shorter survival times than patients with wild-type $p16^{INK4a}$ [80, 81]. These alterations are not found in chronic pancreatitis, indicating this may be a genetic alteration that is specific to PDAC and an attractive target for molecular therapeutics.

The TGF-β Pathway

TGF-β is a cytokine that has been implicated in a diverse range of biological processes. In addition to modulating proliferation, the TGF-β pathway has been implicated in numerous cellular and biological processes including embryogenesis, differentiation, apoptosis, angiogenesis, immunosuppression, and wound healing.

There are three mammalian TGF-β isoforms (TGF-β1, TGF-β2, and TGF-β3) that are encoded by different genes with different expression patterns [82]. TGF-β ligands initiate signaling by acting through two specific cell surface receptors: type I TGF-β receptor (TβRI) and type II TGF-β receptor (TβRII) [83–85]. In the presence of TGF-β ligand and following binding to the TβRII homodimer, TβRII complexes with and phosphorylates TβRI [86]. Phosphorylation results in the activation of TβRI kinase activity and subsequent phosphorylation of TGF-β signal transducers: the Smad family proteins [85, 87–90].

Smad proteins are a family of transcription factors that are divided into three structure/function subcategories: the receptor regulated Smads (R-Smads: Smads2 and 3), the common partner Smad (Co-Smad: Smad4), and the inhibitory Smads (I-Smads: Smads6 and 7). Smad2 and Smad3 are directly phosphorylated by TβRI [91–96]. Smad2 and Smad3 then heterodimerize with Smad4 and the resulting complex translocates to the nucleus. Once in the nucleus, the complex modulates gene transcription in conjunction with coactivators and corepressors such as AP-1, FAST, TFE3, p300/CBP, and Ski [84, 85, 87, 88, 90, 97]. It is the specific interactions of the Smad complex with these nuclear factors that facilitates the specificity and complexity of TGF-β signaling.

The TGF-β signaling pathway is often dysfunctional in human cancers. When TGF-β signaling is aberrant, cells become more invasive, metastatic, and proliferative. TGF-β is a key mediator of pancreatic fibrosis and is thought to contribute to the formation of desmoplastic stroma in both PDAC and chronic pancreatitis [98–101]. Similar to its effects in normal epithelium, TGF-βs act as tumor suppressors in the early stages of pancreatic tumorigenesis, but the expression of TGF-β at later stages of cancer progression fosters a more aggressive phenotype [102–107]. The mechanisms driving this functional "switch" remain to be elucidated.

A number of alterations in the TGF-β signaling occur in PDAC. For instance, the levels of all three TGF-β ligands and TβRII are elevated in PDAC, yet pancreatic cancer cells are often resistant to TGF-β induced growth arrest [108]. *Smad4* mutations are the most frequent TGF-β alteration documented in human PDAC [109], followed by decreased TβRI expression [106, 110], increased TβRII expression, overexpression of Smad6/7 [111, 112], and (rarely) mutations in TβRI/TβRII.

Smad4 is mutated in about half of PDAC cases [109, 113–115]. Immunohistochemical studies suggest that Smad4 is lost late in pancreatic tumorigenesis because it can be detected in PanIN-1 and PanIN-2 lesions, but to a lesser extent in PanIN-3 and in PDAC [116]. PDAC patients that have mutations in Smad4 displayed a shorter survival time after surgery compared to patients with wild-type Smad4 (14.7 mo versus 19.2 mo) [117].

BMPs

Bone morphogenetic proteins (BMPs) are growth factors that are members of the TGF-β superfamily and have been implicated in PDAC. BMPs signal by binding cooperatively to two types of receptors simultaneously: the type-I BMP receptors (BMPR-IA/ALK-3 and BMPR-IB/ALK-6) and the type-II BMP receptor (BMPR-II). Once bound by ligand, BMPR-I becomes activated and phosphorylates Smad1. Activated Smad1 then forms a complex with Smad4 and translocates to the nucleus to modify gene expression as described above for the TGF-β pathway [118, 119].

The BMP pathway is thought to be important for the overactive proliferation of pancreatic cancer cells. Studies with PDAC samples and cell lines have shown increased expression of the BMP-2 ligand, BMPR-IA, and BMPR-II [120, 121]. In addition to its role in the TGF-β signaling pathway, Smad4 has been shown to mediate signaling of the BMP pathway. This crosstalk between growth factor signaling pathways complicates the delineation of the role of individual pathways in PDAC.

CONTRIBUTIONS OF PANCREATIC STROMA AND STROMAL COMPONENTS IN PDAC

The cellular interaction between the stromal elements and epithelial cells is required for normal glandular development. In the pancreas, stromal interactions stimulate the embryonic pancreas to form the exocrine and the endocrine components [122]. When these normal stromal interactions are perturbed, the proliferation and differentiation of adjacent epithelial cells are affected [123–128]. In general, tumor associated pancreatic stroma consists of pancreatic stellate cells and cancer associated fibroblasts, infiltrating inflammatory cells, blood vessels, nerves, and the extracellular matrix (Figure 41.3). Each of these is known to play a role in PDAC progression.

Pancreatic stellate cells are the predominant mesenchymal cells in the pancreatic stroma [129]. Stellate cells are thought to promote the development of a desmoplastic stroma through the production of extracellular matrix proteins (e.g., collagen) and the release of pro-fibrotic inflammatory factors (e.g., TGFβ) [130]. *In vitro*, pancreatic stellate cells can be activated by adjacent pancreatic cells [131]. Recently, stellate cells were shown to contribute to increased tumor growth and metastasis in an orthotopic model [132].

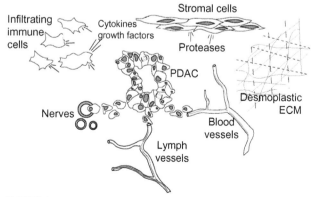

FIGURE 41.3 Role of pancreatic stroma.
Stromal elements are known to participate in pancreatic cancer initiation, progression, and dissemination. These elements include an extensive extracellular matrix (ECM), infiltrating immune cells (e.g., macrophages), stromal cells (including proliferating fibroblasts and pancreatic stellate cells), lymph/blood vessels, and adjacent nerves.

Infiltrating immune cells emigrate within the tumor microenvironment and release cytokines, growth factors, and reactive oxygen species, all of which affect epithelial cell growth. One such family of factors secreted by macrophages is the interleukins (IL). IL-1 promotes the motility and spread of cancer cells [133] while IL-6 activates pancreatic stellate cells and induces fibrotic ECM formation through TGF-β signaling (130). Additionally, fibroblasts in the stromal compartment can release cyclooxygenase 2 (COX-2), a mediator of inflammation and cancer cell invasion [128, 134].

Nerves and blood vessels located in tumor associated stroma provide a conduit for cancer cell dissemination and metastasis. Perineural invasion is frequent in PDAC, allows cancer cells to spread, and is associated with a poorer prognosis [135, 136], whereas angiogenesis provides oxygen, nutrients, and a mode for rapid dissemination. Neo-angiogenesis is initiated by factors secreted by cancer and stromal cells that guide the distant endothelial cells to degrade the extracellular matrix and migrate toward the tumor mass [137]. Additionally, inflammatory cell infiltrates secrete pro-angiogenic factors to induce endothelial cell proliferation in the tumor microenvironment [138]. The expression of VEGF ligand correlates with disease progression [139, 140] and is known to activate mitogenic signaling in some pancreatic cancer cells *in vitro* [94]. In general, marked neo-angiogenesis is a poor prognostic factor in solid tumors, including PDAC [141].

Finally, the extracellular matrix (ECM) is a complex mix of glycoproteins and collagen that form a scaffolding for glandular formation. The tumor associated ECM is desmoplastic, meaning it contains high amounts of fibronectin and collagen [131] and is thought to be important in early neoplastic changes within the pancreas [142]. Type I collagen, which is abundant in PDAC stroma, is associated with increased motility of pancreatic ductal cells [143, 144]. In addition, stromal cells secrete high levels of proteases, such as matrix metalloproteinases (MMPs) [145] and urokinase-type plasminogen activators (uPAs). MMPs degrade the basement membrane and allow cancer cells to escape into adjacent blood vessels and nerves. uPAs stimulate tumor cell invasion and their expression correlates with the presence of metastases in PDAC patients [146]. ECM associated proteins, which promote pancreatic cancer cell invasion and matrix remodeling, are also elevated in PDAC and include thrombospondin, tenascin, vitronectin, and versican [147–151]. Alterations in the extracellular matrix components are also known to contribute to epithelial–mesenchymal transition [152] and to form a hydrostatic barrier against chemotherapy.

Foci of inflammatory cells in the stroma produce chemokines and cytokines while nerve fibers release nerve growth factors [153, 154]. These factors stimulate growth in both fibroblasts and pancreatic stellate cells [155]. Additionally, the stroma is stimulated to proliferate by the abundance of growth factors, including FGF, EGF, TGF-β, and connective tissue growth factor (CTGF) secreted from proliferating cancer cells. The stroma functions in a storage capacity for these growth factors, further potentiating cancer growth [156, 157]. Overall, the tumor associated stroma is an important component in pancreatic tumorigenesis, contributing to chemoresistance and to the high metastatic capacity of PDAC (Figure 41.3).

DIFFERENTIATION

Cellular differentiation is a complex process that involves the coordinated regulation of genes by a multitude of cellular pathways. Differentiation is controlled a number of DNA binding proteins that are aberrantly expressed in PDAC. One group of proteins regulating differentiation is the family of helix-loop-helix (HLH) DNA binding proteins. The HLH motif consists of two conserved amphiphatic α-helix structures separated by a variable loop region [158]. HLH family members form heterodimeric complexes with each other to regulate gene expression. A subgroup of this family, which contain a region of basic amino acids in the DNA binding domain, are called bHLH proteins and are known to affect a number of differentiation specific genes, including immunoglobin genes, neuronal specific genes, muscle specific genes, and insulin related genes [159, 160]. bHLH proteins are well established regulators of pancreatic cell function and differentiation, by affecting genes such as NeuroD/BETA2, E47, Neurogenin3, HES-1 and PTF1-p48, and Pax [57, 59, 161–164].

The Id family of HLH proteins includes four members (Id1 through Id4). These proteins lack the DNA binding region, but affect gene expression by heterodimerizing with bHLH proteins and preventing their associations with DNA [165]. Id family proteins are known to have some effect on pancreatic organogenesis. Id1 and Id2 levels increase during the maturation process of β-cells and inversely correlate with insulin gene activation [166]. Id2 is regulated by the type 1 insulin-like growth factor receptor (IGF-IR) and the insulin receptor substrate-1 (IRS-1), as well as by the mitogen activated protein kinase (MAPK) and phosphatidylinositide 3-kinase (PI3K) pathways [167]. Id expression inhibits insulin production, perhaps by blocking the *cis* regulated insulin control element (ICE) and/or by sequestering protein products that regulate the insulin gene [168].

There is *in vitro* data linking Id expression with cell cycle progression and arrested differentiation in various cell types [169–171]. In the HIT pancreatic cell line, Id proteins regulate p21 activity [172] thereby affecting both cellular proliferation and differentiation. In PANC-1 cells, Id2 mRNA levels are decreased after induction of differentiation [173]. Combined with studies showing overexpression of Id proteins in pancreatic cancer and in dysplastic lesions [174, 175], these data suggest that the Id family proteins may play some role in the early stages of pancreatic cancer by promoting cellular proliferation and inhibiting full differentiation of pancreatic cells.

APOPTOTIC PATHWAYS AND APOPTOTIC RESISTANCE

Programmed cell death, or apoptosis, is a natural mechanism for the essential clearing of diseased cells from an organism. Apoptosis can be initiated in response to various conditions, including exposure to cytotoxic drugs, activation of pro-death receptor pathways (e.g., Fas, TNF, and TRAIL), cellular detachment, and the absence of growth factors [176]. A decreased apoptotic response is known to play an important role in cancer initiation and progression [177].

Human pancreatic cancer cell lines are generally resistant to apoptosis in response to chemotherapy and death receptor signaling [178]. One proapoptotic pathway that is aberrant in pancreatic cancer is the Fas pathway. Normal binding of the Fas receptor by the Fas ligand induces a conformational change in the receptor, which leads to the recruitment and activation of the proapoptotic caspase pathway. Despite normal expression of the Fas receptor and ligand, pancreatic cancer cells display marked resistance to Fas mediated apoptosis [179]. This may be a consequence of somatic mutations leading to mutated Fas protein, enhanced expression of a Fas associated phosphatase that inhibits the effects exerted by Fas [180], or mutations in downstream proteins that perpetuate the Fas signaling. Additionally, Fas expression correlates with Ki-67 staining and poor patient outcome [181, 182].

CLINICAL ASPECTS OF PDAC

Current Therapy

The standard surgical treatments for PDAC consist of three procedures designed to remove the tumor: the Whipple procedure, a distal pancreatectomy, and a total pancreatectomy. If the cancer has progressed beyond the point of resection, palliative surgery can be performed to relieve symptoms associated with tumor blockage of the small intestine, bile duct, or stomach.

Neoadjuvant chemotherapy and radiation are also employed in the treatment of PDAC, depending on the stage of disease at presentation. Such therapy, administered prior to attempts at resection, may occasionally shrink the tumor sufficiently to allow for subsequent surgery [183, 184]. By contrast, radiation therapy given following surgical resection, may be associated with shortened survival [185]. The current standard chemotherapy for PDAC, which received approval by the Food and Drug Administration (FDA) in 1996, is gemcitabine (Gemzar®). This drug is converted intracellularly to active metabolites difluorodeoxycytidine di- and triphosphate (dFdCDP, dFdCTP), which act to both inhibit ribonucleotide reductase and decrease the amount of deoxynucleotide that is available for DNA synthesis. dFdCTP is also incorporated into DNA, resulting in DNA strand termination and apoptosis. It is currently used both as a first-line therapy and as a radiosensitizer in the treatment of pancreatic cancer [186].

Molecular Markers as Prognostic Indicators

Prognostic markers are important for predicting a patient's risk of recurrence and overall survival at time of diagnosis. They may also aid the clinician in performing an accurate diagnosis and associate with a patient's response to anticancer therapy. Proposed prognostic markers in PDAC include the presence of EGFR, TGF-α, p53 mutations, and the overexpression of c-erbB3, uPA, FGF-2, VEGF, or TGF-β [108, 187–190].

New molecular biology techniques enable the detection of K-ras mutations not only in pancreatic tissue sections, but also in pancreatic juice, pancreatic biopsies, and pancreatic duct brushings collected during endoscopic retrograde cholangiopancreatography [191–193]. The rapid and reproducible detection of K-ras mutations may be helpful in the diagnosis of PDAC and eventually may help to select patients that are more responsive to particular anti-tumor therapy or targeted molecular therapy.

Clinical Trials of Molecular Therapeutics

A number of targetable molecular alterations are known to occur in PDAC and several therapies have been devised to target those alterations (Figure 41.4). For example, because matrix metalloproteinases (MMPs) are expressed at high levels in PDAC, Marimastat, an inhibitor of MMPs, was tested in a large phase II clinical trial on 414 patients as a first-line agent in PDAC patients with unresectable disease. Overall, there was no difference in the median survival between patients treated with either marimastat or gemcitabine. This study showed that marimastat had no significant survival advantage when used alone, but suggested it might be good in the adjuvant setting (194). A more recent phase III trial, by Bramhall *et al.* (2002) went on to conclude that marimastat did not improve survival when used in combination to gemcitabine, compared to gemcitabine alone in patients with advanced PDAC [195].

Farnesylation, which is catalyzed by farnesyl-transferase (FT), is a critical step in Ras activation and therefore affects cell growth and differentiation [196]. FT inhibitors were shown to suppress the growth of human pancreatic cells *in vitro* [197, 198]. However these inhibitors lacked efficacy in clinical trials. For example, tipifarnib (Zarnestra), an FT inhibitor, was evaluated in a small phase II clinical trial for patients with advanced PDAC. This study failed to show any benefit for those patients treated with tipifarnib when compared to gemcitabine [199]. A second larger phase III study evaluated the ability of tipifarnib to enhance the therapeutic efficacy of gemcitabine in PDAC. This study concluded that there was no significant difference for

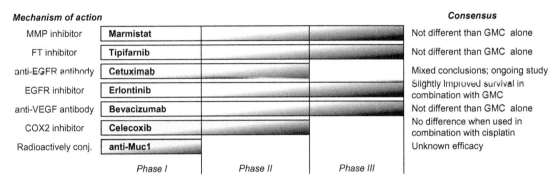

FIGURE 41.4 Overview of several inhibitors used in clinical settings.
As shown, a number of inhibitors that target specific pathways have been used in the treatment of pancreatic cancer. Their mechanisms of action, use in different phase clinical trials, and study outcomes are shown. GMC: gemcitabine.

patients treated with the combination therapy when compared to gemcitabine alone [200]. Ongoing clinical trials are evaluating the efficacy of tipifarnib as a radiosensitizer in the absence or presence of multimodal chemotherapy in patients with unresectable pancreatic cancer.

Another area of targeted therapeutics for pancreatic cancer is in the abrogation of EGF signaling. Two therapeutic approaches have been used: antibody therapy to block EGFR activation or tyrosine kinase inhibitors (TKIs) to inhibit EGFR kinase activity. Cetuximab (Erbitux) is a chimeric human-mouse antibody that binds EGFR with high affinity and prevents stimulation of the receptor by other ligands [201, 202]. Cetuximab was shown to block the G1 phase of the cell cycle, affect the apoptotic response, and inhibit tumor angiogenesis in several human tumor cell lines and xenograft models [203, 204]. It has been approved for treatment of EGFR-positive colorectal cancers and for metastatic head and neck cancer [205, 206]. In a small phase II multicenter trial, cetuximab was shown to have better overall survival when used in combination with gemcitabine, compared to gemcitabine only data from a previous trial [207]. These results were not recapitulated, however, in a large multisite phase II clinical trial evaluating the efficacy of cetuximab in combination with gemcitabine and cisplatin [208]. A number of clinical trials are currently underway to more clearly delineate the potential benefits of cetuximab in PDAC when used in combination with other treatment modalities.

Erlontinib (Tarceva) is a reversible inhibitor of EGFR. *In vitro*, human pancreatic cancer cell lines displayed significant growth inhibition in response to erlontinib [12]. A large multicenter phase III clinical showed that erlontinib could improve the median survival of patients by 0.5 months, when used in combination with gemcitabine [209]. Following these results, a number of clinical trials have been initiated to evaluate the therapeutic benefits of combining erlontinib with chemotherapy and radiation therapy in pancreatic cancer.

As introduced elsewhere in this *Handbook*, vascular endothelial growth factor (VEGF) is another potential target

for molecular therapy. VEGF blockade would theoretically diminish the growth of new blood vessels (neo-angiogenesis). A number of approaches have been used to block VEGF signaling. Bevacizumab is a humanized monoclonal antibody against VEGF and received FDA approval for the treatment of metastatic colon cancer in combination with fluorouracil and leucovorin [210]. In a multicenter phase II clinical trial, PDAC patients treated with a combination of bevacizumab and gemcitabine were found to have better survival statistics than those patients who received gemcitabine alone [211]. Additionally, bevacizumab was found to have benefit in a stage IV pancreatic cancer patient who had previously failed chemotherapy [210]. Based on these results, a larger phase III clinical study was undertaken to assess the efficacy of adding bevacizumab to gemcitabine in patients with advanced PDAC. Unfortunately, the addition of bevacizumab to gemcitabine did not prolong survival in this more robust study [212].

VEGF signaling may also be blocked by inhibiting cyclooxygenase-2 (COX-2), a proinflammatory enzyme that is overexpressed in 70–90 percent of pancreatic tumors. COX-2 overexpression has been linked with tumor proliferation and increased angiogenesis [213, 214]. Celecoxib is a compound that inhibits COX-2 [215, 216]. Celecoxib potentiates gemcitabine induced apoptosis *in vitro* and inhibits angiogenesis and tumor growth *in vivo* [216, 217]. When studied in clinical trials in combination with 5-fluorouracil, celecoxib was initially found to be well tolerated and to produce a durable response in a patient with gemcitabine resistant pancreatic cancer [218]. However, in a phase II clinical study testing the efficacy of celecoxib in combination with gemcitabine and cisplatin, celecoxib did not confer a significant advantage in patients with advanced PDAC [219].

Stromal Targets

As discussed above, the tumor associated stromal compartment is emerging as an important regulator of cancer progression and response to anticancer therapy. In the pancreas, proliferating fibroblasts and stellate cells support the growth of cancer and endothelial cells by activating growth pathways dependent on cell-to-cell contacts, production of collagen, and the release of local mitogenic factors (e.g., EGF, FGF, PDGF, and IGF-1). Molecular therapies could be used to both suppress the proliferation of these stromal components and interfere with the secreted factors. For instance, a VEGF "trap" could be used to bind to and inhibit the pro-angiogenic signals produced by secreted VEGF-A ligand [220]. In a similar manner, expression of a soluble type II TGF-β receptor has been shown to attenuate tumor growth, angiogenesis, and pro-metastatic genes (e.g., plasminogen activator inhibitor 1 and urokinase plasminogen activator) [221, 222], and targeting CTGF has been effective in mouse models [223, 224].

CONCLUSIONS

In spite of tremendous progress in understanding the pathobiology of PDAC, very little progress has been made with respect to impacting the 5-year survival rates of PDAC patients. There are several reasons for this failure. First, as these drugs are designed to target cancers with a particular molecular alteration, it follows that the selection of patients can only be as good as the tools used to characterize the biology of their tumor. While the use of immunohistochemistry (IHC) has become widespread to detect protein alterations, there is still considerable variability in this process. Site-to-site differences in protocol, incomplete characterization of antibody specificity, and even alterations between commercial batches of the "same" antibody contribute to the absence of rigorous reproducibility. Additionally, IHC is basically a snapshot of the protein expression level in a tumor at a particular time and is generally inadequate for assessing functionality. Second, therapeutic failure could be due to the inability of the agent to permeate the tumor, as solid tumors are notorious for having a disordered angiogenic network. This inability to reach the target within the pancreatic tumor mass is compounded by the desmoplastic nature of this malignancy and the resultant increased intra-tumoral hydrostatic pressures. Third, the range of doses used in clinical trials are constrained by the side effect profile of the drugs, and may be insufficient to recapitulate the biological effects observed *in vitro* and in subcutaneous mouse models. Fourth, tumor heterogeneity and crosstalk between signaling pathways may result in cancer cell populations that are resistant to a particular targeted therapy approach, or that acquire resistance to that approach. Finally, in addition to cross talk mechanisms, redundant defects in specific aberrant signaling pathways (e.g., the TGF-β pathway) may contribute to the resistance to targeted therapies [225, 226]. Thus, combinational regimens in conjunction with standard chemotherapy may need to be devised, based on the specific molecular alterations that are delineated in a particular patients cancer. This personalized combinatorial targeted approach may offer a new ray of hope to patients with PDAC.

ACKNOWLEDGEMENT

This work was supported by NIH grant R37-8184.

REFERENCES

1. Hruban RH, Wilentz RE, Kern SE. Genetic Progression in the Pancreatic Ducts. *Am J Pathol* 2000;**156**:1821–5.
2. Gudjonsson B. Cancer of the pancreas. 50 years of surgery. *Cancer* 1987;**60**:2284–303.
3. DiMagno EP, Reber HA, Tempero MA. AGA technical review on the epidemiology, diagnosis, and treatment of pancreatic ductal adenocarcinoma. American Gastroenterological Association. *Gastroenterology* 1999;**117**:1464–84.
4. Ho CK, Kleeff J, Friess H, Buchler MW. Complications of pancreatic surgery. *HPB (Oxford)* 2005;**7**:99–108.
5. Buchler MW, Kleeff J, Friess H. Surgical treatment of pancreatic cancer. *J Am Coll Surg* 2007;**205**:S81–6.
6. Rozenblum E, Schutte M, Goggins M, Hahn SA, Panzer S, Zahurak M, Goodman SN, Sohn TA, Hruban RH, Yeo CJ, Kern SE. Tumor-suppressive Pathways in Pancreatic Carcinoma. *Cancer Res* 1997;**57**:1731–4.
7. Tada M, Ohashi M, Shiratori Y, Okudaira T, Komatsu Y, Kawabe T, Yoshida H, Machinami R, Kishi K, Omata M. Analysis of K-ras gene mutation in hyperplastic duct cells of the pancreas without pancreatic disease. *Gastroenterology* 1996;**110**:227–31.
8. Moskaluk CA, Hruban RH, Kern SE. p16 and K-ras Gene Mutations in the Intraductal Precursors of Human Pancreatic Adenocarcinoma. *Cancer Res* 1997;**57**:2140–3.
9. Terhune PG, Phifer DM, Tosteson TD, Longnecker DS. K-ras mutation in focal proliferative lesions of human pancreas. *Cancer Epidemiol Biomarkers Prev* 1998;**7**:515–21.
10. Day JD, Digiuseppe JA, Yeo C, Lai-Goldman M, Anderson SM, Goodman SN, Kern SE, Hruban RH. Immunohistochemical evaluation of HER-2/neu expression in pancreatic adenocarcinoma and pancreatic intraepithelial neoplasms. *Hum Pathol* 1996;**27**:119–24.
11. Yamano M, Fujii H, Takagaki T, Kadowaki N, Watanabe H, Shirai T. Genetic Progression and Divergence in Pancreatic Carcinoma. *Am J Pathol* 2000;**156**:2123–33.
12. Durkin AJ, Bloomston PM, Rosemurgy AS, Giarelli N, Cojita D, Yeatman TJ, Zervos EE. Defining the role of the epidermal growth factor receptor in pancreatic cancer grown in vitro. *Am J Surg* 2003;**186**:431–6.
13. Yamanaka Y, Friess H, Kobrin MS, Buchler M, Beger HG, Korc M. Coexpression of epidermal growth factor receptor and ligands in human pancreatic cancer is associated with enhanced tumor aggressiveness. *Anticancer Res* 1993;**13**:565–9.
14. Korc M, Friess H, Yamanaka Y, Kobrin MS, Buchler M, Beger HG. Chronic pancreatitis is associated with increased concentrations of epidermal growth factor receptor, transforming growth factor alpha, and phospholipase C gamma. *Gut* 1994;**35**:1468–73.
15. Safran H, Steinhoff M, Mangray S, Rathore R, King TC, Chai L, Berzein K, Moore T, Iannitti D, Reiss P, Pasquariello T, Akerman P, Quirk D, Mass R, Goldstein L, Tantravahi U. Overexpression of the HER-2/neu oncogene in pancreatic adenocarcinoma. *Am J Clin Oncol* 2001;**24**:496–9.
16. Jaskiewicz K, Krige JE, Thomson J. Expression of p53 tumor suppressor gene, oncoprotein c-erbB-2, cellular proliferation and differentiation in malignant and benign pancreatic lesions. *Anticancer Res* 1994;**14**:1919–22.

17. Zhang L, Yuan SZ. Expression of c-erbB-2 oncogene protein, epidermal growth factor receptor, and TGF-beta1 in human pancreatic ductal adenocarcinoma. *Hepatobiliary Pancreat Dis Int* 2002;**1**:620–3.

18. Kolb A, Kleeff J, Arnold N, Giese NA, Giese T, Korc M, Friess H. Expression and differential signaling of heregulins in pancreatic cancer cells. *Int J Cancer* 2007;**120**:514–23.

19. Jiao L, Zhu J, Hassan MM, Evans DB, Abbruzzese JL, Li D. K-ras mutation and p16 and preproenkephalin promoter hypermethylation in plasma DNA of pancreatic cancer patients: in relation to cigarette smoking. *Pancreas* 2007;**34**:55–62.

20. Kim J, Reber HA, Dry SM, Elashoff D, Chen SL, Umetani N, Kitago M, Hines OJ, Kazanjian KK, Hiramatsu S, Bilchik AJ, Yong S, Shoup M, Hoon DS. Unfavourable prognosis associated with K-ras gene mutation in pancreatic cancer surgical margins. *Gut* 2006;**55**:1598–605.

21. Castells A, Puig P, Mora J, Boadas J, Boix L, Urgell E, Sole M. Capella G, Lluis F, Fernandez-Cruz L, Navarro S, Farre A: K-ras mutations in DNA extracted from the plasma of patients with pancreatic carcinoma: diagnostic utility and prognostic significance. *J Clin Oncol* 1999;**17**:578–84.

22. Aguirre AJ, Bardeesy N, Sinha M, Lopez L, Tuveson DA, Horner J, Redston MS, DePinho RA. Activated Kras and Ink4a/Arf deficiency cooperate to produce metastatic pancreatic ductal adenocarcinoma. *Genes Dev* 2003;**17**:3112–26.

23. Ingham PW, McMahon AP. Hedgehog signaling in animal development: paradigms and principles. *Genes Dev* 2001;**15**:3059–87.

24. Ramalho-Santos M, Melton DA, McMahon AP. Hedgehog signals regulate multiple aspects of gastrointestinal development. *Development* 2000;**127**:2763–72.

25. Freeman M. Feedback control of intercellular signalling in development. *Nature* 2000;**408**:313–19.

26. Thayer SP, di Magliano MP, Heiser PW, Nielsen CM, Roberts DJ, Lauwers GY, Qi YP, Gysin S, Fernandez-del Castillo C, Yajnik V, Antoniu B, McMahon M, Warshaw AL, Hebrok M. Hedgehog is an early and late mediator of pancreatic cancer tumorigenesis. *Nature* 2003;**425**:851–6.

27. Pasca di Magliano M, Hebrok M. Hedgehog signalling in cancer formation and maintenance. *Nat Rev Cancer* 2003;**3**:903–11.

28. Pasca di Magliano M, Biankin AV, Heiser PW, Cano DA, Gutierrez PJ, Deramaudt T, Segara D, Dawson AC, Kench JG, Henshall SM, Sutherland RL, Dlugosz A, Rustgi AK, Hebrok M. Common activation of canonical Wnt signaling in pancreatic adenocarcinoma. *PLoS ONE* 2007;**2**:e1155.

29. Cohen Jr MM. The hedgehog signaling network. *Am J Med Genet A* 2003;**123A**:5–28.

30. Gill PS, Rosenblum ND. Control of murine kidney development by sonic hedgehog and its GLI effectors. *Cell Cycle* 2006;**5**:1426–30.

31. Hezel AF, Kimmelman AC, Stanger BZ, Bardeesy N, Depinho RA. Genetics and biology of pancreatic ductal adenocarcinoma. *Genes Dev* 2006;**20**:1218–49.

32. Taipale J, Beachy PA. The Hedgehog and Wnt signalling pathways in cancer. *Nature* 2001;**411**:349–54.

33. Berman DM, Karhadkar SS, Maitra A, Montes De Oca R, Gerstenblith MR, Briggs K, Parker AR, Shimada Y, Eshleman JR, Watkins DN, Beachy PA. Widespread requirement for Hedgehog ligand stimulation in growth of digestive tract tumours. *Nature* 2003;**425**:846–51.

34. Qualtrough D, Buda A, Gaffield W, Williams AC, Paraskeva C. Hedgehog signalling in colorectal tumour cells: induction of apoptosis with cyclopamine treatment. *Int J Cancer* 2004;**110**:831–7.

35. Feldmann G, Dhara S, Fendrich V, Bedja D, Beaty R, Mullendore M, Karikari C, Alvarez H, Iacobuzio-Donahue C, Jimeno A, Gabrielson

KL, Matsui W, Maitra A. Blockade of hedgehog signaling inhibits pancreatic cancer invasion and metastases: a new paradigm for combination therapy in solid cancers. *Cancer Res* 2007;**67**:2187–96.

36. Ji Z, Mei FC, Xie J, Cheng X. Oncogenic KRAS activates hedgehog signaling pathway in pancreatic cancer cells. *J Biol Chem* 2007;**282**:14,048–14,055.

37. Hruban RH, Adsay NV, Albores-Saavedra J, Anver MR, Biankin AV, Boivin GP, Furth EE, Furukawa T, Klein A, Klimstra DS, Kloppel G, Lauwers GY, Longnecker DS, Luttges J, Maitra A, Offerhaus GJA, Perez-Gallego L, Redston M, Tuveson DA. Pathology of Genetically Engineered Mouse Models of Pancreatic Exocrine Cancer: Consensus Report and Recommendations. *Cancer Res* 2006;**66**:95–106.

38. Pasca di Magliano M, Sekine S, Ermilov A, Ferris J, Dlugosz AA, Hebrok M. Hedgehog/Ras interactions regulate early stages of pancreatic cancer. *Genes Dev* 2006;**20**:3161–73.

39. Feldmann G, Habbe N, Dhara S, Bisht S, Alvarez H, Fendrich V, Beaty R, Mullendore M, Karikari C, Bardeesy N, Ouellette MM, Yu W, Maitra A. Hedgehog inhibition prolongs survival in a genetically engineered mouse model of pancreatic cancer. *Gut* 2008;**57**:1420–30.

40. Feldmann G, Fendrich V, McGovern K, Bedja D, Bisht S, Alvarez H, Koorstra JB, Habbe N, Karikari C, Mullendore M, Gabrielson KL, Sharma R, Matsui W, Maitra A. An orally bioavailable small-molecule inhibitor of Hedgehog signaling inhibits tumor initiation and metastasis in pancreatic cancer. *Mol Cancer Ther* 2008;**7**:2725–35.

41. Sjolund J, Manetopoulos C, Stockhausen MT, Axelson H. The Notch pathway in cancer: differentiation gone awry. *Eur J Cancer* 2005;**41**:2620–9.

42. Miele L, Golde T, Osborne B. Notch signaling in cancer. *Curr Mol Med* 2006;**6**:905–18.

43. Brou C, Logeat F, Gupta N, Bessia C, LeBail O, Doedens JR, Cumano A, Roux P, Black RA, Israel A. A novel proteolytic cleavage involved in Notch signaling: the role of the disintegrin-metalloprotease TACE. *Mol Cell* 2000;**5**:207–16.

44. Mumm JS, Schroeter EH, Saxena MT, Griesemer A, Tian X, Pan DJ, Ray WJ, Kopan R. A ligand-induced extracellular cleavage regulates gamma-secretase-like proteolytic activation of Notch1. *Mol Cell* 2000;**5**:197–206.

45. Struhl G, Adachi A. Nuclear access and action of notch in vivo. *Cell* 1998;**93**:649–60.

46. Schroeter EH, Kisslinger JA, Kopan R. Notch-1 signalling requires ligand-induced proteolytic release of intracellular domain. *Nature* 1998;**393**:382–6.

47. Tamura K, Taniguchi Y, Minoguchi S, Sakai T, Tun T, Furukawa T, Honjo T. Physical interaction between a novel domain of the receptor Notch and the transcription factor RBP-J kappa/Su(H). *Curr Biol* 1995;**5**:1416–23.

48. Kato H, Taniguchi Y, Kurooka H, Minoguchi S, Sakai T, Nomura-Okazaki S, Tamura K, Honjo T. Involvement of RBP-J in biological functions of mouse Notch1 and its derivatives. *Development* 1997;**124**:4133–41.

49. Lai EC. Keeping a good pathway down: transcriptional repression of Notch pathway target genes by CSL proteins. *EMBO Rep* 2002;**3**:840–5.

50. Wu L, Aster JC, Blacklow SC, Lake R, Artavanis-Tsakonas S, Griffin JD. MAML1, a human homologue of Drosophila mastermind, is a transcriptional co-activator for NOTCH receptors. *Nat Genet* 2000;**26**:484–9.

51. Fryer CJ, Lamar E, Turbachova I, Kintner C, Jones KA. Mastermind mediates chromatin-specific transcription and turnover of the Notch enhancer complex. *Genes Dev* 2002;**16**:1397–411.

52. Iso T, Kedes L, Hamamori Y. HES and HERP families: multiple effectors of the Notch signaling pathway. *J Cell Physiol* 2003;**194**:237–55.

53. Ronchini C, Capobianco AJ. Induction of cyclin D1 transcription and CDK2 activity by Notch(ic): implication for cell cycle disruption in transformation by Notch(ic). *Mol Cell Biol* 2001;**21**:5925–34.

54. Sarmento LM, Huang H, Limon A, Gordon W, Fernandes J, Tavares MJ, Miele L, Cardoso AA, Classon M, Carlesso N. Notch1 modulates timing of G1-S progression by inducing SKP2 transcription and p27 Kip1 degradation. *J Exp Med* 2005;**202**:157–68.

55. Klinakis A, Szabolcs M, Politi K, Kiaris H, Artavanis-Tsakonas S, Efstratiadis A. Myc is a Notch1 transcriptional target and a requisite for Notch1-induced mammary tumorigenesis in mice. *Proc Natl Acad Sci U S A* 2006;**103**:9262–7.

56. Weng AP, Millholland JM, Yashiro-Ohtani Y, Arcangeli ML, Lau A, Wai C, Del Bianco C, Rodriguez CG, Sai H, Tobias J, Li Y, Wolfe MS, Shachaf C, Felsher D, Blacklow SC, Pear WS, Aster JC. c-Myc is an important direct target of Notch1 in T-cell acute lymphoblastic leukemia/lymphoma. *Genes Dev* 2006;**20**:2096–109.

57. Jensen J, Pedersen EE, Galante P, Hald J, Heller RS, Ishibashi M, Kageyama R, Guillemot F, Serup P, Madsen OD. Control of endodermal endocrine development by Hes-1. *Nat Genet* 2000;**24**:36–44.

58. Apelqvist A, Li H, Sommer L, Beatus P, Anderson DJ, Honjo T, Hrabe de Angelis M, Lendahl U, Edlund H. Notch signalling controls pancreatic cell differentiation. *Nature* 1999;**400**:877–81.

59. Jensen J, Heller RS, Funder-Nielsen T, Pedersen EE, Lindsell C, Weinmaster G, Madsen OD, Serup P. Independent development of pancreatic alpha- and beta-cells from neurogenin3-expressing precursors: a role for the notch pathway in repression of premature differentiation. *Diabetes* 2000;**49**:163–76.

60. Lomberk G, Urrutia R. Primers on molecular pathways: notch. *Pancreatology* 2008;**8**:103–4.

61. Siveke JT, Lubeseder-Martellato C, Lee M, Mazur PK, Nakhai H, Radtke F, Schmid RM. Notch signaling is required for exocrine regeneration after acute pancreatitis. *Gastroenterology* 2008;**134**:544–55.

62. Wang Z, Zhang Y, Li Y, Banerjee S, Liao J, Sarkar FH. Down-regulation of Notch-1 contributes to cell growth inhibition and apoptosis in pancreatic cancer cells. *Mol Cancer Ther* 2006;**5**:483–93.

63. Miyamoto Y, Maitra A, Ghosh B, Zechner U, Argani P, Iacobuzio-Donahue CA, Sriuranpong V, Iso T, Meszoely IM, Wolfe MS, Hruban RH, Ball DW, Schmid RM, Leach SD. Notch mediates TGF alpha-induced changes in epithelial differentiation during pancreatic tumorigenesis. *Cancer Cell* 2003;**3**:565–76.

64. Rizzo P, Osipo C, Foreman K, Golde T, Osborne B, Miele L. Rational targeting of Notch signaling in cancer. *Oncogene* 2008;**27**:5124–31.

65. Noguera-Troise I, Daly C, Papadopoulos NJ, Coetzee S, Boland P, Gale NW, Lin HC, Yancopoulos GD, Thurston G. Blockade of Dll4 inhibits tumour growth by promoting non-productive angiogenesis. *Nature* 2006;**444**:1032–7.

66. Ridgway J, Zhang G, Wu Y, Stawicki S, Liang WC, Chanthery Y, Kowalski J, Watts RJ, Callahan C, Kasman I, Singh M, Chien M, Tan C, Hongo JA, de Sauvage F, Plowman G, Yan M. Inhibition of Dll4 signalling inhibits tumour growth by deregulating angiogenesis. *Nature* 2006;**444**:1083–7.

67. Fan X, Mikolaenko I, Elhassan I, Ni X, Wang Y, Ball D, Brat DJ, Perry A, Eberhart CG. Notch1 and notch2 have opposite effects on embryonal brain tumor growth. *Cancer Res* 2004;**64**:7787–93.

68. Garces C, Ruiz-Hidalgo MJ, Font de Mora J, Park C, Miele L, Goldstein J, Bonvini E, Porras A, Laborda J. Notch-1 controls the expression of fatty acid-activated transcription factors and is required for adipogenesis. *J Biol Chem* 1997;**272**:29,729–29,734.

69. Purow BW, Haque RM, Noel MW, Su Q, Burdick MJ, Lee J, Sundaresan T, Pastorino S, Park JK, Mikolaenko I, Maric D, Eberhart CG, Fine HA. Expression of Notch-1 and its ligands, Delta-like-1 and Jagged-1, is critical for glioma cell survival and proliferation. *Cancer Res* 2005;**65**:2353–63.

70. Nickoloff BJ, Qin JZ, Chaturvedi V, Denning MF, Bonish B, Miele L. Jagged-1 mediated activation of notch signaling induces complete maturation of human keratinocytes through NF-kappaB and PPARgamma. *Cell Death Differ* 2002;**9**:842–55.

71. Talar-Wojnarowska R, Gasiorowska A, Smolarz B, Romanowicz-Makowskal H, Strzelczyk J, Janiak A, Malecka-Panas E. Comparative evaluation of p53 mutation in pancreatic adenocarcinoma and chronic pancreatitis. *Hepatogastroenterology* 2006;**53**:608–12.

72. Lubensky IA, Zhuang Z. Molecular genetic events in gastrointestinal and pancreatic neuroendocrine tumors. *Endocr Pathol* 2007;**18**:156–62.

73. Zhang SY, Ruggeri B, Agarwal P, Sorling AF, Obara T, Ura H, Namiki M, Klein-Szanto AJ. Immunohistochemical analysis of p53 expression in human pancreatic carcinomas. *Arch Pathol Lab Med* 1994;**118**:150–4.

74. Ruggeri BA, Huang L, Berger D, Chang H, Klein-Szanto AJ, Goodrow T, Wood M, Obara T, Heath CW, Lynch H. Molecular pathology of primary and metastatic ductal pancreatic lesions: analyses of mutations and expression of the p53, mdm-2, and p21/WAF-1 genes in sporadic and familial lesions. *Cancer* 1997;**79**:700–16.

75. Kirsch DG, Kastan MB. Tumor-suppressor p53: implications for tumor development and prognosis. *J Clin Oncol* 1998;**16**:3158–68.

76. Hansel DE, Kern SE, Hruban RH. Molecular pathogenesis of pancreatic cancer. *Annu Rev Genomics Hum Genet* 2003;**4**:237–56.

77. Wilentz RE, Geradts J, Maynard R, Offerhaus GJ, Kang M, Goggins M, Yeo CJ, Kern SE, Hruban RH. Inactivation of the p16 (INK4A) tumor-suppressor gene in pancreatic duct lesions: loss of intranuclear expression. *Cancer Res* 1998;**58**:4740–4.

78. Gerdes B, Ramaswamy A, Kersting M, Ernst M, Lang S, Schuermann M, Wild A, Bartsch DK. p16(INK4a) alterations in chronic pancreatitis-indicator for high-risk lesions for pancreatic cancer. *Surgery* 2001;**129**:490–7.

79. Moskaluk CA, Hruban RH, Kern SE. p16 and K-ras gene mutations in the intraductal precursors of human pancreatic adenocarcinoma. *Cancer Res* 1997;**57**:2140–3.

80. Ohtsubo K, Watanabe H, Yamaguchi Y, Hu YX, Motoo Y, Okai T, Sawabu N. Abnormalities of tumor suppressor gene p16 in pancreatic carcinoma: immunohistochemical and genetic findings compared with clinicopathological parameters. *J Gastroenterol* 2003;**38**:663–71.

81. Hu YX, Watanabe H, Ohtsubo K, Yamaguchi Y, Ha A, Okai T, Sawabu N. Frequent loss of p16 expression and its correlation with clinicopathological parameters in pancreatic carcinoma. *Clin Cancer Res* 1997;**3**:1473–7.

82. Derynck R, Akhurst RJ, Balmain A. TGF-beta signaling in tumor suppression and cancer progression. *Nat Genet* 2001;**29**:117–29.

83. Kingsley DM. The TGF-beta superfamily: new members, new receptors, and new genetic tests of function in different organisms. *Genes Dev* 1994;**8**:133–46.

84. Siegel PM, Massague J. Cytostatic and apoptotic actions of TGF-beta in homeostasis and cancer. *Nat Rev Cancer* 2003;**3**:807–21.

85. Shi Y, Massague J. Mechanisms of TGF-beta signaling from cell membrane to the nucleus. *Cell* 2003;**113**:685–700.

86. Derynck R, Feng XH. TGF-beta receptor signaling. *Biochim Biophys Acta* 1997;**1333**:F105–50.

87. Attisano L, Wrana JL. Signal transduction by the TGF-beta superfamily. *Science* 2002;**296**:1646–7.

88. Massague J, Seoane J, Wotton D. Smad transcription factors. *Genes Dev* 2005;**19**:2783–810.

89. Heldin CH, Miyazono K, ten Dijke P. TGF-beta signalling from cell membrane to nucleus through SMAD proteins. *Nature* 1997;**390**:465–71.

90. Derynck R, Zhang Y, Feng XH. Smads: transcriptional activators of TGF-beta responses. *Cell* 1998;**95**:737–40.

91. Shi Y, Wang YF, Jayaraman L, Yang H, Massague J, Pavletich NP. Crystal structure of a Smad MH1 domain bound to DNA: insights on DNA binding in TGF-beta signaling. *Cell* 1998;**94**:585–94.

92. Massague J, Wotton D. Transcriptional control by the TGF-beta/Smad signaling system. *EMBO J* 2000;**19**:1745–54.

93. Murakami M, Nagai E, Mizumoto K, Saimura M, Ohuchida K, Inadome N, Matsumoto K, Nakamura T, Maemondo M, Nukiwa T, Tanaka M. Suppression of metastasis of human pancreatic cancer to the liver by transportal injection of recombinant adenoviral NK4 in nude mice. *Int J Cancer* 2005;**117**:160–5.

94. Korc M. Pathways for aberrant angiogenesis in pancreatic cancer. *Mol Cancer* 2003;**2**:8.

95. Gold LI. The role for transforming growth factor-beta (TGF-beta) in human cancer. *Crit Rev Oncog* 1999;**10**:303–60.

96. Korc M. Role of growth factors in pancreatic cancer. *Surg Oncol Clin N Am* 1998;**7**:25–41.

97. Feng XH, Derynck R. Specificity and versatility in tgf-beta signaling through Smads. *Annu Rev Cell Dev Biol* 2005;**21**:659–93.

98. Yoo BM, Yeo M, Oh TY, Choi JH, Kim WW, Kim JH, Cho SW, Kim SJ, Hahm KB. Amelioration of pancreatic fibrosis in mice with defective TGF-beta signaling. *Pancreas* 2005;**30**:e71–9.

99. Slater SD, Williamson RC, Foster CS. Expression of transforming growth factor-beta 1 in chronic pancreatitis. *Digestion* 1995;**56**:237–41.

100. Fukumura Y, Kumasaka T, Mitani K, Karita K, Suda K. Expression of transforming growth factor beta1, beta2, and beta3 in chronic, cancer-associated, obstructive pancreatitis. *Arch Pathol Lab Med* 2006;**130**:356–61.

101. Lohr M, Schmidt C, Ringel J, Kluth M, Muller P, Nizze H, Jesnowski R. Transforming growth factor-beta1 induces desmoplasia in an experimental model of human pancreatic carcinoma. *Cancer Res* 2001;**61**:550–5.

102. Massague J, Blain SW, Lo RS. TGFbeta signaling in growth control, cancer, and heritable disorders. *Cell* 2000;**103**:295–309.

103. Beauchamp Jr RD, Lyons RM, Yang EY, Coffey RJ, Moses HL. Expression of and response to growth regulatory peptides by two human pancreatic carcinoma cell lines. *Pancreas* 1990;**5**:369–80.

104. Baldwin RL, Korc M. Growth inhibition of human pancreatic carcinoma cells by transforming growth factor beta-1. *Growth Factors* 1993;**8**:23–34.

105. Friess H, Kleeff J, Korc M, Buchler MW. Molecular aspects of pancreatic cancer and future perspectives. *Dig Surg* 1999;**16**:281–90.

106. Wagner M, Kleeff J, Lopez ME, Bockman I, Massaque J, Korc M. Transfection of the type I TGF-beta receptor restores TGF-beta responsiveness in pancreatic cancer. *Int J Cancer* 1998;**78**:255–60.

107. Kleeff J, Korc M. Up-regulation of transforming growth factor (TGF)-beta receptors by TGF-beta1 in COLO-357 cells. *J Biol Chem* 1998;**273**:7495–500.

108. Friess H, Yamanaka Y, Buchler M, Ebert M, Beger HG, Gold LI, Korc M. Enhanced expression of transforming growth factor beta isoforms in pancreatic cancer correlates with decreased survival. *Gastroenterology* 1993;**105**:1846–56.

109. Hahn SA, Schutte M, Hoque AT, Moskaluk CA, da Costa LT, Rozenblum E, Weinstein CL, Fischer A, Yeo CJ, Hruban RH, Kern SE. DPC4, a candidate tumor suppressor gene at human chromosome 18q21.1. *Science* 1996;**271**:350–3.

110. Baldwin RL, Friess H, Yokoyama M, Lopez ME, Kobrin MS, Buchler MW, Korc M. Attenuated ALK5 receptor expression in human pancreatic cancer: correlation with resistance to growth inhibition. *Int J Cancer* 1996;**67**:283–8.

111. Kleeff J, Maruyama H, Friess H, Buchler MW, Falb D, Korc M. Smad6 suppresses TGF-beta-induced growth inhibition in COLO-357 pancreatic cancer cells and is overexpressed in pancreatic cancer. *Biochem Biophys Res Commun* 1999;**255**:268–73.

112. Arnold NB, Ketterer K, Kleeff J, Friess H, Buchler MW, Korc M. Thioredoxin is downstream of Smad7 in a pathway that promotes growth and suppresses cisplatin-induced apoptosis in pancreatic cancer. *Cancer Res* 2004;**64**:3599–606.

113. Wilentz RE, Su GH, Dai JL, Sparks AB, Argani P, Sohn TA, Yeo CJ, Kern SE, Hruban RH. Immunohistochemical labeling for dpc4 mirrors genetic status in pancreatic adenocarcinomas : a new marker of DPC4 inactivation. *Am J Pathol* 2000;**156**:37–43.

114. Tang ZH, Zou SQ, Hao YH, Wang BJ, Yang XP, Chen QQ, Qiu FZ. The relationship between loss expression of DPC4/Smad4 gene and carcinogenesis of pancreatobiliary carcinoma. *Hepatobiliary Pancreat Dis Int* 2002;**1**:624–9.

115. Hua Z, Zhang YC, Hu XM, Jia ZG. Loss of DPC4 expression and its correlation with clinicopathological parameters in pancreatic carcinoma. *World J Gastroenterol* 2003;**9**:2764–7.

116. Wilentz RE, Iacobuzio-Donahue CA, Argani P, McCarthy DM, Parsons JL, Yeo CJ, Kern SE, Hruban RH. Loss of expression of Dpc4 in pancreatic intraepithelial neoplasia: evidence that DPC4 inactivation occurs late in neoplastic progression. *Cancer Res* 2000;**60**:2002–6.

117. Tascilar M, Skinner HG, Rosty C, Sohn T, Wilentz RE, Offerhaus GJ, Adsay V, Abrams RA, Cameron JL, Kern SE, Yeo CJ, Hruban RH, Goggins M. The SMAD4 protein and prognosis of pancreatic ductal adenocarcinoma. *Clin Cancer Res* 2001;**7**:4115–21.

118. Chen D, Zhao M, Mundy GR. Bone morphogenetic proteins. *Growth Factors* 2004;**22**:233–41.

119. Kitisin K, Saha T, Blake T, Golestaneh N, Deng M, Kim C, Tang Y, Shetty K, Mishra B, Mishra L. TGF-beta Signaling in Development. *Sci STKE* 2007;**2007**:cm1.

120. Kleeff J, Maruyama H, Ishiwata T, Sawhney H, Friess H, Buchler MW, Korc M. Bone morphogenetic protein 2 exerts diverse effects on cell growth in vitro and is expressed in human pancreatic cancer in vivo. *Gastroenterology* 1999;**116**:1202–16.

121. Kayed H, Kleeff J, Keleg S, Jiang X, Penzel R, Giese T, Zentgraf H, Buchler MW, Korc M, Friess H. Correlation of glypican-1 expression with TGF-beta, BMP, and activin receptors in pancreatic ductal adenocarcinoma. *Int J Oncol* 2006;**29**:1139–48.

122. Li Z, Manna P, Kobayashi H, Spilde T, Bhatia A, Preuett B, Prasadan K, Hembree M, Gittes GK. Multifaceted pancreatic mesenchymal control of epithelial lineage selection. *Dev Biol* 2004;**269**:252–63.

123. Sadlonova A, Novak Z, Johnson MR, Bowe DB, Gault SR, Page GP, Thottassery JV, Welch DR, Frost AR. Breast fibroblasts modulate epithelial cell proliferation in three-dimensional in vitro co-culture. *Breast Cancer Res* 2005;**7**:R46–59.

124. Barcellos-Hoff MH, Ravani SA. Irradiated mammary gland stroma promotes the expression of tumorigenic potential by unirradiated epithelial cells. *Cancer Res* 2000;**60**:1254–60.

125. Surowiak P, Suchocki S, Gyorffy B, Gansukh T, Wojnar A, Maciejczyk A, Pudelko M, Zabel M. Stromal myofibroblasts in

breast cancer: relations between their occurrence, tumor grade and expression of some tumour markers. *Folia Histochem Cytobiol* 2006;**44**:111–16.

126. Yazhou C, Wenlv S, Weidong Z, Licun W. Clinicopathological significance of stromal myofibroblasts in invasive ductal carcinoma of the breast. *Tumour Biol* 2004;**25**:290–5.

127. Binkley CE, Zhang L, Greenson JK, Giordano TJ, Kuick R, Misek D, Hanash S, Logsdon CD, Simeone DM. The molecular basis of pancreatic fibrosis: common stromal gene expression in chronic pancreatitis and pancreatic adenocarcinoma. *Pancreas* 2004;**29**:254–63.

128. Sato N, Maehara N, Goggins M. Gene expression profiling of tumor-stromal interactions between pancreatic cancer cells and stromal fibroblasts. *Cancer Res* 2004;**64**:6950–6.

129. Apte MV, Park S, Phillips PA, Santucci N, Goldstein D, Kumar RK, Ramm GA, Buchler M, Friess H, McCarroll JA, Keogh G, Merrett N, Pirola R, Wilson JS. Desmoplastic reaction in pancreatic cancer: role of pancreatic stellate cells. *Pancreas* 2004;**29**:179–87.

130. Aoki H, Ohnishi H, Hama K, Shinozaki S, Kita H, Yamamoto H, Osawa H, Sato K, Tamada K, Sugano K. Existence of autocrine loop between interleukin-6 and transforming growth factor-beta1 in activated rat pancreatic stellate cells. *J Cell Biochem* 2006;**99**:221–8.

131. Bachem MG, Schunemann M, Ramadani M, Siech M, Beger H, Buck A, Zhou S, Schmid-Kotsas A, Adler G. Pancreatic carcinoma cells induce fibrosis by stimulating proliferation and matrix synthesis of stellate cells. *Gastroenterology* 2005;**128**:907–21.

132. Hwang RF, Moore T, Arumugam T, Ramachandran V, Amos KD, Rivera A, Ji B, Evans DB, Logsdon CD. Cancer-associated stromal fibroblasts promote pancreatic tumor progression. *Cancer Res* 2008;**68**:918–26.

133. Stefani AL, Basso D, Panozzo MP, Greco E, Mazza S, Zancanaro F, De Franchis G, Plebani M. Cytokines modulate MIA PaCa 2 and CAPAN-1 adhesion to extracellular matrix proteins. *Pancreas* 1999;**19**:362–9.

134. Aoki H, Ohnishi H, Hama K, Shinozaki S, Kita H, Osawa H, Yamamoto H, Sato K, Tamada K, Sugano K. Cyclooxygenase-2 is required for activated pancreatic stellate cells to respond to proinflammatory cytokines. *Am J Physiol Cell Physiol* 2007;**292**:C259–68.

135. Koide N, Yamada T, Shibata R, Mori T, Fukuma M, Yamazaki K, Aiura K, Shimazu M, Hirohashi S, Nimura Y, Sakamoto M. Establishment of perineural invasion models and analysis of gene expression revealed an invariant chain (CD74) as a possible molecule involved in perineural invasion in pancreatic cancer. *Clin Cancer Res* 2006;**12**:2419–26.

136. Meduri F, Diana F, Merenda R, Losacco L, Zuin A, Cecchetto A, Gerunda GE, Neri D, Zangrandi F, Maffei-Faccioli A. Pancreatic cancer and retroperitoneal neural tissue invasion. Its implication for survival following radical surgery. *Zentralbl Pathol* 1994;**140**:277–9.

137. Gasparini G. The rationale and future potential of angiogenesis inhibitors in neoplasia. *Drugs* 1999;**58**:17–38.

138. Esposito I, Menicagli M, Funel N, Bergmann F, Boggi U, Mosca F, Bevilacqua G, Campani D. Inflammatory cells contribute to the generation of an angiogenic phenotype in pancreatic ductal adenocarcinoma. *J Clin Pathol* 2004;**57**:630–6.

139. Beckermann BM, Kallifatidis G, Groth A, Frommhold D, Apel A, Mattern J, Salnikov AV, Moldenhauer G, Wagner W, Diehlmann A, Saffrich R, Schubert M, Ho AD, Giese N, Buchler MW, Friess H, Buchler P, Herr I. VEGF expression by mesenchymal stem cells contributes to angiogenesis in pancreatic carcinoma. *Br J Cancer* 2008;**99**:622–31.

140. Chang YT, Chang MC, Wei SC, Tien YW, Hsu C, Liang PC, Tsao PN, Jan IS, Wong JM. Serum vascular endothelial growth factor/soluble vascular endothelial growth factor receptor 1 ratio is an independent prognostic marker in pancreatic cancer. *Pancreas* 2008;**37**:145–50.

141. Karademir S, Sokmen S, Terzi C, Sagol O, Ozer E, Astarcioglu H, Coker A, Astarcioglu I. Tumor angiogenesis as a prognostic predictor in pancreatic cancer. *J Hepatobiliary Pancreat Surg* 2000;**7**:489–95.

142. Hartel M, Di Mola FF, Gardini A, Zimmermann A, Di Sebastiano P, Guweidhi A, Innocenti P, Giese T, Giese N, Buchler MW, Friess H. Desmoplastic reaction influences pancreatic cancer growth behavior. *World J Surg* 2004;**28**:818–25.

143. Ottaviano AJ, Sun L, Ananthanarayanan V, Munshi HG. Extracellular matrix-mediated membrane-type 1 matrix metalloproteinase expression in pancreatic ductal cells is regulated by transforming growth factor-beta1. *Cancer Res* 2006;**66**:7032–40.

144. Koenig A, Mueller C, Hasel C, Adler G, Menke A. Collagen type I induces disruption of E-cadherin-mediated cell-cell contacts and promotes proliferation of pancreatic carcinoma cells. *Cancer Res* 2006;**66**:4662–71.

145. Maatta M, Soini Y, Liakka A, Autio-Harmainen H. Differential expression of matrix metalloproteinase (MMP)-2, MMP-9, and membrane type 1-MMP in hepatocellular and pancreatic adenocarcinoma: implications for tumor progression and clinical prognosis. *Clin Cancer Res* 2000;**6**:2726–34.

146. Shin SJ, Kim KO, Kim MK, Lee KH, Hyun MS, Kim KJ, Choi JH, Song HS. Expression of E-cadherin and uPA and their association with the prognosis of pancreatic cancer. *Jpn J Clin Oncol* 2005;**35**:342–8.

147. Albo D, Berger DH, Tuszynski GP. The effect of thrombospondin-1 and TGF-beta 1 on pancreatic cancer cell invasion. *J Surg Res* 1998;**76**:86–90.

148. Tobita K, Kijima H, Dowaki S, Oida Y, Kashiwagi H, Ishii M, Sugio Y, Sekka T, Ohtani Y, Tanaka M, Inokuchi S, Makuuchi H. Thrombospondin-1 expression as a prognostic predictor of pancreatic ductal carcinoma. *Int J Oncol* 2002;**21**:1189–95.

149. Linder S, Castanos-Velez E, von Rosen A, Biberfeld P. Immunohistochemical expression of extracellular matrix proteins and adhesion molecules in pancreatic carcinoma. *Hepatogastroenterology* 2001;**48**:1321–7.

150. Lohr M, Trautmann B, Gottler M, Peters S, Zauner I, Maillet B, Kloppel G. Human ductal adenocarcinomas of the pancreas express extracellular matrix proteins. *Br J Cancer* 1994;**69**:144–51.

151. Lohr M, Trautmann B, Gottler M, Peters S, Zauner I, Maier A, Kloppel G, Liebe S, Kreuser ED. Expression and function of receptors for extracellular matrix proteins in human ductal adenocarcinomas of the pancreas. *Pancreas* 1996;**12**:248–59.

152. Thiery JP. Epithelial-mesenchymal transitions in tumour progression. *Nat Rev Cancer* 2002;**2**:442–54.

153. Zhu Z, Kleeff J, Kayed H, Wang L, Korc M, Buchler MW, Friess H. Nerve growth factor and enhancement of proliferation, invasion, and tumorigenicity of pancreatic cancer cells. *Mol Carcinog* 2002;**35**:138–47.

154. Sangai T, Ishii G, Kodama K, Miyamoto S, Aoyagi Y, Ito T, Magae J, Sasaki H, Nagashima T, Miyazaki M, Ochiai A. Effect of differences in cancer cells and tumor growth sites on recruiting bone marrow-derived endothelial cells and myofibroblasts in cancer-induced stroma. *Int J Cancer* 2005;**115**:885–92.

155. Farrow B, Sugiyama Y, Chen A, Uffort E, Nealon W, Mark Evers B. Inflammatory mechanisms contributing to pancreatic cancer development. *Ann Surg* 2004;**239**:763–9. discussion 769–771.

156. Ding K, Lopez-Burks M, Sanchez-Duran JA, Korc M, Lander AD. Growth factor-induced shedding of syndecan-1 confers glypican-1

dependence on mitogenic responses of cancer cells. *J Cell Biol* 2005;**171**:729–38.

157. Korc M. Pancreatic cancer-associated stroma production. *Am J Surg* 2007;**194**:S84–6.

158. Biggs J, Murphy EV, Israel MA. A human Id-like helix-loop-helix protein expressed during early development. *Proc Natl Acad Sci U S A* 1992;**89**:1512–16.

159. Zhao Y, Johansson C, Tran T, Bettencourt R, Itahana Y, Desprez PY, Konieczny SF. Identification of a basic helix-loop-helix transcription factor expressed in mammary gland alveolar cells and required for maintenance of the differentiated state. *Mol Endocrinol* 2006;**20**:2187–98.

160. Gasa R, Mrejen C, Leachman N, Otten M, Barnes M, Wang J, Chakrabarti S, Mirmira R, German M. Proendocrine genes coordinate the pancreatic islet differentiation program in vitro. *Proc Natl Acad Sci U S A* 2004;**101**:13,245–13,250.

161. Qiu Y, Guo M, Huang S, Stein R. Insulin gene transcription is mediated by interactions between the p300 coactivator and PDX-1, BETA2, and E47. *Mol Cell Biol* 2002;**22**:412–20.

162. Lee JC, Smith SB, Watada H, Lin J, Scheel D, Wang J, Mirmira RG, German MS. Regulation of the pancreatic pro-endocrine gene neurogenin3. *Diabetes* 2001;**50**:928–36.

163. Gradwohl G, Dierich A, LeMeur M, Guillemot F. neurogenin3 is required for the development of the four endocrine cell lineages of the pancreas. *Proc Natl Acad Sci U S A* 2000;**97**:1607–11.

164. Krapp A, Knofler M, Ledermann B, Burki K, Berney C, Zoerkler N, Hagenbuchle O, Wellauer PK. The bHLH protein PTF1-p48 is essential for the formation of the exocrine and the correct spatial organization of the endocrine pancreas. *Genes Dev* 1998;**12**:3752–63.

165. Ling MT, Wang X, Zhang X, Wong YC. The multiple roles of Id-1 in cancer progression. *Differentiation* 2006;**74**:481–7.

166. Jensen J, Serup P, Karlsen C, Nielsen TF, Madsen OD. mRNA profiling of rat islet tumors reveals nkx 6.1 as a beta-cell-specific homeodomain transcription factor. *J Biol Chem* 1996;**271**:18,749–18,758,.

167. Prisco M, Peruzzi F, Belletti B, Baserga R. Regulation of Id gene expression by type I insulin-like growth factor: roles of Stat3 and the tyrosine 950 residue of the receptor. *Mol Cell Biol* 2001;**21**:5447–58.

168. Robinson GL, Cordle SR, Henderson E, Weil PA, Teitelman G, Stein R. Isolation and characterization of a novel transcription factor that binds to and activates insulin control element-mediated expression. *Mol Cell Biol* 1994;**14**:6704–14.

169. Moldes M, Lasnier F, Feve B, Pairault J, Djian P. Id3 prevents differentiation of preadipose cells. *Mol Cell Biol* 1997;**17**:1796–804.

170. Coletta RD, Jedlicka P, Gutierrez-Hartmann A, Ford HL. Transcriptional control of the cell cycle in mammary gland development and tumorigenesis. *J Mammary Gland Biol Neoplasia* 2004;**9**:39–53.

171. Zebedee Z, Hara E. : Id proteins in cell cycle control and cellular senescence. *Oncogene* 2001;**20**:8317–25.

172. Prabhu S, Ignatova A, Park ST, Sun XH. Regulation of the expression of cyclin-dependent kinase inhibitor p21 by E2A and Id proteins. *Mol Cell Biol* 1997;**17**:5888–96.

173. Kleeff J, Ishiwata T, Friess H, Buchler MW, Israel MA, Korc M. The helix-loop-helix protein Id2 is overexpressed in human pancreatic cancer. *Cancer Res* 1998;**58**:3769–72.

174. Lee KT, Lee YW, Lee JK, Choi SH, Rhee JC, Paik SS, Kong G. Overexpression of Id-1 is significantly associated with tumour angiogenesis in human pancreas cancers. *Br J Cancer* 2004;**90**:1198–203.

175. Maruyama H, Kleeff J, Wildi S, Friess H, Buchler MW, Israel MA, Korc M. Id-1 and Id-2 are overexpressed in pancreatic cancer and in dysplastic lesions in chronic pancreatitis. *Am J Pathol* 1999;**155**:815–22.

176. Melet A, Song K, Bucur O, Jagani Z, Grassian AR, Khosravi-Far R. Apoptotic pathways in tumor progression and therapy. *Adv Exp Med Biol* 2008;**615**:47–79.

177. Schmitt CA. Senescence, apoptosis and therapy: cutting the lifelines of cancer. *Nat Rev Cancer* 2003;**3**:286–95.

178. Zalatnai A, Molnar J. Review. Molecular background of chemoresistance in pancreatic cancer. *In Vivo* 2007;**21**:339–47.

179. Ibrahim SM, Ringel J, Schmidt C, Ringel B, Muller P, Koczan D, Thiesen HJ, Lohr M. Pancreatic adenocarcinoma cell lines show variable susceptibility to TRAIL-mediated cell death. *Pancreas* 2001;**23**:72–9.

180. Elnemr A, Ohta T, Yachie A, Kayahara M, Kitagawa H, Fujimura T, Ninomiya I, Fushida S, Nishimura GI, Shimizu K, Miwa K. Human pancreatic cancer cells disable function of Fas receptors at several levels in Fas signal transduction pathway. *Int J Oncol* 2001;**18**:311–16.

181. Alo PL, Amini M, Piro F, Pizzuti L, Sebastiani V, Botti C, Murari R, Zotti G, Di Tondo U. Immunohistochemical expression and prognostic significance of fatty acid synthase in pancreatic carcinoma. *Anticancer Res* 2007;**27**:2523–7.

182. Ohta T, Elnemr A, Kitagawa H, Kayahara M, Takamura H, Fujimura T, Nishimura G, Shimizu K, Yi SQ, Miwa K. Fas ligand expression in human pancreatic cancer. *Oncol Rep* 2004;**12**:749–54.

183. Breslin TM, Janjan NA, Lee JE, Pisters PW, Wolff RA, Abbruzzese JL, Evans DB. Neoadjuvant chemoradiation for adenocarcinoma of the pancreas. *Front Biosci* 1998;**3**:E193–203.

184. Tse RV, Dawson LA, Wei A, Moore M. Neoadjuvant treatment for pancreatic cancer: a review. *Crit Rev Oncol Hematol* 2008;**65**:263–74.

185. Neoptolemos JP, Stocken DD, Friess H, Bassi C, Dunn JA, Hickey H, Beger H, Fernandez-Cruz L, Dervenis C, Lacaine F, Falconi M, Pederzoli P, Pap A, Spooner D, Kerr DJ, Buchler MW. the European Study Group for Pancreatic C: A Randomized Trial of Chemoradiotherapy and Chemotherapy after Resection of Pancreatic Cancer. *N Engl J Med* 2004;**350**:1200–10.

186. Crane C, Janjan N, Evans D, Wolff R, Ballo M, Milas L, Mason K, Charnsangavej C, Pisters P, Lee J, Lenzi R, Vauthey J, Wong A, Phan T, Nguyen Q, Abbruzzese J. Toxicity and Efficacy of Concurrent Gemcitabine and Radiotherapy for Locally Advanced Pancreatic Cancer. *Int J Gastrointest Cancer* 2001;**29**:9–18.

187. Cantero D, Friess H, Deflorin J, Zimmermann A, Brundler MA, Riesle E, Korc M, Buchler MW. Enhanced expression of urokinase plasminogen activator and its receptor in pancreatic carcinoma. *Br J Cancer* 1997;**75**:388–95.

188. Friess H, Yamanaka Y, Kobrin MS, Do DA, Buchler MW, Korc M. Enhanced erbB-3 expression in human pancreatic cancer correlates with tumor progression. *Clin Cancer Res* 1995;**1**:1413–20.

189. Yamanaka Y, Friess H, Buchler M, Beger HG, Uchida E, Onda M, Kobrin MS, Korc M. Overexpression of acidic and basic fibroblast growth factors in human pancreatic cancer correlates with advanced tumor stage. *Cancer Res* 1993;**53**:5289–96.

190. Itakura J, Ishiwata T, Friess H, Fujii H, Matsumoto Y, Buchler MW, Korc M. Enhanced expression of vascular endothelial growth factor in human pancreatic cancer correlates with local disease progression. *Clin Cancer Res* 1997;**3**:1309–16.

191. van Heek T, Rader AE, Offerhaus GJ, McCarthy DM, Goggins M, Hruban RH, Wilentz RE. K-ras, p53, and DPC4 (MAD4) alterations

in fine-needle aspirates of the pancreas: a molecular panel correlates with and supplements cytologic diagnosis. *Am J Clin Pathol* 2002;**117**:755–65.

192. Yamaguchi K, Chijiiwa K, Noshiro H, Torata N, Kinoshita M, Tanaka M. Ki-ras codon 12 point mutation and p53 mutation in pancreatic diseases. *Hepatogastroenterology* 1999;**46**:2575–81.

193. Shibata D, Almoguera C, Forrester K, Dunitz J, Martin SE, Cosgrove MM, Perucho M, Arnheim N. Detection of c-K-ras mutations in fine needle aspirates from human pancreatic adenocarcinomas. *Cancer Res* 1990;**50**:1279–83.

194. Bramhall SR, Rosemurgy A, Brown PD, Bowry C, Buckels JA. Marimastat as first-line therapy for patients with unresectable pancreatic cancer: a randomized trial. *J Clin Oncol* 2001;**19**:3447–55.

195. Bramhall SR, Schulz J, Nemunaitis J, Brown PD, Baillet M, Buckels JA. A double-blind placebo-controlled, randomised study comparing gemcitabine and marimastat with gemcitabine and placebo as first line therapy in patients with advanced pancreatic cancer. *Br J Cancer* 2002;**87**:161–7.

196. Cohen LH, Pieterman E, van Leeuwen RE, Overhand M, Burm BE, van der Marel GA, van Boom JH. Inhibitors of prenylation of Ras and other G-proteins and their application as therapeutics. *Biochem Pharmacol* 2000;**60**:1061–8.

197. Song SY, Meszoely IM, Coffey RJ, Pietenpol JA, Leach SD. K-Ras-independent effects of the farnesyl transferase inhibitor L-744,832 on cyclin B1/Cdc2 kinase activity, G2/M cell cycle progression and apoptosis in human pancreatic ductal adenocarcinoma cells. *Neoplasia* 2000;**2**:261–72.

198. Venkatasubbarao K, Choudary A, Freeman JW. Farnesyl transferase inhibitor (R115777)-induced inhibition of STAT3(Tyr705) phosphorylation in human pancreatic cancer cell lines require extracellular signal-regulated kinases. *Cancer Res* 2005;**65**:2861–71.

199. Macdonald 3rd JS, McCoy S, Whitehead RP, Iqbal S, Wade JL, Giguere JK, Abbruzzese JL. A phase II study of farnesyl transferase inhibitor R115777 in pancreatic cancer: a Southwest oncology group (SWOG 9924) study. *Invest New Drugs* 2005;**23**:485–7.

200. Van Cutsem E, van de Velde H, Karasek P, Oettle H, Vervenne WL, Szawlowski A, Schoffski P, Post S, Verslype C, Neumann H, Safran H, Humblet Y, Perez Ruixo J, Ma Y, Von Hoff D. Phase III trial of gemcitabine plus tipifarnib compared with gemcitabine plus placebo in advanced pancreatic cancer. *J Clin Oncol* 2004;**22**:1430–8.

201. Bruns CJ, Harbison MT, Davis DW, Portera CA, Tsan R, McConkey DJ, Evans DB, Abbruzzese JL, Hicklin DJ, Radinsky R. Epidermal growth factor receptor blockade with C225 plus gemcitabine results in regression of human pancreatic carcinoma growing orthotopically in nude mice by antiangiogenic mechanisms. *Clin Cancer Res* 2000;**6**:1936–48.

202. Sclabas GM, Fujioka S, Schmidt C, Fan Z, Evans DB, Chiao PJ. Restoring apoptosis in pancreatic cancer cells by targeting the nuclear factor-kappaB signaling pathway with the anti-epidermal growth factor antibody IMC-C225. *J Gastrointest Surg* 2003;**7**:37–43. discussion 43.

203. Bruns CJ, Solorzano CC, Harbison MT, Ozawa S, Tsan R, Fan D, Abbruzzese J, Traxler P, Buchdunger E, Radinsky R, Fidler IJ. Blockade of the epidermal growth factor receptor signaling by a novel tyrosine kinase inhibitor leads to apoptosis of endothelial cells and therapy of human pancreatic carcinoma. *Cancer Res* 2000;**60**:2926–35.

204. Kim ES, Khuri FR, Herbst RS. Epidermal growth factor receptor biology (IMC-C225). *Curr Opin Oncol* 2001;**13**:506–13.

205. Cunningham D, Humblet Y, Siena S, Khayat D, Bleiberg H, Santoro A, Bets D, Mueser M, Harstrick A, Verslype C, Chau I, Van Cutsem E. Cetuximab monotherapy and cetuximab plus irinotecan in irinotecan-refractory metastatic colorectal cancer. *N Engl J Med* 2004;**351**:337–45.

206. Bonner J, Harari P, Giralt J (2005). Cetuximab improves locoregional control and survival of locoregionally advanced head and neck cancer: independent review of mature data with a median follow-up of 45 months. In: *Annual AACR-NCI-EORTC international conference on molecular targets and cancer therapeutics* Philadelphia, PA:

207. Xiong HQ, Rosenberg A, LoBuglio A, Schmidt W, Wolff RA, Deutsch J, Needle M, Abbruzzese JL. Cetuximab, a monoclonal antibody targeting the epidermal growth factor receptor, in combination with gemcitabine for advanced pancreatic cancer: a multicenter phase II Trial. *J Clin Oncol* 2004;**22**:2610–16.

208. Cascinu S, Berardi R, Labianca R, Siena S, Falcone A, Aitini E, Barni S, Di Costanzo F, Dapretto E, Tonini G, Pierantoni C, Artale S, Rota S, Floriani I, Scartozzi M, Zaniboni A. Cetuximab plus gemcitabine and cisplatin compared with gemcitabine and cisplatin alone in patients with advanced pancreatic cancer: a randomised, multicentre, phase II trial. *Lancet Oncology* 2008;**9**:39–44.

209. Moore MJ. Brief communication: a new combination in the treatment of advanced pancreatic cancer. *Semin Oncol* 2005;**32**:5–6.

210. Hurwitz H, Fehrenbacher L, Novotny W, Cartwright T, Hainsworth J, Heim W, Berlin J, Baron A, Griffing S, Holmgren E, Ferrara N, Fyfe G, Rogers B, Ross R, Kabbinavar F. Bevacizumab plus irinotecan, fluorouracil, and leucovorin for metastatic colorectal cancer. *N Engl J Med* 2004;**350**:2335–42.

211. Kindler HL, Friberg G, Singh DA, Locker G, Nattam S, Kozloff M, Taber DA, Karrison T, Dachman A, Stadler WM, Vokes EE. Phase II Trial of Bevacizumab Plus Gemcitabine in Patients With Advanced Pancreatic Cancer. *J Clin Oncol* 2005;**23**:8033–40.

212. Kindler HL, Niedzwiecki D, Hollis D, Oraefo E, Schrag D, Hurwitz H, McLeod HL, Mulcahy MF, Schilsky RL, Goldberg RM. A double-blind, placebo-controlled, randomized phase III trial of gemcitabine (G) plus bevacizumab (B) versus gemcitabine plus placebo (P) in patients (pts) with advanced pancreatic cancer (PC): A preliminary analysis of Cancer and Leukemia Group B (CALGB) 80303. [Abstract]. In: *2007 Gastrointestinal cancers symposium*. Orlando, FL: American Society of Clinical Oncology; 2007. 212.

213. Okami J, Yamamoto H, Fujiwara Y, Tsujie M, Kondo M, Noura S, Oshima S, Nagano H, Dono K, Umeshita K, Ishikawa O, Sakon M, Matsuura N, Nakamori S, Monden M. Overexpression of cyclooxygenase-2 in carcinoma of the pancreas. *Clin Cancer Res* 1999;**5**:2018–24.

214. Molina MA, Sitja-Arnau M, Lemoine MG, Frazier ML, Sinicrope FA. Increased cyclooxygenase-2 expression in human pancreatic carcinomas and cell lines: growth inhibition by nonsteroidal anti-inflammatory drugs. *Cancer Res* 1999;**59**:4356–62.

215. Eibl G, Takata Y, Boros LG, Liu J, Okada Y, Reber HA, Hines OJ. Growth stimulation of COX-2-negative pancreatic cancer by a selective COX-2 inhibitor. *Cancer Res* 2005;**65**:982–90.

216. Wei D, Wang L, He Y, Xiong HQ, Abbruzzese JL, Xie K. Celecoxib inhibits vascular endothelial growth factor expression in and reduces angiogenesis and metastasis of human pancreatic cancer via suppression of Sp1 transcription factor activity. *Cancer Res* 2004;**64**:2030–8.

217. El-Rayes BF, Ali S, Sarkar FH, Philip PA. Cyclooxygenase-2-dependent and -independent effects of celecoxib in pancreatic cancer cell lines. *Mol Cancer Ther* 2004;**3**:1421–6.

218. Milella M, Gelibter A, Di Cosimo S, Bria E, Ruggeri EM, Carlini P, Malaguti P, Pellicciotta M, Terzoli E, Cognetti F. Pilot study of celecoxib and infusional 5-fluorouracil as second-line treatment for advanced pancreatic carcinoma. *Cancer* 2004;**101**:133–8.

219. El-Rayes BF, Zalupski MM, Shields AF, Ferris AM, Vaishampayan U, Heilbrun LK, Venkatramanamoorthy R, Adsay V, Philip PA. A phase II study of celecoxib, gemcitabine, and cisplatin in advanced pancreatic cancer. *Invest New Drugs* 2005;**23**:583–90.

220. Fukasawa M, Korc M. Vascular endothelial growth factor-trap suppresses tumorigenicity of multiple pancreatic cancer cell lines. *Clin Cancer Res* 2004;**10**:3327–32.

221. Rowland-Goldsmith MA, Maruyama H, Kusama T, Ralli S, Korc M. Soluble type II transforming growth factor-beta (TGF-beta) receptor inhibits TGF-beta signaling in COLO-357 pancreatic cancer cells in vitro and attenuates tumor formation. *Clin Cancer Res* 2001;**7**:2931–40.

222. Rowland-Goldsmith MA, Maruyama H, Matsuda K, Idezawa T, Ralli M, Ralli S, Korc M. Soluble type II transforming growth factor-beta receptor attenuates expression of metastasis-associated genes and suppresses pancreatic cancer cell metastasis. *Mol Cancer Ther* 2002;**1**:161–7.

223. Dornhofer N, Spong S, Bennewith K, Salim A, Klaus S, Kambham N, Wong C, Kaper F, Sutphin P, Nacamuli R, Hockel M, Le Q, Longaker M, Yang G, Koong A, Giaccia A. Connective tissue growth factor-specific monoclonal antibody therapy inhibits pancreatic tumor growth and metastasis. *Cancer Res* 2006;**66**:5816–27.

224. Aikawa T, Gunn J, Spong SM, Klaus SJ, Korc M. Connective tissue growth factor-specific antibody attenuates tumor growth, metastasis, and angiogenesis in an orthotopic mouse model of pancreatic cancer. *Mol Cancer Ther* 2006;**5**:1108–16.

225. Jones S, Zhang X, Parsons DW, Lin JC-H, Leary RJ, Angenendt P, Mankoo P, Carter H, Kamiyama H, Jimeno A, Hong S-M, Fu B, Lin M-T, Calhoun ES, Kamiyama M, Walter K, Nikolskaya T, Nikolsky Y, Hartigan J, Smith DR, Hidalgo M, Leach SD, Klein AP, Jaffee EM, Goggins M, Maitra A, Iacobuzio-Donahue C, Eshleman JR, Kern SE, Hruban RH, Karchin R, Papadopoulos N, Parmigiani G, Vogelstein B, Velculescu VE, Kinzler KW. Core Signaling Pathways in Human Pancreatic Cancers Revealed by Global Genomic Analyses. *Science* 2008;**321**:1801–6.

226. Korc M. Targeted therapeutics in pancreatic cancer: a ray of hope. *Clin Cancer Res* 2005;**11**:410–11.

Chapter 42

Regulatory Signaling in Pancreatic Organogenesis: Implications for Aberrant Signaling in Pancreatic Cancer

Catherine Carrière and Murray Korc

Departments of Medicine, and Pharmacology and Toxicology, Norris Cotton Cancer Center,
Dartmouth-Hitchcock Medical Center, Lebanon, New Hampshire, and Dartmouth Medical School, Hanover, New Hampshire

INTRODUCTION

The pancreas is a complex organ that is essential for life, exerting a critical role in digestion and glucose homeostasis. It is composed of an exocrine component consisting of acini and ductal network, and an endocrine component consisting of hormone-secreting islets. The acinar cells constitute approximately 95 percent of the total gland. They synthesize and secrete digestive enzymes. The duct cells secrete bicarbonate-rich fluid that transports the acinar enzymes to the duodenum. The endocrine islets, also known as the islets of Langherans, make up only 1–2 percent of the gland and are dispersed throughout the exocrine pancreas. Four major peptide hormone-producing cell types constitute the islets: α cells, which secrete glucagons; β cells, which secrete insulin; δ cells, which make somatostatin; and PP cells, which make pancreatic polypeptide.

Both the exocrine and endocrine components of the pancreas have a common origin, the gut endoderm [1]. In the mouse, by embryonic day 8–8.5 (e8–8.5), an area of the endoderm is fated to become pancreas. Pancreatic dorsal and ventral buds will develop between e9.5 and e10.5. In an initial step called the first transition, the pancreatic epithelium goes through an extensive proliferation phase and few cells start differentiating and expressing endocrine markers. The buds expand and grow into a highly branched structure. A second wave of differentiation, called the second transition, occurs around e13.5: more endocrine cells differentiate and islets form; exocrine cells appear and start organizing into acini with the duct cells lining the central epithelium. It is interesting to note that the first population of endocrine cells appears to come from a different progenitor pool than the second population. Between e16 and e17,

the dorsal and ventral buds fuse following the rotation of the stomach and the duodenum (for more details, see [2]).

Pancreas development is highly conserved between human and rodent. It has been extensively studied in the mouse, and numerous transcription factors have been characterized as being important for the embryonic development as well as for regulation of differentiated functions in the adult. Proper expression of these transcription factors depends on the balanced activation of conserved developmental signaling pathways. If insufficient signals are received, organ growth will be deficient. By contrast, excessive unregulated signaling will lead to hyperproliferation, tumor, and metastasis.

In this chapter, we will focus on five major embryonic signaling networks and discuss the role of their aberrant reactivation in the pathobiology of pancreatic cancer.

NOTCH SIGNALING PATHWAY

One of the earliest signaling pathways known to be important in pancreas development is the Notch pathway. Once the transmembrane Notch receptor is activated by binding of its ligands Delta and Serrate (in mammals, Delta-like, Dll, Jagged1 and Jagged2), its intracellular domain translocates in the nucleus where it associates with Recombination Signal Binding Proteins (RBPs), also known as CSL/CBF1/Su(H)/LAG-1 transcription factors. This leads to the activation of transcription on the Hairy Enhancer of Split1 (HES1) promoter and subsequent expression of the HES-1 and HES-related family of transcriptional repressors [3, 4]. In the developing central nervous system where it was originally characterized, Notch signaling prevents the expression of pro-neural bHLH genes, by a mechanism known as lateral inhibition, thereby impeding neuronal differentiation [5].

Several lines of evidence suggest that the Notch pathway has similar role during pancreas development, by preventing endocrine differentiation. Thus, blocking the Notch pathway in early stages of embryonic development by overexpressing a dominant-negative Notch receptor [6] or by inactivating other components of the pathway, as in mice deficient for Delta-like, HES-1, or RBP-j [6, 7], causes a precocious differentiation and excessive formation of endocrine cells in the pancreas. These data suggested that Notch signaling might be critical for the early cell-fate decision to differentiate into either endocrine cells or precursor exocrine cells.

RBP-j is the main partner of Notch during embryogenesis. Its targeted inactivation specifically in the pancreas (to circumvent early embryonic lethality) reduces differentiation into acinar and endocrine cells, and induces the formation of abnormal tubular structures [8]. Conversely, pancreatic overexpression of a constitutively activated form of Notch leads to a reduced number of endocrine cells and, in later stages, branching morphogenesis abnormalities and development of duct-like structures in the absence of acinar differentiation [9, 10]. Taken together, these observations suggest that Notch signaling prevents the premature differentiation of pancreatic progenitor cells into endocrine and ductal cells during early development of the pancreas, and exocrine differentiation in later stages. Thus, one possible role for Notch1 in pancreas development would be to maintain the progenitor cell pool.

RBP-j is also part of the trimeric transcription factor Ptf1, which is essential for early pancreas development as well as differentiation and maintenance of exocrine cells [11]. Ptf1 is composed of p48, a pancreas- and cerebellum-specific factor, a class A bHLH protein, and RBP-j, these latter two being ubiquitously expressed. Mutations in p48 that eliminate its interaction with RBPs are associated with a human genetic disorder characterized by pancreatic and cerebellar agenesis [12]. Part of the phenotypes observed in the acinar population when activated Notch is overexpressed could be due to a competition between Notch itself and p48 for the binding of RBP-j. As Ptf1 is responsible for the activation of numerous acinar-specific genes, disruption of the complex would result in loss of differentiation of the acinar cells [13]. This would suggest that the original pool of pancreatic progenitor cell is converted into a pool of exocrine progenitor cells during the second transition of the pancreas, and Ptf1 would be the main factor for the induction of pancreatic differentiation.

Expression profile analysis, comparing human pancreatic ductal adenocarcinoma (PDAC) with the normal pancreas, reveals an upregulation of multiple Notch pathway components [14]. Epidermal Growth Factor (EGF) signaling, which is highly upregulated in PDAC, has been shown to activate Notch signaling, and the latter is essential for the expansion of a metaplastic ductal epithelium [14].

Moreover, in the most recent mouse models of PDAC, it has been shown that Notch activation represents an early event in tumorigenesis. Thus, Notch activation was detected in the earliest pre-neoplastic lesions. These pancreatic intraepithelial neoplasia (PanIN) lesions are classified into stages I to III, and are believed to represent the precursor lesions for invasive pancreatic cancer [15]. It is not clear, however, whether the original genetic lesion targets a putative progenitor cell that has an active Notch, or whether the Notch pathway is reactivated in PDAC. Activation of the Ras pathway, which is almost invariably found in PDAC [16], would also mediate increases in Notch signaling (14). These observations raise the possibility that Notch activity is an essential component in PDAC progression.

HEDGEHOG SIGNALING PATHWAY

Hedgehog (Hh) signaling appears to play multiple roles in mouse embryonic pancreas development, as well as in the maintenance of endocrine cells in the adult. Two transmembrane receptors Patched 1 and 2 (Ptch 1 and 2) act as receptors of the processed Hh ligands [17]. In the absence of ligand, Ptch receptors repress the activity of the G-protein-coupled-like receptor Smoothened (Smo). Smo repression blocks all downstream signaling events. After binding to Hh ligands, Ptch repression of Smo is alleviated and Smo can initiate a signaling cascade that will result in the activation and/or repression of transcription by the Gli family of transcription factors [18].

Sonic hedgehog (Shh) and its receptors are highly expressed in the embryonic foregut before any budding of the pancreas. By embryonic day 8, Shh expression is downregulated in the region fated to give rise to the dorsal bud of the pancreas. Pdx1/IPF1, a transcription factor that is essential for pancreas development and one of its earliest markers, has initially a broader expression domain, as it is expressed in the future stomach, pancreas, and duodenum [19, 20]. Shh downregulation would be the key event in restricting the zone of Pdx1 expression that will become the pancreas. Gain-of-function studies in which ectopic Shh was expressed under the control of the Pdx1 promoter, during early stages of development, resulted in severely disrupted pancreatic morphogenesis and an abnormal, duodenal-like pancreatic epithelium [21, 22]. Conversely, the pancreas of Shh null mouse embryos is the same size as a wild-type pancreas in spite of the severely decreased size of the embryos [23]. This suggests that pancreas development does not depend on Shh signaling, but on its absence. By contrast, in the mature mouse pancreas, Hh signaling components are present in the ducts and islets. While their function in these cell types *in vivo* is not known, cell culture studies showed that ectopic expression of Shh can promote insulin production, and this could be in part through upregulation of Pdx1 expression [23–25].

As described in rodents, expression of components of the Hh signaling pathway (such as Ptch and Smo) is restricted to the endocrine cells in normal adult human pancreatic tissues [26]. Interestingly, there is a clear upregulation of these molecules in the fibrotic pancreas of chronic pancreatitis patients [26, 27]. Parallel *in vitro* experiments on non-transformed pancreatic duct cells show that treatment with Hh agonists promotes cell proliferation [26]. It is possible, therefore, that Hh signaling upregulation is required for the proliferation of certain cell types in chronic pancreatitis.

The Hh signaling pathway is also dramatically upregulated during the progression from PanINs lesions to PDAC, and PTCH expression is also present in the stromal elements adjacent to the cancer [28]. The possibility that Shh may have a role in the development of PDAC was initially suggested by results obtained through Shh overexpression in the pancreas using the Pdx1 promoter. These transgenic mice exhibit abnormal pancreatic development with mixed populations of duodenal and pancreatic cells and significant intestinal phenotypes [21]. At 3 weeks of age, their pancreata display lesions strongly reminiscent of human PanINs. Interestingly, they also show an upregulation of the EGF receptor and HER2/neu, and harbor a mutated activated form of KRas (G12D), a mutation that seems to occur spontaneously in the lesions. As in human PDAC, Hh signaling components were present in the stroma surrounding the lesions, suggesting that a dysregulated Hh pathway can act in a paracrine as well as an autocrine manner. Human pancreatic cancer cell lines also express high levels of Hh signaling components, and blocking the Hh pathway with a specific inhibitor, cyclopamine, results in attenuated proliferation, increased apoptosis, and reduced tumor growth in *ex vivo* experiments [28]. Thus, activated Hh signaling pathways appear to have a critical role in PDAC and in certain genetically engineered mouse models of PDAC, and Hh pathway activation could potentially be one of the initiating events in the development of this malignancy [29].

TRANSFORMING GROWTH FACTOR-BETA SIGNALING PATHWAY

Transforming growth factor-beta (TGFβ) isoforms are major regulators of pancreatic endocrine and exocrine cell fates, and all three isoforms and their signaling components are expressed in the pancreatic epithelium and surrounding mesenchyme from embryonic to adult stages. The TGFβ superfamily of secreted growth factors consists of several sub-families that include the three mammalian TGFβ isoforms, the activins/inhibins and the Bone Morphogenetic Proteins (BMPs). These signaling pathways include the TGFβ heterotetrameric receptor family that can bind various combinations of ligands with different affinities; antagonist ligands such as follistatin, noggin and gremlin; and

intracellular complexes of Smad proteins that regulate transcription. Smads are divided into three categories: receptor-activated Smads or R-Smads (Smad1, 2, 3, 5, 8), common mediator Smad (Smad4), and inhibitory Smads (Smad6, 7). Ligand binding to the TGFβ receptors is initiated by direct interaction of the ligand with the type II TGFβ receptor (TβRII) homodimer, which then associates with the type I receptor (TβRI) homodimer [30]. TβRI is phosphorylated by the TβRII kinase, leading to the activation of its serine/threonine-kinase activity, thereby allowing it to phosphorylate Smad2 and Smad3. Phosphorylated R-Smads can then form a complex with Smad4 and translocate to the nucleus, where they can interact with multiple classes of co-activator and co-repressor complexes and modulate the expression of TGFβ target genes. Repressor Smads compete with R-Smads for binding to activated TβRI, thereby inhibiting the phosphorylation of R-Smads [31, 32]. TGFβ can also activate alternative signaling pathways, such as mitogen-activated protein kinase (MAPK) or phosphatidylinositol 3-kinase (PI3K) pathways [33].

As indicated earlier, by embryonic day 8–8.5 (e8–8.5) an area of the endoderm is fated to become the mouse pancreas, and pancreatic dorsal and ventral buds will develop between e9.5 and e10.5, ultimately giving rise to endocrine islets, as well as acinar and ductal cells. At e8.5, when pancreas formation is initiated, the notochord secretes numerous growth factors, including activin-B and FGF-2. Tissue culture experiments have shown that both growth factors can mimic the notochord effect, which is to induce transcription of early pancreatic marker genes, such as Pdx1, by repressing Shh in the dorsal pre-pancreatic epithelium. By contrast, in the adjacent future hepatic domain, a combination of FGF-2 and BMP4 secreted by the cardiac mesenchyme, induce Shh and repress Pdx1 [23]. Not surprisingly, mice deficient for the B form of the type II activin receptor (ActRIIB) have severe pancreatic hypoplasia [34, 35].

Altering TGFβ expression in the pancreas during embryogenesis has been shown to affect pancreatic development at multiple levels, from tissue determination, to proliferation, to cell type differentiation via autocrine and paracrine mechanisms. Some components of the pathway, such as activin-B and BMP2, are required for fate determination; others such as follistatin promote the growth of exocrine tissue, while TGFβ1 inhibits acinar growth and favors the development of β cells [36]. This last effect is in fact more complex, as TGFβ1 was shown to be present at low level in acinar tissue. Modest levels of TGFβ1 would then be permissive for acinar development, while increased levels of TGFβ1 later in time would allow for the formation of the islets [37]. Null mutations in the activin signaling pathway are associated with gross pancreatic defects [34, 38]. Mice overexpressing a dominant-negative receptor TβRII, thereby actively preventing activation of the TGFβsignaling pathway, display increased proliferation and severe abnormalities in differentiation of the acinar cells [39], while the specific deletion of this receptor during pancreas development does not affect its formation and function [40]. Another expected major component of the TGFβ signaling pathway, Smad4, was also specifically deleted in the pancreas during embryogenesis and again no developmental abnormalities were observed [41], while the overexpression of a dominant-negative Smad4 in acinar cells seem to result in an increase of islet size [42]. Finally, overexpression of inhibitory Smad, Smad7, during pancreas development results in severe pancreatic malformations and β cell defects [43]. These various phenotypes show the complexity of this particular pathway, and suggest roles for Smad4- and TβRII-independent TGFβ signaling pathways in these developmental processes. It also underscores the potentially non-physiological gain of function effects of ectopically over expressed proteins.

The complex expression patterns of the numerous members of the TGFβ signaling pathway reflect the complexity of their functions [44, 45]. Moreover, in a variety of biological systems, TGFβs present in the circulation and platelets, and/or produced within tissues, regulate cell growth by enhancing the proliferation of cells that are of mesenchymal origin while inhibiting the proliferation of epithelial cell types. This inhibition would be due to the downregulation of cell cycle activators such as Cdk4 and cyclin D1, and upregulation of cell cycle inhibitors such as p21^{Cip1}, p27^{Kip1}, and p15^{Ink4b} [46, 47].

Although epithelial in origin, human pancreatic cancer cells are generally resistant to TGFβ-mediated growth inhibition. Several perturbations in TGFβ signaling pathways have been described in PDAC, including inactivating Smad4 mutations (in 50 percent of PDAC), upregulation of TGFβ1, -2, and -3, BMP2 and its receptors, as well as inhibitory Smads, Smad6 and 7 [48–50].

In vitro experiments using PDAC-derived cell lines have allowed for the elucidation of the mechanisms that contribute to dysregulated TGFβ signaling. One common mechanism is the inactivation of the tumor suppressor gene Smad4, a frequent event in PDAC [50]. Some human pancreatic cancer lines carrying mutations in Smad4 can be growth-stimulated by BMP2, and the reintroduction of wild-type Smad4 in the cells is sufficient to restore growth inhibition by BMP2 [51]. Other studies have shown that restoring Smad4 in one pancreatic cancer cell line led to decreased vascular endothelial growth factor (VEGF) expression, which in turn resulted in smaller tumors with reduced vascular density when compared with sham transfected cells [52]. Conversely, restoration of Smad4 in another pancreatic cancer cell line resulted in a transient attenuation in its capacity to proliferate *in vivo*, and was not associated with attenuated angiogenesis, indicating that Smad4 growth inhibitory actions can be circumvented in later stages of pancreatic tumorigenicity [53]. Thus, Smad4 mutations in PDAC result in complex alterations that are context dependent, and that confer a growth advantage

to pancreatic cancer cells by rendering them resistant to TGFβ-mediated growth inhibition and by allowing, in some circumstances, for the acquisition of a pro-angiogenic profile.

The repressor Smad7 is commonly upregulated in human pancreatic cancer cells. In the pancreatic cell lines that retain TGFβ-mediated growth inhibitory response, the overexpression of Smad7 is sufficient to overcome this growth inhibition. Interestingly, the cells maintain the ability to translocate Smad2/3 in the nucleus and to modulate the expression of several TGFβ target genes [51]. These data show that Smad7 may also act independently from its known repressor function. Indeed, Smad7 overexpression in these cells interferes with TGFβ-mediated attenuation of cyclin A and B levels, inhibition of cdc2 dephosphorylation, and functional inactivation the retinoblastoma protein [54].

A recent report also demonstrated that while epithelial cells with a sustained TGFβ/Smad response are sensitive to TGFβ-induced growth arrest, some pancreatic cell lines that do not harbor mutations in Smad4 have a comparatively more transient nuclear accumulation of active Smad complexes. This would allow pancreatic cancer cells to evade TGFβ-induced growth arrest, while allowing them to maintain other TGFβ responses that can lead to enhanced transformation and malignancy [55].

Finally, TGFβ1 can cooperate with Ras to promote epithelial – mesenchymal transformation (EMT). In this process, epithelial cells acquire mesenchymal markers and lose cell–cell junctional integrity; β-catenin and E-cadherin become mislocalized principally to the nucleus and cytoplasm, respectively, in conjunction with the upregulation of transcription factors such as Twist and Snail, and increased expression of the mesenchymal markers vimentin and N-cadherin [33, 56–58]. The importance of this process rests in the fact that EMT is associated with enhanced cellular invasiveness. In addition, oncogenic Ras cooperates with activated Smad2 to promote EMT and metastasis [59]. Oncogenic KRas can also modulate TGFβ actions by repressing normal TGFβ signaling [60], inducing Smad4 degradation [61], and redirecting TGFβ actions towards tumor promotion [62–64]. Moreover, in mouse keratinocytes, Smad7 (but not Smad6) cooperates with oncogenic Ras to induce carcinoma formation [65], while raf activation upregulates TGFβ expression, which then acts to enhance cancer cell invasion [66]. Together, these alterations allow TGFβ to exert intercellular effects that promote cancer spread and metastasis.

In support of a deleterious intersection between mutated KRas and TGFβ signaling pathways with respect to malignant transformation in the pancreas, studies of various mouse models of PDAC have shown that specific deletions of Smad4 or TβRII, when superimposed on oncogenic KRas activation, cause more aggressive disease when compared to KRas oncogenic activation alone [40, 41, 67]. These results argue for a role of TGFβ signaling pathways in controlling tumor initiation, and subsequently cancer progression as suggested by the frequent mutations of this pathway in PDAC. Despite these observations, the TGFβ signaling pathway has also been implicated in advanced stages of tumorigenesis, underscoring the dual role of the TGF signaling pathway in cancer initiation and progression.

WNT SIGNALING PATHWAY

Wnt proteins play crucial roles in embryogenesis, and this pathway was recently shown to be involved in pancreas development [28]. Like other major signaling pathways, Wnt signaling is involved in proliferation, stem cell maintenance, tissue commitment and differentiation. In most of the cases, it acts via the so-called "canonical" pathway: Wnts are glycoproteins that are normally secreted by specific subpopulations of cells within tissues, which then bind to the Frizzled (Frz) receptors or lipoprotein receptor-related proteins 5 and 6 (LRP5 and 6) on neighboring cells, thereby acting as morphogens or modulators of cell function. Wnt activation of its receptor results in the dissociation of an intracellular complex containing APC (adenomatous polyposis coli), axin and the serine-threonine kinase Gsk3β; this dissociation prevents the phosphorylation and subsequent degradation of cytosolic β-catenin. β-catenin can then translocate in the nucleus where, in collaboration with the Lef/TCF family of transcription factors (and potentially other families), it regulates transcription [68]. Like the TGFβ signaling pathway, the Wnt pathway has numerous components that have highly complex expression patterns. There are at least eight unique Frz genes in the mouse, but little is known about the specificity of Wnt–Frz interactions [69]. Additionally, in the mouse, four secreted Frz-related proteins (sFRPs) have been cloned which have been shown to bind Wnts and antagonize Wnt-mediated signaling [70]. Because multiple Wnts and Frizzleds appear to be expressed in similar patterns, it has been suggested that secreted FRPs may locally regulate which Wnt ligands may interact with a specific receptor to create local morphogen gradients. Finally, an increasing number of Wnt signaling modifiers have been characterized in recent years [68], resulting in multicellular interactive signaling cascades that regulate cell–cell interactions.

Several Wnt pathway components are expressed during pancreas development, and can be detected in the pancreatic endoderm as well as in the surrounding mesenchyme [71–73]. Mice overexpressing Wnt1 under control of the Pdx1 promoter display pancreatic hypoplasia, which suggests that Wnt signaling is important for early proliferation stages and that the maintenance of this signaling prevents further specification/differentiation of the pancreas [72]. Specific disruptions of the Wnt pathway in the pancreas were reported recently. Thus, pancreas-specific deletion

of β-catenin and overexpression of dominant-negative Frz receptor in the pancreas (under the control of the Pdx1 promoter) result in hypoplasia of the exocrine pancreas at birth [73–75]. The endocrine population appears to be relatively intact in these animals, with the endocrine cells exhibiting timely differentiation and the adult mice showing normal glucose homeostasis. These observations suggest that the early pancreatic progenitor pool is not affected and that β-catenin is required solely for the differentiation of exocrine progenitor cells into acini [74].

Activating mutations of β-catenin and loss-of-function mutations of APC were described in non-ductal and non-endocrine tumors of the pancreas such as pancreatoblastomas, acinar cells carcinomas, solid cystic papillary tumors, and solid pseudopapillary tumors (reviewed in [76]), but the Wnt pathway was shown to be aberrantly activated in advanced pancreatic adenocarcinomas independently of β-catenin mutation [77]. However, at this stage of our knowledge regarding the role of Wnt signaling in the pancreas, it is difficult to explain its specific roles in pancreatic cancer development.

FIBROBLAST GROWTH FACTORS SIGNALING PATHWAY

The importance of epithelial–mesenchymal interactions in the control of pancreas development has been well known since the 1960s. In the absence of mesenchyme, it was observed that isolated pancreatic epithelium failed to grow and undergo morphogenesis, and only endocrine cells underwent differentiation. More recent studies have shown that the mesenchyme can in fact regulate the expansion of the pancreas epithelium and the ratio between endocrine and exocrine cells [78]. The mesenchyme secretes numerous growth factors, and among these, Fibroblast Growth Factors (FGFs) seem to play a major role in pancreas formation.

Functionally, FGFs constitute a family of 23 known members in mammals. Signaling is mediated by binding to high-affinity receptors that often require the presence of low-affinity heparan sulfate proteoglycans (HSPGs) for their efficient activation. There are four known receptor genes that generate a larger number of receptors by alternative splicing. Upon ligand binding, the FGF receptors (FGF-R) dimerize, and downstream signaling is mediated through activation of the FGFR substrate FRS and, subsequently, the Ras/MAPK cascade [79].

FGF-10 is expressed in the mesenchyme immediately adjacent to the early dorsal and ventral epithelial buds that give rise to the pancreas between e9.5 and e12.5 [80]. FGF-1 and -7 were also detected [81]. In FGF-10 null mouse embryos, while initial pancreatic bud formation occurs, there is no subsequent growth, differentiation, or branching of the epithelium [80]. This is primarily due to a dramatic reduction in the proliferation of the progenitor cells. Few endocrine and exocrine cells can be detected, showing that the differentiation capacity of the progenitors themselves is not affected [80]. The broad mid-gestational overexpression of a kinase-deficient soluble form of FGFR2 shows a similar result, with a drastic reduction in size of the pancreas as well as the kidney, lung, and other glands, suggesting that FGFR2 has a more general role in the maintenance of progenitor cells in multiple organs [82]. By contrast, the specific targeting of a dominant-negative FGFR to the pancreas with the Pdx1 promoter did not generate any abnormalities during pancreatic organogenesis [83]. One explanation for this absence of effect could be the redundancy of the signaling pathways, as other FGF receptors are expressed during pancreas development [44]. Another explanation could be the difference in targeting of the dominant-negative receptors. The interplay between pancreatic bud and surrounding mesoderm is known to be essential for pancreas morphogenesis. In the case of the dominant-negative FGFR2, which was expressed in the whole organism, this interplay was disrupted, while in the case of the dominant-negative FGFR targeted to the pancreas only, FGF signaling coming from the surrounding mesoderm was not affected.

The ectopic expression of FGF-10 under the Pdx1 promoter causes increased proliferation in pancreatic progenitor cells and suppression of all cell types at a stage when differentiation should have started. Later on, as proliferation slows down, few differentiated cells start to appear. Notch1 and Notch2, as well as their ligands Jagged1 and Jagged2, and their target Hes1 can be detected at a very high level in the proliferating epithelium [84]. A recent study performed on cultured explants of pancreas shows that specific inhibition of Notch processing after treatment with FGF-10 is sufficient to abrogate the effect of FGF-10 on proliferation and maintenance of pancreatic progenitors [85], thereby demonstrating a direct connection between FGF and Notch signaling pathways. Altogether, these data suggest that the mesenchyme signals via FGF-10 to the pancreatic epithelium at early stages of development, and that this signal is integrated by the Notch pathway at the level of the pancreatic progenitor pools. Maintenance of the FGF-10 signaling later in development results in maintenance of an activated Notch pathway and, consequently, maintenance of highly proliferative progenitor pools, which would be sufficient to prevent or at least delay differentiation.

Several FGFs and FGFRs are overexpressed in PDAC [86]. The available data suggest that each FGF may have specific roles in PDAC progression. FGF-1 and -2 are overexpressed principally in the cancer cells, not in the surrounding stroma, and increases in their protein levels correlate with advanced tumor stages [87]. High FGF-2 protein levels also correlate with shorter postoperative patient survival [87]. By contrast, FGF-5, which is also overexpressed in PDAC, was shown to be abundant predominantly in the stromal fibroblasts and infiltrating macrophages adjacent to the cancer cells, as well as in the endocrine islet cells, and only to a lesser extent in the cancer cells [14]. FGF 7/KGF localizes essentially around the acinar and ductal cells adjacent to the cancer cells, but in situ hybridization studies show that it is also present at low level in the cancer cells [88].

In vitro experiments show that several FGFs exert mitogenic effects on multiple PDAC-derived cell lines [89–91]. Moreover, in vitro, pancreatic cancer cells secrete abundant amounts of FGF-5, suggesting that it can participate in autocrine and paracrine regulatory pathways [90]. Studies of the effect of a dominant-negative FGFR-1 (devoid of its intrinsic tyrosine kinase activity) in PDAC-derived cell lines shows a clearly diminished proliferative response to FGF-2 as well as decreased downstream signaling [92]. Conversely, overexpression of the isoform IIIc of FGFR-1 in an immortalized ductal cell line dramatically increased the tumorigenicity of these cells [93]. Taken together, these observations indicate that FGFs contribute to aberrant intercellular signaling pathways that promote pancreatic cancer cell growth.

Finally, the HSPG glypican-1 (GPC1) is a co-receptor for heparin binding growth factors (HBGFs), including FGFs. Cancer-cell derived GPC1 plays a crucial role in PDAC, as evidenced by the observations that down-regulation of GPC1 in pancreatic cancer cells results in decreased anchorage-independent growth, and attenuated tumor growth, angiogenesis, and metastasis [94–96]. It was also recently shown that tumors derived from PDAC cells implanted in athymic mice that are GPC1 null also exhibit decreased tumor angiogenesis and metastasis, indicating that host-derived GPC1 is also important for efficient pancreatic cancer cell metastasis [96]. Inasmuch as GPC1 is expressed in pancreatic cancer cells, cancer-associated fibroblasts, and endothelial cells, it is likely that GPC1 and the associated FGF signaling components contribute to cell autonomous effects in pancreatic cancer cells, as well as to aberrant cancer cell–fibroblast interactions, abnormal cancer cell–endothelial cell interactions, and dysregulated FGF signaling cascades within the pancreatic tumor microenvironment.

CONCLUSION

Studies of signaling pathways in pancreatic embryogenesis point towards complex regulatory interactions between different tissues and/or different cell types that are turned on and off in a carefully orchestrated manner. The fine-tuning of proliferation versus specification and differentiation is under the control of multiple, but clearly defined, signaling pathways that allow for exquisitely sensitive control of pancreatic organogenesis. The mechanisms used by pancreatic cancer cells in their progression toward invasiveness

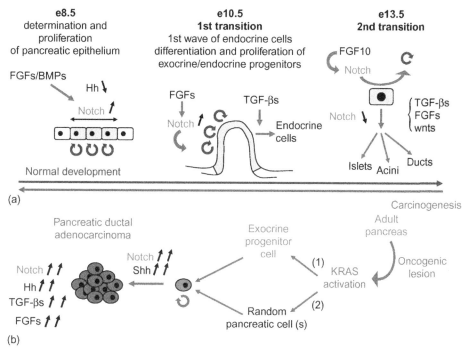

FIGURE 42.1 Regulatory signaling: pancreas embryonic development versus cancer formation.
(a) At mouse embryonic day 8.5 (e8.5), mesenchymal FGFs/BMPs signal to the future pancreatic epithelium: Shh is downregulated and Notch is upregulated. Maintenance of FGF and consequently Notch signaling is required for the proliferation of the pancreatic epithelium. By e10.5, first transition, the pancreatic buds form and under the control of TGFβ signaling, few endocrine cells differentiate. Maintenance of Notch signaling till at least e13.5 is required for the expansion of the endocrine and exocrine progenitor pools. By e13.5, second transition, the endocrine and exocrine cells start differentiating and form ascini, islets, and ducts under the finely tuned control of TGFβ, Wnts, and FGFs signaling pathways. (b) Oncogenic lesions in the adult pancreas, such as Kras activation, can occur (1) in a potential progenitor cell, (2) in any pancreatic cell. From the earliest stages of development of cancer, Notch and Shh signaling are highly activated. At the PDAC stage, there is a dramatic dysregulation of all the main signaling pathways.

are reminiscent of the ones used in normal pancreatic development (see Figure 42.1). However, reactivation of these pathways in the adult pancreas in the context of genetic alterations that include KRas activation, and INK4A/ARF, TRP53, and SMAD/DPC4 tumor suppressor loss of function, subsequently results in the activation of multiple autocrine and paracrine regulatory loops that are not readily controlled. Instead, these aberrant regulatory loops promote cancer cell growth and metastasis by "manipulating" the surrounding stroma, leading to EMT, excessive expression of multiple growth factors, aberrant cell–cell and cell–matrix interactions, and tumor angiogenesis. A better understanding of these pathways during early development may thus provide an improved understanding of the early stages of cancer formation and, potentially, better means of disease detection and prevention. Similarly, an improved understanding of the factors that maintain the equilibrium between proliferation and differentiation will open new therapeutic avenues for suppressing tumor progression and invasion.

ACKNOWLEDGEMENTS

We apologize to our colleagues whose important contributions we could not cite due to limitations in reference number. This study was supported in part by Public Health Service Grant CA10130, awarded by the National Cancer Institute to Murray Korc, and CA127095, awarded by the National Cancer Institute to Catherine Carrière.

REFERENCES

1. Slack JM. Developmental biology of the pancreas. *Development* 1995;**121**(6):1569–80.
2. Edlund H. Pancreatic organogenesis – developmental mechanisms and implications for therapy. *Nat Rev Genet* 2002;**3**(7):524–32.
3. Katoh M. Identification and characterization of human HESL, rat Hesl and rainbow trout hesl genes *in silico*. *Intl J Mol Med* 2004;**14**(4):747–51.
4. Katoh M. Identification and characterization of human HES2, HES3, and HES5 genes in silico. *Intl J Oncol* 2004;**25**(2):529–34.
5. Cornell RA, Eisen JS. Notch in the pathway: the roles of Notch signaling in neural crest development. *Semin Cell Dev Biol* 2005;**16**(6):663–72.
6. Apelqvist A, Li H, Sommer L, Beatus P, Anderson DJ, Honjo T, Hrabe de Angelis M, Lendahl U, Edlund H. Notch signalling controls pancreatic cell differentiation. *Nature* 1999;**400**(6747):877–81.
7. Jensen J, Pedersen EE, Galante P, Hald J, Heller RS, Ishibashi M, Kageyama R, Guillemot F, Serup P, Madsen OD. Control of endodermal endocrine development by Hes-1. *Nat Genet* 2000;**24**(1):36–44.
8. Fujikura J, Hosoda K, Iwakura H, Tomita T, Noguchi M, Masuzaki H, Tanigaki K, Yabe D, Honjo T, Nakao K. Notch/Rbp-j signaling

prevents premature endocrine and ductal cell differentiation in the pancreas. *Cell Metab* 2006;**3**(1):59–65.

9. Hald J, Hjorth JP, German MS, Madsen OD, Serup P, Jensen J. Activated Notch1 prevents differentiation of pancreatic acinar cells and attenuate endocrine development. *Dev Biol* 2003;**260**(2):426–37.

10. Esni F, Ghosh B, Biankin AV, Lin JW, Albert MA, Yu X, MacDonald RJ, Civin CI, Real FX, Pack MA, et al. Notch inhibits Ptf1 function and acinar cell differentiation in developing mouse and zebrafish pancreas. *Development* 2004;**131**(17):4213–24.

11. Krapp A, Knofler M, Ledermann B, Burki K, Berney C, Zoerkler N, Hagenbuchle O, Wellauer PK. The bHLH protein PTF1-p48 is essential for the formation of the exocrine and the correct spatial organization of the endocrine pancreas. *Genes Dev* 1998;**12**(23):3752–63.

12. Sellick GS, Barker KT, Stolte-Dijkstra I, Fleischmann C, Coleman RJ, Garrett C, Gloyn AL, Edghill EL, Hattersley AT, Wellauer PK, et al. Mutations in PTF1A cause pancreatic and cerebellar agenesis. *Nat Genet* 2004;**36**(12):1301–5.

13. Beres TM, Masui T, Swift GH, Shi L, Henke RM, MacDonald RJ. PTF1 is an organ-specific and Notch-independent basic helix–loop–helix complex containing the mammalian suppressor of hairless (RBP-J) or its paralogue, RBP-L. *Mol Cell Biol* 2006;**26**(1):117–30.

14. Miyamoto Y, Maitra A, Ghosh B, Zechner U, Argani P, Iacobuzio-Donahue CA, Sriuranpong V, Iso T, Meszoely IM, Wolfe MS, et al. Notch mediates TGF alpha-induced changes in epithelial differentiation during pancreatic tumorigenesis. *Cancer Cell* 2003;**3**(6):565–76.

15. Leach SD. Epithelial differentiation in pancreatic development and neoplasia: new niches for nestin and Notch. *J Clin Gastroenterol* 2005;**39**(4 Suppl. 2):S78–82.

16. Hruban RH, Wilentz RE, Goggins M, Offerhaus GJ, Yeo CJ, Kern SE. Pathology of incipient pancreatic cancer. *Ann Oncol* 1999;**10**(Suppl. 4):9–11.

17. Carpenter D, Stone DM, Brush J, Ryan A, Armanini M, Frantz G, Rosenthal A, de Sauvage FJ. The Nestin progenitor lineage is the compartment of origin for pancreatic epithelial neoplasia. Characterization of two patched receptors for the vertebrate hedgehog protein family. *Proc Natl Acad Sci USA* 1998;**95**(23):13,630–13,634,.

18. Pasca di Magliano M, Hebrok M. Hedgehog signalling in cancer formation and maintenance. *Nat Rev Cancer* 2003;**3**(12):903–11.

19. Offield MF, Jetton TL, Labosky PA, Ray M, Stein RW, Magnuson MA, Hogan BL, Wright CV. PDX-1 is required for pancreatic outgrowth and differentiation of the rostral duodenum. *Development* 1996;**122**(3):983–95.

20. Jonsson J, Carlsson L, Edlund T, Edlund H. Insulin-promoter-factor 1 is required for pancreas development in mice. *Nature* 1994;**371**(6498):606–9.

21. Apelqvist A, Ahlgren U, Edlund H. Sonic hedgehog directs specialised mesoderm differentiation in the intestine and pancreas. *Curr Biol* 1997;**7**(10):801–4.

22. Kawahira H, Scheel DW, Smith SB, German MS, Hebrok M. Hedgehog signaling regulates expansion of pancreatic epithelial cells. *Dev Biol* 2005;**280**(1):111–21.

23. Hebrok M, Kim SK, St Jacques B, McMahon AP, Melton DA. Regulation of pancreas development by hedgehog signaling. *Development* 2000;**127**(22):4905–13.

24. Thomas MK, Rastalsky N, Lee JH, Habener JF. Hedgehog signaling regulation of insulin production by pancreatic beta-cells. *Diabetes* 2000;**49**(12):2039–47.

25. Kawahira H, Ma NH, Tzanakakis ES, McMahon AP, Chuang PT, Hebrok M. Combined activities of hedgehog signaling inhibitors regulate pancreas development. *Development* 2003;**130**(20):4871–9.

26. Kayed H, Kleeff J, Keleg S, Buchler MW, Friess H. Distribution of Indian hedgehog and its receptors patched and smoothened in human chronic pancreatitis. *J Endocrinol* 2003;**178**(3):467–78.

27. Kayed H, Kleeff J, Esposito I, Giese T, Keleg S, Giese N, Buchler MW, Friess H. Localization of the human hedgehog-interacting protein (Hip) in the normal and diseased pancreas. *Mol Carcinog* **42**(4):183–92.

28. Thayer SP, di Magliano MP, Heiser PW, Nielsen CM, Roberts DJ, Lauwers GY, Qi YP, Gysin S, Fernandez-del Castillo C, Yajnik V, et al. Hedgehog is an early and late mediator of pancreatic cancer tumorigenesis. *Nature* 2003;**425**(6960):851–6.

29. Lau J, Kawahira K, Hebrok M. Hedgehog signaling in pancreas development and disease. *Cell Mol Life Sci* 2006.

30. Zuniga JE, Groppe JC, Cui Y, Hinck CS, Contreras-Shannon V, Pakhomova ON, Yang J, Tang Y, Mendoza V, Lopez-Casillas F, et al. *J Mol Biol* 2005;**354**:1052–68.

31. ten Dijke P, Hill CS. New insights into TGF-beta-Smad signalling. *Trends Biochem Sci* 2004;**29**(5):265–73.

32. Massague J, Seoane J, Wotton D. Smad transcription factors. *Genes Dev* 2005;**19**(23):2783–810.

33. Zavadil J, Bottinger EP. TGF-beta and epithelial-to-mesenchymal transitions. *Oncogene* 2005;**24**(37):5764–74.

34. Kim SK, Hebrok M, Li E, Oh SP, Schrewe H, Harmon EB, Lee JS, Melton DA. Activin receptor patterning of foregut organogenesis. *Genes Dev* 2000;**14**(15):1866–71.

35. Oh SP, Li E. The signaling pathway mediated by the type IIB activin receptor controls axial patterning and lateral asymmetry in the mouse. *Genes Dev* 1997;**11**(14):1812–26.

36. Sanvito F, Nichols A, Herrera PL, Huarte J, Wohlwend A, Vassalli JD, Orci L. TGF-beta 1 overexpression in murine pancreas induces chronic pancreatitis and, together with TNF-alpha, triggers insulin-dependent diabetes. *Biochem Biophys Res Commun* 1995;**217**(3):1279–86.

37. Crisera CA, Maldonado TS, Kadison AS, Li M, Alkasab SL, Longaker MT, Gittes GK, Crisera CA, Maldonado TS, Kadison AS, Li M, Alkasab SL, Longaker MT, Gittes GK. Transforming growth factor-beta 1 in the developing mouse pancreas: a potential regulator of exocrine differentiation. *Differentiation* 2000;**65**(5):255–9.

38. Harmon EB, Apelqvist AA, Smart NG, Gu X, Osborne DH, Kim SK. GDF11 modulates NGN3+ islet progenitor cell number and promotes beta-cell differentiation in pancreas development. *Development* 2004;**131**(24):6163–74.

39. Bottinger EP, Jakubczak JL, Roberts IS, Mumy M, Hemmati P, Bagnall K, Merlino G, Wakefield LM. Expression of a dominant-negative mutant TGF-beta type II receptor in transgenic mice reveals essential roles for TGF-beta in regulation of growth and differentiation in the exocrine pancreas. *EMBO J* 1997;**16**(10):2621–33.

40. Ijichi H, Chytil A, Gorska AE, Aakre ME, Fujitani Y, Fujitani S, Wright CV, Moses HL. Aggressive pancreatic ductal adenocarcinoma in mice caused by pancreas-specific blockade of transforming growth factor-beta signaling in cooperation with active Kras expression. *Genes Dev* 2006;**20**(22):3147–60.

41. Bardeesy N, Cheng KH, Berger JH, Chu GC, Pahler J, Olson P, Hezel AF, Horner J, Lauwers GY, Hanahan D, et al. Smad4 is dispensable for normal pancreas development yet critical in progression and tumor biology of pancreas cancer. *Genes Dev* 2006;**20**(22):3130–46.

42. Simeone DM, Zhang L, Treutelaar MK, Zhang L, Graziano K, Logsdon CD, Burant CF. Islet hypertrophy following pancreatic disruption of Smad4 signaling. *Am J Physiol Endocrinol Metab* 2006;**291**(6):E1305–16.

43. Smart NG, Apelqvist AA, Gu X, Harmon EB, Topper JN, MacDonald RJ, Kim SK. Conditional expression of Smad7 in pancreatic beta cells disrupts TGF-beta signaling and induces reversible diabetes mellitus. *PLoS Biol* 2006;**4**(2):e39.

44. Dichmann DS, Miller CP, Jensen J, Scott Heller R, Serup P. Expression and misexpression of members of the FGF and TGFbeta families of growth factors in the developing mouse pancreas. *Dev Dyn* 2003;**226**(4):663–74.

45. Rane SG, Lee JH, Lin HM. Transforming growth factor-beta pathway: role in pancreas development and pancreatic disease. *Cytokine Growth F R* 2006;**17**(1–2):107–19.

46. Senderowicz AM. Inhibitors of cyclin-dependent kinase modulators for cancer therapy. *Prog Drug Res* 2005;**63**:183–206.

47. Denicourt C, Dowdy SF. Cip/Kip proteins: more than just CDKs inhibitors. *Genes Dev* 2004;**18**(8):851–5.

48. Friess H, Yamanaka Y, Buchler M, Ebert M, Beger HG, Gold LI, Korc M. Enhanced expression of transforming growth factor beta isoforms in pancreatic cancer correlates with decreased survival. *Gastroenterology* 1993;**105**(6):1846–56.

49. Friess H, Yamanaka Y, Buchler M, Berger HG, Kobrin MS, Baldwin RL, Korc M. Enhanced expression of the type II transforming growth factor beta receptor in human pancreatic cancer cells without alteration of type III receptor expression. *Cancer Res* 1993;**53**(12):2704–7.

50. Hahn SA, Schutte M, Hoque AT, Moskaluk CA, da Costa LT, Rozenblum E, Weinstein CL, Fischer A, Yeo CJ, Hruban RH, et al. DPC4, a candidate tumor suppressor gene at human chromosome 18q21.1. *Science* 1996;**271**(5247):350–3.

51. Kleeff J, Ishiwata T, Maruyama H, Friess H, Truong P, Buchler MW, Falb D, Korc M. The TGF-beta signaling inhibitor Smad7 enhances tumorigenicity in pancreatic cancer. *Oncogene* 1999;**18**(39):5363–72.

52. Schwarte-Waldhoff I, Volpert OV, Bouck NP, Sipos B, Hahn SA, Klein-Scory S, Luttges J, Kloppel G, Graeven U, Eilert-Micus C, et al. Smad4/DPC4-mediated tumor suppression through suppression of angiogenesis. *Proc Natl Acad Sci USA* 2000;**97**(17):9624–9.

53. Yasutome M, Gunn J, Korc M. Restoration of Smad4 in BxPC3 pancreatic cancer cells attenuates proliferation without altering angiogenesis. *Clin Exp Metastasis* 2005;**22**(6):461–73.

54. Boyer Arnold N, Korc M. Smad7 abrogates transforming growth factor-beta1-mediated growth inhibition in COLO-357 cells through functional inactivation of the retinoblastoma protein. *J Biol Chem* 2005;**280**(23):21,858–21,866.

55. Nicolas FJ, Hill CS. Attenuation of the TGF-beta-Smad signaling pathway in pancreatic tumor cells confers resistance to TGF-beta-induced growth arrest. *Oncogene* 2003;**22**(24):3698–711.

56. Ellenrieder V, Hendler SF, Boeck W, Seufferlein T, Menke A, Ruhland C, Adler G, Gress TM. Transforming growth factor beta1 treatment leads to an epithelial–mesenchymal transdifferentiation of pancreatic cancer cells requiring extracellular signal-regulated kinase 2 activation. *Cancer Res* 2001;**61**(10):4222–8.

57. Grande M, Franzen A, Karlsson JO, Ericson LE, Heldin NE, Nilsson M. Transforming growth factor-beta and epidermal growth factor synergistically stimulate epithelial to mesenchymal transition (EMT) through a MEK-dependent mechanism in primary cultured pig thyrocytes. *J Cell Sci* 2002;**115**(22):4227–36.

58. Massague J, Gomis RR. The logic of TGFbeta signaling. *FEBS Letts* 2006;**580**(12):2811–20.

59. Oft M, Akhurst RJ, Balmain A. Metastasis is driven by sequential elevation of H-ras and Smad2 levels. *Nat Cell Biol* 2002;**4**(7):487–94.

60. Kretzschmar M, Doody J, Timokhina I, Massague J. A mechanism of repression of TGFbeta/ Smad signaling by oncogenic Ras. *Genes Dev* 1999;**13**(7):804–16.

61. Saha D, Datta PK, Beauchamp RD. Oncogenic ras represses transforming growth factor-beta /Smad signaling by degrading tumor suppressor Smad4. *J Biol Chem* 2001;**276**(31):29,531–29,537.

62. Cui W, Fowlis DJ, Bryson S, Duffie E, Ireland H, Balmain A, Akhurst RJ. TGFbeta1 inhibits the formation of benign skin tumors, but enhances progression to invasive spindle carcinomas in transgenic mice. *Cell* 1996;**86**(4):531–42.

63. Portella G, Cumming SA, Liddell J, Cui W, Ireland H, Akhurst RJ, Balmain A. Transforming growth factor beta is essential for spindle cell conversion of mouse skin carcinoma *in vivo*: implications for tumor invasion. *Cell Growth Differ* 1998;**9**(5):393–404.

64. Korc M. Pancreatic cancer-associated stroma production. *Am J Surg* 2007;**194**(4 Suppl.):S84–6.

65. Liu X, Lee J, Cooley M, Bhogte E, Hartley S, Glick A. Smad7 but not Smad6 cooperates with oncogenic ras to cause malignant conversion in a mouse model for squamous cell carcinoma. *Cancer Res* 2003;**63**(22):7760–8.

66. Lehmann K, Janda E, Pierreux CE, Rytomaa M, Schulze A, McMahon M, Hill CS, Beug H, Downward J. Raf induces TGFbeta production while blocking its apoptotic but not invasive responses: a mechanism leading to increased malignancy in epithelial cells. *Genes Dev* 2000;**14**(20):2610–22.

67. Izeradjene K, Combs C, Best M, Gopinathan A, Wagner A, Grady WM, Deng CX, Hruban RH, Adsay NV, Tuveson DA, et al. Kras(G12D) and Smad4/Dpc4 haploinsufficiency cooperate to induce mucinous cystic neoplasms and invasive adenocarcinoma of the pancreas. *Cancer Cell* 2007;**11**(3):229–43.

68. Logan CY, Nusse R. The Wnt signaling pathway in development and disease. *Annu Rev Cell Dev Biol* 2004;**20**:781–810.

69. Wang HY, Liu T, Malbon CC. Structure–function analysis of Frizzleds. *Cell Signal* 2006.

70. Leimeister C, Bach A, Gessler M. Developmental expression patterns of mouse sFRP genes encoding members of the secreted frizzled related protein family. *Mech Dev* 1998;**75**(1–2):29–42.

71. Heller RS, Klein T, Ling Z, Heimberg H, Katoh M, Madsen OD, Serup P. Expression of Wnt, Frizzled, sFRP, and DKK genes in adult human pancreas. *Gene Expr* 2003;**11**(3–4):141–7.

72. Heller RS, Dichmann DS, Jensen J, Miller C, Wong G, Madsen OD, Serup P. Expression patterns of Wnts, Frizzleds, sFRPs, and misexpression in transgenic mice suggesting a role for Wnts in pancreas and foregut pattern formation. *Dev Dyn* 2002;**225**(3):260–70.

73. Papadopoulou S, Edlund H. Attenuated Wnt signaling perturbs pancreatic growth but not pancreatic function. *Diabetes* 2005;**54**(10):2844–51.

74. Murtaugh LC, Law AC, Dor Y, Melton DA. Beta-catenin is essential for pancreatic acinar but not islet development. *Development* 2005;**132**(21):4663–74.

75. Dessimoz J, Bonnard C, Huelsken J, Grapin-Botton A. Pancreas-specific deletion of beta-catenin reveals Wnt-dependent and Wnt-independent functions during development. *Curr Biol* 2005;**15**(18):1677–83.

76. Dessimoz J, Grapin-Botton A. Pancreas development and cancer: Wnt/beta-catenin at issue. *Cell Cycle* 2006;**5**(1):7–10.

77. Zeng G, Germinaro M, Micsenyi A, Monga NK, Bell A, Sood A, Malhotra V, Sood N, Midda V, Monga DK, et al. Aberrant Wnt/beta-catenin signaling in pancreatic adenocarcinoma. *Neoplasia* 2006;**8**(4):279–89.

78. Miralles F, Czernichow P, Scharfmann R. Follistatin regulates the relative proportions of endocrine versus exocrine tissue during pancreatic development. *Development* 1998;**125**(6):1017–24.

79. Thisse B, Thisse C. Functions and regulations of fibroblast growth factor signaling during embryonic development. *Dev Biol* 2005;**287**(2):390–402.

80. Bhushan A, Itoh N, Kato S, Thiery JP, Czernichow P, Bellusci S, Scharfmann R. Fgf10 is essential for maintaining the proliferative capacity of epithelial progenitor cells during early pancreatic organogenesis. *Development* 2001;**128**(24):5109–17.

81. Miralles F, Czernichow P, Ozaki K, Itoh N, Scharfmann R. Signaling through fibroblast growth factor receptor 2b plays a key role in the development of the exocrine pancreas. *Proc Natl Acad Sci USA* 1999;**96**(11):6267–72.

82. Celli G, LaRochelle WJ, Mackem S, Sharp R, Merlino G. Soluble dominant-negative receptor uncovers essential roles for fibroblast growth factors in multi-organ induction and patterning. *EMBO J* 1998;**17**(6):1642–55.

83. Hart A, Papadopoulou S, Edlund H. Fgf10 maintains notch activation, stimulates proliferation, and blocks differentiation of pancreatic epithelial cells. *Dev Dyn* 2003;**228**(2):185–93.

84. Norgaard GA, Jensen JN, Jensen J. FGF10 signaling maintains the pancreatic progenitor cell state revealing a novel role of Notch in organ development. *Dev Biol* 2003;**264**(2):323–38.

85. Miralles F, Lamotte L, Couton D, Joshi RL. Interplay between FGF10 and Notch signalling is required for the self-renewal of pancreatic progenitors. *Intl J Dev Biol* 2006;**50**(1):17–26.

86. Kornmann M, Beger HG, Korc M. Role of fibroblast growth factors and their receptors in pancreatic cancer and chronic pancreatitis. *Pancreas* 1998;**17**(2):169–75.

87. Yamanaka Y, Friess H, Buchler M, Beger HG, Uchida E, Onda M, Kobrin MS, Korc M. Overexpression of acidic and basic fibroblast growth factors in human pancreatic cancer correlates with advanced tumor stage. *Cancer Res* 1993;**53**(21):5289–96.

88. Ishiwata T, Friess H, Buchler MW, Lopez ME, Korc M. Characterization of keratinocyte growth factor and receptor expression in human pancreatic cancer. *Am J Pathol* 1998;**153**(1):213–22.

89. Leung HY, Gullick WJ, Lemoine NR. Expression and functional activity of fibroblast growth factors and their receptors in human pancreatic cancer. *Int. J Cancer* 1994;**59**(5):667–75.

90. Kornmann M, Ishiwata T, Beger HG, Korc M. Fibroblast growth factor-5 stimulates mitogenic signaling and is overexpressed in human pancreatic cancer: evidence for autocrine and paracrine actions. *Oncogene* 1997;**15**(12):1417–24.

91. Beauchamp Jr RD, Lyons RM, Yang EY, Coffey RJ, Moses HL. Expression of and response to growth regulatory peptides by two human pancreatic carcinoma cell lines. *Pancreas* 1990;**5**(4):369–80.

92. Kleeff J, Kothari NH, Friess H, Fan H, Korc M. Adenovirus-mediated transfer of a truncated fibroblast growth factor (FGF) type I receptor blocks FGF-2 signaling in multiple pancreatic cancer cell lines. *Pancreas* 2004;**28**(1):25–30.

93. Kornmann M, Ishiwata T, Matsuda K, Lopez ME, Fukahi K, Asano G, Beger HG, Korc M. IIIc isoform of fibroblast growth factor receptor 1 is overexpressed in human pancreatic cancer and enhances tumorigenicity of hamster ductal cells. *Gastroenterology* 2002;**123**(1):301–13.

94. Kleeff J, Ishiwata T, Kumbasar A, Friess H, Buchler MW, Lander AD, Korc M. The cell-surface heparan sulfate proteoglycan glypican-1 regulates growth factor action in pancreatic carcinoma cells and is overexpressed in human pancreatic cancer. *J Clin Invest* 1998;**102**(9):1662–73.

95. Kleeff J, Wildi S, Kumbasar A, Friess H, Lander AD, Korc M. Stable transfection of a glypican-1 antisense construct decreases tumorigenicity in PANC-1 pancreatic carcinoma cells. *Pancreas* 1999;**19**(3):281–8.

96. Aikawa T, Whipple CA, Lopez ME, Gunn J, Young A, Lander AD, Korc M. Glypican-1 modulates the angiogenic and metastatic potential of human and mouse cancer cells. *J Clin Invest* 2008;**118**(1):89–99.

Chapter 43

Angiogenesis Signaling Pathways as Targets in Cancer Therapy

Chery A. Whipple and Murray Korc

Departments of Medicine, Pharmacology and Toxicology, Norris Cotton Cancer Center, Dartmouth-Hitchcock Medical Center,
Lebanon, New Hampshire, and Dartmouth Medical School, Hanover, New Hampshire

INTRODUCTION

Under normal conditions blood vessels supply essential nutrients to all tissues through a highly organized network of blood vessels. New blood vessels may form through vasculogenesis, a process whereby the vessels arise as a result of the *de novo* emergence of endothelial cell progenitors from the mesoderm during embryogenesis, or through angiogenesis, which is dependent on the generation of endothelial cells by sprouting from pre-existing vessels [1, 2]. In the adult, physiological angiogenesis is crucial for wound and tissue repair as well as for the recurrent formation of the shed endometrial lining. By contrast, cancer associated angiogenesis is often driven by an abnormal pro-angiogenic profile, a phenomenon that may also be observed in certain inflammatory states, whereas inadequate angiogenesis may be associated with inefficient tissue repair and defective collateral blood vessel formation [2–4]. In solid tumors, which account for more than 85 percent of cancer mortality, angiogenesis is essential for growth and metastasis [5]. Thus, targeting vital pro-angiogenic pathways may prove to be highly effective at inducing tumor regression and enhancing the effectiveness of chemo- and radiotherapy.

OVERVIEW OF ANGIOGENESIS AND ITS ROLE IN TUMOR DEVELOPMENT

Tumor angiogenesis is driven by hypoxia and by a variety of growth factors and chemokines. It requires the recruitment of neighboring host vasculature in order to grow new capillaries toward and into the tumor mass, a phenomenon called cooption, as well as circulating endothelial precursor cells from the bone marrow [4, 6–11]. The delicate balance between angiogenic stimulators and inhibitors is lost and the resulting pro-angiogenic imbalance promotes angiogenesis that supports tumor growth beyond the distance that

oxygen can diffuse (~1–2 mm) [9, 10, 12]. Angiogenesis is further enhanced by the release of angiogenic factors and proteolytic enzymes that initiate and support the proliferation of endothelial cells and facilitate the breakdown of the basement membrane and the extracellular matrix (ECM) [13–15]. The resulting enhanced vascularity increases the likelihood of metastatic spread as well as the growth of the metastatic foci [15]. Additionally, there is evidence that cancer cells may have integrated into the vessel walls in some tumors, and that angiogenesis is enhanced by the recruitment of endothelial precursor cells from the bone marrow [16, 17].

The angiogenic process in tumors can be compartmentalized into three main phases: inflammatory, proliferative, and remodeling. During the inflammatory phase leukocytes and monocytes are recruited to the tumor where they produce pro-angiogenic chemokines and cytokines. During the proliferative phase, in conjunction with the proliferation of endothelial cells there is increased proliferation of fibroblasts, which produce ECM components such as collagen. The proliferation stage is crucial for successful tumor angiogenesis, as it is during this phase that microvascular endothelial cells contribute to the hyperpermeability and local degradation of the basement membrane. This allows for the release of growth factors, heparin, platelet factors, and proteases, which combine to promote the formation of new blood vessels as well as to induce their migration and sprouting into the local stroma [11, 18]. In the remodeling phase, the blood vessels are remodeled, pruned, and allowed to mature [11].

TUMOR VESSEL STRUCTURE

Solid tumors are highly heterogenous and are comprised of cancer cells and a mixture of stromal cells including endothelial cells, pericytes, fibroblasts, myofibroblasts, macrophages,

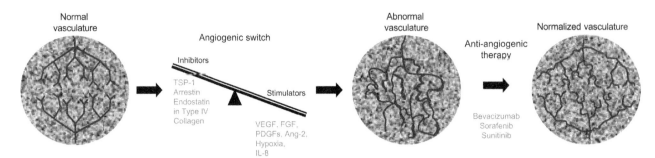

FIGURE 43.1 Schematic representation of tissue and tumor vasculature.
The normal vasculature has a highly organized hierarchy of branching tubes. When the angiogenic balance is tipped in favor of angiogenic stimulators excessive angiogenesis occurs and the tumor vasculature becomes chaotic, leaky, and disorganized. Following anti-angiogenic therapy the tumor vasculature may initially be normalized and during this normalization window additional therapies may become more effective due to improved delivery of therapeutic agents.

inflammatory cells, dendritic cells, and mast cells, all of which are embedded within the ECM [19]. The vessels within solid tumors often exhibit a highly chaotic structure that is devoid of the typical hierarchical branching observed in normal vasculature (Figure 43.1). The resulting loss of equilibrium between vascular growth and cellular demands creates avascular and hypoxic regions, leaky vessels with uneven diameters (due to compression from proliferating tumor cells), a variation in expression of endothelial markers, abnormal or absent lymphatic vessels, and heterogeneous expression of adhesion molecules [1, 20–26].

Additionally, tumor vasculature is continuously undergoing remodeling. Consequently, there is variability in blood flow and permeability between tumors, their metastatic lesions, and within an individual tumor even from one day to the next [19]. Overall perfusion rates (blood flow rate per unit volume) as well as red blood cell velocity are lower in tumors than in normal tissues and tumor blood flow is unevenly distributed and can fluctuate over time and even reverse direction [25, 27]. Circulation within the tumor is also reduced and interstitial fluid pressure is elevated as a consequence of the loss of both vascular hyperpermeability and functional lymphatic vessels that are critical for maintaining interstitial fluid balance [20, 25, 26]. These features, in combination with the leaky blood vessels and high hypoxic conditions, interfere with the efficient delivery of therapeutic agents.

TUMOR STROMA AND THE EXTRACELLULAR MATRIX

The interactions between tumor cells and host stromal cells have a profound influence on tumor cell proliferation, invasion, angiogenesis, and metastasis [28–31]. Potentially the two most critical components of the tumor stroma are the endothelial cells and the fibroblasts [32]. By promoting basement membrane degradation, endothelial cells lead to the release of matrix bound growth factors, such as vascular

endothelial growth factor (VEGF) and fibroblast growth factors (FGFs). In addition, matrix metalloproteinases (MMPs), especially MMP 2, 3, and 9, contribute to ECM degradation and cause the release of pro-angiogenic growth factors. These proteases also generate anti-angiogenic factors such as angiostatin and matrix molecules such as collagen [1], resulting in complex oscillations in the angiogenic switch.

Cancer associated fibroblasts (CAFs) also play a critical role in tumor growth. CAFs are abundant at the host–tumor interface, express VEGF and FGF, and co-localize with the vasculature throughout the tumor [33, 34]. Activated CAFs stimulate the growth of tumor cells [35]; their activation is mediated by platelet derived growth factor (PDGF) and transforming growth factor-beta (TGF-β) isoforms [36, 37]. PDGF also recruits pericytes, which are vascular smooth muscle cells that have a prominent nucleus, scant cytoplasm, and long processes that wrap around blood capillaries and promote the maturation of tumor microvasculature [38]. Pericytes collaborate with endothelial cells and fibroblasts in the production of the basement membrane. Thus, in combination with extracellular proteins fibronectin, collagen, vitronectin, tenascin, and laminin, pericytes contribute to the maintenance of the integrity of the tumor vasculature [1, 39].

THE ROLE OF HYPOXIA IN REGULATING TUMOR ANGIOGENESIS

The abnormal microcirculation within tumors results in areas of hypoxia, which serves to alter the metabolic profile of cancer cells and to further enhance angiogenesis. Typically, there is an associated acidosis, but low pH and hypoxia do not always coincide, inasmuch as tumor vessels may be able to remove waste products such as lactic acid [25, 40]. Hypoxia and acidic pH combine to significantly reduce cancer cells' sensitivity to radiation and chemotherapy, while also interfering with the cellular uptake of certain drugs and with cancer directed immune mechanisms [25, 41, 42]. A hypoxic environment may also induce genetic instability thereby increasing tumor cells invasiveness and metastatic potential [5, 19, 30, 43, 44].

Hypoxia exerts its influence on tumor cells by upregulating glucose transporter 1 (GLUT1) and the expression of a host of growth factors, including VEGF, angiopoietin-2, PDGF, placenta growth factor (PlGF), transforming growth factor-alpha (TGFα), interleukin-8 (IL-8), and hepatocyte growth factor (HGF) [25, 45]. A crucial mediator of the hypoxic response is the transcription factor, hypoxia inducible factor-1 (HIF-1), which is upregulated in many types of human tumors and binds to the hypoxia responsive element (HRE) in the promoter region of responsive genes [45, 46]. HIF-1 is a heterodimeric transcription factor consisting of a constitutive subunit (HIF-1β) and an oxygen responsive subunit (HIF-1α), which undergoes ubiquitin mediated degradation in the presence of oxygen. Under hypoxic condition, HIF-1α becomes stable, allowing HIF-1 to control the expression of numerous genes that mediate developmental and physiological pathways that either deliver oxygen to cells or allow cells to survive hypoxic conditions [47]. HIF-1 is a pro-angiogenic factor that acts upstream of VEGF-A (both increasing its expression and stabilizing the mRNA transcript), thereby rendering cancer cells resistant to hypoxia induced apoptosis while promoting angiogenesis [9, 15, 48–50].

OVERVIEW OF CRITICAL PATHWAYS INVOLVED IN STIMULATING ANGIOGENESIS

Many growth factors and, occasionally, their corresponding high affinity transmembrane receptors, are overexpressed in a variety of cancers. For example, VEGF, TGFα, PDGF, TGFβ, and FGFs are expressed at high levels in many solid tumors. These growth factors have been shown to be essential for the development and progression of cancer, to promote cell proliferation, metastasis, and angiogenesis, and to act synergistically to stimulate endothelial motility and angiogenesis [9, 15, 51–53]. In addition to promoting angiogenesis, PDGF, TGFβ, and the angiopoetins (Ang-1 and -2) in combination with their receptor, Tie2, participate in the regulation of the maturation of nascent vessels into specialized structures [1].

It is generally accepted that VEGF is one of the most crucial growth factors that contributes to tumor angiogenesis. The VEGF family comprises seven secreted homodimeric glycoproteins: VEGF-A, -B, -C, -D, -E, -F, and PlGF [54]. Although VEGF-C and -D are involved in both angiogenesis and lymphangiogenesis and are associated with lymphatic metastasis in a variety of tumor types, VEGF-A appears to be the critical regulator of tumor angiogenesis factor [55]. There are five major VEGF-A isoforms, all of which suppress endothelial cell apoptosis, are vasodilatory, and promote endothelial cell migration and proliferation [19, 56, 57]. These isoforms have 121, 145, 165, 189, and 206 amino acid residues, and arise as a result of alternative splicing from a single gene [58,59]. $VEGF_{121}$ and $VEGF_{145}$ are usually secreted, whereas $VEGF_{189}$ and $VEGF_{206}$ are nearly completely sequestered in the ECM [59]. Interestingly, $VEGF_{165}$ is half secreted and half bound to the cell surface and the ECM [60]. The importance of VEGF-A in embryonic vasculogenesis and angiogenesis is underscored by gene knockout studies in which loss of a single VEGF-A allele in mice resulted in embryonic lethality between day 11 and 12, impaired angiogenesis and blood-island formation, and severe developmental abnormalities [61, 62].

VEGF-A stimulates endothelial cell proliferation following binding to two endothelial cell surface tyrosine kinase receptors, VEGFR-1 (flt-1) and VEGFR-2 (flk-1/KDR) [63–65]. A third high affinity VEGF receptor, termed VEGFR-3 (Flt4), is generally preferentially expressed in lymphatic vessels [66,67]. By contrast, placenta growth factor and VEGF-B bind only VEGFR-1, whereas VEGF-C and VEGF-D interact with both VEGFR-2 and VEGFR-3, and VEGF-E binds only to VEGFR-2 [66]. All three VEGFRs are class III transmembrane protein tyrosine kinases that possess seven immunoglobulin-like sequences in their extracellular domains and a kinase insert in their intracellular domains [63–67]. Their importance in angiogenesis has also been demonstrated in gene knockout studies, which have shown that both VEGFR-1$^{-/-}$ and VEGFR-2$^{-/-}$ mice die *in utero* between day 8.5 and 9.5 [68, 69]. Specifically, in VEGFR-1$^{-/-}$ mice, endothelial cells developed in both embryonic and extra-embryonic sites but failed to organize into normal vascular channels [68]. In VEGFR-2$^{-/-}$ mice, hematopoietic precursors were severely reduced, yolk sac blood islands were absent, and organized blood vessels failed to develop throughout the embryo or the yolk sac [69].

In addition to being upregulated by HIF-1 and hypoxia, VEGF-A expression may be induced by multiple mechanisms, including mutant *ras* and mutant *p53*, transcription factors such as SP1 and the VHL protein, or factors such as FGF-2 and TGFβ [70–73]. Consequently, VEGF-A expression is elevated *in vivo* in many types of tumors including gliomas, breast, colorectal, renal, liver, ovarian, gastric, and pancreatic carcinomas [74, 75]. Moreover, VEGF-A overexpression has been correlated with poor prognosis in many cancers. For example, breast cancer patients with metastatic disease whose tumors exhibit increased angiogenesis have a worse prognosis than the corresponding patients whose tumors do not exhibit increased angiogenesis [76, 77]. Furthermore, suppression of VEGF functions inhibits tumor growth in animal models as demonstrated with a dominant-negative VEGFR-2, soluble VEGFR-1, neutralizing anti-VEGF antibody, VEGF antisense expression, anti-VEGFR-1 or anti-VEGFR-2 ribozymes, tyrosine kinase inhibitors of VEGFR-2, and anti-VEGFR-2 antibodies [78–86].

An additional method of inhibition of VEGF is through VEGF Trap. VEGF Trap is a modified soluble VEGFR that is essentially a chimeric protein consisting of the second immunoglobulin (Ig)-like domain of VEGFR-1 and the third Ig-like domain of VEGFR-2 [87, 88], stabilized by the fusion to a human immunoglobulin heavy chain Fc fragment [89]. VEGF Trap binds VEGF-A with a kD of approximately 1 pM and completely blocks ligand induced phosphorylation of VEGFR-2 [88], the crucial receptor that mediates the mitogenic effects of VEGF-A in endothelial cells. VEGF Trap also blocks the subcutaneous growth and vascularity of tumors deriving from melanoma cells, rat C6 glioma cells, rhabdomyosarcoma cells, and pancreatic cancer cells [88, 90].

VEGF-A interactions with VEGFRs are facilitated by two co-receptors, neuropilin-1 (Np-1) and neuropilin-2 (Np-2). Np-1, originally identified as a mediator of chemorepulsive guidance for axons in the developing nervous system [91, 92], is a transmembrane protein that acts as a co-receptor for VEGF-A [93]. Its extracellular region consists of two complementing binding-like domains (a1 and a2), two coagulation factor V/VIII homology domains (b1 and b2), and a meprin A5 (MAM or c) domain, whereas its intracellular domain consists of a short cytoplasmic tail of about 40 amino acids [91, 92]. Although Np-2 has a similar domain structure, the overall sequence homology between the two genes is only 44 percent [94]. Both Np-1 and Np-2 also act as co-receptors for several, but not necessarily the same, class 3 secreted semaphorins [91, 92]. Semaphorins also bind to plexins, which are transmembrane receptors that are related to semaphorins [95].

VEGF-A activation of VEGFR2 in these cells is facilitated by glypican-1 (GPC1), a heparan sulfate proteoglycan that is attached to the surface of endothelial cells through a glycophophsatidyl inositol anchor [96]. Indeed, endothelial cells isolated form GPC1 knockout mice exhibit an attenuated mitogenic response to VEGF-A [97], whereas targeted deletion of N-acetylgluocsamine N-deacetylase/N-sulfotransferase (Ndst1) leads to decreased N-sulfation of glusoamine residues and attenuated tumor angiogenesis [98]. Once VEGFR2 is phosphorylated on tyrosine residues and becomes activated, there is activation of a cascade of downstream signaling events that includes the Ras-MAPK, src, phosphatidyl-inositol 3'-kinase (PI3-K)/AKT, and eNOS pathways [54, 99]. VEGF-A also induces the mobilization and recruitment of bone marrow derived cells and acts as a survival agent by signaling within the PI3K/Akt signaling pathway to induce expression of the antiapoptotic protein Bcl-2. In addition, VEGF-A renders endothelial cells more radioresistant [100], and promotes the survival of leukemic cells, certain tumor cells, and hematopoietic stem cells [101, 102].

FGF-2 is another potent stimulator of angiogenesis and has been shown to interact synergistically with VEGF in tumor development and growth [103]. There are 23 members in the FGF family, and overexpression of several FGFs has been correlated with tumor invasiveness, angiogenesis, and lymph node metastasis [15, 104, 105]. FGFs bind to their cognate high affinity receptors (FGFRs), which consist of two or three extracellular immunoglobulin-like domains, a single pass transmembrane region, and an intracellular discontinuous tyrosine kinase domain [106]. Of the four known FGFRs, FGFR1 is commonly expressed on endothelial cells. Stable binding of FGF to FGFR1 requires the presence of heparan sulfate proteoglycans (HSPGs) and in particular GPC1 [96]. This interaction leads to an increased cellular response and induces endothelial cell proliferation, migration, and tubulogenesis [15, 103, 107–109].

In addition to binding their high affinity receptors, growth factors rely upon synergistic interactions with major transmembrane cell surface receptors called integrins to mediate efficient signaling between a variety of cell types and the ECM. These interactions lead to alternate mechanisms for the activation of Ras, MAP kinase, focal adhesion kinase (FAK), Src, and PI3-K/AKT signaling pathways [110]. Integrins are often required for optimal signaling via cell adhesion to the ECM with EGF, PDGF, and VEGF. These observations have therapeutic implications, inasmuch as antagonism of integrin α1β1 can block VEGF induced angiogenesis whereas blockade of αVβ3 impedes FGF mediated angiogenesis [11, 111, 112]. Moreover, activation of the endothelial integrin, αVβ3, facilitates the binding of endothelial cells to the ECM and promotes their migration [113, 114].

THE ROLE OF CYTOKINES IN ANGIOGENESIS

Several proinflammatory cytokines, in particular interleukin-6 (IL-6) and interleukin-8 (IL-8), are overexpressed in many cancers and have been associated with increased VEGF expression, especially under hypoxic conditions [15, 115–118]. IL-6 upregulates the expression of several other cytokines that act together to create a tumor environment that favors tumor growth and suppresses cancer directed immunity [115]. For example, in gastric cancer cell lines, transfection with IL-8 resulted in rapidly growing highly vascular tumors [9, 119], whereas neutralizing antibodies against IL-8 and/or VEGF have been shown to attenuate the growth and metastasis of human pancreatic cancer in an orthotopic mouse model [15, 120, 121].

Another family of multifaceted cytokines is the TGFβ superfamily. TGFβs play an essential role in almost every aspect of cellular processes, including cell growth and differentiation, apoptosis, and angiogenesis [122]. Specifically, TGFβs modulate angiogenesis by regulating the proliferation, migration, and differentiation of endothelial cells, promoting capillary tubule formation, and enhancing ECM deposition. As described in more detail elsewhere in the *Handbook*, TGFβs modulate the levels of cyclin dependent kinases (CDK) and CDK inhibitors such as p21 and p27 [15, 123].

TGFβ actions on angiogenesis are complex. While TGFβ is potentially best known for inhibiting growth in many epithelial cell types, TGFβ has also been shown to play an important role in the transcriptional activation of ECM associated genes and their regulatory proteins [124]. Additionally, TGFβ stimulates carcinoma cells to undergo an epithelial-to-mesenchymal transdifferentiation (EMT) leading to enhanced metastasis and invasion [122, 124]. TGFβ promotes tumorigenesis by paracrine stimulation of angiogenesis, which occurs in part by inducing the recruitment of inflammatory cells that aid in the release of VEGF and FGFs [125]. TGFβ also induces the expression of connective tissue growth factor (CTGF), which promotes both stroma formation and angiogenesis, thereby enhancing the pro-angiogenic imbalance in certain cancers [126].

TGFβ actions require the presence of the TGFβ serine-threonine kinase receptors types I (TβRI) and II (TβRII) and betaglycan [127–130]. Betaglycan, which is also known as the type III TGFβ receptor, binds all three TGFβs through its core protein, and facilitates the activation of TβRI by TβRII [130]. A related co-receptor, called endoglin, is predominantly expressed in endothelial cells [131]. After TβRII binds its ligand it is constitutively active as a homodimeric kinase [132,133] and must associate with a TβRI homodimer to initiate a signaling cascade. TβRI is also known as activin-like kinase-5 (ALK5). In addition, endothelial cells express the related receptor ALK1. ALK1 and ALK5 are phosphorylated within their GS region by TβRII, allowing them to phosphorylate receptor activated Smads (R-Smads), Smad2 and Smad3, at their C-terminal SSXS motif [134]. This interaction is facilitated by the co-receptor endoglin [135]. Mice that are null for endoglin, TβRI, ALK1, or TβRII, are all embryonic lethal due to severe vascular defects, underscoring the important role of these receptors and their signaling pathways in vasculogenesis [136–142]. Moreover, endoglin and ALK1 germline mutations are associated with specific vascular abnormalities in humans, termed hereditary hemorrhagic telangiectasia types I and II, respectively [136–142]. It is not surprising therefore, that sequestration of TGFβs by expression of a soluble receptor leads to attenuated tumor growth and angiogenesis in a mouse model of pancreatic cancer [143, 144].

THE ANGIOPOIETINS AND THE ANGIOGENIC SHIFT

Of the four angiopoietins, only Angiopoietin-1 (Ang-1) and angiopoietin-2 (Ang-2) have been implicated in tumor angiogenesis. Both interact with the tyrosine kinase receptor Tie-2. However, Ang-2 antagonizes Ang-1 and prevents Tie-2 activation by competitively binding Tie-2 [15,145–147]. Because Ang-2 is expressed at the site of vascular remodeling, its

antagonism of Ang-1 actions promotes vessel destabilization [11, 148, 149]. These destabilized vessels may undergo regression unless VEGF is present to promote angiogenesis [145]. Conversely, Ang-1 acts as a maturation factor promoting development and stabilization of mature normal vessels *in vivo* by mediating endothelial cell interactions [150, 151]. Activation of Tie-2 by Ang-1 results in downstream activation of the PI3-K/Akt survival pathway, which ultimately leads to endothelial cell migration, tube formation, sprouting, and survival [146, 148], thereby assuring a dynamic balance between vessel regression and growth.

ANGIOGENESIS INHIBITORS

A number of agents and factors act as direct inhibitors of angiogenesis (Table 43.1). The glycoprotein thrombospondin-1 (TSP-1) was the first angiogenic inhibitor discovered that plays a role in cell adhesion, angiogenesis, cell proliferation, cell survival, and the activation of both TGFβ and a variety of proteases [2, 152, 153]. TSP-1 has been shown to prevent VEGF induced angiogenesis by interfering with its ability to bind HSPGs [2, 152]. However, clinical use of TSP-1 is limited due to its large size, limited bioavailability, and general instability [154].

Additional inhibitors that act directly on the endothelial cell include such agents as type IV collagen, arresten, endostatin, angiostatin, and 2-methoxyestradiol. Type IV collagen complexes with other macromolecules including laminin, HSPGs, fibronectin, and entactin, thereby inhibiting capillary endothelial cell proliferation [155]. Arresten is another endogenous inhibitor of endothelial cell tube formation found within the basement membrane [156]. Arrestin may function by blocking the binding of the α1β1 integrin to type I collagen and highly efficient signaling is likely to require HSPGs [111, 157].

TABLE 43.1 Endogenous angiogenesis inhibitors

Inhibitor	Mechanism
Type IV collagen	Binds to other macromolecules to inhibit endothelial cell proliferation
Arrestin	Derived from Type IV collagen and inhibits endothelial cell tube formation
Endostation	Interferes with activity of FGF-2 and VEGF
TSP-1	Interacts with cell adhesion receptors to inhibit neovascularization and endothelial cell migration
Angiostatin	Binds endothelial cell surface ATP synthase and integrins to inhibit endothelial cell migration and proliferation

TABLE 43.2 Anti-angiogenic treatments in clinical trials

Therapeutic agent	Target pathway	Type of chemical
Sunitinib	VEGFR-2 and PDGFR	Kinase inhibitor
Sorafenib (Nexavar)	VEGFR-1, -2, -3, and PDGFR	Kinase inhibitor
Bevacizumab (Avastin™)	VEGF	Monoclonal antibody
DC101	VEGF	Monoclonal antibody
SU11657	VEGF	Monoclonal antibody
Herceptin/ trastuzumab	HER-2	Neutralizing antibody
Vitaxin/ Etaracizumab	$\alpha_v\beta_3$	Humanized monoclonal antibody
VEGF Trap	VEGF-A	Modified soluble VEGFR
SU6668	VEGFR, FGFR, PDGFR	Small molecule kinase inhibitor

Similarly, endostatin is an endogenous collagen derived angiogenesis inhibitor that suppresses angiogenesis, tumor growth, and metastasis [158, 159]. Endostatin interferes with FGF-2 signaling, thereby blocking endothelial cell motility and invasion, inducing apoptosis, and blocking VEGF mediated signaling [160–162]. In addition, several MMPs hydrolyze plasminogen generating angiostatin fragments that also inhibit endothelial cell proliferation and migration [163]. By contrast, 2-methoxyestradiol induces endothelial cell apoptosis by inhibiting microtubule function in proliferating endothelial cells [164].

ANTI-ANGIOGENESIS APPROACHES AND TREATMENTS

A large body or work suggests that anti-angiogenic therapy that restores the angiogenic balance and normalizes the blood vessels may prove to be a highly advantageous approach to inhibit tumor growth and metastasis (Table 43.2). In fact, judiciously applied anti-angiogenic therapy given during the normalization window resulted in more organized vasculature (Figure 43.1), an increase in blood flow to the tumor, and more efficient drug delivery [19, 165]. Moreover, combining anti-angiogenic treatment with radiation therapy during the normalization window

resulted in a synergistic effect on decreased tumor growth [166, 167].

The first anti-VEGF monoclonal antibody to receive U.S. Food and Drug Administration (FDA) approval was bevacizumab (Avastin™) [168]. It is a recombinant humanized antibody that neutralizes VEGF. Its clinical effectiveness was initially established in patients with metastatic colorectal cancer, when given in combination with intravenous 5-fluorouracil [169]. By contrast, in metastatic renal carcinoma, which is often highly vascular, bevacizumab was shown to be effective for first line therapy, alone or in combination with interferon alpha [170]. Bevacizumab was also shown to confer a survival benefit in patients with non-small cell lung cancer when given in combination with paclitaxel and carboplatin [171]. In addition, bevacizumab improved progression free survival when given with taxol in previously untreated metastatic breast cancer patients [172].

The kinase inhibitor sunitinib, which blocks signaling by VEGFR, PDGFR, and the c-kit receptor, recently received FDA approval for treatment of gastrointestinal stromal tumors [173]. A more multifunctional kinase inhibitor, Sorafenib, which inhibits Raf kinase and the kinase activities of VEGFR-1, -2, and -3, PDGFR-β, RET, and c-Kit protein, was shown to prolong progression free survival in patients with advanced clear-cell renal-cell carcinoma in whom previous therapy has failed [170]. Such anti-VEGF treatments not only reduce the size, length, and permeability of tumor associated vessels, but this normalized vasculature also exhibits an increased number of perivascular cells and a more normal basement membrane [166, 167].

Another anti-angiogenic treatment that has shown encouraging results is the targeting of the human epidermal growth factor receptor (HER-2) by the neutralizing antibody trastuzumab (Herceptin). Herceptin, which has been successfully used to treat HER2-neu overexpressing tumor, normalizes tumor vasculature by downregulating growth factors such as VEGF, TGFα, and Ang-1 [174]. One specific way that Herceptin reduces angiogenesis is by targeting VEGF signaling resulting in pruned immature vessels and maturation of other blood vessels, thereby allowing cytotoxic agents to be more efficient [1, 175, 176]. Although VEGF levels are reduced in the cancer cells, the overall levels in the tumor remains the same, which is most likely due to increased VEGF expression by CAFs [174]. Thus, Herceptin may be most effective when used in combination with other anti-angiogenic treatments. It is also plausible that hormone withdrawal from a hormone dependent tumor will be necessary to aid in the reduction of VEGF levels [165]. VEGF levels may also be reduced by utilizing an adenoviral vector that expresses Ang-1 [177] or VEGF Trap, which sequesters VEGF [178].

In some cancers, such as breast cancer, VEGF is the predominant angiogenic growth factor in the early stages of the disease, whereas in later stages tumor growth and angiogenesis is also promoted by additional growth factors such as FGF-1, FGF-2, TGFβ, and PlGF [5, 179]. Therefore it is plausible that a late stage breast tumor may not respond to anti-VEGF treatment and for optimal cancer treatment it will be necessary to target multiple angiogenic pathways specifically tailored to the specific gene and protein profile.

IMPLICATIONS OF TARGETING ANGIOGENESIS–ADVANTAGES AND DISADVANTAGES

While anti-angiogenic therapies can normalize tumor vasculature and the tumor microenvironment, the effect is often transient and is dictated by the type of tumor and its location [25], underscoring the importance of timing and duration of therapy. In addition, tumors can become resistant to a specific anti-angiogenic drug as a consequence of the activation of alternate signaling pathways, as a result of the cancer cells acquiring additional genetic mutations, or through the acquisition by the tumor of alternative methods for sustaining growth [180–184], such as the recruitment of bone marrow derived pro-angiogenic cells, increased pericyte activity and function, and invasion of normal tissue vasculature [39]. Taken together, these observations suggest that it may be important to concomitantly target multiple pro-angiogenic factors and their downstream signaling pathways as well as to simultaneously target the stromal elements and cancer cells [4, 185, 186].

CONCLUSIONS

Anti-angiogenic therapy can effectively normalize the tumor vasculature and attenuate vessel growth for a period of time, known as the normalization window, during which additional therapies such as chemotherapy and radiation become more efficacious. Anti-angiogenic therapy may also cause endothelial cell apoptosis, leading to attenuated tumor growth. To achieve additional advances in anti-angiogenic therapy, it is critical to understand the gene and protein expression profile of each tumor type, determine the timing and extent of treatment based on these profiles, and develop highly accurate tools and biomarkers to measure treatment effectiveness. It is likely that targeting multiple pro-angiogenic factors and their downstream signaling pathways will be necessary to induce tumor regression. However, there is a high risk that such a multitargeted approach will greatly increase health care costs and may increase to unacceptable levels the side effect profile of anti-angiogenic therapy.

ACKNOWLEDGEMENT

Supported by National Cancer Institute grants CA-10130 and CA-102687 awarded to MK.

REFERENCES

1. Jain RK. Molecular regulation of vessel maturation. *Nat Med* 2003; **9**:685–93.
2. Gupta K, Zhang J. Angiogenesis: a curse or cure? *Postgrad Med J* 2005;**81**:236–42.
3. Folkman J. Angiogenesis in cancer, vascular, rheumatoid and other disease. *Nat Med* 1995;**1**:27–31.
4. Pandya NM, Dhalla NS, Santani DD. Angiogenesis: a new target for future therapy. *Vascul Pharmacol* 2006;**44**:265–74.
5. Jain RK. Normalization of tumor vasculature: an emerging concept in antiangiogenic therapy. *Science* 2005;**307**:58–62.
6. Holash J, Maisonpierre PC, Compton D, et al. Vessel cooption, regression, and growth in tumors mediated by angiopoietins and VEGF. *Science* 1999;**284**:1994–8.
7. Asahara T, Takahashi T, Masuda H, et al. VEGF contributes to postnatal neovascularization by mobilizing bone marrow-derived endothelial progenitor cells. *Embo J* 1999;**18**:3964–72.
8. Kerbel RS. Tumor angiogenesis: past, present and the near future. *Carcinogenesis* 2000;**21**:505–15.
9. Garcea G, Lloyd TD, Gescher A, et al. Angiogenesis of gastrointestinal tumours and their metastases: a target for intervention? *Eur J Cancer* 2004;**40**:1302–13.
10. Reinmuth N, Parikh A, Ahamad S, et al. Biology of angiogenesis in tumours of the gastrointestinal tract. *Microsc Res Tech* 2003; **60**:199–207.
11. Li J, Zhang YP, Kirsner RS. Angiogenesis in wound repair: angiogenic growth factors and the extracellular matrix. *Microsc Res Tech* 2003;**60**:107–14.
12. Folkman J. Tumor angiogenesis: therapeutic implications. *N Engl J Med* 1971;**285**:1182–6.
13. John A, Tuszynski G. The role of matrix metalloproteinases in tumor angiogenesis and tumor metastasis. *Pathol Oncol Res* 2001;**7**:14–23.
14. Curran S, Murray GI. Matrix metalloproteinases: molecular aspects of their roles in tumour invasion and metastasis. *Eur J Cancer* 2000; **36**:1621–30.
15. Whipple Chery, Korc M. Targeting angiogenesis in pancreatic cancer: rationale and pitfalls. *Langenbecks Archives of Surgery* 2008; **393**:901–10.
16. Goon PK, Lip GY, Boos CJ, et al. Circulating endothelial cells, endothelial progenitor cells, and endothelial microparticles in cancer. *Neoplasia* 2006;**8**:79–88.
17. Jussila L, Alitalo K. Vascular growth factors and lymphangiogenesis. *Physiol Rev* 2002;**82**:673–700.
18. Marx M, Perlmutter R, Madri JA. Modulation of PDGF-receptor expression in microvascular endothelial cells during in vitro angiogenesis. *J Clin Invest* 1994;**93**:131–9.
19. Fukumura D, Jain RK. Tumor microenvironment abnormalities: causes, consequences, and strategies to normalize. *J Cell Biochem* 2007;**101**:937–49.
20. Padera TP, Kadambi A, di Tomaso E, et al. Lymphatic metastasis in the absence of functional intratumor lymphatics. *Science* 2002; **296**:1883–6.
21. Baish JW, Jain RK. Fractals and cancer. *Cancer Res* 2000;**60**:3683–8.
22. Hashizume H, Baluk P, Morikawa S, et al. Openings between defective endothelial cells explain tumor vessel leakiness. *Am J Pathol* 2000; **156**:1363–80.
23. Helmlinger G, Netti PA, Lichtenbeld HC, et al. Solid stress inhibits the growth of multicellular tumor spheroids. *Nat Biotechnol* 1997;**15**:778–83.

24. Hobbs SK, Monsky WL, Yuan F, et al. Regulation of transport pathways in tumor vessels: role of tumor type and microenvironment. *Proc Natl Acad Sci U S A* 1998;**95**:4607–12.

25. Fukumura D, Jain RK. Tumor microvasculature and microenvironment: targets for anti-angiogenesis and normalization. *Microvasc Res* 2007;**74**:72–84.

26. Leu AJ, Berk DA, Lymboussaki A, et al. Absence of functional lymphatics within a murine sarcoma: a molecular and functional evaluation. *Cancer Res* 2000;**60**:4324–7.

27. Yuan F, Salehi HA, Boucher Y, et al. Vascular permeability and microcirculation of gliomas and mammary carcinomas transplanted in rat and mouse cranial windows. *Cancer Res* 1994;**54**:4564–8.

28. Li G, Satyamoorthy K, Meier F, et al. Function and regulation of melanoma-stromal fibroblast interactions: when seeds meet soil. *Oncogene* 2003;**22**:3162–71.

29. Liotta LA, Kohn EC. The microenvironment of the tumour–host interface. *Nature* 2001;**411**:375–9.

30. Erler JT, Bennewith KL, Nicolau M, et al. Lysyl oxidase is essential for hypoxia-induced metastasis. *Nature* 2006;**440**:1222–6.

31. Elenbaas B, Weinberg RA. Heterotypic signaling between epithelial tumor cells and fibroblasts in carcinoma formation. *Exp Cell Res* 2001;**264**:169–84.

32. Hughes CC. Endothelial-stromal interactions in angiogenesis. *Curr Opin Hematol* 2008;**15**:204–9.

33. Fukumura D, Xavier R, Sugiura T, et al. Tumor induction of VEGF promoter activity in stromal cells. *Cell* 1998;**94**:715–25.

34. Brown EB, Campbell RB, Tsuzuki Y, et al. In vivo measurement of gene expression, angiogenesis and physiological function in tumors using multiphoton laser scanning microscopy. *Nat Med* 2001;**7**:864–8.

35. Micke P, Ostman A. Tumour-stroma interaction: cancer-associated fibroblasts as novel targets in anti-cancer therapy? *Lung Cancer* 2004;**45**:S163–75.

36. Dumont N, Arteaga CL. Transforming growth factor-beta and breast cancer: Tumor promoting effects of transforming growth factor-beta. *Breast Cancer Res* 2000;**2**:125–32.

37. Ronnov-Jessen L, Petersen OW, Bissell MJ. Cellular changes involved in conversion of normal to malignant breast: importance of the stromal reaction. *Physiol Rev* 1996;**76**:69–125.

38. Lindahl P, Johansson BR, Leveen P, Betsholtz C. Pericyte loss and microaneurysm formation in PDGF-B-deficient mice. *Science* 1997;**277**:242–5.

39. Bergers G, Hanahan D. Modes of resistance to anti-angiogenic therapy. *Nat Rev Cancer* 2008;**8**:592–603.

40. Helmlinger G, Sckell A, Dellian M, et al. Acid production in glycolysis-impaired tumors provides new insights into tumor metabolism. *Clin Cancer Res* 2002;**8**:1284–91.

41. Brown JM. The hypoxic cell: a target for selective cancer therapy: eighteenth Bruce F. Cain Memorial Award lecture. *Cancer Res* 1999;**59**:5863–70.

42. Vukovic V, Tannock IF. Influence of low pH on cytotoxicity of paclitaxel, mitoxantrone and topotecan. *Br J Cancer* 1997;**75**:1167–72.

43. Pennacchietti S, Michieli P, Galluzzo M, et al. Hypoxia promotes invasive growth by transcriptional activation of the met protooncogene. *Cancer Cell* 2003;**3**:347–61.

44. Rofstad EK, Mathiesen B, Kindem K, Galappathi K. Acidic extracellular pH promotes experimental metastasis of human melanoma cells in athymic nude mice. *Cancer Res* 2006;**66**:6699–707.

45. Harris AL. Hypoxia: a key regulatory factor in tumour growth. *Nat Rev Cancer* 2002;**2**:38–47.

46. Semenza GL. Targeting HIF-1 for cancer therapy. *Nat Rev Cancer* 2003;**3**:721–32.

47. Semenza GL. Hypoxia-inducible factor 1 (HIF-1) pathway. *Sci STKE* 2007;**2007**:cm8.

48. Wenger RH, Gassman M. Oxygen and the hypoxia-inducible factor 1. *Biol Chem* 1997;**378**:609–16.

49. Buchler P, Reber H, Buchler M, et al. Hypoxia-inducible factor 1 regulates vascular endothelial growth factor expression in pancreatic cancer.. *Pancreas* 2003;**26**:56–64.

50. Akakura N, Kobayashi O, Horiuchi I, et al. Constitutive expression of hypoxia-inducible factor-1 alpha renders pancreatic cancer cells resistant to apoptosis induced by hypoxia and nutrient deprivation. *Cancer Res* 2001;**61**:6548–54.

51. Bachem MG, Schünemann M, Ramadani M, et al. Pancreatic carcinoma cells induce fibrosis by stimulating proliferation and matrix synthesis of stellate cells. *Gastroenterology* 2005;**128**:907–21.

52. Gaspar NJ, Li L, Kapoun AM, et al. Inhibition of transforming growth factor beta signaling reduces pancreatic adenocarcinoma growth and invasiveness. *Mol Pharmacol* 2007;**72**:152–61.

53. Takahashi Y, Bucana C, Akagi Y, et al. Significance of platelet-derived endothelial cell growth factor in the angiogenesis of human gastric cancer. *Clin Cancer Res* 1998;**4**:429–34.

54. Otrock ZK, Makarem JA, Shamseddine AI. Vascular endothelial growth factor family of ligands and receptors: review. *Blood Cells Mol Dis* 2007;**38**:258–68.

55. Alitalo K, Tammela T, Petrova TV. Lymphangiogenesis in development and human disease. *Nature* 2005;**438**:946–53.

56. Dvorak HF. Vascular permeability factor/vascular endothelial growth factor: a critical cytokine in tumor angiogenesis and a potential target for diagnosis and therapy. *J Clin Oncol* 2002;**20**:4368–80.

57. Rafii S, Lyden D, Benezra R, et al. Vascular and haematopoietic stem cells: novel targets for anti-angiogenesis therapy? *Nat Rev Cancer* 2002;**2**:826–35.

58. Houck KA, Ferrara N, Winer J, et al. The vascular endothelial growth factor family: identification of a fourth molecular species and characterization of alternative splicing of RNA. *Mol Endocrinol* 1991;**5**:1806–14.

59. Poltorak Z, Cohen T, Sivan R, et al. VEGF145, a secreted vascular endothelial growth factor isoform that binds to extracellular matrix. *J Biol Chem* 1997;**272**:7151–8.

60. Park JE, Keller GA, Ferrara N. The vascular endothelial growth factor (VEGF) isoforms: differential deposition into the subepithelial extracellular matrix and bioactivity of extracellular matrix-bound VEGF. *Mol Biol Cell* 1993;**4**:1317–26.

61. Berman DM, Karhadkar SS, Maitra A, et al. Widespread requirement for Hedgehog ligand stimulation in growth of digestive tract tumours. *Nature* 2003;**425**:846–51.

62. Ferrara N, Carver-Moore K, Chen H, et al. Heterozygous embryonic lethality induced by targeted inactivation of the VEGF gene. *Nature* 1996;**380**:439–42.

63. Shibuya M. Structure and function of VEGF/VEGF-receptor system involved in angiogenesis. *Cell Struct Funct* 2001;**26**:25–35.

64. Neufeld G, Cohen T, Gengrinovitch S, Poltorak Z. Vascular endothelial growth factor (VEGF) and its receptors. *FASEB J* 1999;**13**:9–22.

65. Veikkola T, Karkkainen M, Claesson-Welsh L, Alitalo K. Regulation of angiogenesis via vascular endothelial growth factor receptors. *Cancer Res* 2000;**60**:203–12.

66. Kukk E, Lymboussaki A, Taira S, et al. VEGF-C receptor binding and pattern of expression with VEGFR-3 suggests a role in lymphatic vascular development. *Development* 1996;**122**:3829–37.

67. Iljin K, Karkkainen MJ, Lawrence EC, et al. VEGFR3 gene structure, regulatory region, and sequence polymorphisms. *FASEB J* 2001;**15**:1028–36.

68. Fong GH, Rossant J, Gertsenstein M, Breitman ML. Role of the Flt-1 receptor tyrosine kinase in regulating the assembly of vascular endothelium. *Nature* 1995;**376**:66–70.

69. Shalaby F, Rossant J, Yamaguchi TP, et al. Failure of blood-island formation and vasculogenesis in Flk-1-deficient mice. *Nature* 1995;**376**:62–6.

70. Korc M. Pathways for aberrant angiogenesis in pancreatic cancer. *Mol Cancer* 2003;**2**:8.

71. Blancher C, Moore JW, Robertson N, Harris AL. Effects of ras and von Hippel-Lindau (VHL) gene mutations on hypoxia-inducible factor (HIF)-1alpha, HIF-2alpha, and vascular endothelial growth factor expression and their regulation by the phosphatidylinositol 3′-kinase/Akt signaling pathway. *Cancer Res* 2001;**61**:7349–55.

72. Meadows KN, Bryant P, Pumiglia K. Vascular endothelial growth factor induction of the angiogenic phenotype requires Ras activation. *J Biol Chem* 2001;**276**:49,289–98.

73. Okada F, Rak JW, Croix BS, et al. Impact of oncogenes in tumor angiogenesis: mutant K-ras up-regulation of vascular endothelial growth factor/vascular permeability factor is necessary, but not sufficient for tumorigenicity of human colorectal carcinoma cells. *Proc Natl Acad Sci U S A* 1998;**95**:3609–14.

74. Ferrara N. Molecular and biological properties of vascular endothelial growth factor. *J Mol Med* 1999;**77**:527–43.

75. Itakura J, Ishiwata T, Friess H, et al. Enhanced expression of vascular endothelial growth factor in human pancreatic cancer correlates with local disease progression. *Clin Cancer Res* 1997;**3**:1309–16.

76. Arora R, Joshi K, Nijhawan R, et al. Angiogenesis as an independent prognostic indicator in node-negative breast cancer. *Anal Quant Cytol Histol* 2002;**24**:228–33.

77. Sledge Jr. GW. Vascular endothelial growth factor in breast cancer: biologic and therapeutic aspects. *Semin Oncol* 2002;**29**:104–10.

78. Millauer B, Shawver LK, Plate KH, et al. Glioblastoma growth inhibited in vivo by a dominant-negative Flk-1 mutant. *Nature* 1994;**367**:576–9.

79. Kong HL, Hecht D, Song W, et al. Regional suppression of tumor growth by in vivo transfer of a cDNA encoding a secreted form of the extracellular domain of the flt-1 vascular endothelial growth factor receptor. *Hum Gene Ther* 1998;**9**:823–33.

80. Goldman CK, Kendall RL, Cabrera G, et al. Paracrine expression of a native soluble vascular endothelial growth factor receptor inhibits tumor growth, metastasis, and mortality rate. *Proc Natl Acad Sci USA* 1998;**95**:8795–800.

81. Kim KJ, Li B, Winer J, et al. Inhibition of vascular endothelial growth factor-induced angiogenesis suppresses tumour growth in vivo. *Nature* 1993;**362**:841–4.

82. Cheng SY, Huang HJ, Nagane M, et al. Suppression of glioblastoma angiogenicity and tumorigenicity by inhibition of endogenous expression of vascular endothelial growth factor. *Proc Natl Acad Sci USA* 1996;**93**:8502–7.

83. Saleh M, Stacker SA, Wilks AF. Inhibition of growth of C6 glioma cells in vivo by expression of antisense vascular endothelial growth factor sequence. *Cancer Res* 1996;**56**:393–401.

84. Pavco PA, Bouhana KS, Gallegos AM, et al. Antitumor and antimetastatic activity of ribozymes targeting the messenger RNA of vascular endothelial growth factor receptors. *Clin Cancer Res* 2000;**6**:2094–103.

85. Fong TA, Shawver LK, Sun L, et al. SU5416 is a potent and selective inhibitor of the vascular endothelial growth factor receptor (Flk-1/KDR) that inhibits tyrosine kinase catalysis, tumor vascularization, and growth of multiple tumor types. *Cancer Res* 1999;**59**:99–106.

86. Witte L, Hicklin DJ, Zhu Z, et al. Monoclonal antibodies targeting the VEGF receptor-2 (Flk1/KDR) as an anti-angiogenic therapeutic strategy. *Cancer Metastasis Rev* 1998;**17**:155–61.

87. Kim ES, Serur A, Huang J, et al. Potent VEGF blockade causes regression of coopted vessels in a model of neuroblastoma. *Proc Natl Acad Sci U S A* 2002;**99**:11,399–404.

88. Holash J, Davis S, Papadopoulos N, Croll SD, et al. VEGF-Trap: a VEGF blocker with potent antitumor effects. *Proc Natl Acad Sci U S A* 2002;**99**:11,393–8.

89. Komesli S, Vivien D, Dutartre P. Chimeric extracellular domain type II transforming growth factor (TGF)-beta receptor fused to the Fc region of human immunoglobulin as a TGF-beta antagonist. *Eur J Biochem* 1998;**254**:505–13.

90. Fukasawa M, Korc M. Vascular endothelial growth factor-trap suppresses tumorigenicity of multiple pancreatic cancer cell lines. *Clin Cancer Res* 2004;**10**:3327–32.

91. He Z, Tessier-Lavigne M. Neuropilin is a receptor for the axonal chemorepellent Semaphorin III. *Cell* 1997;**90**:739–51.

92. Neufeld G, Cohen T, Shraga N, et al. The neuropilins: multifunctional semaphorin and VEGF receptors that modulate axon guidance and angiogenesis. *Trends Cardiovasc Med* 2002;**12**:13–9.

93. Soker S, Takashima S, Miao HQ, et al. Neuropilin-1 is expressed by endothelial and tumor cells as an isoform-specific receptor for vascular endothelial growth factor. *Cell* 1998;**92**:735–45.

94. Kolodkin AL, Levengood DV, Rowe EG, et al. Neuropilin is a semaphorin III receptor. *Cell* 1997;**90**:753–62.

95. Tamagnone L, Comoglio PM. Signalling by semaphorin receptors: cell guidance and beyond. *Trends Cell Biol* 2000;**10**:377–83.

96. Kleeff J, Ishiwata T, Kumbasar A, et al. The cell-surface heparan sulfate proteoglycan glypican-1 regulates growth factor action in pancreatic carcinoma cells and is overexpressed in human pancreatic cancer. *J Clin Invest* 1998;**102**:1662–73.

97. Aikawa T, Whipple CA, Lopez ME, et al. Glypican-1 Modulates the Angiogenic and Metastatic Potential of Cancer Cells. *J Clin Invest* 2007;**118**:89–99.

98. Fuster MM, Wang L, Castagnola J, et al. Genetic alteration of endothelial heparan sulfate selectively inhibits tumor angiogenesis. *J Cell Biol* 2007;**177**:539–49.

99. Xu J, Liu X, Jiang Y, et al. MAPK/ERK signaling mediates VEGF-induced bone marrow stem cell differentiation into endothelial cell. *J Cell Mol Med* 2008;**12**:2395–406.

100. Gupta VK, Jaskowiak NT, Beckett MA, et al. Vascular endothelial growth factor enhances endothelial cell survival and tumor radioresistance. *Cancer J* 2002;**8**:47–54.

101. Harmey JH, Bouchier-Hayes D. Vascular endothelial growth factor (VEGF), a survival factor for tumour cells: implications for anti-angiogenic therapy. *Bioessays* 2002;**24**:280–3.

102. Dias S, Shmelkov SV, Lam G, Rafii S. VEGF(165) promotes survival of leukemic cells by Hsp90-mediated induction of Bcl-2 expression and apoptosis inhibition. *Blood* 2002;**99**:2532–40.

103. Qiao D, Meyer K, Mundhenke C, et al. Heparan sulfate proteoglycans as regulators of fibroblast growth factor-2 signaling in brain endothelial cells. Specific role for glypican-1 in glioma angiogenesis. *J Biol Chem* 2003;**278**:16,045–53.

104. Noda M, Hattori T, Kimura T, Naitoh H, et al. Expression of fibroblast growth factor 2 mRNA in early and advanced gastric cancer. *Acta Oncol* 1997;**36**:695–700.

105. Ueki T, Koji T, Tamiya S, Nakane PK, Tsuneyoshi M. Expression of basic fibroblast growth factor and fibroblast growth factor receptor in advanced gastric carcinoma. *J Pathol* 1995;**177**:353–61.

106. Klint P, Claesson-Welsh L. Signal transduction by fibroblast growth factor receptors. *Front Biosci* 1999;**4**:D165–77.

107. Sperinde GV, Nugent MA. Heparan sulfate proteoglycans control intracellular processing of bFGF in vascular smooth muscle cells. *Biochemistry* 1998;**37**:13,153–64.

108. Sperinde GV, Nugent MA. Mechanisms of fibroblast growth factor 2 intracellular processing: a kinetic analysis of the role of heparan sulfate proteoglycans. *Biochemistry* 2000;**39**:3788–96.

109. Javerzat S, Auguste P, Bikfalvi A. The role of fibroblast growth factors in vascular development. *Trends Mol Med* 2002;**8**:483–9.

110. Eliceiri BP, Cheresh DA. Adhesion events in angiogenesis. *Curr Opin Cell Biol* 2001;**13**:563–8.

111. Senger DR, Claffey KP, Benes JE, et al. Angiogenesis promoted by vascular endothelial growth factor: regulation through alpha-1beta1 and alpha2beta1 integrins. *Proc Natl Acad Sci U S A* 1997;**94**:13,612–17.

112. Brooks PC, Clark RA, Cheresh DA. Requirement of vascular integrin alpha v beta 3 for angiogenesis. *Science* 1994;**264**:569–71.

113. Brooks PC, Stromblad S, Sanders LC, et al. Localization of matrix metalloproteinase MMP-2 to the surface of invasive cells by interaction with integrin alpha v beta 3. *Cell* 1996;**85**:683–93.

114. Sepp NT, Li LJ, Lee KH, et al. Basic fibroblast growth factor increases expression of the alpha v beta 3 integrin complex on human microvascular endothelial cells. *J Invest Dermatol* 1994;**103**:295–9.

115. Feurino LW, Zhang Y, Bharadwaj U, et al. IL-6 Stimulates Th2 Type Cytokine Secretion and Upregulates VEGF and NRP-1 Expression in Pancreatic Cancer Cells. *Cancer Biol Ther* 2007;**6**:1096–100.

116. Hedin KE. Chemokines: new, key players in the pathobiology of pancreatic cancer. *Int J Gastrointest Cancer* 2002;**31**:23–9.

117. Wente MN, Keane MP, Burdick MD, et al. Blockade of the chemokine receptor CXCR2 inhibits pancreatic cancer cell-induced angiogenesis. *Cancer Lett* 2006;**241**:221–7.

118. Shi Q, Abbruzzese JL, Huang S, et al. Constitutive and inducible interleukin 8 expression by hypoxia and acidosis renders human pancreatic cancer cells more tumorigenic and metastatic. *Clin Cancer Res* 1999;**5**:3711–21.

119. Kitadai Y, Takahashi Y, Haruma K, et al. Transfection of interlukin-8 increases angiogenesis and tumorigenesis of human gastric carcinoma cell s in nude mice. *Br J Cancer* 1998;**81**:647–53.

120. Xie K, Wei D, Shi Q, Huang S. Constitutive and inducible expression and regulation of vascular endothelial growth factor. *Cytokine Growth Factor Rev* 2004;**15**:297–324.

121. Xie K, Wei D, Huang S. Transcriptional anti-angiogenesis therapy of human pancreatic cancer. *Cytokine Growth Factor Rev* 2006;**17**:147–56.

122. Yue J, Mulder KM. Transforming growth factor-beta signal transduction in epithelial cells. *Pharmacol Ther* 2001;**91**:1–34.

123. Boyer Arnold N, Korc M. Smad7 abrogates transforming growth factor-beta1-mediated growth inhibition in COLO-357 cells through functional inactivation of the retinoblastoma protein. *J Biol Chem* 2005;**280**:21,858–66.

124. Huang SS, Huang JS. TGF-beta control of cell proliferation. *J Cell Biochem* 2005;**96**:447–62.

125. Bachman KE, Park BH. Duel nature of TGF-beta signaling: tumor suppressor vs. tumor promoter. *Curr Opin Oncol* 2005;**17**:49–54.

126. Aikawa T, Gunn J, Spong SM, et al. Connective tissue growth factor-specific antibody attenuates tumor growth, metastasis, and angiogenesis in an orthotopic mouse model of pancreatic cancer. *Mol Cancer Ther* 2006;**5**:1108–16.

127. Kingsley DM. The TGF-beta superfamily: new members, new receptors, and new genetic tests of function in different organisms. *Genes Dev* 1994;**8**:133–46.

128. Lin HY, Wang XF, Ng-Eaton E, et al. Expression cloning of the TGF-beta type II receptor, a functional transmembrane serine/threonine kinase. *Cell* 1992;**68**:775–85.

129. Franzen P, ten Dijke P, Ichijo H, et al. Cloning of a TGF beta type I receptor that forms a heteromeric complex with the TGF beta type II receptor. *Cell* 1993;**75**:681–92.

130. Wang XF, Lin HY, Ng-Eaton E, et al. Expression cloning and characterization of the TGF-beta type III receptor. *Cell* 1991;**67**:797–805.

131. Cheifetz S, Bellon T, Cales C, et al. Endoglin is a component of the transforming growth factor-beta receptor system in human endothelial cells. *J Biol Chem* 1992;**267**:19,027–30.

132. Attisano L, Wrana JL. Signal transduction by the TGF-beta superfamily. *Science* 2002;**296**:1646–7.

133. Shi Y, Massague J. Mechanisms of TGF-beta signaling from cell membrane to the nucleus. *Cell* 2003;**113**:685–700.

134. Wieser R, Wrana JL, Massague J. GS domain mutations that constitutively activate T beta R-I, the downstream signaling component in the TGF-beta receptor complex. *EMBO J* 1995;**14**:2199–208.

135. Lee NY, Ray BN, How T, Blobe GC. Endoglin promotes transforming growth factor-beta-mediated smad 1/5/8 signaling and inhibits endothelial cell migration through its association with GIPC. *J Biol Chem* 2008;**283**:32,527–3.

136. Bourdeau A, Dumont DJ, Letarte M. A murine model of hereditary hemorrhagic telangiectasia. *J Clin Invest* 1999;**104**:1343–51.

137. Derynck R, Akhurst RJ. Differentiation plasticity regulated by TGF-beta family proteins in development and disease. *Nat Cell Biol* 2007;**9**:1000–4.

138. Li DY, Sorensen LK, Brooke BS, et al. Defective angiogenesis in mice lacking endoglin. *Science* 1999;**284**:1534–7.

139. Massague J, Gomis RR. The logic of TGFbeta signaling. *FEBS Lett* 2006;**580**:2811–20.

140. Oh SP, Seki T, Goss KA, et al. Activin receptor-like kinase 1 modulates transforming growth factor-beta 1 signaling in the regulation of angiogenesis. *Proc Natl Acad Sci U S A* 2000;**97**:2626–31.

141. Yang X, Castilla LH, Xu X, et al. Angiogenesis defects and mesenchymal apoptosis in mice lacking SMAD5. *Development* 1999;**126**:1571–80.

142. Fonsatti E, Altomonte M, Nicotra MR, et al. Endoglin (CD105): a powerful therapeutic target on tumor-associated angiogenetic blood vessels. *Oncogene* 2003;**22**:6557–63.

143. Rowland-Goldsmith MA, Maruyama H, Kusama T, et al. Soluble type II transforming growth factor-beta (TGF-beta) receptor inhibits TGF-beta signaling in COLO-357 pancreatic cancer cells in vitro and attenuates tumor formation. *Clin Cancer Res* 2001;**7**:2931–40.

144. Rowland-Goldsmith MA, Maruyama H, Matsuda K, et al. Soluble type II transforming growth factor-beta receptor attenuates expression of metastasis-associated genes and suppresses pancreatic cancer cell metastasis. *Mol Cancer Ther* 2002;**1**:161–7.

145. Caine GJ, Blann AD, Stonelake PS, et al. Plasma angiopoietin-1, angiopoietin-2 and Tie-2 in breast and prostate cancer: a comparison with VEGF and Flt-1. *Eur J Clin Invest* 2003;**33**:883–90.

146. Niedzwiecki S, Stepien T, Kopec K, et al. Angiopoietin 1 (Ang-1), angiopoietin 2 (Ang-2) and Tie-2 (a receptor tyrosine kinase) concentrations in peripheral blood of patients with thyroid cancers. *Cytokine* 2006;**36**:291–5.

147. Asahara T, Chen D, Takahashi T, et al. Tie2 receptor ligands, angiopoietin-1 and angiopoietin-2, modulate VEGF-induced postnatal neovascularization. *Circ Res* 1998;**83**:233–40.

148. Bach F, Uddin FJ, Burke D. Angiopoietins in malignancy. *Eur J Surg Oncol* 2007;**33**:7–15.

149. Tait CR, Jones PF. Angiopoietins in tumours: the angiogenic switch. *J Pathol* 2004;**204**:1–10.

150. Suri C, McClain J, Thurston G, et al. Increased vascularization in mice overexpressing angiopoietin-1. *Science* 1998;**282**:468–71.

151. Thurston G, Suri C, Smith K, et al. Leakage resistant blood vessels in mice transgenically overexpressing angiopoietin-1. *Science* 1999;**286**:2511–14.

152. Gupta K, Gupta P, Wild R, et al. Binding and displacement of vascular endothelial growth factor (VEGF) by thrombospondin: effect on human microvascular endothelial cell proliferation and angiogenesis. *Angiogenesis* 1999;**3**:147–58.

153. Chen H, Herndon ME, Lawler J. The cell biology of thrombospondin-1. *Matrix Biol* 2000;**19**:597–614.

154. Gupta K, Radotra BD, Banerjee AK, Nijhawan R. Quantitation of angiogenesis and its correlation with vascular endothelial growth factor expression in astrocytic tumors. *Anal Quant Cytol Histol* 2004;**26**:223–9.

155. Timpl R. Macromolecular organization of basement membranes. *Curr Opin Cell Biol* 1996;**8**:618–24.

156. Colorado PC, Torre A, Kamphaus G, et al. Anti-angiogenic cues from vascular basement membrane collagen. *Cancer Res* 2000;**60**:2520–6.

157. Pozzi A, Moberg PE, Miles LA, et al. Elevated matrix metalloprotease and angiostatin levels in integrin alpha 1 knockout mice cause reduced tumor vascularization.. *Proc Natl Acad Sci U S A* 2000;**97**:2202–7.

158. O'Reilly MS, Boehm T, Shing Y, et al. Endostatin: an endogenous inhibitor of angiogenesis and tumor growth. *Cell* 1997;**88**:277–85.

159. Marneros AG, Olsen BR. The role of collagen-derived proteolytic fragments in angiogenesis. *Matrix Biol* 2001;**20**:337–45.

160. Dixelius J, Cross M, Matsumoto T, et al. Endostatin regulates endothelial cell adhesion and cytoskeletal organization. *Cancer Res* 2002;**62**:1944–7.

161. Dhanabal M, Volk R, Ramchandran R, et al. Cloning, expression, and in vitro activity of human endostatin. *Biochem Biophys Res Commun* 1999;**258**:345–52.

162. Hanai J, Dhanabal M, Karumanchi SA, et al. Endostatin causes G1 arrest of endothelial cells through inhibition of cyclin D1. *J Biol Chem* 2002;**277**:16,464–69.

163. O'Reilly MS, Holmgren L, Shing Y, et al. Angiostatin: a novel angiogenesis inhibitor that mediates the suppression of metastases by a Lewis lung carcinoma. *Cell* 1994;**79**:315–28.

164. Yue TL, Wang X, Louden CS, et al. 2-Methoxyestradiol, an endogenous estrogen metabolite, induces apoptosis in endothelial cells and inhibits angiogenesis: possible role for stress-activated protein kinase signaling pathway and Fas expression. *Mol Pharmacol* 1997;**51**:951–62.

165. Jain RK. Normalizing tumor vasculature with anti-angiogenic therapy: a new paradigm for combination therapy. *Nat Med* 2001;**7**:987–9.

166. Winkler F, Kozin SV, Tong RT, et al. Kinetics of vascular normalization by VEGFR2 blockade governs brain tumor response to radiation: role of oxygenation, angiopoietin-1, and matrix metalloproteinases. *Cancer Cell* 2004;**6**:553–63.

167. Tong RT, Boucher Y, Kozin SV, et al. Vascular normalization by vascular endothelial growth factor receptor 2 blockade induces a pressure gradient across the vasculature and improves drug penetration in tumors. *Cancer Res* 2004;**64**:3731–6.

168. Ranieri G, Patruno R, Ruggieri E, et al. Vascular endothelial growth factor (VEGF) as a target of bevacizumab in cancer: from the biology to the clinic. *Curr Med Chem* 2006;**13**:1845–57.

169. Giantonio BJ, Catalano PJ, Meropol NJ, et al. Bevacizumab in combination with oxaliplatin, fluorouracil, and leucovorin (FOLFOX4) for previously treated metastatic colorectal cancer: results from the Eastern Cooperative Oncology Group Study E3200. *J Clin Oncol* 2007;**25**:1539–44.

170. Escudier B, Pluzanska A, Koralewski P, et al. Bevacizumab plus interferon alfa-2a for treatment of metastatic renal cell carcinoma: a randomised, double-blind phase III trial. *Lancet* 2007;**370**:2103–11.

171. Sandler A, Gray R, Perry MC, et al. Paclitaxel-carboplatin alone or with bevacizumab for non-small-cell lung cancer. *N Engl J Med* 2006;**355**:2542–50.

172. Miller K, Wang M, Gralow J, et al. Paclitaxel plus bevacizumab versus paclitaxel alone for metastatic breast cancer. *N Engl J Med* 2007;**357**:2666–76.

173. Motzer RJ, Hutson TE, Tomczak P, et al. Sunitinib versus interferon alfa in metastatic renal-cell carcinoma. *N Engl J Med* 2007;**356**:115–24.

174. Izumi Y, Xu L, di Tomaso E, et al. Tumour biology: herceptin acts as an anti-angiogenic cocktail. *Nature* 2002;**416**:279–80.

175. Kadambi A, Mouta Carreira C, Yun CO, et al. Vascular endothelial growth factor (VEGF)-C differentially affects tumor vascular function and leukocyte recruitment: role of VEGF-receptor 2 and host VEGF-A. *Cancer Res* 2001;**61**:2404–8.

176. Benjamin LE, Golijanin D, Itin A, et al. Selective ablation of immature blood vessels in established human tumors follows vascular endothelial growth factor withdrawal. *J Clin Invest* 1999;**103**:159–65.

177. Thurston G, Rudge JS, Ioffe E, et al. Angiopoietin-1 protects the adult vasculature against plasma leakage. *Nat Med* 2000;**6**:460–3.

178. Dupont J, Bienvenu B, Aghajanian C, et al. Phase I and pharmacokinetic study of the novel oral cell-cycle inhibitor Ro 31-7453 in patients with advanced solid tumors. *J Clin Oncol* 2004;**22**:3366–74.

179. Relf M, LeJeune S, Scott PA, et al. Expression of the angiogenic factors vascular endothelial cell growth factor, acidic and basic fibroblast growth factor, tumor growth factor beta-1, platelet-derived endothelial cell growth factor, placenta growth factor, and pleiotrophin in human primary breast cancer and its relation to angiogenesis. *Cancer Res* 1997;**57**:963–9.

180. Casanovas O, Hicklin DJ, Bergers G, Hanahan D. Drug resistance by evasion of antiangiogenic targeting of VEGF signaling in late-stage pancreatic islet tumors. *Cancer Cell* 2005;**8**:299–309.

181. Kerbel RS. Therapeutic implications of intrinsic or induced angiogenic growth factor redundancy in tumors revealed. *Cancer Cell* 2005;**8**:269–71.

182. Blouw B, Song H, Tihan T, et al. The hypoxic response of tumors is dependent on their microenvironment. *Cancer Cell* 2003;**4**:133–46.

183. Rubenstein JL, Kim J, Ozawa T, et al. Anti-VEGF antibody treatment of glioblastoma prolongs survival but results in increased vascular cooption. *Neoplasia* 2000;**2**:306–14.

184. Fernando NT, Koch M, Rothrock C, et al. Tumor escape from endogenous, extracellular matrix-associated angiogenesis inhibitors by up-regulation of multiple proangiogenic factors. *Clin Cancer Res* 2008;**14**:1529–39.

185. Yu JL, Rak JW, Coomber BL, et al. Effect of p53 status on tumor response to antiangiogenic therapy. *Science* 2002;**295**:1526–8.

186. Marx J. Cancer research. Obstacle for promising cancer therapy. *Science* 2002;**295**:1444.

Chapter 44

Clinical Applications of Kinase Inhibitors in Solid Tumors

William Pao[1] and Nicolas Girard[2]

[1]Vanderbilt-Ingram Cancer Center, Nashville, Tennessee

[2]Human Oncology and Pathogenesis Program, Memorial Sloan-Kettering Cancer Center, New York, New York

INTRODUCTION

The treatment of cancer has traditionally involved three major therapeutic modalities: surgery, radiotherapy, and systemic chemotherapy. Within the latter category, therapies could be broadly categorized as cytotoxic chemotherapies (e.g., alkylating agents, DNA damaging agents, antimetabolites, topoisomerase interactive agents, and antimicrotubule agents) and biotherapeutics (e.g., interferons, interleukins, hormonal therapies, differentiation agents, and monoclonal antibodies) [1]. Recently, a newer class of "targeted therapies," involving selective kinase inhibitors, has been developed. The development of such agents has been the direct result of major progress made in the elucidation of molecular signaling events that lead to and sustain cancer. Remarkably, it was only in 1979 that a cancer-causing gene (i.e., the *v-src* oncogene), was first found to act biochemically as a kinase [2].

That kinase inhibitors have effective and sustained antitumor activity is best exemplified by the use of imatinib mesylate (STI-571, Gleevec™ (US), Glivec™ (Europe); Novartis, Basel, Switzerland) to treat patients with chronic myelegenous leukemia (CML). Imatinib, a 2-phenylaminopyrimidine derivative (4-4-methylpiperazin-1-yl)-methyl]-N-[4-methyl-3-4-pyridin-3-ylpyrimidin-2-yl)-amino]-phenyl]-benzamide; Figure 44.1), is an oral small molecule selective inhibitor of the ABL kinase. Patients with CML contain tumor cells driven by a constitutively activated ABL kinase (due to the pathognemonic *BCR-ABL* translocation), and imatinib can be used effectively to treat CML. In May 2001 imatinib became the first kinase inhibitor approved by the FDA to treat a human cancer [3], ushering in a new paradigm in cancer drug development.

Some skeptics thought that imatinib may represent an outlier class of drug with only a small niche in hematologic malignancies. However, developments over the past few years have demonstrated that kinase inhibitors also have extensive utility in treating solid tumors from a variety of tissue origins. This chapter will highlight the clinical applications of kinase inhibitors in the first three specific solid tumors for which kinase inhibitors were approved: gastrointestinal stromal tumor (GISTs), non-small cell lung cancer (NSCLC), and renal cell carcinoma (RCC) (Table 44.1). In one case (GIST), understanding of the molecular etiology of the disease led to rational drug development; in another case (NSCLC), clinical drug development led to rational dissection of the underlying biology; and in a third case (RCC), an existing body of molecular knowledge has provided a framework to begin to understand why a certain class of drugs may be effective. As more detailed kinase biology is discussed in other chapters, we will only discuss relevant translational studies. For detailed information on (1) clinical aspects of these diseases, (2) pharmacologic properties of the kinase inhibitors used in these diseases, or (3) clinical trial design, the reader is referred to other references (for example, [4]).

RATIONALE FOR KINASE INHIBITION IN THE TREATMENT OF SOLID TUMORS

Cancers arise from the progressive accumulation of activating mutations in growth-enhancing genes (oncogenes) and inactivating mutations in growth-inhibitory genes (tumor suppressor genes). For most cancers, it is currently not known whether a genetic lesion that is necessary for the initial development or progression of a specific tumor

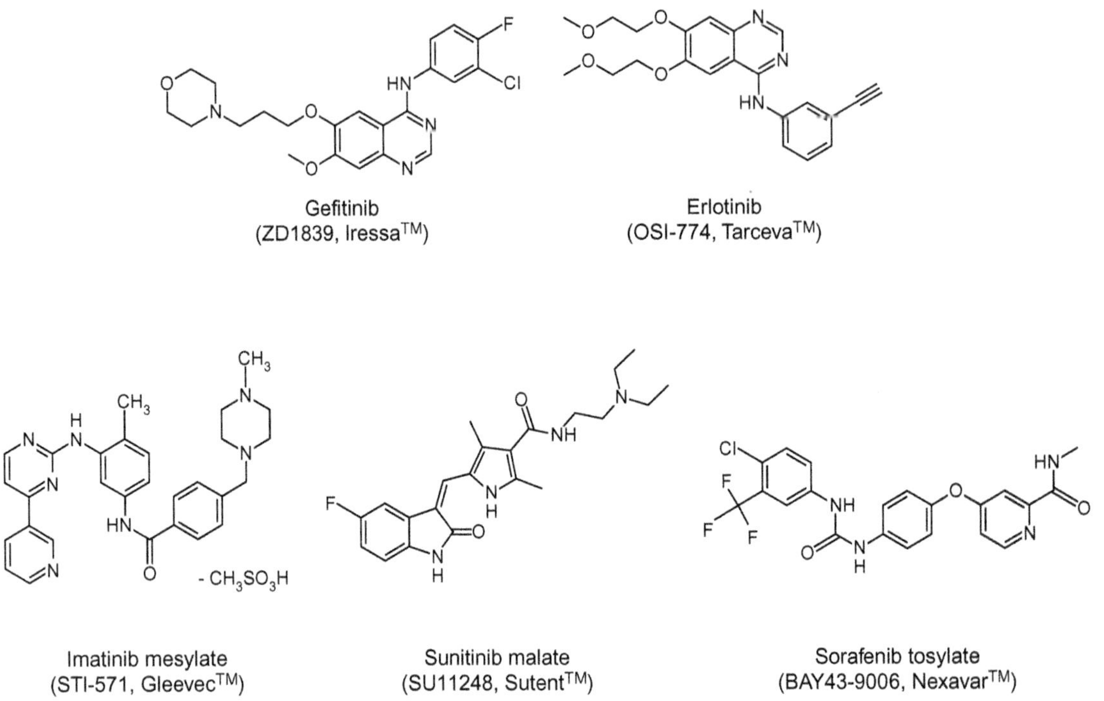

FIGURE 44.1 Chemical structures of erlotinib, gefitinib, imatinib, sorafenib, and sunitinib.

TABLE 44.1 Small molecule tyrosine kinase inhibitors approved by the US FDA for use in solid tumors

Disease	Drug	Relevant target(s)	FDA approval		Reference
			Indication	Date	
Gastrointestinal stromal tumor	Imatinib	CKIT/PDGFR-α (mutant)	First-line	2002	24
	Sunitinib	CKIT/PDGFR-α (mutant)	Second-line, after imatinib failure	2006	37
Non-small cell lung cancer	Gefitinib	EGFR	Third-line, restricted to patients benefiting from gefitinib	2003 2005	48, 49 50
	Erlotinib	EGFR	Second-line	2004	52
Renal cell carcinoma	Sorafenib	VEGFR?	Second-line	2005	127
	Sunitinib	VEGFR?	First-line	2007	131
Breast cancer	Lapatinib	HER2 (amplified)	Third/Fourth-line, in association with capecitabine, in HER2 overexpressing tumors, after previous anthracycline, taxane, and trastuzumab	2007	134
Hepatocellular carcinoma	Sorafenib	?	First-line	2007	135
Thyroid carcinoma	Vandetinib	RET (mutant)	First-line, orphan drug designation	2004	136

is also required for the maintenance of that tumor's survival. However, multiple mouse models of cancer involving inducible transgenes have demonstrated that tumors, remarkably, can be dependent upon the expression of single oncogenes for survival, even in the absence of tumor suppressor genes [5-7]. Furthermore, the success of targeting the constitutively activated BCR-ABL tyrosine kinase with the oral kinase inhibitor, imatinib, in patients with CML shows that human malignancies can also rely upon single oncogenes for survival [3]. This concept of tumor maintenance has also been called "oncogene addiction" [8]. Studies supportive of this notion offer an optimistic view on new approaches for treating cancer; they suggest that tumors harboring complex genetic lesions nevertheless have an "Achilles heel" that just needs to be systematically identified and exploited.

Although tumor suppressor genes (for example, *p53*) may also play a role in tumor maintenance [9], the most clinically relevant examples of "oncogene addiction" involve aberrant kinases. The human genome encodes approximately 518 kinases (1.7 percent of all human genes), classified phylogenetically into seven major groups [10], including:

1. Sixty-three cyclic nucleotide-regulated kinases (containing PKA, PKG, and PKC families)
2. Seventy-four calcium/calmodulin-dependent kinases (CaMK)
3. Twelve casein kinases (CK1)
4. Sixty-one cyclin-dependent kinases (containing CDK, MAPK, GSK3, CLK families)
5. Forty-seven STE kinases (homologs of yeast Sterile 7, Sterile 11, and Sterile 20 kinases)
6. Ninety tyrosine kinases
7. Forty-three tyrosine-like kinases.

The remaining 128 kinases fall into "atypical" protein kinase families. Among these, thus far, mutant tyrosine kinases have proven to be the most effective targets for drug therapy for at least four major reasons. First, malignant transformation is often the result of deregulated tyrosine kinase activity [11]. Second, simple screening methods (e.g., antiproliferative and/or apoptotic activity) can be used to identify and select novel inhibitors for clinical development [12]. Third, elucidation of tyrosine kinase crystal structures has facilitated structure-based drug design of ATP-competitive analogs [13]. Finally, aberrant kinases are often inhibited at lower doses of drug versus their wild-type counterparts, providing a therapeutic "window of opportunity" that maximizes specificity and minimizes side effects. Thus far, the most successful kinase inhibitors have been agents that block enzymatic activity by competing with ATP in the ATP binding pocket of target kinases.

Notably, kinase inhibitors are very different from conventional cytotoxic chemotherapies and/or biotherapeutics, most of which are administered intravenously in outpatient oncology clinics. All available kinase inhibitors are orally administered drugs with a once- or twice-daily schedule. While they are not completely devoid of side effects, in general kinase inhibitors are well-tolerated, with relatively fewer life-threatening complications. These considerations make a big difference in the lives of patients who want to balance quality of life with the therapeutic benefits of various treatment regimens.

KINASE INHIBITION IN GASTROINTESTINAL STROMAL TUMORS

GIST was the first human solid tumor in which specific inhibition of a tyrosine kinase was shown be a clinically useful therapeutic intervention. Thus, this disease has served as a paradigm for the development of kinase inhibitors in solid tumors.

GIST Background

Soft tissue sarcomas are a heterogenous group of neoplasms, and represent 1 percent of adult malignancies. GISTs are a subset of soft tissue sarcomas that arise from mesenchymal precursors that normally give rise to connective-tissue cells of the gastrointestinal tract [14]. The annual incidence in the United States is estimated at 3800 [15]. Approximately 70 percent of GISTs are found in the stomach, 20-30 percent in the small intestine, and less than 10 percent elsewhere in the gastrointestinal tract [16]. Most cases are sporadic, although familial GISTs also occur [17]. Surgery is the only curative option. In the twentieth century, for patients with unresectable or recurrent disease, no systemic treatments were shown to have meaningful clinical activity against GISTs [18]. The median duration of survival for patients with metastatic GIST was 20 months, and for patients with local recurrence it was 9-12 months [18, 19].

Classification of these neoplasms used to be based upon histological assessments, and was historically controversial. However, key observations in 1998 helped to clarify the nature of GISTs. First, GISTs were shown to strongly express the receptor tyrosine kinase KIT (CD117) [20]. This marker, which is rarely present on other spindle cell tumors occurring in the abdominal cavity, is now accepted as the most specific immunohistochemical marker for GISTs. Second, GISTs were shown to harbor mutations in the juxtamembrane domain (exon 11) of the *KIT* gene (Figure 44.2a) [20]. These mutations led to constitutive activation of the KIT kinase in the absence of ligand (stem cell factor; SCF). Finally, based upon IHC staining of cell differentiation markers, GISTs were proposed to arise from the interstitial cells of Cajal (ICC). These cells reside in and near the circular muscle layer of the gastrointestinal tract, and are thought to serve as pacemakers that trigger gut contraction [14, 21]. Myenteric plexus ICC fail to develop

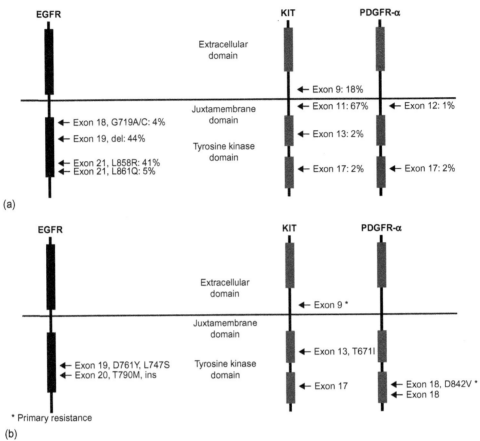

FIGURE 44.2 Spectrum of drug-sensitive (a) and -resistant (b) mutations found in EGFR in NSCLC, and KIT/PDGFRA in GIST, respectively. Percentages indicate the relative proportion of each mutation.

in mice lacking KIT or SCF, indicating that the KIT-SCF axis is essential to development of these cells [21].

Development of Kinase Inhibitors in GIST

The treatment of GIST was revolutionized based on two key preclinical observations. First, the kinase inhibitor, imatinib, was shown to block the *in vitro* kinase activity of both wild-type KIT and a mutant KIT isoform commonly found in GISTs [22]. Second, imatinib was shown *in vitro* using cells from GIST patients harboring activating KIT mutations to arrest cell proliferation and induce apoptosis [22]. As discussed above, imatinib is an oral small molecule kinase inhibitor that targets the BCR-ABL fusion protein in CML, but it also inhibits the kinase activity in the nanomolar range of both KIT and platelet-derived growth factor receptor α (PDGFRA). Imatinib was subsequently shown to induce a rapid, substantial, and durable response in a patient with chemotherapy-resistant GIST [23]. Within a year, a phase II multi-national trial enrolling patients with advanced CD117-positive GIST demonstrated efficacy and safety of the drug. Among 147 patients, 54 percent (95% CI 45.3-62.0) demonstrated partial, sustained, clinically significant responses [24]. Based on this trial, the FDA approved imatinib, in February 2002, for the treatment of

patients with KIT (CD117)-positive unresectable and/or metastatic malignant GISTs.

Correlative translational studies have demonstrated that KIT expression is exhibited by 95 percent of GISTs and is related to *KIT* activating mutations in a majority of cases [25]. *KIT* mutations occur predominantly within exons 9, 11, 13, and 17 (Figure 44.2a) [25]. Exon 9 mutations (duplication/insertions) occur in the extracellular domain; the mechanism of action remains to be determined, but it is hypothesized that they disrupt an antidimerization motif. Exon 11 mutations (deletions, insertions, point mutations, and internal tandem duplications) occur in the juxtamembrane domain, which normally functions to inhibit receptor dimerization in the absence of ligand. Exon 13 mutations (point mutations) occur in the kinase I domain, and may lead to spontaneous receptor homodimerization. Exon 17 mutations occur in the activation loop; the mechanism of activation is unclear.

Among the minority of GISTs that lack detectable *KIT* mutations, nearly half harbor intragenic activating mutations in the gene encoding the related receptor tyrosine kinase, *PDGFRA*. These were identified after a series of elegant immunoprecipation and immunoblotting experiments revealed that PDGFRA was activated in *KIT*-wild-type GIST tumor lysates [26]. Mutations occur in exons 12 (juxtamembrane region) or 18 (activation loop), and lead to constitutive activation of the receptor. Tumors expressing KIT or PDGFRA oncoproteins appear to be indistinguishable with respect to activation of downstream signaling intermediates and cytogenetic changes associated with tumor progression [25, 26].

In terms of the prevalence of the various *KIT* and *PDGFRA* mutations, one comprehensive analysis of 127 patients with GIST enrolled onto a phase II clinical study of imatinib found *KIT* and *PDGFRA* mutations in 88.2 percent and 4.7 percent of GISTs, respectively [27]. In total, 66.9 percent of cases harbored exon 11 *KIT* mutations, 18.1 percent exon 9 *KIT* mutations, and 3.9 percent *PDGFRA* 18 mutations. The remaining mutations occurred in *KIT* exons 13 or 17 or *PDGFRA* exon 12; 7.1 percent of patients had no detectable mutations in either gene. Interestingly, disease response rates to imatinib vary according to tumor mutation type, ranging from 65–87 percent in tumors bearing *KIT* exon 11 mutations vs 34-48 percent in tumors with *KIT* exon 9 mutations [27]. These differences translate into significantly longer event-free and overall survival rates for patients harboring exon 11 mutations.

Resistance to Kinase Inhibitors in GIST

Primary Resistance

Despite harboring *KIT* or *PDGFRA* mutations, some GISTs are resistant to treatment with imatinib. These are more likely to harbor *KIT* exon 9 mutations or *PDGFRA* D842V mutations (Figure 44.2b) [28]. Additional molecular mechanisms underlying primary resistance remain to be elucidated.

Acquired Resistance

In patients whose disease demonstrates initial response, secondary or acquired resistance invariably develops after a median time of about 2 years [29]. Analyses of tumor specimens biopsied after disease response and progression have revealed that resistant tumor cells harbor second-site kinase domain mutations in exons 13, 14, or 17 of *KIT*, or exon 18 of *PDGFRA*, respectively (Figure 44.2b) [28, 30, 31]. About one-fourth of secondary KIT mutations result in the same amino acid substitution, changing a threonine to isoleucine at position 670 (encoded by exon 14). This mutation is analogous to a mutation that affects the corresponding residue (T315I) in ABL, found in CML patients who develop acquired resistance to imatinib [32]. Replacement of this threonine residue by isoleucine is thought to lead to bulky steric clash with the kinase inhibitor but not ATP [33]. Because of its key location at the entrance to a hydrophobic pocket in the back of the ATP binding cleft, this threonine is an important determinant of inhibitor specificity in protein kinases and has been therefore called a "gatekeeper" residue. The remaining mutations in exons 13 and 17 are predicted to destabilize the

inactive conformation of KIT, thus preventing proper binding of imatinib [34].

More recently, an alternative mechanism of acquired resistance has been identified, involving a "kinase switch" from dependency on KIT to a different tyrosine kinase, AXL [35]. This mechanism was elucidated by the study of imatinib-sensitive GIST cell lines selected *in vitro* over time for resistance to the drug. A "kinase switch" has also been observed in acquired resistance to EGFR inhibitors in lung cancer (see below), suggesting that this may be a common mechanism by which tumor cells adapt to inhibition of a kinase upon which they are dependent for survival.

Importantly, soon after second-site kinase domain mutations were found to mediate acquired resistance to imatinib in GIST, sunitinib malate (SU11248, Sutent ™; Pfizer, New-York, NY) was identified as a potential agent to overcome resistance. Sunitinib is an oral quinazoline (*N*-[2-diethylamino)ethyl]-5-*Z*)-(5-fluoro-1,2-dihydro-2-oxo-3*H*-indol-3-ylidine)methyl]-2,4-dimethyl-1*H*-pyrrole-3-carboxamide; Figure 44.1) that inhibits KIT, PDGFR, and all three isoforms of the vascular endothelial growth factor receptor (VEGFR-1, VEGFR-2, and VEGFR-3) in the nanomolar range (Figure 44.3) [36]. Although sunitinib binds the same ATP binding site as imatinib, both on KIT and PDGFR, these two molecules are members of different chemical classes and bear different binding characteristics with different affinities for the respective receptors. As second-line treatment after failure to imatinib, a phase III randomized, double-blind, placebo-controlled, multicenter international trial assessed the tolerability and anticancer efficacy of sunitinib in patients (*n*=312) with advanced GIST who were resistant to or intolerant of previous treatment with imatinib. Patients were randomized in a 2 : 1 ratio to receive sunitinib or placebo; the trial was unblinded early when a planned interim analysis showed significantly longer time to tumor progression with sunitinib (27.3 weeks vs 6.4 weeks, HR 0.33; 95% CI 0.23-0.47; p< 0.0001) [36]. Further studies showed that exon 13 and 14 secondary *KIT* mutations were sensitive *in vitro* to sunitinib [37]. Based on this trial, in January 2006 the FDA approved sunitinib for this indication. Response rates to sunitinib were significantly better in patients whose tumors bore exon 9 mutations (37 percent vs 5 percent in case of exon 11 mutations), the opposite result to that with imatinib. This suggests that GISTs that arise from different types of mutations may best be treated with different kinase inhibitors [28]. The additional inhibition of VEGFRs by sunitinib may also play a significant role in tumor responses, as GISTs are highly vascularized tumors [16].

KINASE INHIBITORS IN NON-SMALL CELL LUNG CANCER

As discussed above, small molecule kinase inhibitors were rationally developed for use in GIST, based upon an

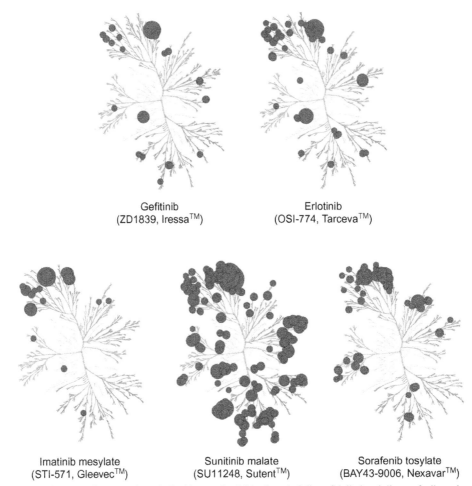

Gefitinib
(ZD1839, Iressa™)

Erlotinib
(OSI-774, Tarceva™)

Imatinib mesylate
(STI-571, Gleevec™)

Sunitinib malate
(SU11248, Sutent™)

Sorafenib tosylate
(BAY43-9006, Nexavar™)

FIGURE 44.3 Kinase interaction maps demonstrating relative kinase selectivities for erlotinib, gefitinib, imatinib, sorafenib, and sunitinib (see [137] for details).

understanding of the underlying pathophysiology of the disease. By contrast, the development of kinase inhibitors in NSCLC revealed kinase biology that was highly unexpected.

Lung Cancer Background

Cancer of the lung is the most frequent cause of cancer-related death worldwide, accounting for more than 1 million deaths per year [38]. Despite recent advances in the treatment of this disease, the overall 5-year survival in the United States remains only 15 percent, highlighting the need for novel treatment strategies.

Lung cancers are currently classified into two major groups depending on histology: small cell lung cancer, and non-small cell lung cancer (NSCLC). The latter is comprised of three different subtypes: adenocarcinoma, squamous cell carcinoma, and large cell carcinoma. The incidence of the adenocarcinoma subtype has been rising, and now accounts for >50 percent of all cases of lung cancer

[39]. The main risk factor for lung cancer is cigarette smoking; however, nearly half of lung cancers occur in patients who quit smoking more than a year prior to diagnosis (i.e., in "former" as opposed to "current" smokers) and 10 percent of lung cancers occur in "never smokers" (i.e., individuals who smoked less than 100 cigarettes in a lifetime). Sixty percent of patients are diagnosed at incurable stages.

Standard treatment for metastatic lung cancer involves empiric cytotoxic chemotherapy, which appears to have reached a plateau in terms of efficacy. In a landmark clinical trial involving 1207 patients randomized to one of four treatment arms using "modern doublets" (i.e., cisplatin/gemcitabine, cisplatin/paclitaxel, cisplatin/docetaxel, carboplatin/paclitaxel), the median overall survival was 8 months regardless of the doublet used [40]. Recently, addition to chemotherapy of the "angiogenesis inhibitor" bevacizumab – a monoclonal antibody against the vascular endothelial growth factor – conferred an additional 2-month survival benefit to patients with non-squamous NSCLC over those receiving chemotherapy alone [41].

Development of EGFR Tyrosine Kinase Inhibitors in Lung Cancer

Contrary to the development of kinase inhibitors in GISTs, EGFR inhibitors were developed in lung cancer originally along clinical lines. The dysregulation of EGFR activity in some cancers coupled with the limited activity of cytotoxic chemotherapies in general spurred the development in the 1990s of small molecule agents to inhibit EGFR kinase activity. Investigation of the structure and catalytic mechanism of the wild-type EGFR tyrosine kinase led to the identification of a new structural class of anilinoquinazoline tyrosine kinase inhibitors [42]. One structural derivative, 4-quinazolinamine, N-(3-chloro-4-fluorophenylamino)-7-methoxy-6-3-4-morpholinyl)-propoxy], molecular weight 446.9, eventually became gefitinib (ZD1839; Iressa ™; AstraZeneca, Macclefields, UK; Figure 44.1). A different quinazoline derivative - [6,7-bis (2-methoxy-ethoxy)-quinazolin-4-yl]-[3-ethylphenyl]-amine - became erlotinib (OSI-774 ; Tarveva ™ ; OSI Pharmaceuticals, Melville, NY, and Genentech, South San Francisco, CA; Figure 44.1). Both agents are orally available, reversible, ATP-competitive inhibitors of the EGFR tyrosine kinase (Figure 44.3). Gefitinib was developed first.

In preclinical studies of cell lines and human tumor xenografts, gefitinib produced growth inhibition in a variety of solid tumor types, including lung, prostate, breast, colon, and ovarian cancers [43]. Subsequently, in four phase I studies of gefitinib in unselected patients, durable clinical benefit was seen in advanced cancers of multiple tumor types [44–47]. Although no complete responses were seen in 221 patients studied, unanticipated dramatic radiographic regressions (within several days to weeks) were noted in 10 of 100 heavily pretreated NSCLC patients.

The promising activity seen in NSCLC led to two multi-center phase II trials: the Iressa Dose Evaluation in Advanced Lung Cancer (IDEAL)-1 trial in Japan and Europe with 210 patients, and the IDEAL-2 trial in the US with 221 patients [48, 49]. In both trials, patients were required to have received prior platinum-containing chemotherapy regimens. Overall response rates in IDEAL-1 and -2 were 18 percent and 10 percent, respectively. Based upon these data involving "surrogate endpoints", gefitinib received approval in Japan and South Korea in July 2002 as second-line chemotherapy for advanced NSCLC. In the US, the FDA approved gefitinib in May 2003, under the Agency's accelerated approval program, for the treatment of patients with NSCLC who had failed two or more courses of chemotherapy.

After the approval of gefitinib in 2003, AstraZeneca conducted an FDA-required study (IRESSA Survival Evaluation in Lung cancer (ISEL)) to compare the efficacy of gefitinib vs placebo in patients ($n=1692$) with refractory NSCLC [50]. A statistically significant difference in survival was not observed (5.6 vs 5.1 months; HR 0.89, p=0.11). For this reason, and because alternatives

to gefitinib by then existed for patients with NSCLC (e.g., docetaxel, pemetrexed, and erlotinib; see below), the FDA withdrew gefitinib from the US market in June 2005. However, preplanned subset analyses did demonstrate a significant survival benefit in patients who had never smoked, and in patients of Asian origin [51]. The drug remains widely used in East Asia.

While the phase III trial of gefitinib failed to show a survival benefit, the phase III trial of erlotinib (NCIC BR.21) did [52]. This trial randomized 731 patients to either drug or placebo after first- or second-line chemotherapy. The overall response to erlotinib was 9 percent, and the overall survival was 6.7 months for erlotinib versus 4.7 months for placebo (p=0.001). Erlotinib received approval by the FDA in November 2004. Why gefitinib failed and erlotinib succeeded in their respective phase III trials remains a subject of intense debate. Major explanations include differences in (1) drug potency (gefitinib was dosed at one-third its maximum tolerated dose (MTD), while erlotinib was dosed at its MTD), (2) drug pharmacokinetics, (3) clinical trial design (ISEL patients had more treatment-refractory patients than BR.21), and (4) potential off-target activity of the respective kinase inhibitors [50].

Preclinical models demonstrated synergy between EGFR TKIs and cytotoxic chemotherapy [53, 54]. Thus, both gefitinib and erlotinib were also evaluated as first-line therapy for NSCLC, given continuously with cytotoxic chemotherapy in four separate phase III trials: Iressa NSCLC Trials Assessing Combination Therapy (INTACT)-1 ($n=1093$) and -2 ($n=1037$) [55, 56], and TRIBUTE ($n=1059$) and TALENT ($n=1172$) (for Tarceva) [57, 58]. No survival benefit or improvements in response rates were noted with the addition of TKI to chemotherapy in either trial. Thus, concurrent use of an EGFR TKI with chemotherapy in the first-line setting remains investigational.

EGFR MUTATIONS IN LUNG CANCER

When gefitinib received FDA approval in May 2003, erlotinib studies were still ongoing, and the specific target(s) of these drugs in human tumors were unknown. Three groups took different approaches leading to the identification of EGFR kinase domain mutations in NSCLC. All groups were motivated by the observation that tumors could be "addicted" to signaling from aberrant TKs [59, 60] and other oncogenes [61–63], as discussed above. One approach hypothesized that patients who had striking responses to gefitinib had somatic mutations in EGFR that would indicate the essential role of the EGFR signaling pathway in the tumor [64]. Another approach involved high-throughput re-sequencing in lung tumors of exons encoding the kinase domains of receptor tyrosine kinases, to determine whether mutations in specific kinases played a causal role in NSCLC [65]. This group followed that of Bardelli and colleagues, who showed the feasibility of determining the mutational status of all TKs within a tumor type [66]. Similar to Bardelli and colleagues, we initiated an effort to perform mutational profiling on all tyrosine kinases, choosing to start with EGFR first, since it was the putative target of gefitinib and erlotinib [67]. We also reasoned that further careful study of tumors from patients with clinical characteristics predictive of response – a strategy of "clinical enrichment" – would accelerate the identification of either the "true" molecular targets of gefitinib and erlotinib or other molecular predictors of response [68]. Thus, we chose initially to focus on tumors from "never smokers" that demonstrated marked clinical or radiographic responses on drug and who had adenocarcinoma histology, particularly a subtype called bronchioloalveolar carcinoma (BAC) [69]. All three studies demonstrated that somatic mutations in the EGFR kinase domain were present in a subset of NSCLCs, and that such mutations were associated with increased sensitivity to gefitinib or erlotinib.

EGFR kinase domain mutations are almost exclusively found in NSCLCs [70–74]. Many types of mutations have been reported, but there thus far are only four types of drug-sensitive mutations, validated from *in vitro* studies [75] and/or from actual tumor responses in human patients. These are point mutations in exons 18 (G719A/C) and 21 (L858R and L861Q), and in-frame deletions in exon 19, most of which eliminate four amino acids, LREA, just downstream of a critical lysine residue at position 745 (Figure 44.2a). (Note that EGFR has two numbering systems. The first denotes the initiating methionine in the signal sequence as amino acid -24; the second, used here, denotes the methionine as amino acid +1.) The most common of these four drug-sensitive mutations are exon 19 deletions and the exon 21 L858R substitution, together representing 85–90 percent of *EGFR* mutations in NSCLC (Figure 44.2a) [76]. Thus far, four kinase domain mutations are associated with drug resistance: two exon 19 point mutations (L747S and D761Y), an exon 20 point mutation (T790M) (see section on acquired resistance) and exon 20 insertions (e.g., D770_N771insNPG). Within lung cancers, *EGFR* kinase domain mutations are more common in adenocarcinomas, East Asians, women, and never smokers (reviewed in [76]). Mutations in *EGFR* may be more common in women, since the majority of never smokers are women [77]. These characteristics had been previously noted as clinical predictors of response to gefitinib and erlotinib [48, 49, 68]. Kinase domain mutations are usually somatic, although a family with rare germ-line mutations in *EGFR* has been identified [78]. Mutations outside the exons encoding the kinase domain are uncommon in lung cancers [64, 65, 67], but the EGFRvIII mutations found in gliomas have been identified in some NSCLCs, especially in those with squamous cell histology [79, 80].

EGFR exon 19 deletion and L858R mutants confer ligand-independent activation and prolonged receptor kinase activity after ligand stimulation [64, 65, 81]. Such mutations are also sufficient for oncogenic transformation. *In vitro* work has shown that selected mutations in *EGFR* (exon 18 G719S, exon 19 deletion, exon 21 L858R, and exon 20 insertion) can transform both fibroblasts and lung epithelial cells [82, 83]. Additionally, tetracycline-regulatable mouse model systems indicate that expression of either *EGFR* exon 19 deletions or *L858R* alleles in mouse lung epithelia leads to formation of tumors analogous to human lung cancers [84, 85].

Kinetic analysis of the purified intracellular domains of the L858R mutant and a deletion mutant reveals that both mutants exhibit a higher K_m for ATP (i.e. a decreased affinity for ATP) and a lower K_i for erlotinib (i.e. increased affinity for drug) relative to wild-type receptor [83]. Separate *in vitro* kinase activity assays show that the catalytic efficiency (k_{cat}/K_M) of the L858R mutant form of the kinase domain is ~20-fold higher than that for the wild-type kinase domain, suggesting that while the wild-type kinase domain is autoinhibited, the L858R mutant is constitutively active, probably because the L→R amino acid substitution destabilizes the inactive EGFR conformation [86, 87]. The structural basis for the enhanced sensitivity of the deletion mutants is not apparent from previously published reports of crystal structure data of EGFR TKIs with the kinase domain of EGFR [88, 89], and no crystal structures have yet been elucidated.

Similar to GIST, where patients whose tumors bear exon 11 and 9 mutations have different clinical outcomes to imatinib, different mutations in *EGFR* may confer different tumor activation profiles which lead to variations in both natural history and clinical course after treatment with erlotinib or gefitinib. In one study, in NSCLC patients treated with surgery alone, patients with *EGFR* point mutations ($n=31$) had a prolonged overall survival when compared to patients with exon 19 deletions ($n=31$) [90]. In contrast, retrospective data from our group [91] and others [92] suggest that after treatment with gefitinib or erlotinib, patients with *EGFR* exon 19 deletions have a longer overall survival when compared to patients with *EGFR* L858R (34 months vs 8 months, log-rank p=0.01). The molecular basis for this observation remains to be elucidated, although kinetic analyses of EGFR mutant proteins suggest that the off-rate for erlotinib may be slower for the deletion mutant, as compared to the L858R mutant, thus prolonging the duration of erlotinib binding to the deletion mutant [83]. Remarkably, a 34-month median survival far exceeds the norm for patients treated with standard chemotherapy (~10-12 months). Patients with *EGFR* mutant tumors may also have a better prognosis than those with wild-type tumors [93].

Drug sensitivity may be associated with both *EGFR* mutation and amplification. Several groups have investigated the predictive value of *EGFR* amplification in patients treated with gefitinib or erlotinib on clinical trials

[94, 95]. In these studies, patients with amplification or polysomy of *EGFR* were more likely to respond to erlotinib or gefitinib as compared to patients with normal *EGFR* copy number. Patients with amplification or high polysomy also had longer median time to progression and overall survival. However, confounding these results, in most studies, amplification of *EGFR* has been associated with somatic mutation in *EGFR* [96]. Thus, whether amplified wild-type *EGFR* alone contributes to lung cancer oncogenesis and susceptibility to erlotinib and gefitinib remains to be established. In the absence of ligand, wild-type *EGFR* is not transforming in mouse fibroblasts or bronchial epithelial cells [75, 97].

Resistance to EGFR Kinase Inhibitors in NSCLC

Primary Resistance

Dramatic radiographic responses to gefitinib and erlotinib are only observed in a minority of patients. Since the majority of patients do not respond dramatically to gefitinib or erlotinib, and since some patients whose tumors are reported to have wild-type EGFR do respond to treatment, we and others investigated whether we could find molecular markers predictive of primary resistance to these drugs. As a first step, we examined the status of *KRAS* (exon 2) mutations in drug-sensitive vs drug-refractory tumors, since (1) KRAS is a known downstream signaling molecule in the EGFR pathway, and (2) *KRAS* mutations had previously been reported to occur in 15–30 percent of lung adenocarcinomas. Our analysis of 60 tumors revealed that *EGFR* mutations were found only in sensitive tumors, while *KRAS* mutations were found only in resistant tumors [98]. Subsequent prospective studies have further verified these findings [99, 100]. Since *EGFR* and *KRAS* mutations are, with some rare exceptions, mutually exclusive [98], *KRAS* mutation testing in conjunction with *EGFR* mutation testing can be used to help guide treatment decisions regarding the use of gefitinib or erlotinib.

Acquired Resistance

Despite an initial response to *EGFR* tyrosine kinase inhibitors, patients whose tumors harbor drug-sensitive *EGFR* mutations rarely achieve a complete radiographic or pathologic response, and their disease usually begins to grow again after about 12 months. Analyses of tumor tissue from patients with such acquired resistance have revealed the presence of additional second-site *EGFR* mutations (Figure 44.2b) [101–103]. The major lesion identified to date is an *EGFR* T790M mutation (exon 20), which has been reported in about half of patient tumors after disease progression [101–105], and is analogous to mutations seen in acquired resistance to imatinib in GIST and CML (Figure

44.2b). Like T670I in *KIT*, the T790M mutation in *EGFR* was predicted to block binding of erlotinib or gefitinib to the kinase ATP binding pocket. However, recently, it has been suggested that the amino acid substitution causes drug resistance by increasing the affinity for ATP [106]. Other second-site mutations implicated in acquired resistance include L747S (exon 19) [107] and D761Y (exon 19) (Figure 44.2b) [108].

Another mechanism of acquired resistance may involve a "kinase switch", as amplification of the gene encoding the MET receptor tyrosine kinase has been implicated in two recent separate studies. One group isolated gefitinib-resistant clones from an *EGFR* mutant lung cancer cell line and found that the resistant cells displayed amplification of *MET* and maintained activation of ERBB3/PI3K/Akt signaling in the presence of gefitinib. Inhibition of MET signaling in these cells restored their sensitivity to gefitinib [109]. Our group used array-based comparative genomic hybridization (aCGH) to compare genomic profiles of *EGFR* mutant tumors from untreated patients with those from patients with acquired resistance [110]. Among three loci demonstrating copy number alterations specific to the acquired resistance set, one contained the *MET* proto-oncogene. Collectively, *MET* amplification has been found in about 20 percent of samples from patients with acquired resistance, with or without EGFR T790M mutations. Notably, *MET* encodes a heterodimeric transmembrane receptor tyrosine kinase composed of an extracellular α-chain disulfide bonded to a membrane-spanning β chain [111, 112]. Binding of the receptor to its ligand, hepatocyte growth factor/scatter factor, induces receptor dimerization, triggering conformational changes that activate MET tyrosine kinase activity. MET activation can have profound effects on cell growth, survival, motility, invasion, and angiogenesis [113].

Understanding the basis for acquired resistance has led to the identification of agents that may overcome acquired resistance. Currently, a number of "second-generation" EGFR inhibitors are under clinical investigation specifically for patients with secondary resistance (HKI-272, Wyeth, Pearl River, NY; XL647, Exelixis, San Francisco, CA; BIBW2992, Boehringer-Ingelheim, Ingelheim, Germany). MET kinase inhibitors are in phase I/II development, and are also being considered for treatment of this cohort of patients.

KINASE INHIBITORS IN RENAL CELL CARCINOMA

The success of kinase inhibitors in RCC was the culmination of a third paradigm of drug development. Here, while the exact mechanisms underlying tumor responses remain under investigation, an existing body of molecular knowledge rapidly provided a framework to comprehend why a certain class of drugs was effective.

Renal Cell Carcinoma Background

Cancer of the kidney and renal pelvis is the seventh leading malignant condition among men and the ninth among women, accounting for ~3 percent of all cancers in the US annually [38]. Although more than 75 percent of patients initially present with early-stage resectable tumours, RCC recurs in about one-third of patients treated for localized disease. Median survival for patients with metastatic disease is about 13 months [114].

Renal carcinomas arise from the proximal renal tubular epithelium. At the histological level, RCCs are comprised of five main cellular types: clear cell, papillary, chromophobe, oncocytoma, and collecting duct. Clear-cell carcinomas constitute 85 percent of kidney cancers. Papillary renal carcinoma makes up approximately 10 percent of all kidney cancers, with the remainder comprised of chromophobe, collecting duct, and miscellaneous histologic types. Renal oncocytoma is considered to be a predominantly benign lesion [114].

Renal carcinoma occurs in both sporadic and hereditary forms. At least four forms of hereditary renal carcinoma are recognized: von Hippel-Landau Syndrome, hereditary papillary renal carcinoma, Birt-Hogg-Dubé Syndrome, and hereditary clear-cell renal carcinoma. VHL patients harbor germline mutations in the *VHL* tumor suppressor gene and develop clear-cell RCC [115]. Sporadic (non-hereditary) clear-cell RCCs also harbor VHL gene mutations [116]. Papillary RCC is attributed to germline mutations in the tyrosine kinase domain of the *MET* gene [117]. Genes responsible for the other hereditary forms of RCC remain to be identified.

Development of Kinase Inhibitors in RCC

In the twentieth century, three decades of extensive clinical investigation identified only two agents – non-cytotoxic chemotherapies – that result in improved survival for patients with metastatic renal cell carcinoma [118]. High-dose intravenous interleukin-2 administered in an intensive care unit setting demonstrated a 14 percent partial or complete response rate [119]. Two phase III trials comparing interferon-α therapy with vinblastine or medroxyprogesterone showed a modest survival benefit for interferon α [120, 121]. Thus, standard first-line treatment involved cytokine therapy. No effective treatments were available for patients whose disease progressed after initial response, or who did not respond to cytokine treatment.

Concurrent with the systematic testing of various systemic agents against RCC, studies of RCCs associated with VHL syndrome provided mechanistic insights into the etiology of the disease. An important observation was that mutations or deletions in the *VHL* gene (found in both hereditary and sporadic forms) stimulate hypoxia by preventing degradation of the hypoxia-inducible factor 1α (HIF-1A) [122, 123]. This in turn leads to unregulated expression of hypoxia-inducible genes, including *VEGF* and *PDGF*

[122]. These secreted growth factors then bind to VEGFRs and PDGFRs on the surface of endothelial cells and vascular pericytes, respectively, promoting angiogenesis and stimulating cell migration, proliferation, and survival. In 2003, VEGF was validated as a clinically relevant target in RCC in a randomized, double-blind, phase II trial comparing the effect of placebo with the anti-VEGF antibody, bevacizumab (rhuMAb-VEGF, Avastin ™; Genentech, San Francisco, CA) in patients with clear-cell RCC previously treated or intolerant of IL-2 therapy. Bevacizumab significantly prolonged time to progression [124].

Subsequently, multiple kinase inhibitors that target receptors for VEGF and PDGF were found to have antitumor activity in RCC. However, during the drug development process for most of these agents, RCC was not the first tumor type considered. For example, sorafenib tosylate (BAY43-9006, Nexavar ™; Bayer, West Haven, CT) is an oral kinase inhibitor (Figure 44.1), originally developed to target RAF1 but found to inhibit a number of kinases in the nanomolar range, including wild-type BRAF, mutant BRAF^{V600E}, VEGFR-1, VEGFR-2, VEGFR-3, PDGFR, FLT-3, and KIT (reviewed in [125]; Figure 44.3). Based on phase I data showing that the primary benefit of sorafenib was disease stabilization, a phase II study was conducted using a novel randomized discontinuation design in which all patients received study drug (sorafenib) for an initial run-in period, followed by random assignment of patients with disease stabilization to either the study drug or placebo [126]. The trial was initially designed to focus on patients with colorectal cancer, while allowing patients with other tumor types to enroll. When responses were observed in RCC but not colorectal cancer, recruitment was directed to RCC. A total of 502 patients were enrolled, including 202 patients with metastatic refractory RCC. In this population, sorafenib demonstrated significant disease-stabilizing activity [126]. Subsequently, in a phase III randomized, double-blind, placebo-controlled trial of sorafenib in patients (n=903) with clear-cell RCC in whom previous therapy had failed, the drug prolonged progression-free survival (5.5 months vs 2.8 months, HR 0.44; 95% CI 0.35-0.55; p<0.01) [127]. Sorafenib was approved by the FDA in December 2005 for the second-line treatment of advanced clear-cell RCC.

Sunitinib, as discussed above, is another highly potent, selective competitive inhibitor of VEGFR1-3 and PDGFR-α/β (Figure 44.3) [125]. In a phase I trial, the drug demonstrated evidence of antitumor activity in a subset of patients with metastatic RCC [128]. The follow-up phase II study, for patients (n=63) with metastatic RCC and progression on first-line cytokine therapy, demonstrated a 40 percent response rate, with a median time to progression of 8.7 months [129]. Based upon this study, sunitinib received a priority review by the FDA and received "accelerated approval" in less than 6 months in January 2006. A separate phase II trial involving another 106 patients specifically with clear-cell RCC [130] further validated the original findings. Soon thereafter, in a phase III trial involving 750 patients with clear-cell RCC, sunitinib was compared with interferon-α in the first-line setting [131]. Progression-free survival was longer (11 vs 5 months; HR 0.42, 95% CI 0.32-0.54, p<0.001), and response rates were higher (31 percent vs 6 percent, p<0.0001) in patients who received sunitinib than in those receiving interferon-α. Based on this trial, the FDA converted the approval of sunitinib from accelerated approval to regular approval for first-line treatment of advanced RCC in February 2007.

Currently, the relative benefit of sorafenib and sunitinib for patients with metastatic RCC has not been compared directly. Moreover, despite the preclinical data linking *VHL* mutations and RCC to angiogenesis, the actual molecular basis for the activity of these agents in RCC is unclear. For example, investigators have been unable to correlate disease responses with specific types of *VHL* mutations [132] or blood levels of VEGF [133]. Not surprisingly, then, the molecular basis of acquired resistance to these agents is not yet understood. Since sunitinib and sorafenib inhibit multiple kinases, including ones potentially involved in the tumor microenvironment rather than within tumor cells itself, it is currently unclear whether RCCs are "addicted" to specific aberrant kinases for survival in the same manner as GISTs and *EGFR* mutant NSCLCs.

KINASE INHIBITORS IN OTHER SOLID TUMORS

Above, we have highlighted the first three solid tumors treated effectively with small molecule kinase inhibitors. However, already in the past several years, kinase inhibitors have shown promise in other solid tumors (Table 44.1). For example, in anti-HER2 antibody (trastuzumab)-resistant, metastatic HER2-positive breast tumors [134], the dual EGFR/HER2 inhibitor lapatinib (GW572016, Tykerb™, Glaxo-Wellcome, London, UK) has been shown, compared to placebo, to delay time to disease progression. Notably, the drug was given concurrently with the oral antimetabolite, capecitabine. The multi-kinase inhibitor sorafenib was approved by the FDA in 2007 for the treatment of unresectable hepatocellular carcinoma (HCC), following a phase III trial that showed a 44 percent improvement in overall survival compared to placebo [135]. Similar to RCC, the exact mechanism(s) of action is/are unclear. Finally, in medullary thyroid carcinomas, which often harbor mutations in the gene encoding the RET tyrosine kinase, the oral anilinoquinazoline RET inhibitor vandetanib (ZD6474, Zactima™, Astra-Zeneca) has shown promising antitumor activity [136] and was designated as an orphan drug by FDA in 2004 for treatment of this disease. Collectively, all of these data demonstrate that kinase inhibitors are likely to play an increasingly important role in the treatment of human solid tumor malignancies.

LESSONS LEARNED?

The clinical development of kinase inhibitors in GIST, NSCLC, and RCC provides three distinct examples of how this class of agents can be developed to treat solid tumors. Can lessons be drawn from these case histories?

Good Science Can Lead to Rapid and Rational Development of a Drug

The development of imatinib for GIST was based upon detailed understanding of the pathophysiology of the disease. The observation that *KIT* mutations caused GIST led to preclinical assessment of a KIT inhibitor (imatinib) against mutant KIT, which led to testing of the drug in a patient, and ultimately a clinical trial proving the drug's efficacy over existing therapies. The observation that some *KIT* wild-type patients responded to therapy spurred identification of *PDGFRA* mutations as a cause of GIST. Finally, an understanding of acquired resistance mechanisms led to the rapid development of a second kinase inhibitor for that setting. This experience indicates that clinical trials of promising new kinase inhibitors should, when possible, enrich for cohorts of patients likely to benefit using criteria based upon molecular mechanisms of disease.

Despite What We Think We Know, We Don't Understand Everything

Despite the best laid plans, clinical trials can often unexpectedly reveal biological insights heretofore unrecognized. This notion is exemplified by the surprising dramatic responses to EGFR inhibitors seen in NSCLC patients across four phase I trials, and the subsequent discovery of *EGFR* mutations in a subset of patients with NSCLC after multiple advanced trials had already been completed. A similar theme was experienced with the demonstration that sorafenib offers significant benefit in RCC, while the drug was originally developed for colon cancer. These occurrences suggest that while we should apply molecular criteria as best we can to enrich for patients likely to benefit from a kinase inhibitor, early trials should still consider enrolling a broad spectrum of patients with various diseases. Clinical researchers then need to make astute observations to steer the development of promising agents in the right direction.

Kinase Inhibitors Have Off-Target Activities Which Can be Exploited

In the early phase of kinase inhibitor development, only handfuls of kinases were tested using *in vitro* assays (see, for example, [137]). However, while large-scale *in vitro* kinase assays have not yet been established for all

518 kinases encoded in the human genome, recent high-throughput methods have begun to make it easier to assess a larger spectrum of drug target activity against kinases [137]. Now, for example, we know that gefitinib and erlotinib demonstrate relatively high specificity, while sunitinib and sorafenib do not (Figure 44.3). "Off-target activity" may lead to more side effects, but can also lead to alternative uses of drugs, just as imatinib was originally developed for use in CML to target ABL and then was developed in GIST to target mutant KIT and PDGFRA. In the future, investigators will need to understand the full spectrum of kinase inhibition of various compounds in order to mix and match the appropriate ones to the appropriate patients.

Solid Tumors Refractory to Conventional Treatments Nonetheless Harbor Novel Targets for Therapy

By the end of the twentieth century, treatments for GIST, NSCLC, and RCC appeared to have reached a therapeutic plateau with regard to existing therapies. However, for each disease, novel targeted agents were identified and shown to have proven clinical benefit over standard treatments, even in patients with heavily pre-treated disease. These studies indeed support an optimistic message with respect to new approaches for treating cancer, because they do demonstrate that tumors harboring complex genetic lesions nevertheless have an "Achilles heel" that can be identified and exploited [138]. In the future, as large-scale efforts to define the role of the kinome in human solid tumors come to fruition, the tasks will be to determine how many actual cancers have a kinase-based Achilles heel, and to identify the appropriate kinase inhibitors to treat them.

GLOSSARY

Hazard ratio

The hazard ratio in *survival analysis* is the effect of an exploratory? variable on the hazard or risk of an event. Hazard ratio can be considered as an estimate of *relative risk*, which is the risk of an event (or of developing a disease) relative to exposure. Relative risk is a ratio of the probability of the event occurring in the exposed group versus the control (non-exposed) group.

95% Confidence Interval

A confidence interval is an interval estimate of a population parameter. Instead of estimating the parameter by a single value, an interval of likely estimates is given.

Phase I trial

Phase I trials are conducted in a small number of patients, primarily to determine the appropriate and safe dose of a new drug or drug combination. Antitumor activity is usually a secondary endpoint of a phase I trial.

Phase II trial

Phase II trials are conducted in a larger number of patients to determine if a new drug or drug combination has efficacy in a given population.

Phase III trial

Phase III trials compare a new treatment (as determined from phase II trials) versus the standard of care. These trials usually involve hundreds of patients in order to determine with statistical significance if the new treatment is indeed superior to the standard of care.

ACKNOWLEDGEMENTS

N. Girard is recipient of the Lavoisier Grant from the French Ministry of Foreign and European Affairs. W. Pao acknowledges funding from the Doris Duke Charitable Foundation, the Thomas G. Labrecque Foundation, Joan's Legacy: The Joan Scarangello Foundation to Conquer Lung Cancer (WP), the NIH (K08-CA097980, R01-CA121210, and P01 CA129243), the Jodi Spiegel Fisher Cancer Foundation, The Carmel Hill Fund, and funds from the Miner Family. We thank Ross Levine for critical reading of the manuscript.

REFERENCES

1. De Vita JT, Hellman SP, Rosenberg SA, editors. *Cancer: Principles and Practice of Oncology.* 7th edn Philadelphia, PA: Lippincott Williams and Wilkins; 2005. p. 307–421.

2. Sefton BM, Hunter T, Beemon K. Product of in vitro translation of the Rous sarcoma virus src gene has protein kinase activity. *J Virol* 1979;**30**:311–18.

3. Druker BJ, Talpaz M, Resta DJ, Peng B, Buchdunger E, Ford JM, Lydon NB, Kantarjian H, Capdeville R, Ohno-Jones S, Sawyers CL. Efficacy and safety of a specific inhibitor of the BCR-ABL tyrosine kinase in chronic myeloid leukemia. *N Engl J Med* 2001;**344**:1031–7.

4. De Vita JT, Hellman S, Rosenberg SA, editors. *Cancer: Principles and Practice of Oncology.* 7th edn. Philadelphia, PA: Lippincott Williams and Wilkins; 2005.

5. Felsher DW, Denis KA, Weiss D, Ando DT, Braun J. A murine model for B-cell lymphomagenesis in immunocompromised hosts: c-myc-rearranged B-cell lines with a premalignant phenotype. *Cancer Res* 1990;**50**:7042–9.

6. Chin L. Modeling malignant melanoma in mice: pathogenesis and maintenance. *Oncogene* 1999;**18**:5304–10.

7. Fisher GH, Wellen SL, Klimstra D, Lenczowski JM, Tichelaar JW, Lizak MJ, Whitsett JA, Koretsky A, Varmus HE. Induction and apoptotic regression of lung adenocarcinomas by regulation of a K-Ras transgene in the presence and absence of tumor suppressor genes. *Genes Dev* 2001;**15**:3249–62.

8. Weinstein IB. Cancer. Addiction to oncogenes – the Achilles heel of cancer. *Science* 2002;**297**:63–4.

9. Ventura A, Kirsch DG, McLaughlin ME, Tuveson DA, Grimm J, Lintault L, Newman J, Reczek EE, Weissleder R, Jacks T. Restoration of p53 function leads to tumour regression in vivo. *Nature* 2007;**445**:661–5.

10. Manning G, Whyte DB, Martinez R, Hunter T, Sudarsanam S. The protein kinase complement of the human genome. *Science* 2002;**298**:1912–34.

11. Blume-Jensen P, Hunter T. Oncogenic kinase signalling. *Nature* 2001;**411**:355–65.

12. Yaish P, Gazit A, Gilon C, Levitzki A. Blocking of EGF-dependent cell proliferation by EGF receptor kinase inhibitors. *Science* 1988;**242**:933–5.

13. Huse M, Kuriyan J. The conformational plasticity of protein kinases. *Cell* 2002;**109**:275–82.

14. Wang L, Vargas H, French SW. Cellular origin of gastrointestinal stromal tumors: a study of 27 cases. *Arch Pathol Lab Med* 2000;**124**:1471–5.

15. Goettsch WG, Bos SD, Breekveldt-Postma N, Casparie M, Herings RMC, Hogendoorn PCW. Incidence of gastrointestinal stromal tumours is underestimated: Results of a nation-wide study. *Eur J Cancer* 2005;**41**:2868–72.

16. Mazur MT, Clark HB. Gastric stromal tumors. Reappraisal of histogenesis. *Am J Surg Pathol* 1983;**7**:507–19.

17. Beghini A, Tibiletti MG, Roversi G, Chiaravalli AM, Serio G, Capella C, Larizza L. Germline mutation in the juxtamembrane domain of the kit gene in a family with gastrointestinal stromal tumors and urticaria pigmentosa. *Cancer* 2001;**92**:657–62.

18. Dematteo RP, Heinrich MC, El-Rifai WM, Demetri G. Clinical management of gastrointestinal stromal tumors: before and after STI-571. *Hum Pathol* 2002;**33**:466–77.

19. Edmonson JH, Marks RS, Buckner JC, Mahoney MR. Contrast of response to dacarbazine, mitomycin, doxorubicin, and cisplatin (DMAP) plus GM-CSF between patients with advanced malignant gastrointestinal stromal tumors and patients with other advanced leiomyosarcomas. *Cancer Invest* 2002;**20**:605–12.

20. Hirota S, Isozaki K, Moriyama Y, Hashimoto K, Nishida T, Ishiguro S, Kawano K, Hanada M, Kurata A, Takeda M, Muhammad Tunio G, Matsuzawa Y, Kanakura Y, Shinomura Y, Kitamura Y. Gain-of-function mutations of c-kit in human gastrointestinal stromal tumors. *Science* 1998;**279**:577–80.

21. Huizinga JD, Thuneberg L, Klüppel M, Malysz J, Mikkelsen HB, Bernstein A. W/kit gene required for interstitial cells of Cajal and for intestinal pacemaker activity. *Nature* 1995;**373**:347–9.

22. Heinrich MC, Griffith DJ, Druker BJ, Wait CL, Ott KA, Zigler AJ. Inhibition of c-kit receptor tyrosine kinase activity by STI 571, a selective tyrosine kinase inhibitor. *Blood* 2000;**96**:925–32.

23. Joensuu H, Roberts PJ, Sarlomo-Rikala M, Andersson LC, Tervahartiala P, Tuveson D, Silberman S, Capdeville R, Dimitrijevic S, Druker B, Demetri GD. Effect of the tyrosine kinase inhibitor STI571 in a patient with a metastatic gastrointestinal stromal tumor. *N Engl J Med* 2001;**344**:1052–6.

24. Demetri GD, von Mehren M, Blanke CD, Van den Abbeele AD, Eisenberg B, Roberts PJ, Heinrich MC, Tuveson DA, Singer S, Janicek M, Fletcher JA, Silverman SG, Silberman SL, Capdeville R, Kiese B, Peng B, Dimitrijevic S, Druker BJ, Corless C, Fletcher CD, Joensuu H. Efficacy and safety of imatinib mesylate in advanced gastrointestinal stromal tumors. *N Engl J Med* 2002;**347**:472–80.

25. Corless CL, Fletcher JA, Heinrich MC. Biology of gastrointestinal stromal tumors. *J Clin Oncol* 2004;**22**:3813–25.

26. Heinrich MC, Corless CL, Duensing A, McGreevey L, Chen CJ, Joseph N, Singer S, Griffith DJ, Haley A, Town A, Demetri GD, Fletcher CD, Fletcher JA. PDGFRA activating mutations in gastrointestinal stromal tumors. *Science* 2003;**299**:708–10.

27. Heinrich MC, Corless CL, Demetri GD, Blanke CD, von Mehren M, Joensuu H, McGreevey LS, Chen CJ, Van den Abbeele AD, Druker BJ, Kiese B, Eisenberg B, Roberts PJ, Singer S, Fletcher CD, Silberman S, Dimitrijevic S, Fletcher JA. Kinase mutations and imatinib response in patients with metastatic gastrointestinal stromal tumor. *J Clin Oncol* 2003;**21**:4342–9.

28. Heinrich MC, Corless CL, Blanke CD, Demetri GD, Joensuu H, Roberts PJ, Eisenberg BL, von Mehren M, Fletcher CD, Sandau K, McDougall K, Ou WB, Chen CJ, Fletcher JA. Molecular correlates of imatinib resistance in gastrointestinal stromal tumors. *J Clin Oncol* 2006;**24**:4764–74.

29. Van Glabbeke M, Verweij J, Casali PG, Le Cesne A, Hohenberger P, Ray-Coquard I, Schlemmer M, van Oosterom AT, Goldstein D, Sciot R, Hogendoorn PC, Brown M, Bertulli R, Judson IR. Initial and late resistance to imatinib in advanced gastrointestinal stromal tumors are predicted by different prognostic factors: a European Organisation for Research and Treatment of Cancer-Italian Sarcoma Group-Australasian Gastrointestinal Trials Group study. *J Clin Oncol* 2005;**23**:5795–804.

30. Antonescu CR, Besmer P, Guo T, Arkun K, Hom G, Koryotowski B, Leversha MA, Jeffrey PD, Desantis D, Singer S, Brennan MF, Maki RG, DeMatteo RP. Acquired resistance to imatinib in gastrointestinal stromal tumor occurs through secondary gene mutation. *Clin Cancer Res* 2005;**11**:4182–90.

31. Wardelmann E, Thomas N, Merkelbach-Bruse S, Pauls K, Speidel N, Büttner R, Bihl H, Leutner CC, Heinicke T, Hohenberger P. Acquired resistance to imatinib in gastrointestinal stromal tumours caused by multiple KIT mutations. *Lancet Oncol* 2005;**6**:249–51.

32. Shah NP, Nicoll JM, Nagar B, Gorre ME, Paquette RL, Kuriyan J, Sawyers CL. Multiple BCR-ABL kinase domain mutations confer polyclonal resistance to the tyrosine kinase inhibitor imatinib (STI571) in chronic phase and blast crisis chronic myeloid leukemia. *Cancer Cell* 2002;**2**:117–25.

33. Modugno M, Casale E, Soncini C, Rosettani P, Colombo R, Lupi R, Rusconi L, Fancelli D, Carpinelli P, Cameron AD, Isacchi A, Moll J. Crystal structure of the T315I Abl mutant in complex with the aurora kinases inhibitor PHA-739358. *Cancer Res* 2007;**67**:7987–90.

34. Tamborini E, Pricl S, Negri T, Lagonigro MS, Miselli F, Greco A, Gronchi A, Casali PG, Ferrone M, Fermeglia M, Carbone A, Pierotti MA, Pilotti S. Functional analyses and molecular modeling of two c-Kit mutations responsible for imatinib secondary resistance in GIST patients. *Oncogene* 2006;**25**:6140–6.

35. Mahadevan D, Cooke L, Riley C, Swart R, Simons B, Della Croce K, Wisner L, Iorio M, Shakalya K, Garewal H, Nagle R, Bearss D. A novel tyrosine kinase switch is a mechanism of imatinib resistance in gastrointestinal stromal tumors. *Oncogene* 2007;**26**:3909–19.

36. Demetri GD, van Oosterom AT, Garrett CR, Blackstein ME, Shah MH, Verweij J, McArthur G, Judson IR, Heinrich MC, Morgan JA, Desai J, Fletcher CD, George S, Bello CL, Huang X, Baum CM, Casali PG. Efficacy and safety of sunitinib in patients with advanced gastrointestinal stromal tumour after failure of imatinib: a randomised controlled trial. *Lancet* 2006;**368**:1329–38.

37. Heinrich MC, Maki RG, Corless CL, Antonescu CR, Fletcher JA, Fletcher CD, Huang X, Baum CM, Demetri GD. Sunitinib (SU) response in imatinib-resistant (IM-R) GIST correlates with KIT and PDGFRA mutation status. *Proc Am Soc Clin Oncol* 2006. Abstract 9502.

38. American Cancer Society. *Cancer Facts and Figures 2007*. Atlanta, GA: American Cancer Society; 2007.

39. Gabrielson E. Worldwide trends in lung cancer pathology. *Respirology* 2006;**11**:533–8.

40. Schiller JH, Harrington D, Belani CP, Langer C, Sandler A, Krook J, Zhu J, Johnson DHEastern Cooperative Oncology Group. Comparison

of four chemotherapy regimens for advanced non-small-cell lung cancer. *N Engl J Med* 2002;**346**:92–8.

41. Sandler A, Gray R, Perry MC, Brahmer J, Schiller JH, Dowlati A, Lilenbaum R, Johnson DH. Paclitaxel-carboplatin alone or with bevacizumab for non-small-cell lung cancer. *N Engl J Med* 2006;**355**:2542–50.

42. Ward WH, Cook PN, Slater AM, Davies DH, Holdgate GA, Green LR. Epidermal growth factor receptor tyrosine kinase. Investigation of catalytic mechanism, structure-based searching and discovery of a potent inhibitor. *Biochem Pharmacol* 1994;**48**:659–66.

43. Sirotnak FM. Studies with ZD1839 in preclinical models. *Semin Oncol* 2003;**30**:12–20.

44. Baselga J, Rischin D, Ranson M, Calvert H, Raymond E, Kieback DG, Kaye SB, Gianni L, Harris A, Bjork T, Averbuch SD, Feyereislova A, Swaisland H, Rojo F, Albanell J. Phase I safety, pharmacokinetic, and pharmacodynamic trial of ZD1839, a selective oral epidermal growth factor receptor tyrosine kinase inhibitor, in patients with five selected solid tumor types. *J Clin Oncol* 2002;**20**:4292–302.

45. Herbst RS, Maddox AM, Rothenberg ML, Small EJ, Rubin EH, Baselga J, Rojo F, Hong WK, Swaisland H, Averbuch SD, Ochs J, LoRusso PM. Selective oral epidermal growth factor receptor tyrosine kinase inhibitor ZD1839 is generally well-tolerated and has activity in non-small-cell lung cancer and other solid tumors: results of a phase I trial. *J Clin Oncol* 2002;**20**:3815–25.

46. Ranson M, Hammond LA, Ferry D, Kris M, Tullo A, Murray PI, Miller V, Averbuch S, Ochs J, Morris C, Feyereislova A, Swaisland H, Rowinsky EK. ZD1839, a selective oral epidermal growth factor receptor-tyrosine kinase inhibitor, is well tolerated and active in patients with solid, malignant tumors: results of a phase I trial. *J Clin Oncol* 2002;**20**:2240–50.

47. Nakagawa K, Tamura T, Negoro S, Kudoh S, Yamamoto N, Yamamoto N, Takeda K, Swaisland H, Nakatani I, Hirose M, Dong RP, Fukuoka M. Phase I pharmacokinetic trial of the selective oral epidermal growth factor receptor tyrosine kinase inhibitor gefitinib ("Iressa", ZD1839) in Japanese patients with solid malignant tumors. *Ann Oncol* 2003;**14**:922–30.

48. Fukuoka M, Yano S, Giaccone G, Tamura T, Nakagawa K, Douillard JY, Nishiwaki Y, Vansteenkiste J, Kudoh S, Rischin D, Eek R, Horai T, Noda K, Takata I, Smit E, Averbuch S, Macleod A, Feyereislova A, Dong RP, Baselga J. Multi-institutional randomized phase II trial of gefitinib for previously treated patients with advanced non-small-cell lung cancer (The IDEAL 1 Trial). *J Clin Oncol* 2003;**21**:2237–46.

49. Kris Jr MG, Natale RB, Herbst RS, Lynch TJ, Prager D, Belani CP, Schiller JH, Kelly K, Spiridonidis H, Sandler A, Albain KS, Cella D, Wolf MK, Averbuch SD, Ochs JJ, Kay AC. Efficacy of gefitinib, an inhibitor of the epidermal growth factor receptor tyrosine kinase, in symptomatic patients with non-small cell lung cancer: a randomized trial. *J Am Med Assoc* 2003;**290**:2149–58.

50. Thatcher N, Chang A, Parikh P, Rodrigues Pereira J, Ciuleanu T, von Pawel J, Thongprasert S, Tan EH, Pemberton K, Archer V, Carroll K. Gefitinib plus best supportive care in previously treated patients with refractory advanced non-small-cell lung cancer: results from a randomised, placebo-controlled, multicentre study (Iressa Survival Evaluation in Lung Cancer). *Lancet* 2005;**366**:1527–37.

51. Chang A, Parikh P, Thongprasert S, Tan EH, Perng RP, Ganzon D, Yang CH, Tsao CJ, Watkins C, Botwood N, Thatcher N. Gefitinib (IRESSA) in patients of Asian origin with refractory advanced non-small cell lung cancer: subset analysis from the ISEL study. *J Thorac Oncol* 2006;**1**:847–55.

52. Shepherd FA, Rodrigues Pereira J, Ciuleanu T, Tan EH, Hirsh V, Thongprasert S, Campos D, Maoleekoonpiroj S, Smylie M, Martins R, van Kooten M, Dediu M, Findlay B, Tu D, Johnston D, Bezjak A, Clark G, Santabárbara P, Seymour L. National Cancer Institute of Canada Clinical Trials Group. Erlotinib in previously treated non-small-cell lung cancer. *N Engl J Med* 2005;**353**:123–32.

53. Sirotnak F, Zakowski M, Miller V, Scher H, Kris M. Efficacy of cytotoxic agents against human tumor xenografts is markedly enhanced by co-administration of ZD1839 (Iressa), an inhibitor of EGFR tyrosine kinase. *Clin Can Res* 2000;**6**:4885–92.

54. Ciardiello F, Caputo R, Bianco R, Damiano V, Pomatico G, De Placido S, Bianco AR, Tortora G. Antitumor effect and potentiation of cytotoxic drugs activity in human cancer cells by ZD-1839 (Iressa), an epidermal growth factor-selective tyrosine kiinase inhibitor. *Clin Cancer Res* 2000;**6**:2053–63.

55. Giaccone G, Herbst RS, Manegold C, Scagliotti G, Rosell R, Miller V, Natale RB, Schiller JH, Von Pawel J, Pluzanska A, Gatzemeier U, Grous J, Ochs JS, Averbuch SD, Wolf MK, Rennie P, Fandi A, Johnson DH. Gefitinib in combination with gemcitabine and cisplatin in advanced non-small-cell lung cancer: a phase III trial - INTACT 1. *J Clin Oncol* 2004;**22**:777–84.

56. Herbst RS, Giaccone G, Schiller JH, Natale RB, Miller V, Manegold C, Scagliotti G, Rosell R, Oliff I, Reeves JA, Wolf MK, Krebs AD, Averbuch SD, Ochs JS, Grous J, Fandi A, Johnson DH. Gefitinib in combination with paclitaxel and carboplatin in advanced non-small-cell lung cancer: a phase III trial - INTACT 2. *J Clin Oncol* 2004;**22**:785–94.

57. Herbst RS, Prager D, Hermann R, Fehrenbacher L, Johnson BE, Sandler A, Kris MG, Tran HT, Klein P, Li X, Ramies D, Johnson DH, Miller VA. TRIBUTE Investigator Group. TRIBUTE: a phase III trial of erlotinib hydrochloride (OSI-774) combined with carboplatin and paclitaxel chemotherapy in advanced non-small-cell lung cancer. *J Clin Oncol* 2005;**23**:5892–9.

58. Gatzemeier U, Pluzanska A, Szczesna A, Kaukel E, Roubec J, De Rosa F, Milanowski J, Karnicka-Mlodkowski H, Pesek M, Serwatowski P, Ramlau R, Janaskova T, Vansteenkiste J, Strausz J, Manikhas GM, Von Pawel J. Phase III study of erlotinib in combination with cisplatin and gemcitabine in advanced non-small-cell lung cancer: the Tarceva Lung Cancer Investigation Trial. *J Clin Oncol* 2007;**25**:1545–52.

59. Moody SE, Sarkisian CJ, Hahn KT, Gunther EJ, Pickup S, Dugan KD, Innocent N, Cardiff RD, Schnall MD, Chodosh LA. Conditional activation of Neu in the mammary epithelium of transgenic mice results in reversible pulmonary metastasis. *Cancer Cell* 2002;**2**:451–61.

60. Huettner CS, Zhang P, Van Etten RA, Tenen DG. Reversibility of acute B-cell leukaemia induced by BCR-ABL1. *Nat Genet* 2000;**24**:57–60.

61. Fisher GH, Wellen SL, Klimstra D, Lenczowski JM, Tichelaar JW, Lizak MJ, Whitsett JA, Koretsky A, Varmus HE. Induction and apoptotic regression of lung adenocarcinomas by regulation of a K-Ras transgene in the presence and absence of tumor suppressor genes. *Genes Dev* 2001;**15**:3249–62.

62. Felsher DW, Bishop JM. Reversible tumorigenesis by MYC in hematopoietic lineages. *Mol Cell* 1999;**4**:199–207.

63. Chin 2nd L, Tam A, Pomerantz J, Wong M, Holash J, Bardeesy N, Shen Q, O'Hagan R, Pantginis J, Zhou H, Horner JW, Cordon-Cardo C, Yancopoulos GD, DePinho RA. Essential role for oncogenic Ras in tumour maintenance. *Nature* 1999;**400**:468–72.

64. Lynch TJ, Bell DW, Sordella R, Gurubhagavatula S, Okimoto RA, Brannigan BW, Harris PL, Haserlat SM, Supko JG, Haluska FG, Louis DN, Christiani DC, Settleman J, Haber DA. Activating

mutations in the epidermal growth factor receptor underlying responsiveness of non-small-cell lung cancer to gefitinib. *N Engl J Med* 2004;**350**:2129–39.

65. Paez JG, Jänne PA, Lee JC, Tracy S, Greulich H, Gabriel S, Herman P, Kaye FJ, Lindeman N, Boggon TJ, Naoki K, Sasaki H, Fujii Y, Eck MJ, Sellers WR, Johnson BE, Meyerson M. EGFR mutations in lung cancer: correlation with clinical response to gefitinib therapy. *Science* 2004;**304**:1497–500.

66. Bardelli A, Parsons DW, Silliman N, Ptak J, Szabo S, Saha S, Markowitz S, Willson JK, Parmigiani G, Kinzler KW, Vogelstein B, Velculescu VE. Mutational analysis of the tyrosine kinome in colorectal cancers. *Science* 2003;**300**:949.

67. Pao W, Miller V, Zakowski M, Doherty J, Politi K, Sarkaria I, Singh B, Heelan R, Rusch V, Fulton L, Mardis E, Kupfer D, Wilson R, Kris M, Varmus H. EGF receptor gene mutations are common in lung cancers from "never smokers" and are associated with sensitivity of tumors to gefitinib and erlotinib. *Proc Natl Acad Sci USA* 2004;**101**:13,306–13,311,

68. Miller VA, Kris MG, Shah N, Patel J, Azzoli C, Gomez J, Krug LM, Pao W, Rizvi N, Pizzo B, Tyson L, Venkatraman E, Ben-Porat L, Memoli N, Zakowski M, Rusch V, Heelan RT. Bronchioloalveolar pathologic subtype and smoking history predict sensitivity to gefitinib in advanced non-small-cell lung cancer. *J Clin Oncol* 2004;**22**:1103–9.

69. West HL, Franklin WA, McCoy J, Gumerlock PH, Vance R, Lau DH, Chansky K, Crowley JJ, Gandara DR. Gefitinib therapy in advanced bronchioloalveolar carcinoma: Southwest Oncology Group Study S0126. *J Clin Oncol* 2006;**24**:1807–13.

70. Guo M, Liu S, Lu F. Gefitinib-sensitizing mutations in esophageal carcinoma. *N Engl J Med* 2006;**354**:2193–4.

71. Gwak GY, Yoon JH, Shin CM, Ahn YJ, Chung JK, Kim YA, Kim TY, Lee HS. Detection of response-predicting mutations in the kinase domain of the epidermal growth factor receptor gene in cholangiocarcinomas. *J Cancer Res Clin Oncol* 2005;**131**:649–52.

72. Lee JW, Soung YH, Kim SY, Park WS, Nam SW, Lee JY, Yoo NJ, Lee SH. Absence of EGFR mutation in the kinase domain in common human cancers besides non-small cell lung cancer. *Int J Cancer* 2005;**113**:510–11.

73. Nagahara H, Mimori K, Ohta M, Utsunomiya T, Inoue H, Barnard GF, Ohira M, Hirakawa K, Mori M. Somatic mutations of epidermal growth factor receptor in colorectal carcinoma. *Clin Cancer Res* 2005;**11**:1368–71.

74. Kwak EL, Jankowski J, Thayer SP, Lauwers GY, Brannigan BW, Harris PL, Okimoto RA, Haserlat SM, Driscoll DR, Ferry D, Muir B, Settleman J, Fuchs CS, Kulke MH, Ryan DP, Clark JW, Sgroi DC, Haber DA, Bell DW. Epidermal growth factor receptor kinase domain mutations in esophageal and pancreatic adenocarcinomas. *Clin Cancer Res* 2006;**12**:4283–7.

75. Greulich H, Chen TH, Feng W, Jänne PA, Alvarez JV, Zappaterra M, Bulmer SE, Frank DA, Hahn WC, Sellers WR, Meyerson M. Oncogenic transformation by inhibitor-sensitive and -resistant EGFR mutants. *PLoS Med* 2005;**2**:e313.

76. Shigematsu H, Gazdar AF. Somatic mutations of epidermal growth factor receptor signaling pathway in lung cancers. *Intl J Cancer* 2006;**118**:257–62.

77. Zang EA, Wynder EL. Differences in lung cancer risk between men and women: examination of the evidence. *J Natl Cancer Inst* 1996;**88**:183–92.

78. Bell DW, Gore I, Okimoto RA, Godin-Heymann N, Sordella R, Mulloy R, Sharma SV, Brannigan BW, Mohapatra G, Settleman J, Haber DA. Inherited susceptibility to lung cancer may be associated

with the T790M drug resistance mutation in EGFR. *Nat Genet* 2005;**37**:1315–16.

79. Mellinghoff IK, Wang MY, Vivanco I, Haas-Kogan DA, Zhu S, Dia EQ, Lu KV, Yoshimoto K, Huang JH, Chute DJ, Riggs BL, Horvath S, Liau LM, Cavenee WK, Rao PN, Beroukhim R, Peck TC, Lee JC, Sellers WR, Stokoe D, Prados M, Cloughesy TF, Sawyers CL, Mischel PS. Molecular determinants of the response of glioblastomas to EGFR kinase inhibitors. *N Engl J Med* 2005;**353**:2012–24.

80. Ji H, Zhao X, Yuza Y, Shimamura T, Li D, Protopopov A, Jung BL, McNamara K, Xia H, Glatt KA, Thomas RK, Sasaki H, Horner JW, Eck M, Mitchell A, Sun Y, Al-Hashem R, Bronson RT, Rabindran SK, Discafani CM, Maher E, Shapiro GI, Meyerson M, Wong KK. Epidermal growth factor receptor variant III mutations in lung tumorigenesis and sensitivity to tyrosine kinase inhibitors. *Proc Natl Acad Sci USA* 2006;**103**:7817–22.

81. Amann J, Kalyankrishna S, Massion PP, Ohm JE, Girard L, Shigematsu H, Peyton M, Juroske D, Huang Y, Stuart SalmonJ, Kim YH, Pollack JR, Yanagisawa K, Gazdar A, Minna JD, Kurie JM, Carbone DP. Aberrant epidermal growth factor receptor signaling and enhanced sensitivity to EGFR inhibitors in lung cancer. *Cancer Res* 2005;**65**:226–35.

82. Greulich H, Chen TH, Feng W, Jänne PA, Alvarez JV, Zappaterra M, Bulmer SE, Frank DA, Hahn WC, Sellers WR, Meyerson M. Oncogenic transformation by inhibitor-sensitive and -resistant EGFR mutants. *PLoS Med* 2005;**2**:e313.

83. Carey KD, Garton AJ, Romero MS, Kahler J, Thomson S, Ross S, Park F, Haley JD, Gibson N, Sliwkowski MX. Kinetic analysis of epidermal growth factor receptor somatic mutant proteins shows increased sensitivity to the epidermal growth factor receptor tyrosine kinase inhibitor, erlotinib. *Cancer Res* 2006;**66**:8163–71.

84. Ji H, Li D, Chen L, Shimamura T, Kobayashi S, McNamara K, Mahmood U, Mitchell A, Sun Y, Al-Hashem R, Chirieac LR, Padera R, Bronson RT, Kim W, Jänne PA, Shapiro GI, Tenen D, Johnson BE, Weissleder R, Sharpless NE, Wong KK. The impact of human EGFR kinase domain mutations on lung tumorigenesis and *in vivo* sensitivity to EGFR-targeted therapies. *Cancer Cell* 2006;**9**:485–95.

85. Politi K, Zakowski MF, Fan PD, Schonfeld EA, Pao W, Varmus HE. Lung adenocarcinomas induced in mice by mutant EGF receptors found in human lung cancers respond to a tyrosine kinase inhibitor or to down-regulation of the receptors. *Genes Dev* 2006;**20**:1496–510.

86. Zhang X, Gureasko J, Shen K, Cole PA, Kuriyan J. An allosteric mechanism for activation of the kinase domain of epidermal growth factor receptor. *Cell* 2006;**125**:1137–49.

87. Yun CH, Boggon TJ, Li Y, Woo MS, Greulich H, Meyerson M, Eck MJ. Structures of lung cancer-derived EGFR mutants and inhibitor complexes: mechanism of activation and insights into differential inhibitor sensitivity. *Cancer Cell* 2007;**11**:217–27.

88. Stamos J, Sliwkowski MX, Eigenbrot C. Structure of the epidermal growth factor receptor kinase domain alone and in complex with a 4-anilinoquinazoline inhibitor. *J Biol Chem* 2002;**277**:46,265–46,272.

89. Carey KD, Garton AJ, Romero MS, Kahler J, Thomson S, Ross S, Park F, Haley JD, Gibson N, Sliwkowski MX. A unique structure for epidermal growth factor receptor bound to GW572016 (Lapatinib): relationships among protein conformation, inhibitor off-rate, and receptor activity in tumor cells. *Cancer Res* 2004;**64**:6652–9.

90. Shigematsu H, Lin L, Takahashi T, Nomura M, Suzuki M, Wistuba II, Fong KM, Lee H, Toyooka S, Shimizu N, Fujisawa T, Feng Z, Roth JA, Herz J, Minna JD, Gazdar AF. Clinical and biological features associated with epidermal growth factor receptor gene mutations in lung cancers. *J Natl Cancer Inst* 2005;**97**:339–46.

91. Riely GJ, Pao W, Pham D, Li AR, Rizvi N, Venkatraman ES, Zakowski MF, Kris MG, Ladanyi M, Miller VA. Clinical course of patients with non-small cell lung cancer and epidermal growth factor receptor exon 19 and exon 21 mutations treated with gefitinib or erlotinib. Clin Cancer Res 2006;12:839–44.

92. Jackman DM, Yeap BY, Sequist LV, Lindeman N, Holmes AJ, Joshi VA, Bell DW, Huberman MS, Halmos B, Rabin MS, Haber DA, Lynch TJ, Meyerson M, Johnson BE, Jänne PA. Exon 19 deletion mutations of epidermal growth factor receptor are associated with prolonged survival in non-small cell lung cancer patients treated with gefitinib or erlotinib. Clin Cancer Res 2006;12:3908–14.

93. Marks JL, Broderick S, Zhou Q, Chitale D, Li AR, Zakowski MF, Kris MG, Rusch VW, Azzoli CG, Seshan VE, Ladanyi M, Pao W. Prognostic and therapeutic implications of EGFR and KRAS mutations in resected lung adenocarcinoma. J Thorac Oncol 2008. in press.

94. Cappuzzo Jr. F, Hirsch FR, Rossi E, Bartolini S, Ceresoli GL, Bemis L, Haney J, Witta S, Danenberg K, Domenichini I, Ludovini V, Magrini E, Gregorc V, Doglioni C, Sidoni A, Tonato M, Franklin WA, Crino L, Bunn PA, Varella-Garcia M Epidermal growth factor receptor gene and protein and gefitinib sensitivity in non-small-cell lung cancer. J Natl Cancer Inst 2005;97:643–55.

95. Tsao MS, Sakurada A, Cutz JC, Zhu CQ, Kamel-Reid S, Squire J, Lorimer I, Zhang T, Liu N, Daneshmand M, Marrano P, da Cunha Santos G, Lagarde A, Richardson F, Seymour L, Whitehead M, Ding K, Pater J, Shepherd FA. Erlotinib in lung cancer - molecular and clinical predictors of outcome. N Engl J Med 2005;353:133–44.

96. Kaye FJ. A curious link between epidermal growth factor receptor amplification and survival: effect of "allele dilution" on gefitinib sensitivity? J Natl Cancer Inst 2005;97:621–3.

97. Velu TJ, Beguinot L, Vass WC, Willingham MC, Merlino GT, Pastan I, Lowy DR. Epidermal-growth-factor-dependent transformation by a human EGF receptor proto-oncogene. Science 1987;238:1408–10.

98. Pao W, Wang TY, Riely GJ, Miller VA, Pan Q, Ladanyi M, Zakowski MF, Heelan RT, Kris MG, Varmus HE. KRAS mutations and primary resistance of lung adenocarcinomas to gefitinib or erlotinib. PLoS Med 2005;2:e17.

99. Miller VA, Riely GJ, Zakowski MF, Li AR, Patel JD, Heelan RT, Kris MG, Sandler AB, Carbone DP, Tsao A, Herbst RS, Heller G, Ladanyi M, Pao W, Johnson DH. Molecular characteristics of bronchioloalveolar carcinoma and adenocarcinoma, bronchioloalveolar carcinoma subtype, predict response to erlotinib. J Clin Oncol 2008;26:1472–8.

100. Jackman DM, Yeap BY, Lindeman NI, Fidias P, Rabin MS, Temel J, Skarin AT, Meyerson M, Holmes AJ, Borras AM, Freidlin B, Ostler PA, Lucca J, Lynch TJ, Johnson BE, Jänne PA. Phase II clinical trial of chemotherapy-naive patients>or =70 years of age treated with erlotinib for advanced non-small cell lung cancer. J Clin Oncol 2007;25:760–6.

101. Kobayashi S, Ji H, Yuza Y, Meyerson M, Wong KK, Tenen DG, Halmos B. An alternative inhibitor overcomes resistance caused by a mutation of the epidermal growth factor receptor. Cancer Res 2005;65:7096–101.

102. Kwak EL, Sordella R, Bell DW, Godin-Heymann N, Okimoto RA, Brannigan BW, Harris PL, Driscoll DR, Fidias P, Lynch TJ, Rabindran SK, McGinnis JP, Wissner A, Sharma SV, Isselbacher KJ, Settleman J, Haber DA. Irreversible inhibitors of the EGF receptor may circumvent acquired resistance to gefitinib. Proc Natl Acad Sci USA 2005;102:7665–70.

103. Pao W, Miller VA, Politi KA, Riely GJ, Somwar R, Zakowski MF, Kris MG, Varmus H. Acquired resistance of lung adenocarcinomas to gefitinib or erlotinib is associated with a second mutation in the EGFR kinase domain. PLoS Med 2005;2:e7.

104. Gow CH, Shih JY, Chang YL, Yu CJ. Acquired gefitinib-resistant mutation of EGFR in a chemonaive lung adenocarcinoma harboring gefitinib-sensitive mutation L858R. PLoS Med 2005;2:e269.

105. Jain A, Tindell CA, Laux I, Hunter JB, Curran J, Galkin A, Afar DE, Aronson N, Shak S, Natale RB, Agus DB. Epithelial membrane protein-1 is a biomarker of gefitinib resistance. Proc Natl Acad Sci USA 2005;102:11858–11863.

106. Yun CH, Mengwasser KE, Toms AV, Woo MS, Greulich H, Wong KK, Meyerson M, Eck MJ. The T790M mutation in EGFR kinase causes drug resistance by increasing the affinity for ATP. Proc Natl Acad Sci USA 2008. in press.

107. Costa DB, Halmos B, Kumar A, Schumer ST, Huberman MS, Boggon TJ, Tenen DG, Kobayashi S. BIM mediates EGFR tyrosine kinase inhibitor-induced apoptosis in lung cancers with oncogenic EGFR mutations. PLoS Med 2007;4:1669–79.

108. Balak MN, Gong Y, Riely GJ, Somwar R, Li AR, Zakowski MF, Chiang A, Yang G, Ouerfelli O, Kris MG, Ladanyi M, Miller VA, Pao W. Novel D761Y and common secondary T790M mutations in epidermal growth factor receptor-mutant lung adenocarcinomas with acquired resistance to kinase inhibitors. Clin Cancer Res 2006;12:6494–501.

109. Engelman JA, Zejnullahu K, Mitsudomi T, Song Y, Hyland C, Park JO, Lindeman N, Gale CM, Zhao X, Christensen J, Kosaka T, Holmes AJ, Rogers AM, Cappuzzo F, Mok T, Lee C, Johnson BE, Cantley LC, Jänne PA. MET amplification leads to gefitinib resistance in lung cancer by activating ERBB3 signaling. Science 2007;316:1039–43.

110. Bean J, Brennan C, Shih JY, Riely G, Viale A, Wang L, Chitale D, Motoi N, Szoke J, Broderick S, Balak M, Chang WC, Yu CJ, Gazdar A, Pass H, Rusch V, Gerald W, Huang SF, Yang PC, Miller V, Ladanyi M, Yang CH, Pao W. MET amplification occurs with or without T790M mutations in EGFR mutant lung tumors with acquired resistance to gefitinib or erlotinib. Proc Natl Acad Sci USA 2007;104:20,932–20,937.

111. Giordano S, Ponzetto C, Di Renzo MF, Cooper CS, Comoglio PM. Tyrosine kinase receptor indistinguishable from the c-met protein. Nature 1989;339:155–6.

112. Park M, Dean M, Kaul K, Braun MJ, Gonda MA, Vande Woude G. Sequence of MET protooncogene cDNA has features characteristic of the tyrosine kinase family of growth-factor receptors. Proc Natl Acad Sci USA 1987;84:6379–83.

113. Birchmeier C, Birchmeier W, Gherardi E, Vande Woude GF. Met, metastasis, motility and more. Nat Rev Mol Cell Biol 2003;4:915–25.

114. Linehan WM, Bates SE, Yang JC. Cancer of the kidney. In: De Vita JT, Hellman S, Rosenberg SA, editors. Cancer: Principles and Practice of Oncology. 7th edn Philadelphia, PA: Lippincott Williams and Wilkins; 2005. p. 1139–65.

115. Zbar B, Kishida T, Chen F, Schmidt L, Maher ER, Richards FM, Crossey PA, Webster AR, Affara NA, Ferguson-Smith MA, Brauch H, Glavac D, Neumann HP, Tisherman S, Mulvihill JJ, Gross DJ, Shuin T, Whaley J, Seizinger B, Kley N, Olschwang S, Boisson C, Richard S, Lips CH, Linehan WM, Lerman M. Germline mutations in the Von Hippel-Lindau disease (VHL) gene in families from North America, Europe, and Japan. Hum Mutat 1996;8:348–57.

116. Suzuki H, Ueda T, Komiya A, Okano T, Isaka S, Shimazaki J, Ito H. Mutational state of von Hippel-Lindau and adenomatous polyposis coli genes in renal tumors. *Oncology* 1997;**54**:252–7.

117. Schmidt L, Duh FM, Chen F, Kishida T, Glenn G, Choyke P, Scherer SW, Zhuang Z, Lubensky I, Dean M, Allikmets R, Chidambaram A, Bergerheim UR, Feltis JT, Casadevall C, Zamarron A, Bernues M, Richard S, Lips CJ, Walther MM, Tsui LC, Geil L, Orcutt ML, Stackhouse T, Lipan J, Slife L, Brauch H, Decker J, Niehans G, Hughson MD, Moch H, Storkel S, Lerman MI, Linehan WM, Zbar B. Germline and somatic mutations in the tyrosine kinase domain of the MET proto-oncogene in papillary renal carcinomas. *Nat Genet* 1997;**16**:68–73.

118. Motzer RJ, Russo P. Systemic therapy for renal cell carcinoma. *J Urology* 2000;**163**:408–17.

119. Fyfe GA, Fisher RI, Rosenberg SA, Sznol M, Parkinson DR, Louie AC. Long-term response data for 255 patients with metastatic renal cell carcinoma treated with high-dose recombinant interleukin-2 therapy. *J Clin Oncol* 1996;**14**:2410–11.

120. Pyrhönen S, Salminen E, Ruutu M, Lehtonen T, Nurmi M, Tammela T, Juusela H, Rintala E, Hietanen P, Kellokumpu-Lehtinen PL. Prospective randomized trial of interferon alfa-2a plus vinblastine versus vinblastine alone in patients with advanced renal cell cancer. *J Clin Oncol* 1999;**17**:2859–67.

121. Medical Research Council and Collaborators. Interferon-alfa and survival in metastatic renal carcinoma: Early results of a randomised controlled trial. *Lancet* 1999;**353**:14–17.

122. Turner KJ, Moore JW, Jones A, Taylor CF, Cuthbert-Heavens D, Han C, Leek RD, Gatter KC, Maxwell PH, Ratcliffe PJ, Cranston D, Harris AL. Expression of hypoxia-inducible factors in human renal cancer: relationship to angiogenesis and to the von Hippel-Lindau gene mutation. *Cancer Res* 2002;**62**:2957–61.

123. Kaelin WG. The von Hippel-Lindau tumor suppressor protein: roles in cancer and oxygen sensing. *Cold Spring Harb Symp Quant Biol* 2005;**70**:159–66.

124. Yang JC, Haworth L, Sherry RM, Hwu P, Schwartzentruber DJ, Topalian SL, Steinberg SM, Chen HX, Rosenberg SA. A randomized trial of bevacizumab, an anti-vascular endothelial growth factor antibody, for metastatic renal cancer. *N Engl J Med* 2003;**349**:427–34.

125. Stein MN, Flaherty KT. CCR drug updates: sorafenib and sunitinib in renal cell carcinoma. *Clin Cancer Res* 2007;**13**:3765–70.

126. Ratain MJ, Eisen T, Stadler WM, Flaherty KT, Kaye SB, Rosner GL, Gore M, Desai AA, Patnaik A, Xiong HQ, Rowinsky E, Abbruzzese JL, Xia C, Simantov R, Schwartz B, O'Dwyer PJ. Phase II placebo-controlled randomized discontinuation trial of sorafenib in patients with metastatic renal cell carcinoma. *J Clin Oncol* 2006;**24**:2505–12.

127. Escudier B, Eisen T, Stadler WM, Szczylik C, Oudard S, Siebels M, Negrier S, Chevreau C, Solska E, Desai AA, Rolland F, Demkow T, Hutson TE, Gore M, Freeman S, Schwartz B, Shan M, Simantov R, Bukowski RM. TARGET Study Group. Sorafenib in advanced clear-cell renal-cell carcinoma. *N Engl J Med* 2007;**356**:125–34.

128. Faivre S, Delbaldo C, Vera K, Robert C, Lozahic S, Lassau N, Bello C, Deprimo S, Brega N, Massimini G, Armand JP, Scigalla P, Raymond E. Safety, pharmacokinetic, and antitumor activity of SU11248, a novel oral multitarget tyrosine kinase inhibitor, in patients with cancer. *J Clin Oncol* 2006;**24**:25–35.

129. Motzer RJ, Michaelson MD, Redman BG, Hudes GR, Wilding G, Figlin RA, Ginsberg MS, Kim ST, Baum CM, DePrimo SE, Li JZ, Bello CL, Theuer CP, George DJ, Rini BI. Activity of SU11248, a multitargeted inhibitor of vascular endothelial growth factor receptor and platelet-derived growth factor receptor, in patients with metastatic renal cell carcinoma. *J Clin Oncol* 2006;**24**:16–24.

130. Motzer RJ, Rini BI, Bukowski RM, Curti BD, George DJ, Hudes GR, Redman BG, Margolin KA, Merchan JR, Wilding G, Ginsberg MS, Bacik J, Kim ST, Baum CM, Michaelson MD. Sunitinib in patients with metastatic renal cell carcinoma. *J Am Med Assoc* 2006;**295**:2516–24.

131. Motzer RJ, Hutson TE, Tomczak P, Michaelson MD, Bukowski RM, Rixe O, Oudard S, Negrier S, Szczylik C, Kim ST, Chen I, Bycott PW, Baum CM, Figlin RA. Sunitinib versus interferon alfa in metastatic renal-cell carcinoma. *N Engl J Med* 2007;**356**:115–24.

132. Rini BI, Jaeger E, Weinberg V, Sein N, Chew K, Fong K, Simko J, Small EJ, Waldman FM. Clinical response to therapy targeted at vascular endothelial growth factor in metastatic renal cell carcinoma: impact of patient characteristics and Von Hippel-Lindau gene status. *BJU Intl* 2006;**98**:756–62.

133. Deprimo SE, Bello CL, Smeraglia J, Baum CM, Spinella D, Rini BI, Michaelson MD, Motzer RJ. Circulating protein biomarkers of pharmacodynamic activity of sunitinib in patients with metastatic renal cell carcinoma: modulation of VEGF and VEGF-related proteins. *J Transl Med* 2007;**5**:32–43.

134. Geyer CE, Forster J, Lindquist D, Chan S, Romieu CG, Pienkowski T, Jagiello-Gruszfeld A, Crown J, Chan A, Kaufman B, Skarlos D, Campone M, Davidson N, Berger M, Oliva C, Rubin SD, Stein S, Cameron D. Lapatinib plus capecitabine for HER2-positive advanced breast cancer. *N Engl J Med* 2006;**355**:2733–43.

135. Llovet J, Ricci S, Mazzaferro V, Hilgard P, Raoul J, Zeuzem S, Poulin-Costello M, Moscovici M, Voliotis D, Bruix J. For the SHARP Investigators Study Group. Sorafenib improves survival in advanced hepatocellular carcinoma (HCC): results of a phase III randomized placebo-controlled trial (SHARP trial). *Proc Am Soc Clin Oncol* 2007. Abstract LBA1.

136. Wells SA, Gosnell JE, Gagel RF, Moley JF, Pfister DG, Sosa JA, Skinner M, Krebs A, Hou J, Schlumberger M. Vandetanib in metastatic hereditary medullary thyroid cancer: Follow-up results of an open-label phase II trial. *Proc Am Soc Clin Oncol* 2007. Abstract 6018.

137. Karaman MW, Herrgard S, Treiber DK, Gallant P, Atteridge CE, Campbell BT, Chan KW, Ciceri P, Davis MI, Edeen PT, Faraoni R, Floyd M, Hunt JP, Lockhart DJ, Milanov ZV, Morrison MJ, Pallares G, Patel HK, Pritchard S, Wodicka LM, Zarrinkar PP. A quantitative analysis of kinase inhibitor selectivity. *Nat Biotechnol* 2008;**26**:127–32.

138. Pao W, Miller VA, Kris MG. "Targeting" the epidermal growth factor receptor tyrosine kinase with gefitinib (Iressa) in non-small cell lung cancer (NSCLC). *Semin Cancer Biol* 2004;**14**:33–40.

Chapter 45

Adipokine Signaling: Implications for Obesity

Rexford S. Ahima and Gladys M. Varela

University of Pennsylvania School of Medicine, Department of Medicine, Division of Endocrinology, Diabetes and Metabolism, Philadelphia, Pennsylvania

ADIPOSE TISSUE AND ITS RELATION TO OBESITY

Obesity has reached epidemic levels globally, affecting both developed and developing countries [1]. More than 1.6 billion adults worldwide are overweight (body mass index; BMI > 25) and 400 million are obese (BMI > 30). Obesity is also increasing rapidly among children [1]. Obesity poses enormous challenges to health care systems because it is associated with increased risk of Type 2 diabetes, hypertension, coronary artery disease, sleep apnea, cancer, and various ailments [1]. The past decade has seen major advances in our understanding of the pathogenesis of obesity [2]. Fundamentally obesity is the result of an imbalance between energy intake and expenditure. The current epidemic of obesity and related diseases is attributed mainly to increased intake of energy-dense foods rich in fat and sugar, and sedentary lifestyle [1, 2].

White adipose tissue contains lipid filled cells (adipocytes), preadipocytes, and immune cells, and has a rich vascular supply and innervation. The distribution of adipose tissue is influenced by sex hormones, such that women have greater amounts of subcutaneous adipose tissue partly due to estrogen, while accumulation of visceral adiposity in males and postmenopausal women is promoted by androgens [2]. The stored triglycerides in white adipocytes provides an enormous reservoir of metabolic fuel. During fasting, lipolysis is stimulated in adipocytes under the influence of catecholamines, hormone sensitive lipase, and adipose tissue triglyceride lipase [3]. Non-esterified fatty acids are released into the circulation and transported to various organs, in particular skeletal muscle, to be oxidized for energy. Unlike rodents, the capacity for *de novo* lipogenesis in adipocytes is very low in humans. Instead, in the fed state, triglycerides derived from the diet or synthesized in the liver are transported as lipoprotein particles, hydrolyzed by lipoprotein lipase into fatty acids, and then taken up by adipocytes for triglyceride synthesis. Excessive food consumption leads to major alterations in the structure and function of adipose tissue to accommodate the increased demand for triglyceride storage. Adipocytes undergo hyperplasia and hypertrophy, the extracellular matrix expands, and angiogenesis and macrophage infiltration are enhanced in obesity [2–6]. Accumulation of adipose tissue in obesity, in particular visceral adipose tissue, results in greater rate of lipolysis, ectopic lipid accumulation in the liver, muscle, and pancreatic islets, insulin resistance, diabetes, hypertension, dyslipidemia, and cardiovascular morbidity [2, 7].

In addition to fatty acids, adipose tissue is the source of several circulating peptides including leptin, adiponectin, retinol binding protein-4, proinflammatory cytokines, and complement vasoactive and procoagulant factors, some of which act through autocrine and paracrine mechanisms to control the growth, and metabolic and storage functions of adipose tissue [2]. Other adipokines act in the brain and peripheral organs to modulate energy balance, glucose and lipid metabolism, and neuroendocrine and immune systems [2]. The following sections will provide insights into adipokine signaling under normal physiological conditions, obesity, and associated diseases.

LEPTIN

The discovery of leptin was a major milestone in the concept of adipose tissue as "an endocrine organ" [2, 3, 8]. Leptin is secreted mainly by white adipocytes, but small amounts are produced in the gastric fundus, intestine, and muscle. The levels of leptin in adipose tissue and plasma are proportional to fat stores, hence leptin is increased in obesity and reduced in lean individuals. Leptin rises several hours after feeding and falls rapidly in response to fasting [9, 10]. These changes are mediated, at least in part, by insulin. Leptin levels are higher in females than males because of increased synthesis in subcutaneous adipose tissue, stimulation by estrogen and suppression by androgens [9, 11]. Leptin is increased in response to chronic glucocorticoid exposure and acutely by tumor necrosis factor-α [9]. On the other hand, cold exposure and β3 adrenergic stimulation decrease leptin levels [9].

Leptin acts in the brain to control feeding, energy balance, and the neuroendocrine axis. The fall in leptin during fasting results in suppression of reproduction, immunity, thyroid and growth hormones and energy expenditure, and stimulation of appetite [9–15]. These responses are attenuated by exogenously administered leptin that signals to the brain that energy stores are sufficient [9–15]. Similar changes in energy balance, immune responses, and hormones reversible by leptin treatment occur in lipodystrophy [16–18]. Moreover, abrogation of leptin signaling due to mutation of leptin or the leptin receptor genes results in hyperphagia, reduction in energy expenditure, hypothyroidism and hypogonadism, decreased growth, and immunosuppression [9]. Together, these findings demonstrate a critical role of leptin as a signal for energy deficiency.

Multiple leptin receptor (LR) isoforms derived from alternative splicing of the *Lepr* gene product have been described [19]. Short-form LRs (LRa, LRc, LRd, and LRf in mice) and the long-form LR (LRb in mice) share identical extracellular and transmembrane domains and the first 29 intracellular amino acids. However, only LRb has the intracellular domain necessary for leptin signaling. LRa is the most abundantly expressed isoform, well conserved among species, and is proposed to transport leptin across the blood–brain barrier. Secreted forms of LR, e.g., murine LRe, comprising only the extracellular domain bind leptin in the plasma thereby controlling bioavailability. LRb is critical for leptin action, as evidenced by identical obese phenotypes of *db/db* mice lacking LRb due to a mutation that causes missplicing of the LRb mRNA, *db3j/db3j* mice lacking all LR isoforms, and leptin deficient *ob/ob* mice [9, 20, 21].

LRb is mainly present in the brain, with the highest levels in hypothalamus nuclei, including the arcuate (Arc), and dorsomedial, ventromedial, and ventral premammillary nuclei [9]. One population of LRb neurons in the Arc expresses neuropeptide Y (NPY) and agouti related peptide (AGRP) (Figure 45.1). Other LRb neurons synthesize pro-opiomelanocortin (POMC), precursor of α-melanocyte stimulating hormone (αMSH) (Figure 45.1). NPY/AGRP neurons project to the paraventricular nucleus (PVN) to stimulate feeding, while α-MSH inhibits feeding by activating the melanocortin-4 receptor (MC4R) and the melanocortin-3 receptor (MC3R) [9]. The binding of leptin to LRb exerts an anorexigenic action by activating LRb/POMC, and stimulating the synthesis of and secretion of α-MSH [9] (Figure 45.1). AGRP normally blocks α-MSH/MC4R signaling. In the fed state, leptin acts as a negative feedback signal via LRb to suppress the expression and secretion of NPY and AGRP and inhibit feeding [9]. Conversely, the reduction in leptin levels during fasting or lack of leptin signaling in *ob/ob* and *db/db* mice, decreases α-MSH and increases NPY and AGRP, resulting in hyperphagia and weight gain [9].

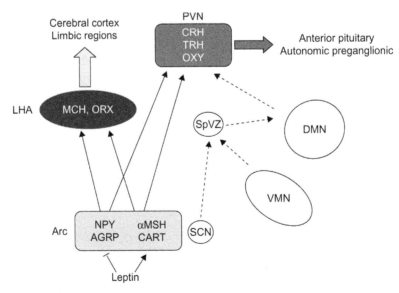

FIGURE 45.1 Neuronal targets of leptin in the hypothalamus.
Leptin stimulates POMC and CART, and suppresses AGRP and NPY in the Arc. Arc neurons project to the paraventricular (PVN) and lateral hypothalamic area (LHA) to inhibit feeding and increase energy expenditure, through the regulation of corticotropin releasing hormone (CRH), thyrotropin releasing hormone (TRH), oxytocin (OXY), orexins (ORX), and melanin concentrating hormone (MCH). Leptin responsive neurons in the PVN and LHA control hormonal and autonomic functions, and feeding behavior via projections to the median eminence, cerebral cortex, limbic regions, and brainstem. Leptin also affects circadian rhythms and glucose metabolism, likely through indirect interactions with the suprachiasmatic (SCN), ventromedial (VMN) and dorsomedial nuclei (DMN), and subparaventricular zone (SpVZ).

Leptin signaling has been analyzed *in vitro* and in mice [19]. The binding of leptin to LRb enables the transphosphorylation and activation of the intracellular LRb associated janus kinase-2 (Jak2), which then phosphorylates other tyrosine residues within the LRb/Jak2 complex (Figure 45.2). Three conserved tyrosine residues present on the intracellular domain of LRb, Tyr985, Tyr1077, and Tyr1138, are phosphorylated and contribute to leptin signaling. The phosphorylation of Tyr985 creates a binding site for the specialized phosphotyrosine binding (SH2) domain of the tyrosine phosphatase SHP-2, which leads to activation of p21ras and ERK signaling. Phosphorylation of Tyr1138 recruits STAT3 to the LRb/Jak2 complex, resulting in the tyrosine phosphorylation and nuclear translocation of STAT3 to mediate transcriptional regulation (Figure 45.2). Among the STAT3 regulated genes is SOCS3 (suppressor of cytokine signaling 3), which binds to Tyr985 of LRb and terminates LRb-STAT3 signaling (Figure 45.2). Tyr1077 activates STAT5 (signal transducer and activator of transcription 5) phosphorylation and transcriptional regulation by leptin.

The roles of leptin signaling molecules have been studied in mice [22–28]. Deletion of LRb or STAT3 recapitulated obesity, thermoregulatory and neuroendocrine deficits, and diabetes seen in *db/db* mice [22, 23]. A homologously targeted knockin mouse model in which LRb was replaced by a mutant molecule (LRbS1138) containing a substitution mutation of Tyr1138 (the STAT3 binding site), failed to mediate STAT3 activation in response to leptin, resulting in hyperphagia, decreased energy expenditure, and obesity [24]. LRbS1138 mutation caused hyperleptinemia, leptin resistance, and central hypothyroidism similar to *db/db* mice [24]. However, LRbS1138 mice showed improved glucose tolerance and fertility, and increased linear growth and immunity, in contrast to *db/db* mice [24]. Hypothalamic POMC expression was reduced in both LRbS1138 and *db/db* mice, whereas NPY and AGRP were suppressed in LRbS1138 but elevated in *db/db* mice. Together, these data reveal different roles of LRb-Tyr1138-STAT3 signaling in the pathogenesis of obesity, hormonal control, and diabetes.

A mutation of Tyr985 prevented phosphorylation of the site and blocked recruitment of SHP-2/SOCS3, and decreased food intake and adiposity [28]. In this model, hypothalamic NPY and AGRP were suppressed, basal STAT3 activation was increased, and sensitivity to leptin treatment was increased, indicating Tyr985 plays a key role in inhibiting LRb signaling [28].

Although leptin deficiency offers important lessons on LRb signaling, obesity is often associated with elevated levels of leptin [2]. The failure of high leptin levels to prevent obesity has given rise to the notion of "leptin resistance," akin to hyperinsulinemia and insulin resistance [2]. A number of mechanisms are proposed to explain leptin resistance in diet induced obesity (DIO). The transport of leptin across the blood–brain barrier is impaired in DIO [29]. Although there is no obvious mutation of LRb to explain leptin resistance in DIO, LRb signaling, particularly in the Arc, is abnormal [30, 31]. Leptin's ability to suppress food intake is attenuated in DIO, and this is related to decreased STAT3 phosphorylation and impaired neuropeptide release [30–32].

As noted earlier, SOCS3 binds to LRb Tyr985 and Jak2, leading to termination of LRb signaling. Obesity is

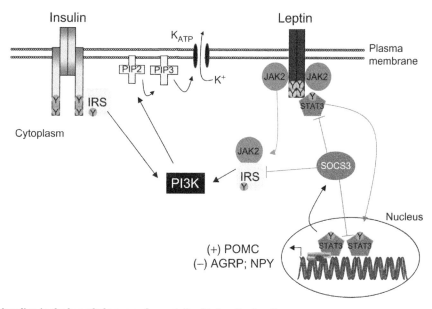

FIGURE 45.2 Leptin signaling in the hypothalamus and crosstalk with insulin signaling.
Leptin binding to LRb activates Janus kinase (JAK), leading to nuclear translocation of STAT3 to activate transcription of neuropeptides and suppressor of cytokine signaling-3 (SOCS3). SOCS3 terminates leptin signaling. Leptin and insulin both activate phosphatidylinositol-3 kinase (PI3K) and inhibit feeding.

characterized by hyperleptinemia, decreased LRb Tyr985 mediated phosphorylation of STAT3, and induction of SOCS3 expression in the hypothalamus [26, 27]. Tyr985 and SOCS3 contribute to leptin resistance, as evidenced by increased leptin sensitivity and leanness when SOCS3 was deleted in neurons, specifically Arc POMC neurons [26, 27]. Potentially, the absence of a blood–brain barrier in the median eminence of the hypothalamus allows free access of leptin and other circulating factors to the Arc, while other hypothalamic sites are protected by the blood–brain barrier [33]. Another possible mediator of leptin resistance is the tyrosine phosphatase PTP1B that dephosphorylates Jak2 and blunts LRb signaling [34, 35]. Neuron specific deletion of PTP1B increased leptin sensitivity and protected against obesity, whereas PTP1B deletion in adipocytes and liver did not decrease weight [36].

The robust responses signaled by leptin deficiency is consistent with the evolutionary pressure to maximize feeding and energy storage against the threat of starvation [2], but is there a physiological need to attenuate leptin signaling? In seasonal animals, such as hibernators, the leptin-LRb feedback mechanism may work in concert with other processes to increase food intake and promote energy storage [37]. The feedback inhibition of LRb signaling has also been proposed as a means of maintaining high food consumption and energy storage to meet the demands of pregnancy and lactation [38, 39].

LRb is expressed in various regions of the brain, including the nucleus of the solitary tract (NTS), lateral parabrachial nucleus, and ventral tegmental area [9]. Studies have shown a crucial role of leptin in the feeding reward circuitry, through induction of STAT3 phosphorylation in dopamine and GABA (γ-amino butyric acid) neurons of the ventral tegmental area and mesoaccumbens [40, 41]. AMP activated protein kinase (AMPK) is another leptin target in the brain [42, 43]. AMPK is phosphorylated and activated in response to energy deficits during cellular stress or fasting, leading to increased fatty acid oxidation and inhibition of anabolic pathways [43]. AMPK is co-localized with STAT3, NPY, and other hypothalamic neuropeptides. Leptin inhibits AMPK in the hypothalamus, resulting in appetite suppression and weight loss [42]. Studies have also revealed a crosstalk between leptin and insulin signaling in the hypothalamus mediated through Jak2, PI3K (phosphoinositide 3-kinase) and IRS1 and IRS2 (insulin receptor substrate 1 and 2), resulting in inhibition of feeding [44] (Figure 45.2).

Leptin affects neurotransmission in a manner that cannot be explained by Jak-STAT signaling. For example, leptin depolarizes POMC neurons in the Arc and decreases the inhibitory tone of GABA on POMC neurons [45]. Rising leptin levels also hyperpolarize and inactivate NPY neurons in the Arc [45]. Conversely, low leptin levels during fasting activate NPY/AGRP neurons and stimulate feeding [46]. Leptin also hyperpolarizes glucose responsive neurons in

the hypothalamus by opening KATP channels, and this has been linked to inhibition of feeding and weight reduction [47] (Figure 45.2).

Leptin restores synaptic density in NPY and POMC neurons in the hypothalamus within a few hours, and this is thought to mediate leptin induced satiety [48]. Congenital leptin deficiency is associated with obesity and reduced brain size due to neuronal loss and impaired myelination [49, 50]. These deficits are reversed by leptin treatment in concert with inhibition of appetite and weight loss [49, 50]. Leptin also stimulates the development of axonal projections from Arc to PVN, and this trophic action is attenuated in DIO [51, 52].

Leptin has profound effects on human brain activity [53–55]. In one study, obese patients were food restricted, maintained at 90 percent of their initial weight, and received replacement doses of leptin or placebo [53]. Brain activity responses to visual food and non-food visual stimuli were monitored using functional magnetic resonance imaging. Leptin levels fell during weight loss and increased brain activity in areas involved in emotional, cognitive, and sensory control of food intake. Leptin replacement maintained weight loss and reversed the changes in brain activity, confirming leptin is a critical factor linking reduced energy stores to eating behavior [53]. Leptin treatment attenuated the desire to eat in patients with congenital leptin deficiency, and this was related to suppression of activity in the striatum, a region involved in the pleasure and reward responses to food [54]. In another study, leptin suppressed brain activity in regions related to hunger, and increased activity in areas linked to satiety [55]. Leptin also stimulates hippocampal activity cognitive function [56, 57].

Leptin resistance in obesity has been linked to steatosis, lipotoxicity, and organ dysfunction [7]. Although the brain is the major site of leptin action, low levels of LRb are expressed in peripheral tissues and involved in metabolism [7]. Deletion of LRb from pancreatic β-cells increased islet mass, and impaired glucose stimulated insulin release and glucose tolerance [58]. LRb is also expressed by CD34+ hematopoietic bone marrow precursors, monocytes and macrophages, and T and B cells [59]. Leptin promotes innate immunity through activation of monocytes/macrophages, neutrophils, and natural killer cells [59].

Obesity protects against osteoporosis suggesting a connection between energy metabolism and regulation of bone [60]. Leptin regulates bone mass through the sympathetic nervous system and activation of cocaine and amphetamine regulated transcript (CART) neurons [60]. Deletion of neuronal LRb increased bone formation and resorption, resulting in a high bone mass [61]. Enhancement of leptin signaling through a Y985L substitution in LRb decreased bone mass without changing feeding or energy expenditure [61]. Furthermore, leptin decreased the levels of osteocalcin, leading to attenuation of insulin release [62].

Thus, leptin is an important signal linking adipose tissue and skeleton to glucose homeostasis.

ADIPONECTIN

Adiponectin is produced exclusively by white adipocytes and composed of an N-terminal sequence, hypervariable domain, 15 collagenous repeats, and a C-terminal domain [63]. A trimeric form of adiponectin is secreted by adipocytes and gives rise to hexamers (low molecular weight; LMW) and six trimers (18 mers, high molecular weight; HMW) through non-covalent bonding [63]. HMW adiponectin is thought to be the bioactive form in plasma, while trimeric and hexameric adiponectin is predominant in the cerebrospinal fluid [64, 65]. Adiponectin also undergoes post-translational modifications including glycosylation [63]. Total and HMW adiponectin are more abundant in females, partly due to suppression of adiponectin by androgens in males. In contrast to leptin, adiponectin is reduced in obesity and rises with prolonged fasting and severe weight reduction. Adiponectin, particularly HMW, is increased by thiazolidinediones (TZD) and mediates the insulin sensitizing effect of TZD [64].

Adiponectin deficiency in rodents increases hepatic insulin resistance, which is reversed by adiponectin treatment [66–68]. Hypoadiponectinemia in humans is also strongly associated with obesity, hepatic steatosis, insulin resistance, inflammation, dyslipidemia, and cardiovascular morbidity [63] (Figure 45.3). Insulin resistance is associated with impaired skeletal muscle oxidation capacity and reduced mitochondrial number and function [69]. Individuals with a family history of Type 2 diabetes display skeletal muscle insulin resistance and impaired mitochondrial function strongly associated with adiponectin deficiency [69]. Adiponectin treatment of human myotubes in primary culture induced mitochondrial biogenesis, and fatty acid oxidation and citrate synthase activity, suppressed reactive oxygen species production, and improved glucose uptake [69].

Reduced adiponectin levels in obesity and insulin resistant states contribute to the excess cardiovascular risk observed in these conditions [63]. Adiponectin ameliorates the progression of vascular injury and atherosclerosis in rodents, consistent with its association with improved vascular outcomes in epidemiological studies [63]. In endothelial cells, adiponectin stimulates production of nitric oxide, suppresses reactive oxygen species, and protects against inflammation resulting from exposure to hyperglycemia or inflammatory cytokines, by activating cyclic AMP dependent protein kinase (AMPK) [63]. Adiponectin protects against ischemic-reperfusion injury in the heart via cyclo-oxygenase-2 mediated suppression of TNF signaling, inhibition of apoptosis by AMPK, and inhibition of peroxynitrite induced oxidative stress [70]. Adiponectin inhibits

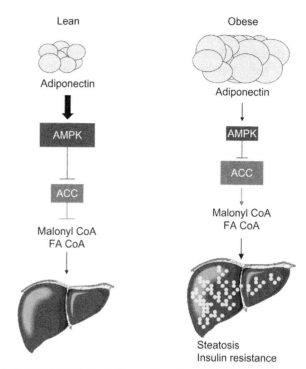

FIGURE 45.3 Effects of adiponectin on hepatic lipid metabolism. In lean individuals, adiponectin activates AMPK and inhibits ACC, resulting in diminution of malonyl-CoA, enhanced fatty acid oxidation and reduced lipogenesis. Adiponectin is decreased in obesity, and this results in attenuation of AMPK activity, increased ACC activity, elevated malonyl-CoA levels, reduced fatty acid oxidation, and increased lipogenesis. Hepatic steatosis in obesity has been associated with insulin resistance.

monocyte adhesion, macrophage transformation, and proliferation and migration of vascular smooth muscle cells, by activating AMPK and inhibiting NFκB (nuclear factor κB) [63]. Putative adiponectin receptors, AdipoR1 and AdipoR2, containing seven transmembrane domains with an internal N-terminus and an external C-terminus, mediate the signaling of adiponectin [71]. In contrast to G protein coupled receptors, activation of AdipoRs does not affect cAMP levels. AdipoR1 is highly expressed in skeletal muscle, while AdipoR2 is mainly expressed in liver. AdipoR1 has a higher affinity for globular adiponectin than for full length adiponectin. AdipoR2 has an intermediate affinity for both globular and full length adiponectin [71].

Adiponectin stimulates glucose uptake and fatty acid oxidation in skeletal muscle [68 ,69, 71], through interaction between AdipoR1 and adaptor protein containing pleckstrin homology domain, phosphotyrosine domain, and leucine zipper domain (APPL) [72]. Once bound to AdipoR1, adiponectin stimulates APPL binding to the intracellular region of AdipoR1, which activates Rab5, a small GTPase known to facilitate the membrane translocation of glucose transporter-4. APPL also interacts with PI3K and Akt, suggesting a crosstalk between adiponectin and insulin signaling [72]. Association of APPL and AdipoR1 stimulates phosphorylation and activation of AMPK, leading to inactivation of ACC [72]. Normally, ACC catalyzes the reaction that produces malonyl-CoA, an inhibitor of fatty acid oxidation. Therefore, AdipoR mediated activation of AMPK has a net effect to enhance fatty acid oxidation. AMPK activation also mediates adiponectin's ability to increase glucose uptake and lactate production in muscle and suppress gluconeogenesis [72].

The function of AdipoRs has been studied in mice. Adenovirus mediated expression of AdipoR1 and R2 in the liver of *db/db* mice activated AMPK and PPARα signaling, decreased gluconeogenesis, and enhanced fatty acid oxidation [73]. Conversely, ablation of AdipoR1 attenuated adiponectin induced AMPK activation, while ablation of AdipoR2 decreased PPARα signaling. Disruption of both AdipoR1 and AdipoR2 abolished adiponectin binding, induced lipid accumulation in liver and muscle, and induced inflammation, oxidative stress, and insulin resistance [73].

Adiponectin also plays a role in energy balance via neuronal targets in the brain. Adiponectin is present in the cerebrospinal fluid (CSF) in rodents and humans, and CSF adiponectin is increased following peripheral adiponectin administration [65, 74, 75]. This indicates adiponectin can cross the blood–brain barrier [65, 74, 75]. Central administration of adiponectin stimulated energy expenditure and decreased weight and fat content in mice [74]. Adiponectin also enhanced AMPK activity in the Arc through AdipoR1, stimulated food intake, and decreased energy expenditure [75]. In contrast, adiponectin knockout mice showed decreased AMPK phosphorylation in the Arc, decreased food intake, increased energy expenditure, and resistance to obesity [75]. Serum and cerebrospinal fluid levels of adiponectin and AdipoR1 expression in the Arc are increased during fasting, and this has led to the proposal that adiponectin is a major signal for the physiological adaptation to fasting [75].

Contrary to these findings, intracerebroventricular injection of adiponectin suppressed food intake, and this effect was associated with activation of IRS1/2, ERK, Akt, FOXO1, Jak2, and STAT3, via AdipoR1 in the Arc and LHA [76]. Thus, adiponectin shares similar signaling pathways with leptin and insulin in the hypothalamus. Another study revealed opposing effects of AdipoR deletion on energy balance [77]. AdipoR1 deficient mice had increased adiposity associated with decreased glucose tolerance, locomotor activity, and energy expenditure. In contrast, AdipoR2 knockout mice were lean, and had improved glucose tolerance, higher locomotor activity and energy expenditure, and reduced plasma cholesterol levels [77].

As with leptin, adiponectin exerts electrophysiological actions in the brain [78–81]. The area postrema (AP) in the brainstem lacks a blood–brain barrier and is a critical homeostatic integrator for humoral and neural signals. AP neurons expressing both AdipoR1 and AdipoR2 were depolarized by adiponectin, and direct injection of

adiponectin into AP increased blood pressure [78]. In contrast, adiponectin decreased blood pressure by modulating the excitability of NPY neurons in the NTS [79].

Adiponectin also depolarized neuroendocrine corticotropin releasing hormone neurons in the PVN, and increased plasma ACTH levels [80]. In contrast, adiponectin did not affect thyrotropin releasing hormone (TRH) neurons in the PVN [81]. Instead, adiponectin depolarized both preautonomic TRH and oxytocin neurons, indicating the existence of distinct populations of PVN neurons involved in the neuroendocrine and autonomic functions of adiponectin [81].

Excessive caloric intake leads to accumulation of lipids not only in adipose tissue, but also in non-adipose tissue. Ectopic lipid accumulation has been linked to insulin resistance in liver and muscle and pancreatic β-cell failure [7]. Some studies have suggested that ectopic fat in the liver rather than fat accumulation in adipose tissue plays a critical role in the development of abnormal metabolism in obesity. Indeed, removal of subcutaneous adipose tissue through liposuction did not alter plasma concentrations of adiponectin or improve obesity associated metabolic abnormalities [82]. Scherer and colleagues explored the connection between adiponectin and metabolic alterations in obesity [83]. A modest overexpression of adiponectin in *ob/ob* mice resulted in marked expansion of subcutaneous adipose tissue. However, hepatic steatosis, insulin resistance, and islet function, were all improved in these massively obese mice [83]. Whether adiponectin acts directly to promote adipogenesis and lipid storage in adipose tissue is unclear [83]. Nonetheless, this is a novel example of metabolically benign obesity [83].

PROINFLAMMATORY CYTOKINES

TNFα is expressed by adipocytes, stromovascular cells, and macrophages in adipose tissue [2, 84, 85]. TNFα induces the expression of genes involved in cholesterol and fatty acid synthesis while suppressing the expression of genes involved in fatty acid oxidation and glucose uptake in liver [86]. Obesity is associated with increased TNFα expression, insulin resistance, and hyperlipidemia [85]. Conversely, deletion of TNFα or its receptors improved insulin sensitivity and reduced the levels of circulating free fatty acids in obese mice [87]. In rodents, TNFα attenuates insulin signaling in part through activation of the NFκB pathway [88]. Overexpression of IKKβ impairs insulin signaling, while *ob/ob* mice heterozygous for IKKβ are protected from insulin resistance [88]. TNFα also induces insulin resistance through activation of the Jun N-terminal kinase (JNK) family of serine/threonine protein kinases. JNK phosphorylates IRS-1/IRS-2 on serine residues, making these poor substrates for the insulin receptor kinase and decreasing their affinity for PI3K. Insulin resistance in obesity is associated with increased JNK activity in liver, muscle, and adipose tissue [89].

TNFα in the circulation is generated from cleavage of plasma membrane bound TNF by TNF converting enzyme (TACE). Mice with TACE haploinsufficiency were protected from obesity and insulin resistance associated with increased uncoupling protein-1 and GLUT4 expression in white adipose tissue [90].

Interleukin-6 (IL-6) is another proinflammatory cytokine that is increased in obesity [2, 84]. Adipocytes and stromal cells express IL-6 and its receptor (IL-6R), which belongs to the same cytokine receptor family as LRb. IL-6 binding to IL-6R and gp130 results in activation of Jak/STAT3 signaling pathway [91]. Increased serum concentrations of IL-6 parallel the development of insulin resistance and cardiovascular disease in humans [2]. Administration of IL-6 inhibits insulin signaling in hepatocytes by decreasing tyrosine phosphorylation of the insulin receptor, association with PI3K to IRS-1, and activation of Akt [91]. IL-6 also induces the expression of SOCS3, which binds to the insulin receptor and decreases its autophosphorylation [92]. Injection of IL-6 in the brain decreases body fat and increases energy expenditure in rodents. Conversely, mice lacking IL-6 develop late onset obesity that is partly reversed by IL-6 treatment [93].

OTHER ADIPOKINES RELATED TO OBESITY

Resistin belongs to a family of cysteine-rich C-terminal domain proteins called RELMs (resistin-like molecules) [94, 95]. Initial studies revealed that resistin was suppressed by thiazolidinediones (TZD) and induced insulin resistance when administered in rodents [94]. Resistin deficiency decreases glucose and enhances insulin sensitivity in mice [96], while transgenic overexpression of resistin or infusion of recombinant resistin induces insulin resistance [97, 98]. The resistin receptor is not known; however, studies in rodents suggest resistin inhibits the phosphorylation and activation of AMPK, and induces SOCS3 [96, 98].

Unlike rodents where resistin is synthesized and secreted from adipocytes, the source of resistin in humans is mononuclear cells and activated macrophages [99]. Resistin has been associated with obesity, insulin resistance, vascular inflammation, and atherogenesis, but whether this is clinically relevant is unknown [100–102].

Resistin is present in cerebrospinal fluid, and inhibits the release of dopamine and norepinephrine from hypothalamic synaptosomes [103, 104]. Central administration of resistin in rodents induced insulin resistance in the liver [105, 106]. This action was partly explained by activation of neuropeptide Y in the hypothalamus and an increase in TNFα, IL-6, and SOCS3 expression in liver [105, 106].

Retinol binding protein 4 (RBP4) was discovered in mice lacking GLUT-4 in (glucose transport)-4 in adipose tissue, and shown to be elevated in insulin resistant mice and obese and diabetic patients [107, 108]. Transgenic overexpression of human RBP4 or injection of recombinant RBP4 induced insulin resistance in mice, while deletion of *Rbp4* enhanced insulin sensitivity [107]. However, the role of RPB4 in humans is controversial [109, 110]. Some studies have reported an association between serum RBP4 levels, obesity, and cardiovascular risk, but others have not consistently observed a relationship between RBP4 and glucose and lipid metabolism [109, 110].

CONCLUDING REMARKS

Adipose tissue has gained recognition not only as the main energy storage organ, but also as a source of secreted peptides. This review highlights the roles of leptin, adiponectin, and proinflammatory cytokines in obesity, diabetes, and related disorders. Current research areas include the origin of adipose tissue, and specific functions of subcutaneous and visceral adipose tissue, and how they relate to normal physiology and disease. Our knowledge of adipokine signaling has benefited immensely from animal models, but there are potential pitfalls, e.g., differences in the sources of adipokines and target tissues. Moreover, important differences exist between rodent and human circadian rhythms, thermoregulation, immune function, and glucose and lipid metabolism. Thus, it is necessary to confirm discoveries about adipokine signaling in humans under normal health and disease states.

ACKNOWLEDGEMENT

This work was supported by grant RO1-DK62348 and PO1-DK49210 from the National Institutes of Health.

REFERENCES

1. James WP. The epidemiology of obesity: the size of the problem. *J Intern Med* 2008;**263**:336–52.
2. Flier JS. Obesity wars: molecular progress confronts an expanding epidemic. *Cell* 2004;**116**:337–50.
3. Ahima RS. Leptin and the neuroendocrinology of fasting. *Front Horm Res* 2000;**26**:42–6.
4. Rupnick MA, Panigrahy D, Zhang CY, et al. Adipose tissue mass can be regulated through the vasculature. *Proc Natl Acad Sci U S A* 2002;**99**:10,730–5.
5. Xu H, Barnes GT, Yang Q, et al. Chronic inflammation in fat plays a crucial role in the development of obesity-related insulin resistance. *J Clin Invest* 2003;**112**:1821–30.
6. Weisberg SPD, Hunter R, Huber., et al. Obesity is associated with macrophage accumulation in adipose tissue. *J Clin Invest* 2003;**112**:1796–8.
7. Unger RH. The physiology of cellular liporegulation. *Annu Rev Physiol* 2003;**65**:333–47.

8. Zhang Y, Proenca R, Maffei M, et al. Positional cloning of the mouse obese gene and its human homologue. *Nature* 1994;**372**:425–32.

9. Ahima RS, Saper CB, Flier JS, Elmquist JK. Leptin regulation of neuroendocrine systems. *Front Neuroendocrinol* 2000;**21**:263–307.

10. Ahima RS, Prabakaran D, Mantzoros C, et al. Role of leptin in the neuroendocrine response to fasting. *Nature* 1996;**382**:250–2.

11. Licinio J, Negrão AB, Mantzoros C, et al. Sex differences in circulating human leptin pulse amplitude: clinical implications. *J Clin Endocrinol Metab* 1998;**83**:4140–7.

12. Howard JK, Lord GM, Matarese G, et al. Leptin protects mice from starvation-induced lymphoid atrophy and increases thymic cellularity in ob/ob mice. *J Clin Invest* 1999;**104**:1051–9.

13. Farooqi IS, Matarese G, Lord GM, et al. Beneficial effects of leptin on obesity, T cell hyporesponsiveness, and neuroendocrine/metabolic dysfunction of human congenital leptin deficiency. *J Clin Invest* 2002;**110**:1093–103.

14. Chan JLK, Heist AM, DePaoli., et al. The role of falling leptin levels in the neuroendocrine and metabolic adaptation to short-term starvation in healthy men. *J Clin Invest* 2003;**111**:1409–21.

15. Welt CK, Chan JL, Bullen J, et al. Recombinant human leptin in women with hypothalamic amenorrhea. *N Engl J Med* 2004;**351**:987–97.

16. McDuffie JR, Riggs PA, Calis KA, et al. Effects of exogenous leptin on satiety and satiation in patients with lipodystrophy and leptin insufficiency. *J Clin Endocrinol Metab* 2004;**89**:4258–63.

17. Javor ED, Cochran EK, Musso C, et al. Long-term efficacy of leptin replacement in patients with generalized lipodystrophy. *Diabetes* 2005;**54**:1994–2002.

18. Musso C, Cochran E, Javor E, et al. The long-term effect of recombinant methionyl human leptin therapy on hyperandrogenism and menstrual function in female and pituitary function in male and female hypoleptinemic lipodystrophic patients. *Metabolism* 2005;**54**:255–63.

19. Robertson SA, Leininger GM, Myers Jr. MG Molecular and neural mediators of leptin action. *Physiol Behav* 2008;**94**:637–42.

20. Tartaglia LA, Dembski M, Weng X, et al. Identification and expression cloning of a leptin receptor, OB-R. *Cell* 1995;**83**:1263–71.

21. Lee GH, Proenca R, Montez JM, et al. Abnormal splicing of the leptin receptor in diabetic mice. *Nature* 1996;**379**:632–5.

22. Cohen P, Zhao C, Cai X, et al. Selective deletion of leptin receptor in neurons leads to obesity. *J Clin Invest* 2001;**108**:1113–21.

23. Gao Q, Wolfgang MJ, Neschen S, et al. Disruption of neural signal transducer and activator of transcription 3 causes obesity, diabetes, infertility, and thermal dysregulation. *Proc Natl Acad Sci U S A* 2004;**101**:4661–6.

24. Bates SH, Stearns WH, Dundon TA, et al. STAT3 signalling is required for leptin regulation of energy balance but not reproduction. *Nature* 2003;**421**:856–9.

25. Balthasar N, Coppari R, McMinn J, et al. Leptin receptor signaling in POMC neurons is required for normal body weight homeostasis. *Neuron* 2004;**42**:983–91.

26. Mori H, Hanada R, Hanada T, et al. Socs3 deficiency in the brain elevates leptin sensitivity and confers resistance to diet-induced obesity. *Nat Med* 2004;**10**:739–43.

27. Kievit P, Howard JK, Badman MK, et al. Enhanced leptin sensitivity and improved glucose homeostasis in mice lacking suppressor of cytokine signaling-3 in POMC-expressing cells. *Cell Metab* 2006;**4**:123–32.

28. Björnholm M, Münzberg H, Leshan RL, et al. Mice lacking inhibitory leptin receptor signals are lean with normal endocrine function. *J Clin Invest* 2007;**117**:1354–60.

29. Banks WA, Farrell CL. Impaired transport of leptin across the blood-brain barrier in obesity is acquired and reversible. *Am J Physiol Endocrinol Metab* 2003;**285**:E10–5.

30. El-Haschimi K, Pierroz DD, Hileman SM, et al. Two defects contribute to hypothalamic leptin resistance in mice with diet-induced obesity. *J Clin Invest* 2000;**105**:1827–32.

31. Levin BE, Dunn-Meynell AA, Banks WA. Obesity-prone rats have normal blood-brain barrier transport but defective central leptin signaling before obesity onset. *Am J Physiol Regul Integr Comp Physiol* 2004;**286**:R143–50.

32. Enriori PJ, Evans AE, Sinnayah P, et al. Diet-induced obesity causes severe but reversible leptin resistance in arcuate melanocortin neurons. *Cell Metab* 2007;**5**:181–94.

33. Munzberg H, Flier JS, Bjorbaek C. Region-specific leptin resistance within the hypothalamus of diet-induced obese mice. *Endocrinology* 2004;**145**:4880–9.

34. Zabolotny JM, Bence-Hanulec KK, Stricker-Krongrad A, et al. PTP1B regulates leptin signal transduction in vivo. *Dev Cell* 2002;**2**:489–95.

35. Kaszubska W, Falls HD, Schaefer VG, et al. Protein tyrosine phosphatase 1B negatively regulates leptin signaling in a hypothalamic cell line. *Mol Cell Endocrinol* 2002;**195**:109–18.

36. Bence KK, Delibegovic M, Xue B, et al. Neuronal PTP1B regulates body weight, adiposity and leptin action. *Nat Med* 2006;**12**:917–24.

37. John D. Annual lipid cycles in hibernators: integration of physiology and behavior. *Annu Rev Nutr* 2005;**25**:469–97.

38. Grattan DR, Ladyman SR, Augustine RA. Hormonal induction of leptin resistance during pregnancy. *Physiol Behav* 2007;**91**:366–74.

39. Augustine RA, Ladyman SR, Grattan DR. From feeding one to feeding many: hormone-induced changes in bodyweight homeostasis during pregnancy. *J Physiol* 2008;**586**:387–97.

40. Fulton S, Pissios P, Manchon RP, et al. Leptin regulation of the mesoaccumbens dopamine pathway. *Neuron* 2006;**51**:811–22.

41. Hommel JD, Trinko R, Sears RM, et al. Leptin receptor signaling in midbrain dopamine neurons regulates feeding. *Neuron* 2006;**51**:801–10.

42. Minokoshi Y, Kim YB, Peroni OD, et al. Leptin stimulates fatty-acid oxidation by activating AMP-activated protein kinase. *Nature* 2002;**415**:339–43.

43. Kahn BB, Alquier T, Carling D, Hardie DG. AMP-activated protein kinase: ancient energy gauge provides clues to modern understanding of metabolism. *Cell Metab* 2005;**1**:15–25.

44. Niswender KD, Baskin DG, Schwartz MW. Insulin and its evolving partnership with leptin in the hypothalamic control of energy homeostasis. *Trends Endocrinol Metab* 2004;**15**:362–9.

45. Cowley MA, Smart JL, Rubinstein M, et al. Leptin activates anorexigenic POMC neurons through a neural network in the arcuate nucleus. *Nature* 2001;**411**:480–4.

46. Takahashi KA, Cone RD. Fasting induces a large, leptin-dependent increase in the intrinsic action potential frequency of orexigenic arcuate nucleus neuropeptide Y/Agouti-related protein neurons. *Endocrinology* 2005;**146**:1043–7.

47. Spanswick D, Smith MA, Mirshamsi S, et al. Insulin activates ATP-sensitive K+ channels in hypothalamic neurons of lean, but not obese rats. *Nat Neurosci* 2000;**3**:757–8.

48. Pinto S, Roseberry AG, Liu H, et al. Rapid rewiring of arcuate nucleus feeding circuits by leptin. *Science* 2004;**304**:110–5.

49. Ahima RS, Bjorbaek C, Osei S, Flier JS. Regulation of neuronal and glial proteins by leptin: implications for brain development. *Endocrinology* 1999;**140**:2755–62.

50. Matochik JA, London ED, Yildiz BO, et al. Effect of leptin replacement on brain structure in genetically leptin-deficient adults. *J Clin Endocrinol Metab* 2005;**90**:2851–4.

51. Bouret SG, Draper SJ, Simerly RB. Trophic action of leptin on hypothalamic neurons that regulate feeding. *Science* 2004;**304**:108–10.

52. Bouret SG, Gorski JN, Patterson CM, et al. Hypothalamic neural projections are permanently disrupted in diet-induced obese rats. *Cell Metab* 2008;**7**:179–85.

53. Rosenbaum M, Sy M, Pavlovich K, et al. Leptin reverses weight loss-induced changes in regional neural activity responses to visual food stimuli. *J Clin Invest* 2008;**118**:2583–91.

54. Farooqi IS, Bullmore E, Keogh J, et al. Leptin regulates striatal regions and human eating behavior. *Science* 2007;**317**:1355.

55. Baicy K, London ED, Monterosso J, et al. Leptin replacement alters brain response to food cues in genetically leptin-deficient adults. *Proc Natl Acad Sci U S A* 2007;**104**:18,276–9.

56. Harvey J. Leptin regulation of neuronal excitability and cognitive function. *Curr Opin Pharmacol* 2007;**7**:643–7.

57. Paz-Filho GJ, Babikian T, Asarnow R, et al. Leptin replacement improves cognitive development. *PLoS ONE* 2008;**3**:e3098.

58. Covey SD, Wideman RD, McDonald C, et al. The pancreatic beta cell is a key site for mediating the effects of leptin on glucose homeostasis. *Cell Metab* 2006;**4**:291–302.

59. De Rosa V, Procaccini C, Calì G, et al. A key role of leptin in the control of regulatory T cell proliferation. *Immunity* 2007;**26**:241–55.

60. Karsenty G. Convergence between bone and energy homeostases: leptin regulation of bone mass. *Cell Metab* 2006;**4**:341–8.

61. Shi Y, Yadav VK, Suda N, et al. Dissociation of the neuronal regulation of bone mass and energy metabolism by leptin in vivo. *Proc Natl Acad Sci U S A* 2008;**105**:20,529–33.

62. Hinoi E, Gao N, Jung DY, et al. The sympathetic tone mediates leptin's inhibition of insulin secretion by modulating osteocalcin bioactivity. *J Cell Biol* 2008;**183**:1235–42.

63. Kadowaki T, Yamauchi T, Kubota N, et al. Adiponectin and adiponectin receptors in insulin resistance, diabetes, and the metabolic syndrome. *J Clin Invest* 2006;**116**:1784–92.

64. Nawrocki AR, Rajala MW, Tomas E, et al. Mice lacking adiponectin show decreased hepatic insulin sensitivity and reduced responsiveness to peroxisome proliferator-activated receptor gamma agonists. *J Biol Chem* 2006;**281**:2654–60.

65. Kusminski CM, McTernan PG, Schraw T, et al. Adiponectin complexes in human cerebrospinal fluid: distinct complex distribution from serum. *Diabetologia* 2007;**50**:634–42.

66. Maeda N, Shimomura I, Kishida K, et al. Diet-induced insulin resistance in mice lacking adiponectin/ACRP30. *Nat Med* 2002;**8**:731–7.

67. Berg AH, Combs TP, Du X, et al. The adipocyte-secreted protein Acrp30 enhances hepatic insulin action. *Nat Med* 2001;**7**:947–53.

68. Yamauchi T, Kamon J, Minokoshi Y, et al. Adiponectin stimulates glucose utilization and fatty-acid oxidation by activating AMP-activated protein kinase. *Nat Med* 2002;**8**:1288–95.

69. Civitarese AE, Ukropcova B, Carling S, et al. Role of adiponectin in human skeletal muscle bioenergetics. *Cell Metab* 2006;**4**:75–87.

70. Shibata R, Sato K, Pimentel DR, et al. Adiponectin protects against myocardial ischemia-reperfusion injury through AMPK- and COX-2-dependent mechanisms. *Nat Med* 2005;**11**:1096–103.

71. Yamauchi T, Kamon J, Ito Y, et al. Cloning of adiponectin receptors that mediate antidiabetic metabolic effects. *Nature* 2003;**423**:762–9. Erratum in: Nature, 2004. 431: p. 1123.

72. Mao X, Kikani CK, Riojas RA, et al. APPL1 binds to adiponectin receptors and mediates adiponectin signalling and function. *Nat Cell Biol* 2006;**8**:516–23.

73. Yamauchi T, Kikani CK, Riojas RA, et al. Targeted disruption of AdipoR1 and AdipoR2 causes abrogation of adiponectin binding and metabolic actions. *Nat Med* 2007;**13**:332–9.

74. Qi Y, Takahashi N, Hileman SM, et al. Adiponectin acts in the brain to decrease body weight. *Nat Med* 2004;**10**:524–9. Erratum in: Nat Med, 2004.10: p. 649.

75. Kubota N, Yano W, Kubota T, et al. Adiponectin stimulates AMP-activated protein kinase in the ypothalamus and increases food intake. *Cell Metab* 2007;**6**:55–68.

76. Coope A, Milanski M, Araújo EP, et al. AdipoR1 mediates the anorexigenic and insulin/leptin-like actions of adiponectin in the hypothalamus. *FEBS Lett* 2008;**582**:1471–6.

77. Bjursell M, Ahnmark A, Bohlooly-Y M, et al. Opposing effects of adiponectin receptors 1 and 2 on energy metabolism. *Diabetes* 2007;**56**:583–93.

78. Fry M, Smith PM, Hoyda TD, et al. Area postrema neurons are modulated by the adipocyte hormone adiponectin. *J Neurosci* 2006;**26**:9695–702.

79. Hoyda TD, Smith PM, Ferguson AV. Adiponectin acts in the nucleus of the solitary tract to decrease blood pressure by modulating the excitability of neuropeptide Y neurons. *Brain Res* 2008;**1256**:76–84.

80. Hoyda TD, Fry M, Ahima RS, Ferguson AV. Adiponectin selectively inhibits oxytocin neurons of the araventricular nucleus of the hypothalamus. *J Physiol* 2007;**585**:805–16.

81. Hoyda TD, Samson WK, Ferguson AV. Adiponectin depolarizes parvocellular paraventricular nucleus neurons controlling neuroendocrine and autonomic function. *Endocrinology* 2008;**150**:832–40.

82. Klein S, Fontana L, Young VL, et al. Absence of an effect of liposuction on insulin action and risk factors for coronary heart disease. *N Engl J Med* 2004;**350**:2549–57.

83. Kim JY, van de Wall., Laplante M, et al. Obesity-associated improvements in metabolic profile through expansion of adipose tissue. *J Clin Invest* 2007;**117**:2621–37.

84. Tilg H, Moschen AR. Adipocytokines: mediators linking adipose tissue, inflammation and immunity. *Nat Rev Immunol* 2006;**6**:772–83.

85. Hotamisligil GS, Shargill NS, Spiegelman BM. Adipose expression of tumor necrosis factor-alpha: direct role in obesity-linked insulin resistance. *Science* 1993;**259**:87–91.

86. Ruan H, Miles PD, Ladd CM, et al. Profiling gene transcription in vivo reveals adipose tissue as an immediate target of tumor necrosis factor-alpha: implications for insulin resistance. *Diabetes* 2002;**51**:3176–88.

87. Uysal KT, Wiesbrock SM, Marino MW, Hotamisligil GS. Protection from obesity-induced insulin resistance in mice lacking TNF-alpha function. *Nature* 1997;**389**:610–4.

88. Yuan M, Konstantopoulos N, Lee J, et al. Reversal of obesity- and diet-induced insulin resistance with salicylates or targeted disruption of Ikkbeta. *Science* 2001;**293**:1673–7.

89. Hirosumi J, Tuncman G, Chang L, et al. A central role for JNK in obesity and insulin resistance. *Nature* 2002;**420**:333–6.

90. Serino M, Menghini R, Fiorentino L, et al. Mice heterozygous for tumor necrosis factor-alpha converting enzyme are protected from obesity-induced insulin resistance and diabetes. *Diabetes* 2007;**56**:2541–6.

91. Senn JJ, Menghini R, Fiorentino L, et al. Interleukin-6 induces cellular insulin resistance in hepatocytes. *Diabetes* 2002;**51**:3391–9.

92. Senn JJ, Klover PJ, Nowak IA, et al. Suppressor of cytokine signaling-3 (SOCS-3), a potential mediator of interleukin-6-dependent insulin resistance in hepatocytes. *J Biol Chem* 2003;**278**:13,740–6.

93. Wallenius V, Wallenius K, Ahrén B, et al. Interleukin-6-deficient mice develop mature-onset obesity. *Nat Med* 2002;**8**:75–9.

94. Steppan CM, Bailey ST, Bhat S, et al. The hormone resistin links obesity to diabetes. *Nature* 2001;**409**:307–12.

95. Steppan CM, Brown EJ, Wright CM, et al. A family of tissue-specific resistin-like molecules. *Proc Natl Acad Sci U S A* 2001;**98**:502–6.

96. Banerjee RR, Rangwala SM, Shapiro JS, et al. Regulation of fasted blood glucose by resistin. *Science* 2004;**303**:1195–8.

97. Satoh H, Nguyen MT, Miles PD, et al. Adenovirus-mediated chronic "hyper-resistinemia" leads to in vivo insulin resistance in normal rats. *J Clin Invest* 2004;**114**:224–31.

98. Rangwala SM, Rich AS, Rhoades B, et al. Abnormal glucose homeostasis due to chronic hyperresistinemia. *Diabetes* 2004;**53**:1937–41.

99. Savage DB, Sewter CP, Klenk ES, et al. Resistin/Fizz3 expression in relation to obesity and peroxisome proliferator-activated receptor-gamma action in humans. *Diabetes* 2001;**50**:2199–202.

100. Azuma K, Katsukawa F, Oguchi S, et al. Correlation between serum resistin level and adiposity in obese individuals. *Obes Res* 2003;**11**:997–1001.

101. Kunnari A, Ukkola O, Päivänsalo M, Kesäniemi YA. High plasma resistin level is associated with enhanced highly sensitive C-reactive protein and leukocytes. *J Clin Endocrinol Metab* 2006;**91**:2755–60.

102. Reilly MP, Lehrke M, Wolfe ML, et al. Resistin is an inflammatory marker of atherosclerosis in humans. *Circulation* 2005;**111**:932–9.

103. Kos K, Harte AL, da Silva NF, et al. Adiponectin and resistin in human cerebrospinal fluid and expression of adiponectin receptors in the human hypothalamus. *J Clin Endocrinol Metab* 2007;**92**:1129–36.

104. Brunetti L, Orlando G, Recinella L, et al. Resistin, but not adiponectin, inhibits dopamine and norepinephrine release in the hypothalamus. *Eur J Pharmacol* 2004;**493**:41–4.

105. Muse ED, Lam TK, Scherer PE, Rossetti L. Hypothalamic resistin induces hepatic insulin resistance. *J Clin Invest* 2007;**117**:1670–8.

106. Singhal NS, Lazar MA, Ahima RS. Central resistin induces hepatic insulin resistance via neuropeptide Y. *J Neurosci* 2007;**27**:12,924–32.

107. Yang Q, Graham TE, Mody N, et al. Serum retinol binding protein 4 contributes to insulin resistance in obesity and type 2 diabetes. *Nature* 2005;**436**:356–62.

108. Graham TE, Yang Q, Blüher M, et al. Retinol-binding protein 4 and insulin resistance in lean, obese, and diabetic subjects. *N Engl J Med* 2006;**354**:2552–63.

109. Janke J, Engeli S, Boschmann M, et al. Retinol-binding protein 4 in human obesity. *Diabetes* 2006;**55**:2805–10.

110. Graham TE, Wason CJ, Blüher M, Kahn BB. Shortcomings in methodology complicate measurements of serum retinol binding protein (RBP4) in insulin-resistant human subjects. *Diabetologia* 2007;**50**:814–23.

Chapter 46

CXC Chemokine Signaling in Interstitial Lung Diseases

Borna Mehrad and Robert M. Strieter

Division of Pulmonary and Critical Care Medicine, University of Virginia, Charlottesville, Virginia

INTRODUCTION

Interstitial lung diseases are a heterogeneous group of parenchymal lung diseases characterized by varying degrees of lung inflammation and fibrosis. Some interstitial lung diseases occur in response to known environmental insults and autoimmune mechanisms; the other idiopathic interstitial lung diseases are classified as several distinct syndromes on the basis of clinical, radiographic, and histopathologic features [1]. The prototypic interstitial lung disease, idiopathic pulmonary fibrosis (IPF), is a progressive fibrotic illness with no effective therapy that typically results in respiratory failure a median of 3 years after diagnosis [2–4].

Usual interstitial pneumonia (UIP) is the term used to describe the histologic pattern shared between IPF and several interstitial lung diseases of known etiology. The hallmark of UIP, described as temporal heterogeneity, is the juxtaposition of relatively normal areas of lung with areas of leukocyte infiltration and other areas with advanced fibrosis and architectural distortion within a given low power field [5–7]. Another histologic feature of UIP is focal collections of fibroblasts and myofibroblasts, referred to as fibroblastic foci, which are thought to represent focal aggregates within an organized reticulum of fibroblasts that courses through the entire lung [8]. UIP has been hypothesized to represent the result of repeated epithelial injury and repair and vascular remodeling in the absence of intact basement membrane [9–11].

Members of the CXC chemokine family were originally described for their role in recruiting leukocytes in the context of inflammation. In this chapter, we review the data on the role of this family of chemokine ligands and receptors in mediating vascular remodeling and in orchestrating the recruitment of circulating fibroblast progenitors to the lung in the context of interstitial lung disease.

CHEMOKINE REGULATION OF ANGIOGENESIS IN PULMONARY FIBROSIS: RECIPROCAL ROLES OF CXCR2 AND CXCR3

Angiogenesis, defined as formation and remodeling of new capillaries, is a critical biological process that is intimately connected in the pathogenesis of chronic fibroproliferative diseases, malignancy, and compensatory revascularization of ischemic tissues. In human lung disease, increase in vascular resistance in the pulmonary circulation (arising from the right ventricle) results in new blood vessel formation in the bronchial circulation (which, as part of the systemic circulation, arises from the left ventricle); similar vascular remodeling between the systemic and pulmonary circulations occur in mouse models of increased pulmonary vascular resistance [12–16]. In his context, vascular remodeling is well documented in pathological studies of human interstitial lung diseases [17–20].

The process of angiogenesis is regulated by a number of secreted mediators, which include the CXC chemokines. The defining feature of chemokine ligands is four highly conserved cysteine amino acid residues. In the CXC family, the first two of these cysteines near the amino terminus are separated by a non-conserved residue, resulting in the CXC motif. The CXC family is further classified based on the presence or absence of another amino acid sequence (glutamic acid-leucine-arginine or the ELR motif) immediately upstream of the CXC sequence [21, 22]. The ELR containing CXC chemokines, which include CXCL1/Gro-α, CXCL2/Gro-β, CXCL3/Gro-γ, CXCL5/ENA-78, CXCL6/GCP-2, CXCL7/NAP-2, and CXCL8/IL-8, were originally discovered for their neutrophil chemoattractant properties but are also potent inducers of angiogenesis [22]. The ELR+ chemokine ligands signal exclusively via the

CXCR2 receptor in mice. In humans, all angiogenic human chemokines bind and signal via CXCR2, whereas CXCL6 and CXCL8 also signal via CXCR1. Although endothelial cells express both CXCR1 and CXCR2, ligand mediated angiogenesis is dependent on CXCR2 but not CXCR1, both in the context of *in vitro* chemotaxis assays and *in vivo* angiogenesis models [23–26].

A subset of the ELR-negative chemokines, CXCL9, CXCL10, CXCL11, as well as CXCL4 are potent inhibitors of angiogenesis via their interaction with the receptor, CXCR3 [21, 22]. Among these angiostatic chemokines, CXCL9-11 (but not CXCL4) are potently induced by both interferon-α/β and interferon-γ [27]; as such, Th-1 mediated immune responses in the context of interstitial lung disease may be associated with reduced angiogenesis [28]. The angiostatic chemokine receptor, CXCR3, exists in at least three variants generated by alternative mRNA splicing and exon skipping, designated CXCR3A, CXCR3B, and CXCR3alt [29, 30]. CXCR3A is primarily responsible for mediating leukocyte influx whereas CXCR3B is expressed on endothelial cells and mediates the angiostatic properties of the receptor via the p38 MAP kinase pathway [31–34]; the functional role of CXCR3alt in regulation of angiogenesis remains to be established.

In the context of human interstitial lung disease, bronchoalveolar lavage fluid from patients with IPF is potently angiogenic in the rat corneal micropocket model and contains high levels of the angiogenic chemokines CXCL5 and CXCL8 and low levels of angiostatic chemokines CXCL10 and CXCL11. Furthermore, neutralization of CXCL5 and CXCL8 inhibits the angiogenic properties of the fluid [28, 35–38]. In the context of the mouse model of bleomycin induced pulmonary fibrosis, the induction of the angiogenic chemokines CXCL1/KC and CXCL2-3/MIP-2 in the lungs correlated with the degree of tissue angiogenic activity, whereas the expression of the angiostatic chemokines CXCL10 and CXCL11 were suppressed [39, 40]. In this model, the immunoneutralization of CXCL1/KC or CXCL2-3/MIP-2 resulted in attenuation of lung angiogenic activity as well as the degree of fibrosis [38–40]. Conversely, exogenous administration of the angiostatic chemokines CXCL10 or CXCL11 resulted in both reduced lung angiogenic activity and reduced lung fibrosis [38–40]. In addition, while the chemokine CXCL11 can bind and signal via CXCR7 in addition to CXCR3, the angiostatic effect of exogenously administered CXCL11 in this model system was entirely CXCR3 dependent [38–40].

MESENCHYMAL PROGENITORS IN PULMONARY FIBROSIS: ROLE OF CXCR4

Among the many cells that have been implicated in the pathogenesis of interstitial lung diseases, cells of the fibroblast-myofibroblast lineage play pivotal roles in the generation

of the extracellular matrix and generation of fibrosis. The classical hypothesis regarding the source of these cells in the fibrotic lung is that tissue injury induces activation and proliferation of resident interstitial fibroblasts that subsequently migrate into the alveolar spaces, expresses constituents of the extracellular matrix, and differentiates into myofibroblasts [41–43]. A second hypothetical mechanism involves trans-differentiation of epithelial cells into fibroblasts and myofibroblasts as a result of changes in the lung microenvironment [43–45]. Finally, a circulating bone marrow derived progenitor cell, the fibrocyte, can home to sites of lung injury, differentiate to myofibroblasts, proliferate, and contribute to the generation of extracellular matrix [27, 46–48]. Fibrocytes, defined as collagen-1 expressing circulating leukocytes, comprise approximately 0.5 percent of nucleated cells in peripheral blood of healthy humans [46, 48, 49], and can be cultured from a CD14+ cell population and have a monocytic morphology [47]. Fibrocytes express fibroblast markers (vimentin, fibronectin, collagens I and III), the common leukocyte antigen (CD45RO), the pan-myeloid antigen (CD13), HLA-DR, and the hematopoietic stem cell marker, CD34 [27, 46–50], as well as the adhesion molecules CD11b and CD18 [27, 47–49]. In contrast, fibrocytes do not express markers of monocyte/macrophage lineage, lymphocyte markers, or surface markers for epithelial and endothelial cells myofibroblasts [27, 46–50]. In culture, fibrocytes lose expression of CD34 and CD45 and spontaneously express alpha-smooth muscle actin (α-SMA); α-SMA is upregulated in the presence of TGFβ compatible with differentiation into myofibroblasts [27, 46–51]. Human and mouse fibrocytes both express the chemokine receptors CCR7 and CXCR4 but differ in that humans express CCR3 and CCR5 whereas mouse fibrocytes express CCR2 [46, 47, 49, 52]. CXCR4 and its ligand, CXCL12, are critical to the homing of both hematopoietic and non-hematopoietic progenitor cells, including fibrocytes [46, 53].

In the mouse model of bleomycin induced pulmonary fibrosis, the fibrocyte pool progressively expands in the bone marrow and blood early after exposure to bleomycin, and fibrocytes accumulated in the lungs and correlated with deposition of collagen. Similarly, human cultured fibrocytes administered intravenously to immunocompromised mice challenged with intrapulmonary bleomycin home to the lungs. This accumulation corresponded with expression of CXCL12 in the lungs after administration of bleomycin and neutralization of CXCL12 resulted in both reduced lung numbers of fibrocytes and attenuated lung fibrosis without influencing other lung leukocyte populations [46] suggesting that CXCR4 mediated homing of fibrocytes to the lung is important to the pathophysiology of pulmonary fibrosis. A smaller pool of CCR7+ fibrocytes were also found to accumulate in the lung in response to bleomycin, potentially indicating that a CXCR4 independent mechanism may be relevant homing of this smaller population of fibrocytes to the lungs.

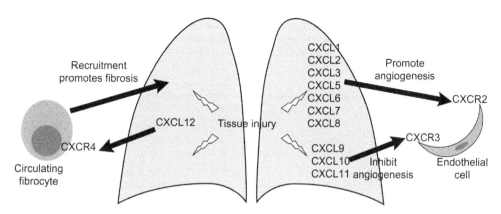

FIGURE 46.1 Diagrammatic representation of chemokine signaling in fibrotic interstitial lung diseases.

Fibrocytes were also found to accumulate in the lung in another mouse model of fluorescein induced pulmonary fibrosis [52]. These cells were found to express CXCR4, CCR5, CCR7, and CCR2, migrated in response to the CCR2 ligands, CCL2 and CCL12, and lost expression of CCR2 when cultured [52]; in addition, CCR2 deficient mice challenged with intrapulmonary fluorescein had fewer number of lung fibrocytes and attenuated fibrosis [52]. The effects of CCR2 in this model were subsequently shown to be via the ligand CCL12 rather than CCL2 [54]. Since only a small subset of human fibrocytes express CCR2 and the mouse ligand CCL12 does not have a human counterpart, the relevance of this observation in human interstitial lung disease remains to be determined. Another group has reported fibrocyte influx to the lung in the bleomycin model to be mediated by the CCL3-CCR5 ligand receptor pair [55]. Interestingly, this work also found parallel attenuation of CXCL12 expression in the lungs of CCL3- and CCR5 deficient animals, suggesting that the effect of CCL3/CCR5 may be mediated via the regulation of CXCL12 expression.

The relevance of chemokines and fibrocytes has also been investigated in the context of human interstitial lung diseases. Patients with fibrotic interstitial lung disease were found to have elevated lung and blood levels of CXCL12 that correlated with a log-fold elevated concentration of circulating fibrocytes in the peripheral blood, which consisted mostly of CXCR4 expressing cells [56]. In addition, fibrocytes were detectable by immunofluorescent microscopy in the lungs of patients with IPF but not in normal lungs and their number correlated with the number of fibroblastic foci and lung CXCL12 levels [57].

POTENTIAL THERAPEUTIC APPLICATIONS OF CHEMOKINE SIGNALING

Therapeutic manipulation of chemokine ligands or receptors has not, to our knowledge, been applied to patients with interstitial lung disease. A small molecule CXCR2 antagonist

has been developed by Glaxo-SmithKline for use in COPD [58, 59] and a phase I safety study of the compound has recently been completed in healthy adults (Trials.gov identifier NCT00504439). With regards to CXCR4, several small molecule antagonists have been developed with the aim of inhibiting cellular entry of CXCR4-trophic strains of HIV and for mobilization of bone marrow stem cells [60–65]. Phase I safety trials have been published for one such agent, AMD070, in healthy volunteers [66, 67], and for another agent, AMD3100, in a small number of individuals with HIV [61]. As of the time of writing, Trials.gov lists nine recruiting and six completed trials of CXCR4 antagonists in several cancers, HIV, and for mobilization of progenitor cells.

CONCLUSIONS

The CXC chemokine family contributes to the pathogenesis of interstitial lung diseases via several distinct mechanisms, including the regulation of vascular modeling and mediating the traffic of bone marrow derived progenitor cells to the lungs. In the context of animal models, manipulation of these mechanisms results in measurable alteration of disease severity, suggesting that CXC chemokines might represent novel therapeutic targets in interstitial lung diseases (Figure 46.1).

ACKNOWLEDGEMENT

This work was supported by NIH grants HL73848 and an American Lung Association Career Investigator Award (BM) and CA87879 and HL66027 (RMS).

REFERENCES

1. American Thoracic Society/European Respiratory Society International Multidisciplinary Consensus Classification of the

Idiopathic Interstitial Pneumonias. This joint statement of the American Thoracic Society (ATS), and the European Respiratory Society (ERS) was adopted by the ATS board of directors, June 2001 and by the ERS Executive Committee, June 2001. *Am J Respir Crit Care Med* 2002;**165**:277–304.

2. Bjoraker JA, Ryu JH, Edwin MK, et al. Prognostic significance of histopathologic subsets in idiopathic pulmonary fibrosis. *Am J Respir Crit Care Med* 1998;**157**:199–203.

3. Flaherty KR, Travis WD, Colby TV, et al. Histopathologic variability in usual and nonspecific interstitial pneumonias. *Am J Respir Crit Care Med* 2001;**164**:1722–7.

4. Nicholson AG, Colby TV, du Bois RM, et al. The prognostic significance of the histologic pattern of interstitial pneumonia in patients presenting with the clinical entity of cryptogenic fibrosing alveolitis. *Am J Respir Crit Care Med* 2000;**162**:2213–17.

5. American Thoracic Society. Idiopathic pulmonary fibrosis: diagnosis and treatment. International consensus statement. *Am J Respir Crit Care Med* 2000;**161**:646–64. American Thoracic Society (ATS), and the European Respiratory Society (ERS).

6. Katzenstein AL, Myers JL. Idiopathic pulmonary fibrosis: clinical relevance of pathologic classification. *Am J Respir Crit Care Med* 1998;**157**:1301–15.

7. Katzenstein AL, Zisman DA, Litzky LA, et al. Usual interstitial pneumonia: histologic study of biopsy and explant specimens. *Am J Surg Pathol* 2002;**26**:1567–77.

8. Cool CD, Groshong SD, Rai PR, et al. Fibroblast foci are not discrete sites of lung injury or repair: the fibroblast reticulum. *Am J Respir Crit Care Med* 2006;**174**:654–8.

9. Maher TM, Wells AU, Laurent GJ. Idiopathic pulmonary fibrosis: multiple causes and multiple mechanisms? *Eur Respir J* 2007;**30**:835–9.

10. Strieter RM. What differentiates normal lung repair and fibrosis? Inflammation, resolution of repair, and fibrosis. *Proc Am Thorac Soc* 2008;**5**:305–10.

11. Selman M, Pardo A. Idiopathic pulmonary fibrosis: an epithelial/fibroblastic cross-talk disorder. *Respir Res* 2002;**3**:3.

12. Mitzner W, Lee W, Georgakopoulos D, Wagner E. Angiogenesis in the mouse lung. *Am J Pathol* 2000;**157**:93–101.

13. Charan NB, Carvalho P. Angiogenesis in bronchial circulatory system after unilateral pulmonary artery obstruction. *J Appl Physiol* 1997;**82**:284–91.

14. Schlaepfer K. Ligation of the pulmonary artery of one lung with and without resection of the phrenic nerve. *Arch Surg* 1924;**9**:25–94.

15. Michel RP, Hakim TS, Petsikas D. Segmental vascular resistance in postobstructive pulmonary vasculopathy. *J Appl Physiol* 1990;**69**:1022–32.

16. Dutly AE, Andrade CF, Verkaik R, et al. A novel model for posttransplant obliterative airway disease reveals angiogenesis from the pulmonary circulation. *Am J Transplant* 2005;**5**:248–54.

17. Turner-Warwick M. Precapillary systemic-pulmonary anastomoses. *Thorax* 1963;**18**:225–37.

18. Renzoni EA, Walsh DA, Salmon M, et al. Interstitial vascularity in fibrosing alveolitis. *Am J Respir Crit Care Med* 2003;**167**:438–43.

19. Cosgrove GP, Brown KK, Schiemann WP, et al. Pigment epithelium-derived factor in idiopathic pulmonary fibrosis: a role in aberrant angiogenesis. *Am J Respir Crit Care Med* 2004;**170**:242–51.

20. Keane MP, Belperio JA, Xue YY, et al. Depletion of CXCR2 inhibits tumor growth and angiogenesis in a murine model of lung cancer. *J Immunol* 2004;**172**:2853–60.

21. Belperio JA, Keane MP, Arenberg DA, et al. CXC chemokines in angiogenesis. *J Leukoc Biol* 2000;**68**:1–8.

22. Strieter RM, Polverini PJ, Kunkel SL, et al. The functional role of the ELR motif in CXC chemokine-mediated angiogenesis. *J Biol Chem* 1995;**270**:27,348–57.

23. Addison CL, Daniel TO, Burdick MD, et al. The CXC chemokine receptor 2, CXCR2, is the putative receptor for ELR+ CXC chemokine-induced angiogenic activity. *J Immunol* 2000;**165**:5269–77.

24. Murdoch C, Monk PN, Finn A. Cxc chemokine receptor expression on human endothelial cells. *Cytokine* 1999;**11**:704–12.

25. Schraufstatter IU, Trieu K, Zhao M, et al. IL-8-mediated cell migration in endothelial cells depends on cathepsin B activity and transactivation of the epidermal growth factor receptor. *J Immunol* 2003;**171**:6714–22.

26. Heidemann J, Ogawa H, Dwinell MB, et al. Angiogenic effects of interleukin 8 (CXCL8) in human intestinal microvascular endothelial cells are mediated by CXCR2. *J Biol Chem* 2003;**278**:8508–15.

27. Bucala R, Spiegel LA, Chesney J, et al. Circulating fibrocytes define a new leukocyte subpopulation that mediates tissue repair. *Mol Med* 1994;**1**:71–81.

28. Antoniou KM, Tzouvelekis A, Alexandrakis MG, et al. Different angiogenic activity in pulmonary sarcoidosis and idiopathic pulmonary fibrosis. *Chest* 2006;**130**:982–8.

29. Loetscher M, Loetscher P, Brass N, et al. Lymphocyte-specific chemokine receptor CXCR3: regulation, chemokine binding and gene localization. *Eur J Immunol* 1998;**28**:3696–705.

30. Ehlert JE, Addison CA, Burdick MD, et al. Identification and partial characterization of a variant of human CXCR3 generated by posttranscriptional exon skipping. *J Immunol* 2004;**173**:6234–40.

31. Romagnani P, Annunziato F, Lasagni L, et al. Cell cycle-dependent expression of CXC chemokine receptor 3 by endothelial cells mediates angiostatic activity. *J Clin Invest* 2001;**107**:53–63.

32. Salcedo R, Resau JH, Halverson D, et al. Differential expression and responsiveness of chemokine receptors (CXCR1-3) by human microvascular endothelial cells and umbilical vein endothelial cells. *Faseb J* 2000;**14**:2055–64.

33. Lasagni L, Francalanci M, Annunziato F, et al. An alternatively spliced variant of CXCR3 mediates the inhibition of endothelial cell growth induced by IP-10, Mig, and I-TAC, and acts as functional receptor for platelet factor 4. *J Exp Med* 2003;**197**:1537–49.

34. Petrai I, Rombouts K, Lasagni L, et al. Activation of p38(MAPK) mediates the angiostatic effect of the chemokine receptor CXCR3-B. *Int J Biochem Cell Biol* 2008;**40**:1764–74.

35. Keane MP, Arenberg DA, Lynch 3rd JP, et al. The CXC chemokines, IL-8 and IP-10, regulate angiogenic activity in idiopathic pulmonary fibrosis. *J Immunol* 1997;**159**:1437–43.

36. Keane MP, Belperio JA, Burdick MD, et al. ENA-78 is an important angiogenic factor in idiopathic pulmonary fibrosis. *Am J Respir Crit Care Med* 2001;**164**:2239–42.

37. Belperio JA, Keane MP, Burdick MD, et al. Role of CXCL9/CXCR3 chemokine biology during pathogenesis of acute lung allograft rejection. *J Immunol* 2003;**171**:4844–52.

38. Burdick MD, Murray LA, Keane MP, et al. CXCL11 attenuates bleomycin-induced pulmonary fibrosis via inhibition of vascular remodeling. *Am J Respir Crit Care Med* 2005;**171**:261–8.

39. Keane MP, Belperio JA, Arenberg DA, et al. IFN-gamma-inducible protein-10 attenuates bleomycin-induced pulmonary fibrosis via inhibition of angiogenesis. *J Immunol* 1999;**163**:5686–92.

40. Keane MP, Belperio JA, Moore TA, et al. Neutralization of the CXC chemokine, macrophage inflammatory protein-2, attenuates bleomycin-induced pulmonary fibrosis. *J Immunol* 1999;**162**:5511–18.

41. Fukuda Y, Ishizaki M, Masuda Y, et al. The role of intraalveolar fibrosis in the process of pulmonary structural remodeling in patients with diffuse alveolar damage. *Am J Pathol* 1987;**126**:171–82.

42. Marshall R, Bellingan G, Laurent G. The acute respiratory distress syndrome: fibrosis in the fast lane. *Thorax* 1998;**53**:815–7.

43. Kalluri R, Neilson EG. Epithelial-mesenchymal transition and its implications for fibrosis. *J Clin Invest* 2003;**112**:1776–84.

44. Kim KK, Kugler MC, Wolters PJ, et al. Alveolar epithelial cell mesenchymal transition develops in vivo during pulmonary fibrosis and is regulated by the extracellular matrix. *Proc Natl Acad Sci U S A* 2006;**103**:13,180–5.

45. Iwano M, Plieth D, Danoff TM, et al. Evidence that fibroblasts derive from epithelium during tissue fibrosis. *J Clin Invest* 2002;**110**:341–50.

46. Phillips RJ, Burdick MD, Hong K, et al. Circulating fibrocytes traffic to the lungs in response to CXCL12 and mediate fibrosis. *J Clin Invest* 2004;**114**:438–46.

47. Abe R, Donnelly SC, Peng T, et al. Peripheral blood fibrocytes: differentiation pathway and migration to wound sites. *J Immunol* 2001;**166**:7556–62.

48. Metz CN. Fibrocytes: a unique cell population implicated in wound healing. *Cell Mol Life Sci* 2003;**60**:1342–50.

49. Quan TE, Cowper S, Wu SP, et al. Circulating fibrocytes: collagen-secreting cells of the peripheral blood. *Int J Biochem Cell Biol* 2004;**36**:598–606.

50. Schmidt M, Sun G, Stacey MA, et al. Identification of circulating fibrocytes as precursors of bronchial myofibroblasts in asthma. *J Immunol* 2003;**171**:380–9.

51. Chauhan H, Abraham A, Phillips JR, et al. There is more than one kind of myofibroblast: analysis of CD34 expression in benign, in situ, and invasive breast lesions. *J Clin Pathol* 2003;**56**:271–6.

52. Moore BB, Kolodsick JE, Thannickal VJ, et al. CCR2-mediated recruitment of fibrocytes to the alveolar space after fibrotic injury. *Am J Pathol* 2005;**166**:675–84.

53. Murdoch C. CXCR4: chemokine receptor extraordinaire. *Immunol Rev* 2000;**177**:175–84.

54. Moore BB, Murray L, Das A, et al. The role of CCL12 in the recruitment of fibrocytes and lung fibrosis. *Am J Respir Cell Mol Biol* 2006;**35**:175–81.

55. Ishida Y, Kimura A, Kondo T, et al. Essential roles of the CC chemokine ligand 3-CC chemokine receptor 5 axis in bleomycin-induced pulmonary fibrosis through regulation of macrophage and fibrocyte infiltration. *Am J Pathol* 2007;**170**:843–54.

56. Mehrad B, Burdick MD, Zisman DA, et al. Circulating peripheral blood fibrocytes in human fibrotic interstitial lung disease. *Biochem Biophys Res Commun* 2007;**353**:104–8.

57. Andersson-Sjoland A, de Alba CG, Nihlberg K, et al. Fibrocytes are a potential source of lung fibroblasts in idiopathic pulmonary fibrosis. *Int J Biochem Cell Biol* 2008;**40**:2129–40.

58. Nicholson GC, Tennant RC, Carpenter DC, et al. A novel flow cytometric assay of human whole blood neutrophil and monocyte CD11b levels: upregulation by chemokines is related to receptor expression, comparison with neutrophil shape change, and effects of a chemokine receptor (CXCR2) antagonist. *Pulm Pharmacol Ther* 2007;**20**:52–9.

59. Podolin PL, Bolognese BJ, Foley JJ, et al. A potent and selective non-peptide antagonist of CXCR2 inhibits acute and chronic models of arthritis in the rabbit. *J Immunol* 2002;**169**:6435–44.

60. Devine SM, Flomenberg N, Vesole DH, et al. Rapid mobilization of CD34+ cells following administration of the CXCR4 antagonist AMD3100 to patients with multiple myeloma and non-Hodgkin's lymphoma. *J Clin Oncol* 2004;**22**:1095–102.

61. Fransen S, Bridger G, Whitcomb JM, et al. Suppression of dualtropic human immunodeficiency virus type 1 by the CXCR4 antagonist AMD3100 is associated with efficiency of CXCR4 use and baseline virus composition. *Antimicrob Agents Chemother* 2008;**52**:2608–15.

62. Holtan SG, Porrata LF, Micallef IN, et al. AMD3100 affects autograft lymphocyte collection and progression-free survival after autologous stem cell transplantation in non-Hodgkin lymphoma. *Clin Lymphoma Myeloma* 2007;**7**:315–18.

63. Lack NA, Green B, Dale DC, et al. A pharmacokinetic-pharmacodynamic model for the mobilization of CD34+ hematopoietic progenitor cells by AMD3100. *Clin Pharmacol Ther* 2005;**77**:427–36.

64. Liles WC, Rodger E, Broxmeyer HE, et al. Augmented mobilization and collection of CD34+ hematopoietic cells from normal human volunteers stimulated with granulocyte-colony-stimulating factor by single-dose administration of AMD3100, a CXCR4 antagonist. *Transfusion* 2005;**45**:295–300.

65. Shepherd RM, Capoccia BJ, Devine SM, et al. Angiogenic cells can be rapidly mobilized and efficiently harvested from the blood following treatment with AMD3100. *Blood* 2006;**108**:3662–7.

66. Cao YJ, Flexner CW, Dunaway S, et al. Effect of low-dose ritonavir on the pharmacokinetics of the CXCR4 antagonist AMD070 in healthy volunteers. *Antimicrob Agents Chemother* 2008;**52**:1630–4.

67. Stone ND, Dunaway SB, Flexner C, et al. Multiple-dose escalation study of the safety, pharmacokinetics, and biologic activity of oral AMD070, a selective CXCR4 receptor inhibitor, in human subjects. *Antimicrob Agents Chemother* 2007;**51**:2351–8.

Chapter 47

ER and Oxidative Stress: Implications in Disease

Jyoti D. Malhotra[1] and Randal J. Kaufman[1,2,3]

[1]*Department of Biological Chemistry, University of Michigan Medical School, Ann Arbor, Michigan*

[2]*Department of Internal Medicine, University of Michigan Medical School, Ann Arbor, Michigan*

[3]*Department of Howard Hughes Medical Institute, University of Michigan Medical School, Ann Arbor, Michigan*

INTRODUCTION

The endoplasmic reticulum (ER) is a membranous network extending throughout the cytoplasm of the eukaryotic cell and is contiguous with the nuclear envelope. The ER is the site of synthesis of sterols, lipids, core-asparagine linked oligosaccharides, and membrane and secreted proteins biosynthesis. The ER has evolved as a protein folding machine and a major intracellular signaling organelle. Numerous posttranslational modification reactions occur at the ER and many of these are required for proteins to attain their final-folded functional conformation. The quality of protein folding is strictly monitored by protein chaperones that prevent aberrant folding and aggregation. These chaperones permit only properly folded proteins to exit the ER, a process termed as *quality control*. Quality control is a surveillance mechanism that permits only properly folded proteins to exit the ER *en route* to other intracellular organelles and the cell surface. Misfolded proteins are retained within the ER lumen in complex with molecular chaperones or are directed toward degradation through the 26S proteasome in a process called ER associated protein degradation (ERAD) or through autophagy.

The ER provides a unique environment for protein folding, assembly, and disulfide bond formation prior to transit to Golgi compartment. ER function is perturbed when unfolded or misfolded proteins exceed the folding capacity of the ER. The high concentration of partially folded and unfolded proteins predisposes partially folded proteins to aggregation. Polypeptide binding proteins, such as BiP and GRP94, act to slow protein folding reactions and prevent aberrant interactions and aggregation. The ER lumen is an oxidizing environment so disulfide bonds form. As a consequence, cells have evolved sophisticated machinery composed of many protein disulfide isomerases (PDIs) that

are required to ensure proper disulfide bond formation and prevent formation of illegitimate disulfide bonds. Protein folding in the ER requires extensive amounts of energy and depletion of energy stores prevents proper protein folding. The ER is also the primary Ca^{2+} storage organelle in the cell. Both protein folding reactions and protein chaperone functions require high levels of ER intralumenal calcium. All these processes are highly sensitive to alterations in the ER luminal environment. As a consequence, innumerable environmental insults alter protein folding reactions in the ER through mechanisms that include depletion of ER calcium, alteration in the redox status, and energy (sugar/glucose) deprivation. The accumulation of unfolded proteins signals activation of an adaptive process known as the unfolded protein response (UPR). Appropriate adaptation to misfolded protein accumulation in the ER lumen requires regulation at all levels of gene expression including transcription, translation, translocation into the ER lumen, and ERAD. Coordinate regulation of all these processes is required to restore proper protein folding and ER homeostasis [1–6]. Conversely, if the protein folding defect is not resolved, the UPR is chronically activated to signal an apoptotic (programmed cell death) response.

UPR SIGNALING

Upon accumulation of unfolded or misfolded proteins in the ER lumen three ER localized transmembrane signal transducers are activated to initiate adaptive responses. These transducers are two protein kinases IRE1 (inositol requiring kinase 1) [7, 8], and PERK (double-stranded RNA activated protein kinase-like ER kinase) [9] and the transcription factor ATF6 (activating transcription factor 6) [8, 10] (Figure 47.1).

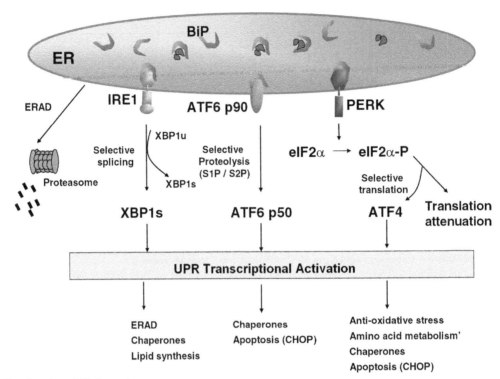

FIGURE 47.1 Signaling the unfolded protein response.
Three proximal sensors, IRE1, PERK, and ATF6, act in concert to regulate the UPR through their respective signaling cascade and collectively referred to as tripartite signaling in the ER. The protein chaperone BiP is the master regulator and negatively regulates these pathways. Under non-stressed conditions, BiP binds to the lumenal domains of IRE1 and PERK to prevent their dimerization. With the accumulation of the unfolded proteins, BiP is released from IRE1 permits dimerization to activate its kinase and RNase activities to initiate XBP1 mRNA splicing thereby creating a potent transcriptional activator. Primary targets that require IRE1/XBP1 pathway for induction are genes encoding functions in ERAD. Similarly BiP release from ATF6 permits transport to the Golgi compartment where ATF6 is cleaved by SIP and S2P proteases to yield cytosolic fragment that migrates to the nucleus to further activate transcription of UPR responsive genes. Finally BiP release permits PERK dimerization and activation to phosphorylate eIF2α on Ser 51, which leads to general attenuation of translational initiation. eIF2α / phosphorylation preferentially induces ATF4 mRNA and also recent evidence has shown that PERK/eIF2α/ATF4 regulatory axis induces expression of antioxidative stress response genes pathway and also promotes expression of proapoptotic transcription factor CHOP.

IRE1 SIGNALING

The first component in the UPR pathway was identified in the budding yeast *Saccharomyces cerevisiae* in the early 1990s using a genetic screen to identify mutants in UPR signaling. Two independent groups identified Ire1p/Ern1p as an ER transmembrane protein kinase that acts as a proximal sensor in the yeast UPR [3, 11]. Subsequently, it was discovered that Ire1p is a bifunctional protein that also has a site specific endoribonuclease (RNase) activity. When the cells are not stressed, Ire1p protein kinase is maintained in an inactive monomeric form through interactions with the protein chaperone Kar2p/BiP. Under conditions of ER stress, Ire1p is released from Kar2p/BiP and undergoes homodimerization and trans-autophosphorylation to activate its RNase activity. The RNase activity of Ire1p cleaves a 252-base intron from mRNA encoding the basic leucine zipper (bZIP) containing transcription factor Hac1p [12]. The protein encoded by spliced *HAC1* mRNA binds and activates transcription from the UPR element (UPRE, minimal motif TGACGTG(C/A)) upstream of many UPR target genes [2, 13]. In *S. cerevisiae*, the UPR activates transcription of approximately 381 genes [14].

Two mammalian homologs of yeast IRE1 have been identified; IRE1α [15] and IRE1β [16]. IRE1α is expressed in most cells and tissues, with highest levels of expression in the pancreas and placenta [15]. IRE1β expression is prominent only in intestinal epithelial cells [16]. The cleavage specificities of IRE1α and IRE1β are quite similar, thereby suggesting that they do not recognize distinct substrates but rather confer temporal and tissue specific expression [17].

Analysis of promoter regions of UPR inducible genes in mammals, such as BiP, Grp94, and calreticulin, identified a mammalian ER stress response element (ERSE, CCAAT(N9)CCACG) that is necessary and sufficient for UPR gene activation [18]. Subsequently, Yoshida *et al.* [18] used a yeast one hybrid screen to identify the bZIP containing transcription factor XBP1 (X-box binding protein) as an ERSE binding protein. Several groups demonstrated that *XBP1* mRNA is a substrate for the endoribonuclease activity of mammalian IRE1 [8, 19–21]. On activation of the UPR, IRE1 RNase cleaves *XBP1* mRNA to remove 26 nucleotide intron. This splicing reaction creates a translational frame shift to produce a larger form of XBP1 that contains a novel transcriptional activation domain in its C-terminus. Spliced Xbp1 is a transcriptional activator that plays a key role activation of wide variety of UPR target genes. Some of the genes identified that require the IRE1/XBP1 pathway are those that encode functions involved in ERAD, such as EDEM. Indeed, cells that are deficient in either IRE1 or XBP1 are defective in ERAD.

Deletion of *Ire1α* or *Xbp1* in mice creates an embryonic lethality at E11.5–E14 [20, 22]. Mice with heterozygous Xbp1 deletion appear normal but develop insulin resistance when fed a high fat diet [23]. Thus, it was proposed that the UPR might be important in insulin signaling (see below). In addition, both Ire1 and Xbp 1 have critical roles for B cell differentiation. Antigenic stimulation of mature B lymphocytes activates the UPR and signaling through IRE1 mediated XBP1 mRNA splicing is required to drive cells to differentiate into plasma cells [19, 24–26]. These studies suggest that the IRE1/XBP1 subpathway of the UPR might be required for differentiation of cell types that secrete high levels of protein [27].

PERK SIGNALING

In response to ER stress there is an immediate transient attenuation of mRNA translation, thereby preventing continued influx of newly synthesized polypeptides into the stressed ER lumen [28]. This translational attenuation is signaled through PERK mediated phosphorylation of the eukaryotic translation initiation factor 2 on the alpha subunit (eIF2α) at Ser51. eIF2α phosphorylation inhibits the guanine nucleotide exchange factor eIF2B that recycles the eIF2 complex to its active GTP bound form. The formation of the ternary translation initiation complex eIF2-GTP-tRNAMet is required for AUG initiation codon recognition and joining of the 60S ribosomal subunit that occurs during initiation phase of polypeptide chain synthesis. Lower levels of active ternary complex result in lower levels of translation initiation [9, 29, 30]. PERK is an ER associated transmembrane serine/threonine protein kinase. Upon accumulation of unfolded proteins in the ER lumen, PERK dimerization and trans-autophosphorylation leads to activation of the eIF2α kinase function [9, 31]. In addition to translational attenuation, activation of PERK also contributes to transcriptional induction of the majority of the UPR genes [29, 30, 32, 33]. Although phosphorylation of eIF2α inhibits general translation initiation, it is required for the selective translation of several mRNAs. One fundamental transcription factor for which translation is activated upon PERK mediated phosphorylation of eIF2α, is the activating transcription factor 4 (ATF4). Expression profiling identified that genes encoding amino acid biosynthesis and transport functions, anti-oxidative stress responses,

and apoptosis, such as growth arrest and DNA damage 34 (GADD34) and CAAT/enhancer binding protein (C/EBP) homologous protein (CHOP) [31, 34] require PERK, eIF2α phosphorylation, and ATF4 [29, 30, 32, 33].

ATF6 SIGNALING

The bZiP containing activating transcription factor 6 (ATF6) was identified as another regulatory protein that, like XBP1, binds the ERSE1 element in the promoters of UPR responsive genes [18]. There are two alleles of ATF6, ATF6α and ATF6β, both synthesized in all cell types as ER transmembrane proteins. In the unstressed state ATF6 is localized at the ER membrane and bound to BiP. In response to ER stress, BiP dissociation leads to transport of ATF6 to the Golgi complex. In the Golgi complex, ATF6 is sequentially cleaved by two proteases. The serine protease site-1 protease (S1P) cleaves ATF6 in the luminal domain. The N-terminal portion is cleaved by the metalloprotease site-2 protease S2P [35]. The processed forms of ATF6α and ATF6β translocate to the nucleus and bind to the ATF/cAMP response element (CRE) and to the ER stress responsive element (ERSE-1) to activate target genes [36]. ATF6α (90 kda) and ATF6β (110 kDa) both require the presence of the transcription factor CBF (also called NF-Y) to bind ERSEI [36–38]. Recently, ATF6α and ATF6β have been deleted in the mouse. Although deletion of either alone produce no significant phenotype, combined deletion is an early embryonic lethal. Where ATF6α contributes significantly to UPR induced gene expression, no genes were identified that are regulated through ATF6β.

Recently, CREBH was identified as a liver specific bZiP transcription factor of the CREB/ATF family with a transmembrane domain directs localization to the ER. Pro-inflammatory cytokines IL6, 1L-1β, and TNFα increase transcription of CREBH to produce an inert protein that is localized to the ER. Upon ER stress, CREBH transits to the Golgi compartment where it is cleaved by S1P and S2P processing enzymes. However, cleaved CREBH does not activate transcription of UPR genes but, rather, induces transcription of many acute phase response genes, such as C-reactive protein and murine Serum Amyloid P component (SAP) in hepatocytes. These studies have identified CREBH as a novel ER localized transcription factor that has an essential role in induction of innate immune response genes and links for the first time ER stress to inflammatory responses [39].

Biochemical studies have demonstrated that in the unstressed state, the luminal domains of IRE1, PERK, and ATF6 are bound to the protein chaperone BiP. As unfolded proteins accumulate, they bind to BiP, thereby promoting BiP release from the UPR signal transducers. When these sensors are bound to BiP, they are maintained in an inactive state. This model for negative regulation of the UPR by BiP is also supported by the observation that BiP overexpression prevented activation of the UPR upon ER stress. However, recently, based on the crystal structure of the ycast Ire1p luminal domain, Credle et al. identified that a deep, long MHC1-type groove exists in an Ire1p dimer and proposed that unfolded polypeptides directly bind Ire1p to mediate its dimerization [40]. However, although analysis of the human IRE1 indicated a similar structure, the MHC1-type groove was not solvent accessible [41]. In addition, the luminal domain was shown to form dimers in the absence of added polypeptide, bringing into question the requirement for peptide binding to promote dimerization. Future studies should resolve this issue.

ER STRESS INDUCED APOPTOSIS

If the efforts to correct the protein folding defect fail, apoptosis is activated. Both mitochondrial dependent and independent cell death pathways likely mediate apoptosis in response to ER stress. The ER might actually serve as a site where apoptotic signals are generated and integrated to elicit the death response. Several mechanisms have been proposed

by which apoptotic signals at the ER are generated. These include Bak/Bax regulated Ca^{2+} release from the ER, cleavage and activation of procaspase-12, IRE1 mediated activation of ASK1 (apoptosis signal regulating kinase 1)/JNK (c-Jun amino terminal kinase), and PERK/eIF2α dependent induction of the proapoptotic transcription factor CHOP (Figure 47.2).

Upon ER stress, Bak and Bax in the ER membrane undergo a conformational change to permit Ca^{2+} efflux, which activates Ca^{2+} dependent processes such as protein kinase C and calcineurin. It was also proposed that Ca^{2+} activates m-Calpain in the cytoplasm to cleave and activate ER resident procaspase 12. However, the significance of caspase 12 activation remains in question since a functional caspase 12 is not conserved in humans. The Ca^{2+} efflux also activates mitochondria dependent apoptotic pathways. Activated IRE1 binds to c-Jun N-terminal inhibitory kinase (JIK) and recruits TRAF2, which leads to the activation of ASK1/JNK. The Ca^{2+} released from the ER enters mitochondria leading to depolarization of the inner membrane, cytochrome c release, and activation of Apaf-1 (apoptosis protease activating factor 1)/procaspase-9 regulated apoptosis. CHOP (CEBP homologous protein) is a downstream UPR transcriptional target. CHOP is a basic leucine zipper

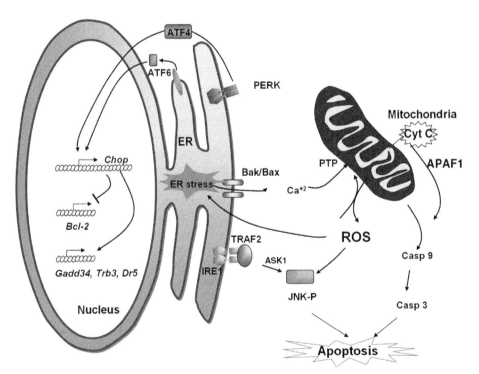

FIGURE 47.2 UPR regulated programmed cell death.
ER stress leads to several redundant pathways for caspase activation that involve mitochondrial dependent or independent pathways. Activated IRE1 recruits TRAF2 that leads to activation of JNK phosphorylation. A second death signaling pathway activated by ER stress is mediated by transcriptional activation of genes encoding proapoptotic functions. Activation of PERK, ATF6, and possibly IRE1, lead to transcriptional activation of CHOP that potentiates apoptosis possibly through regulation of Gadd 34, Dr5, and Trb3, or by inhibiting the antiapoptotic protein Bcl2. ER stress can also lead to ROS production and this can also occur subsequent to accumulation of unfolded protein in the ER. Mitochondrial ROS can also be generated as a result of ER stress induced Ca^{+2} release by depolarizing the inner mitochondrial membrane (PTP), cytochrome C release, and activation of Apaf-1/procaspase-9 regulated apoptosis. Thus oxidative stress in association of unresolved ER stress contributes to multiple pathways of cell death.

containing transcription factor that inhibits the expression of Bcl-2, and activates transcription of several genes that encode apoptotic functions including GADD34, DR5, and TRB3.

Analysis of gene deleted mice has provided insight into ER stress induced apoptosis. Cells from Apaf-1 deficient mice are susceptible to ER stress induced apoptosis, indicating that non-mitochondrial death pathways exist. Bak/Bax double knockout, *Caspase-12–/–*, and *Chop–/–* MEFs all show partial resistance to ER stress induced apoptosis, further supporting the idea that they facilitate the apoptotic response upon ER stress. Although, caspase 12 deficient and CHOP deficient mice show no developmental defects, they display protection to genetically imposed or environmentally imposed ER stress.

ER–MITOCHONDRIAL INTERACTIONS

ER and mitochondria interact both physiologically and functionally and the most critical aspect is the calcium signaling. The key processes linking the apoptosis to the ER–mitochondria interactions is the discovery that a massive and/or a prolonged influx of calcium into mitochondria can lead to the opening of a large conductance pore in the PTP swelling of the organelle, breakage of OMM and release into the cytosol of a series of proapoptotic proteins such as cytochrome c, apoptosis inducing factor (AIF), and smac/Diablo [42]. Antiapoptotic Bcl-2 members (Bcl2 and Bcl2-XL) can affect ER Ca^{2+} storage. Overexpression of Bcl2 results in decrease of ER luminal Ca^{2+} and this effect has been attributed to an increase in the Ca^{2+} leak from the organelle [43–46]. Bax and Bak knockdown increases the interaction of Bcl2 with type-1 IP3s and promotes both the phosphorylation of the IP3R and constitutive Ca^{2+} leak through the IP3Rs [47–49]. A reduced ER Ca^{2+} content as in the case of Bcl2 overexpression or knockdown of Bax/Bak reduces the amount of Ca^{2+} that can be released from the ER owing to an apoptotic stimulus and thus decreases the probability of a Ca^{2+} dependent PTP opening [50–53].

OXIDATIVE PROTEIN FOLDING IN THE ER

There is accumulating evidence to suggest that protein folding and production of ROS are closely linked events; however, this area of ER stress is not well explored. In eukaryotes, oxidative protein folding occurs in the ER. A growing family of ER oxidoreductases, including PDI (protein disulphide isomerase), ERp57, ERp72, PDIR, PDIp, and P5, catalyze these protein folding reactions in mammalian cells. PDI is a multi-functional protein capable of catalyzing the formation, isomerization, and reduction of disulfide bonds *in vitro* as well as being an essential subunit for the enzymes prolyl 4-hydroxylase

and microsomal triacylglycerol transfer protein [54]. When disulfide bond formation occurs, cysteine residues within the PDI active site [-C-X-X-C-] accept two electrons from the polypeptide chain substrate. This electron transfer results in the oxidation of the substrate and the reduction of the PDI active site. Despite the ability of PDI to enhance the rate of disulfide linked folding, the mechanisms by which the ER disposes of electrons as a result of the oxidative disulfide bond formation reaction have remained enigmatic. A number of different factors have been proposed to maintain the oxidizing environment of the ER, including the preferential secretion of reduced thiols and uptake of oxidized thiols, and a variety of different redox enzymes and small molecule oxidants. However, there is a lack of genetic evidence demonstrating that these factors are physiologically important [55–57]. It was believed for many years that the low molecular mass thiol glutathione is responsible for oxidizing the PDI active sites. This was contrary to observations in yeast where depletion of glutathione did not interfere with disulfide bond formation [58, 59]

One fundamental unanswered question is whether the presence of an unfolded protein in the ER lumen is sufficient to activate oxidative stress. It has been estimated that approximately 25 percent of the ROS generated in a cell may result from formation of disulfide bonds in the ER during oxidative protein folding. During formation of disulfide bonds, ROS are a by-product formed as ERO1 and PDI act in concert to transfer electrons from thiol groups in proteins to molecular oxygen. In addition, ROS may be formed as a consequence of the glutathione depletion that occurs as glutathione reduces unstable and improper disulfide bonds. The consumption of GSH would return thiols involved in non-native disulfide bonds to their reduced form so they may again interact with ERO1/PDI1 to be reoxidized. This would generate a futile cycle of disulfide bond formation and breakage in which each cycle would generate ROS and consume GSH. As a consequence, it is expected that proteins that have multiple disulfide bonds may be more prone to generating oxidative stress. It is presently unknown whether misfolding of a protein that has no disulfide bonds would generate oxidative stress.

ER STRESS AND OXIDATIVE STRESS: IMPLICATIONS IN HUMAN DISEASE

The UPR evolved as a complex homeostatic mechanism to balance the load of newly synthesized proteins with the capacity for chaperone assisted protein folding in the lumen of the ER. Increasing evidence suggests that protein misfolding in the ER lumen and alterations in the UPR play important roles in numerous disease states, including diabetes mellitus, atherosclerosis, neoplasia, and neurodegenerative diseases.

UPR AND OXIDATIVE STRESS IN METABOLIC DISEASE

The development of Type 2 diabetes is associated with a combination of insulin resistance in fat, muscle, and liver and a failure of pancreatic beta cells to adequately compensate by increased insulin production [60, 61]. There is also evidence that oxidative damage is associated with development of the diabetic state [62–64].

The requirement for the UPR in beta cell function was first suggested through the identification of *PERK* as the gene defective in the human disease Wolcott-Rallison syndrome (WRS) [65]. Individuals with WRS, as well as *Perk–/–* mice, develop beta cell apoptosis with infantile onset insulin dependent diabetes [65, 66]. In addition, mice with homozygous Ser51Ala mutation at the PERK phosphorylation site in eIF2α display even greater beta cell loss that appears *in utero* [29]. There are several mechanisms that may explain the unique requirement beta cells display for PERK/eIF2α. First, beta cells may require PERK/eIF2α signaling because they are sensitive to physiological fluctuations in blood glucose. In beta cells the generation of ATP fluctuates with blood glucose because glycolysis is controlled by glucokinase that has a low affinity for glucose. Periodic decreases blood glucose level would decrease the ATP:ADP ratio and compromise protein folding in the ER so that UPR may be frequently activated. Through this mechanism, PERK/eIF2α signaling would couple protein synthesis with energy available for protein folding reactions in the ER lumen. Alternatively, glucose stimulates insulin transcription, translation, and secretion. PERK phosphorylation of eIF2α may be required for beta cells to attenuate protein synthesis so that insulin production does not exceed folding capacity of the ER. Finally, it is also possible that beta cells require PERK/eIF2α to minimize oxidative stress. There is increasing evidence that suggests oxidative stress contributes to the beta cell failure in diabetes. Beta cells express low levels of catalase and glutathione peroxidase, two enzymes that protect from ROS. Recent evidence indicates the PERK/eIF2α pathway prevents oxidative stress [67]. In addition, the PERK/eIF2α/ATF4 pathway induces expression of antioxidative stress response genes [30, 68]. Thus, defects in PERK/eIF2α signaling might generate oxidative stress.

Antioxidants have been reported to preserve glucose stimulated insulin secretion, prevent apoptosis, and expand beta cell mass, without significantly affecting cell proliferation. For example, treatment of Zucker diabetic fatty (ZDF) rats with the antioxidants N-acetyl-L-cysteine or aminoguanidine prevented hyperglycemia, improved insulin secretion, and increased PDX1 binding to the insulin promoter [69]. Although the mechanism by which antioxidants improve beta cell function is not known, there is evidence to support the idea that oxidative stress activates JNK. Activated JNK can phosphorylate PDX1 to suppress PDX1 binding to specific promoters by preventing its translocation to the nucleus [70]. Significantly, JNK inhibition protects beta cells from oxidative stress, prevents apoptosis, and improves islet graft function [71]. In conclusion, evidence supports the notion that oxidative stress and ER stress play central roles in the pathogenesis of Type 2 diabetes. The findings support the notion that targeted therapy to intervene to prevent JNK activation may reduce progression of insulin resistance to diabetes.

NEURODEGENERATIVE DISEASES

Neurodegenerative diseases, such as Alzhiemer's disease and Parkinson's disease, represent a large class of conformational diseases associated with accumulation of abnormal protein aggregates in and around affected neurons. Oxidative stress and protein misfolding play critical roles in the pathogenesis of these neurodegenerative diseases [72]. These diseases are characterized by fibrillar aggregates that are composed of misfolded proteins [73]. At the cellular level, neuronal death or apoptosis may be mediated by oxidative stress and/or ER stress. Upregulation of ER stress markers has been demonstrated in postmortem brain tissues and cell culture models of many neurodegenerative disorders, including Parkinson's disease, Alzheimer's disease, amyotropic lateral sclerosis (ALS), and expanded polyglutamine diseases such as Huntingtons's disease and spinocerebellar ataxias [74]. The significance of ER dysfunction that occurs by direct action of oligomeric, a potentially toxic species that in turn generates oxidative stress and cell death is not clear. However several *in vivo* studies suggest that at least in some cases ER stress may have a significant correlation with neurodegeneration [75].

The mechanisms underlying the accumulation of abnormal protein aggregates in and around neurons in Alzheimer's and Parkinson's disease has bewildered both researchers and clinicians. How these protein aggregates disturb ER function directly is not understood, but *in vitro* studies suggest these can inhibit the proteasome and ERAD. For example, in disease Machado-Joseph syndrome, the polyglutamine repeats present in spinocerebrocellular atrophy protein (SCA3) form cytosolic aggregates that can inhibit the proteasome. Proteasome inhibition in the cytosol can interfere with ERAD to elicit UPR activation, caspase 12 activation, and apoptosis [76, 77]. In humans, mutations in SIL1, which encodes an adenine nucleotide exchange factor of BiP cause Marinesco-Sjögren syndrome, a rare disease associated with cerebellar ataxia, progressive myopathy, and cataracts [78]. In mice, homozygous mutations in *Sil1* cause cerebellar Purkinje cell degeneration and subsequent ataxia [75]. Analysis of Sil-1 deficient mice demonstrated that the mutant Purkinje cells have ubiquitinated nuclear and ER associated protein aggregates and also exhibit upregulation of several ER stress markers BiP, CHOP, and ORP150 [75]. As the protein chaperone BiP requires ATP/ADP exchange during its peptide binding and release cycle, it is possible that SIL1 is essential to preserve proper protein folding in the ER and prevent UPR activation [79].

Recent studies have also implicated oxidative stress in the pathogenesis of neurodegenerative diseases. The pathology of a group of neurodegenerative diseases, including Alzheimer's disease, are characterized by the deposition of intracellular aggregates containing abnormally phosphorylated forms of the microtubule binding protein tau [80]. Using a *Drosophila* model relevant to human neurodegenerative diseases, it was demonstrated that oxidative stress plays a casual role in neurotoxicity and promotes tau-phosphorylation. In this model, activation of the JNK pathway correlated with the degree of tau induced neurodegeneration [81]. Although oxidative stress and ER stress have been linked to neurodegenerative diseases, in most instances these events may not be the primary cause of neuron death. However, these stresses may modify the progression and severity of these complex diseases.

FUTURE DIRECTIONS

There has been significant progress in understanding the mechanisms underlying the cause of ER stress over the past few decades. New insights have been gained into how the cells respond to ER stress to mediate survival as well as cell death responses. Each of the signaling pathways that constitute the UPR contributes to activation of different subsets of UPR genes. There is a close link between ER stress and oxidative stress but the mechanisms linking both are not very well understood. Future studies are required to understand how these stresses affect protein folding, misfolding, and secretion *in vivo*. Further studies should identify under what physiological and pathological states these pathways are activated *in vivo* and how they influence the disease outcome. Studies in this important area will aid in comprehending how interactions between ER stress and oxidative stress are integrated into other signaling pathways. A greater exploration of the understanding of complex interaction between protein misfolding and oxidative stress may lead to development of more specific therapeutic agents targeted for diseases associated with ER/oxidative stress.

ACKNOWLEDGEMENTS

We thank Janet Mitchell for assistance with manuscript preparation and the members of the Kaufman laboratory for critical input. This work was supported in part by DK042394, HL052173, and HL057346 (RJK). RJK is an Investigator of the Howard Hughes Medical Institute.

REFERENCES

1. Kaufman RJ. Orchestrating the unfolded protein response in health and disease. *J Clin Invest* 2002;**110**:1389–98.
2. Patil C, Walter P. Intracellular signaling from the endoplasmic reticulum to the nucleus: the unfolded protein response in yeast and mammals. *Curr Opin Cell Biol* 2001;**13**:349–55.
3. Mori K, Ma W, Gething MJ, Sambrook J. A transmembrane protein with a cdc2+/CDC28-related kinase activity is required for signaling from the ER to the nucleus. *Cell* 1993;**74**:743–56.
4. Harding HP, Calfon M, Urano F, Novoa I, Ron D. Transcriptional and translational control in the Mammalian unfolded protein response. *Annu Rev Cell Dev Biol* 2002;**18**:575–99.
5. Schroder M, Kaufman RJ. The mammalian unfolded protein response. *Annu Rev Biochem* 2005;**74**:739–89.
6. Wu J, Kaufman RJ. From acute ER stress to physiological roles of the Unfolded Protein Response. *Cell Death Differ* 2006;**13**:374–84.
7. Sidrauski C, Walter P. The transmembrane kinase Ire1p is a site-specific endonuclease that initiates mRNA splicing in the unfolded protein response. *Cell* 1997;**90**:1031–9.
8. Yoshida H, Matsui T, Yamamoto A, Okada T, Mori K. XBP1 mRNA is induced by ATF6 and spliced by IRE1 in response to ER stress to produce a highly active transcription factor. *Cell* 2001;**107**:881–91.
9. Harding HP, Zhang Y, Bertolotti A, Zeng H, Ron D. Perk is essential for translational regulation and cell survival during the unfolded protein response. *Mol Cell* 2000;**5**:897–904.
10. Yoshida H, Okada T, Haze K, Yanagi H, Yura T, Negishi M, Mori K. ATF6 activated by proteolysis binds in the presence of NF-Y (CBF) directly to the cis-acting element responsible for the mammalian unfolded protein response. *Mol Cell Biol* 2000;**20**:6755–67.
11. Cox JS, Walter P. A novel mechanism for regulating activity of a transcription factor that controls the unfolded protein response. *Cell* 1996;**87**:391–404.
12. Cox JS, Shamu CE, Walter P. Transcriptional induction of genes encoding endoplasmic reticulum resident proteins requires a transmembrane protein kinase. *Cell* 1993;**73**:1197–206.
13. Mori K, Ogawa N, Kawahara T, Yanagi H, Yura T. mRNA splicing-mediated C-terminal replacement of transcription factor Hac1p is required for efficient activation of the unfolded protein response. *Proc Natl Acad Sci USA* 2000;**97**:4660–5.
14. Travers KJ, Patil CK, Wodicka L, Lockhart DJ, Weissman JS, Walter P. Functional and genomic analyses reveal an essential coordination between the unfolded protein response and ER-associated degradation. *Cell* 2000;**101**:249–58.
15. Tirasophon W, Welihinda AA, Kaufman RJ. A stress response pathway from the endoplasmic reticulum to the nucleus requires a novel bifunctional protein kinase/endoribonuclease (Ire1p) in mammalian cells. *Genes Dev* 1998;**12**:1812–24.
16. Wang XZ, Harding HP, Zhang Y, Jolicoeur EM, Kuroda M, Ron D. Cloning of mammalian Ire1 reveals diversity in the ER stress responses. *Embo J* 1998;**17**:5708–17.
17. Niwa M, Sidrauski C, Kaufman RJ, Walter P. A role for presenilin-1 in nuclear accumulation of Ire1 fragments and induction of the mammalian unfolded protein response. *Cell* 1999;**99**:691–702.
18. Yoshida H, Haze K, Yanagi H, Yura T, Mori K. Identification of the cis-acting endoplasmic reticulum stress response element responsible for transcriptional induction of mammalian glucose-regulated proteins. Involvement of basic leucine zipper transcription factors. *J Biol Chem* 1998;**273**:33,741–33,749.

19. Calfon M, Zeng H, Urano F, Till JH, Hubbard SR, Harding HP, Clark SG, Ron D. IRE1 couples endoplasmic reticulum load to secretory capacity by processing the XBP-1 mRNA. *Nature* 2002;**415**:92–6.

20. Lee K, Tirasophon W, Shen X, Michalak M, Prywes R, Okada T, Yoshida H, Mori K, Kaufman RJ. IRE1-mediated unconventional mRNA splicing and S2P-mediated ATF6 cleavage merge to regulate XBP1 in signaling the unfolded protein response. *Genes Dev* 2002;**16**:452–66.

21. Shen X, Ellis RE, Lee K, Liu CY, Yang K, Solomon A, Yoshida H, Morimoto R, Kurnit DM, Mori K, Kaufman RJ. Complementary signaling pathways regulate the unfolded protein response and are required for C. elegans development. *Cell* 2001;**107**:893–903.

22. Reimold AM, Etkin A, Clauss I, Perkins A, Friend DS, Zhang J, Horton HF, Scott A, Orkin SH, Byrne MC, Grusby MJ, Glimcher LH. An essential role in liver development for transcription factor XBP-1. *Genes Dev* 2000;**14**:152–7.

23. Ozcan U, Cao Q, Yilmaz E, Lee AH, Iwakoshi NN, Ozdelen E, Tuncman G, Gorgun C, Glimcher LH, Hotamisligil GS. Endoplasmic reticulum stress links obesity, insulin action, and type 2 diabetes. *Science* 2004;**306**:457–61.

24. Reimold AM, Iwakoshi NN, Manis J, Vallabhajosyula P, Szomolanyi-Tsuda E, Gravallese EM, Friend D, Grusby MJ, Alt F, Glimcher LH. Plasma cell differentiation requires the transcription factor XBP-1. *Nature* 2001;**412**:300–7.

25. Iwakoshi NN, Lee AH, Vallabhajosyula P, Otipoby KL, Rajewsky K, Glimcher LH. Plasma cell differentiation and the unfolded protein response intersect at the transcription factor XBP-1. *Nat Immunol* 2003;**4**:321–9.

26. Zhang K, Wong HN, Song B, Miller CN, Scheuner D, Kaufman RJ. The unfolded protein response sensor IRE1alpha is required at 2 distinct steps in B cell lymphopoiesis. *J Clin Invest* 2005;**115**:268–81.

27. Lee AH, Chu GC, Iwakoshi NN, Glimcher LH. XBP-1 is required for biogenesis of cellular secretory machinery of exocrine glands. *Embo J* 2005;**24**:4368–80.

28. Kaufman RJ. Regulation of mRNA translation by protein folding in the endoplasmic reticulum. *Trends Biochem Sci* 2004;**29**:152–8.

29. Scheuner D, Song B, McEwen E, Liu C, Laybutt R, Gillespie P, Saunders T, Bonner-Weir S, Kaufman RJ. Translational control is required for the unfolded protein response and *in vivo* glucose homeostasis. *Mol Cell* 2001;**7**:1165–76.

30. Harding HP, Zhang Y, Zeng H, Novoa I, Lu PD, Calfon M, Sadri N, Yun C, Popko B, Paules R, Stojdl DF, Bell JC, Hettmann T, Leiden JM, Ron D. An integrated stress response regulates amino acid metabolism and resistance to oxidative stress. *Mol Cell* 2003;**11**:619–33.

31. Harding HP, Novoa I, Zhang Y, Zeng H, Wek R, Schapira M, Ron D. Regulated translation initiation controls stress-induced gene expression in mammalian cells. *Mol Cell* 2000;**6**:1099–108.

32. Ron D. Translational control in the endoplasmic reticulum stress response. *J Clin Invest* 2002;**110**:1383–8.

33. Harding HP, Zhang Y, Ron D. Protein translation and folding are coupled by an endoplasmic-reticulum-resident kinase. *Nature* 1999;**397**:271–4.

34. Ma Y, Brewer JW, Diehl JA, Hendershot LM. Two distinct stress signaling pathways converge upon the CHOP promoter during the mammalian unfolded protein response. *J Mol Biol* 2002;**318**:1351–65.

35. Ye J, Rawson RB, Komuro R, Chen X, Dave UP, Prywes R, Brown MS, Goldstein JL. ER stress induces cleavage of membrane-bound ATF6 by the same proteases that process SREBPs. *Mol Cell* 2000;**6**:1355–64.

36. Yoshida H, Okada T, Haze K, Yanagi H, Yura T, Negishi M, Mori K. Endoplasmic reticulum stress-induced formation of transcription fac-

tor complex ERSF including NF-Y (CBF) and activating transcription factors 6alpha and 6beta that activates the mammalian unfolded protein response. *Mol Cell Biol* 2001;**21**:1239–48.

37. Haze K, Yoshida H, Yanagi H, Yura T, Mori K. Mammalian transcription factor ATF6 is synthesized as a transmembrane protein and activated by proteolysis in response to endoplasmic reticulum stress. *Mol Biol Cell* 1999;**10**:3787–99.

38. Li M, Baumeister P, Roy B, Phan T, Foti D, Luo S, Lee AS. ATF6 as a transcription activator of the endoplasmic reticulum stress element: thapsigargin stress-induced changes and synergistic interactions with NF-Y and YY1. *Mol Cell Biol* 2000;**20**:5096–106.

39. Zhang K, Shen X, Wu J, Sakaki K, Saunders T, Rutkowski DT, Back SH, Kaufman RJ. Endoplasmic reticulum stress activates cleavage of CREBH to induce a systemic inflammatory response. *Cell* 2006;**124**:587–99.

40. Credle JJ, Finer-Moore JS, Papa FR, Stroud RM, Walter P. On the mechanism of sensing unfolded protein in the endoplasmic reticulum. *Proc Natl Acad Sci USA* 2005;**102**:18,773–18,784.

41. Zhou J, Liu CY, Back SH, Clark RL, Peisach D, Xu Z, Kaufman RJ. The crystal structure of human IRE1 luminal domain reveals a conserved dimerization interface required for activation of the unfolded protein response. *Proc Natl Acad Sci USA* 2006;**103**:14,343–14,348.

42. Bernardi P, Petronilli V, Di Lisa F, Forte M. A mitochondrial perspective on cell death. *Trends Biochem Sci* 2001;**26**:112–17.

43. Pinton P, Ferrari D, Magalhaes P, Schulze-Osthoff K, Di Virgilio F, Pozzan T, Rizzuto R. Reduced loading of intracellular Ca(2+) stores and downregulation of capacitative Ca(2+) influx in Bcl-2-overexpressing cells. *J Cell Biol* 2000;**148**:857–62.

44. Foyouzi-Youssefi R, Arnaudeau S, Borner C, Kelley WL, Tschopp J, Lew DP, Demaurex N, Krause KH. Bcl-2 decreases the free Ca2+ concentration within the endoplasmic reticulum. *Proc Natl Acad Sci USA* 2000;**97**:5723–8.

45. Palmer AE, Jin C, Reed JC, Tsien RY. Bcl-2-mediated alterations in endoplasmic reticulum Ca2+ analyzed with an improved genetically encoded fluorescent sensor. *Proc Natl Acad Sci USA* 2004;**101**:17,404–17,409.

46. White C, Li C, Yang J, Petrenko NB, Madesh M, Thompson CB, Foskett JK. The endoplasmic reticulum gateway to apoptosis by Bcl-X(L) modulation of the InsP3R. *Nat Cell Biol* 2005;**7**:1021–8.

47. Scorrano L, Oakes SA, Opferman JT, Cheng EH, Sorcinelli MD, Pozzan T, Korsmeyer SJ. BAX and BAK regulation of endoplasmic reticulum Ca2+: a control point for apoptosis. *Science* 2003;**300**:135–9.

48. Oakes SA, Opferman JT, Pozzan T, Korsmeyer SJ, Scorrano L. Regulation of endoplasmic reticulum Ca2+ dynamics by proapoptotic BCL-2 family members. *Biochem Pharmacol* 2003;**66**:1335–40.

49. Oakes SA, Scorrano L, Opferman JT, Bassik MC, Nishino M, Pozzan T, Korsmeyer SJ. Proapoptotic BAX and BAK regulate the type 1 inositol trisphosphate receptor and calcium leak from the endoplasmic reticulum. *Proc Natl Acad Sci USA* 2005;**102**:105–10.

50. Szalai G, Krishnamurthy R, Hajnoczky G. Apoptosis driven by IP(3)-linked mitochondrial calcium signals. *Embo J* 1999;**18**:6349–61.

51. Oakes SA, Lin SS, Bassik MC. The control of endoplasmic reticulum-initiated apoptosis by the BCL-2 family of proteins. *Curr Mol Med* 2006;**6**:99–109.

52. Pinton P, Rizzuto R. Bcl-2 and Ca2+ homeostasis in the endoplasmic reticulum. *Cell Death Differ* 2006;**13**:1409–18.

53. Giacomello M, Drago I, Pizzo P, Pozzan T. Mitochondrial Ca2+ as a key regulator of cell life and death. *Cell Death Differ* 2007;**14**:1267–74.

54. Ferrari DM, Soling HD. The protein disulphide-isomerase family: unravelling a string of folds. *Biochem J* 1999;**339**:1–10.

55. Carelli S, Ceriotti A, Cabibbo A, Fassina G, Ruvo M, Sitia R. Cysteine and glutathione secretion in response to protein disulfide bond formation in the ER. *Science* 1997;**277**:1681–4.

56. Frand AR, Kaiser CA. Two pairs of conserved cysteines are required for the oxidative activity of Ero1p in protein disulfide bond formation in the endoplasmic reticulum. *Mol Biol Cell* 2000;**11**:2833–43.

57. Frand AR, Kaiser CA. The ERO1 gene of yeast is required for oxidation of protein dithiols in the endoplasmic reticulum. *Mol Cell* 1998;**1**:161–70.

58. Jessop CE, Chakravarthi S, Watkins RH, Bulleid NJ. Oxidative protein folding in the mammalian endoplasmic reticulum. *Biochem Soc Trans* 2004;**32**:655–8.

59. Cuozzo JW, Kaiser CA. Competition between glutathione and protein thiols for disulphide-bond formation. *Nat Cell Biol* 1999;**1**:130–5.

60. Saltiel AR, Kahn CR. Insulin signalling and the regulation of glucose and lipid metabolism. *Nature* 2001;**414**:799–806.

61. Shulman GI. Cellular mechanisms of insulin resistance. *J Clin Invest* 2000;**106**:171–6.

62. Dandona P, Thusu K, Cook S, Snyder B, Makowski J, Armstrong D, Nicotera T. Oxidative damage to DNA in diabetes mellitus. *Lancet* 1996;**347**:444–5.

63. Nishikawa T, Edelstein D, Du XL, Yamagishi S, Matsumura T, Kaneda Y, Yorek MA, Beebe D, Oates PJ, Hammes HP, Giardino I, Brownlee M. Normalizing mitochondrial superoxide production blocks three pathways of hyperglycaemic damage. *Nature* 2000;**404**:787–90.

64. Gorogawa S, Kajimoto Y, Umayahara Y, Kaneto H, Watada H, Kuroda A, Kawamori D, Yasuda T, Matsuhisa M, Yamasaki Y, Hori M. Probucol preserves pancreatic beta-cell function through reduction of oxidative stress in type 2 diabetes. *Diabetes Res Clin Pract* 2002;**57**:1–10.

65. Delepine M, Nicolino M, Barrett T, Golamaully M, Lathrop GM, Julier C. EIF2AK3, encoding translation initiation factor 2-alpha kinase 3, is mutated in patients with Wolcott-Rallison syndrome. *Nat Genet* 2000;**25**:406–9.

66. Bertolotti A, Zhang Y, Hendershot LM, Harding HP, Ron D. Dynamic interaction of BiP and ER stress transducers in the unfolded-protein response. *Nat Cell Biol* 2000;**2**:326–32.

67. Song B, Scheuner D, Ron D, Pennathur S, Kaufman RJ. CHOP deleton reduces oxidative stress, improves beta cell function and promotes cell survival in multiple mouse models of diabetes. *J Clin Invest* 2008;**118**:3378–89.

68. Marciniak SJ, Yun CY, Oyadomari S, Novoa I, Zhang Y, Jungreis R, Nagata K, Harding HP, Ron D. CHOP induces death by promoting protein synthesis and oxidation in the stressed endoplasmic reticulum. *Genes Dev* 2004;**18**:3066–77.

69. Tanaka Y, Tamoto H, Tozuka Z, Sato A, Kimura T. Metabolism and degradation products of recombinant human insulin-like growth factor-I in lysosomes of rat kidney. *Xenobiotica* 1999;**29**:281–95.

70. Kawamori D, Kajimoto Y, Kaneto H, Umayahara Y, Fujitani Y, Miyatsuka T, Watada H, Leibiger IB, Yamasaki Y, Hori M. Oxidative stress induces nucleo-cytoplasmic translocation of pancreatic transcription factor PDX-1 through activation of c-Jun NH(2)-terminal kinase. *Diabetes* 2003;**52**:2896–904.

71. Noguchi H, Nakai Y, Matsumoto S, Kawaguchi M, Ueda M, Okitsu T, Iwanaga Y, Yonekawa Y, Nagata H, Minami K, Masui Y, Futaki S, Tanaka K. Cell permeable peptide of JNK inhibitor prevents islet apoptosis immediately after isolation and improves islet graft function. *Am J Transplant* 2005;**5**:1848–55.

72. Forman MS, Lee VM, Trojanowski JQ. "Unfolding" pathways in neurodegenerative disease. *Trends Neurosci* 2003;**26**:407–10.

73. Selkoe DJ. Folding proteins in fatal ways. *Nature* 2003;**426**:900–4.

74. Lindholm D, Wootz H, Korhonen L. ER stress and neurodegenerative diseases. *Cell Death Differ* 2006;**13**:385–92.

75. Zhao L, Longo-Guess C, Harris BS, Lee JW, Ackerman SL. Protein accumulation and neurodegeneration in the woozy mutant mouse is caused by disruption of SIL1, a cochaperone of BiP. *Nat Genet* 2005;**37**:974–9.

76. Nishitoh H, Matsuzawa A, Tobiume K, Saegusa K, Takeda K, Inoue K, Hori S, Kakizuka A, Ichijo H. ASK1 is essential for endoplasmic reticulum stress-induced neuronal cell death triggered by expanded polyglutamine repeats. *Genes Dev* 2002;**16**:1345–55.

77. Bence NF, Sampat RM, Kopito RR. Impairment of the ubiquitin-proteasome system by protein aggregation. *Science* 2001;**292**:1552–5.

78. Anttonen AK, Mahjneh I, Hamalainen RH, Lagier-Tourenne C, Kopra O, Waris L, Anttonen M, Joensuu T, Kalimo H, Paetau A, Tranebjaerg L, Chaigne D, Koenig M, Eeg-Olofsson O, Udd B, Somer M, Somer H, Lehesjoki AE. The gene disrupted in Marinesco-Sjogren syndrome encodes SIL1, an HSPA5 cochaperone. *Nat Genet* 2005;**37**:1309–11.

79. Gething MJ. Role and regulation of the ER chaperone BiP. *Semin Cell Dev Biol* 1999;**10**:465–72.

80. Lee VM, Goedert M, Trojanowski JQ. Neurodegenerative tauopathies. *Annu Rev Neurosci* 2001;**24**:1121–59.

81. Dias-Santagata D, Fulga TA, Duttaroy A, Feany MB. Oxidative stress mediates tau-induced neurodegeneration in Drosophila. *J Clin Invest* 2007;**117**:236–45.

Chapter 48

Protein Serine/Threonine Phosphatase Inhibitors and Human Disease

Shirish Shenolikar [1,2] and Matthew H. Brush[1]

[1]*Department of Pharmacology and Cancer Biology, Duke University Medical Center, Durham, North Carolina*
[2]*Signature Research Program in Neuroscience and Behavioral Disorders, Duke-NUS Graduate Medical School Singapore.*

INTRODUCTION

Coordinating the activity of protein kinases and phosphatases allows cells to amplify and accelerate hormonal signals and elicit a robust and decisive physiological response. Hormonal signals have been shown to modulate the expression or activity of endogenous phosphatase inhibitor proteins (described in Chapter 105 of Handbook of Cell Signaling, Second Edition) that dampen the functions of specific protein phosphatases and enhance the phosphorylation of their cellular substrates. This paradigm is best exemplified by protein serine/threonine phosphatases, many of which are targeted by an increasing number of endogenous protein inhibitors. By suppressing the functions of the relatively few phosphatase catalytic subunits that catalyze serine/threonine dephosphorylation in eukaryotic cells, these inhibitor proteins have the potential to modulate a broad range of physiological events and orchestrate an integrated physiological response that coordinates such diverse processes as metabolism, migration, gene transcription, and growth. Emerging studies suggest that some phosphatase inhibitors physically associate with cellular phosphatase complexes comprised of a catalytic subunit bound to one or more regulatory proteins that restrict the actions of the phosphatases to control specific cellular events. The functions of these phosphatase inhibitors are in turn controlled by their expression, reversible phosphorylation, and dynamic association with target protein phosphatases. In this manner, endogenous serine/threonine phosphatase inhibitors achieve both spatial and temporal control of phosphoproteins that regulate normal cell physiology, and growing evidence suggests that aberrant expression and/or activity of phosphatase inhibitors may be associated with many human diseases. This chapter focuses on new information on the mode of action of phosphatase inhibitors and discusses their potential contributions to the pathophysiology of human disease.

ENVIRONMENTAL TOXINS AS PHOSPHATASE INHIBITORS

Okadaic acid, responsible for human diarrhetic shellfish poisoning, was first identified as a toxin concentrated by marine sponges, and functions as a potent inhibitor of several eukaryotic protein serine/threonine phosphatases. By inhibiting one or more phosphatases, okadaic acid enhances smooth muscle myosin phosphorylation and the contractility of the gut [1]. Subsequent studies identified cyclic peptides like microcystin and nodularin in freshwater ponds and lakes infested with cyanobacteria, which cause severe hepatotoxicity that is often lethal to animals [2]. Like okadaic acid, these compounds also inhibited the major mammalian serine/threonine phosphatases. To date, more than a dozen environmental toxins or natural products have been shown to inhibit protein serine/threonine phosphatases. Almost all of these compounds have the potential to promote cell growth and transformation, and induce a variety of cancers in experimental animals. However, at least one report hinted at other cellular actions of these toxins, specifically that okadaic acid impaired DNA repair mechanisms and elicited DNA damage through an unknown mechanism independent of phosphatase inhibition [3]. In any case, a number of these cell-permeable compounds are now commercially available and, by enhancing cellular protein phosphorylation, have become valuable experimental tools in elucidating the physiological functions of phosphoproteins. Some compounds contain a phosphate moiety and yet others possess an acidic residue, which serves as a phosphomimetic and directly interacts with the metals in the catalytic site of serine/threonine phosphatases [4]. Although compounds like okadaic acid and calyculin-A show differing selectivity for the inhibition of protein serine/threonine phosphatases (e.g., type 1 versus type 2) at the exposures that promote cell growth and transformation, they most likely inhibit as many as five of the seven families of mammalian serine/threonine phosphatases to elicit their broad effects on cell physiology. By comparison, the deletion of individual genes encoding protein serine/threonine phosphatases in the model eukaryotes (yeast and flies) suggests that several phosphatases play critical and non-overlapping roles in eukaryotic biology such that the loss of function of any of these enzymes inhibits cell growth and reduces cell viability [5]. In this regard, the enhanced growth and transformation of mammalian cells elicited by xenobiotics most likely reflects an incomplete or partial inhibition of protein phosphatases that increases or prolongs the phosphorylation of proteins involved in cell division and growth. However, higher concentrations of these compounds associated with greater inhibition of cellular phosphatases are generally cytotoxic, severely impairing cell growth and often inducing programmed cell death. Due to their broad actions, these compounds have not provided a clear-cut link between the inhibition of individual protein phosphatases and the development of human disease. However, it is also worth stressing that compounds like fostriecin, which inhibits several type 2 phosphatases but has no discernable activity against type 1 enzymes, have been extensively analyzed as potential anticancer drugs [6]. Indeed, the two natural products, cyclosporin A and FK-506, are known to be potent and selective inhibitors of calcineurin (protein phosphatase-2B), and are among the most effective immunosuppressive drugs that are widely used to prevent rejection of transplanted organs [7]. These drugs are also clinically approved for the treatment of psoriasis, rheumatoid arthritis, aplastic anemia, nephritic syndrome, and atopic dermatitis, and the list may soon be extended to include other autoimmune diseases. However, these drugs are costly and have some serious side effects. For example, the inhibition of calcineurin has also been implicated in their renal toxicity. Thus, the potential of natural products to be either cytotoxic or valuable therapeutic agents may depend on their phosphatase specificity, degree of enzyme inhibition and physiological roles of the targeted enzymes in mammalian tissues. The availability of natural product inhibitors has also provided scientists with valuable tools to affinity-isolate and -analyze a wide variety of cellular phosphatase complexes [8].

NEW INSIGHTS IN CELLULAR PHOSPHATASE INHIBITORS

As reviewed in Chapter 105 of Handbook of Cell Signaling, Second Edition, endogenous inhibitor proteins are much more selective than the environmental toxins discussed above, targeting selected members of the protein serine/threonine phosphatase family and, at the highest concentrations found in mammalian tissues, having little or no effect on other phosphatases [9]. The largest group of known mammalian protein inhibitors targets type 1 phosphatases (PP1). An emerging theme from studies of PP1 inhibitors is that they can exist in unique cellular complexes that contain not only the PP1 catalytic subunit but also selected regulatory or targeting subunits that dictate PP1's localization and substrate specificity. Thus, some PP1 inhibitors appear to target highly specific physiological events.

The first suggestion that cells utilized PP1 inhibitors to target certain physiological events came from the finding that inhibitor-1 (I-1), a PKA-activated PP1 inhibitor, bound both PP1 and its regulator, GADD34. Later experiments showed that the I-1 C-terminus, which directs its interaction with GADD34, was essential for the efficient transduction of cyclic AMP signals that inhibit PP1 activity and promote the phosphorylation of the eukaryotic translation initiation factor, eIF2α, a substrate of the GADD34/PP1 complex [10]. This raised the possibility that the heterotrimeric complex of PP1/GADD34/I-1 tranduces hormonal signals that inhibit protein translation in some mammalian tissues. Moreover, the region of I-1 mRNA encoding its C-terminus that binds GADD34 is alternately spliced, and is also not conserved in protein products of other predicted human genes encoding I-1 isoforms. This also suggests that mammalian tissues may contain multiple I-1 isoforms, only one of which regulates protein translation, while the other I-1 polypeptides transduce cAMP signals that control other physiological events [11]. It is also noteworthy that I-1 expression and alternate splicing may be developmentally controlled, peaking after birth to reach much lower steady-state levels in many adult tissues.

The recent co-crystallization of PP1 with inhibitor-2 (I-2) provided new insights into the mechanism by which this inhibitor protein regulates PP1 activity [12]. Acute modulation of PP1 most likely occurs by I-2's binding at an allosteric site shared with many PP1 inhibitors, including I-1, which lies some distance from the PP1 catalytic site. In addition, I-2 may chronically suppress phosphatase activity by a direct interaction with the PP1 active site, resulting in the displacement of one of the two catalytic metals and leading to a more prolonged inactivation of PP1. These and other studies demonstrated that I-2 displays two modalities of PP1 regulation, one short-term or transient, and the other prolonged or more persistent. Following the identification of several mammalian PP1 complexes containing I-2, it was speculated that I-2 also targets selected cellular PP1 pools. Thus, I-2 may control the duration of protein phosphorylation at the actin cytoskeleton through its binding to the actin- and PP1-binding protein, neurabin [13], or at mitotic spindles through its association with the kinase, Nek2 [14]. It should be noted that the persistent inactivation of the PP1/I-2 complex is reversed *in vitro* by the phosphorylation of I-2 (threonine-72) by GSK-3, which also requires the prior phosphorylation of I-2 at several serines by casein kinase-II [15]. The precise physiological circumstances under which GSK-3 and casein kinase-II phosphorylate I-2 and reactivate the latent PP1 complexes in mammalian cells are still unclear.

Most recently, a novel complex containing the PP1 catalytic subunit, the nuclear regulatory subunit, Sds22, and inhibitor-3 (I-3) was identified in yeast and mammalian cells [16,17]. While the precise function of the PP1/Sds22/I-3 complex remains to be investigated, the subcellular localization of this complex suggests that this complex controls nucleolar phosphorylation events. Yet other studies have shown that a PKC-activated PP1 inhibitor, CPI-17, and its structural relatives, the PHI proteins, target a PP1 catalytic subunit bound to the regulatory subunit, MYPT1. This in turn targets the heterotrimeric PP1/MYPT1/CPI-17 complex to smooth muscle myosin, and transduces signals from PKC and other kinases to enhance the calcium sensitivity of muscle contraction and elicit rapid and robust contraction of smooth muscle tissue [18]. Together, these studies begin to highlight a critical difference in the actions of endogenous phosphatase inhibitors from those of environmental toxins, distinguishing them on the basis of their selectivity for phosphatase catalytic subunits and their ability to target defined pools of phosphatases within all eukaryotic cells.

Recent work shows that cytokines such as IL-6 downregulate CPI-17 expression in smooth muscle cells [19], while the structurally related PKC-activated PP1 inhibitor, KEPI, is upregulated in neurons in response to opioids [20]. Prior studies also show that levels of I-2 mRNA, and protein, as well as the localization of I-2 in the nucleus, are regulated during the cell division cycle [21]. This highlights a critical mode of modulating the signaling capacity of mammalian cells by rapid changes in cellular content and subcellular distribution of phosphatase inhibitors.

CELLULAR PHOSPHATASE INHIBITORS AND HUMAN DISEASE

Reduced expression of phosphatase inhibitors and human disease

A widely-held view in cell signaling is that in some tissues, under unstimulated or basal conditions, cellular protein phosphatase activity effectively antagonizes or clamps the functions of protein kinases, especially those displaying significant basal activity. This eliminates leaky or unwarranted signaling, but it also means that, following cell stimulation, intracellular signals that are transduced by these protein kinases are severely blunted or sluggish. Thus, cells require mechanisms to activate endogenous phosphatase inhibitors to remove the "brake" imposed by their target phosphatases in parallel with the activation of the protein kinases to achieve speedy and effective signal transduction. Such a "necessary" role for phosphatase inhibitors in hormone signaling may be limited to some tissues and

processes, where speed and accuracy of signaling is vital. Thus far, the best evidence for a necessary or essential role for phosphatase inhibitors in cell signaling was obtained by the genetic deletion of mouse genes encoding I-1 and its structural homologue, DARPP-32.

DARPP-32 is primarily expressed in dopaminergic neurons. The remarkable finding in the DARPP-32 null mice was a dramatic diminution or near complete loss in many aspects of dopamine (D1 receptor) signaling [22]. Thus, the mutant mice demonstrate an altered response to neurotransmitters and drugs of abuse as well as a variety of other behavioral defects [23]. This led to the hypothesis that a loss of DARPP-32 function may also contribute to human neurological disease, specifically schizophrenia. Indeed, several studies reported reduced levels of DARPP-32 protein in brain samples from individuals with schizophrenia [24]. However, analysis of the human DARPP-32 gene in the *post mortem* brain samples provided no insight into the molecular mechanism underlying DARPP-32 reduction in schizophrenia patients, and a role for DARPP-32 deficiency in human neurological disease remains to be established. Interestingly, a search for candidate genes from mouse models of neurological disease identified DARPP-32 as a potential disease-causing gene [25], but more work is needed to establish a clear link between DARPP-32 and bipolar disease and other neurological disorders.

In contrast to DARPP-32, I-1 is widely expressed in mammalian tissues, with the highest content of I-1 protein found in brain and muscle. While the I-1 null mice showed defects in excitatory neurotransmission in some areas of the brain, no major behavioral abnormalities were noted in the mutant mice [26]. This argued for the presence of redundant mechanisms, possibly DARPP-32 or other PKA-regulated PP1-binding proteins, for amplifying cAMP signaling in the mammalian brain. The I-1 null mice did, however, display defects in cardiac contractility, similar to those previously reported in transgenic mice that overexpressed PP1 catalytic subunit in the heart [27]. Indeed, levels of PP1 inhibitors, both I-1 and I-2, were diminished in experimental models of heart failure [28]. While these data pointed to changes in PP1 inhibitors as a contributing factor in heart disease, direct evidence that PP1 inhibitors improved cardiac function and alleviated or delayed heart failure was obtained from the generation of transgenic mice that overexpressed I-1 [29] or I-2 [30] in the heart. These animals showed significantly improved β-adrenergic signaling and enhanced calcium cycling which results from the phosphorylation of the sarcoplasmic reticulum protein, phospholamban, a known PP1 substrate. Additional support for a specific role of I-1 in human heart disease came from the observation that I-1 levels were greatly reduced in the myocardium from failing human hearts [31]. Together, the above studies make a compelling case for I-1 as a major contributor in effective hormonal signaling and cardiac function in the normal human heart, and suggest that errors in I-1 expression and/or activity play an important role in heart disease.

Increased Expression of Phosphatase Inhibitors

While some tissues express high levels of specific phosphatase inhibitors, no inhibitor proteins are detected in other tissues. For example, compared to muscle and brain, mammalian liver expresses very low levels of I-1. Similarly, no measurable DARRP-32 protein can be seen in most non-neural tissues, and significantly less CPI-17 is expressed in tissues other than smooth muscle. The transcriptional mechanisms that dictate the tissue-specific expression of phosphatase inhibitors are just being understood. In addition, there are undoubtedly mechanisms that can also downregulate these proteins. Thus, it comes as no surprise that errors in the regulatory processes that control the phosphatase inhibitors can result in their aberrant expression, and this in turn may contribute to the pathophysiology of many human diseases.

Indeed, a growing number of publications have linked an elevated expression of I-1 with human hepatic cancers [32], DARPP-32 overexpression in gastric, esophageal, gastrointestinal, and colorectal cancers [33,34], and I-2 in prostrate cancer [35,36]. As the tumor-promoting activity of environmental toxins that inhibit protein serine/threonine phosphatases is well documented, the development of cancer associated with abnormally elevated expression of phosphatase inhibitors simply points to the ability of protein serine/threonine phosphatases to regulate critical events in the mammalian cell division cycle and phosphorylation of substrates that promote cell proliferation. For example, the PP1/MYPT1 complex, which dephosphorylates smooth muscle myosin, also targets members of the ERM family of actin-binding proteins, including merlin, the product of the human neurofibromatosis NF2 gene. By inhibiting the PP1/MYPT1 complex in mammalian cells, CPI-17 can elevate merlin phosphorylation and lead to activation of the oncogene, Ras, and cell transformation [37]. Elevated expression of the putative PP2A inhibitor, SET, has also been linked with cell growth and transformation, but direct evidence has not been obtained to show that PP2A activity is inhibited in human cancers [38]. However, emerging studies show that a growing number of cell-cycle regulators, such as Cdc25, are subject to regulation by dephosphorylation events catalyzed by both PP1 and PP2A [39,40]. In conclusion, the overall concept that abnormal levels of PP1 inhibitors contribute to rapid growth and metastasis of human cancers is perhaps not surprising. However, the cause-or-effect relationship between cellular changes in phosphatase inhibitors and the disease process needs further clarification.

The product of the Down Syndrome Critical Region 1 or *DSCR1* gene (also known as *Adapt78*, a gene induced by oxidative stress) is elevated in Down syndrome patients, who also display neural pathology similar to that of early onset Alzheimer's disease [41]. Studies of *post mortem* tissue from Alzheimer's patients also showed that increased expression of *DSCR1* correlated with the severity of disease and the abundance of neurofibrillar tangles. *DSCR1* encodes RCAN1 (regulator of calcineurin 1), a protein previously termed calcipressin 1, Rcn1 (the yeast regulator of calcineurin) and MCIP1 (myocyte-enriched calcineurin-interacting protein), and is highly conserved in all eukaryotes. RCAN proteins share a signature sequence also found in NFAT (nuclear factor of activated T cells), the calcineurin substrate that directly binds the phosphatase. Biochemical and genetic studies indicate some similarities between calcineurin regulation by RCAN1 and PP1 regulation by I-2. For example, several factors, including oxidative stress, elevated intracellular calcium, and amyloid (Aβ) peptide induce RCAN1 expression. RCAN1 binds and inhibits calcineurin or PP2B activity, specifically towards the substrate, NFAT. Like I-2, RCAN1 may also function as a possible chaperone to elevate cellular functions of its target phosphatase, and, like I-2, phosphatase inhibition by RCAN1 appears to be subject to regulation by GSK-3-mediated phosphorylation [42]. While RCAN1 and calcineurin are highly expressed in the brain, calcium and calcineurin also play a critical role in cardiac output and promote cardiac hypertrophy. Genetic studies in mice suggest that excessive NFAT dephosphorylation by calcineurin promotes cardiac hypertrophy, and transgenic mice with elevated RCAN1 levels in the heart tissue are protected from hypertrophy [43]. However, these animals show defects in heart valve formation. Thus, either reduced inhibition of calcineurin or excessive levels of its inhibitor, RCAN1, may contribute to different aspects of cardiac disease. Other studies suggest that that aberrant RCAN1 function may also contribute to diabetes, immunological diseases, and skin disorders. However, more work is needed to establish the relevance of these findings to human disease.

CONCLUDING REMARKS

The availability of protein and non-protein inhibitors of protein serine/threonine phosphatases has provided researchers with an extensive toolkit to study the role of phosphatases and phosphatase inhibitors in modulating the mammalian physiology. Rapid progress is also being made in the understanding of cellular phosphatase complexes and their functions. In addition, the mechanisms that regulate the expression and activity of phosphatase inhibitors and their mode of action in controlling specific pools of cellular protein phosphatases are being actively investigated. Many studies have correlated changes in phosphatase inhibitors with human diseases, including cancer,

Alzheimer's, and cardiomyopathy. Thus, it is anticipated that it will not be long before experimental evidence clearly demonstrates the mechanism or mode of action of phosphatase inhibitors in the genesis and progression of human disease.

REFERENCES

1. Bialojan C, Takai A. Inhibitory effects of marine sponge toxin on protein phosphatases: specificity and kinetics. *Biochem J* 1988;**256**:283–90.

2. Herfindal L, Selheim F. Microcystin produces disparate effects on liver cells in a dose dependent manner. *Mini Rev Med Chem* 2006;**6**:279–85.

3. Nakagama H, Kaneko S, Shima H, Inamori H, Fukuda H, Kominami R, Sugimura T, Nagao M. Induction of minisatellite mutation in NIH3T3 cells by treatment with the tumor promoter okadaic acid. *Proc Natl Acad Sci USA* 1997;**94**:10,813–10,816.

4. Holmes CF, Maynes JT, Perreault KR, Dawson JF, James MN. Molecular enzymology underlying regulation of protein phosphatase-1 by natural toxins. *Curr Med Chem* 2002;**9**:1981–9.

5. Stark MJ, Black S, Sneddon AA, Andrews PD. Genetic analyses of yeast protein serine/threonine phosphatases. *FEMS Microbiol Letts* 1994;**117**:121–30.

6. Roberge M, Tudan C, Hung SMF, Harder KW, Jirik FR, Anderson H. Antitumor drug Fostriecin inhibits the mitotic entry checkpoint and protein phosphatases 1 and 2A. *Cancer Res* 1994;**54**:6115–21.

7. Keown PA. Emerging indications for the use of cyclosporin in organ transplantation and autoimmunity. *Drugs* 1990;**40**:315–25.

8. Moorhead G, MacKintosh RW, Morrice N, Gallagher T, Mackintosh C. Purification of type 1 protein (serine/threonine) phosphatases by microcystin-Sepharose affinity chromatography. *FEBS Letts* 1994;**356**:46–50.

9. Oliver CJ, Shenolikar S. Physiological importance of protein phosphatase inhibitors. *Front Biosci* 1998;**3**:961–72.

10. Brush MH, Weiser DC, Shenolikar S. The growth arrest and DNA-damage-inducible gene product, GADD34, targets protein phosphatase-1 to endoplasmic reticulum and promotes the dephosphorylation of α-subunit of the eukaryotic translation initiation factor 2. *Mol Cell Biol* 2003;**23**:1292–303.

11. Weiser DC, Sikes S, Li S, Shenolikar S. Inhibitor-1 C-terminus facilitates hormonal regulation of cellular protein phosphatase-1: functional implications for inhibitor-1 isoforms. *J Biol Chem* 2004;**279**:48,904–48,914.

12. Hurley TD, Yang J, Zhang L, Goodwin KD, Zou Q, Cortese M, Dunker AK, DePaoli-Roach AA. Structural basis for regulation of protein phosphatase 1 by inhibitor-2. *J Biol Chem* 2007;**282**:28,874–28,883.

13. Eto M, Elliott E, Prickett TD, Brautigan DL. Inhibitor-2 regulates protein phosphatase-1 complexed with NimA-related kinase to induce centrosome separation. *J Biol Chem* 2002;**277**:44,013–44,020.

14. Terry-Lorenzo RT, Elliot E, Weiser DC, Prickett TD, Brautigan DL, Shenolikar S. Neurabins recruit protein phosphatase-1 and inhibitor-2 to actin cytoskeleton. *J Biol Chem* 2002;**277**:46,535–46,543.

15. DePaoli-Roach AA. Synergistic phosphorylation and activation of ATP-Mg-dependent phosphoprotein phosphatase by Fa/GSK-3 and casein kinase II (PC0.7). *J Biol Chem* 1984;**259**:12,144–12,152.

16. Pedelini L, Marquina M, Joaquin Ariño J, Casamayor A, Sanz L, Bollen M, Sanz P, Garcia-Gimeno MA. YPI1 and SDS22 proteins

17. Lesage B, Beullens M, Pedelini L, Garcia-Gimeno MA, Waelkens E, Sanz P, Bollen M. A complex of catalytically inactive protein phosphatase-1 sandwiched between Sds22 and inhibitor-3. *Biochemistry* 2007;**46**:8909–19.

18. Ito M, Nakano T, Erdodi F, Hartshorne DJ. Myosin phosphatase: structure, regulation and function. *Mol Cell Biochem* 2004;**259**:197–209.

19. Ohama T, Hori M, Sato K, Ozaki H, Karaki H. Chronic treatment with interleukin-1 attenuates contractions by decreasing the activities of CPI-17 and MYPT-1 in intestinal smooth muscle. *J Biol Chem* 2003;**278**:48,794–48,804.

20. Liu QR, Zhang PW, Zhen Q, Walther D, Wang XB, Uhl GR. KEPI, a PKC-dependent protein phosphatase 1 inhibitor regulated by morphine. *J Biol Chem* 2002;**277**:13,312–13,320.

21. Brautigan DL, Sunwoo J, Labbe JC, Fernandez A, Lamb NJ. Cell cycle oscillation of phosphatase inhibitor-2 in rat fibroblasts coincident with p34cdc2 restriction. *Nature* 1990;**344**:74–8.

22. Greengard P. The neurobiology of slow synaptic transmission. *Science* 2001;**294**:1024–30.

23. N Hiroi H, Fienberg AA, Haile CN, Alburges M, Hanson GR, Greengard P, Nestler EJ. Neuronal and behavioural abnormalities in striatal function in DARPP-32-mutant mice. *Eur J Neurosci* 1999;**11**:1114–18.

24. Manji HK, Gottesman II, Gould TD. Signal transduction and genes-to-behaviors pathways in psychiatric diseases. *Science STKE* 2003;**2003**:49.

25. Sunkin SM, Hohmann JG. Insights from spatially mapped gene expression in the mouse brain. *Hum Mol Genet* 2007;**16**:209–19.

26. Allen PB, Havlby O, Jensen V, Errington ML, Ramsay M, Chaudhry FA, Bliss TV, Storm-Mathisen J, Morris RG, Andersen P, Greengard P. Protein phosphatase-1 regulation in the induction of long-term potentiation: heterogeneous molecular mechanisms. *J Neurosci* 2000;**20**:3537–43.

27. Carr AN, Schmidt AG, Suzuki Y, del Monte F, Sato Y, Lanner C, Breeden K, Jing SL, Allen PB, Greengard P, Yatani A, Hoit BD, Grupp IL, Hajjar RJ, DePaoli-Roach AA, Kranias EG. Type-1 phosphatase, a negative regulator of cardiac function. *Mol Cell Biol* 2002;**22**:4124–35.

28. Gupta RC, Mishra S, Yang XP, Sabbah HN. Reduced inhibitor-1 and -2 activity is associated with increased protein phosphatase type 1 activity in left ventricular myocardium of one-kidney, one-clip hypertensive rats. *Mol Cell Biochem* 2005;**269**:49–57.

29. Pathak A, del Monte F, Zhao W, Schultz JE, Lorenz JN, Bodi I, Weiser D, Hahn H, Carr AN, Syed F, Mavila N, Jha L, Mareez Y, Chen G, McGraw DW, Heist EK, Guerrero L, DePaoli-Roach AA, Hajjar RJ, Kranias EG. Enhancement of cardiac function and suppression of heart failure progression by inhibition of protein phosphatase-1. *Circ Res* 2005;**96**:756–66.

30. Kirchhefer U, Baba HA, Boknik P, Breeden KM, Mavila N, Bruchert N, Justus I, Matus M, Schmitz W, DePaoli-Roach AA, Neumann J. Enhanced cardiac function in mice overexpressing protein phosphatase inhibitor-2. *Cardiovasc Res* 2005;**68**:98–108.

31. El-Armouche A, Pamminger T, Ditz D, Zolk O, Eschenhagen T. Decreased protein and phosphorylation level of the protein phosphatase inhibitor-1 in failing human hearts. *Cardiovasc Res* 2004;**61**:87–93.

32. Aleem E, Flohr T, Thielman HW, Bannasch P, Mayer D. Protein phosphatase inhibitor-1 mRNA expression correlates with neoplastic transformation in epithelial liver cells and progression of hepatocellular carcinomas. *Intl J Oncol* 2004;**24**:869–77.

33. El-Rafai W, Smith MF, Li G, Beckler A, Carl VS, Montgomery E, Knuutila S, Moskaluk CA, Frierson HF, Powell SM. Gastric cancers overexpress DARPP-32 and a novel isoform, t-DARPP. *Cancer Res* 2002;**62**:4061–4.

34. Wang MS, Pan Y, Liu N, Guo C, Hong L, Fan D. Overexpression of DARPP-32 in colorectal adenocarcinoma. *Intl J Clin Pract* 2005;**59**:58–61.

35. Dhanasekaran SM, Barrette TR, Ghosh D, Shah R, Varambally S, Kurachi K, Pienta KJ, Rubin MA, Chinnaiyan AM. Delineation of prognostic biomarkers in prostate cancer. *Nature* 2001;**412**:822–6.

36. Lapointe J, Li C, Higgins JP, van de Rijn M, Bair E, Montgomery K, Ferrari M, Egevad L, Rayford W, Bergerheim U, Ekman P, DeMarzo AM, Tibshirani R, Botstein D, Brown PO, Brooks JD, Pollack JR. Gene expression profiling identifies clinically relevant subtypes of prostate cancer. *Proc Natl Acad Sci USA* 2004;**101**:811–16.

37. Jin H, Sperka T, Herrlich P, Morrison H. Tumorigenic transformation by CPI-17 through inhibition of a merlin phosphatase. *Nature* 2006;**442**:576–9.

38. Li M, Makkinje A, Damuni Z. The myeloid leukemia-associated protein, SET is a potent inhibitor of protein phosphatase-2A. *J Biol Chem* 1996;**271**:11,059–11,062,.

39. Margolis SS, Walsh S, Weiser DC, Yoshida M, Shenolikar S, Kornbluth S. PP1 control of M-phase entry exerted through 14-3-3-regulated Cdc25 dephosphorylation. *EMBO J* 2003;**22**:5734–45.

40. Margolis SS, Perry J, Forester C, Nutt LK, Guo X, Jardim MJ, Thomenius MJ, Freel CD, Darbandi R, Ahn JH, Aroyo JD, Wang XF, Shenolikar S, Nairn AC, Dunphy WG, Han WC, Virshup DM, Kornbluth S. Role for the PP2A/B56δphosphatase in regulating 14-3-3 release from Cdc25 to control mitosis. *Cell* 2006;**127**:759–73.

41. Harris CD, Ermak G, Davies KJA. Multiple roles of the DSCR1 (Adapt78 or RCAN1) gene and its protein product Calcipressin 1 (or RCAN1) in disease. *Cell Mol Life Sci* 2005;**62**:2477–86.

42. Hilioti Z, Gallagher DA, Low-Nam ST, Ramaswamy P, Gajer P, Kingsbury TJ, Birchwood CJ, Levchenko A, Cunningham KW. GSK-3 kinases enhance calcineurin signaling by phosphorylation of RCNs. *Genes Dev* 2004;**18**:35–47.

43. Rothermel BA, McKinsey TA, Vega RB, Nicol RL, Mammen P, Yang J, Antos CL, Shelton JM, Bassel-Duby R, Olson EN, Williams RS. Myocyte-enriched calcineurin-interacting protein, MCIP1, inhibits cardiac hypertrophy *in vivo*. *Proc Natl Acad Sci USA* 2001;**98**:3328–33.

Chapter 49

Signal Transduction in Rheumatoid Arthritis and Systemic Lupus Erythematosus

Thomas Dörner[1] and Peter E. Lipsky[2]

[1]Department of Medicine, Rheumatology, and Clinical Immunology, Charite University Medicine Berlin, Germany
[2]NIAMS, National Institutes of Health, Bethesda, MD

INTRODUCTION

Rheumatoid arthritis (RA) is the most common cause of adult inflammatory arthritis and causes substantial morbidity and mortality. T and B lymphocytes together with macrophages and synovial fibroblasts play a central role in the immunopathogenesis of inflammatory joint diseases, in particular RA.

Studies of twins identified a genetic contribution to disease susceptibility [1] and the siblings of patients with seropositive, erosive rheumatoid arthritis have an estimated risk of developing the disease of between 5 and 10 times that of the general population [2].

The introduction of new therapies, including the blockade of TNF, IL-1, and IL-6, as well B cell depletion and co-stimulation blockade using CTLA4-Ig has improved the signs and symptoms, disability and radiologic progression in patients with severe RA [3]. A major lesson learned by this therapeutic success is that immune intervention has clearly gained clinical value beyond what was initially expected. On the other hand, there is promise that deeper understanding of immune mechanisms, including genetic predispositions and intracellular signaling processes can improve immune targeting and even achieve ultimate goals of remission or prevention of the disease, with fewer adverse events.

The highly polymorphic HLA region is a major contributor to genetic risk of rheumatoid arthritis [4]. Several HLA associated and other non-HLA genes associated with a modest risk for RA have recently been identified, including the Arg620Trp variant of the intracellular phosphatase gene PTPN22 [5] that elevates the risk for RA, whereas an IL4-R gene polymorphism primarily enhances the risk for erosive disease [6]. Other candidate genes associated with RA have been reported in the literature (e.g., CTLA4 and PADI4), but had only a modest statistical association [7–9]. Reassuring associations between RA and loci in and around HLA-DRB1 and PTPN22 have been repeatedly implicated as genetic risk factors in persons of European descent. Earlier large scale linkage disequilibrium studies implicated a variant of PADI4 as a risk factor for RA [9]. This variant would seem to have a more potent effect in Asian populations than in those of European descent. Variants of these genes are believed to confer a risk for the development of RA by affecting the presentation of autoantigens (in the case of HLA-DRB1), T cell receptor signal transduction (in the case of PTPN22), and the citrullination of proteins, the targets of anti-CCP antibodies (in the case of PADI4). The definitive identification of risk genes outside the HLA region has been challenging. In spite of the likely involvement of specific genes in RA, there is no clear evidence for an RA specific abnormality in intracellular signaling rather than disturbances related to generally enhanced inflammation as also seen in other autoimmune disorders.

Communication between plasma membrane receptors, cytoplasmic events, and the nucleus allows cells to respond appropriately to environmental danger signals. Rapid and adequate transduction of this information is critical for appropriate cell reactions and survival. Intracellular messengers are important for the interaction between the extracellular and intracellular milieus and the genes inside the nucleus. A number of these pathways are suspected to be involved in RA pathogenesis mainly based in the identification of non-HLA SNPs linked to RA itself or RA severity.

STAT4 RS7574865 ALLELE AND THE RISK OF RA

It is known that STAT4 is a latent cytosolic factor and phosphorylated after activation by cytokines and then accumulates in the nucleus. Of note, STAT4 is a transcription factor that transmits signals induced by several key cytokines, such as interleukin-12, type 1 interferons, and interleukin-23 [10]. Activated STAT4 induces transcription of specific genes including interferon-γ, a key indicator of T cell differentiation into type 1 helper T (Th1) cells. Therefore, STAT4 dependent signaling by interleukin-12 receptors plays a critical role in the development of a Th1-type T cell response [11,12]. STAT4 has also been shown to be involved in differentiation of Th17 cells dependent in part on the activity of interleukin-23, a cytokine related to interleukin-12 [13]. These proinflammatory Th17 cells can play an important, if not predominant, role in chronic inflammatory disorders [14]. Thus, STAT4 represents a key player in both TH1 and TH17 lineages as evidenced by experimental models of autoimmunity. STAT4 deficient mice are generally resistant to models of autoimmune disease, including arthritis [15]. In addition, it has been shown that specific blockade of STAT4 signaling by inhibitory oligodeoxynucleotides or antisense oligonucleotides ameliorated the disease in arthritis models [16] suggesting the utility of STAT4 as a therapeutic target.

In a recent large genetic study, a variant allele of STAT4 was identified to confer an enhanced risk for RA [16]. A linkage of genes located on the long (q) arm of chromosome 2 was found previously to be associated with RA in 642 families of European ancestry [17] collected by the North American Rheumatoid Arthritis Consortium (NARAC) [18]. The detailed study cited above [16], tested single nucleotide polymorphisms (SNPs) in and around 13 candidate genes within the previously linked chromosome 2q region and found the STAT1-STAT4 region in 1620 RA patients versus 2635 controls to be linked to RA. The relation was replicated in 1529 RA and 881 controls from Sweden. In detail, a SNP haplotype in the third intron of STAT4 was found to be associated with susceptibility to RA but also SLE. The odds ratio for having the risk allele in chromosomes of RA patients vs. those of controls was 1.32, whereas for SLE patients the odds ratio was calculated 1.55. Homozygosity of the risk allele was associated with a more than doubled risk for SLE and a 60 percent increased risk for RA. Among anti-CCP positive NARAC patients (81 percent positive for anti-cyclic citrullinated peptide antibody), the rs7574865 minor allele frequency did not differ significantly ($P > 0.05$) in the subgroup that was positive for the antibody (0.28) and the subgroup that was negative for the antibody (0.27). Interestingly, logistic-regression analysis after accounting for the rs7574865 genotype showed that this one SNP could explain the signal across the STAT1-STAT4 region. Even after accounting

for the CTLA4 SNP associated with RA (rs3087243), the result for the STAT4 rs7574865 remained significant.

Since STAT4 rs7574865 was also identified in SLE patients, it was concluded that common risk genes apparently underlie multiple autoimmune disorders and likely the involvement of common pathways of pathogenesis among these different diseases, similarly as seen with PNTP22 alleles [19]. Although the association of STAT4 rs7574865 genotype has been identified in American, Swedish, and a Korean [16] population, whether the polymorphism affects STAT4 expression or function is not known. Moreover, the specific role of STAT4 in RA and other autoimmune disorders needs to be delineated as well as influences of allelic variation on certain disease subgroups.

TNFR1 AND C5 (RS376147 AND RS2900180) AND THE RISK OF RA

Another study addressing risk factors for RA outside the HLA region [20] identified an association of a 100 kb region on chromosome 9 containing the TRAF1 and C5 genes with disease in anti-CCP positive patients with RA. Since the most highly associated SNPs (rs376147 and rs2900180) are in linkage disequilibrium with both genes, it cannot be dissected whether the causal alleles or group of alleles influences TRAF1 or C5 (or both) to increase susceptibility for RA.

It is known that the TRAF1 gene encodes an intracellular protein that mediates signal transduction through tumor necrosis factor (TNF) receptors. TNF has been considered as a critical cytokine in the pathogenesis of RA and blocking TNF is effective in the treatment of RA. TRAF1 knockout mice have exaggerated T cell proliferation and activation in response to TNF or when stimulated through the T cell receptor complex, suggesting that TRAF1 acts as a negative regulator of these signaling pathways [21]. TRAF1 binds several intracellular proteins, including the nuclear factor-(kappa)B inhibitory protein TNFAIP3 [22].

The clinical and biologic data for C5 for the involvement in RA are also very compelling. The complement pathway has been implicated in the pathogenesis of RA for more than 30 years [23]. Complement activation leading to significant depletion of complement components has been shown in synovial fluid of patients with RA. C5 cleavage generates the proinflammatory anaphylatoxin C5a as well as C5b, which initiates the generation of the membrane attack complex. C5 deficient mice are resistant to inflammatory arthritis in models with dominant humoral immunity [24–26]. It cannot be excluded that the allele identified acts through C5 by amplifying complement activation in the joints of patients with RA.

Collectively, it is important to note that despite the association of genetic abnormalities and RA there is no known or obvious functional allele that explains these associations although intracellular and extracellular consequences are conceivable. Moreover, it is possible that the identified SNPs need the presence of certain other genetic abnormalities in order to express the disease.

CHALLENGES TO THE APPLICATION OF FINDINGS FROM GENETIC MAPPING STUDIES TO CLINICAL CONSEQUENCES

The impact of SNPs associated with an enhanced risk for RA and their potential functional differences are usually smaller than those of mutations responsible for monogenic disorders. Therefore it is difficult to show that such variants have a relevant biologic effect and most association studies lack the provision of functional data. Thus, sensitive and sophisticated methods in molecular biology and translational immunology are required for the investigation of functionally important variants to understand their clinical relevance.

Another critical issue is related to SNP associations across populations of various ancestries. For example, PADI4 associations were identified in Japanese and Korean RA patients [9] but could not be consistently confirmed in European populations. Along this line, one study [8] observed weaker odds ratios and P values for TRAF1-C5 and RA in the Swedish population compared to North American RA patients, whereas the TRAF1-C5 locus did not surface in the Wellcome Trust Case Control Consortium study among British RA patients [27]. Similarly, the PTPN22 allele associated with RA in Caucasians does not exist in Asians. It is commonly accepted that polymorphisms identified in a signaling pathway in any population implies a role for that pathway in disease pathogenesis. It is possible, therefore, that polymorphisms in other genes in that signaling pathway may explain the disease risk in other populations.

MAPKp38

Mitogen activated protein kinase p38 (p38MAPK) is an important intracellular kinase activated by cellular stress that links inflammatory as well as environmental stress to transcription factors [28, 29]. Signal transduction is accomplished by a cascade of activation steps involving sequential kinases linking the plasma membrane level with the transcription factor level binding to DNA. P38MAPK is ultimate downstream signaling step before the transcription factor level. Whereas, p38MAPK and c-Jun N-terminal kinase (JNK) are mainly regulated by extracellular stress factors, the third pathway, extracellular signal related kinases (ERK) is preferentially a target for mitogenic stimuli.

Induction of p38MAPK Pathway

P38MAPK comprises four different isoforms termed p38MAPK-alpha, beta, gamma, and delta. Importantly, all isoforms are serine-threonine protein kinases that share the common phosphorylation motif TGY. Upon activation p38MAPK undergoes dual phosphorylation at threonine 180 and tyrosine 182 [30]. Inflammatory stimuli, such as lipopolysaccharide (LPS), tumor necrosis factor (TNF) and interleukin-1 (IL-1) are the major inducers of p38MAPK activation. Initially, it was found that LPS induced p38MAPK [30] and research focused on the role of p38MAPK in septic shock and LPS mediated induction of inflammatory cytokines, such as TNF [30, 31]. Moreover, TNF itself also activates p38MAPK by engaging type I TNF receptor [32, 33]. Downstream activation of p38MAPK then allows TNF to transduce its inflammatory message to the target organ, e.g., the synovial membrane. TNF mediated activation is also relevant *in vivo* because systemic TNF overexpression in mice leads to activation of p38MAPK in the inflamed joints [33]. P38MAPK does not only integrate inflammatory stimuli but also signals heat stress, osmotic shock, ultraviolet light, and cytotoxic chemicals [28, 29]. The activity of p38MAPK is tightly regulated by phosphatases such as mitogen activated protein kinase phosphatase-1 (MKP-1), dephosphorylating p38MAPK [34]. Interestingly MKP-1 is strongly upregulated by glucocorticoids, suggesting that part of the anti-inflammatory properties of these drugs are base on p38MAPK inhibition [35] as it has been shown for the regulation of MKP-1 by glucocorticoids in synovial fibroblasts from RA patients [36].

Since several different stress factors are present in the rheumatoid synovium, activation of the p38MAPK pathway is conceivable and may be a target for therapeutic approaches. In this context, proinflammatory cytokines are chronically elevated in RA, which appears to be critical for p38MAPK activation [37].

A detailed description of regulation of the activation cascade via MAPKKK, MAPKK, and finally resulting into p38MAPK induction can be found elsewhere [33, 38]. It should be emphasized that a particular abnormality in RA with regard to p38MAPK signaling has not been identified so far.

Role of p38MAPK in Synovial Inflammation

The mechanisms of gene regulation are complex and can include both transcriptional and translational events as well as alterations in mRNA stability. One major group of target genes for p38MAPK activation are proinflammatory cytokines such as TNF, IL-1, and IL-6 [30, 31]. Regulation of these inflammatory mediators is thus a major function of p38MAPK, which influences the balance of pro- and anti-inflammatory mechanisms. Selective blockade of TNF, IL-1, or IL-6 can control disease by inhibiting joint inflammation and structural damage. Since p38MAPK plays a central role in intracellular signaling, it was considered as an important target for inhibition. Such compounds indeed have anti-inflammatory properties in experimental animal models of arthritis but failed early clinical trials because of toxicity [39–44]. Inhibition of p38MAPK in animal models led to reduced inflammation and correlated with reduced expression of cytokines such as IL-1, IL-6, and also RANKL [44]. Upon inhibition of p38MAPK, formation of synovial inflammatory tissue is generally reduced along with a decrease in infiltrating cells. It is interesting to note that pharmacological inhibition of p38MAPK affects more than one isoform of p38MAPK, whereas the selective genetic deletion of the β-isoform is not sufficient to inhibit experimental arthritis [45].

Role of p38MAPK in Cartilage Damage

Cartilage damage is a hallmark of RA. The expression of matrix metalloproteinases by synovial tissue appears to be a key prerequisite for synovial tissue to invade and destroy cartilage [46, 47]. Synthesis of MMPs is regulated through multiple MAPK families, including p38MAPK, suggesting that blockade of p38MAPK may have an effect in arthritis [48, 49]. This could be confirmed by animal models of arthritis that suggested that p38MAPK activation might be especially important for the destructive features of arthritis since p38 inhibitors strongly reduce cartilage degradation [44]. It is uncertain, whether this effect is directly through p38MAPK dependent regulation of MMP expression or an indirect effect related to lower expression of proinflammatory cytokines, especially IL-1, which is a key inducer of MMPs.

Role of p38MAPK in Inflammatory Bone Loss

Inflammatory bone destruction is another central component of RA explaining local bone erosions in RA leading to structural damage, changed joint architecture, and loss of joint function. Bone damage in arthritis results from osteoclasts that are derived from monocyte precursors and are able to damage bone [50–52]. Interestingly, the formation of osteoclasts is modulated by the stimulation of hematopoietic precursor cells with cytokines, i.e., RANKL and TNF, which ultimately results in p38MAPK phosphorylation [51, 53]. Evidence for the central role of p38MAPK in this process comes from the substantial reduction of mature osteoclasts and osteoclast precursors in the synovial tissue of experimental arthritis after p38MAPK blockade [44]. As a consequence tissue invasion into juxta-articular bone can be effectively blocked by the use of p38MAPK inhibitors. In contrast, deregulation of p38MAPK signaling such as found in $CD44^{-/-}$ mice when challenged with

TNF leads to increased osteoclast formation and increased bone resorption [54]. Resulting osteoclasts are increased in number and size caused by a decreased expression of MKP-1, which is a major regulatory molecule of the p38MAPK [54].

Other Functions

It is notable, that activation of p38MAPK in synovial microvessels may represent the result of autocrine, paracrine, or endocrine activation by proinflammatory cytokines. VEGF also uses p38MAPK to communicate mitogenic stimuli to endothelial cells, which are essential for the synovial microvessel proliferation into the newly formed inflammatory tissue [55]. Activation of p38MAPK is critical for several different functions of endothelial cells, such as (i) chemoattraction, (ii) vasodilation, and (iii) angiogenesis, all of which are important for RA. Chemoattraction or chemotaxis is a particularly critical event in synovial inflammation, since a majority of cells migrate in the inflamed synovium from the blood stream and have passed the endothelial barrier. P38MAPK regulates adhesion molecule expression, including E-selectin and VCAM-1 on endothelial cells, which regulate rolling and adhesion of leukocytes on the endothelium before transmigrating to the inflamed tissue, respectively [56, 57]. Further, molecules involved in chemoattraction such as MCP-1 are regulated by p38MAPK [58].

Collectively, P38MAPK are important stress kinases that are involved in a variety of processes of synovial inflammation and bone resorption but also appear to be active under a number of other conditions as well. So far, their inhibition in clinical trials also encountered toxicity problems that halted further development.

PTPN22

The Arg620Trp variant of the intracellular phosphatase gene PTPN22 [5, 59] has been found as a SNP conferring a higher risk for the susceptibility for RA. It is interesting to note that PTPN22 is involved in modulating the strength of the T cell receptor signal and therefore able to translate a functional consequence. Very similar to the identification of STAT4, this appears to be a common predisposition gene for both lupus and RA [16], since other studies reported broad associations of the intracellular phosphatase PTPN22 with RA and SLE and with other autoimmune diseases [59] such as type 1 diabetes mellitus [60], autoimmune thyroid disease [61], and myasthenia gravis [62]. It is notable that the R620W mutation of PTPN22 conveys a gain of function that may truncate the T cell receptor response, possibly resulting in diminished activation induced T cell apoptosis and persistence of autoreactive T cells.

CTLA4

Activation of T cells requires two distinct signals. The first is an antigen specific interaction between the T cell receptor and antigen presented in the context of the MHC on the surface of an antigen presenting cell. The second signal may be provided through a number of potential co-stimulatory molecules, of which CD28 may be the most important. Co-stimulation is especially important for the initial T cell response, and its effects are mediated by promoting proliferation and survival. One of the most prominent T cell co-stimulatory signals is mediated through the CD28-CD80/86 pathway, which regulates interleukin-2 production and the expression of antiapoptotic molecules, such as Bcl-x_L [63, 64]. CD28 is present on most T cells and it binds to both CD80 (B7-1) and CD86 (B7-2), which are present on antigen presenting cells, including dendritic cells, B cells, and macrophages. These ligands are also expressed on activated T cells and are present on T cells obtained from the RA joint, suggesting a self-sustaining mechanism for T cell activation [65]. Engagement with these ligands provides the second signal required for maximal T cell activation, and the absence of a co-stimulatory signal may result in anergy and apoptotic cell death. Cytotoxic T lymphocyte associated antigen (CTLA)4 (CD152), which is upregulated on T cells following their activation, also interacts with CD80 and CD86, providing an important mechanism for downregulating T cell function [66, 67]. Not only does CTLA4 permit interruption of the activating CD28 pathway but it may also provide important negative signals that permit long term tolerance. CD28/B7 interactions are critical for the generation of CD4$^+$, CD25$^+$, CTLA4$^+$ T regulatory cells, and signaling through CTLA4 may promote the release of immunoregulatory cytokines such as TGFβ [64, 68]. Of interest, CTLA4 is expressed on T cells in the RA joint [6], supporting the potential importance of this pathway in regulating T cell activation in RA.

The regulatory effects of interrupting CD28 interactions with CD80/86 have been harnessed in recombinant molecules (CTLA4-immunoglobulin (Ig), abatacept) that combine the extracellular domain of human CTLA4 with a portion of the Fc domain of IgG$_1$ [69], which has obtained approval for the treatment of RA.

IL4R VARIANTS I50V AND Q551R AND RA

Identification of non-*HLA* loci associated with RA has also been extended to the 16p12 region [70, 71], which included *IL4R*, the gene coding for a specific receptor subunit of the Th2 cytokine interleukin-4 (IL-4). IL-4 and its receptor IL-4R play an important modifying role in the pathogenesis of RA, since diminished production of IL-4 is believed to contribute to the characteristic Th1 mediated autoimmune rheumatoid inflammation [72, 73]. Within the coding region of the *IL4R* gene, 2 SNPs, the I50V, and the Q551R polymorphisms reside within sequences coding for functionally important regions of the *IL4R* molecule. A recent study [6] identified two *IL4R* SNPs (I50V and Q551R) with RA susceptibility and severity in an association study of 371 controls and 471 well characterized RA patients with erosive disease. Although no association between I50V and Q551R *IL4R* SNPs and disease susceptibility was identified, the I50V SNP was strongly associated with rapid development of joint erosions. The predictive power of the I50V SNP for early erosive disease was higher than that of the auto-antibody, rheumatoid factor and the *HLA-DR* shared epitope (SE), thus identifying the I50V *IL4R* SNP as a novel genetic marker in RA with a high predictive value for early joint erosion. Interestingly, the functional role of the I50V SNP was analyzed and it could be shown that the response of CD4 T cells from subjects homozygous for the V50 allele had a significantly lower responsiveness to IL-4 than did cells from subjects homozygous for I50, as assessed by STAT-6 phosphorylation, GATA-3 induction, and IL-12R2 downmodulation. This study provided evidence for a possible mechanism that might underlie the newly identified association of the I50V *IL4R* SNP with early erosions in RA and confirms that subtle but predictable functional consequences that occur as a result of change of function polymorphisms can contribute to specific aspects of disease pathogenesis.

TLR SIGNALING AND ARTHRITIS AND AUTOIMMUNITY

Recognition of invading microorganisms by pattern recognition receptoirs (PRR), such as TLRs, results in the activation of genes encoding proinflammatory cytokines and chemokines, which induce local inflammatory reactions. Moreover, TLR signaling leads to upregulation of co-stimulatory molecules on antigen presenting cells, facilitating the subsequent activation of the adaptive immune system. Finally, engagement of TLRs on B Cells along with ligation of the B cell receptor can lead to increased stimulation of specific B cells and enhanced production of antibody [74].

The innate immune system is therefore able to influence adaptive immune responses via the provision of a second signal to T cell and B cell stimulation. Because of this important role of the innate immune system, it has been postulated that dysregulation of innate immune recognition of pathogens may be associated with autoimmunity. In this regard, an association of a TLR4 polymorphism (Asp299Gly) with Crohn's disease and ulcerative colitis has been identified [75]. However, a study in patients with rheumatoid arthritis and systemic lupus erythematosus (SLE) revealed no evidence for an association with TLR2 or -4 polymorphisms [76], whereas another report described a decreased susceptibility to rheumatoid arthritis in individuals with the TLR4 variant Asp299Gly [77].

Classical animal models of arthritis such as adjuvant arthritis or streptococcal cell wall arthritis are dependent on the activation of the innate immune system by TLR ligands as confirmed in mice deficient in the adaptor molecule MyD88; such mice did not develop arthritis [78]. The availability of TLR ligands might be sufficient to initiate arthritis in a susceptible host but require maintenance by adaptive immune mechanisms, although it remains unclear whether chronic stimulation of the innate immune system by TLRs is required. Heat shock proteins, fibrinogen, and hyaluronan are commonly found in inflamed joints and can bind to TLR4. A study conducted by Marshak-Rothstein and colleagues [74] has demonstrated in a transgenic mouse model, that chromatin containing immune complexes can activate B cells to produce rheumatoid factor auto-antibodies by synergistic engagement of the B cell receptor and TLR9. There is considerable evidence for the presence of exogenous as well as endogenous TLR ligands in autoimmune disease, although it is unclear whether activation of TLR signaling pathways is present or required in arthritis.

There is also evidence for a role for TLRs in the development of murine lupus. The autoimmunity susceptibility locus, *yaa*, that resides on the Y chromosome and accounts for the susceptibility to lupus in the BXSB strain of mice, contains a duplication of the TLR7 gene [79]. Moreover, genetic disruption of the TLR7 gene in BXSB mice prevents some but not all disease manifestations [80]. The complexity of TLR function in lupus is further emphasized by the finding that disruption of TLR7 in the MRL lp/lpr strain of mice blocked the development of auto-antibodies to RNA containing autoantigens and also disease development [81]. In contrast, disruption of TLR9 inhibits the production of some anti-histone antibodies, but actually causes the disease to worsen [81]. Efforts to control human lupus with antagonistic TLR ligands are currently in progress.

NFκB SIGNALING IN ARTHRITIS AND INFLAMMATION

Activation of the nuclear factor-κB (NFκB) pathway in the inflamed RA synovium appears to play a central role and results in the transactivation of a number of responsive elements (p105/p50, p100/p52, p65, RelB, c-Rel) that are intimately involved in the inflammatory reactions, including TNF, release of matrix metalloproteinases by synovial fibroblasts, and the production of proinflammatory chemokines. As a consequence, immune cells are recruited to the inflamed joints. Of central importance, the activation of the so-called "canonical" NFκB pathway leads to the formation of heterodimers of p50/p65. Although murine studies provide very compelling information on the role of NFκB in inflammation and autoimmunity [82], important differences exist in the signaling networks between human and murine immune cells and immortalized cell lines, and

TABLE 49.1 Potential B cell abnormalities leading to autoimmunity

V(D)J recombination

Entry of B cells into the immune repertoire

Survival of B cells by altered apoptosis

Selection

Somatic hypermutation

Receptor editing/revision

Differentiation of plasma cells

Extrinsic:

T cells, cytokines, APC, autoantigens

Altered activation threshold

require further studies. Although inhibition of the NFκB is attempting pharmacologic approach, it can result in activation and simultaneous exacerbation of inflammation [82]. A more detailed knowledge about the impact of NFκB inhibition is necessary before its safe potential use in the clinic is contemplated.

B CELL SIGNALING IN AUTOIMMUNITY

Historically, B cells have not been thought of as playing a central role in the immunopathogenesis of inflammatory joint diseases, in particular RA. Despite this, it has been accepted that auto-antibodies and immune complexes in the inflamed joints play an important amplifying role in synovitis. Recent data show that deleting B cells using an anti-CD20 antibody, rituximab, provides an effective therapy with an acceptable safety profile [7, 83].

Currently it is not clear whether intrinsic abnormalities of B cell function (Table 49.1) and/or their interaction with other immune cells appear to be of central importance of RA [7, 84]. Loss of precise regulatory influences stabilizing B cell homeostasis can result into autoimmunity or immune deficiency or sometimes both, although the predisposition and/or mechanisms remain unclear.

B Cell Signal Transduction Pathways and their Implications for Autoimmunity

The immune system is maintained by a fine balance between activation and inhibition (Table 49.2). On the one hand, it must possess adequate reactivity to generate an

TABLE 49.2 Immune functions of B cells

Precursors of (auto)antibody secreting plasma cells

Essential functions of B cells in regulating immune responses:

a. antigen presenting cells

b. differentiation of follicular dendritic cells in secondary lymphoid organs

c. essential role in lymphoid organogenesis as well as in the initiation and regulation of T and B cell responses

d. development of effective lymphoid architecture (antigen presenting M cells)

e. activated B cells express co-stimulatory molecules and may differentiate into polarized cytokine producing effector cells that can be essential for the evolution of T effector cells

f. differentiation of T effector cells

g. immunoregulatory functions by IL-10 positive B cells

h. cytokine production by activated B cells may influence the function of antigen presenting dendritic cells

effective immune response to target non-self molecules, whereas the emergence of autoimmunity must be avoided on the other hand. Essential to this process is the ability to control the timing and site of B cell activation and to limit the extent of activation precisely. All of this is sufficiently regulated by a number of extrinsic and intrinsic mechanisms in a normal immune system to avoid pathogenic autoimmunity. It is currently believed that failure to maintain this balance of activating and inhibiting factors, receptors, and pathways could result in either immunodeficiency or autoimmunity. Since B cells represent a unique crossroad between the innate and adaptive immune system, especially since they can be directly activated by toll-like receptors (TLRs), it becomes important that simple structures, such as methylated bacterial DNA, are able to activate B cells resulting in the production of auto-antibodies, such as rheumatoid factor [74].

Multiple checkpoints permit both positive and negative selection of B cells, both centrally in the bone marrow and in the peripheral lymphoid tissues, such as the spleen and lymph nodes. These checkpoints are necessary to produce a diverse population of B cells capable of generating high affinity effector antibodies in the absence of pathologic autoreactivity.

Disturbances that Alter B Cell Survival and Lead to Autoimmunity

A major process involved in B cell decisions in germinal centers (GC) is apoptosis, which is centrally involved in the selection of high affinity variants. Transgenic mice expressing

the genes bcl-2 and bcl-x_L, the products of which inhibit certain forms of apoptosis, provided evidence for the role of apoptosis in GC differentiation of B cells. A classical example of dysregulated apoptotic regulatory genes leading to autoimmunity was identified in a bcl-2 transgenic mouse model. Enhanced bcl-2 expression allows inappropriate survival of autoreactive B cell clones [85]. Bcl-2 transgenic mice develop antinuclear antibodies and glomerulonephritis caused by immune complex deposition. Primary immunization of bcl-2 transgenic mice resulted in 20-fold more memory B cells than controls. Furthermore, a high proportion of these bcl-2 transgenic memory B cells, despite having a typical memory phenotype, retained their V_H genes in a low affinity configuration. Thus, these memory B cells showed no evidence of having undergone affinity maturation.

An additional bcl-x_L transgenic mouse also preserved low affinity cells in the GC, although in this case the low affinity cells were not germline variants of the dominant clonotype but rather were B cells using V_H genes that usually appear only during the early stages of the response [86]. In both cases, blocking apoptosis in the GC had the effect of promoting the survival of low affinity variants that, in the case of the bcl-2 transgenic mice, entered the memory compartment. This demonstrates the role of apoptosis in selecting B cells that enter the memory compartment, including a certain frequency of autoreactive cells. In contrast with the alterations of the memory compartment, bcl-2 transgenic mice showed no perturbation of selection of bone marrow antibody secreting cells. This compartment remained predominantly composed of high affinity antibody producers. These findings suggest that resistance to apoptosis in the GC is an important prerequisite to allow differentiation of a GC cell into a memory cell, whereas the differentiation to long lived plasma cells requires a more stringent affinity based signal. How this translates into human autoimmunity is still a matter of debate.

Another example of autoimmunity developing as a result of alterations in lymphocyte apoptosis is provided by MRL mice homozygous for mutations in the Fas gene, a death inducing receptor required for normal regulation of B cell and T cell survival. MRL$^{lpr/lpr}$ mice develop a spectrum of autoreactivity resembling that found in human SLE and other autoimmune diseases. Thus, enhanced lymphocyte survival caused by inappropriate expression of apoptotic and/or antiapoptotic signals can promote the emergence of autoimmunity.

In addition to intrinsic defects that can lead to increased B cell survival, external signals permit autoreactive B cells to escape deletion. A signal that is particularly important in B cell growth, differentiation, and survival is BAFF (also known as BlyS, TALL-1, THANK, and zTNF4) [87]. BAFF is a member of the TNF family of cytokines that is produced by myeloid cells, such as dendritic cells, monocytes, and macrophages in inflamed tissue. It has remote effects and induces immature B cell survival as well as growth of mature B cells within peripheral lymphoid tissues. BAFF binds three receptors; BCMA (B cell maturation antigen), TACI (transmembrane activator and calcium modulator and cyclophilin ligand interactor), and BAFF receptor. Through these receptors, BAFF acts as a potent co-stimulator for B cell survival when coupled with B cell antigen receptor ligation. In this regard, it was reported that BAFF ligation increased bcl-2 expression and increased activation of NF-b, both of which increase B cell survival [88]. BCMA or BAFF transgenic mice display mature B cell hyperplasia and develop an SLE-like disease, with anti-DNA antibodies, elevated serum IgM, vasculitis, and glomerulonephritis [88]. Moreover, BAFF expression is elevated in MRL[lpr/lpr] mice and lupus prone (NZW×NZB)F1 hybrid mice and correlates with disease progression. In this regard, elevated BAFF levels were found in the serum of some patients with SLE, Sjögren's syndrome and idiopathic thrombocytopenic purpura patients [89] as well as in the synovial fluid of RA patients. Conversely, BAFF deficient mice show a complete loss of follicular and marginal zone B lymphocytes.

Studies of knockout mice have shown that BCMA, TACI, and BAFF-R are not directly equivalent in function [90]. Mice lacking BCMA show normal B cell development and antibody responses [91], whereas TACI deficient mice were shown to be deficient only in T cell independent antibody responses [92]. Paradoxically, mice lacking TACI show increased B cell proliferation and accumulation suggesting an inhibitory role for TACI in B cell homeostasis. Recently, BCMA has been identified as being involved in the generation of long lived plasma cells [93]. Thus far, gene targeted mice lacking BAFF-R have not been reported, but the natural mouse mutant, A/WySnJ, has a disruption of the intracellular domain of BAFF-R. A/WySnJ mice display a phenotype that is similar to BAFF[−/−] mice, although follicular and marginal zone B cells are not completely abolished [94]. A/WySnJ mice are impaired only in T cell dependent antibody responses, in contrast to the comprehensive defect of BAFF deficient mice. These results suggest that, while BAFF-R may be the major receptor of BAFF mediated signals for B cell survival, redundancy in function may be provided by the other two receptors, especially by TACI.

An interesting role for BAFF was recently shown when it was reported that BAFF regulated the survival of both marginal zone and follicular B cells in mice treated with anti-CD20 antibody [95]. The increased levels of BAFF found in some subjects with SLE or Sjögren's syndrome [87-89] may limit the therapeutic potential of this B cell depleting antibody.

Altered Thresholds for B Cellular Activation can Lead to Autoimmunity

Signals generated through the B cell antigen receptor (BCR) are critical for B cell development and survival [96] as well

as responses to antigen. The BCR is non-covalently associated with the signal transduction elements, Igα (CD79a) and Igβ (CD79b) (Table 49.2). The cytoplasmic domains of Igα and β contain highly conserved motifs that are the sites of Src family kinase docking and tyrosine phosphorylation, termed the immunoreceptor tyrosine based activation motifs (ITAM). Phosphorylation of tyrosines within these motifs is mediated by Src family kinases, including Lyn, Fyn, or Blk. This phosphorylation cascade promotes BCR recruitment of another tyrosine kinase, Syk, which facilitates receptor phosphorylation and initiates downstream signaling cascades that promote B cell activation [97].

The generation and maintenance of self-reactive B cells is regulated by autoantigen signaling through the BCR complex. These responses are further influenced by other cell surface signal transduction molecules, including CD19, CD21, and CD22, which function as response regulators to amplify or inhibit BCR signaling. CD19, CD21, and CD22 modulate BCR mediated signals by altering intrinsic intracellular signal transduction thresholds and thereby adjusting the strength of signal needed to initiate BCR mediated activation [97]. Intracellular regulatory molecules that also control the BCR signaling intensity include Lyn, Btk, Vav, and the SHP1 protein tyrosine phosphatase [97]. Notably, CD19, CD21, CD22, Lyn, Vav, and SHP1 are functionally linked in a common signaling pathway as summarized in Figure 49.1.

Inhibitory Receptors of B Cells

Currently, two major classes of inhibitory receptors have been described that share a number of structural and functional similarities. Each inhibitory receptor contains one or more immunoreceptor tyrosine based inhibitory motifs (ITIMs) within its cytoplasmic domain essential for generation and transduction of inhibitory signals. Ligation of the inhibitory receptor to an immunoreceptor tyrosine based activating motif (ITAM) containing activating molecule, results in tyrosine kinase phosphorylation of the tyrosine residue within the ITIM [98] by lyn [99], which allows it to bind and activate phosphatases containing an src homology 2 (SH2) domain. Two classes of SH2 containing inhibitory phosphatases have been identified: (i) the protein tyrosine phosphatases SHP-1 and SHP-2, and (ii) the phosphoinositol phosphatases, SHIP and SHIP2. These classes have separate downstream signaling pathways through which they modulate cellular inhibition. In general, each class of phosphatase interacts with the ITIMs of different inhibitory receptors, but each inhibitory receptor acts predominantly through only one class of phosphatase [99]. The surface molecules FcRII, CD22, and PD1 are inhibitory surface receptors and experimental evidence suggests that defective regulation by B cell inhibitory receptors may be of importance in autoimmunity.

B cell receptor signaling is critically dependent on
fine-tuning by activatory/inhibitory signals

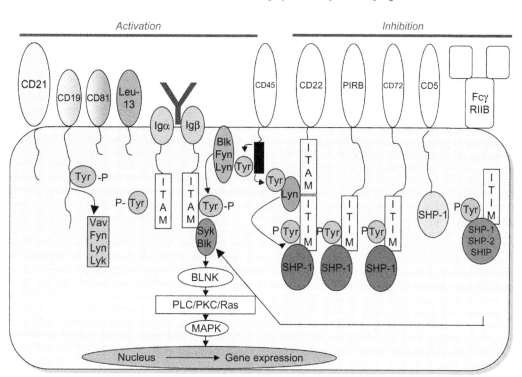

FIGURE 49.1 Activation and inhibitory surface molecules and subsequent signaling pathways of B cells that modulate the strength of the BCR signal.

Fc Receptors FcRIIb

Three classes of FcR have been described in humans, FcRI, FcRII, and FcRIII. FcRII and III are further divided into a and b forms. FcRI, IIa and IIIa are activating receptors, whereas FcRIIb is considered as inhibitory receptor. The function of FcRIIIb, which lacks an intracellular domain, is still unknown. Coordinate expression of FcR has been implicated in various diseases involving immune complexes, such as insulin dependent diabetes mellitus, SLE, RA, multiple sclerosis, and autoimmune anemia. FcRIIb is a member of the Ig superfamily and represents a single chain, low affinity receptor for the Fc portion of IgG. It is a 40 kDa protein that consists of two extracellular Ig-like domains, a transmembrane domain and an intracytoplasmic domain that contains a single ITIM. It binds IgG either complexed to multivalent soluble antigens as immune complexes or bound to cell membranes [100]. The isoform on B cells is unique in containing an intracytoplasmic motif that prevents its internalization [101].

In B cells, which do not express any Fc receptors other than FcRIIb, it acts to inhibit signaling through the B cell receptor (BCR), whereas in myeloid cells, FcRIIb inhibits activation through activating Fc receptors. Co-ligation of FcRIIb to the BCR leads to tyrosine phosphorylation of the ITIM by the tyrosine kinase lyn, recruitment of SHIP, and inhibition of Ca^{2+} flux and proliferation. The precise

mechanism by which SHIP prevents B cell proliferation is uncertain [101].

Evidence of a role for defective FcRIIb inhibition in the pathogenesis of autoimmunity is provided by studies of FcRII deficient mice, murine models of autoimmune disease, and human SLE as well as RA. FcRIIb deficiency renders normally resistant strains of mice susceptible to collagen induced arthritis and Goodpasture's syndrome. $FcRIIb^{-/-}$ mice derived on a C57BL/6 but not a Balb/c background produce auto-antibodies and develop immune complex mediated autoimmune disease resembling SLE, including immune complex mediated glomerulonephritis and renal failure. The *Fcr2b* and its polymorphisms represent a candidate gene of an inhibitory receptor that is likely to be involved in human autoimmune disease, in particular SLE. However, there has been no association reported for RA so far.

CD22

CD22 is a B cell specific glycoprotein that first appears intracellularly during the late pro-B cell stage of ontogeny. Subsequently, CD22 shifts to the plasma membrane with B cell maturation until plasma cell differentiation. Plasma cells do not express the molecule. CD22 has seven extracytoplasmic Ig-like domains and belongs to the Ig superfamily. It serves as a receptor for carbohydrate determinants on a wide variety of cell surface and soluble molecules *in vivo*. In contrast to CD19, CD22 can act as an antagonist to B cell activation most likely by enhancing the threshold of BCR induced signals. Following BCR engagement, CD22 is predominantly phosphorylated within ITIMs present in its cytoplasmic domain. Phosphorylation is predominantly mediated by Lyn, downstream of the CD19 dependent Lyn kinase amplification loop. If phosphorylated by Lyn, CD22 recruits the SHP-1 and SHIP phosphatases, leading to activation of a CD22/SHP-1/SHIP regulatory pathway that downregulates CD19 phosphorylation and BCR mediated signal transduction. Thus, CD19 and CD22 together define signaling thresholds critical for expansion of the peripheral B cell pool [101]. Ligation of CD22 to the BCR, and subsequent SHP-1 activation inhibits B cell activation by inhibiting the MAP kinases ERK2, JNK, and p38, and dephosphorylating molecules involved in the early events of BCR mediated activation. These include the BCR itself, tyrosine kinases activated by phosphorylation of Igαβ (such as syk) and the targets of these kinases (including the adaptor protein BLNK and PLC). Since co-ligation of CD22 to the BCR reduces B cell activation, the interaction of CD22 with its ligand may be involved in downregulating B cell activation [101].

CD22 has been linked genetically to autoimmune disease indicating a possible role for defects in CD22 and subsequent signaling pathways in the development of autoimmunity. It has been shown that CD22 deficient mice have an expanded B1 cell population and develop increased serum IgM. The B cells of these mice are hyper-responsive to stimulation through the BCR [102] suggesting that CD22 is an important inhibitory receptor in BCR dependent B cell activation. Importantly, these mice develop high affinity auto-antibodies to dsDNA, myeloperoxidase, and cardiolipin, although they do not develop autoimmune disease. Interestingly, CD22 has been successfully used as a target of biological therapies in patients with non-Hodgkin lymphomas as well as in Sjögren's syndrome [103] and SLE [104] by employing a humanized monoclonal antibody to CD22. To what extent this monoclonal antibody exerts its effects by depletion, inducing apoptosis, or by inhibition of B cell activation via CD22 remains to be shown.

PD-1

PD-1 is a 55 kDa highly conserved inhibitory receptor of the Ig superfamily [105]. It is expressed on resting B cells, T cells, and macrophages and is induced on activation of these cells [102, 105]. PD-1 is composed of a single extracellular Ig-like domain, a transmembrane region and has two tyrosine residues in the cytoplasmic tail, one of which forms part of an ITIM. Two PD-1 ligands (PD-Ls) have been identified and are constitutively expressed on dendritic cells and on heart, lung, thymus, and kidney and also on monocytes after IFN stimulation. *In vitro* studies on a B cell lymphoma line using a chimeric molecule with the FcRII extracellular domain and the PD-1 cytoplasmic domain have shown that ligation of the PD-1 cytoplasmic domain to the BCR can inhibit BCR mediated signaling. This inhibition prevented BCR mediated proliferation, Ca^{2+} mobilization, and tyrosine phosphorylation of molecules, including CD79beta, syk, PLC2, and ERK1/2. The physiological role of PD-1 in B cells remains unclear, but it may play a role in maintaining peripheral tolerance by limiting activation of autoreactive B cells by crosslinking PD-1 during interactions with PD-L expressing cells. PD-1 knockout mice develop either a lupus-like syndrome or autoimmune myocarditis, depending on the genetic background.

Collectively, preclinical studies in animal models, genetic analysis as well as linkage studies, indicate that the three inhibitory receptors PD-1, CD22, and FcRIIb candidate as being involved in autoimmunity. Whether these potential defects are involved in initiation or the maintenance of autoimmunity needs to be addressed.

Inhibitory Receptor Pathways and Autoimmunity

It needs emphasis that the large number of inhibitory receptors on the surface of B cells is subserved by remarkably similar intracellular signaling pathways. To date, Lyn is the only tyrosine kinase that has been identified as phosphorylating ITIMs on the B cell inhibitory receptors. Most of these ITIMs then associate with SHP-1 or SHIP (Figure 49.1).

Inhibitory receptors control the activation threshold of many immune cells, including B cells. There are many similarities in the signaling pathways of these inhibitory receptors. Inhibitory receptors also have specific effects, as they bind different ligands, including the activation of different phosphatases. Consistent with a role of defects in inhibitory receptor function in autoimmunity is the finding that B cells from inhibitory receptor deficient mice have similarities in phenotype and lowered thresholds for activation to that reported for B cells from SLE patients.

SHP-1

SHP-1 is a protein tyrosine phosphatase and is similar in structure to SHP-2. SHP-1 is the phosphatase that is utilized most widely in the inhibitory receptor signaling pathways. SHP-1 plays the predominant role in regulating through ITIMs, whereas increasing evidence suggests that SHP-2 may well have an additional activating role. Obviously, these molecules have an important role in regulation of a normal immune system. Defects in SHP-1 expression have been associated with SLE in humans; reduced levels of SHP-1 and Lyn are found in the lymphocytes of patients with inactive SLE, suggesting a potential role in pathogenesis.

SHIP

SHIP is a highly conserved SH2 containing inositol phosphatase related to SHIP-2 [104]. Both share a conserved N-terminal SH2 catalytic domain. SHIP acts predominantly on the FcRIIb signaling pathway. The molecule is expressed in myeloid and lymphoid lineages, including B cells [106]. Genetic studies in humans have identified susceptibility loci for both diabetes and SLE mapping to the region of the genome containing SHIP, although direct evidence for abnormal SHIP function in human disease is lacking.

Recently, the K/BxN mouse has generated particular interest. In this model, spontaneous arthritis occurs in mice that express both the transgene encoded KRN T cell receptor and the IAg^7 MHC class II allele [107]. The transgenic T cells have a specificity for glucose-6-phosphate isomerase (G6PI) and are able to break tolerance in the B cell compartment resulting in the production of auto-antibodies to G6PI. Affinity purified anti-G6PI Ig from these mice can transfer joint specific inflammation to healthy recipients. A mechanism for joint specific disease arising from autoimmunity to G6PI has been suggested recently. G6PI bound to the surface of cartilage serves as the target for anti-G6PI binding and subsequent complement mediated damage. In this model, the inciting event is the expression of an autoreactive T cell receptor in the periphery. However, joint destruction is delegated by the adaptive response to innate immune mechanisms and can be transferred to animals that lack B and T cells [108]. Whereas these animal studies are very compelling, anti-G6PI antibodies do not frequently occur in the serum of RA patients.

SUMMARY

Although a number of associations of genetic abnormalities and RA suggest a functional involvement in this disease, so far none has been proven to be effective in disease susceptibility or disease activity. That extracellular and intracellular factors involved in signaling pathways clearly play a role in RA with p38MAPK, has been characterized most intensively. With regard to systemic autoimmune diseases, much attention has been focused on interaction of immune cells (CD40/CD40L, PD-1) and intracellular signaling. Because of the central role of B cells in these entities, inhibitory receptors and their signaling pathways appear to be central in the regulation of activation thresholds. Although still at early stages of investigation, identification of distinct signaling pathways have the promise for new therapeutic avenues.

REFERENCES

1. MacGregor AJ, Snieder H, Rigby AS, Koskenvuo M, Kaprio J, Aho K, et al. Characterizing the quantitative genetic contribution to rheumatoid arthritis using data from twins. *Arthritis Rheum* 2000;**43**(1):30–7.

2. Seldin MF, Amos CI, Ward R, Gregersen PK. The genetics revolution and the assault on rheumatoid arthritis. *Arthritis Rheum* 1999;**42**(6):1071–9.

3. Smolen JS, Aletaha D, Koeller M, Weisman MH, Emery P. New therapies for treatment of rheumatoid arthritis. *Lancet* 2007;**370**(9602):1861–74.

4. Wordsworth BP, Bell JI. The immunogenetics of rheumatoid-arthritis. *Springer Semin Immunopathol* 1992;**14**(1):59–78.

5. Begovich AB, Carlton VEH, Honigberg LA, Schrodi SJ, Chokkalingam AP, Alexander HC, et al. A missense single-nucleotide polymorphism in a gene encoding a protein tyrosine phosphatase (PTPN22) is associated with rheumatoid arthritis. *Am J Hum Genet* 2004;**75**(2):330–7.

6. Prots I, Skapenko A, Wendler J, Mattyasovszky S, Yone CL, Spriewald B, et al. Association of the IL4R single-nucleotide polymorphism I50V with rapidly erosive rheumatoid arthritis. *Arthritis Rheum* 2006;**54**(5):1491–500.

7. Hoppe B, Haupl T, Gruber R, Kiesewetter H, Burmester GR, Salama A, et al. Detailed analysis of the variability of peptidylarginine deiminase type 4 in German patients with rheumatoid arthritis: a case control study.. *Arthritis Res Ther* 2006;**8**(2):R34.

8. Plenge RM, Padyukov L, Remmers EF, Purcell S, Lee AT, Karlson EW, et al. Replication of putative candidate-gene associations with rheumatoid arthritis in >4,000 samples from North America and Sweden: Association of susceptibility with PTPN22, CTLA4, and PADI4. *Am J Hum Genet* 2005;**77**(6):1044–60.

9. Suzuki A, Yamada R, Chang XT, Tokuhiro S, Sawada T, Suzuki M, et al. Functional haplotypes of PADI4, encoding citrullinating enzyme peptidylarginine deiminase 4, are associated with rheumatoid arthritis. *Nat Genet* 2003;**34**(4):395–402.

10. Watford WT, Hissong BD, Bream JH, Kanno Y, Muul L, O'Shea JJ. Signaling by IL-12 and IL-23 and the immunoregulatory roles of STAT4. *Immunol Rev* 2004;**202**:139–56.

11. Morinobu A, Gadina M, Strober W, Visconti R, Fornace A, Montagna C, et al. STAT4 serine phosphorylation is critical for IL-12-induced IFN-gamma production but not for cell proliferation. *Proc Natl Acad Sci U S A* 2002;**99**(19):12,281–12,286,.

12. Nishikomori R, Usui T, Wu CY, Morinobu A, O'Shea JJ, Strober W. Activated STAT4 has an essential role in Th1 differentiation and proliferation that is independent of its role in the maintenance of IL-12R beta 2 chain expression and signaling. *J Immunol* 2002;**169**(8):4388–98.

13. Mathur AN, Chang HC, Zisoulis DG, Stritesky GL, Yu Q, O'Malley JT, et al. Stat3 and Stat4 direct development of IL-17-secreting Th cells. *J Immunol* 2007;**178**(8):4901–7.

14. Bettelli E, Oukka M, Kuchroo VK. T-H-17 cells in the circle of immunity and autoimmunity. *Nat Immunol* 2007;**8**(4):345–50.

15. Finnegan A, Grusby MJ, Kaplan CD, O'Neill SK, Eibel H, Koreny T, et al. IL-4 and IL-12 regulate proteoglycan-induced arthritis through stat-dependent mechanisms. *J Immunol* 2002;**169**(6):3345–52.

16. Remmers EF, Plenge RM, Lee AT, Graham RR, Hom G, Behrens TW, et al. STAT4 and the risk of rheumatoid arthritis and systemic lupus erythematosus. *N Eng J Med* 2007;**357**(10):977–86.

17. Amos CI, Chen WV, Lee A, Li W, Kern M, Lundsten R, et al. High-density SNP analysis of 642 Caucasian families with rheumatoid arthritis identifies two new linkage regions on 11p12 and 2q33. *Genes Immun* 2006;**7**(4):277–86.

18. Jawaheer D, Seldin MF, Amos CI, Chen WV, Shigeta R, Etzel C, et al. Screening the rheumatoid arthritis genome for susceptibility genes: A replication study and combined analysis of 512 multicase families. *Arthritis Rheum* 2003;**48**(4):906–16.

19. Gregersen PK, Behrens TW. Genetics of autoimmune diseases: disorders of immune homeostasis. *Nat Rev Genet* 2006;**7**(12):917–28.

20. Plenge RM, Seielstad M, Padyukov L, Lee AT, Remmers EF, Ding B, et al. TRAF1-C5 as a risk locus for rheumatoid arthritis: a genomewide study. *N Eng J Med* 2007;**357**(12):1199–209.

21. Tsitsikov EN, Laouini D, Dunn IF, Sannikova TY, Davidson L, Alt FW, et al. TRAF1 is a negative regulator of TNF signaling: Enhanced TNF signaling in TRAF1-deficient mice. *Immunity* 2001;**15**(4):647–57.

22. Bradley JR, Pober JS. Tumor necrosis factor receptor-associated factors (TRAFs). *Oncogene* 2001;**20**(44):6482–91.

23. Cooke TD, Hurd ER, Jasin HE, Bienenstock J, Ziff M. Identification of immunoglobulins and complement in rheumatoid articular collagenous tissues. *Arthritis Rheum* 1975;**18**(6):541–51.

24. Ji H, Ohmura K, Mahmood U, Lee DM, Hofhuis FMA, Boackle SA, et al. Arthritis critically dependent on innate immune system players. *Immunity* 2002;**16**(2):157–68.

25. Wang Y, Rollins S, Madri J, Matis L. Anti-C5 monoclonal-antibody therapy parents collagen-induced arthritis and ameliorates established disease. *Arthritis Rheum* 1995;**38**(9):1316.

26. Wang Y, Kristan J, Hao LM, Lenkoski CS, Shen YM, Matis LA. A role for complement in antibody-mediated inflammation: C5-deficient DBA/1 mice are resistant to collagen-induced arthritis. *J Immunol* 2000;**164**(8):4340–7.

27. Anon. Genome-wide association study of 14,000 cases of seven common diseases and 3,000 shared controls. *Nature* 2007;**447**(7145):661–7.

28. Cohen DM. Mitogen-activated protein kinase cascades and the signaling of hyperosmotic stress to immediate early genes. *Comparative Biochemistry and Physiology A-Physiology* 1997;**117**(3):291–9.

29. Johnson GL, Lapadat R. Mitogen-activated protein kinase pathways mediated by ERK, JNK, and p38 protein kinases. *Science* 2002;**298**(5600):1911–12.

30. Lee JC, Laydon JT, Mcdonnell PC, Gallagher TF, Kumar S, Green D, et al. A Protein-Kinase Involved in the Regulation of Inflammatory Cytokine Biosynthesis. *Nature* 1994;**372**(6508):739–46.

31. Beyaert R, Cuenda A, VandenBerghe W, Plaisance S, Lee JC, Haegeman G, et al. The p38/RK mitogen-activated protein kinase pathway regulates interleukin-6 synthesis in response to tumour necrosis factor. *Embo J* 1996;**15**(8):1914–23.

32. Ferrero E, Zocchi MR, Magni E, Panzeri MC, Curnis F, Rugarli C, et al. Roles of tumor necrosis factor p55 and p75 receptors in TNF-alpha-induced vascular permeability. *Am J Physiol Cell Physiol* 2001;**281**(4):C1173–9.

33. Gortz B, Hayer S, Tuerck B, Zwerina J, Smolen JS, Schett G. Tumour necrosis factor activates the mitogen-activated protein kinases p38 alpha and ERK in the synovial membrane in vivo. *Arthritis Res Ther* 2005;**7**(5):R1140–7.

34. Theodosiou A, Ashworth A. Differential effects of stress stimuli on a JNK-inactivating phosphatase. *Oncogene* 2002;**21**(15):2387–97.

35. Kassel O, Sancono A, Kratzschmar J, Kreft B, Stassen M, Cato ACB. Glucocorticoids inhibit MAP kinase via increased expression and decreased degradation of MKP-1. *Embo J* 2001;**20**(24):7108–16.

36. Toh ML, Yang Y, Leech M, Santos L, Morand EF. Expression of mitogen-activated protein kinase phosphatase 1, a negative regulator of the mitogen-activated protein kinases, in rheumatoid arthritis: Up-regulation by interleukin-1 beta and glucocorticoids. *Arthritis Rheum* 2004;**50**(10):3118–28.

37. Feldmann M, Maini RN, Bondeson J, Taylor P, Foxwell BMJ, Brennan FM. Cytokine blockade in rheumatoid arthritis. *Mechanisms of Lymphocyte Activation and Immune Regulation Viii: Autoimmunity 2000 and Beyond* 2001;**490**:119–27.

38. Sweeney SE, Firestein GS. Signal transduction in rheumatoid arthritis. *Curr Opin Rheumatol* 2004;**16**(3):231–7.

39. Badger AM, Griswold DE, Kapadia R, Blake S, Swift BA, Hoffman SJ, et al. Disease-modifying activity of SB 242235, a selective inhibitor of p38 mitogen-activated protein kinase, in rat adjuvant-induced arthritis. *Arthritis Rheum* 2000;**43**(1):175–83.

40. Medicherla S, Ma JY, Mangadu R, Jiang YB, Zhao JJ, Almirez R, et al. A selective p38 alpha mitogen-activated protein kinase inhibitor reverses cartilage and bone destruction in mice with collagen-induced arthritis. *J Pharmacol Exp Ther* 2006;**318**(1):132–41.

41. Nishikawa M, Myoui A, Tomita T, Takahi K, Nampei A, Yoshikawa H. Prevention of the onset and progression of collagen-induced arthritis in rats by the potent p38 mitogen-activated protein kinase inhibitor FR167653. *Arthritis Rheum* 2003;**48**(9):2670–81.

42. Patten C, Bush K, Rioja I, Morgan R, Wooley P, Trill J, et al. Characterization of pristane-induced arthritis, a murine model of chronic disease: response to antirheumatic agents, expression of joint cytokines, and immunopathology. *Arthritis Rheum* 2004;**50**(10):3334–45.

43. Revesz L, Blum E, Di Padova FE, Buhl T, Feifel R, Gram H, et al. Novel p38 inhibitors with potent oral efficacy in several models of rheumatoid arthritis.. *Bioorg Med Chem* 2004;**14**(13):3595–9.

44. Zwerina J, Hayer S, Redlich K, Bobacz K, Smolen JS, Schett G. Activation of p38MAPK is a key step in TNF-Mediated inflammatory bone destruction. *Arthritis Rheum* 2005;**52**(9):S160.

45. Beardmore VA, Hinton HJ, Eftychi C, Apostolaki M, Armaka M, Darragh J, et al. Generation and characterization of p38 beta (MAPK11) gene-targeted mice. *Mol Cell Biol* 2005;**25**(23):10,454–10,464,.

46. Muller-Ladner U, Gay S. The SCID mouse: A novel experimental model for gene therapy in human rheumatoid arthritis. *Drugs Today* 1999;**35**(4–5):379–88.

47. Pap T, Aupperle KR, Gay S, Firestein GS, Gay RE. Invasiveness of synovial fibroblasts is regulated by p53 in the SCID mouse in vivo model of cartilage invasion. *Arthritis Rheum* 2001;**44**(3):676–81.

48. Liacini A, Sylvester J, Li WQ, Huang WS, Dehnade F, Ahmad M, et al. Induction of matrix metalloproteinase-13 gene expression by TNF-alpha is mediated by MAP kinases, AP-1, and NF-kappa B transcription factors in articular chondrocytes. *Exp Cell Res* 2003;**288**(1):208–17.

49. Suzuki M, Tetsuka T, Yoshida S, Watanabe N, Kobayashi M, Matsui N, et al. The role of p38 mitogen-activated protein kinase in IL-6 and IL-8 production from the TNF-alpha- or IL-1 beta-stimulated rheumatoid synovial fibroblasts. *Febs Lett* 2000;**465**(1):23–7.

50. Gravallese EM, Harada Y, Wang JT, Gorn AH, Thornhill TS, Goldring SR. Identification of cell types responsible for bone resorption in rheumatoid arthritis and juvenile rheumatoid arthritis. *Am J Pathol* 1998;**152**(4):943–51.

51. Li XT, Udagawa N, Itoh K, Suda K, Murase Y, Nishihara T, et al. p38 MAPK-mediated signals are required for inducing osteoclast differentiation but not for osteoclast function. *Endocrinology* 2002;**143**(8):3105–13.

52. Redlich K, Hayer S, Ricci R, David JP, Tohidast-Akrad M, Kollias G, et al. Osteoclasts are essential for TNF-alpha-mediated joint destruction. *J Clin Invest* 2002;**110**(10):1419–27.

53. Matsumoto M, Sudo T, Maruyama M, Osada H, Tsujimoto M. Activation of p38 mitogen-activated protein kinase is crucial in osteoclastogenesis induced by tumor necrosis factor. *Febs Lett* 2000;**486**(1):23–8.

54. Hayer S, Steiner G, Gortz B, Reiter E, Tohidast-Akrad M, Amling M, et al. CD44 is a determinant of inflammatory bone loss. *J Exp Med* 2005;**201**(6):903–14.

55. Rousseau S, Houle F, Landry J, Huot J. p38 MAP kinase activation by vascular endothelial growth factor mediates actin reorganization and cell migration in human endothelial cells. *Oncogene* 1997;**15**(18):2169–77.

56. Pietersma A, Tilly BC, Gaestel N, deJong N, Lee JC, Koster JF, et al. P38 mitogen activated protein kinase regulates endothelial VCAM-1 expression at the post-transcriptional level. *Biochem Biophys Res Commun* 1997;**230**(1):44–8.

57. Read MA, Whitley MZ, Gupta S, Pierce JW, Best J, Davis RJ, et al. Tumor necrosis factor alpha-induced E-selectin expression is activated by the nuclear factor-kappa B and c-JUN N-terminal kinase/p38 mitogen-activated protein kinase pathways. *J Biol Chem* 1997;**272**(5):2753–61.

58. Goebeler M, Kilian K, Gillitzer R, Kunz M, Yoshimura T, Brocker EB, et al. The MKK6/p38 stress kinase cascade is critical for tumor necrosis factor-alpha-induced expression of monocyte-chemoattractant protein-1 in endothelial cells. *Blood* 1999;**93**(3):857–65.

59. Gregersen PK, Lee HS, Batliwalla F, Begovich AB. PTPN22: Setting thresholds for autoimmunity. *Semin Immunol* 2006;**18**(4):214–23.

60. Bottini N, Musumeci L, Alonso A, Rahmouni S, Nika K, Rostamkhani M, et al. A functional variant of lymphoid tyrosine phosphatase is associated with type I diabetes. *Nat Genet* 2004;**36**(4):337–8.

61. Criswell LA, Pfeiffer KA, Lum RF, Gonzales B, Novitzke J, Moser KL, et al. Analysis of families in the multiple autoimmune disease genetics consortium (MADGC) collection: the PTPN22 620W allele associates with multiple autoimmune phenotypes. *Am J Hum Genet* 2005;**76**(4):561–71.

62. Vandiedonck C, Capdevielle C, Giraud M, Krumeich S, Jais JP, Eymard B, et al. Association of the PTPN22*R620W polymorphism with autoimmune myasthenia gravis. *Ann Neurol* 2006;**59**(2):404–7.

63. Jenkins MK, Taylor PS, Norton SD, Urdahl KB. Cd28 delivers a costimulatory signal involved in antigen-specific Il-2 production by human T-Cells. *J Immunol* 1991;**147**(8):2461–6.

64. Salomon B, Bluestone JA. Complexities of CD28/B7: CTLA-4 costimulatory pathways in autoimmunity and transplantation. *Annu Rev Immunol* 2001;**19**:225–52.

65. Verwilghen J, Lovis R, Deboer M, Linsley PS, Haines GK, Koch AE, et al. Expression of functional B7 and Ctla4 on rheumatoid synovial T-Cells. *J Immunol* 1994;**153**(3):1378–85.

66. Karandikar NJ, Vanderlugt CL, Walunas TL, Miller SD, Bluestone JA. CTLA-4: a negative regulator of autoimmune disease. *J Exp Med* 1996;**184**(2):783–8.

67. Walunas TL, Bakker CY, Bluestone JA. CTLA-4 ligation blocks CD28-dependent T cell activation. *J Exp Med* 1996;**183**(6):2541–50.

68. Tang QZ, Henriksen KJ, Boden EK, Tooley AJ, Ye JQ, Subudhi SK, et al. Cutting edge: CD28 controls peripheral homeostasis of CD4(+)CD25(+). *J Immunol* 2003;**171**(7):3348–52.

69. Linsley PS, Brady W, Urnes M, Grosmaire LS, Damle NK, Ledbetter JA. Ctla-4 is a 2Nd receptor for the B-Cell activation antigen-B7. *J Exp Med* 1991;**174**(3):561–9.

70. Cornelis F, Faure S, Martinez M, Prud'Homme JF, Fritz P, Dib C, et al. New susceptibility locus for rheumatoid arthritis suggested by a genome-wide linkage study. *Proc Natl Acad Sci U S A* 1998;**95**(18):10,746–10,750,.

71. Jawaheer D, Seldin MF, Amos CI, Chen WV, Shigeta R, Monteiro J, et al. A genomewide screen in multiplex rheumatoid arthritis families suggests genetic overlap with other autoimmune diseases. *Am J Hum Genet* 2001;**68**(4):927–36.

72. Simon AK, Seipelt E, Sieper J. Divergent T-Cell cytokine patterns in inflammatory arthritis. *Proc Natl Acad Sci U S A* 1994;**91**(18):8562–6.

73. Skapenko A, Wendler J, Lipsky PE, Kalden JR, Schulze-Koops H. Altered memory T cell differentiation in patients with early rheumatoid arthritis. *J Immunol* 1999;**163**(1):491–9.

74. Leadbetter EA, Rifkin IR, Hohlbaum AM, Beaudette BC, Shlomchik MJ, Marshak-Rothstein A. Chromatin-IgG complexes activate B cells by dual engagement of IgM and Toll-like receptors. *Nature* 2002;**416**(6881):603–7.

75. Brentano F, Kyburz D, Schorr O, Gay R, Gay S. The role of toll-like receptor signalling in the pathogenesis of arthritis. *Cell Immunol* 2005;**233**(2):90–6.

76. Sanchez E, Jimenez-Alonso J, Raya E, Martin J. Polymorphisms of Toll-like receptor 2 and 4 genes in rheumatoid arthritis and systemic lupus erythematosus. *Genes Immun* 2004;**5**:S34.

77. Radstake TRDJ, Franke B, Hanssen S, Netea MG, Welsing P, Barrera P, et al. The toll-like receptor 4 Asp299Gly functional variant is associated with decreased rheumatoid arthritis disease susceptibility but does not influence disease severity and/or outcome. *Arthritis Rheum* 2004;**50**(3):999–1001.

78. Joosten LAB, Koenders MI, Smeets RL, Heuvelmans-Jacobs M, Helsen MMA, Takeda K, et al. Toll-like receptor 2 pathway drives streptococcal cell wall-induced joint inflammation: Critical role of myeloid differentiation factor 88. *J Immunol* 2003;**171**(11):6145–53.

79. Subramanian S, Tus K, Li QZ, Wang A, Tian XH, Zhou J, et al. A Tlr7 translocation accelerates systemic autoimmunity in murine lupus. *Proc Natl Acad Sci U S A* 2006;**103**(26):9970–5.

80. Santiago-Raber ML, Kikuchi S, Borel P, Uematsu S, Akira S, Kotzin BL, et al. Evidence for genes in addition to Tlr7 in the Yaa translocation linked with acceleration of systemic lupus erythematosus. *J Immunol* 2008;**181**(2):1556–62.

81. Christensen SR, Shlomchik MJ. Regulation of lupus-related autoantibody production and clinical disease by Toll-like receptors. *Semin Immunol* 2007;**19**(1):11–23.

82. Simmonds RE, Foxwell BM. Signalling, inflammation and arthritis: NF-kappa B and its relevance to arthritis and inflammation. *Rheumatology* 2008;**47**(5):584–90.

83. Cohen SB, Emery P, Greenwald MW, Dougados M, Furie RA, Genovese MC, et al. Rituximab for rheumatoid arthritis refractory to anti-tumor necrosis factor therapy: results of a multicenter, randomized, double-blind, placebo-controlled, phase III trial evaluating primary efficacy and safety at twenty-four weeks. *Arthritis Rheum* 2006;**54**(9):2793–806.

84. Dorner T, Lipsky PE. Signalling pathways in B cells: implications for autoimmunity. *Curr Top Microbiol Immunol* 2006;**305**:213–40.

85. Strasser A, Harris AW, Cory S. Bcl-2 transgene inhibits T-Cell death and perturbs thymic self-censorship. *Cell* 1991;**67**(5):889–99.

86. Takahashi Y, Cerasoli DM, Dal Porto JM, Shimoda M, Freund R, Fang W, et al. Relaxed negative selection in germinal centers and impaired affinity maturation in bcl-x(L) transgenic mice. *J Exp Med* 1999;**190**(3):399–409.

87. Dorner T, Putterman C. B cells, BAFF/zTNF4, TACl, and systemic lupus erythematosus. *Arthritis Res* 2001;**3**(4):197–9.

88. Mackay F, Tangye SG. The role of the BAFF/APRIL system in B cell homeostasis and lymphoid cancers. *Curr Opin Pharmacol* 2004;**4**(4):347–54.

89. Groom J, Kalled SL, Cutler AH, Olson C, Woodcock SA, Schneider P, et al. Association of BAFF/BLyS overexpression and altered B cell differentiation with Sjogren's syndrome. *J Clin Invest* 2002;**109**(1):59–68.

90. Thompson JS, Bixler SA, Qian F, Vora K, Scott ML, Cachero TG, et al. BAFF-R, a newly identified TNF receptor that specifically interacts with BAFF. *Science* 2001;**293**(5537):2108–11.

91. Yan MH, Brady JR, Chan B, Lee WP, Hsu B, Harless S, et al. Identification of a novel receptor for B lymphocyte stimulator that is mutated in a mouse strain with severe B cell deficiency. *Curr Biol* 2001;**11**(19):1547–52.

92. Carsetti R, Köhler G, Lamers MC. Transitional B cells are the target of negative selection in the B cell compartment. *J Exp Med* 1995;**181**(6):2129–40.

93. Schiemann B, Gommerman JL, Vora K, Cachero TG, Shulga-Morskaya S, Dobles M, et al. An essential role for BAFF in the normal development of B cells through a BCMA-independent pathway. *Science* 2001;**293**(5537):2111–14.

94. Benschop RJ, Cambier JC. B cell development: signal transduction by antigen receptors and their surrogates. *Curr Opin Immunol* 1999;**11**(2):143–51.

95. Gong Q, Ou Q, Ye S, Lee WP, Cornelius J, Diehl L, Lin WY, Hu Z, Lu Y, Chen Y, Wu Y, Meng YG, Gribling P, Lin Z, Nguyen K, Tran T, Zhang Y, Rosen H, Martin F, Chan AC. Importance of cellular microenvironment and circulatory dynamics in B cell immunotherapy. *J Immunol* 2005;**174**(2):817–26.

96. Maruyama M, Lam KP, Rajewsky K. Memory B-cell persistence is independent of persisting immunizing antigen. *Nature* 2000;**407**(6804):636–42.

97. Tedder TF, Tuscano J, Sato S, Kehrl JH. CD22, a B lymphocyte-specific adhesion molecule that regulates antigen receptor signaling. *Annu Rev Immunol* 1997;**15**:481–504.

98. Fong DC, Brauweiler A, Minskoff SA, Bruhns P, Tamir I, Mellman I, et al. Mutational analysis reveals multiple distinct sites within Fc gamma receptor IIB that function in inhibitory signaling. *J Immunol* 2000;**165**(8):4453–62.

99. Daeron M, Lesourne R. Negative signaling in Fc receptor complexes. *Adv Immunol* 2006;**89**:39–86.

100. Latour S, Fridman WH, Daeron M. Identification, molecular cloning, biologic properties, and tissue distribution of a novel isoform of murine low-affinity IgG receptor homologous to human Fc gamma RIIB1. *J Immunol* 1996;**157**(1):189–97.

101. Pritchard NR, Smith KGC. B cell inhibitory receptors and autoimmunity. *Immunology* 2003;**108**(3):263–73.

102. Okazaki T, Maeda A, Nishimura H, Kurosaki T, Honjo T. PD-1 immunoreceptor inhibits B cell receptor-mediated signaling by recruiting src homology 2-domain-containing tyrosine phosphatase 2 to phosphotyrosine. *Proc Natl Acad Sci U S A* 2001;**98**(24):13,866–13,871,.

103. Steinfeld SD, Rommes S, Tant L, Song I, Burmester GR, Wegener WA, et al. Initial clinical study of immunotherapy in primary Sjogren's syndrome with humanized anti-CD22 antibody epratuzumab. *Ann Rheum Dis* 2005;**64**:311.

104. Dorner T, Kaufmann J, Wegener WA, Teoh N, Goldenberg DM, Burmester GR. Initial clinical trial of epratuzumab (humanized anti-CD22 antibody) for immunotherapy of systemic lupus erythematosus. *Arthritis Res Ther* 2006;**8**(3):R74.

105. Nishimura H, Honjo T. PD-1: an inhibitory immunoreceptor involved in peripheral tolerance. *Trends Immunol* 2001;**22**(5):265–8.

106. Helgason CD, Damen JE, Rosten PM, Grewal R, Sorensen P, Chappel SM, et al. Targeted disruption of ship leads to hemopoietic perturbations, lung pathology, and a shortened lifespan. *Exp Hematol* 1998;**26**(8):815.

107. Matsumoto I, Staub A, Benoist C, Mathis D. Arthritis provoked by linked T and B cell recognition of a glycolytic enzyme. *Science* 1999;**286**(5445):1732–5.

108. Korganow AS, Ji H, Mangialaio S, Duchatelle V, Pelanda R, Martin T, et al. From systemic T cell self-reactivity to organ-specific autoimmune disease via immunoglobulins. *Immunity* 1999;**10**(4):451–61.

Chapter 50

Translational Concepts in Vasculitis

Daniel A. Albert and David B. Talmadge

Dartmouth Medical School and Dartmouth-Hitchcock Medical Center, Lebanon, New Hampshire

INTRODUCTION

Delivery of oxygen and nutrients to the end organs and the removal of by-products from bodily tissues are major functions of the vascular tree. These vital processes may be severely perturbed by the vasculitides, a varied group of inflammatory diseases that have as their common denominator inflammation of blood vessels, decreased blood flow, and subsequent tissue damage through ischemic and/or hemorrhagic mechanisms. Organized classification schemes devised in the last 20 years have been very helpful in characterizing the multitude of vasculitic syndromes. These rubrics have allowed basic science researchers and clinicians alike to gain a better understanding of the core pathological processes at work in vasculitis. Though the etiology of vasculitis remains relatively obscure, our understanding of the pathophysiology of these diseases has grown through the study of the vascular endothelium and antiendothelial cell antibodies (AECAs), immune complex formation and deposition, antineutrophil cytoplasmic antibodies (ANCAs), and animal models. Translational research involving signal transduction and resulting clinical application is now gaining a foothold in the vasculitis scientific community, and these efforts promise to usher in a new revolution of therapies targeted toward a group of diseases that generate significant morbidity and mortality.

MECHANISMS IN THE PATHOGENESIS OF VASCULITIS

The Vascular Endothelium

The endothelial layer of blood vessels constitutes a dynamic and physiologically active layer of cells that line blood vessels. The endothelial cell (EC) layer is not a static structure, but instead a highly specialized single layer of cells that is able to react to the microenvironment and alter the physiology of the end organs they serve. ECs have a number of functions, some of which include hemostasis, neovascularization, control of local blood flow, and coordination of the transport and recruitment of leukocytes. Through these mechanisms and their interface with the blood stream, ECs are involved in the normal physiology of the end organ, but also in the pathophysiology as it relates to vasculitic processes and other deleterious insults. The characteristics and functions of the endothelial cells vary depending on the size and location of the blood vessel studied, and there are both microvascular and macrovascular ECs. Much of our understanding of EC signaling comes from the *in vitro* study of human umbilical vein endothelial cells (HUVECs), though researchers are now starting to incorporate tenets from more organ specific endothelial cell subtypes in the study of vasculitis. For example, human kidney microvascular endothelial cells (HKMECs) are found in Wegener's granulomatosis (WG) [1]. Effector cells of the immune system interact with the endothelium to evoke an immune response. This chapter will encompass the signaling mechanisms by which these processes occur in vasculitis.

Endothelial cells are activated by inflammatory processes, either local or systemic. Trauma, infection, and vasculitis are three examples of inflammatory insults that can trigger the activation of ECs. Interleukin-1 (IL-1) and tumor necrosis factor-alpha (TNFα) are two very potent inflammatory cytokines produced by dendritic cells (DCs), fibroblasts, and macrophages that can activate the EC. An overview of EC adhesion pathways is illustrated in Figure 50.1.

Through a cascade of intracellular signaling and enzymatic modifications nuclear factor kappa beta (NFκβ) is generated, which leads to a whole host of downstream events. These include the production of inflammatory and recruitment cytokines Interleukin-6 (IL-6) and Interleukin-8 (IL-8), respectively, upregulation of the adhesion molecules intercellular adhesion molecule-1 (ICAM-1) and vascular cell adhesion molecule-1 (VCAM-1) among others. Cyclo-oxygenase-2 (COX-2) and inducible nitric oxide

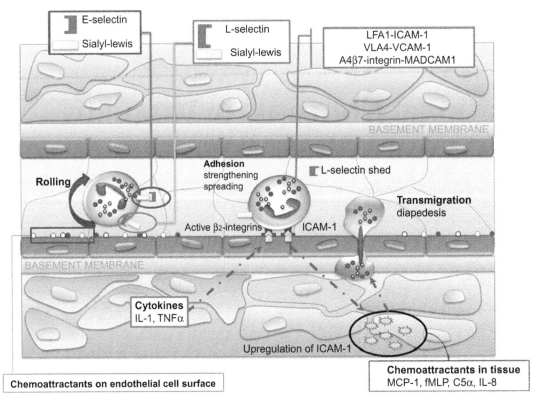

FIGURE 50.1 Leukocyte adhesion cascade.
The inflammatory process that results in vasculitis is initiated by a series of molecular events between leukocytes and the vascular endothelium. The process involves the recruitment of leukocytes to inflammation sites, due, in part, to the actions of the indicated cytokines.

synthase (iNOS) enzymes are also produced. The net effect is recruitment of further immune cells to the inflammatory site, with egress of these effector cells into the surrounding tissue [2], as shown in Figure 50.2 [3].

Through an understanding of these signaling mechanisms, translational research has revealed potential therapeutic targets in the treatment of vasculitis [4].

Antiendothelial Cell Antibodies

Antiendothelial cell antibodies (AECAs) are a heterogeneous group of auto-antibodies without well defined antigens that have been implicated in the pathogenesis of several autoimmune connective tissue diseases and vasculitides [5–7]. A host of pathogenetic mechanisms have been posited in an effort to link AECAs in the chain of events leading to tissue damage in vasculitis. Some of these include activation of thrombosis, increased leukocyte adhesion, direct cytotoxicity, and induction of apoptosis. The clinical utility of AECAs is unclear at the present time [8], though several authors have speculated that AECAs may be a worthwhile assay to pursue in patients with vasculitis, in particular, Wegener's granulomatosis [9–12]. Perhaps the most compelling support for AECA involvement in the pathogenesis of vasculitis results from the work of Damianovich *et al.* who described a murine model of AECA induced vasculitis

[13]. In this model, BALB/c mice were immunized with IgG AECA from a patient with WG, which triggered murine AECA production and resultant pulmonary and renal vasculitic lesions. Other evidence for the possible role of AECAs in vasculitis comes from AECA titer studies [14], *in vitro* effects of AECA on endothelial cells [15], and the upregulation of vascular endothelial adherence molecules with increased leukocyte activation in patients with WG who have been exposed to AECA IgG [16]. A recent study by Holmén *et al.* lends some support to AECA signaling through the stress activated protein kinase (SAPK)/c-Jun N-terminal Kinase (JNK) and NFκβ pathways, which has generated some interest in translational efforts for modulation of vascular adhesion protein-1 (VAP-1) in the treatment of WG [17].

However, several groups have questioned the methodology of AECA detection and quantification [18–20] resulting in a call for more robust data by combining at least two molecular techniques [21] in the identification and measurement of AECAs. One of the more promising biochemical techniques used in the last few years to better identify putative AECA antigens is proteomics [22]. In review, ongoing and future investigations should provide a clearer view into the role of AECAs in the pathogenesis of vasculitides and putative signaling mechanisms that may affect inflammation and tissue damage.

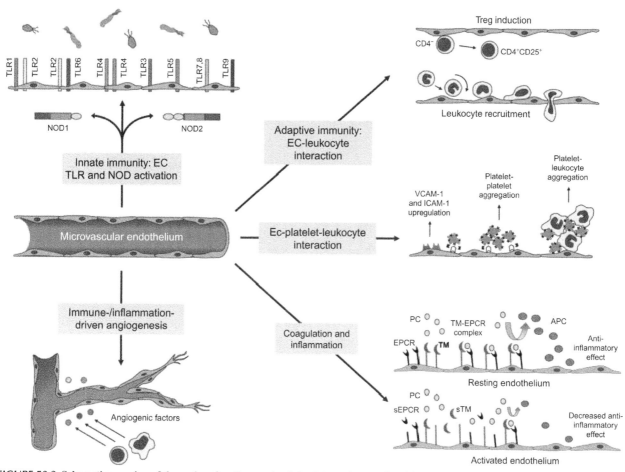

FIGURE 50.2 Schematic overview of the various functions and cellular interactions mediated by microvascular ECs that impact on innate and adaptive immunity, coagulation, and inflammation.
PC, Protein C; APC, activated protein C; sEPCR, soluble EPCR; sTM, soluble TM.P 64. Both the innate and adaptive immune systems are involved in the pathogenesis of vasculitis.

Immune Complex Mediated Vasculitic Syndromes

The immunological mechanisms involved in the immune complex mediated vasculitides can be broadly considered humoral in nature with examples including cryoglobulinemic vasculitis, hypersensitivity vasculitis, and Henoch-Schonlein Purpura (HSP). An immune complex (IC) comprises the formation of antigen-antibody conglomerates. The IC is an important function of the immune system in that the foreign antigens are more readily removed when detected by the immune system as an immune complex. These ICs are normally rather innocuous and do not stimulate the immune system to any great degree. In addition, they do not normally deposit in the walls of blood vessels. However, in the immune complex mediated vasculitides, ICs are deposited in the walls of the blood vessels in organs such as the skin and kidneys. This often triggers a pathological cascade of deleterious pathways that include complement activation, recruitment of lymphocytes, and resultant perpetuation of inflammation (Figure 50.3). This

cycle leads to endothelial damage of the vessel, decreased blood flow to vital organs through ischemia and/or hemorrhage, and the widespread clinical manifestations seen in the immune complex mediated vasculitic syndromes.

The complement system is a cascade of tightly regulated plasma proteins that functions to convert pathogen recognition into host defense [23]. It has multiple other related roles, for example, defense against bacterial infection, linkage of innate and adaptive immunity, and removal of ICs. The early events in the complement cascade are a series of proenzymatic cleavage reactions that result in activation of downstream complement components. It can be activated in three ways: directly by the pathogen (the classical pathway); activated C3 binding in plasma to the surface of a pathogen (the alternative pathway); and by binding of mannose binding lectin (the MB-lectin pathway). These three pathways converge to produce the critical C3 convertases that function to cleave C3 into C3b. C3b then acts as an opsonizing molecule, and it also binds to C3 convertase to form C5 convertase. C5a, an important peptide mediator of inflammation, is produced from C3b binding to C3 convertase,

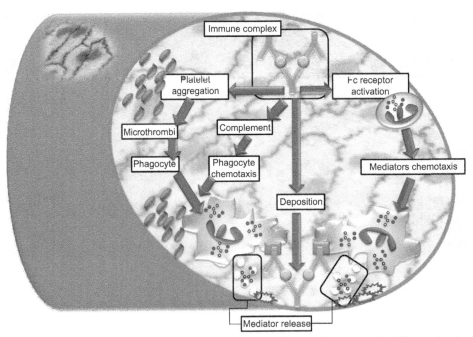

FIGURE 50.3 Immune complexes (IC) result in vasculitis primarily by activating the complement cascade. Inflammation is often triggered by circulating ICs entering tissue.

and the larger cleavage product, C5b, then triggers the downstream events of the complement cascade, with the production of the membrane attack complex (C5b-9 complex). This results in the creation of C5a, further chemotaxis and leukocyte recruitment, and the initiation of a destructive cycle that may involve the membrane attack complex in both the skin and the kidney [24]. More recently, the MB-lectin and alternative pathways of complement activation, in addition to the classical complement pathway, have been implicated in the renal pathogenesis of immune complex disease, and others have suggested that monitoring certain levels of complement may be useful in monitoring HSP disease activity [25]. Alternatively, C4 complement deficiencies have also been linked to immune complex disease, perhaps reflecting the importance of the complement system in the clearance of Ics [26, 27]. There are a number of regulating pathways that motivate the activation.

ANCA: Implications in Vasculitis

Antineutrophil cytoplasmic antibodies (ANCA) are auto-antibodies that have activity against intracytoplasmic enzymes found in neutrophils, namely proteinase 3 (PR3) and myeloperoxidase (MPO) [28,29]. ANCAs are recognized based on their characteristic staining patterns on direct immunofluorescence, with C-ANCA denoting cytoplasmic staining and P-ANCA illustrating a predominant perinuclear staining pattern. Vasculitic processes that are associated with ANCAs include WG, MPA, and CSS, and these entities often have a preponderance of either PR3 or MPO on enzyme-linked immunosorbent assay (ELISA).

PR3-ANCA is strongly associated with WG, while MPO-ANCA is more commonly found in MPA and CSS. There is overlap between the vasculitides, however, with respect to their antigen specificities. In contrast to the immune complex mediated vasculitic processes, the small vessel vasculitides associated with ANCAs are considered to be pauci-immune, that is, they have very little immune complex deposition in the walls of the blood vessel. In addition, they tend to be more granulomatous in nature, thus implicating predominantly cell mediated immune mechanisms rather than humoral involvement.

In vitro data supporting the pathogenic link between ANCA and vasculitis began to emerge over the last 10 years, with several studies illustrating the ability of MPO-ANCA and PR3-ANCA to stimulate neutrophils to produce reactive oxygen radicals, release lytic enzymes, and increase the neutrophil adherence to the endothelium; these mechanisms ultimately result in lysis of the endothelial cells with subsequent tissue damage [30,31]. Neutrophil activation is thought to be mediated by Fc and Fab'2 binding, with signaling pathways possibly distinct from those used in the immune complex mediated vasculitides [32]. The Fc receptor signaling mechanisms are further explored below. Figure 50.4 illustrates the proposed pathways in the pathogenesis of ANCA mediated vasculitis.

The group of Fcγ receptors (FcγRs) deserve special mention, as they are finding an increasingly significant role in the field of translational research in immunology and autoimmune disease [33–35]. The FcγRs are a family of glycoproteins that are intimately involved in the homeostasis of the immune system, as they have both activation and

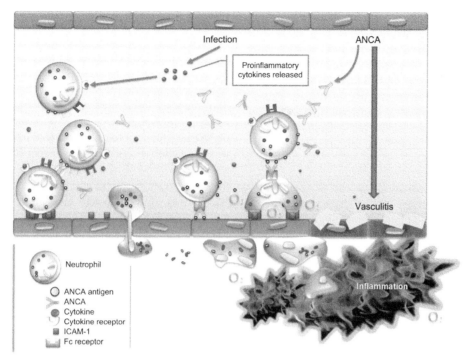

FIGURE 50.4 Neutrophil activation resulting in lysosomal enzyme release and subsequent damage to the vascular endothelium.

inhibitory functions. They have roles in antigen presentation, effector cell activation, B cell activation, and dendritic cell maturation. Through these functions, the FcγRs modulate the intensity of the immune response and serve as a link between the innate and adaptive immune systems. Their signaling pathways are mediated through phosphorylation of immunoreceptor tyrosine based activation motifs (ITAMs) by the SRC family of kinases, with subsequent recruitment of SYK family (cytosolic tyrosine kinase) kinases [36, 37]. Downstream events involve activation of Bruton's tyrosine kinase (BTK) and phospholipase Cγ (PLCγ), with resultant calcium influx, in addition to the activation of RAS-RAF-MAPK pathways [38]. Work by Hewins *et al.* has shown that Syk phosphorylation is induced during ANCA mediated neutrophil activation, and that phosphorylation is abrogated by blockade of FcγR [39]. The precise involvement of the various members of the FcγR family in ANCA associated vasculitis is still being investigated, but the number of possible therapeutic targets is staggering in these complex pathways, and translational research is underway to best determine viable targets, especially now that there are FcγR humanized mice models [40, 41].

Clinical evidence supporting the pathogenic role of ANCA lies not only in the strong association of PR3-ANCA and MPO-ANCA with WG, CSS, and MPA, but also in clinical case reports of vasculitis in human patients. As illustrated in Table 50.2, the CHCC incorporated the diagnostic value of ANCA into their vasculitis rubric, and further studies have corroborated the usefulness of ANCA in the diagnosis of CSS, MPA, and WG [42,43]. Furthermore, because of the strong association of ANCA

with MPA, CSS, and WG, ANCA has been implicated in the pathogenesis of these vasculitides [44]. Some have advocated following titers of ANCA in an effort to predict relapse of disease and guide treatment [45]. As in the case of AECAs, however, the utility of serial ANCA measurements has been questioned by others [46].

In 2004 and 2005, a case of possible vertical transmission of MPO-ANCA IgG was reported [47]. This newborn child developed pulmonary hemorrhage and glomerulonephritis after delivery. The child's mother had active MPA that was associated with MPO-ANCA, and the causal link between the MPO-ANCA IgG in the infant's blood and the resulting clinical manifestations has been postulated. Speculation has also arisen regarding a possible link between propylthiouracil and hydralazine and the induction of ANCA associated disease manifesting as pauci-immune glomerulonephritis [48].

There are several lines of evidence that support the role of ANCA in the pathogenesis of vasculitis; however, it should be stressed that the preponderance of the data thus far do not provide a direct link, and indeed provide a pathogenic model more by inference than by direct causality. Whether or not the ANCA is playing a direct role in the pathogenesis of these vasculitic syndromes, or is an epiphenomenon of immunological activation, remains speculation at this time.

Animal Models of Vasculitis

There are several animal models that investigate the possible pathogenetic relationships of ANCA and infectious agents with vasculitic syndromes. They include knockout murine models, rat studies, and antigen transfer studies, all of which have shed some light on the specter of the role of infection and ANCA in the pathogenesis of vasculitis. It should be noted that these are animal models, however, and they do not perfectly describe the human disease phenotype. However, they do serve an important purpose in that they function to illustrate possible pathogenetic pathways by which the vasculitides perturb the vascular tree.

ANCA + Vasculitis

Investigations from the 1990s hinted at the possibility of ANCA pathogenicity in animals [49]. These initial studies have been further substantiated by work from Xiao, Jennette, Huugen, and others at the University of North Carolina at Chapel Hill, and a possible role of the complement system in the ANCA associated vasculitides has also been postulated. Xiao's model utilizes MPO deficient mice that are immunized with murine MPO; this immunization initiates an MPO directed immunological reaction. When splenocytes or IgG are passively transferred into recipient wild-type or immunodeficient mice without T or B cells, leukocytoclastic angiitis and pulmonary capillaritis ensues in some of the mice. All of them develop a focal necrotizing glomerulonephritis, which is exacerbated by injection of lipopolysaccharide [50–52]. Microscopic immunofluorescent analysis has revealed this to be a pauci-immune process, w/o significant complement or immunoglobulin deposition. The role of anti-MPO T cells in this pathological process is less clear.

A rat model has been developed by Little and others that implicates MPO-ANCA in the pathogenesis of vasculitis [53]. In this model, rats are immunized with human MPO, which creates anti-MPO antibodies that cross react with both rat MPO and human MPO. The ability of anti-MPO to interfere with leukocyte migration and adherence, in addition to the generation of a focal pauci-immune glomerulonephritis and pulmonary capillaritis, *in vivo*, was shown in this study.

The evidence for ANCA pathogenicity is less convincing with respect to the PR3 antigen and the possible link with WG in animal models. A model developed by Pfister and others illustrates the potential for anti-PR3 pathogenicity, however [54]. In this mouse model, PR3 and elastase genes are knocked out, and the mice are immunized with recombinant murine PR3. This results in anti-PR3 autoantibodies against the cytoplasm of the neutrophils. Interestingly, these mice did not develop full blown renal or pulmonary disease, only the earliest renal and pulmonary lesions. They did develop cutaneous inflammation, however, when the mice received both a skin injection of tumor necrosis factor-alpha (TNFα) and an intravenous injection of anti-PR3 antibodies.

Kawasaki Disease

Perhaps the most exciting mouse model of vasculitis, with respect to translational applications, comes from Kawasaki disease (KD). KD is a disease of young persons and is characterized by prolonged fever, rash, conjunctival injection, mucosal inflammation, and enlarged cervical lymph nodes [55]. Duong and others have created a murine model of KD in which inbred mice receive intraperitoneal injections of lactobacillus cell wall extract [56]. The histology and the time course of the coronary artery lesions in these mice is very similar to those found in children affected with KD, by way of immune cells invading myocardial tissue as early as day 3 post-injection and formation of aneurisms at day 42. Intravenous immunoglobulin therapy prevents the vascular lesions in this mouse model, just as it does in children with KD [57,58]. Upregulation of leukocyte recruitment proteins is thought to be mediated by TNFα, and blockade of TNFα by entanercept completely ameliorates the elastin breakdown in the coronary arteries, which is the major defect in the aneurysm formation. This type of exciting basic science research has led to translational applications of anti-TNFα agents to treat KD.

TRANSLATIONAL APPROACHES TO THERAPEUTICS IN VASCULITIS

The current treatment approach to life or organ threatening systemic vasculitides (with the exception of Kawasaki disease) involve the administration of high dose corticosteroids and Cyclophosphamide (CYC), to put the diseases into remission. Though effective, their use is associated with high morbidity and, occasionally, mortality. After the severe manifestations of the vasculitis are controlled, alternative, less toxic agents such as methotrexate, mycophenolate, or azathioprine are used as maintenance therapies. As discussed above, intensive basic science research has illustrated many of the pathways involved in the pathogenesis of vasculitis, and this has resulted in the development of new agents more specifically targeted at the underlying mechanisms. Two classes of medications, TNFα inhibitors and anti-B cell agents, have been studied in prospective trials, and though the results have been mixed, they provide good examples of "bench to bedside to practice" methodology.

TNFα is found in vasculitic lesions, and levels of the cytokine have been shown to correlate with disease activity, in addition to its putative role in neutrophil priming. There is also evidence in an animal model of TNFα blockade efficacy [59]. The Wegener's granulomatosis entanercept trial (WGET) trial was initiated on this basic science foundation [60]. Over 100 patients were randomized to receive "standard therapy," which consisted of CYC and corticosteroids plus placebo or entanercept, a TNFα fusion protein receptor blocker, plus placebo. This was a negative study, in that no differences were found between the two groups with respect to remission rates.

Rituximab (RTX) is a monoclonal IgG antibody directed against the CD20 molecule on B cells that result in a rapid and sustained depletion of most subsets of B cells. It has found widespread use in hematology with the treatment of Non-Hodgkin Lymphoma, and it has found rheumatological applications in the treatment of rheumatoid arthritis. Given that B cells produce auto-antibodies and cytokines and that ANCAs are posited to be pathogenic in vasculitis, there is biological plausibility for the use of RTX in the vasculitides. Two recent studies in patients with WG have illustrated this proof of principle. Keogh *et al.* examined 11 patients with refractory vasculitis despite the use of CYC or in those who had a contraindication to its use. RTX induced a remission in all patients, with concomitant decreases in peripheral B cells and ANCA titers [61]. In the other study, 10 patients (seven of whom had renal involvement) went into remission after therapy with RTX [62]. The results of these promising preliminary studies need to be validated by larger randomized trials, and the ongoing rituximab for ANCA associated vasculitis (RAVE) trial hopes to address these unanswered questions.

CLASSIFICATION OF VASCULITIC SYNDROMES

Correct classification of clinical syndromes allows the practitioner to better characterize the disease entity and to direct his or her therapies against the pathological process. In addition, appropriate classification of patients allows clinical and basic science researchers to better understand the effectiveness of treatments in clinical trials and the pathophysiological mechanisms underlying the diseases. Until the 1990s, the vasculitides were lacking such organizational rubrics. Table 50.1 depicts the classification criteria of systemic vasculitis developed by the American College of Rheumatology (ACR) in 1990, which helped bring some order to the spectrum of vasculitic disease [63].

Approximately 4 years later, experts from the field of vasculitis met at the Chapel Hill Consensus Conference (CHCC) on the Nomenclature of Systemic Vasculitis in an effort to create clearer definitions of vasculitis and to develop diagnostic terminology related to the vasculitic syndromes [64] as depicted in Table 50.2.

More recently, the European League against Rheumatism (EULAR) and the Paediatric Rheumatology European Society (PReS) developed classification criteria for the vasculitic syndromes found in children. Table 50.3 illustrates the EULAR/PReS rubric.

These classification schemes take into account the most common clinical and pathologic presentations of the individual disorders, though they somewhat artificially divide

TABLE 50.1 1990 ACR classification of systemic vasculitis

Dominant vessel	Primary	Secondary
Large arteries	Giant cell arteritis	Aortitis associated with RA
	Takayasu's arteritis	Infection (e.g., syphilis, TB)
Medium arteries	Classical PAN	Hepatitis B associated PAN
	Kawasaki disease	
Small vessels and medium arteries	Wegener's granulomatosis[1]	Vasculitis secondary to RA, SLE, Sjögren's syndrome
	Churg-Strauss syndrome[1]	
	Microscopic polyangiitis[1]	Drugs Infection (e.g., HIV)
Small vessels (leukocytoclastic)	Henoch-Schönlein purpura	Drugs[2]
	Cryoglobulinemia	Hepatitis C associated
	Cutaneous leukocytoclastic angiitis	Infection

PAN - polyarteritis nodosa; RA - rheumatoid arteritis; TB – tuberculosis
[1]*Diseases most commonly associated with ANCA – significant risk of renal involvement, most responsive to immunosuppression with cyclophosphamide.*
[2]*For example: sulfonamides, penicillins, thiazide diuretics, and many others.*

TABLE 50.2 Names and definitions of vasculitis adopted by the Chapel Hill Consensus Conference on the nomenclature of systemic vasculitis

Name	Definition
Polyarteritis nodosa (PAN)	Necrotizing inflammation of medium-sized or small arteries without glomerulonephritis or vasculitis in arterioles, capillaries, or venules.
Wegener's granulomatosis (WG)	Granulomatous inflammation involving the respiratory tract, and necrotizing vasculitis affecting small to medium-sized vessels, e.g., capillaries, venules, arterioles, and arteries. *Necrotizing glomerulonephritis is common.*
Churg-Strauss syndrome (CSS)	Eosinophil-rich and granulomatous inflammation involving the respiratory tract and necrotizing vasculitis affecting small- to medium-sized vessels, and associated with asthma and blood eosinophilia.
Microscopic polyangiitis (MSA)	Necrotizing vasculitis with few or no immune deposits affecting small vessels, i.e., capillaries, venules, or arterioles. *Necrotizing arteritis of small- and medium-sized arteries may be present. Necrotizing glomerulonephritis is very common. Pulmonary capillaritis often occurs.*

TABLE 50.3 New classification of childhood vasculitis

1. Predominantly large vessel vasculitis
   - Takayasu arteritis
2. Predominantly medium-sized vessel vasculitis
   - Childhood polyarteritis nodosa
   - Cutaneous polyarteritis
   - Kawasaki disease
3. Predominantly small vessels vasculitis
   a. Granulomatous
      - Wegener's granulomatosis
      - Churg-Strauss syndrome
   b. Non-granulomatous
      - Microscopic polyangiitis
      - Henoch-Schönlein purpura
      - Isolated cutaneous leucocytoclastic vasculitis
      - Hypocomplementic urticarial vasculitis
4. Other vasculitides
   - Behçet disease
   - Vasculitis secondary to infection (including hepatitis B associated polyarteritis nodosa), malignancies, and drugs, including hypersensitivity vasculitis
   - Vasculitis associated with connective tissue diseases
   - Isolated vasculitis of the central nervous system
   - Cogan syndrome
   - Unclassified

them according to the size of blood vessel affected by the vasculitic process. Several of the vasculitic syndromes affect more than one size of blood vessel, and this overlap often leads to the protean manifestations of these diseases, which can result in diagnostic dilemmas. These classification motifs are not all inclusive, as there are a multitude of rare vasculitic syndromes that are not listed in the ACR, CHCC, or EULAR/PreS classification schemes. A second way to group the vasculitides is according to their relationship with other diseases; thus, they may be considered primary or secondary processes. Secondary vasculitic syndromes occur as a result of, or in association with, other systemic processes or insults. Examples of secondary vasculitides include those related to malignancy, substance abuse, and radiation therapy. While not perfect by any means, the aforementioned sets of classification criteria serve to gain a better sense of the clinical, pathological, and diagnostic characteristics of the vasculitic syndromes.

CONCLUSIONS

The vasculitides describe a very heterogeneous group of diseases that all have a common thread, that is, inflammation

TABLE 50.4 Pathogenic mechanisms

Disease	Mechanisms
Kawasaki	1. Mouse model: lactobacillus casei cell-wall extract induction of toll-like receptors (TLR2) through myeloid differentiation factor 88 (MyD88) and nuclear factor-kappaβ [65] 2. Superantigen (toxic shock syndrome toxin 1) or (streptococcal pyrogenic exotoxin A) or (streptococcal pyrogenic exotoxin C) or (Lactobacillus casei cell-wall extract) [66] 3. Metalloproteinase-2 activation with elastolytic activity [66] 4. Anti-alpha-enolase antibody shock protein [67]
Wegener's granulomatosis Microscopic polyangiitis	1. Anti-myeloperoxidase antibodies binding myeloperoxidase on surface of leukocytes inducing leukocyte adhesion [68] to endothelial cells through CD11a/CD18, CD11b/CD18 beta2 integrins, and chemokine receptors CXCR1 and CXCR2 [69] 2. Antibodies to HSP60 [70] 3. ANCA activation of alternative pathway complement [71]
Takayasu's arteritis	1. Antibodies to HSP60 [72, 73] 2. Antiendothelial antibodies [65] 3. Antibodies to HSP65 [74]
Behçet's disease	1. HLA-B51 binds MIC-A [75] (non-classical) HLA Class I antigen which is unregulated by bacterial infection TH1 and 2 and TBet cells are involved
Mixed cryoglobulinemia	1. Reactivity to HSP70, HSP 90 [76] 2. Activation of matrix metalloproteinases (MMP-1, MMP-7, MMP-9) and Il1 β
Henoch-Schönlein purpura	1. Abnormal glycosylated IgA1 [70] 2. IgA containing immune complexes bind Fcγ receptors activate SRC family of kinases
Churg-Strauss syndrome	1. IL10 polymorphisms [77]

TABLE 50.5 Mechanism summarization

Mechanism	Signal transduction pathways	Disease(s)
Immune complex	1. Bind Fcγ receptors activate SRC family of kinases 2. Activate complement	Cryoglobulinemic vasculitis, Henoch-Schönlein purpura, Leukocytoclastic vasculitis Polyarteritis nodosa
ANCA	Neutrophil degranulation by binding to PR3 or MPO on surface of activated neutrophils or through FcγR mechanism (see above)	Microscopic polyangiitis Wegener's granulomatosis
Antiendothelial antibodies	1. Mechanism unknown 2. Stress activated protein kinases 3. c-Jun N-terminal kinases 4. NFκβ	Scleroderma Wegener's granulomatosis

of blood vessel walls with subsequent downstream tissue damage. They may affect small, medium, and/or large vascular structures, and they may be primary in nature or secondarily associated with other disease processes. They manifest in protean phenotypes with a great deal of overlap, which can make diagnosis, especially at the onset of disease, quite challenging. However, *in vitro*, *in vivo*, and animal model studies are beginning to unlock many of the previous mysteries that have been associated with the pathogenesis of vasculitis. The study of immune complex deposition and the biochemical and clinical relevance of AECAs and ANCAs are bringing to light the importance of overlap between basic science and clinical as well as translational medicine. The molecular mechanisms that underlie the pathogenesis of vasculitis are now being teased out by basic science researchers around the world, and the signal transduction pathways are starting to find translational implications in the realm of therapeutics. The pathogenic mechanisms are summarized in Tables 50.4 and 50.5.

We can only hope that continued translational efforts in the field of vasculitis research will be able to shed further light on a group of diseases that, even in present day technologically advanced medicine, persist in producing significant morbidity and mortality.

REFERENCES

1. Holmén C, Christensson M, Pettersson E, Bratt J, Stjärne P, Karrar A, Sumitran-Holgersson S. Wegener's granulomatosis is associated with organ-specific antiendothelial cell antibodies. *Kidney Int* 2004;**66**(3):1049–60.

2. McIntyre TM, Prescott SM, Weyrich AS, Zimmerman GA. Cell-cell interactions: leukocyte-endothelial interactions. *Curr Opin Hematol* 2003;**10**(2):150–8.

3. Danese S, Dejana E, Fiocchi C. Immune regulation by microvascular endothelial cells: directing innate and adaptive immunity, coagulation, and inflammation. *J Immunol* 2007;**178**(10):6017–22.

4. Ulbrich H, Eriksson EE, Lindbom L. Leukocyte and endothelial cell adhesion molecules as targets for therapeutic interventions in inflammatory disease. *Trends Pharmacol Sci* 2003;**24**(12):640–7.

5. Renaudineau Y, Grunebaum E, Krause I, Praprotnik S, Revelen R, Youinou P, Blanks M, Gilburd B, Sherer Y, Luderschmidt C, Eldor A, Weksler B, Gershwin EM, Shoenfeld Y. Anti-endothelial cell antibodies (AECA) in systemic sclerosis: increased sensitivity using different endothelial cell substrates and association with other autoantibodies. *Autoimmunity* 2001;**33**(3):171–9.

6. Bordron A, Révélen R, D'Arbonneau F, Dueymes M, Renaudineau Y, Jamin C, Youinou P. Functional heterogeneity of anti-endothelial cell antibodies. *Clin Exp Immunol* 2001;**124**(3):492–501.

7. Cid MC, Segarra M, García-Martínez A, Hernández-Rodríguez J. Endothelial cells, antineutrophil cytoplasmic antibodies, and cytokines in the pathogenesis of systemic vasculitis. *Curr Rheumatol Rep* 2004;**6**(3):184–94.

8. Sebastian JK, Mahr AD, Ahmed SS, Stone JH, Romay-Penabad Z, Davis JC, Hoffman GS, McCune WJ, St Clair EW, Specks U, Spiera R, Pierangeli S, Merkel PA. Antiendothelial cell antibodies in patients with Wegener's granulomatosis: prevalence and correlation with disease activity and manifestations. *J Rheumatol* 2007;**34**(5):1027–31.

9. Ferraro G, Meroni PL, Tincani A, Sinico A, Barcellini W, Radice A, Gregorini G, Froldi M, Borghi MO, Balestrieri G. Anti-endothelial cell antibodies in patients with Wegener's granulomatosis and micropolyarteritis. *Clin Exp Immunol* 1990;**79**(1):47–53.

10. Frampton G, Jayne DR, Perry GJ, Lockwood CM, Cameron JS. Autoantibodies to endothelial cells and neutrophil cytoplasmic antigens in systemic vasculitis. *Clin Exp Immunol* 1990;**82**(2):227–32.

11. Chan TM, Frampton G, Jayne DR, Perry GJ, Lockwood CM, Cameron JS. Clinical significance of anti-endothelial cell antibodies in systemic vasculitis: a longitudinal study comparing anti-endothelial cell antibodies and anti-neutrophil cytoplasm antibodies. *Am J Kidney Dis* 1993;**22**(3):387–92.

12. Göbel U, Eichhorn J, Kettritz R, Briedigkeit L, Sima D, Lindschau C, Haller H, Luft FC. Disease activity and autoantibodies to endothelial cells in patients with Wegener's granulomatosis. *Am J Kidney Dis* 1996;**28**(2):186–94.

13. Damianovich M, Gilburd B, George J, Del Papa N, Afek A, Goldberg I, Kopolovic Y, Roth D, Barkai G, Meroni PL, Shoenfeld Y. Pathogenic role of anti-endothelial cell antibodies in vasculitis. An idiotypic experimental model. *J Immunol* 1996;**156**(12):4946–51.

14. Woywodt A, Streiber F, de Groot K, Regelsberger H, Haller H, Haubitz M. Circulating endothelial cells as markers for ANCA-associated small-vessel vasculitis. *Lancet* 2003;**361**(9353):206–10.

15. Praprotnik S, Blank M, Meroni PL, Rozman B, Eldor A, Shoenfeld Y. Classification of anti-endothelial cell antibodies into antibodies against microvascular and macrovascular endothelial cells: the pathogenic and diagnostic implications. *Arthritis Rheum* 2001;**44**(7):1484–94.

16. Del Papa N, Guidali L, Sironi M, Shoenfeld Y, Mantovani A, Tincani A, Balestrieri G, Radice A, Sinico RA, Meroni PL. Anti-endothelial cell

IgG antibodies from patients with Wegener's granulomatosis bind to human endothelial cells in vitro and induce adhesion molecule expression and cytokine secretion. *Arthritis Rheum* 1996;**39**(5):758–66.

17. Holmén C, Elsheikh E, Christensson M, Liu J, Johansson AS, Qureshi AR, Jalkanen S, Sumitran-Holgersson S. Anti endothelial cell autoantibodies selectively activate SAPK/JNK signalling in Wegener's granulomatosis. *J Am Soc Nephrol* 2007;**18**(9):2497–508.

18. Chanseaud Y, García de la Peña-Lefebvre P, Guilpain P, Mahr A, Tamby MC, Uzan M, Guillevin L, Boissier MC, Mouthon L. IgM and IgG autoantibodies from microscopic polyangiitis patients but not those with other small- and medium-sized vessel vasculitides recognize multiple endothelial cell antigens. *Clin Immunol* 2003;**109**(2):165–78.

19. Youinou P, Le Dantec C, Bendaoud B, Renaudineau Y, Pers JO, Jamin C. Endothelium, a target for immune-mediated assault in connective tissue disease. *Autoimmun Rev* 2006;**5**(3):222–8.

20. Chan TM, Cheng IK. Identification of endothelial cell membrane proteins that bind anti-DNA antibodies from patients with systemic lupus erythematosus by direct or indirect mechanisms. *J Autoimmun* 1997;**10**(5):433–9.

21. Révélen R, D'Arbonneau F, Guillevin L, Bordron A, Youinou P, Dueymes M. Comparison of cell-ELISA, flow cytometry and Western blotting for the detection of antiendothelial cell antibodies. *Clin Exp Rheumatol* 2002;**20**(1):19–26.

22. Servettaz A, Guilpain P, Camoin L, Mayeux P, Broussard C, Tamby MC, Tamas N. Kaveri SV, Guillevin L, Mouthon L. Identification of target antigens of antiendothelial cell antibodies in healthy individuals: A proteomic approach. *Proteomics* 2008;**8**(5):1000–8.

23. Walport MJ. Complement. Second of Two Parts. *N Engl J Med* 2001;**344**(15):1140–4.

24. Kawana S, Nishiyama S. Serum SC5b-9 (terminal complement complex) level, a sensitive indicator of disease activity in patients with Henoch-Schönlein purpura. *Dermatology* 1992;**184**(3):171–6.

25. Motoyama O, Iitaka K. Henoch-Schonlein purpura with hypocomplementemia in children. *Pediatr Int* 2005;**47**(1):39–42.

26. Stefansson Thors V, Kolka R, Sigurdardottir SL, Edvardsson VO, Arason G, Haraldsson A. Increased frequency of C4B*Q0 alleles in patients with Henoch-Schönlein purpura. *Scand J Immunol* 2005;**61**(3):274–8.

27. Ault III. BH, Stapleton FB, Rivas ML, Waldo FB, Roy S, McLean RH, Bin JA, Wyatt RJ Association of Henoch-Schönlein purpura glomerulonephritis with C4B deficiency. *J Pediatr* 1990;**117**(5):753–5.

28. Falk RJ, Jennette JC. Anti-neutrophil cytoplasmic autoantibodies with specificity for myeloperoxidase in patients with systemic vasculitis and idiopathic necrotizing and crescentic glomerulonephritis. *N Engl J Med* 1988;**318**(25):1651–7.

29. Jennette JC, Hoidal JR, Falk RJ. Specificity of anti-neutrophil cytoplasmic autoantibodies for proteinase 3. *Blood* 1990;**75**(11):2263–4.

30. Falk RJ, Terrell RS, Charles LA, Jennette JC. Anti-neutrophil cytoplasmic autoantibodies induce neutrophils to degranulate and produce oxygen radicals in vitro. *Proc Natl Acad Sci U S A* 1990;**87**(11):4115–19.

31. Harper L, Savage CO. Leukocyte-endothelial interactions in antineutrophil cytoplasmic antibody-associated systemic vasculitis. *Rheum Dis Clin North Am* 2001;**27**(4):887–903.

32. Ben-Smith A, Dove SK, Martin A, Wakelam MJ, Savage CO. Antineutrophil cytoplasm autoantibodies from patients with systemic vasculitis activate neutrophils through distinct signaling cascades: comparison with conventional Fcgamma receptor ligation. *Blood* 2001;**98**(5):1448–55.

33. Nimmerjahn F. Activating and inhibitory FcgammaRs in autoimmune disorders. *Springer Semin Immunopathol* 2006;**28**(4):305–19.

34. Nimmerjahn F, Ravetch JV. Fcgamma receptors: old friends and new family members. *Immunity* 2006;**24**(1):19–28.

35. Nimmerjahn F, Ravetch JV. Fcgamma receptors as regulators of immune responses. *Nat Rev Immunol* 2008;**8**(1):34–47.

36. Wang AV, Scholl PR, Geha RS. Physical and functional association of the high affinity immunoglobulin G receptor (Fc gamma RI) with the kinases Hck and Lyn. *J Exp Med* 1994;**180**(3):1165–70.

37. Odin JA, Edberg JC, Painter CJ, Kimberly RP, Unkeless JC. Regulation of phagocytosis and [Ca2+]i flux by distinct regions of an Fc receptor. *Science* 1991;**254**(5039):1785–8.

38. Liao F, Shin HS, Rhee SG. Tyrosine phosphorylation of phospholipase C-gamma 1 induced by cross-linking of the high-affinity or low-affinity Fc receptor for IgG in U937 cells. *Proc Natl Acad Sci U S A* 1992;**89**(8):3659–63.

39. Hewins P, Williams JM, Wakelam MJ, Savage CO. Activation of Syk in neutrophils by antineutrophil cytoplasm antibodies occurs via Fcgamma receptors and CD18. *J Am Soc Nephrol* 2004;**15**(3):796–808.

40. Hazenbos WL, Gessner JE, Hofhuis FM, Kuipers H, Meyer D, Heijnen IA, Schmidt RE, Sandor M, Capel PJ, Daëron M, van de Winkel JG, Verbeek JS. Impaired IgG-dependent anaphylaxis and Arthus reaction in Fc gamma RIII (CD16) deficient mice. *Immunity* 1996;**5**(2):181–8.

41. Li M, Wirthmueller U, Ravetch JV. Reconstitution of human Fc gamma RIII cell type specificity in transgenic mice. *J Exp Med* 1996;**183**(3):1259–63.

42. Hoffman GS, Specks U. Antineutrophil cytoplasmic antibodies. *Arthritis Rheum* 1998;**41**(9):1521–37.

43. Savige J, Davies D, Falk RJ, Jennette JC, Wiik A. Antineutrophil cytoplasmic antibodies and associated diseases: a review of the clinical and laboratory features. *Kidney Int* 2000;**57**(3):846–62.

44. Bartunková J, Tesar V, Sedivá A. Diagnostic and pathogenetic role of antineutrophil cytoplasmic autoantibodies. *Clin Immunol* 2003;**106**(2):73–82.

45. Boomsma MM, Stegeman CA, van der Leij MJ, Oost W, Hermans J, Kallenberg CG, Limburg PC, Tervaert JW. Prediction of relapses in Wegener's granulomatosis by measurement of antineutrophil cytoplasmic antibody levels: a prospective study. *Arthritis Rheum* 2000;**43**(9):2025–33.

46. Finkielman JD, Merkel PA, Schroeder D, Hoffman GS, Spiera R, St Clair EW, Davis Jr. JC, McCune WJ, Lears AK, Ytterberg SR, Hummel AM, Viss MA, Peikert T, Stone JH, Specks U WGET Research Group. Antiproteinase 3 antineutrophil cytoplasmic antibodies and disease activity in Wegener granulomatosis. *Ann Intern Med* 2007;**147**(9):611–19.

47. Schlieben DJ, Korbet SM, Kimura RE, Schwartz MM, Lewis EJ. Pulmonary-renal syndrome in a newborn with placental transmission of ANCAs. *Am J Kidney Dis* 2005;**45**(4):758–61.

48. Morita S, Ueda Y, Eguchi K. Anti-thyroid drug-induced ANCA-associated vasculitis: a case report and review of the literature. *Endocr J* 2000;**47**(4):467–70.

49. Heeringa P, Brouwer E, Tervaert JW, Weening JJ, Kallenberg CG. Animal models of anti-neutrophil cytoplasmic antibody associated vasculitis. *Kidney Int* 1998;**53**(2):253–63.

50. Xiao H, Heeringa P, Hu P, Liu Z, Zhao M, Aratani Y, Maeda N, Falk RJ, Jennette JC. Antineutrophil cytoplasmic autoantibodies specific for myeloperoxidase cause glomerulonephritis and vasculitis in mice. *J Clin Invest* 2002;**110**(7):955–63.

51. Huugen D, Xiao H, van Esch A, Falk RJ, Peutz-Kootstra CJ, Buurman WA, Tervaert JW, Jennette JC, Heeringa P. Aggravation of

anti-myeloperoxidase antibody-induced glomerulonephritis by bacterial lipopolysaccharide: role of tumor necrosis factor-alpha. *Am J Pathol* 2005;**167**(1):47–58.

52. Xiao H, Schreiber A, Heeringa P, Falk RJ, Jennette JC. Alternative complement pathway in the pathogenesis of disease mediated by anti-neutrophil cytoplasmic autoantibodies. *Am J Pathol* 2007;**170**(1):52–64.

53. Little MA, Smyth CL, Yadav R, Ambrose L, Cook HT, Nourshargh S, Pusey CD. Antineutrophil cytoplasm antibodies directed against myeloperoxidase augment leukocyte-microvascular interactions in vivo. *Blood* 2005;**106**(6):2050–8.

54. Pfister H, Ollert M, Fröhlich LF, Quintanilla-Martinez L, Colby TV, Specks U, Jenne DE. Antineutrophil cytoplasmic autoantibodies against the murine homolog of proteinase 3 (Wegener autoantigen) are pathogenic in vivo. *Blood* 2004;**104**(5):1411–18.

55. Ozen S, Ruperto N, Dillon MJ, Bagga A, Barron K, Davin JC, Kawasaki T, Lindsley C, Petty RE, Prieur AM, Ravelli A, Woo P. EULAR/PReS endorsed consensus criteria for the classification of childhood vasculitides. *Ann Rheum Dis* 2006;**65**(7):936–41.

56. Duong TT, Silverman ED, Bissessar MV, Yeung RS. Superantigenic activity is responsible for induction of coronary arteritis in mice: an animal model of Kawasaki disease. *Int Immunol* 2003;**15**(1):79–89.

57. Chan WC, Duong TT, Yeung RS. Presence of IFN-gamma does not indicate its necessity for induction of coronary arteritis in an animal model of Kawasaki disease. *J Immunol* 2004;**173**(5):3492–503.

58. Hui-Yuen JS, Duong TT, Yeung RS. TNF-alpha is necessary for induction of coronary artery inflammation and aneurysm formation in an animal model of Kawasaki disease. *J Immunol* 2006;**176**(10):6294–301.

59. Jayne K. What place for the new biologics in the treatment of necrotising vasculitides. *Clin Exp Rheumatol* 2006;**24**(2 Suppl 41):S1–5.

60. WGET Research Group. Design of the wegener's granulomatosis etanercept trial (WGET). *Control Clin Trials* 2002;**23**(4):450–68.

61. Keogh KA, Wylam ME, Stone JH, Specks U. Induction of remission by B lymphocyte depletion in eleven patients with refractory antineutrophil cytoplasmic antibody-associated vasculitis. *Arthritis Rheum* 2005;**52**(1):262–8.

62. Keogh KA, Ytterberg SR, Fervenza FC, Carlson KA, Schroeder DR, Specks U. Rituximab for refractory Wegener's granulomatosis: report of a prospective, open-label pilot trial. *Am J Respir Crit Care Med* 2006;**173**(2):180–7.

63. Hunder GG, Arend WP, Bloch DA, Calabrese LH, Fauci AS, Fries JF, et al. The American College of Rheumatology 1990 criteria for the classification of vasculitis. Introduction. *Arthritis Rheum* 1990;**33**:1065–7.

64. Jennette JC, Falk RJ, Andrassy K, Bacon PA, Churg J, Gross WL, Hagen EC, Hoffman GS, Hunder GG, Kallenberg CG, McCluskey RT, Sinico RA, Rees AJ, van Es LA, Waldherr R, Wiik A. Nomenclature of systemic vasculitides. Proposal of an international consensus conference. *Arthritis Rheum* 1994;**37**:187–92.

65. Rosenkranz ME, Schulte DJ, Agle LMA, Wong MH, Zhang W, Ivashkiv L, Doherty TM, Fishbein MC, Lehman TJA, Michelsen KS, Arditi M. TLR2 and MyD88 contribute to Lactobacillus casei extract-induced focal coronary arteritis in a mouse model of Kawasaki disease. *Circulation* 2005;**112**(19):2966–73.

66. Yeung RSM. Pathogenesis and treatment of Kawasaki's disease. *Curr Opin Rheumatol* 2008;**17**:617–23.

67. Chun JK, Lee TJ, Choi KM, Lee KH, Kim DS. Elevated anti-alpha-enolase antibody levels in Kawasaki disease. *Scand J Rheumatol* 2008;**37**(1):48–52.

68. Nolan SL, Kalia N, Nash GB, Kamel D, Heeringa P, Savage CO. Mechanisms of ANCA-mediated leukocyte-endothelial cell interactions in vivo. *J Am Soc Nephrol* 2008;**19**(5):973–84.

69. Calderwood JW, Williams JM, Morgan MD, Nash GB, Savage CO. ANCA induces beta2 integrin and CXC chemokine-dependent neutrophil-endothelial cell interactions that mimic those of highly cytokine-activated endothelium. *J Leukoc Biol* 2005;**77**(1):33–43.

70. Lau KK, Wyatt RJ, Moldoveanu Z, Tomana M. Julian BA, Hogg RJ, Lee JY, Huang WQ, Mestecky J, Novak J. Serum levels of galactose-deficient IgA in children with IgA nephropathy and Henoch-Schönlein purpura. *Pediatr Nephrol* 2007;**22**(12):2067–72.

71. Jennette JC, Falk RJ. New insight into the pathogenesis of vasculitis associated with antineutrophil cytoplasmic autoantibodies. *Curr Opin Rheumatol* 2008;**20**(1):55–60.

72. Alard JE, Dueymes M, Youinou P, Jamin C. HSP60 and Anti-HSP60 antibodies in vasculitis: they are two of a kind. *Clin Rev Allergy Immunol* 2008;**35**(1–2):66–71.

73. Chauhan SK, Singh M, Nityanand S. Reactivity of gamma/delta T cells to human 60-kd heat-shock protein and their cytotoxicity to aortic endothelial cells in Takayasu arteritis. *Arthritis Rheum* 2007;**56**(8):2798–802.

74. Arnaud L, Kahn JE, Girszyn N, Piette AM, Bletry O. Takayasu's arteritis: An update on physiopathology. *Eur J Intern Med* 2006;**17**(4):241–6.

75. Rajendram R, Rao NA. Molecular mechanisms in Behet's disease. *BR J Ophthalmol* 2003;**87**:119–20.

76. Saadoun D, Bieche I, Authier F-J, Laurendeau I, Jambou F, Piette JC, Vidaud M, Maisonobe T, Cacoub P. Role of matrix metalloproteinases, proinflammatory cytokines, and oxidative stress-derived molecules in hepatitis C virus-associated mixed cryoglobulinemia vasculitis neuropathy. *Arthritis Rheum* 2007;**56**(4):1315–24.

77. Wieczorek S, Hellmich B, Arning L, Moosig F, Lamprecht P, Gross WL, Epplen JT. Functionally relevant variations of the interleukin-10 gene associated with antineutrophil cytoplasmic antibody-negative Churg-Strauss syndrome, but not with Wegener's granulomatosis. *Arthritis Rheum* 2008;**58**(6):1839–48.

Chapter 51

Translational Implications of Proteomics

Sam Hanash

Fred Hutchinson Cancer Research Center, Seattle, Washington

INTRODUCTION

Biomarkers are commonly defined as indicators that inform about events in biological samples or systems. In addition to their diagnostic utility, biomarkers offer the promise of assessing disease risk and more efficient discovery and development of novel therapies and individualized disease treatment [1]. Biomarker proteins detectable in serum and plasma are the basis of commonly relied upon tests to screen and monitor prostate cancer through the measurement of PSA, to monitor ovarian, pancreatic, and colon cancer response to therapy and disease recurrence through the measurement of CA125, CA19.9, and carcinoembryonic antigen respectively, among others [2]. While such protein biomarkers were not identified through proteomics, but mostly through antibody based searches for tumor antigens, they do occur in serum at concentrations that are within the reach of current proteomic profiling technologies. Therefore it is reasonable to assume that many more cancer markers with similar concentrations in serum and plasma may be identified through systematic proteomic searches. While there is currently an intense effort to apply proteomics to biomarker discovery, the field is challenged by a requirement for biomarkers with high sensitivity and specificity in the face of substantial heterogeneity among human subjects and disease processes. The search for such biomarkers has been quite diversified from the point of view of sources of biomarkers investigated (Table 51.1) and the particular approaches being followed.

PROFILING OF TISSUES TO IDENTIFY POTENTIAL CIRCULATING MARKERS

Given the wide dynamic range of protein abundance in serum and plasma and the likely occurrence of tumor tissue derived proteins in circulation at the lower end of this range, there is merit in profiling tumor tissue to identify proteins that potentially may be released into the circulation

---

TABLE 51.1 Sources of biomarkers for discovery

Serum, plasma, and other biological fluids

- comprehensive analysis of intact proteins
- comprehensive analysis of protein digests
- enriched protein and peptide subsets e.g., glycoproteins and glycopeptides

Isolated fresh cells and cultured cell lines

- whole lysates
- subcellular compartments e.g., cell surface, secreted/shed proteins, phosphoproteins, glycoproteins

Tissue sources

- whole tissue lysates
- microdissected tissue
- vascular endothelium
- infiltrating cells

Animal models

- engineered mouse models
- xenotransplants

---

and therefore may serve as blood based biomarkers. Potential circulating protein markers may be deduced from proteomic and also from transcriptomic profiling data for tissues. However, the ability to predict from tissue sources potential circulating proteins is complicated by issues of tissue specificity of identified proteins. A comparison of tumor expression profiles relative to healthy or unaffected tissues from the same organ type may identify protein differences. However, the extent to which the contributions of tumor tissue proteins to the overall circulating levels for these proteins, relative to the contribution of other unaffected organ types that may be releasing the same proteins into the circulation, is difficult to predict and requires exhaustive tissue analysis. Additional issues include rates of release and clearance from the circulation, which are not predictable from analyses at the tissue level. Additional factors that may contribute to discrepancies between tissue and plasma findings include heterogeneity within tumors not represented in samples subjected to profiling and difficulty in defining and obtaining adequate controls from confounding conditions that inform about disease specificity. It should also be pointed out that levels of a protein in tumor tissue may be unchanged compared to unaffected tissues, but because of altered processing, increased turnover, and cell breakdown, the protein may occur at increased concentrations in circulation [3]. Therefore analysis of tumor tissue for the purpose of identifying circulating markers has numerous caveats that need to be taken into consideration.

PROTEOMIC PROFILING OF PROXIMAL BIOLOGICAL FLUIDS

Because circulating proteins largely represent a subset of tissue proteins that are primarily secreted or shed from cells, biological fluids, and disease related effusions that are proximal to the disease tissue may represent useful sources for discovery of circulating biomarkers. Furthermore, given their proximity to disease tissue, proximal fluids potentially contain higher concentrations of proteins released from tumor tissue into extracellular fluids through secretion or cell and tissue breakdown [4]. Such fluids include pleural effusions in the case of intrathoracic tumors, ascites fluid for intra-abdominal tumors, seminal fluid for prostate cancer, breast ductal fluid for breast cancer, and cerebrospinal fluid for central nervous system tumors. Proximal biological fluids also have some challenges as exemplified by nipple aspirates and ductal lavages to identify potential breast cancer markers. Procedures for obtaining fluid are difficult to standardize. Controls may not be adequate even when obtained from a contra-lateral presumably unaffected breast, due to disparities in the amount of fluid obtained and occurrence of micro-hemorrhages among others. Additional heterogeneity may be introduced due to varying cell and tissue admixture and breakdown.

PROFILING OF TUMOR CELL POPULATIONS

The use of isolated cell populations and cell lines to identify candidate markers requires that the cell material utilized adequately reflects the tumor type for which biomarkers are being sought. While a single cell line is unlikely to be representative of the spectrum of changes that may occur in a disease, analysis of a set of well chosen cell populations and cell lines may be quite informative. This is illustrated in a study of ovarian cancer cells from the author's laboratory [5]. The repertoire of proteins expressed in three

ovarian adenocarcinoma cell lines, OVCAR3, CaOV3, and ES2, as well as in ovarian cancer cells enriched from ascites fluid has been substantially elucidated [6]. Separate mass spectrometric analysis of proteins released into culture media and of cell surface proteins were done, using total cell lysates as a reference (Figure 51.1). Some 6400 proteins were identified with high confidence. Similar or complementary data have been published for ovarian cell lines and ovarian ascites fluid [7, 8] with substantial concordance. Included among secreted proteins identified are WFDC2 (HE4), MUC16 (mucin 16, CA125), IGFBP3 (insulin-like growth factor binding protein 3), MDK (midkine), PROS1(vitamin k dependent protein s), and SLPI (secretory leukoprotease inhibitor) for which prior associations with ovarian cancer have been made. Numerous additional candidates have been identified that require further testing to confirm their relevance to ovarian cancer.

Important features of secreted or shed proteins that inform with respect to their potential as biomarkers include a high secretion rate specifically by tumor cells and a low baseline concentration in normal plasma. A recent mathematical model has derived estimates of the balance between tumor biomarker secretion into and removal out of the intravascular compartment that take into account protein secretion rates by tumor cells [9]. The model was primed by the authors based on publicly available data for the ovarian marker CA125 leading to estimate the minimal tumor size required to detect elevated levels of CA125 in circulation in ovarian cancer. Applying this model to TIMP1, a secreted protein identified in ovarian cancer cell profiling, the secretion rate of which was found to be ~3 ng/million cells/h) [6], would lead to an estimate that a tumor ~2 cm in diameter would result in a detectable 50 percent increase in the level of TIMP1. Assumptions made for the calculation are: (i) a half-life in blood of 48 h, which is significantly below 151 h calculated for CA125; (ii) 50 percent of the secreted protein reaches the blood. These favorable predictions led to the testing of TIMP1 levels in serum from newly diagnosed patients with ovarian cancer and demonstration of increased levels in ovarian cancer including in early stage disease (Hanash *et al.* unpublished).

PROFILING THE PLASMA PROTEOME FOR CANCER BIOMARKER IDENTIFICATION

Undoubtedly, in principle, the objective of identifying plasma (or serum) based biomarkers is best accomplished through profiling of plasma rather than other indirect sources from which inferences need to be made and that may not reflect protein content in plasma. However, in addition to the technological challenges related to the vast dynamic range of protein concentrations, profiling plasma for cancer biomarker identification is challenged in other ways; perhaps most important are the numerous sources of

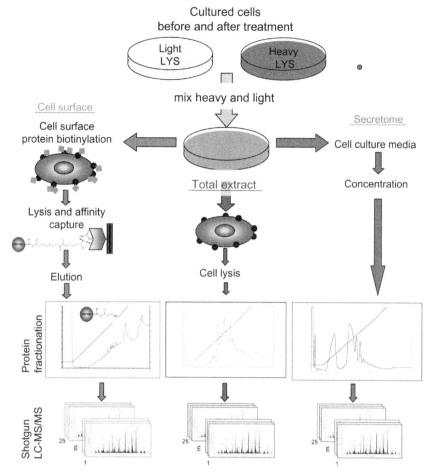

FIGURE 51.1 Profiling the secretome and cell surface proteome of cancer cells.

variability that may contribute to false discovery and that have to be taken into consideration as part of the experimental design and data interpretation. These include subject variability, and sample processing variability unrelated to the disease state being investigated. Thus, artifacts or non-specific disease related changes in plasma proteins need to be distinguished from potentially specific markers. A case in point is cancer serum protein profiling by MALDI mass spectrometry, which has uncovered mass profiles for unidentified proteins that were proposed to be diagnostic for several common types of cancer but whose validity was subsequently questioned [10].

Currently the workhorse for proteomic profiling is mass spectrometry, which has evolved from a tool to identify and characterize isolated proteins or for mass peak profiling as in the application of MALDI to clinical samples, to a platform for interrogating complex proteomes by matching mass spectra to sequence databases to derive protein identifications [11]. However, even with substantial improvements in sensitivity and mass accuracy, the complexities of plasma far exceeds the current capabilities of mass spectrometry to fully resolve the full complement of individual

protein and peptide constituents in a single analysis. Current strategies to achieve in-depth coverage require sample fractionation followed by separate analyses of individual fractions as illustrated in Figure 51.2, or capture of protein or peptide subsets such as glycosylated proteins or phosphopeptides that are analyzed exclusively [12]. The trade-off is limited throughput with extensive fractionation or limited coverage of the proteome if a protein subset is singled out for analysis. Quantitative analysis of protein and peptide constituents is achieved by means of isotopic labeling of proteins and peptides or label free quantification of derived mass spectra.

Cancer specificity of circulating markers may not be required for certain applications, e.g., for delineating signaling pathways or for predicting response to therapy and monitoring disease progression. As cancer is increasingly defined based on deregulated pathways, relevant markers may cut across tumor types without exhibiting tissue specificity. In this regard, there is substantial interest in defining proteomic signatures that predict response to therapy. A case in point is identifying subjects who are likely to benefit from treatment with EGFR tyrosine kinase (TK) inhibitors [13].

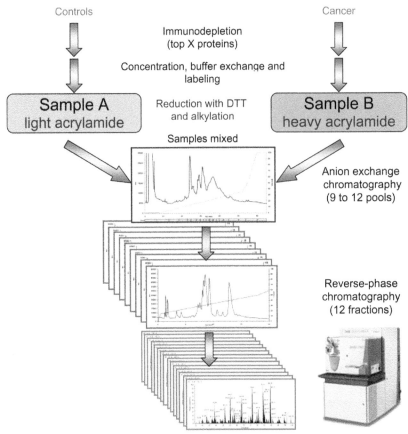

FIGURE 51.2 Quantitative profiling of plasma proteins.
Control and cancer plasmas are first immunodepleted to remove abundant proteins and then labeled with acrylamide isotopes to distinguish cancer from control. Plasmas are then mixed and subjected to intact protein fractionation by anion exchange, followed by reverse phase chromatography. After tryptic digestion, samples are subjected to high resolution mass spectrometry and shotgun LC-MS/MS analysis for protein identification and quantitation. Following statistical analysis and data mining, upregulated and downregulated proteins in cancer plasma relative controls are selected for further investigations and validation as candidate cancer markers.

POSTTRANSLATIONAL MODIFICATIONS AS A SOURCE OF CANCER BIOMARKERS

The study of posttranslational modifications (PTMs) as a source of biomarkers in cancer or other disease is still at a relatively early stage. Aside from phosphorylation analysis, which can be rather readily accomplished using a variety of methods from mass spectrometry to application of phosphospecific antibodies, most other modifications have not been investigated as a source of biomarkers, perhaps with the exception of altered glycosylation [14]. Glycan modifications of proteins are primarily Asn linked (N-linked glycans) or Ser or Thr linked (O-linked glycans). Glycoproteins with complex glycans are membrane bound or secreted. Proteins that are predominantly nuclear or cytoplasmic frequently exhibit glycosylation with the monosaccharide O-linked N-acetylglucosamine (O-GlcNAc) at serine residues, a site of protein phosphorylation. Research going back several decades has yielded evidence that cancer cells exhibit altered glycans relative to normal cells [15]. Interestingly, many of the initial studies with naturally occurring and hybridoma derived monoclonal antibodies that were targeted against tumor antigens yielded evidence of reactivity that were directed against carbohydrate epitopes as in the case of so-called oncofetal antigens [16].

Some effort within the field of proteomics has focused on glycoproteins because of their biological significance and because of their importance as sources of biomarkers. One approach is to capture and enrich for glycoproteins, followed by their identification through mass spectrometric analysis after their deglycosylation and therefore without elucidating their glycan modifications. A more glycan informative approach has been to enrich glycoproteins based on their lectin affinity, followed by mass spectrometry analysis. Glycoproteins are then classified based on their lectin binding properties. Alternative approaches other than mass spectrometry for glycoprotein profiling include array based analyses. Glycoproteins in complex mixtures are subclassified based on their binding to lectin arrays, followed by their individual quantitative analysis based on their recognition by antibodies that recognize particular proteins [17]. Glycomic centric approaches have largely been focused on analysis of individual targeted glycoproteins obtained through preparative approaches followed by detailed analysis of their glycan composition and structure. Some effort at profiling cleaved glycan from serum or tissue to identify biomarkers without their corresponding proteins have yielded potential glycan markers [18]. Increased fucosylation of serum ribonuclease 1 and altered glycosylation of other proteins have been reported in pancreatic cancer [19,20].

MICROARRAY BASED APPROACHES FOR BIOMARKER IDENTIFICATION

Microarrays, using various formats that contain proteins, lysates, or affinity capture agents, are increasingly relied upon for investigations of signaling processes in cells and tissues and for identification of biomarkers [21]. Additionally microarrays have been particularly useful for the identification of tumor antigens that induce an antibody response. For example, recombinant protein microarrays were utilized to screen for auto-antibodies in ovarian cancer [22]. Sera from 30 ovarian cancer patients and 30 healthy individuals were used to probe microarrays containing 5005 human proteins for immunoglobulin reactivity. A total of 94 antigens were identified that exhibited enhanced reactivity with cancer patient sera relative to control sera. Reactive antigens tested using tissue microarrays were found to exhibit increased expression in ovarian cancer tissue relative to controls providing support for their immune recognition as aberrantly expressed proteins. In blinded validation studies that relied on natural protein containing microarrays, anti-annexin antibodies were detected in sera collected a year before a diagnosis of lung cancer was made. Annexin auto-antibodies together with auto-antibodies to PGP9.5 and 14-3-3- theta proteins gave a sensitivity of 55 percent at 95 percent specificity in discriminating lung cancer at the preclinical stage from matched controls [23]. Microarrays containing the repertoire of natural proteins expressed in tumor cells have the potential to substantially accelerate the pace of discovery of tumor antigens and could provide a molecular signature for immune responses in different types of cancer [24].

VALIDATION STRATEGIES FOR DISCOVERED PROTEIN BIOMARKERS

Following an initial discovery, further progress toward biomarker development requires validation studies to be conducted. The biomarker validation process has been conceptually divided into five phases [25]. Assay development for validation studies remains a major hurdle and the number of samples needed for validation increases with advanced validation phases and so does the need to implement assays

that are applicable in the clinical laboratory and that can provide the needed throughput. The most relied upon approach for validation remains the Enzyme Linked Immuno Sorbent Assay (ELISA), which provides the requisite specificity and ready application in the clinical laboratory. For further optimization and increases in sensitivity, antibodies have been coupled to a wide variety of fluorescent molecules, such as quantum dots [27]. The limits of detection have also been increased with amplification strategies taking advantage of the polymerase chain reaction with rolling circle amplification [28] or proximity ligation [29]. Technological developments are also improving the limits of detection for antibody–antigen complexes, as with Surface Plasmon Resonance [30], or with nanomechanical resonators that accurately measure the mass of molecules [31].

Of interest to the field of proteomics, and eventually more broadly, are mass spectrometry based approaches that do not require antibodies. The use of multiple reaction monitoring (MRM) to monitor fragments of specific peptides to directly quantify corresponding proteins in serum, using stable isotope labeled peptides as internal standards, presents an attractive alternative [32]. Other approaches namely peptide [33], RNA, or DNA [34] aptamers also have a potential to overcome the need for antibodies.

REFERENCES

1. Hanash SM, Pitteri SJ, Faca VM. Mining the plasma proteome for cancer biomarkers. *Nature* 2008;**452**(7187):571–9.
2. Ludwig JA, Weinstein JN. Biomarkers in cancer staging, prognosis and treatment selection. *Nat Rev Cancer* 2005;**5**(11):845–56.
3. Chignard N, Shang S, Wang H, et al. Cleavage of endoplasmic reticulum proteins in hepatocellular carcinoma: Detection of generated fragments in patient sera. *Gastroenterology* 2006;**130**(7):2010–22.
4. Hu S, Loo JA, Wong DT. Human body fluid proteome analysis. *Proteomics* 2006;**6**(23):6326–53.
5. Faca V, Hanash S. In-depth proteomics to define the cell surface and secretome of ovarian cancer cells and processes of protein shedding. *Cancer Res* 2008;**69**(3):728–30.
6. Faca VM, Ventura AP, Fitzgibbon MP, et al. Proteomic analysis of ovarian cancer cells reveals dynamic processes of protein secretion and shedding of extra-cellular domains. *PLoS ONE* 2008;**3**(6):e2425.
7. Gortzak-Uzan L, Ignatchenko A, Evangelou AI, et al. A proteome resource of ovarian cancer ascites: integrated proteomic and bioinformatic analyses to identify putative biomarkers. *J Proteome Res* 2008;**7**(1):339–51.
8. Sodek KL, Evangelou AI, Ignatchenko A, et al. Identification of pathways associated with invasive behavior by ovarian cancer cells using multidimensional protein identification technology (MudPIT). *Mol Biosyst* 2008;**4**(7):762–73.
9. Lutz AM, Willmann JK, Cochran FV, et al. A mathematical model relating secreted blood biomarker levels to tumor sizes. *PLoS Med* 2008;**19**:e170.
10. Ransohoff DF. Bias as a threat to the validity of cancer molecular-marker research. *Nat Rev Cancer* 2005;**5**(2):142–9.
11. Cox J, Mann M. Is proteomics the new genomics? *Cell* 2007;**130**(3):395–8.

12. Wu SL, Kim J, Bandle RW, et al. Dynamic profiling of the post-translational modifications and interaction partners of epidermal growth factor receptor signaling after stimulation by epidermal growth factor using Extended Range Proteomic Analysis (ERPA). *Mol Cell Proteomics* 2006;**5**(9):1610–27.

13. Taguchi F, Solomon B, Gregorc V, et al. Mass spectrometry to classify non-small-cell lung cancer patients for clinical outcome after treatment with epidermal growth factor receptor tyrosine kinase inhibitors: a multi-cohort cross-institutional study. *J Natl Cancer Inst* 2007;**99**(11):838–46.

14. Kirmiz C, Li B, An HJ, et al. A serum glycomics approach to breast cancer biomarkers. *Mol Cell Proteomics* 2007;**6**(1):43–55.

15. Dube DH, Bertozzi CR. Glycans in cancer and inflammation: potential for therapeutics and diagnostics. *Nat Rev Drug Discov* 2005;**4**(6):477–88.

16. Feizi T. Carbohydrate antigens in human cancer. *Cancer Surv* 1985;**4**(1):245–69.

17. Chen S, LaRoche T, Hamelinck D, et al. Multiplexed analysis of glycan variation on native proteins captured by antibody microarrays. *Nat Methods* 2007;**4**(5):437–44.

18. Kyselova Z, Mechref Y, Al Bataineh MM, et al. Alterations in the serum glycome due to metastatic prostate cancer. *J Proteome Res* 2007;**6**(5):1822–32.

19. Barrabes S, Pages-Pons L, Radcliffe CM, et al. Glycosylation of serum ribonuclease 1 indicates a major endothelial origin and reveals an increase in core fucosylation in pancreatic cancer. *Glycobiology* 2007;**17**(4):388–400.

20. Zhao J, Qiu W, Simeone DM, Lubman DM. N-linked glycosylation profiling of pancreatic cancer serum using capillary liquid phase separation coupled with mass spectrometric analysis. *J Proteome Res* 2007;**6**(3):1126–38.

21. Wingren C, Borrebaeck CA. Antibody microarray analysis of directly labelled complex proteomes. *Curr Opin Biotechnol* 2008;**19**(1):55–61.

22. Hudson ME, Pozdnyakova I, Haines K, et al. Identification of differentially expressed proteins in ovarian cancer using high-density protein microarrays. *Proc Natl Acad Sci U S A* 2007;**104**(44):17,494–99.

23. Pereira-Faca SR, Kuick R, Puravs E, et al. Identification of 14-3-3 theta as an antigen that induces a humoral response in lung cancer. *Cancer Res* 2007;**67**(24):12,000–6.

24. Madoz-Gurpide J, Kuick R, Wang H, et al. Integral protein microarrays for the identification of lung cancer antigens in sera that induce a humoral immune response. *Mol Cell Proteomics* 2007;**7**(2).268–81.

25. Yasui Y, Pepe M, Thompson ML, et al. A data-analytic strategy for protein biomarker discovery: profiling of high-dimensional proteomic data for cancer detection. *Biostatistics* 2003;**4**(3):449–63.

26. Deragatis LR, Melisaratos N. The brief symptom inventory: An introductory report. *Psychol Med* 1983;**13**:595–605.

27. Tada H, Higuchi H, Wanatabe TM, Ohuchi N. In vivo real-time tracking of single quantum dots conjugated with monoclonal anti-HER2 antibody in tumors of mice. *Cancer Res* 2007;**67**(3):1138–44.

28. Shafer MW, Mangold L, Partin AW, Haab BB. Antibody array profiling reveals serum TSP-1 as a marker to distinguish benign from malignant prostatic disease. *Prostate* 2007;**67**(3):255–67.

29. Fredriksson S, Gullberg M, Jarvius J, et al. Protein detection using proximity-dependent DNA ligation assays. *Nat Biotechnol* 2002;**20**(5):473–7.

30. Kato K, Ishimuro T, Arima Y, et al. High-Throughput Immunophenotyping by Surface Plasmon Resonance Imaging. *Anal Chem* 2007;**79**(22):8616–23.

31. Burg TP, Godin M, Knudsen SM, et al. Weighing of biomolecules, single cells and single nanoparticles in fluid. *Nature* 2007;**446**(7139):1066–9.

32. Kuhn E, Wu J, Karl J, et al. Quantification of C-reactive protein in the serum of patients with rheumatoid arthritis using multiple reaction monitoring mass spectrometry and 13C-labeled peptide standards. *Proteomics* 2004;**4**(4):1175–86.

33. Hoppe-Seyler F, Butz K. Peptide aptamers: powerful new tools for molecular medicine. *J Mol Med* 2000;**78**(8):426–30.

34. Cerchia L, Hamm J, Libri D, et al. Nucleic acid aptamers in cancer medicine. *FEBS Lett* 2002;**528**(1–3):12–6.

Chapter 52

Translational Implications of MicroRNAs in Clinical Diagnostics and Therapeutics

Lorenzo F. Sempere[1] and Sakari Kauppinen[2]

[1]Department of Medicine, Norris Cotton Cancer Center, Lebanon, New Hampshire

[2]Santaris Pharma, Hørsholm, Denmark, and Wilhelm Johannsen Centre for Functional Genome Research, Department of Cellular and Molecular Medicine, University of Copenhagen, Copenhagen, Denmark

BIOGENESIS AND FUNCTIONS OF ANIMAL MICRORNAS

The founding members of the miRNA class, lin-4 [1] and let-7 [2], were identified by forward genetics as central players of the heterochronic gene pathway in the nematode *C. elegans* and first deemed as curious oddities of this little worm [3]. However, three studies published in 2001 reported on miRNA discovery in *C. elegans*, *Drosophila*, and human HeLa cells [4–6]. More than 1000 miRNA genes are estimated to be encoded in the human genome [7, 8]. The accumulating number of miRNA genes along with their highly diverse expression patterns and the multiple target genes (>100) thought to be regulated by individual miRNAs imply that miRNAs play important roles in a wide variety of physiological processes and pathological conditions.

The functional mature miRNA is released after sequential enzymatic cleavage of a pri-pre-miRNA molecule (Figure 52.1). First, a long capped and polyadenylated primary transcript is cleaved in the nucleus by Drosha and associated proteins of the Microprocessor to release a canonical 70 nt precursor hairpin, which is exported via the exportin 5 pathway into the cytoplasm where it is further cleaved by the Dicer/Argonaute multiprotein complex (miRISC for miRNA induced silencing complex) [9, 10]. The ~18–25 nt mature miRNA, which is loaded in the miRISC, guides it to the 3'-untranslated region (3' UTR) of target mRNAs thereby triggering translational repression and/or degradation of the target mRNA [11] (Figure 52.1). In some cases, miRNAs bind to promoter regions and can stimulate transcription [12]. The exact mechanisms of the translational repression are not well understood and may vary depending on the cell type and the composition of auxiliary regulatory proteins and sequence features of the target gene. A series of studies probing for the mechanistic details have proposed that miRNAs affect mRNA accessibility to the translational machinery when stored in P-bodies, ribosome assembly, initiation, and elongation steps. The shortcomings of some of these experimental designs have been thoroughly and critically reviewed [13, 14]. As for mRNA degradation, miRNAs are thought to stimulate decapping and deadenylation permitting access to endonucleases or they may act as RNA guides to direct cleavage of target mRNAs, reminiscent of the siRNA triggered RNAi mechanism [14].

Prediction and Validation of Target Genes

The identification of the miRNA/target mRNA interactions should shed light on their mode of action. Initial computational target prediction algorithms were based on simulation of a few known miRNA/mRNA interactions that lack unifying features and consequently yielded a high rate of false positives (for review, see [15]). Experimental approaches have established the importance of the 5' seed sequence (2–8 nucleotides) of the miRNA in directing the interaction with its target mRNA [11]. A possible caveat of these experiments was that they relied on detectable changes of mRNA levels as the principal mode of action or a collateral effect of translational repression. This potential bias was recently circumvented using other strategies. Co-immunoprecipitation protocols were developed to capture the physical association of miRNA/mRNA molecules. The miRISC and interacting mRNAs were pulled down with antibodies against Ago2 or directly against a biotinylated miRNA mimic, which was followed by high throughput mRNA analysis for target discovery or by specific primer RT-PCR for target validation

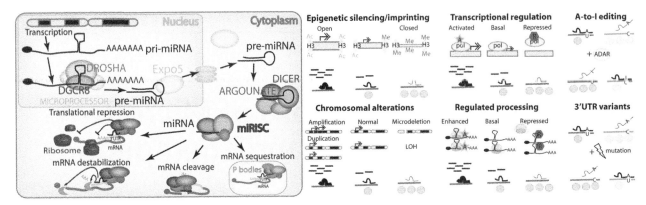

FIGURE 52.1 miRNA biogenesis, regulation, and functions.
Left panel, schematic depiction of stepwise processing of a miRNA molecule from transcription and nuclear processing to export and cytoplasmic processing. Once loaded in the miRISC, the miRNA can control gene expression via different mechanisms. Right panel, different regulatory mechanisms that affect miRNA levels and/or binding interaction with the cognate target mRNAs. Higher levels of mature miRNA increase inhibitory effect on target genes, while lower levels or absence of miRNA relieve target mRNA repression. Abbreviations: H3-Ac, acetylated lysine 9 on histone 3; H3-Me, tri-methylated lysine 9 on histone 3; Me, methylated DNA; LOH, loss of heterozygosity; transcriptional firing (arrow); activator (star); repressor (hexagon).

[16–18]. Two recent reports utilized a high throughput mass spectrometry approach to identify changes of differentially labeled proteins obtained after manipulation of specific miRNA activity [19, 20]. An emerging trend suggested that the higher the levels of protein repression (>fourfold) the more likely it was for mRNA levels to be significantly diminished as well, yet several targets were mainly affected at the translational level [19, 20]. Of the identified target genes, ~60–70 percent could be ascribed to the seed directed binding category, whereas the remaining targets did not present a discernable seed motif, suggesting that other sequences or structural features are important and can mediate the interaction with miRISC loaded miRNA [19, 20]. Together, these experimental approaches should assist in refining and improving computer predictions. Accurate target prediction and experimental validation should further our understanding of the specific genes and pathways regulated by miRNAs.

Regulatory Mechanisms of miRNA Expression and Activity

Several studies have demonstrated that miRNAs are generally transcribed by RNA polymerase II and follow a processing that is similar to mRNAs (but see [21] for RNA pol III transcription). Indeed, more than 50 percent of the human miRNAs reside in exonic regions of transcriptional units without discernable protein encoding potential or in introns of host genes [22–24]. High throughput profiling by cloning, microarray, and qRT-PCR analyses have been extensively used to characterize and classify miRNA expression in a wide variety of tissues and organs at different stages during development, in cell lines, and in animal models upon administration of growth factors or other chemical compounds to stimulate a cellular response, as well as in pathological specimens. These massive efforts have identified tissue specific miRNAs or linked small subsets of miRNAs

to a given disease. However, little is known about the transcriptional regulation of miRNAs, even though they should be subject to the same transcription factor associated regulatory networks as other RNA pol II transcribed genes. In some instances, restricted expression of a miRNA to a specific cell type or tissue has allowed to test transcriptional regulation of miRNAs by master gene regulator(s) such as MyoD in skeletal muscle (see below). The fact that the location of the miRNA hairpin does not inform as to where the transcription starts (this can be up to 20–50 kb upstream), has hampered a more global analysis of transcription factor binding sites. Genome-wide chromatin immunoprecipitation analyses have shed light on features of miRNA promoter and other regulatory regions, such as CpG islands. Using tri-methylated lysine 4 on Histone 3 as a landmark of the transcriptional start site, Marson et al. 2008 found that about 20 percent of the miRNA promoter regions were occupied by Oct4/Sox2/Nanog/TCF3 core embryonic stem (ES) cell transcription factors and were actively engaged in the production of primary transcripts in murine ES cells [25]. Promoter binding by Myc has been shown to repress transcription of multiple miRNAs with tumor suppressive functions [26], while selectively activating transcription of the proto-oncogenic miR-17~miR-92 cluster [27], suggesting that Myc can alter miRNA mediated processes to enhance malignant transformation.

Besides transcriptional control, miRNA levels and activity can be subject to other types of regulation that affect target gene expression (Figure 52.1). Epigenetic silencing of the mir-127 locus has been observed in several cancer cell lines. For example, treatment with chromatin modifying drugs 5-aza-20-deoxycytidine and 4-phenylbutyric acid restored miR-127 expression via usage of a cryptic promoter [28]. In a similar study, chromatin silencing by hypermethylation of the tumor suppressive mir-9, mir-34b/c, and mir-148 genes in cancer cells correlated with metastatic potential [29]. Chromosomal deletion of tumor suppressive miRNAs and increased gene copy number of proto-oncogenic miRNAs were frequently detected in ovarian and breast cancer as well as in melanomas [30]. In human vascular smooth muscle cells, pri-miR-21 processing by Drosha is regulated by a TGFβ mediated mechanism, which involves direct interaction of a SMAD protein complex with the p68 helicase component of the Microprocessor [31]. All these modes of regulation affect the steady state levels of the miRNA and hence increase or decrease miRNA activity. Conversely, miRNA editing of adenosine to inosine, catalyzed by the adenosine deaminases ADAR1 and ADAR2, can alter the affinity for target sites as exemplified by miR-376 mediated regulation of the phosphoribosyl pyrophosphate synthetase (involved in purine and uric acid metabolism) in murine cortex, but not in the liver [32]. Accordingly, miR-376 is preferentially edited in the brain, while other miRNAs present a different editing pattern, such as miR-144 and miR-451, which are exclusively edited in the spleen and testis [33]. Thus, A-to-I editing may have a profound effect in miRNA target recognition to the extent that almost non-overlapping sets of regulated mRNAs may be controlled by the same miRNA in different cell types or under different physiological or pathological conditions. Similarly, single nucleotide polymorphisms (SNP) or other mutations in the 3'UTR of target genes that reduce or increase miRNA binding affinity provide an *in trans* mechanism to alter miRNA regulated processes. Reduced myostatin levels contribute to muscular hypertrophy, a single point mutation (6723G→A transition) in the 3'UTR of myostatin in the Texel sheep creates a *de novo* binding site for the muscle specific miR-1 and miR-206, which may account for the renowned meatiness of this strain [34]. In contrast, a single point mutation (829C→T transition) in 3'UTR of human dihydrofolate reductase (DHFR) mRNA abolishes miR-24 mediated regulation; elevated DHFR expression increased cellular resistance to the anti-tumoral drug methotrexate [35].

MICRORNAS IN PHYSIOLOGICAL AND METABOLIC PROCESSES

The RNase III enzyme Dicer cleaves the miRNA hairpin precursor to render the mature miRNA duplex. In mice and humans, Dicer enzyme is encoded by a single gene *Dicer1*, whereas other species such as *Drosophila* contain several Dicer isoforms specialized in the processing of miRNAs or siRNAs generated during an RNAi response. Loss of function (lf) Dicer mutants in plants, the nematode *C. elegans*, the fruit fly *Drosophila*, zebrafish, and mouse lead to blockage of maturation for all miRNAs. Phenotypes of Dicer deficient organisms are consistent with the loss of miRNA functions. In *C. elegans*, *dcr-1 KO* animals exhibit developmental defects in the temporal specification of cell fates

reminiscent of the lf-mutant phenotypes of *lin-4* and *let-7* miRNAs [36]. In zebrafish, brain morphological defects of Dicer deficient animals can be rescued by injection of miR-430 [37]. In the mouse, *Dicer1* deficient animals do not survive past embryonic day 7.5, inferring an important developmental role for Dicer [38]. To circumvent this early lethality, Harfe *et al.* 2005, generated a conditional lf-allele of *Dicer1* gene by flanking exon 23 with *LoxP* sites ("floxed" allele; "flox") [38]. Upon Cre mediated recombination, exon 23 is excised rendering a non-functional *Dicer1* gene product. In the original report of the *Dicer1^{flox/flox}* mouse, expression of Cre recombinase was driven by *Prx-1* promoter in the mesoderm of budding limbs; Dicer deficient animals exhibited a retarded development and had smaller limbs compared to age matched littermates [38]. In a follow-up study, the authors observed in *Prx-1::Cre;Dicer1^{flox/flox}* animals a relief of an unknown inhibitory activity that prevented retinoic acid mediated activation in the hindlimb. This led to the discovery of miR-196 as a negative regulator of Hoxb8 gene expression in the hindlimb [39]. Other groups have employed similar strategies to conditionally inactivate Dicer by crossing *Dicer1^{flox/flox}* animals with different driver strains that express Cre recombinase in specific cell types and study the consequences of impaired miRNA functions in different physiological and developmental contexts. Dicer deficient animals exhibit defects in lung morphogenesis (*Sonic hedgehog::Cre*) [40]; in skin morphogenesis (*Cytokeratin-14::Cre*) [41]; in skeletal muscle development (*MyoD::Cre*) [42]; in postnatal angiogenesis (*Tie-2::Cre* or *Vascular endothelial Cadherin::Cre-ER^{T2}*) [43]; in proliferation and differentiation of chondrocytes during skeletal development (*Col2a1::Cre*) [44]; in cerebellar neurodegeneration (*Purkinje cell specific::Cre*) [45]; in forebrain morphogenesis (α-Calmodulin kinase II::Cre) [46]; in cytoskeletal dynamics and develop glomerular disease (*Podocyte-specific::Cre*) [47, 48]; in homeostasis and function of regulatory T lymphocytes and develop aggressive autoimmune disease (*Foxp3::Cre*) [49,50]; in the development and function of the female reproductive tract (*Anti Müllerian hormone receptor type 2::Cre*) [51]; in structural and functional retinal neurodegeneration (*Chx10::Cre*) [52]; and in increased premature senescence of cultured murine embryonic fibroblast (*Cagg::Cre-ER^{T2}*) [53]. These proxy studies were refined by generation of targeted "knockout" (KO) chromosomal deletion or forced expression of specific miRNAs in transgenic mouse models; and complemented by gain and loss of function approaches in cell culture systems.

MicroRNAs in Brain Development

Profiling experiments in P19 and NTera2 cell lines during retinoic acid induced neural differentiation and at different stages of rodent brain development uncovered a subset of miRNAs that were brain enriched or brain specific, whose expression dynamically changed in close association with morphological and differentiation hallmarks of neural development [22, 54, 55]. Some of these miRNAs are neuronal specific, such as miR-124, others are enriched in glial cells, and yet others (e.g., miR-9) exhibit a broader but brain specific pattern. Unveiled roles for these miRNAs are consistent with their restricted expression pattern. During brain development, miR-9 promotes progression of neurogenesis in the midbrain–hindbrain regions via regulation of the fibroblast growth factor (FGF) signaling pathway in the midbrain–hindbrain boundary [56]. In the midbrain, miR-133b is involved in maturation and homeostasis of dopaminergic neurons through a negative feedback loop with the paired-like homeodomain transcriptional factor Pitx3 [57]. miR-124 functions to initiate and maintain neuronal cell identity; miR-124 directly represses expression of non-neuronal mRNAs and indirectly favors expression of neuronal specific mRNAs via inhibition of PTBP1, which interferes with neuronal specific alternative splicing forms [58]. Thus, miR-124 globally fine-tunes neuronal gene expression. miR-134 localizes to the synaptodendritic compartment, where it locally and dynamically controls expression of the Limk1 kinase [59]. The miR-134/Limk1 complex modulates dendritic spine morphogenesis and synaptic plasticity. Translational regulation by miR-34 and other microRNAs at synapses is thought to be implicated in processes of memory consolidation and retrieval (for review, see [60]). Interestingly, several miRNAs have been associated with neurodegenerative disease and cancer. In Alzheimer's disease, a marked decrease of miR-29 [61] and miR-107 [62] was observed in afflicted individuals and the β-site amyloid precursor protein cleaving enzyme 1 (BACE1)/β secretase was suggested as key target of these miRNAs. Decreased levels of miR-133b were observed in the midbrain of patients with Parkinson's disease [57]. Altered expression of miR-7, miR-9, miR-21, miR-124, miR-137, and miR-181 was documented in gliomas and other forms of brain cancer [63–68].

MicroRNAs in Cardiac and Skeletal Muscle Development

Restricted expression of miR-1, miR-133, miR-206, and miR-208 to skeletal and cardiac muscle is associated by direct regulation of the myogenic transcriptional factors SRF/myocardin and MyoD/MEF2, respectively [69, 70]. KO experiments in *Drosophila* and mouse have revealed a crucial role for miR-1 in muscular integrity [71, 72]. In the mouse, Hand2 has been proposed as a key miR-1 target in cardiac development [71], while the *Drosophila* miR-1 (analogous to miR-124 in the brain) appears to act as a fine-tuner of muscle specific gene expression by actively repressing non-muscle mRNAs, mostly neuronal mRNAs [72]. A mouse KO of the myocardium specific miR-208 revealed that miR-208 was required to protect cardiomyocyte growth under stress conditions and to safeguard cardiomyocyte

identity by preventing ectopic expression of skeletal muscle specific and other mRNAs [73]. Altered expression and/or function of miR-1, miR-133, miR-206, and/or miR-208 has also been linked to muscular atrophy, cardiac hypertrophy, and other medical conditions ([74], for review, see [75, 76]).

MicroRNAs in the Immune System

Specific subsets of miRNAs are expressed in lymphoid and myeloid lineages derived from hematopoietic stem cells. These miRNAs are important players not only in differentiation programs, but also in highly specialized processes, such as innate immunity, antigen presentation, and T cell activation. The miR-17~miR-92 cluster, miR-150, and miR-181 participate in differentiation of B cells, miR-142, miR-181, and miR-223 in T cells, miR-223 in granulocytes (for review, see [77, 78]), and miR-451 in erythrocytes [79]. Independent studies using mouse KO strains have revealed that miR-155 is required for normal functioning of B and T cells [80–82]. miR-155 null animals failed to mount an adaptive immune response against *Salmonella* challenge after vaccination with attenuated strain; fewer class switched antibodies after immunization (B cell function) and reduced IL-2 and IFN-γ production (T cell function) were observed compared to control animals [81]. miR-155 was also required to control the germinal center reaction after pathogen challenge [82] and it was suggested that this occurred in part by regulating activation induced cytidine deaminase [83, 84].

miR-132, miR-146, and miR-155 have been implicated in the innate immune response in monocytes and macrophages. Lipopolysaccharide (LPS) triggers an inflammatory response via toll-like receptor 4 (TLR-4). *In vitro* treatment with LPS induced expression of miR-132, miR-146 in a monocytic cell line [85]. In addition, miR-146 expression was also stimulated by TNFα and IL-1β treatment. LPS, TNFα, and IL-1β stimuli converged in the activation of the NFκB pathway. Thus, miR-146 was proposed to act in an NFκB dependent manner as an attenuator of toll-like receptor and cytokine signaling via repression of IRAK1 and TRAF6, which are key adapter molecules in these signaling cascades that lead to activation of NFκB and AP-1 pathways [85].

MicroRNAs in the Liver

miR-122 is a highly abundant miRNA with restricted expression to hepatocytes (>50,000 copies/cell) (for review, see [86]). miR-122 regulates expression of the cationic amino acid transporter protein (CAT-1), which is involved in the import of the essential amino acids lysine and arginine [87]. miR-122 is also implicated in the regulation of cholesterol, fatty acid, and lipid metabolism. Recent studies have reported on profound and prolonged lowering of total plasma cholesterol levels after silencing of miR-122 *in vivo* in mice and non-human primates [88–92]. Several target mRNAs of miR-122 have been experimentally validated, including Aldolase A (ALDOA), Cyclin G1 (CCNG1), Glycogen Synthase 1 (GYS1), prolyl 4-hydroxylase alpha subunit (P4HA1), branched chain α-ketoacid dehydrogenase kinase (Bckdk), CD320 antigen-putative VLDL receptor, and N-Myc downstream regulated gene 3 (Ndrg3) [88, 90, 91].

miR-122 may also play an important role in liver neoplasia. Decreased levels of miR-122 were detected in hepatocellular carcinomas [93, 94]. Bcl-w, an antiapoptotic Bcl-2 family member, and promitogenic CCNG1 were suggested as miR-122 targets and may explain the need of liver cancer cells to dismantle this negative regulation [94, 95]. Contrary to the aforementioned findings, a recent report showed that miR-122 is upregulated in hepatitis C virus associated hepatocellular carcinomas [96].

MICRORNAS IN HUMAN DISEASE

Human diseases have been associated with mutations in components of the multiprotein machineries that process miRNAs or assist in miRNA mediated regulation. Although a dysfunctional machinery could lead to global alterations of miRNA functions, specific miRNAs and their regulated targets may be more sensitive to dose effects. DGCR8 is a key component of the nuclear microprocessor complex. DGCR8 recognizes the structural features of the primiRNA transcript and binds to the basal region of the stem loop, correctly positioning Drosha for productive cleavage of the pre-miRNA hairpin [97]. DGCR8 (for DiGeorge chromosomal region gene 8) was cloned from the DiGeorge syndrome region on human chromosome 22q11.2 [98]. DiGeorge syndrome affects multiple organs and afflicted individuals present with cardiac malformations, craniofacial, limb, and digit anomalies, as well as with learning, language, behavioral, and other mental disorders. The haploinsufficiency of this syndrome suggests that dose reduction of some miRNAs may interfere with proper regulation of gene expression. Experiments on an engineered mouse strain to recapitulate syntenic 22q11.2 microdeletion, which deletes one copy of DGCR8 locus, uncovered the reduced processing of brain specific miRNAs, which partially explained the observed defects [99]. Fragile X syndrome, one of the most common forms of inherited mental retardation, is caused by the functional loss of fragile X mental retardation protein (FMRP). FMRP is an RNA binding protein with known roles in translational control. Several lines of evidence have shown that FMRP genetically interacts and/or physically associates with protein components of the (mi)RISC complexes (for review, see [100]). Although there is no direct evidence that FMRP affects functions of specific miRNAs, it is possible that recruitment of FMRP to the miRISC complex potentiates inhibitory activity of brain expressed miRNAs, since even slight leakiness of key target genes under tight miRNA mediated regulation may have deleterious effects. In contrast, trisomic dosage overexpression of miRNAs located on chromosome 21 has been suggested as an etiological factor in Down's syndrome (or Trisomy 21), which is the most common genetic cause of cognitive impairment and congenital heart defects [101]. This exposes the importance of adequate miRNA mediated control of gene expression, where too much repression may distort the balance in biological processes.

Genome-wide SNP analyses have linked allele variants with increased risk of different diseases. Some of these variants map to 3' UTRs and causality for altered miRNA binding and regulation of the affected gene has been probed for and found (for review, see [102]). Briefly, SNPs in the 3' UTRs of the following genes have been implicated in disease: (i) the candidate gene for Tourette's syndrome SLITRK1 was associated with miR-189 regulation in brain tissue of afflicted patients; (ii) the fibroblast growth factor 20 (FGF20) with miR-433 in Parkinson's disease; (iii) the angiotensin receptor-1 (AGTR1) with miR-155 in hypertension; (iv) the IGF-II receptor gene (IGF2R) with miR-657 in Type 2 diabetes; (v) and the HLA-G gene with miR-148/miR-152 in asthma.

MICRORNAS IN HUMAN CANCER

During the past few years, numerous seminal studies on the role of miRNAs in cancer have greatly contributed to the understanding of tumorigenic processes. The following will present an overview of this rapidly expanding field of cancer biology.

microRNA Signatures in Liquid and Solid Cancers

Chromosomal deletion and/or downregulation of miR-15a and miR-16-1 expression in B cell chronic lymphocytic leukemia (CLL) was the first association between miRNAs and cancer [103]. Now a unique signature affecting 13 miRNA genes has been linked to prognosis and progression of this disease [104]. High throughput technologies have been applied for detection of miRNA in fresh, frozen, and archived specimens from liquid and solid tumors, including leukemias, lymphomas, and carcinomas of the breast, colon, ovaries, pancreas, liver, lung, and thyroid, as well as brain cancer and melanoma (for review, see [105]). These profiling efforts have uncovered specific miRNA signatures associated with different cancer types, while a few miRNAs are consistently detected at higher (e.g., miR-21 and miR-155) and lower levels (e.g., let-7, miR-34, and miR-145), respectively, across several cancer types. Importantly, miRNA signatures can more robustly distinguish normal from tumor tissues compared with expression patterns for protein encoding genes [106, 107]. For example, miRNA based classification of poorly differentiated gastro-intestinal tumors was more accurate than mRNA profiles, when applied to the same specimens [106].

The utility of miRNAs as markers for detection, diagnosis, and treatment selection (i.e., drug resistance profile [108]), prognostic indicators of progression and recurrence, as well as molecular surrogate markers to monitor treatment response has been clinically assessed or suggested. Higher levels of miR-21, miR-155, and miR-221 in activated B cell-like compared with germinal center B cell-like could separate these subtypes of diffuse large B cell lymphoma [109]. In lung adenocarcinomas, elevated miR-155 detection and low let-7 levels correlated with a poor clinical outcome [110, 111]. High levels of miR-21 correlated with poor prognosis in breast, colon, lung, and pancreatic cancer as well as glioblastomas, but not in gastric cancer [112]. In breast cancer, hypoxia induced miR-210 was suggested as an independent prognostic indicator [113]. In colonic adenocarcinomas, miR-145 was consistently detected at lower levels [114–116], while high levels of miR-320 and miR-498 correlated with the probability of recurrence free survival [115]. In hepatocellular carcinomas, a 20-miRNA metastasis signature confidently predicted which primary tumors were associated with venous metastases or were metastasis free [117]. In conclusion, miRNAs hold great promise as novel biomarkers to assist in multiple aspects of cancer management.

miRNA Detection in Archived Specimens, Fine Needle Aspirates, and Blood Samples

Differential detection of miRNA levels (or other molecules) in whole tissue biopsies should be cautiously interpreted. It is generally assumed that an elevated signal of a miRNA corresponds to its upregulation within cancer cells, whereas low or no signal is associated with downregulation of miRNA expression. However, differential detection could merely reflect cell type heterogeneity among normal and tumor tissues and/or the recruitment of reactive stroma and infiltrating lymphocytes to the cancerous lesions. Differential detection of an miRNA (or miRNA signature) that can distinguish patients with favorable (e.g., low miRNA-X) or poor prognosis (e.g., high miRNA-X) will be clinically relevant independent of the underlying biological cause; for example miR-X is upregulated in cancer cells and increases aggressiveness or miR-X is expressed in lymphocytes and/or macrophages and elevated signal correlates with inflammation. To gain further insight into the differential miRNA expression in cancer, an *in situ* hybridization (ISH) protocol has been developed and implemented to visualize the spatial distribution of miRNA expression in archived formalin fixed paraffin embedded (FFPE) sections [118, 119]. This is an innovative method due to the technical limitations associated with detection of small RNAs, such as miRNAs using conventional methods [120]. The success of this ISH method relies primarily on the design and use of LNA modified oligonucleotide detection probes with high binding affinity to their cognate miRNAs.

Locked nucleic acids (LNAs) are a class of bicyclic RNA analogs with unprecedented affinity against their complementary RNA molecules [121]. Differential detection of miR-124 in oligodendrogliomas suggested a potential etiological contribution, but an LNA based ISH assay conducted on FFPE sections of normal brain and oligodendrogliomas revealed a neuronal specific expression of miR-124; this explained the differential detection since neurons are underrepresented in oligodendroglioma lesions [119]. In breast cancer, decreased detection of miR-451 merely reflected changes of the tumor associated vasculature; as miR-451 was predominantly expressed in mature erythrocytes as determined by LNA-ISH [118], and independently confirmed by functional assays of hematopoietic cell differentiation [79,122]. Nonetheless, the ISH approaches have clinically validated differential expression of some miRNAs: the brain specific miR-9 is upregulated in glioma lesions [119], miR-145 and miR-205 are predominantly expressed in the mammary myoepithelial cell layer, while miR-21 is upregulated in tumor associated fibroblasts and in breast cancer cells (Figure 52.2 and [118]). ISH is not as sensitive as quantitative RT-PCR or microarray profiling. Thus, alternative or complementary techniques, such as laser capture microdissection to enrich for stromal cells or cancer cells followed by miRNA expression analysis should be employed to clinically validate differential expression and prioritize further studies of the most etiologically relevant miRNAs. For example, enrichment of pancreatic cancer lesions by physically drilling with a 2 mm bore through FFPE tissue blocks identified a different miRNA signature [123] as compared to whole tissue biopsies [124, 125].

Mature miRNAs have a very long half-life and their small size makes them less sensitive than mRNAs to degradation often associated with processing of tissue for formalin fixation and paraffin embedding, cell preparations obtained by endoscopic ultrasound guided (EUS) fine needle aspiration (FNA), or blood based samples. Thus, miRNAs may be more suitable as biomarkers than mRNAs for expression profiling in these biological materials. Indeed, miRNA expression analysis of pancreatic cell preparations obtained by EUS-FNA suggested that the ratio of miR-196a/miR-217 could assist in separating benign cases and chronic pancreatitis from pancreatic ductal adenocarcinoma [126]. miR-217 is highly enriched in the acinar cell compartment [125], hence cell type heterogeneity may at least in part explain these findings. If confirmed in a larger patient cohort, this miRNA assay could represent an important breakthrough for early detection of pancreatic cancer, which is one of the most aggressive and deadliest cancers. Remarkably, circulating cell free miRNAs have been detected in blood plasma and serum, which could provide a non-invasive method of diagnosis (Figure 52.2). The placental miRNAs, miR-526a, miR-527, and miR-520d-5p, are detected at higher levels in blood serum of pregnant women [127]. Elevated detection of the tumor associated miR-21, miR-155, and miR-210 in serum could separate patients with diffuse large B cell lymphoma from healthy volunteers [128]. Similarly, elevated detection of miR-141 in blood plasma has been suggested as an independent marker for detection of prostate cancer, which moderately correlates with prostate specific antigen (PSA) levels [129]. PSA can be elevated in benign prostate conditions, thus miR-141 may assist in refining PSA based diagnosis. miR-141 is also differentially detected in breast and lung cancer, and thus it will be interesting to determine whether circulating miR-141 can detect patients afflicted with other carcinomas.

Contribution of miRNAs to Cancer Initiation, Progression, and Metastasis

Experiments in cell culture and in mouse models have provided detailed molecular mechanisms of miRNA mediated regulation on cancer progression and metastasis. These studies clearly indicate that specific miRNAs can act as oncogenes (OG) or tumor suppressor genes (TSG). The following examples demonstrate that altered miRNA functions confer a growth advantage to cancer cells, either by reducing activities that repress mitogenesis and/or induce apoptosis, or by increasing activities that favor malignant transformation (Figure 52.3).

Overexpression of miR-155 by a B cell specific promoter caused rapid and aggressive B cell lymphoma in mice [130]. Similarly, overexpression of the polycistronic miRNA cluster, miR-17~miR-92, accelerated tumor development in a c-Myc induced mouse model of B cell lymphoma [131]. In contrast, overexpression of miR-15 and miR-16 increased cell death in the MEG-01 leukemic cell line via targeting of the antiapoptotic gene Bcl-2 [132]. Frequent loss of miR-15 and miR-16 in CLL cases may in part contribute to the reduced apoptotic potential of malignant cells.

The miR-17~miR-92 cluster consists of six tightly linked miRNAs: miR-17, miR-18a, miR-19a, miR-20a, miR-19b-1, and miR-92-1, that are co-transcribed as a long pri-miRNA, which was previously identified in B cell lymphomas as the C13orf25 transcript (for review, see [133]). Myc stimulates transcription of the *mir-17~mir-92* gene cluster and concomitant elevation of Myc, and these miRNAs were detected in B cell lymphomas and lung cancer [133]. Moreover, miR-17-5p and miR-20 overexpression protected prostate and lung cancer cell lines from undergoing apoptosis [134, 135]. This is consistent with the notion that these miRNAs mediate Myc driven processes to promote tumor growth. However,

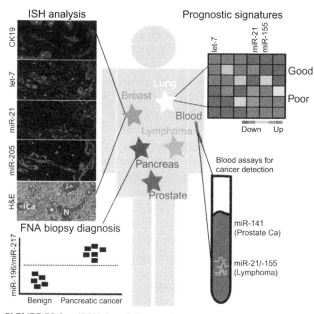

FIGURE 52.2 miRNA based diagnostics.
RNA extracted from fresh and archived tissue biopsies, FNA cell preparations and blood collections can be used to determine changes of miRNA expression and to obtain clinically valuable information for early detection, diagnosis, and prognosis. Different technological platforms have been employed to detect miRNA levels such as ISH in archived breast FFPE, microarray analysis in frozen lung tissue, quantitative RT-PCR in pancreatic FNA, and blood samples.

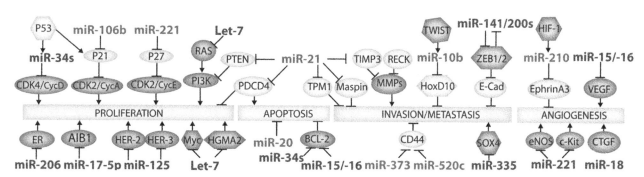

FIGURE 52.3 miRNA mediated mechanisms in cancer biology.
Schematic illustration of representative tumor suppressive and oncogenic miRNAs and their regulatory interactions with protein encoding genes. miRNAs and proteins whose functions promote malignant transformation are indicated in bold print, whereas those that act as tumor suppressors are shown in regular print. The actions of these miRNAs may be contextual and/or cell type specific as exemplified by miR-221. → indicates positive and ⊣ negative effect that may be direct or require intermediary players. ⬡ denotes a protein that acts as a transcription factor.

the miR-17~miR-92 cluster can also exert tumor suppressive functions. miR-17-5p and miR-20 are engaged in regulatory loops with Myc and the E2F family members as well as with Cyclin D1 (CycD1). While Myc and E2Fs stimulate the transcriptional output of each other and that of the miRNA cluster, miR-17-5p and miR-20 repress expression of E2Fs, dampening post-transcriptionally the magnitude of transcriptional activation [133]. Similarly, in the MCF7 breast cancer cell line, CycD1 stimulated expression of this miRNA cluster, while miR-17-5p and miR-20 inhibited CycD1 expression via binding to its 3'UTR [136]. Thus, in breast and other cancers with low levels of the miR-17~miR-92 cluster, members of this cluster may act as TSGs by limiting CycD1 and E2Fs proliferative signals, while in B cell lymphomas and colon cancer that are associated with Myc and miR-17~miR-92 overexpression the miRNA cluster members may promote tumorigenesis by dissociating proliferative Myc activities from the proapoptotic functions of E2F1 and other proteins.

One of the first miRNAs discovered, let-7, targets the oncogenic HMGA2, Myc, N-Ras, and K-Ras [137, 138]. Poor prognosis of lung adenocarcinoma was associated with low levels of let-7 and elevated Ras activity [137]. In a complementary study, global decrease of miRNA activity by loss of Dicer (or other enzymes required for miRNA maturation) enhanced the tumorigenic potential of a lung adenocarcinoma cell line [139]. Inactivation of Dicer in combination with constitutively active K-Ras^{G12D} by adenoviral delivery of Cre recombinase into the respiratory tract accelerated tumor progression in a mouse model of lung cancer [139]. let-7 mediated regulation of K-Ras and c-Myc was an important tumor suppressive mechanism impaired in the Dicer deficient animals [139]. Forced expression of let-7 in mouse models of lung cancer reduces tumor burden [140, 141]. In Burkitt's lymphoma cell lines, replenishing let-7 activity with a synthetic let-7 mimic repressed Myc expression and had a profound antiproliferative effect [142].

The miR-34 family members (miR-34a, b, c) are transcriptionally activated by p53 in response to DNA damage and mitogenic signals (for review, see [143]). miR-34 inhibits translation of several cell cycle and/or survival promoting genes including cyclin dependent kinase 4 (CDK-4), CDK-6, CycD1, CycE2, and BCL-2 [143]. Conversely, a group of proto-oncogenic miRNAs represses inhibitory cell cycle regulators to promote rapid tumor growth. The miR-221/-222 family members have been shown to repress the CDK inhibitor p27(kip1) expression in breast, thyroid, liver, prostate, melanoma, and glioblastoma cells [144–149], as well as the CDK inhibitor p57 in liver cancer cells [145], while miR-17-5p and miR-106 were shown to repress the CDK inhibitor p21 expression [116, 150].

miR-21, which is commonly upregulated in both liquid and solid cancers, is considered as a master oncogenic miRNA. Validated mRNA targets of miR-21 include: phosphatase and tensin homolog deleted on chromosome 10 (PTEN), which is a TSG and natural attenuator of the Ras/PI3K signaling pathway [151,152]; programmed cell death 4 gene (PDCD4), which is a TSG that inhibits translation and interferes with the transactivation of the NFκ and AP-1 transcriptional factors and their neoplastic programs [153–156]; tropomyosin (TPM1) and maspin [153, 157], and membrane anchored matrix metalloproteinase regulator (RECK) and tissue inhibitors of matrix metalloproteinases (TIMP3), which are TSGs and inhibitors of cell migration and invasion [65]. Thus, miR-21 orchestrates a coordinated disassembly of anti-mitogenic, proapoptotic, and anti-metastatic mechanisms, thereby contributing to aggressiveness of malignant cells.

Loss of E-cadherin weakens epithelial cell-to-cell junctions and facilitates an epithelial to mesenchymal transition (EMT), which leads to cell invasion and metastasis. The miR-141/miR-200 family members, miR-205 and miR-489, maintain expression of E-cadherin by post-transcriptional repression of ZEB1 and ZEB2, which are zinc finger transcriptional repressors of E-cadherin [158–162]. ZEB1 and ZEB2 also repress transcription of the miR-141/miR-200 genes via binding to E boxes [158]. Thus, the ZEBs and these miRNAs are engaged in a reciprocal negative feedback loop [158]. miR-335 suppresses breast cancer metastasis by inhibiting the pro-metastatic transcriptional factor SOX4 and the extracellular matrix component tenascin C [115], while miR-373 and miR-520c promote cell migration and invasion [163] via repression of anti-metastatic CD44. miR-10b is a key player of a regulatory transcriptional/post-transcriptional cascade that promotes metastatic programs in part by stimulating an EMT [164]. Twist drives transcription of miR-10b, which in turn inhibits Homeobox D10 expression, releasing HoxD10 mediated transcriptional repression of pro-metastatic gene RhoC [164].

miRNAs have also been shown to participate in cancer cell response to hypoxia and angiogenesis. miR-210 expression is upregulated by hypoxia inducible factor (HIF-1) under hypoxic conditions in breast cancer and endothelial cell lines (for review, see [165]). A connection between hypoxia and the need of blood supplies suggest that hypoxic induction may stimulate blood vessel formation. Indeed, overexpression of miR-210 in the HUVEC cell line promoted tubulogenesis and cell migration by repressing EphrinA3 [166]. Similar observations *in vitro* and *in vivo* demonstrated a pro-angiogenic role of the miR-17~miR-92 cluster, miR-130b, and miR-378 and an anti-angiogenic role of miR-15/-16 via VEGF and Bcl-2 repression, and miR-221/-222 via c-Kit and eNOS [165]. None of these miRNAs is specifically expressed in endothelial cells and, thus, functional analysis of known miRNAs with an enriched or restricted expression to endothelial cells could uncover other regulatory mechanisms of blood vessel formation. This approach has been undertaken to study the endothelial specific miR-126. miR-126 null animals exhibit defects in endothelial cell proliferation, migration, and angiogenesis [167]. Under physiological conditions, miR-126 is required to relay VEGF and FGF signaling to activate angiogenic programs, by opposing the inhibitory action of SPRED-1 [167]. Interestingly, miR-126 was detected at lower levels in tumor tissue of breast, lung, and other organs, suggesting that miR-126 is not required or perhaps interferes with tumor induced angiogenesis. This warrants further investigations given the potential therapeutic implication of blocking tumor access to blood supply via miR-126 mediated mechanisms.

MICRORNAS AND VIRAL LIFE CYCLES

Viral encoded miRNAs were first demonstrated in the Epstein–Barr virus (EBV) and have since been identified in several human pathogenic herpesviruses, including Kaposi's Sarcoma associated herpesvirus, cytomegalovirus, and herpes simplex virus type 1 (HSV-1) (for review, see [168]). Herpesviruses possess long dsDNA genomes (>120kb), which may have enhanced their evolutionary potential to encode miRNAs. Indeed, viral encoded miRNAs have primarily been detected in dsDNA virus types (adenoviruses, herpesviruses, polyomaviruses) whereas attempts to clone miRNAs from retroviral and ssRNA viruses have failed to identify miRNAs [168]. Intriguingly, the structured RNA transactivation responsive (TAR) element, which is located at the 5' end of all transcripts derived from retroviral human immunodeficiency virus type 1 (HIV-1), is recognized as a Dicer substrate and yields processed and functional miRNAs [169,170]. This opens the possibility that non-dsDNA viruses may have devised alternative mechanisms to encode and utilize miRNAs. Viruses have not only acquired viral encoded miRNAs during evolution that control expression of viral and host gene transcripts, but have also recruited and exploited host miRNAs to regulate viral replication and latency [168].

Herpesvirus Latency

During latency, the only viral product of HSV-1 is the non-coding RNA transcript (LAT). Four viral miRNAs are processed from LAT (a pri-miRNA) and one of these, miR-H2-3p, post-transcriptionally represses expression of ICP0 [171]. ICP0 is a viral immediate early transcriptional activator required for replication and re-entry into the lytic cycle. In the related virus HSV-2, LAT also functions as a pri-miRNA from which a single miRNA, miR-I, is processed [172]. It has been postulated that miR-I modulates HSV-2 virulence by controlling expression of ICP34.5 [172]. In EBV, latently expressed miR-BART2 directs the mRNA cleavage of BALF5, the viral polymerase required during the lytic cycle [173]. These examples illustrate the evolution of miRNA mediated mechanisms to actively maintain latency.

HIV-1 Latency

HIV-1 escapes the immune system by latently "hiding" in resting CD4$^+$ lymphocytes. The 3' region of HIV-1 RNA contains binding sites for multiple miRNAs whose expression is enriched in resting CD4$^+$ cells [174]. Cooperative binding of these host miRNAs (miR-28, miR-125b, miR-150, miR-223, and miR-382) attenuate viral protein production [174] and hence maintain latency. Treatment with inhibitors of the host miRNAs *in vitro* was shown to stimulate viral replication, even though viral transcriptional activation was also required [174]. In addition, HIV-1 encodes viral miRNAs that also contribute to maintenance of latency [169,170]. A TAR derived miRNA directs binding of histone deacetylase HDAC-1 to the HIV-1 Long Term Repeat region, which triggers chromatin silencing of HIV expression [170].

Hepatitis C Virus Replication

Several studies have shown that the liver expressed miR-122 is required for replication of the positive-strand RNA Hepatitis C virus (HCV) (for review, see [86]). miR-122 is recruited via two binding sites in the 5' end of the HCV genome [175]. The 5' location of these sites is crucial for the positive miR-122 effect as placing this motif in the 3' region inhibits expression [175]. Interestingly, transfection of Huh-7 cells harboring the HCV-N replicon NNeo/C-5B with a 2'-O-methyl antimiR oligonucleotide complementary to miR-122 resulted in significant reduction of viral RNA underscoring the role of miR-122 in HCV replication [175,176]. In addition, miR-196, miR-296, miR-351, miR-431, and miR-448, whose expression is induced by interferon-β (natural cellular response and first line of therapeutic treatment against viral infections), have been reported to inhibit HCV viral production [177].

FUTURE MICRORNA BASED THERAPEUTIC STRATEGIES

Several biopharmaceutical companies are currently developing miRNA based therapeutics. These include: Santaris Pharma (Denmark), Regulus Therapeutics (USA), Miragen Therapeutics (USA), and Mirna Therapeutics (USA). In May 2008, Santaris Pharma initiated a Phase I human volunteer trial of the world's first miRNA medicine to be tested in man. Santaris Pharma's LNA based antagonist of miR-122 (SPC3649) is being developed as a potential new approach to the treatment for Hepatitis C infection. The Phase I clinical trial is a placebo controlled, double-blind, randomized, single dose, dose escalating safety study in a total of 48 healthy male volunteers. If successful, the company is planning to test SPC3649 in patients with hepatitis C virus (HCV) infection.

miRNA based therapeutics exhibits several features that may provide patient benefits unobtainable by other therapeutic approaches (Figures 52.4, 52.5). (i) The ability of miRNAs to coordinately modulate the expression of multiple target genes and thereby affect several output pathways in disease through key target gene(s) may have a major etiological contribution. Such a combinatorial effect is difficult to achieve

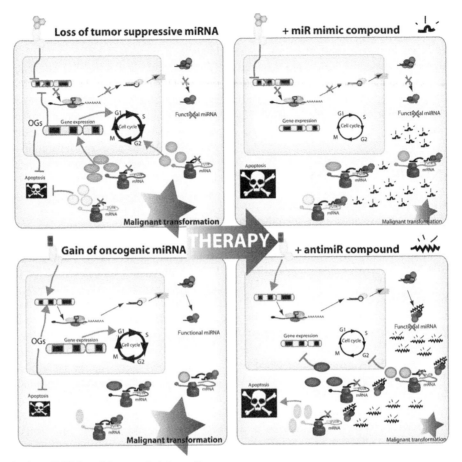

FIGURE 52.4 Strategies for miRNA based therapeutic intervention.
Reduced levels/activity of a miRNA with tumor suppressive functions (upper panel) or increased levels/activity of a miRNA with tumor promoting functions (lower panel) by altered regulation of their expression results in increased activity of proteins that accelerate cell cycle progression, inhibit cell death, and promote neoplastic programs, or that inhibit cell cycle progression, enable cell death, and maintain cell identify programs, respectively. Together, these changes confer a higher tumorigenic potential and facilitate malignant transformation. Administration of a miR mimic compound replenishes miRNA activity and concomitantly restores negative regulation on multiple target genes (upper panel) whereas an antimiR compound antagonizes miRNA activity and thereby relieves inhibition of target genes (lower panel), effectively reducing aggressiveness and malignancy of cancer cells.

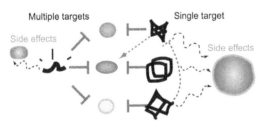

FIGURE 52.5 Potential benefits of miRNA based therapeutics.
The use of miRNA based therapeutics enables simultaneous targeting of multiple mRNAs in disease pathways, whereas this cannot be achieved using conventional drug compounds. Furthermore, single agent miRNA based medications may exhibit milder, more predictable side effects and adverse drug–drug interactions compared with conventional medications.

with other therapeutic approaches, which most often aim at interfering with the function of one or a few selected proteins. (ii) miRNA based therapeutics could be used to replace a given miRNA activity using synthetic miRNA mimics, while replacing a protein encoding gene requires viral delivery and

production (transcription/translation) by the endogenous cellular machinery. (iii) The flexible design and the ability to scale-up the synthesis of chemically modified oligonucleotide compounds provide the basis for development of therapeutic approaches that either mimic an miRNA activity (miRNA mimics) or effectively bind the mature miRNA, thereby antagonizing its function (antimiR).

Modulation of miRNA Activity

Synthetic miRNA compounds are commercially available from various companies. An miRNA mimic approach, commercialized by Ambion as "pre-miR," comprises a double-stranded RNA molecule with proprietary modified nucleosides and with asymmetric features to favor unwinding of the miRNA guide (sense) strand. Other approaches have also used double-stranded RNA, chemically identical to siRNAs or with modifications, to obtain a similar effect (Dharmacon; [19, 178]). Neither of these compounds

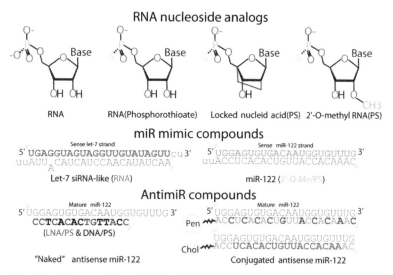

FIGURE 52.6 Synthetic RNA analogs for miRNA based therapeutics.
Chemical structures of different nucleoside modifications used to design miRNA mimics or antimiR oligonucleotides. Depicted miR mimic and antimiR compounds are based on: Selbach *et al.*, 2008 [19], RNA let-7 mimic; Shan *et al.*, 2007 [178], 2-O-methyl miR-122 mimic; Elmen *et al.*, 2008 [90], unconjugated 15-mer LNA-antimiR-122; Krutzfeldt *et al.*, 2005 [88], cholesterol (chol)-conjugated 2'-O-methyl/RNA antimiR-122; Fabani *et al.*, 2008 [89], penetratin (pen)-conjugated 2'-O-methyl/LNA antimiR-122.

exactly mimics a hairpin RNA precursor, which may affect their mode of action. These siRNA-like molecules are processed by Dicer and loaded into miRISC (for miRNAs) and/or RISC (for siRNAs); sorting of RNA guide strand into miRISC or RISC depends on the structural features of precursor molecule [179]. Despite the similarity between miRNAs and siRNAs with respect to their processing machinery, they are biologically and functionally different: miRNAs are endogenous and processed from a hairpin precursor, while siRNAs are typically exogenously administered and processed from a double-stranded molecule. Thus, the use of double-stranded compounds may not fully recapitulate endogenous miRNA functions, if they are preferentially loaded in RISC rather than miRISC. The miRNA mimics have been utilized in cell culture systems, and hence additional modifications and/or conjugation chemistries will be needed for stabilization and efficient uptake in animals.

Since binding of the miRNA to its cognate target mRNA is guided by Watson-Crick base pairing, miRNA function can be effectively blocked using an antisense oligonucleotide, termed here as antimiR, that binds to the miRNA in competition with cellular target mRNAs. Indeed, different chemically modified oligonucleotides have proven efficient in functional inhibition of miRNAs in invertebrate and mammalian cell culture and *in vivo* in animals including worms, zebrafish, and rodents. These include morpholinos [180], 2'-O-methyl [88], 2'-O-methoxyethyl (2'-MOE) [91], and LNA modified antimiR oligonucleotides [90, 92] (Figure 52.6). Several studies have reported on miR-122 silencing in mice using antimiR oligonucleotides delivered either as unconjugated, phosphorothiolated compounds or conjugated to cholesterol using different dosing regimens [88, 90, 91]. In a recent study, Elmen *et al.*

2008 [92] used a high affinity LNA-antimiR to silence miR-122 in non-human primates. Systemic delivery of PBS-formulated LNA-antimiR at doses ranging from 3×1 to 3×10 mg/kg with three intravenous infusions over 5 days resulted in dose dependent and long-lasting decrease of total plasma cholesterol in African green monkeys. Consistent with the data from miR-122 antagonized mice [88,90–92], miR-122 silencing in primates was reversible as the levels of cholesterol returned to normal over a 3-month period after antimiR treatment. The LNA mediated miR-122 silencing approach was well tolerated in primates as no acute or subchronic toxicities were observed in this study. Importantly, no hepatotoxicity, renal toxicity, or abnormalities in histopathological investigations of primate liver biopsies were detected in the LNA-antimiR treated primates. Considered together, reports on efficient miRNA silencing in rodents and non-human primates using high affinity targeting by chemically modified antimiR oligonucleotides underscore the potential of such compounds in the future development of miRNA based therapeutics.

Delivery

Efficient *in vivo* delivery of therapeutic compounds is a critical factor for the development of successful miRNA based treatment modalities. Many peripheral tissues can be effectively targeted by systemically delivered oligonucleotides containing phosphorothioate backbone modifications, which provide good pharmacokinetic properties and tissue uptake along with good biostability *in vivo* [181]. Unconjugated ("naked"), phosphorothiolated 2'-MOE, and LNA-antimiR oligonucleotides have been used to silence miR-122 *in vivo*, whereas the use of cholesterol-conjugated

2'-O-methyl oligonucleotides (termed as antagomirs) represents another delivery approach to silence miRNA function [88,90–92]. A number of different strategies for efficient siRNA delivery are being tested and these could also be applied to miRNA based therapies. For example, ligands for specific cell surface receptors capable of being internalized can be conjugated to oligonucleotides thereby facilitating both cellular uptake and cell type specific delivery. The use of supramolecular nanocarriers, such as liposomes, and polymeric nanoparticles represents another strategy for delivering antisense oligonucleotides (for review, see [182, 183]). The field of nanoparticle delivery is advancing at a fast pace, and different formulations are already in preclinical trials for targeted delivery to ovarian cancer cells using poly(beta-amino ester) derivatives and to epithelial cancer cells with folic acid coated dendrimers [184]. Recent findings that siRNAs are recognized by cell surface toll-like receptor 3 (TLR3) and elicit non-target extracellular induction of interferon-γ and interleukin-12 via TLR3 and its adaptor TRIF [185] should be taken into consideration when refining miRNA mimic formulations.

Therapeutic Intervention

Accumulating evidence implies that miR-21 and miR-155 could be potential targets for antimiR based cancer therapeutics. Several studies have shown that miR-21 is overexpressed in glioblastomas and in breast, lung, colon, prostate, and stomach cancers [65, 153–157], suggesting that miR-21 acts as an oncogene. Recent data suggest that miR-21 plays a role in malignancy by repressing matrix metalloprotease (MMP) inhibitors, which, in turn, leads to activation of MMPs thereby accelerating glioma cell invasiveness [65]. The oncogenic activity of miR-155, which is overexpressed in Burkitt's and Hodgkin's lymphomas, has been confirmed in transgenic mice where miR-155 overexpression leads to aggressive B cell malignancy mice [130]. However, miR-155 has also been implicated as a central regulator of the immune system [78, 165], whereas miR-21 is detected in normal fibroblasts and monocyte derived dendritic cells and mast cells [118, 186]. Thus, preclinical studies in mouse cancer models using chemically modified antimiR oligonucleotides will be necessary to address the therapeutic value of targeting miR-21 and miR-155. Furthermore, spatial localization of miR-21 and miR-155 accumulation in tumor lesions will provide useful information for future development of targeted drug delivery into the cancer cells, the stromal compartment, and/or infiltrating immune cells.

Given the prevalence of activated K-Ras mutations in colon, pancreas, and lung cancer, replenishing let-7 activity with a let-7 mimic may provide a viable avenue for development of let-7 based cancer therapeutics. Enabling global let-7 suppressive mechanisms against K-Ras, c-Myc, and other oncogenic targets in conjunction with current chemotherapy may bring hopes for a cure, where other drugs such as farnesyltransferase inhibitors and small molecule inhibitors of K-Ras or downstream effector pathways have failed so far. Similar strategies could be envisioned for replenishing miR-34 activity. As a mediator of the p53 regulated processes, a miR-34 mimic could partially restore p53 functions and render cancer cells more vulnerable to p53 dependent cytotoxic compounds.

Antiviral Agents

Most antiviral agents act primarily on the replication active virus. For example, HIV is refractory in resting CD4$^+$ lymphocytes to suppressive highly active antiretroviral therapy (HAART) [187]. AntimiRs against viral miRNAs that control and maintain latency in combination with antiviral agents may provide an effective therapy to eradicate viral infection (e.g., herpesviruses and HIV-1). However, careful preclinical studies should evaluate protocol safety for a controlled entrance in the lytic cycle, as antimiR compounds that specifically target viral sequences should have minimized side effects on host cells. Interestingly, miRNA based strategies to eliminate HIV-1 via inhibition of host miRNAs have been successfully tested in cell line systems [174]. Thus, the development of therapeutic approaches that would allow targeted delivery of antimiR compounds against miR-28, miR-125b, miR-150, miR-223, and miR-382 to resting CD4$^+$ cells could be used in combination with HAART to improve treatment of HIV patients. The impediments of applying host miRNA based therapies have been thoroughly discussed [187].

A high proportion of HCV infections become chronic and lead to HCC, which is one of the most common cancers in the world. Due to the suboptimal efficacy and safety profiles of currently employed HCV therapies there is a high unmet medical need for novel HCV targeted therapeutics. Since miR-122 is required for HCV replication it represents a potential therapeutic target for treatment of HCV infection, which, in turn, necessitates the development of effective and safe approaches for miR-122 silencing *in vivo*.

Passive Regulators of Gene Therapy

There are several highly abundant miRNAs with organ/cell type specific expression (e.g., miR-122 in hepatocytes, miR-1 in muscle cells, miR-124 in neurons). Although the clinical application of gene therapy is currently halted and under serious scrutiny, the use of tissue specific miRNAs as passive regulators of transgene expression may provide a reliable safeguard mechanism to effect a precise delivery and functioning of gene therapy in the intended target cells. Proof-of-principle experiments were recently conducted in preclinical animal models. Reduced hepatotoxicity of suicide gene therapy was accomplished by engineering the thymidine kinase (HSVtk) and ganciclovir (GCV) transgenes to contain four perfectly complementary sites for miR-122 in their 3'UTR [188]. Using miR-122 as an endogenous RNA guide to cleave the HSVtk and GCV mRNAs, adenovirus delivered vectors were expressed more than 100-fold lower in hepatocytes compared to intended target tumor cells [188].

CONCLUSIONS

miRNAs were presented to the scientific community in 2001 and have emerged as a new paradigm for the control of gene expression. This miRNA mediated regulatory layer of gene expression has been repeatedly and convincingly shown to play an important role in developmental, physiological, and pathological processes. Indeed, after 7 years of miRNA research and with over 3400 miRNA related publications in the Pubmed database, very few fields of life sciences and biomedicine have been left unshaken. The significance of this breakthrough discovery has already been acknowledged with prestigious awards including the Lasker Award in 2008.

The field of cancer biology has been significantly transformed by the discovery of miRNAs and is expected to have a major impact on translational research. miRNAs have already shed light on the molecular mechanisms of carcinogenesis, provided potential new biomarker tools for cancer management and suggested novel approaches for therapeutic intervention. Furthermore, our understanding of the etiological contribution of miRNAs to other human diseases is rapidly advancing. Hence, we foresee an impending biomedical revolution based on the clinical application of miRNAs and miRNA mediated pathways for diagnostic and therapeutic purposes.

ACKNOWLEDGEMENTS

We would like to thank Elena Bryleva, Murray Korc, and Haoxu Ouyang for their valuable comments and help during the preparation of this manuscript. This work was supported by: Junior Investigator Award from Department of Medicine, Dartmouth-Hitchcock Medical Center, and Pancreatic Cancer Action Network-AACR 08-20-25-SEMP to LFS; and by grants from the Danish National Advanced Technology Foundation, Danish Medical Research Council and the Lundbeck Foundation to SK.

REFERENCES

1. Lee RC, Feinbaum RL, Ambros V. The C. elegans heterochronic gene lin-4 encodes small RNAs with antisense complementarity to lin-14. *Cell* 1993;**75**:843–54.
2. Reinhart BJ, Slack FJ, Basson M, et al. The 21-nucleotide let-7 RNA regulates developmental timing in Caenorhabditis elegans. *Nature* 2000;**403**:901–6.
3. Moss EG. Heterochronic genes and the nature of developmental time. *Curr Biol* 2007;**17**:R425–34.
4. Lee RC, Ambros V. An extensive class of small RNAs in Caenorhabditis elegans. *Science* 2001;**294**:862–4.
5. Lau NC, Lim LP, Weinstein EG, Bartel DP. An abundant class of tiny RNAs with probable regulatory roles in Caenorhabditis elegans. *Science* 2001;**294**:858–62.
6. Lagos-Quintana M, Rauhut R, Lendeckel W, Tuschl T. Identification of novel genes coding for small expressed RNAs. *Science* 2001;**294**: 853–8.
7. Bentwich I, Avniel A, Karov Y, et al. Identification of hundreds of conserved and nonconserved human microRNAs. *Nat Genet* 2005;**37**:766–70.
8. Griffiths-Jones S, Saini HK, van DS, Enright AJ. miRBase: tools for microRNA genomics. *Nucleic Acids Res* 2008;**36**:D154–8.
9. Lee Y, Jeon K, Lee JT, et al. MicroRNA maturation: stepwise processing and subcellular localization. *EMBO J* 2002;**21**:4663–70.
10. Lee Y, Ahn C, Han J, et al. The nuclear RNase III Drosha initiates microRNA processing. *Nature* 2003;**425**:415–9.
11. Lim LP, Lau NC, Garrett-Engele P, et al. Microarray analysis shows that some microRNAs downregulate large numbers of target mRNAs. *Nature* 2005;**433**:769–73.
12. Place RF, Li LC, Pookot D, et al. MicroRNA-373 induces expression of genes with complementary promoter sequences. *Proc Natl Acad Sci U S A* 2008;**105**:1608–13.
13. Kozak M. Faulty old ideas about translational regulation paved the way for current confusion about how microRNAs function. *Gene* 2008;**423**:108–15.
14. Filipowicz W, Bhattacharyya SN, Sonenberg N. Mechanisms of post-transcriptional regulation by microRNAs: are the answers in sight? *Nat Rev Genet* 2008;**9**:102–14.
15. Gusev Y. Computational methods for analysis of cellular functions and pathways collectively targeted by differentially expressed microRNA. *Methods* 2008;**44**:61–72.
16. Karginov FV, Conaco C, Xuan Z, et al. A biochemical approach to identifying microRNA targets. *Proc Natl Acad Sci U S A* 2007;**104**: 19,291–6.
17. Hock J, Weinmann L, Ender C, et al. Proteomic and functional analysis of Argonaute-containing mRNA-protein complexes in human cells. *EMBO Rep* 2007;**8**:1052–60.
18. Orom UA, Lund AH. Isolation of microRNA targets using biotinylated synthetic microRNAs. *Methods* 2007;**43**:162–5.
19. Selbach M, Schwanhausser B, Thierfelder N, et al. Widespread changes in protein synthesis induced by microRNAs. *Nature* 2008; **455**:58–63.
20. Baek D, Villen J, Shin C, et al. The impact of microRNAs on protein output. *Nature* 2008;**455**:64–71.
21. Borchert GM, Lanier W, Davidson BL. RNA polymerase III transcribes human microRNAs. *Nat Struct Mol Biol* 2006;**13**:1097–101.
22. Sempere LF, Freemantle S, Pitha-Rowe I, et al. Expression profiling of mammalian microRNAs uncovers a subset of brain-expressed microRNAs with possible roles in murine and human neuronal differentiation. *Genome Biol* 2004;**5**:R13.
23. Baskerville S, Bartel DP. Microarray profiling of microRNAs reveals frequent coexpression with neighboring miRNAs and host genes. *RNA* 2005;**11**:241–7.
24. Rodriguez A, Griffiths-Jones S, Ashurst JL, Bradley A. Identification of mammalian microRNA host genes and transcription units. *Genome Res* 2004;**14**:1902–10.
25. Marson A, Levine SS, Cole MF, et al. Connecting microRNA genes to the core transcriptional regulatory circuitry of embryonic stem cells. *Cell* 2008;**134**:521–33.
26. Chang TC, Yu D, Lee YS, et al. Widespread microRNA repression by Myc contributes to tumorigenesis. *Nat Genet* 2008;**40**:43–50.

27. O'Donnell KA, Wentzel EA, Zeller KI, et al. c-Myc-regulated micro-RNAs modulate E2F1 expression. *Nature* 2005;**435**:839–43.

28. Saito Y, Liang G, Egger G, et al. Specific activation of microRNA-127 with downregulation of the proto-oncogene BCL6 by chromatin-modifying drugs in human cancer cells. *Cancer Cell* 2006;**9**:435–43.

29. Lujambio A, Esteller M. CpG island hypermethylation of tumor suppressor microRNAs in human cancer. *Cell Cycle* 2007;**6**:1455–9.

30. Zhang L, Huang J, Yang N, et al. microRNAs exhibit high frequency genomic alterations in human cancer. *Proc Natl Acad Sci U S A* 2006; **103**:9136–41.

31. Davis BN, Hilyard AC, Lagna G, Hata A. SMAD proteins control DROSHA-mediated microRNA maturation. *Nature* 2008;**454**:56–61.

32. Kawahara Y, Zinshteyn B, Sethupathy P, et al. Redirection of silencing targets by adenosine-to-inosine editing of miRNAs. *Science* 2007;**315**:1137–40.

33. Blow MJ, Grocock RJ, van DS, et al. RNA editing of human micro-RNAs. *Genome Biol* 2006;**7**:R27.

34. Clop A, Marcq F, Takeda H, et al. A mutation creating a potential illegitimate microRNA target site in the myostatin gene affects muscularity in sheep. *Nat Genet* 2006;**38**:813–8.

35. Mishra PJ, Humeniuk R, Mishra PJ, et al. A miR-24 microRNA binding-site polymorphism in dihydrofolate reductase gene leads to methotrexate resistance. *Proc Natl Acad Sci U S A* 2007;**104**: 13,513–8.

36. Grishok A, Pasquinelli AE, Conte D, et al. Genes and mechanisms related to RNA interference regulate expression of the small temporal RNAs that control C. elegans developmental timing. *Cell* 2001; **106**:23–34.

37. Giraldez AJ, Cinalli RM, Glasner ME, et al. MicroRNAs regulate brain morphogenesis in zebrafish. *Science* 2005;**308**:833–8.

38. Harfe BD, McManus MT, Mansfield JH, et al. The RNaseIII enzyme Dicer is required for morphogenesis but not patterning of the vertebrate limb. *Proc Natl Acad Sci U S A* 2005;**102**:10,898–903.

39. Hornstein E, Mansfield JH, Yekta S, et al. The microRNA miR-196 acts upstream of Hoxb8 and Shh in limb development. *Nature* 2005;**438**:671–4.

40. Harris KS, Zhang Z, McManus MT, et al. Dicer function is essential for lung epithelium morphogenesis. *Proc Natl Acad Sci U S A* 2006;**103**:2208–13.

41. Yi R, O'Carroll D, Pasolli HA, et al. Morphogenesis in skin is governed by discrete sets of differentially expressed microRNAs. *Nat Genet* 2006;**38**:356–62.

42. O'Rourke JR, Georges SA, Seay HR, et al. Essential role for Dicer during skeletal muscle development. *Dev Biol* 2007;**311**:359–68.

43. Suarez Y, Fernandez-Hernando C, Yu J, et al. Dicer-dependent endothelial microRNAs are necessary for postnatal angiogenesis. *Proc Natl Acad Sci U S A* 2008;**105**:14,082–7.

44. Kobayashi T, Lu J, Cobb BS, et al. Dicer-dependent pathways regulate chondrocyte proliferation and differentiation. *Proc Natl Acad Sci U S A* 2008;**105**:1949–54.

45. Schaefer A, O'Carroll D, Tan CL, et al. Cerebellar neurodegeneration in the absence of microRNAs. *J Exp Med* 2007;**204**:1553–8.

46. Davis TH, Cuellar TL, Koch SM, et al. Conditional loss of Dicer disrupts cellular and tissue morphogenesis in the cortex and hippocampus. *J Neurosci* 2008;**28**:4322–30.

47. Harvey SJ, Jarad G, Cunningham J, et al. Podocyte-specific deletion of Dicer alters cytoskeletal dynamics and causes glomerular disease. *J Am Soc Nephrol* 2008;**19**:2150–8.

48. Shi S, Yu L, Chiu C, et al. Podocyte-selective deletion of Dicer induces proteinuria and glomerulosclerosis. *J Am Soc Nephrol* 2008;**19**:2159–69.

49. Chong MM, Rasmussen JP, Rundensky AY, Littman DR. The RNAseIII enzyme Drosha is critical in T cells for preventing lethal inflammatory disease. *J Exp Med* 2008;**205**:2005–17.

50. Liston A, Lu LF, O'Carroll D, et al. Dicer-dependent microRNA pathway safeguards regulatory T cell function. *J Exp Med* 2008;**205**:1993–2004.

51. Nagaraja AK, Andreu-Vieyra C, Franco HL, et al. Deletion of Dicer in somatic cells of the female reproductive tract causes sterility. *Mol Endocrinol* 2008;**22**:2336–52.

52. Damiani D, Alexander JJ, O'Rourke JR, et al. Dicer inactivation leads to progressive functional and structural degeneration of the mouse retina. *J Neurosci* 2008;**28**:4878–87.

53. Mudhasani R, Zhu Z, Hutvagner G, et al. Loss of miRNA biogenesis induces p19Arf-p53 signaling and senescence in primary cells. *J Cell Biol* 2008;**181**:1055–63.

54. Krichevsky AM, King KS, Donahue CP, et al. A MicroRNA array reveals extensive regulation of microRNAs during brain development. *RNA* 2003;**9**:1274–81.

55. Krichevsky AM, Sonntag KC, Isacson O, Kosik KS. Specific MicroRNAs modulate embryonic stem cell-derived neurogenesis. *Stem Cells* 2006;**24**:857–64.

56. Leucht C, Stigloher C, Wizenmann A, et al. MicroRNA-9 directs late organizer activity of the midbrain-hindbrain boundary. *Nat Neurosci* 2008;**11**:641–8.

57. Kim J, Inoue K, Ishii J, et al. A MicroRNA feedback circuit in midbrain dopamine neurons. *Science* 2007;**317**:1220–4.

58. Makeyev EV, Zhang J, Carrasco MA, Maniatis T. The MicroRNA miR-124 promotes neuronal differentiation by triggering brain-specific alternative pre-mRNA splicing. *Mol Cell* 2007;**27**:435–48.

59. Schratt GM, Tuebing F, Nigh EA, et al. A brain-specific microRNA regulates dendritic spine development. *Nature* 2006;**439**:283–9.

60. Fiore R, Siegel G, Schratt G. MicroRNA function in neuronal development, plasticity and disease. *Biochim Biophys Acta* 2008;**1779**:471–8.

61. Hebert SS, Horre K, Nicolai L, et al. Loss of microRNA cluster miR-29a/b-1 in sporadic Alzheimer's disease correlates with increased BACE1/beta-secretase expression. *Proc Natl Acad Sci U S A* 2008;**105**:6415–20.

62. Wang WX, Rajeev BW, Stromberg AJ, et al. The expression of microRNA miR-107 decreases early in Alzheimer's disease and may accelerate disease progression through regulation of beta-site amyloid precursor protein-cleaving enzyme 1. *J Neurosci* 2008;**28**:1213–23.

63. Shi L, Cheng Z, Zhang J, et al. hsa-mir-181a and hsa-mir-181b function as tumor suppressors in human glioma cells. *Brain Res* 2008;**1236**:185–93.

64. Chao TF, Zhang Y, Yan XQ, et al. [MiR-9 regulates the expression of CBX7 in human glioma]. *Zhongguo Yi Xue Ke Xue Yuan Xue Bao* 2008;**30**:268–74.

65. Gabriely G, Wurdinger T, Kesari S, et al. MicroRNA 21 promotes glioma invasion by targeting matrix metalloproteinase regulators. *Mol Cell Biol* 2008;**28**:5369–80.

66. Silber J, Lim DA, Petritsch C, et al. miR-124 and miR-137 inhibit proliferation of glioblastoma multiforme cells and induce differentiation of brain tumor stem cells. *BMC Med* 2008;**6**:14.

67. Kefas B, Godlewski J, Comeau L, et al. microRNA-7 inhibits the epidermal growth factor receptor and the Akt pathway and is down-regulated in glioblastoma. *Cancer Res* 2008;**68**:3566–72.

68. Chan JA, Krichevsky AM, Kosik KS. MicroRNA-21 is an antiapoptotic factor in human glioblastoma cells. *Cancer Res* 2005;**65**:6029–33.

69. Liu N, Williams AH, Kim Y, et al. An intragenic MEF2-dependent enhancer directs muscle-specific expression of microRNAs 1 and 133. *Proc Natl Acad Sci U S A* 2007;**104**:20,844–9.

70. Sweetman D, Goljanek K, Rathjen T, et al. Specific requirements of MRFs for the expression of muscle specific microRNAs, miR-1, miR-206 and miR-133. *Dev Biol* 2008;**321**:491–9.

71. Zhao Y, Samal E, Srivastava D. Serum response factor regulates a muscle-specific microRNA that targets Hand2 during cardiogenesis. *Nature* 2005;**436**:214–20.

72. Sokol NS, Ambros V. Mesodermally expressed Drosophila microRNA-1 is regulated by Twist and is required in muscles during larval growth. *Genes Dev* 2005;**19**:2343–54.

73. van RE, Sutherland LB, Qi X, et al. Control of stress-dependent cardiac growth and gene expression by a microRNA. *Science* 2007;**316**:575–9.

74. Luo X, Lin H, Pan Z, et al. Down-regulation of miR-1/miR-133 contributes to re-expression of pacemaker channel genes HCN2 and HCN4 in hypertrophic heart. *J Biol Chem* 2008;**283**:20,045–52.

75. Callis TE, Wang DZ. Taking microRNAs to heart. *Trends Mol Med* 2008;**14**:254–60.

76. Callis TE, Deng Z, Chen JF, Wang DZ. Muscling through the microRNA world. *Exp Biol Med (Maywood)* 2008;**233**:131–8.

77. Lindsay MA. microRNAs and the immune response. *Trends Immunol* 2008;**29**:343–51.

78. Sonkoly E, Stahle M, Pivarcsi A. MicroRNAs and immunity: novel players in the regulation of normal immune function and inflammation. *Semin Cancer Biol* 2008;**18**:131–40.

79. Zhan M, Miller CP, Papayannopoulou T, et al. MicroRNA expression dynamics during murine and human erythroid differentiation. *Exp Hematol* 2007;**35**:1015–25.

80. Turner M, Vigorito E. Regulation of B- and T-cell differentiation by a single microRNA. *Biochem Soc Trans* 2008;**36**:531–3.

81. Vigorito E, Perks KL, breu-Goodger C, et al. microRNA-155 regulates the generation of immunoglobulin class-switched plasma cells. *Immunity* 2007;**27**:847–59.

82. Thai TH, Calado DP, Casola S, et al. Regulation of the germinal center response by microRNA-155. *Science* 2007;**316**:604–8.

83. Dorsett Y, McBride KM, Jankovic M, et al. MicroRNA-155 suppresses activation-induced cytidine deaminase-mediated Myc-Igh translocation. *Immunity* 2008;**28**:630–8.

84. Teng G, Hakimpour P, Landgraf P, et al. MicroRNA-155 is a negative regulator of activation-induced cytidine deaminase. *Immunity* 2008;**28**:621–9.

85. Taganov KD, Boldin MP, Chang KJ, Baltimore D. NF-kappaB-dependent induction of microRNA miR-146, an inhibitor targeted to signaling proteins of innate immune responses. *Proc Natl Acad Sci U S A* 2006;**103**:12,481–6.

86. Girard M, Jacquemin E, Munnich A, et al. miR-122, a paradigm for the role of microRNAs in the liver. *J Hepatol* 2008;**48**:648–56.

87. Chang J, Nicolas E, Marks D, et al. miR-122, a mammalian liver-specific microRNA, is processed from hcr mRNA and may downregulate the high affinity cationic amino acid transporter CAT-1. *RNA Biol* 2004;**1**:106–13.

88. Krutzfeldt J, Rajewsky N, Braich R, et al. Silencing of microRNAs in vivo with "antagomirs". *Nature* 2005;**438**:685–9.

89. Fabani MM, Gait MJ. miR-122 targeting with LNA/2'-O-methyl oligonucleotide mixmers, peptide nucleic acids (PNA), and PNA-peptide conjugates. *RNA* 2008;**14**:336–46.

90. Elmen J, Lindow M, Silahtaroglu A, et al. Antagonism of microRNA-122 in mice by systemically administered LNA-antimiR leads to up-regulation of a large set of predicted target mRNAs in the liver. *Nucleic Acids Res* 2008;**36**:1153–62.

91. Esau C, Davis S, Murray SF, et al. miR-122 regulation of lipid metabolism revealed by in vivo antisense targeting. *Cell Metab* 2006;**3**:87–98.

92. Elmen J, Lindow M, Schutz S, et al. LNA-mediated microRNA silencing in non-human primates. *Nature* 2008;**452**:896–9.

93. Kutay H, Bai S, Datta J, et al. Downregulation of miR-122 in the rodent and human hepatocellular carcinomas. *J Cell Biochem* 2006;**99**:671–8.

94. Gramantieri L, Ferracin M, Fornari F, et al. Cyclin G1 is a target of miR-122a, a microRNA frequently down-regulated in human hepatocellular carcinoma. *Cancer Res* 2007;**67**:6092–9.

95. Lin CJ, Gong HY, Tseng HC, et al. miR-122 targets an anti-apoptotic gene, Bcl-w, in human hepatocellular carcinoma cell lines. *Biochem Biophys Res Commun* 2008;**375**:315–20.

96. Varnholt H, Drebber U, Schulze F, et al. MicroRNA gene expression profile of hepatitis C virus-associated hepatocellular carcinoma. *Hepatology* 2008;**47**:1223–32.

97. Han J, Lee Y, Yeom KH, et al. Molecular basis for the recognition of primary microRNAs by the Drosha-DGCR8 complex. *Cell* 2006;**125**:887–901.

98. Shiohama A, Sasaki T, Noda S, et al. Molecular cloning and expression analysis of a novel gene DGCR8 located in the DiGeorge syndrome chromosomal region. *Biochem Biophys Res Commun* 2003;**304**:184–90.

99. Stark KL, Xu B, Bagchi A, et al. Altered brain microRNA biogenesis contributes to phenotypic deficits in a 22q11-deletion mouse model. *Nat Genet* 2008;**40**:751–60.

100. Li Y, Lin L, Jin P. The microRNA pathway and fragile X mental retardation protein. *Biochim Biophys Acta* 2008;**1779**:702–5.

101. Kuhn DE, Nuovo GJ, Martin MM, et al. Human chromosome 21-derived miRNAs are overexpressed in Down syndrome brains and hearts. *Biochem Biophys Res Commun* 2008;**370**:473–7.

102. Borel C, Antonarakis SE. Functional genetic variation of human miRNAs and phenotypic consequences. *Mamm Genome* 2008;**19**:503–9.

103. Calin GA, Dumitru CD, Shimizu M, et al. Frequent deletions and down-regulation of micro- RNA genes miR15 and miR16 at 13q14 in chronic lymphocytic leukemia. *Proc Natl Acad Sci U S A* 2002;**99**:15,524–9.

104. Calin GA, Ferracin M, Cimmino A, et al. A MicroRNA signature associated with prognosis and progression in chronic lymphocytic leukemia. *N Engl J Med* 2005;**353**:1793–801.

105. Barbarotto E, Schmittgen TD, Calin GA. MicroRNAs and cancer: profile, profile, profile. *Int J Cancer* 2008;**122**:969–77.

106. Lu J, Getz G, Miska EA, et al. MicroRNA expression profiles classify human cancers. *Nature* 2005;**435**:834–8.

107. Yanaihara N, Caplen N, Bowman E, et al. Unique microRNA molecular profiles in lung cancer diagnosis and prognosis. *Cancer Cell* 2006;**9**:189–98.

108. Salter KH, Acharya CR, Walters KS, et al. An integrated approach to the prediction of chemotherapeutic response in patients with breast cancer. *PLoS ONE* 2008;**3**:e1908.

109. Lawrie CH, Soneji S, Marafioti T, et al. MicroRNA expression distinguishes between germinal center B cell-like and activated B cell-like subtypes of diffuse large B cell lymphoma. *Int J Cancer* 2007;**121**:1156–61.

110. Takamizawa J, Konishi H, Yanagisawa K, et al. Reduced expression of the let-7 microRNAs in human lung cancers in association with shortened postoperative survival. *Cancer Res* 2004;**64**:3753–6.

111. Karube Y, Tanaka H, Osada H, et al. Reduced expression of Dicer associated with poor prognosis in lung cancer patients. *Cancer Sci* 2005;**96**:111–5.

112. Chan SH, Wu CW, Li AF, et al. miR-21 microRNA expression in human gastric carcinomas and its clinical association. *Anticancer Res* 2008;**28**:907–11.

113. Camps C, Buffa FM, Colella S, et al. hsa-miR-210 Is induced by hypoxia and is an independent prognostic factor in breast cancer. *Clin Cancer Res* 2008;**14**:1340–8.

114. Michael MZ, O' Connor SM, Holst Pellekaan NG, et al. Reduced accumulation of specific microRNAs in colorectal neoplasia. *Mol Cancer Res* 2003;**1**:882–91.

115. Tavazoie SF, Alarcon C, Oskarsson T, et al. Endogenous human microRNAs that suppress breast cancer metastasis. *Nature* 2008;**451**:147–52.

116. Ivanovska I, Ball AS, Diaz RL, et al. MicroRNAs in the miR-106b family regulate p21/CDKN1A and promote cell cycle progression. *Mol Cell Biol* 2008;**28**:2167–74.

117. Budhu A, Jia HL, Forgues M, et al. Identification of metastasis-related microRNAs in hepatocellular carcinoma. *Hepatology* 2008;**47**:897–907.

118. Sempere LF, Christensen M, Silahtaroglu A, et al. Altered MicroRNA expression confined to specific epithelial cell subpopulations in breast cancer. *Cancer Res* 2007;**67**:11,612–20.

119. Nelson PT, Baldwin DA, Kloosterman WP, et al. RAKE and LNA-ISH reveal microRNA expression and localization in archival human brain. *RNA* 2006;**12**:187–91.

120. Stenvang J, Silahtaroglu AN, Lindow M, et al. The utility of LNA in microRNA-based cancer diagnostics and therapeutics. *Semin Cancer Biol* 2008;**18**:89–102.

121. Kauppinen S, Vester B, Wengel J. Locked nucleic acid: high-affinity targeting of complementary RNA for RNomics. *Handb Exp Pharmacol* 2006;**173**:405–22.

122. Rathjen T, Nicol C, McConkey G, Dalmay T. Analysis of short RNAs in the malaria parasite and its red blood cell host. *FEBS Lett* 2006;**580**:5185–8.

123. Bloomston M, Frankel WL, Petrocca F, et al. MicroRNA expression patterns to differentiate pancreatic adenocarcinoma from normal pancreas and chronic pancreatitis. *JAMA* 2007;**297**:1901–8.

124. Lee EJ, Gusev Y, Jiang J, et al. Expression profiling identifies microRNA signature in pancreatic cancer. *Int J Cancer* 2007;**120**:1046–54.

125. Szafranska AE, Davison TS, John J, et al. MicroRNA expression alterations are linked to tumorigenesis and non-neoplastic processes in pancreatic ductal adenocarcinoma. *Oncogene* 2007;**26**:4442–52.

126. Szafranska AE, Doleshal M, Edmunds HS, et al. Analysis of MicroRNAs in pancreatic fine-needle aspirates can classify benign and malignant tissues. *Clin Chem* 2008;**54**:1716–24.

127. Gilad S, Meiri E, Yogev Y, et al. Serum microRNAs are promising novel biomarkers. *PLoS ONE* 2008;**3**:e3148.

128. Lawrie CH, Gal S, Dunlop HM, et al. Detection of elevated levels of tumour-associated microRNAs in serum of patients with diffuse large B-cell lymphoma. *Br J Haematol* 2008;**141**:672–5.

129. Mitchell PS, Parkin RK, Kroh EM, et al. Circulating microRNAs as stable blood-based markers for cancer detection. *Proc Natl Acad Sci U S A* 2008;**105**:10,513–8.

130. Costinean S, Zanesi N, Pekarsky Y, et al. Pre-B cell proliferation and lymphoblastic leukemia/high-grade lymphoma in Emicro-miR155 transgenic mice. *Proc Natl Acad Sci U S A* 2006;**103**:7024–9.

131. He L, Thomson JM, Hemann MT, et al. A microRNA polycistron as a potential human oncogene. *Nature* 2005;**435**:828–33.

132. Cimmino A, Calin GA, Fabbri M, et al. miR-15 and miR-16 induce apoptosis by targeting BCL2. *Proc Natl Acad Sci U S A* 2005;**102**:13,944–9.

133. Coller HA, Forman JJ, Legesse-Miller A. "Myc'ed messages": myc induces transcription of E2F1 while inhibiting its translation via a microRNA polycistron. *PLoS Genet* 2007;**3**:e146.

134. Sylvestre Y, De GV, Querido E, et al. An E2F/miR-20a autoregulatory feedback loop. *J Biol Chem* 2007;**282**:2135–43.

135. Matsubara H, Takeuchi T, Nishikawa E, et al. Apoptosis induction by antisense oligonucleotides against miR-17-5p and miR-20a in lung cancers overexpressing miR-17-92. *Oncogene* 2007;**26**:6099–105.

136. Yu Z, Wang C, Wang M, et al. A cyclin D1/microRNA 17/20 regulatory feedback loop in control of breast cancer cell proliferation. *J Cell Biol* 2008;**182**:509–17.

137. Johnson SM, Grosshans H, Shingara J, et al. RAS is regulated by the let-7 microRNA family. *Cell* 2005;**120**:635–47.

138. Lee YS, Dutta A. The tumor suppressor microRNA let-7 represses the HMGA2 oncogene. *Genes Dev* 2007;**21**:1025–30.

139. Kumar MS, Lu J, Mercer KL, et al. Impaired microRNA processing enhances cellular transformation and tumorigenesis. *Nat Genet* 2007;**39**:673–7.

140. Esquela-Kerscher A, Trang P, Wiggins JF, et al. The let-7 microRNA reduces tumor growth in mouse models of lung cancer. *Cell Cycle* 2008;**7**:759–64.

141. Kumar MS, Erkeland SJ, Pester RE, et al. Suppression of non-small cell lung tumor development by the let-7 microRNA family. *Proc Natl Acad Sci U S A* 2008;**105**:3903–8.

142. Sampson VB, Rong NH, Han J, et al. MicroRNA let-7a down-regulates MYC and reverts MYC-induced growth in Burkitt lymphoma cells. *Cancer Res* 2007;**67**:9762–70.

143. He L, He X, Lowe SW, Hannon GJ. microRNAs join the p53 network: another piece in the tumour-suppression puzzle. *Nat Rev Cancer* 2007;**7**:819–22.

144. Miller TE, Ghoshal K, Ramaswamy B, et al. MicroRNA-221/222 confers tamoxifen resistance in breast cancer by targeting p27(Kip1). *J Biol Chem* 2008;**283**:29,897–903.

145. Fornari F, Gramantieri L, Ferracin M, et al. MiR-221 controls CDKN1C/p57 and CDKN1B/p27 expression in human hepatocellular carcinoma. *Oncogene* 2008;**27**:5651–61.

146. Medina R, Zaidi SK, Liu CG, et al. MicroRNAs 221 and 222 bypass quiescence and compromise cell survival. *Cancer Res* 2008;**68**:2773–80.

147. Visone R, Russo L, Pallante P, et al. MicroRNAs (miR)-221 and miR-222, both overexpressed in human thyroid papillary carcinomas, regulate p27Kip1 protein levels and cell cycle. *Endocr Relat Cancer* 2007;**14**:791–8.

148. Sage C, Nagel R, Egan DA, et al. Regulation of the p27(Kip1) tumor suppressor by miR-221 and miR-222 promotes cancer cell proliferation. *EMBO J* 2007;**26**:3699–708.

149. Galardi S, Mercatelli N, Giorda E, et al. miR-221 and miR-222 expression affects the proliferation potential of human prostate carcinoma cell lines by targeting p27Kip1. *J Biol Chem* 2007;**282**:23,716–24.

150. Fontana L, Fiori ME, Albini S, et al. Antagomir-17-5p abolishes the growth of therapy-resistant neuroblastoma through p21 and BIM. *PLoS ONE* 2008;**3**:e2236.

151. Meng F, Henson R, Wehbe-Janek H, et al. MicroRNA-21 regulates expression of the PTEN tumor suppressor gene in human hepatocellular cancer. *Gastroenterology* 2007;**133**:647–58.

152. Meng F, Henson R, Lang M, et al. Involvement of human microRNA in growth and response to chemotherapy in human cholangiocarcinoma cell lines. *Gastroenterology* 2006;**130**:2113–29.

153. Zhu S, Wu H, Wu F, et al. MicroRNA-21 targets tumor suppressor genes in invasion and metastasis. *Cell Res* 2008;**18**:350–9.

154. Lu Z, Liu M, Stribinskis V, et al. MicroRNA-21 promotes cell transformation by targeting the programmed cell death 4 gene. *Oncogene* 2008;**27**:4373–9.

155. Frankel LB, Christoffersen NR, Jacobsen A, et al. Programmed cell death 4 (PDCD4) is an important functional target of the microRNA miR-21 in breast cancer cells. *J Biol Chem* 2008;**283**:1026–33.

156. Asangani IA, Rasheed SA, Nikolova DA, et al. MicroRNA-21 (miR-21) post-transcriptionally downregulates tumor suppressor Pdcd4 and stimulates invasion, intravasation and metastasis in colorectal cancer. *Oncogene* 2008;**27**:2128–36.

157. Zhu S, Si ML, Wu H, Mo YY. MicroRNA-21 targets the tumor suppressor gene tropomyosin 1 (TPM1). *J Biol Chem* 2007;**282**:14,328–36.

158. Burk U, Schubert J, Wellner U, et al. A reciprocal repression between ZEB1 and members of the miR-200 family promotes EMT and invasion in cancer cells. *EMBO Rep* 2008;**9**:582–9.

159. Gregory PA, Bert AG, Paterson EL, et al. The miR-200 family and miR-205 regulate epithelial to mesenchymal transition by targeting ZEB1 and SIP1. *Nat Cell Biol* 2008;**10**:593–601.

160. Korpal M, Lee ES, Hu G, Kang Y. The miR-200 family inhibits epithelial-mesenchymal transition and cancer cell migration by direct targeting of E-cadherin transcriptional repressors ZEB1 and ZEB2. *J Biol Chem* 2008;**283**:14,910–4.

161. Park SM, Gaur AB, Lengyel E, Peter ME. The miR-200 family determines the epithelial phenotype of cancer cells by targeting the E-cadherin repressors ZEB1 and ZEB2. *Genes Dev* 2008;**22**:894–907.

162. Hurteau GJ, Carlson JA, Spivack SD, Brock GJ. Overexpression of the microRNA hsa-miR-200c leads to reduced expression of transcription factor 8 and increased expression of E-cadherin. *Cancer Res* 2007;**67**:7972–6.

163. Huang Q, Gumireddy K, Schrier M, et al. The microRNAs miR-373 and miR-520c promote tumour invasion and metastasis. *Nat Cell Biol* 2008;**10**:202–10.

164. Ma L, Teruya-Feldstein J, Weinberg RA. Tumour invasion and metastasis initiated by microRNA-10b in breast cancer. *Nature* 2007;**449**:682–8.

165. Urbich C, Kuehbacher A, Dimmeler S. Role of microRNAs in vascular diseases, inflammation, and angiogenesis. *Cardiovasc Res* 2008;**79**:581–8.

166. Fasanaro P, D'Alessandra Y, Di S, et al. MicroRNA-210 modulates endothelial cell response to hypoxia and inhibits the receptor tyrosine kinase ligand Ephrin-A3. *J Biol Chem* 2008;**283**:15,878–83.

167. Wang S, Aurora AB, Johnson BA, et al. The endothelial-specific microRNA miR-126 governs vascular integrity and angiogenesis. *Dev Cell* 2008;**15**:261–71.

168. Gottwein E, Cullen BR. Viral and cellular microRNAs as determinants of viral pathogenesis and immunity. *Cell Host Microbe* 2008;**3**:375–87.

169. Ouellet DL, Plante I, Landry P, et al. Identification of functional microRNAs released through asymmetrical processing of HIV-1 TAR element. *Nucleic Acids Res* 2008;**36**:2353–65.

170. Klase Z, Kale P, Winograd R, et al. HIV-1 TAR element is processed by Dicer to yield a viral micro-RNA involved in chromatin remodeling of the viral LTR. *BMC Mol Biol* 2007;**8**:63.

171. Umbach JL, Kramer MF, Jurak I, et al. MicroRNAs expressed by herpes simplex virus 1 during latent infection regulate viral mRNAs. *Nature* 2008;**454**:780–3.

172. Tang S, Bertke AS, Patel A, et al. An acutely and latently expressed herpes simplex virus 2 viral microRNA inhibits expression of ICP34.5, a viral neurovirulence factor. *Proc Natl Acad Sci U S A* 2008;**105**:10,931–6.

173. Barth S, Pfuhl T, Mamiani A, et al. Epstein-Barr virus-encoded microRNA miR-BART2 down-regulates the viral DNA polymerase BALF5. *Nucleic Acids Res* 2008;**36**:666–75.

174. Huang J, Wang F, Argyris E, et al. Cellular microRNAs contribute to HIV-1 latency in resting primary CD4+ T lymphocytes. *Nat Med* 2007;**13**:1241–7.

175. Jopling CL, Schutz S, Sarnow P. Position-dependent function for a tandem microRNA miR-122-binding site located in the hepatitis C virus RNA genome. *Cell Host Microbe* 2008;**4**:77–85.

176. Jopling CL, Yi M, Lancaster AM, et al. Modulation of hepatitis C virus RNA abundance by a liver-specific MicroRNA. *Science* 2005;**309**:1577–81.

177. Pedersen IM, Cheng G, Wieland S, et al. Interferon modulation of cellular microRNAs as an antiviral mechanism. *Nature* 2007;**449**:919–22.

178. Shan Y, Zheng J, Lambrecht RW, Bonkovsky HL. Reciprocal effects of micro-RNA-122 on expression of heme oxygenase-1 and hepatitis C virus genes in human hepatocytes. *Gastroenterology* 2007;**133**:1166–74.

179. Tang G. siRNA and miRNA: an insight into RISCs. *Trends Biochem Sci* 2005;**30**:106–14.

180. Flynt AS, Li N, Thatcher EJ, et al. Zebrafish miR-214 modulates Hedgehog signaling to specify muscle cell fate. *Nat Genet* 2007;**39**:259–63.

181. Crooke ST. Progress in antisense technology: the end of the beginning. *Methods Enzymol* 2000;**313**:3–45.

182. Juliano R, Alam MR, Dixit V, Kang H. Mechanisms and strategies for effective delivery of antisense and siRNA oligonucleotides. *Nucleic Acids Res* 2008;**36**:4158–71.

183. De PD, Bentley MV, Mahato RI. Hydrophobization and bioconjugation for enhanced siRNA delivery and targeting. *RNA* 2007;**13**:431–56.

184. Zhang L, Gu FX, Chan JM, et al. Nanoparticles in medicine: therapeutic applications and developments. *Clin Pharmacol Ther* 2008;**83**:761–9.

185. Kleinman ME, Yamada K, Takeda A, et al. Sequence- and target-independent angiogenesis suppression by siRNA via TLR3. *Nature* 2008;**452**:591–7.

186. Sonkoly E, Wei T, Janson PC, et al. MicroRNAs: novel regulators involved in the pathogenesis of Psoriasis? *PLoS ONE* 2007;**2**:e610.

187. Zhang H. Reversal of HIV-1 latency with anti-microRNA inhibitors. *Int J Biochem Cell Biol* 2008;**41**:451–4.

188. Suzuki T, Sakurai F, Nakamura SI, et al. miR-122a-regulated expression of a suicide gene prevents hepatotoxicity without altering antitumor effects in suicide gene therapy. *Mol Ther* 2008;**16**:1719–26.

Printed and bound by CPI Group (UK) Ltd, Croydon, CR0 4YY

03/10/2024

01040319-0015